IMMS' GENERAL TEXTBOOK OF ENTOMOLOGY

TENTH EDITION

Volume 2: Classification and Biology

O. W. RICHARDS
M.A., D.Sc., F.R.S.
(deceased)
Emeritus Professor of Zoology and Applied Entomology, Imperial College, University of London

and

R. G. DAVIES
M.Sc.
Professor of Entomology, Imperial Colleg University of London

AF769165

CHAPMAN & HALL
London · Glasgow · Weinheim · New York · Tokyo · Melbourne · Madras

Published by Chapman & Hall, 2-6 Boundary Row, London SE1 8HN, UK

Chapman & Hall, 2-6 Boundary Row, London SE1 8HN, UK

Blackie Academic & Professional, Wester Cleddens Road, Bishopbriggs, Glasgow G64 2NZ, UK

Chapman & Hall GmbH, Pappelallee 3, 69469 Weinheim, Germany

Chapman & Hall USA., One Penn Plaza, 41st Floor, New York, NY10119, USA

Chapman & Hall Japan, ITP - Japan, Kyowa Building, 3F, 2-2-1 Hirakawacho, Chiyoda-ku, Tokyo 102, Japan

Chapman & Hall Australia, Thomas Nelson Australia, 102 Dodds Street, South Melbourne, Victoria 3205, Australia

Chapman & Hall India, R. Seshadri, 32 Second Main Road, CIT East, Madras 600 035, India

First edition 1925
Second edition 1930
Third edition, revised and enlarged 1934
Fourth edition 1938
Fifth edition 1942
Sixth edition 1974
Seventh edition 1948
Eighth edition 1951
Ninth edition, revised by O. W. Richards and R. G. Davies 1957
Tenth edition published in two volumes in 1977 by Chapman & Hall Ltd
Reprinted 1994

© 1977 O. W. Richards and R. G. Davies

ISBN 978-94-017-0474-8

Apart from any fair dealing for the purposes of research or private study, or criticism or review, as permitted under the UK Copyright Designs and Patents Act, 1988, this publication may not be reproduced, stored, or transmitted, in any form or by any means, without the prior permission in writing of the publishers, or in the case of reprographic reproduction only in accordance with the terms of the licences issued by the Copyright Licensing Agency in the UK, or in accordance with the terms of licences issued by the appropriate Reproduction Rights Organization outside the UK. Enquiries concerning reproduction outside the terms stated here should be sent to the publishers at the London address printed on this page.

The publisher makes no representation, express or implied, with regard to the accuracy of the information contained in this book and cannot accept any legal responsibility or liability for any errors or omissions that may be made.

A Catalogue record for this book is available from the British Library

Library of Congress Cataloging-in-Publication Data available

CONTENTS

VOLUME I

PART I. Anatomy and Physiology

PART II. Development and Metamorphosis

VOLUME II

PART III. The Orders of Insects

FOREWORD

The continuing demand for *Imms' General Textbook of Entomology* enables me to respond to the publisher's suggestion that I write a few words on the evolutionary history of a book which, in one form or another, has now been available for almost seventy years. It was, in fact, on March 12th 1925 that *A General Textbook of Entomology* by A. D. Imms first appeared under the imprint of Methuen & Co. Augustus Daniel Imms was born in 1880 in what is now a part of Birmingham. Following a boyhood interest in insect collecting and natural history, he studied zoology at Mason College – the forerunner of the University of Birmingham – and then at Cambridge under A. E. Shipley and G. H. F. Nuttall. During a professional career extending over forty years Imms occupied five successive posts: Professor of Zoology at the University of Allahabad, Forest Entomologist to the Government of India at Dehra Dun, Reader in Agricultural Zoology at Manchester, head of the Entomology Department at Rothamsted Experimental Station, and finally Reader in Entomology at Cambridge. In each of these positions he faced and overcame the formidable challenge of starting and building up a new department from virtually nothing. Despite the onerous tasks of organization and administration that this entailed, Imms not only published a creditable volume of original research, mainly on insect morphology and biology, but also found time to write five books on various aspects of entomology. His *General Textbook*, originally written while he was at Rothamsted, set a new standard for such works and was characterised by three distinctive features: the advanced level of treatment, with numerous references to the primary literature, the world-wide selection of illustrative examples, and – most of all – the judicious balance between all aspects of the subject. Structure, function, development, classification and biology all received their appropriate emphasis. And, as Sir Vincent Wigglesworth once wrote, '. . . the amount of information and its accessibility in this textbook is a constant surprise to any one who makes regular use of it.'

The *General Textbook* went through six further editions during Imms's lifetime and when he died in 1949 he was still gathering material for a further revision. By 1950, however, there was clearly a need for more extensive revision and with this in mind the publishers approached O. W. Richards, then Reader in Entomology at Imperial College and one of the most distinguished entomologists of his generation, with strong interests in systematics and ecology. As a result Richards and I collaborated to produce in 1957 the revised and substantially rewritten ninth edition. This preserved the balance and general treatment established earlier by Imms and was sufficiently well received for a separate reprint, sponsored and subsidized by the English Language Book Society for distribution in the developing countries, to appear in 1965. The general sections of the ninth edition (amounting to some nineteen chapters) were also given wider circulation in India when they were translated into Hindi and published as a separate book in 1975. Meanwhile, however, the progress of entomology had continued actively and some of this was reflected in a tenth edition, again substantially revised and rewritten by Professor Richards and myself, which was published by Chapman & Hall in 1977. This edition too was given wider currency overseas when it was translated into Spanish and appeared as *Tratado de Entomologia Imms* in 1983.

The scope and depth of entomology have increased enormously since Imms first wrote. The subject has also been invaded progressively by biologists of all kinds, who have recognised the favourable material which insects offer for investigating some of the fundamental problems of physiology, ecology, genetics, evolution and, most recently, of molecular biology. It may be hoped that *Imms' General Textbook of Entomology* will continue to provide much of the basic information needed by those who work in entomology and in these other fields.

London R. G. DAVIES
February 1994

PREFACE TO THE TENTH EDITION

In the twenty years that have elapsed since our last complete revision of this textbook, entomology has developed greatly, both in extent and depth. There are now over 8000 publications on the subject each year (excluding the applied literature) and the difficulty of incorporating even a fraction of the more important new results has occupied us considerably. We have nevertheless retained the original plan of the book, especially as it has the merit of familiarity for many readers, but we have made a number of appreciable changes in the text as well as innumerable smaller alterations. We have decided, with some reluctance, to dispense with the keys to families that were formerly given for most of the orders of insects. These are increasingly difficult to construct because specialists tend to recognize ever larger numbers of families, often based on regional revisions and therefore applicable with difficulty, if at all, to the world fauna. Our revision of the text has also entailed extensive changes in the bibliographies, which have been brought more or less up to date. In doing this we have had to be rigorously selective and we have tended to give some emphasis to review articles or recent papers at the expense of older works. We recognize that this has sometimes done less than justice to the contributions of earlier authorities, but the immense volume of literature left little alternative and we apologize to those who feel our choice of references has sometimes been almost arbitrary.

Every chapter has been revised in detail, many of them include new sections, and some have been extensively rewritten. In a few groups such as the Plecoptera and Heteroptera the higher classification has been recast; more often we have made smaller amendments in the number and arrangement of families so as to bring the scheme into broad but conservative agreement with modern views. The general chapters now include some information on ultrastructure and we have retained and tried to modernize the physiological sections; as non-specialists in this field we owe a great debt to the textbooks of Wigglesworth and of Rockstein. Inevitably the book has grown in size with the development of the subject. It may, indeed, be argued that the day of the general textbook has passed and that it must be replaced by a series of special monographs. We believe, however, that there are some advantages in a more unified viewpoint and it is our hope that the new balance we have reached between the various aspects of entomology will seem as appropriate now as the original balance was when Dr A. D. Imms' textbook was first published over fifty years ago.

There are 35 new figures, all based on published illustrations, the sources of which are acknowledged in the captions. We are grateful to the authors concerned and also to Miss K. Priest of Messrs Chapman & Hall, who saved us from many errors and omissions, and to Mrs R. G. Davies for substantial help in preparing the bibliographies and checking references.

London O.W.R.
May 1976 R.G.D.

Part III

THE ORDERS OF INSECTS

THE CLASSIFICATION AND PHYLOGENY OF INSECTS

The classification of insects has passed through many changes and with the growth of detailed knowledge an increasing number of orders has come to be recognized. Handlirsch (1908) and Wilson and Doner (1937) have reviewed the earlier attempts at classification, among which the schemes of Brauer (1885), Sharp (1899) and Börner (1904) did much to define the more distinctive recent orders. In 1908 Handlirsch published a more revolutionary system, incorporating recent and fossil forms, which gave the Collembola, Thysanura and Diplura the status of three independent Arthropodan classes and considered as separate orders such groups as the Sialoidea, Raphidioidea, Heteroptera and Homoptera. He also split up the old order Orthoptera, gave its components ordinal rank and regrouped them with some of the other orders into a subclass Orthopteroidea and another subclass Blattaeformia. This system was modified somewhat by Handlirsch (1926) and as such has influenced Brues, Melander and Carpenter (1954), Weber (1933), Martynov (1938) and Jeannel (1949). More recently, general classifications of the insects have been set out by Beier (1969), Hennig (1953, 1969) and Mackerras (1970).

Most classifications of the insects have tended to reflect their authors' opinions on the evolutionary relationships of the major groups. Unfortunately, the palaeontological record does not provide a very satisfactory basis on which to reconstruct insect phylogeny, while attempts to infer it from the comparative morphology of recent species are also subject to many uncertainties and qualifications. Until there is wider agreement on phylogenetic issues, therefore, it seems best not to insist on a taxonomic scheme whose detailed structure depends on too many evolutionary hypotheses, however interesting these may be in their own right. We have therefore simply enumerated the 29 orders of insects recognized in this book without grouping them in a formal hierarchical system of superorders, subclasses and so on. The various alternative arrangements of these orders are then discussed in connection with the palaeontological data and some of the more probable phylogenetic interpretations. In addition to the detailed studies cited below there are modern accounts of the general phylogeny of the insects by Wille

(1960), Rohdendorf (1969*b*) and especially by Hennig (1953, 1969) and Kristensen (1974). The fossil groups have been reviewed by Handlirsch (1937–39), Martynov (1938), Laurentiaux (1953), Martynova (1961) and Rohdendorf (1962, 1969*a*) as well as in many special studies, some of which are listed on pp. 427–31. Details of the fossil record are tabulated by Crowson *et al.* (1967).

The first four orders, the Apterygote insects, are primitively apterous forms with only a slight metamorphosis; they usually moult several times after attaining sexual maturity and the adults have one or more pairs of pregenital appendages. The mandibles usually articulate with the head-capsule at a single point.

Order 1. THYSANURA
Order 2. DIPLURA
Order 3. PROTURA
Order 4. COLLEMBOLA

The remaining 25 orders are the Pterygote insects and their adults are winged or secondarily apterous. Their metamorphosis is varied, the adults do not moult and they have no pregenital abdominal appendages. Unless highly modified, the mandibles articulate with the head-capsule at two points.

The Pterygotes fall into two sections. Orders 5 to 20 below constitute the Exopterygotes, normally with a simple, incomplete (hemimetabolous) metamorphosis. There is usually no pupal instar, the wings develop externally, and the immature stages (known either as larvae or nymphs) usually resemble the adults in structure and habits.

Order 5. EPHEMEROPTERA	Palaeopteran orders
Order 6. ODONATA	Palaeopteran orders
Order 7. PLECOPTERA	Orthopteroid orders
Order 8. GRYLLOBLATTODEA	Orthopteroid orders
Order 9. ORTHOPTERA	Orthopteroid orders
Order 10. PHASMIDA	Orthopteroid orders
Order 11. DERMAPTERA	Orthopteroid orders
Order 12. EMBIOPTERA	Orthopteroid orders
Order 13. DICTYOPTERA	Orthopteroid orders
Order 14. ISOPTERA	Orthopteroid orders
Order 15. ZORAPTERA	Orthopteroid orders
Order 16. PSOCOPTERA	Hemipteroid orders
Order 17. MALLOPHAGA	Hemipteroid orders
Order 18. SIPHUNCULATA	Hemipteroid orders
Order 19. HEMIPTERA	Hemipteroid orders
Order 20. THYSANOPTERA	Hemipteroid orders

The 9 remaining orders are the Endopterygotes, with a complete (holometabolous) metamorphosis, accompanied by a pupal instar. The wings develop internally and the larvae differ from the adults in structure and habits.

Order 21. NEUROPTERA
Order 22. COLEOPTERA
Order 23. STREPSIPTERA
Order 24. MECOPTERA
Order 25. SIPHONAPTERA
Order 26. DIPTERA — Panorpoid orders (Orders 24–28)
Order 27. LEPIDOPTERA
Order 28. TRICHOPTERA
Order 29. HYMENOPTERA

The Apterygote orders are a rather diverse assemblage which seem to represent more than one evolutionary line and which should probably not be grouped together as a single subclass, the Apterygota, as was done in older classifications. Manton (1964, 1972, 1973) adopts an extreme position in believing that all four Apterygote orders and the Pterygotes evolved their hexapod condition independently from more primitive Myriapod-like stock, thus supporting those entomologists who, since the time of Handlirsch (1908), have treated them as separate subclasses or classes. Certainly the Thysanura, Diplura, Protura and Collembola are very different from one another, but the Lepismatidae resemble the Pterygota in several apparently fundamental respects (p. 439) and the Diplura, Protura and Collembola all have entognathous mouthparts. Manton also regards entognathy as a convergent feature but Hennig (1969), Tuxen (1959, 1970), Lauterbach (1972) and Kristensen (1975) unite the three orders into one supposedly monophyletic group, the Entognatha, which they contrast with the Ectognatha (= Thysanura + Pterygota). Associated with these differences of opinion are variations in the use of the names Insecta and Hexapoda and in the taxonomic status accorded to the groups so named. Among recent authors, Kristensen (1975) restricts the name Insecta to the ectognathous orders, treating them and the Entognatha as the two components of a larger monophyletic unit, the Hexapoda. Beier (1969) retains the comprehensive class Insecta with subclasses Entognatha, Ectognatha (= Thysanura) and Pterygota. Manton (1973) refers to the Hexapoda as a subphylum of her Uniramia (p. 5), but presumably regards it as a polyphyletic group. To all these discussions on the mutual relationships of the Apterygote insects, palaeontology has contributed little. The earliest known insect, the Devonian *Rhyniella praecursor*, seems to be a normal Collembolan and therefore does little to elucidate their relationships (Scourfield, 1940). Two species of another Apterygote, *Dasyleptus*, from the Upper Carboniferous and Lower Permian, are also somewhat uninformative; though treated as the sole representatives of a distinct order, the Monura, by Sharov (1957), they seem to be not far removed from the Machilidae (Hennig, 1969), as is also the Upper Triassic *Triassomachilis uralensis* (Sharov, 1948). Whether the Thysanura (*sens. lat.*) should be divided into two groups, each of ordinal status, is discussed on p. 439.

Of the origin of the winged insects, nothing certain is known. The

Devonian *Eopterum* and *Eopteridium*, formerly regarded as the oldest Pterygote fossils, are now known to be Crustacean remains (Rohdendorf, 1972). In the lower part of the Upper Carboniferous, however, there occur a few fossils belonging to the Odonata (*Erasipteron larischi*), the Auchenorrhynchan Homoptera (*Protoprosbole straeleni*) and the extinct order Palaeodictyoptera (Kukalova, 1969–70), as well as some others of uncertain position such as *Ampeliptera* and the Miomoptera. The Pterygotes had thus already achieved appreciable diversity by an early geological stage and the presence of characteristic Pterygote features in the mouthparts, tentorium, ovipositor, tracheal system and embryonic membranes of the Lepismatidae suggests strongly that they and the Pterygotes had a common ancestry (Kristensen, 1975). This is the basis for recognizing a large monophyletic taxon composed of all or some of the Thysanura plus the Pterygotes. Martynov (1925, 1938) has emphasized that the Ephemeroptera and Odonata, unlike other Recent winged insects, are unable to flex their wings over the abdomen into a position of repose. They also have an atypical wing articulation, and they retain an anterior median vein and the primitive alternation of concave and convex veins. Martynov formalized these distinctions by placing the Ephemeroptera and Odonata in one taxonomic section, the Palaeoptera, with all other Recent Pterygote orders forming a second, much larger section, the Neoptera. An inability to flex the wings seems also to have characterized the extinct Palaeozoic orders Palaeodictyoptera, Protephemeroptera, Archodonata, Protodonata (Carpenter, 1961), Megasecoptera (Kukalova–Peck, 1974) and Campylopterodea. The mayflies and dragonflies are thus the survivors of a large and varied Palaeozoic fauna, but though the two orders share the Palaeopteran features mentioned above they differ considerably in many other respects. Indeed, Kristensen (1975) prefers to regard the Odonata + Neoptera as a monophyletic group, to be contrasted with the Ephemeroptera.

Among the Neopteran insects one finds Carboniferous representatives of an Orthopteroid group of orders, sometimes collectively known as the Polyneoptera or Paurometabola. As conceived by Sharov (1968), this complex contains four purely fossil orders, the Protoblattoidea (Carboniferous to Jurassic), the Protorthoptera (Carboniferous), the Protelytroptera (Permian; see Carpenter and Kukalova, 1964; Kukalova, 1966) and the Titanoptera (Triassic). The remaining nine Orthopteroid orders are the extant Plecoptera, Grylloblattodea, Orthoptera, Phasmida, Dermaptera, Embioptera, Dictyoptera, Isoptera and Zoraptera. Viewed morphologically they are characterized by, or easily derivable from forms characterized by: (i) unmodified mandibulate mouthparts; (ii) presence of a large anal lobe in the hind wing; (iii) presence of cerci; (iv) presence of numerous Malpighian tubules and (v) presence of several separate ganglia in the ventral nerve-cord.

While the Orthopteroid orders almost certainly make up a monophyletic group, their mutual relationships have proved difficult to unravel

(Kristensen, 1975; Carpenter, 1966). The Dictyoptera include the cockroaches (Blattaria) which probably had a common ancestry with the Protoblattoidea and though the latter died out in the Upper Carboniferous, the rich Palaeozoic Blattoid fauna gave rise to modern descendants. The anatomy of Recent Mantids shows that they are undoubtedly close relatives of the cockroaches, but as the earliest indubitable fossil Mantids date only to the early Tertiary the details of their origin remain obscure. Comparative anatomy also makes it very probable that the Isoptera were originally little more than cockroaches which adopted a peculiar mode of social organization but the fact that no Isopteran fossils are known before the Cretaceous Hodotermitid *Cretatermes carpenteri* (Emerson, 1967) makes their precise origin uncertain. Tillyard (1937) has shown, however, that the mode of folding of the hind wing of *Mastotermes* resembles that of Palaeozoic cockroaches rather than Recent ones (see also McKittrick, 1965). There is some justification for including the cockroaches, termites and mantids all in a single taxon, though whether this should be ordinal or supraordinal is not agreed.

The saltatorial Orthoptera can be traced back to the fossil Sthenaropodidae (Lower Permian) and Oedischiidae (Carboniferous), but their relationship with the Protorthoptera is not well established (Carpenter, 1966). Also unclear are the affinities of the Phasmida; they have a poor fossil record and *Timema*, perhaps their most primitive living member, needs further study. Sharov (1968) relates them and the Titanoptera to a family of Permian Orthoptera, the Tcholmanvisiidae. The Grylloblattodea combine primitive and specialized features; they have been regarded as the remnants of a Protorthopteran stock (Zeuner, 1939) and as close to the ancestor of both the Blattoid and Orthopteran lines of descent, but neither hypothesis is well supported. There are many resemblances between the Grylloblattodea and the Dermaptera (Giles, 1963); the suggestion of a close phylogenetic relationship between the two orders has been challenged (Kristensen, 1975) but it is perhaps more convincing than various alternative theories which ally the Dermaptera with the cockroaches, phasmids, Plecoptera, Embioptera, or even the beetles. The Plecoptera and Embioptera are, in fact, two very isolated Orthopteroid orders despite the repeated suggestions that they are quite closely related to each other (Zeuner, 1936, and others). The Plecoptera have been associated with the fossil Paraplecoptera (see, e.g., Illies, 1965); their uncertain affinities with other Recent orders has led to a variety of proposals for the taxonomic subdivision of the Orthopteroid complex.

The remaining Exopterygote orders alive today form, on anatomical grounds, what might be called an Hemipteroid group (Königsmann, 1960), also known as the Paraneoptera. They may be defined as those which (i) possess specialized mandibulate or suctorial mouthparts; (ii) lack a large anal lobe in the hind wing; (iii) lack cerci; (iv) possess only a few Malpighian tubules and (v) show a more or less highly concentrated group of ganglia in the ventral nervous system. The distinction between the Hemipteroid and Orthopteroid groups is not sharp because the Zoraptera have a reduced wing-venation, few Malpighian tubules and a somewhat concentrated nervous system. The Zoraptera have therefore been considered a primitive Hemipteroid order though the specializations just mentioned may have been acquired convergently. Both anatomically and on palaeontological grounds the Psocoptera may be regarded as a generalized Hemipteroid stock, appearing first in the Lower Permian. Directly connected with them, though fossil lice are lacking, are the Mallophaga, which share with the Psocoptera a unique type of hypopharynx. The Siphunculata, in turn, are probably closely related to some Mallophaga, which they resemble not only in many features of external and internal anatomy and in habits, but also in spiracular structure and in the mode of hatching from the egg. Indeed, it may be more satisfactory to group the biting and sucking lice together in a single order, the Phthiraptera, comprising three main sections (Königsmann, 1960; Clay, 1970; and see p. 664). The Protoprosbolidae from the Upper Carboniferous are probably to be regarded as the earliest known Hemiptera. In the Lower Permian other Homopteran wings similar to the Permian Psocoptera have been found, as well as *Paraknightia*, which seems to be the earliest Heteropteran (Evans, 1963, 1964). Many insect classifications assign the Homoptera and Heteroptera to separate orders but there is no doubt that they together form a monophyletic group, though it is one whose further natural classification presents problems (Hennig, 1969; Kristensen, 1975; and see p. 702). The affinities of the remaining Hemipteroid order, the Thysanoptera, are obscure, though alleged representatives are known from the Permian.

Despite the isolation of the Coleoptera, Strepsiptera and Hymenoptera, there is little doubt that the Endopterygote insects are a monophyletic group. Its origin is not known and the claim for ancestors among the fossils assigned to the Protoperlaria is not very strong (Adams, 1958). Apart from the three orders mentioned above, the Endopterygotes have been regarded since the classical work of Tillyard (1918–20, 1935) as forming a Panorpoid complex of orders. This is centred on the Mecoptera with the Neuroptera forming a somewhat distinct branch that is often separated from the Panorpoid orders *sensu stricto* (Hinton, 1958; Mickoleit, 1969). Among the Neuroptera there is little doubt that the Megaloptera (Lower Permian-Recent) include some of the most primitive Endopterygotes; the interrelationships of the Sialoidea, Raphidioidea and Plannipennia are not altogether clear (Achtelig and Kristensen, 1973) but there seems no need at present to give each of these groups ordinal status. The Mecoptera are known first from the Lower Permian, where members of the suborder Protomecoptera are found. The Boreidae are a distinctive family which Hinton (1958) places in an order of its own, the Neomecoptera, though this has not yet been widely accepted. From early Panorpoid stock there probably arose on the one hand the Diptera and Siphonaptera, and on the other hand the Trichoptera and Lepidoptera. The position of the Micropterigidae is interesting in this connection since although traditionally classed with the Lepidoptera, its members are actually more primitive than any other known Lepidoptera or Trichoptera and Hinton (1946; 1958) has therefore urged that it be given the rank of a distinct order (Zeugloptera). There has been considerable discussion of the phylogenetic relationships of the lower Lepidoptera (e.g. Friese, 1970; Niculescu, 1970; Kristensen, 1971; Common, 1975) but several adult features favour the retention of the Micropterigidae in the Lepidoptera. The Siphonaptera are very distinct in their imaginal structure but the larvae are not unlike those of some Nematoceran Diptera (Mycetophilidae) and it is likely that if not of early Dipteran origin they are at least derived from a Panorpoid stock.

With the Hymenoptera, Coleoptera and Strepsiptera one reaches unsolved phylogenetic problems, to which palaeontology has contributed very little since the few Mesozoic Symphytan wings are apparently Xyelids (Riek, 1955) and therefore already relatively specialized while the fossil remains of the earliest (Lower Permian) beetles are mostly fragmentary elytra, impossible to relate to more generalized orders. Handlirsch attempted to derive the Hymenoptera from Protorthopteran stock, but his arguments are not convincing and the similarity of Symphytan larvae to those of Panorpoid insects, together with the fact that the wing-venation of the Symphyta can, without great difficulty, be derived from a Megalopteran pattern (Ross, 1936), inclines many to regard the Hymenoptera as having had a common ancestry with the Neuroptera and other Panorpoid orders. The Coleoptera have also been thought to have arisen independently of the other Endopterygote insects (either from a Protoblattoid-like stock or some earlier group) but there is little real evidence for this and most authorities now favour a derivation from Neuropteran-like ancestors (Crowson, 1960; Mickoleit, 1973). The Strepsiptera (with no pre-Tertiary fossils) are generally considered to be related to the Coleoptera though this is debatable (Crowson, 1968; Kinzelbach, 1971).

Literature on Classification and Phylogeny

ACHTELIG, M. AND KRISTENSEN, N. P. (1973), A re-examination of the relationships of the Raphidioptera (Insecta), *Z. zool. Syst. Evolutionsforsch.*, 11, 268–274.

ADAMS, P. A. (1958), The relationship of the Protoperlaria and the Endopterygota, *Psyche*, 65, 115–127.

BEIER, M. (1969), Klassifikation, In: Helmcke, J. G., Starck D. and Wermuth, H. (eds), *Handbuch der Zoologie*, 4 (1), Lfg. 9, 1–17.

BÖRNER, C. (1904), Zur Systematik der Hexapoden, *Zool. Anz.*, 27, 511–533.

BRAUER, F. (1885), Systematisch-zoologische Studien, *S. B. Akad. Wiss. Wien*, 91, 237–413.

BRUES, C. T., MELANDER, A. L. AND CARPENTER, F. M. (1954), *Classification of Insects*, Harvard Univ. Press, Cambridge, Mass., 2nd edn, 917 pp.

CARPENTER, F. M. (1961), Studies of North American Carboniferous insects, 1. The Protodonata, *Psyche*, **67** (1960), 98–110.

—— (1962–64), Studies on Carboniferous insects of Commentry, France. III, IV, VI, *Psyche*, **68**, 145–153; **70**, 120–128; **71**, 104–116.

—— (1966), The lower Permian insects of Kansas. Part 11. The orders Protorthoptera and Orthoptera, *Psyche*, **73**, 46–88.

CARPENTER, F. M. AND KUKALOVA, J. (1964), The structure of the Protelytroptera, with description of a new genus from Permian strata of Moravia, *Psyche*, **71**, 183–197.

CARPENTER, F. M. AND RICHARDSON, E. S. (1968), Megasecopterous nymphs in Pennsylvanian concretions from Illinois, *Psyche*, **75**, 295–309.

—— (1971), Additional insects in Pennsylvanian concretions from Illinois, *Psyche*, **78**, 267–295.

CLAY, T. (1970), The Amblycera (Phthiraptera: Insecta), *Bull. Br. Mus. nat. Hist.* (*Ent.*), **25**, 73–98.

COMMON, I. F. B. (1975), Evolution and classification of the Lepidoptera, *A. Rev. Ent.*, **20**, 183–203.

CROWSÓN, R. A. (1960), The phylogeny of Coleoptera, *A. Rev. Ent.*, **5**, 111–134.

—— (1968), *A Natural Classification of the families of Coleoptera*, Nathaniel Lloyd, London, 195 pp.

CROWSON, R. A., ROLFE, W. D. I., SMART, J., WATERSTON, C. D., WILLEY, E. C. AND WOOTTON, R. J. (1967), Arthropoda: Chelicerata, Pycnogonida, Palaeoisopus, Myriapoda and Insecta, In: Harland, W. B. *et al.*, (eds), *The Fossil Record*, London, pp. 499–534.

EMERSON, A. E. (1967), Cretaceous insects from Labrador. 3. A new genus and species of termite (Isoptera: Hodotermitidae), *Psyche*, **74**, 276–289.

EVANS, J. W. (1963), The phylogeny of the Homoptera, *A. Rev. Ent.*, **8**, 77–94.

—— (1964), The periods of origin and diversification of the superfamilies of the Homoptera-Auchenorrhyncha as determined by a study of the wings of Palaeozoic and Mesozoic fossils, *Proc. Linn. Soc. Lond.*, **175**, 171–181.

FRIESE, G. (1970), Zur Phylogenie der älteren Teilgruppen der Lepidopteren, *Ber. 10. Wanderversamml. dt. Ent.*, 203–222.

GILES, E. T. (1963), The comparative external morphology and the affinities of the Dermaptera, *Trans. R. ent. Soc. Lond.*, **115**, 95–164.

HANDLIRSCH, A. (1908), *Die fossilen Insekten und die Phylogenie der rezenten Formen*, Engelmann, Leipzig, 1430 pp.

—— (1926), Vierter Unterstamm des Stammes der Arthropoda. Insecta = Insekten, In: Kükenthal, W. and Krumbach, T. (eds), *Handbuch der Zoologie*, **4**, 403–592.

—— (1937–39), Neue Untersuchungen über die fossilen Insekten mit Ergänzungen und Nachträgen sowie Ausblicken auf phylogenetische, paläogeographische und allgemein biologische Probleme. I, II, *Annln naturh. Mus., Wien*, **48**, 1–140; **49**, 1–240.

HENNIG, W. (1935), Kritische Bemerkungen zum phylogenetischen System der Insekten, *Beitr. Ent.*, **3**, 1–85.

—— (1969), *Die Stammesgeschichte der Insekten*, Kramer, Frankfurt a. M., 436 pp.

HINTON, H. E. (1946), On the homology and nomenclature of the setae of Lepidopterous larvae, with some notes on the phylogeny of the Lepidoptera, *Trans. R. ent. Soc. Lond.*, **97**, 1–37.

—— (1958), The phylogeny of the Panorpoid orders, *A. Rev. Ent.*, **3**, 181–206.

ILLIES, I. (1965), Phylogeny and zoogeography of the Plecoptera, *A. Rev. Ent.*, **10**, 117–140.

JEANNEL, R. (1949), Classification et phylogénie des insectes, In: Grassé, P. P. (ed.), *Traité de Zoologie*, **9**, 1–110.

KINZELBACH, R. K. (1971), Morphologische Befunde an Fächerflüglern und ihre phylogenetische Bedeutung (Insecta: Strepsiptera), *Zoologica, Stuttgart*, **41**, (119/126), 1–256.

KÖNIGSMANN, E. (1960), Zur Phylogenie der Parametabola, unter besonderer Berücksichtigung der Phthiraptera, *Beitr. Ent.*, **10**, 705–744.

KRISTENSEN, N. P. (1971), The systematic position of the Zeugloptera in the light of recent anatomical investigations, *Proc. 13th int. Congr. Ent., Moscow*, **1**, 261.

—— (1975), The phylogeny of hexapod 'orders'. A critical review of recent accounts, *Z. zool. Syst. Evolutionsforsch.*, **13**, 1–44.

KUKALOVA, J. (1966), Protelytroptera from the Upper Permian of Australia, with a discussion of the Protocoleoptera and Paracoleoptera, *Psyche*, **73**, 89–111.

—— (1968), Permian mayfly nymphs, *Psyche*, **75**, 310–327.

—— (1969–70), Revisional study of the order Palaeodictyoptera in the Upper Carboniferous shales of Commentry, France. I–III, *Psyche*, **76**, 163–215; **76**, 439–486; **77**, 1–44.

KUKALOVA-PECK, J. (1973), Unusual structures in the Palaeozoic insect orders Megasecoptera and Palaeodictyoptera, with a description of a new family, *Psyche*, **79**, 243–268.

—— (1974), Wing-folding in the Palaeozoic order Diaphanopterodea, with a description of new representatives of Elmoidae (Insecta, Paleoptera), *Psyche*, **81**, 315–333.

—— (1975), Megasecoptera from the lower Permian of Moravia, *Psyche*, **82**, 1–79.

LAURENTIAUX, D. (1953), Classe des insectes, In: Piveteau, J. (ed.), *Traité de Paléontologie*, **3**, 397–527.

LAUTERBACH, K.-E. (1972), Die morphologischen Grundlagen für die Entstehung der Entognathie bei den apterygoten Insekten in phylogenetischer Sicht, *Zool. Beitr.*, **18**, 25–69.

MACKERRAS, I. M. (ed.) (1970), *The Insects of Australia*, Melbourne Univ. Press, Melbourne, 1029 pp.

MANTON, S. M. (1964), Mandibular mechanisms and the evolution of arthropods, *Phil. Trans. R. Soc. Ser. B*, **427**, 1–183.

—— (1972), The evolution of arthropodan locomotory mechanisms. Part 10. Locomotory habits, morphology and evolution of the hexapod classes, *J. Linn. Soc. (Zool.)*, **51**, 203–400.

—— (1973), Arthropod phylogeny – a modern synthesis, *J. Zool., Lond.*, **171**, 111–130.

MARTYNOV, A. V. (1925), Ueber zwei Grundtypen der Flügel bei den Insekten und ihre Evolution, *Z. Morph. Ökol. Tiere*, **4**, 465–501.

—— (1938), Études sur l'histoire géologique et de phylogénie des ordres des insectes (Pterygota). 1e partie. Palaeoptera et Neoptera-Polyneoptera, *Trav. Inst. paléont. Acad. Sci. U.R.S.S.*, **7** (4), 150 pp.

MARTYNOVA, O. (1961), Palaeoentomology, *Ann. Rev. Ent.*, **6**, 285–294.

MCKITTRICK, F. A. (1965), A contribution to the understanding of cockroach-termite affinities, *Ann. ent. Soc. Am.*, **58**, 18–22.

MICKOLEIT, G. (1969), Vergleichend-anatomische Untersuchungen an der pterothorakalen Pleurotergalmuskulatur der Neuropteria und Mecopteria (Insecta, Holometabola), *Z. Morph. Tiere*, **64**, 151–178.

—— (1973), Über den Ovipositor der Neuropteroidea und Coleoptera und seine phylogenetische Bedeutung (Insecta, Holometabola), *Z. Morph. Ökol. Tiere*, **74**, 37–64.

NICULESCU, E. V. (1970), Aperçu critique sur la systematique et la phylogénie des Lépidoptères, *Bull. Soc. ent. Mulhouse*, **1970**, 1–16.

RIEK, E. F. (1955), Fossil insects from the Triassic beds at Mt. Crosby, Queensland, *Aust. J. Zool.*, **3**, 654–691.

ROHDENDORF, B. B. (ed.) (1961), Palaeozoic insects of the Kusnetzk Basin, *Trav. Inst. paléont. Acad. Sci. U.R.S.S.*, **85**, 1–705.

—— (ed.) (1962), *Principles of Palaeontology: Arthropoda, Tracheata & Chelicerata*, Acad. Sci. U.S.S.R., Moscow, 560 pp.

—— (1969*a*), Phylogenie, In: Helmcke, J.-G., Starck, D. and Wermuth, H. (eds), *Handbuch der Zoologie*, **4** (1), Lfg 9, 1–28.

—— (1969*b*), Paläontologie, In: Helmcke, J.-G., Starck, D. and Wermuth, H. (eds), *Handbuch der Zoologie*, **4** (2), Lfg 9, 1–27.

—— (1972), Devonian Eopterids were not insects but eumalacostracans (Crustacea), *Ent. Obozr.*, **51**, 96–97.

ROSS, H. H. (1936), The ancestry and wing venation of the Hymenoptera, *Ann. ent. Soc. Am.*, **29**, 99–111.

SCOURFIELD, D. J. (1940), The oldest known fossil insect (*Rhyniella praecursor*, Hirst and Maulik) – further details from additional specimens, *Proc. Linn. Soc. Lond.*, **152**, 113–131.

SHAROV, A. G. (1948), Triassic Thysanura from the Ural foreland, *C. r. Acad. Sci. U.R.S.S. (N.S.)*, **61**, 517–519.

—— (1957), Peculiar Palaeozoic wingless insects belonging to a new order Monura (Insecta, Apterygota), *Dokl. Akad. Nauk. S.S.S.R.*, **115**, 795–798.

—— (1968), The phylogeny of the Orthopteroidea, *Trud. paleont. Inst. Akad. Nauk. U.S.S.R.*, **118**, 1–213. [English translation, 1971.]

SHARP, D. (1899), Some points in the classification of the Insecta Hexapoda, *Congr. int. Zool.*, **4**, 246–249.

SMART, J. AND HUGHES, N. F. (1972), The insect and the plant: progressive palaeoecological integration, In: Van Emden, H. F. (ed.), *Insect/Plant Relationships*, pp. 143–155. Symposium No. 6 (1972), London, Royal Entomological Society of London.

TILLYARD, R. J. (1918–20), The Panorpoid complex. A study of the phylogeny of the Holometabolous insects with special reference to the sub-classes Panorpoidea and Neuropteroidea, *Proc. Linn. Soc. N.S.W.*, **43**, 265–284, 395–408, 626–657; **44**, 533–718; **45**, 214–217.

—— (1935), The evolution of the scorpion flies and their derivatives, *Ann. ent. Soc. Am.*, **28**, 1–45.

—— (1937), Kansas Permian insects. Part 20. The Cockroaches or order Blattaria. Pt. II, *Am. J. Sci.*, **34**, 249–276.

TUXEN, S. L. (1959), The phylogenetic significance of entognathy in entognathous apterygotes, *Smithson. misc. Collns*, **137**, 379–416.

—— (1970), The systematic position of entognathous apterygotes, *Anals Esc. nac. Cienc. biol. Méx.*, **17**, 65–79.

WEBER, H. (1933), *Lehrbuch der Entomologie*, Fischer, Jena, 726 pp.

WILLE, A. (1960), The phylogeny and relationships between the insect orders, *Rev. Biol. trop.*, 8 (1), 93–123.

WILSON, H. F. AND DONER, M. H. (1937), *The Historical Development of Insect Classification*, Univ. Wisconsin, Wisconsin, 133 pp.

ZEUNER, F. E. (1936), Das erste Protoperlar aus europäischem Perm und die Abstammung der Embien, *Jahrb. preuss. geol. Landesanst.*, 56, 266–273.

—— (1939), *Fossil Orthoptera Ensifera*, Brit. Mus. (Nat. Hist.), London, 2 vols. 321 pp.

Order 1

THYSANURA (BRISTLE-TAILS; SILVERFISH)

Apterygota with ectognathous mouthparts, adapted for biting. Antennae many-segmented, but only the basal segment provided with intrinsic muscles. Compound eyes present or absent. Tarsi with 2–5, commonly 3, segments. Abdomen 11-segmented, with a variable number of lateral, styliform, pregenital appendages, a pair of many-segmented cerci, and ending in a segmented median process. Tracheal system and Malpighian tubules present. Metamorphosis slight or wanting.

This order includes some of the most primitive insects and is very widely distributed; about 9 species have been found in the British Isles out of a total of over 550. Its members live a concealed life in the soil, in rotting wood, under stones, or in the leaf-deposits of forest floors; a considerable number occur in the nests of ants and termites. Unlike many Collembola they are not usually found among living herbage. The 'silverfish', *Lepisma saccharina* (Fig. 204) and *Ctenolepisma lineata* and *C. longicaudata* occur in buildings in many parts of the world, where they are destructive to paper, book-bindings, etc., while *Thermobia domestica* frequents warm buildings. *Petrobius brevistylis* and *P. maritimus* (Fig. 205) inhabit rocky coasts, close to the sea. Although the order includes a number of minute species, the majority are larger than the Collembola, though they do not exceed 2 cm in length. Most species are brownish, grey or white, and the scaled forms exhibit a metallic sheen.

External Anatomy – The body is more or less spindle-shaped in outline, depressed in the Lepismatoidea and somewhat compressed in the Machiloidea. It is clothed with scales except in some Meinertellidae and Nicoletiidae. The antennae are long and filiform, often consisting of 30 or more segments, the absence of muscles in the flagellar segments distinguishing them from those of the Diplura (Imms, 1939; Slifer and Sekhon, 1970). Compound eyes are well developed in the Machiloidea where they are approximated or contiguous dorsally, but in the Lepismatoidea they are considerably reduced (Brandenburg, 1960; Elofsson, 1970). The latter group also usually lacks ocelli but in the Machilidae median and paired ocelli occur. They are variable in form with the retinal cells in small groups

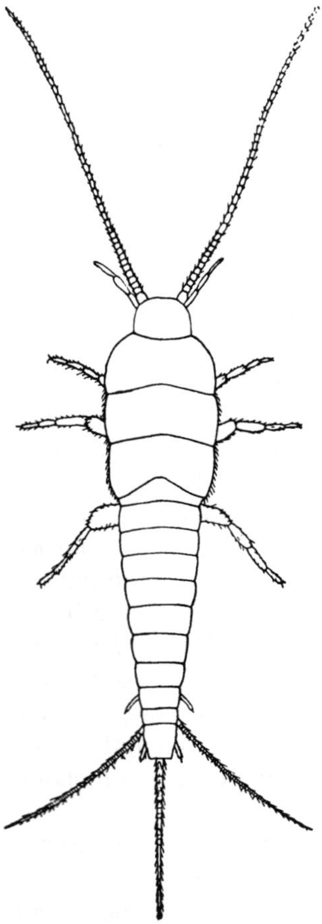

FIG. 204
Lepisma saccharina (magnified). Britain
After Lubbock.

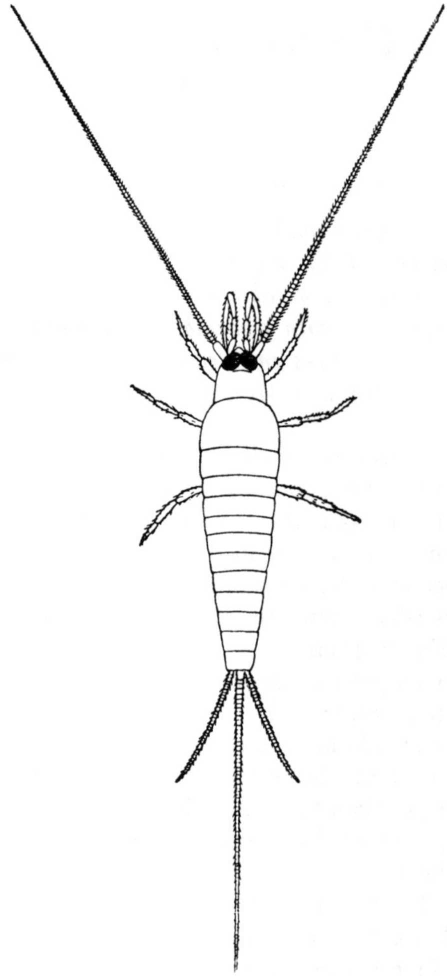

FIG. 205
Petrobius maritimus (magnified). Britain
After Lubbock.

surrounding a rhabdom-like structure (Hanström, 1940; Marlier, 1941). The head (Chaudonneret, 1951; Bitsch, 1963) often exhibits the 'epicranial suture' and both the labrum and clypeus are well developed. The mouthparts (O'Harra and Adams, 1942; Chaudonneret, 1949) are normal and exserted and in *Petrobius* and other Machilids they are relatively generalized structures from which the mouthparts of all the other Apterygote orders and the Pterygotes might well have evolved (Manton, 1964). The mandibles are long and pointed organs, with a transverse strengthening ridge and a well-defined projecting molar area. The superlinguae are exceptionally well developed;

each organ is attached by membrane to the base of the hypopharynx, and exhibits differentiation into two lobes together with a small palp-like process. The maxillae are composed of the typical sclerites and their palps are 7-segmented. In the labium the mentum and submentum are broad plates, the prementum is paired, and the palps are 3-segmented. Paired glossae and paraglossae are present and the latter are longitudinally subdivided into three lobes. In *Lepisma* the mandibles are each provided with two cephalic articulations instead of the one found in Machilidae and in the labium the glossae and paraglossae are single organs on either side. The cephalic endoskeleton comprises anterior and posterior tentorial arms which, in some Lepismatids, are united into a structure essentially similar to that of the Pterygote insects (Snodgrass, 1952).

The thorax (Barlet, 1951–54, 1967; Manton, 1972) consists of three well-defined segments. The prothorax is somewhat narrower than the other two in Machilids and in some Lepismatoidea the terga are produced laterally into paranotal lobes, tracheated in a way that recalls a small Pterygote wing-pad (p. 67). Functionally, the skeletomuscular system of the Thysanuran thorax differs from that of the Pterygotes and the other Apterygote orders and may perhaps have evolved independently (Manton, 1972).

The legs have 3 tarsal segments in the Machiloidea and 2–5 in the Lepismatoidea; paired pretarsal claws are always present. In some Machilidae (Fig. 206) the coxae of the 2nd and 3rd pairs of legs each bear a small, movable, unsegmented style, but in other members of this family they are absent or occur only on the posterior pair of legs.

The abdomen is composed of 11 segments (Bitsch, 1973–74; Rousset, 1973; Birket-Smith, 1974). The 10th segment is reduced and bears no appendages while the 11th is also small but carries the cerci and its tergum is prolonged into the median cerciform appendage. The abdominal sterna (Fig. 206) are exhibited best on the pregenital segments (1–7 inclusive) where each may be divided transversely, as in *Nicoletia*, or be composed of a triangular sternum with, in some cases, a pair of laterosternites. Typically, each segment possesses a pair of laterally placed appendages made up of a basal, plate-like coxite and a small terminal style. These appendages are probably serial homologues of the thoracic legs but the extent to which they are developed varies considerably in different genera. In the Machilidae coxites and styles are present on segments 2–9 inclusive but styles do not occur on the 1st segment. In the Lepismatoidea styles are found on segments 2–9 in *Nicoletia* but other genera have fewer, *Lepisma* and its allies rarely having more than 3 pairs (segments 7–9). In some cases the coxites are fused with the sternal plate to form a compound coxosternum. Each appendage of the pregenital segments may be provided medially with one or two *eversible vesicles* (Fig. 206). The latter can be extended by blood-pressure and retracted by special muscles: their function is uncertain but they may be concerned in respiration or water-uptake. The vesicles are absent from many Lepismatidae but *Nicoletia* has 6 pairs (segments 2–7). They are always present in Machilidae (almost invariably on segments 1–7) and in some genera (e.g. *Machilis* and *Petrobius*) segments 2–5 each carry two pairs.

The 8th and 9th abdominal segments are modified through the development of external *genitalia* (Figs. 45 and 48). In the female, both segments bear a pair of coxites and styles and articulated basally with each coxite is a long, annulated gonapophysis. The four gonapophyses fit together to form the ovipositor; a gonangulum is present in the Lepismatoidea but not in the

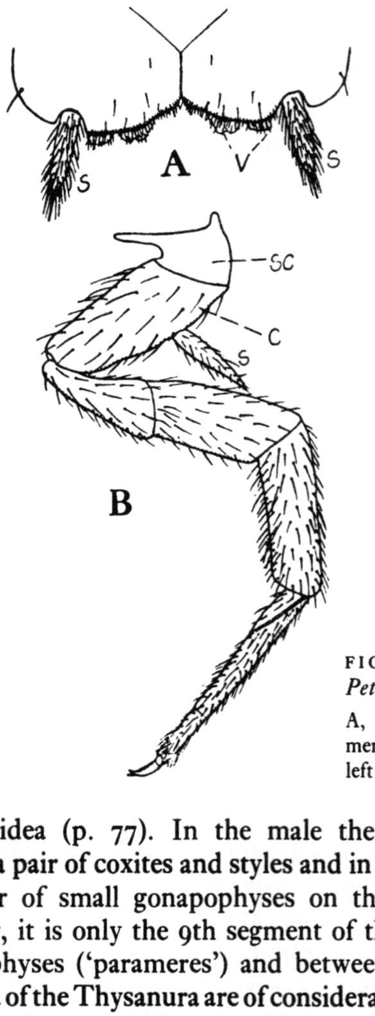

FIG. 206
Petrobius maritimus. Britain
A, hind margin of 5th abdominal segment showing vesicles *v* and styles *s*. B, left leg of 3rd pair; *c*, coxa; *s*, style.

Machiloidea (p. 77). In the male the genital segments likewise usually possess a pair of coxites and styles and in a few Machilids (e.g. *Machilis*) there is a pair of small gonapophyses on the 8th segment. In all other cases, however, it is only the 9th segment of the male which bears a pair of small gonapophyses ('parameres') and between them lies the median penis. The genitalia of the Thysanura are of considerable morphological interest because of the light they are believed to throw on the homologies of these organs in Pterygote insects (Gustafson, 1950; Scudder, 1971).

Internal Anatomy (Barnhart, 1961) – The *alimentary canal* (Fig. 207) is usually a simple straight tube, but in *Lepisma* the hind intestine presents a single convolution. There is a large gizzard in *Lepisma* and in this genus and in *Machilis* enteric caeca are present. Salivary glands are generally present (Philiptschenko,

1907–08) while the Malpighian tubules are well developed and number 12 to 20 in the Machilidae, and 4 to 8 in the Lepismatidae. The *nervous system* (Hilton, 1917; Watson, 1963; Rousset, 1975) is generalized with 3 thoracic and 8 abdominal ganglia, and double longitudinal connectives throughout (Fig. 66A). The *tracheal system* exhibits differences in the two best-known families. In the Machilidae there are 9 pairs of spiracles: the 1st pair is located between the pro- and mesothorax, the 2nd pair is placed near the hinder border of the mesothorax, and the remaining pairs are placed on the 2nd to 8th abdominal segments. The tracheae associated with the abdominal spiracles remain unconnected with

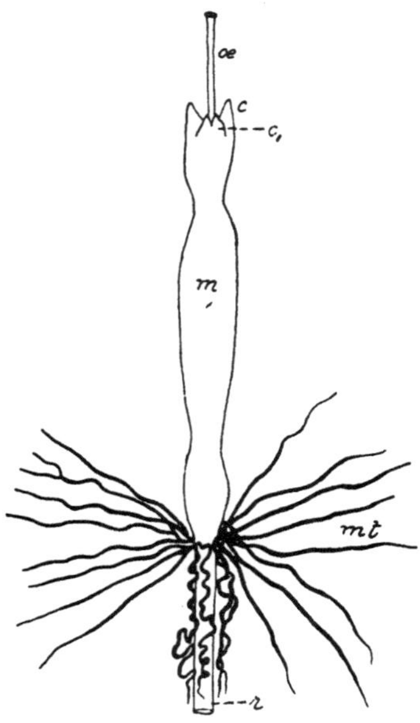

FIG. 207
Alimentary canal of *Petrobius*

oe, oesophagus; *c*, large caeca; c_1, smaller caeca; *m*, mid intestine; *mt*, Malpighian tubules; *r*, rectum. *After* Oudemans.

those of adjacent segments (Stobbart, 1956). In the Lepismatidae there are 10 pairs of spiracles which belong to the 2nd and 3rd thoracic and the first 8 abdominal segments. In this family the tracheal system is relatively highly developed; there is a common longitudinal tracheal trunk passing down either side of the body, and there is a transverse trunk in each segment uniting the tracheae of opposite sides (Šulc, 1927).

The *heart* in *Machilis* (Bär, 1912, Barth, 1963) extends from the 10th abdominal segment into the mesothorax, passing anteriorly into the aorta. There are 11 pairs of dorsally situated ostioles and, in the 8th and 9th segments, two pairs of ventral ones. 11 pairs of alary muscles are said to be present. See also Rousset (1974).

The *reproductive system* (Fig. 208) exhibits some differences in the two main families. In the female, the panoistic ovarioles number 5 on each side in Lepismatidae and 7 in Machilidae. In the latter family the ovarioles join a lateral oviduct one behind the other so as to present a metameric appearance, though a strict segmental disposition does not seem to be preserved. The two oviducts join to form a short vagina which, according to Gustafson (1950), opens behind the 7th abdominal segment in *Lepisma* and *Neomachilis* and behind the 8th in *Nicoletia*. In the Lepismatidae a spermatheca and a pair of accessory glands (unpaired in *Nicoletia*) are present, but the

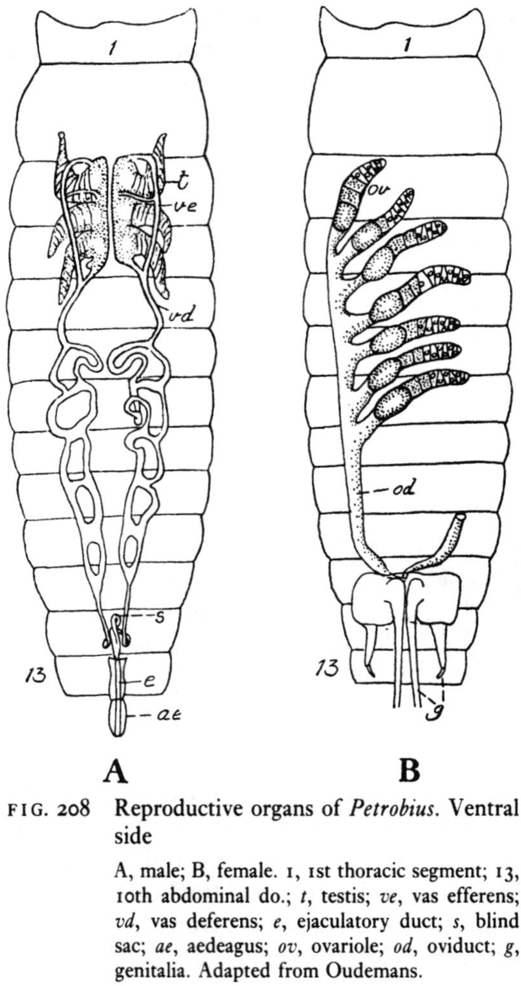

FIG. 208 Reproductive organs of *Petrobius*. Ventral side

A, male; B, female. 1, 1st thoracic segment; 13, 10th abdominal do.; *t*, testis; *ve*, vas efferens; *vd*, vas deferens; *e*, ejaculatory duct; *s*, blind sac; *ae*, aedeagus; *ov*, ovariole; *od*, oviduct; *g*, genitalia. Adapted from Oudemans.

Machilidae do not possess them. In the male the testes comprise a group of lobes, each apparently made up of several follicles. The number of lobes varies considerably (many in *Nicoletia*, 6 in *Lepisma*, 3 in *Petrobius*). The vasa deferentia are more or less convoluted and in *Petrobius* each is double throughout the greater part of its length, the two canals thus formed being united by a series of 5 transverse connecting tubes (Fig. 208, A). According to Gustafson (1950) in *Lepisma* and *Ctenolepisma* the vasa deferentia of opposite sides do not join but extend separately into the penis on which they open by a pair of gonopores. In other cases there is a short median ejaculatory duct. For further anatomical and histological information see Bitsch (1968*a*, *b*, *c*), Barth (1962), Torgerson and Akre (1969) and Wygodzinsky (1959).

Biology and Postembryonic Development – Despite their well-developed external genitalia the Thysanura do not copulate but transfer sperm indirectly after a more or less complicated courtship. The male spins a thread or threads with the aedeagus and deposits sperm droplets on them; the female then encounters the thread and takes up the sperm into her reproductive tract (Sturm, 1955–56). The absence or rarity of males in some species suggests that parthenogenesis may occur (e.g. Janetschek, 1954*b*). The eggs of Thysanura are relatively large, somewhat variable in shape and the number laid in one season does not exceed about 30. The growth and postembryonic development of several species has been summarized by Delany (1957). Hatching takes place with the aid of a cephalic spine and the early instars may lack scales and styles on the coxae and abdomen. External genital rudiments first become apparent at the fourth instar in *Petrobius brevistylis* and the eighth in *Ctenolepisma longicaudata*, and sexual maturity is reached after a further five or six moults. The changes which occur during development are slight and the different instars can sometimes only be recognized biometrically. Moulting continues into the adult stage and the total number of moults may be considerable – from 25 to 66 have been recorded in *Ctenolepisma* and from 19 to 58 in *Thermobia*. For further details of postembryonic development see Delany (1959, 1961), Sweetman (1952), Lindsay (1940), Sahrhage (1953), Bitsch (1964) and Larink (1969). Longevity depends on the species and the environmental conditions, but most investigated species live from one to four years; moulting and growth cease during the winter in temperate zones. At each adult moult the cuticular lining of the spermatheca is lost, together with its contents, so that copulation has to occur in each adult instar in order that fertile eggs may continued to be laid (but see Sahrhage, 1953).

Affinities – The Lepismatoidea and the Pterygota share a number of specialized features which do not occur in the Machiloidea. Hennig (1953, 1969) and Kristensen (1975) have therefore argued that the Thysanura (in the wide sense adopted above) should be divided between two monophyletic groups. One of these is the Machiloidea, to which the ordinal name Microcoryphia (= Archaeognatha) is sometimes applied. The other consists of the Lepismatoidea (order Zygentoma or Thysanura *s.str.*) plus all the Pterygote orders. The specialized features shared by the Lepismatoids and the Pterygotes include: (i) presence of two mandibular articulations; (ii) presence of a gonangulum; (iii) origin of ventral mandibular and stipital adductors on the tentorium; (iv) absence of 'fulturae' in the cephalic endoskeleton; (v) presence of longitudinal and transverse tracheal trunks in the abdomen; and (vi) the closed amniotic cavity of the embryo. While some of these features probably deserve fuller scrutiny, there seems little doubt that the Lepismatoids resemble the Pterygotes far more than do the Machiloids. This important distinction may be adequately recognized for the present by placing the Lepismatoidea and Machiloidea in distinct suborders of the Thysanura.

Classification – The higher classification summarized below is based on Remington (1954). For taxonomic keys or monographs on smaller groups or regional faunas, see especially Delany (1954), Escherich (1904), Janetschek (1954*a*, *b*), Paclt (1963, 1967), Palissa (1964), Womersley (1939) and Wygodzinsky (1941, 1963, 1970, 1972).

Suborder MICROCORYPHIA

Eyes large, contiguous, with many ommatidia; median and paired lateral ocelli present; mandible with one articulation, distinct molar area and long, pointed incisor process; maxillary palp 7-segmented; paraglossae 3-lobed; meso- and metathoracic coxae often with styles; eversible vesicles usually on 2nd to 7th abdominal sterna; no gonangulum.

Superfamily **Machiloidea (Bristletails)**
(With characters of the suborder)

FAM. MACHILIDAE. *Sterna large, triangular; at least one and often two pairs of eversible vesicles on 2nd to 6th abdominal segments; scales present at least on legs and base of antenna.* Mainly a Northern hemisphere group. *Dilta, Petrobius, Machilis* and others.

FAM. MEINERTELLIDAE. *Sterna small; never more than one pair of abdominal vesicles on abdominal segments; antennae and legs unscaled.* Mainly from the Southern hemisphere, e.g. *Allomachilis, Nesomachilis.*

Suborder ZYGENTOMA

Eyes small and separate or absent; ocelli absent; mandibles with anterior and posterior articulations, molar and incisor areas confluent; maxillary palps 5-segmented; paraglossae simple; coxae without styles; eversible vesicles absent or one pair present on some segments.

Superfamily **Lepismatoidea**
(With characters of the suborder)

FAM. LEPIDOTRICHIDAE. *Tricholepidion gertschi*, described by Wygodzinsky (1961) from beneath bark and rotting logs of Douglas Fir in California, is the only living representative of this very primitive family. The body is unscaled, eyes and ocelli are present, the abdominal sterna and coxites are well-developed, with styles present on the 2nd to 9th segments and eversible vesicles on the 2nd to 6th. In some respects, therefore, it provides exceptions to the subordinal characters listed above.

FAM. NICOLETIIDAE. *Eyes and ocelli absent; scales usually absent; male gonapophyses long; subterranean, myrmecophilous or termitophilous species. Nicoletia* is widely distributed. *Atelura formicarius* is a small myrmecophilous form from Central Europe.

FAM. LEPISMATIDAE (SILVERFISH). *Eyes present but no ocelli; scales usually present; male gonapophyses short; rarely myrmecophilous, usually free-living or domestic.* A large family with representatives in all regions. Most of the species live outdoors but a few are closely associated with houses, greenhouses or stores. *Lepisma saccharina* and *Thermobia domestica* are the best known but species of *Ctenolepisma*, *Acrotelsa* and *Peliolepisma* are domestic insects in various parts of the world; their natural distribution has been widely extended by commerce.

Literature on the Thysanura

BÄR, H. (1912), Beiträge zur Kenntnis der Thysanuren, *Jena. Z. Naturwiss.*, **48**, 1–92.

BARLET, J. (1951–54), Morphologie du thorax de *Lepisma saccharina* L. (Aptérygote, Thysanoure). I–III, *Bull. Ann. Soc. ent. Belg.*, **87**, 253–271; **89**, 215–236; **90**, 299–321.

—— (1952), Ressemblances entre le thorax de *Nicoletia* (Thysanoure Lepismatidae) et celui d'autres Aptérygotes, *Bull. Inst. Sci. nat. Belg.*, **28** (54), 8 pp.

—— (1967), Squelette et musculature thoraciques de *Lepismachilis y-signata* Kratochvil (Thysanoures), *Bull. Ann. Soc. r. ent. Belg.*, **103**, 110–157.

BARNHART, C. S. (1961), The internal anatomy of the Silverfish *Ctenolepisma campbelli* and *Lepisma saccharinum* (Thysanura: Lepismatidae), *Ann. ent. Soc. Am.*, **54**, 177–195.

BARTH, R. (1962), Die Drüsen des weiblichen Genitalapparats einer Machilide (Thysanura), *An. Acad. bras. Cienc.*, **34**, 553–562.

—— (1963), Ueber das Zirkulationssystem einer Machilide (Thysanura), *Mem. Inst. Osw. Cruz.*, **61**, 371–400.

BIRKET-SMITH, S. J. R. (1974), On the abdominal morphology of Thysanura (Archaeognatha and Thysanura s. str.), *Ent. Scand., Suppl.*, **6**, 5–67.

BITSCH, J. (1963), Morphologie céphalique des Machilides (Insecta, Thysanura), *Annls Sci. nat.* (*Zool.*) (12) **5**, 501–584.

—— (1964), Observations sur le développement postembryonnaire des Machilides (Insecta, Thysanura), *Trav. Lab. Zool. Stn aquic. Grimaldi*, **54**, 1–17.

—— (1968*a*), Sur les phénomènes de résorption d'oocytes chez les *Machilis* (Insectes, Thysanoures), *Bull. Soc. zool. Fr.*, **93**, 385–395.

—— (1968*b*), Données histologiques sur l'oogenèse des *Machilis* (Insectes, Thysanura), *Annls Sci. nat.* (*Zool.*), (12) **10**, 267–290.

—— (1968*c*), Anatomie et histologie de l'appareil génital mâle du genre *Machilinus* (Thysanura, Meinertellidae), *Bull. Soc. ent. Fr.*, **73**, 100–113.

—— (1973–74), Morphologie abdominale des Machilides (Insecta: Thysanura) I–III, *Annls Sci. nat.* (*Zool.*), **15**, 173–200; *Int. J. Insect Morphol. and Embryol.*, **3**, 101–120; 203–224.

BRANDENBURG, J. (1960), Die Feinstruktur des Seitenauges von *Lepisma saccharina* L., *Zool. Beitr.*, (*N.F.*), **5**, 291–300.

CHAUDONNERET, J. (1949), Le labium des Thysanoures, *Annls Sci. nat.* (*Zool.*), **10**, 1–26.

—— (1951), La morphologie céphalique de *Thermobia domestica* (Packard) (Insecte Aptérygote Thysanoure), *Annls Sci. nat.* (*Zool.*), **12**, 145–302.

DELANY, M. J. (1954), Thysanura and Diplura, *R. ent. Soc. Lond., Handb. Ident. Brit. Ins.*, **1** (2), 7 pp.

DELANY, M. J. (1957), Life histories in the Thysanura, *Acta Zool. Cracoviensia*, **2** (3), 61–90.

—— (1959), The life histories and ecology of two species of *Petrobius* Leach, *P. brevistylis* and *P. maritimus, Trans. Roy. Soc. Edinb.*, **63**, 501–533.

—— (1961), A study of the post-embryonic development of *Machiloides delanyi* Wygodzinsky (Thysanura: Meinertellidae), *Proc. R. ent. Soc. Lond. Ser. A*, **36**, 81–87.

ELOFSSON, R. (1970), Brain and eyes of Zygentoma (Thys., Lepismatidae), *Ent. Scand.*, **1**, 1–20.

ESCHERICH, K. (1904), Das System der Lepismatiden, *Zoologica (Stuttgart)*, **43**, 164 pp.

GUSTAFSON, J. F. (1950), The origin and evolution of the genitalia of the Insecta, *Microentomology*, **15**, 35–67.

HANSTRÖM, B. (1940), Inkretorische Organe, Sinnesorgane und Nervensystem des Kopfes einiger niederer Insektenordnungen, *K. svenska VetenskAkad. Handl.*, **18**, 266 pp.

HENNIG, W. (1953), Kritische Bemerkungen zum phylogenetischen System der Insekten, *Beitr. Ent.*, **3**, 1–85.

—— (1969), *Die Stammesgeschichte der Insekten*, Frankfurt a.M., 436 pp.

HILTON, W. A. (1917), The nervous system of Thysanura, *Ann. ent. Soc. Am.*, **10**, 303–313.

IMMS, A. D. (1939), On the antennal musculature in insects and other Arthropods, *Q. Jl microsc. Sci.*, **81**, 273–320.

JANETSCHEK, H. (1954*a*), Ueber Felsenspringer der Mittelmeerländer (Thysanura, Machilidae), *Eos*, **30**, 163–314.

—— (1954*b*), Ueber mitteleuropäische Felsenspringer (Ins. Thysanura), *Öst. zool. Z.*, **5**, 281–328.

KRISTENSEN, N. P. (1975), The phylogeny of hexapod 'orders'. A critical review of recent accounts, *Z. zool. Syst. Evolutionsforsch.*, **13**, 1–44.

LARINK, O. (1969), Zur Entwicklungsgeschichte von *Petrobius brevistylis* (Thysanura: Insecta), *Helgol. wiss. Meeres.*, **19**, 111–155.

LINDSAY, E. (1940), The biology of the silverfish *Ctenolepisma longicaudata* Esch., with particular reference to its feeding habits, *Proc. R. Soc. Vict. (N.S.)*, **52**, 35–83.

MANTON, S. M. (1964), Mandibular mechanisms and the evolution of arthropods, *Phil. Trans. R. Soc. Ser. B*, **247**, 1–183, 1 pl.

—— (1972), The evolution of arthropodan locomotory mechanisms. Part 10. Locomotory habits, morphology and evolution of the hexapod classes, *Zool. J. Linn. Soc.*, **51**, 203–400.

MARLIER, G. (1941), Recherches sur les organes photorécepteurs des insectes aptilotes, *Annls Soc. r. zool. Belg.*, **72**, 204–236.

O'HARRA, R. AND ADAMS, J. (1942), The mouthparts of the firebrat *Thermobia domestica* (Packard) (Thysanura), *Proc. Iowa Acad. Sci.*, **49**, 507–516.

PACLT, J. (1963), Thysanura fam. Nicoletiidae, *Genera Insect.*, Fasc. **216e**, 58 pp.

—— (1967), Thysanura fam. Lepidotrichidae, Maindroniidae, Lepismatidae, *Genera Insect.*, Fasc. **218e**, 1–86.

PALISSA, A. (1964), Apterygota – Urinsekten, *Tierwelt Mitteleuropas*, **4** (1), 407 pp.

PHILIPTSCHENKO, J. (1907–08), Beiträge zur Kenntnis der Apterygoten, I, II, *Z. wiss. Zool.*, **88**, 99–116; **91**, 93–111.

REMINGTON, C. L. (1954), The suprageneric classification of the order Thysanura (Insecta), *Ann. ent. Soc. Am.*, **47**, 277–286.

ROUSSET, A. (1973), Squelette et musculature des régions génitales et postgénitales de la femelle de *Thermobia domestica* (Packard). Comparaison avec la région génitale de *Nicoletia* sp. (Insecta: Aptérygota: Lepismatida), *Int. J. Insect Morphol. & Embryol.*, **2**, 55–80.

—— (1974), Les différenciations postérieures du vaisseau dorsal de *Thermobia domestica* (Packard), Anatomie et innervation, *C. r. Acad. Sci. Paris*, **278**, 2449–2452.

—— (1975), Innervation sensorielle et motrice des régions abdominales postérieures de la femelle de *Thermobia domestica* (Packard) (Insecta: Apterygota: Lepismatida), *Int. J. Insect Morphol. & Embryol.*, **4**, 61–76.

SAHRHAGE, D. (1953), Ökologische Untersuchungen an *Thermobia domestica* (Packard) und *Lepisma saccharina* L., *Z. wiss. Zool.*, **157**, 77–168.

SCUDDER, G. G. E. (1971), Comparative morphology of insect genitalia, *A. Rev. Ent.*, **16**, 379–406.

SLIFER, E. H. AND SEKHON, S. S. (1970), Sense-organs of a Thysanuran, *Ctenolepisma lineata pilifera*, with special reference to those on the antennal flagellum (Lepismatidae), *J. Morph.*, **132**, 1–26.

SNODGRASS, R. E. (1952), *A Textbook of Arthropod Anatomy*, Ithaca, N.Y., 363 pp.

STOBBART, R. H. (1956), A note on the tracheal system of the Machilidae, *Proc. R. ent. Soc. Lond. Ser. A*, **31**, 34–36.

STURM, H. (1955), Beiträge zur Ethologie einiger mitteldeutscher Machiliden, *Z. Tierpsychol.*, **12**, 337–363.

—— (1956), Die Paarung beim Silberfisch *Lepisma saccharina*, *Z. Tierpsychol.*, **13**, 1–12.

ŠULC, K. (1927), Das Tracheensystem von *Lepisma* (Thysanura) und Phylogenie der Pterygogenea, *Acta Soc. Sci. nat. Moravia*, **4**, 227–344.

SWEETMAN, H. L. (1952), The number of instars among the Thysanura as influenced by environment, *Proc. 9th int. Congr. Ent.*, **1**, 411–414.

TORGERSON, R. L. AND AKRE, R. D. (1969), Reproductive morphology and behavior of a Thysanuran, *Trichatelura manni*, associated with army ants, *Ann. ent. Soc. Am.*, **62**, 1367–1374.

WATSON, J. A. L. (1963), The cephalic endocrine system in the Thysanura, *J. Morph.*, **113**, 359–369.

WOMERSLEY, H. (1939), *Primitive Insects of South Australia. Silverfish, Springtails and their Allies*, Government Printer, Adelaide, 322 pp.

WYGODZINSKY, P. (1941), Beiträge zur Kenntnis der Dipluren und Thysanuren der Schweiz, *Denkschr. Schweiznaturf. Ges.*, **74**, 113–227.

—— (1959), Beobachtungen an Spermatolophiden und Spermatophoren bei Nicoletiidae (Thysanura, Insecta), *Zool. Anz.*, **161** (1958), 280–287.

—— (1961), On a surviving representative of the Lepidotrichidae (Thysanura), *Ann. ent. Soc. Am.*, **54**, 621–627.

—— (1963), On J. Paclt's Nicoletiidae (Thysanura) in the 'Genera Insectorum', *Ann. Mag. nat. Hist.*, (13) **6**, 265–269.

—— (1970), Thysanura associated with termites in southern Africa (Insecta), *Bull. Am. Mus. nat. Hist.*, **142**, 213–254.

—— (1972), A review of the silverfish (Lepismatidae, Thysanura) of the United States and Caribbean area, *Am. Mus. Novit.*, **2481**, 1–26.

Order 2

DIPLURA

Apterygota with entognathous mouthparts. Antennae many-segmented, flagellar segments provided with muscles. Compound eyes and ocelli absent. Tarsi 1-segmented. Abdomen with lateral styliform appendages on most or all of the pregenital segments and ending in paired cerci of variable form. Terminal median filament absent. Tracheal system present; Malpighian tubules vestigial or absent.

Like the Thysanura, with which they were formerly united in a single order, the Diplura are a group of widely distributed insects living in concealed situations under stones, in dead wood, among fallen leaves or in soil (Paclt, 1957). The Campodeidae are well represented in the Holarctic region while the other families occur mainly in the tropics and subtropics of all regions. About 600 species are known, of which 11 species of *Campodea* (Fig. 209) are British (Delany, 1954). The Diplura are usually small insects, the largest forms occurring in the genus *Heterojapyx* (Fig. 210) where *H. soulei*, for example, measures up to 50 mm in length.

External Anatomy – The integument is generally thin and pale and scales occur only in a few Campodeidae (e.g. *Lepidocampa*; see Bareth, 1963*a*). The head-capsule (François, 1970) is oval or quadrangular in outline and is subdivided in some forms by the ecdysial cleavage line and the postoccipital sulcus while the clypeus and labrum are distinct sclerites. The antennae are more or less elongate structures with 20 to 40 or more segments, all except the last being provided with intrinsic muscles (Imms, 1939). The reduced mouthparts (Fig. 211) are partially sunk into the head-capsule and resemble those of the Collembola (Snodgrass, 1952; Manton, 1964). The mandibles are elongate, apically toothed structures with a conical base fitting loosely into a corresponding socket on the head-capsule and in the Campodeidae and Projapygidae a prostheca is present. In the maxilla the cardo is small, the stipes rather elongate and a small 1- or 2-segmented palp may also occur. A lacinia and galea can be recognized, the former often serrated. The labium is sub-divided into pre- and postmentum and bears a pair of small papilla-like palps (absent in *Parajapyx*) and a ligula which is divided into glossae and paraglossae. Lateral to the prementum is a pair of sclerites of uncertain homologies – the admental plates of Silvestri (1933). A well-developed hypopharynx occurs with large superlinguae. As in the

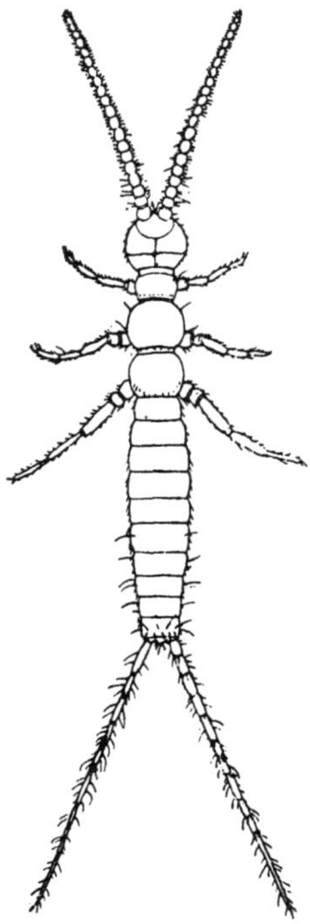

FIG. 209
Campodea (× *circa* 15). Britain
After Lubbock.

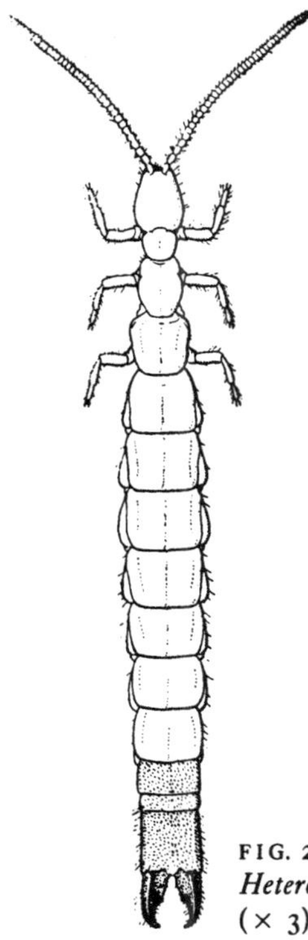

FIG. 210
Heterojapyx sp. Australia (× 3)

Collembola, the cephalic endoskeleton is not fully understood. Manton (1964) regards it as composed of transverse mandibular and maxillary 'tendons' and the posterior tentorial arms, but others believe that the arm-like structures supporting the hypopharynx ('fulturae' or 'intermaxillary brachia') are not really tentorial (Snodgrass, 1952; François, 1970). It is also a matter of controversy whether the entognathous condition has been evolved independently in the Diplura, Collembola and Protura (see p. 32).

The 3 thoracic segments are clearly separated, the prothorax being the smallest (Carpentier and Barlet, 1951; Barlet and Carpentier, 1962). The sterna and sometimes also the terga are subdivided by transverse sutures but there is no general agreement on the homologies of the sclerites so delimited. The pleural sclerites are reduced but indications of the primitive anapleural and coxopleural arcs have been claimed to occur in some genera. Manton (1972), however, considers that the thoracic sclerites of the Diplura evolved independently of those in the three other Apterygote orders and in the Pterygotes and that it is misleading to seek homologies between the various groups. The 3 pairs of legs differ little, the tarsi are 1-segmented and there are usually 2 pretarsal claws, though an additional median claw-like appendage occurs in *Japyx, Anajapyx* and *Lepidocampa*. There is a single sternal articulation between coxa and pleuron.

The abdomen is composed of 10 well-developed segments and a small 11th segment which bears the cerci. The abdominal sterna, at the front of which narrow transverse presternites are sometimes delimited, bear a variable number of lateral styliform appendages. These occur on segments 1–7 in the Japygidae and Projapygidae and on 2–7 in the Campodeidae, the 1st segment in the latter family being provided with a pair of larger, lobe-like appendages which may show sexual differences. Paired eversible vesicles

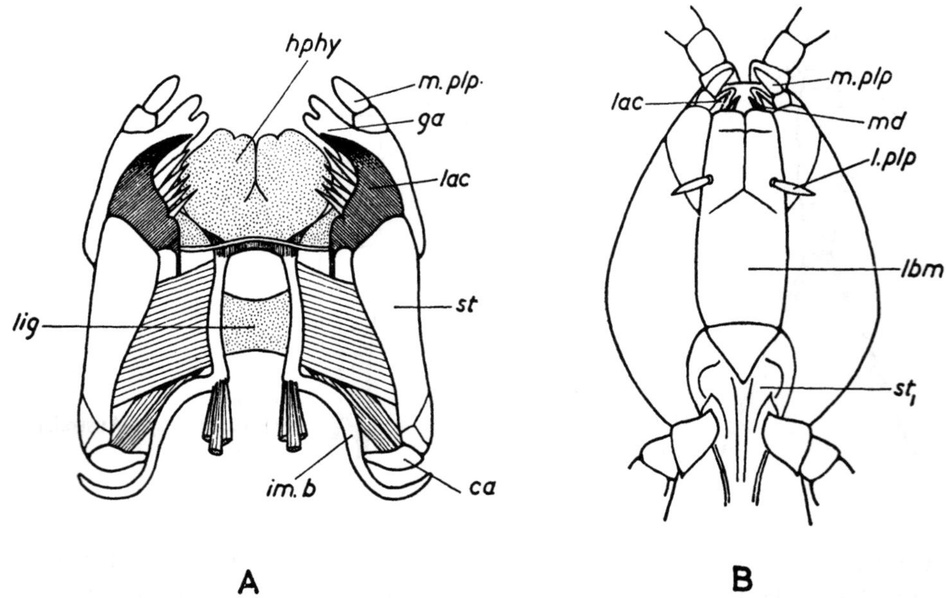

FIG. 211 Mouthparts of *Heterojapyx* (*after* Snodgrass, 1952). A. Maxilla and Hypopharynx. B. Ventral view of head.

ca, cardo; *ga*, galea; *hphy*, hypopharynx; *im.b*, intermaxillary brachium; *lac*, lacinia; *lbm*, labium: *l.plp*, labial palp; *lig*, interbrachial ligament; *md*, mandible; *m.plp*, maxillary palp; st_1, first thoracic sternum; *st*, stipes.

resembling those of the Thysanura are found on the 2nd to 7th abdominal segments of the Campodeidae and Anajapygidae, on the 2nd and 3rd sterna of Parajapygids and on the 1st to 7th in the Japygids; in the Projapygidae there are no vesicles. The vesicles of *Campodea* absorb solutions of methylene blue experimentally and may therefore be sites of water uptake (Drummond, 1953). In some Japygidae the cuticle of the 1st abdominal segment is differentiated just medial to the styles to form a pair of setose 'subcoxal organs' between which may lie an unpaired glandular structure. The eighth abdominal segment bears small papillae associated with the gonopore (Pagés, 1962). The cerci assume very different forms in the three families. In the Campodeidae they are long, many-segmented, antenna-like structures and the Projapygidae also retain segmented cerci though they are here short, robust organs perforated apically by the opening of a gland. In the Japygidae the cerci are represented by stout, strongly sclerotized forceps which the insects use in catching their prey (Kosaroff, 1935).

Internal Anatomy – The *alimentary canal* is a straight tube. In *Anajapyx* the mid gut is very short but the oesophagus is of great length, extending into the 4th abdominal segment. *Campodea* has a large rectum which can be extended by dilator muscles and the epithelium of its mid gut undergoes partial or complete degeneration and renewal at intervals during postembryonic development (Bareth, 1969). The head contains several paired exocrine glands, seven of which have been distinguished in *Campodea* by Bareth (1968*a*). Three pairs open to the exterior while the other four discharge into the pre-oral food cavity. Of the latter, one pair is long and tubular, with a terminal sac, and is perhaps excretory ('labial nephridia'), while another is bilobed and apparently salivary. Malpighian tubules are represented in some Diplura by small papillae of which *Campodea* has 16, *Projapyx* 5 and *Anajapyx* 6, but in *Japyx* these structures are entirely absent.

The nervous system shows little concentration, the ventral cord including eight abdominal ganglia in the Japygidae and Parajapygidae and seven in the other families. Neurosecretory cells are present in the brain and ventral ganglia and the stomatogastric system, corpora cardiaca and corpora allata are like those of the Collembola (Cazal, 1948; Bareth, 1968*b*). The *respiratory* system shows several unusual features. In *Campodea* it is poorly developed and opens by 3 pairs of thoracic spiracles (2 on the mesothorax and 1 on the metathorax). The tracheae from each spiracle remain unconnected with those from the others, the tracheal intima lacks spiral thickenings (Marten, 1939) and the whole system is said to be absent from newly-hatched nymphs. Abdominal spiracles are absent in *Campodea*. In *Heterojapyx* and *Japyx solifugus* (Fig. 101) there are 11 pairs of spiracles, of which 4 are thoracic and 7 abdominal. The 1st, 2nd and 4th pairs correspond with the 3 pairs of thoracic spiracles of *Campodea*: the 3rd pair is situated on the metathorax in front of the 4th pair. A longitudinal trunk unites the tracheae on either side of the body into a single system, but there is only a single delicate transverse commissure which is situated near the junction of the 9th and 10th abdominal segments. In *Parajapyx isabellae* there are 9 pairs of spiracles; those homologous with the 2nd and 4th pairs of *Japyx solifugus* being unrepresented. In *Projapyx* there are 10 pairs – 3 thoracic and 7 abdominal: in *Anajapyx* there are 9 pairs of which the 1st and 2nd correspond with the 1st and 3rd of *J. solifugus*.

The *dorsal vessel* is notable because the heart extends into the mesothorax. In *Japyx* it is composed of 10 chambers and in *Campodea* Marten (1939) found 9 pairs of ostia. A pair of *posterior glands*, possibly repugnatorial, opens at the apices of the cerci in the Projapygidae and may be homologous with similar glands in the Symphyla and Diplopoda. In *Campodea* other epidermal glands open by setae on the labium and the first abdominal sternum (Bareth, 1962).

The reproductive organs (Fig. 212) differ considerably within the

Diplura. In the females, *Campodea* has a single pair of large polytrophic ovarioles but in other members of the order they are panoistic. *Japyx* has 7 metamerically arranged ovarioles on each side (1st to 7th abdominal segments) while *Anajapyx* has 2 pairs. In all cases the vagina is extremely short and the two oviducts combine immediately before opening by the gonopore

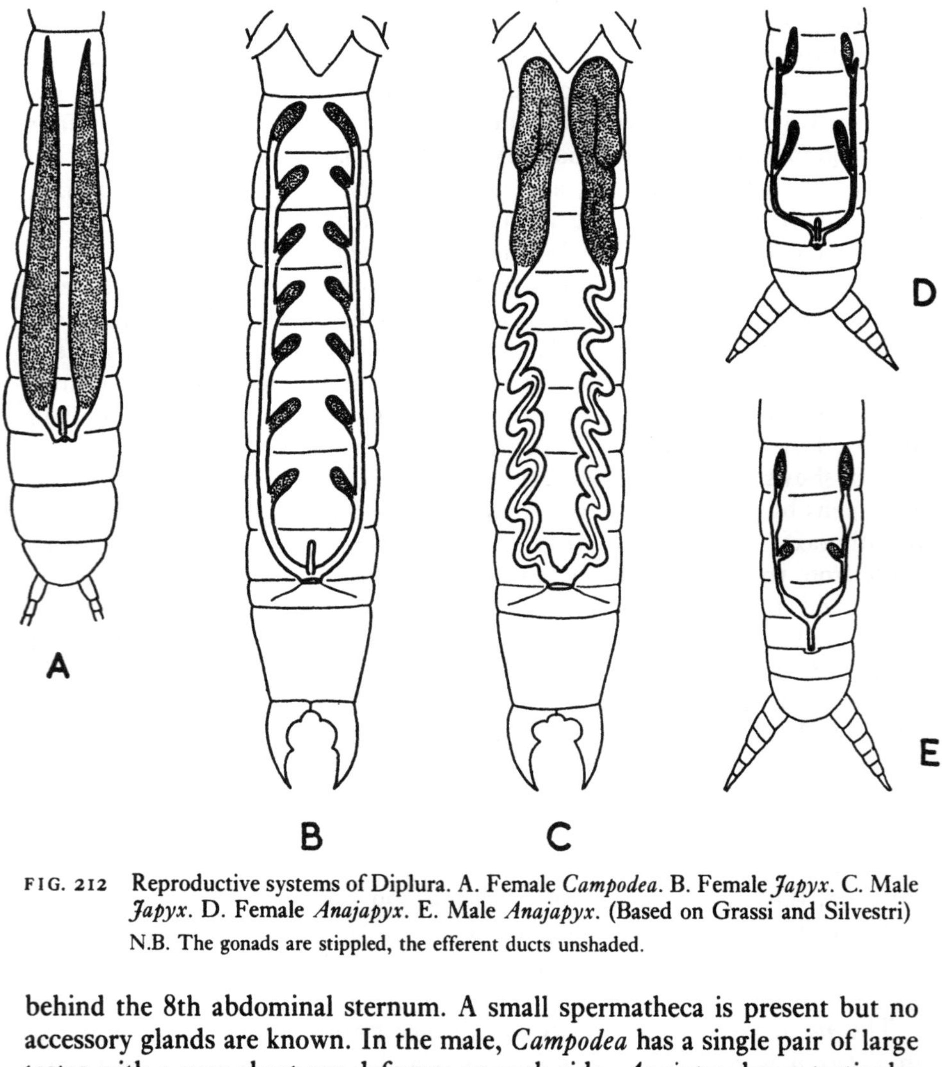

FIG. 212 Reproductive systems of Diplura. A. Female *Campodea*. B. Female *Japyx*. C. Male *Japyx*. D. Female *Anajapyx*. E. Male *Anajapyx*. (Based on Grassi and Silvestri)
N.B. The gonads are stippled, the efferent ducts unshaded.

behind the 8th abdominal sternum. A small spermatheca is present but no accessory glands are known. In the male, *Campodea* has a single pair of large testes with a very short vas deferens on each side. *Anajapyx* has 2 testicular lobes on each side, arranged as are the ovarioles of this genus, while *Japyx* has 1 pair of testes with long, convoluted vasa deferentia. The ductus ejaculatorius, when present, is always short; in *Campodea* it is surrounded by masses of glandular tissue that secrete the spermatophore material (Bareth, 1968*b*) but otherwise accessory glands are not known. The male gonopore, like that of the female, opens behind the eighth abdominal sternum.

Biology and Postembryonic Development – Sperm transfer has been observed only in the Campodeidae where it is indirect; the male deposits stalked spermatophores from which the female takes up the terminal sperm-droplet without courtship (von Orelli, 1956; Bareth, 1966, 1968*b*). The eggs of Campodeidae and Japygidae are normally laid in a mass of up to 40, suspended in a cavity in the soil on a common stalk (Bareth, 1963*b*; von Orelli, 1956; Gyger, 1960); in *Parajapyx*, however, each egg is individually

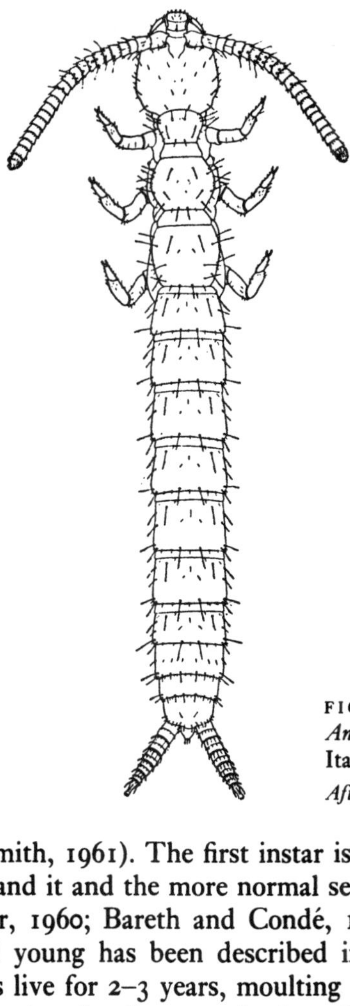

FIG. 213
Anajapyx vesiculosus (magnified)
Italy
After Silvestri, 1905.

stalked (Smith, 1961). The first instar is a short-lived, incompletely formed 'prelarva' and it and the more normal second instar live on the yolk in their gut (Gyger, 1960; Bareth and Condé, 1965). Simple maternal care of the aggregated young has been described in the Japygidae. In *Campodea* the individuals live for 2–3 years, moulting up to 20 times a year and attaining sexual maturity after 8–11 moults (von Orelli, *l.c.*). Japygids may have fewer instars but they also moult after reaching the adult state. For further information on the general biology and development of Diplura see also Condé (1956), Paclt (1956; 1957) and Pagés (1952, 1967).

Affinities – The Diplura were formerly included in the order Thysanura as a suborder (Entognatha). Their elevation to separate ordinal status by most modern authorities is based mainly on the distinctive mouthparts and leg movements, the intrinsic flagellar musculature of the antennae and the atypical arrangement of thoracic spiracles, the latter feature being unique among the Insecta. Although among the Apterygotes it is the Thysanura which approach most closely to the Pterygote insects, the Diplura are of considerable phylogenetic interest because of their resemblance to the Symphyla (Imms, 1936; Smith, 1960). Thus *Anajapyx* and the Symphyla agree in the possession of abdominal styli, eversible vesicles and anal glands opening on the cerci as well as in the structure of the mouthparts, antennal musculature and legs. The fact that the Symphyla are progoneate is not considered by Imms (1936) or Tiegs (1945) to preclude the likelihood of the Diplura, and with them the other insects, having originated from a primitive Symphylan-like stock. Manton (1964, 1972), however, by emphasizing the unique features of the Dipluran mouthparts and locomotor mechanisms, concludes that they – like each of the other Apterygote orders – form an isolated hexapodan group (pp. 4–6).

Classification – The monograph by Paclt (1957) lists earlier taxonomic works. Among later papers are those by Palissa (1964) on the European species; and by Smith (1960), Condé and Thomas (1957) and Bareth and Condé (1958) on N. American forms. A classification of the Diplura into two suborders and six familes, due to Pagés (1959), is summarized below.

Suborder I. RHABDURA

Mandible with prostheca; abdominal styli soft, setose; abdominal segments 8–10 of same consistency as preceding ones; many-segmented cerci; 7 abdominal ganglia in ventral nerve cord.

Superfamily Projapygoidea

Antennae with trichobothria from 4th or 5th segment onwards; styles on first seven abdominal segments; 7 pairs of abdominal spiracles; cerci with gland-openings apically

FAM. ANAJAPYGIDAE. *Antennae with trichobothria on segments 5-12; eversible vesicles on 2nd to 7th abdominal segments; 16 rudimentary Malpighian tubules. Anajapyx* has a few species from Italy, Africa and N. America.

FAM. PROJAPYGIDAE. *Antennae with trichobothria on segments 4–22; no eversible vesicles; no Malpighian tubules. Projapyx* from Brazil and W. Africa; *Symphylurinus*, with a wide distribution.

Superfamily Campodeoidea

Antennae with trichobothria from 3rd segment onwards; styles on 2nd to 7th abdominal segments; no abdominal spiracles; cerci closed apically.

FAM. PROCAMPODEIDAE. *Trichobothria on antennal segments 3 to 7; 2 pairs thoracic spiracles; 1st abdominal segment without appendages; cerci with mass of glandular cells near base. Procampodea* (Italy and California).

FAM. CAMPODEIDAE. *Trichobothria on antennal segments 3 to 6; 3 pairs thoracic spiracles; 1st abdominal segment with one pair lobate appendages; cerci without associated glands.* *Campodea* is a large, cosmopolitan genus with over 150 species; *Plusiocampa* from Europe, China and Mexico; *Lepidocampa*, tropicopolitan.

Suborder 2. DICELLURATA

Mandible without prostheca; abdominal styli spiniform, with few setae; abdominal segments 8 to 10 strongly sclerotized; cerci 1-segmented, forcipate; 8 abdominal ganglia in ventral nerve cord.

Superfamily Japygoidea
(With characters of the suborder)

FAM. JAPYGIDAE. *Antennae with trichobothria at least on segments 4 to 6; 4 pairs thoracic spiracles; very small eversible vesicles on first seven abdominal segments; cerci without proximal glandular orifices.* Contains several large, widely distributed genera such as *Japyx, Indjapyx, Metajapyx* and *Burmjapyx*. *Evalljapyx* is American.

FAM. PARAJAPYGIDAE. *Antennae without trichobothria; 2 pairs thoracic spiracles; two large eversible vesicles on 2nd and 3rd abdominal segments; cerci with glandular orifices proximally.* *Parajapyx* is widely distributed, with about 40 species.

Literature on the Diplura

BARETH, C. (1962), Histologie de quelques glandes tégumentaires chez *Campodea* (*C.*) *remyi* Denis, *Bull. Soc. zool. Fr.*, **87**, 280–288.

—— (1963*a*), Étude morphologique et histologique de quelques formations tégumentaires des Diploures Campodéidés, *Bull. Mus. Hist. nat., Paris*, **35**, 370–380.

—— (1963*b*), À propos des pontes et des éclosions chez *Campodea* (*C.*) *remyi* Denis, *Bull. Soc. zool. Fr.*, **88**, 663–671.

—— (1966), Étude comparative des spermatophores chez les Campodéidés, *C.r. Acad. Sci., Paris*, **262**, D, 2055–2058.

—— (1968*a*), Les glandes exocrines céphaliques de *Campodea remyi*, *Bull. Soc. zool. Fr.*, **93**, 629–646.

—— (1968*b*), Biologie sexuelle et formations endocrines de *Campodea remyi* Denis (Diploures Campodéidés), *Rev. Ecol. Biol. Sol.*, **3**, 303–426.

—— (1969), Structure et évolution de l'intestin moyen de *Campodea remyi* Denis en fonction de l'intermue, *Annls Spéléol.*, **24**, 603–612.

BARETH, C. AND CONDÉ, B. (1958), Campodéidés endogés de l'Ouest des États-Unis (Washington, Oregon, Californie, Arizona), *Bull. mens. Soc. linn. Lyon*, **27**, 226–248; 265–276; 297–304.

—— (1965), La prélarve de *Campodea* (*C.*) *remyi*. *Rev. Ecol. Biol. Sol.*, **2**, 397–401.

BARLET, J. AND CARPENTIER, F. (1962), Le thorax des Japygides, *Bull. A. Soc. R. ent. Belg.*, **98**, 95–123.

CARPENTIER, F. AND BARLET, J. (1951), Les sclérites pleuraux du thorax de *Campodea* (Insectes Aptérygotes), *Mém. Inst. r. Sci. nat. Belg.*, **27** (47), 1–7.

CAZAL, P. (1948), Les glandes endocrines rétrocérébrales des insectes, *Bull. biol. Fr. Belg., Suppl.*, **32**, 227 pp.

CONDÉ, B. (1956), Matériaux pour une monographie des Diploures Campodéidés, *Mém. Mus. natn. Hist. nat., Paris*, **12**, (1955), 1–201.

CONDÉ, B. AND THOMAS, J. (1957), Contribution à la faune des Campodéidés de Californie (Insectes Diploures), *Bull. mens. Soc. linn. Lyon*, **26**, 81–96; 118–127; 142–155.

DELANY, M. J. (1954), Thysanura and Diplura, *R. ent. Soc. Handb. Ident. Brit. Insects*, **1** (2), 7 pp.

DRUMMOND, F. H. (1953), The eversible vesicles of *Campodea* (Thysanura), *Proc. R. ent. Soc. Lond.*, (A), **28**, 145–148.

FRANÇOIS, J. (1970), Squelette et musculature céphalique de *Campodes chardardi* Condé (Diplura: Campodeidae), *Zool. Jb.* (*Anat.*), **87**, 331–376.

GYGER, H. (1960), Untersuchungen zur postembryonalen Entwicklung von *Dipljapyx humberti* (Grassi), *Verh. naturf. Ges. Basel*, **71**, 29–95.

IMMS, A. D. (1936), The ancestry of insects, *Trans. Soc. Br. Ent.*, **3**, 1–32.

—— (1939), On the antennal musculature in insects and other Arthropods, *Q. Jl microsc. Sci.*, **81**, 273–320.

KOSAROFF, G. (1935), Beobachtungen über die Ernährung der Japygiden, *Mitt. naturw. Inst. Sophia*, **8**, 181–185.

MANTON, S. M. (1964), Mandibular mechanisms and the evolution of arthropods, *Phil. Trans. R. Soc. Ser. B*, **247**, 1–183.

—— (1972), The evolution of arthropodan locomotory mechanisms. Part 10. Locomotory habits, morphology and evolution of the hexapod classes, *Zool. J. Linn. Soc.*, **51**, 203–400.

MARTEN, W. (1939), Zur Kenntnis von *Campodea*, *Z. Morph. Ökol. Tiere*, **36**, 40–88.

ORELLI, M. VON (1956), Untersuchungen zur postembryonalen Entwicklung von *Campodea* (Insecta, Apterygota), *Verh. naturf. Ges. Basel*, **67**, 501–574.

PACLT, J. (1956), *Biologie der primär flügellosen Insekten*, Fischer, Jena, 258 pp.

—— (1957), Diplura, *Genera Insectorum*, Fasc. **212**, 123 pp.

PAGÉS, J. (1952), Parajapyginae (Insecta, Entotrophi, Japyginae) de l'Angola, *Publ. Cultur. Comp. Diamantes de Angola*, **13**, 53–96.

—— (1959), Remarques sur la classification des Diploures, *Trav. Lab. Zool. Stn aquic. Grimaldi*, **26**, 1–25.

—— (1962), Comparaison et interprétation des papilles génitales femelles des Diploures, *C.r. Acad. Sci., Paris*, **252**, 2001–2003.

—— (1967), Données sur la biologie de *Dipljapyx humberti* (Grassi), *Rev. Écol. Biol. Sol.*, **4**, 187–281.

PALISSA, A. (1964), Apterygota – Urinsekten, *Tierwelt Mitteleuropas*, **4** (1), 407 pp.

SILVESTRI, F. (1933), Sulle appendici del capo degli Japygidae (Thysanura Entotropha) e rispettivo confronto quelle dei Chilopodi, dei Diplopodi e dei Crostacei, *5e Congr. int. Ent., Paris*, **2**, 329–343.

SMITH, L. M. (1960), The family Projapygidae and Anajapygidae (Diplura) in North America, *Ann. ent. Soc. Am.*, **53**, 575–583.

—— (1961), Japygidae of North America. 8. Postembryonic development of Parajapyginae and Evalljapyginae (Insecta, Diplura), *Ann. ent. Soc. Am.*, **54**, 437–441.

SNODGRASS, R. E. (1952), *A Textbook of Arthropod Anatomy*, Comstock Publ. Ass., Ithaca, N.Y., 363 pp.

TIEGS, O. W. (1945), The postembryonic development of *Hanseniella agilis* (Symphyla), *Q. Jl microsc. Sci.*, **85**, 191–238.

WYGODZINSKY, P. W. (1941), Beiträge zur Kenntnis der Dipluren und Thysanuren der Schweiz, *Denkschr. Schweiznaturfr. Ges.*, **74**, 113–227.

Order 3

PROTURA

Minute insects with entognathous, piercing mouthparts; antennae and eyes absent. Anterior legs sensory, usually held forward and little used in walking; all tarsi one-segmented, with a single claw. Abdomen of 11 segments and a well-developed telson; first three segments each with a pair of small appendages; cerci absent. External genitalia associated with eleventh abdominal segment, male with a pair of gonopores. Tracheal system present or absent. Malpighian tubules represented by papillae. Metamorphosis slight, accompanied by an increase in the number of abdominal segments.

The Protura are minute, whitish insects 0·5 to 2·5 mm long, occurring in all zoogeographical regions. Because of their small size they are easily overlooked, but they are numerous in certain types of moist soil, peat, woodland litter and turf (Strenzke, 1942; Raw, 1956); they are also encountered under stones and beneath bark. The order was first recognized by Silvestri in 1907 from Italy, then studied in considerable anatomical detail by Berlese (1909). For modern general accounts see especially Tuxen (1931, 1964) and Janetschek (1970).

External Anatomy – The head is prognathous, pyriform and narrowing anteriorly, with lateral oral folds produced downwards to enclose the mandibles and maxillae in a pair of gnathal pouches (François, 1959, 1969). The sclerotized parts of the folds do not quite meet ventrally, thus leaving a median *linea ventralis* comparable to that of the Collembola. Otherwise the sutures of the head-capsule are peculiar to the order and have no general morphological significance. There are neither compound eyes nor ocelli but on each side of the head is a small *pseudoculus* (Bedini and Tongiorgi, 1971; Haupt, 1972). This is a circular dome of thin, perforated cuticle surmounting a group of 2–6 neurons with ciliary dendritic processes and it may well be a chemoreceptor.

The *labrum* is pointed or vestigial, while the mandibles and maxillae are withdrawn into the head (Fig. 215). The former appendages are stylet-like, adapted for piercing and are each attached to the head by a single articulation. The maxillae are divided into an outer and an inner lobe, and the palps are 3- or 4-segmented; either the inner or both lobes are modified into piercing organs (Fig. 216). The labium comprises an elongate, basal submentum and paired mental and premental sclerites while the ligula is com-

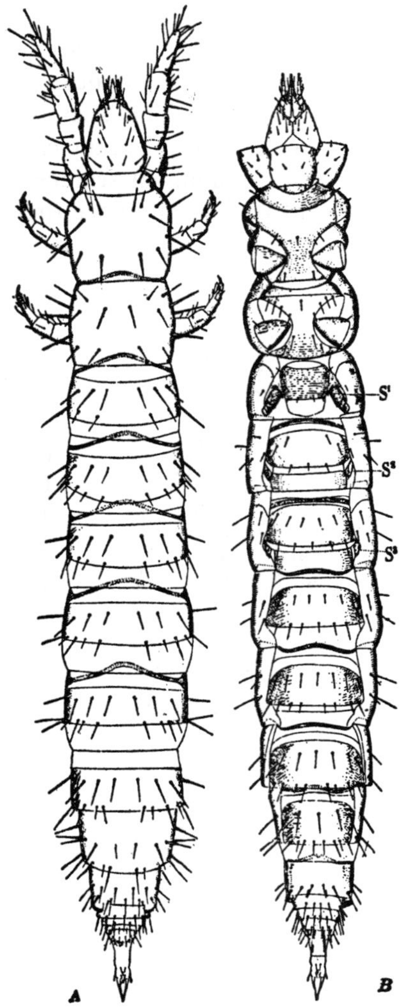

FIG. 214
Acerentomon doderoi (highly magnified). Europe
A, dorsal; B, ventral. *S*, styli. *After* Silvestri, 1907.

posed of a pair of pointed structures. The labial palpi are short and 2- or 3-segmented. Superlinguae appear to be absent, but from the hypopharyngeal region and the front of the head arises a complicated X-shaped tentorium-like structure resembling that of the Collembola. Little is known of the feeding habits of the Protura, but *Acerentomon doderoi* and *Eosentomon transitorium* have been observed feeding on the external mycorhiza associated with the roots of oak and beech (Sturm, 1959); the mouthparts pierce the fungal hyphae and the mid gut undergoes peristaltic movements like those of a sucking pump.

The thorax is clearly defined with the first segment reduced; the meso- and metaterga and all three sterna are well-developed plates but the pleura consist of several small sclerites, some arcuate and others less strongly sclerotized (Prell, 1913; François, 1964). The legs are long with 1-segmented tarsi, each terminated by a single claw and an empodial appendage which may be absent. The abdomen is very long and slender; in the newly-hatched insect it is composed of 8 segments and a telson and three more segments develop postembryonically between the telson and the last segment. This anamorphosis, or increase in the number of segments after emergence from the egg, is a generalized Arthropod character. The first three abdominal segments each carry a pair of small appendages (Fig. 214); in the Eosentomidae they

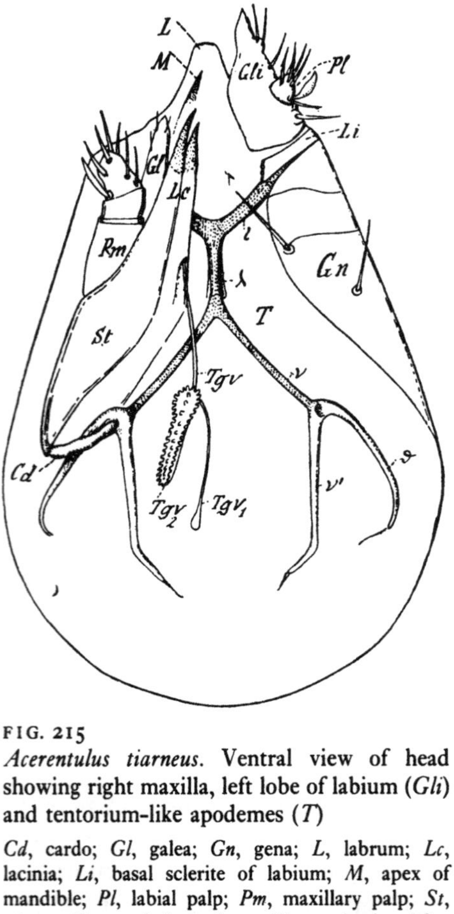

FIG. 215
Acerentulus tiarneus. Ventral view of head showing right maxilla, left lobe of labium (*Gli*) and tentorium-like apodemes (*T*)

Cd, cardo; *Gl*, galea; *Gn*, gena; *L*, labrum; *Lc*, lacinia; *Li*, basal sclerite of labium; *M*, apex of mandible; *Pl*, labial palp; *Pm*, maxillary palp; *St*, stipes; *Tgv*, tubules of maxillary gland. *After* Berlese, 1909.

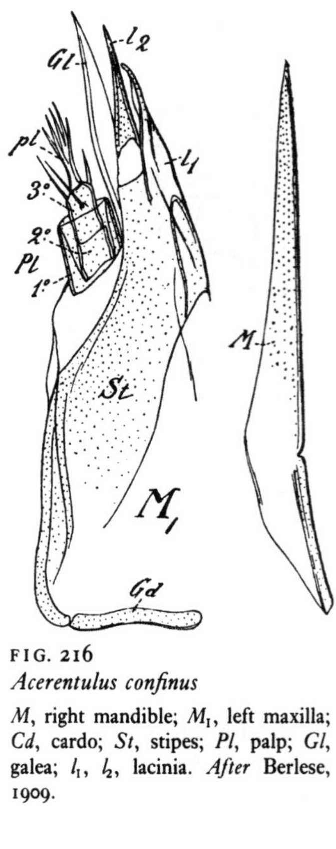

FIG. 216
Acerentulus confinus

M, right mandible; M_I, left maxilla; *Cd*, cardo; *St*, stipes; *Pl*, palp; *Gl*, galea; l_1, l_2, lacinia. *After* Berlese, 1909.

are 2-segmented, the second segment being reduced and provided with a protrusible vesicle. In other genera either the first pair or first two pairs are 2-segmented and the others consist of a single minute lobe. Cerci are absent in the order, and the name Protura is derived from the simple telson. Small eversible external genitalia occur behind the eleventh abdominal segment. The male has a pair of two-jointed style-like processes, each carrying a preapical gonopore on its long, distal segment; in females the basal segments of

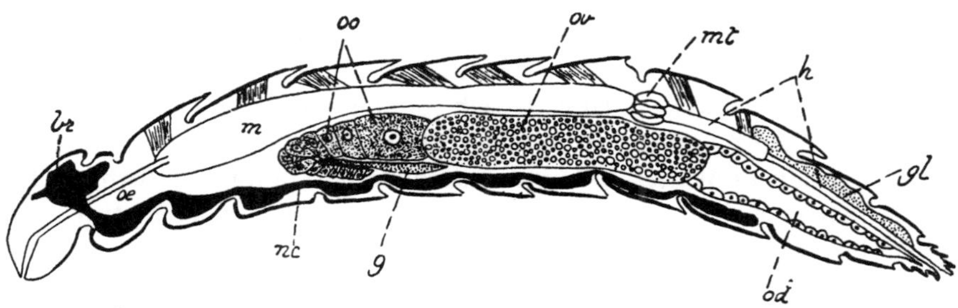

FIG. 217 *Acerentulus confinus*, female: general anatomy

br, brain; *g*, germarium; *gl*, abdominal gland; *h*, hind gut; *m*, mid gut; *mt*, excretory papillae; *nc*, nerve cord; *od*, oviduct; *oe*, fore gut; *oo*, oocytes; *ov*, mature ovum. Adapted from Berlese, 1909.

the styles are relatively much larger and broader and the single gonopore opens between them proximally.

Internal Anatomy (Fig. 217) – The alimentary canal is a simple straight tube and its most extensive region is the large cylindrical mid gut. Two pairs of maxillary glands and a pair of labial (salivary) glands are present. The Malpighian tubes are represented by six uni- or bicellular papillae disposed in two groups of three. The nervous system consists of the brain and fused suboesophageal and prothoracic ganglia, while there are separate ganglia in the remaining thoracic and the first six abdominal segments. The connectives throughout are double. The terminal ganglion is larger than those preceding it and there is a supplementary ganglion on each pedal nerve at the bases of the legs. Beneath and behind the brain lies a median corpus cardiacum and paired corpora allata (François, 1965), near which is the anterior end of the dorsal vessel (Aubertot, 1939). In the Eosentomidae and Sinentomidae, which alone possess a *tracheal system* (Fig. 218), the

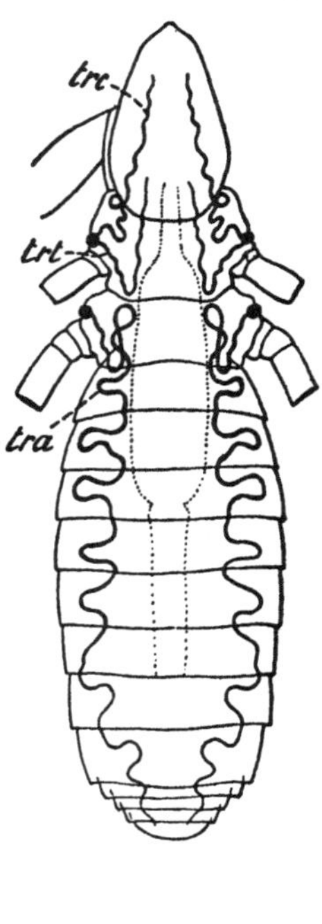

FIG. 218
Tracheal system of *Eosentomon* (much retracted)

trc, cephalic trachea; *trt*, thoracic do.; *tra*, abdominal do. *After* Berlese, 1909.

latter communicates with the exterior by two pairs of spiracles – one pair on the mesothorax and the other on the metathorax. There is no communication between the tracheae associated with each spiracle and the tracheae are absent from the first-stage nymph (Tuxen, 1949). The *reproductive system* in the female consists of a pair of ovaries and oviducts; the latter combine to form a short vagina which opens by a median pore. Each ovary extends, when mature, from the metathorax into the 9th abdominal segment and is homologous with a single panoistic ovariole of other insects. The germarium is situated in the reflexed apex of the ovary and from it is derived a single chain of egg-cells. In the male the testes are a pair of elongate sacs, united anteriorly about the level of the mesothorax. The vasa deferentia are closely coiled tubes which enter the genitalia separately. The germarium is apical and the remainder of the testis contains spermatozoa in various stages of development.

Postembryonic Development – Tuxen (1949) described the five nymphal stages of *Acerella danica* (= *Acerentulus danicus*) and *Eosentomon transitorium* (= *armatum*). These differ in chaetotaxy and have 8, 8, 9, 11 and 11 abdominal segments respectively. Similar instars have now been described in other species (François, 1960; Tuxen, 1961; Imadaté, 1966); in no case is there evidence that the adult stage moults as it does in the other Apterygote groups.

Affinities – The affinities of the Protura are still unsettled. Berlese (1909) placed them in a class of their own, the Myrientomata, and an isolated position is also advocated for them by Manton (1973) who has brought forward morphological evidence that hexapody and the entognathous condition arose independently in the Protura, Diplura and Collembola. On the other hand, Tuxen (1970) and others have emphasized the features which the Protura share with the Collembola: entognathy and the *linea ventralis*, a reduced tracheal system, the absence of cerci and Malpighian tubules, and the specialized pretarsus. The embryology of the Protura remains unknown; it may perhaps eventually help to solve the problem of their relationships.

Classification – The monograph by Tuxen (1964) supersedes much previous work and is the basis of modern taxonomic studies. Other present-day authorities include Condé (1951, etc.), Imadaté (1964–65; 1966) and Yin (1965) who recently added the family Sinentomidae. Classification depends greatly on the abdominal chaetotaxy and the distribution of setae and sensilla on the fore tarsus. There are over 200 described species arranged in four families. The first two of these form the superfamily Eosentomoidea and the others make up the Acerentomoidea.

FAM. EOSENTOMIDAE. *Tracheae and spiracles present; all three pairs of abdominal appendages 2-segmented.* Contains a single large genus, *Eosentomon*, with over 100 species from all regions.

FAM. SINENTOMIDAE. *Tracheae and spiracles present; only the first pair of abdominal appendages 2-segmented.* Contains only the primitive *Sinentomon erythranum* from China.

FAM. PROTENTOMIDAE. *Without tracheae and spiracles; at least the first two pairs of abdominal appendages 2-segmented. Hesperentomon*, *Proturentomon* and *Protentomon*, each with several species and wide distributions, are the most important genera.

FAM. ACERENTOMIDAE. *Without tracheae and spiracles; only the first pair of abdominal appendages 2-segmented.* A diverse family with some 10 genera, of which *Acerentulus*, *Acerella*, *Acerentomon* and *Berberentulus* contain the majority of species.

Literature on Protura

Tuxen (1964) and Janetschek (1970) give good bibliographies.

AUBERTOT, M. (1939), Présence d'un vaisseau dorsal contractile chez les Protoures du genre *Acerentomon*, *C.r. Acad. Sci., Paris*, **208**, 120–123.

BEDINI, C. AND TONGIORGI, P. (1971), The fine structure of the pseudoculus of acerentomid Protura (Insecta Apterygota), *Monit. zool., Italia*, **5**, 25–38.

BERLESE, A. (1909), Monografia dei Myrientomata, *Redia*, **6**, 1–182.

CONDÉ, B. (1951), Les grandes divisions de l'ordre des Protoures, *Bull. Mus. natn. Hist. nat.*, **23**, 121–125.

FRANÇOIS, J. (1959), Squelette et musculature céphaliques d'*Acerentomon propinquum* (Condé) (Ins. Protoures), *Trav. Lab. zool. Dijon*, **29**, 57 pp.

—— (1960), Développement postembryonnaire d'un Protoure du genre *Acerentomon* Silv., *Trav. Lab. zool. Dijon*, **33**, 1–11.

—— (1964), Le squelette thoracique des Protoures, *Trav. Lab. zool. Sta. Aqu. Grimaldi, Fac. Sci. Dijon*, **55**, 1–18.

—— (1965), Sur la présence de glandes neurendocrines retrocérébrales chez les Protoures (Insectes Apterygotes), *C.r. Acad. Sci., Paris*, **1965**, 2307–2309.

—— (1969), Anatomie et morphologie céphalique des Protoures (Insecta Apterygota), *Mém. Mus. natn. Hist. nat., Paris, (N.S.), A, (Zoologie)*, **59** (1), 144 pp.

HAUPT, J. (1972), Ultrastruktur des Pseudoculus von *Eosentomon* (Protura, Insecta), *Z. Zellforsch.*, **135**, 539–551.

IMADATÉ, G. (1964–65), Taxonomic arrangement of Japanese Protura. I–III, *Bull. nat. Sci. Mus. Tokyo*, **7**, 37–81; **7**, 263–293; **8**, 23–69.

—— (1966), Taxonomic arrangement of Japanese Protura. (IV) The Proturan chaetotaxy and its meaning to phylogeny, *Bull. nat. Sci. Mus. Tokyo*, **9**, 277–315.

JANETSCHEK, H. (1970), Ordnung Protura (Beintastler), In: *Handbuch der Zoologie*, Bd. **4**, 2. Hälfte, Lfg. 14, Beitr. 3, 1–72.

MANTON, S. M. (1973), Arthropod phylogeny – a modern synthesis, *J. Zool.*, **171**, 111–130.

PRELL, H. (1913), Das Chitinskelett von *Eosentomon*, ein Beitrag zur Morphologie des Insektenkörpers, *Zoologica, Stuttgart*, **25**, 1–58.

RAW, F. (1956), The abundance and distribution of Protura in grassland, *J. Anim. Ecol.*, **25**, 15–21.

SILVESTRI, F. (1907), Descrizione di novo genera d'insetti Apterigoti representante di un novo ordine, *Bull. Lab. Zool., Portici*, **1**, 296–311.

STRENZKE, K. (1942), Norddeutsche Proturen, *Zool. Jb. (Syst.)*, **75**, 73–102.

STURM, H. (1959), Die Nahrung der Proturen. (Beobachtungen an *Acerentomon doderoi* Silv., und *Eosentomon transitorium* Berl.), *Naturwissenschaften*, **46**, 90–91.

TUXEN, S. L. (1931), Monographie der Proturen. I. Morphologie nebst Bemerkungen über Systematik und Oekologie, *Z. Morph. Ökol. Tiere*, **22**, 671–720.

—— (1949), Über den Lebenszyklus und die postembryonale Entwicklung zweier dänischer Proturengattungen, *K. danske Vidensk. Selsk. Skr.*, **6**, 49 pp.

—— (1961), Die Variabilität einer Proturen-Art (*Acerentomon gallicum* Ion.) nebst deren postembryonaler Entwicklung, *Zool. Anz.*, **167**, 58–69.

—— (1964), *The Protura*, Hermann, Paris, 360 pp.

—— (1970), The systematic position of entognathous Apterygotes, *An.Esc. nac. Cienc. biol. Méx.*, **17**, 65–79.

YIN, W. (1965), Studies on Chinese Protura II, *Acta ent. sin.*, **14**, 186–195.

Order 4

COLLEMBOLA (Spring-tails)

Mouthparts entognathous, principally adapted for biting; antennae usually 4-segmented, the first 3 segments with intrinsic muscles; compound eyes absent. Abdomen 6-segmented, usually with three pairs of appendages, i.e. a ventral tube on segment I, a minute retinaculum on III, and a forked springing organ on IV. A tracheal system is usually absent and there are no Malpighian tubules. Metamorphosis slight.

Collembola are small insects rarely exceeding 5 mm in length, and occurring in almost all situations (Schaller, 1970; Paclt, 1956; Christiansen, 1964; Butcher *et al.*, 1971). They are found in the soil, in decaying vegetable matter, among herbage, and under bark of trees. A few species frequent the nests of ants and termites, others occur on the surface of fresh water and several are littoral or marine: *Anurida maritima*, for example, is daily submerged by each tide. Cavernicolous species have evolved a characteristic facies by convergence (Christiansen, 1961, 1965). The only condition which seems essential for their welfare is a certain amount of moisture, for they are rare in very dry situations. Various works on their biology, especially of the soil-inhabiting species, are among those listed on pp. 471–5. The order is world-wide and is remarkable for the extensive distribution of many of its genera and species (Salmon, 1949).

Collembola vary very much in coloration. Many are of a uniform dull blue-black, as in *Anurida*; others are green or yellowish with irregular patches of a darker colour; a few species are banded, some are all white, one or two are bright red while metallic forms are not infrequent. In habits they are saprophagous or phytophagous, pollen grains and fungal spores or mycelium often being eaten (MacNamara, 1924; Poole, 1959).

External Anatomy (Figs. 219–222) – In most species the body is clothed with hairs but some genera, notably *Tomocerus* and *Lepidocyrtus*, are scaled (Maiwald, 1972). The hairs vary in shape, often on different regions of the body; they may be simple and tapering, clavate, flattened and partially resembling scales, or plumose. The head (Denis, 1928; Bruckmoser, 1965) is pro- or hypognathous with the labrum distinctly marked off and sometimes also the clypeus and frons. The cephalic endoskeleton is not fully understood. Manton (1972) describes anterior and posterior tentorial arms and transverse segmental 'tendons' between the mandibles and maxillae, but others have denied that the cephalic apodemes constitute a true tentorium (Denis, 1928; François, 1972). The antennae vary greatly in length and the distal segments may be secondarily annulated. They are typically 4-segmented; the maximum number of six is found in *Orchesella*. In the Neelidae the antennae may be shorter than the head, while in some of the Entomobryidae they are longer than the whole body. Sensory organs of varied types are usually present on the last two segments and take the form of cones, rods, pits or papillae. In some Sminthuridae the second and third segments of the male antenna form a gripping organ armed with bristles and other cuticular processes (Massoud and Betsch, 1972). A variable number of ocelli is generally present on each side of the head behind the antennae; there are never more than eight to a side and often many fewer. Each ocellus typically has 8 retinular cells, a crystalline cone, corneagen cells and a simple or subdivided rhabdom (Barra, 1971; Paulus, 1972). Marlier (1941) has described additional ocelliform structures on the head of some genera. Immediately behind the antennae of some Collembola is a very characteristic structure known as the postantennal organ (Dallai, 1971; Karuhize, 1971; Altner *et al.*, 1970–71). This occurs in most Poduroidea and some Entomobryoidea but not in a typical form in the Symphypleona. It assumes a great variety of forms among different genera, being simple and ring-like in *Isotoma*, in the form of a rosette in *Anurida*, while in *Onychiurus* it attains considerable complexity of structure. Histologically, it comprises a sense-cell and several enveloping cells and the way in which dendrites of the sense-cell pass to porous regions of the cuticle suggests that it may be a chemoreceptor. The *mouthparts* (Manton, 1964; Goto, 1972) are deeply withdrawn into the head and are greatly elongated, which allows of their freedom of movement when protruded. Their deeply seated position is a secondary acquisition and has been brought about in the following manner. In the embryo, the sides of the head develop from a pair of lateral evaginations of the germ band. These evaginations eventually fuse with the developing labrum and labium and, in this way, form a kind of enclosing box which, by further growth, comes to surround the remaining mouthparts. The mouth-cavity is roofed over by the labrum and clypeus. The mandibles (Fig. 219) are slender organs usually with toothed extremities; they undergo rotary movements and are apparently provided with only a single main articulation with the head-capsule. They are rarely absent, as in *Brachystomella*. The maxillae each consist of a complex apical portion or 'head' which possibly represents a lacinia. In some species a digit-like palpifer carries a vestigial palp and the galea. The cardo and stipes are variable in form and sometimes rod-like. The superlinguae are well developed lamellate structures overlying the hypopharynx: as a rule they are undivided but in *Isotomurus palustris* they are bilobed. The labium is very much reduced and, although it exhibits evidences of a paired structure, neither glossa nor paraglossae are separately developed. Labial palps have been detected in the early embryo but as a rule they subsequently atrophy. In *Neanura* and its allies the mouthparts are specialized for sucking and piercing:

the labrum and labium together form a conical tube enclosing the rest of the mouthparts, the latter being modified into stylets (Wolter, 1963).

The thorax of the more generalized forms consists of three very similar segments but in the Entomobryoidea the prothorax is greatly reduced, and

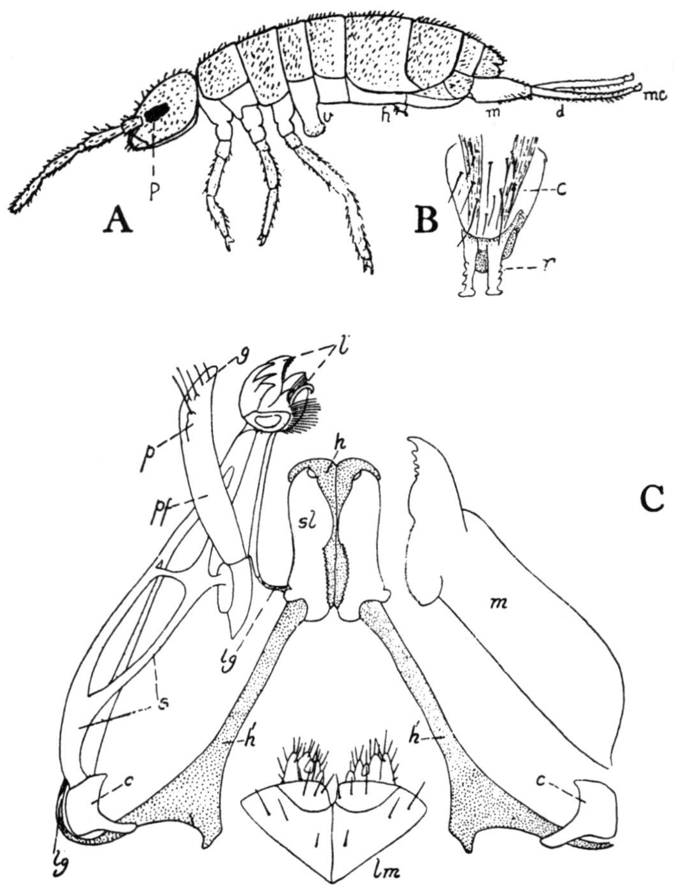

FIG. 219 Structural details of Collembola

A, *Axelsonia*. *p*, pigment surrounding eyes; *v*, ventral tube; *h*, hamula; *m*, manubrium; *d*, left dens and mucro *mc*. (Adapted from Carpenter.)
B, *Tomocerus*, retinaculum. *h*, hamula; *c*, corpus; *r*, ramus. (*After* Willem, 1900.)
C, Mouthparts of *Orchesella*, dorsal view. *c*, cardo; *g*, galea; *h*, hypopharynx; *h'*, posterior tentorial apodemes; *l*, lacinia; *lg*, anterior apodemes; *lm*, labium; *m*, right mandible; *p*, maxillary palp; *pf*, palpifer; *s*, stipes; *sl*, superlingua. (Partly after Folsom.)

its tergum is fused with that of the mesothorax (Manton, 1972). In the Symphypleona the thorax becomes intimately fused with the abdomen and its segmentation is largely obsolete. The legs have no separate tarsal segments and the tibiotarsi generally terminate in a pair of claws, an upper and a lower, though the latter, which possibly represents a modified empodium rather than a true claw, may be vestigial or wanting. A group of short setae on the trochanter forms the so-called trochanteral organ, a structure of some taxonomic importance. The abdomen is composed of six segments only; in this respect Collembola differ from all other insects and at no stage in development are there more than that number present. In some Arthropleona the 4th and 5th, or 4th to 6th segments undergo fusion, while in the Symphypleona the first four segments are almost entirely undifferentiated. On the ventral aspect of the first segment, in all Collembola, there is a bilobed structure known as the *ventral tube* (Rüppel, 1953; Sedlag, 1952). It is formed by the union of the first pair of embryonic abdominal appendages, and consists of a basal column containing a pair of protrusible vesicles. The

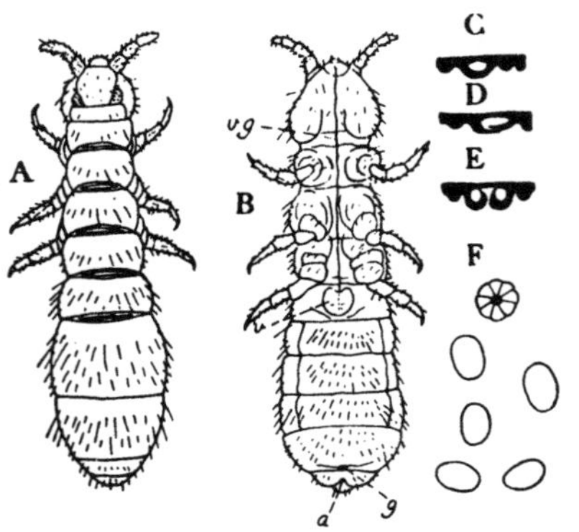

FIG. 220
Anurida maritima

A, dorsal. B, ventral. *a*, anus; *g*, genital pore, *v*, ventral tube; *vg*, ventral groove. C, D, E: transverse sections of ventral groove in regions of the head, prothorax and metathorax respectively. F, eyes and postantennal organ, right side.

latter are commonly shallow sacs though in some genera they are long and tubular. The cavity of the ventral tube communicates freely with that of the body and contains blood; the vesicles are everted by blood-pressure and withdrawn by the contraction of special muscles. Many divergent opinions have been expressed with respect to the function of the ventral tube. It has been supposed by some to be respiratory, and Noble-Nesbitt (1963) has shown that it is a site of water absorption. It also acts as an adhesive organ, enabling the insect to walk over smooth or steep surfaces; according to some, the surface of the vesicles is moistened by the secretion of cephalic glands which is discharged into the commencement of the *ventral groove* (Fig. 220). This is a cuticular channel passing down the middle ventral line of the body; it arises just behind the labium and terminates on the anterior aspect of the ventral tube. The ultrastructure of the epidermis of the tube also suggests an osmoregulatory function (Eisenbeis and Wichard, 1975). Many Collembola retain a minute pair of appendages on the 3rd abdominal segment. They are fused proximally to form a basal piece or *corpus*, while their distal portions remain free and are termed the *rami*. The organ thus formed is variously known as the *retinaculum* or *hamula*, and it serves to retain the furca in position, when the latter is stowed away under the abdomen while not in use. Most Collembola carry a jumping organ or *furca*, associated with the fourth abdominal segment. This is usually thought to represent a highly modified pair of segmental appendages, though studies of the musculature and of

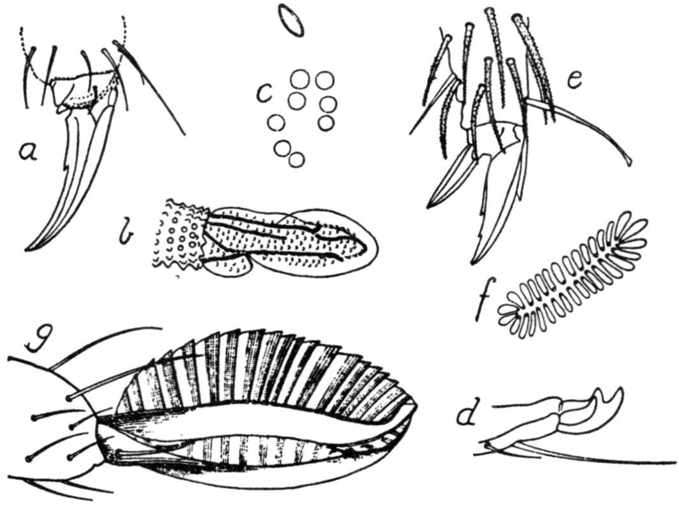

FIG. 221 Structural details of Collembola

a, *Podura*, claw of left leg; *b*, left mucro; *c*, *Isotoma*, eyes and postantennal organ; *d*, left mucro; *e*, *Lepidocyrtus*, claws of left leg; *f*, *Onychiurus*, right postantennal organ; *g*, *Sminthurides*, left mucro. Adapted from Folsom.

embryonic development do not fully confirm the homology (Pistor, 1955; Bretfeld, 1963). The mechanism whereby the furca is suddenly moved downwards and backwards to strike the substrate and so propel the insect into the air is also not fully understood. Earlier anatomical studies suggested that the furca was moved from its resting position by extensor muscles (Pistor, Bretfeld, *loc. cit.*). Manton (1972), however, has given detailed reasons for believing that the so-called extensor muscles have only a stabilizing function; the jumping movement is thought to result from a rapid increase in the hydrostatic pressure of the blood when the body is com-

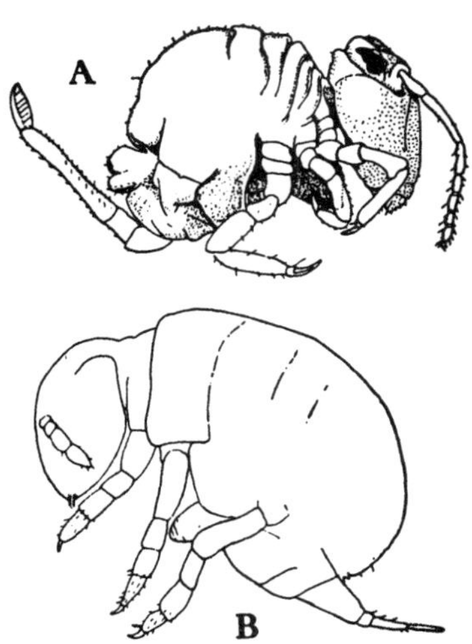

FIG. 222
Collembola, Symphypleona (magnified)

A, *Sminthurides aquaticus*. *After* Willem.
B, *Neelus folsomi*. *After* Caroli.

pressed through contraction of thoracic and abdominal muscles. The common basal piece of the furca is termed the *manubrium*; it carries a pair of distal arms or *dentes*, each bearing a very variably shaped claw-like process or *mucro*. The furca varies greatly in size; in *Entomobrya*, for example, it extends at rest to beyond the ventral tube; in *Hypogastrura* it is often very short, while in *Neanura* and *Anurida* it is wanting. Males and females are similar in Collembola, there being no external genitalia although the gonopore and the region around it may differ slightly in the sexes (Agrell, 1937). The genital aperture is near the hind margin of the 5th sternum and the anus on the 6th sternum.

Internal Anatomy (Fig. 223) – The *alimentary canal* (Boelitz, 1933; Tóth, 1942) is usually a simple straight tube, passing from mouth to anus

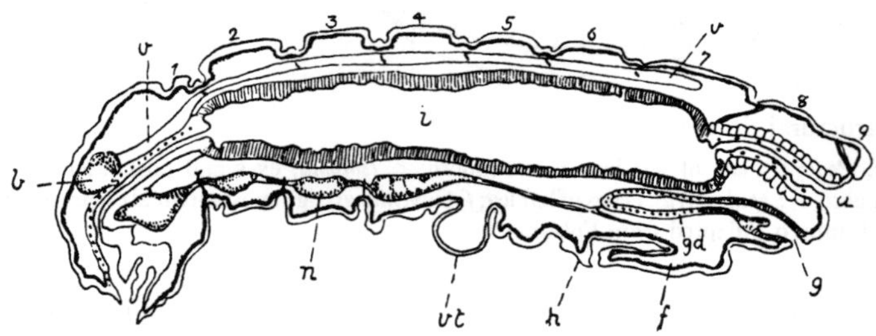

FIG. 223 *Hypogastrura viatica*, longitudinal section

a, anus; *b*, brain; *f*, furca; *g*, genital pore; *gd*, gonoduct; *h*, hamula; *i*, mid intestine; *n*, nerve-cord; *v*, dorsal vessel; *vt*, ventral tube. *After* Willem, 1900.

without convolutions. The greater portion is formed by the extensive mid intestine which in *Neelus* forms four subequal chambers. The head contains one or two pairs of salivary glands whose secretion enters the pre-oral food cavity and which, in *Bilobella massoudi*, contain some cells with giant polytene chromosomes (Cassagnau, 1968). There is also a pair of tubular 'labial nephridia' whose terminal sac can accumulate experimentally injected dyestuffs selectively and which may play some excretory role (Feustel, 1958; Altner, 1968). The central *nervous system* is considerably specialized and consists of the cerebral ganglia and a ventral nerve-cord composed of four ganglionic centres – the suboesophageal and three thoracic ganglia, which are united by double connectives. There are no separate abdominal ganglia, the nerve centres of that region having fused with the metathoracic ganglion. In the Sminthuridae the ventral ganglia are closely merged together and there are no intervening connectives. The stomatogastric nervous system and associated endocrine structures are not unlike those of the Pterygote insects, but the corpora allata are innervated from the suboesophageal ganglion (Cassagnau and Juberthie, 1967–68). The *heart*, in the more generalized forms, consists of six chambers with paired lateral ostia and alary muscles at each constriction. Anteriorly, it is prolonged into the aorta and in *Anurida* the latter surrounds the fore gut as a cylinder which opens in the head beneath the brain. There are no Malpighian tubules and *excretion* is chiefly performed by the fat-body (Feustel, 1958). This contains numerous concretions of ammonium urate which increase in size with the age of the individual and are not eliminated from the insect. The epithelium of the mid gut also performs an excretory function. Its cells contain concretions of a similar nature to those found in the fat-body. These congregate in the inner halves of the cells, which divide off from the remainder, and are periodically discharged into the lumen of the gut. They are removed from the body during each ecdysis, and a regeneration of the epithelium takes place (Folsom and Welles, 1906; Humbert, 1974). According to Boelitz (1933) the urate concretions of the mid gut are actually transported to this position from the fat-body by phagocytes (see also Ichikawa, 1931; Jura, 1958; Thibaud, 1968).

Respiration in the majority of Collembola is cutaneous (Rüppel, 1953) but tracheae are present in *Actaletes* and the Sminthuridae while vestiges of a tracheal system are said to occur in *Dicyrtoma fusca*. They are well developed in *Sminthurus* where there is a single pair of simple spiracular openings between the head and prothorax (Fig. 224). Tracheal branches are distributed to the head, legs and abdomen, but no anastomosis takes place between the tracheae of opposite sides of the body (Davies, W. M., 1927).

The *reproductive system* is extremely simple; the gonads consist of a pair of large sacs, their ducts are extremely short and they unite to form the vagina or ejaculatory canal as the case may be. The ovaries contain groups of vitellogenous cells and developing eggs but there is no arrangement into ovarioles, and the testes are filled with dense masses of developing spermatozoa and associated secretions. Unlike other insects, the germarium in both the ovaries and testes is lateral and not apical in position. There are no well-defined accessory glands but the ejaculatory duct of the male is

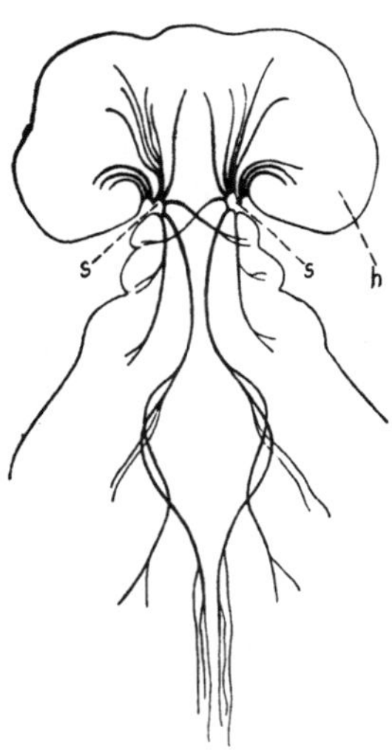

FIG. 224
Sminthurus fuscus. Tracheal system
h, head; *s*, spiracle. *After* Willem, 1900.

modified in order to produce the characteristic spermatophore (Schliwa, 1965).

Postembryonic Development and Biology – Sperm-transfer is indirect (p. 301); the male deposits spermatophores on the substrate, each consisting of a stalk surmounted by a droplet of spermatozoa that is taken up into the reproductive tract of the female when she moves over it (Schaller, 1953; Mayer, 1957). There may be a loose behavioural association between the sexes at this time (Bretfeld, 1971) and in the closer forms of epigamic behaviour (e.g. in *Heterosminthurus*) the male grips the female with his modified antennae. The eggs of Collembola are smooth and spherical, usually cream-coloured, and are deposited in small groups. The newly-hatched insects are white excepting for an area of dark pigment surrounding the ocelli. Agrell (1949) finds that the number of moults which occur before the maximum size is attained is six in *Arrhopalites pygmaeus*, seven in *Hypogastrura sahlbergi* and eight in *Folsomia 4-oculata*, the external changes which accompany postembryonic development being relatively slight (e.g. secondary annulation of antennal segments and differentiation of the trochanteral organ, the teeth of the claws and the genital region). Sexual maturity is attained before maximum size – after five moults, for example, in *Folsomia 4-oculata*. For further information see, for example, Hale (1965*a*, *b*), Lindenmann (1950), Milne (1960), and Thibaud (1967). Some species show appreciable structural differences according to the environment in which they have developed, an effect referred to as ecomorphosis (e.g. Cassagnau and de Izarra, 1969, but cf. Willson, 1960).

Collembola make up a major component of the soil fauna (Butcher *et al.*, 1971) and some species play an active part in breaking down dead vegetation to humus (Schaller, 1950; Dunger, 1956). Different soils may be characterized by quantitative and qualitative features of the Collembolan fauna and characteristic Collembolan associations have been defined (Gisin, 1943; Strenzke, 1949; Hüther, 1961). In an interesting example, Gisin (1955) found that in soils most suitable for viticulture the dominant Collembolan was *Tullbergia krausbaueri* while second-grade soils contained associations of other species not found elsewhere.

The species *Sminthurus viridis* has been more fully studied than any other (MacLagan, 1932; Davidson, 1934; Wallace, 1967–68). It feeds on the leaves of various plants, especially Leguminosae and, according to MacLagan, the optimum conditions for growth are a humidity near saturation and soil with a pH of about 6·5. Under these conditions, and a temperature of about 13° C., the maximum number of eggs laid per individual is about 120; they are coated with soil voided through the anus. The incubation period is near 26 days and about 48 days elapse from eclosion to sexual maturity. There are eight instars and the last ecdysis takes place after attaining sexual maturity. As an adult the insect lives for about 15 days; the life span, from the egg onward, is approximately 2½ months and there are five generations in the year.

A very characteristic feature of Collembola is a tendency for enormous numbers of individuals, comprising both adult and immature forms, to mass together. This behaviour has been noted in various countries (Turk, 1932; Paclt, 1956) but its significance is unknown. In some cases it may be connected with an abundance of a particular food or in others with migration.

Affinities – The systematic position of the Collembola is uncertain. Although Tillyard (1930) regarded them as the modern representatives of a primitive group possessing few postcephalic segments and from which the other Insecta originated, other authorities such as Handlirsch (1908) emphasize their isolated position by giving them the status of a separate Arthropodan class. Certainly they differ from other insects in the possession of only six abdominal segments, gonads with a lateral germarium, a postantennal organ which resembles most closely the organ of Tömösvary in Chilopoda and Diplopoda, and eggs which undergo total cleavage during embryonic development. Their inclusion among the Insecta depends on their being opisthogoneate hexapods with three gnathal head segments. Kristensen and others (p. 423) see them and the Protura as the specialized 'sister-group' of the Diplura, the three orders together forming the supposedly monophyletic Entognatha. Manton (1964, 1972) believes, however, that entognathy and the hexapod condition have both evolved convergently. She has emphasized the role of unique hydrostatic mechanisms in Collembolan locomotion and regards the group as an isolated section of the Uniramian complex, equivalent in status to the Myriapods and the Pterygote insects. A final decision requires closer agreement on the morphological interpretation of cephalic structures and on the principles to be followed in reconstructing animal phylogeny. The Devonian fossil *Rhyniella praecursor* (pp. 399, 423) is probably allied to *Neanura* and is therefore a specialized Collembolan that provides little indication of the group's affinities (Massoud, 1967). Immunologically, the Collembola are most like the Diplura, Thysanura and lower Pterygotes and show no Myriapodan affinities (Parisi, 1962).

Classification – There are over 1500 species of Collembola, mostly listed by Salmon (1964). The classification below follows the traditional division into two suborders, Arthropleona and Symphypleona, though Salmon (loc. cit.) has divided the Arthropleonan families into three subordinal groups and Massoud (1971) has created a new suborder for the family Neelidae. Recent major systematic works include those by Gisin (1944, 1960) on the Holarctic and European species and Salmon (1964) on the genera of the world, all with extensive bibliographies. Stach (1947–63), Bonet (1947) and Richards, W. R. (1968) have monographed selected families while Palissa (1964) and Scott (1961) deal respectively with the Collembola of Central Europe and North America. Ellis and Bellinger (1973) list the genera of the world and their type species.

Suborder 1. ARTHROPLEONA

Body more or less elongate; thoracic and abdominal segments distinctly separated, except sometimes for the last 2–3 abdominal segments; head almost always prognathous; tracheae absent (except *Actaletes*).

Superfamily **Poduroidea**

Pronotum distinct, well-developed, setose and visible from above; body without scales; cuticle usually granular or tuberculate; commonly with postantennal organ; furca sometimes reduced or absent.

FAM. PODURIDAE. *Head hypognathous; mandible with well-developed molar plate; 8 ocelli on each side; no postantennal organ; well-developed furca with distally annulate dens.* Includes only *Podura*. The European *P. aquatica* occurs on the surface of fresh and occasionally brackish waters, overwintering in damp soil.

FAM. HYPOGASTRURIDAE. *Head obliquely prognathous; biting mouthparts, mandible with molar plate; without pseudocelli. Hypogastrura* is a large genus with species from various habitats; the cosmopolitan *H. viatica* is littoral and *H. armata* sometimes occurs in large numbers on snow. The family also includes *Xenylla, Willemia* and other genera.

FAM. NEANURIDAE. *Mouthparts more or less adapted for piercing; mandibles without plate or absent; without pseudocelli.* These are somewhat specialized Collembola. The European *Anurida maritima, A. denisi* and allied forms are found on the sea-shore (Imms, 1906); *Neanura muscorum* occurs in moss, under bark, in leaf litter and similar habitats. Other genera include *Friesea, Pseudachorutes* and *Brachystomella*.

FAM. ONYCHIURIDAE. *Usually with pseudocelli on thoracic and abdominal terga; biting mouthparts; complicated sense-organ on third antennal segment; postantennal organ present. Onychiurus* is a large genus, commonly represented by the cosmopolitan *O. armatus*. Other genera are *Tullbergia* and *Tetrodontophora*, the latter with a relatively large, woodhouse-like montane species, *T. bielanensis*, from C. Europe.

Superfamily **Entomobryoidea**

Pronotum reduced or apparently absent, not setose; cuticle smooth, setose or scaled; usually without postantennal organ; furca rarely absent.

FAM. ISOTOMIDAE. *Body setose, without scales; 3rd and 4th abdominal segments about equal; legs without trochanteral organ; furca usually well developed.* This important family includes the large genera *Folsomia, Isotoma* and others. *F. candida* is a widely distributed, blind, facultatively parthenogenetic species, sometimes found in caves. *Isotomurus palustris* occurs at pond margins and other wet places. *Isotoma saltans* is the Alpine 'Gletscherfloh', living amid snow and ice, where it feeds on pollen and organic detritus.

FAM. ACTALETIDAE. Sometimes included in the preceding family, *Actaletes neptuni* is unusual in having a tracheal system that opens by a pair of cephalic spiracles.

FAM. ENTOMOBRYIDAE. *Body usually scaled; antennae short, 4–6 segments; 4th abdominal segment usually much longer than 3rd; hind legs with trochanteral organ; furca well developed. Entomobrya, Orchesella, Lepidocyrtus* and *Pseudosinella* are some of the important genera. Species of *Entomobrya* are morphologically very similar and must often be separated by their colour patterns, even though these are subject to considerable variation.

FAM. TOMOCERIDAE. *Body scaled; antennae long, 4-segmented; 3rd abdominal segment as long or longer than 4th.* The principal genus is *Tomocerus.* The members of this family are sometimes included in the Entomobryidae, as are also those of the **Cyphoderidae** and **Oncopoduridae**. *Cyphoderus* and its allies include most of the myrmecophilous and termitophilous Collembola (Delamare-Deboutteville, 1948).

Suborder 2. SYMPHYPLEONA

Body subglobose, thoracic and first four abdominal segments fused together; head hypognathous; tracheae often present; rarely soil-dwellers.

FAM. NEELIDAE. *Antennae shorter than head; eyes absent; minute forms without tracheae. Neelus* and *Megalothorax* are the most important genera.

FAM. SMINTHURIDAE. *Antennae at least as long as head, more or less geniculate between 3rd and 4th segments; eyes present; abdomen with bothriotrichia.* This is the largest Symphypleonan family, including such genera as *Sminthurus, Bourletiella* and *Sminthurides. Sminthurides aquaticus* is found on the surface of ponds, feeding on duckweed (*Lemna*).

FAM. DICYRTOMIDAE. These differ from the Sminthuridae in having the antennae elbowed between the 2nd and 3rd segments. *Dicyrtoma.*

Literature on the Collembola

AGRELL, I. (1937), Der Sexualdimorphismus der äusseren Genitalien bei den Collembolen, nebst Bemerkungen über Verschiedenheiten in Grösse und Frequenz der Geschlechter bei denselben, *Opusc. ent.*, 1 (1936), 119–127.

—— (1941), Zur Oekologie der Collembolen. Untersuchungen im schwedischen Lappland, *Opusc. ent., Suppl.* 3, 1–236.

—— (1949), Studies on the postembryonic development of Collemboles, *Ark. Zool.*, 41A (12), 1–35.

ALTNER, H. (1968), Die Ultrastruktur der Labialnephridien von *Onychiurus quadriocellatus* (Collembola), *J. Ultrastruct. Res.*, **24**, 349–366.

ALTNER, H., ERNST, K. D. AND KARUHIZE, G. (1970), Untersuchungen am Postantennalorgan der Collembolen (Apterygota). I, II, *Z. Zellforsch.*, **111**, 263–285; *Rev. Écol. Biol. Sol.*, 8, 31–35.

BARRA, J.-A. (1971), Les photorécepteurs des Collemboles, étude ultrastructurale. 1. L'appareil dioptrique, *Z. Zellforsch.*, **117**, 322–353.

BOELITZ, E. (1933), Beiträge zur Anatomie und Histologie der Collembolen. Darmkanal und Mitteldarmepithelregeneration bei *Tomocerus vulgaris* Tullb. und *Sinella caeca* Schött, *Zool. Jb.* (*Anat.*), **57**, 375–432.

BONET, F. (1947), Monografía de la familia Neelidae (Collembola, *Rev. Soc. Mexicana Hist. nat.*, 8, 131–192.

BRETFELD, G. (1963), Zur Anatomie und Embryologie der Rumpfmuskulatur und der abdominalen Anhänge der Collembolen, *Zool. Jb.* (*Anat.*), **80**, 309–384.

—— (1971), Das Paarungsverhalten europäischer Bourletiellinae (Sminthuridae), *Rev. Ecol. Biol. Sol.*, 8, 145–153.

BRUCKMOSER, P. (1965), Embryologische Untersuchungen über den Kopfbau der Collembole *Orchesella villosa* L, *Zool. Jb.* (*Anat.*), **82**, 299–364.

BUTCHER, J. W., SNIDER, R. AND SNIDER, R. J. (1971), Bioecology of edaphic Collembola and Acarina, *A. Rev. Ent.*, **16**, 249–288.

CARPENTIER, F. (1947), Quelques remarques concernant la morphologie thoracique des Collemboles (Aptérygotes), *Bull.* (*Ann.*) *Soc. ent. Belg.*, **83**, 297–303.

CASSAGNAU, P. (1968), Sur la structure des chromosomes salivaires de *Bilobella massoudi* Cassagnau (Collembola: Neanuridae), *Chromosoma*, **24**, 42–58.

CASSAGNAU, P. AND DE IZARRA, D. C. (1969), Contribution à l'étude des écomorphoses. 4, *Bull. Soc. zool. Fr.*, **94**, 243–250.

CASSAGNAU, P. AND JUBERTHIE, C. (1967–68), Structures nerveuses, neurosécrétion et organes endocrines chez les Collemboles. I–III, *Bull. Soc. Hist. nat. Toulouse*, **103**, 178–222; *Gen. comp. Endocr.*, 8, 489–502; **10**, 61–69.

CHRISTIANSEN, K. A. (1961), Convergence and parallelism in cave Entomobryinae, *Evolution*, **15**, 288–301.

—— (1964), Bionomics of Collembola, *A. Rev. Ent.*, **9**, 147–178.

—— (1965), Behavior and form in the evolution of cave Collembola, *Evolution*, **19** (4), 529–537.

DALLAI, R. (1971), First data on the ultrastructure of the postantennal organ of Collembola, *Rev. Écol. Biol. Sol.*, 8, 11–29.

DAVIDSON, J. (1934), The 'lucerne flea' *Smynthurus viridis* L. (Collembola) in Australia, *Bull. Counc. Sci. ind. Res. Aust.*, **79**, 1–66.

DAVIES, W. M. (1927), On the tracheal system of Collembola with special reference to that of *Sminthurus viridis* Lubbock, *Q. Jl microsc. Sci.*, **71**, 15–30.

DELAMARE-DEBOUTTEVILLE, C. (1948), Recherches sur les Collemboles termitophiles et myrmecophiles (Écologie, éthologie, systématique), *Archs Zool. exp. gén.*, **85**, 261–425.

DENIS, J. R. (1928), Études sur l'anatomie de la tête de quelques Collemboles, suivies de considérations sur la morphologie de la tête des insectes, *Archs Zool. exp. gén.*, **68**, 1–291.

DUNGER, W. (1956), Untersuchungen über Laubstreuzersetzung durch Collembolen, *Zool. Jb.* (*Syst.*), **84**, 75–98.

EISENBEIS, G. AND WICHARD, W. (1975), Histochemischer Chloridnachweis im Transportepithel am Ventraltubus arthropleoner Collembolen, *J. Insect Physiol.*, **21**, 231–236.

ELLIS, W. M. AND BELLINGER, P. F. (1973), An annotated list of the generic names of Collembola (Insecta) and their type species, *Monogr. Nederlandse Ent. Vereniging*, **7**, 74 pp.

FEUSTEL, H. (1958), Untersuchungen über die Exkretion bei Collembolen, *Z. wiss. Zool.*, **161** (1/2), 209–238.

FOLSOM, J. W. AND WELLES, M. U. (1906), Epithelial degeneration, regeneration and secretion in the mid-intestine of Collembola, *Univ. Ill. Bull.*, **4**, 5–32.

FRANÇOIS, J. (1972), Contribution à l'anatomie des Collemboles: les formations endosquelettiques céphaliques, *Rev. Ecol. Biol. Sol.*, 8, 45–48.

GISIN, H. (1943), Ökologie und Lebensgemeinschaften der Collembolen im schweizerischen Exkursionsgebiet Basels, *Revue suisse Zool.*, **50**, 131–224.

—— (1944), Hilfstabellen zum Bestimmen der holarktischen Collembolen, *Verh. naturf. Ges. Basel*, **55**, 1–130.

—— (1955), Recherches sur la relation entre la fauna endogée de Collemboles et les qualités agrologiques de sols viticoles, *Revue suisse Zool.*, **62**, 601–648.

—— (1960), *Collembolenfauna Europas*, Geneva, 312 pp.

GLASGOW, J. P. (1939), A population study of subterranean soil Collembola, *J. Anim. Ecol.*, **8**, 323–353.

GOTO, H. E. (1972), On the structure and function of the mouthparts of the soil-inhabiting collembolan *Folsomia candida, Biol. J. Linn. Soc.*, **4**, 147–168.

HALE, W. G. (1965*a*), Observations on the breeding biology of Collembola. I, II, *Pedobiologia*, **5**, 146–152; 161–177.

—— (1965*b*), Postembryonic development of some species of Collembola, *Pedobiologia*, **5**, 228–243.

HAMMER, M. (1944), Studies on the Oribatids and Collemboles of Greenland, *Medd. Grønland*, **141** (3), 1–210.

HANDLIRSCH, A. (1908), *Die fossilen Insekten und die Phylogenie der rezenten Formen*, Wien, 2 vols., 1430 pp.

HUMBERT, W. (1974), Localisation, structure et genèse des concrétions minérales dans le mésentéron des Collemboles Tomoceridae (Insecta, Collembola), *Z. Morph. Tiere*, **78**, 93–109.

HÜTHER, W. (1961), Ökologische Untersuchungen über die Fauna pfälzischer Weinbergsböden mit besonderer Berücksichtigung der Collembolen und Milben, *Zool. Jb. (Syst.)*, **89**, 243–368.

ICHIKAWA, M. (1931), On the renewal of the mid-intestinal epithelium of Collembola, *Mem. Coll. Sci. Kyoto*, (8), **7**, 135–141.

IMMS, A. D. (1906), *Anurida, L.M.B.C. Memoirs*, **13**, 1–99.

JURA, C. (1958), The alimentary canal of *Tetrodontophora bielanensis* Waga (Collembola) and the regeneration of the midgut epithelium, *Bull. ent. Pologne*, **22**, 85–89.

KARUHIZE, G. R. (1971), The structure of the postantennal organ in *Onychiurus* (Insecta: Collembola) and its connection to the central nervous system, *Z. Zellforsch.*, **118**, 263–282.

LINDENMANN, W. (1950), Untersuchungen zur postembryonalen Entwicklung schweizerischer Orchesellen, *Revue suisse Zool.*, **57**, 353–428.

MACLAGAN, D. S. (1932), An ecological study of the 'Lucerne flea' (*Sminthurus viridis* Linn.), *Bull. ent. Res.*, **23**, 101–145; 151–190.

MACNAMARA, C. (1924), The food of Collembola, *Can. Ent.*, **56**, 99–104.

MAIWALD, M. (1972), Die Feinstruktur der Körperschuppen von zwei Collembolengattungen und einer Thysanurenart, *Rev. Écol. Biol. Sol.*, **9**, 145–154.

MANTON, S. M. (1964), Mandibular mechanisms and the evolution of arthropods, *Phil. Trans. R. Soc. Ser. B*, **247**, 1–183.

—— (1972), The evolution of arthropodan locomotory mechanisms. Part 10. Locomotory habits, morphology and evolution of the hexapod classes, *Zool. J. Linn. Soc.*, **51**, 203–400.

MARLIER, G. (1941), Recherches sur les organes photorécepteurs des insectes aptilotes, *Annls Soc. r. zool. Belg.*, **72**, 204–236.

MASSOUD, Z. (1967), Contribution à l'étude de *Rhyniella praecursor* Hirst et Maulik 1926, Collembole fossile du Devonien, *Rev. Ecol. Biol. Sol.*, **4**, 497–505.

—— (1971), Contribution à la connaissance morphologique et systématique des Collemboles Neelidae, *Rev. Écol. Biol. Sol.*, **8**, 195–198.

MASSOUD, Z. AND BETSCH, J.-M. (1972), Étude sur les insectes Collemboles. II – Les caractères sexuels secondaires des antennes des Symphypléones, *Rev. Écol. Biol. Sol.*, **9**, 55–97.

MAYER, H. (1957), Zur Biologie und Ethologie einheimischer Collembolen, *Zool. Jb. (Syst.)*, **85**, 501–570.

MILLS, H. B. (1934), *A Monograph of the Collembola of Iowa*, Iowa Collegiate Press, Ames, Iowa, 143 pp.

MILNE, S. (1960), Studies on the life histories of various species of Arthropleone Collembola, *Proc. R. ent. Soc. Lond.*, **35**, 133–140.

MÜLLER, G. (1959), Untersuchungen über das Nahrungswahlvermögen einiger im Ackerboden häufig vorkommenden Collembolen und Milben, *Zool. Jb. (Syst.)*, **87**, 231–256.

NOBLE-NESBITT, J. (1963), A site of water and ionic exchange with the medium in *Podura aquatica* L. (Collembola, Isotomidae), *J. exp. Biol.*, **40**, 701–711.

PACLT, J. (1956), *Biologie der primär flügellosen Insekten*, Fischer, Jena, 258 pp.

PALISSA, A. (1964), Apterygota (Urinsekten), In : Brohmer, Ehrmann and Ulmer, *Die Tierwelt Mitteleuropas*, **4**, 1–407.

PARISI, V. (1962), Le affinita dei Collemboli (Insecta, Apterigota) studiate con metodo immunologico, *Monit. zool. ital.*, **69**, 202–210.

PAULUS, H. F. (1972), Zum Feinbau der Komplexaugen einiger Collembolen. Eine vergleichend-anatomische Untersuchung. (Insecta, Apterygota), *Zool. Jb. (Anat.)*, **89**, 1–116.

PISTOR, D. (1955), Die Sprungmuskulatur der Collembolen, *Zool. Jb. (Syst.)*, 83, 511–540.

POOLE, T. B. (1959), Studies on the food of Collembola in a Douglas Fir plantation, *Proc. zool. Soc. Lond.*, **132**, 71–82.

RICHARDS, W. R. (1968), Generic classification, evolution and biogeography of the Sminthuridae of the world (Collembola), *Mem. ent. Soc. Canada*, **53**, 1–53.

RÜPPEL, H. (1953), Physiologische Untersuchungen über die Bedeutung des Ventraltubus und die Atmung der Collembolen, *Zool. Jb. (Allg. Zool.)*, **64**, 429–469.

SALMON, J. T. (1949), The zoogeography of the Collembola, *Brit. Sci. News*, **2** (19), 196–198.

—— (1964), An Index to the Collembola, *Bull. R. Soc. New Zealand*, **7**, 644 pp.

SCHALLER, F. (1950), Biologische Beobachtungen an humusbildenden Bodentieren, insbesondere an Collembolen, *Zool. Jb. (Syst.)*, **78**, 506–525.

—— (1953), Untersuchungen zur Fortpflanzungsbiologie arthropleoner Collembolen, *Z. Morph. Ökol. Tiere*, **41**, 265–277.

—— (1970), Collembola (Springschwänze), In: Helmcke, J.-G., Starck, D. and Wermuth, H. (eds), *Handbuch der Zoologie*, **4** (2) Lf. 12, 1–72.

SCHLIWA, W. (1965), Vergleichend anatomisch-histologische Untersuchungen über die Spermatophorenbildung bei Collembolen (mit Berücksichtigung der Dipluren und Oribatiden), *Zool. Jb. (Anat.)*, **82**, 445–520.

SCOTT, H. G. (1961), Collembola: Pictorial keys to the Nearctic genera, *Ann. ent. Soc. Am.*, **54**, 104–113.

SEDLAG, U. (1952), Untersuchungen über den Ventraltubus der Collembolen, *Wiss. Z. Univ. Halle*, **1**, 93–127.

STACH, J. (1947–63), The Apterygotan fauna of Poland in relation to the world-fauna of this group of insects, *Acta monograph. Mus. hist. nat. Polska Akad. Umiej.*, **1947**, 488 pp. [Isotomidae]; **1949**, 341 pp. [Neogastruridae and Brachystomellidae]; **1949**, 122 pp. [Anuridae and Pseudachorutidae]; **1941**, 100 pp. [Bilobidae]; *Polska Akademia Nauk, Cracow*, **1954**, 219 pp. [Onychiuridae; **1956**, 287 pp. [Sminthuridae]; **1957**, 113 pp. [Neelidae and Dicyrtomidae]; **1960**, 151 pp. [Orchesellini]; **1963**, 126 pp. [Entomobryini].

STREBEL, O. (1932), Beiträge zur Biologie, Oekologie und Physiologie einheimischer Collembolen, *Z. Morph. Ökol. Tiere*, **25**, 31–153.

STRENZKE, K. (1949), Ökologische Studien über die Collembolengesellschaften feuchter Böden Ost-Holsteins, *Arch. Hydrobiol.*, **43**, 201–303.

THIBAUD, J. M. (1967), Contributions à l'étude du développement postembryonnaire chez les Collemboles Hypogastruridae épigés et cavernicoles, *Annls Spéléol.*, **22**, 167–198.

—— (1968), Cycle du tube digestif lors de l'intermue chez les Hypogastruridae (Collemboles) épigés et cavernicoles, *Rev. Écol. Biol. Sol.*, **4**, 647–655.

—— (1970), Biologie et écologie des Collemboles Hypogastruridae édaphiques et cavernicoles, *Mém. Mus. natn. Hist. nat., Paris*, **61**, 35–201.

TILLYARD, R. J. (1930), The evolution of the class Insecta, *Pap. Proc. R. Soc. Tasmania*, **1930**, 1–89.

TÓTH, L. (1942), Der Darmkanal der Collembolen, *Arb. Ung. biol. Forsch.-Inst.*, **14**, 397–400.

TURK, F. A. (1932), Swarming of Collembola in England, *Nature*, **129**, 830–831.

WALLACE, M. M. H. (1967–68), The ecology of *Sminthurus viridis* (L.) (Collembola). I, II, *Aust. J. Zool.*, **15**, 1173–1206; **16**, 871–883.

WEIS-FOGH, T. (1948), Ecological investigations on mites and collemboles in the soil, *Natura Jutlandicae*, **1**, 135–270.

WILLSON, M. F. (1960), The effect of temperature and light upon the phenotypes of some Collembola, *Iowa Acad. Sci.*, **67**, 598–601.

WOLTER, H. (1963), Vergleichende Untersuchungen zur Anatomie und Funktionsmorphologie der stechend-saugenden Mundwerkzeuge der Collembolen, *Zool. Jb. (Anat.)*, **81**, 27–100.

WOMERSLEY, H. (1939), *Primitive Insects of South Australia: Silverfish, Springtails and their Allies*, Adelaide, 322 pp.

ZINKLER, D. (1966), Vergleichende Untersuchungen zur Atmungsphysiologie von Collembolen (Apterygota) und anderen Bodenkleinarthropoden, *Z. vergl. Physiol.*, **52**, 99–144.

Order 5

EPHEMEROPTERA (PLECTOPTERA: MAYFLIES)

Soft-bodied insects with short setaceous antennae and vestigial mouthparts derived from a biting type. Wings membranous, held vertically upwards when at rest, the hind pair considerably reduced; 'intercalary' veins and numerous cross-veins usually present. Abdomen terminated by very long cerci with or without a similar median caudal prolongation. Metamorphosis hemimetabolous: nymphs aquatic, campodeiform, usually with long cerci and a median caudal filament; lamellate or plumose, metameric, tracheal gills present. Adult preceded by a subimaginal winged instar.

Existing mayflies are the remnants of a formerly extensive order. They are familiar insects near lakes, streams and rivers, and their association with the Ephemerides of Grecian mythology expresses their brief life above water which, in species such as *Ephoron, Campsurus* and *Palingenia*, lasts only a few hours. In their nymphal stages, however, they are at least as long-lived as most insects and some may live three years.

When a mayfly is about to emerge the nymph usually floats to the surface of the water; the dorsal cuticle splits, and a winged insect flies away in the course of a few seconds. Less frequently the nymph moults under water. In either case the resulting winged form is known as the *subimago*, and it differs from the mature imago in several features. In their general form the two stages are alike, the wings are fully expanded and spiracular respiration is established. The subimago may be recognized by its duller appearance, and by its somewhat translucent wings which are usually margined by prominent fringes of hairs. The passage from the subimago to the imago is marked by an ecdysis which is unique among insects; the subimago casts a delicate cuticle from its whole body, including the wings, and then issues as a fully formed imago (Edmunds, 1956; Taylor & Richards, A. G., 1963). In the adult condition the insect presents a shiny appearance and has assumed its full coloration, the wings become transparent, while the eyes and legs attain their complete development. Among certain short-lived species the sub-imaginal cuticle is partially or completely persistent in one or other sex. The males of *Oligoneuria*, for example, retain it on the wings, while the females of

Palingenia, Ephoron and Campsurus do not appear to shed it at all (Spieth, 1940). The subimaginal stage is of variable duration but there is some correspondence between it and the length of life of the imago. The short-lived species are often night fliers; species of *Palingenia, Oligoneuria, Ephemera, Hexagenia* and *Caenis* emerge about sundown in vast swarms and are attracted in large numbers to lights near the waterside. The males of some species carry out familiar 'dances' in the air, a fluttering swift ascent being followed by a passive leisurely fall many times repeated.

FIG. 225
Ephemera vulgata
(Reproduced by permission of the Trustees of the British Museum.)

Mayflies are eagerly devoured by fishes and most of the 'duns', 'spinners' and several of the 'drakes', of the fly-fisher, are made up to represent various species (Harris, 1952; Kimmins, 1972).

For general accounts of the order see Needham, Traver and Hsu (1935), Verrier (1956), Illies (1968) and Peters and Peters (1973).

The Adult

External Anatomy – Anatomical accounts of a few species are given by Drenkelfort (1910), Heiner (1914) and Grandi (1940–1947). The head (Fig. 226) is free with the antennae short, and composed of two basal segments, followed by a multi-articulate setaceous flagellum. The compound eyes are largest in the males and, in some genera, the upper portion of each has larger facets than the lower. In the Baetidae the upper divisions are mounted upon pillar-like outgrowths of the head (Verrier, 1940) and are used to recognize the female in the nuptial flight (Spieth, 1940). Between the compound eyes there are three ocelli. The mouthparts are degenerate; degeneration begins in the late nymph, it is externally complete in the subimago and complete in the imago (Sternefeld, 1907; Murphy, 1922). Individual parts do not undergo equal degrees of atrophy and the various genera differ very much in this respect. Mandibles are vestigial or wanting, and the maxillae, though greatly reduced, usually retain their palps; in *Ephemera* the labium is represented by the postmentum and a pair of distal lobes with small palps. In some genera the mouthparts have atrophied so much that they are scarcely recognizable and in no mayflies do the adults feed.

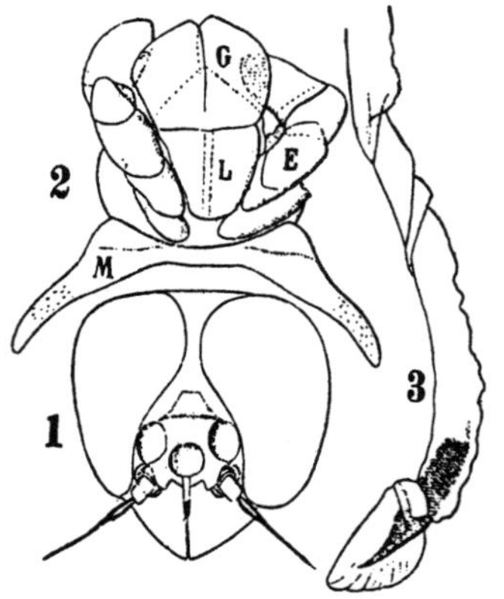

FIG. 226
Ephemera. 1, head viewed from above. 2, maxillae and labium. 3, apex of tibia with tarsus

E, maxilla; *G*, ligula; *L*, prementum; *M*, postmentum. Adapted from Silvestri.

The thorax (Knox, 1935; Matsuda, 1956; Tsui and Peters, 1972) is mainly notable for the predominance of the mesothorax, the other two segments being small. In *Hexagenia* the meso- and metanota are subdivided and postnota are present. The pleural sutures are reduced or absent and the pleura of the pterothorax are fused ventrally with the sternum to form pre- and postcoxal bridges and dorsally with the postnotum to form postalar bridges. The wings are markedly triangular and fragile. The axillary sclerites are atypical (Grandi, 1947) and, as in other Palaeopteran insects, the mayflies cannot flex the wings over the abdomen at rest but hold them in a vertically raised position (see also Kukalova-Peck, 1975). The hind wings are more or less reduced and in some genera, such as *Cloeon* and *Caenis*, they are absent. The fore wings are longitudinally fluted or corrugated but they do not fold along the flutes, except in those species where the female oviposits under water. The fluting, which strengthens the wing in flight (Edmunds and Traver, 1954*a*), is associated with the presence of characteristic intercalary veins, which are not often absent in the Ephemeroptera. Each vein, whether a main vein or an intercalary, either follows the crest of a ridge, when it is called a convex vein, or the bottom of a furrow, when it is known as a concave vein. The intercalaries appear to be branches which have lost their basal connections with the remaining veins but are united with the latter by a greatly developed system of cross-veins. The wing-venation, which is of considerable taxonomic value (Spieth, 1933), has been the subject of several conflicting interpretations (Needham *et al.*, 1935) but that of Tillyard (1932) is followed here. It shows the primitive feature of a media retaining its anterior (MA) and posterior (M) divisions in both pairs of wings. R_s is attached basally to MA. The legs are not used for walking and are sometimes greatly reduced though in males the fore legs are usually much longer and grip the female from below in the mating flight. In the American *Campsurus* the four posterior legs are mere stumps. There are primitively five tarsal segments but one or two of the basal segments are fused with the tibia in more specialized forms and in degenerate legs there may be only one or two tarsal segments. Two pretarsal claws are present, of which one is commonly degenerate and blunt.

The abdomen (Birket-Smith, 1970) is evidently 10-segmented with an 11th segment reduced and fused to the 10th. In the female the two oviducts open by separate apertures between the 7th and 8th sterna. An appendicular ovipositor is always absent though Morrison (1919) has shown that posterior extensions of the 7th and 8th sterna form a functional ovipositor in some Leptophlebiidae. The male genitalia (Spieth, 1933; Levy, 1948; Snodgrass, 1957; Qadri, 1940; Grandi, 1962) include a pair of claspers borne on the posterior margin of the 9th sternum, consisting each of a coxite (partially or completely fused with that of the opposite side) and a secondarily annulated style. Between the claspers lies a pair of penes which are more or less fused basally, bear variously shaped outgrowths and may represent appendages of the 10th abdominal segment (Brinck, 1957).

Internal Anatomy – The most characteristic internal feature is the modification of the alimentary canal for aerostatic purposes (Sternefeld, 1907; Pickles, 1931; Grandi, 1950). This region no longer functions as a digestive tract, but has assumed an entirely new role, and has undergone certain structural changes in consequence. In the nymph the oesophagus is wide, but in the imago it becomes an extremely narrow tube and there is a complicated apparatus of dilator muscles which appears to regulate the air-content of the gut. Air is taken in or expelled through the mouth, and the mid gut is modified into a kind of storage balloon; its epithelium is no longer secretory, but is converted into a pavement type, and the muscular coat has disappeared. The Malpighian tubes number up to 140, and the first portion of the hind gut is modified to form a complex valve which prevents the escape of air from the mid gut. In these short-lived insects it is more important that their specific gravity should be lessened to facilitate the mating flight, rather than that they should feed and live longer. The reproductive organs of mayflies are remarkable for their primitive nature (Brinck, 1957; Grandi, 1955). There are no accessory glands, and the gonoducts are paired in both sexes, each duct opening to the exterior separately.

In the male the testes are ovoid sacs, and the two vasa deferentia each communicate with a separate penis of its side. Each ovary is composed of a large number of small panoistic ovarioles, disposed along a common tube which is continued posteriorly as the oviduct. The respiratory system is well developed and opens to the exterior by ten pairs of spiracles, two pairs are thoracic and eight abdominal. The heart, which extends throughout the abdomen, comprises ten chambers and continues as the aorta into the head.

Oviposition and Postembryonic Development – Mating swarms of mayflies (Spieth, 1940; Brinck, 1957) consist normally of males. Females enter the swarm and leave with a male, copulation (which lasts only a very few minutes) occurring in flight, after which oviposition occurs. The eggs are very variable in form and structure (Smith, 1935; Koss, 1968) and though the type of egg is quite constant within a species or genus in some families, there is no clear relation between egg-structure and the type of habitat in which they are laid. The differences involve variations in chorionic sculpturing; in the polar cap; in the presence or absence of the micropylar apparatus; and in the form and occurrence of special anchoring filaments. A few examples in illustration of these facts may be cited. The egg of *Heptagenia interpunctata* is provided at each pole with a skein of fine yellow threads, which unravel in the water and serve to anchor it by becoming entangled with foreign objects. The egg of *Ephemerella excrucians* is white and slightly dumb-bell shaped, with a strongly sculptured chorion, but with no anchoring filaments or micropylar apparatus. That of *E. rotunda* is yellowish and oval, with a smooth chorion, and a prominent mushroom-shaped micropylar apparatus; there are four anchoring filaments each being terminated by a knob-like structure. The ovoid eggs of *Ecdyonurus* are provided with numerous short coiled filaments; after they have been in the water a little while each coil unwinds with a sudden spring, when it is seen to be terminated by a minute viscid button-like cap. The number of eggs laid by different species varies from several hundred up to about 4000. Some short-lived species discharge their eggs *en masse* as a pair of clusters which are laid on the water; these soon disintegrate and the eggs scatter over the river-bed. The longer lived species lay them in smaller numbers at a time, either alighting on the surface for the purpose or descending beneath the water and depositing their eggs under stones, etc.; the insects float up again and fly away to repeat the process, or die without reappearing. According to Heymons (1896) the eggs of *Ephemera vulgata* hatch in 10–11 days at 20–25 °C. In many species they require a much longer period for their development which may extend to a diapause lasting several months. Ovoviviparity has been recorded in *Cloeon dipterum* and some species of *Callibaetis* and seems to be correlated with a long imaginal life in the female – 8 to 21 days have been recorded (Berner, 1941). For further information on mating flights and reproduction see Tjønneland (1960) and Degrange (1960).

Immature Stages

The nymphs of many species have been figured and described by Vayssière (1882), Morgan (1913), Lestage (1917–25), Needham *et al.* (1935) and in most of the larger taxonomic works and keys (e.g. Bertrand, 1954; Edmunds *et al.*, 1963; and others listed on pp. 488–493). Berner (1959) gives a tabular summary of the biology of N. American mayfly nymphs. The exact number of moults is difficult to determine, but it is always high – from 23 to 45 instars have been recorded in the few species studied (e.g. Ide, 1935). In ecological work the nymphs are usually divided into size-classes or grouped into categories based on a few of the more obvious external features (e.g. Macan, 1970; Pleskot, 1958). Mayfly nymphs are essentially herbivorous, feeding on plant detritus, algal growths, fragments of higher plants and so on, but some species will eat small arthropods under experimental conditions and a few species are apparently predacious (e.g. *Prosopistoma, Pseudiron, Anepeorus*). The nymphal stages frequent a great variety of aquatic situations; many live concealed in the banks, some burrow in mud, while others hide beneath stones in lakes, streams and rivers. Certain genera occur among water plants and are active swimmers, others live in swift currents or near waterfalls, and there are some species found among decaying vegetation at the bottoms of ponds or ditches. This wide range of habitat is accompanied by a diversity of adaptive modifications greater than that found in other aquatic insects. The general shape of the body is very variable, but all are campodeiform with evident antennae, and usually elongate multi-articulate cerci. Compound eyes and ocelli are well developed, and most species possess seven pairs of plate-like or filamentous abdominal tracheal gills. These are responsible for an appreciable proportion of the total oxygen uptake of the nymph in *Hexagenia recurvata* (Morgan and Grierson, 1932) and *Ephemera vulgata*, but they do not appear to play an important respiratory role in *Baetis* and *Cloeon* (Wingfield, 1939). The gills are capable of co-ordinated movements by special muscles (Eastham, 1936–58) and even if they are not always major respiratory organs they may perform an accessory function in providing a flow of water over the general respiratory surface of the body. In this way, some species can regulate their oxygen uptake, at least when living in the appropriate substrate (Eriksen, 1963). The nymphs of *Ephemera* and *Hexagenia* burrow in mud or in the banks of streams; they have elongate bodies, with strong fossorial legs. The first pair of gills is vestigial, and the remainder are biramous, consisting of a pair of lamellae fringed with long filaments which are penetrated by tracheoles. When necessary the gills are carried reflexed over the back and are thus protected from abrasion. In *Iron, Epeorus* and *Heptagenia* (Fig. 227) the body and appendages are flattened, and the gills are lamellate with a basal tuft of branchial filaments. Such nymphs are adapted to life in rapidly flowing streams, though the flattened body may be a device that enables them to live in crevices rather than to resist the water current directly (Stuart, 1958). *Cloeon* and

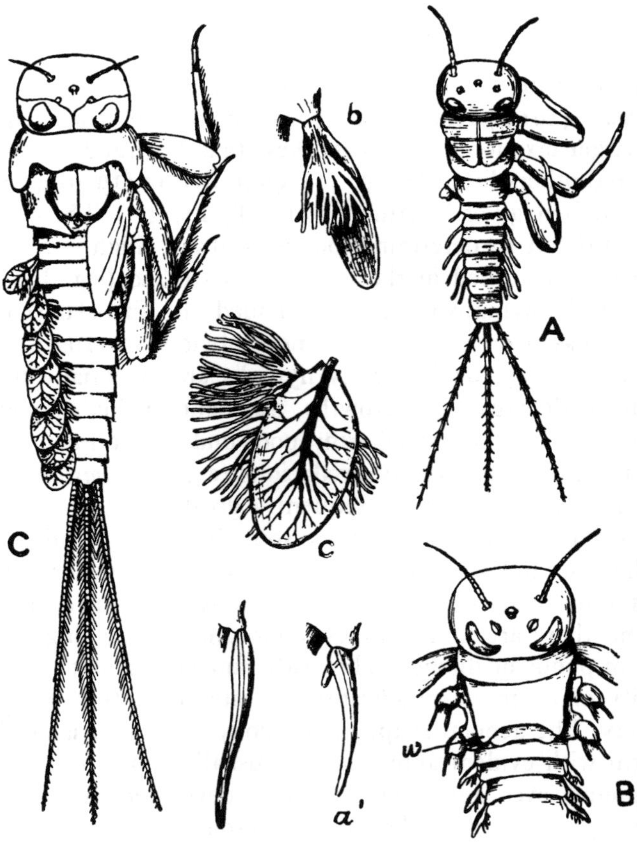

FIG. 227 Nymphal instars of *Heptagenia*

A, 3rd instar × 16. *a*, abdominal appendage (gill-rudiment); a^1, do. of later instar × 45. B, 7th instar, anterior region with wing-rudiments *w*, × 12. *b*, abdominal gill. C, 8th instar with prominent wing-rudiments (on the right), × 4. *c*, abdominal gill. From Carpenter *after* Vayssière.

Siphlonurus have seven pairs of simple lamellae which project from the sides of the body; the three caudal filaments are fringed with setae and function as a kind of tail. They are active swimming nymphs living among water plants, etc. In *Caenis, Tricorythus* and others the nymphs live in an environment of mud and sand; there are six pairs of gills and the upper lamellae of the second pair form opercula concealing the gills behind. The branchial chamber thus formed is guarded by fringes of setae which prevent the entrance of mud or sand particles held in suspension by the inhalant current. In *Oligoneuria* six pairs of dorsal gills are present on segments 2 to 7; each gill consists of a small, thick, scale-like, non-respiratory lamina with a bunch of gill-filaments at its base. A pair of similar ventral gills occurs on the first segment and a tuft of gill-filaments at the base of each maxillary palp. Similar adventitious gills occur on the thorax and mouthparts of a few other

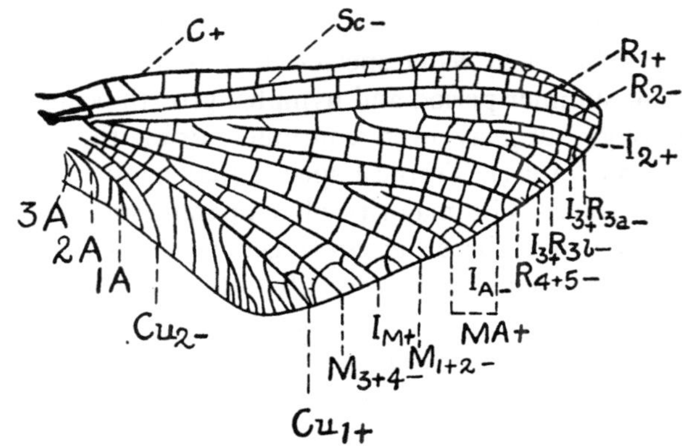

FIG. 228 Fore wing of *Chironetes albimanicatus*

I_2, intercalary branch of R_2. I_3, intercalary branches of R_3. I_A, intercalary branch of *MA*. I_M, intercalary branch of media. *MA*, fork of anterior branch of media. (Convex veins +, concave do. −.) Venation *after* Morgan, modified.

species. *Prosopistoma* (Vayssière, 1890) has a most highly modified nymph which uses its body as a kind of sucker, attaching itself to stones in flowing water; it can also swim rapidly with its fan-like caudal filaments (Fig. 229). In this genus there are five pairs of gills in a branchial chamber roofed over by a carapace that is formed from the greatly developed pro- and mesothoracic terga fused with the anterior wing-pads. The sides of the chamber are formed by the posterior wing-sheaths, and the floor by the combined terga of the metathorax and first six abdominal segments. Water enters this very perfect type of branchial chamber by a pair of lateral apertures and the exhalant stream passes through a median opening.

The Ephemeroptera also show a number of other interesting biological features. In temperate regions some species, such as *Ephemerella ignita* in Europe (Macan, 1957*b*), are summer forms, with nymphs present in the

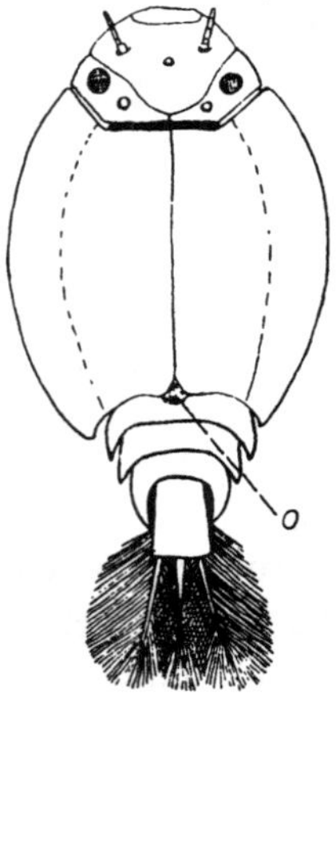

FIG. 229 Nymph of *Prosopistoma*

o, exhalant aperture of branchial chamber. *After* Vayssière, *loc. cit.*

warmer months only and a long period of dormancy as the diapausing egg. In others, the so-called cold water species such as *Rhithrogena semicolorata*, there is a dormant summer period and nymphal growth occurs mainly in the autumn and spring, but may even continue during the winter (see also Maxwell and Benson, 1963). Tropical species may show no seasonal effects (e.g. Tjønneland, 1960). The majority of temperate species are univoltine, but some may have two or three successive generations a year and a few take 2–3 years to develop. Some are especially interesting in that while each individual takes a year to complete its life-cycle, there may at a given time be two or three broods at different stages of development (e.g. Lyman, 1955). The behaviour of mayfly nymphs is influenced by many environmental factors (e.g. Hughes, 1966) and there is often an endogenous diurnal rhythm of activity. *Baetis rhodani, Ecdyonurus venosus* and other species are more active at night, when they move to the upper side of the stones under which they shelter by day (Elliott, 1968; see also Harker, 1953*a*). Like other invertebrates from running waters, mayfly nymphs may drift downstream with the current, an effect which does not seem to be counteracted by any tendency for females to fly upstream before oviposition, as occurs in the Plecoptera (Müller, 1954; Elliott, 1967; Hynes, 1970; Waters, 1972). Nymphs of some species enter into phoretic associations with Chironomid and Simuliid larvae or act as intermediate hosts of digenetic trematodes (Crosskey, 1965; Arvy and Peters, 1973). In addition to papers cited above, there are accounts of nymphal biology by Bretschko (1965), Grandi (1940–43), Harker (1952, 1953*b*), Hynes (1941, 1967), Macan (1957*a*) and Pleskot (1953).

The morphology of mayfly nymphs has been studied in detail by Vayssière (1882), Strenger (1953), Landa (1948, 1969*a*) and others and certain of their more important anatomical features may be enumerated. Gills are commonly undeveloped in the newly-hatched nymphs; in *Ephemera* Heymons states that they arise as integumental outgrowths about the fourth day. The researches of Snodgrass (1931) and others into their development and musculature indicate that they may be serially homologous with legs and should, therefore, be regarded as abdominal appendages which have become adapted for respiratory needs. They also bear specialized 'chloride cells' that are probably concerned in osmoregulation (Wichard *et al.*, 1972; Komnick *et al.*, 1972).

The mouthparts (e.g. Brown, 1961) are well-developed, with large mandibles, and maxillae bearing a single endite and a 3- or 4-segmented palp. The labial palps are generally 3-segmented, the ligula is four-lobed, and the prominent hypopharynx bears a pair of large superlinguae. *Arthroplea* feeds by straining out particles from the water, using long hairs on the palps and its mandibles are inconspicuous (Froehlich, 1964).

The digestive system is characterized by the great size of the mid gut and the large number of Malpighian tubes; the latter differ greatly in character among various genera (Landa 1969*a*). They may either open directly into the hind gut, or combine in groups, each group discharging into a separate duct which, in its turn, communicates with the intestine. The circulatory system is very well developed and easily observed. The dorsal vessel consists of one chamber for each abdominal segment and in the metathorax it is continued forwards as the aorta. In *Cloeon* it gives off a definite vessel into each of the caudal filaments and the terminal chamber acts as a pump driving the blood into these organs; here it escapes by orifices in the walls of the vessels and flows into the cavity of each filament, probably absorbing oxygen from the surrounding water.

The ganglia of the nervous system present varying degrees of fusion. The brain is small but, correlated with the presence of compound eyes, the optic ganglia are well developed. In *Tricorythus* there are three thoracic and seven abdominal ganglia; in *Oniscigaster* the abdominal ganglia are reduced to six and the last two centres are closely united. In *Baetisca* and *Prosopistoma* the ganglia are fused into a common thoracico-abdominal centre (Landa, *loc. cit.*).

Affinities and Phylogeny – The earliest known fossil mayfly is *Triplosoba pulchella* from the Upper Carboniferous, placed by Demoulin (1958) and Tshernova (1970) in the suborder Protephemeroptera, though another Carboniferous genus, the Palaeodictyopteran *Lithoneura*, is venationally closer to the later mayflies (Edmunds, 1972). Apart from a probable relationship between *Triplosoba* and the so-called Archodonata (Demoulin, 1958), there is little to be said on the origin of the order. The paired gonopores are probably primitive and the unique subimaginal instar may be the relic of a formerly more extensive system of imaginal moults such as occurs in the Apterygote insects and which may have been found in some other early Pterygotes. Support for this idea comes from the Permian fossil Paraplecopteran *Atactophlebia termitoides* which seems to have adults of three different sizes, thus leading Sharov (1957) to infer that they moulted twice. Later Permian mayflies of the family Protereismatidae and some related groups made up over three per cent of the insect fauna of that period. These show various primitive features such as the almost equal thoracic segments, wings and legs; the basal attachment of the intercalaries and of R_s to R_1; and the nymphs with 8 or 9 pairs of lateral abdominal gills. Since that time the order has declined in importance: it now constitutes only about 0·2 per cent of the described species of insects. Early Tertiary Ephemeroptera from amber deposits belong to Recent families and these differ from the Permian forms in the characters listed above: the hind wings are smaller, the intercalaries are detached basally, R_s is connected to MA and the nymphs have 7 or fewer pairs of gills.

Classification – There is no modern taxonomic monograph or catalogue of the 2000 or so species in this order, but a number of regional works and faunal lists are available. See especially Berner (1950), Burks (1953) and Edmunds and Allen (1957) for the American fauna; Demoulin (1970) for African species; Kimmins (1972) and Macan (1970) for the 47 British species; and Bogoescu (1958), Grandi (1960), Illies (1967) and Landa (1969*b*) for other parts of Europe. There is considerable agreement on the family limits in the Ephemeroptera but the interrelations of the families and their grouping into higher taxa are still debatable. The classification of the

Ephemeroptera given below is based on Edmunds and Traver (1954*b*) and Illies (1968). It distinguishes 19 families in 5 superfamilies and excludes a few small groups mentioned by Edmunds (1972). The classification differs appreciably from the other modern ones by Demoulin (1958) and Tshernova (1970), who rely extensively on palaeontological evidence. Families occurring in Britain are marked with an asterisk.

Superfamily **Heptagenioidea**

A variable group, combining generalized venational and nymphal characteristics with specialized features in some families.

***FAM. SIPHLONURIDAE.** *Copious wing venation;* Cu_1 *with curved cross-veins joining it posteriorly to the wing margin; male eyes divided; nymphs variable.* A widely distributed family, mainly adapted to cool waters and containing some relatively primitive genera. *Siphlonurus* and *Ameletus.*

FAM. SIPHLAENIGMATIDAE. Represented by the recently discovered New Zealand species *Siphlaenigma janae* (Penniket, 1962); allied to the Siphlonuridae and Baetidae.

***FAM. BAETIDAE.** *Wing venation reduced, with free marginal intercalaries; hind wing reduced or absent; males with turbinate eyes; hind legs with 3 tarsal segments; nymphs with 6–7 pairs of lamellate gills.* Widely distributed. *Baetis*, *Callibaetis*, *Cloeon* and *Centroptilum* are examples.

FAM. OLIGONEURIIDAE. *A specialized family with large, triangular fore wings and reduced venation; legs often reduced and functionless; nymphs with simple gills.* Holarctic, Neotropical, Ethiopian and Oriental regions. *Oligoneuriella rhenana* is from C. Europe.

***FAM. HEPTAGENIIDAE (= ECDYONURIDAE).** *Two pairs of intercalaries between* Cu_1 *and* Cu_2, *the anterior pair shorter; male eyes undivided; hind legs with 5 tarsal segments; nymphs with flattened head and body; gills with lamellate and tufted portions.* Characteristically associated with running water, especially mountain streams, this family occurs in all zoogeographical regions except the Australasian. *Rhithrogena, Heptagenia, Ecdyonurus.*

FAM. AMETROPODIDAE. *Venation like the Heptageniidae, but anterior pair of intercalaries longer; nymphs with subcylindrical body and long claws on mid and hind leg.* A Holarctic family. *Ametropus.*

Superfamily **Leptophlebioidea**

Wing venation generalized; hind leg with only four tarsal segments; fore legs long in both sexes, with unequal claws; male eyes divided; usually three tail filaments; nymphs variable.

***FAM. LEPTOPHLEBIIDAE.** *Fore wing with* Cu_2 *curved gently to hind margin; wing margin without free intercalaries; nymphs with 5–7 lamellate or plumose gills.* A large, cosmopolitan family. *Leptophlebia*, *Paraleptophlebia*, *Habrophlebia* and *Choroterpes* occur in Europe and elsewhere. Over half the Australian mayflies belong to this family.

***FAM. EPHEMERELLIDAE.** *Fore wing with* Cu_2 *bent almost at right angles to reach the hind margin; some free intercalaries present at outer margin; nymphs robust, with dorsal rows of spines or tubercles.* Widely distributed. *Ephemerella.*

FAM. TRICORYTHIDAE. *Fore wing with very few cross-veins; hind wings absent; nymphs sometimes with concealed gills.* The resemblance to the Caenidae is probably the result of convergence. Africa, N. and S. America, Oriental.

Superfamily **Ephemeroidea**

Wings usually with copious venation; posterior branch of MP strongly curved at the base and running to margin parallel with Cu_I; hind legs with 4 freely movable tarsal segments; nymphs usually burrowing, with plumose gills and long mandibular processes.

FAM. BEHNINGIIDAE. *Venation highly specialized with main longitudinal veins arranged in closely parallel pairs; legs reduced; highly modified burrowing nymphs with fore and mid legs fossorial and hind legs adapted to protect gills.* Holarctic. *Behningia.*

***FAM. POTAMANTHIDAE.** *Generalized wing venation, with many rectangular cells; 1A forked; nymph with divided, plumose gills.* A poorly known Holarctic and Oriental family. *Potamanthus luteus* is European, occurring in large streams and small rivers.

FAM. EUTHYPLOCIIDAE. *Wing venation like Polymitarcidae (q.v.); male claspers 1-segmented; nymphs with long maxillary palps and setose mandibular processes.* Tropical.

***FAM. EPHEMERIDAE.** *Fore wing with 1A unforked but connected to hind margin by many intercalaries; nymphs with gills held dorsally and long, pointed mandibular processes.* A cosmopolitan family; *Ephemera* is associated with lakes and slow rivers. *Hexagenia.*

FAM. POLYMITARCIDAE. *Fore wing with many intercalaries between* Cu_I *and* Cu_2*; fore legs long, especially in male; nymphs similar to Ephemeridae.* Widely distributed. Nymphs of the tropical *Povilla* and related genera bore into wood, with enlarged mandibles (Corbet, 1957; Hartland-Rowe, 1958). *Campsurus.*

FAM. PALINGENIIDAE. *Fore wing with distal half of Sc concealed in a fold; legs of female reduced; no median filament; claspers of male three-segmented; nymphs burrowing, with dorsal gills and externally toothed mandibles.* Palaearctic, Oriental and Madagascar. *Palingenia longicauda*, the largest European mayfly, was once common in all the large rivers of C. Europe but is now confined to the lower Danube and Theiss.

Superfamily **Caenoidea**

Recognized especially from the specialized nymphs, in which the gills of the first abdominal segment are absent and those of the second segment form a protective operculum concealing the more normal posterior gills.

FAM. NEOEPHEMERIDAE. *Large forms; wing venation like Ephemeroidea.* Holarctic.

***FAM. CAENIDAE.** *Small species; normal longitudinal veins but few cross-veins and no short, free intercalaries at margin; all legs with 5-segmented tarsi; nymphs flattened, with posterolateral angles of abdominal segments pointed and produced.* All regions except New Zealand and Madagascar. *Caenis, Brachycercus.*

Superfamily **Prosopistomatoidea**

Larvae with a characteristic 'carapace' formed by extensions of mesonotum to cover thorax and half or more of abdomen. Probably related to the Caenoidea.

FAM. BAETISCIDAE. *Generalized venation, with many cross-veins; larval carapace covering thorax and half abdomen. Baetisca*, the only genus, is found in eastern N. America.

FAM. PROSOPISTOMATIDAE. *Venation without cross-veins; legs reduced, functionless; females do not moult from subimago; nymphal carapace covering almost the whole body. Prosopistoma*, with a relict Old World distribution but mainly African, is the only genus. Its adult shows fewer Baetiscid affinities than the nymph (Gillies, 1954).

Literature on the Ephemeroptera

ARVY, L. AND GABE, M. (1953), Données histophysiologiques sur la neurosécrétion chez les Paléoptères (Ephéméroptères et Odonates), *Z. Zellforsch.*, **38**, 591–610.

ARVY, L. AND PETERS, W. L. (1973), Phorésies, biocoenoses et thanatocoenoses chez les Ephéméroptères, *Proc. 1st int. Congr. Ephemeroptera*, 254–311.

BERNER, L. (1941), Ovoviviparous mayflies in Florida, *Florida Ent.*, **24**, 32–34.

—— (1950), *The Mayflies of Florida*, University of Florida Press, Gainesville, 267 pp.

—— (1959), Tabular summary of the biology of North American Mayfly nymphs, *Bull. Fla. St. Mus.*, **4**, 1–58.

BERTRAND, H. (1954), Les insectes aquatiques d'Europe (Genres: larves, nymphes, imagos). I. Introduction: Collemboles, Hemiptères, Odonates, Plécoptères, Ephéméroptères, Megaloptères, Plannipennes, Coleoptères, *Encyc. ent.*, (A) **30**, 1–556.

BIRKET-SMITH, J. (1970), The abdominal morphology of *Povilla adusta* Navas (Polymitarcidae) and of Ephemeroptera in general, *Ent. Scand.*, **2**, 139–160.

BOGOESCU, C. (1958), Ephemeroptera, *Fauna Republicii Populare Romine. Insecta*, 7 (3), 1–187.

BRETSCHKO, G. (1965), Zur Larvalentwicklung von *Cloeon dipterum, Cloeon simile, Centroptilum luteum* und *Baëtis rhodani*, *Z. wiss. Zool.*, **172**, 17–36.

BRINCK, P. (1957), Reproductive system and mating in Ephemeroptera, *Opusc. ent.*, **22**, 1–37.

BROWN, D. S. (1961), The morphology and functioning of the mouthparts of *Chloeon dipterum* L. and *Baetis rhodani* Pictet (Insecta, Ephemeroptera), *Proc. zool. Soc. Lond.*, **136**, 147–176.

BURKS, B. D. (1953), The mayflies or Ephemeroptera of Illinois, *Bull. Ill. nat. Hist. Surv.*, **26**, 1–216.

CORBET, P. S. (1957), Duration of the aquatic stages of *Povilla adusta* Navás (Ephemeroptera: Polymitarcidae), *Bull. ent. Res.*, **48**, 243–250.

CROSSKEY, R. W. (1965), The identification of African Simuliidae (Diptera) living in phoresis with nymphal Ephemeroptera, with special reference to *Simulium berneri* Freeman, *Proc. R. ent. Soc. Lond.*, **40**, 118–124.

DEGRANGE, C. (1960), Recherches sur la reproduction des Ephéméroptères, *Trav. Lab. Hydrobiol. Piscicult., Grenoble*, **50–51**, 7–193.

DEMOULIN, G. (1958), Nouveau schéma de classification des Archodonates et des Ephéméroptères, *Bull. Inst. r. Sci. nat. Belg.*, **34** (27), 1–19.

—— (1970), Ephemeroptera des faunes éthiopiennes et Malgache, *S. Afr. anim. Life*, **14**, 24–170.

DRENKELFORT, H. (1910), Neuer Beitrag zur Kenntnis der Biologie und Anatomie von *Siphlurus lacustris* Etn., *Zool. Jb.* (*Anat.*), **29**, 527–617.

EASTHAM, L. E. S. (1936), The rhythmical movements of the gills of nymphal *Leptophlebia marginata* (Ephemeroptera) and the currents produced by them in water, *J. exp. Biol.*, **13**, 443–449.

—— (1937), The gill movements of nymphal *Ecdyonurus venosus* (Ephemeroptera) and the currents produced by them in water, *J. exp. Biol.*, **14**, 219–229.

—— (1939), Gill movements of nymphal *Ephemera danica* (Ephemeroptera) and the water-currents caused by them, *J. exp. Biol.*, **16**, 18–33.

—— (1958), The abdominal musculature of nymphal *Chloeon dipterum* L. (Insecta: Ephemeroptera) in relation to gill movement and swimming, *Proc. zool. Soc. Lond.*, **131**, 279–291.

EDMUNDS, G. F. (1956), Exuviation of subimaginal mosquitoes in flight, *Ent. News*, **67**, 91–93.

—— (1972), Biogeography and evolution of Ephemeroptera, *A. Rev. Ent.*, **17**, 21–42.

EDMUNDS, G. F. AND ALLEN, R. K. (1957), A check list of the Ephemeroptera of North America north of Mexico, *Ann. ent. Soc. Am.*, **50**, 317–324.

EDMUNDS, G. F., ALLEN, R. K. AND PETERS, W. L. (1963), An annotated key to the nymphs of the families and subfamilies of mayflies, *Univ. Utah Biol. Ser.*, **13** (1), 1–49.

EDMUNDS, G. F. AND TRAVER, J. R. (1954*a*), The flight mechanics and evolution of the wings of Ephemeroptera, with notes on the archetype insect wing, *J. Wash. Acad. Sci.*, **44**, 390–400.

—— (1954*b*), An outline of a reclassification of the Ephemeroptera, *Proc. ent. Soc. Wash.*, **56**, 236–240.

ELLIOTT, J. M. (1967), The life histories and drifting of the Plecoptera and Ephemeroptera in a Dartmoor stream, *J. Anim. Ecol.*, **36**, 343–362.

—— (1968), The daily activity patterns of mayfly nymphs (Ephemeroptera), *J. Zool.*, **155**, 201–221.

ERIKSEN, C. H. (1963), Respiratory regulation in *Ephemera simulans* Walker and *Hexagenia limbata* (Serville) (Ephemeroptera), *J. exp. Biol.*, **40**, 455–467.

FROEHLICH, C. G. (1964), The feeding apparatus of the nymph of *Arthroplea congener* Bengtsson (Ephemeroptera), *Opusc. ent.*, **29**, 188–208.

GILLIES, M. T. (1954), The adult stages of *Prosopistoma* Latreille (Ephemeroptera) with descriptions of two new species from Africa, *Trans. R. ent. Soc. Lond.*, **105**, 355–372.

GRANDI, M. (1940–43), Contributi allo studio degli Efemerotteri italiani. I–III, V, *Boll. Ist. Ent. Univ. Bologna*, **12**, 1–2; 179–205; **13**, 29–71; **14**, 114–130.

—— (1947), Contributi allo studio degli Efemerotteri italiani. VIII. Gli scleriti ascellari (pseudopteralia) degli Efemeroidei, loro morfologia e miologia comparata, *Boll. Ist. Ent. Univ. Bologna*, **16**, 85–114.

—— (1950), Contributi allo studio degli Efemerotteri italiani. XIV. Morfologia ed istologia dell'apparato digerente degli standi preimmaginali, subimmaginali ed immaginali di veri generi et specie, *Boll. Ist. Ent. Univ. Bologna*, **18**, 58–92.

—— (1955), I gonodotti femminili degli Efemeroidei, loro comportamento e loro sbocco. Studio anatomico comparato. (Contributo allo studio degli Efemeroidei italiani. XIX), *Boll. Ist. Ent. Univ. Bologna*, **21**, 9–41.

—— (1960), Ephemeroidea, *Fauna d'Italia*, **3**, 474 pp.

—— (1962), Contributi allo studio degli Efemeroidei italiani. XXIII. Gli organi genitali esterni maschili degli Efemeroidei, *Boll. Ist. Ent. Univ. Bologna*, **24** (1960), 67–120.

HARKER, J. E. (1952), A study of the life cycles and growth-rates of four species of mayflies, *Proc. R. ent. Soc. Lond.*, (A), **27**, 77–85.

—— (1953*a*), The diurnal rhythm of activity of mayfly nymphs, *J. exp. Biol.*, **30**, 525–533.

—— (1953*b*), An investigation of the distribution of the mayfly fauna of a Lancashire stream, *J. Anim. Ecol.*, **22**, 1–13.

HARRIS, J. R. (1952), *An Angler's Entomology*, Collins, London, 268 pp.

HARTLAND-ROWE, R. (1958), The biology of a tropical mayfly *Povilla adusta* Navás (Ephemeroptera Polymitarcidae) with special reference to the lunar rhythm of emergence, *Rev. Zool. Bot. afr.*, **58**, 185–202.

HEINER, H. (1914), Zur Biologie und Anatomie von *Cloëon dipterum* L., *Baetis binoculatus* L., und *Habrophlebia fusca* Curt., *Jena. Z. Naturw.*, **53**, 5–56.

HEYMONS, R. (1896), Über die Lebensweise und Entwicklung von *Ephemera vulgata* L., *S. B. Ges. naturf. Fr. Berlin*, **1896**, 82–96.

HUGHES, D. A. (1966), The role of responses to light in the selection and maintenance of microhabitat by the nymphs of two species of mayfly, *Anim. Behav.*, **14**, 17–33.

HYNES, H. B. N. (1941), The invertebrate fauna of a Welsh mountain stream, *Arch. Hydrobiol.*, **57**, 344–388.

—— (1967), Further studies on the invertebrate fauna of a Welsh mountain stream, *Arch. Hydrobiol.*, **65**, 360–379.

—— (1970), Ecology of stream insects, *A. Rev. Ent.*, **15**, 25–42.

IDE, F. P. (1935), Post-embryological development of Ephemeroptera (mayflies), external characters only, *Can. J. Res.*, **12**, 433–478.

ILLIES, J. (1967), *Limnofauna Europaea*, Fischer, Stuttgart, (Ephemeroptera, pp. 220–229).

—— (1968), Ephemeroptera (Eintagsfliegen), In: Helmcke, J.-G., Starck, D. and Wermuth, H. (eds), *Handbuch der Zoologie*, **4** (2), Lief. 7, 63 pp.

KIMMINS, D. E. (1972), A revised key to the adults of the British species of Ephemeroptera with notes on their ecology, *Scient. Publs Freshwat. biol. Ass.*, **15** (2nd rev. edn), 75 pp.

KNOX, V. (1935), The body wall of the thorax, In: Needham, J. G. *et al.* (eds), *The Biology of Mayflies*, Comstock Publ. Co., Ithaca, N.Y., 759 pp.

KOMNICK, H., RHEES, R. W. AND ABEL, J. H. (1972), The function of ephemerid chloride cells. Histochemical, autoradiographic and physiological studies with radioactive chloride on *Callibaetis*, *Cytobiologie*, **5**, 65–82.

KOSS, R. W. (1968), Morphology and taxonomic use of Ephemeroptera eggs, *Ann. ent. Soc. Am.*, **61**, 696–721.

KUKALOVA-PECK, J. (1975), The pteralia of Palaeozoic Palaeopterous insects (Megasecoptera, Palaeodictyoptera, Diaphanopterodea) compared to those of Ephemeroptera, *Psyche*, 81, 416–430.

LANDA, V. (1948), Contribution to the anatomy of Ephemerid larvae. I. Topography and anatomy of tracheal system, *Vestn. Čsl. zool. spol., Prag*, **21**, 25–82.

—— (1959), Problems of internal anatomy of Ephemeroptera and their relation to the phylogeny and systematics of the order, *Proc. XVth int. Congr. Zool.*, (1958), 113–115.

—— (1969*a*), Comparative anatomy of mayfly larvae (Ephemeroptera), *Acta ent. bohemoslov.*, **66**, 289–316.

—— (1969*b*), Jepice – Ephemeroptera, *Fauna Č.S.S.R.*, **18**, 1–347.

LESTAGE, J. A. (1917–25), Contribution à l'étude des larves des Ephémères paléarctiques, *Ann. Biol. lacustr.*, **8**, 213–458; **9**, 79–182; **13**, 237–302.

LEVY, H. A. (1948), The male genitalia of Ephemerida (Mayflies), *Jl N.Y. ent. Soc.*, **56**, 25–41.

LYMAN, F. E. (1955), Seasonal distribution and life cycles of Ephemeroptera, *Ann. ent. Soc. Am.*, **48**, 380–391.

MACAN, T. T. (1957*a*), The Ephemeroptera of a stony stream, *J. Anim. Ecol.*, **26**, 317–342.

—— (1957*b*), The life histories and migrations of the Ephemeroptera in a stony stream, *Trans. Soc. Br. Ent.*, **12**, 129–156.

—— (1970), A key to the nymphs of British species of Ephemeroptera with notes on their ecology, *Scient. Publs Freshwat. biol. Ass.*, **20** (2nd rev. edn), 68 pp.

MATSUDA, R. (1956), Morphology of the thoracic exoskeleton and musculature of a mayfly *Siphlonurus columbianus* McDunnough (Siphlonuridae, Ephemeroptera). A contribution to the subcoxal theory of the insect thorax, *J. Kansas ent. Soc.*, **29**, 92–113.

MAXWELL, G. R. AND BENSON, A. (1963), Wing pad and tergite growth of mayfly nymphs in winter, *Am. Midl. Nat.*, **69**, 224–230.

MORGAN, A. H. (1913), A contribution to the biology of mayflies, *Ann. ent. Soc. Am.*, **6**, 371–426.

MORGAN, A. H. AND GRIERSON, M. C. (1932), The function of the gills in burrowing mayflies (*Hexagenia recurvata*), *Physiol. Zool.*, **5**, 230–245.

MORRISON, E. R. (1919), The mayfly ovipositor, with notes on *Leptophlebia* and *Hagenulus*, *Can. Ent.*, 51, 139–146.

MÜLLER, K. (1954), Investigations on the organic drift in North Swedish streams, *Rep. Inst. Freshwater Res. Drottningholm, Lund*, 35, 133–148.

MURPHY, H. E. (1922), Notes on the biology of some of our North American species of mayflies. I. The metamorphosis of mayfly mouthparts, *Bull. Lloyd Library ent. Ser.*, 22 (2), 1–39; II. Notes on the biology of mayflies of the genus *Baetis*. *Ibid.*, 40–46.

NEEDHAM, J. G., TRAVER, J. R. AND HSU, Y. (1935), *The Biology of Mayflies*, Comstock Publ. Co., Ithaca, 759 pp.

PENNIKET, J. G. (1962), Notes on New Zealand Ephemeroptera. III, *Rec. Canterbury Mus.*, 7, 389–398.

PETERS, W. L. AND PETERS, J. G. (eds) (1973), *Proceedings of the First International Congress on Ephemeroptera. 1970*, London, 312 pp.

PICKLES, A. (1931), On the metamorphosis of the alimentary canal in certain Ephemeroptera, *Trans. R. ent. Soc. Lond.*, 79, 263–276.

PLESKOT, G. (1953), Zur Oekologie der Leptophlebiiden (Ins., Ephemeroptera), *Öst. zool. Z.*, 4, 45–107.

—— (1958), Die Periodizität einiger Ephemeropteren der Schwechat, *Wasser u. Abwässer*, 1958, 1–32.

QADRI, M. A. H. (1940), On the development of the genitalia and their ducts of Orthopteroid insects, *Trans. R. ent. Soc. Lond.*, 90, 121–175.

RAWLINSON, R. (1939), Studies on the life-history and breeding of *Ecdyonurus venosus* (Ephemeroptera), *Proc. zool. Soc. Lond.* (B), 109, 377–450.

SCHOENEMUND, E. (1930), Eintagsfliegen oder Ephemeroptera, *Tierwelt Deutschlands*, 11, 1–106.

SHAROV, A. G. (1957), Types of insect metamorphosis and their relationship, *Ent. Obozr.*, 36, 569–576. (In Russian.)

SMITH, O. R. (1935), The eggs and egg-laying habits of North American Mayflies, in: Needham *et al.* (1935), pp. 67–89.

SNODGRASS, R. E. (1931), Morphology of the insect abdomen. Part 1. General structure of the abdomen and its appendages, *Smithson. misc. Collns*, 85 (6), 1–128.

—— (1957), A revised interpretation of the external reproductive organs of male insects, *Smithson. misc. Collns*, 135 (6), 60 pp.

SPIETH, H. T. (1933), The phylogeny of some mayfly genera, *J. New York ent. Soc.*, 41, 55–86; 327–390.

—— (1940), Studies on the biology of the Ephemeroptera. II. The nuptial flight, *J. New York ent. Soc.*, 48, 379–390.

STERNEFELD, R. (1907), Die Verkümmerung der Mundteile und der Funktionswechsel des Darms bei Ephemeriden, *Zool. Jb.* (*Anat.*), 24, 415–430.

STRENGER, A. (1953), Zur Kopfmorphologie der Ephemeridenlarven. Erster Teil. *Ecdyonurus und Rhithrogenia*, *Öst. zool. Z.*, 4, 191–228.

STUART, A. M. (1958), The efficiency of adaptive structures in the nymph of *Rhithrogenia semicolorata* (Curtis) (Ephemeroptera), *J. exp. Biol.*, 35, 27–38.

TAYLOR, R. L. AND RICHARDS, A. G. (1963), The sub-imaginal cuticle of the mayfly *Callibaetis* sp. (Ephemeroptera), *Ann. ent. Soc. Am.*, 56, 418–426.

TILLYARD, R. J. (1932), Kansas Permian Insects. Part 15. The order Plectoptera, *Am. J. Sci.*, 23, 97–134; 237–272.

TJØNNELAND, A. (1960), The flight of mayflies as expressed in some East African species, *Arbok Univ. Bergen, mat.-naturv. ser.*, 1, 1–88.

TSHERNOVA, O. A. (1970), On the classification of fossil and recent Ephemeroptera, *Ent. Rev.*, **49**, 71–81.

TSUI, P. T. P. AND PETERS, W. L. (1972), The comparative morphology of the thorax of selected genera of the Leptophlebiidae (Ephemeroptera), *Proc. zool. Soc. Lond.*, **168**, 309–368.

VAYSSIÈRE, A. (1882), Organisation des larves des Ephémérines, *Ann. Sci. nat. Zool.*, Sér. 6, 13–14, 1–137.

—— (1890), Monographie zoologique et anatomique du genre *Prosopistoma* Latr., *Ann. Sci. nat. Zool.*, Sér. 7, **9**, 19–87.

VERRIER, M. L. (1940), Recherches sur les yeux et la vision des Arthropodes. I, *Bull. biol. Fr. Belg.*, **74**, 309–326.

—— (1956), *Biologie des Ephémères*, Colin, Paris, pp. 216.

WATERS, T. F. (1972), The drift of stream insects, *A. Rev. Ent.*, **17**, 253–272.

WICHARD, W., KOMNICK, H. AND ABEL, J. H. (1972), Typology of Ephemerid chloride cells, *Z. Zellforsch.*, **132**, 533–551.

WINGFIELD, C. A. (1939), The function of the gills of mayfly nymphs from different habitats, *J. exp. Biol.*, **16**, 363–373.

Order 6

ODONATA (DRAGONFLIES)

Predacious insects with biting mouthparts and two equal or subequal pairs of elongate, membranous wings; each wing with a complex reticulation of small cross-veins and usually a conspicuous stigma. Eyes very large and prominent; antennae very short and filiform. Abdomen elongate, often extremely slender; male accessory genital armature developed on 2nd and 3rd abdominal sterna. Nymphs aquatic, hemimetabolous; labium modified into a prehensile organ; respiration by means of rectal or caudal gills.

Rather more than 5000 species of these elegant insects have been described, included in over 500 genera. They attain their greatest abundance in the Oriental and Neotropical regions and, except for Japan, no part of the Palaearctic zone contains an abundant or striking dragonfly fauna. Apart from the Hawaiian *Megalagrion oahuense* and a few others they are aquatic in their early stages. The adults, however, are not confined to the proximity of

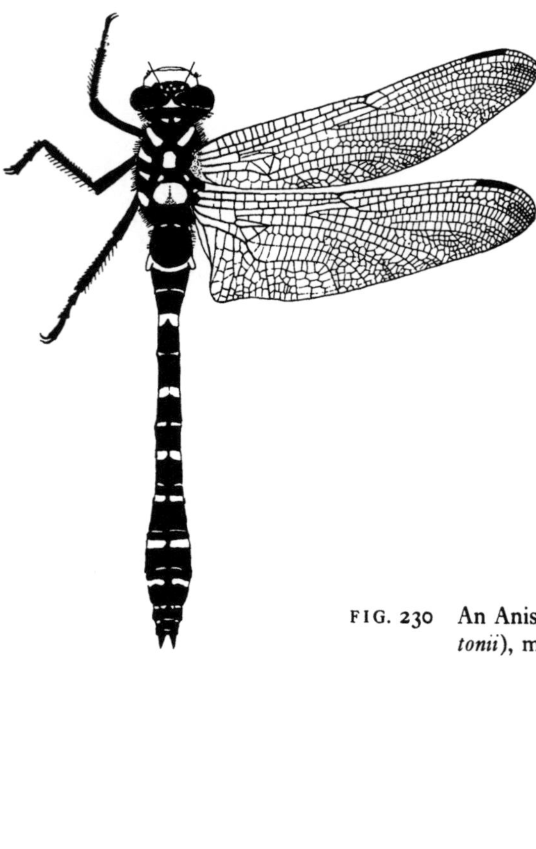

FIG. 230 An Anisopteran dragonfly (*Cordulegaster boltonii*), male. Britain

water, and the females of many groups seldom fly there except for pairing or oviposition. They are essentially sun-loving insects though some oriental species are only known to fly at night. Many are very swift on the wing and according to Tillyard *Austrophlebia* can fly at nearly 60 miles per hour; other species, particularly those of *Coenagrion* and *Agrion*, possess feeble powers of flight and may easily be caught. Although no existing member of the order can compare in size with the Upper Carboniferous *Meganeura*, which had a wing expanse of over 600 mm, the females of *Megaloprepus coerulatus* measure about 190 mm across the wings.

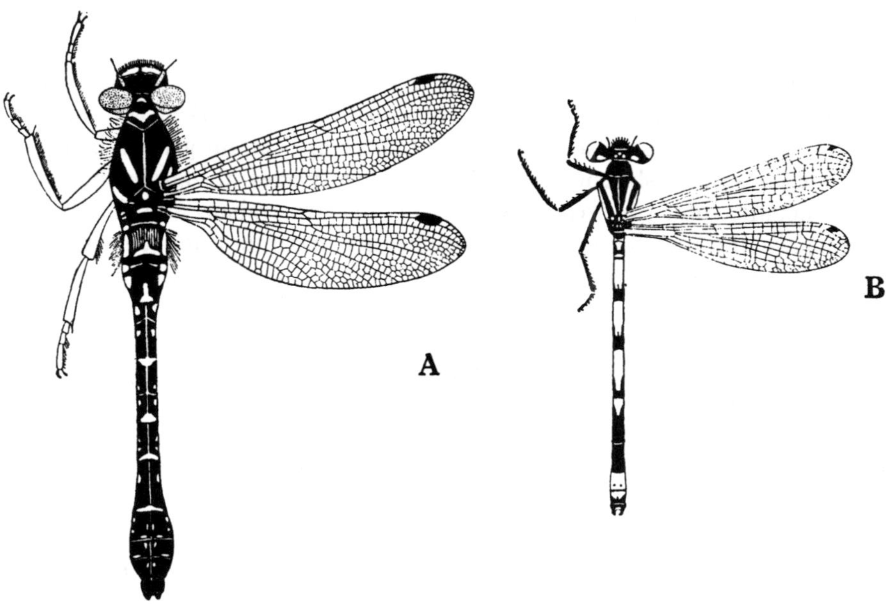

FIG. 231 A. *Epiophlebia superstes*, female. Japan
B. Zygopteran dragonfly (*Coenagrion puella*), male. Britain
Adapted from de Selys with wing-venation after Munz.

Adult dragonflies are generalized predators, usually feeding on almost any suitably sized insects (Hobby, 1934). They capture their prey in flight or while it is resting, using their forwardly directed legs to hold and transfer it to the mouthparts. Most species feed during the day but some species, when feeding on swarms of Culicidae, Chironomidae, or other Diptera, may do so at dusk or dawn (Corbet, 1962). In many species the mature males exhibit territorial behaviour, establishing themselves along stretches of water which they defend against other males and in which mating and oviposition take place (e.g. Moore, 1952; Jacobs, 1955; Kormondy, 1961; Johnson, 1962; Heymer, 1972). This behaviour tends to control the density of the species in the most suitable habitats, reduces disturbance to mating and egg-laying, and results in the dispersal of sexually mature males to new areas (Moore,

1953, 1964). Apart from local movements of this sort, some species such as *Libellula quadrimaculata* undertake longer migratory flights, sometimes in swarms (Williams, 1958).

Odonata are noted for the beauty and brilliance of their coloration; in addition to pigmentary and structural colours, a whitish or bluish pruinescence is often present, especially in the males. It appears to be associated with the maturation of the gonads and is exuded through fine cuticular pores. Among dragonflies a bicolorous pattern is the most primitive; many unicolorous males have bicolorous females, and newly emerged males often exhibit traces of an original bicolorous marking. Among the Coenagriidae the females are sometimes dimorphic, and one or other colour form may closely resemble the male. In the common *Ischnura elegans*, for example, the predominant colour form of the female is very like that of the male whereas the rarer or 'heteromorphic' females are conspicuously marked with orange. Most Odonata possess hyaline wings but there are certain groups in which they are conspicuously coloured. Thus among species of *Agrion* the males have metallic blue or green wings. In the Australian and E. Indian *Rhinocypha* the metallic coloration reaches its maximum and consists of a combination of glistening reds, mauves, purples, bronzes and greens. In *Rhyothemis* the wings are also exquisitely coloured with metallic green, purple or bronze.

General works on the order include those by Tillyard (1917*a*), Robert (1959), Corbet *et al.* (1960), St Quentin and Beier (1968) and especially the account of their biology by Corbet (1962).

The Imago

External Anatomy – In addition to special accounts mentioned below there are monographs of the external anatomy of *Sympetrum* (Winkelmann, 1973), *Onychogomphus* (Chao, 1953) and *Epiophlebia* (Asahina, 1954). The *head* (Lew, 1934; Hakim, 1964; Short, 1955) has become modified in association with the great development of the *eyes* (Fig. 232). The latter, in many Anisoptera, meet mid-dorsally and compose by far the largest part of the cephalic region; in the Zygoptera the eyes are much smaller and button-like, but their range of vision is increased by the transversely elongate head. In the Anisoptera, occiput, vertex and frons are distinct sclerites but in the Zygoptera the sutures of the head-capsule in these regions are less distinct or absent. The frons is divided into two regions, the ante- and post-frons, by a transfrontal sulcus and its posterior limit is marked by a sulcus rather than the remnants of an ecdysial cleavage line. A characteristic Π-shaped tentorium is present (Hudson, 1948) and there are three ocelli (Ruck and Edwards, 1964; Munchberg, 1967).

The *antennae* are always very short and inconspicuous; they are composed of three to seven segments, the latter number being usual, and bear numerous coeloconic sensilla that are thought to act as chemoreceptors

(Slifer and Sekhon, 1972). The reduced condition of the antennae is correlated with the increased power of the compound eyes. The *mouthparts* (Fig. 233) are entirely of a biting and masticatory type (Mathur, 1962). The *mandibles* are stout with very powerful teeth, and the *maxillae* each carry a lobe-like unsegmented palp and a dentate mala, the latter probably representing the fused lacinia and galea. The morphology of the *labium* has given rise to considerable controversy (Fig. 233). The prementum is expanded by the development of sidepieces or *squamae* and each squama carries the *lateral lobe* of its side. The inner border of each lobe terminates in an

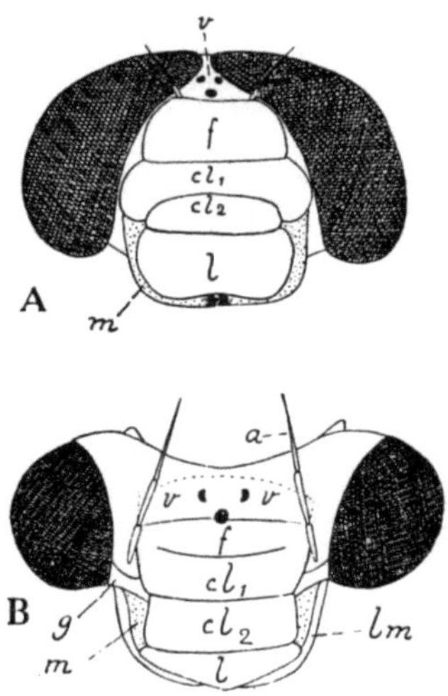

FIG. 232
Head of A, Anisoptera: B, Zygoptera

a, antenna; cl_1, post-clypeus; cl_2, ante-clypeus; *f*, frons; *g*, gena; *l*, labrum; *lm*, labium; *m*, mandible; *v*, vertex.

end-hook, slightly external to which is a small *movable hook*. The lateral lobes and hooks are probably modified labial palps. The prementum carries a single distal lobe or *ligula* which is often medially cleft. In the Libelluloidea the movable hook is wanting, the end-hook and median lobe are vestigial, and the two lateral lobes are greatly developed. The head is exceptionally mobile and attached to an exceedingly small slender neck region which is supported on either side by four cervical sclerites.

The prothorax, though greatly reduced, remains a distinct segment, but the meso- and metathorax are intimately fused together and modified to meet the requirements of the legs and wings (Fig. 234). The legs have shifted their attachments anteriorly and the sterna have migrated with them. The wings and terga, on the other hand, have moved posteriorly. Although the sterna and terga of these segments are reduced their pleura are very greatly developed. The mesepisterna extend forwards and dorsalwards to meet in front of the mesotergum and form the dorsal carina; by this means the terga are pushed backwards and lie between the wing bases. The metepimera on the other hand have grown downwards and backwards, usually fusing ventrally behind the metasternum. In this way the sterna are pushed forwards and the legs lie close behind the mouth to hold the prey.

The legs are unfitted for walking but are of some value for climbing, and the tarsi are 3-segmented with paired claws.

The Odonata are accomplished fliers and the skeletomuscular system of the thorax is functionally modified to enable them to move swiftly and manoeuvre skilfully (Clark, 1940; Hatch, 1966). Each wing has two fulcra and six functional connections with the tergum, and can move independently of the other wings. The propulsive downstroke is brought about mainly by one basalar and two subalar muscles and the upstroke by a tergosternal and two coxo-alar muscles. The large size and importance of the direct flight

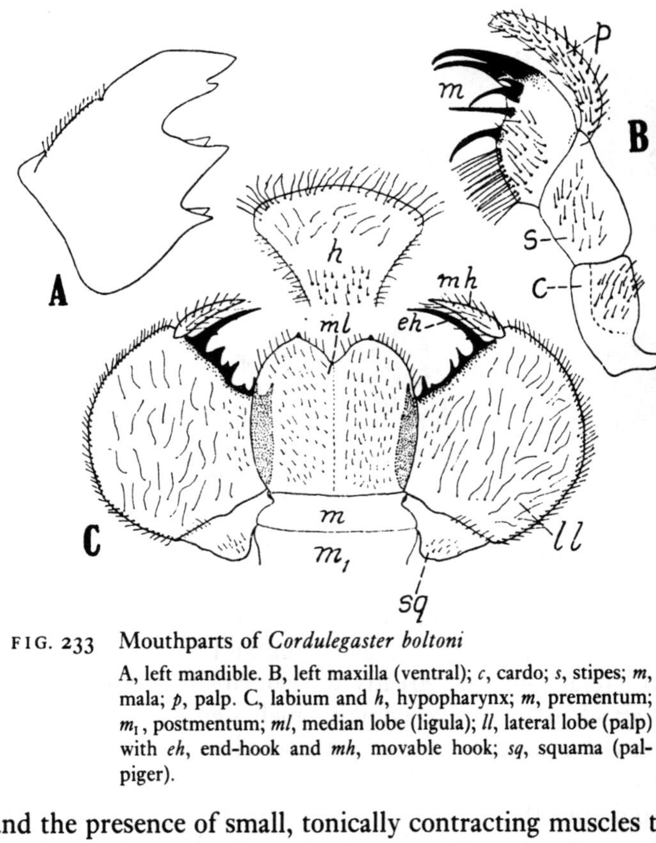

FIG. 233 Mouthparts of *Cordulegaster boltoni*

A, left mandible. B, left maxilla (ventral); *c*, cardo; *s*, stipes; *m*, mala; *p*, palp. C, labium and *h*, hypopharynx; *m*, prementum; m_1, postmentum; *ml*, median lobe (ligula); *ll*, lateral lobe (palp) with *eh*, end-hook and *mh*, movable hook; *sq*, squama (palpiger).

muscles and the presence of small, tonically contracting muscles that modify the flight movements, are of special interest. See also Russenberger and Russenberger (1963) and Tannert (1958).

The two pairs of wings are almost identical in the Zygoptera, but in Anisoptera the hind wings are broader basally and there are minor venational differences (Fig. 235). The veinlets are developed to a remarkable degree and form a complex reticulum of often minute cells. In a single wing of *Neurothemis* according to Tillyard there are over 3000 cells. The stigma, a thickening of the wing-membrane between C and R, is a very characteristic feature. In the Petaluridae it is very elongate while in the Pseudostigmatidae it may be absent or abnormal.

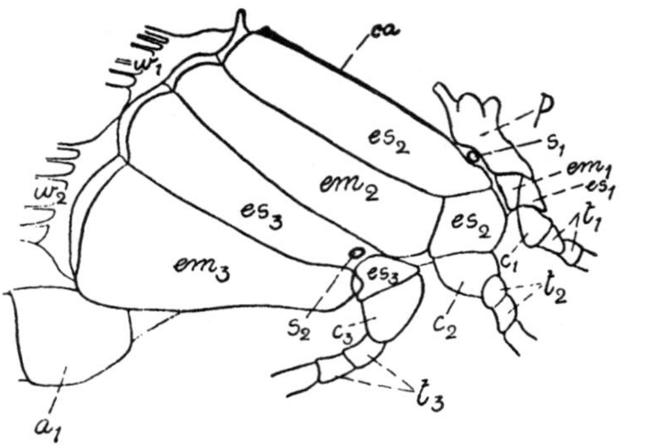

FIG. 234
Lateral view of thorax of *Agrion virgo*
a_1, 1st abdominal segment; *c*, coxa; *ca*, dorsal carina; *em*, epimeron; *es*, episternum (es_2 and es_3 are divided into an- and katepisternum); *p*, pronotum; *s*, spiracle; *t*, trochanter (double); *w*, wing.

The homologies of the main veins have not been unequivocally decided. Needham (1951) and others rely much on the pattern of nymphal tracheation while Tillyard (1928) and Tillyard and Fraser (1938–40) – who are followed here – base their interpretation on fossil forms. Sc lies a little behind the costal margin and appears to end at the thickened cross-vein

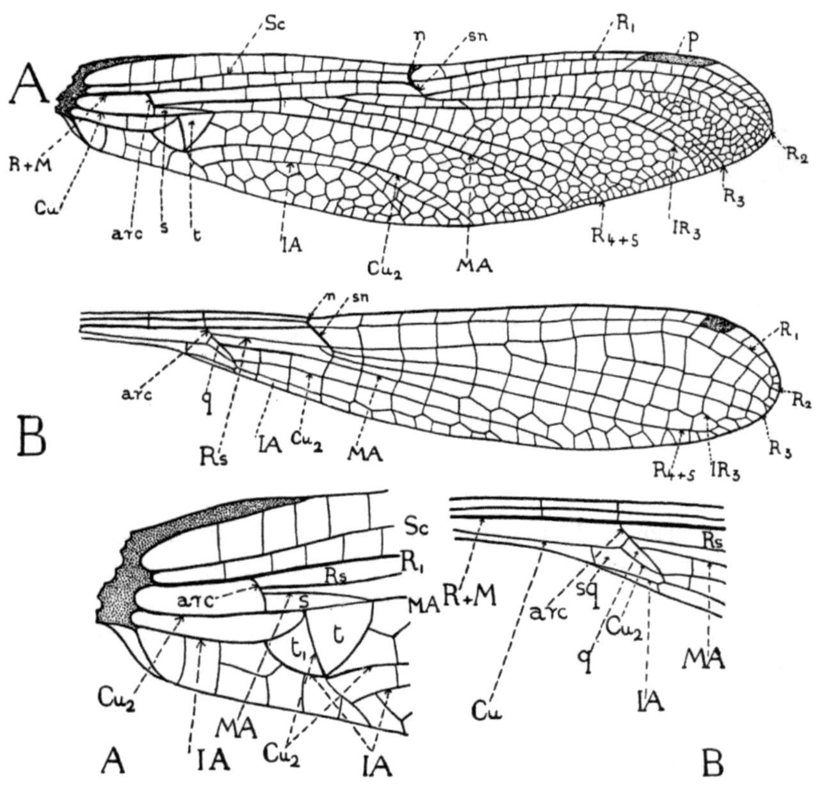

FIG. 235 A, venation of an Anisopteran dragonfly (chief features only). B, of a Zygopteran dragonfly. Basal area of fore wing of C, an Anisopteran. D, of a Zygopteran
arc, arculus; IR_3, intercalary vein; *n*, nodus; *p*, pterostigma; *q*, quadrilateral; *s*, supratriangle; *sn*, subnodus; *t*, triangle; *sq*, subquadrangle; t_1, subtriangle.

which constitutes the joint-like *nodus*. The radius and media arise basally as a single vein from which R_1 is soon given off as an unbranched vein running to the wing apex and joined to the costal margin by a series of *ante-* and *postnodal* cross-veins. The backwardly bent stalk of $R_s + M$, together with the cross-vein behind it, forms the *arculus*, just distal to which is a conspicuous *discoidal cell*. In the Zygoptera this is known from its shape as the *quadrilateral*, but in the Anisoptera it is divided by a cross-vein into the *triangle* and *supratriangle*, both of which may be further subdivided by thin cross-veins. R_s is divided into three branches between which secondary intercalary or supplementary veins may occur. The media is represented

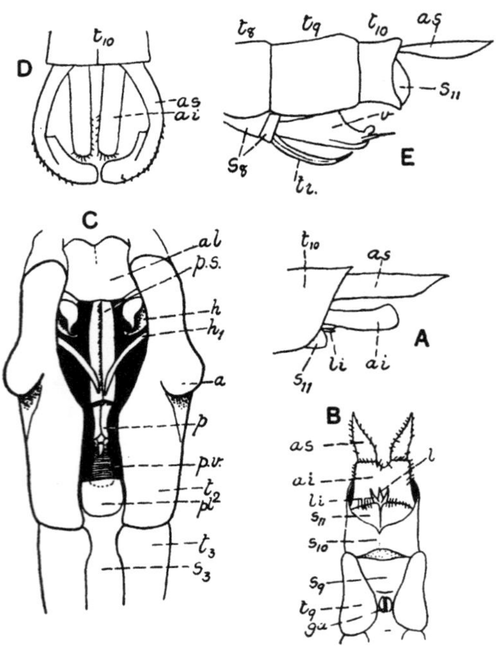

FIG. 236
Cordulegaster. A, lateral view of male genitalia. B, ventral view of terminal abdominal segments of male. C, ventral view of male copulatory apparatus. D, *Agrion*, dorsal view of male genitalia. E, *Aeshna*, lateral view of terminal abdominal segments of female with left parts of ovipositor slightly separated

a, oreillet; *ai*, inferior, and *as*, superior anal appendages; *al*, *pl*, anterior and posterior sternal laminae; *ga*, genital aperture; *h*, h_1, hamuli; *l*, lamina supra-analis; *li*, lamina infra-analis; *p*, penis with *ps*, its sheath; *pv*, penis vesicle; *s*, sternum; *t*, tergum; *tr*, terebra; *v*, valve.

only by the anterior media (MA) and the cubitus only by its posterior branch (Cu_2), though basal traces of the missing MP and Cu_1 are found in the fossil Meganeuridae. The anal vein (1A) arises independently and runs almost parallel to Cu_2 (Fraser, 1938) but is coincident with the hind margin of the wing basally in many Zygoptera. 1A and Cu_2 are connected basally by a cross-vein, the *anal crossing* (Ac). The venation provides many important taxonomic characters, an indication of some being given in the diagnoses of systematic groups on pp. 511–514).

The abdomen (Fig. 236) is always greatly elongate in proportion to its breadth, and in extreme cases it is so attenuated as to be scarcely thicker than a stout bristle (Whedon, 1918). Ten complete segments are evident and parts of the 11th segment and the telson are also recognizable. In males of the Anisoptera and a few Zygoptera the second abdominal tergite bears lateral, spinose processes, the auricles or oreillets, which apparently act as guides, enabling the end of the female abdomen to make effective contact with the secondary copulatory organ of the male (Fraser, 1943). The interpretation of the terminal abdominal structures has been disputed, but the views of Snodgrass (1954) are followed here. Behind the tenth segment is a pair of *superior anal appendages*, apparently modified cerci, which are well-developed in the male but reduced or vestigial in females. The 11th segment comprises a tergite and a small divided sternite. The tergite (epiproct) is produced into the *median inferior anal appendage* in Anisopteran males but is otherwise rudimentary. The sternites (paraprocts) form the *paired inferior anal appendages* of male Zygoptera but are absent or vestigial in other members of the order. The telson is probably represented by three small processes immediately surrounding the anus: a median dorsal *lamina supra-analis* and paired lateroventral *laminae infra-anales*. During pairing, which often occurs in flight, the female is grasped by the anal appendages of the male, the superior pair establishing a firm grip in the region of the neck (among Anisoptera) or prothorax (among Zygoptera) while the inferior appendage of the Anisopteran male is pressed down upon the occiput. In the Zygoptera the inferior pair is usually too short to reach the head. Mechanical factors involved in pairing seem to play a part in the sexual isolation of some species (e.g. Paulson, 1974). Males do not always distinguish visually between females of their own species and other dragonflies, but when they try to mate with heterospecific females their abdominal appendages are unable to secure a firm grip.

The secondary copulatory organ of the male (Fig. 236) is unique among insects, being developed from the second and third abdominal sterna, though the true genital aperture opens on the 9th segment. A spermatophore must therefore be transferred from the gonopore to the secondary copulatory organ before mating. Great variation is exhibited by the copulatory organ, which therefore has some taxonomic value; for details see Schmidt (1916) and Poonawalla (1966). The precise functions of the various parts are difficult to make out and are discussed by Pfau (1971) and Johnson (1972). On the 2nd sternum is a depression or *genital fossa* in which the copulatory organs are housed and walls of which are supported by a complex sclerotized framework. The fossa communicates posteriorly with a small sac – the *penis vesicle*, which is developed from the anterior part of the 3rd sternum. The intromittent organ appears to be a somewhat different structure in each of the three suborders (Pfau, *loc. cit.*). In the Anisoptera the penis is a complex, jointed organ, partly covered anteriorly by a sheath-like ligula, and provided with two orifices (one to receive the sperm and one from which it is discharged). In the Zygoptera it is the ligula which acts as the intromittent organ and in *Epiophlebia* the posterior hamuli carry out the function. Otherwise the hamuli appear to guide and retain the female genital structures during copulation. The posterior pair of hamuli is universal, but the anterior pair occurs only in the Aeshnoidea. In some species sexual isolation may depend on physical incompatibility of the secondary copulatory organ and the female genitalia (Watson, 1966).

In the female the external genitalia consist typically, as in the Zygoptera, of three pairs of appendages which make up the ovipositor (Fig. 236, E). An anterior pair, developed from the gonapophyses of the 8th abdominal segment, and an inner pair from those of the 9th segment are slender structures adapted for cutting and, together, constitute the terebra. A dorsal pair of valves, also arising from the 9th segment and representing the gonoplac and style, are broad, lamellate organs each terminating in a hard, pointed structure which may be tactile in function. A similar type of ovipositor is found among the Anisoptera in the Aeshnidae and Petaluridae, but among the other Anisopteran families either the lateral valves alone are vestigial (Cordulegasteridae) or all three pairs are vestigial or absent. The different types of ovipositor are correlated with different modes of oviposition (St. Quentin, 1962).

Internal Anatomy – Most of the internal organs are greatly elongated in conformity with the length of the body in these insects. The *alimentary canal* is an unconvoluted tube throughout its course and salivary glands are present (Oka, 1930). The oesophagus is long and slender, expanding into a crop at the commencement of the abdomen. A rudimentary gizzard is present but its armature of denticles is very weak or absent. The mid gut is the largest division of the gut and extends through the greater part of the abdomen; it is devoid of enteric caeca and is followed by a very short hind gut. Attached to the latter are 50 to 70 Malpighian tubes which unite in groups of five or six, each group discharging into the gut by a narrow common duct. Six longitudinal rectal papillae are usually present. The *nervous system* is well developed and exhibits comparatively little concentration. The brain is transversely elongated and is characterized by the great development of the optic ganglia, which is associated with the large size of the eyes. The ventral nerve-cord consists of three thoracic ganglia and seven evident ganglia (2nd to 8th) in the abdomen, the 1st abdominal ganglion being amalgamated with that of the metathorax. A well-developed stomatogastric system is present (Cazal, 1948). The circulatory system has not been studied in any detail but appears to be very similar to that of the nymph, except that a ventral blood sinus is present in the imago in close relation with the main nerve-cord. The tracheal system (Wolf, 1935) consists of three pairs of principal longitudinal trunks which give off segmental branches. It communicates with the exterior by ten pairs of spiracles on the last two thoracic and the first eight abdominal segments. The *male reproductive organs* consist of a pair of very elongate *testes* extending, in *Aeshna*, from the 4th to the 8th abdominal segments: each organ is composed of a large number of spherical lobules in which the spermatozoa develop. The vasa deferentia are rather short narrow tubes which enter a common duct just above the genital aperture. The common passage is dilated dorsally to form a conspicuous sperm-sac. The spermatozoa adhere in a radiating fashion forming rounded spermatophores, each apparently derived from a single lobule of the testis. The spermatophores are somewhat mucilaginous externally and are adapted for transfer from the 9th to the 2nd segment before copulation. The *female reproductive organs* are characterized by the great size and length of the ovaries which extend from the base of the abdomen down to the 7th segment. Each ovary is composed of a large number of longitudinally arranged panoistic ovarioles. The two oviducts are very short and open into a large pouch-like spermatheca in the 8th segment. A pair of accessory glands communicates by a common duct with the dorsal side of the spermatheca.

Structure and Biology of the Developmental Instars

Oviposition in dragonflies may be either endophytic or exophytic. In the latter case the eggs are rounded and are either dropped freely into the water or attached superficially to aquatic plants. This method is the rule among the Anisoptera, with the exceptions mentioned below. In *Sympetrum* and *Tetragoneuria* the eggs are laid in gelatinous strings attached to submerged twigs. Endophytic oviposition is characteristic of the Zygoptera and the Anisopteran families Aeshnidae and Petaluridae. Dragonflies adopting this method have elongate eggs which they insert into slits cut by the ovipositor in the stems and leaves of plants or other objects, near or beneath the water. In some cases the female (alone, or accompanied by the male) descends below the water-surface for the purpose.

Before the nymph emerges from the egg it swallows amniotic fluid, the associated contractions of stomodaeal musculature being visible through the shell (Grieve, 1937). Pressure of the head of the embryo against the chorion is the immediate cause of hatching, since it forces open the lid-like anterior extremity of the egg. The newly-hatched insect is known as the pronymph: at this stage it exhibits a more or less embryonic appearance, the whole body and appendages being invested by a delicate cuticular sheath. The pronymph is of extremely brief duration, lasting only a few seconds in *Anax* (Tillyard), for two or three minutes in *Coenagrion* (Balfour-Browne, 1909), but up to 30 minutes in *Sympetrum striolatum* (Gardner, 1951). At this stage the pulsations of the stomodaeum increase in frequency and the pronymphal cuticle is ruptured. The insect which emerges is in its second instar and is now a free nymph fully equipped for its future life. The nymphs of the Odonata are campodeiform and may be divided into two main types – the Anisopteran and the Zygopteran. In the former the body ends in three small processes, a median epiproct or appendix dorsalis and a pair of lateral paraprocts: when closed they form a pyramid which conceals the anus (Fig. 239). Respiration takes place by means of concealed rectal tracheal gills. In the Zygopteran type the three terminal processes are greatly developed to form caudal gills, and rectal tracheal gills are wanting (Fig. 239). *Megalagrion oahuense* nymphs are terrestrial, living among moist debris on the floor of Hawaiian forests and some other species of *Megalagrion* spend much time crawling out of streams in a water-film on rocks (Williams, 1936). A few other terrestrial or semi-terrestrial nymphs are known (e.g. Wolfe, 1953; Willey, 1955) but otherwise the immature stages of the Odonata are exclusively aquatic, living in various situations in fresh water. Many remain hidden in sand or mud and are homogeneously coloured without any pattern. Those which live on the river-bottom or among weed exhibit a cryptic pattern which tends to conceal them from enemies and prey. A few live in the water which accumulates at the leaf-bases of Bromeliads and other tropical plants. Certain species cling to rocks and tend to simulate the colour of the surface which they frequent. Dragonfly nymphs are also able to

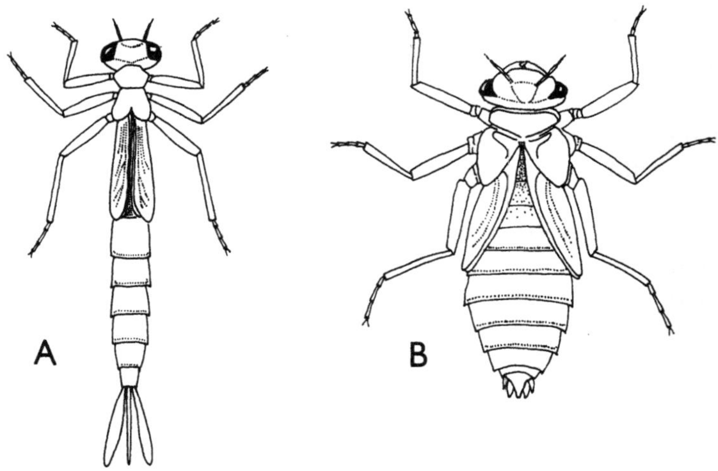

FIG. 237 Larvae of Odonata. A, *Ischnura* (Zygoptera: Coenagriidae); B, *Synthemis* (Anisoptera: Synthemidae). (*After Insects of Australia*, C.S.I.R.O., 1970)

change their general coloration according to differences in their environment. Without exception all the species are predacious, feeding on various aquatic animals, the nature of the food depending upon the age of the nymphs; older nymphs often feed mainly on Ephemeropteran nymphs and Culicid larvae as well as nymphs of their own and other species of Odonata. The larger Aeshnid nymphs will also attack tadpoles and occasionally small fish.

The number of instars that intervene between the egg and the imago varies in different species and also among individuals of the same species. It ranges between about ten and fifteen (Munchberg, 1938; Gardner, 1951) and the whole nymphal period may be passed through within a year as in most Zygoptera, or occupy two years as in *Aeshna*, or may even last from three to five years. In temperate species there may be a diapause, such as occurs in the last-instar nymph of *Anax imperator* and which ensures the synchronized emergence of adults in the following spring (Corbet, 1956*a*, 1957*a*). For further accounts of the life-cycles and nymphal biology of dragonflies see Corbet (1962) and various papers listed on pp. 515–520.

The principal external changes involved during metamorphosis (e.g. Snodgrass, 1954; Calvert, 1934) include an increase in the size of the compound eyes, and during the last few instars ocelli become evident. The antennal segments increase in number, and the wing-rudiments change so that the developing hind wings overlap the anterior pair; the wing-bearing segments increase in size, and changes take place in the caudal gills of Zygoptera. The internal changes which occur in the fully grown nymph just before metamorphosis is complete have been less fully studied, but they include alterations in the histological structure of the gut (Straub, 1943) and the disintegration of the musculature of the nymphal labium with its replacement by a newly developed set of imaginal muscles (Munscheid, 1933). The optic lobes of the brain increase in size and complexity and new ommatidia differentiate in the eye from an adjacent area of epidermal cells (Ando, 1957; Mouze, 1972). The secondary copulatory organ of the male develops largely in the last instar nymph and the pharate adult (Defossez, 1973) and the skeletomuscular system of the thorax and abdomen undergo considerable changes (Maloeuf, 1935). Altogether the metamorphosis of the Odonata is more profound than that of most Exopterygote insects (Snodgrass, 1954).

When the time for emergence of the imago is approaching the nymph ceases to feed and appears tense and swollen. The thorax in particular becomes noticeably inflated and the wing-sheaths become sub-erect. The gills are no longer functional and at the same time the thoracic spiracles are brought into use, the nymph partially protruding itself from the water in order to breathe air. When the internal changes are complete the nymph climbs up some suitable object out of the water and fixes its claws so firmly in position that the exuviae remain tightly adherent to the support long after the imago has flown away. The nymph remains stationary and sooner or later the cuticle splits along the mid-dorsal line of the thorax, the fracture extending forwards to the head. The imago then withdraws its head and thorax through the opening and the legs and wings become free, but the abdomen is not yet fully drawn out from the exuviae. The insect usually hangs head downwards until the legs attain strength and freedom of movement. The withdrawal of the abdomen forms the final act, and the insect crawls away to rest until the wings and abdomen are fully extended (Fig. 193). A variable period elapses before the imaginal colour pattern is fully acquired and teneral forms, or individuals which have not yet developed their mature coloration, are often seen on the wing.

The main difference between the head of the nymph and that of the imago is found in the labium. In the nymph this is modified for prehensile purposes and is known as the *mask* (Fig. 238) from the fact that it conceals the other mouthparts (Butler, 1904). The prementum and postmentum are markedly lengthened, and there is a great freedom of movement between the two parts. The ligula is undivided and represented by a median lobe which is fused with the prementum. The labial palps are modified to form lateral lobes, each of which carries on its outer side a movable hook. The nymph uses its mask entirely for the capture of prey. At rest the postmentum is reflexed between the bases of the legs with the prementum hinged upon it ventrally. When seizing prey the mask is extended very rapidly and the prey impaled on the movable hooks. In the Anisoptera extension of the labium occurs hydraulically; the body cavity is divided by a transverse diaphragm in the abdomen and contraction of segmental muscles near this leads to an increase in the blood pressure which forces the labium forwards and causes its movable hooks to diverge (Olesen, 1972). The return of the labium with the prey occurs

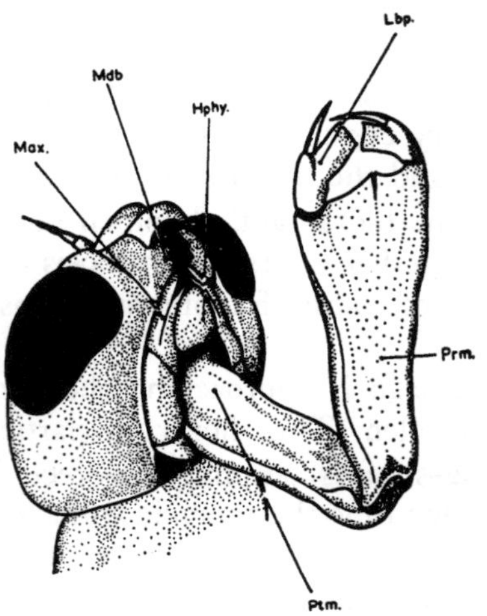

FIG. 238
Latero-ventral view of head and mouthparts of an Odonate nymph (*after* Weber, 1933)

Hphy, hypopharynx; *Lbp*, labial pulp; *Max*, maxilla; *Mdb*, mandible; *Prm*, prementum; *Ptm*, postmentum.

more slowly through the action of retractor muscles. Pritchard (1965) distinguishes between two types of predatory behaviour: nymphs of some species climb among aquatic vegetation and use their large compound eyes to locate the prey (see also Baldus, 1926); others live sprawled on the bottom and catch their prey through tactile perception followed by a rapid strike with the labium.

The prothorax in the nymph is always longer than in the imago; in advanced nymphs the meso- and metathorax are closely fused. The legs are considerably longer than those of the imago and the femoro-trochanteral articulation is modified to form a breaking joint. By a sudden contraction of the trochanteral muscles the intervening membrane can be ruptured and the limb discarded should it be seized by a predator. Eleven segments are recognizable in the abdomen, the 11th being modified differently in the two main suborders (Snodgrass, 1954). In the Zygoptera the median caudal gill is a process of the 11th tergite (epiproct) and the two lateral gills are developed from the divided 11th sternum (paraprocts). In the Anisoptera the paraprocts and the epiproct for the three anal 'appendages' that fit together to form the anal pyramid (Fig. 239). The cerci, sometimes referred to as 'cercoids', are a pair of small, unsegmented structures lying one on each side of the epiproct. At the final metamorphosis the terminal abdominal appendages of the adult are formed in relation to the corresponding nymphal structures.

The alimentary canal of the nymph (Sadones, 1896; Straub, 1943) differs from that of the imago in several features. The gizzard, for example, is a very highly specialized organ provided with internal denticle-bearing longitudinal ridges; these are either four or some multiple of four in number among different groups. The mid gut is considerably shorter than in the imago, and the Malpighian tubes at first number only three but gradually increase at each instar until the full complement is acquired. The nervous system has

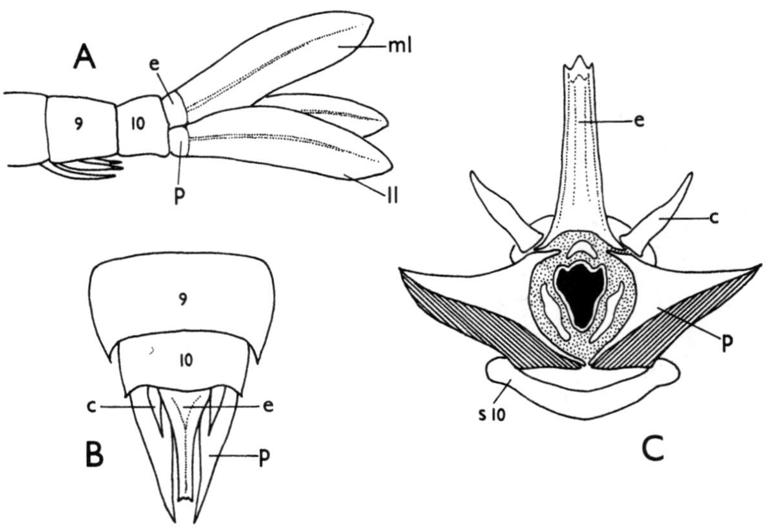

FIG. 239 Abdominal terminalia of Odonate larvae (*after* Snodgrass, 1954). A, *Archilestes grandis* (Zygoptera), end segments and gills, lateral view; B, *Anax junius* (Anisoptera), end segments and apical lobes, dorsal view; C, the same, with apical structures spread apart, showing three small circum-anal valves

c, cercus; *e*, epiproct (= 11th tergum); *ll*, lateral gill lobe; *ml*, median gill lobe; *p*, paraproct (= half of divided 11th sternum); *s10*, tenth sternum; *9*, *10*, ninth and tenth terga.

eight abdominal ganglia (Mill, 1965), the first being quite distinct from the metathoracic ganglion although fused with it in the imago. The circulatory system has been studied by Zawarsin (1911) in *Aeshna*. The heart consists of eight chambers corresponding with the 2nd to 9th abdominal segments, in which they lie; alary muscles are only present in relation with the two hindmost chambers. The respiratory system presents features of exceptional interest and has been investigated more particularly by Sadones (1896), Tillyard (1916) and Wolf (1935). Spiracles are present on the meso- and metathorax, but only the mesothoracic pair is well developed and is functional when the larva leaves the water. The metathoracic and abdominal spiracles are small and usually non-functional. Special respiratory organs in the form of tracheal gills are present in the nymphs of all dragonflies. In the

Anisoptera they take the form of rectal gills which form an elaborate and beautiful apparatus known as the branchial basket. In most Zygoptera the respiratory organs are caudal gills, while in a few rare cases (Polythoridae, Epallagidae) lateral abdominal 'gills' are also present though their respiratory function is doubtful. These three types are treated separately below.

(1) *The Branchial Basket.* This structure is formed by the expanded anterior two-thirds of the rectum, which assumes the form of a barrel-like chamber (Fig. 240). The gills are primarily developed as six longitudinal folds of the rectal wall and are homologous with the six rectal papillae. They are covered with an extremely delicate

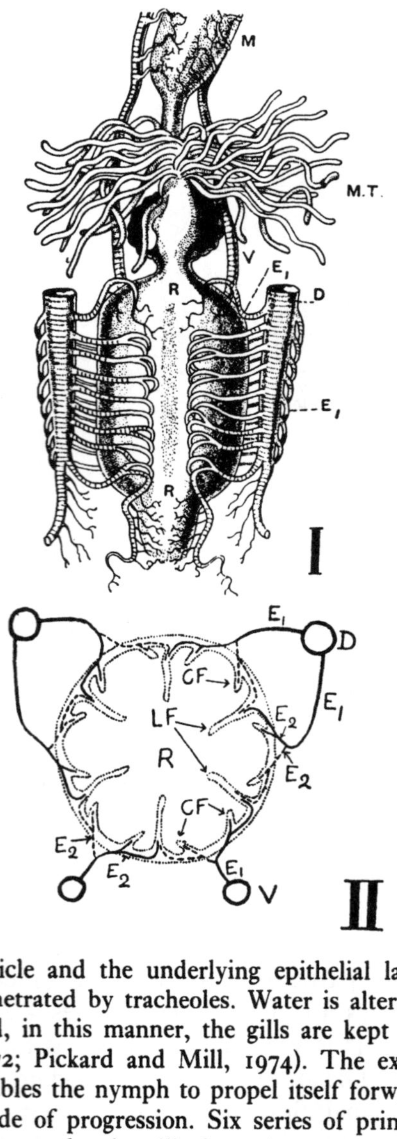

FIG. 240
I. Hind gut of a nymph of *Aeshna* showing tracheal supply. *After* Oustalet. II. Diagrammatic transverse section of the rectum of the nymph of *Austrogomphus.* Adapted from Tillyard

CF, cross-fold; *D*, dorsal tracheal trunk; E_1, primary efferent trachea; E_2, secondary do.; *LF*, longitudinal fold; *M*, mid gut; *MT*, Malpighian tubes; *R*, rectum; *V*, visceral tracheal trunk.

cuticle and the underlying epithelial layer is modified to form a syncytial core penetrated by tracheoles. Water is alternately taken into the rectum and expelled and, in this manner, the gills are kept aerated (Tonner, 1936; Mill and Pickard, 1972; Pickard and Mill, 1974). The expulsion of the water, when forcible, also enables the nymph to propel itself forward by a series of jerks, which is its usual mode of progression. Six series of primary efferent tracheae convey the oxygen, taken up by the gills from the water, to the main longitudinal trunks of the body.

Each primary efferent trachea divides into two secondary efferents which give off a very large number of tracheoles to the gills. Each tracheole forms a complete loop within the gill, returning to the same secondary efferent from which it arose. The gill system may be either simplex or duplex in character (Fig. 241). In the *simplex system* there are six principal longitudinal gill folds supported right and left by a double series of cross-folds. The simplex system is divisible into two types, the undulate and the papillate. In the undulate type the free edge of each gill-fold is undulated or wavy in character. This is the primary type of gill which persists throughout life in the more archaic groups (Cordulegasteridae, Petaluridae and in *Austrogomphus*). In most of the Gomphidae all the gill-folds are broken up into elongate filaments forming what is termed the papillate type. This specialization brings about greater respiratory efficiency since each filament is bathed on all sides by the water. The *duplex system* is a secondary development and differs in that the main longitudinal folds are either non-functional or wanting, the gills being entirely

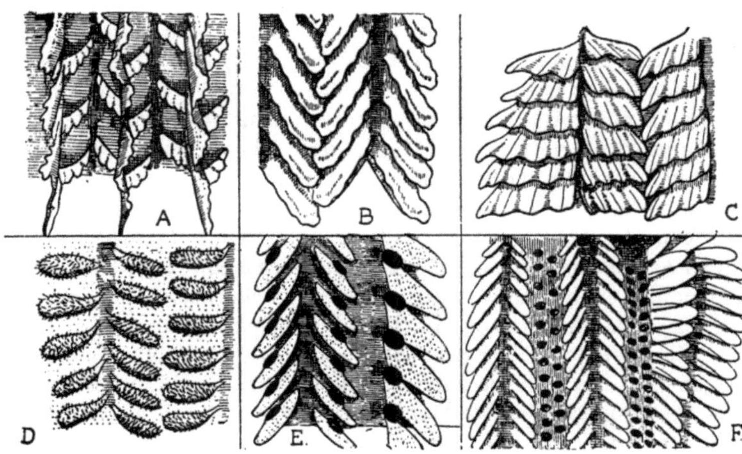

FIG. 241 Portions of the freshly-opened branchial basket, to show form of gills
A, undulate simplex; B, implicate; C, foliate; D, papillo-foliate; E, F, lamellate. *After* Tillyard, *Biology of Dragonflies*.

formed from the double series of cross-folds. Three main types are recognizable in this system and depend upon the form assumed by the gills. The implicate type occurs in the tribe Brachytronini of the family Aeshnidae. The gills resemble a series of obliquely placed concave tiles slightly overlapping one another. In the foliate type, which is found in the Aeshnini, each gill is basally constricted and leaf-like in form. The lamellate type occurs in the family Libellulidae: the gills appear as flat plates projecting into the cavity of the rectum and are attached by broad bases. For full details concerning the types of rectal gills and the differences in their tracheal supply reference should be made to papers by Tillyard (1916) and Rich (1918). Ultrastructural evidence suggests that ionic absorption also occurs in the gills (Greven and Rudolph, 1973).

(2) *The Caudal Gills.* Nearly all Zygopteran nymphs possess three external tracheal gills at the hinder extremity (Fig. 239), for details of which see Tillyard (1917*b*) and MacNeill (1960). In the young nymph the caudal gills are filamentous and hairy, but they soon acquire a triquetral form (i.e triangular in cross-section). The triquetral gill is retained throughout life in a few cases, as for example in the lateral

gills of *Agrion*. In most instances it either becomes swollen (saccoid gill) or flattened (lamellate gill). Internally the gills contain one or more large tracheal branches ending in an anastomosing tracheolar system, nerves and definite blood channels while the remaining space is filled by a peculiar alveolar tissue. Reduction of the gills and their tracheae occurs in the less fully aquatic species of *Megalagrion*.

The problem of respiration in Zygopteran nymphs is reviewed by Calvert (1915) but needs further study. Pennak and McColl (1944), working with nymphs of *Enallagma* spp., found that large nymphs from which the caudal appendages had been removed lived for over two weeks in well-aerated water, apparently obtaining an adequate supply of oxygen through the general body-surface. They found, however, that intact nymphs could extract rather more oxygen from water in closed containers than could those without gills. See also Harnisch (1958) and Zahner (1959). Although young nymphs periodically take water into the rectum and expel it there is no experimental evidence that this is a respiratory process, nor have the rectal papillae of the Zygoptera any special tracheal supply.

(3) *The Lateral Abdominal Gills.* These occur on either side of the 2nd to 7th or 8th abdominal segments in a few genera (e.g. *Cora, Anisopleura, Pseudophaea*). They are attached towards the ventral surface, are filamentous in form and may possibly be true abdominal appendages; their respiratory role has not been studied.

Affinities and Phylogeny – The Odonata have a better fossil record than most insect groups and when the extinct forms are taken into account the order includes a further four suborders, for whose interrelationships see Fraser (1957*a*). To one of these suborders, the Meganisoptera, belongs the earliest known Odonate, the Carboniferous *Erasipteron larischi*, as well as the Meganeuridae (which include the gigantic dragonflies of the genus *Meganeura*). Some Meganisoptera were formerly included in the order Protodonata, but the latter has now been restricted in scope. Nevertheless the Protodonata are probably related to the earliest Odonata and are also connected with the Ephemeroptera through forms like *Permothemis* which is placed in the Protodonate suborder Archodonata (see p. 485). From Meganisopteran ancestors there probably evolved a further two extinct suborders of dragonflies, the Protanisoptera and Protozygoptera. The latter gave rise to the Zygoptera (first known from the Upper Permian) and from early Zygoptera there probably arose the Anisozygoptera (Triassic to Recent). In turn, the Anisozygoptera included the probable ancestors of the Anisoptera. The morphological study of recent forms has not solved the question of relationships between the Ephemeroptera, Odonata and the remaining (Neopteran) Pterygotes. Hennig (1969) supports the orthodox view that the Ephemeroptera and Odonata form a monophyletic group of Palaeopteran orders, whereas Kristensen (1975) regards the resemblances as due to convergence and treats the Odonata as more closely related to the Neopteran insects.

Classification – There is no general catalogue or monograph, but many regional studies, among which are accounts of the dragonflies of North America (Needham and Westfall, 1954; Walker, 1953–58); India (Fraser, 1933–36); Australia (Fraser, 1960); China (Needham, 1930); Africa (Pinhey, 1951, 1962) and Madagascar (Fraser, 1957*b*; Lieftinck, 1965). The 45 British species have been dealt with by Lucas (1900, 1930), Longfield (1949), Fraser (1956) and Gardner (1954) while for other parts of Europe see Cirdei and Bulimar (1965), Conci and Nielsen (1956), May (1933), and Aguesse (1968).

The higher classification summarized below is due to Fraser (1957*a*) but omits the fossil groups that he also considers. It depends largely on venational characters and recognizes 26 families, of which the 9 represented in Britain are denoted by an asterisk.

Suborder I. **ZYGOPTERA**

Fore and hind wings closely similar in form and venation, more or less petiolate basally, in repose usually held vertically above the abdomen; nodus usually before middle of wing; discoidal cell similar in fore and hind wings, very rarely open basally, never divided into triangle and supratriangle though sometimes crossed by veins. Eyes projecting laterally and separated by a space greater than their dorsal diameter. Labium with middle lobe deeply cleft. Male with two superior and two inferior anal appendages; penis not distinctly jointed. Female with superior appendages only and complete ovipositor. Nymphs with 3 caudal gills and slender, elongate abdomen; gizzard with 8–16 radially symmetrical folds.

Superfamily **Coenagrioidea**

Narrow petiolate wings; nodus well before middle of costal margin; boundary between meso- and metathorax extending only to spiracle; nymphal labium with entire margin.

FAM. PLATYSTICTIDAE. *Small or medium-sized species from Neotropical and Oriental regions; arculus distal to second antenodal; anal vein and sometimes also* Cu_2 *greatly reduced; supplementary postcubital cross-veins present basally, joining* Cu_2 *to hind margin.*

FAM. PROTONEURIDAE. *Like the previous family but abdomen often very long; no supplementary postcubitals;* widely distributed.

*FAM. PLATYCNEMIDAE. *Anal vein and* Cu_2 *normal; discoidal cell elongate;* a somewhat heterogeneous Old World group. *Platycnemis pennipes* occurs in Europe and is usually associated with running water.

*FAM. COENAGRIIDAE (= AGRIONIDAE). A large and successful Zygopteran family with a worldwide distribution. Differs from the Platycnemidae in the short discoidal cell, the anterior side being much shorter than the posterior side. *Ischnura* and *Enallagma* are cosmopolitan, *Coenagrion* is mainly Palaearctic; *Pyrrhosoma* is found in Europe and *Erythromma* occurs there and in N. America.

FAM. PSEUDOSTIGMATIDAE. A small, rather isolated family; wings with very many cross-veins; nodus near wing-base and pterostigma reduced or absent.

FAM. MEGAPODAGRIIDAE. A peculiar group of rare species, having affinities also with the Lestidae and their allies. Australia and Madagascar have many species but others occur in the Neotropical and Oriental regions.

Superfamily **Hemiphlebioidea**

Postnodals not aligned with the cross-veins behind them; discoidal cell open basally.

FAM. HEMIPHLEBIIDAE. Contains only the very local Australian *Hemiphlebia mirabilis*.

Superfamily **Lestinoidea**

Anterior hamuli of secondary male copulatory apparatus elongate; venation varied.

FAM. PERILESTIDAE. *Small species, with long, slender abdomen; discoidal cell approaching or impinging upon the hind margin of the wing near the point at which 1A arises from the margin; pterostigma more or less quadrate or oval;* occurring in the Western Hemisphere, Africa and Australia.

FAM. CHLOROLESTIDAE. *Larger, more robust species, with more elongate pterostigma.* Oriental region, Africa and Australia.

*FAM. LESTIDAE. *Moderately robust species, variable in size; discoidal cell well separated from wing margin. Lestes* is cosmopolitan and *Sympecma* is Palaearctic; *S. fusca* and *S. paedisca* are the only European dragonflies to overwinter as the imago (Prenn, 1928). Most of the other genera are Oriental.

Superfamily **Agrioidea (= Calopterygoidea)**

Medium or large Zygoptera with broad, often coloured, wings; nodus towards middle of wing; origin of R_{4+5} and IR_3 nearer to arculus than to nodus; boundary between meso- and metathorax usually complete; nymphal labium incised medially.

FAM. PSEUDOLESTIDAE. A small family of 7 genera with a scattered distribution, annectant between the Coenagrioidea and Agrioidea.

FAM. AMPHIPTERYGIDAE. *Another small family of medium-sized species; wings moderately petiolate; few secondary antenodals; some species hold the wings open at rest like Anisoptera.* Occur in Australia, Africa and the Oriental and Neotropical regions.

FAM. CHLOROCYPHIDAE. A large family, whose distribution extends from West Africa to the Pacific. *Head narrow; eyes large; clypeus produced, snout-like; females usually with plain wings, males often having them metallic and coloured; many secondary antenodals.*

FAM. HELIOCHARITIDAE. *Wings uncoloured, long, narrow and petiolate; discoidal cell with one cross-vein; many secondary antenodals.* Confined to South America.

FAM. POLYTHORIDAE. *Also S. American and probably derived from the preceding family; arculus absent; base of discoidal cell connected directly to* R_s. The nymphs of *Cora* have lateral abdominal 'gills' which apparently function as organs of attachment.

FAM. EPALLAGIDAE. *Medium-sized with moderately petiolate wings; discoidal cell short and traversed by few or no cross-veins; nymphs robust with saccoid or triquetral terminal gills and also lateral abdominal 'gills'.*

***FAM. AGRIIDAE (= CALOPTERYGIDAE).** *A large family, with non-petiolate, densely reticulate wings, with enormous numbers of cross-veins and many supplementary longitudinal veins; nymphs usually elongate, spidery, with long caudal gills, usually in fast-running streams.* Well represented in the Tropics; *Hetaerina* is N. American and *Agrion* is Holarctic, with two attractive European species – *A. virgo* and *A. splendens.*

Suborder II. ANISOZYGOPTERA

It is not easy to frame a definition of this suborder to embrace its many extinct Mesozoic forms, but the recent fauna includes only two species of *Epiophlebia* from Japan and India (Fig. 231, A). Like the fossil forms these combine the characteristics of the two other suborders: the imago has a Zygopteran venation and the general body form of an Anisopteran while the nymph resembles the Anisoptera in general facies, labium and anal appendages.

FAM. EPIOPHLEBIIDAE. (With the characters of the suborder.) *Epiophlebia superstes* from Japan has been monographed by Asahina (1954), who also summarizes its life-history (Asahina, 1950). *E. laidlawi*, from the Himalayas, is known only by its nymph.

Suborder III. ANISOPTERA

Fore and hind wings not petiolate, dissimilar in form and venation, the hind wings broadened basally; held horizontally or depressed in repose (except *Cordulephya*); discoidal cell differentiated into triangle and supratriangle. Eyes large, never separated by more than their dorsal diameter and often contiguous dorsally. Labium variable. Male with 2 superior and 1 inferior anal appendages; penis jointed. Female with superior appendages only and ovipositor normal or atrophied.

Larvae with rectal gills and anus closed by a pyramid made up of a median dorsal appendage and the two cerci. Gizzard with 4–8 folds.

Superfamily Aeshnoidea

Eyes separate or broadly confluent above; nymphal labium flattened and without setae, its lateral lobes narrow and with long, movable hooks.

*FAM. GOMPHIDAE. *A large family, mainly of black and yellow forms with hyaline wings; eyes separate; ovipositor incomplete; abdomen variable but posterior segments sometimes dilated or foliate ('club-tails'); nymphs variable in form and habits;* found in all zoogeographical regions, *Gomphus vulgatissimus* is the only British species. Some Gomphids lay their eggs at the water surface or above it and they are provided with adhesive egg jelly or filaments.

FAM. PETALURIDAE. *A small, archaic family of only 9 very large species with a relict distribution. Eyes separate; ovipositor complete; wings narrow, highly reticulate.* The nymphs are unique in penetrating swampy or marshy soil to a foot or more, emerging from their burrows at night to find their prey (Wolfe, 1953).

*FAM. AESHNIDAE. *Large, robust species; eyes contiguous, ovipositor complete; nymphs living among weed or crawling on bottom or among debris.* A large family, widely distributed. *Anax* and *Aeshna* are cosmopolitan.

Superfamily **Cordulegasteroidea**

Large, robust forms; head transversely elongate; eyes separate or just meeting above; wings closely reticulate; ovipositor variably reduced; nymphs elongate, fusiform, with clypeus produced into a ridge-like structure for delving in mud; nymphal labium spoon-shaped.

*FAM. CORDULEGASTERIDAE. Holarctic and Oriental species, often with wings extensively coloured. The Oriental *Chlorogomphus papilio* is one of the largest and most magnificent dragonflies. The black and yellow *Cordulegaster boltoni* is European.

Superfamily **Libelluloidea**

A large, dominant group, whose families seem to form a sequence of increasing venational specialization. Eyes meeting or confluent; ovipositor reduced; abdomen variable: fusiform or depressed; nymphal labium spoon-shaped and setose.

FAM. SYNTHEMIDAE. *Medium or slender Australasian species with uncoloured wings; triangle and subtriangle not sub-divided by cross-veins; primary antenodals thickened.* Some species have a secondary ovipositor. The nymphs are robust and hairy, resembling those of the Cordulegasteridae.

*FAM. CORDULIIDAE. *Variable in size; wings rarely coloured; nodus placed about two-thirds along costa from base of wing; nymphs variable.* Well represented in Australia but occurring elsewhere. *Cordulia* and *Somatochlora* are Holarctic.

FAM. MACRODIPLACTIDAE. A small, annectant, mainly Old World family whose members are sometimes placed in the Corduliidae or in the Libellulidae.

*FAM. LIBELLULIDAE. A very large, successful and variable family, breeding mainly in still water. They differ from the Corduliidae in being rarely metallic, in the rounded base of the hind wing, and in the absence of keels on the male tibiae. *Libellula*, *Leucorrhinia* and *Sympetrum* are Holarctic; *Orthetrum* is cosmopolitan.

Literature on the Odonata

AGUESSE, P. (1968), Les Odonates de l'Europe occidentale du nord de l'Afrique et des Iles Atlantiques, *Faune de l'Europe et du bassin méditerranéen*, **4**, 1–255.

ANDO, H. (1957), A comparative study on the development of ommatidia in Odonata, *Sci. Rep. Tokyo Kyoiku Daig* (*B*), **8**, 174–216.

ASAHINA, S. (1950), On the life history of *Epiophlebia superstes* (Odonata, Anisozygoptera), *Proc. 8th int. Congr. Ent.*, 337–341.

—— (1954), *A Morphological Study of a Relic Dragonfly* Epiophlebia superstes, *Jap. Soc. Promotion Sci.*, Tokyo, 153 pp.

—— (1958), On a rediscovery of the larva of *Epiophlebia laidlawi* Tillyard from the Himalayas (Odonata, Anisozygoptera), *Tombo*, **1**, 1–2.

BALDUS, K. (1926), Experimentelle Untersuchungen über die Entfernungslokalisation der Libellen (*Aeschna cyanea*), *Z. vergl. Physiol.*, **3**, 475–505.

BALFOUR-BROWNE, F. (1909), The life-history of the Agrionid dragonfly, *Proc. zool. Soc. Lond.*, **1909**, 253–285.

BUTLER, H. (1904), The labium of the Odonata, *Trans. Am. ent. Soc.*, **30**, 111–134.

CALVERT, P. P. (1901–08), Odonata, *Biol. cent.-Amer.*, 17–420.

—— (1915), Studies on Costa Rican Odonata, VII. Internal organs of larva, and the respiration and rectal tracheation of Zygopterous larvae in general, *Ent. News*, **26**, 385–395; 435–447.

—— (1934), The rates of growth, larval development and seasonal distribution of the genus *Anax* (Odonata: Aeshnidae), *Proc. Am. phil. Soc.*, **73**, 1–70.

CAZAL, P. (1948), Recherches sur les glandes endocrines rétrocérébrales des insectes II – Odonates, *Arch. Zool. exp. gén.*, **85**, 55–82.

CHAO, HSIU-FU (1953), The external morphology of the dragonfly *Onychogomphus ardens* Needham, *Smithson. misc. Collns*, **122** (6), 56 pp.

CIRDEI, F., AND BULIMAR, F. (1965), Odonata, *Fauna Repub. pop. rom.*, **7** (5), 1–274.

CLARK, H. W. (1940), The adult musculature of the anisopterous dragonfly thorax (Odonata, Anisoptera), *J. Morph.*, **67**, 523–565.

CONCI, C. AND NIELSEN, C. (1956), Odonata, *Fauna d'Italia*, **1**, 295 pp.

CORBET, P. S. (1955), The larval stages of *Coenagrion mercuriale* (Charp.) (Odonata, Coenagriidae), *Proc. R. ent. Soc. Lond.*, (A), **30**, 115–126.

—— (1956*a*), Environmental factors influencing the induction and termination of diapause in the Emperor Dragonfly, *Anax imperator* Leach. (Odonata, Aeshnidae), *J. exp. Biol.*, **33**, 1–14.

—— (1956*b*), The life-histories of *Lestes sponsa* (Hansemann) and *Sympetrum striolatum* (Charp.) (Odonata), *Tijdschr. Ent.*, **99**, 217–229.

—— (1956*c*), The influence of temperature on diapause development in the dragonfly *Lestes sponsa* (Hansemann) (Odonata, Lestidae), *Proc. R. ent. Soc. Lond.*, (A), **31**, 45–48.

—— (1957*a*), The life-history of the Emperor Dragonfly *Anax imperator* Leach. (Odon., Aeshnidae), *J. Anim. Ecol.*, **26**, 1–69.

CORBET, P. S. (1957*b*), The life-histories of two summer species of dragonfly (Odonata, Coenagriidae), *Proc. zool. Soc. Lond.*, **128**, 403–418.

—— (1957*c*), The life histories of two spring species of dragonfly (Odonata, Zygoptera), *Ent. Gaz.*, **8**, 79–89.

—— (1962), *A Biology of Dragonflies*, Witherby, London, 247 pp.

CORBET, P. S., LONGFIELD, C. AND MOORE, N. W. (1960), *Dragonflies*, Collins, London, 260 pp.

DEFOSSEZ, A. (1973), Développement de l'appareil copulateur mâle au cours de la métamorphose des Aeschnidae (Odonata), *Int. J. Insect Morphol. & Embryol.*, **2**, 153–167.

FRASER, F. C. (1933–36), Odonata, *Fauna of British India*, **1**, 1–423; **2**, 1–398; **3**, 1–461.

—— (1938), A note on the fallaciousness of the theory of pretracheation in the venation of the Odonata, *Proc. R. ent. Soc. Lond.*, (A), **13**, 60–70.

—— (1943), The function and comparative anatomy of the oreillets in the Odonata, *Proc. R. ent. Soc. Lond.*, (A), **18**, 50–56.

—— (1956), Odonata, *Handb. R. ent. Soc. Ident. Brit. Ins.*, **1** (10), 1–49.

—— (1957*a*), *A Reclassification of the Order Odonata*, R. Zool. Soc. N.S.W., Sydney, 133 pp.

—— (1957*b*), *Faune de Madagascar I. Insectes Odonates Anisoptères*, Inst. Rech. Sci., Tananarive, 125 pp.

—— (1960), *A Handbook of the Dragonflies of Australia with keys for the identification of all species*, R. Zool. Soc. N.S.W., Sydney, 167 pp.

GARDNER, A. E. (1951), The early stages of Odonata, *Proc. Trans. S. London ent. nat. Hist. Soc.*, **1951**, 83–88.

—— (1954), A key to the larvae of the British Odonata. I, II, *Ent. Gaz.*, **5**, 157–171; 193–213.

GREVEN, H. AND RUDOLPH, R. (1973), Histologie und Feinstruktur der larvalen Kiemenkammer von *Aeshna cyanea* Müller (Odonata: Anisoptera), *Z. Morph. Tiere*, **76**, 209–226.

GRIEVE, E. G. (1937), Studies on the biology of the damselfly *Ischnura verticalis*, with notes on certain parasites, *Entomologica am.*, **17**, 121–152.

HAKIM, Z. M. (1964), Comparative anatomy of the head capsules of adult Odonata, *Ann. ent. Soc. Am.*, **57**, 267–278.

HARNISCH, O. (1958), Untersuchungen an den Analkiemen der Larve von *Agrion*, *Biol. Zbl.*, **77**, 30–310.

HATCH, G. (1966), Structure and mechanics of the dragonfly pterothorax, *Ann. ent. Soc. Am.*, **59**, 702–714.

HENNIG, W. (1969), *Die Stammesgeschichte der Insekten*, Kramer, Frankfurt a. M., 436 pp.

HEYMER, A. (1972), Comportements social et territorial des Calopterygidae (Odon. Zygoptera), *Annls Soc. ent. Fr.*, (N.S.), **8**, 3–53.

HOBBY, B. M. (1934), The prey of British dragonflies, *Trans. ent. Soc. S. England*, **8**, 65–76.

HUDSON, G. B. (1948), Studies in the comparative anatomy and systematic importance of the Hexapod tentorium. III, *J. ent. Soc. sth. Afr.*, **11**, 38–49.

JACOBS, M. E. (1955), Studies on territorialism and sexual selection in dragonflies, *Ecology*, **36**, 566–586.

JOHNSON, C. (1962), A description of territorial behavior and a quantitative study of its function in males of *Hetaerina americana* (Fabricius) (Odonata: Agriidae), *Can. Ent.*, **94**, 178–192.

—— (1972), Tandem linkage, sperm translocation and copulation in the dragonfly, *Hagenius brevistylus* (Odonata, Gomphidae), *Am. Midl. Nat.*, **88**, 131–149.

KORMONDY, E. J. (1961), Territoriality and dispersal in dragonflies (Odonata), *Jl N. Y. ent. Soc.*, **69**, 42–52.

KRISTENSEN, N. P. (1975), The phylogeny of hexapod 'orders'. A critical review of recent accounts, *Z. zool. Syst. Evolutionsforsch.*, **13**, 1–44.

LAMB, L. (1925), A tabular account of the differences between the earlier instars of *Pantala flavescens* (Odonata: Libellulidae), *Trans. Am. ent. Soc.*, **50**, 289–311.

LEVINE, H. R. (1957), Anatomy and taxonomy of the mature naiads of the Dragonfly genus *Plathemis*, *Smithson. misc. Collns*, **134** (6), 28 pp.

LEW, G. T. (1934), Head characters of the Odonata, with special reference to the development of the compound eye, *Ent. Amer.*, **14**, 41–96.

LIEFTINCK, M. A. (1965), Notes on Odonata of Madagascar with special reference to the Zygoptera and with comparative notes on other faunal regions, *Verh. naturf. Ges. Basel*, **76**, 229–256.

LONGFIELD, C. (1949), *The Dragonflies of the British Isles*, Warne, London, 2nd edn, 256 pp.

LUCAS, W. J. (1900), *British Dragonflies (Odonata)*, London (Ray Society), 356 pp.

—— (1930), *The Aquatic (Naiad) Stage of the British Dragonflies*, London (Ray Society), 132 pp.

MacNEILL, N. (1960), A study of the caudal gills of dragonfly larvae of the sub-order Zygoptera, *Proc. R. Ir. Acad.*, **61** (B), 115–140.

MALOEUF, N. S. (1935), The postembryonic history of the somatic musculature of the dragonfly thorax, *J. Morph.*, **58**, 87–115.

MATHUR, K. C. (1962), The musculature of the head capsule and mouth parts of adult *Pantala flavescens* (Fabricius) (Odonata, Anisoptera, Libellulidae, Libellulinae), *J. Anim. Morph. Physiol.*, **9**, 18–31.

MAY, E. (1933), Libellen oder Wasserjungfern (Odonata), *Tierwelt Deutschlands*, **27**, 1–124.

MILL, P. J. (1965), An anatomical study of the abdominal nervous and muscular systems of dragonfly (Aeshnidae) nymphs, *Proc. zool. Soc. Lond.*, **145**, 57–73.

MILL, P. J. AND PICKARD, R. S. (1972), A review of the types of ventilation and their control in aeshnid larvae, *Odonatologica*, **1**, 41–50.

MOORE, N. W. (1952), On the so-called 'territories' of dragonflies, *Behaviour*, **4**, 85–100.

—— (1953), Population density in adult dragonflies (Odonata – Anisoptera), *J. Anim. Ecol.*, **22**, 344–359.

—— (1964), Intra- and interspecific competition among dragonflies (Odonata), *J. Anim. Ecol.*, **33**, 49–71.

MOUZE, M. (1972), Croissance et métamorphose de l'appareil visuel des Aeschnidae (Odonata), *Int. J. Insect Morphol. & Embryol.*, **1**, 181–200.

MUNCHBERG, P. (1938), Über die Entwicklung und die Larve der Libelle *Sympetrum pedemontanum* Allioni, zugleich ein Beitrag über die Anzahl der Häutungen der Odonatenlarven, *Arch. Naturgesch.*, **7**, 559–568.

—— (1967), Zum morphologischen Bau und zur funktionellen Bedeutung der Ocellen der Libellen, *Beitr. Ent.*, **16**, 221–249.

MUNSCHEID, L. (1933), Die Metamorphose des Labiums der Odonaten, *Z. wiss. Zool.*, **143**, 201–240.

MUNZ, P. A. (1919), A venational study of the suborder Zygoptera, with keys for the identification of genera, *Trans. Am. ent. Soc.*, **3**, 1–78.

NEEDHAM, J. G. (1930), A Manual of the dragonflies of China, *Zool. Sinica*, (A), **11** (1), 1–355.

—— (1951), Prodrome for a manual of the dragonflies of North America, with extended comments on wing venation systems, *Trans. Am. ent. Soc.*, **77**, 21–62.

NEEDHAM, J. G. AND WESTFALL, M. J. (1954), *A Manual of the Dragonflies of North America (Anisoptera)*, Univ. Calif. Press, Berkeley, 615 pp.

OKA, H. (1930), Untersuchungen über die Speicheldrüsen der Libellen, *Z. Morph. Ökol. Tiere*, **17**, 275–301.

OLESEN, J. (1972), The hydraulic mechanism of labial extension and jet propulsion in dragonfly nymphs, *J. comp. Physiol.*, **81**, 53–55.

PAULSON, D. R. (1974), Reproductive isolation in damselflies, *Syst. Zool.*, **23**, 40–69.

PENNAK, R. W. AND MCCOLL, C. M. (1944), An experimental study of oxygen absorption in some damselfly naiads, *J. cell. comp. Physiol.*, **23**, 1–10.

PFAU, H. K. (1971), Struktur und Funktion des sekundären Kopulationsapparates der Odonaten (Insecta Palaeoptera), ihre Wandlung in der Stammesgeschichte und Bedeutung für die adaptive Entfaltung der Ordnung, *Z. Morph. Tiere*, **70**, 281–371.

PICKARD, R. S. AND MILL, P. J. (1974), Ventilatory movements of the abdomen and branchial apparatus in dragonfly larvae (Odonata: Anisoptera), *J. Zool., Lond.*, **174**, 23–40.

PINHEY, E. C. G. (1951), The Dragonflies of Southern Africa, *Mem. Mus. Transvaal*, **5**, 335 pp.

—— (1962), A descriptive catalogue of the Odonata of the African Continent. I, II, *Publ. cult. Cia Diamantes, Angola:* pp. 1–322.

POONAWALLA, Z. T. (1966), The structure and musculature of the secondary male genitalia of the Odonata and the functional significance of the muscle disposition, *Ann. ent. Soc. Am.*, **59**, 810–818.

PRENN, F. (1928), Zur Biologie von *Sympycna (Sympecma) paedisca* Br., *Verh. zool.-bot. Ges. Wien*, **78**, 19–28.

PRITCHARD, G. (1965), Prey capture by dragonfly larvae (Odonata: Anisoptera), *Can. J. Zool.*, **43**, 271–289.

RICH, S. G. (1918), The gill chamber of dragonfly nymphs, *J. Morph.*, **31**, 317–349.

ROBERT, P. A. (1959), *Les Libellules (Odonates)*, Delachaux & Niestlé, Paris, 364 pp.

RUCK, P. AND EDWARDS, G. A. (1964), The structure of the insect dorsal ocellus. I. General organisation of the ocellus in dragonflies, *J. Morph.*, **115**, 1–26.

RUSSENBERGER, H. AND RUSSENBERGER, M. (1963), Bau und Wirkungsweise des Flugapparates von Libellen, *Mitt. naturf. Ges. Schaffhausen*, **27**, 1–88.

SADONES, J. (1896), L'appareil digestif et respiration larvaire des Odonates, *Cellule*, **11**, 273–324.

ST. QUENTIN, D. (1962), Der Eilegeapparat der Odonaten, *Z. Morph. Ökol. Tiere*, **51**, 165–189.

ST. QUENTIN, D. AND BEIER, M. (1968), Odonata (Libellen), In: Helmcke, J.-G., Starck, D. and Wermuth, H., *Handbuch der Zoologie*, **4** (2), Lief. 3, 39 pp.

SCHAFFER, G. D. (1923), The growth of dragonfly nymphs at the moult and between moults, *Stanford Univ. Publ. Biol. Sci.*, 3, 307–337.

SCHALLER, F. (1960), Étude du développement post-embryonnaire d'*Aeschna cyanea* Müll., *Annls Sci. nat.* (*Zool.*), (12) **2**, 751–868.

SCHMIDT, E. (1916), Vergleichende Morphologie des 2. und 3. Abdominalsegments bei männlichen Libellen, *Zool. Jb.* (*Anat.*), **34**, 87–200.

SHORT, J. R. T. (1955), The morphology of the head capsule of *Aeschna cyanea* (Odonata, Anisoptera), *Trans. R. ent. Soc. Lond.*, **106**, 197–211.

SLIFER, E. H. AND SEKHON, S. S. (1972), Sense organs on the antennal flagella of damselflies and dragonflies (Odonata), *Int. J. Insect. Morphol. & Embryol.*, **1**, 289–300.

SNODGRASS, R. E. (1954), The dragonfly larva, *Smithson. misc. Collns*, **123** (2), 1–38.

STRAUB, E. (1943), Stadien und Darmkanal der Odonaten in Metamorphose und Häutung, sowie die Bedeutung des Schlupfaktes für die systematische Biologie, *Arch. Naturgesch.*, **12**, 1–93.

TANNERT, W. (1958), Die Flügelgelenkung bei Odonaten, *Dt. ent. Z.*, (N.F.), **5**, 394–455.

TILLYARD, R. J. (1916), A study of the rectal breathing-apparatus in the larvae of Anisopterid dragonflies, *J. Linn. Soc. Zool.*, **33**, 127–196.

—— (1917*a*), *The Biology of Dragonflies*, Cambridge Univ. Press, Cambridge, 396 pp.

—— (1917*b*), On the morphology of the caudal gills of the larvae of Zygopterid dragonflies, *Proc. Linn. Soc., N.S. Wales*, **42**, 31–112.

—— (1921), On an Anisozygopterous larva from the Himalayas (Order Odonata), *Rec. Ind. Mus.*, **22**, 93–107.

—— (1928), The evolution of the order Odonata. Part 1. Introduction and early history of the order, *Ibid.*, **30**, 151–172.

TILLYARD, R. J. AND FRASER, F. C. (1938–40), A reclassification of the order Odonata, based on some new interpretations of the venation of the dragonfly wing, *Austr. Zool.*, **9**, 125–169; 195–221; 359–396.

TONNER, F. (1936), Mechanik und Koordination der Atem- und Schwimm-bewegung bei Libellen-Larven, *Z. wiss. Zool.*, **147**, 433–454.

WALKER, E. M. (1953), *The Odonata of Canada and Alaska*, Univ. Toronto Press, Toronto, **1**, 292 pp.; **2** (3), 318 pp.

WATSON, J. A. L. (1966), Genital structure as an isolating mechanism in Odonata, *Proc. R. ent. Soc. Lond.*, (A), **41**, 171–174.

WHEDON, A. D. (1918), The comparative morphology and possible adaptation of the abdomen in the Odonata, *Trans. Am. ent. Soc.*, **44**, 373–437.

WILLEY, R. L. (1955), A terrestrial damselfly nymph (Megapodagrionidae) from New Caledonia, *Psyche*, **62**, 137–144.

WILLIAMS, C. B. (1958), *Insect Migration*, Collins, London, 235 pp.

WILLIAMS, F. X. (1936), Biological studies in Hawaiian water-loving insects. I, II, *Proc. Hawaii. ent. Soc.*, **9**, 235–349.

WINKELMANN, F. (1973), *Sympetrum vulgatum*, Heidelibelle, *Grosses zoologisches Praktikum*, **14d**, 100 pp.

WOLF, H. (1935), Das larvale und imaginale Tracheensystem der Odonaten und seine Metamorphose, *Z. wiss. Zool.*, **146**, 591–620.

WOLFE, L. S. (1953), A study of the genus *Uropetala* Selys (Order Odonata) from New Zealand, *Trans. R. Soc. N.Z.*, **80**, 245–275.

ZAHNER, R. (1959–60), Über die Bindung der mitteleuropäischen *Calopteryx*-Arten (Odonata, Zygoptera) an den Lebensraum des strömenden Wassers. I, II, *Int. Revue ges. Hydrobiol. Hydrogr.*, **44**, 51–130; **45**, 101–123.
ZAWARSIN, A. (1911–12), Histologische Studien über Insekten. I, II, *Z. wiss. Zool.*, **97**, 481–510; **100**, 245–286.

Order 7

PLECOPTERA (PERLARIA: STONEFLIES)

Soft-bodied insects of moderate to rather large size with elongate, setaceous antennae. Mouthparts weak, of the biting type; mandibles normal or vestigial, ligula 4-lobed. Wings membranous, held flat over the back in repose, hind pair usually the larger, with well-developed anal lobes. Venation variable, often considerably specialized; vein M 2-branched. Tarsi 3-segmented. Abdomen usually terminated by long multi-articulate cerci; ovipositor wanting. Metamorphosis hemimetabolous; nymphs aquatic, campodeiform, with the antennae and usually the cerci elongate; tracheal gills, which are variable in position, commonly present.

The Plecoptera are a small order, whose members are of certain interest on account of the archaic features in their structure, and the aquatic habits of their nymphs (Hynes, 1976). The adults share some features with the Orthoptera, but the mouthparts are weaker, there is never more than a slight difference in texture between the fore and hind wings, and the coxae are small. They are poor fliers, and do not wander far from the margins of streams and lakes: they are commonly found resting upon stones, tree-trunks or palings near the water's edge, while the green forms frequent herbage. The larger species are well known to anglers as a bait for trout. The nymphs are almost exclusively aquatic, living beneath stones in clear water, particularly in streams with stony beds, and places where there are waterfalls, or where the water is otherwise well aerated. They do not live in polluted streams and few species occur in standing water. The adults of many of the species with well-developed mandibles feed on lichens and unicellular algae (Frison, 1935; Hynes, 1942; Brinck, 1949).

External Anatomy – There is no detailed account of the anatomy of the order as a whole, but the works of Schoenemund (1912), Wu (1923), Clark (1934), Hanson (1946), Grandi (1948–50) and Nelson and Hanson (1969, 1971) are useful. The head-capsule (Hoke, 1924; Chisholm, 1962; Moulins, 1968) resembles that of the Orthoptera but is prognathous, the 'epicranial suture' may be reduced or even absent and the frontoclypeal sulcus is wanting in many species. The antennae are long and setaceous, with a large number of small segments. Compound eyes are well developed, and there are three (more rarely two) ocelli. The mouthparts (Fig. 243), although completely formed, are sometimes weak structures: the mandibles are normally developed in most families, but in the Perlidae and related families they are vestigial flexible lamellae. The maxillae consist of the typical sclerites and their palps are 5-segmented. In the labium, the submentum is large, the prementum is sometimes divided, and both glossae and paraglossae are evident; the labial palps are 3-segmented. The whole of the trunk is somewhat flattened, none of the parts are strongly sclerotized, and much

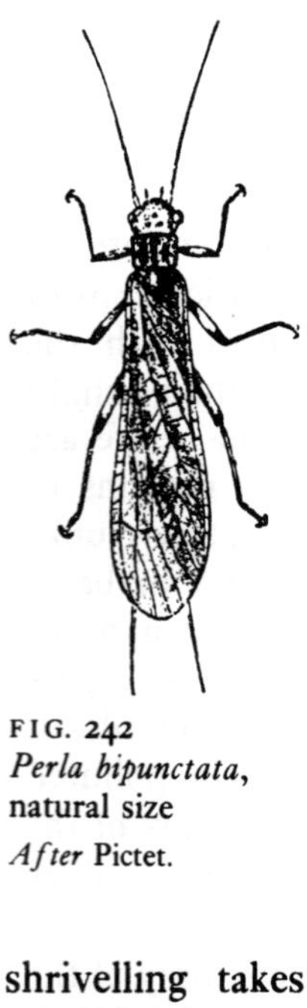

FIG. 242
Perla bipunctata, natural size
After Pictet.

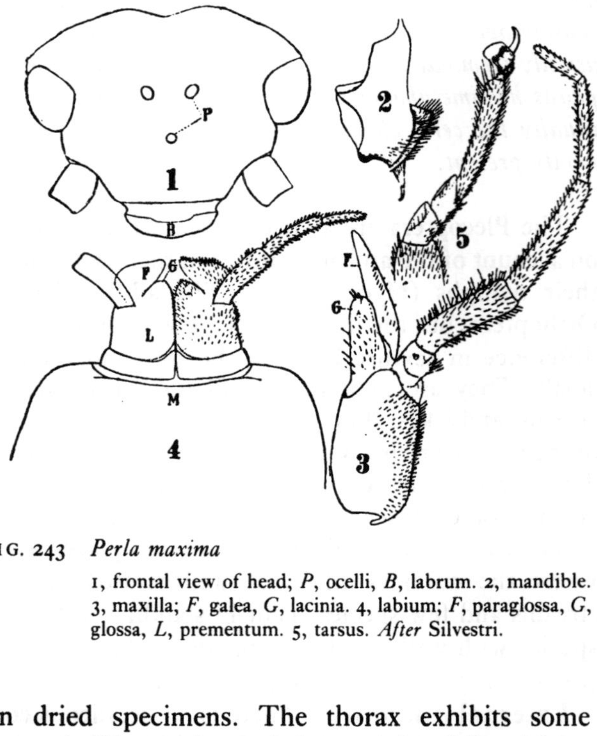

FIG. 243 *Perla maxima*
1, frontal view of head; *P*, ocelli, *B*, labrum. 2, mandible. 3, maxilla; *F*, galea, *G*, lacinia. 4, labium; *F*, paraglossa, *G*, glossa, *L*, prementum. 5, tarsus. *After* Silvestri.

shrivelling takes place in dried specimens. The thorax exhibits some primitive features (Wittig, 1955). The prothorax is large and mobile with an undivided notum and the pleuron is not differentiated into episternum and epimeron though Snodgrass (1935) recognizes in it the anapleurite and coxopleurite of more primitive forms. The meso- and metathorax are subequal segments, each composed dorsally of prescutum, scutum, scutellum and postnotum while the pleura show the usual division into episternum and epimeron. The sterna exhibit the full complement of sternites except that the spinasternum is missing from the metathorax while in the Capniidae the pro- and mesosternum have an additional sclerite lying between sternum and sternellum. On each sternum a pair of furcal arms arises from pits lying between the legs.

The wings are membranous; the hind wings are almost always considerably larger than the anterior pair, and a coupling-apparatus is not developed though some primitive families have a row of hooked hairs along part of the hind margin of the fore wing that may perhaps function as one. The anal lobe is folded fanwise against the body when in repose. The tracheation in the nymphs has been studied in several genera; it closely resembles the hypothetical type in the absence of the transverse basal trachea (Fig. 31). The fully developed wings exhibit great instability of the subordinate veins, and individuals are frequently unlike with respect to the wing-venation of the two sides of the body. Apterous forms are few but several species exhibit a brachypterous condition. These include sexually

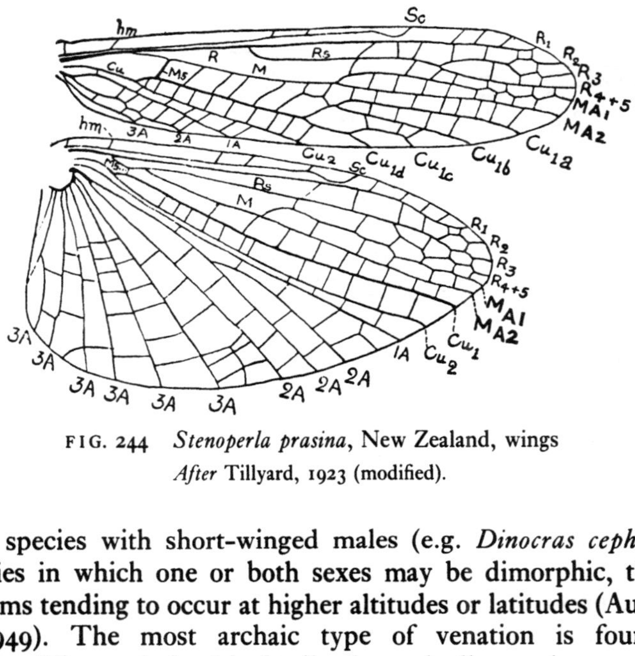

FIG. 244 *Stenoperla prasina*, New Zealand, wings
After Tillyard, 1923 (modified).

dimorphic species with short-winged males (e.g. *Dinocras cephalotes*) and other species in which one or both sexes may be dimorphic, the brachypterous forms tending to occur at higher altitudes or latitudes (Aubert, 1945; Brinck, 1949). The most archaic type of venation is found in the Eustheniidae (Fig. 244); in this family the archedictyon is present over all parts of the wings, R_s exhibits three or more branches, and there is a large fan-like anal lobe to the hind wings carrying a number of anal veins. M is 2-branched and Tillyard (1928), on palaeontological grounds, claims that this vein corresponds to the anterior media of primitive fossil insects, the posterior media having been lost. Sharov (1961), however, believes that MP has become fused with Cu_I rather than lost. Various transitional groups (e.g. Fig. 245) lead to more specialized forms like the Capniidae in which the close network of cross-veins has disappeared, R_s is two-branched in the fore wing and unbranched in the hind wing, while the latter has lost the anal lobe and vein 1A.

The abdomen is composed of ten evident segments, together with vestiges of an 11th segment. There is no ovipositor, the female gonopore opening on or, much more frequently, behind the 8th abdominal sternum while the 7th,

8th or 9th sternite is usually modified to form a subgenital plate. The male, in which the gonopore opens behind the 9th sternum, lacks an appendicular copulatory organ though a secondary structure is developed from an eversible genital chamber and in the Perlidae and Chloroperlidae it incorporates paired penis rudiments (Brinck, 1956, 1970). The 10th segment is usually a complete annulus in females but in males the sternum is generally reduced and membranous. The 11th segment is represented by a tergum (epiproct or supra-anal lobe) and a divided sternum (paraprocts or subanal lobes), all of which may be modified in males to subserve a copulatory function. A peculiar feature of males of the Nemouridae, Taeniopterygidae and some

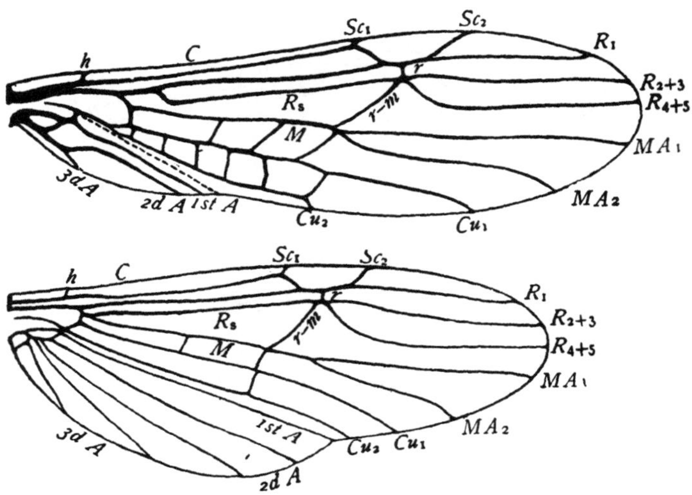

FIG. 245 *Nemoura*, wings
After Comstock, *Wings of Insects* (modified).

others is the presence on the enlarged 9th abdominal sternum of a drumming lobe. This bears mechanoreceptor hairs and probably plays a part in controlling the repeated drumming on the ground that stoneflies of both sexes carry out before mating (Rupprecht, 1968; Gnatzy and Rupprecht, 1972). Cerci are characteristic of the order and are usually long multi-articulate structures of a primitive type; in the Nemouridae, however, where they are sometimes used in copulation, they are small and composed of a single segment with, in some, a rudimentary 2nd segment.

Internal Anatomy – Much of the scattered information on internal anatomy has been summarized and supplemented by Zwick (1973). The oesophagus is very long and in *Pteronarcys* it extends into the 4th abdominal segment; the gizzard is wanting or rudimentary, and the mid gut is small. In *Perla* there are ten anterior enteric caeca, the lateral pair being the largest. The hind intestine is short and the Malpighian tubes vary between about twenty and sixty. A pair of salivary glands is present. The brain and suboesophageal ganglia are small and the stomatogastric system is normal (Arvy and

Gabe, 1954). In *Pteronarcys* there are three thoracic and eight abdominal ganglia, but in *Perla* and *Capnia* certain of the latter have undergone coalescence, with the result that there are only six evident ganglia in the abdomen (Illies, 1962). There are some variations in the structure of the reproductive system (Brinck, 1956). In the male, *Leuctra* has a pair of testes, each composed of several follicles but in most stoneflies the testis of each side is joined to its fellow, the two forming an arch-like structure. The paired vasa deferentia communicate with a pair of seminal vesicles (one in *Leuctra*) and in *Taeniopteryx* an accessory gland is also found. The reproductive system opens to the exterior by a median ejaculatory duct (sometimes very long) or, in some cases, by ducts from the seminal vesicles which remain paired up to the gonopore. The ovaries are composed of many panoistic ovarioles and are either separate (*Leuctra, Capnia*) or, more often, the ovarioles come off from a common duct joining the oviducts of each side. There is a spermatheca of variable form (absent in *Capnia*) and, in *Nemoura*, a bursa copulatrix. The rudimentary, non-functional hermaphroditism of *Perla marginata* is discussed on p. 307. The tracheal system opens to the exterior by two pairs of thoracic and eight pairs of abdominal spiracles.

Oviposition and Postembryonic Development – Claassen (1931), Frison (1935), Kuhtreiber (1934), Hynes (1941), Brinck (1949) and Schwarz (1970) give much valuable information on the nymphal stages and on the biology of the order. Mating does not occur in flight, the male mounting the back of the female on the ground and curving his abdomen down at one side of hers (Brinck, 1956). In the Perlidae the eggs are fully developed when mating occurs and are laid soon after, but in other families maturation of the eggs in the adult takes several weeks. Very many eggs are laid, probably in several masses (e.g. Miller (1939) found that *Pteronarcys proteus* lays 500–1000 eggs in batches of about 150 over a period of three weeks), the eggs of each mass being held together by a sticky slime which dissolves in water so that the eggs separate. Brachypterous females and some of the larger Perlidae crawl on stones etc. near the water's edge or move over its surface to lay their eggs, but the remainder oviposit while flying over the water, occasionally dipping the abdomen beneath its surface. According to Frison (1935) the females of some species of *Allocapnia* crawl into the water to lay their eggs and McLachlan (1864) says that females of *Leuctra* carry the eggs on their back before depositing them. The eggs of several species are described by Brinck (1949), Degrange (1957) and Knight *et al.* (1965). Those of the Perlidae are ovoidal or tetrahedral with a gelatinous outer membrane bearing adhesive structures which attach the egg to a substratum while the eggs of the other families are spherical and though they have a gelatinous coat they lack the adhesive bodies. *Allocapnia vivipara* and *Capnia nigra* are ovoviviparous.

The nymphs (Claassen, 1931; Hynes, 1941) resemble the adults very closely in their general form. Apart from the absence of fully developed wings, the characters which differentiate the nymphs from the adults are adaptive in nature, fitting them for an aquatic existence; only *Megandiperla kuscheli* nymphs live out of water but even they inhabit very rainy areas in Chile and respire through gills (Illies, 1960*a*). Plecopteran nymphs are characterized by their long multi-articulate antennae, and their similar elongate cerci (Fig. 246). The head may carry both ocelli and compound eyes; the legs are long, laterally fringed with natatory hairs, and end in paired claws. The tracheal system is apneustic, and respiration is either cutaneous or branchial. A primitive type of nymph is found in the Eustheniidae in which there are five or six pairs of lateral abdominal appendages which function as gills. In other families the nymphs may breathe by secondary tracheal gills which, according to the species, occur on various parts of the body – submentum, neck, thorax, coxae, anterior 2–3 abdominal segments or the anal region.

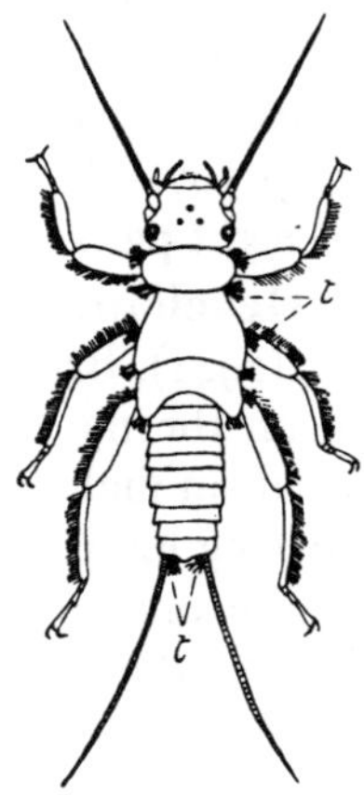

FIG. 246
Perla nymph
t, tracheal gills.

The gills may either be few and finger-like or in copious tufts. Nymphs of several Plecoptera such as *Isoperla, Leuctra* and *Capnia* are without gills. Plecoptera are also remarkable for the fact that the branchiae may persist in a somewhat shrivelled, non-functional condition in the imagines (Eggert, 1937). Perlid nymphs are largely carnivorous, preying chiefly on Ephemeropteran nymphs and Chironomid larvae, but those of other families are mainly herbivorous, feeding on diatoms, algae and mosses; *Pteronarcys, Isogenus* and others are polyphagous (Richardson and Gaufin, 1971). Correlated with these dietary distinctions there are differences in the structure of the mouthparts. The herbivorous nymphs have large mandibles with a well-developed prostheca, stout maxillae and long glossae, while in carnivorous forms the mandibles are slender and sharp without a prostheca, the maxillae are weaker and the glossae reduced.

The time occupied in development appears to range from about a year in the smaller forms up to 3½ or 4 years in the larger species. According to Wu (1923) a species of *Nemoura* passed through 22 instars and the same number was observed by Samal (1923) in *Perla burmeisteriana*. In *Dinocras cephalotes* Schoenemund (1912) recorded 33 ecdyses during a period of three years and in *Pteronarcys proteus* there are 12 nymphal instars occupying two years (Holdsworth, 1941). Last-stage nymphs move to the water's edge and may crawl some distance on land before the adult emerges. Adults of different species emerge at characteristic times over a shorter or longer period (e.g. Harper and Pilon, 1970). Some species undergo a nymphal diapause that enables them to survive the summer, when streams become too warm or dry up (Harper and Hynes, 1970). In other cases diapausing eggs are laid, as in certain populations of *Diura bicaudata* that have a two-year cycle, of which more than nine months is spent in the egg (Khoo, 1968). Plecopteran nymphs tend to drift downstream but stable population levels are maintained either by active nymphal migration upstream or by the upstream flight of adults (e.g. Schwarz, 1970).

Affinities and Phylogeny – Anatomical studies of Recent Plecoptera serve mainly to emphasize their isolated position, though it is not unreasonable to regard them as Orthopteroid and perhaps to contrast them with all the other Orthopteroid orders. The earliest Plecoptera occur in the Permian (Eustheniidae, Taeniopterygidae and some extinct families) and Tillyard (1928) supposed that they evolved from forms referred to the order Protoperlaria. More recently Sharov (1961) has included these in another extinct order, the Paraplecoptera, one family of which (the Carboniferous Narkemidae) shows venational features that suggest an ancestor of the Plecoptera. So far as Recent stoneflies are concerned, there is little doubt that the Eustheniidae and Diamphipnoidae are primitive families and that among the other two suborders the Austroperlidae and Pteronarcidae retain some generalized features. More detailed reconstructions of phylogeny, as inferred from the anatomy of Recent forms, have been presented by Illies (1965) and Zwick (1973) but differ in many respects.

Classification – The catalogues of Illies (1966) and Zwick (1973) list over 1700 species and include extensive bibliographies. The North American species have been dealt with by Needham and Claassen (1925), Frison (1935, 1942), Ricker (1943, 1952), Jewett (1959, 1960) and others. Kawai (1967) has monographed the Japanese species and the 34 British species can be identified from Hynes (1967). For the Plecoptera of other parts of Europe see Despax (1951), Illies (1955), Aubert (1964) and Steinmann (1968). Modern attempts at the phylogenetic classification of the Plecoptera owe much to Tillyard (1921), Frison (1935) and Ricker (1950, 1952). More recently Illies (1960*b*, 1965) has extended the subject and there is substantial agreement on the 14 or so families to be recognized. Their arrangement is, however, still open to discussion. The classification summarized below follows Illies, but Zwick (1973) has proposed alternative groupings based on the evaluation of a wide range of adult and nymphal characters; his views are outlined briefly at the end of the classification on p. 530. Families occurring in Britain are indicated by an asterisk.

Suborder 1. ARCHIPERLARIA

Primitive, highly coloured forms from Australasia and S. America; abundant cross-veins in both wings; glossae almost as long as paraglossae; maxillary palps filiform; mandibles well developed; nymphs with paired lateral gills on the first 4–6 abdominal segments (Illies, 1960*b*).

FAM. EUSTHENIIDAE. *Hind margin of hind wing uniformly rounded from* R_s *to last anal vein;* R_s *usually branched only once; nymphs predacious, with more or less tubular gills on the first 5 or 6 abdominal segments.* A primitive family of about 17 species with a characteristic 'amphinotic' distribution (Australia, Tasmania, New Zealand and S. America). *Eusthenia, Stenoperla.*

FAM. DIAMPHIPNOIDAE. *Hind margin of hind wing indented at end of* Cu_2*;* R_s *always branched several times in fore wing; nymphs with first four abdominal segments bearing tufted, paired gills.* There are 5 species, all from Chile, in *Diamphipnoa* and *Diamphipnopsis.*

Suborder 2. FILIPALPIA

Cross-veins generally much reduced; glossae about as long as paraglossae; maxillary palps filiform; mandibles well developed; nymphs without segmentally arranged gills.

Superfamily Leptoperloidea

A somewhat archaic, heterogeneous group, usually with numerous cross-veins and nymphs with tubular anal gills.

FAM. AUSTROPERLIDAE (= PENTUROPERLIDAE). *The most archaic Filipalpian family, with many cross-veins (except in anal lobe of hind wing).* This family also has an amphinotic distribution (Australia, Tasmania, New Zealand and Chile). There are 12 species monographed by Illies (1969); e.g. *Tasmanoperla, Klapopteryx.*

FAM. SCOPURIDAE. Represented only by the wingless *Scopura longa* from Japan (Ueno, 1938); the larva has suggested a relationship with the Austroperlidae, though Zwick puts it near the Nemuroidea (*q.v.*).

FAM. PELTOPERLIDAE. Another family of uncertain affinities, with 35 Holarctic and Oriental species. Zwick places it near the Perlids and their allies.

FAM. GRIPOPTERYGIDAE. About 120 species, distributed over Australasia and S. America; related to the Austroperlidae, from which they differ in the arrangement of anal veins. Over half the Australian stoneflies belong to this family, *Trinotaperla, Dinotaperla* and *Leptoperla* having the most species there (Kimmins, 1951).

Superfamily Nemouroidea

A more homogeneous group of smaller stoneflies, with more or less reduced cross-veins, anal lobe and cerci; nymphal gills not anal, sometimes on head or thorax.

***FAM. TAENIOPTERYGIDAE.** *Relatively copious venation; large anal lobe in hind wing; cerci usually 5- or 6-segmented.* About 70 Northern Hemisphere species. *Taeniopteryx* is Holarctic and *Rhabdiopteryx* and *Brachyptera* are Palaearctic. All three genera are represented in Britain.

***FAM. LEUCTRIDAE.** *A moderately large Northern Hemisphere family, with about 170 rather small species with somewhat rolled wings. Leuctra* is the largest genus, with species in Europe, N. Africa, Asia and N. America.

***FAM CAPNIIDAE.** *Wing venation and anal lobe of hind wing reduced; cerci long.* Another large Northern Hemisphere family, with over 200 species. *Capnia* is the largest genus, and it and *Allocapnia* have a number of N. American forms (see Ross and Ricker, 1971). *Notonemura* and its allies are S. Hemisphere forms that are sometimes placed in a family of their own, the **Notonemuridae**.

***FAM. NEMOURIDAE.** *Wing venation uniform and rather generalized; cerci reduced to one segment.* The largest Filipalpian family, with about 350 small or medium-sized species from the Holarctic and Oriental regions, a few of which are unusual among stoneflies in developing in standing water. *Amphinemura, Nemoura* and *Protonemura* are large genera; the first two occur in the Holarctic and the third in the Palaearctic.

Suborder 3. SETIPALPIA

Cross-veins usually reduced; glossae usually shorter than paraglossae; maxillary palps subuliform; mandibles usually weakly developed; nymphs without segmentally arranged abdominal gills.

Superfamily Pteronarcoidea

FAM. PTERONARCIDAE. *The smallest and most generalized of the Setipalpian families, with numerous cross-veins; mandibles relatively well developed; nymphs with branched gills on thorax and anterior part of abdomen.* About 12 species from N. America and E. Asia. *Pteronarcys dorsata*, the Giant Stonefly of N. America, is up to 60 mms long.

Superfamily Perloidea (=Subulipalpia)

Glossae reduced; mandibles reduced and functionless; first abdominal sternum reduced; first and second tarsal segments very small.

***FAM. PERLIDAE.** The largest family of stoneflies, with over 350 species from all regions except Australasia. *Dinocras cephalotes* and *Perla bipunctata* are the only British species; both genera are Palaearctic.

***FAM. PERLODIDAE.** A large and somewhat heterogeneous family of about 220 Northern Hemisphere species. *Isogenus, Perlodes, Diura* and *Isoperla* are among the European forms, some of which have short-winged males.

***FAM. CHLOROPERLIDAE.** *Small, specialized stoneflies with reduced venation and anal lobe.* About 110 species from N. America, Europe and Palaearctic Asia.

In Zwick's classification (Zwick, 1973), the above families and the Notonemuridae are rather differently grouped. The Eustheniidae, Diamphipnoidae, Austroperlidae and Gripopterygidae form one suborder (the Southern Hemisphere Antarctoperlaria), while the remaining families make up the other suborder, Arctoperlaria, within which are two sections. One of these, the Systellognatha, includes the Perloidea, Peltoperlidae and Pteronarcidae, while the other one, the Euholognatha, comprises the Nemouroidea and Scopuridae.

Literature on the Plecoptera

ARVY, L. AND GABE, M. (1954), The intercerebralis-cardiacum-allatum system of some Plecoptera, *Biol. Bull. mar. biol. Lab., Woods Hole*, **106**, 1–14.

AUBERT, J. (1945), Le microptérisme chez les Plécoptères (Perlariés), *Revue suisse Zool.*, **52**, 395–399.

—— (1964), Plecoptera, *Insecta Helvetica*, **1**, 140 pp.

BRINCK, P. (1949), Studies on Swedish stoneflies, *Opusc. ent., Suppl.*, **11**, 250 pp.

—— (1956), Reproductive system and mating in Plecoptera, *Opusc. ent.*, **21**, 57–127.

—— (1970), Plecoptera, In: *Taxonomist's Glossary of Genitalia in Insects* (Tuxen, S. L., ed.), pp. 50–55, Munksgaard, Copenhagen.

CHISHOLM, P. J. (1962), The anatomy in relation to feeding habits of *Perla cephalotes* Curtis (Plecoptera, Perlidae) and other Plecoptera, *Trans. Soc. Br. Ent.*, **15**, 55–101.

CLAASSEN, P. W. (1931), *Plecoptera Nymphs of North America*, Thomas Say Foundation, Springfield, 199 pp.

CLARK, R. L. (1934), The external morphology of *Acroneura evoluta* Klapálek (Perlidae: Plecoptera), *Ohio J. Sci.*, **34**, 121–128.

DEGRANGE, C. (1957), L'oeuf et le mode d'éclosion de quelques Plécoptères, *Trav. Lab. Hydrobiol. Piscic. Grenoble*, **48–49**, 37–49.

DESPAX, R. (1951), Plécoptères, *Faune de France*, **55**, 1–280.

EGGERT, B. (1937), Die imaginalen Tracheenkiemen von *Protonemura nitida* Pict. (Plecopt.), *Zool. Anz.*, **118**, 113–117.

FRISON, T. H. (1935), The stoneflies or Plecoptera of Illinois, *Bull. Illinois nat. Hist. Surv.*, **20**, 281–471.

—— (1942), Studies on North American Plecoptera with special reference to the fauna of Illinois, *Bull. Illinois nat. Hist. Surv.*, **22**, 235–355.

GNATZY, W. AND RUPPRECHT, R. (1972), Die Bauchblase von *Nemurella picteti* Klapálek (Insecta, Plecoptera), *Z. Morph. Tiere*, **73**, 325–342.

GRANDI, M. (1948), Contributi allo studio dei Plecotteri. I. Morfologia e miologia thoracica di *Perla marginata* Panz., *Boll. Ist. Univ. Padua*, **17**, 130–157.

—— (1950), Contributi allo studio dei Plecotteri. II. Morfologia comparata, *Boll. Ist. Ent. Univ. Bologna*, **18**, 30–57.

HANSON, J. F. (1946), Comparative morphology and taxonomy of the Capniidae, *Am. Midl. Nat.*, **35**, 193–249.

HARPER, P. P. AND HYNES, H. B. N. (1970), Diapause in the nymphs of Canadian winter stoneflies, *Ecology*, **51**, 925–927.

HARPER, P. P. AND PILON, J.-G. (1970), Annual patterns of emergence of some Quebec stoneflies (Insecta: Plecoptera), *Can. J. Zool.*, **48**, 681–694.

HOKE, G. (1924), The anatomy of the head and mouthparts of Plecoptera, *J. Morph.*, **38**, 347–373.

HOLDSWORTH, R. P. (1941), The life-history and growth of *Pteronarcys proteus* Newman, *Ann. ent. Soc. Am.*, **34**, 495–502.

HYNES, H. B. N. (1941), The taxonomy and ecology of the nymphs of British Plecoptera with notes on the adults and eggs, *Trans. R. ent. Soc. Lond.*, **91**, 459–557.

—— (1942), A study of the feeding of adult stoneflies (Plecoptera), *Proc. R. ent. Soc. Lond.*, (A), **17**, 81–82.

—— (1967), A key to the adults and nymphs of British stoneflies (Plecoptera), *Scient. Publs Freshwat. biol. Ass.*, **17**, 1–90.

—— (1976), Biology of the Plecoptera, *A. Rev. Ent.* **21**, 135–153.

ILLIES, J. (1955), Steinfliegen oder Plecopteren, *Tierwelt Deutschlands*, **43**, 150 pp.

—— (1960*a*), Archiperlaria, eine neue Unterordnung der Plecopteren, *Beitr. Ent.*, **10**, 661–697.

—— (1960*b*), Die erste auch im Larvenstadium terrestrische Plecoptere, *Mitt. schweiz. ent. Ges.*, **33**, 161–168.

—— (1962), Das abdominale Zentralnervensystem der Insekten und seine Bedeutung für Phylogenie und Systematik der Plecopteren, *Wanderversamml. dt. Ent.*, **9** (45), 139–152.

—— (1965), Phylogeny and zoogeography of the Plecoptera, *A. Rev. Ent.*, **10**, 117–140.

—— (1966), Katalog der rezenten Plecoptera, *Tierreich*, Lief. **82**, 632 pp.

—— (1969), Revision der Plecopterenfamilie Austroperlidae, *Ent. Tidskr.*, **90**, 19–51.

JEWETT, S. G. (1959), The stoneflies (Plecoptera) of the Pacific Northwest, *Corvallis Oregon State Monogr.*, **3**, 95 pp.

—— (1960), The stoneflies (Plecoptera) of California, *Bull. Calif. Ins. Survey*, **6**, 125–177.

KAWAI, T. (1967), Plecoptera (Insecta), *Fauna japon.*, **1967**, 211 pp.

KHOO, S. G. (1968), Experimental studies on diapause in stoneflies. I–III, *Proc. R. ent. Soc. Lond.*, (A), **43**, 40–48; 49–56; 141–146.

KIMMINS, D. E. (1951), A revision of the Australian and Tasmanian Gripopterygidae and Nemouridae (Plecoptera), *Bull. Br. Mus. nat. Hist.* (*Ent.*), **2**, 45–93.

KNIGHT, A. W., NEBEKER, A. V. AND GAUFIN, A. R. (1965), Descriptions of the eggs of common Plecoptera of western United States, *Ent. News*, **76**, 105–111; 233–239.

KUHTREIBER, J. (1934), Die Plekopterenfauna Nordtirols, *Ber. naturw.-med. Ver. Innsbruck*, **43–44**, 1–219.

MCLACHLAN, R. (1864), Note on the manner in which the females of the genus *Leuctra* carry their eggs, *Entomologist's monthly Mag.*, **1**, 216.

MILLER, A. (1939), The egg and early development of the stonefly, *Pteronarcys proteus* Newman (Plecoptera), *J. Morph.*, **64**, 555–609.

MOULINS, M. (1968), Contribution à la connaissance anatomique des Plécoptères: la région céphalique de la larve de *Nemoura cinerea* (Nemouridae), *Annls Soc. ent. Fr.* (N.S.), **4**, 91–143.

NEEDHAM, J. G. AND CLAASSEN, P. W. (1925), *A Monograph of the Plecoptera or Stoneflies of America north of Mexico*, Thomas Say Foundation, Lafayette, 397 pp.

NELSON, C. H. AND HANSON, J. F. (1969), The external anatomy of *Pteronarcys* (*Allonarcys*) *proteus* Newman and *Pteronarcys* (*Allonarcys*) *biloba* Newman (Plecoptera: Pteronarcidae), *Trans. Am. ent. Soc.*, **94**, 429–472.

—— (1971), Contribution to the anatomy and phylogeny of the family Pteronarcidae (Plecoptera), *Trans. Am. ent. Soc.*, **97**, 123–200.

RICHARDSON, J. W. AND GAUFIN, A. R. (1971), Food habits of some Western stonefly nymphs, *Trans. Am. ent. Soc.*, **97**, 91–121.

RICKER, W. E. (1943), Stoneflies of southwestern British Columbia, *Indiana Univ. Publ. Sci. Ser.*, **12**, 145 pp.

—— (1950), Some evolutionary trends in Plecoptera, *Proc. Indiana Acad. Sci.*, **59**, 197–209.

—— (1952), Systematic studies in Plecoptera, *Indiana Univ. Pub. Sci. Ser.*, **18**, 1–200.

ROSS, H. H. AND RICKER, W. E. (1971), The classifiation, evolution and dispersal of the winter stonefly genus *Allocapnia*, *Illinois biol. Monogr.*, **45**, 166 pp.

RUPPRECHT, R. (1968), Das Trommeln der Plecopteren, *Z. vergl. Physiol.*, **59**, 38–71.

SAMAL, J. (1923), Étude morphologique et biologique de *Perla abdominalis* Burm. (Plécoptère), *Annls Biol. lacustr.*, **12**, 229–272.

SCHOENEMUND, E. (1912), Zur Biologie und Morphologie einiger *Perla*-Arten, *Zool. Jb.* (*Anat.*), **34**, 1–56.

SCHWARZ, P. (1970), Autökologische Untersuchungen zum Lebenszyklus von Setipalpia-Arten (Plecoptera), *Arch. Hydrobiol.*, **67**, 103–140; 141–172.

SHAROV, A. G. (1961), The origin of the order Plecoptera, *Verh. XI. int. Kongr. Ent.*, **1**, 296–298.

SNODGRASS, R. E. (1935), *Principles of Insect Morphology*, McGraw-Hill, New York, 667 pp.

STEINMANN, H. (1968), Álkérészek – Plecoptera, *Faun. Hung. Entom.*, **92** (8), 1–185.

TILLYARD, R. J. (1921), A new classification of the order Perlaria, *Can. Ent.*, **53**, 35–44.

—— (1928), Kansas Permian insects. Part II. Order Protoperlaria: family Lemmatophoridae (continued), *Am. J. Sci.*, **16**, 313–348.

UENO, M. (1938) Scopuridae, an aberrant family of the order Plecoptera, *Insecta matsum.*, **12**, 154–159.

WITTIG, G. (1955), Untersuchungen am Thorax von *Perla abdominalis* Burm. (Larve und Imago), *Zool. Jb.* (*Anat.*), **74**, 491–570.

WU, C. F. (1923), Morphology, anatomy and ethology of *Nemoura*, *Bull. Lloyd Library*, **23**, *Ent. Ser.* 3, 81 pp.

ZWICK, P. (1973), Insecta: Plecoptera. Phylogenetisches System und Katalog, *Das Tierreich*, Lfg. **94**, 465 pp.

Order 8

GRYLLOBLATTODEA

Apterous, with eyes reduced or absent and no ocelli. Antennae moderately long and filiform. Mouthparts mandibulate. Legs approximately similar to each other; tarsi 5-segmented. Female with well-developed ovipositor; neither 7th nor 8th abdominal sternum enlarged to form subgenital plate. Male genitalia asymmetrical. Cerci long, 8-segmented.

This order, the first representative of which (*Grylloblatta campodeiformis*) was discovered by Walker (1914) in the Canadian Rockies, contains only sixteen species. Its members exhibit many primitive features and are therefore of considerable phylogenetic interest.

External Anatomy – Walker (1931) describes the *head-capsule* of *Grylloblatta campodeiformis* as flattened and prognathous. The epicranial, frontoclypeal, subgenal, occipital and postoccipital sulci are present, together with a 'parietal' suture which runs from the antennal socket to the foramen. Of the head sclerites the labrum is well developed, the clypeus large and subdivided into ante- and post-clypeus while the gena and postocciput are small. There is a complete tentorium and the eyes are either absent (as in *Galloisiana notabilis*) or reduced to about 60 ommatidia on each side. There are no ocelli.

The *antennae* in this order are moderately long and filiform, comprising about 28 to over 40 segments according to the species. The *mandibles* are well developed, toothed apically and near the base but without a molar region. The *maxillae* are composed of the typical parts, the lacinia being provided with two teeth and the subdivided galea less heavily sclerotized. The palp consists of five segments and there is no palpifer. In the *labium* a prementum, mentum and submentum are readily distinguished. The prementum bears a pair of 3-segmented palps, a pair of paraglossae and a pair of slightly smaller glossae. The *hypopharynx* is a flattened, broadly oval lobe, beneath which the salivary duct opens.

The three *thoracic terga* are subequal, undivided and carry no internal phragmata (Walker, 1938). This simple condition may well be related to the loss of the wings. The *pleura* are well-developed, each being divided into episternum and epimeron while the propleuron also reveals traces of a subdivision into what have been regarded as the anapleural and coxopleural arcs of the primitive insect (see p. 43). A pair of large trochantins occurs in each segment and broad precoxal bridges are found in both meso- and metathorax. The *sternal region* is partly membranous, the basisternum being most heavily sclerotized. A pair of apophyseal pits is found on each segment and an unusually primitive condition is seen in the retention of a spina in each of the three thoracic segments. The legs are cursorial, differing little in size and with large coxae but no mera. The 5-segmented tarsi are provided with ventral pads and end in a pair of claws, the pretarsus lacking an arolium and pulvilli.

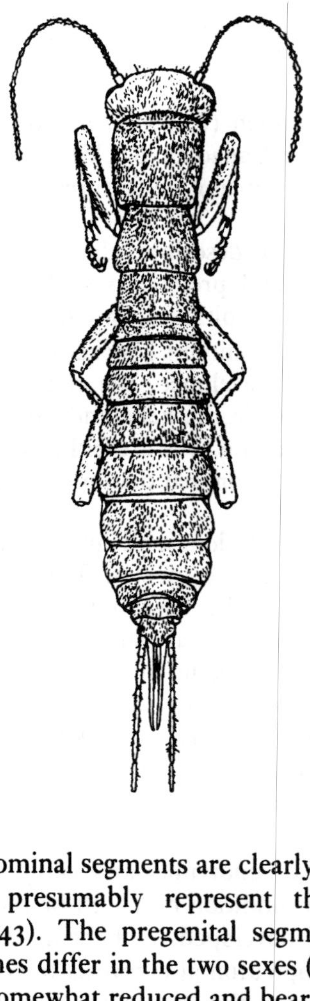

FIG. 247
Grylloblatta, female
After Walker, *Canad. Ent.*, 1914.

Ten abdominal segments are clearly evident while the epiproct and paired paraprocts presumably represent the 11th segment (Crampton, 1927; Walker, 1943). The pregenital segments are all fully developed but the posterior ones differ in the two sexes (Scudder, 1970). In the female the 8th sternite is somewhat reduced and bears the anterior pair of ovipositor valves. The 9th sternite is greatly reduced but bears the remaining two pairs of ovipositor valves; a gonangulum is also present. In the male (Walker, 1919; 1943), the 9th sternite is large and bears an asymmetrical pair of coxites, each with a small terminal style. Immediately behind the coxites is the male copulatory organ, composed of a pair of lobes; the right one bears irregular sclerotizations while the left is membranous and carries an eversible sac of unknown function. The gonopore lies on the inner side of the right lobe. Both sexes of all species carry flexible, 8-segmented cerci.

Internal Anatomy – The alimentary canal (Walker, 1949) begins with the pharynx, oesophagus and thin-walled capacious crop. This leads into a muscular proventriculus which opens into the short wide mid gut, the latter being produced anteriorly into two ill-defined gastric caeca. The hind gut is partly looped, receives 12–24 Malpighian tubules and the rectum bears six rectal pads. The compact salivary glands surround the oesophagus and lack a reservoir. The ventral nerve-cord includes seven free abdominal ganglia, the one supplying the 1st abdominal segment having fused with the metathoracic ganglion (Nesbit, 1958). The tracheae are very delicate and open by ten pairs of spiracles – two are thoracic and eight abdominal.

Biology – Ford (1926), Walker (1937) and Mills and Pepper (1937) have contributed short accounts of the biology of *Grylloblatta campodeiformis*. The insects are found beneath stones etc., at altitudes of 1500–6500 ft. They are apparently omnivorous and nocturnal, with a low temperature preference of about 1° C. (Henson, 1957; see also Edwards and Nutting, 1959). The black eggs are deposited singly in the soil or among moss when the adult female is about a year old. There is an incubation period of about a year and eight nymphal instars which together occupy about five years. Some other species are cavernicolous (Kamp, 1963, 1970).

Affinities – The affinities of this group have been the subject of much discussion (see Walker, 1933; 1937; 1943; Crampton, 1933; Zeuner, 1939; Kristensen, 1975). On the one hand, the 5-segmented tarsi, multi-articulate cerci, large coxae and asymmetrical male genitalia suggest Blattoid affinities, while on the other the absence of a meron, the well-developed ovipositor and the structure of the tentorium indicate connexions with the Orthoptera (*s. str.*). The Grylloblattodea are perhaps to be regarded as the only living remnants of a primitive stock from which both Blattoids and Orthoptera evolved, Zeuner going so far as to call them recent Protorthoptera. The reduced eyes, absence of ocelli and wings and the simplified hypopharynx are probably specializations which preclude their being directly ancestral to any of the other Orthopteroid orders.

Classification – Gurney (1937–61) has extended and summarized most of the taxonomic work, only a few species having been added later (Kamp, 1963; Asahina, 1961). There are 16 species in three genera: *Grylloblatta* occurs in the mountains and caves of western N. America and has 9 species; *Galloisiana* contains 6 species from Japan; and *Grylloblattina djakonovi* occurs in Siberia (Beĭ-Bienko, 1951).

Literature on the Grylloblattodea

ASAHINA, S. (1961), A new *Galloisiana* from Hokkaido (Grylloblattoidea), *Kontyû*, **29**, 85–87.

BEĬ-BIENKO, G. Y. (1951), A new representative of the Orthopteroid insects of the group Grylloblattoidea (Orthoptera) in the fauna of the U.S.S.R., *Ent. obozr.*, **31**, 506–509. [In Russian.]

CRAMPTON, G. C. (1927), The abdominal structures of the Orthopteroid family Grylloblattidae and the relationships of the group, *Pan-Pacific Ent.*, **3**, 115–135.

—— (1933), The affinities of the archaic Orthopteroid family Grylloblattidae and its position in the general phylogenetic scheme, *J. New York ent. Soc.*, **41**, 127–166.

EDWARDS, G. A. AND NUTTING, W. L. (1959), The influence of temperature upon the respiration and heart activity of *Thermobia* and *Grylloblatta*, *Psyche*, **57**, 33–44.

FORD, N. (1926), On the behaviour of *Grylloblatta, Can. Ent.*, **58**, 66–70.

GURNEY, A. B. (1937), Synopsis of the Grylloblattidae with the description of a new species from Oregon (Orthoptera), *Pan-Pacific Ent.*, **13**, 159–171.

—— (1948), The taxonomy and distribution of the Grylloblattidae (Orthoptera), *Proc. ent. Soc. Washington*, **50**, 86–110.

—— (1953), Recent advances in the taxonomy and distribution of *Grylloblatta* (Orthoptera: Grylloblattidae), *J. Washington Acad. Sci.*, **43**, 325–332.

—— (1961), Further advances in the taxonomy and distribution of the Grylloblattidae, *Proc. biol. Soc. Washington*, **74**, 67–76.

HENSON, W. R. (1957), Temperature preference of *Grylloblatta campodeiformis* (Walker), *Nature*, **179**, 637.

KAMP, J. W. (1963), New species of Grylloblattodea with an interpretation of their geographical distribution, *Ann. ent. Soc. Am.*, **56**, 53–68.

—— (1970), The cavernicolous Grylloblattodea of the western United States (1), *Ann. spéléol.*, **25**, 223–230.

KRISTENSEN, N. P. (1975), The phylogeny of hexapod orders. A critical review of recent accounts, *Z. zool. syst. Evolutionsforsch.*, **13**, 1–44.

MILLS, H. B. AND PEPPER, J. H. (1937), Observations on *Grylloblatta campodeiformis* Walker, *Ann. ent. Soc. Am.*, **30**, 269–274.

NESBITT, H. H. (1958), Contributions to the anatomy of *Grylloblatta campodeiformis* Walker. 6. The nervous system, *Proc. 10th int. Congr. Ent.*, **1**, 525–529.

SCUDDER, G. G. E. (1970), Grylloblattodea, In: *Taxonomist's Glossary of Genitalia in Insects*, Tuxen, S. L. (ed.), pp. 55–58.

WALKER, E. M. (1914), A new species of Orthoptera forming a new genus and family, *Can. Ent.*, **46**, 93–99.

—— (1919), On the male and immature state of *Grylloblatta campodeiformis* Walker, *Can. Ent.*, **51**, 131–139.

—— (1931), On the anatomy of *Grylloblatta campodeiformis* Walker. I. Exoskeleton and musculature of the head, *Ann. ent. Soc. Am.*, **24**, 519–536.

—— (1933), Ditto. II. Comparisons of the head with those of other Orthopteroid insects, *Ann. ent Soc. Am.*, **26**, 309–337.

—— (1937), *Grylloblatta*, a living fossil, *Trans. R. Soc. Can.*, **31**, 1–10.

—— (1938), Ditto. III. Exoskeleton and musculature of the neck and thorax, *Ann. ent. Soc. Am.*, **31**, 588–640.

—— (1943), Ditto. IV. Exoskeleton and musculature of the abdomen, *Ann. ent. Soc. Am.*, **36**, 681–706.

—— (1949), Ditto. V. The organs of digestion, *Canad. J. Res.*, (D), **27**, 309–344.

ZEUNER, F. E. (1939), *Fossil Orthoptera Ensifera*, Brit. Mus. (Nat. Hist.) London, 2 vols., 321 pp.

ORTHOPTERA (GRASSHOPPERS, LOCUSTS, CRICKETS, ETC.)

Usually medium- or large-sized insects; winged, brachypterous or apterous. Mouthparts mandibulate. Prothorax large. Hind legs usually enlarged and modified for jumping; coxae small and somewhat widely separated; tarsi 3- or 4-segmented, rarely with 5 or fewer than 3 segments. Fore wings forming more or less thickened tegmina with submarginal costal vein. Wing-pads of nymph undergo reversal during development. Female generally with well-developed ovipositor, not concealed by 7th or 8th abdominal sterna. Male external genitalia symmetrical, concealed at rest by enlarged 9th abdominal sternum which may or may not bear a pair of styles. Cerci usually short and almost invariably unsegmented. Specialized auditory and stridulatory organs frequently developed. Metamorphosis slight.

This large order, with over 17 000 described species, was formerly held to include the groups here treated as the separate orders Grylloblattodea, Dictyoptera and Phasmida. As restricted by the above definition, the Orthoptera includes not only such familiar forms as the grasshoppers, locusts and crickets but also the mole-crickets and grouse-locusts together with the wetas and king-crickets of Australasia and many others. The order is best represented in the tropics though members occur in all but the coldest zones. They are almost all terrestrial and, though usually capable of jumping actively, relatively few strong fliers are known, all belonging to the Acrididae.

The large literature on the order is best approached through the general accounts of Beier (1972) and Chopard (1938; 1949).

External Anatomy – Among other anatomical works may be mentioned those of Albrecht (1953; 1956), Ander (1939), Blackith and Blackith (1966), Carbonell (1959), Davis (1927), Jannone (1939*a*, *b*), Kramer (1944), Maskell (1927), Snodgrass (1929; 1935; 1937) and Zolessi (1968).

The head-capsule (Yuasa, 1920; Strenger, 1942) is hypognathous or occasionally prognathous and exhibits most of the sulci and sclerites in a relatively primitive condition though the ecdysial cleavage line is not always fully developed. The frontoclypeal sulcus is distinct and a trans-clypeal sulcus is usually present. The tentorium is well-developed, X-shaped and without a central aperture (Hudson, 1945). The compound eyes are usually large, but are reduced in some Stenopelmatids and the Cylindrachetidae. Ocelli are absent in apterous species but in winged forms there are usually three, though some Tettigoniidae have only two. The frons bears wind receptors in Acridoids (Camhi, 1969).

The antennae of the suborder Ensifera are elongate, filiform structures, often greatly exceeding the length of the body and composed of a large number of small segments. In the Caelifera they are shorter, with fewer than 30 segments (Mason, 1954) and though generally filiform are sometimes

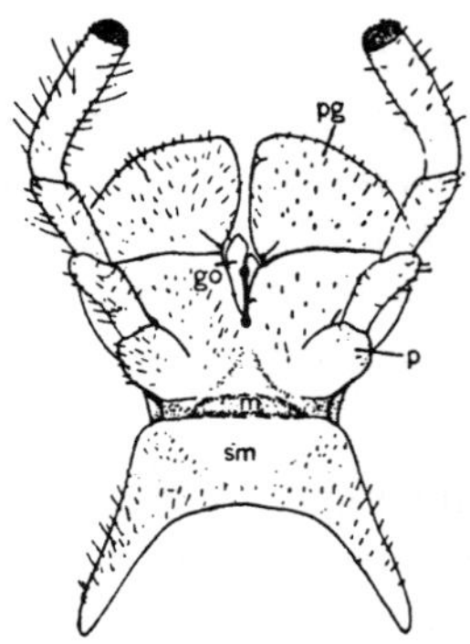

FIG. 248
Labium of *Melanoplus differentialis*, ventral aspect

pg, paraglossa; *go*, glossa; *p*, palpiger; *m*, mentum; *sm*, submentum. *After* Yuasa, *J. Morph.*, 33.

ensiform, clavate, serrate or pectinate. The mouthparts are mandibulate and relatively primitive. The mandibles are well developed and bear a series of strong grinding ridges in the phytophagous Acridoidea (Iseley, 1944; Chapman, 1964) but are more elongate and apically pointed in omnivorous or carnivorous species. The males of some Stenopelmatidae have greatly enlarged, tusk-like mandibles. The maxillae are typically developed with a pair of 5-segmented palps and the laciniae apically bidentate. The labium (Fig. 248) is divided into submentum, mentum and prementum, the latter bearing 3-segmented palps, large paraglossae and more or less reduced glossae; like the maxilla, it bears numerous chemoreceptor sensilla (Thomas, 1966; Louveaux, 1972, 1975). The hypopharynx shows no specially interesting features except in the Gryllidae, where it is a large dilatable structure crossed by numerous cuticular grooves (Rietschel, 1953).

The thoracic structures are most fully developed in alate species and are best known in the Acrididae (Snodgrass, 1929). The prothorax is large and its notum is extended laterally so as to conceal a great deal of the small propleuron (Ander, 1939) while the meso- and metathorax are closely associated to form the pterothorax. The pterothoracic nota are usually divided into prescutum, scutum and scutellum, while the intersegmental sclerites are represented by an acrotergite (precosta) on the mesothorax and both acrotergite and postnotum on the metathorax. The pterothoracic pleura are clearly divided into episternum and epimeron. Ventrally, the broad sternum of each pterothoracic segment is made up largely of the basisternum; widely separated apophyseal pits are present and sternellar regions may usually be discerned. Spinasterna occur only on the pro- and mesothorax. In apterous species the thoracic structure is simplified by the loss of many sutures and in the Tridactyloidea the prosternum and pronotum are directly connected by a precoxal bridge. The legs are usually unequally developed. The hind pair is typically adapted for jumping with enlarged femora which accommodate the powerful tibial levator muscles (Heitler, 1974). In the Gryllotalpidae, Pneumoridae, Cylindrachetidae and some others the hind legs are secondarily reduced to a more normal appearance. The fore legs are strongly fossorial in the Gryllotalpidae (Fig. 20, E) and Cylindrachetidae, the tibia being particularly broad and provided with large teeth. Four tarsal segments are found in the Tettigonioidea, three in the Grylloidea and Acridoidea and one or two in the Tridactyloidea. The legs may also bear structures concerned in stridulation or sound-perception (see below).

The degree of development of the wings varies considerably; in some species both sexes may include normal and brachypterous forms while in others the short-winged condition is characteristic of the female (Atzinger, 1957; Knetsch, 1939). Certain species or larger groups are exclusively brachypterous or apterous and in the Tetrigidae the fore wings are always vestigial while the hind wings may be polymorphic. When fully developed, the wings have a relatively complete venation with numerous cross-veins (Ragge, 1955*a*). In the fore wing – which is sclerotized to form a tegmen – the costa is submarginal, the radial sector, media and first cubitus possess several branches and Cu_2 is a straight vein delimiting a long, narrow anal region (Fig. 249). Extensive modifications of the cubito-anal area occur in males of the Tettigoniidae and Gryllidae (q.v.) in connexion with the stridulatory apparatus. The hind wing is membranous and notable for its enlarged anal lobe supported by numerous anal veins; the costa is marginal.

Eleven segments are recognizable in the abdomen, though the first sternum is reduced and the terminal segments modified in connexion with the genitalia (Ander, 1957). In the female (Qadri, 1940; Snodgrass, 1933; 1935) there is usually a well-developed ovipositor which, in its most complete form (Tettigonioidea), consists of three pairs of long valves held together by tongue and groove joints (Figs. 46, 47). The anterior (ventral) valves are derived from the 8th abdominal segment, a basal sclerite representing the coxite while the large valve corresponds with the gonapophysis. The inner and posterior (dorsal) valves originate from the reduced 9th abdominal sternum, the former pair being gonapophyses while the latter are modified coxites. A discrete gonangulum is present, with the normal articulations (p. 77). In the Gryllidae the ovipositor is long and needle-like, with vestigial inner valves, but in the Gryllotalpidae it has been lost completely. The Acridoidea (Agarwala, 1952) have three pairs of valves but the inner ones are reduced and the others are short, stout structures adapted for boring into the soil where the eggs are laid (Fig. 255). In the Tridactyloidea the ovipositor of *Rhipipteryx* resembles that of the Acrididae, but the Cylindrachetidae lack one. The male genitalia (Snodgrass, 1957; Qadri, 1940) are mostly concealed by the enlarged 9th abdominal sternum which in the Tettigonioidea bears a pair of styles. The aedeagus is a complex structure without parameres and its homologies are uncertain. Qadri (*l.c.*) and Else (1934) consider it to be derived by fusion of the appendages of the 10th abdominal segment. Its variation is of considerable taxonomic importance in some groups, such as

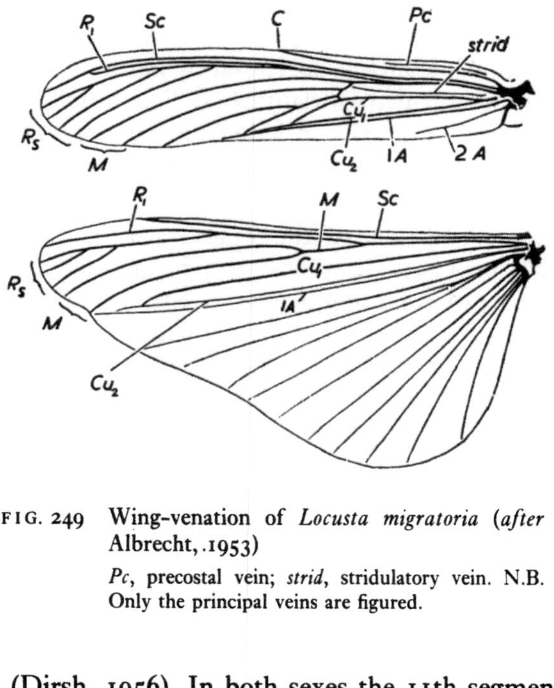

FIG. 249 Wing-venation of *Locusta migratoria* (*after* Albrecht, 1953)

Pc, precostal vein; *strid*, stridulatory vein. N.B. Only the principal veins are figured.

the Acridoidea (Dirsh, 1956). In both sexes the 11th segment consists of a median dorsal epiproct, representing the tergum (and united with the 10th tergum in Gryllidae), and a pair of lateral paraprocts which correspond to the divided sternum. The cerci are almost invariably unsegmented (the Stenopelmatid *Lezina* being an exception) and though usually short they are elongate in the Grylloidea and some others.

Among the most characteristic features of the Orthoptera are the stridulatory organs. With some exceptions (Fulton, 1933; Ragge, 1955*b*) it is the males alone which stridulate and the organs concerned are of two main types: (*a*) alary and (*b*) femoro-alary. The first type occurs in a simple form in the Prophalangopsidae and reaches its highest development in the Tettigoniidae and Grylloidea; it is described in detail below (pp. 546, 547). The femoro-alary type is found in most Acridoidea and is discussed on p. 550. In a few groups (e.g. the Pneumoridae) there is a modified zone on each side of the 2nd or 3rd abdominal segment which, when rubbed by the inner face of the hind femur, produces a sound. A somewhat similar organ in the Pamphagidae is unlikely to be stridulatory (Uvarov, 1943). Though the significance of all aspects of stridulation is not clear, in most Orthoptera it leads to aggregation or pair-formation and plays a part in courtship and in the copulatory and post-copulatory behaviour that follow (Alexander, 1967). The patterns of sound emitted are relatively complex (Pierce, 1948) and characteristic sounds are produced by each species in many cases (Faber, 1953; Jacobs, 1953; Busnel, 1955, 1963). In addition to the normal sounds, special songs are associated with characteristic phases of epigamic behaviour. Correlated with the presence of stridulatory organs is the occurrence of auditory organs which are found in both sexes (p. 131 ff.). Each organ consists of a thin cuticular membrane or tympanum whose vibrations are transmitted to the scolopidia and thence to the central nervous system. In the Acridoidea the tympanal organs are seen one on either side of the first abdominal segment (Mason, 1969). In Tettigoniidae and Gryllidae these organs are different and consist of a pair of tympana near the proximal end of each fore tibia. In many genera these tympana are freely exposed, but in some they are largely concealed; each is then covered by an integumental fold, so that it lies in a cavity which only communicates with the exterior by an elongate slit-like opening (see also p. 133).

Internal Anatomy – The *alimentary canal* (Bordas, 1897; Anadon, 1949) is almost straight in the Acridoidea but a little convoluted in the Ensifera (Fig. 95). The narrow oesophagus opens into a capacious crop which is followed by the gizzard (Judd, 1948). This is not very strongly developed in the Acridoidea but is provided with a strong, sclerotized armature in the Ensifera. The mid gut is produced anteriorly into a number of gastric caeca – two simple outgrowths occur in the Tettigonioidea and Grylloidea, but the Acridoidea possess six caeca, each composed of a forwardly directed part and a shorter posterior diverticulum (Hodge, 1936–43). The Malpighian tubules are numerous and enter the gut separately (Acridoidea) or in groups at the end of small ampullae (Tettigoniidae) or by uniting to form a common ureter (Grylloidea, Fig. 95). The hind gut ends in the rectum which bears six rectal papillae. Salivary glands are not very strongly developed in the Acridoidea, where they lack a reservoir, but they are large and provided with one in the Ensifera.

The *central nervous system* (Nesbitt, 1941; Hanström, 1940; Ewer, 1953–57) is generalized with three thoracic and usually five or six free abdominal ganglia. The ganglia supplying the anteriormost one or two abdominal segments are fused with the metathoracic centre while fusion of the ganglia of segments 7–11 or 8–11 always occurs and there may also be fusion of the ganglia supplying segments two and three. *Gryllotalpa* is exceptional in having only four free abdominal ganglia. The *stomatogastric nervous system* is well developed (Cazal, 1948) with a frontal ganglion, a pair of partially fused corpora cardiaca, a hypocerebral ganglion and a pair of oesophageal nerves; the paired spherical corpora allata are each connected to the corpora cardiaca by a short nerve. The *tracheal system* (Vinal, 1919; Carpentier, 1927) communicates with the exterior by two pairs of thoracic spiracles and eight abdominal pairs. In many Acridoidea there is a highly developed system of segmentally arranged air-sacs whose main function is probably to increase the efficiency of tracheal ventilation. In the *circulatory* system (Nutting, 1951), the heart extends throughout the abdomen into the mesothorax and is flanked by 10–12 pairs of alary muscles. Nine pairs of abdominal incurrent ostioles are present and two or three pairs of thoracic ones. In addition there may be up to two pairs of excurrent ostia in the thorax and up to five pairs in the abdomen.

The male *reproductive organs* of several species have been described by Snodgrass (1937). The testes are a pair of compact bodies composed of a variable, usually large, number of follicles enclosed in a peritoneal sheath. In the Acridoidea the testes are closely pressed together in the mid-line (Laird, 1943) while in the Gryllacrididae they are united in a common sheath. The vas deferentia may be simple (Acridoidea) or much convoluted near the testis to form an epididymis-like organ, especially conspicuous in the Gryllotalpidae. After looping round the cercal nerves the vasa join the mesodermal part of the ductus ejaculatorius from which arise the numerous tubular accessory glands and, when they are present, the paired seminal vesicles (Cantacuzène, 1967; Odhiambo, 1969–71). Finally the ectodermal part of the ductus runs into the aedeagus after giving off, in the Ensifera, a pair of globular vesicles (prostate glands) of unknown function. The female gonads consist of a pair of ovaries made up of a more or less large number of panoistic ovarioles (Voy, 1949; Waloff, 1954; Kraft, 1960) which may be arranged pectinately along the lateral oviducts or arise from them in a cluster. A spermatheca (Dirsh, 1957) is always present, opening independently of and behind the gonopore while accessory glands of variable form also occur. In the Acridoidea the latter consist of a tubular structure arising from the anterior end of each lateral oviduct and their secretion makes up the pod in which the eggs are laid (Lauverjat, 1965; Baccetti, 1967). Some Acridids also possess a pair of small Comstock-Kellogg glands at the sides of the vagina. In some Ensifera accessory glands open at the base of the ovipositor independently of the gonopore and spermatheca.

Postembryonic Development – Parthenogenesis is rare in the Orthoptera and viviparity is unknown. The epigamic behaviour often involves stridulation but otherwise is not very striking though the males of a few species have special glands on the dorsum of the thorax (Oecanthine crickets) or abdomen (*Troglophilus* etc.), the secretion of which is attractive to the female. Copulation takes place in different ways in the various species (Boldyrev, 1915; 1929) and a spermatophore (Khalifa, 1949) is always found. Oviposition usually takes place either in or on the ground (most Gryllidae, some Stenopelmatidae, almost all Acridoidea, Fig. 255) or in plant tissues (e.g. the Oecanthine crickets; many Tettigoniidae, notably the Phaneropterinae). The eggs of the Ensifera are almost always laid separately (either in clusters or widely spaced), but those of most Acridoids are enclosed in a cylindrical 'pod' (Zimin, 1938; Waloff, 1950; Chapman, 1958) made of the hardened secretion of the accessory glands sometimes mixed with particles of soil or debris. The eggs of Orthoptera are usually somewhat elongate and ovoidal or are flattened (Phaneropterinae) with one or a few micropyles on the ventral side near the anterior end. The first instar in most, if not all, Orthoptera, is the so-called 'vermiform larva' (pronymph) with a loose cuticle which envelops the appendages in such a way that they are pressed to the sides of the body and their segmentation is indistinct. This stage is of very short duration and is succeeded by instars of more normal appearance; these are usually 4–7 in number, but ten or more moults have been recorded in Grylloidea and Gryllacridoidea and up to 15 occur in the order (Ramsay, 1964). In many Orthoptera the soft cervical membrane plays an important part during hatching: it is capable of being distended, by the influx of blood, into a swollen dorsal ampulla which protrudes immediately behind the head. According to Künckel d'Herculais (1890) in *Dociostaurus* a turgid condition is maintained by the accumulation of air in the crop which lies beneath the ampulla and by means of the pressure thus exerted the insect is able to rupture the old cuticle. The cervical ampulla also plays an important part in the escape of the insects from the egg pod. Six or seven young insects combine their efforts and force open the lid of the capsule by means of their ampullae, thereby effecting their exit (see also Bernays, 1972).

In apterous forms postembryonic growth consists mainly of an increase in size, and in the further differentiation of the appendages and genital segments; in other words metamorphosis is slight and the young closely resemble their parents. In the winged forms a slight but gradual metamorphosis is also evident, and the wing-pads usually appear in the third instar. In *Melanoplus* and *Oecanthus*, which pass through six nymphal instars, the wing-rudiments arise as slight extensions of the meso- and metanota in the second instar, becoming clearly evident after the subsequent ecdysis; for the Acridoidea in general see Dirsh, 1968. The position assumed by the wings in the saltatorial Orthoptera during their later nymphal instars is different from that found in the adults. In the immature forms the wings have undergone torsion with the result that their surfaces and margins are inverted; the costal margin thus assumes a dorsal position and the hind wings are placed outside the fore wings. At the last moult the wings are untwisted into the normal positions of the adult. This type of wing development appears to occur elsewhere only in the Odonata.

Affinities and Phylogeny – Views on the origin and phylogeny of the various Orthopteran groups have changed with the accumulation of fossil material, though even now they cannot be regarded as finally settled. Despite claims to the contrary, the order (in its restricted sense) is probably monophyletic but no clear relationships with the other Recent Orthopteroid orders can be made out. Sharov (1968) considered that the Orthoptera arose from Protorthopteran stock in the Upper Carboniferous, where they are represented by an extinct, primitive Ensiferan family, the Oedischiidae. Not only did these give rise to several other purely fossil families of Ensifera (mainly from the Mesozoic) but they also led to the Locustopsidae, a primitive Caeliferan family from which the Recent Caelifera subsequently diverged. The Oedischiidae were also thought to have been ancestral to the Prophalangopsidae, from which the Gryllacridoidea and Grylloidea arose, though Sharov thought that the Tettigonioidea had an independent Oedischiid origin. These views differ in several respects from the earlier palaeontological conclusions of Zeuner (1939) and from the morphological inferences of Ander (1939), and it seems premature to accept any detailed scheme of Orthopteran phylogeny yet. Contributions to the subject from serology (Leone, 1947), cytology (White, 1951) and numerical taxonomy (Blackith and Blackith, 1968) are very interesting but not yet sufficiently conclusive.

Classification – The Orthoptera have been catalogued by Beier (1962–70) but most modern taxonomic works are restricted to the Orthopteran fauna of relatively small regions or to particular taxonomic groups within the order. Some of the latter are mentioned below, but there are general works on the N. American Orthoptera by Blatchley (1920) and Rehn and Grant (1961) and on various parts of Europe by Chopard (1951), Götz (1965) and Harz (1957–75); Ragge (1965, 1973) deals with the classification and biology of the 30 British species. See also Beĭ-Bienko (1966) and Beĭ-Bienko and Mishchenko (1951) for the Russian fauna. Phylogenetic aspects of higher classification are discussed by Ander (1939), Zeuner (1939) and Sharov (1968), the last two from a palaeontological standpoint. While the main outlines of the classification are now widely agreed, there are differences of opinion on the taxonomic rank to be given to many groups that are variously treated as families or subfamilies; some of these are indicated below. Beier (1955, 1972) has good general accounts.

Suborder 1. ENSIFERA

This suborder includes the long-horned grasshoppers or bush-crickets, the crickets and mole-crickets, and a number of less well-known groups. Antennae about as long as body, with many segments; tympanal organs, when present, on fore tibiae; stridulatory organs, when present, usually tegminal; ovipositor, when present, usually more or less elongate. Three superfamilies are recognized below.

Superfamily **Gryllacridoidea**

Stridulatory organs (tegminal or femoro-abdominal) rarely present; usually no tympanal organ on fore tibia; four tarsal segments, usually compressed; ovipositor seldom reduced, often very long.

FAM. STENOPELMATIDAE. *Fossorial, usually wingless, with fore legs sometimes strongly spinose; tarsi compressed or sub-cylindrical, with tarsal pulvilli.* The 200 or so species are usually some shade of brown and live in the soil, in rotting wood or under bark, feeding nocturnally on plant and animal matter. *Australostoma* is one of the Australian king-crickets and the family also includes the tree and ground wetas of New Zealand; *Deinacrida*, a giant weta, stridulates in both sexes and passes through 9–10 nymphal instars (Richards, A. M., 1973). Various species of *Stenopelmatus* occur in N. America and the anatomy of one has been studied by Davis (1927). For other anatomical and taxonomic work see Maskell (1927) and Karny (1937).

FAM. GRYLLACRIDIDAE. *Fore tibiae without tympanal organs; tarsi depressed; wings usually well developed but without stridulatory organs; ovipositor complete, long and thin.* Like the Stenopelmatidae, this is another relatively primitive family of saltatorial Orthoptera. It is a predominantly tropical group of over 550 species (Karny, 1937), its members being usually dark brown in colour and mostly living in trees where some construct shelters of rolled leaves with the aid of an oral secretion. They appear to be mainly predacious.

FAM. SCHIZODACTYLIDAE. This group has a discontinuous Old-World distribution and includes only three genera (Ramme, 1931; Karny, 1937). *Schizodactylus*, the only alate genus, has long wings whose apices are coiled spirally at rest. These insects are able to burrow into the soil, spending the daylight hours in the holes they have dug and seeking their prey at night.

FAM. RHAPHIDOPHORIDAE. *Wingless species, with long hind legs; no tympanal organs; tarsi strongly compressed, usually without tarsal pulvilli.* This family of about 300 species is closely allied to the Stenopelmatidae. Its members are known as camel crickets in North America, where it is represented by *Ceuthophilus* and other genera. It includes various cavernicolous forms in such genera as *Troglophilus* and *Dolichopoda* (Chopard, 1933; Varricchio, 1936). *Tachycines asynamorus* has been introduced from E. Asia into heated greenhouses in Europe, where it is nocturnal, feeding mainly on dead animal matter, but occasionally reported as attacking plants. *Daihinia brevipes* is a minor agricultural pest in Oklahoma (Whitehead and Miner, 1944). The New Zealand cave wetas also belong here and for their biology see Richards, A. M., (1970 and earlier papers).

Superfamily **Tettigonioidea**

Tarsi 4-segmented; fore wings usually with stridulatory apparatus; fore tibiae usually with tympanal organ; ovipositor with three pairs of valves usually well developed.

FAM. PROPHALANGOPSIDAE (= HAGLIDAE). Only three recent species are placed here – the Indian *Prophalangopsis obscura*, which is unusual in having only three tarsal segments, and two N. American species of *Cyphoderris*. Zeuner (1939), however, has referred a number of Mesozoic fossils to the family and considers it of great phylogenetic importance, being ancestral to the Tettigoniidae and Grylloidea. A simple stridulatory apparatus is present in the male, similar on both fore wings, which are used interchangeably (Spooner, 1973).

FAM. TETTIGONIIDAE (Locustidae: Bush-crickets, Long-horned Grasshoppers, Katydids). The Tettigoniidae (Fig. 250) are a large, predominantly tropical group of over 5000 species, divided into 19 or so sections which, following

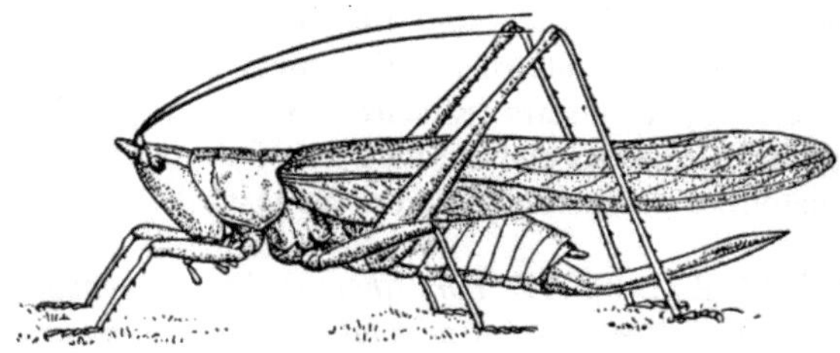

FIG. 250 A long-horned grasshopper *Neoconocephalus palustris*
After Blatchley.

Beier (1960–62) and older authorities (Caudell, 1908–16; Karny, 1912–13; Zeuner, 1936) we regard here as subfamilies, though they are sometimes treated as families. Apterous forms are common and include some of the largest species. When winged, the left tegmen usually overlaps the right one and in males the cubito-anal regions of the tegmina are modified asymmetrically for stridulation. Typically, each such tegmen has a specialized, approximately circular area delimited by branches of Cu_1 and behind this lies the stridulatory vein, Cu_2. The circular area is best developed on the right tegmen, where it is known as the 'mirror', while the stridulatory vein on the left tegmen bears a row of teeth and is known as the 'file'. Stridulation occurs through the file being scraped by the edge of the right tegmen, the mirror acting as a resonator (see papers cited on p. 180). In some groups the hind wings are absent and the tegmina reduced to the stridulating areas only, as in the Ephippigerinae and Bradyporinae where the females are also able to stridulate, though they apparently only do so infrequently. Winged Tettigoniidae are predominantly green and live amidst herbage, particularly bushes and trees, where the Pseudophyllinae (Beier, 1960–62) simulate leaves. The wingless forms live nearer the ground but some are agile climbers and reach the tree-tops. The eggs of Tettigoniidae are not enclosed in pods, and the ovipositor frequently attains a great length, even exceeding that of the body (Fig. 46). In some cases it is used for depositing the eggs in the earth, but usually they are laid in plant-tissues of various

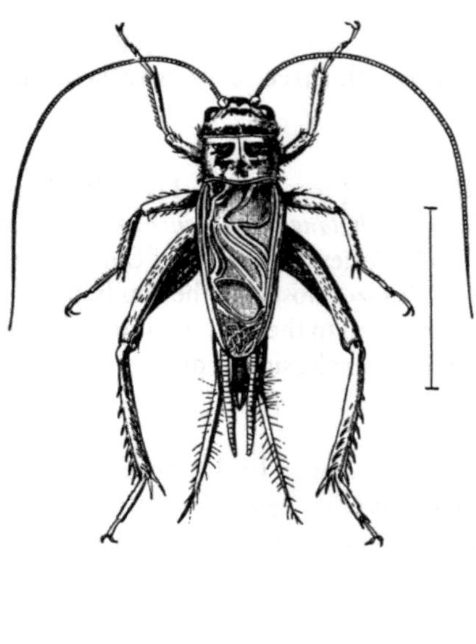

FIG. 251
Acheta domesticus, male
After Sharp, *Camb. Nat. Hist.*

kinds, often in neat longitudinal rows. Five or six ecdyses are prevalent and the members of this family are less predominantly herbivorous than the Acrididae; some forms (Saginae) are notably carnivorous while others appear to be omnivorous. *Phasmodes ranatriformis*, the only member of the Australian subfamily Phasmodinae, has a striking superficial resemblance to an elongate, wingless Phasmid.

Only ten species extend into the British Isles, the largest being *Tettigonia viridissima* which occurs in the southern half of England where it attracts attention from its strident notes; *Pholidoptera griseoaptera* has vestigial wings, and its range in England is very similar. The biology of several N. American species is described by Iseley (1941).

Superfamily **Grylloidea**

Tarsi 3-segmented; male fore wing with stridulatory apparatus; fore tibiae with tympanal organs; ovipositor needle-like with vestigial inner valves.

FAM. GRYLLIDAE (Crickets). Like the Tettigoniidae, the members of this large family (over 2300 species) stridulate by friction of the tegmina and possess tibial auditory organs (Fig. 75). The stridulatory apparatus of the male tegmen occupies a larger area than in the Tettigoniidae and the two tegmina are similarly modified. The mirror, bounded by branches of Cu_1, is displaced distally in comparison with the long-horned grasshoppers and a large space between Cu_1 and Cu_2 is traversed by cross-veins and forms the harp. Cu_2 is the file and is provided with a row of teeth while a small zone on the hind margin of each tegmen forms the scraper. During stridulation the tegmina are elevated at an angle of about 45° to the abdomen and moved backwards and forwards laterally. Although the right and left tegmina are similar to each other it is said that the sound is always produced by the scraper of the left tegmen rubbing against the file of the right one, so throwing the mirrors into vibration (Stärk, 1958). The sound-producing powers of the Gryllidae are well exemplified in the house cricket; *Brachytrypes megacephalus* is stated to make a noise so penetrating that it can be heard at the distance of a mile. The calls are characteristic of the species, thus playing an important role in the maintenance of reproductive isolation and enabling taxonomists to distinguish between sibling species (see, for example, Alexander, 1957, 1962; Leroy, 1966; Walker, 1969; and references cited on p. 181). The auditory organs differ from those of the Tettigoniidae in that the pair on each fore leg differ from each other, the outer organ being larger than the inner one. Many crickets are entirely devoid of tegmina and wings: in *Trigonidium* the tegmina are arched and horny, and impart to these insects the appearance of Coleoptera. The ovipositor is slender and cylindrical, being more or less acicular, and there is a pair of exceptionally long unsegmented cerci.

The eggs of most species are laid singly in the ground; a few of the subterranean forms deposit them in masses in underground chambers, while some Oecanthinae place them in a single uniform row in the pith of twigs (Fig. 252). There are five ecdyses in the latter subfamily but more among other Gryllidae. The life-cycle of temperate species is adapted in various ways to the climatic conditions. In Britain, for example, *Gryllus campestris* hibernates in subterranean burrows in the penultimate (10th) nymphal instar, the adults emerging in the following spring (Ragge, 1965); but in N. America other species of *Gryllus* may overwinter as the egg, or as nymphs, or even – in *Gryllus assimilis* – as eggs, nymphs and adults together (Alexander, 1968). Crickets are mostly omnivorous and frequent hot dry places, or live in holes or burrows, under logs, or among dead leaves, while the Oecanthinae occur on trees and bushes. The Gryllinae include the typical crickets and form the largest subfamily though there are only three indigenous British species, viz. the Ground Cricket *Nemobius sylvestris*, the Field Cricket *Gryllus campestris* and the

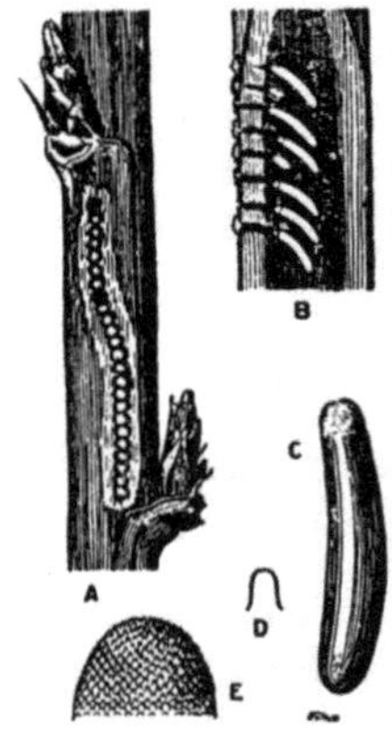

FIG. 252
Oecanthus nigricornis. A, egg punctures in stem of raspberry. B, longitudinal section. C, egg, magnified. D, projection of egg-cap. E, egg-cap
After Fulton, *N.Y. Agric. exp. Sta. Tech. Bull.*, 42.

House Cricket *Acheta domesticus* (Fig. 251). The Myrmecophilinae (Schimmer, 1909) are very small subspherical apterous crickets that live in association with ants and occur in Europe, Asia and America (Fig. 253). *Mogoplistes* and its allies are covered with minute scales and make up another subfamily while the Oecanthinae (Fulton, 1915) are a large group of pale-coloured tree crickets. The Eneopterinae include the larger brown bush crickets mostly found in the Old World.

FAM. GRYLLOTALPIDAE (Mole-crickets). This well-defined group of about 50 species (Fig. 254) is distinguished by its adaptations to a subterranean habit, the fore legs being greatly expanded and armed with strong teeth to assist

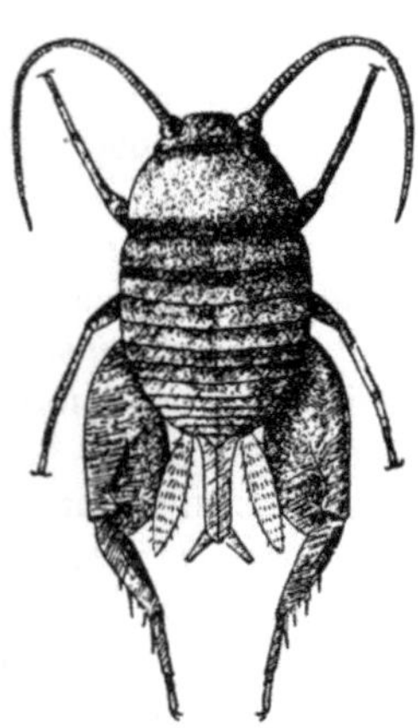

FIG. 253
Myrmecophila acervorum, female × 5
After Chopard, *Faune de France.*

FIG. 254 *Gryllotalpa gryllotalpa*, with eggs and nymphs
After Curtis.

digging, the eyes reduced and the ovipositor vestigial. When fully winged, the mole-crickets are able to fly, but brachypterous and apterous species are also known. *Gryllotalpa gryllotalpa* is rare in Britain but common in continental Europe and has been introduced into N. America. Like other species, it is sometimes injurious to cultivated plants though it is said also to eat insects and worms. The eggs are laid in a sort of nest, 10–15 cm below the surface, there are probably 10 nymphal instars and the life-cycle takes two years to complete. *G. gryllotalpa* may be separated from the very similar *G. vineae* by its song; both species construct horn-shaped, double-mouthed burrows which concentrate the sound they emit (Bennet-Clark, 1970). There is evidence that *Gryllotalpa gryllotalpa* comprises several chromosomally distinct populations which may represent sibling species (White, 1973). Hayslip (1943) describes the biology of *Scapteriscus* in Florida and Tindale (1928) that of some Australian species.

Suborder 2. CAELIFERA

This order includes the Short-horned Grasshoppers and locusts, as well as some less familiar groups like the Grouse-locusts (Tetrigidae) and the Pigmy Mole-crickets (Tridactylidae). Antennae shorter than body, with fewer than 30 segments; tympanal organs, when present, at base of abdomen; stridulatory apparatus varied or absent, but typically femoro-alary; ovipositor, when present, short and robust, with inner valves reduced. Two superfamilies are recognized below though the Tetrigidae and Proscopiidae are sufficiently distinct to have been given the rank of superfamilies by some taxonomists.

Superfamily Acridoidea

Hind tarsi, and usually also fore and mid tarsi, with three segments; pretarsus usually with arolium; abdomen of female with eight normal sterna; ovipositor present. The delimitation and subdivision of this group, which contains more than half the species of Orthoptera, presents some difficulties. The Eumastacidae, Proscopiidae and Tetrigidae (Grouse-locusts) are sufficiently distinct for each to form a family, or even a superfamily, of their own. The remainder are sometimes treated as subfamilies of a single family, the Acrididae (*s.l.*) but the modern tendency is to raise each of the fourteen or so groups to family rank (Dirsh, 1961, 1975). This procedure is followed here, but many of the resulting families contain only a few species and are not listed below. Even in this modern classification the large majority of species belong to the Acrididae (*s. str.*).

FAM. EUMASTACIDAE. *Over 300 species of small Acridoids of varied facies; head usually flattened anteriorly; prosternum without a process; wings often reduced or absent; tympanal and stridulatory organs absent; hind femur with Brunner's organ.* See Bolivar y Pieltain, 1930; Rehn, 1948. These are relatively primitive forms, mainly tropical and subtropical and living among bushes. They include the 200 species of elongate, apterous Morabinae, endemic to Australia (Blackith and Blackith, 1966).

FAM. PNEUMORIDAE. A small group of distinctive S. African species remarkable for the inflated abdomen of the male, which bears stridulatory ridges on the second tergum, and for the relatively small hind legs (Rehn, 1941).

FAM. PAMPHAGIDAE. *Usually rather large, often short-winged or apterous; tympanum present or absent according to development of wings; pronotum often keeled; Brunner's organ present and usually also Krauss's organ.* About 200 species, including those of *Lamarckiana* (Africa).

FAM. PYRGOMORPHIDAE. *About 400 species, often with rather bright, aposematic coloration; without stridulatory apparatus or Krauss's organ; head somewhat conical; otherwise not unlike Pamphagidae.* Mainly from arid regions of Africa and S. Europe and in mountains and semi-deserts of Asia. See Kevan (1966) and Rehn (1952–57).

FAM. ACRIDIDAE (Short-horned Grasshoppers; Locusts). The Acrididae, with about 9000 species, is the largest Orthopteran family and, though found predominantly in the hotter regions, includes the familiar grasshoppers of the temperate countryside as well as the notoriously destructive locusts. Stridulation takes place in several ways, only three of which will be mentioned here (Faber, 1953; Jacobs, 1953; Busnel, 1955). The best-known method, seen for example in the subfamilies Gomphocerinae and Truxalinae, is by a ridge bearing many small peg-like projections on the inner side of each hind femur which is rubbed against the hardened radial vein of the closed tegmen, thus producing a low, buzzing sound. The males stridulate by day and females of some species can also produce a sound though they possess a somewhat reduced stridulatory apparatus. In the Oedipodinae, stridulation is again femoro-alary but the row of peg-like projections lies on a secondary vein near the base of the tegmen and the femur bears a simple longitudinal ridge. Some other Acridids, mostly Oedipodines, are also able to stridulate during flight, apparently by friction between the hind wings and the under surface of the tegmina. A crackling sound results, which has been compared to that of burning stubble. The auditory organs are located one on each side of the basal segment of the abdomen (Mason, 1969). The ovipositor is not conspicuous and its

FIG. 255 Locusts in the act of oviposition
After Riley.

valves are short and curved; with them the female excavates a hole in the ground or more rarely in decaying wood. The eggs are then deposited (Federov, 1927; Agarwala, 1952–54) until they form a mass of 30–100 or more and, during the process, a glutinous fluid is discharged around them which hardens to form the egg-pod (Zimin, 1938; Waloff, 1950; Chapman, 1958) – a waterproof protection, corresponding to the more perfect ootheca of the Dictyoptera. Several of these masses are usually deposited by each female and the oviposition period in *Melanoplus* extends, according to Riley, over a period of two months. There appear to be four to eight ecdyses in the life of a species and commonly one or two generations in the year. These insects are voracious devourers of vegetation during both their young and adult stages. Iseley (1938) has shown that the Acridinae and Oedipodinae are primarily grass-feeders while the Catantopinae eat broad-leaved plants and are more selective in their feeding habits (see also Mulkern, 1967).

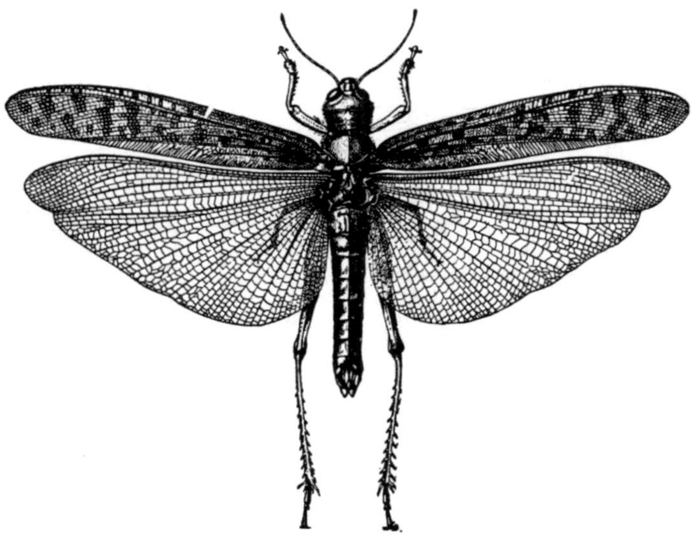

FIG. 256 A typical locust, *Schistocerca gregaria*
Reproduced by permission of the Trustees of the British Museum.

The Acrididae are usually divided into nineteen subfamilies (Uvarov, 1966; Dirsh, 1961) and with the exception of *Mecostethus grossus*, all eleven British species are referred to the Truxalinae. For their biology, see Clark (1948) and Richards and Waloff (1954). The Oedipodinae or band-winged grasshoppers are more or less brightly-coloured insects, often with blue, yellow or red hind wings crossed by a characteristic black fascia. The tegmina, however, are sombrely coloured and when closed the insect harmonizes very closely with its environment.

The term 'locust' is correctly given to a few species of Acridids which are capable under certain conditions not fully understood of forming large swarms which move over wide areas causing great devastation of natural and cultivated vegetation where they feed (Fig. 256). Uvarov (1921; 1966) has proposed a theory that each species of locust can exist in two main forms ('phases') which differ structurally and biologically. These are the gregarious phase (*phasis gregaria*) and the solitary phase (*phasis solitaria*) and the two are often so distinct as to have been regarded by earlier taxonomists as separate species. Intermediates (*phasis transiens*) also occur during the transition of a population from one extreme to the other. The solitary phase is characterized in its nymphal instars by being variable in colour, green, grey or brown and similar to the colour of its normal environment; in the adult state, the pronotum is longer and crested (Fig. 257), while the hind femur is relatively long compared with the fore wing. In *gregaria* forms, the nymphal coloration is a bold pattern mainly of black and yellow or orange and the adult has a shorter, saddle-shaped pronotum and a relatively shorter hind femur. Biologically, the most important difference between the phases is the higher activity and gregarious tendencies of the *gregaria* phase. This is manifested in the nymphs by their habit of living in large bands which, during the hotter part of the day, march from place to place. In adults,

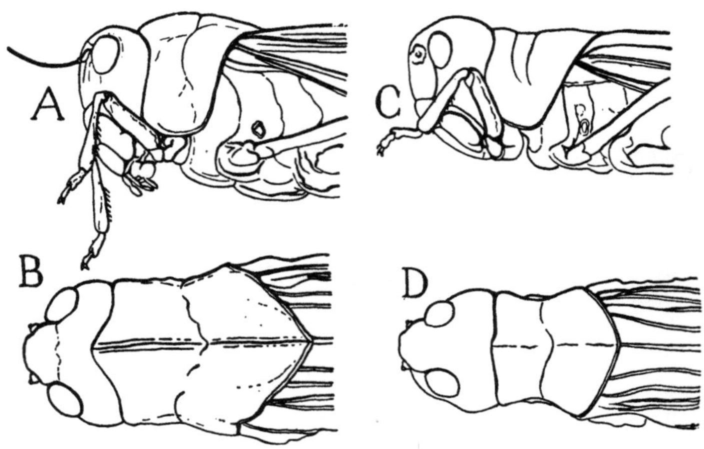

FIG. 257 Form of thorax in phases of the locust, *Locusta migratoria migratorioides*

A, B, *phasis solitaria*; C, D, *phasis gregaria*. *After* Faure. N.B. The tarsi should be represented with 3 segments.

the *gregaria* forms occur in large, more or less dense swarms which may fly over great distances under the influence of winds until environmental conditions (e.g. a fall of temperature) cause them to settle. Laboratory studies have shown that nymphs reared in isolation are of the *solitaria* phase while the crowding together of many young nymphs results in increased activity which, in turn, is associated with the development of black pigment and usually of other characteristic attributes of the *gregaria* form. Further, the high level of activity of the latter is promoted by the higher internal body temperature of the black *gregaria* nymphs which absorb more radiant heat than do the green or brownish *solitaria* nymphs. The natural conditions which induce crowding and therefore cause 'gregarization' and lead to locust plagues are not completely clear, but in general it appears that the process takes place in restricted regions ('outbreak areas') where, as a result of flooding or variable rainfall, the habitats favourable to breeding are liable to irregularly periodic contractions in size. Swarms originating in the outbreak areas invade large regions and, when they breed there, may give rise to *solitaria* or *gregaria* forms according to local conditions. After a few years, however, the area affected by swarms becomes smaller and the locust plague ends. New plagues apparently originate only in the more or less permanent outbreak areas and measures designed to prevent gregarization in these places are therefore fundamental to locust control.

The most important locusts of the Old World are (i) *Locusta migratoria*, with several subspecies, of which the African one has an outbreak area in the flood-plains of the Middle Niger; (ii) *Nomadacris septemfasciata* (the Red Locust) with outbreak areas in E. Africa and (iii) *Schistocerca gregaria* (the Desert Locust), the outbreak areas of which vary with the rainfall in Pakistan, Arabia and Africa. Other economically important locusts include the S. American *Schistocerca paranensis*, the Mediterranean *Dociostaurus maroccanus*, the Indomalaysian *Patanga succincta* and the Australian Plague Locust, *Chortoicetes terminifera*. Other injurious Acridids that do not show such pronounced phase polymorphism include *Zonocerus variegatus* and *Oedaleus senegalensis* from Africa and the Australian *Austroicetes cruciata*.

There is a very large biological and applied literature on Acrididae, much of it summarized by Uvarov (1966, 1976), Albrecht (1967), Dempster (1963), Fraser-Rowell (1971), Gunn (1960), Hemming and Taylor (1972), and Kennedy (1956); see also such publications as *Anti-Locust Bulletin*, *Anti-Locust Memoirs* and *Acrida*. For taxonomic work see, among others, Bigelow (1967), Bruner (1900–09), Dirsh (1961, 1965, 1968, 1974, 1975), Johnston (1956, 1968), Key (1954) and Rehn (1952–57).

FAM. PROSCOPIIDAE. This family comprises about 120 species endemic to S. America (Mello-Leitão, 1939). They are superficially similar to elongate Phasmids, enjoying a protective resemblance to the vegetation on which they live; almost all are apterous and they rarely jump. See Zolessi, 1968.

FAM. TETRIGIDAE (Tettigidae, Acrydiidae: Grouse-locusts) – This well-defined group consists of over 1000 species (Hancock, 1906; Günther, 1938; Podgornaya, 1969) and its members are readily distinguished from other Acridoids by the backwardly directed, sometimes grotesquely shaped, process of the pronotum (Fig. 258) which covers the abdomen and conceals the hind wings, the tegmina

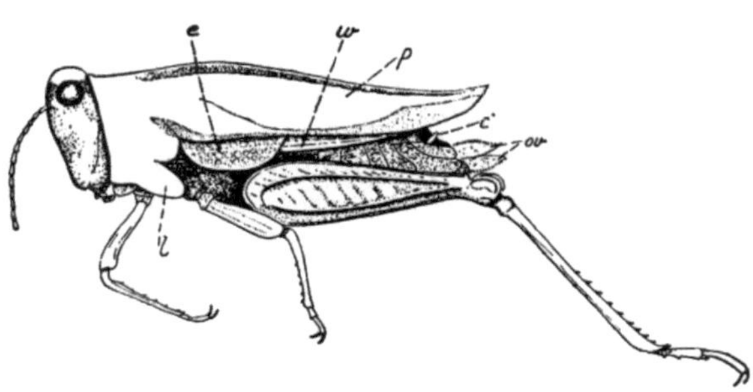

FIG. 258 *Tetrix undulata* × 5
c, cercus; *e*, tegmen; *ov*, ovipositor; *p*, pronotum; *l*, lateral lobe; *w*, wing.

being reduced to small scales; the fore and mid tarsi have 2 segments and there are no arolia. Many species exhibit polymorphism in the development of the hind wings and pronotum. Though the Tetrigidae are best represented in the tropics, they are relatively numerous in cooler regions, over 50 species being Palaearctic (three British). They tend to frequent moist places and some are semi-aquatic. Their biology is not very well known but at least some of the temperate species pass the winter both as adults and partly grown nymphs (see Hodgson, 1963). Their eggs are laid in the ground in a loosely associated mass and they feed on the algae and mosses growing in moist soil or at the sides of lakes and streams.

Superfamily Tridactyloidea

Hind tarsi 1-segmented; fore and mid tarsi 1- or 2-segmented; arolium absent; abdomen with 9 sterna in female; ovipositor usually absent.

FAM. TRIDACTYLIDAE. About 100 species of these small insects are known, four occurring in the Mediterranean region. They possess short antennae and very large hind femora while each hind tibia, in addition to a number of subapical articulated spines, bears a pair of longer apical appendages between which lies the vestigial, unsegmented hind tarsus. The anterior tarsi are 2-segmented. The Tridactylidae live near water, where they burrow into sandy ground, apparently feeding on fragments of vegetable matter in the soil. They are also capable of moving on or beneath the water-surface. For biology, see Urquhart (1937).

FAM. CYLINDRACHETIDAE. – Though formerly associated with the Gryllotalpidae, this small family is now known to be allied to the Tridactylidae (Ander, 1934; Carpentier, 1936). Its members occur in Australia, New Guinea and Patagonia and are highly adapted to a burrowing subterranean life, being apterous with an elongate, cylindrical body, fossorial fore legs, short antennae and reduced ocelliform eyes; the hind legs are not saltatorial. See Tindale (1928).

Literature on the Orthoptera

AGARWALA, S. B. D. (1952–54), A comparative study of the ovipositor in Acrididae, *Indian J. Ent.*, **13**, 147–181; **14**, 61–75; **15**, 53–69, 299–318.

ALBRECHT, F. O. (1953), *The Anatomy of the Migratory Locust*, Athlone Press, London, 118 pp.

—— (1956), The anatomy of the red locust (*Nomadacris septemfasciata* Serville), *Anti-Locust Bull.*, **23**, 9 pp.

—— (1967), *Polymorphisme Phasaire et Biologie des Acridiens Migrateurs*, Masson et Cie., Paris, 194 pp.

ALEXANDER, R. D. (1957), The taxonomy of the field crickets of the eastern United States (Orthoptera: Gryllidae), *Ann. ent. Soc. Am.*, **50**, 584–602.

—— (1962), The role of behavioural study in cricket classification, *Syst. Zool.*, **11**, 53–72.

—— (1967), Acoustical communication in arthropods, *A. Rev. Ent.*, **12**, 495–526.

—— (1968), Life cycle origins, speciation and related phenomena in crickets, *Q. Rev. Biol.*, **43**, 1–41.

ANADON, E. (1949), Estudios sobre el aparato digestivo de los Ephippigerinos, *Trab. Inst. Cienc. nat. Madrid*, **2**, 95–241.

ANDER, K. J. (1934), Über die Gattung *Cylindracheta* und ihre systematische Stellung, *Ark. Zool.*, **26** (21), 1–16.

—— (1939), Vergleichend-anatomische und phylogenetische Studien über die Ensifera (Saltatoria), *Opusc. ent., Suppl.*, **2**, 306 pp.

—— (1957), Zur Morphologie des Hinterleibsendes der Ensiferen, *Ent. Tidskr.*, **78**, 89–100.

ATZINGER, L. (1957), Vergleichende Untersuchungen über die Beziehungen zwischen Ausbildung der Flügel, der Flugmuskulatur und des Flugvermögen bei Feldheuschrecken, *Zool. Jb. (Anat.)*, **76**, 199–222.

BACCETTI, B. (1967), L'ultrastruttura delle ghiandole della ooteca in Ortotteri Acridoidei, Blattoidei e Mantoidei, *Z. Zellforsch.*, **77**, 64–79.

BEĬ-BIENKO, G. Y. (1966), Orthoptera: Vol. 2, No. 2. Tettigonioidea, Phaneropterinae, *Fauna U.S.S.R. (N.S.)*, **59**, 1–381.

BEĬ-BIENKO, G. Y. AND MISHCHENKO, L. L. (1951), Acridoidea of the fauna of the U.S.S.R. and neighbouring countries. Parts 1 and 2, *Opred. Faun. S.S.S.R.*, **38**, 378 pp.; **40**, 667 pp.

BEIER, M. (1955), Orthopteroidea, *Bronn's Kl. Ord. Tierreich*, 5/III/6, 31–304.

—— (1960, 1962), Pseudophyllinae. I, II, *Tierreich*, **73**, 1–434; **74**, 1–396.

—— (ed.) (1962–70), *Orthopterorum Catalogus*, The Hague. Partes 1–14, 2160 pp.

—— (1972), Saltatoria (Grillen und Heuschrecken), In: Helmcke, J.-G., Starck, D. and Wermuth, H. (eds), *Handbuch der Zoologie*, **4** (2), Lfg. 17, 1–217.

BENNET-CLARK, H. C. (1970), The mechanism and efficiency of sound production in mole crickets, *J. exp. Biol.*, **52**, 619–652.

BERNAYS, E. A. (1972*a*), The muscles of newly-hatched *Schistocerca gregaria* larvae and their possible functions in hatching, digging and ecdysial movements (Insecta: Acrididae), *J. Zool.*, **166**, 14–158.

—— (1972*b*), The intermediate moult (first ecdysis) of *Schistocerca gregaria* (Forskål) (Insecta, Orthoptera), *Z. Morph. Tiere*, **71**, 160–179.

BIGELOW, R. S. (1967), *The Grasshoppers (Acrididae) of New Zealand*, University of Canterbury, Christchurch, 111 pp.

BLACKITH, R. E. AND BLACKITH, R. M. (1966–67), The anatomy and physiology of the morabine grasshoppers: I–III, *Aust. J. Zool.*, **14**, 31–48; 1035–1071; **15**, 961–998.

——(1968), A numerical taxonomy of Orthopteroid insects, *Aust. J. Zool.*, **16**, 111–31.

BLATCHLEY, W. S. (1920), *The Orthoptera of North Eastern America*, Nature Publ. Co., Indianapolis, 784 pp.

BOLDYREV, B. T. (1915), Contributions á l'étude de la structure des spermatophores et des particularités de la copulation chez Locustodea et Gryllodea, *Hor. Soc. ent. Ross.*, **41**, 1–245.

—— (1929), Spermatophore fertilization in the Migratory Locust (*Locusta migratoria*, L.), *Izv. prikl. Ent., Leningrad*, **4** (1), 189–218.

BOLIVAR Y PIELTAIN, C. (1930), Monografia de los Eumastacidos (Orth. Acrid.), *Trab. mus. nac. Ciencias nat.*, **46**, 380 pp.

BORDAS, L. (1897), L'appareil digestif des Orthoptères. (Études morphologiques, histologiques et physiologiques de cet organe et son importance pour la classification des Orthoptères.) *Annls Sci. nat. (Zool.)*, **5**, 1–208.

BRUNER, L. (1900–1909), Orthoptera: Acrididae, *Biol. centr.-Amer.*, **2**, 412 pp.

BUSNEL, R. G. (ed.) (1955), Colloque sur l'acoustique des Orthoptères, *Annls Epiphyt.*, (C), **5** num. hors série, 1–448.

—— (ed.) (1963), *Acoustic Behaviour of Animals*, London and Amsterdam, 933 pp.

CAMHI, J. M. (1969), Locust wind receptors I–III, *J. exp. Biol.*, **50**, 335–348; 349–362; 363–373.

CANTACUZÈNE, A. M. (1967), Recherches morphologiques et physiologiques sur les glandes annexes mâles des Orthoptères. I, III, *Bull. Soc. zool. Fr.*, **92**, 725–738; *Z. Zellforsch.*, **90**, 113–126.

CARBONELL, C. S. (1959), The external anatomy of the South American semiaquatic grasshopper *Marellia remipes* Uvarov (Acridoidea, Pauliniidae), *Smithson. misc. Collns*, **137**, 61–97.

CARPENTIER, F. (1927), Sur les trachées de la base des pattes et des ailes de la sautenelle verte (*Phasgoneura viridissimus, L.*), *Ann. Soc. Sci., Bruxelles*, (B), **47**, 63–86.

—— (1936), Le thorax et ses appendices chez les vraies et les faux Gryllotalpides, *Mem. Mus. Hist. nat. Belg.*, (2), **4**, 86 pp.

CAUDELL, A. N. (1908–16), Locustidae, *Genera Insectorum*, **72**, 43 pp., **120**, 7 pp., **138**, 25 pp., **140**, 10 pp., **167**, 9 pp., **168**, 13 pp., **171**, 29 pp.

CAZAL, P. (1948), Les glandes endocrines rétrocérébrales des insectes (étude morphologie), *Bull. biol. Fr. Belg., Suppl.*, **32**, 227 pp., **33**, 9–18.

CHAPMAN, R. F. (1958), The egg pods of some tropical African grasshoppers. I, II, *J. ent. Soc. sth. Afr.*, **21**, 85–112; **24**, 259–284.

—— (1964), The structure and wear of mandibles of some African grasshoppers, *Proc. zool. Soc. Lond.*, **142**, 107–121.

CHOPARD, L. (1933), Biospeologica LVIII. Les Orthoptères cavernicoles de la faune paléarctique, *Arch. Zool. exp. gén.*, **74**, 263–286.

—— (1938), La biologie des Orthoptères, *Encycl. Ent., Paris*, **20**, 541 pp.

—— (1949), Ordre des Orthoptères, In Grassé's *Traité de Zoologie*, **9**, 617–722.

—— (1951), Orthoptéroïdes, *Faune de France*, **56**, 359 pp.

CLARK, E. J. (1948), Studies in the ecology of British grasshoppers, *Trans. R. ent. Soc. Lond.*, **99**, 173–222.

DAVIS, A. C. (1927), Studies of the anatomy and histology of *Stenopelmatus fuscus* Hald, *Univ. Calif. Publs Ent.*, **4**, 159–208.

DEMPSTER, J. P. (1963), The population dynamics of grasshoppers and locusts, *Biol. Rev.*, **38**, 490–529.

DIRSH, V. M. (1956), The phallic complex in Acridoidea in relation to taxonomy, *Trans. R. ent. Soc. Lond.*, **108**, 223–356.

—— (1957), The spermatheca as a taxonomic character in Acridoidea, *Proc. R. ent. Soc. Lond.*, (A), **32**, 107–114.

—— (1961), A preliminary revision of the families and subfamilies of Acridoidea, *Bull. Br. Mus. nat. Hist., Ent.*, **10**, 350–419.

—— (1965), *The African genera of Acridoidea*, Cambridge, 634 pp.

—— (1968), The post-embryonic ontogeny of Acridomorpha (Orthoptera), *Eos*, **43**, 413–514.

—— (1974), Genus *Schistocerca, Series Entomologica*, **10**, 240 pp.

—— (1975), *Classification of the Acridomorphoid Insects*, Classey, Faringdon, 184 pp.

ELSE, F. L. (1934), The developmental anatomy of male genitalia in *Melanoplus differentialis*, *J. Morph.*, **55**, 577–610.

EWER, D. W. (1953–57), The anatomy of the nervous system of the tree locust, *Acanthacris ruficornis* (Fab.) I–V, *Ann. Natal Mus.*, **12**, 367–381; *J. ent. Soc. sth. Afr.*, **17**, 27–37; 232–236; **20**, 195–216.

FABER, A. (1953), Laut- und Gebärdensprache bei Insekten. Orthoptera (Geradflügler). Teil 1, *Mitt. Mus. Naturk. Stuttgart*, **287**, 198 pp.

FEDEROV, S. M. (1927), Studies in the copulation and oviposition of *Anacridium aegyptium, L.*, (Orthoptera, Acrididae), *Trans. ent. Soc. Lond.*, **75**, 53–61.

FRASER-ROWELL, C. H. (1971), The variable coloration of the Acridoid grasshoppers, *Adv. Insect Physiol.*, **8**, 145–198.

FULTON, B. B. (1915), The tree crickets of New York: life history and bionomics, *Tech. Bull. agric. Exp. Stn, New York*, **42**, 1–47.

—— (1933), Stridulatory organs of female Tettigoniidae (Orthoptera), *Ent. News*, **44**, 270–275.

GÖTZ, W. (1965), Orthoptera, Geradflügler, *Tierwel Mitteleuropas*, Bd. 4, Lf. 2., 71 pp.

GUNN, D. L. (1960), The biological background of locust control, *A. Rev. Ent.*, **5**, 279–300.

GÜNTHER, K. (1938), Revision der Acrydiinae. 1, 2, *Mitt. zool. Mus. Berlin*, **23**, 299–437; *Stettin. ent. Ztg.*, **99**, 117–148; 161–220.

HANCOCK, J. L. (1906), Acrididae, Tetrigidae, *Genera Insectorum, Fasc.* **48**, 79 pp.

HANSTRÖM, B. (1940), Inkretorische Organe, Sinnesorgane und Nervensystem des Kopfes einiger niederer Insektenordnungen, *K. svenska Vetensk.Akad. Handl.*, **18**, 1–266.

HARTLEY, J. C. AND WARNE, A. C. (1972), The developmental biology of the egg stage of Western European Tettigoniidae (Orthoptera), *Proc. zool. Soc. Lond.*, **168**, 267–298.

HARZ, K. (1957), *Die Geradflügler Mitteleuropas*, Fischer, Jena, 494 pp.

—— (1969), (1975), *Die Orthopteren Europas*, Junk, The Hague, Vol. I. (1969) 750 pp; Vol. II. (1975) 939 pp.

HAYSLIP, N. C. (1943), Notes on biological studies of mole crickets at Plant City, Florida, *Florida Ent.*, **26**, 33–46.

HEITLER, W. J. (1974), The locust jump. Specializations of the metathoracic femoral-tibial joint, *J. comp. Physiol.*, **89**, 93–104.

HEMMING, C. F. AND TAYLOR, T. H. C. (1972), *Proceedings of the International Study Conference on the current and future problems of Acridology*, London (C.O.P.R.) 533 pp.

HODGE, C. (1936), The anatomy and histology of the alimentary tract of the grasshopper *Melanoplus differentialis* Thomas, *J. Morph.*, **59**, 423–434.

—— (1939), The anatomy and histology of the alimentary tract of *Locusta migratoria* L, *J. Morph.*, **64**, 375–400.

—— (1940), The anatomy and histology of the alimentary tract of *Radenotatum carinatum* var. *peninsulare* Rehn and Hebard, *J. Morph.*, **66**, 581–604.

—— (1943), The internal anatomy of *Leptysma marginicollis* (Serv.) and of *Opshomala vitreipennis* (Marsch.), *J. Morph.*, **72**, 87–124.

HODGSON, C. J. (1963), Some observations on the habits and life history of *Tetrix undulata* (Swrb.) (Orthoptera: Tetrigidae), *Proc. R. ent. Soc. Lond.*, (A), **38**, 200–205.

HUDSON, G. B. (1945), A study of the tentorium in some Orthopteroid Hexapoda, *J. ent. Soc. sth. Afr.*, **8**, 71–90.

ISELEY, F. B. (1938), The relations of Texas Acrididae to plants and soils, *Ecol. Monogr.*, **8**, 551–604.

—— (1941), Researches concerning Texas Tettigoniidae, *Ecol. Monog.*, **11**, 457–475.

—— (1944), Correlation between mandibular morphology and food specificity in grasshoppers, *Ann. ent. Soc. Am.*, **37**, 47–67.

JACOBS, W. (1953), Verhaltensbiologische Studien an Feldheuschrecken, *Z. Tierpsychol.*, Beih. **1**, 228 pp.

JANNONE, G. (1939*a*), Studio morfologica, anatomica e istologica del *Dociostaurus maroccanus* Thunb. nelle sue fasi *transiens, congregans, gregaria* e *solitario* (terzo contributo), *Boll. Lab. Ent. agr. Portici*, **4**, 1–443.

JANNONE, G. (1939*b*), Contributi alla conoscenza dell'ortotterofauna italica. III, *Boll. Lab. Zool. Portici*, **31**, 201–217.

JOHNSTON, H. B. (1956), *Annotated Catalogue of African Grasshoppers*, Cambridge Univ. Press, Cambridge, 833 pp.

—— (1968), *Annotated Catalogue of African Grasshoppers, Supplement*, Cambridge Univ. Press, Cambridge, 448 pp.

JUDD, W. W. (1948), A comparative study of the proventriculus of Orthopteroid insects with reference to its use in taxonomy, *Can. J. Res.*, (D), **26**, 93–161.

KARNY, H. (1912–13), Locustidae, *Genera Insect.*, Fasc. **131**, 20 pp; Fasc. **135**, 17 pp; Fasc. **139**, 50 pp; Fasc. **141**, 47 pp.

—— (1937) Gryllacrididae, *Genera Insect.*, Fasc. **206**, 317 pp.

KENNEDY, J. S. (1956), Phase transformation in locust biology, *Biol. Rev.*, **31**, 349–370.

KEVAN, D. K. MCE. (1966), The Pyrgomorphidae of South Africa (Orthoptera: Acridoidea), *Trans. Am. ent. Soc.*, **92**, 557–584.

KEY, K. H. L. (1954), Taxonomy, phases and distribution of *Chortoicetes* and *Austroicetes* (Orthoptera), Canberra (C.S.I.R.O.), 237 pp.

KHALIFA, A. (1949), The mechanism of insemination and the mode of action of the spermatophore in *Gryllus domesticus*, *Q. Jl microsc. Sci.*, **9**, 281–292.

KNETSCH, H. (1939), Die Korrelation in der Ausbildung der Tympanalorgane, der Flügel, der Stridulationsapparate und anderer Organsysteme bei den Orthopteren, *Arch. Naturgesch.*, (N.F.) **8**, 1–69.

KRAFT, A. (1960), Entwicklungsgeschichtliche und histochemische Untersuchungen zur Oogenese von *Tachycines*. I, II, *Zool. Jb.* (*Anat.*), **78**, 457–484; 485–558.

KRAMER, S. (1944), The external morphology of the oblong-winged katydid *Amblycorypha oblongifolia* (DeGeer) (Orthoptera, Tettigoniidae), *Ann. ent. Soc. Am.*, **37**, 167–192.

KÜNCKEL d'HERCULAIS, J. (1890), Mécanisme physiologique de l'éclosion, des mues et de la métamorphose chez les insectes Orthoptères de la famille des Acridides, *C.r. Acad. Sci., Paris*, **110**, 657–659.

LAIRD, A. K. (1943), A study of the types of male gonads found in the Acrididae (Orthoptera), *J. Morph.*, **72**, 477–490.

LAUVERJAT, S. (1965), Données histologiques et histochimiques sur les voies génitales femelles et sur la sécrétion de l'ootheque chez quelques Acridiens, *Annls Soc. ent. Fr.*, (N.S.), **1**, 879–935.

LEONE, C. A. (1947), A serological study of some Orthoptera, *Ann. ent. Soc. Am.*, **40**, 417–433.

LEROY, Y. (1966), Signaux acoustiques, comportement et systématique de quelques espèces de Gryllides, *Bull. biol. Fr. Belg.*, **100**, 3–134.

LOUVEAUX, A. (1972), Équipement sensoriel et système nerveux périphérique des pieces buccales de *Locusta migratoria* L, *Insectes soc.*, **19**, 359–368.

—— (1975), Étude de l'innervation sensorielle de l'hypopharynx de larves de *Locusta migratoria migratorioides* R. et F. (Orthoptère, Acrididae), *Insectes soc.*, **22**, 3–11.

MASKELL, F. G. (1927), The anatomy of *Hemideina thoracica*, *Trans. Proc. N. Zealand Inst.*, **57**, 637–670.

MASON, J. B. (1954), Numbers of antennal segments in adult Acrididae, *Proc. R. ent. Soc. Lond.*, (B), **23**, 228–238.

—— (1969), The tympanal organs of Acridomorpha, *Eos*, **44**, 267–355.

MELLO-LEITÃO, C. de (1939), Estudio monografico de los Proscopidos, *Rev. Mus. La Plata Secc. Zool.*, **1**, 279–448.

MULKERN, G. B. (1967), Food selection by grasshoppers, *A. Rev. Ent.*, **12**, 59–78.

NESBITT, H. H. J. (1941), A comparative morphological study of the nervous system of the Orthoptera and related orders, *Ann. ent. Soc. Am.*, **34**, 51–71.

NUTTING, W. L. (1951), A comparative anatomical study of the heart and accessory structures of the Orthopteroid insects, *J. Morph.*, **89**, 501–597.

ODHIAMBO, T. R. (1969), The architecture of the accessory reproductive glands of the desert locust, *Phil. Trans. R. Soc. Ser. B*, **256**, 85–114.

—— (1969–71), The architecture of the accessory reproductive glands of the male desert locust. I, II, V, *Tissue and Cell*, **1**, 155–182; 325–340; **3**, 309–324.

PIERCE, G. W. (1948), *The Songs of Insects*, Harvard Univ. Press, Cambridge, Mass., 329 pp.

PODGORNAYA, L. I. (1969), Some anatomical peculiarities of the family Tetrigidae (Orthoptera), *Ent. Obozr.*, **48**, 255–262. [*Ent. Rev.*, **48**, 149–153.]

QADRI, M. A. H. (1940), On the development of the genitalia and their ducts of Orthopteroid insects, *Trans. R. ent. Soc. Lond.*, **90**, 121–175.

RAGGE, D. R. (1955*a*), *The Wing-venation of the Orthoptera Saltatoria*, (Brit. Mus. (Nat. Hist.)), London, 159 pp.

—— (1955*b*), La problème de la stridulation des femelles Acridinae (Orthoptera, Acrididae), In: *Colloque sur l'Acoustique des Orthoptères.* (Busnel, R. G., ed.), **5** num. hors série des *Annls Epiphyt.*, 448 pp.

—— (1965), *Grasshoppers, crickets and cockroaches of the British Isles*, Warne, London, 299 pp.

—— (1973), The British Orthoptera: a supplement, *Ent. Gaz.*, **24**, 227–245.

RAMME, W. (1931), Systematisches, Verbreitung und Morphobiologisches aus der Gryllacrididen-Unterfamilie Schizodactylinae (Orth.), *Z. Morph. Ökol. Tiere*, **22**, 163–172.

RAMSAY, G. W. (1964), Moult number in Orthoptera (Insecta), *N. Z. Jl Sci.*, **7**, 644–666.

REHN, J. A. G. (1941), On new and previously known species of Pneumoridae (Orthoptera, Acridoidea), *Trans. Am. ent. Soc.*, **67**, 137–159.

—— (1948), The Acridoid family Eumastacidae (Orthoptera). A review of our knowledge of its components, features and systematics, with a suggested new classification of its major groups, *Proc. Acad. nat. Sci. Philad.*, **100**, 77–139.

—— (1952–57), *The Grasshoppers and Locusts (Acridoidea) of Australia* I–III, C.S.I.R.O., Melbourne, 326 pp.; 270 pp.; 273 pp.

REHN, J. A. G. AND GRANT, H. J. (1961), A monograph of the Orthoptera of North America (north of Mexico). Vol. I, *Monog. Acad. nat. Sci. Philad.*, **12**, 282 pp.

RICHARDS, A. M. (1970), Observations on the biology of *Pallidotettix nullarborensis* Richards (Rhaphidophoridae: Orthoptera) from the Nullarbor Plain, *Proc. Linn. Soc. N.S.W.*, **94**, 195–206.

—— (1973), A comparative study of the biology of the giant wetas *Deinacrida heteracantha* and *D. fallai* (Orthoptera: Henicidae) from New Zealand, *J. Zool. Lond.*, **169**, 195–236.

RICHARDS, O. W. AND WALOFF, N. (1954), Studies on the biology and population dynamics of British grasshoppers, *Anti-Locust Bull.*, **17**, 182 pp.

RIETSCHEL, P. (1953), Der Hypopharynx von *Gryllus* und anderen Gryllinae, ein dem Fliegenrüssel analoges Organ, *Z. Morph. Ökol. Tiere*, **41**, 386–410.

SCHIMMER, F. (1909), Beitrag zu einer Monographie der Gryllodeengattung *Myrmecophila* Latr., *Z. wiss. Zool.*, **93**, 409–534.

SHAROV, A. G. (1968), The phylogeny of the Orthopteroidea, *Trud. paleont. Inst. Akad. Nauk. USSR.*, **118**, 1–213. (In Russian; English translation, 1971.)

SNODGRASS, R. E. (1929), The thoracic mechanism of a grasshopper and its antecedents, *Smithson, misc. Collns*, **93**, 409–534.

—— (1933), Morphology of the insect abdomen. II. The genital ducts and the ovipositor, *Smithson. misc. Collns*, **89**, 148 pp.

—— (1935), The abdominal mechanisms of a grasshopper, *Smithson, misc. Collns*, **94**, 89 pp.

—— (1937), The male genitalia of Orthopteroid insects, *Smithson. misc. Collns*, **96**, 107 pp.

SPOONER, J. D. (1973), Sound production in *Cyphoderris monstrosa* (Orthoptera: Prophalangopsidae), *Ann. ent. Soc. Am.*, **66**, 4–5.

STÄRK, A. A. (1958), Untersuchungen am Lautorgan einiger Grillen- und Laubheuschrecken-Arten, zugleich ein Beitrag zum Recht-Links-Problem, *Zool. Jb.* (*Anat.*), **77**, 9–50.

STRENGER, A. (1942), Funktionelle Analyse des Orthopterenkopfes, eine systematisch-funktionsanalytische Studie, *Zool. Jb.* (*Syst.*), **75**, 1–72.

THOMAS, J. G. (1966), The sense organs on the mouthparts of the Desert Locust (*Schistocerca gregaria*), *J. Zool., Lond.*, **148**, 420–448.

TINDALE, N. B. (1928), Australasian mole-crickets of the family Gryllotalpidae, *Rec. S. Aust. Mus., Adelaide*, **4**, 1–42.

URQUHART, F. A. (1937), Some notes on the sand cricket (*Tridactylus apicalis* Say), *Canad. Fld. Nat.*, **51**, 28–29.

UVAROV, B. P. (1921), A revision of the genus *Locusta*, L. (=*Pachytelus* Fieb.), with a new theory as to the periodicity and migrations of locusts, *Bull. ent. Res.*, **12**, 135–163.

—— (1943), The tribe Thrinchini of the subfamily Pamphaginae and the interrelations of the Acridid subfamilies, *Trans. R. ent. Soc. Lond.*, **93**, 1–72.

—— (1966, 1977), *Grasshoppers and Locusts: A Handbook of General Acridology*, I, II. Cambridge Univ. Press, Cambridge, 481 pp., 597 pp.

VARRICCHIO, P. (1936), Note sulla morfologia e lo sviluppo postembrionale della *Dolichopoda palpata* Sulzer, *Ann. Mus. zool. Univ. Napoli*, (*N.S.*), **6** (19), 20 pp.

VINAL, S. C. (1919), The respiratory system of the Carolina Locust, *Dissosteira carolina*, *J. New York. ent. Soc.*, **27**, 19–32.

VOY, A. (1949), Contribution à l'étude anatomique et histologique des organes accessoires de l'appareil génital femelle chez quelques espèces d'Orthoptéroides, *Ann. Sci. nat. Paris*, **11**, 269–345.

WALKER, T. J. (1964), Cryptic species among sound-producing ensiferan Orthoptera (Gryllidae and Tettigoniidae), *Quart. Rev. Biol.*, **39**, 345–355.

—— (1969), Systematics and acoustic behavior of United States crickets of the genus *Orocharis* (Orthoptera: Gryllidae), *Ann. ent. Soc. Am.*, **62**, 752–762.

WALOFF, N. (1950), The egg pods of British short-horned grasshoppers (Acrididae), *Proc. R. ent. Soc. Lond.*, (A), **25**, 115–126.

—— (1954), The number and development of ovarioles of some Acridoidea in relation to climate, *Physiol. comp. Oecol.*, **3**, 370–390.

WHITE, M. J. D. (1951), Cytogenetics of Orthopteroid insects, *Adv. Genetics*, **4**, 268–330.

—— (1973), *Animal Cytology and Evolution*, Cambridge Univ. Press, Cambridge, 3rd edn, 961 pp.

WHITEHEAD, F. E. AND MINER, F. D. (1944), The biology and control of the camel cricket, *J. econ. Ent.*, **37**, 573–581.

YUASA, H. (1920), The head and mouthparts of Orthoptera and Euplexoptera, *J. Morph.*, **33**, 251–290.

ZEUNER, F. E. (1936), The subfamilies of Tettigoniidae (Orthoptera), *Proc. R. ent. Soc. Lond.*, (B), **5**, 103–109.

—— (1939), *Fossil Orthoptera Ensifera*, Brit. Mus. (Nat. Hist.), London, 2 vols., 321 pp.

ZIMIN, L. S. (1938), Les pontes des Acridiens. Morphologie, classification et écologie. Tableaux analytiques, *Faune U.R.S.S.*, **23**, 84 pp.

ZOLESSI, L. C. de (1968), Morphologie, endosquelette et musculature d'un Acridien aptère (Orthoptera: Proscopiidae), *Trans. R. ent. Soc. Lond.*, **120**, 55–113.

PHASMIDA (STICK- AND LEAF-INSECTS)

Large, apterous or winged insects, frequently of elongate, cylindrical form, more rarely depressed and leaf-like. Mouthparts mandibulate. Prothorax short; meso- and metathorax usually elongate, the latter closely associated with 1st abdominal segment. Legs similar to each other; coxae small and rather widely separated; tarsi almost always 5-segmented. Fore wings, when present, usually small and with submarginal costa. Wing-pads do not undergo reversal during development. Ovipositor small and mostly concealed by enlarged 8th abdominal sternum. Male external genitalia variable and asymmetrical, concealed by 9th abdominal segment. Cerci short, unsegmented. Specialized auditory and stridulatory organs absent. Eggs deposited singly. Metamorphosis slight.

The Phasmida is a group of predominantly tropical insects (Günther, 1953; Beier, 1957) remarkable for their close protective resemblance to the foliage or, more frequently, twigs, of the vegetation on which they occur and feed. Though formerly classified with the Dictyoptera as Cursorial Orthoptera, they show some similarities to the Saltatoria and because of their uncertain affinities they are here given separate ordinal status. For general accounts, see Beier (1968) and Chopard (1938; 1949).

External Anatomy – In addition to the references cited below, the following papers deal with the anatomy of various members of this order: Heymons (1897), de Sinety (1901), Marshall and Severin (1906), Leuzinger *et al.* (1926), Littig (1942) and Bauchhenss (1971). Two main types of bodily structure occur in the Phasmida: elongate, often apterous, cylindrical, rod-like forms which closely resemble twigs in colour and appearance (Fig. 260) and the less frequent depressed, leaf-like, winged species (Phylliidae) with lamellate expansions of the legs (Fig. 259). A few more thick-set forms, superficially similar to saltatorial Orthoptera, are also known and are probably more primitive than the plant-simulating species.

The cuticle of the Phasmids is frequently beset with spines and other cuticular prominences. The head is prognathous with a well-developed frontoclypeal sulcus, but the ecdysial cleavage line and occipital sulcus are not evident and the transverse division of the clypeus is indistinct. The labrum is large and cleft apically and the tentorium (Hudson, 1945) is imperforate, with long anterior arms and a small body. Compound eyes are always present, but ocelli (usually 2) are found only in some of the winged species. The antennae vary considerably in length, being filiform or moniliform, with from 8 to over 100 segments. The biting mouthparts have strong mandibles and each maxilla possesses a 5-segmented palp, a 2-segmented galea and a lacinia which is spinose on its inner face. The labium is subdivided into prementum, mentum and submentum; the labial palps are 3-segmented, the paraglossae are well developed and the glossae narrower. A simple hypopharynx is also present.

The prothorax is always short, while the meso- and metathorax are longer, the latter being closely attached to the 1st abdominal segment (median segment), the suture between them sometimes being obliterated. In the

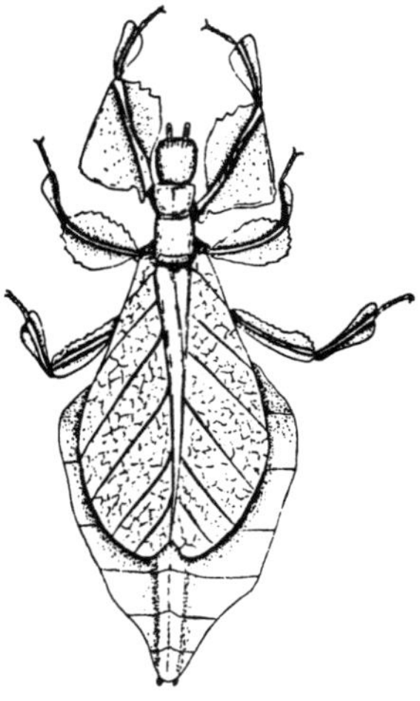

FIG. 259
Phyllium crurifolium, half natural size
Oriental region.

elongate, cylindrical, apterous forms, the terga and sterna form a tube-like structure, the pleura being greatly reduced. The ventral side of the thorax is dominated by the large basisterna, but a sternellum is also found on each segment and a spinasternum occurs in the meso- and metathorax. The prothoracic pleura are typical in appearance, but the epimera of the other two segments are very small. Many Phasmids are completely apterous and many others show varying degrees of brachyptery. When wings are well developed, the fore wings are rather strongly sclerotized to form tegmina and are commonly reduced in size. The hind wings possess an anterior sclerotized region, correlated with the presence of small tegmina and a membranous posterior part made up largely of the anal lobe. The venation has been studied by Ragge (1955). It is rather uniform and not unlike the Orthoptera but in the leaf-like forms that of the fore wing is considerably modified to imitate the veins of a leaf. The costa of the fore wing is set back from the front margin and the other veins run longitudinally with little or no branching. The hind wing also has a simple venation with a large number of straight anal veins supporting the large anal lobe, which can be folded at rest in a fan-like fashion. The legs are similar to each other, the coxae rather small and widely separated and, in the Phylliidae, the femora and tibiae are provided with lamellate expansions. There are five tarsal segments, except in *Timema*, which has three, and in regenerated legs where at most four are present.

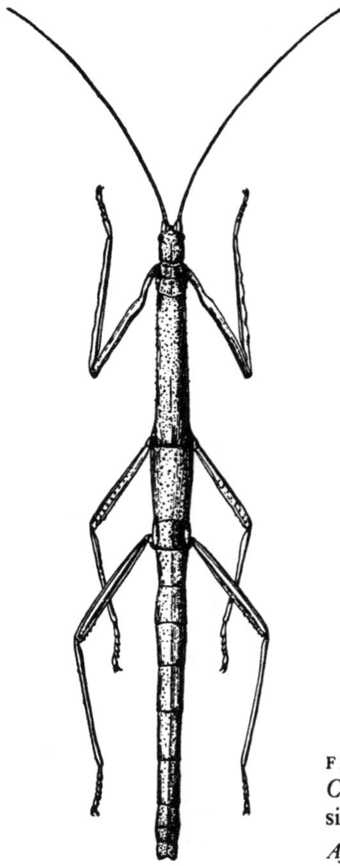

FIG. 260
Carausius morosus, natural size
After Ling Roth.

Eleven segments may be recognized in the abdomen. The 1st segment is closely associated with the metathorax (though the tergum is free in *Timema*) and the first sternum is reduced. The genital segments differ according to the sex, but the 10th segment is always well developed and the 11th is represented by a small epiproct (tergum), a pair of paraprocts (divided sternum) and the cerci. The ovipositor (Günther, 1933; Maki, 1935*a*) is composed of three pairs of small valves, the first pair being appendages of the 8th abdominal segment while the two posterior pairs are derived from the 9th segment. The gonangulum is fused with the first valvifer (coxite of segment 8) except in *Timema* (Kristensen, 1975). The whole ovipositor is concealed within the enlarged 8th sternum (operculum). In the male (Pantel, 1915; Snodgrass, 1937) the terminal abdominal segments and aedeagus are rather variable in form. The latter is asymmetrical and made up of a number of lobes, which are sometimes more or less completely united into a single structure. The 9th abdominal sternum is prolonged beneath the genitalia and is usually divided into an anterior part and a distinct subgenital plate. The 10th segment in the male may be a simple plate or is sometimes desclerotized except for a conspicuous anterior lobe, the so-called vomer. The cerci, which may have a prehensile appearance in some males, are always unsegmented.

Internal Anatomy – The alimentary canal (Bordas, 1897) is without convolutions and the crop is normal but the gizzard is atrophied and there are no gastric caeca. The anterior part of the mid gut is provided with thick bands of circular muscles while the posterior part bears numerous glandular papillae on its outer surface, each produced into a terminal filament. There are numerous Malpighian tubules and a pair of large, bilobed salivary glands. In the central nervous system there are three thoracic ganglia and the first 1–3 abdominal ganglia are united with the metathoracic nerve-centre so that 5–7 free abdominal ganglia occur (see also Marquardt, 1940). The stomatogastric nervous system (Pflugfelder, 1937; Hanström, 1940) is well developed with a frontal ganglion, recurrent nerve, hypocerebral ganglion, partially fused corpora cardiaca and a single oesophageal nerve. A pair of asymmetrically arranged corpora allata is also found (Boisson, 1949). The dorsal vessel (Nutting, 1951) extends forwards from the 9th abdominal segment into the metathorax as the heart and is then continued into the head in the form of the aorta. There are 9–11 pairs of alary muscles and 9 abdominally situated pairs of incurrent ostia while one pair of excurrent ostia is found in the metathorax and another pair in each of the first two abdominal segments. In the male reproductive system (Snodgrass, 1937), the testes of *Timema* are each composed of a longitudinally arranged series of ovoidal follicles, but in other Phasmids the testes are a pair of elongate, tubular structures not divided into follicles. The short vasa deferentia are not convoluted and open into the ductus ejaculatorius at the point where a variable number of tubular accessory glands arise. Distinct seminal vesicles have not usually been reported but de Sinety (1901) claims that two tubular structures very similar to the accessory glands act as sperm reservoirs in *Leptyniella*. In the female, each ovary comprises a variable number of panoistic ovarioles widely spaced along the median side of the lateral oviduct. The common oviduct opens behind the 8th sternite and dorsal to it there lies a large bursa copulatrix which opens near the gonopore. The single or paired spermatheca opens into or just behind the bursa and a pair of accessory glands is also present (Heberdey, 1931; Günther, 1933).

In many Phasmids, a pair of long tubular repugnatorial glands is found in the prothorax; they open separately in front of each fore coxa (Moreno, 1940; Wegner, 1955).

Postembryonic Development and Biology – Mating (Stockard, 1908) occurs by the male mounting the back of the female and curving his abdomen down at the side of, or behind (*Phyllium*), that of the female. Copulation sometimes lasts several hours and a spermatophore is probably formed in most species (Lefevre, 1939). The eggs are laid singly, usually falling to the ground, and they often resemble seeds very closely – those of *Aplopus mayeri*, for example, being similar to the seeds of *Suriana maritima*, the plant on which it feeds (Stockard, *l.c.*). Each egg has a rather complex structure, with a distinct operculum and a thick shell which, in *Bacillus libanicus* (Moscona, 1950), for instance, comprises a many-layered, hard, brittle exochorion whose organic framework is partly impregnated with calcium salts, a double-layered, membranous endochorion and a thin vitelline membrane (Fig. 261). The egg is often retained in the genital chamber of the female for some time before laying and many months may elapse before it hatches. The postembryonic development has been studied in only a few species. Grimpe (1921) found that in *Phyllium bioculatum* females there are usually six nymphal instars which are completed in about 125 days at 20–24° C. Males develop more rapidly and have one or two less instars. They also have a shorter adult life than the females which live for about 90 days and start oviposition 14–20 days after copulation, producing over 100 eggs before they die. *Carausius morosus* has 5 nymphal instars in the male and six in the female, which lays up to 800 or so eggs, though 1–200 is more usual. For further details see Favrelle (1938), Ling Roth (1917), Talbot (1920), Voy (1954) and Stringer (1970).

The biology of the Phasmids shows many interesting features, some of which are reviewed by Chopard (1938). Some species of *Bacillus, Clonopsis* and *Carausius* are almost exclusively parthenogenetic, while in other Phasmids the males are uncommon and facultative parthenogenesis probably occurs, only females developing from unfertilized eggs (Cappe de Baillon and Vichet, 1940; Bergerard, 1958). Another peculiarity is the occurrence, on stimulation or injury, of autotomy of the nymphal legs, the affected limb breaking off at a specialized point between trochanter and femur and being capable of rapid regeneration (though with only four tarsal segments) in subsequent instars. Rapid 'physiological' colour changes have been studied in a few Phasmids, notably *Carausius morosus*. Various colour-forms of this species are known, varying from brown to green and in all except the green variety a change in the intensity of the colour can be produced quite rapidly by the migration of pigment-granules within the epidermal cells, probably under hormonal influence (Dupont-Raabe, 1957; Mothes, 1960). High humidity, low temperature and low light intensities cause darkening, while pallor is induced by the opposite conditions. Under normal influences a rhythmical colour-change occurs diurnally (dark at night, light in the day) and this rhythm persists for some time if the insects are subjected to continuous darkness. The Australian Phasmids *Podocanthus wilkinsoni, Didymuria violescens* and *Ctenomorphodes tessulata* may sometimes occur in very large

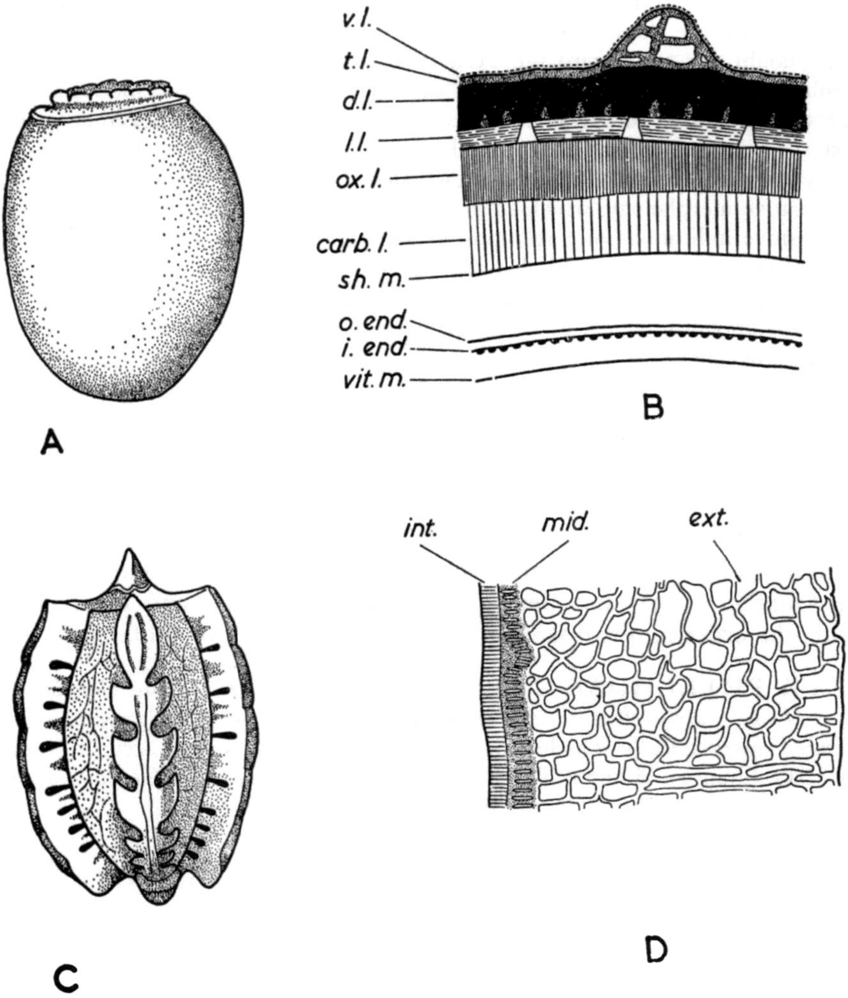

FIG. 261 Eggs of Phasmida. A, B. *Bacillus*. C, D. *Phyllium* (B *after* Moscona, 1950; C & D *after* Henneguy, 1904)

carb.l, carbonate layer; *d.l*, dark layer; *ext*, external layer; *i.end*, inner endochorion; *int*, internal layer; *l.l*, lamellar layer; *mid*, intermediate layer; *o.end*, outer endochorion; *ox.l*, oxalate layer; *sh.m*, shell membrane; *t.l*, tubercular layer; *vit.m*, vitelline membrane; *v.l*, varnish layer.

numbers on eucalyptus trees, which they damage. These species show phase differences analogous to the type of polymorphism found in locusts, differing in colour and biometric features according to whether they develop in crowds or in isolation (Key, 1957). Catalepsy (Godden, 1974) and well-defined diurnal rhythms in activity and behaviour (Eidmann, 1956) are two further phenomena found in some species of Phasmida.

Affinities and Phylogeny – The Triassic Xiphopteridae, Aeroplanidae and Chresmodidae are thought by some to be true Phasmids (Sharov, 1968), though others doubt this and regard the Tertiary amber fossils as the earliest undoubted members of the order. Sharov further considers that the Triassic species and the representatives of his order Titanoptera (also Triassic) both arose together from an Ensiferan-like Orthopteran group, the Permian Tcholmanvissiidae. An Orthopteran theory of Phasmid origins is the traditional view, but it is not very well supported by comparative morphology, which suggests that they form an isolated order showing few positive indications of a relationship with any particular Orthopteroid group. Kristensen (1975) has argued that the peculiar N. American genus *Timema* is a distinctive relict of a primitive stage in Phasmid evolution, but until more is known of the anatomy of the more generalized Phasmids it seems unprofitable to speculate on their phylogeny.

Classification – About 2500 species of Phasmida are known, the greatest number occurring in the Oriental region. Only a small number of species are found outside the tropics though two New Zealand species (*Acanthoxyla prasina* and *Clitarchus laeviusculus*) have now established themselves as members of the British fauna in the Scilly Isles (Uvarov, 1950). Taxonomic works include those of Brunner von Wattenwyl and Redtenbacher (1906–08), Günther (1953) and Beier (1957). Günther (1953) and Beier (1968) divide the order into two families, each with a number of subfamilies.

FAM. PHYLLIIDAE. *Usually rather thick-set species, occasionally flattened and leaf-like, rarely elongate and stick-like; tibiae with a small, triangular area delimited ventro-apically.* The leaf-like genus *Phyllium* and its allies occur in tropical Asia, New Guinea and Papua. Other genera include *Clonopsis, Bacillus, Timema* and numerous genera in the subfamily Pseudophasmatinae.

FAM. PHASMATIDAE. *Elongate, often extremely so, only rarely thick-set and never leaf-like; tibiae without a triangular apical area.* Includes *Carausius, Lonchodes, Megacrania, Clitumnus* and many other genera.

Literature on the Phasmida

BAUCHHENSS, E. (1971), *Carausius morosus* Br. Stabheuschrecke, *Grosses Zoologisches Praktikum*, **140**, 53 pp.

BEIER, M. (1957), Arthropoda. Insecta. Orthopteroidea: Ordnung Cheleutoptera Crampton 1915 (Phasmida Leach 1815), In: Bronn's *Klassen u. Ordnungen der Tiere*, **5**, Abt. iii (6, 2), 305–454.

—— (1968), Phasmida (Stab- oder Gespenstheuschrecken), In: *Handbuch der Zoologie*, (Helmcke, J.-G., Starck, D. and Wermuth, H. eds), **4** (2), Lief. 6: 56 pp.

BERGERARD, J. (1958), Étude de la parthénogénèse facultative de *Clitumnus extradentatus* Br., *Bull. biol. Fr. Belg.*, **92**, 87–182.

BOISSON, C. (1949), Recherches histologiques sur la complexe allatocardiaque de *Bacillus rossii* Fabr., *Bull. biol. Fr. Belg., Suppl.*, **34**, 1–92.

BORDAS, L. (1897), Considérations générales sur l'appareil digestif des Phasmidae, *Bull. Mus. nat. Hist. nat., Paris*, **2**, 378–379.

BRUNNER V. WATTENWYL, K. AND REDTENBACHER, J. (1906–08), *Die Insektenfamilie der Phasmiden*, Leipzig, 589 pp.

CAPPE DE BAILLON, P. AND VICHET, G. (1940), La parthénogénèse des espèces du genre *Leptynia* Pant., *Bull. biol. Fr. Belg.*, **74**, 43–57.

CHOPARD, L. (1938), *La Biologie des Orthoptères*, Lechevalier, Paris, 541 pp.

—— (1949), Ordre des Chéleutoptères, In: *Traité de Zoologie*, (Grassé, P. P.), **9**, 594–616.

DUPONT-RAABE, M. (1957), Les mécanismes de l'adaptation chromatiques chez les insectes, *Arch. Zool. exp. gén.*, **94**, 61–293.

EIDMANN, H. (1956), Über rhythmische Erscheinungen bei der Stabheuschrecke *Carausius morosus* Br., *Z. vergl. Physiol.*, **38**, 370–390.

FAVRELLE, M. (1938), Étude du *Phalces longiscaphus* (Orthopt. Phasmidae), *Ann. Soc. ent. Fr.*, **107**, 197–211.

GODDEN, D. H. (1974), The physiological mechanism of catalepsy in the stick insect *Carausius morosus* Br., *J. comp. Physiol.*, **90**, 251–274.

GRIMPE, G. (1921), Beiträge zur Biologie von *Phyllium bioculatum* G. R. Gray (Phasmidae), *Zool. Jb. (Syst.)*, **44**, 227–266.

GÜNTHER, K. (1928), Die Phasmoiden der Deutschen Kaiserin-Augusta-Fluss Expedition 1912/13. Ein Beitrag zur Kenntnis der Phasmoidenfauna Neuguineas, *Mitt. zool. Mus. Berlin*, **14**, 599–746.

—— (1933), Funktionell-anatomische Untersuchungen über die Bursa copulatrix, den Ovipositor und den männlichen Kopulationsapparat bei Phasmiden, *Jena. Z. Naturwiss.*, **68**, 403–462.

—— (1938), Beitrag zur Kenntnis der Fortpflanzungsbiologie der Stabheuschrecke *Orxines macklotti* de Haan, *Verh. 7. int. Kongr. Ent.*, 1156–1169.

—— (1953), Über die taxonomische Gliederung und die geographische Verbreitung der Insektenordnung der Phasmatodea, *Beitr. Ent.*, **3**, 541–563.

GUSTAFSON, J. F. (1966), Biological observations on *Timema californica* (Phasmoidea: Phasmidae), *Ann. ent. Soc. Am.*, **59**, 59–61.

HADLINGTON, P. AND HOSCHKE, F. (1959), Observations on the ecology of the Phasmatid *Ctenomorphodes tessulata* (Gray), *Proc. Linn. Soc. N.S.W.*, **84**, 46–159.

HANSTRÖM, B. (1940), Inkretorische Organe, Sinnesorgane und Nervensystem des Kopfes einiger niederer Insektenordnungen, *K. svenska VetenskAkad. Handl.*, **18** (8), 1–266.

HEBERDEY, R. F. (1931), Zur Entwicklungsgeschichte, vergleichenden Anatomie und Physiologie der weiblichen Geschlechtsausführwege der Insekten, *Z. Morph. Ökol. Tiere*, **22**, 416–586.

HENRY, L. M. (1937), Biological notes on *Timema californica* Scudder, *Pan Pacific Ent.*, **13**, 137–141.

HEYMONS, R. (1897), Ueber die Organisation und Entwicklung von *Bacillus Rossii* Fabr., *S.B. preuss. Akad. Wiss.*, **1897**, 363–373.

HUDSON, G. B. (1945), A study of the tentorium in some Orthopteroid Hexapoda, *J. ent. Soc. sthn. Africa*, **8**, 71–90.

HUGHES-SCHRADER, S. (1959), On the cytotaxonomy of Phasmids, *Chromosoma*, **10**, 268–277.

KEY, K. H. L. (1957), Kentromorphic phases in three species of Phasmatodea, *Aust. J. Zool.*, **5**, 247–284.

KRISTENSEN, N. P. (1975), The phylogeny of the hexapod 'orders'. A critical review of recent accounts, *Z. zool. Syst. Evolutionsforsch.*, **13**, 1–44.

LEFEVRE, W. P. (1939), A phasmid with spermatophore, *Proc. R. ent. Soc. Lond.*, A, **14**, 24.

LEUZINGER, H., WIESMANN, R. AND LEHMANN, F. E. (1926), *Zur Kenntnis der Anatomie und Entwicklungsgeschichte der Stabheuschrecke* Carausius morosus *Br.*, Fischer, Jena, 414 pp.

LING ROTH, H. (1917), Observations on the growth and habits of the stick insect, *Carausius morosus* Br., *Trans. ent. Soc. Lond.*, **1916**, 345–386.

LITTIG, K. S. (1942), External anatomy of the Florida walking stick *Anisomorpha buprestoides* Stoll, *Florida Ent.*, **25**, 33–41.

MAKI, T. (1935*a*), The abdominal structures of *Megacrania tsudai* Shiraki and their development in the late nymphal stage, *Trans. nat. Hist. Soc. Formosa*, **25**, 184–196.

—— (1935*b*), A study of the musculature of the Phasmid *Megacrania tsudai* Shiraki, *Mem. Fac. Sci. Agric. Taihoku imp. Univ.*, **12**, 181–279.

MARQUARDT, F. (1940), Beiträge zur Anatomie der Muskulatur und der peripheren Nerven von *Carausius* (*Dixippus*) *morosus* Br., *Zool. Jb.* (*Anat.*), **66**, 63–128.

MARSHALL, W. AND SEVERIN, H. (1906), Ueber die Anatomie der Gespenstheuschrecke *Diapheromera femorata* Say., *Arch. Biontol., Berlin*, **1**, 215–244.

MORENO, A. (1940), Glandulas odoriferas en *Paradoxomorpha, Notas Mus. La Plata*, **5** (*Zool.*), 319–323.

MOSCONA, A. (1950), Studies of the egg of *Bacillus libanicus* (Orthoptera, Phasmidae). Part 1. The egg envelopes, *Q. Jl microsc. Sci.*, **91**, 183–193.

MOTHES, G. (1960), Weitere Untersuchungen über den physiologischen Farbwechsel von *Carausius morosus* (Br.), *Zool. Jb.* (*Physiol.*), **69**, 133–162.

NUTTING, W. L. (1951), A comparative anatomical study of the heart and accessory structures of the orthopteroid insects, *J. Morph.*, **89**, 501–597.

PANTEL, J. (1915), Notes orthoptérologiques. VI. Le 'vomer sous-anal' n'est pas le 'titillateur'; études des segments abdominaux et principalement du segment terminal des mâles chez les Phasmides, *Ann. Soc. ent. Fr.*, **84**, 173–243.

PEHANI, H. (1925), Die Geschlechtszellen der Phasmiden, zugleich ein Beitrag zur Fortpflanzungsbiologie der Phasmiden, *Z. wiss. Zool.*, **125**, 167–238.

PFLUGFELDER, O. (1937), Bau, Entwicklung und Funktion der Corpora allata und cardiaca von *Dixippus morosus* Br., *Z. wiss. Zool.*, **149**, 477–512.

RAGGE, D. R. (1955), The wing-venation of the order Phasmida, *Trans. R. ent. Soc. Lond.*, **106**, 375–392.

SHAROV, A. G. (1968), Phylogeny of the Orthopteroidea, *Trud. paleont. Inst. Akad. Nauk. USSR*, **118**, 1–213.

SINETY, R. de (1901), Recherches sur la biologie et l'anatomie des Phasmes, *La Cellule*, **19**, 117–278.

SNODGRASS, R. E. (1937), The male genitalia of Orthopteroid insects, *Smithson. misc. Collns*, **96**, 107 pp.

STOCKARD, C. R. (1908), Habits, reactions and mating instincts of the 'walking stick', *Aplopus mayeri. Publ. Carnegie Inst.*, No. 103, **2**, 43–59.

STRINGER, I. A. N. (1970), The nymphal and imaginal stages of the bisexual stick insect *Clitarchus hookeri* (Phasmidae: Phasminae), *New Zealand Ent.*, **4**, 85–95.

TALBOT, G. (1920), A contribution to our knowledge of the life-history of the Stick Insect. *Carausius morosus* Br., *Trans. ent. Soc. Lond.*, **1920**, 285–304.

TEISSIER, G. (1956), Analyse factorielle de la variabilité de *Dixippus morosus* aux differents stades de son developpement, *Proc. 14th int. Congr. Zool. Kopenhagen*, 250–251.

UVAROV, B. P. (1950), A second New Zealand stick-insect (Phasmatodea) established in the British Isles, *Proc. R. ent. Soc. Lond.*, (B), **19**, 174–175.

VOY, A. (1954), Biologie et croissance chez le phasme femelle (*Clonopsis gallica*, Charp.), *Bull. biol.*, **88**, 101–129.

WEGNER, A. M. R. (1955), Biological notes on *Megacrania wegneri* Willemse und *M. alpheus* Westwood (Orthoptera, Phasmidae), *Treubia*, **23**, 47–52.

DERMAPTERA (EARWIGS)

Elongate insects with typical biting mouthparts; superlinguae distinct; ligula 2-lobed. Fore wings modified into very short leathery tegmina devoid of veins; hind wings semicircular, membranous, with the veins highly modified and disposed radially. Apterous forms common. Tarsi 3-segmented. Cerci unjointed and almost always modified into heavily sclerotized forceps; ovipositor reduced or absent. Metamorphosis slight.

The general appearance of these insects is well exemplified by the common European earwig, *Forficula auricularia* (Fig. 262), which also occurs in other parts of the Palaearctic region and has been introduced into N. and S. America, Australia, New Zealand and S. Africa. About 1200 species of Dermaptera are known and the majority show little marked variation in structure and habits. They are mostly nocturnal and many tropical species are attracted to light; during the day they hide away in the soil, under bark and stones, or among herbage. Except for *Labia minor* the European species rarely fly, even when (as in *Forficula*) they have well-developed wings. Two small sections of the order, represented by *Hemimerus* and *Arixenia*, have tended to adopt parasitic habits and differ somewhat from the majority. There are general accounts of the order by Beier (1958) and Günther and Herter (1974).

External Anatomy – There is a detailed comparative account by Giles (1963). The head (Henson, 1950; Strenger, 1950*a*) is more or less prognathous with a distinct gular sclerite and remnants of an ecdysial cleavage line, along part of which inflection has occurred to form a coronal and two post-frontal sulci. Epistomal, occipital and postoccipital sulci are present and the clypeus is divided into a sclerotized postclypeus and a membranous anteclypeus. There is a well-developed, imperforate tentorium (Hudson, 1947). Compound eyes are usually well-developed but ocelli are absent; the antennae have from 10 to 50 segments and bear chemoreceptor sensilla (Slifer, 1967). The mouthparts are of a generalized biting pattern with long, complete maxillae, each bearing a 5-segmented palp. The labium has been variously interpreted but the characteristic two-lobed ligula seems to consist entirely of the paraglossae. The hypopharynx is two-lobed with 3 paired suspensory sclerites (Moulins, 1969). Popham (1959) recognizes ten major types of head and mouthparts within the order, depending on the degree of prognathism, the kind of food eaten and the associated modifications of the mandibles and maxillae. The thorax (Henson, 1953) has a large pronotum, the meso- and metatergum are normally subdivided in winged species and the metathorax has a postnotum. The metapleura are almost horizontally arranged and the sterna are broad, flat plates with well-defined apophyseal pits. Tegmina and wings are absent in *Anisolabis*, the Brachylabini and in *Arixenia* and *Hemimerus*, while the wings vary greatly in development in other members of the order. The tegmina are short, truncated structures devoid of veins, and meet along the median line, thus resembling the elytra of the Staphylinidae. The large semicircular wings are almost entirely com-

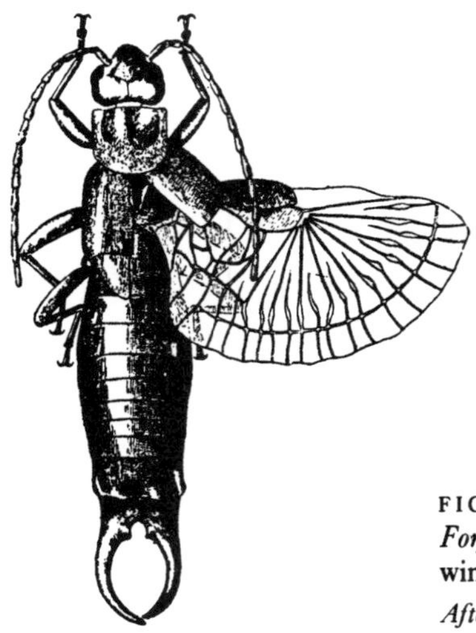

FIG. 262
Forficula auricularia, male with right wing extended
After Chopard, *Faune de France.*

posed of the greatly extended anal lobe: the pre-anal portion of the wing is sclerotized and contains two reduced longitudinal veins (R and Cu). The greater part of the wing is supported by a series of secondarily developed radially disposed branches. The wings are folded longitudinally in a fan-like manner, accompanied by two folds in a transverse direction and, in this way, they are stowed beneath the small tegmina (Verhoeff, 1917; Kleinow, 1966). The legs are subequal with three tarsal segments.

The *abdomen* is 11-segmented; the 1st tergum is fused with the metathorax and the 11th is represented by the epiproct and paraprocts. In the females of the Forficulina and Hemimeridae the 8th and 9th terga are greatly reduced and invisible without dissection. In the Arixeniidae, on the other hand, there is only very slight reduction of the 8th and 9th terga in the female. The 1st sternum is always wanting, while sterna 2 to 9 in the male and 2 to 7 in the female are clearly visible. The 9th sternum in the male largely overlies the 10th, the latter being represented in both sexes by a pair of plates at the base of the cerci. In the female the 7th sternum completely conceals the 8th and 9th (Fig. 263). The epiproct is divided transversely to form the 'opisthomeres' of taxonomists; in their most complete condition there are three such sclerites, the so-called pygidium, metapygidium and telson. The females of some primitive Forficulina (Pygidicranidae) possess a reduced ovipositor composed of a pair of valves developed from the 8th abdominal segment and another pair from the 9th but the more specialized families lack an ovipositor. The male genitalia (Burr, 1915–16; Popham, 1965) are of considerable taxonomic value: the Pygidicranidae, Carcinophoridae and Labiduridae have two penes but one of these is greatly reduced or absent in the other families. In all the Forficulina the cerci are modified into unjointed forceps (Lhoste, 1942; Strenger, 1950*b*). The latter

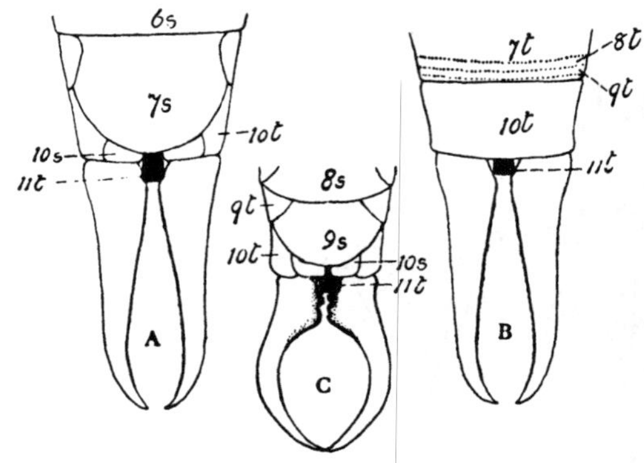

FIG. 263 *Forficula*, terminal abdominal segments
A, female (ventral). B, female (dorsal). C, male (ventral). *s*, sterna; *t*, terga.

present great diversity of form among different species, and are often variable within the limits of a single species as in *Forficula auricularia* (Bateson and Brindley, 1892). In the females of almost all earwigs they are shorter than in the males, being as a rule straight and unarmed. In *Hemimerus* the cerci are represented by hairy unjointed styliform appendages but those of *Arixenia* are bowed inwards in the male, and bear a closer resemblance to the forceps of true earwigs. The forceps are used in defence, to hold and catch prey, and play a part in courtship.

Internal Anatomy – The literature on internal anatomy is sparse and scattered but Lhoste (1957) includes much general and histological information. The *alimentary canal* (Fig. 264) is of a very uniform structure throughout the order. The oesophagus leads into the crop which is followed by a small globular gizzard. The mid gut has no enteric caeca and is slightly coiled posteriorly, but in *Arixenia* it forms nearly two complete coils. The peritrophic membrane is well-developed (Giles, 1965). The Malpighian tubules vary from eight or ten to about twenty, and are grouped in bundles. In *Forficula* there are four groups of five tubules (Henson, 1946). The hind gut presents a partial or, in *Arixenia*, a complete convolution, and there are six rectal papillae. The *nervous system* (Fig. 59) appears to be very constant and, in addition to the two cephalic centres, there are three thoracic and six abdominal ganglia. The *tracheal system* communicates with the exterior by ten pairs of spiracles. The *female reproductive organs* (Fig. 149) are divisible into two types. In *Forficula* there are three rows of numerous, very short, polytrophic ovarioles, distributed at regular intervals along the greater part of the length of each oviduct. In *Labidura riparia, Arixenia* and *Hemimerus* the ovarioles are much fewer, and are disposed in a single series. For *Anisolabis* see Giles (1961*b*) and Bonhag (1956). In *Labidura* there are five

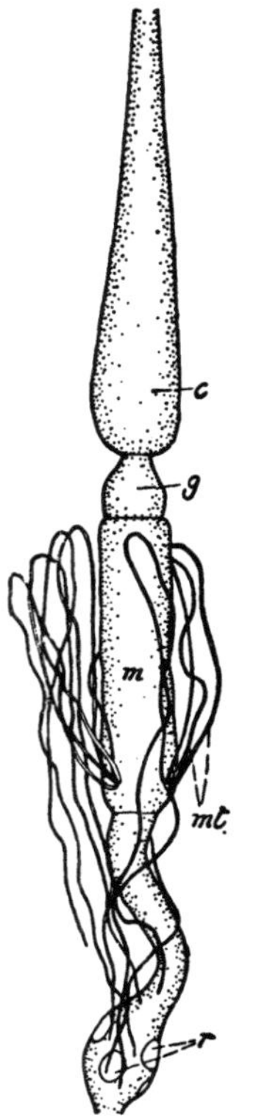

FIG. 264
Forficula, alimentary canal

c, crop; *g*, gizzard; *m*, mid gut; *mt*, Malpighian tubes; *r*, rectal papillae.

elongate ovarioles: in *Hemimerus* there are eight (Jordan) or 10–12 (Heymons), while in *Arixenia* there are fewer (Jordan). In the two last-mentioned genera the ovarioles are very short, each containing a single egg, and viviparous reproduction occurs. According to Heymons (1912) a maternal placenta is present in *Hemimerus* and envelops the embryo. At its anterior extremity the placenta forms a large cell-mass and, lying beneath it, is a foetal placenta which is developed as a proliferation of the amnion and serosa in that region. The whole placental organ thus formed is in direct connection with the body of the embryo by means of a diverticulum of the head-cavity, known as the cephalic vesicle. The embryos, about six at a time, are nourished *in situ* within their respective ovarioles, until they develop into young insects and are ready for birth. The *male reproductive organs* (Fig. 146) exhibit considerable differences among various genera. In *Forficula* and

Anisolabis (Giles, 1961*a*) the testes each consist of a pair of elongate closely apposed follicles; in *Hemimerus* the follicles are likewise paired, but are filiform and tightly coiled; in *Arixenia* the testes are compact and globular, each consisting of sixteen short follicles. The vasa deferentia are very slender, and in *Anisolabis* and *Hemimerus* they dilate posteriorly to form vesiculae seminales; the latter open, in *Hemimerus*, into a small vesicle which communicates with the base of the penis. Contrary to previous reports, *Hemimerus* has a single, median ejaculatory duct (Davies, R. G., 1966*b*). In the Forficulina, however, there is a secondary bifurcation to give two terminal ducts, each opening on a separate penis-lobe; one of these ducts then undergoes partial or almost complete degeneration in the more specialized (monandric) forms that have lost a penis-lobe.

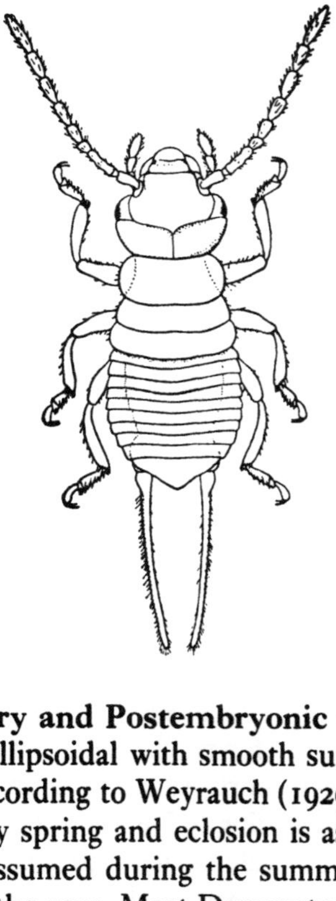

FIG. 265
Forficula, newly hatched nymph, × 15

Life-History and Postembryonic Growth – The eggs of Dermaptera are pale and ellipsoidal with smooth surfaces. In Europe *F. auricularia* lays 21–80 eggs according to Weyrauch (1929). The eggs have been found during winter or early spring and eclosion is assisted by an egg-burster. The adult condition is assumed during the summer, and there appears to be a single generation in the year. Most Dermaptera pass through four nymphal instars but some species of *Euborellia* and *Anisolabis* have five. The young nymphs resemble their parents in general form, except that they have fewer antennal segments and the forceps are simple and more or less styliform (Fig. 265). In *Diplatys* and *Karschiella* the forceps are preceded by segmented cerci in the nymphal stages. According to Green, *D. gerstaeckeri* nymphs 2·5 mm long bear 14-segmented cerci which are equal in length to the body. During subsequent instars the number of segments increases up to 45, and the cerci attain a length nearly double that of the body. In the preimaginal instar they become abruptly curtailed to a single segment, within which the future forceps can be made out. In almost all other earwigs the forceps are not preceded by segmented cerci, and no trace of segmentation in these organs has been detected in the few embryos that have been examined. Several species of Forficulina, including *Cheluridella*, *Anechura* and *Forficula*, display simple maternal care; the eggs are laid in the soil in a group and the female rests over and around them and the resulting young stages, tending them until they can look after themselves. The majority of Dermaptera seem to be omnivorous though some, like *Forficula*, prefer plant material, while others such as the Labiduridae and Pygidicranidae favour small arthropods and other animal food. *Chelisoches morio* has a somewhat specialized diet of aphids, leaf-hoppers and leaf-mining Hispid larvae.

Affinities and Phylogeny – A few Jurassic Dermaptera are known, of which the somewhat problematic *Protodiplatys fortis* is especially interesting since it has indications of tegminal venation, five tarsal segments and multiarticulate cerci; it is perhaps annectant between the Dermaptera and the Permian Protelytroptera. Giles (1963) found that the Recent Dermaptera share far more features with the Grylloblattodea than with any other Orthopteroid order, but the resemblance may be due to the common possession of certain primitive characters and does not necessarily provide conclusive evidence of a close phylogenetic relationship. Also debatable is the position of the Hemimerina and Arixenina. Popham (1961, 1962) has argued, largely from their feeding habits, that *Hemimerus* represents a distinct order of insects (as some older taxonomists also held) but that *Arixenia* is a Labioid earwig. Giles (1963, 1974) has brought forward other evidence that *Hemimerus* and *Arixenia* are both Dermaptera but that they are not closely related to the Forficulina or to each other. The cytological evidence of White (1971, 1972) confirms this view, which is followed in the classification below.

Classification – The Dermaptera have been monographed by Burr (1911), there is a catalogue by Sakai (1970–73), and regional works by Burr (1910), Brindle (1966), Blatchley (1920), Chopard (1951), Harz (1957) and others. For the 6 British species see Lucas (1920) and Hincks (1956). Zacher (1911), Burr (1911–16), Hincks (1955–59) and others have contributed to the development of the higher classification. Popham (1961, 1965) has recently proposed substantial changes, based mainly on his study of the male external genitalia. His views are discussed by Günther and Herter (1974), whose modified version of the classification is summarized below. It recognizes the traditional three suborders, with the Forficulina divided into three superfamilies and six families.

Suborder 1. FORFICULINA

Eyes well developed; cerci forcipate; wings present or absent; 1st tarsal segment longer than 2nd, 3rd not longer than the other two together; body without strong bristles; free-living, usually oviparous.

Superfamily **Pygidicranoidea**

Usually larger species, pubescent or tomentose; usually winged; opisthomeres comprising metapygidium, pygidium and telson; forceps of male often asymmetrical; aedeagus usually with two penis lobes flexed forwards.

FAM. PYGIDICRANIDAE. A relatively primitive tropical family (Hincks, 1955–59). Typical genera are *Echinosoma, Cranopygia, Diplatys* and *Karschiella*; the last two, with their allies, have been placed in separate families of their own by Popham (1965).

Superfamily **Labioidea**

Body not pubescent; forceps of male usually symmetrical; penis lobes variable, single or with only one directed forward.

FAM. CARCINOPHORIDAE. *Usually apterous or without hind wings; aedeagus with paired penis lobes, one directed forward the other backward; Euborellia* and *Anisolabis* are large, cosmopolitan genera. The Australian *Titanolabis colossea* is the largest Dermapteran (55 mm).

FAM. LABIIDAE. *Small species, usually winged; aedeagus with a single median penis lobe.* Predominantly an Old World group. *Spongovostox, Chaetospania* and *Labia* are large genera; *Labia minor* is cosmopolitan.

Superfamily **Forficuloidea**

2nd tarsal segment bilobed or extending more or less beneath 3rd segment; aedeagus with long, simple, distally rounded parameres; penis lobes paired or simple.

FAM. LABIDURIDAE. *Medium-sized species; usually winged; males usually with one penis lobe directed forward and one backward. Labidura riparia* is cosmopolitan; *Apachyus.*

FAM. CHELISOCHIDAE. *Single, median penis lobe, parameres acuminate; 2nd tarsal segment extending somewhat under 3rd. Chelisoches, Proreus.*

FAM. FORFICULIDAE. *Single median penis lobe; parameres rounded; 2nd tarsal segment bilobed. Forficula, Anechura.*

Suborder 2. ARIXENIINA

Ectoparasitic; eyes vestigial; apterous; cerci not horny but arched and hairy.

FAM. ARIXENIIDAE. This small group contains only *Arixenia* with two species, *A. esau* from Sarawak and *A. jacobsoni* from Java (Fig. 266). The first species was found in the breast-pouch of the bat *Cheiromeles torquatus*; the second occurs in large numbers on guano in a cave much resorted to by bats (Medway, 1958). *Arixenia* is apterous and viviparous, the eyes are greatly reduced, and the mandibles are strongly flattened with their inner edges rounded and clothed with bristles. The cerci are feebly sclerotized and hairy; they are unjointed and somewhat bowed, which gives them the appearance of incipient forceps. For further information on the genus see Jordan (1909*a*), Burr and Jordan (1913), Cloudsley-Thompson (1957) and Giles (1961*c*).

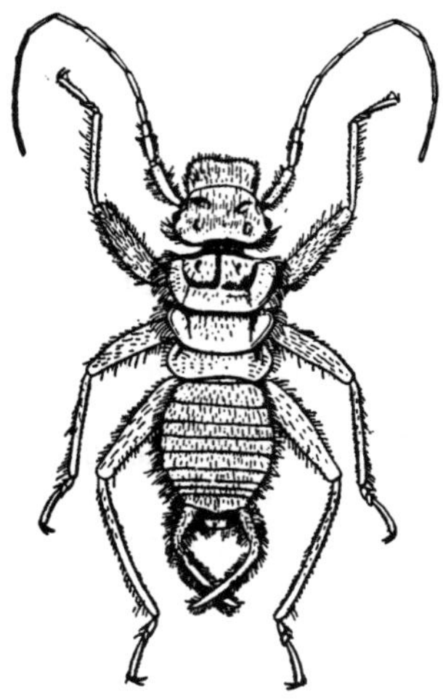

FIG. 266
Arixenia jacobsoni, male
After Burr, *Ent. Month. Mag.*, 1912.

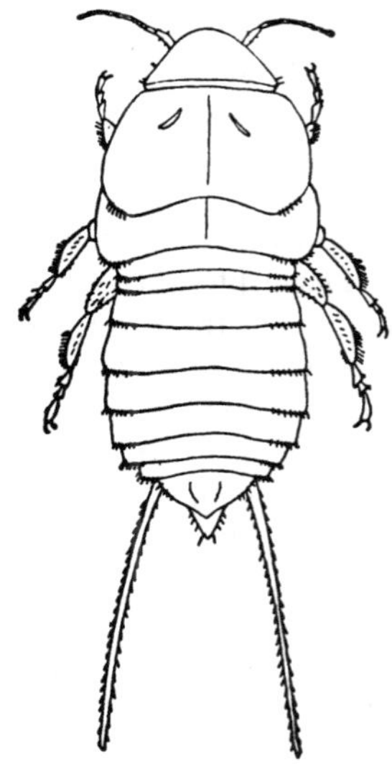

FIG. 267
Hemimerus talpoides
Adapted from Hansen.

Suborder 3. HEMIMERINA

Ectoparasitic; eyes absent; apterous; cerci long, straight and unsegmented.

FAM. HEMIMERIDAE. This family is composed of the 10 or so species of *Hemimerus* (Rehn and Rehn, 1936). They are ectoparasitic on different species or subspecies of rats of the genus *Cricetomys* from Africa between about 10° N. and 20° S. and feed upon the epidermal products of their hosts (Ashford, 1970; Popham, 1962); *H. morrisi* from Malawi lives in the food-stores of the pouched rat *Beamys major* (Hanney, 1963). *Hemimerus* is apterous and viviparous, devoid of eyes, and the cerci are feebly sclerotized, unjointed, hairy appendages which resemble those of the Gryllidae. The structure of these remarkable insects has been investigated by Hansen (1894), Jordan (1909*b*), Heymons (1912) and Deoras (1941*a*; 1941*b*); there are four nymphal instars (Davies, 1966*a*)

Literature on the Dermaptera

ASHFORD, R. W. (1970), Observations on the biology of *Hemimerus talpoides* (Insecta: Dermaptera), *J. Zool.*, **162**, 413–418.

BATESON, W. AND BRINDLEY, H. H. (1892), On some causes of variation in secondary sexual characters statistically examined, *Proc. zool. Soc. Lond.*, **1892**, 585–594.

BEĬ-BIENKO, G. J. (1936), Insectes Dermaptères, *Faune U.R.S.S.* (N.S.), (5), **10**, 1–240.

BEIER, M. (1958), Dermaptera, In: Bronn's *Klassen u. Ordnungen des Tierreichs*, 5 (Buch 6, Lief. 3), 455–585.

BLATCHLEY, W. S. (1920), *The Orthoptera of North Eastern America*, Nature Publ. Co., Indianapolis, 784 pp.

BONHAG, P. F. (1956), The origin and distribution of periodic acid-Schiff-positive substances in the oocyte of the earwig, *Anisolabis maritima* (Gené), *J. Morph.*, **99**, 433–464.

BRINDLE, A. (1966), The Dermaptera of Madagascar, *Trans. R. ent. Soc. Lond.*, **118**, 221–259.

BURR, M. (1910), Dermaptera, *Fauna Brit, India,* 217 pp.

—— (1911), Dermaptera, *Genera Insect.*, Fasc. **122**, 1–112.

—— (1915–16), On the male genital armature of the Dermaptera I–III, *J. roy. microsc. Soc.*, **1915**, 413–447; 521–546; **1916**, 1–18.

BURR, M. AND JORDAN, K. (1913), On Arixenima Burr, a sub-order of Dermaptera, *Trans. 2nd int. Congr. Ent.*, 398–421.

CAUSSANEL, C. (1966), Étude du développement larvaire de *Labidura riparia* (Derm., Labiduridae), *Ann. Soc. ent. Fr.*, (N.S.), **2**, 469–498.

CHOPARD, L. (1951), Orthoptéroïdes, *Faune de France*, **56**, 359 pp.

CLOUDSLEY-THOMPSON, J. L. (1957), On the habits and growth stages of *Arixenia esau* Jordan and *A. jacobsoni* Burr (Dermaptera, Arixenioidea), with descriptions of the hitherto unknown adults of the former, *Proc. R. ent. Soc. Lond.*, (A), **32**, 1–12.

CRUMB, E. S., EIDE, P. M. AND BONN, A. E. (1941), The European Earwig, *Tech. Bull. U.S. Dept. Agric.*, **766**, 1–76.

DAVIES, R. G. (1966*a*), The postembryonic development of *Hemimerus vicinus* Rehn & Rehn (Dermaptera: Hemimeridae), *Proc. R. ent. Soc. Lond.*, (A), **41**, 67–77.

—— (1966*b*), The male of *Hemimerus vicinus* Rehn & Rehn (Dermaptera: Hemimeridae), with notes on the structure of the aedeagus, *Proc. R. ent. Soc. Lond.*, (B), **35**, 61–64.

DEORAS, P. J. (1941*a*), Structure of *Hemimerus deceptus* Rehn var. *ovatus* an external parasite of *Cricetomys gambiense, Parasitology*, **33**, 172–185.

—— (1941*b*), The internal anatomy and description of *Hemimerus deceptus* var. *ovatus* Deoras (Dermaptera) with remarks on the systematic position of the Hemimeridae, *Indian J. Ent.*, **3**, 321–333.

GILES, E. T. (1953), The biology of *Anisolabis littorea* (White) (Dermaptera: Labiduridae), *Trans. roy. Soc. N.Z.*, **80**, 383–398.

—— (1961*a*), The male reproductive organs and genitalia of *Anisolabis littorea* (White) (Dermaptera: Labiduridae), *Trans. roy. Soc. N.Z.*, (*Zool.*), **1**, 203–213.

—— (1961*b*), The female reproductive system and genital segments of *Anisolabis littorea* (White) (Dermaptera: Labiduridae), *Trans. roy. Soc. N.Z.*, (*Zool.*), **1**, 293–302.

—— (1961*c*), Further studies on the growth stages of *Arixenia esau* Jordan and *Arixenia jacobsoni* Burr (Dermaptera, Arixeniidae), with a note on the first instar antennae of *Hemimerus talpoides* Walker (Dermaptera, Hemimeridae), *Proc. R. ent. Soc. Lond.*, (A), **36**, 21–26.

—— (1963), The comparative external morphology and affinities of the Dermaptera, *Trans. R. ent. Soc. Lond.*, **115**, 95–164.

—— (1965), The alimentary canal of *Anisolabis littorea* (White) (Dermaptera: Labiduridae), with special reference to the peritrophic membrane, *Trans. roy. Soc. N.Z., (Zool.)*, 6, 87–101.
—— (1974), The relationship between the Hemimerina and the other Dermaptera: a case for reinstating the Hemimerina within the Dermaptera, based upon a numerical procedure, *Trans. R. ent. Soc. Lond.*, **126**, 129–206.
GÜNTHER, K. AND HERTER, K. (1974), 11. Ordnung Dermaptera (Ohrwürmer), In: Helmcke, J.-G., Starck, D. and Wermuth, H. (eds), *Handbuch der Zoologie*, 4 (2), Lfg. 23, 1–158.
HANNEY, P. W. (1963), A new species of *Hemimerus* (Dermaptera) from Nyassaland and its association with the genus *Beamys* Thomas, *Proc. R. ent. Soc. Lond.*, (B), **32**, 38–40.
HARZ, K. (1957), *Die Geradflügler Mitteleuropas*, Fischer, Jena, 494 pp.
HANSEN, H. J. (1894), Beiträge zur Kenntnis der Insektenfauna von Kamerun. 3. On the structure and habits of *Hemimerus talpoides* Walker, *Ent. Tidskr.*, **15**, 63–93.
HENSON, H. E. (1946), On the Malpighian tubules of *Forficula auricularia* (Dermaptera), *Proc. R. ent. Soc. Lond.*, (A), **21**, 29–39.
—— (1950), On the head capsule and mouthparts of *Forficula auricularia* Linn. (Dermaptera), *Proc. R. ent. Soc. Lond.*, (A), **25**, 10–18.
—— (1951), The wings of *Forficula auricularia* L. (Dermaptera), *Proc. R. ent. Soc. Lond.*, (A), **26**, 135–142.
—— (1953), On the external morphology of the neck and thorax of *Forficula auricularia* L., *Trans. R. ent. Soc. Lond.*, **104**, 25–37.
HERTER, K. (1943), Zur Fortpflanzungsbiologie eines lebendgebärenden Ohrwurmes, *Z. Morph. Ökol. Tiere*, **40**, 158–180.
—— (1960), Zur Fortpflanzungsbiologie des Meerstrand-Ohrwurmes *Anisolabis maritima* (Géné), *Zool. Beitr.*, (N.F.), **5**, 199–239.
—— (1963), Zur Fortpflanzungsbiologie des Sand- oder Ufer-Ohrwurmes *Labidura riparia* Pall., *Zool. Beitr.* (N.F.), **8**, 297–329.
—— (1964), Zur Fortpflanzungsbiologie des Ohrwurmes *Forficula pubescens* (Géné), *Zool. Beitr.*, (N.F.), **10**, 1–28.
—— (1965), Die Fortpflanzungsbiologie des Ohrwurmes *Forficula auricularia* L., *Zool. Jb. (Syst.)*, **92**, 405–466.
—— (1967), Weiteres zur Fortpflanzungsbiologie des Ohrwurmes *Forficula auricularia* L., *Zool. Beitr.* (N.F.), **13**, 213–244.
HEYMONS, R. (1912), Über den Genitalapparat und die Entwicklung von *Hemimerus talpoides* Walk., *Zool. Jb. Suppl.*, **15** (2), 141–184.
HINCKS, W. D. (1955–59), *A Systematic Monograph of the Dermaptera of the World.* I, II, Brit. Mus. (Nat. Hist.), London, 132 pp.; 218 pp. [Pygidicranidae only.]
—— (1956), Dermaptera and Orthoptera, *R. ent. Soc. Handb. Ident. Brit. Ins.*, **1** (5), 24 pp.
HUDSON, G. B. (1947), Studies in the comparative anatomy and systematic importance of the hexapod tentorium. 2. Dermaptera, Embioptera and Isoptera, *J. ent. Soc. sthn. Afr.*, **9**, 99–110.
JORDAN, K. (1909*a*), Description of a new kind of apterous earwig, apparently parasitic on a bat, *Novit. zool.*, **16**, 313–326.
—— (1909*b*), Notes on the anatomy of *Hemimerus talpoides*, *Novit. zool.*, **16**, 327–330.

KLEINOW, W. (1966), Untersuchungen zum Flügelmechanismus der Dermapteren, *Z. Morph. Ökol. Tiere*, **56**, 363–416.
LHOSTE, J. (1941), Aperçu anatomique et histologique du tube digestif de *Forficula auricularia* L., *Bull. Soc. ent. Fr.*, **46**, 43–46.
—— (1942), Les cerques des Dermaptères, *Bull. biol. Fr. Belg.*, **76**, 192–201.
—— (1957), Données anatomiques et histophysiologiques sur *Forficula auricularia* L. (Dermaptera), *Arch. Zool. exp. gén.*, **95**, 75–252.
LUCAS, W. J. (1920), *A Monograph of the British Orthoptera*, London (Ray Society), 264 pp.
MEDWAY, Lord (1958), On the habits of *Arixenia esau* Jordan (Dermaptera), *Proc. R. ent. Soc. Lond.*, (A), **33**, 191–195.
MOULINS, M. (1969), Étude anatomique de l'hypopharynx de *Forficula auricularia* L. (Insecte, Dermaptère): Teguments, musculature, organes sensoriels et innervations. Interprétation morphologique, *Zool. Jb.* (*Anat.*), **86**, 1–27.
POPHAM, E. J. (1959), The anatomy in relation to feeding habits of *Forficula auricularia* and other Dermaptera, *Proc. zool. Soc. Lond.*, **133**, 251–300.
—— (1961), On the systematic position of *Hemimerus* Walker – a case for ordinal status, *Proc. R. ent. Soc. Lond.*, (B), **30**, 19–25.
—— (1962), The anatomy related to the feeding habits of *Arixenia* and *Hemimerus* (Dermaptera), *Proc. zool. Soc. Lond.*, **139**, 429–450.
—— (1965), The functional morphology of the reproductive organs of the common earwig (*Forficula auricularia*) and other Dermaptera with reference to the natural classification of the order, *J. Zool.*, **146**, 1–43.
REHN, J. A. G. AND REHN, J. W. H. (1936), A study of the genus *Hemimerus* (Dermaptera, Hemimerina, Hemimeridae), *Proc. Acad. nat. Sci. Philadelphia*, **87**, 457–508.
SAKAI, S. (1970–73), *Dermapterorum Catalogus Preliminaris*. Parts I–VII, Daito Bunka Univ., Tokyo, 1–91; 1–177; 1–68; 1–14; 1–162; 1–265; 1–357.
SLIFER, E. H. (1957), Sense organs on the antennal flagella of earwigs (Dermaptera) with special reference to those of *Forficula auricularia*, *J. Morph.*, **122**, 63–80.
STRENGER, A. (1950*a*), Funktionstudie des Kopfes von *Forficula auricularia*, *Zool. Jb.* (*Anat.*), **70**, 557–575.
—— (1950*b*), Eine funktionsanatomische Untersuchung einiger Dermapterencerci, *Zool. Jb.* (*Anat.*), **70**, 576–600.
VERHOEFF, K. W. (1917), Über Bau und Faltung der Flügel von *Forficula auricularia*, *Arch. Naturg.*, (A), **83**, 1–23.
WEYRAUCH, W. K. (1929), Experimentelle Analyse der Brutpflege des Ohrwurmes *Forficula auricularia* L., *Biol. Zbl.*, **49**, 543–558.
WHITE, M. J. D. (1971), The chromosomes of *Hemimerus bouvieri* Chopard (Dermaptera), *Chromosoma*, **34**, 183–189.
—— (1972), The chromosomes of *Arixenia esau* Jordan (Dermaptera), *Chromosoma*, **36**, 338–342.
ZACHER, F. (1911), Studien über das System der Protodermapteren, *Zool. Jber.*, **30**, 303–400.

Order 12

EMBIOPTERA

Gregarious insects living in silken tunnels. Mouthparts mandibulate, ligula 4-lobed. Tarsi 3-segmented; 1st segment of anterior pair greatly inflated. Females apterous, males usually with 2 pairs of similar wings; radius greatly thickened, remaining veins often reduced or vestigial. Cerci 2-segmented, generally asymmetrical in the male. Metamorphosis gradual in the male, less pronounced in the female.

The Embioptera are a small group of fragile insects with a soft thin cuticle and weak powers of flight. All are sombre coloured, being either brown or yellowish brown, with smoky wings. In their habits these insects generally prefer warm, damp conditions and avoid daylight, living beneath stones, or under bark, etc. The females are much more rarely met with than males, the latter not infrequently being attracted to a light. Sexual dimorphism is a marked characteristic of the order, the males usually being winged and the females apterous (Figs. 268, 269). In some species, however, notably *Anisembia texana* and *Embia tyrrhenica*, the males may be winged or apterous while several groups have independently evolved completely apterous males (e.g. *Metoligotoma* and all species of Australembiidae). For general accounts of the order see Beier (1955), Kaltenbach (1968) and Ross (1970).

The most striking feature in the biology of the Embioptera is their habit of constructing silken tunnels, a life to which they show many structural adaptations (Ross, 1970). The tunnels are anchored by silken threads to the substrate and their walls may contain fragments of plant material, frass and exuviae. Their main function is probably to protect the inhabitants against the attacks of Arthropod predators and they communicate with special chambers for mating, moulting and overwintering. The insects can move backwards or forwards in the tunnels with equal facility and spend much of the day within them, emerging at night to forage for food. In *Embia ramburi* and *Oligotoma nigra* each tunnel contains only one individual or a female and her young (Friedrichs, 1934). Mature males live only a few weeks, entering a female's tunnel to mate and then leaving (Ananthasubramanian, 1957). Many contiguous tunnels form a 'colony' (sometimes composed of hundreds of individuals) but the Embioptera are not social or even subsocial insects. During the construction of the tunnels the fore legs are in constant activity, crossing and recrossing one another repeatedly. The faculty of weaving the

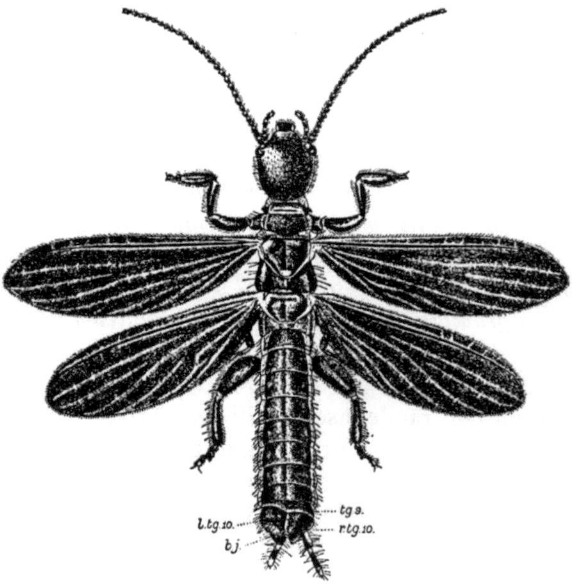

FIG. 268 *Embia major*, male. India
Tg9, 9th tergum; *tg10*, right and left plates of 10th tergum; *bj*, basal segment of left cercus. From Imms, 1913.

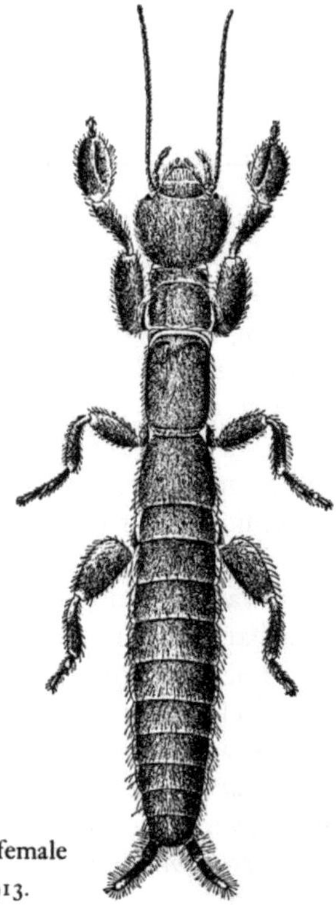

FIG. 269
Embia major, female
From Imms, 1913.

tunnels is possessed equally by both sexes and also by the nymphs. Newly hatched nymphs, removed from the proximity of the mother, weave fine tunnels on their own. The silk glands are situated in the fore tarsi (Barth, 1954). On the plantar surface of the 1st and 2nd tarsal segments of the fore legs are a number of hollow cuticular processes (silk ejectors) which are probably not modified macrotrichia and which communicate, each by a fine duct, with a small glandular chamber. The chambers are situated on the lower area of the enlarged 1st tarsal segment; each is bounded by a single layer of epithelium which encloses a central space filled with a viscid secretion (Fig. 270). In *Embolyntha batesi* over 200 chambers are present in the whole segment. Since a fine thread is emitted from each ejector a number are available simultaneously, which accounts for the rapidity with which these insects weave their tunnels.

Female and nymphal Embioptera feed almost exclusively on fresh or dry plant material: flowers, leaves, roots and even woody structures. Adult males probably do not feed and in some species their mouthparts are used mainly to hold the female by the head during copulation (Stefani, 1953; Lacombe, 1958).

The eggs are urn-shaped or ovoidal with a conspicuous operculum towards one end. They are laid in small groups along the silken tunnels and the females exhibit simple parental care in some species: the eggs are licked and may be moved around in the nest and a few forms provide their newly-hatched young with finely chewed plant material. Reproduction is normally bisexual but a few species show occasional parthenogenesis and others have parthenogenetic 'races' or exhibit obligatory parthenogenesis (Ross, 1960; Stefani, 1956, 1961).

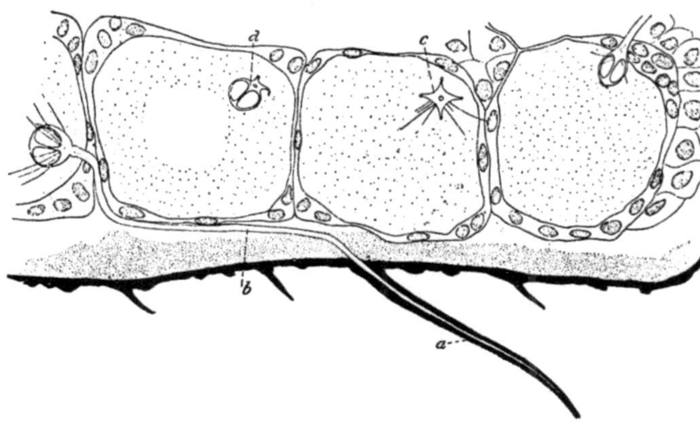

FIG. 270 *Anisembia texana*, section of portion of 1st tarsal segment, showing silk glands *a*, silk ejector; *b*, duct of silk gland; *c*, *d*, ampullae at bases of ducts. *After* Melander, *Biol. Bull.*, 1902.

Embioptera are tropicopolitan but extend their range into the warmer temperate zones. They are best represented in South America but occur in all zoogeographical regions including Australia; species are also found in Madagascar, New Zealand, Ceylon and various smaller islands. The Mediterranean region has 23 species from the genera *Embia, Cleomia, Oligotoma* and *Haploembia* (Ross, 1966). *Oligotoma saundersi* and *O. humbertiana* have been distributed through the trade in tropical plants and survive in greenhouses outside their original areas of distribution. *O. saundersi, O. michaeli, Clothoda urichi* and others injure cultivated orchids by feeding on their roots.

External Anatomy – The head is prognathous with a ventral gula and with epistomal, trans-clypeal, clypeolabral and postoccipital sulci; a normal tentorium is present but no ocelli (Lacombe, 1958; Rähle, 1970). The compound eyes are elliptical or reniform and smaller in the female. The antennae are shorter than the head, filiform, and made up of 15–32 segments with tactile hairs and chemoreceptor sensilla (Slifer and Sekhon, 1972). The mouthparts (Fig. 271) are Orthopteroid and show sexual dimorphism, mainly in the mandibles of the male being more slender with fewer teeth. The maxillary palps are 5-segmented, the galea is membranous and the lacinia sclerotized and provided with a pair of apical teeth. Both cardo and stipes are well developed. In the labium the ligula consists of a pair of rather fleshy paraglossae and between them the very small pointed glossae; the labial palps are 3-segmented. The hypopharynx is large and its dorsal surface is covered with minute pectinate scales.

The prothorax (Rähle, 1970; Bitsch and Ramond, 1970) is narrower than the head, and a deep transverse sulcus cuts off the anterior portion of the pronotum from the remainder. The meso- and metathorax are subequal in size and broader than long in the male, but elongate and narrower in the female. The fore legs are stout, the middle pair is reduced in size, and the hind pair has swollen femora accommodating the large tibial depressor muscles; these enable the insects to move backwards quickly in their tunnels.

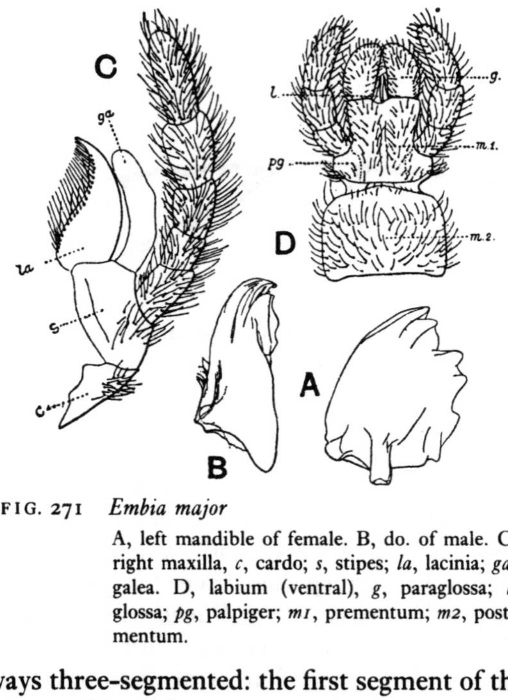

FIG. 271 *Embia major*

A, left mandible of female. B, do. of male. C, right maxilla, *c*, cardo; *s*, stipes; *la*, lacinia; *ga*, galea. D, labium (ventral), *g*, paraglossa; *l*, glossa; *pg*, palpiger; *m1*, prementum; *m2*, postmentum.

The tarsi are always three-segmented: the first segment of the anterior pair is inflated in all stages and both sexes.

The two pairs of wings (Fig. 272) are almost identical in size and shape and only differ in unimportant details of venation. The wing-membrane is smoky in colour, with narrow hyaline areas running in a longitudinal manner between the principal veins, giving the wings a very characteristic appearance. The surface of the wings is clothed with microtrichia, together with macrotrichia distributed along and between the vins. The radial vein is a broad tube containing a well-developed blood sinus which enables the wing to be stiffened by blood pressure for effective flight. When the wings are folded the sinuses collapse and, because the other veins are greatly reduced, the whole wing becomes flexible and so less liable to damage when the insect moves backwards in its tunnel. The venation is seen in a generalized condition in *Donaconethis*, but even in this genus reduction is evident, as R_s is only 3-branched and M is represented by a single fork. In the Oligotomidae and some other families the venation is greatly reduced and markedly degenerate; R_{4+5} is represented by a mere spur, M has practically disappeared and Cu is unbranched. Traces of a former more complete venation are evident as slight thickenings of the wing-membrane.

The abdomen is composed of 10 evident terga; in the females and immature forms of both sexes the 10th tergum is entire but in the adult males it is divided into a pair of asymmetrical plates (hemitergites). One or both of these plates is drawn out into a sclerotized process of variable form. In

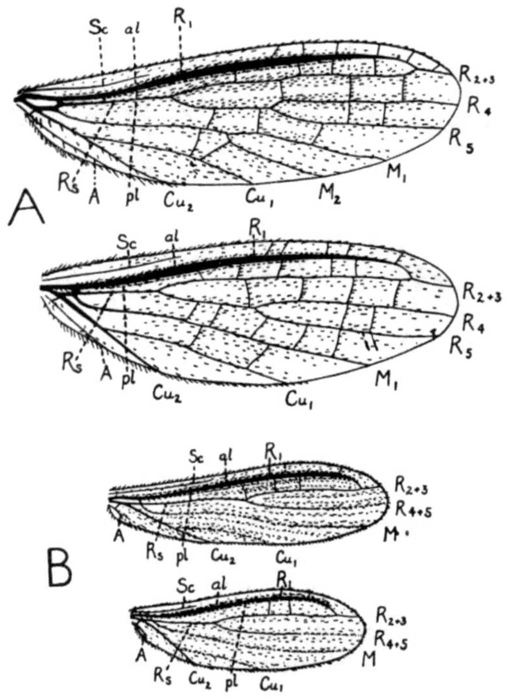

FIG. 272 Right wings of A, *Embia major*; B, *Oligotoma saundersi*

al, *pl*, anterior and posterior radial lines.

Clothoda the 10th tergum of the male is entire and symmetrical as in the female (Fig. 273). A pair of 2-segmented cerci is present at the apex of the abdomen; as a rule the left cercus in the male is modified basally and the pair is asymmetrical in consequence (Fig. 273). In *Clothoda*, however, the cerci are unmodified and they show only slight asymmetry in the Oligotomidae. Each cercus is borne on a basal plate which may represent vestiges of an 11th segment. Ten sterna are usually present though in some the 1st sternum of the female is largely aborted. In the immature forms of both sexes, and the females, the 10th sternum is divided into two symmetrical plates. In the males the 9th sternum is usually asymmetrical and the 10th sternum is

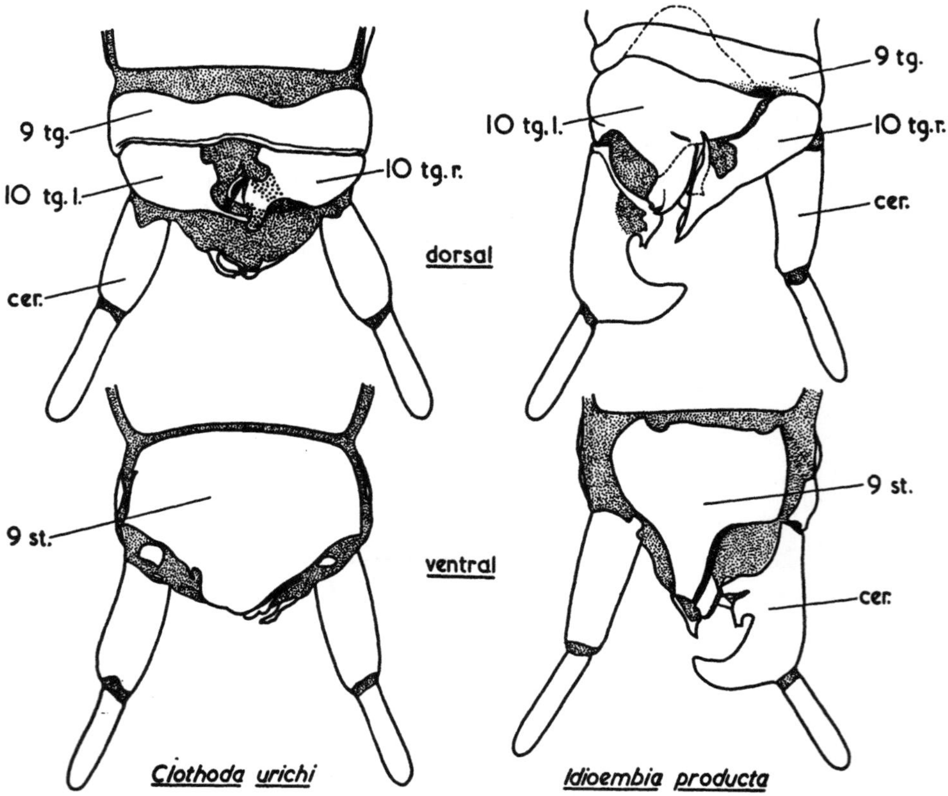

FIG. 273 Abdominal terminalia of male Embioptera (*after* Ross, 1944)
cer, cercus; 9 *st*, 9th abdominal sternite; 9 *tg*, 9th abdominal tergite; 10 *tg.l*, 10 *tg.r*, left and right hemitergites of 10th abdominal segment.

possibly represented by a pair of minute sclerites sometimes fused with the 9th sternum. The female genital aperture lies near the hind border of the 8th sternum, the latter functioning as the subgenital plate. In the male the 9th sternum is the subgenital plate and an aedeagus is absent or little developed.

Internal Anatomy – The internal organization of the Embioptera needs fuller investigation as its general features are mainly known from incomplete accounts of a few species by Grassi and Sandias (1896–97), Melander (1903), Mukerji (1927), Barth and Lacombe (1955) and Lacombe (1958–65). The *alimentary canal* is an almost straight tube from the mouth to the anus. The mouth leads into a small buccal cavity which is lined with backwardly directed denticles. This is succeeded by a narrow pharynx, and the remainder of the fore gut consists of a large dilated oesophagus and crop. The mid gut is a long tubular chamber which narrows somewhat posteriorly. The hind gut consists of a slightly coiled ileum, a very short colon and a dilated rectum, the latter provided with six cushion-like rectal papillae. In males some parts of the gut are histologically degenerate. *Malpighian tubules* are variable in number: in adults there are about 14–30 of these organs. A pair of small *salivary glands* and reservoirs are present in the thorax, and their ducts unite anteriorly to form a common canal which opens beneath the hypopharynx. The *nervous system* consists of a rather small brain (Rähle, 1970), a suboesophageal ganglion, and a chain of three thoracic and seven abdominal ganglia, which are united throughout by double connectives; the stomatogastric system is also well developed and there is a single bilobed corpus allatum. The *tracheal system* opens by ten pairs of spiracles which belong to the meso- and metathorax, and first eight abdominal segments respectively. The tracheae anastomose by longitudinal and transverse branches. In females each ovary consists of five panoistic ovarioles which open at intervals along the lateral oviduct; there is a short vagina which receives the aperture of a large spermatheca. In males the numerous follicles of each testis are disposed in five lobes, arranged successively along the vas deferens. The latter dilates posteriorly to form a seminal vesicle and ultimately unites with its fellow to form a common ejaculatory duct; two pairs of tubular accessory glands are also present.

Postembryonic Development – There are four nymphal instars, of which the last two bear wing pads in species with alate males. The characteristic asymmetry of the terminal abdominal segments appears only with the last moult and otherwise there are few metamorphic changes beyond alterations in the proportions of the body, an increase in the number of antennal segments and the development of the reproductive system (Imms, 1913; Mills, 1932).

Affinities and Phylogeny – The first undoubted Embiopteran, *Electroembia antiqua*, occurs in Oligocene amber (Ross, 1956). Like other Tertiary Embioptera it belongs to an extant family, in this case the Embiidae, and provides no clue to the origin of the order. The Embioptera are certainly Orthopteroids and have often been linked with the Plecoptera (e.g. Ross, 1970). A morphological study by Rähle (1970) gives little support to this relationship and suggests that they may be more like the Phasmida, though neither hypothesis is yet substantiated.

Classification – Our knowledge of the Embioptera has been greatly extended in recent years through the work of Davis (1939–40*b*) and, more particularly, of Ross (1944, 1963, 1966, 1970), whose publications replace the older taxonomic monographs (Krauss, 1911; Enderlein, 1912). Some 300 species have been described and many more await description. The main taxonomic characters are provided by the wing venation and male abdominal terminalia, so that unassociated females may be difficult to place. Ross (1970) has indicated a classification into three suborders and 13 families; some of these have been redefined and the following may be mentioned here. The **CLOTHODIDAE** are the most generalized Embioptera with almost symmetrical male terminalia and well-developed venation; there are three species of *Clothoda* from the tropical forests of northern South America. The **EMBIIDAE** is the largest family, with many genera and species, mainly from savanna and woodland habitats in Africa, India and the Mediterranean area as well as some more generalized species from South America. The

NOTOLIGOTOMIDAE include a few Asian and Australian species and the **EMBONYCHIDAE** is represented only by *Embonycha interrupta* from North Vietnam. The **ANISEMBIIDAE** is a relatively large and diverse New World family, ranging from the southwestern United States to Argentina (e.g. *Saussurembia*, *Chelicerca*) while two archaic Australian genera form the **AUSTRALEMBIIDAE**. The **OLIGOTOMIDAE** include several common webspinners from Asia and Australia, with *Haploembia* extending into the Mediterranean region and a few species more widely distributed by commerce. The **TERATEMBIIDAE** (= **OLIGEMBIIDAE**) may eventually prove the largest Embiopteran family, including small or diminutive species mainly from the New World and Africa.

Literature on Embioptera

ANANTHASUBRAMANIAN, K. S. (1957), Biology of *Oligotoma humbertiana* (Sauss.) (Oligotomidae, Embioptera), *Indian J. Ent.*, **18**, 226–232.

BARTH, R. (1954), Untersuchungen an den Tarsaldrüsen von *Embolyntha batesi* MacLachlan, 1877 (Embioidea), *Zool. Jb.* (*Anat.*), **74**, 172–188.

BARTH, R. AND LACOMBE, D. (1955), II. Estudos anatômicos e histológicos do ducto intestinal de *Embolyntha batesi* MacLachlan, 1877 (Embiidina), *Mem. Inst. Oswaldo Cruz,* **53**, 67–86.

BEIER, M. (1955), Embioidea und Orthopteroidea, In: Bronns *Klassen Ordn. Tiere*, **5**' (Abt. III, Buch 6), 1–30.

BITSCH, J. AND RAMOND, S. (1970), Étude du squelette et de la musculature prothoracique d'*Embia ramburi* R.-K. (Insecta Embioptera). Comparaison avec la structure du prothorax des autres Polynéoptères et des Aptérygotes, *Zool. Jb.* (*Anat.*), **87**, 63–93.

DAVIS, C. (1939–40), Taxonomic notes on the order Embioptera. I–XIX. *Proc. Linn. Soc. N.S.W.*, **64**, 181–190; 217–222; 369–372; 373–380; 381–384; 474–495, **65**, 171–191; 323–352; 362–387; 525–532.

—— (1940*a*), Taxonomic notes on the order Embioptera. XX. The distribution and comparative morphology of the order Embioptera, *Proc. Linn. Soc. N.S.W.*, **65**, 533–542.

—— (1940*b*), Family classification of the order Embioptera, *Ann. ent. Soc. Am.*, **33**, 677–682.

ENDERLEIN, G. (1912), Embiidinen, *Coll. zool. Selys Longchamps*, **3**, 1–121.

FRIEDRICHS, K. (1934), Das Gemeinschaftleben der Embiiden, *Arch. Naturgesch.*, **3**, 405–444.

GRASSI, B. AND SANDIAS, A. (1896–97), The constitution and development of the society of termites, *Q. Jl microsc. Sci.*, **39**, 245–322; **40**, 1–82.

IMMS, A. D. (1913), On *Embia major* n. sp. from the Himalayas, *Trans. Linn. Soc. Lond., Zool.*, **11**, 167–195.

KALTENBACH, A. (1968), Embiodea (Spinnfüsser), In: Helmcke, J.-G., Starck, D., and Wermuth, H. (eds), *Handbuch der Zoologie*, **4** (2), 2/8, 29 pp.

KRAUSS, H. A. (1911), Monographie der Embien, *Zoologica, Stuttgart*, **32**, 1–78.

LACOMBE, D. (1958–65), Contribução ao estudo dos Embiidae. III, IV, VI–VIII, *Stud. Ent., Petropolis*, **1**, 177–195; (1958); *Mem. Inst. Oswaldo Cruz*, **56**, 655–

683, (1958); *Bol. Mus. Nac., Rio de Janeiro (N.S.) Zool.*, **219**, 1–16, (1960); *An. Acad. brasil.*, **35**, 393–411, (1963); *An. Acad. brasil.*, **37**, 503–517, (1965).

MELANDER, A. L. (1903), Notes on the structure and development of *Embia texana, Biol. Bull.*, **4**, 99–118.

MILLS, H. B. (1932), The life history and thoracic development of *Oligotoma texana* (Mel.), *Ann. ent. Soc. Am.*, **35**, 648–652.

MUKERJI, S. (1927), On the morphology and bionomics of *Embia minor* sp. nov., with special reference to its spinning organ, *Rec. Ind. Mus.*, **29**, 253–282.

RÄHLE, W. (1970), Untersuchungen am Kopf und Prothorax von *Embia ramburi* Rimsky-Korsakow 1906 (Embioptera, Embiidae), *Zool. Jb. (Anat.)*, **87**, 248–330.

ROSS, E. S. (1944), A revision of the Embioptera or web-spinners of the New World, *Proc. U.S. nat. Mus.*, **94**, 401–504.

—— (1956), A new genus of Embioptera from Baltic amber, *Mitt. geol. Staatsinst. Hamburg*, **25**, 76–81.

—— (1960), Parthenogenetic African Embioptera, *Wasmann J. Biol.*, **18**, 297–304.

—— (1963), The families of Australian Embioptera, with descriptions of a new family, genus and species, *Wasmann J. Biol.*, **21**, 121–136.

—— (1966), The Embioptera of Europe and the Mediterranean region, *Bull. Brit. Mus. (Nat. Hist.) Entomol.*, **17** (7), 273–326.

—— (1970), Biosystematics of the Embioptera, *A. Rev. Ent.*, **15**, 157–172.

SLIFER, E. H. AND SEKHON, S. S. (1972), Sense organs on the antennal flagellum of two species of Embioptera (Insecta), *J. Morph.*, **139**, 211–226.

STEFANI, R. (1953), Un particolare modo di accoppiamento negli Insetti Embiotteri, *Rend. Acad. Naz. Lincei (Cl. Sci. fis., mat. nat.)*, (8) **14**, 544–549.

—— (1956), Il problema della partenogenesi in *Haploembia solieri* Ramb. (Embioptera, Oligotomidae), *Atti Accad. Naz. Lincei, Mem. (Cl. Sci. fis., mat. nat.)*, (8) **5**, 127–200.

—— (1961), La citologia della partenogenesi in due nuovi Embiotteri dell'Africa tropicale, *Rend. Accad. Naz. Lincei (Cl. Sci. fis., mat. nat.)*, (8), **30**, 254–256.

DICTYOPTERA (COCKROACHES AND MANTIDS)

Antennae almost invariably filiform, with numerous segments. Mouthparts mandibulate. Legs similar to each other or fore legs raptorial; coxae large and rather closely approximated; tarsi almost always 5-segmented. Fore wings modified into more or less thickened tegmina and with marginal costal vein. Wing-pads of nymph do not undergo reversal during development. Female with reduced ovipositor concealed by enlarged 7th abdominal sternum. Male genitalia complex, asymmetrical and largely concealed by 9th abdominal sternum which bears a pair of styles. Cerci many-segmented. Specialized stridulatory and auditory organs absent. Eggs contained in ootheca.

Though formerly associated with the Stick-Insects, Saltatorial Orthoptera and *Grylloblatta* in the Orthoptera, the Dictyoptera are here regarded as a separate order, with the distinctive characters enumerated above. They are medium or large-sized insects and include two distinct, homogeneous groups: the Cockroaches (suborder Blattaria) and the Mantids (suborder Mantodea). Though a very few semi-aquatic cockroaches are known (Shelford, 1909), the Dictyoptera are essentially terrestrial forms occurring predominantly in tropical and subtropical regions. They do not fly well and the wings of many species are reduced or absent, more especially in the female. For general accounts of these insects see Chopard (1938, 1949) and Beier (1961–74). The biology of *Periplaneta* and some other domestic cockroaches is dealt with in books by Guthrie and Tindall (1968) and Cornwell (1968).

External Anatomy – The skeletal structure of several species of Dictyoptera has been described in detail, e.g. *Periplaneta americana* (Snodgrass, 1952), *Blaberus giganteus* (Pradl, 1971) and *Stagmomantis carolina* (Leverault, 1937–38).

The head, which is usually hypognathous, is relatively primitive in structure with most of the typical sutures and sclerites well-defined, though the arms of the ecdysial cleavage line are not always present in the adult. The frons is well-developed, both clypeus and labrum are large and the tentorium is characterized by the presence of an aperture in its central part (Hudson, 1945). Compound eyes are usually well-developed, especially in Mantids, but in a few cavernicolous or myrmecophilous cockroaches they are reduced or absent. Three ocelli occur in the Mantodea, being larger in males than in females; in a few cockroaches two ocelli are present but in most members of the latter group these structures are represented by the so-called fenestrae – a pair of pale-coloured areas, each with a nervous connection to the brain and a histological structure reminiscent of a degenerate ocellus. The mouth-parts (Bugnion, 1920; Yuasa, 1920, Popham, 1961) are complete and adapted for biting (Fig. 8). The mandibles are strong and toothed, bearing a prostheca in the Blattaria. In the maxilla the cardo is subdivided, the stipes elongate and partly differentiated to form a subgalea; the palp is 5-segmented and borne on a small palpifer. The galea is a relatively soft structure, but the lacinia is toothed apically and bears bristles or teeth on its inner face. The labium comprises a large submentum, a small mentum and a prementum bearing a pair of 3-segmented labial palps, each on a palpiger. Glossae and paraglossae are both well-developed. The hypopharynx is large, with lateral suspensory sclerites and a median basal depression – the sitophore (Moulins, 1971).

In the neck region the cervical sclerites are well-developed (Storch, 1968). The prothorax bears dorsally a large pronotum, shield-like and usually overlapping the head in cockroaches, usually elongate and not covering the head in Mantids. The meso- and metaterga are similar, each subdivided into acrotergite, prescutum, scutum and scutellum, no postnota being present. The pleural region is normal, that of the prothorax not being concealed by lateral expansions of the pronotum (cf. Orthoptera). The sternal region of the Blattaria has undergone extensive desclerotization, especially in the pterothorax where the basisternum is reduced to a small pair of plates and the sternellum is even more reduced. In the Mantids, the sterna are more fully sclerotized and are dominated by the large basisterna (Adam and Lepointe, 1948; La Greca and Raucci, 1949). In all Dictyoptera there are well-developed furcal arms and a spina is present in the pro- and mesothorax.

The legs of the Blattaria are unmodified with large, closely approximated coxae and 5-segmented tarsi (Urvoy, 1963). The middle and hind legs of the Mantids are similarly unmodified but the fore legs are highly adapted for catching the small insects on which they prey. The coxae are elongate and mobile while the femora are thickly spinose and grooved along their lower side. The tibiae, which are also spinose, can fit into the groove along the femur and just before their apex – which is usually produced into a hook – there is inserted a reduced 5-segmented tarsus. When present, the two pairs of wings differ markedly. The front pair, known as tegmina, are rather strongly sclerotized and serve mainly to protect the membranous hind wings. The wing-venation has been studied in detail by Tillyard (1937), Rehn (1951) and Ragge (1955). The costa is marginal, its subcosta short and Rs possesses numerous anterior pectinate branches in the Blattaria. M and Cu_1 occupy a large part of the tegmen and in cockroaches the short, curved Cu_2 cuts off a distinctively shaped clavus. The hind wings possess a large anal lobe which at rest is folded beneath the anterior part between Cu_2 and 1A and is then usually further folded in a fan-like manner (Fig. 274). The tegmina of a few cockroaches (e.g. *Diploptera*) are greatly thickened and resemble the elytra of beetles.

The abdomen consists of 10 evident segments represented by their terga, and a reduced 11th segment. The first sternum is small and the 7th (females) or 9th (males) sterna are enlarged to form the subgenital plate which conceals most of the terminal abdominal structures. The ovipositor (Marks and Lawson, 1962; McKittrick, 1964) consists of three pairs of small valves which are probably homologous with those of the Orthoptera (q.v.) and are covered by the 7th sternum. The 8th, 9th and 10th sterna of the female are

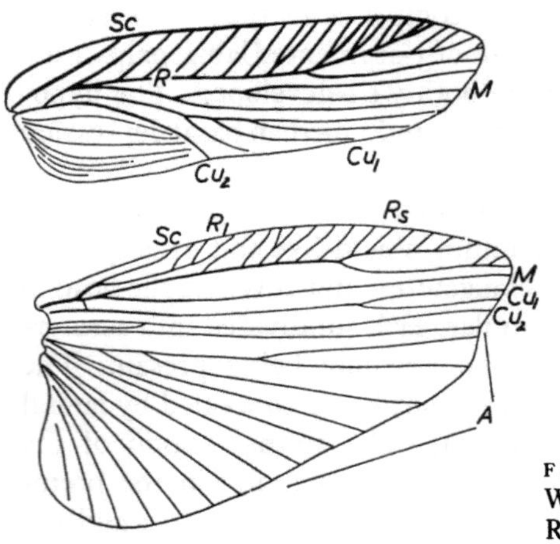

FIG. 274
Wings of *Blattella* (*after* Comstock, 1918, and Rehn, 1951)

reduced and partly membranous. In the male, the external genitalia comprise a group of complex, asymmetrical sclerites, between which lies the gonopore (Snodgrass, 1937; La Greca, 1953–56; McKittrick, 1964; Roth, 1969–73). Khalifa (1950) describes the role which these complicated structures play in copulation; at rest they are largely concealed by the 9th sternum which bears a pair of styles, one or both of which are missing in a few Blattaria. The 11th segment is represented in both sexes by its tergum (epiproct), its subdivided sternum (paraprocts) and its appendages – a pair of cerci, which are relatively short and many-segmented except in the cockroach *Panesthia* where they consist of a single segment.

In cockroaches the abdominal terga are modified somewhat. The 8th and 9th are small and concealed beneath the 7th in both sexes. The males of many species also possess tergal glands which secrete pheromones attractive to the female and which she licks while mating (Roth, 1969). They commonly open into a median setose depression of the tergum and show their greatest variation and development in the Blatellinae, occurring most often on the 7th tergum, though they can be found on as many as five segments and may involve any of the first ten terga.

Internal Anatomy – The alimentary canal (Bordas, 1897) is usually long and sinuous, but in some Mantids it is straight. The crop is well developed and the gizzard either rather poorly developed, as in the Mantids, or provided with a powerful masticatory armature in most Blattaria (Miller and Fisk, 1971). The mid gut (Fig. 95) bears eight tubular enteric caeca and at its junction with the hind gut there arise about 80–100 Malpighian tubules (Henson, 1944). The hind gut bears 6 rectal papillae (Oschman and Wall, 1969). Large salivary glands are present and are provided with conspicuous reservoirs in the cockroaches and some Mantids (Fig. 137). The nervous system (Nesbitt, 1941) is relatively generalized. The circumoesophageal connectives pass through the aperture in the tentorium and 1–3 of the anterior abdominal ganglia are fused with the metathoracic ganglion, so resulting in the presence of only 4–6 separate abdominal ganglia. The stomatogastric nervous system is well developed (Willey, 1961) with a recurrent nerve, hypocerebral ganglion, paired corpora cardiaca and a single oesophageal nerve. The tracheal system (Fig. 114) communicates with the exterior by ten pairs of spiracles – two thoracic and eight abdominal. The circulatory system has been described by Nutting (1951). The heart (Fig. 118) occupies most of the thorax and the first nine abdominal segments; it is flanked by twelve pairs of alary muscles and provided with three thoracic and nine abdominal pairs of incurrent ostia. Six pairs of segmental vessels (two thoracic, four abdominal) leave the heart in the Blattaria, but in the Mantids there are either only four abdominal pairs or the latter are replaced by four pairs of excurrent ostia. In the male reproductive system (Snodgrass, 1937; van Wyk, 1952) the testes each consist of four or more follicles usually enclosed in a common peritoneal sheath. The vasa deferentia run back with few or no convolutions and then loop forward round the cercal nerves to join the ductus ejaculatorius. Arising near the anterior end of the latter are one or more pairs of small ovoidal seminal vesicles and a large number of tubular accessory glands of various lengths. These glands secrete the material from which the spermatophore is constructed (Khalifa, 1950; Graves, 1969) and are apparently mesodermal in origin, developing in the nymph from an ampulla at the end of the rudiment of each vas deferens. In addition, male cockroaches possess an unpaired 'conglobate gland' of variable form which lies beneath the accessory glands and opens separately between the lobes of the phallus. In the female reproductive system (Voy, 1949; van Wyk, 1952, Bonhag, 1959; Roth, 1968*b*), each ovary comprises a number of panoistic ovarioles (6 in *Diploptera*, 15–20 in *Leucophaea*) leading by short ducts into the common oviduct which opens on the reduced 8th abdominal sternum into the large genital chamber, the floor of which is formed of the 7th sternite. Between the 8th and 9th abdominal sterna – i.e. in the dorsal wall of the genital chamber – there opens the duct of the spermatheca (Gupta and Smith, 1969), which is accompanied in some cockroaches by a small spermathecal gland and in a few species (e.g. *Blattella germanica*) is paired, with separate ducts. The female accessory glands ('colleterial glands') are a pair of large structures which also open separately into the genital chamber and secrete the materials from which the ootheca is formed in the chamber. They have been studied in detail in *Periplaneta*: the left gland produces a protein which becomes shaped to form the ootheca while the right one secretes a diphenolic substance which is converted into a quinonoid tanning agent used to harden the oothecal material (Brunet, 1951–52; Brunet and Kent, 1955; Mercer and Brunet, 1959). A similar pair of glands produces the ootheca of *Sphodromantis* (Kenchington and Flower, 1969).

Biology and Postembryonic Development – The Blattaria are a group of predominantly tropical forms although the few species occurring in temperate areas are sometimes numerically abundant under artificial conditions. Several species are readily distributed by human agency and have become established in all areas – e.g. such well-known domestic pests as *Blatta orientalis, Periplaneta americana* and

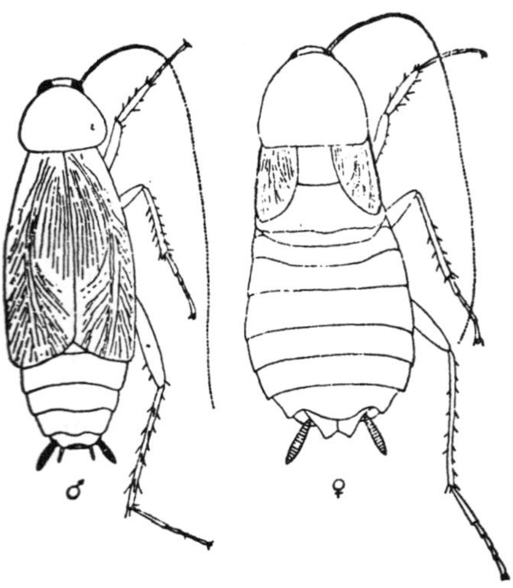

FIG. 275 *Blatta orientalis*, slightly enlarged

Blattella germanica. Although usually of a testaceous or dark mahogany colour, there are tropical species which are more brightly coloured, *Panchlora*, for instance, being pale green. Apterous and brachypterous forms are not uncommon, particularly in the females. Cockroaches are usually found in or on the ground and among low vegetation and debris, but some inhabit caves (e.g. *Nocticola*) and a few, such as *Attaphila*, are myrmecophilous. Some species of *Rhicnoda* and *Epilampra* live near water and can withstand short periods of submergence. On the other hand, a few species live in deserts and the females are provided with spinose fore tibiae which enable them to burrow into the ground. The domestic species are omnivorous, with a partiality for sugary and starchy foods but Brunner considered the food of most wild species to be composed largely of dead animal matter. *Panesthia* and *Cryptocercus* (Cleveland, 1934) feed on dead wood, digesting the cellulose through the assistance of symbiotic bacteria and Protozoa respectively (cf. Isoptera). *Cryptocercus*, incidentally, exhibits subsocial behaviour as does also *Polyzosteria*. For all these and other aspects of cockroach biology see Roth and Willis (1960).

The Mantids are exclusively carnivorous, occurring in all the warmer parts of the world (Beier, 1939). They are easily recognizable by the peculiar form of their front legs and, armed in this way, mantids often sit motionless for long periods at a time, with the head upraised upon the elongate and sub-erect prothorax. The powerful raptorial fore legs are raised together in front, their pincers being partially opened to seize any suitable prey which ventures within range (Rilling *et al.*, 1959; Mittelstaedt, 1957; Holling, 1966). This curious attitude, which suggests one of supplication, has earned for its possessors the name of 'praying mantids'. They feed voraciously on flies, grasshoppers, caterpillars, etc., and are very pugnacious, the larger forms attacking the smaller, and females the males; nymphs of *Pseudocreobotra* may be territorial (MacKinnon, 1970). Some of the larger S. American species have been recorded as even attacking small birds, lizards and frogs. Mantids are extremely variable in form and are assimilated in a remarkable manner to their surroundings, perhaps more to deceive their prey than to protect themselves (but see Edmunds, 1972). The green colour serves this purpose admirably while those that simulate flowers have the advantage of attracting flower-haunting insects within their reach. Certain tropical species possess foliaceous expansions on the prothorax and limbs, while *Pyrgomantis* is so attenuated as to resemble a Phasmid.

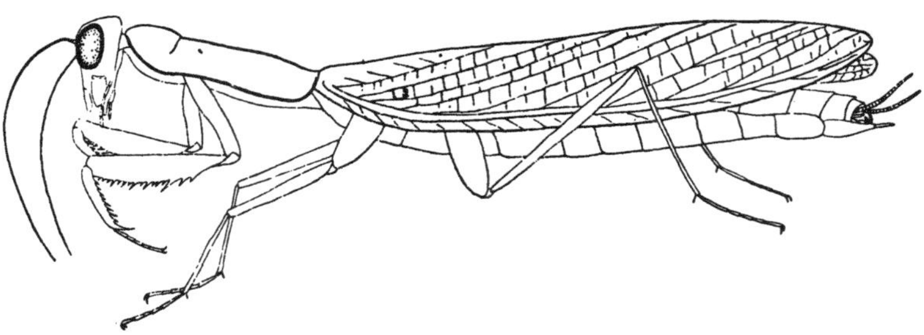

FIG. 276 *Mantis religiosa*, male. S. Europe

Cockroaches mate after a relatively simple courtship (Roth and Willis, 1952; Roth and Barth, 1967; Roth, 1970) and a spermatophore is formed (Khalifa, 1950; Graves, 1969). Three principal modes of reproduction have been distinguished (Roth, 1967, 1968*a*, 1970; Roth and Willis, 1957): (i) Oviparity, in which the eggs develop outside the reproductive tract of the female in a sclerotized ootheca which may be dropped soon after it is formed or which may be carried, protruding from the female's body, for much of the incubation period, as in *Blattella* (Fig. 277). (ii) Ovoviviparity, in which the ootheca (which may be thin and membranous) is first extruded and then retracted into the brood sac, an expansion of the genital cavity of the female parent. Here it remains while the embryos are being nourished from the egg-yolk, though they absorb water from the mother. When they are ready to hatch, the ootheca is extruded completely and the young drop free from the egg-shells and oothecal covering. (iii) Viviparity, confined to *Diploptera punctata* (Roth and Willis, 1955), which resembles ovoviviparity except that the embryos derive both water and food from the parent. Typically the ootheca is divided into two rows of pockets by a longitudinal partition and each egg (16 in *Blatta*, 40 in *Blattella*) occupies a pocket. When the nymphs are ready to emerge, the ootheca splits along its dorsal edge, the two halves separate and the young struggle out. The first-stage nymph in *Blatta orientalis* (Qadri, 1938) is a 'pronymph' with the instar of very short duration and its appendages not fully formed and apparently soldered down so as to give it a ver-

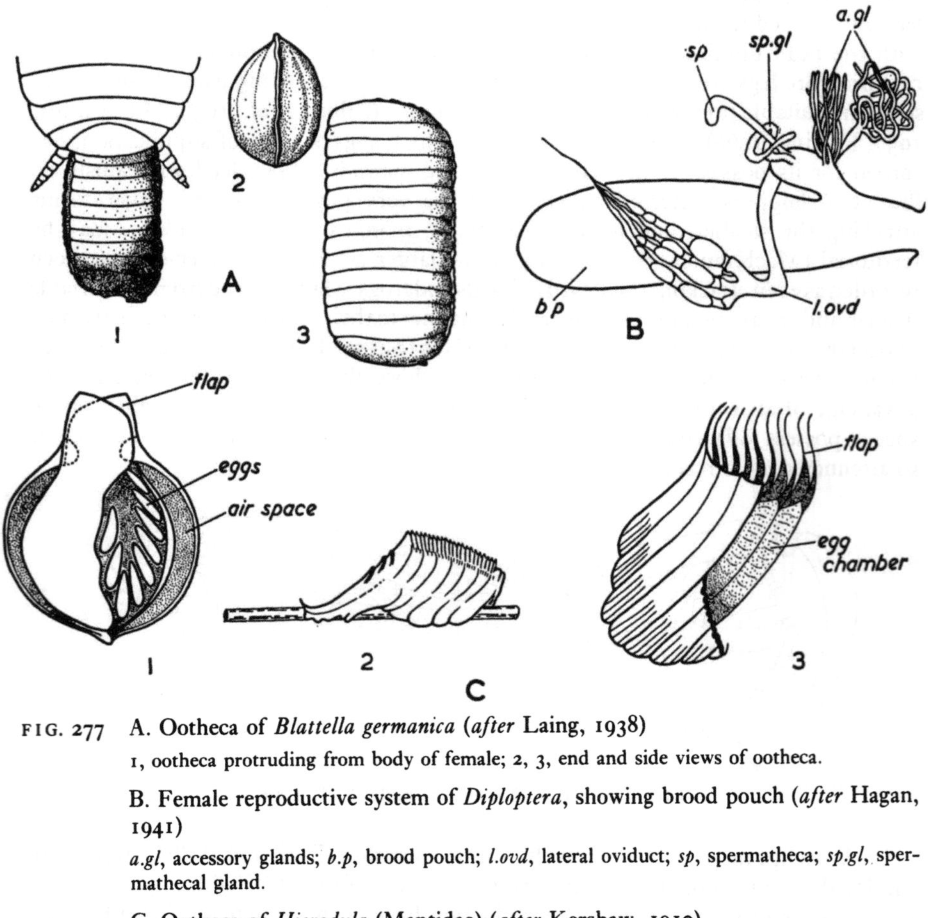

FIG. 277 A. Ootheca of *Blattella germanica* (*after* Laing, 1938)

1, ootheca protruding from body of female; 2, 3, end and side views of ootheca.

B. Female reproductive system of *Diploptera*, showing brood pouch (*after* Hagan, 1941)

a.gl, accessory glands; *b.p*, brood pouch; *l.ovd*, lateral oviduct; *sp*, spermatheca; *sp.gl*, spermathecal gland.

C. Ootheca of *Hierodula* (Mantidae) (*after* Kershaw, 1910)

1, transverse section; 2, entire ootheca; 3, details in lateral view.

miform appearance. The six subsequent nymphal instars of this species can be distinguished from one another by differences in the developing genital segments, number of cercal segments and size of wing-pads and take 279 days to complete their development at 27·5° C. *Periplaneta americana*, according to Griffiths and Tauber (1942), has 11 nymphal instars in the female and 12 in the male and takes about 250–270 days to develop at 29° C. See also Gould and Deay (1940). In Britain, *Ectobius lapponicus* has a two-year life cycle (Brown, 1973). It overwinters either as the ootheca (with an obligatory diapause) or as the 2nd, 3rd or 4th instar nymph.

Mating in the mantids has often been described in rather spectacular terms on account of a tendency for the female to attack and start eating the male before or during copulation. In an experimental study of mating in *Mantis religiosa*, Roeder (1935) suggests that this attack occurs through the failure of the female to recognize the male, which is therefore treated as legitimate prey. If the male is able to secure the correct grip on the female's body from the start of copulation he remains unmolested. Even when the male is attacked and his head eaten, this does not interfere with effective mating since the copulatory movements are under the control of the last abdominal ganglion and are even stimulated by the destruction of inhibitory centres in the suboesophageal ganglion. The eggs of mantids are laid in oothecae (Fig. 277) which are attached to twigs, bark, walls and other objects (Arora and Singh, 1957). Each female makes a number of these cases (twenty-two have been recorded in *Miomantis* by Adair, 1913) and their type of construction varies according to the species. In the Indian *Gongylus*, for example, Williams (1904) states that the ootheca consists of a more or less frothy secretion which hardens into a firm, spongy substance. Within this envelope is a layer of about forty egg-chambers arranged four abreast: they are constructed of a viscid material which very rapidly hardens to the consistency of horn. In the Chinese *Hierodula saussurii* Kershaw (1910) mentions that the oothecae are about an inch long, and the egg-chambers number about twenty-four, arranged in two longitudinal rows. An air space is left between the layer of eggs and the outer covering, and the latter is composed of overlapping strips of extremely hard, tough material. Other types of oothecae are described by Breland and Dobson (1948) and others. Though it would appear that the eggs of Mantids are admirably protected by these cases, the latter are ineffective in warding off the attacks of insect enemies, judging from the frequency with which parasitic Hymenoptera are bred out from them. It is possible, however, that they serve to protect the eggs from birds and lizards. As in the cockroaches, the first-stage nymph is a 'pronymph' and is rapidly succeeded by a more normal second-stage form (Williams and Buxton, 1916). The pronymph was observed by Williams and Buxton to hang suspended from the ootheca by means of silken threads secreted from a pair of papillae on the 10th abdominal sternum, but no silk is produced in later instars. The number of nymphal stages in the Mantidae is not constant and from accounts given by different observers it varies from three to twelve, the whole life-cycle taking about a year to complete.

Affinities and Phylogeny – The anatomy of Recent species leaves little doubt that the Dictyoptera are a natural group, distinct from the saltatorial Orthoptera with which they were formerly associated. Their closest modern relatives are the Isoptera (*q.v.*). The Blattaria are well represented from the Upper Carboniferous and later Palaeozoic by extinct families like the Archimylacridae. Their descendants have persisted with relatively few important structural changes, though the group declined in importance after the Palaeozoic. By the Tertiary, representatives of the Polyphaginae, Blattinae, Blatellinae, Blaberinae, Ectobiinae and several other subfamilies had evolved. It seems likely that the Blattaria arose from the so-called Protoblattoidea (Carboniferous to Jurassic), themselves allied to or part of the Protorthoptera. The Stenoneuridae (Carboniferous) are perhaps the ancestral stock from which the Blattaria took their origin. The fossil record of the Mantodea, on the other hand, is unusually poor. The first undoubted mantids occur in Oligocene amber from which members of the bark-haunting Liturgusinae are known. The relationship of these insects to the Blattaria is therefore based solely on the anatomy of Recent species.

Classification – The cockroaches and mantids are both homogeneous, natural groups which are treated here as suborders. For the Dictyopteran fauna of restricted regions see especially: Russia (Beĭ-Bienko, 1950); Europe (Harz, 1957; Kaltenbach, 1963; Chopard, 1951; Princis, 1965; Ragge, 1965);

N. America (Blatchley, 1920; Gurney, 1951); Malaysia (Bruijning, 1948); Africa (Ragge and Roy, 1967; Rehn, 1931–37). There are about 4000 species of Blattaria and about 2000 of Mantodea.

Suborder 1. BLATTARIA

Dictyoptera with head nearly or completely covered from above by the large, shield-like pronotum; two ocelli usually represented by fenestrae; fore legs unmodified; proventriculus generally with powerful masticatory armature.

The higher classification of the Blattaria (reviewed by Princis, 1960) is not finally settled. Among modern arrangements of families Rehn (1951) has relied mainly on the wing-venation and McKittrick (1964) has proposed a scheme based on a comparative study of the genitalia, proventriculus and oviposition behaviour of 85 genera. This has found support from further work on reproduction (Roth, 1970) and gizzard structure (Miller and Fisk, 1971) and from a detailed numerical analysis of biometric data on the nymphs and adults of 35 genera (Huber, 1974). McKittrick's classification is not, however, as fully worked out as the one developed on more traditional lines by Princis (1960; 1962–71). This authority regards the cockroaches as an order, dividing them into 4 suborders and 28 families. These taxa have mostly been given family and subfamily rank in other classifications and this has been adopted in the following summary of Princis' classification set out below.

FAM. POLYPHAGIDAE. *Under-side of mid and hind femora not spinose, rarely 1–4 small spines on fore or hind margin; anal area of hind wing flat, without fan-like folding at rest; if apterous, then with enlarged postclypeus; 9th sternite of male rounded posteriorly and with long, symmetrical styles.* Widely distributed. The Polyphaginae include some Mediterranean species; there are two myrmecophilous subfamilies, the Neotropical Atticolinae and the American Attaphilinae.

FAM. BLABERIDAE. *Femora as Polyphagidae; anal area of hind wing folded in a fan-like manner at rest; 9th sternite of male variable.* A large and diverse group, including such genera as *Perisphaeria*, *Pycnoscelus*, *Panesthia*, *Oulopteryx* and *Diploptera*.

FAM. BLATTIDAE. *Under-side of mid and hind femora spinose; hind wing folded fan-like at rest; 9th sternite of male unspecialized, symmetrical, with 2 styles; 7th sternum of female produced posteriorly and divided longitudinally.* The Blattinae include about 550 species in many genera, e.g., *Blatta*, *Periplaneta*, *Eurycotis*, *Polyzosteria*; the African and Oriental Nocticolinae are cavernicolous.

FAM. EPILAMPRIDAE. *Femora spinose; hind wing folded, fan-like; male with more specialized 9th sternite; female with undivided 7th sternum.* The largest group is the Blattellinae, with almost 1300 species (e.g., *Blattella*, *Parcoblatta*). The Ectobiinae are largely Old World, including some temperate species of *Ectobius*.

Suborder 2. MANTODEA

Head not covered by pronotum; three ocelli usually present; fore legs raptorial; proventriculus not powerfully armed.

A natural suprageneric classification of the Mantids has proved difficult to establish, as indicated by the work of Giglio-Tos (1913–27), Chopard (1949) and Beier (1934, 1964, 1968). The subdivision of the group into 8 families by Beier is followed here. Of these the **Chaeteessidae**, **Metallycidae** and **Mantoididae** are each represented by one small genus; the characters of the remaining five families are indicated below.

FAM. AMORPHOSCELIDAE. *Small species from Australia and the Old World tropics; prothorax short, often quadrate; fore tibiae not, or only very weakly, spinose; cerci long, often with enlarged terminal segment. Amorphoescelus, Paraoxypilus.*

FAM. EREMIAPHILIDAE. *Thick-set, brachypterous, desert species; prothorax more or less quadrate; fore tibiae strongly spinose; mid and hind legs long; tarsal segmentation 5:5:5 or 4:3:3.* Only two genera: *Eremiaphila* and *Heteronutarsus.*

FAM. HYMENOPODIDAE. *Small or medium-sized; head often with process on vertex; pronotum often expanded laterally; tegmina often with bicoloured transverse band or spiral marking; females sometimes brachypterous; inner ventral spines on fore femur alternately long and short, the outer ventral spines more or less decumbent.* Circumtropical, mainly African and Oriental. *Oxypilus*, *Acromantis*, *Hymenopus* and others.

FAM. MANTIDAE. *Small or large forms of variable facies; inner ventral spines of fore femur alternately long and short, the outer spines erect or oblique; never with bicoloured markings on tegmina.* The largest and most diverse family; among the many genera are *Mantis* (including the European *M. religiosa*, also introduced into N. America), *Pyrgomantis, Miomantis, Hierodula* and *Stagmomantis.*

FAM. EMPUSIDAE. *Large, more or less elongate, often bizarre forms; inner ventral spines of fore tibiae alternately 1 long and 3–4 short.* Old World species, e.g., *Empusa*, extending into S. Europe; *Gongylus.*

Literature on the Dictyoptera

ADAIR, E. V. (1913), Notes sur la ponte et l'éclosion de *Miomantis savignyi* (Sauss.), *Bull. Soc. ent. Egypte*, 3, 117–127.

ADAM, J. P. AND LEPOINTE, J. (1948), Recherches sur le morphologie des sternites et des pleurites des Mantes, *Bull. Mus. natn. Hist. nat. Paris*, (2), **20**, 169–173.

ARORA, G. L. AND SINGH, I. (1957), Mantodean oothecae, *Res. Bull. Punjab Univ.* (*Zoology*), **105**, 261–267.

BEĬ-BIENKO, G. Y. (1950), Fn. USSR. Insects: Blattodea, *Inst. Zool. Acad. Sci. URSS* (N.S.), **40**, 343 pp.

BEIER, M. (1934–35), Mantodea, *Genera Insect.*, **196**, 1–36; **197**, 1–10; **198**, 1–9; **200**, 1–32; **201**, 1–9; **203**, 1–146.

BEIER, M. (1939), Die geographische Verbreitung der Mantodeen, *Verh. 7. int. Kongr. Ent.*, **1**, 5–15.

—— (1961), Arthropoda. Insecta. Überordnung: Blattopteroidea Martynov 1938, Ordnung Blattodea Brunner 1882, In: Bronn's *Klassen u. Ordnungen des Tierreichs*, **5**, III. Abt., 6. Buch, 3. Lief., 587–848.

—— (1964), Blattopteroidea Mantodea, In: Bronn's *Klassen u. Ordnungen des Tierreichs*, **5**, 3. Abt., 5. Lief., 849–970.

—— (1968), Mantodea (Fangheuschrecken), In: Helmcke, J.-G., Starck, D. and Wermuth, H., *Handbuch der Zoologie*, **4** (2), Lief., 4: 47 pp.

—— (1974), 13. Ordnung Blattariae (Schaben), In: Helmcke, J.-G., Starck, D. and Wermuth, H., *Handbuch der Zoologie*, **4** (2), Lief., 23, 1–127.

BLATCHLEY, W. S. (1920), *The Orthoptera of Northeastern America*, Nature Publ. Co., Indianopolis, 784 pp.

BONHAG, P. F. (1959), Histological and histochemical studies on the ovary of the American cockroach *Periplaneta americana* (L.), *Univ. Calif. Publs Ent.*, **16**, 81–124.

BORDAS, L. (1897), L'appareil digestif des Orthoptères (Études morphologiques, histologiques et physiologiques de cet organe et son importance pour la classification des Orthoptères), *Ann. Sci. nat.*, **5**, 1–208.

BRELAND, O. P. AND DOBSON, J. W. (1948), Specificity of mantid oothecae (Orthoptera, Mantidae), *Ann. ent. Soc. Am.*, **40**, 557–575.

BROWN, V. K. (1973), The overwintering stages of *Ectobius lapponicus* (L.) (Dictyoptera: Blattidae), *J. Ent.*, (A), **48**, 11–24.

—— (1975), Development of the male genitalia in *Ectobius* spp. Stephens (Dictyoptera: Blattidae), *Int. J. Insect Morph. Embryol.*, **4**, 49–59.

BRUIJNING, C. F. A. (1948), Studies on Malayan Blattidae, *Zool. Meded.*, **29**, 1–174.

BRUNET, P. C. J. (1951–52), The formation of the ootheca by *Periplaneta americana*. I, II, *Q. Jl. microsc. Sci.*, **92**, 113–127; **93**, 47–69.

BRUNET, P. C. J. AND KENT, P. W. (1955), Observations on the mechanism of a tanning reaction in *Periplaneta* and *Blatta*, *Proc. roy. Soc.* (B), **144**, 259–274.

BUGNION, E. (1920), Les parties buccales de la Blatte et les muscles qui servent à les mouvoir, *Ann. Sci. nat.*, **10**, 41–108.

CHOPARD, L. (1938), *La Biologie des Orthoptères*, Lechevalier, Paris, 541 pp.

—— (1949), Ordre des Dictyoptères, In: Grassé, P. P. (ed.), *Traité de Zoologie*, **9**, 355–407.

—— (1951), Orthoptéroïdes, *Faune de France*, **56**, 359 pp.

CLEVELAND, L. R. (1934), The wood-feeding roach *Cryptocercus*, its Protozoa and the symbiosis between Protozoa and roach, *Mem. Am. Acad. Arts. Sci.*, **17**, 185–342.

COHEN, S. AND ROTH, M. (1970), Chromosome numbers of the Blattaria, *Ann. ent. Soc. Am.*, **63**, 1520–1547.

CORNWELL, P. B. (1968), *The Cockroach*, London, 391 pp.

DAY, M. F. (1950), The histology of a very large insect, *Macropanesthia rhinocerus* Sauss. (Blattidae), *Aust. J. sci. Res.*, B, **3**, 61–75.

EDMUNDS, M. (1972), Defensive behaviour in Ghanaian praying mantids, *Zool. J. Linn. Soc.*, **51**, 1–32.

GIGLIO-TOS, E. (1913–21), Mantidae, *Genera Insect.*, Fasc. **144**, 13 pp.; **177**, 36 pp.

—— (1927), Mantidae, *Tierreich*, Lfg. **50**, 707 pp.

GOULD, G. E. AND DEAY, H. O. (1940), The biology of six cockroaches which inhabit buildings, *Bull. Purdue Univ. agric. exp. Sta.*, **451**, 31 pp.

GRAVES, P. N. (1969), Spermatophores of the Blattaria, *Ann. ent. Soc. Am.*, **62**, 595–602.

GRIFFITHS, J. T. AND TAUBER, O. E. (1942), The nymphal development of the roach, *Periplaneta americana* L., *J. New York ent. Soc.*, **50**, 263–272.

GURNEY, A. B. (1951), Praying mantids of the United States, native and introduced, *Rep. Smithson. Inst., Washington*, **1950**, 339–362.

GUPTA, B. L. AND SMITH, D. S. (1969), Fine structural organization of the spermatheca in the cockroach *Periplaneta americana*, *Tissue & Cell*, **1**, 295–324.

GUTHRIE, D. M. AND TINDALL, A. R. (1968), *The Biology of the Cockroach*, London, 416 pp.

HARZ, K. (1957), *Die Geradflügler Mitteleuropas*, Fischer, Jena, 494 pp.

HENSON, H. E. (1944), The development of the Malpighian tubules of *Blatta orientalis* (Orthoptera), *Proc. R. ent. Soc. Lond.*, (A), **19**, 73–91.

HOLLING, C. S. (1966), The functional response of invertebrate predators to prey density, *Mem. ent. Soc. Canada*, **48**, 1–86.

HUBER, I. (1974), Taxonomic and ontogenetic studies of cockroaches, *Univ. Kansas Sci. Bull.*, **50** (6), 235–331.

HUDSON, G. B. (1945), A study of the tentorium in some Orthopteroid insects, *J. ent. Soc. sthn Africa*, 8, 71–90.

KALTENBACH, A. (1963), Kritische Untersuchungen zur Systematik, Biologie und Verbreitung der europäischen Fangheuschrecken, *Zool. Jb.* (*Anat.*), **90**, 521–598.

KENCHINGTON, W. AND FLOWER, N. E. (1969), Studies on insect fibrous proteins: the structural protein of the ootheca in the praying mantis, *Sphodromantis centralis* Rehn, *J. Microscopy*, **89**, 263–281.

KERSHAW, J. C. (1910), The formation of the ootheca of a Chinese mantid, *Hierodula saussurii*, *Psyche*, **17**, 136–141.

KHALIFA, A. (1950), Spermatophore production in *Blatella germanica* L. (Orthoptera: Blattidae), *Proc. R. ent. Soc. Lond.*, (A), **25**, 53–61.

LA GRECA, M. (1953–54), Sulla struttura morfologica dell'apparato copulatore dei Mantodei, *Ann. Ist. sup. Sci. Lett., Napoli*, **1953–54**, 1–28.

—— (1956), Sulla struttura morfologica dell'apparato copulatore dei Mantodei, *Ann. Ist. Sci. Lett. S. Chiara, Napoli*, **1956** (6), 293–317.

LA GRECA, M. AND RAINONE, A. (1949), Il dermascheletro e la musculatura dell'addome di *Mantis religiosa*, *Ann. Ist. Mus. zool. Univ. Napoli*, **1** (5), 43 pp.

LA GRECA, M. AND RAUCCI, A. (1949), Il dermascheletro e la musculatura del torace di *Mantis religiosa*, *Ann. Ist. Mus. zool. Univ. Napoli*, **1** (3), 1–41.

LEVERAULT, P. (1937–38), The morphology of the Carolina mantis. I, II, *Univ. Kansas Sci. Bull.*, **24**, 205–259.

MACKINNON, J. (1970), Indications of territoriality in mantids, *Z. Tierpsychol.*, **27**, 150–155.

MARKS, E. P. AND LAWSON, F. A. (1962), A comparative study of the Dictyopteran ovipositor, *J. Morph.*, **111**, 139–158.

MCKITTRICK, F. A. (1964), Evolutionary studies of cockroaches, *Mem. Cornell Univ. agric. exp. Sta.*, **389**, 197 pp.

MERCER, E. H. AND BRUNET, P. C. J. (1959), The electron microscopy of the left colleterial gland of the cockroach, *J. Biophys. Biochem. Cytol.*, **5**, 257–262.

MILLER, H. K. AND FISK, F. W. (1971), Taxonomic implications of the comparative morphology of cockroach proventriculi, *Ann. ent. Soc. Am.*, **64**, 671–687.

MITTELSTAEDT, H. (1957), Prey capture in mantids, In: Scheer, B. T., (ed.), *Recent Advances in Invertebrate Physiology*, Univ. Oregon Press, Eugene, Oregon, pp. 51–71.

MOULINS, M. (1971), La cavité préorale de *Blaberus craniifer* Burm. (Insecte, Dictyoptère) et son innervation: étude anatomo-histologique de l'epipharynx et l'hypopharynx, *Zool. Jb. (Anat.)*, **88**, 527–586.

NESBITT, H. H. J. (1941), A comparative morphological study of the nervous system of the Orthoptera and related orders, *Ann. ent. Soc. Am.*, **34**, 51–71.

NUTTING, W. L. (1951), A comparative anatomical study of the heart and accessory structures of the Orthopteroid insects, *J. Morph.*, **89**, 501–597.

OSCHMAN, J. L. AND WALL, B. J. (1969), The structure of the rectal pads of *Periplaneta americana* L. with regard to fluid transport, *J. Morph.*, **127**, 475–510.

POPHAM, E. J. (1961), The functional morphology of the mouthparts of the cockroach *Periplaneta americana* L., *Entomologist*, **94**, 185–192.

PRADL, W.-D. (1971), *Blaberus giganteus*, Schaben. *Grosses Zoologisches Praktikum*, **14b**, 48 pp.

PRINCIS, K. (1960), Zur Systematik der Blattaria, *Eos*, **36**, 427–449.

—— (1962–71), Blattariae, In: Beier, M. (ed.), *Orthopterorum Catalogus*, Partes 3–14, 1224 pp.

—— (1965), Ordnung Blattariae (Schaben), *Bestimmungstabellen zur Bodenfauna Europas*, Lfg. 3, 50 pp.

QADRI, M. A. H. (1938), The life history and growth of the cockroach, *Blatta orientalis*, L., *Bull. ent. Res.*, **29**, 263–276.

RAGGE, D. R. (1965), *The Wing Venation of the Orthoptera Saltatoria with Notes on Dictyopteran Wing Venation*, Brit. Mus. (Nat. Hist.), London, 159 pp.

—— (1965), *Grasshoppers, Crickets and Cockroaches of the British Isles*, Collins, London, 299 pp.

RAGGE, D. R. AND ROY, R. (1967), A review of the praying mantises of Ghana (Dictyoptera, Mantodea), *Bull. Inst. fr. Afr. noire (Sér. A)*, **29**, 586–644.

REHN, J. A. G. (1931–37), African and Malagasy Blattidae (Orthoptera). I–III, *Proc. Acad. nat. Sci., Philadelphia*, **83**, 305–387; **84**, 405–511; **89**, 17–123.

REHN, J. W. H. (1951), Classification of the Blattaria as indicated by their wings, *Mem. Amer. ent. Soc.*, **14**, 134 pp.

RILLING, S., MITTELSTAEDT, H. AND ROEDER, K. (1959), Prey recognition in the praying mantis, *Behaviour*, **14**, 164–184.

ROEDER, K. D. (1935), An experimental analysis of the sexual behaviour of the praying mantis (*Mantis religiosa*, L.), *Biol. Bull.*, **69**, 203–220.

ROTH, L. M. (1967), Sexual isolation in parthenogenetic *Pycnoscelus surinamensis* and application of the name *Pycnoscelus indicus* to its bisexual relative (Dictyoptera: Blattaria: Blaberidae: Pycnoscelinae), *Ann. ent. Soc. Am.*, **60**, 774–779.

—— (1968*a*), Oothecae of the Blattaria, *Ann. ent. Soc. Am.*, **61**, 83–111.

—— (1968*b*), Ovarioles of the Blattaria, *Ann. ent. Soc. Am.*, **61**, 132–140.

—— (1969), Evolution of male tergal glands in the Blattaria, *Ann. ent. Soc. Am.*, **62**, 176–208.

—— (1969–73), The male genitalia of Blattaria, *Psyche*, **76**, 217–250; **77**, 104–119; 217–236; 308–342; 436–480; **78**, 84–106; 180–192; 296–305; **80**, 249–264; 305–348.

—— (1970), Evolution and taxonomic significance of reproduction in Blattaria, *Ann. Rev. Ent.*, **15**, 75–96.

ROTH, L. M. AND BARTH, R. (1967), The sense organs employed by cockroaches in mating behaviour, *Behaviour*, **28**, 58–94.

ROTH, L. M. AND WILLIS, E. R. (1952), A study of cockroach behaviour, *Am. midl. Nat.*, **47**, 66–129.

—— —— (1955), Intra-uterine nutrition of the 'beetle roach' *Diploptera dytiscoides* (Serv.) during embryogenesis, with notes on its biology in the laboratory, *Psyche*, **62**, 55–68.

—— —— (1957), An analysis of oviparity and viviparity in the Blattaria, *Trans. Am. ent. Soc.*, **83**, 221–238.

—— —— (1960), The biotic associations of cockroaches, *Smithson. misc. Collns*, **141**, 470 pp.

SHELFORD, R. (1909), Notes on some amphibious cockroaches, *Rec. Indian Mus.*, **3**, 125–127.

SNODGRASS, R. E. (1937), The male genitalia of Orthopteroid insects, *Smithson. misc. Collns*, **96**, 107 pp.

—— (1952), *A Textbook of Arthropod Anatomy*, Comstock Publ. Co., Ithaca, N.Y., 363 pp.

STORCH, R. H. (1968), The adult cervicothoracic musculature of the cockroach, *Nauphoeta cinerea* (Olivier), *J. Morph.*, **126**, 107–122.

TILLYARD, R. J. (1937), Kansas Permian insects, Pt. 20. The cockroaches or order Blattaria. *Am. J. Sci.*, **34**, 169–202; 249–276.

URVOY, J. (1963), Étude anatomo-functionelle de la patte et de l'antenne de la blatte *Blabera craniifer* Burmeister, *Ann. Sci. nat., Paris (Zool.)*, (12), **5**, 287–413.

VOY, A. (1949), Contribution à l'étude anatomique et histologique des organes accessoires de l'appareil génital femelle chez quelques espèces d'Orthopteroïdes, *Ann. Sci. nat.*, **11**, 269–345.

WILLEY, R. B. (1961), The morphology of the stomodeal nervous system in *Periplaneta americana* (L.) and other Blattaria, *J. Morph.*, **108**, 219–261.

WILLIAMS, C. B. AND BUXTON, P. A. (1916), On the biology of *Sphodromantis guttata*, *Trans. ent. Soc. Lond.*, **1916**, 86–100.

WILLIAMS, C. E. (1904), Notes on the life-history of *Gongylus gongyloides*, a mantis of the tribe Empusides and floral simulator, *Trans. ent. Soc. Lond.*, **1904**, 125–137.

WYK, L. E. VAN (1952), The morphology and histology of the genital organs of *Leucophaea maderae* (Fabr.) (Blattidae, Orthoptera), *J. ent. Soc. sthn Africa*, **15**, 3–62.

YUASA, H. (1920), The anatomy of the head and mouthparts of Orthoptera and Euplexoptera, *J. Morph.*, **33**, 251–290.

Order 14

ISOPTERA (TERMITES OR WHITE ANTS)

Social and polymorphic species living in large communities composed of reproductive forms together with numerous apterous, sterile soldiers and workers. Mouthparts of the typical biting type: ligula 4-lobed. Wings very similar, elongate and membranous, superposed flat over the back when at rest and capable of being shed by means of basal fractures: anterior veins strongly sclerotized, regular cross-veins wanting and an archedictyon often present. Tarsi almost always 4-segmented. Cerci short or very short: genitalia usually wanting or rudimentary in both sexes. Metamorphosis slight or absent.

The Isoptera are usually known as termites or 'white ants'. The latter name is unfortunate since they are only distantly related to the true ants or Formicidae, though the two groups offer striking analogies in structure and habits (Wilson, 1971). Termites abound throughout the tropics and also occur in most warm temperate countries. Only two species, *Kalotermes flavicollis* and *Reticulitermes lucifugus*, are common in Europe but their range does not extend into the British Isles. Termite communities inhabit nests or termitaria of various kinds and consist of four main castes or types of individuals, two of which are reproductive forms and two sterile. The reproductive castes comprise (*a*) sclerotized, macropterous *primary reproductives* which found new colonies (Fig. 278); (*b*) less sclerotized *supplementary reproductives* with more or less reduced wings (Fig. 291). A termite colony usually contains a royal pair, the king and queen, which are commonly primary reproductives that have lost their wings after founding the colony originally (Fig. 290). The sterile castes consist of (*a*) *soldiers* and (*b*) *workers*, both of which may be apterous males and females adapted to perform special non-reproductive functions. Every colony also contains numerous immature individuals of various ages undergoing development into some or all of the above castes. In addition the colony usually contains symbionts and inquilines from other orders of insects or even different groups of animals. It is evident, therefore, that the study of termite associations involves problems of wide biological significance. General reviews of the order are given by Grassé (1949), Schmidt (1955), Krishna and Weesner (1969–70), Howse (1970), Weidner (1970), Harris (1971), Lee and Wood (1971) and the relevant chapters of Wilson (1971). There is a classified bibliography by Snyder (1956–68).

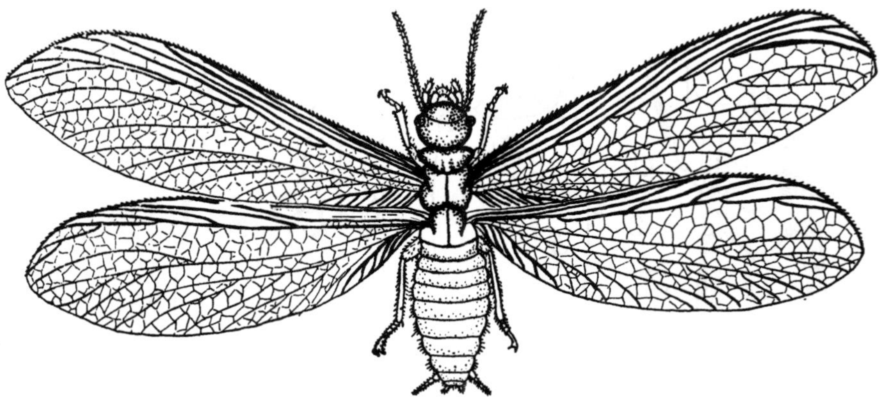

FIG. 278 A winged termite (*Archotermopsis*), male

External Anatomy

In addition to works cited below, there are accounts of *Archotermopsis*, *Stolotermes* and *Anacanthotermes* by Imms (1919), Morgan (1959) and Gupta (1960). The cuticle is more completely sclerotized in the primary reproductives than in other castes; in workers and soldiers it is usually only the head which is sclerotized though species which forage above ground in daylight may be more darkly coloured than subterranean forms. In workers and reproductive castes the head-capsule is ovoid or rounded; in soldiers it is often oblong or pyriform and may be larger than the rest of the body. Remnants of the ecdysial cleavage line are variably developed and the tentorium is perforate like that of the Dictyoptera (Hudson, 1946). Compound eyes are present in macropterous forms and in all castes of the Hodotermitidae but are otherwise degenerate or absent. Paired dorsal ocelli often accompany the compound eyes. The moniliform antennae comprise from 9 to 30 segments depending on the species, caste and instar; macropterous individuals have the highest number. The well-developed, variably shaped labrum overlies the bases of the mandibles and is hinged posteriorly to the clypeus, itself divisible into a sclerotized postclypeus, firmly fused to the frons, and a membranous anteclypeus.

The mouthparts (Figs 280, 281) are Orthopteroid in their general features. The mandibles are similar in reproductives and workers and present few striking differences in form, though their detailed structure is taxonomically important (Ahmad, 1950). The mandibles of soldiers are very variable in the different genera, often attaining a great size accompanied by curious anomalies of shape (Fig. 280); only in the nasute soldiers (*q.v.*) are they vestigial. The *maxillae* only differ in points of detail throughout the order. The galea is hood-like and commonly 2-segmented; the lacinia is strongly sclerotized and powerfully toothed distally, becoming more or less laminate basally, and is armed with stout setae along its inner margin; the palps are 5-segmented. The labium consists of pre- and postmentum with

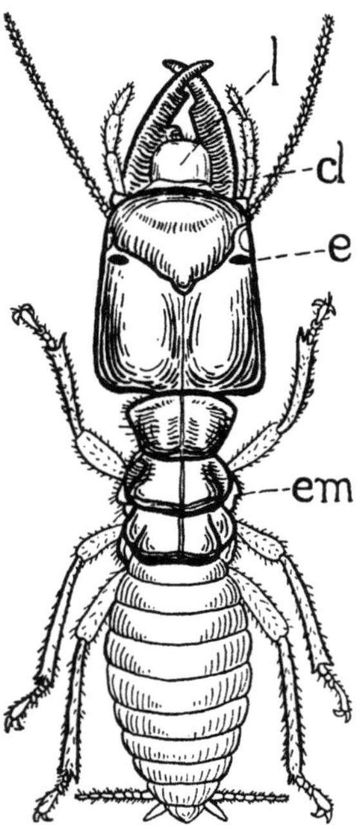

FIG. 279
A soldier termite (*Archotermopsis*), male
cl, clypeus; *e*, eye; *em*, epimeron; *l*, labrum.

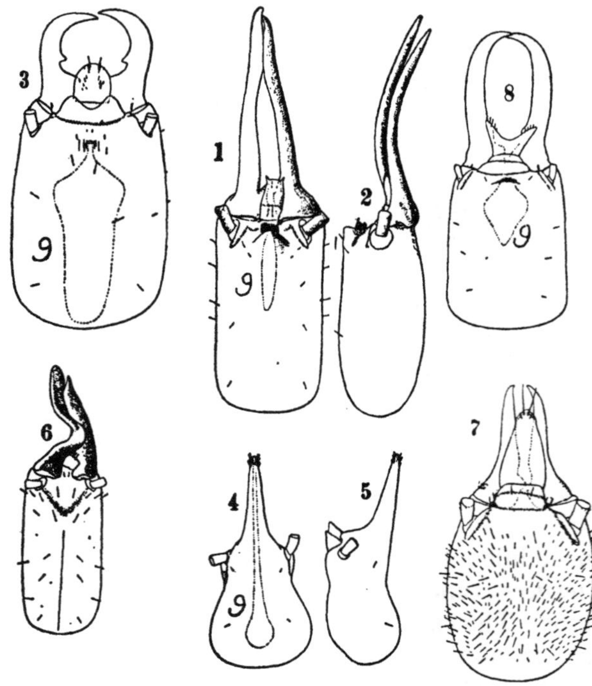

FIG. 280 Heads of soldiers of African termites
1, 2, *Mirotermes*. 3, *Hamitermes*. 4, 5, *Eutermes*. 6, *Pericapritermes*. 7, *Microtermes*. 8, *Cubitermes*; *g*, frontal gland. *After* various figures by Silvestri, *Boll. Lab. Zool. Portici*, 9.

glossae, paraglossae and 3-segmented labial palps. The hypopharynx is large and without superlinguae. On each side of the neck are two lateral cervical sclerites placed at right angles to each other; small dorsal and ventral cervical sclerites are also sometimes present.

In the **thorax** (Fuller, 1924) the terga are well developed. The pronotum shows many taxonomically important variations in shape: it may be flat and shield-like, heart-shaped, laterally lobed or saddle-shaped. The meso- and metanota are subequal and less variable. On the ventral side the prosternum is reduced, the mesosternum large and broad (mainly basisternum) and the metasternum somewhat smaller and V-shaped. The pleura comprise approximately equal episterna and epimera separated by a more or less straight pleural sulcus; a trochantin is also present. The legs are similar to each other and the coxae large and broad. In the middle and hind pairs a *meron* is marked off from the rest of the coxa by a deep sulcus. The tibiae are long and slender; among the most primitive genera they are armed with both terminal and lateral spines, but in the majority lateral spines are wanting. The tarsi are typically 4-segmented; the only exception is *Mastotermes* which has 5 complete segments. In *Archotermopsis, Zootermopsis* and *Hodotermopsis*, the tarsi are imperfectly 5-segmented, the 2nd segment being reduced. In the winged imagines of *Mastotermes* and some Kalotermitidae, an *empodium* is present between the claws of the feet; in other families this structure is wanting.

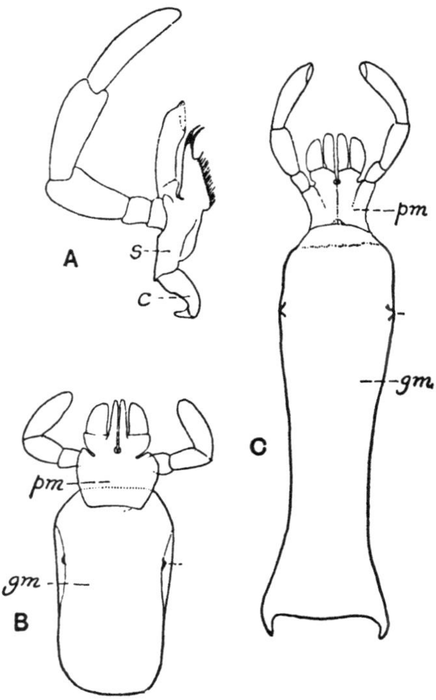

FIG. 281
Archotermopsis

A, maxilla of soldier. B, labium of macropterous form. C, labium of soldier; *c*, cardo; *gm*, postmentum; *pm*, prementum; *s*, stipes.

The two pairs of wings are essentially similar in size, form and venation (Figs 282, 283). There is a striking absence of regular cross-veins though in more primitive genera the membrane is stiffened by an irregular sclerotized network between the main veins. The venation (Tillyard, 1931; Fuller, 1919) is primitive in a few genera (*Mastotermes, Archotermopsis, Zootermopsis*), but in the remainder of the order specialization by reduction is evident, affecting more particularly the radial and median veins. In the fore wing of *Mastotermes* (Fig. 300) there is no true costal vein: Sc is 2-branched, and R_{1-5} are recognizable as separate branches. Both M and Cu are well developed, Cu_2 forms a curved *vena dividens* and there are no distinct anal veins. In the hind wing Sc is unbranched and M arises from the stem of R_{4+5}. Three anal veins are present and support a well-developed anal lobe which recalls the Blattarian hind wing and is a primitive character found in no other termites (Tillyard, 1937). *Archotermopsis* and *Zootermopsis* exhibit

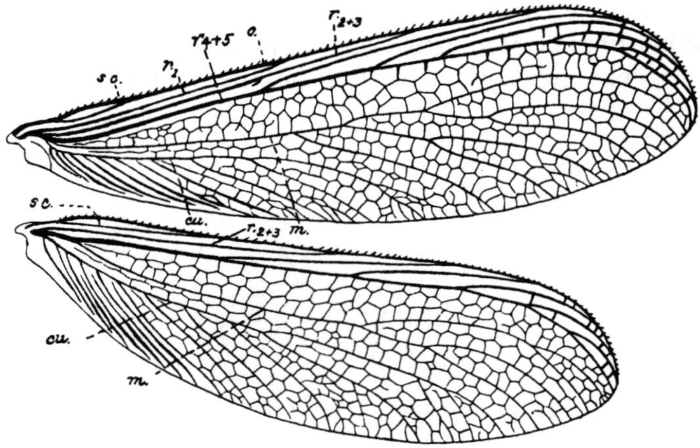

FIG. 282 Fore and hind wings of *Archotermopsis*

the first stage in reduction, R_{2+3} being undivided in the fore wing, and the anal lobe of the hind wing vestigial. In the higher Isoptera the costal margin is greatly thickened through the fusion of certain anterior veins and R is represented by a single stem, possibly R_{4+5}. M usually retains one or more branches and the remainder of the wing is occupied by the accessory branches of Cu. One of the most striking features of the termite wing is the presence of the *basal* or *humeral suture* which is a line of weakness along which fracture and shedding of the wings takes place after swarming. *Mastotermes* is exceptional in that the hind wing has no fracture-line and is torn off irregularly. The stump of the wing, or that portion which lies between the humeral suture and the thorax, persists throughout life and is commonly termed the *scale*.

The **abdomen** has 10 visible terga, the 11th tergum probably being fused with the 10th, while the 11th sternum is represented by a pair of paraprocts. The 1st sternum is atrophied and the sternal plates differ markedly in the two sexes of the reproductive forms. In the males of many termites all the sterna are entire but in certain higher forms the 9th sternum is divided. In the females the 7th sternum is greatly enlarged forming the subgenital plate which overlies the succeeding sterna. The terminal segment of the abdomen

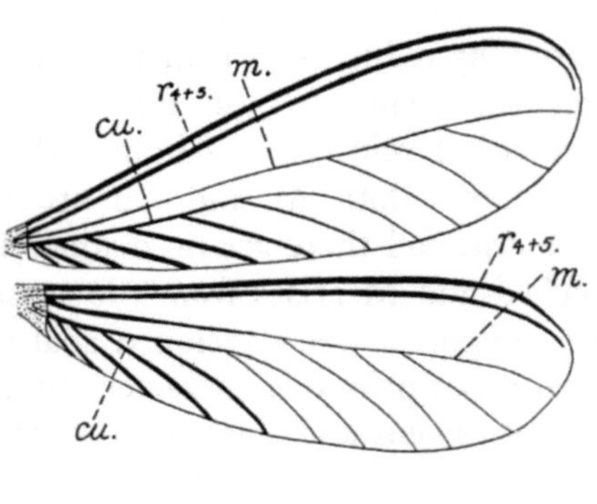

FIG. 283
Fore and hind wings of *Eutermes* sp.

carries a pair of short *cerci* which are present in all castes. In *Archotermopsis* they are composed of 6–8 segments, in *Mastotermes* and *Zootermopsis* of 5 segments, *Hodotermopsis* 3–6 segments, while in the family Termitidae they are, for the most part, reduced to the condition of 1- or 2-segmented tubercles. On the hind border of the 9th sternum a pair of small, unjointed *styles* is frequently present. They occur in both sexes of the soldiers and workers and in the nymphs of all castes; in the reproductive forms, with rare exceptions, they are present in the males only. External sexual differentiation is clearly evident in the soldiers and workers of *Mastotermes*, *Archotermopsis*, and a few other primitive forms. A reduced ovipositor of the Blattarian type is present in *Mastotermes* and is still further reduced in the other families (Browman, 1935; McKittrick, 1965). The male has a membranous median copulatory organ.

Internal Anatomy

The **alimentary canal** (Figs. 284, 285) is a coiled tube of moderate length and exhibits comparatively few important variations in structure (Noirot and Kovoor, 1958; Kovoor, 1968–71). The long, narrow *oesophagus expands distally to form the crop*. The latter organ is seldom capacious and is often only slightly emphasized. It is followed by the *gizzard*, provided with an armature of cuticular denticles; this organ is simple and ring-like in certain of the more primitive forms, becoming more pronounced among other termites (see McKittrick, 1965). Beyond the gizzard the fore intestine protrudes into the cavity of the mid gut forming a large *oesophageal valve*. The mid gut is tubular, of uniform calibre throughout and often completely encircles the hind gut (Noirot-Timothée and Noirot, 1965; Noirot and Noirot-Timothée, 1967). At the junction of the mid and hind gut are the *Malpighian tubules*; these are variable in number, eight usually being present in the Kalotermitidae and two to four among the Termitidae. In some Hodotermitidae four or five *enteric caeca* arise as outgrowths from the anterior end of the mid gut. At its commencement, the *hind intestine* is a short narrow tube (the *ileum*), often separated by a valve from the *colon*. The latter is usually an extensive chamber (the so-called rectal pouch) which, in the wood-feeding termites, is frequently distended through the presence of large numbers of Protozoa. The *rectum* is a narrow tube of very variable length and terminates in an ovoid or spherical chamber opening to the exterior by the anus. A peritrophic membrane is present (Platania, 1938). Racemose salivary glands are present, each with a salivary reservoir. The saliva of workers is used in nest construction and is fed to other members of the colony; soldiers of *Pseudacanthotermes* produce sticky, defensive saliva.

The **circulatory system** has been very little investigated; the *heart* consists of 8–10 chambers and is prolonged anteriorly as the *aorta* which communicates with the cephalic blood space just behind the brain. The fat-body is more extensively developed in the reproductive forms than in the soldiers or workers. Feytaud (1912) states that in the kings and queens this tissue undergoes a complete change several years after swarming. Migratory adipocytes enter it in large numbers and undergo division, gradually building up a new fat-body at the expense of the old. Intracellular micro-organisms, presumed to be symbiotic, occur in the fat-body of *Mastotermes* and are similar to those found in the Blattaria (Koch, 1938).

In the **central nervous system** the brain and optic lobes vary considerably, depending on the size of the compound eyes in the different castes

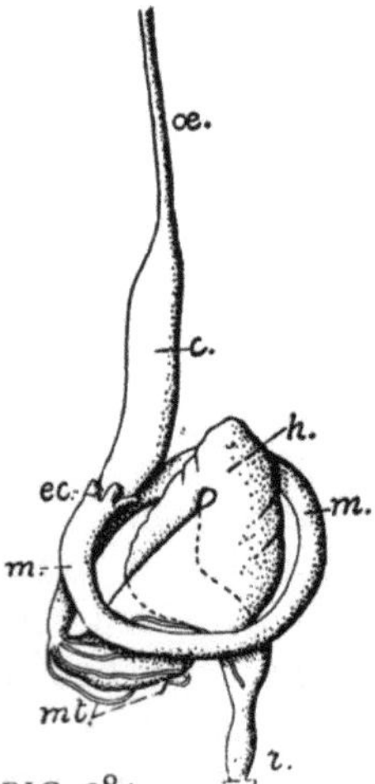

FIG. 284
Archotermopsis, alimentary canal of soldier

oe, oesophagus; *c*, crop; *ec*, enteric caeca; *m*, mid gut; *mt*, Malpighian tubes; *h*, hind gut; *r*, rectum.

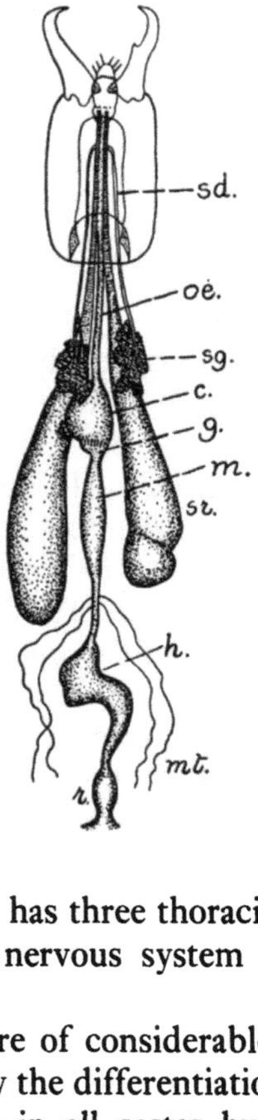

FIG. 285
Odontotermes ceylonicus, alimentary canal of soldier

g, gizzard; *sd*, salivary duct with gland *sg* and reservoir *sr*. Other lettering as in Fig. 284. *After* Bugnion, *Rev. Suisse Zool.*, 1911.

(Zuberi, 1963). The ventral nerve cord has three thoracic and six abdominal ganglia and a normal stomatogastric nervous system is present (Hecker, 1966; Zuberi and Peters, 1964).

Two **epidermal gland systems** are of considerable importance in the termites. The *frontal gland* is formed by the differentiation of a group of cells beneath the frons. It may be present in all castes but usually attains its greatest development in the soldiers (Holmgren, 1909; Feytaud, 1912; Thompson, 1916). In its completely developed condition it is a sac-like gland which opens to the exterior by a frontal pore in a shallow depression on the surface of the head where the cuticle is pale-coloured and forms the *fontanelle*. The gland attains its greatest development in the soldiers of *Prorhinotermes* and *Coptotermes*; in these genera it is an extensive sac, reaching to the extremity of the abdomen (Fig. 293), and discharges a milky latex-like secretion through an enlarged frontal pore. In the soldiers of *Termes* the gland opens at the apex of a prominent *frontal tubercle*, and in the nasute soldiers of *Nasutitermes* the tubercle is prolonged into an elongate rostrum, through which the duct of the gland passes. At least in some genera the secretions of the frontal gland are used defensively by the soldiers. In some Rhinotermitidae and Termitidae a white suspension of lipids in an aqueous mucopolysaccharide solution is sprayed out of the frontal pore and thickens into an adhesive material. In the Nasutitermitinae the same function is performed by a mixture of terpenes that is sprayed out of the apex of the

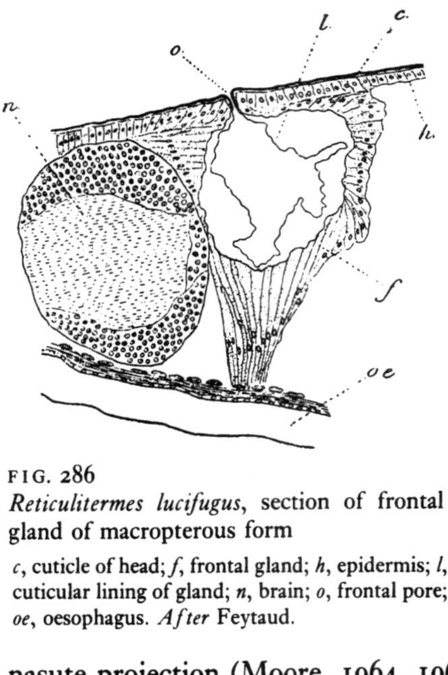

FIG. 286
Reticulitermes lucifugus, section of frontal gland of macropterous form

c, cuticle of head; *f*, frontal gland; *h*, epidermis; *l*, cuticular lining of gland; *n*, brain; *o*, frontal pore; *oe*, oesophagus. *After* Feytaud.

FIG. 287
Archotermopsis, reproductive organs of winged female

ov, ovary; *od*, oviduct; *cg*, colleterial gland; *s*, spermatheca.

nasute projection (Moore, 1964, 1968). The *sternal glands* of termites occur beneath the sterna of the third, fourth and fifth abdominal segments of *Mastotermes*, the fourth segment of the Hodotermitidae and Termopsidae and the fifth segment in other Isoptera (Noirot and Noirot-Timothée, 1965; Stuart and Satir, 1968). The gland consists of secretory and columnar cells and has no reservoir; its secretion apparently passes out through fine cuticular pores. The secretion is used to form trails that can be followed by other termites when foraging or moving to repair breaches in the nest. Such odour trails are more or less specific and consist of a variety of relatively complex hydrocarbons or alcohols (e.g. Hummel and Karlson, 1968; Matsumura *et al.*, 1968).

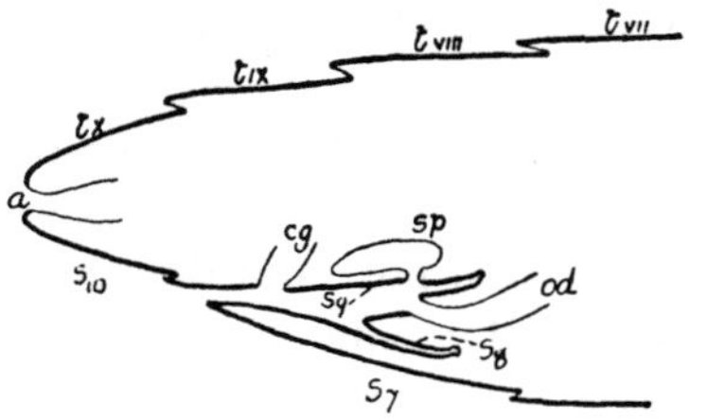

FIG. 288
Archotermopsis, diagrammatic section of the apex of the abdomen of a female soldier

tvii–tx, terga; s_7–s_{10}, sterna; *a*, anus; *cg*, opening of colleterial glands; *sp*, spermatheca; *od*, oviduct.

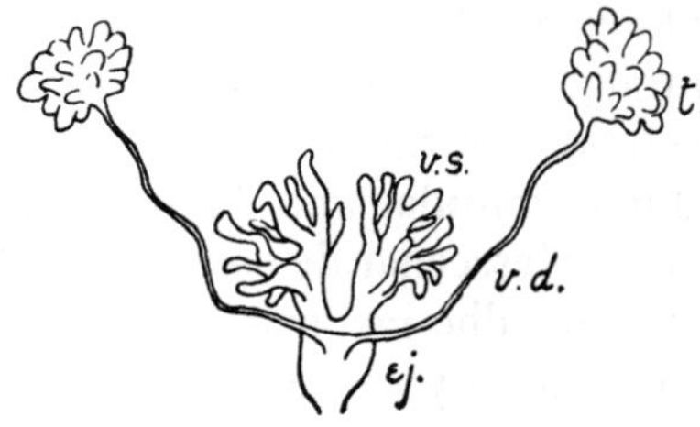

FIG. 289
Archotermopsis, reproductive organs of winged male

t, testis; *vs*, vesicula seminalis; *vd*, vas deferens; *ej*, ejaculatory duct.

The **reproductive system** attains its complete functional development in the primary and supplementary reproductive castes. In workers and soldiers it is always more or less reduced and various degrees of degeneration can be traced in the soldiers from forms like *Archotermopsis*, with fully developed organs, to those like *Hospitalitermes* where no trace of the reproductive system can be found (Weyer, 1935; Stella, 1938–39). In the reproductive castes the testes are simple and usually consist of about 8–10 short, digitate lobes in or near the 8th abdominal segment (Fig. 289). The vasa deferentia are short and converge to form the muscular ejaculatory duct. At their point of union is a pair of seminal vesicles, each consisting of a small group of tubules. The spermatozoa of termites are usually said to be aberrant, non-motile cells without a flagellum, but in *Kalotermes flavicollis* they have a body 10–15 μm long, provided with two anterior flagella and several smaller backwardly directed processes, and they are motile (Truckenbrodt, 1964). In the females each ovary consists of a variable number of panoistic ovarioles which open separately into the lateral oviduct. The number of ovarioles varies greatly, being lower in the more primitive forms and enormous in some Termitidae: 30–45 occur in *Archotermopsis*, 3000 in *Odontotermes* (Truckenbrodt, 1973). The two oviducts communicate by a common aperture with the genital pouch whose floor is formed by the enlarged 7th sternum (Fig. 288). The dorsal wall of the pouch receives the openings of the *spermatheca* and the common duct of the colleterial glands. The latter consist of a large number of elongate and much convoluted tubules, whose function has not been ascertained.

The Castes of Termites

It has been mentioned previously that termites are polymorphic and live in large communities. The following four castes are structurally and functionally distinct terminal forms incapable of further transformation.

Primary Reproductives – The members of this caste have two pairs of large membranous wings which are shed after they have swarmed in the process of founding new colonies. Their body is well sclerotized and more or less dark brown in colour, compound eyes are fully developed and there are often paired ocelli. The brain and optic ganglia are well developed, as is also the frontal gland (when it is present), and the reproductive organs attain a greater size than in any of the other castes. Primary reproductives are adapted for a short aerial life, after which they pair and form the king and queen of a new colony.

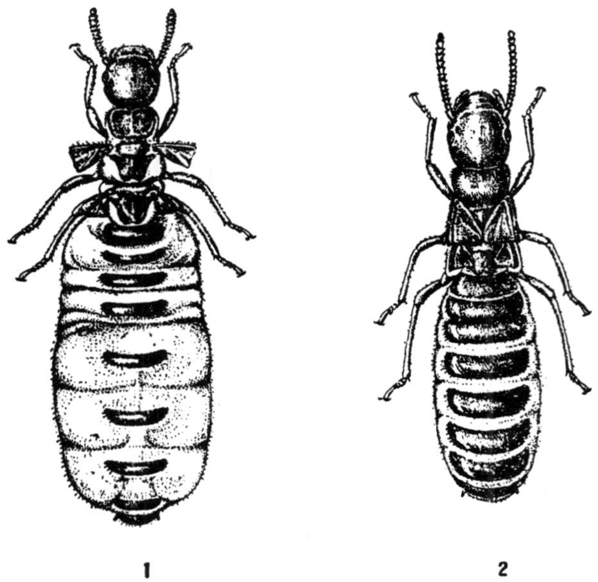

FIG. 290 *Reticulitermes flavipes.* Deälated queen (1) and king (2) of primary reproductive form
After Banks and Snyder, *U.S. Nat. Mus. Bull.* 108.

Supplementary Reproductives (Fig. 291) – The members of this caste have no aerial life, the body is usually much less sclerotized and pigmented than in the primary reproductives and the compound eyes are generally reduced. Growth of the wings is inhibited to varying degrees and they usually resemble wing-pads, though with some indications of venation. The brain, frontal glands and reproductive system are also somewhat reduced in size. Some authorities distinguish three kinds of supplementary reproductives: adultoid supplementaries very similar in appearance to the primaries, nymphoid supplementaries with short, pad-like wings, and ergatoid supplementaries without any wing rudiments. Weyer (1930), however, has found in *Microcerotermes* that a series of transitional forms occurs varying from apterous specimens morphologically very similar to workers up to those which resemble primary reproductives in pigmentation, sclerotization and

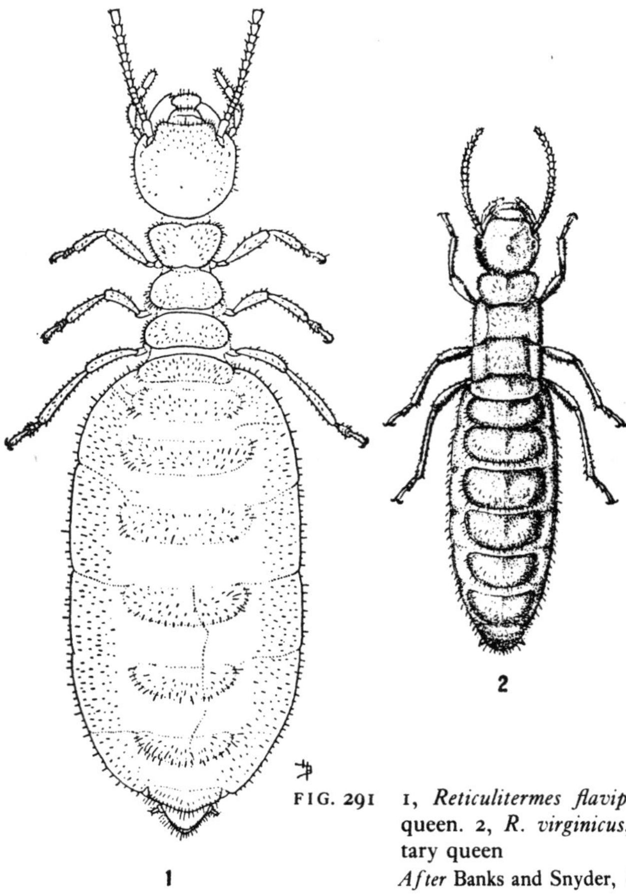

FIG. 291 1, *Reticulitermes flavipes*, apterous supplementary queen. 2, *R. virginicus*, brachypterous supplementary queen
After Banks and Snyder, *U.S. Nat. Mus. Bull.* 108.

eye-development, but with shorter wings (which, of course, are not shed). See also Noirot (1955–56). The so-called 'pseudo-imagines' of *Kalotermes flavicollis* and *Reticulitermes lucifugus* (Grassé, 1949) which occasionally appear in colonies recently deprived of their queen, resemble primary reproductives but are not pigmented and, after tearing off their wings irregularly with the mandibles, become sexually functional without leaving the colony. They are probably to be regarded as extreme examples of supplementary reproductives.

Supplementary reproductives are not normally found in colonies headed by the original primary reproductives. They appear after a lapse of several weeks should one or both of these die and replace them so that the colony can continue to develop. There do not seem to be any general rules governing the number and sex of supplementary reproductives which develop after loss of the primary forms. When one of the primaries dies, supplementaries of that sex alone or of both sexes may develop and though many supplementaries frequently appear (e.g. 4–100 in the colonies of *Microcerotermes* studied by Weyer, 1930), Grassé and Noirot (1946*a*) report that in artificial colonies of *Kalotermes flavicollis* one pair of supplementaries sometimes attains dominance, the others disappearing (possibly through cannibalism). Further details of the role of supplementary reproductives and their differentiation from nymphal stages will be found below.

The reproductive castes exhibit remarkable post-metamorphic growth which is least evident in the primitive families and most pronounced in the fertilized females of the Termitidae (Fig. 296). These physogastric queens attain a length of 5 to 9 cm or more and are largest when derived from the primary females (easily recognizable by the persistent wing-scales; see Fig. 290). The great increase in size is due to the distension of the abdomen through the growth of the ovaries, gut and fat-body. The head and thorax do not grow nor do the tergal and sternal sclerites of the abdomen, which therefore remain as small sclerotized islands set in the expanded intersegmental membrane that forms the main abdominal covering. The principal changes that occur during postmetamorphic growth may be summarized as follows (Feytaud, 1912; Bugnion and Popoff, 1912). The wing-muscles, which occupy the greater part of the thoracic cavity, degenerate and are broken down, partly by phagocytic action. The original fat-body, as mentioned earlier, undergoes complete transformation, being replaced by a new tissue. Certain changes supervene in the digestive system in conformity with an alteration in diet. The queen no longer partakes of woody or other hard matter but is nourished upon saliva or, in the fungus-growing species, upon fungal hyphae in combination with that secretion. The jaw-muscles in consequence become reduced in size and power; the mid gut undergoes correlated changes, both structural and functional; the Malpighian tubules increase in length, while the hind intestine suffers marked curtailment and the intestinal Protozoa are lost. The volume of the blood is greatly increased, while the nervous system and dorsal vessel undergo elongation in conformity with the general extension of the abdomen. The tracheae increase in size but not in complexity as the viscera of the young female are already more richly tracheated than the male (Bordereau, 1971*a, b*). There is also hypertrophy of the corpora allata with histological signs of increased secretory activity (Pflugfelder, 1938). The most striking changes are exhibited in the reproductive system which monopolizes, as it were, the greater part of the abdomen and converts the queen into a vast, almost inert, egg-laying mechanism. The changes involved are those of size and, in its general morphology, the reproductive system of the mature queen does not differ from that of the same individual when in the winged stage though a postmetamorphic increase in the number of functional ovarioles takes place. In *Odontotermes badius* only the anteriormost 5–7 ovarioles are functional in the young queen, but after several years the remaining 3000 or so have also become capable of producing eggs. (Truckenbrodt, 1973).

Workers – True workers are absent from *Mastotermes*, the Kalotermitidae and Termopsidae, their place being taken by nymphal stages or pseudergates (see below). Where they occur, however, workers are numerically the most important members of the community. They are usually pale in colour, with the cuticle only slightly sclerotized, and they look more like nymphs than the adults of the other castes. External sexual characters are hardly perceptible though workers may be genetically male or female. The head is directed downwards; it is relatively wider than in the reproductive castes but never as large as in soldiers. Compound eyes are reasonably well-developed in the workers of the Hodotermitidae that forage above ground in daylight, but are otherwise vestigial or absent. The thorax resembles that of soldiers rather than reproductives. Specific characters are not clearly expressed and unassociated worker termites may be difficult or impossible to identify. In some species the workers are dimorphic, being divisible into major and minor forms, as in *Macrotermes estherae, Odonototermes obesus, O. redemanni, O. horni* and many species of *Trinervitermes*. In *Nasutitermes costalis* large workers come more readily to repair broken nest surfaces, while small workers are more abundant elsewhere (McMahan, 1970). In the Macrotermitinae large workers are male and small ones are female. Although taking no part in reproduction, and seldom any part in the defence of the community to which they belong, practically all other duties devolve upon the members of this caste. They exhibit marked care for the eggs and young and in times of danger may remove them to safer situations. They also feed and tend the queens, forage for food, often at a distance from the nest and, in the fungus-growing species, attend to the cultivation of the 'gardens'. In lignicolous species the workers excavate the galleries and tunnels which serve for the nest; in mound-building forms they construct the termitarium, and repair damage to it. It is the workers which destroy crops, timber, woodwork and other materials.

Soldiers – These are structurally the most specialized members of the community, but their occurrence in virtually all genera and certain features of intercastes (see below) suggest that they are the primitive sterile caste. They are wanting only in several genera allied to *Anoplotermes* (Sands, 1972) and may readily be recognized by their large, strongly sclerotized heads. Four well-defined types of soldiers can be distinguished: (*a*) mandibulate soldiers, the most frequent forms, with large, powerful mandibles that may assume striking or grotesque shapes; (*b*) nasute soldiers (Fig. 292), in which the mandibles are vestigial but there is a long median frontal nasus or rostrum, at the end of which the frontal gland opens; they occur only in the Nasutitermitinae; (*c*) nasutoid soldiers, found among the Rhinotermitidae, in which the mandibles vary in size but the frons, clypeus and labrum are produced into a snout-like structure bearing a dorsal channel along which the secretion of the frontal gland can flow; (*d*) phragmotic soldiers, found in the Kalotermitidae, whose head is a strongly sclerotized, more or less plug-like structure that can be used to block the nest-openings and whose mandibles are reduced. In addition, the soldiers of a species are sometimes divisible into two or three size categories, though in other cases the large and small forms are connected by intermediates. Like the workers, soldiers

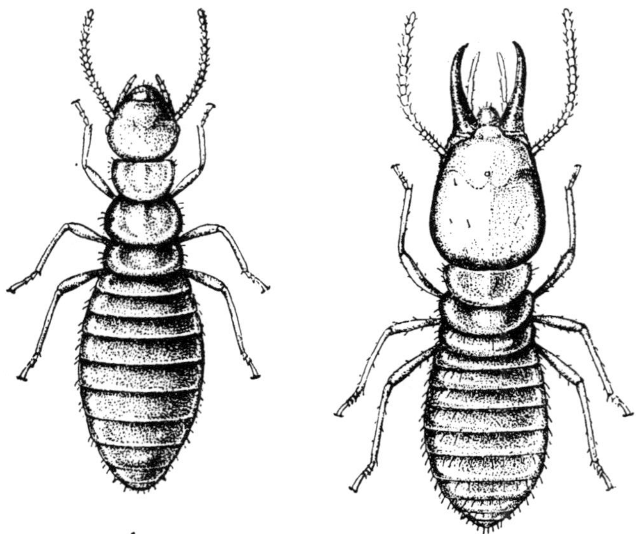

FIG. 292 1, worker, 2, soldier, both of *Prorhinotermes simplex*. *After* Banks and Snyder, *U.S. nat. Mus. Bull.* 108.

usually consist of genetic males and females, but in some Termitidae caste and sex-determination seem to be associated: most Nasutitermitinae have only male soldiers while in the Macrotermitinae and Termitinae they are female. Except in *Mastotermes, Archotermopsis* and some Kalotermitidae, external sexual characters are slight and the sex of the individual can be ascertained best through examination of the reduced gonads. Tolerably well-developed faceted eyes occur in the soldiers of *Hodotermes*, and vestigial eyes are found in those of *Archotermopsis, Kalotermes* and other genera, but more often than not visual organs are totally wanting; a pair of reduced ocelli is sometimes also present. The antennae usually consist of one or several segments less than in the reproductive castes.

Soldiers are specialized in structure and behaviour for the defence of the colony, mainly against other insects – especially ants – and vertebrate predators. The typical soldiers attack with their mandibles, which can pierce or slice, and even the large, twisted, asymmetrical mandibles of such genera as *Capritermes* and *Pericapritermes* (Fig. 281) can be used defensively with a peculiar 'snapping' action, as when one snaps one's fingers (Kaiser, 1954; Deligne, 1965). Other soldiers attack insect intruders with the adhesive secretions of the salivary or frontal glands (p. 612). Both of these may be enormously enlarged and occupy much of the body cavity (Fig. 293). Salivary secretions are discharged through the mouth, frontal gland secretions from the rostrum of nasute or nasutoid soldiers which may direct a fine spray of sticky material against intruders (Ernst, 1959).

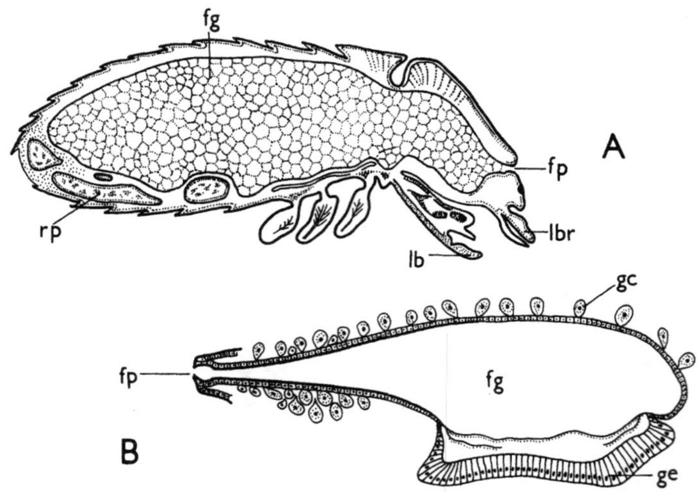

FIG. 293 Frontal gland of Isoptera. A, Sagittal section of worker of *Coptotermes travians* (*after* Bugnion and Popoff). B, Longitudinal section of frontal gland of nasute soldier of *Nasutitermes arborum* (*after* Bonneville)

fg, frontal gland; *fp*, frontal pore; *gc*, glandular cell; *ge*, glandular epithelium; *lb*, labium; *lbr*, labrum; *rp*, rectal pouch.

Other Forms – The above four castes are terminal forms, incapable of further moulting. In the lower termites, however, the workers are absent and their place is taken by 'pseudergates' or by the older nymphs. Grassé and Noirot (1947) designated as pseudergates the large, blind, apterous forms in *Kalotermes flavicollis* which carry out the functions exercised by workers in the higher termites. They consider that in a normal stable colony the pseudergates do not moult again, but under experimental conditions they have moulted into supplementary reproductives so that they can hardly be regarded as a true caste. It is now known that pseudergates occur in the Mastotermitidae (Watson, 1971), Kalotermitidae and Termopsidae, all of which appear to lack a true worker caste. In some cases the pseudergates share their duties with the older nymphs in the colony, in others it is the nymphs alone which perform the duties of a worker. In *Zootermopsis nevadensis*, for example, it is the 6th instar nymphs that behave like workers in most respects, though earlier instars can carry out some digging and nest-building (Howse, 1968).

Also present occasionally, as a result of abnormal conditions, are two other forms, *achrestogonimes* and *intercastes*:

Achrestogonimes. The alate imagines produced in a colony normally all leave when swarming occurs but sometimes a small number remains; these lose their wings and their gonads atrophy. They remain xylophagous, retain their protozoal fauna and though they play no part in the maintenance of the colony are tolerated by its other members. Grassé and Bonneville (1935) refer to such forms as achrestogonimes and have found them in several species.

Intercastes. Rather rarely, individuals are found which are morphologically intermediate between castes. Soldiers with wing-pads and forms intermediate between workers and soldiers are known, but worker-sexual intercastes have never been reported (Adamson, 1940). Though infection with parasites is apparently sometimes the cause of intercaste formation, other cases are readily intelligible in the light of what is known of the postembryonic differentiation of castes under the influence of morphogenetic pheromones.

Postembryonic Development and Caste Differentiation

The study of the postembryonic development of termites is unusually difficult. Direct observation of the numbers of instars and of the course of caste differentiation is impossible in natural colonies and the plasticity shown in the development of the castes from nymphs of different stages makes it difficult to be certain how far some of the developmental processes revealed in artificial colonies are a regular feature of natural development. It is therefore not surprising that reasonably complete accounts of the process have been given for only a few species. Summaries are available in Krishna and Weesner (1969–70) and other general works; there are many diverse modes of postembryonic development and caste-differentiation and care must be taken not to generalize from the few examples given here.

The eggs are normally deposited singly but *Mastotermes* lays clusters of 16–24 eggs cemented together with a gelatinous secretion and recalling a simple type of ootheca (Hill, 1925). The incubation period is rather long (periods of 24 to 90 days have been recorded for various species, while some eggs may overwinter in cooler regions) and the eggs are often tended by workers or nymphs. Postembryonic development is slow, e.g. Pickens (1932) records that workers of *Reticulitermes hesperus* in the U.S.A. take about 32 months to attain full development while Harvey (in Kofoid *et al.*, 1934) found that reproductives of *Kalotermes minor* took from 6 to 14 months to develop at 21° C. and 83 per cent relative humidity. Such figures have little general significance, however, since the duration of development and the number of nymphal instars varies not only with the species, caste and the usual environmental factors, but also with the age, size and caste-composition of the colony. From four to ten nymphal stages have been recorded, the younger colonies producing terminal forms (soldiers, workers or supplementary reproductives) after a smaller number of moults. There are commonly seven nymphal instars among the members of established colonies of the more primitive families while the primary reproductives seem almost invariably to have at least this number of developmental stages. Ecdysis is accompanied by cessation of feeding, loss of most or all of the intestinal contents, shedding of the mid gut epithelium and its renewal from regenerative cells (Weyer, 1935) and by a quiescent period which may last for several days (Snyder, 1913). This quiescent period is most pronounced in the moult giving rise to soldiers and primary reproductives and in the Macrotermitinae (Termitidae) where the insect lies on its side with the head flexed upon the ventral aspect of the thorax, while the limbs and other parts remain immobile (Fig. 294). Such a state presents some analogies with the pupal instar of endopterygote insects, but there is no reason to regard the resemblance as having a phylogenetic significance.

Caste differences appear during the course of postembryonic development. Thus, typically, the primary reproductives and the brachypterous supplementary reproductives are preceded by instars possessing wing-pads, the soldiers develop from apterous nymphs which have larger heads than those destined to become workers, while in species with polymorphic workers and soldiers, the major and minor forms may be distinguished among the preceding nymphal stages. Cases are known, however, where regressive

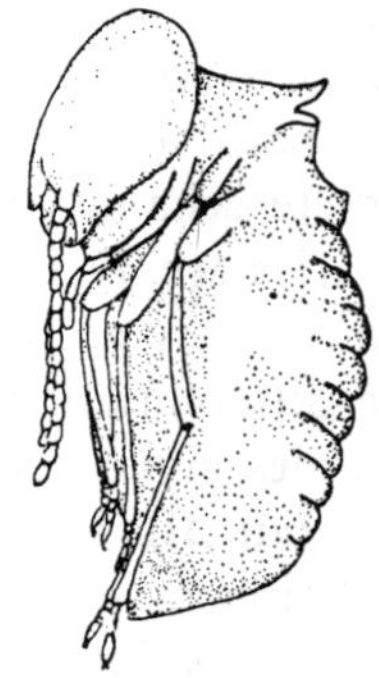

FIG. 294
Quiescent nymphal phase of a termite (*Rhinotermes*)
After Holmgren, *Zool. Jb., Syst.*, 23.

moults occur, e.g. Grassé and Noirot (1947) found that pseudergates could develop in *Kalotermes flavicollis* either from apterous fourth-instar nymphs or from sixth- or seventh-instar nymphs by loss of the wing-pads; Miller (1942) working with *Prorhinotermes* and Grassé and Noirot (1946*b*) with *Kalotermes* have found that soldiers can develop in the same way from nymphs with wing-pads. The instar at which caste differences become apparent varies again with species, caste and colony age. Thus, in *Kalotermes* the first three instars show no differences in young colonies, while in older colonies caste distinctions appear only in sixth or later instars. On the other hand, in *Macrotermes gilvus* Bathellier (1927) finds that the second-instar nymphs are already differentiated into large and small apterous nymphs (destined to produce major and minor sterile castes respectively) and nymphs with wing-pads which give rise to primary reproductives. The variety of conditions occurring in postembryonic development are best represented diagrammatically as in the two examples given in Fig. 295; and see also Noirot (1955, 1956) and Buchli (1958). It is hardly surprising with the large number of developmental courses open to undifferentiated or partially differentiated nymphs that intercastes should occasionally arise.

Origins of Polymorphism

Several theories have been proposed to account for the production of the different castes, especially for the distinction between reproductive and sterile forms. At least in the lower families of termites, it is now agreed that there is little reason to believe that caste is genetically determined. Critical examination of newly hatched nymphs has shown that they do not display caste differences (Heath, 1927; Hare, 1934;

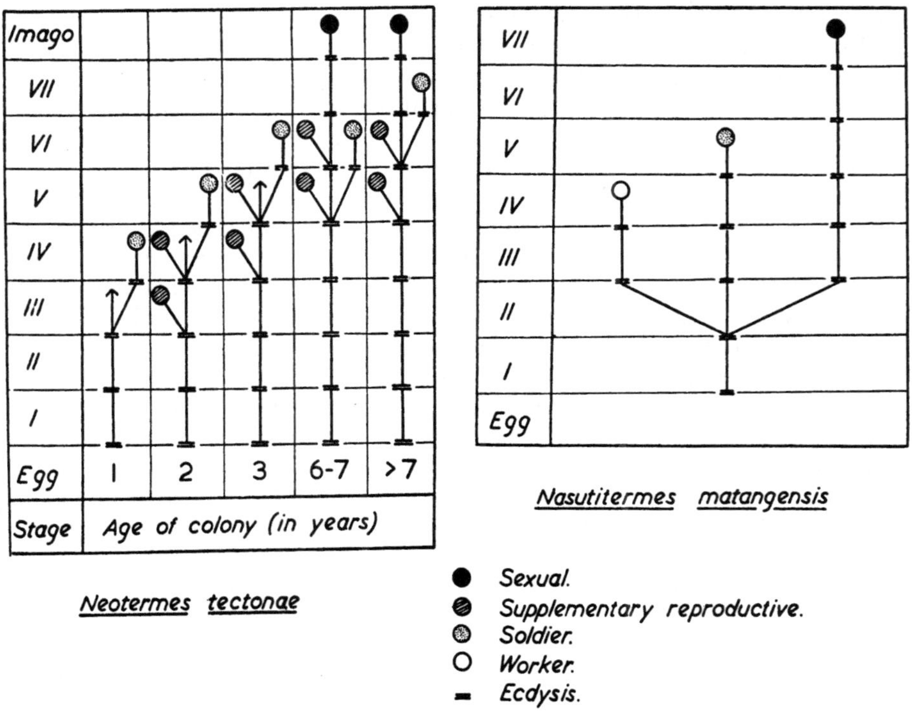

FIG. 295 Diagrammatic representation of caste-differentiation in two species of Isoptera (based on figures in section Isoptères of *Traité de Zoologie*, 9 (1949), under direction of P. P. Grassé, Masson & Cie, Éditeurs)

Grassé and Noirot, 1946*a*). Further, it has been demonstrated by several workers that nymphs which would normally have developed into sterile forms can give rise to supplementary reproductives in experimental colonies lacking these forms. Thus, Light and Illg (1945), working with *Zootermopsis* and Grassé and Noirot (1946*a*) with *Kalotermes flavicollis* showed that all nymphs of the fourth or later instars can differentiate into supplementary reproductives in artificial colonies from which the reproductives are removed as they appear. These results have been extended by later workers (e.g. Nagin, 1972) and comparable experiments with soldiers have shown that these, too, can differentiate from nymphs which would otherwise have developed into other castes (e.g. Light and Weesner, 1955; Grassé and Noirot, 1956). For these reasons it is considered that the newly hatched nymph of the lower families of termites is potentially capable of giving rise to any of the castes, the exact line it follows being determined by environmental factors within the colony. Pickens (1932) first suggested that the sexual forms secreted some substance (an 'ectohormone') which was absorbed, possibly by feeding stomodaeally, proctodaeally or on the products of exudatory glands, by the developing nymphs and inhibited their development into further sexuals. Similarly it was suggested that soldiers produced a substance inhibiting the differentiation of further soldiers. Light (1944), Keene and Light (1944) and Light and Weesner (1951) have reported extensive experiments on the effects of feeding extracts of supplementary reproductives to nymphal colonies of *Zootermopsis* and although the results were far from decisive, they concluded that the balance of evidence favoured the view that extracts of female supplementaries tended to inhibit the development of further supplementary reproductives. Morphogenetic substances of this kind are now known as pheromones (p. 142) and the work of Lüscher (1961) and others has shown that caste differentiation in *Kalotermes flavicollis* requires the combined action of several pheromones concerned in the differentiation of the royal pair from pseudergates. Not only do the king and queen produce inhibitory pheromones that prevent the development of further members of their own sex but another male-produced substance induces female pseudergates to develop into queens, while supernumerary royal males and females recognize each other through further pheromones and fight when they meet. The inhibitory pheromones pass directly from reproductive to pseudergate and circulate further through proctodaeal feeding (see below). The chemical nature of these pheromones is not yet known, but something can be said concerning the endocrine mechanisms that probably mediate their morphogenetic effects. Experimental injection of ecdysone into nymphs promotes their differentiation into sexual forms. This action is opposed by juvenile hormone or its analogues or by the implantation of active corpora allata, which lead either to stationary moults without further differentiation or to the production of soldiers or soldier-sexual intercastes (Lüscher, 1969; Lüscher and Springhetti, 1960; Lebrun, 1970; Wanyonyi, 1974). It is assumed that the inhibitory pheromones act on the endocrine system through the brain and its neuroendocrine cells.

The above results relate to *Kalotermes* and other relatively primitive forms. The situation in the Termitidae is not yet understood; supplementary reproductives seem, at least in some cases, to appear in colonies from which the royal pair is eliminated, but there is no evidence that the mechanism by which they are produced involves pheromones nor that it is subject to endocrine control. There is even evidence that reproductive and sterile castes may be distinct during the first instar in some Termitidae (Kaiser, 1956).

The Termite Community and its Biology

In order to describe the biology of the termite community it is convenient to begin with an account of the founding and development of colonies and then to discuss some aspects of the organization and maintenance of established colonies. Much of the extensive literature on this subject has been summarized in the general works cited on p. 606. Other important papers dealing with the biology of special groups include those of Emerson (1925, 1939), Bathellier (1927), Kemner (1934), Kofoid *et al.* (1934), Grassé (1937, 1944–45), Grassé and Noirot (1961), Bouillon (1964), Sands (1965), Bodot (1969) and Stuart (1972).

Colony-founding – The typical method of founding new colonies is by the emission of swarms of primary reproductives (Fuller, 1915; Grassé, 1942; Light and Weesner, 1948; Weesner, 1956; Williams, 1959; Becker, 1961; Nutting, 1966). Individuals of this caste appear in established colonies, usually at certain seasons, and accumulate in the colony for a short period before departing on their colonizing flight. During this period the gonads are not fully mature, especially in the female, the alates exhibit no sexual behaviour and in the primitive families they remain xylophagous. Prior to the flight the workers make exit holes where necessary in the walls of the termitarium and numerous members of this caste, along with soldiers, congregate around and often just outside these apertures while swarming is in progress. Males and females leave the nest in about equal numbers and may emerge in a continuous stream or a series of smaller batches. Swarming may be diurnal or nocturnal according to the species, those which swarm at night being frequently attracted to light. Swarming is often a seasonal occurrence of limited duration or may occur at times over a rather long period – e.g. *Reticulitermes lucifugus* swarms in S. France between May and August; many tropical species do so after the first rains of the rainy season. Many nests in the same area may produce swarms simultaneously but it is thought that most individuals eventually mate with another from their own nest. After leaving the nest the alates fly weakly for a short distance (Nutting, 1966); *Incisitermes minor* is said to be able to fly about 2 km but in many species less than 200 m are covered unless a light wind assists their flight. Attacks by birds, lizards and small mammals cause a high mortality among swarming alates but the survivors, on completing their flight, cast their wings and on encountering a deälate of the opposite sex show some epigamic behaviour before walking in tandem fashion (the male following the female) to seek a site where, in wood or the ground, they excavate a small nuptial chamber in which copulation takes place. At least in the higher termites, the whole series of acts beginning with the emission of the swarm and ending with the establishment of the royal pair in the nuptial chamber are closely linked together – e.g. alates captured before the end of the swarming flight fail to shed their wings and pair – but no detailed experimental analysis of the series has been made. External factors such as temperature and humidity probably exert some effect in deciding the time of swarming but conditions within the society are doubtless also of great importance.

Other methods of colony-founding have been described but are probably not very common. In *Reticulitermes*, which constructs a very diffuse nest, supplementary reproductives may develop among groups of nymphs remote, though apparently not completely isolated, from the parent couple and so give rise to secondary colonies (Pickens, 1932). It is said that in Italy *Reticulitermes lucifugus* forms new colonies only in this way, the alate forms never reproducing after swarming (Jucci, 1924). Grassé and Noirot (1951) have described the formation of new colonies of *Anoplotermes* and *Trinervitermes* through a large fraction of a colony, including all castes with even the primary founders, leaving the nest and breaking up into smaller groups which, if they lack the royal pair, develop supplementary reproductives. Finally, in *Microcerotermes amboinensis* and *Subulitermes undecimus*, Weyer (1930) has found that two or more pairs of primary reproductives may collaborate in starting a colony and Pickens (1932) says that the first offspring of several primary reproductives may be collected under one of the pairs in *Reticulitermes hesperus*.

Growth of the Colony – Quantitative data on the growth and size of termite populations are given by Bodenheimer (1937). Egg-production by the primary queen is low at first – 15–50 eggs may be laid in the first season, of which some are eaten by the parents. Later, the fecundity increases and mature queens of the Termitidae lay several thousand eggs per day (at least over short periods) though far fewer are produced by queens of the primitive families, e.g. *Kalotermes flavicollis* females reach their full egg-laying capacity after 3–4 years and then produce 4–6 eggs per day. The size attained by a mature colony is much smaller in the primitive genera, though *Mastotermes* sometimes forms exceptionally large societies. In *Kalotermes flavicollis* there are 15–20 members after one year and a maximum of 600–1000 individuals is reached after several years. On the other hand, colonies of the Termitidae may contain over a million individuals, all derived from a single royal pair. The individuals produced in the early stages of a colony are always of the sterile castes; alate reproductives develop later, e.g. after about four years in *Zootermopsis* (Heath, 1927) when the colony comprises about 450 individuals. During the growth of the colony in higher termites the king and queen remain in the royal cell – usually deep in the termitarium – where the queen undergoes the postmetamorphic growth described above and is attended by numerous workers who feed her and the king and devour the secretions produced by the exudate glands which she develops and those issuing from the gonopore and anus. The king lives as long as the queen and copulation takes place frequently. Colonies containing two pairs of primary reproductives or 2–3 queens with one king have been described, but are exceptional. In *Nasutitermes amboinensis* Weyer (1930*b*) found colonies containing primary queens of different ages but this species appears to be unusual in the readiness with which colonies adopt foreign queens and Weyer was able to produce colonies with several queens artificially.

Members of the sterile castes probably live for 2–4 years. The longevity of the reproductive forms is not known accurately; it may be 15–50 years in the higher forms but is shorter in primitive families. The long life of the individuals in a colony is obviously an important factor in permitting the close relationship between successive generations which is the basis of insect social organization. Since the death of the primary queen is normally followed by the development of one or many supplementary reproductives, colonies would seem to be potentially immortal, and colonies of some Termitidae have certainly been estimated to be 40–100 years old. However, Kalshoven (1930) considered that large colonies of *Neotermes tectonae* died out at the age of 15–16 years, having attained a maximum population of nearly 3000 individuals eight years previously, while smaller colonies died after 6–7 years. Newly developed supplementary queens lay more eggs than young primaries but never achieve the fecundity of mature primaries, though the development of many supplementaries to replace a lost primary may compensate for this. There are several records (e.g. in Hill, 1942) of primary and supplementary reproductives being found together in the same colony.

Few quantitative data are available on the proportions of the different castes present in a colony. On general grounds and from the very limited experimental work on artificial colonies (e.g. Miller, 1942; Grassé and Noirot, 1947) one would expect considerable powers of social regulation leading to approximately constant proportions of the various castes in well-established colonies. Emerson (1939) states that the proportion of soldiers varies from 3 to 16 per cent of the sterile castes and Bodot (1969) has demonstrated seasonal variations in the numbers of different castes.

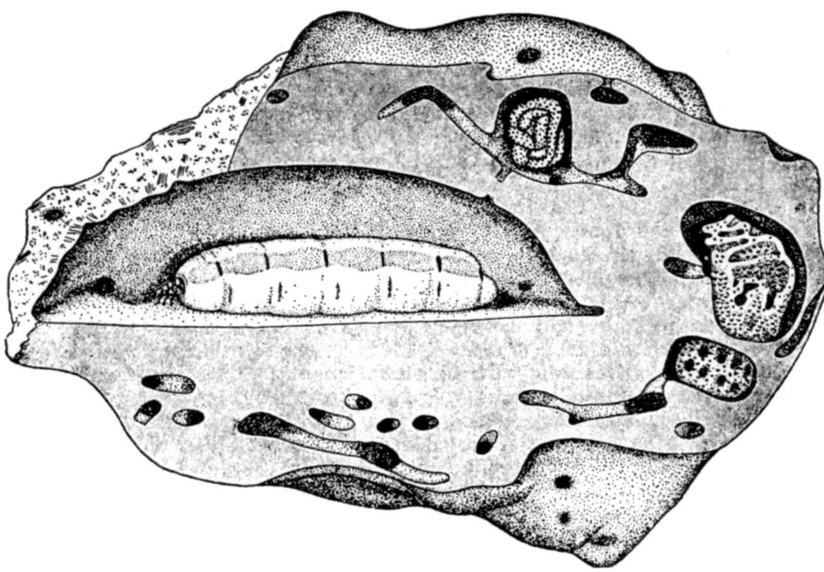

FIG. 296 Section of royal cell with queen of *Odontotermes transvaalensis*, × ½. On the right chambers with fungus gardens
After Sjöstedt.

Feeding Habits and Nutrition – The food requirements and feeding habits of termites display numerous interesting features but the fact that the different castes and developmental stages may have widely different alimentary regimes makes the subject rather complex (Cleveland, 1925*a*; Grassé and Noirot, 1945). Two main types of feeding habit may be distinguished: (*a*) feeding on sound or decaying wood or other plant materials such as humus, grass, fungi, etc. This is practised by the workers of all families which possess them, the young reproductive forms of the Kalotermitidae and the older nymphal stages of most termites, (*b*) feeding on a diet prepared by other members of the colony. This includes (i) stomodaeal feeding, in which a mixture of varying proportions of salivary secretions and regurgitated intestinal contents is received from another insect and (ii) proctodaeal feeding, in which a drop of the contents of the rectal pouch is obtained from the anus of another insect in response to tactile stimulation by the soliciting termite. While the forms feeding primarily on wood etc. may also consume these prepared foods, the latter are the sole diet of soldiers, all young nymphal stages, the entire brood of the subfamily Macrotermitinae (Termitidae), the older sexual forms of the primitive families and the reproductives of all ages in the Termitidae, the food being passed on to them by workers or old nymphs.

In all families except the Termitidae, those forms whose diet includes much cellulose-containing material harbour a rich fauna of Protozoa belong to the flagellate groups Polymastigina and Hypermastigina (Kirby, 1937; Steinhaus, 1946). In most cases all members of a given species of termite are associated with a characteristic protozoal fauna, although experiments to decide whether there is resistance on the part of the insect to infection by Protozoa normally found only in other species appear to have given conflicting results. Usually each species of termite contains only a single species of flagellate but some harbour two or three and occasionally ten or more may be found. At each nymphal moult the flagellates are lost and this is probably also so at the final moult which gives rise to the adult, though Grassé and Noirot (1945) claim that in *Reticulitermes* and possibly some others the flagellates are retained in the rectal pouch though its other contents are lost. After the moult the fauna is rapidly re-acquired, probably through proctodaeal feeding (Andrew, 1930).

It was shown by Cleveland (1924; 1925*b*) that the Protozoa could be removed artificially from the termites by a period of starvation, exposure to temperatures of about 36° C or, best of all, by exposure to a pressure of 3–4 atmospheres of oxygen. Experimentally defaunated termites lose weight and eventually die if they are prevented from acquiring a new fauna. The absence of a cellulase from the digestive enzymes of the insect coupled with the facts that some species of the protozoa can be seen to ingest wood particles and are capable of digesting cellulose *in vitro* indicate that they are essential because of their ability to supply the insects with the end-products of cellulose decomposition. Earlier workers considered that the main product of cellulose digestion by the Protozoa was glucose but the later researches of Hungate (1939; 1943) showed that *in vitro* degradation of cellulose by the Protozoa of *Zootermopsis* led to the appearance of carbon dioxide, hydrogen and simple organic acids (mainly acetic acid) and he considers that it is probably the acetic acid which is absorbed and metabolized by the termite. It should, however, be pointed out that preliminary experiments by Cook (1943) showed that the sodium salts of lactic and acetic acids killed the Protozoa in termites to which they were fed.

Nutrition in the Termitidae has been less well studied and it is not satisfactorily established how they are able to digest the cellulose in their diet. Anaerobic cellulose-digesting bacteria have been isolated from the gut of several Termitidae (Hungate, 1946; Pochon *et al.*, 1959; Misra and Ranganathan, 1954) though it is doubted whether they play as important a role as do the flagellates of the lower termites. Potts and Hewitt (1973) found in workers of *Trinervitermes* that most of the cellulase activity occurred in the mid gut and that some 40 per cent of this was located in the mid gut wall, thus suggesting that the cellulase is synthesized by the termites rather than by the bacteria which are also present. It should also be noted that the food of Termitidae often includes a relatively high proportion of fungal material or partially decomposed organic matter in the form of humus, so that cellulose may not be as important a component of their diet as it is of other families.

The habit of foraging outside the nest is found in various species of Termitidae and in the Hodotermitidae. The workers and soldiers of species of the latter family possess well-developed compound eyes, and exhibit the unusual habit of foraging above ground during daylight. Sorties are made from the nest to collect grass, pine needles, etc., which are cut into short lengths, and carried to the mouth of the burrow. Here the material is either taken directly in, or allowed to accumulate to form a mound whose contents are subsequently removed into the nest. The foraging habits of *Odontotermes latericus* and *Trinervitermes trinervius* in S. Africa are described by Fuller (1915). In the former species there are special cells or granaries within the nest, and lengths of green grass, together with large quantities of seeds are collected. *Nasutitermes triodiae*, in Australia, stores dried grass in chambers which are situated in the walls of the termitarium from the ground to the summit. Bugnion (1914) describes the habits of the 'black termite' (*Hospitalitermes monoceros*) of Ceylon. Long dense files of workers of this species set out about sunset, with the soldiers lined up on guard on either side of the procession. The object of these expeditions is to gather fragments of lichens which serve to nourish the young. Having found a suitable tree,

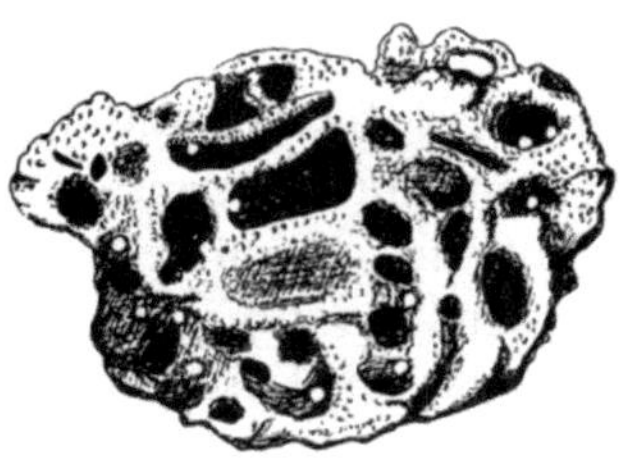

FIG. 297
Fungus bed of *Odontotermes*, India: the small white fungal spheres are seen growing on the substratum. × 2

they remain foraging the whole night and return the following morning. Bugnion estimated that there are 1000 termites to each metre of the moving column and about 300 000 termites involved in the procession.

The habitations of the Macrotermitinae (Termitidae) contain what are commonly termed 'fungus gardens' (Grassé, 1944–45; Sands, 1960). These beds are composed of a spongy dark reddish-brown coral-like 'comb' which is constructed by workers from small balls of comminuted vegetable matter, probably faecal pellets, aggregated into masses varying in size from about 2 to 20 cm across and providing a substrate on which fungal hyphae grow (Fig. 297). The chambers containing the fungus gardens are scattered throughout the nest, sometimes more or less concentrated near the royal cell. Two genera of fungi occur in these gardens – the Ascomycete *Xylaria* which, though not confined to termitaria, constantly occurs there but does not produce fructifications in inhabited nests, and a Basidiomycete, *Termitomyces*, which is exclusively termitophile and produces on the surface of the bed small white spheres composed of bundles of hyphae ending in swollen cells (Heim, 1940; 1942). These spheres form part of the food of workers and are mixed with the stomodaeal food fed to nymphs. It seems likely that they serve as a source of vitamins and organic nitrogen; they do not constitute the principal food.

Termitophiles – In addition to the normal occupants of a termite nest, there is also a very extensive termitophilous fauna consisting of various insects, and other arthropods, which are represented in almost every community by one or more species. The relations between these guests and their hosts are similar to those between myrmecophilous species and ants. The termitophilous forms include true guests or symphiles, indifferently tolerated guests or synoeketes, and synechthrans which are scavengers or predators; but see Delamare-Debouteville (1948) for a somewhat different grouping. The largest number of termitophilous insects belong to the Coleoptera. The Carabidae are principally represented by the larvae of *Orthogonius*; the Staphylinidae include such genera as *Corotoca*, *Spirachtha*, *Termitobia*, *Termitomimus* and *Doryloxenus*, while the Pselaphidae, Scarabaeidae, Tenebrionidae and other families have sundry representatives. Among the Diptera are certain remarkable Phoridae including *Termitoxenia*, *Termitomyia* and *Ptochomyia*; also the equally remarkable Psychodid *Termitomastus*, and several genera of larval Anthomyidae. The Thysanura include a large number of termitophilous forms, there are also about 50 species of Collembola, several larval Tineids and, among the Hemiptera, the anomalous genus *Termitaphis*. In addition to insects the list includes Acarina, Diplopoda and Chilopoda. The literature on termitophilous arthropods is extensive and is principally comprised in numerous papers by Wasmann (1894 onwards; 1934), Silvestri (1903; 1905; 1914–20), Seevers (1957) and others listed in the select bibliography of Wilson (1971). Termite nests also afford shelter to lizards, snakes and scorpions, while in Trinidad a parakeet (*Forpus passerinus*) usually nests in them.

It is noteworthy that more than one species of termite may live in the same habitation and that a kind of social symbiosis exists in consequence. Thus some Apicotermitinae which have no soldiers, are often associated with species of other genera. In S. America five species of termites, belonging to as many different genera, are recorded by Holmgren as sharing a habitation of *Syntermes dirus*, while no less than eight different species are mentioned by Escherich as living amicably with *S. chaquimayensis*. Certain members of the genus *Nasutitermes* particularly exhibit this habit of guest species. It should be noted, however, that termite colonies are generally hostile to individuals of the same or different species, that very few cases are known of termites found only in association with other species and that where two or more species occur together they may occupy gallery systems which are distinct. Termites and ants have often been recorded as inhabiting the same log, or other object, where they may occupy contiguous galleries or even intermingle. Under ordinary circumstances the relations between the two kinds of insects are friendly, unless the nest be disturbed, when the ants soon attack and carry off the termites.

The Habitations of Termites

General accounts of termite nests are given by Emerson (1938) and the general works cited on p. 606. The simplest kind of habitation is found in the wood-feeding species, which usually lack the worker caste and include the most primitive members of the order. *Archotermopsis* and *Zootermopsis*, for example, live in moist decaying trunks and logs of conifers. Their abodes consist of nothing more than a series of galleries, excavated in the wood, without any external manifestation of their presence. Other genera such as *Mastotermes, Kalotermes, Neotermes* and *Cryptotermes* include species which bore into dry wood, often selecting posts and other structures, or furniture in buildings, as the seat of their habitations. *Neotermes militaris* and *N. greeni* are

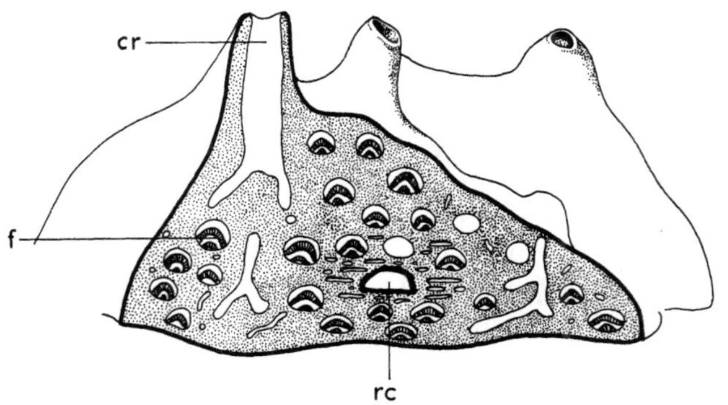

FIG. 298 Vertical section through nest of *Protermes minutus*, measuring 2 m across (*after* Grassé, 1949)

cr, crater-like opening; *f*, chamber containing fungal comb; *rc*, royal cell.

destructive to tea in Ceylon where they burrow in the stems of the bushes. *Rhinotermes, Reticulitermes* and *Coptotermes* live in the ground and infest wood indirectly through the soil; they are exceedingly injurious to any woodwork of buildings in contact with the ground and often issue above ground to obtain access to woodwork. With this object in view they construct covered passage-ways of earth or faecal matter, which enable them to work concealed from the light and enemies, while surrounded by the requisite humidity. They are able, by means of these runways to leave their underground chambers and reach the upper storeys of buildings or ascend lofty trees.

In other cases very extensive structures known as termitaria (Figs 298, 299) are constructed, particularly by the African and Australian Termitidae. These termite mounds are built of earth excavated in making subterranean chambers and were perhaps originally only a convenient method of disposing of this material. The outer walls, passages and royal cells are composed of earth particles cemented together to form a hard brick-like substance. The agglutinating fluid appears to consist of saliva and excrement. The inner galleries, where the brood is contained, are of a softer consistency, and are composed of woody or other comminuted material which has passed through the alimentary canal. Some of the most remarkable of all termitaria are the lofty steeple-like structures constructed by *Nasutitermes triodiae* in Northern Australia. They and the nests of *Macrotermes* exceed in size those of any other termites, and may measure over 8 m in height. The greater bulk of the earth and sand used in their formation is collected on the surface, and not mined from below. The interior of such a termitarium presents a maze of irregular chambers and passages, and its walls are so resistant that it is difficult to make any impression upon them even with a sharp pick. The

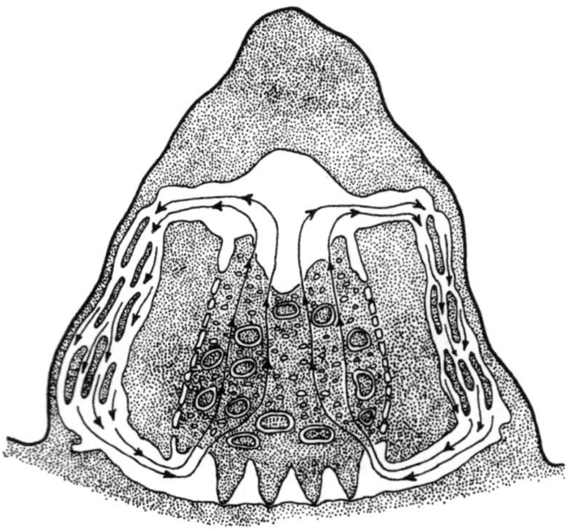

FIG. 299
Vertical section through a large nest of the African *Macrotermes natalensis* to illustrate regulation of temperature and carbon dioxide content of the atmosphere (*after* Wilson and Lüscher). The colony occupies the central part of the nest and air rises by convection through these galleries, reaching a temperature of 30° C. and a CO_2 content of 2·7 per cent. In the space above the colony the air flows radially outwards, then descends through flat channels near the surface of the nest, where it becomes cooled to 25° C. by radiation and where the CO_2 content is reduced by diffusion to less than 1 per cent. The cooler, fresher air then passes inwards towards the bottom of the colony area to be recirculated.

'compass' or 'meridional' termite (*Omitermes meridionalis*) is widely distributed in Australia. Its nests may attain a height of over 3 m and are flattened from side to side in such a manner that the broad sides face east and west, and the narrow ends north and south. It has been suggested that they are built this way to minimize heat-gain at mid-day or to secure the maximum of desiccation, and to allow the repairs, which are made during the wet season, to dry and harden as speedily as possible.

Other species of termites live in the ground, without constructing termitaria above the surface, or only forming small mound-like structures (Fig. 299). Many termites which exhibit this habit are injurious to the roots of grass, field crops, and other vegetation. Although the type of habitation may be very constant for a particular genus or species, in other cases considerable variation occurs. *Odontotermes*, for example, includes both mound builders and subterranean forms, and the two habits may be exhibited in the same species as in the common Indian termite *O. obesus*. Some species of *Nasutitermes* construct gigantic termitaria of the type already referred to while others form arboreal habitations often more or less spherical in form.

The material used in constructing the latter appears to be comminuted wood, and the nest is composed of an outer envelope enclosing a comb-like mass of internal chambers. Such habitations bear a superficial resemblance to the carton nests of arboreal Vespidae. In many cases they are connected by covered passage-ways with subterranean abodes.

Subterranean and mound-building termites exert various complex influences on the soil and plant growth, some beneficial to agriculture and some adverse (Lee and Wood, 1971). Actively occupied mounds and the areas around their base may be without vegetation, though a characteristic flora can develop on old, eroded mounds. Some species promote soil erosion by removing plant cover and perhaps by reducing soil organic matter.

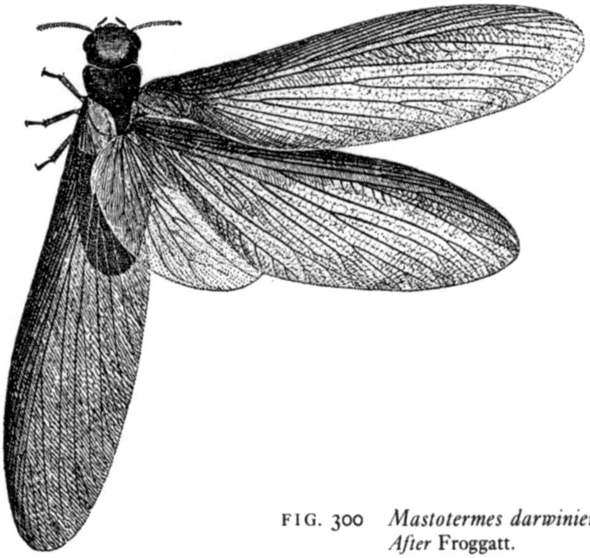

FIG. 300 *Mastotermes darwiniensis* male, right wings extended *After* Froggatt.

Wood-eating species are important agents in decomposing branches, logs and tree-stumps, though the role of termites in cycling organic matter and soil nutrients has not been sufficiently studied; it may be important locally under less intensive systems of agricultural management. The presence of subterranean termite galleries increases soil porosity but this effect is counteracted by the way in which soil particles may be cemented into large structures that are very resistant to erosion or destruction and prevent the penetration of water to the underlying soil.

Affinities and Phylogeny – The earliest known fossil termite is the Cretaceous *Cretatermes carpenteri* (Emerson, 1967). This is a Hodotermitid but the Tertiary termite fauna includes many Kalotermitidae and almost all the known Mastotermitidae (Emerson, 1965, 1969). It has long been known that *Mastotermes darwiniensis* shows many striking similarities to the Blattaria. These include the perforate tentorium, the wing-venation, the ootheca-like egg-mass, mycetocytes and the structure of the proventriculus and ovipositor. In particular, the Blattarian

Cryptocercus is in some respects structurally closer to *Mastotermes* than to other cockroaches (McKittrick, 1965). *Cryptocercus* also harbours symbiotic flagellates very similar to those of termites (Cleveland *et al.*, 1934; see also Nutting, 1956). There is therefore little doubt that the Isoptera arose from cockroach-like ancestors which had begun to adopt a social mode of life. Within the Isoptera the major evolutionary steps have been the development of a worker caste and large colonies in the Rhinotermitidae, Hodotermitidae and Termitidae. The last two families in particular have radiated into a wide variety of habitats to exploit the most diverse sources of cellulose-containing plant material.

Classification

The foundations of the modern classification of termites were laid by Desneux (1904), Silvestri (1909) and particularly Holmgren (1909–13). About 1900 species are now recognized, most of them catalogued by Snyder (1949). Among the more recent taxonomic works are those by Banks and Snyder (1920), Emerson (1925, 1965, 1969), Sjöstedt (1926), Bathellier (1927), Kemner (1934), Grassé (1937), Hill (1942), Ahmad (1950), Krishna (1961) and Sands (1965, 1972). The families listed below are those recognized by Weidner (1970); some authorities include the Termopsidae in the Hodotermitidae.

FAM. MASTOTERMITIDAE. *Hind wing with large anal lobe; tarsi 5-segmented; fontanelle absent; ocelli present; pseudergates, but no true workers.* An archaic family with only one Recent species, *Mastotermes darwiniensis* from N. Australia (Fig. 300) and some fossil forms (Emerson, 1965).

FAM. KALOTERMITIDAE (Dry-wood termites). *Hind wing without anal lobe; tarsi 4-segmented; fontanelle absent; ocelli present; pronotum flat, usually broader than head; anterior wing-scales large; pseudergates, but no true workers.* Over 250 species. *Neotermes, Kalotermes* and *Glyptotermes* are important genera. See Emerson (1969), Krishna (1961).

FAM. TERMOPSIDAE (Damp-wood termites). *Hind wings without anal lobe; tarsi apparently 4-segmented when viewed from above; fontanelle absent; ocelli absent; pronotum flat, narrower than head; anterior wing-scales large; pseudergates, but no true workers.* A relict group of about 15 species. *Zootermopsis, Stolotermes, Porotermes.*

FAM. HODOTERMITIDAE (Harvester termites). *Hind wing without anal lobe; tarsi 4-segmented; fontanelle and ocelli absent; all castes with compound eyes; pronotum saddle-shaped, narrow; anterior wing-scales short; workers present.* A small group of subterranean species. *Hodotermes, Microhodotermes, Anacanthotermes.*

FAM. RHINOTERMITIDAE. *Hind wing without anal lobe; tarsi 4-segmented; fontanelle present; ocelli present; pronotum flat; wings often reticulate, without hairs; anterior wing-scales usually long; workers present; almost all subterranean.* Representative genera: *Psammotermes, Reticulitermes, Coptotermes, Termitogeton, Rhinotermes.*

FAM. SERRITERMITIDAE. Includes only *Serritermes serrifer* from Brazil, with unusual falcate mandibles in adult and worker. This species has been variously placed in the Kalotermitidae, Rhinotermitidae and Termitidae.

FAM. TERMITIDAE. *Hind wing without anal lobe; venation reduced; tarsi 4-segmented; fontanelle present; ocelli present; pronotum of workers and soldiers narrow, with raised anterior lobe; wings not or only slightly reticulate; wing membrane more or less hairy; anterior wing-scales short; workers present; all ground dwelling with wide range of food habits and colony structure.* Includes about three-quarters of all Recent species of Isoptera. Representative genera: *Anoplotermes, Termes, Capritermes, Macrotermes, Odontotermes, Microtermes, Nasutitermes, Trinervitermes.*

Literature on the Isoptera

ADAMSON, A. M. (1940), New termite intercastes, *Proc. roy. Soc.*, (B), **129**, 35–53.

AHMAD, M. (1950), The phylogeny of termite genera based on imago-worker mandibles, *Bull. Am. Mus. nat. Hist.*, **95**, 39–86.

ALIBERT-BERTHOT, J. (1969), La trophallaxie chez le termite à cou jaune, *Calotermes flavicollis* Fabr., etudiée à l'aide de radio-éléments, I, *Ann. Sci. nat.* (*Zool.*), **11** (2), 235–325.

ANDREW, B. J. (1930), Method and rate of protozoan refaunation in the termite *Termopsis angusticollis* Hagen, *Univ. Calif. Publs Zool.*, **33**, 449–470.

BANKS, N. AND SNYDER, T. E. (1920), A revision of the Nearctic termites, *Bull. U.S. nat. Mus.*, **108**, 1–228.

BATHELLIER, J. (1927), Contribution à l'étude systématique et biologique des termites de l'Indochine, *Faun. Colon. fr.*, **1**, 125–365.

BECKER, G. (1961), Beobachtungen und Versuche über den Beginn der Kolonie-Entwicklung von *Nasutitermes ephratae* Holmgren (Isoptera), *Z. angew. Ent.*, **49**, 78–93.

BODENHEIMER, F. S. (1937), Population problems of social insects, *Biol. Rev.*, **12**, 393–403.

BODOT, P. (1969), Composition des colonies de termites: ses fluctuations au cours de temps, *Insectes soc.*, **26**, 39–54.

BORDEREAU, C. (1971*a*), Dimorphisme sexuel du système trachéen chez les imagos ailés de *Bellicositermes natalensis* Haviland (Isoptera, Termitidae): rapports avec la physogastrie de la reine, *Arch. Zool. exp. gén.*, **112**, 33–54.

—— (1971*b*), Le système trachéen de la reine physogastre et du roi chez *Bellicositermes natalensis* Haviland, (Isoptera, Termitidae), *Arch. Zool. exp. gén.*, **112**, 747–760.

BOUILLON, A. (ed.) (1964), *Études sur les Termites Africains*, Masson et Cie., Paris, 414 pp.

BROWMAN, L. G. (1935), The chitinous structures in the posterior abdominal segments of certain female termites, *J. Morph.*, **57**, 113–129.

BUCHLI, H. (1958), L'origine des castes et les potentialités ontogéniques des termites européens du genre *Reticulitermes* Holmgren, *Ann. Sci. nat.* (*Zool.*), (11) **20**, 263–429.

BUGNION, E. (1914), La biologie des termites de Ceylan, *Bull. Mus. Hist. nat. Paris*, **1914**, 170–204.

BUGNION, E. AND POPOFF, N. (1912), Anatomie de la reine et du roi termite (*Termes redemanni, obscuriceps* et *horni*), *Mém. Soc. zool. Fr.*, **25**, 210–231.

CLEVELAND, L. R. (1924), The physiological and symbiotic relationship between the intestinal protozoa of termites and their host, with special reference to *Reticulitermes flavipes* Kollar., *Biol. Bull.*, **46**, 203–227.

—— (1925*a*), The feeding habits of termite castes and its relationship to their intestinal flagellates, *Biol. Bull.*, **48**, 295–306.

—— (1925*b*), The effects of oxygenation and starvation on the symbiosis between the termite, *Termopsis*, and its intestinal flagellates, *Biol. Bull.*, **48**, 309–325.

CLEVELAND, L. R., HALL, S. R., SANDERS, E. P. AND COLLIER, J. (1934), The wood-feeding roach, *Cryptocercus*, its Protozoa, and the symbiosis between Protozoa and roach, *Mem. Am. Acad. Art Sci.*, **17**, 185–342.

COOK, S. F. (1943), Non-symbiotic utilization of carbohydrate by the termite *Zootermopsis angusticollis, Physiol. Zool.*, **16**, 123–128.

DELAMARE-DEBOUTTEVILLE, C. (1948), Recherches sur les collemboles termitophiles et myrmecophiles (écologie, éthologie, systématique), *Archs Zool. exp. gén.*, **85**, 261–425.

DELIGNE, J. (1965), Morphologie et fonctionnement des mandibules chez les soldats des termites, *Biol. Gabon.*, **1**, 179–186.

DESNEUX, J. (1904), Isoptera. Fam. Termitidae, *Genera Insect.*, Fasc. **25**, 1–52.

EMERSON, A. E. (1925), The termites of Kartabo, Bartica District, British Guiana, *Zoologica, N.Y.*, **6**, 191–459.

—— (1938), Termite nests – a study of the phylogeny of behavior, *Ecol. Monogr.*, **8**, 247–284.

—— (1939), Populations of social insects, *Ecol. Monogr.*, **9**, 287–300.

—— (1965), A review of the Mastotermitidae (Isoptera) including a new fossil genus from Brazil, *Am. Mus. Novit.*, **2236**, 1–46.

—— (1967), Cretaceous insects from Labrador. 3. A new genus and species of termite (Isoptera: Hodotermitidae), *Psyche*, **74**, 276–289.

—— (1969), A revision of the Tertiary fossil species of the Kalotermitidae (Isoptera), *Am. Mus. Novit.*, **2359**, 1–57.

ERNST, E. (1959), Beobachtungen beim Spritzakt der *Nasutitermes*-Soldaten, *Rev. suisse Zool.*, **66**, 289–295.

FEYTAUD, J. (1912), Contribution à l'étude du termite lucifuge (anatomie, fondation des colonies nouvelles), *Archs Anat. microsc. Morph. exp.*, **13**, 481–607.

FULLER, C. (1915), Observations on some South African termites, *Ann. Natal Mus.*, **3**, 329–504.

—— (1919), The wing venation and respiratory system of certain South African termites, *Ann. Natal Mus.*, **4**, 19–102.

—— (1924), The thorax and abdomen of winged termites with special reference to the sclerites and muscles of the thorax, *Union S. Afr. Dept. Agric. Ent. Mem.*, **2**, 49–78.

GRASSÉ, P. P. (1937), Recherches sur la systématique et la biologie des termites de l'Afrique et la biologie des termites de l'Afrique occidentale française, *Ann. Soc. ent. Fr.*, **106**, 1–100.

—— (1942), L'essaimage des termites. Essai d'analyse causale d'un complexe instinctif, *Bull. biol. Fr. Belg.*, **76**, 347–382.

—— (1944–45), Recherches sur la biologie des termites champignonnistes (Macrotermitinae), *Ann. Sci. nat.*, **6**, 97–171; **7**, 115–146.

—— (1949), Ordre des Isoptères ou termites, In: Grassé's *Traité de Zoologie*, **9**, 408–544.

GRASSÉ, P. P. AND BONNEVILLE, P. (1935), Les sexués inutilisés ou achrestogonimes des Protermitides, *Bull. biol. Fr. Belg.*, **69**, 474–491.

GRASSÉ, P. P. AND NOIROT, C. (1945), La transmission des flagellés symbiotiques et les aliments des termites, *Bull. Soc. biol. Fr. Belg.*, **79**, 273–292.

—— (1946*a*), La production des sexués néoténiques chez le termite à cou jaune (*Calotermes flavicollis* F.): inhibition germinale et inhibition somatique, *C.r. Acad. Sci. Paris*, **223**, 569–571.

—— (1946*b*), Le polymorphisme social du termite à cou jaune (*Calotermes flavicollis* F.), La production des soldats, *C.r. Acad. Sci. Paris*, **223**, 929–931.

—— (1947), Le polymorphisme social du termite à cou jaune (*Calotermes flavicollis* F.), Les faux ouvriers ou pseudergates et les mues regressives, *C.r. Acad. Sci. Paris*, **224**, 219–221.

—— (1951), La sociotomie: migration et fragmentation de la termitière chez les *Anoplotermes* et les *Trinervitermes*, *Behaviour*, **3**, 146–166.

—— (1956), *Apicotermes arquieri* (Isoptère): ses constructions, sa biologie. Considérations générales sur la sous-famille des Apicotermitinae nov., *Annls Sci. nat.* (*Zool.*), (11), **16**, 345–388.

—— (1961), Nouvelles recherches sur la systématique et l'éthologie des termites champignonnistes du genre *Bellicostitermes* Emerson, *Insectes soc.*, **8**, 311–359.

GUPTA, S. D. (1960), Morphology of the primitive termite *Anacanthotermes macrocephalus* (Desneux) (Isoptera: Hodotermitidae), Parts I, II, *Rec. Indian Mus.*, **58**, 169–194; 195–222.

HARE, L. (1934), Caste determination and differentiation with special reference to the genus *Reticulitermes* (Isoptera), *J. Morph.*, **56**, 267–294.

HARRIS, W. V. (1971), *Termites: their Recognition and Control*, Longmans, London, 2nd edn, 187 pp.

HEATH, H. (1927), Caste formation in the termite genus *Termopsis*, *J. Morph.*, **43**, 387–419.

HECKER, H. (1966), Das Zentralnervensystem des Kopfes und seine postembryonale Entwicklung bei *Bellicositermes bellicosus* (Smeath.) (Isoptera), *Acta Trop.*, **23**, 297–352.

HEIM, R. (1940), Études descriptives et expérimentales sur les Agarics termitophiles d'Afrique tropicale, *Mém. Acad. Sci. Inst. Fr.*, **64**, 1–74.

—— (1942), Nouvelles études descriptives sur les agarics termitophiles d'Afrique tropicale, *Archs Mus. natn. Hist. nat. Paris*, **18**, 107–166.

HEWITT, P. H., NEL, J. J. C. AND CONRADIE, S. (1969*a*), The rôle of chemicals in communication in the harvester termites *Hodotermes mossambicus* (Hagen) and *Trinervitermes trinervoides* (Sjöstedt), *Insectes soc.*, **16**, 79–86.

—— (1969*b*), Preliminary studies on the control of caste formation in the harvester termite *Hodotermes mossambicus* (Hagen), *Insectes soc.*, **16**, 159–172.

HILL, G. F. (1925), Notes on *Mastotermes darwiniensis* Froggatt (Isoptera), *Proc. R. Soc. Vict.*, **37**, 119–124.

—— (1942), *Termites (Isoptera) from the Australian Region*, Melbourne, 479 pp.

HOLMGREN, N. (1909–13), Termitenstudien I–IV, *K. svenska VetenskAkad. Handl.*, **44**, 1–215; **46**, 1–88; **48**, 1–166; **50**, 1–276.

HOWSE, P. E. (1968), On the division of labour in the primitive termite *Zootermopsis nevadensis* (Hagen), *Insectes soc.*, **15**, 45–50.

HOWSE, P. E. (1970), *Termites: a study in social behaviour*, London, 150 pp.

HUDSON, G. B. (1946), A study of the tentorium in some Orthopteroid Hexapoda, *J. ent. Soc. sthn Afr.*, **8**, 71–90.

HUMMEL, H. AND KARLSON, P. (1968), Hexansäure als Bestandteil des Spurpheromons der Termite *Zootermopsis nevadensis* Hagen, *Hoppe-Seyler's Z. physiol. Chem.*, **349**, 725–727.

HUNGATE, R. E. (1936–39), Studies on the nutrition of *Zootermopsis*, I–III, *Zentbl. Bakt. ParasitKde*, **94**, 240–249; *Ecology*, **19**, 1–25; **20**, 230–245.

—— (1943), Quantitative analyses on the cellulose fermentation by termite Protozoa, *Ann. ent. Soc. Am.*, **36**, 730–739.

—— (1946), The symbiotic utilization of cellulose, *J. Elisha Mitchell scient. Soc.*, **62**, 9–24.

IMMS, A. D. (1919), On the structure and biology of *Archotermopsis*, together with descriptions of new species of intestinal Protozoa and general observations on the Isoptera, *Phil. Trans. R. Soc. Ser. B*, **209**, 75–180.

JUCCI, C. (1924), Su la differenziazione de la caste nella società dei Termitidi, *Mem. Accad. Lincei*, (5), **14**, 269–500.

KAISER, P. (1954), Über die Funktion der Mandibeln bei den Soldaten von *Neocapritermes opacus* (Hagen), *Zool. Anz.*, **152**, 228–234.

—— (1956), Die Hormonalorgane der Termiten im Zusammenhang mit der Entstehung ihrer Kasten, *Mitt. zool. StInst. Hamb.*, **54**, 129–178.

KALSHOVEN, L. G. E. (1930), Bionomics of *Kalotermes tectonae* Damm. as a basis for its control, *Meded. Inst. Plantenziekten*, **76**, 154 pp.

KEENE, S. A. AND LIGHT, S. F. (1944), Results of feeding ether extracts of male supplementary reproductives to groups of nymphal termites, *Univ. Calif. Publs Zool.*, **49**, 283–290.

KEMNER, M. A. (1934), Systematische und biologische Studien über die Termiten Javas und Celebes, *K. svenska VetenskAkad. Handl.*, (3), **13** (4), 241 pp.

KIRBY, H. (1937), Host-parasite relations in the distribution of protozoa in termites, *Univ. Calif. Publs Zool.*, **41**, 189–212.

KOCH, A. (1938), Symbiosestudien III: Die intracellulare Bakteriensymbiose von *Mastotermes darwiniensis* Froggatt (Isoptera), *Z. Morph. Ökol. Tiere*, **34**, 584–609.

KOFOID, C. A., *et al.* (1934), *Termites and Termite Control*, Univ. Calif. Press, Berkeley, 734 pp.

KOVOOR, J. (1968), L'intestin d'un termite supérieur (*Microcerotermes edentatus* Wasmann, Amitermitinae). Histophysiologie et flore bactérienne symbiotique, *Bull. biol. Fr. Belg.*, **102**, 45–84.

—— (1969, 1971), Anatomie comparée du tube digestif des termites. II, III, *Insectes soc.*, **16**, 195–234; **18**, 49–70.

KRISHNA, K. (1961), A generic revision and phylogenetic study of the family Kalotermitidae (Isoptera), *Bull. Am. Mus. nat. Hist.*, **122**, 303–408.

KRISHNA, K. AND WEESNER, F. M. (1969–70), *Biology of Termites*, New York, 2 vols. Vol. **I**, 1969, 598 pp.; Vol. **II**, 1970, 643 pp.

LEBRUN, D. (1970), Intercastes expérimentaux de *Calotermes flavicollis* Fabr., *Insectes soc.*, **17**, 159–176.

LEE, K. E. AND WOOD, T. G. (1971), *Termites and Soils*, Academic Press, London and New York, 252 pp.

LIGHT, S. E. (1942–43), The determination of the castes of social insects, *Quart. Rev. Biol.*, **17**, 312–326; **18**, 46–63.

—— (1944), Parthenogenesis in termites of the genus *Zootermopsis, Univ. Calif. Publs Zool.*, **43**, 405–412.

—— (1944*a*), Experimental studies on ectohormonal control of the development of supplementary reproductives in the termite genus *Zootermopsis* (formerly *Termopsis*), *Univ. Calif. Publs Zool.*, **43**, 413–454.

LIGHT, S. F. AND ILLG, P. L. (1945), Rate and extent of development of neotenic reproductives in groups of nymphs of the termite genus *Zootermopsis, Univ. Calif. Publs Zool.*, **53**, 1–40.

LIGHT, S. F. AND WEESNER, F. M. (1948), Biology of Arizona termites with emphasis on swarming, *Pan-pacif. Ent.*, **24**, 54–68.

—— (1951), Further studies in the production of supplementary reproductives in *Zootermopsis* (Isoptera), *J. exp. Zool.*, **117**, 397–414.

—— (1955), The production and replacement of soldiers in incipient colonies of *Tenuirostritermes tenuirostris* (Desneux), *Insectes soc.*, **2**, 135–146.

LÜSCHER, M. (1961), Social control of polymorphism in termites, *Symp. R. ent. Soc. Lond.*, **1**, 57–67.

—— (1969), Die Bedeutung des Juvenilhormons für die Differenzierung der Soldaten bei der Termite *Kalotermes flavicollis, Proc. 6th Congr. Int. Union Study Social Insects*, **1969**, 165–170.

LÜSCHER, M. AND SPRINGHETTI, A. (1960), Untersuchungen über die Bedeutung der Corpora Allata für die Differenzierung der Kasten bei der Termite *Kalotermes flavicollis* F., *J. Insect Physiol.*, **5**, 190–212.

MATSUMURA, F., COPPEL, H. C. AND TAI, A. (1968), Isolation and identification of termite trail-following pheromone, *Nature*, **219**, 963–964.

MCKITTRICK, F. A. (1965), A contribution to the understanding of cockroach-termite affinities, *Ann. ent. Soc. Am.*, **58**, 18–22.

MCMAHAN, E. A. (1970), Polyethism in workers of *Nasutitermes costalis* (Holmgren), *Insectes soc.*, **17**, 113–120.

MILLER, E. M. (1942), The problem of castes and caste differentiation in *Prorhinotermes simplex* Hag., *Bull. Univ. Miami*, **15**, 1–27.

MISRA, J. N. AND RANGANATHAN, V. (1954), Digestion of cellulose by the mound building termite *Termes* (*Cyclotermes*) *obesus* (Rambur), *Proc. Indian Acad. Sci.*, **39**, 100–113.

MOORE, B. P. (1964), Volatile terpenes from *Nasutitermes* soldiers (Isoptera, Termitidae), *J. Insect Physiol.*, **10**, 371–375.

—— (1968), Studies on the chemical composition and function of the cephalic gland secretion in Australian termites, *J. Insect Physiol.*, **14**, 33–49.

MORGAN, F. D. (1959), The ecology and external morphology of *Stolotermes ruficeps* Brauer (Isoptera: Hodotermitidae), *Trans. R. Soc. N.Z.*, **86**, 155–195.

NAGIN, R. (1972), Caste determination in *Neotermes jouteli* (Banks), *Insectes soc.*, **19**, 39–61.

NOIROT, C. (1955), Recherches sur le polymorphisme des termites supérieurs (Termitidae), *Annls Sci. nat.* (*Zool.*), (2), **17**, 400–595.

—— (1956), Les sexués de remplacement chez les termites supérieurs (Termitidae), *Insectes soc.*, **3**, 145–158.

NOIROT, C. AND KOVOOR, J. (1958), Anatomie comparée du tube digestif des termites, I, *Insectes soc.*, **5**, 439–471.

NOIROT, C. AND NOIROT-TIMOTHÉE, C. (1965), La glande sternale dans l'évolution des termites, *Insectes soc.*, **12**, 265–272.

—— (1967), L'épithélium absorbant de la panse d'un termite supérieur, Ultrastructure et rapport avec la symbiose bactérienne, *Annls Soc. ent. Fr.*, 3, (N.S.), 577–592.

—— (1969), La cuticule proctodéale des Insectes. 1. Ultrastructure comparée, *Z. Zellforsch.*, **101**, 477–509.

NOIROT-TIMOTHÉE, C. AND NOIROT, C. (1965), L'intestin moyen chez la reine des termites supérieurs. Étude au microscope électronique, *Annls Sci. nat.*, 7, 185–208.

NUTTING, W. L. (1956), Reciprocal transformation between the roach, *Cryptocercus*, and the termite, *Zootermopsis*, *Biol. Bull. mar. biol. Lab., Woods Hole*, **110**, 83–90.

—— (1966), Colonizing flights and associated activities of termites. I. The desert damp-wood termite *Paraneotermes simplicicornis* (Kalotermitidae), *Psyche*, 73, 131–149.

PFLUGFELDER, O. (1938), Untersuchungen über die histologischen Veränderungen und das Kernwachstum der Corpora allata von Termiten, *Z. wiss. Zool.*, **150**, 451–467.

PICKENS, A. L. (1932), Observations on the genus *Reticulitermes* Holmgren, *Pan-Pacif. Ent.*, **8**, 178–180.

PLATANIA, R. (1938), Richerche sulla struttura del tubo digirente di *Reticulitermes lucifugus* (Rossi) con particolare riguardo alla natura, origine e funzione della peritrofica, *Archo zool. ital.*, **25**, 297–328.

POCHON, J., BARJAC, H. de AND ROCHE, A. (1959), Recherches sur la digestion de la cellulose chez le termite *Sphaerotermes sphaerothorax*, *Annls Inst. Pasteur, Paris*, **96**, 352–355.

POTTS, R. C. AND HEWITT, P. H. (1973), The distribution of intestinal bacteria and cellulase activity in the harvester termite *Trinervitermes trinervoides* (Nasutitermitinae), *Insectes soc.*, **20**, 215–220.

RATCLIFFE, F. N., GAY, F. J. AND GREAVES, T. (1952), *Australian Termites. The Biology, Recognition and Economic Importance of the Common Species*, C.S.I.R.O., Melbourne, 124 pp.

SANDS, W. A. (1960), The initiation of fungus comb construction in laboratory colonies of *Ancistrotermes guineensis* (Silvestri), *Insectes soc.*, 7, 251–263.

—— (1965*a*), A revision of the subfamily Nasutitermitinae from the Ethiopian Region, *Bull. Br. Mus. nat. Hist.(Ent.)*, Suppl. **4**, 172 pp.

—— (1965*b*), Mound population movements and fluctuations in *Trinervitermes ebenerianus* Sjöstedt (Isoptera, Termitidae, Nasutitermitinae), *Insectes soc.*, **12**, 49–58.

—— (1972), The soldierless termites of Africa (Isoptera: Termitidae), *Bull. Br. Mus. nat. Hist. (Ent.)*, Suppl. **18**, 1–244.

SCHMIDT, H. (ed.) (1955), *Die Termiten. Ihre Erkennungsmerkmale und wirtschaftliche Bedeutung*, Geest & Partig, Leipzig, 309 pp.

SCHMIDT, R. S. (1955), The evolution of nest-building behaviour in *Apicotermes* (Isoptera), *Evolution*, **9**, 157–181.

SEEVERS, C. H. (1957), A monograph on the termitophilous Staphylinidae (Col.), *Fieldiana, Zool.*, **40**, 1–334.

SILVESTRI, F. (1903), Contribuzione alla conoscenza dei Termitidi e Termitofili dell'America meridionale, *Redia*, **1**, 1–234.
—— (1905), Contribuzione alla conoscenza dei Termitidi e Termitofili dell' Eritrea, *Redia*, **3**, 341–359.
—— (1909), Isoptera, In: Michaelsen and Hartmeyer's *Die Fauna Südwest-Australiens*, **2** (17), 279–314.
—— (1914–20), Contribuzione alla conoscenza dei Termitidi e Termitofili dell' Africa occidentale, *Boll. Lab. Zool. Portici*, **9**, 3–146; **12**, 287–346; **14**, 265–319.
SJÖSTEDT, Y. (1926), Revision der Termiten Afrikas, 3. Monographie, *K. svenska VetenskAkad. Handl.*, (3), **3**, 1–419.
SNYDER, T. E., (1913), Changes during quiescent stages in the metamorphoses of termites, *Science*, **38**, 487–488.
—— (1949), Catalog of the termites (Isoptera) of the world, *Smithson. misc. Coll.*, **112**, 1–490.
—— (1956–68), Annotated subject-heading bibliography of termites 1350 B.C. to A.D. 1954 [with supplements], *Smithson. misc. Coll.*, **130**, 305 pp.; **143** (3), 137 pp.; **152** (3), 188 pp.
STEINHAUS, E. A. (1946), *Insect Microbiology*, Comstock Publ. Co., Ithaca, N.Y., 763 pp.
STELLA, E. (1938), Richerche citologiche sui neutri e sui reproduttori delle Termiti italiane (*Calotermes flavicollis* e *Reticulitermes lucifugus*), *Mem. Accad. Lincei*, (6) **7**, 1–30.
—— (1939), Studi sulle Termiti: 4°-Alcuni dati citologici sulle gonadi di soldati di *Bellicositermes bellicosus* (Smeathm.), *Riv. Biol. Colon.*, **2**, 255–262.
STUART, A. M. (1972), Behavioral regulatory mechanisms in the social homeostasis of termites (Isoptera), *Am. Zool.*, **12**, 589–594.
STUART, A. M. AND SATIR, P. (1968), Morphological and functional aspects of an insect epidermal gland, *J. Cell Biol.*, **36**, 527–549.
THOMPSON, C. B. (1916), The brain and frontal gland of the castes of the 'White Ant' *Leucotermes flavipes* Kollar, *J. Comp. Neurol.*, **26**, 553–603.
—— (1919), The development of the castes of nine genera and thirteen species of termites, *Biol. Bull.*, **36**, 379–398.
—— (1922), The castes of *Termopsis*, *J. Morph.*, **36**, 495–535.
TILLYARD, R. J. (1931), The wing-venation of the order Isoptera, I, Introduction and the family Mastotermitidae, *Proc. Linn. Soc. N.S.W.*, **56**, 371–390.
TRUCKENBRODT, W. (1964), Zytologische und entwicklungsphysiologische Untersuchungen am besamten und am parthenogenetischen Ei von *Kalotermes flavicollis* Fabr. Reifung, Furchungsablauf und Bildung der Keimanlage, *Zool. Jb., Anat.*, **81**, 359–434.
—— (1973), Ueber die imaginale Ovarvergrösserung im Zusammenhang mit der Physogastrie bei *Odontotermes badius* Haviland (Insecta, Isoptera), *Insectes soc.*, **20**, 21–40.
WANYONYI, K. (1974), The influence of the juvenile hormone analogue ZR512 (Zoecon) on caste development in *Zootermopsis nevadensis* (Hagen) (Isoptera), *Insectes soc.*, **21**, 35–44.
WASMANN, E. (1894), *Kritisches Verzeichnis der myrmekophilen und termitophilen Arthropoden. Mit Angabe der Lebensweise und mit Beschreibung neuer Arten*, Dames, Berlin, 231 pp.
—— (1934), *Die Ameisen, die Termiten und ihre Gäste*, Manz, Regensburg, 148 pp.

WATSON, J. A. L. (1971), The development of 'workers' and reproductives in *Mastotermes darwiniensis* Froggatt (Isoptera), *Insectes soc.*, **18**, 173–176.

—— (1973), The worker caste of the Hodotermitid harvester termites, *Insectes soc.*, **20**, 1–20.

WEESNER, F. M. (1956), The biology of colony foundation in *Reticulitermes hesperus* Banks, *Univ. Calif. Publs Zool.*, **61**, 253–306.

—— (1965), *The Termites of the United States: a Handbook*, Nat. Pest. Control Assoc., Elizabeth, New Jersey, 67 pp.

WEIDNER, H. (1970), Isoptera (Termiten), In: Helmcke, J.-G., Starck, D. and Wermuth, H., *Handbuch der Zoologie*, **4** (2), Lfg. 13, Beitr. 14, 147 pp.

WEYER, F. (1930*a*), Ueber Ersatzgeschlechtstiere bei Termiten, *Z. Morph. Ökol. Tiere*, **19**, 364–380.

—— (1930*b*), Beobachtung über die Enstehung neuer Kolonien bei tropischen Termiten, *Zool. Jb., Syst.*, **60**, 327–380.

—— (1935), Epithelneuerung im Mitteldarm der Termiten während der Häutung, *Z. Morph. Ökol. Tiere*, **30**, 648–672.

WILLIAMS, R. M. C. (1959), Flight and colony foundation in two *Cubitermes* species (Isoptera: Termitidae), *Insectes soc.*, **6** (2), 203–218.

WILSON, E. O. (1971), *The Insect Societies*, Harvard Univ. Press, Cambridge, Mass., 548 pp.

ZUBERI, H. A. (1963), L'anatomie comparée du cerveau chez les termites en rapport avec le polymorphisme, *Bull. biol. Fr. Belg.*, **97**, 147–207.

ZUBERI, H. A. AND PETERS, P. (1964), A study of the neurosecretory cells and the endocrine glands of *Calotermes exiguus* Mathot (Isoptera, Termitinae), In: Bouillon, A. (ed.), *Études sur les termites Africains*, Leopoldville, 87–105.

Order 15

ZORAPTERA

Winged or apterous insects with 9-segmented moniliform antennae. Y-shaped ecdysial cleavage line present. Normal maxillae, 3-segmented labial palps. Wings, when present, capable of being shed by means of basal fractures; venation specialized by reduction. Prothorax well developed. Tarsi 2-segmented. Cerci very short, 1-segmented. Ovipositor absent; male genitalia specialized, sometimes asymmetrical. Metamorphosis slight.

The first Zoraptera were described by Silvestri in 1913, among insects obtained from W. Africa, Ceylon and Java. Weidner (1969, 1970) recognizes 22 species from all zoogeographical regions except the Palaearctic. The known species belong to the genus *Zorotypus* which constitutes the family Zorotypidae; they are minute insects, less than 3 mm long, and the alate forms have a wing-expanse of about 7 mm. They occur under bark, in decaying wood, humus, etc., and are sometimes found near the galleries of termites. Though the alate forms of some species have not yet been described, two distinct types can be recognized in most species of the Zoraptera. The commoner form is apterous, only slightly pigmented and without compound eyes or ocelli while the rarer alates are darker, with eyes and ocelli and differ in details of thoracic structure (Delamare-Deboutteville, 1948*a*). The causes of this dimorphism are not known, but it is not a caste-difference since each form is made up of sexually functional males and females.

The mouthparts of the Zoraptera are of a generalized type (Fig. 303). The mandibles are more or less quadrangular and adapted for mastication; the maxillae do not call for special mention and their palps are 5-segmented; the labium is characterized by the completely divided prementum, and 3-segmented palps. The wings are capable of being shed as in termites, but the fractures are not very definitely located though they are situated near the bases of the veins. The wing-stumps persist in dealated individuals as in termites. The venation (Fig. 302) is greatly specialized by reduction and according to Crampton (1922) it approaches that of some Psocoptera, probably through convergent evolution. The abdomen is 11-segmented and genitalia are wanting in the female; in the male, genitalia are present but their homologies are not known (Snodgrass, 1937). There are ten pairs of spiracles, two being thoracic and the remainder abdominal in position. The

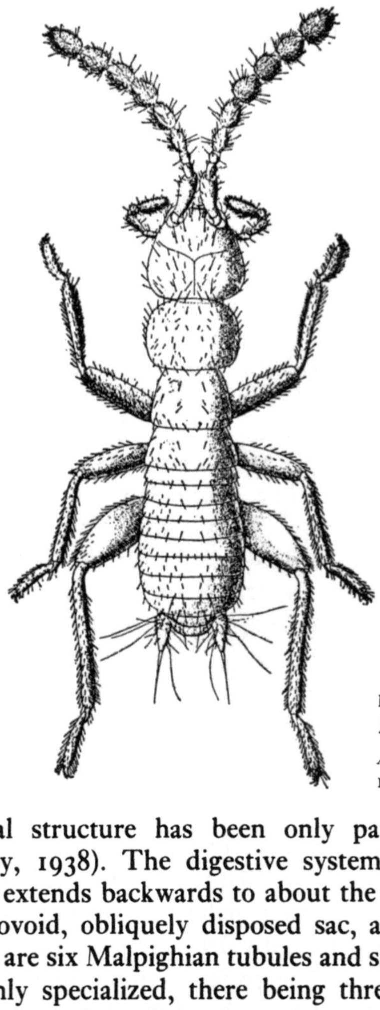

FIG. 301
Zorotypus guineensis, Africa
After Silvestri, *Boll. Lab. Zool. Portici*, 1913.

internal structure has been only partially investigated (Silvestri, 1913; Gurney, 1938). The digestive system is characterized by the large crop which extends backwards to about the 5th abdominal segment; the mid gut is an ovoid, obliquely disposed sac, and the hind intestine is convoluted. There are six Malpighian tubules and six rectal papillae. The nervous system is highly specialized, there being three thoracic and only two abdominal ganglia, the first of the latter located in the thorax. The testes are ovoid

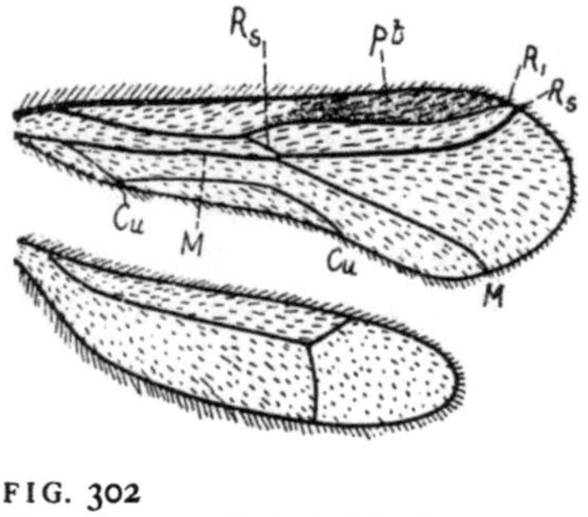

FIG. 302
Zorotypus snyderi, right wings
pt, pterostigma.

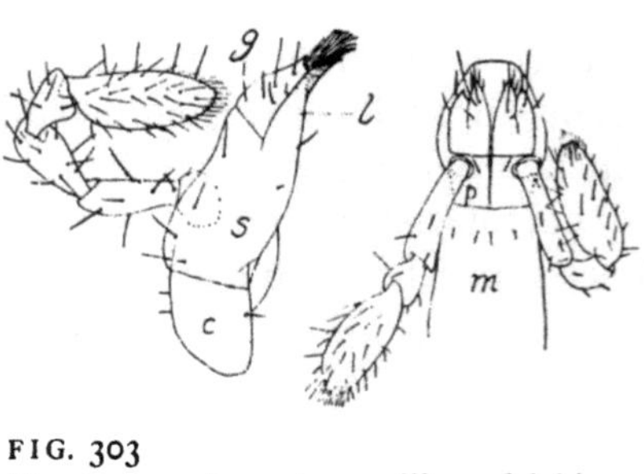

FIG. 303
Zorotypus guineensis, maxilla and labium
c, cardo; *s*, stipes; *g*, galea; *l*, lacinia; *p*, prementum; *m*, postmentum. *After* Silvestri, 1913.

paired bodies communicating by slender vasa deferentia with a large seminal vesicle. From this runs a long ejaculatory duct at the base of which is a pair of accessory glands. The female reproductive system consists of 4–5 panoistic ovarioles on each side and a spermatheca connected with the region of the genital chamber by a long slender duct.

The eggs are simple ovoidal structures and eclosion is assisted by an egg-burster on the head of the embryonic cuticle which is cast off as the young nymph emerges. The number of nymphal instars is unknown but two types of nymphs, corresponding to the adult forms, occur. Though the Zoraptera are gregarious there is no evidence of any form of social organization. Fungal spores and the remains of mites have been found in the alimentary canal.

The affinities of the Zoraptera are uncertain. The concentrated nervous system and few Malpighian tubules are suggestive of the Psocoptera but the presence of cerci and the structure of the head-capsule, mouthparts and thorax indicate an Orthopteroid ancestry perhaps near to the point at which the Psocoptera also originated.

Literature on Zoraptera

CAUDELL, A. N. (1920), Zoraptera not an apterous order, *Proc. ent. Soc. Washington*, **22**, 84–97.

CRAMPTON, G. C. (1920), Some anatomical details of the remarkable winged Zorapteron, *Zorotypus hubbardi* Caudell with notes on its relationships, *Proc. ent. Soc. Washington*, **22**, 98–106.

—— (1922), Evidences of relationship indicated by the venation of the forewings of certain insects with especial reference to the Hemiptera-Homoptera, *Psyche*, **29**, 23–41.

DELAMARE-DEBOUTTEVILLE, C. (1948*a*), Sur la morphologie des adultes aptères et ailés des Zoraptères, *Annls Sci. nat.* (*Zool.*), **9**, 145–154.

—— (1948*b*), Observations sur l'écologie et l'éthologie des Zoraptères. La question de leur vie sociale et de leurs prétendus rapports avec les termites, *Rev. ent.*, **19**, 347–352.

DENIS, R. (1949), Ordre des Zoraptères, In: Grassé's *Traité de Zoologie*, **9**, 545–555.

GURNEY, A. B. (1938), A synopsis of the order Zoraptera with notes on the biology of *Zorotypus hubbardi* Caudell, *Proc. ent. Soc. Washington*, **40**, 57–87.

SILVESTRI, F. (1913), Descrizione di un nuovo ordine di insetti, *Boll. Lab. Zool., Portici*, **7**, 193–209.

SNODGRASS, R. E. (1937), The male genitalia of Orthopteroid insects, *Smithson. misc. Collns*, **96** (5), 1–107.

WEIDNER, H. (1969), Die Ordnung Zoraptera oder Bodenläuse, *Ent. Z.*, **79**, 29–51.

—— (1970), 15. Ordnung Zoraptera (Bodenläuse), In: Helmcke, J.-G., Starck, D. and Wermuth, H. (eds), *Handbuch der Zoologie*, **4** (2), Lfg. 13, 1–12.

Order 16

PSOCOPTERA
(COPEOGNATHA, CORRODENTIA: Booklice or Psocids and their allies)

Hemipteroid insects with long, filiform antennae of 12–50 segments, sometimes secondarily annulate. Head with Y*-shaped 'epicranial suture' usually present; post-clypeus inflated. Maxilla with 4-segmented palp and a rod-like lacinia which is partially sunk into head-capsule; labial palps much reduced, 1- or 2-segmented. Prothorax generally small; wing-venation simple, with few cross-veins, but often with characteristic anastomoses between* R_s *and* M *and between* M *and* Cu_{1a}. *Tarsi 2- or 3-segmented. External genitalia inconspicuous, cerci absent.*

The Psocoptera are small or minute insects with rather soft, stout bodies and usually delicate, membranous wings. Brachypterous, micropterous and apterous forms are characteristic of the female or of both sexes of some species. Several apterous species of *Liposcelis* are often found among accumulations of books and papers. They are commonly known as booklice and feed on fragments of animal and vegetable matter or the paste of bookbindings. The order includes a number of domestic species which may feed on stored food products, natural history specimens, straw and chaff in barns and warehouses, or occur in thatches and haystacks. The majority of Psocoptera, however, occur outdoors on foliage, tree trunks, under bark, on weathered fences and palings, on fungi and among growths of algae and lichens. A few inhabit the nests of birds or mammals and are sometimes found among their plumage or fur (Mockford, 1967*a*, 1971). In general Psocids live on fragments of animal or vegetable matter, particularly on fungi, unicellular algae and lichens. Although sometimes stated to eat paper, certain species actually feed upon moulds growing there and in this way reveal the injury done to the paper. Most Psocids carry foreign matter entangled among their body-hairs and in this way disseminate fungal spores. Many live gregariously and clusters of individuals of various ages are sometimes met with on bark, each colony covered by a canopy of fine silken threads. The winged forms are curiously reluctant to take to flight but sometimes fly in considerable numbers and drift through the air after the manner of winged aphids. They are occasionally recorded as occurring in buildings in large swarms, the commonest species concerned being *Lachesilla pedicularia*.

General accounts of the Psocoptera are given by Badonnel (1951) and Weidner (1972) and various aspects of their biology are discussed by Jentsch (1939), Medem (1951), Schneider (1955), Broadhead and Thornton (1955), Broadhead (1958) and New (1969, 1971).

External Anatomy – The head (Fig. 304) is large and very mobile, with the 'epicranial sutures' more or less distinct. The compound eyes are markedly convex and protrude from the surface of the head; in apterous

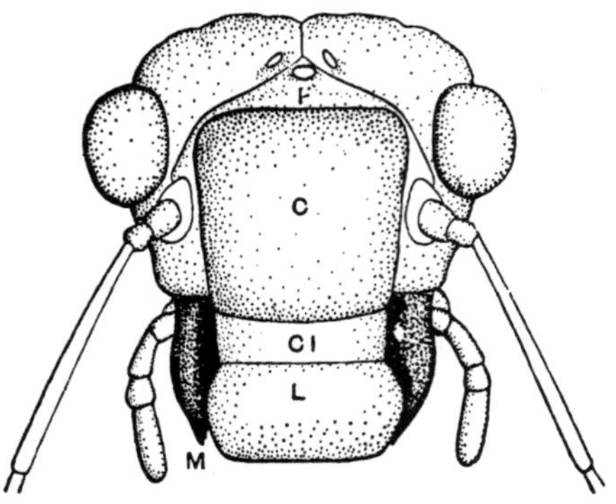

FIG. 304
Frontal view of the head of a psocid

F, frons. *C*, post-clypeus. *C1*, ante-clypeus. *L*, labrum. *M*, mandible.

Liposcelidae and some other families they are vestigial and reduced to two small groups of ommatidia. Three ocelli are present in the winged species but are wanting in the apterous forms. The *labrum* is well developed and attached to the *ante-clypeus*; the *post-clypeus* (prefrons of some authors) is a conspicuous sclerite often presenting an inflated appearance. The *antennae* are long and filiform; they are frequently 13-segmented, but the number of segments is variable and may be as high as fifty. The mouthparts (Fig. 305) have been investigated by many authors, especially Badonnel (1934), Weber (1936), Cope (1940) and von Kéler (1966). The *mandibles* are relatively large and strong, each with a broad striated molar area and a denticulate cutting edge. The *maxillae* are considerably modified. The cardo and stipes are not clearly separated, but the 4-segmented maxillary palp is well developed. The galea is a large, fleshy lobe, medial to which is a strongly sclerotized rod (the 'pick'), the basal half of which is sunk into the head. The innervation of the small internal cavity of the pick by a branch of the maxillary nerve, and the attachment to it of retractor muscles comparable with the cranial flexors of the lacinia in other insects, identify it as a highly modified lacinia. The pick is said to be used for scraping food from the substrate but Pearman (1936) was unable to confirm this, the mandibles alone being used in the feeding Psocids which he studied; New (1974*a*) suggests the picks may prop up the head while feeding. In the *labium*, the mentum is oblong, the prementum is divided and the ligula carries a pair of membranous paraglossae. The inner lobes or glossae are represented by a pair of minute structures forming the external conduit of the labial (silk) glands. The labial palps are reduced to the condition of single or, rarely, 2-segmented lobes. The *hypopharynx* is well-developed and its complex structure has received diverse explanations. The free extremity of the lingua bears a pair of small delicate lobes, the *superlinguae*, while its ventral surface is thickened locally to form a pair of oval *lingual sclerites*. The latter are connected by a branched or paired sclerotized filament (perhaps hollow) to the conspicuous *sitophore sclerite* (sometimes misleadingly called the oesophageal sclerite). This, following Badonnel (1934) and Snodgrass (1944), is to be regarded as the basal part of the cibarial surface of the hypopharynx. Directly opposite the sitophore sclerite, the dorsal wall of the cibarium bears a sclerotized knob-like process which Weber (1936) suggests can be moved against the sitophore so that the two act like a mortar and pestle and help to break up food particles. The sitophore sclerite, lingual sclerites and sclerotized filament correspond to similar structures in the Mallophaga.

The *thorax* of the winged forms has a reduced prothorax, which is largely concealed between the head and mesothorax. The meso- and metathorax are very similar, the nota being subdivided clearly into scutum and scutellum, followed in each case by a phragma-bearing postnotum. The pleura consist of the typical elements. The sterna are reduced to narrow strips lying between the legs, each being subdivided into a sternum and sternellum and provided with well-developed furcal arms. In apterous forms the prothorax is larger and in some (Liposcelidae) the tergites and sternites of the meso- and metathorax are united into a continuous shield. The *wings* (Fig. 306) are membranous with prominent though reduced venation; the anterior pair is considerably larger and the wings when not in use are folded flat or, more

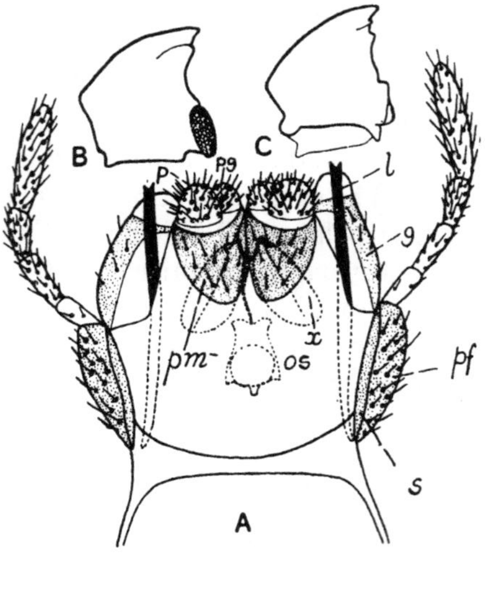

FIG. 305
A, ventral view of the head of a Psocid. B, right mandible (ventral). C, left mandible (dorsal)

g, galea; *l*, lacinia; *os*, sitophore sclerite; *p*, labial palp; *pf*, palpifer; *pg*, paraglossa; *pm*, prementum; *s*, stipes; *x*, lingual sclerite.

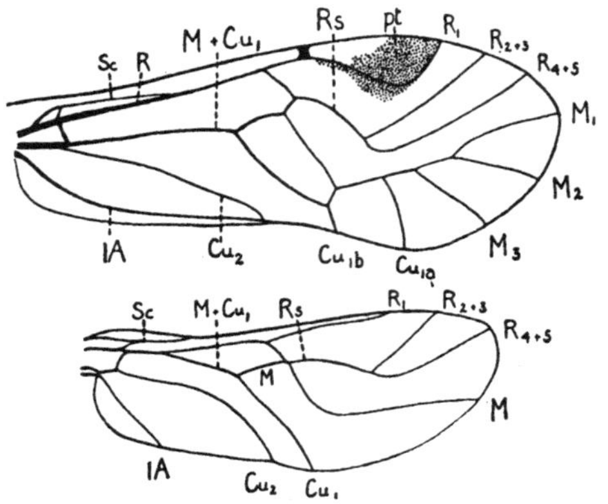

FIG. 306
Right wings of *Amphigerontia bifasciata*
pt, pterostigma.

often, steeply inclined in a roof-like manner over the body, with the hind margins uppermost. In some species the wings are conspicuously marbled and may, together with the body and appendages, bear scales of varied form not unlike those of Lepidoptera. The venation is discussed in detail by Enderlein (1903), Tillyard (1926) and in taxonomic works; a pterostigma is present on the fore wing, and there is a reduction in the branches of the principal veins. Sc is unbranched, R and M are each normally 3-branched and there is a fusion of the main stems of M and Cu. M may also be associated with the basal part of R_s. Cu_I usually branches distally into Cu_{Ia} and Cu_{Ib}, the cell enclosed by the fork (the so-called *areola postica*) being either free, connected with M by a cross-vein or fused with it as in Fig. 306. In the hind wing reduction is carried still further, M being represented as a rule by a single branch. In the Psocidae the wings are effectively braced through the somewhat tortuous courses of the veins and there is a striking absence of cross-veins, though these occur in other members of the order. In many species there is a small hook at the border of the fore wing near Cu_2 which engages the hind costa to couple the wings in flight; a separate projection from the under side of the pterostigma may also engage the hind costa when the wings are folded in repose (New, 1974*b*).

The legs show no special adaptive modifications. In many species there occurs on the inner face of each posterior coxa a structure known as Pearman's organ. This is believed to be stridulatory and consists typically of a sculptured prominence near which lies an area of thin cuticle – the mirror or tympan (Pearman, 1928). According to Cope (1940) the Psocoptera share with the Mallophaga, and with them alone, the peculiarity of a ventral articulation between the anterior coxae and the prosternum, the middle and hind coxae having only the normal pleural articulation. The pretarsus bears two claws, beneath which occur pulvilli of varied form, but there is no empodium.

The morphology of the abdomen, and particularly of its terminal segments, is not entirely clear. Nine terga are generally recognizable, the 10th is presumably fused with the 9th, and the abdomen ends in an epiproct and a pair of paraprocts. These appear to represent the tergum and the divided sternum of the 11th segment and the paraprocts each bear a cluster of trichobothria in many species. The small ovipositor is partly concealed by the enlarged sternum of the 8th abdominal segment and consists, when fully developed, of three pairs of valves of which the anterior pair arises from the 8th sternum and two pairs from the 9th sternum. Atrophy of all the valves may occur. The male genitalia, which are concealed by the enlarged 9th sternum, consist of a pair of parameres between which lies the aedeagus. The homologies of the genitalia are not established but the terminalia provide valuable taxonomic characters.

There are two pairs of thoracic spiracles and seven or eight abdominal pairs.

Internal Anatomy – The principal recent works on the internal anatomy of the order are those of Badonnel (1934), Weber (1936) and Finlayson (1949) who also review earlier work.

In the *digestive system* the oesophagus is elongate and extends into the abdomen, the mid gut is sharply curved and U-shaped and leads into a very short unconvoluted hind gut bearing six rectal papillae; the Malpighian tubules are four in number. There are two pairs of labial glands which extend into the abdomen and whose ducts open at the base of the labium. One pair, normally dorsal in position, is composed of cells with an acidophil cytoplasm and, in all but a few species, secretes silk, while the other pair, made up of basophil cells, is salivary and is usually ventral to the silk glands (Weber, 1938). The spinning glands provide silken threads which form the webs often associated with colonies of these insects. The *nervous system* is highly concentrated: in addition to the brain (Jentsch, 1940) and suboesophageal ganglion there are only three other ganglionic centres. The first of these belongs to the prothorax, the meso- and metathoracic ganglia are fused into a common second centre, and the single abdominal ganglion has shifted forwards so as to lie partly in the thorax. The connectives are extremely short but are double throughout. A pair of large abdominal nerves extend to the posterior end of the body. The *reproductive organs* are relatively simple but accord well with the taxonomic subdivisions of the order (Klier, 1956; Kai and Thornton, 1968; Wong, 1971). Each ovary consists of from three to five polytrophic ovarioles, the oviducts are very short and a small globular spermatheca opens into the dorsal aspect of the vagina. A peculiar type of accessory gland was described by Nitzsch in *Clothilla* many years ago; it contains from one to four small sacs each opening by a narrow canal into a common duct. The *male reproductive system* consists of a pair of testes – which may be simple pyriform organs or three-lobed – leading by short, narrow vasa deferentia into the large seminal vesicles. The latter are somewhat complex structures in some species, consisting of two chambers and responsible for secreting material used in the formation of the spermatophore, a detailed account of which is given by Finlayson for *Lepinotus*. The seminal vesicles open into a short ejaculatory duct.

Development – Females – except in viviparous forms – lay 20–100 eggs which are ellipsoidal or bluntly rounded at one end. They may be scattered singly or laid in groups and are sometimes covered by an incrustation of bark, algae, etc., or by a silken web (Fig. 307). Eclosion is assisted by an egg-burster on the head of the embryonic cuticle, which is shed before emergence is complete. There are usually six subsequent nymphal instars differing from the adult in possessing a smaller number of antennal segments,

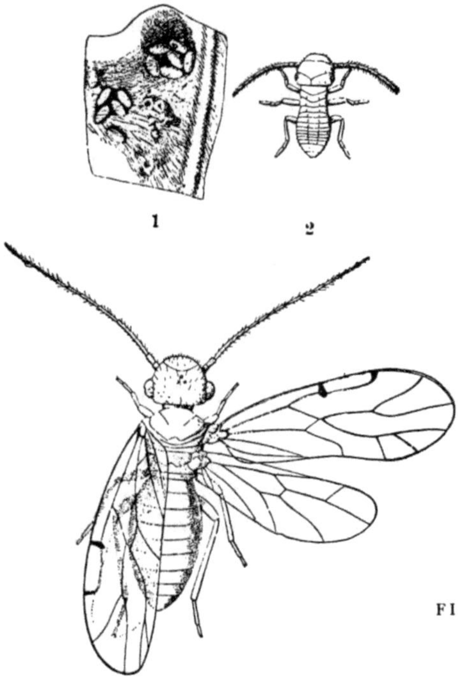

FIG. 307 *Peripsocus phaeopterus*. 1, portion of leaf with eggs beneath silken threads; 2, young nymph; 3, imago
After Silvestri.

fewer ommatidia, two tarsal segments and the gradually developing wing-pads and genitalia (Weber, 1931; Söfner, 1941; Broadhead, 1947; Finlayson, 1949; and other references cited on p. 655 ff.). Facultative parthenogenesis is not infrequent and males of some species are unknown. The Psocoptera occurring in N.W. Europe usually have one, two or three generations per year but some species of *Caecilius* have several generations while the domestic species breed continuously. Overwintering usually occurs in the egg but New (1969*a*) has found that *Ectopsocus briggsi* also overwinters as nymphs and adults while *Graphopsocus cruciatus* does so as eggs and adults.

Affinities and Phylogeny – Several families of fossil Psocoptera, mostly from the Permian, have been grouped into the suborder Permopsocida and are known mainly from the work of Tillyard (1926, 1935) and Carpenter (1932). The Dichentomidae (Lower Permian) had a rather generalized venation with 2 or 3 branches in R_S, 4 branches in M, and a forked Cu_1. The fore and hind wings were similar in size and venation and the tarsi were 4-segmented. There is no palaeontological evidence bearing on the origin of these primitive Psocoptera but it seems likely that they, together with the other Hemipteroid orders, shared a common ancestry with the Zoraptera, from which they differ principally in the loss of cerci, reduction of the first abdominal sternum and specialization of the mouthparts (Königsmann, 1960). It is also likely on morphological grounds that the Mallophaga originated from a Psocopteran-like stock (see p. 426). Within the Psocoptera the probable phyletic relations of the families and other major taxa must be inferred mainly from the morphology of Recent species. A major contribution to this field has come from Smithers (1972), whose paper – together with that of Mockford (1967*b*) – should be consulted for further details.

Classification – A modern classification of the Psocoptera was outlined by Pearman (1936) and has been extended and modified by Roesler (1944), Badonnel (1951), Smithers (1967, 1972) and others. There are nearly 2000 species in about 226 genera (New, 1974*a*) but the order is not well known and many new species await discovery. About 80 per cent of those described belong to the suborder Psocomorpha and some 90 species are British. For the European species see Badonnel (1943), Günther (1974) and New (1974*a*); for those of other regions the following works may be consulted: Russia: Vishniakova (1967); Madagascar: Badonnel (1967); Africa: Badonnel (1946–49; 1955–69); N. America: Chapman (1930); and New Zealand: Tillyard (1923).

The scheme summarized below is due to Badonnel (1951); a later phylogenetic classification has been proposed by Smithers (1972) as a basis for further discussion and improvement. The families marked with an asterisk occur in Britain.

Suborder 1. TROGIOMORPHA

Antennae with more than 20 segments, not secondarily annulate; tarsi 3-segmented; labial palps 2-segmented; winged forms with pterostigma not thickened; hypopharynx with sclerotized filaments separated for their entire length.

***FAM. LEPIDOPSOCIDAE.** *Legs, body and wings usually clothed in scales.* A mainly tropical group of about 100 species, frequenting bark, leaf-litter and occasionally found in buildings.

***FAM. TROGIIDAE.** About 20 species of apterous or brachypterous insects, mostly associated with buildings and widely distributed, e.g. *Trogium pulsatorium.*

***FAM. PSOQUILLIDAE.** About 20 species, mainly from the Old World tropics, but with domestic species that have become widely distributed by commerce.

***FAM. PSYLLIPSOCIDAE.** *Delicate species with the wings often reduced or showing various degrees of alary polymorphism.* The 20 or so species are mostly cavernicolous or domestic. *Psyllipsocus ramburi* is a widely distributed domestic species.

FAM. PRIONOGLARIDAE. Two species from Europe and Afghanistan with extremely long antennae. At least the immature stages are cavernicolous. This and the preceding family are sometimes placed in the suborder Troctomorpha.

Suborder 2. TROCTOMORPHA

Antennae 11- to 17-segmented, the flagellum secondarily annulate; tarsi 3-segmented; labial palps 1- or 2-segmented; pterostigma of winged forms not thickened; hypopharynx with sclerotized filaments separate distally.

FAM. AMPHIENTOMIDAE. *Antennae 15-segmented; 1A and 2A separate; body and wings scaled.* About 50 Ethiopian and Oriental species. Closely allied to and sometimes included in the Amphientomidae are four small families, nearly all the species of which come from C. America: **Troctopsocidae** (= **Plaumanniidae**), **Manicapsocidae**, **Compsocidae** and **Musapsocidae**.

***FAM. LIPOSCELIDAE.** *Small, flattened, oval body; short antennae; winged or apterous, the latter with fused meso- and metanotum, reduced eyes and no ocelli. Embidopsocus* and its allies are mostly African and S. American, under bark. *Liposcelis* is apterous with some 60 species from the nests of ants or birds, or in stored foodstuffs, or in other domestic situations (Broadhead, 1950).

***FAM. PACHYTROCTIDAE.** *Body not flattened, abdomen more or less conical; antennae long; winged or apterous.* About 40 species from various regions in litter, nests and domestic habitats.

***FAM. SPHAEROPSOCIDAE.** *Apterous or with elytron-like fore wings, no hind wings and reduced eyes.* A small, mainly southern hemisphere family.

Suborder 3. PSOCOMORPHA

Antennae with 13 or fewer segments; never secondarily annulate; labial palp 1- or 2-segmented; tarsi 2- or 3-segmented; pterostigma more strongly sclerotized than wing membrane; sclerotized filament of hypopharynx only forked anteriorly.

***FAM. EPIPSOCIDAE.** *Tarsi 2- or 3-segmented; labial palps 1-segmented; areola postica free,* Cu_{1a} *not arched forwards. Epipsocus*, with 40 species from most regions, is the only large genus. The Neotropical **Ptiloneuridae**, with a many-branched media, and the **Callistopteridae**, with one species from New Guinea, are related to the Epipsocidae.

***FAM. CAECILIIDAE.** *Tarsi 2-segmented; areola postica free or joined to M by a cross-vein; claws without teeth.* A large family of foliage-haunting Psocoptera with a world-wide distribution. Most of the species belong to *Caecilius*. The small families **Calopsocidae** (= **Neurosemidae**), with many cells in the venation, and the S. American **Polypsocidae** are allied to the Caeciliidae.

*FAM. STENOPSOCIDAE. *Differing from the Caeciliidae (of which it is sometimes regarded as a subfamily) in having the pterostigma linked to* R_s *by a cross-vein.* Widely distributed, with about 40 species, of which *Graphopsocus cruciatus* is almost cosmopolitan.

*FAM. AMPHIPSOCIDAE. *Usually differing from the Caeciliidae by the incomplete cross-vein running from the pterostigma towards* R_s. About 35 species, of which *Kolbea quisquiliarum* is European.

*FAM. LACHESILLIDAE (= **Pterodelidae**). Contains only two genera of which *Lachesilla* has very many species, about half of them Holarctic. The males are long-winged with hook-like paraprocts, the females long-winged or brachypterous.

*FAM. PERIPSOCIDAE. *Tarsi 2-segmented; areola postica absent.* The two subfamilies, typified by *Ectopsocus* and *Peripsocus* are now often treated as distinct families. Allied to them are the **Hemipsocidae**, with a few species from Australia, New Guinea and the Oriental Region.

FAM. PSEUDOCAECILIIDAE. *Venation like Caeciliidae, but veins bearing several rows of setae; pterostigma and areola postica elongate. Pseudocaecilius*, with a mainly Old World distribution, is the largest genus. The ***Trichopsocidae** are a very small related family and the **Archipsocidae** include 2 mainly tropical genera with setose wings and indistinct venation.

*FAM. ELIPSOCIDAE. *Tarsi usually with 3 segments (2 in brachypterous forms); antennae with 13 or fewer segments; pterostigma not connected with* R_s *by a cross-vein;* R_s *and M anastomosed or linked by a cross-vein; areola postica usually free but sometimes joined to M or absent.* A widely distributed family of about 70 species, e.g. *Nepiomorpha, Propsocus, Pseudopsocus, Elipsocus* and *Lesneia*.

*FAM. PHILOTARSIDAE. *Tarsi 3-segmented; fore wing margin and veins strongly setose; areola postica free; female sub-genital plate one-lobed. Philotarsus* is widely distributed.

*FAM. MESOPSOCIDAE. *Tarsi 3-segmented; wings glabrous; areola postica free, with* Cu_{1a} *strongly arched forwards; females often apterous.* About 20 species of *Mesopsocus* occur in Africa and the Holarctic region. Allied to this family is the **Psoculidae**, containing only *Psoculus neglectus* from France and Germany.

*FAM. PSOCIDAE. *Tarsi 2-segmented; fore wings glabrous; areola postica with* Cu_{1a} *usually anastomosed more or less completely with M.* A large family, distributed throughout the world, with many bark-frequenting species. *Amphigerontia, Psocus, Psococerastis* and *Trichadenotecnum* are among the important genera.

FAM. THYRSOPHORIDAE. Includes about 20 large, Neotropical species with 2-segmented tarsi and the 3rd and 4th antennal segments larger and more thickly pubescent than the others.

FAM. MYOPSOCIDAE. *Tarsi 3-segmented; areola postica with* Cu_{1a} *anastomosed with M; wings more or less regularly spotted.* Over 50 species, mainly from the Old World tropics. The **Psilopsocidae**, with a few Australasian species, are closely related.

Literature on the Psocoptera

BADONNEL, A. (1934), Recherches sur l'anatomie des psoques, *Bull. biol. Fr. Belg., Suppl.*, **18**, 1–241.
—— (1943), Psocoptères, *Faune de France*, **42**, 1–64.
—— (1946–49), Psocoptères du Congo Belge. I–III, *Rev. Zool. Bot. Afr.*, **39**, 137–196; **40**, 266–322; *Bull. Inst. Sci. nat. Belg.*, **25** (11), 1–64.
—— (1951), Ordre des Psocoptères, In: Grassé, P. P. (ed.), *Traité de Zoologie*, **10**, 1301–1340.
—— (1955–69), Psocoptères de l'Angola, *Publ. Cult. Comp. Diam. Angola*, **26**, 11–267; **79**, 1–152.
—— (1962), Psocoptères, *Biol. Amer. australe*, **1**, 185–229.
—— (1967), Insectes psocoptères, *Faune Madagascar*, **23**, 1–235.
BROADHEAD, E. (1947), The life-history of *Embidopsocus enderleini* (Ribaga) (Corrodentia, Liposcelidae), *Ent. mon. Mag.*, **83**, 200–203.
—— (1950), A revision of the genus *Liposcelis* Motschulsky with notes on the position of this genus in the order Corrodentia and on the variability of the *Liposcelis* species, *Trans. R. ent. Soc. Lond.*, **101**, 335–388.
—— (1958), The psocid fauna of larch trees in northern England – an ecological study of mixed species populations exploiting a common resource, *J. anim. Ecol.*, **27**, 217–263.
BROADHEAD, E. AND THORNTON, I. W. B. (1955), An ecological study of three closely related psocid species, *Oikos*, **6**, 1–50.
CARPENTER, F. M. (1932), The Lower Permian insects of Kansas. 5. Psocoptera and additions to the Homoptera, *Am. J. Sci.*, (5), **24**, 1–22.
CHAPMAN, P. J. (1930), Corrodentia of the U.S.A., I. Suborder Isotecnomera, *J. N. Y. ent. Soc.*, **38**, 319–402.
COPE, O. B. (1940), The morphology of *Psocus confraternus* Banks, *Microentomology*, **5**, 91–115.
ENDERLEIN, G. (1903), Über die Morphologie, Gruppierung und systematische Stellung der Corrodentia, *Zool. Anz.*, **26**, 423–437.
FINLAYSON, L. H. (1949), The life history and anatomy of *Lepinotus patruelis* Pearman (Psocoptera – Atropidae), *Proc. zool. Soc. Lond.*, **119**, 301–323.
GÜNTHER, K. K. (1974), Staubläuse, Psocoptera, *Tierwelt Deutschlands*, **61**, 314 pp.
GOSS, R. J. (1954). Ovarian development and oogenesis in the book louse, *Liposcelis divergens* Badonnel (Psocoptera, Liposcelidae), *Ann. ent. Soc. Am.*, **47**, 190–207.
JENTSCH, S. (1939), Beiträge zur Kenntnis der Überordnung Pscoidea. 7. Vergleichend entwicklungsbiologische und ökologische Untersuchungen an einheimischen Psocopteren unter besonderer Berücksichtigung der Art *Hyperetes guestfalicus* Kolbe 1880, *Zool. Jb, Syst.*, **73**, 1–46.
—— (1940), Zur Morphologie des Gehirns und der Lichtsinnesorgane der Psocopteren, *Zool. Jb., Anat.*, **66**, 403–446.
KAI, W. S. AND THORNTON, I. W. B. (1968), The internal morphology of the reproductive system of some Psocid species, *Proc. R. ent. Soc. Lond.*, A, **43**, 1–12.
KÉLER, S. VON (1966), Zur Mechanik der Nahrungsaufnahme bei Corrodentien, *Z. Parasitenk.*, **27**, 64–79.
KLIER, E. (1956), Zur Konstruktionsmorphologie des männlichen Geschlechtsapparates der Psocopteren, *Zool. Jb.* (*Anat.*), **75**, 207–286.
KÖNIGSMANN, E. (1960), Die Phylogenie der Parametabola, *Beitr. Ent.*, **10**, 705–744.

MEDEM, F. (1951), Biologische Beobachtungen an Psocopteren, *Zool. Jb. (Syst.)*, **79**, 591–613.

MOCKFORD, E. L. (1967*a*), Some Psocoptera from plumage of birds, *Proc. ent. Soc. Washington*, **69**, 307–309.

—— (1967*b*), The Electrotomoid Psocids (Psocoptera), *Psyche*, **74**, 118–165.

—— (1971), Psocoptera from sleeping nests of the dusky-footed wood rat in southern California (Psocoptera: Atropidae, Psoquillidae, Liposcelidae), *Pan Pacific Ent.*, **47**, 127–140.

NEW, T. R. (1969*a*), The early stages and life histories of some British foliage-frequenting Psocoptera, with notes on the overwintering species of British arboreal Psocoptera, *Trans. R. ent. Soc. Lond.*, **121**, 59–77.

—— (1969*b*), Aerial dispersal of some British Psocoptera, as indicated by suction trap catches, *Proc. R. ent. Soc. Lond.*, (A) **44**, 49–61.

—— (1969*c*), Observations on the biology of Psocoptera found in leaf litter in Southern England, *Trans. Soc. Brit. Ent.*, **18**, 169–180.

—— (1971), An introduction to the natural history of the British Psocoptera, *Entomologist*, **104**, 59–97.

—— (1974*a*), Psocoptera, *R. ent. Soc. Handb. Ident. Brit. Insects*, **1** (7), 102 pp.

—— (1974*b*), Structural variation in Psocopteran wing-coupling mechanisms, *Int. J. Insect. Morphol. Embryol.*, **3**, 193–201.

PEARMAN, J. V. (1928), On sound-production in the Psocoptera and on a presumed stridulatory organ, *Ent. mon. Mag.*, **64**, 179–186.

—— (1936), The taxonomy of the Psocoptera: a preliminary sketch, *Proc. R. ent. Soc. Lond.*, (B), **5**, 58–62.

ROESLER, R. (1944), Die Gattungen der Copeognathen, *Stettin. ent. Ztg.*, **105**, 117–166.

SCHNEIDER, H. (1955), Vergleichende Untersuchungen über Parthenogenese und Entwicklungsrhythmen bei einheimischen Psocopteren, *Biol. Zbl.*, **74**, 273–310.

SMITHERS, C. N. (1965), A bibliography of the Psocoptera (Insecta), *Aust. Zool.*, **13**, 137–209.

—— (1967), A catalogue of the Psocoptera of the World, *Aust. Zool.*, **14**, 1–145.

—— (1972), The classification and phylogeny of the Psocoptera, *Mem. Aust. Mus., Sydney*, **14**, 351 pp.

SNODGRASS, R. E. (1944), The feeding apparatus of biting and sucking insects affecting man and animals, *Smithson. Misc. Collns*, **104** (7), 113 pp.

SÖFNER, L. (1941), Zur Entwicklungsbiologie und Oekologie der einheimischen Psocopteren-Arten *Ectopsocus meridionalis* (Ribaga 1904) und *Ectopsocus briggsi* McLach. 1899, *Zool. Jb. (Syst.)*, **74**, 325–360.

THORNTON, I. W. B. AND WONG, S. K. (1967), A numerical taxonomic analysis of the Peripsocidae of the Oriental region and the Pacific basin, *Syst. Zool.*, **16**, 217–240.

—— (1968), The peripsocid fauna (Psocoptera) of the Oriental region and the Pacific, *Pacific Insects Monogr.*, **19**, 1–158.

TILLYARD, R. J. (1923), A monograph of the Psocoptera, or Copeognatha, of New Zealand, *Trans. N.Z. Inst., Wellington*, **54**, 170–196.

—— (1926), Kansas Permian insects. 8: The order Copeognatha, *Am. J. Sci.*, **11**, 315–349.

—— (1935), Upper Permian insects of New South Wales, III. The order Copeognatha, *Proc. Linn. Soc. N.S. Wales*, **60**, 265–279.

VISHNIAKOVA, V. N. (1967), Psocoptera, In: Bei-Bienko, G. Y. (ed.), *Keys to the Insects of the European part of the USSR*, **1**, 362–384.
WEBER, H. (1931), Die Lebensgeschichte von *Ectopsocus parvulus* (Kolbe), *Z. wiss. Zool.*, **138**, 457–486.
——(1936), Copeognathen, In: Schulze, P. (ed.), *Biologie der Tiere Deutschlands*, 27.1–27.50.
——(1938), Beiträge zur Kenntnis der Überordnung Psocoidea. I. Die Labialdrüsen der Copeognathen, *Zool. Jb.* (*Anat.*), **64**, 243–286.
WEIDNER, H. (1972), 16. Ordnung Copeognatha (Staubläuse), In: Helmcke, J.-G., Starck, D. and Wermuth, H. (eds), *Handbuch der Zoologie*, **4** (2), Lfg. 18, 1–94.
WONG, S. K. (1971), The study of internal genital systems and the classification of Psocoptera, *N.Z. Ent.*, **4** (4), 66–71.

Order 17

MALLOPHAGA (BITING LICE OR BIRD LICE)

Apterous insects living as ectoparasites mainly of birds, less frequently of mammals. Eyes reduced. No ocelli. Antennae 3- to 5-segmented. Mouthparts of a modified biting type; maxillary palpi 4-segmented or wanting; ligula undivided or 2-lobed, labial palps rudimentary. Prothorax evident, free; meso- and metathorax often imperfectly separated; tarsi 1- or 2-segmented, terminated by single or paired claws. Thoracic spiracles ventral. Cerci absent. Metamorphosis slight.

The Mallophaga are very small or small (0·5 to 10 mm long), flat-bodied, active insects entirely adapted for an ectoparasitic life; for general accounts see Eichler (1963) and von Kéler (1969). The majority of the species infest birds and a smaller number occur on mammals. Most Mallophaga feed on fragments of feathers, hair and other epidermal products, but Ewing (1924) says that *Gyropus ovalis* and *Gliricola porcelli* obtain sebum and possibly serum by probing into the hair follicles of their hosts while Crutchfield and Hixson (1943) find that *Menacanthus* spp. feed habitually on blood in addition to feathers. Other species may imbibe blood from wounds, as, for instance, when a bird is shot. Hosts seem able to withstand the usual degree of infestation without obvious ill-effects but with exceptionally heavy infestations there may be some loss of plumage and deterioration in condition due to irritation. Dust-baths and 'anting' in birds may possibly be attempts to rid themselves of these parasites. Mallophaga tend frequently to be restricted to special areas of the host's body. For example, on the pigeon, *Columbicola columbae* is found mainly on the remiges of the wing while *Goniocotes bidentatus* occurs on the small feathers of the neck (Beier, 1936). In populations of *Bovicola bovis* on cattle, Craufurd-Benson (1941) found that there was some segregation into breeding clusters and nymphal colonies though this was complicated by seasonal changes in the distribution of infestation over the host's body. After the death of the host, Mallophaga are capable of survival for only a short period (a few hours to three days). Migration from one host to another probably occurs mainly through bodily contact of the hosts though recorded instances of Mallophaga clinging to Culicid and Hippoboscid flies shows that phoresy may play some part in their spread (Clay and Meinertzhagen, 1943; Keirans, 1975). There appears to be a definite relation between the evolution of the lice and that of their hosts so that groups of closely related host species tend to be infested by similar Mallophaga (Hopkins, 1942; 1949). Several authors have suggested that this type of host-association may aid in elucidating the phylogeny of the hosts. Thus, the flamingoes (Phoenicopteridae) have four Mallophagan genera in common with the ducks (Anatidae) but only one with the storks (Ciconiidae), thus indicating a closer relationship with the former. The principle needs, however, to be used with care; for general discussions see

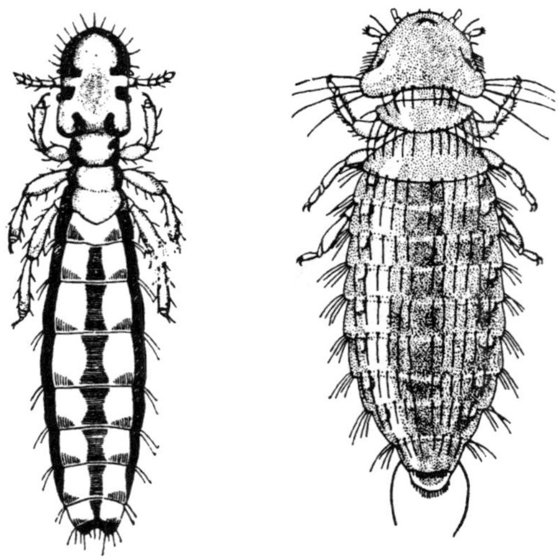

A B

FIG. 308 Mallophaga of domestic fowl

A, *Lipeurus caponis* L. *After* Denny. B, *Menopon pallidum* Nitz. *After* Bishopp and Wood, *U.S. Farmer's Bull.*, 801.

Clay (1949*a*), Timmerman (1957, 1965) and Zlotorzycka (1965, 1968). Many Mallophaga show an interesting symbiotic association with bacteria which, in nymphs and males, occur in specialized mycetocytes distributed among the fat-body. In adult females the bacteria accumulate in the ovarial ampullae, whence they pass into the eggs before they are laid and are so transmitted to the progeny (Ries, 1931). Lice from which the symbionts have been eliminated soon die. The symbionts tend to occur in those species which normally or occasionally imbibe blood and it has been suggested that one factor helping to determine the host-specificity of biting lice is the inability of the symbionts to flourish in a louse feeding on the blood of an abnormal host species.

One of the most injurious members of the order is the Common Chicken-Louse *Menopon gallinae*. Ducks are infested by several species, among which a common form is *Anatoecus dentatus*. Pigeons are almost always infested by an elongate and very slender louse, *Columbicola columbae*. The species living on domestic mammals belong to a few Trichodectid genera: thus the dog is often infested by *Trichodectes canis* and cats by *Felicola subrostratus*. Horses and donkeys harbour several species while *Bovicola bovis* troubles cattle all over the world. The host-relations of the Trichodectidae are discussed by Hopkins (1949).

External Anatomy – The body is usually very much flattened dorsoventrally with the integument well sclerotized (Neuffer, 1954). Over the abdomen the tergal, pleural and sternal regions are separated by very distinct areas of membrane. The *head-capsule* is large and horizontal and many

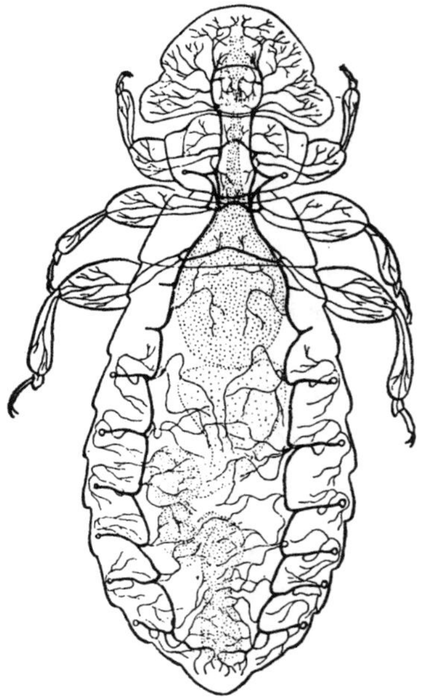

FIG. 309
Tracheal system of *Myrsidea cucularis*
After Harrison, *Parasitology*, 8.

of its sutures have been obliterated though the labrum is distinct and a clypeofrontal suture may be retained; the tentorium is often reduced (Symmons, 1952, and accounts of individual species by Stöwe, 1943; Risler, 1951; Buckup, 1959 and Haub, 1971). The eyes are vestigial (Wundrig, 1936).

The *antennae* differ very markedly in the two suborders: in the Amblycera they are generally capitate, and concealed in deep fossae, while among the Ischnocera they are filiform, exserted and may be modified as clasping organs in the male. The *mouthparts* (Risler and Geisinger, 1965; Haub, 1972–73) are of the biting type with large dentate *mandibles*, which differ in their insertion in the two suborders (Fig. 310). Among the Amblycera they lie parallel with the ventral surface of the head, so that each condyle is ventral and the ginglymus dorsal. In the Ischnocera each mandible is inserted more or less at right angles to the head, the condyle being posterior and the ginglymus anterior. The *maxillae* are single-lobed and lack differentiation into the usual sclerites; they are attached to the lateral margins of the labium. In certain genera a pair of minute forked rods have been described, but are very fragile and easily overlooked. They are evidently homologous with the similar, but more prominent, laciniae of the Psocoptera. The maxillary palps are 4-segmented in the Amblycera and wanting in the Ischnocera. The *labium* is composed of a submentum, mentum and prementum; the palps are reduced to small lobes or absent, and the ligula is either entire or represented by a pair of fleshy processes, probably homologous with paraglossae. The mouthparts of *Haematomyzus* are highly modified (Weber, 1969). The head is produced into a long rostrum at the end of which are a pair of mandibles with laterally developed biting edges which work

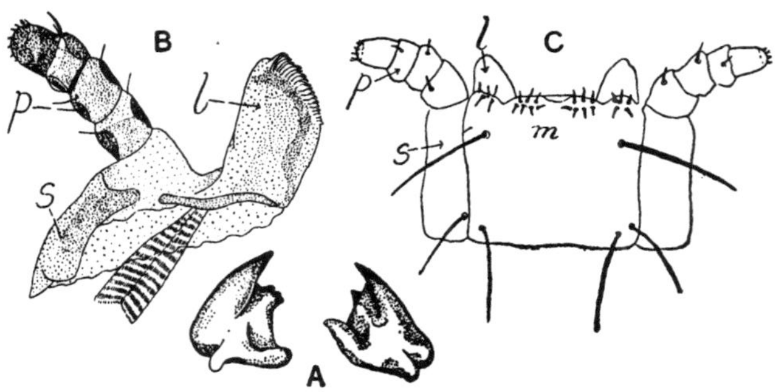

FIG. 310 Mouthparts of Mallophaga. A, mandibles. B, maxilla (*Laemobothrion*). C, labium and maxillae (*Ancistrona*)
l, galea; *m*, mentum; *p*, maxillary palp; *s*, stipes. Adapted from Snodgrass, 1899, 1905.

outwards. Small structures which may be vestigial maxillae and labium are also present and a long tubular food meatus traverses the rostrum to open posteriorly into a cibarial sucking pump. The Mallophagan *hypopharynx* shows some variations in structure (Haub, 1972, 1973), but basically it bears a very strong resemblance to that of the Psocoptera (q.v.). Posteriorly there is a strongly developed sitophore sclerite (the 'oesophageal sclerite' of some authors) from which arises a hollow sclerotized filament which runs forwards and divides into two, the branches uniting on each side with well-developed oval lingual sclerites. The latter have been referred to as lingual glands but there is no evidence that they are associated with secretory cells. The sitophore and lingual sclerites are absent in a few cases. In *Trochilocoetes*, Clay (1949*b*) has found that the hypopharynx is produced into three stylet-like structures which appear to be adapted for piercing.

In the *thorax* (Mayer, 1954), the prothorax is well developed and almost invariably free. Cope (1940; 1941) – who has given a detailed, though in places unorthodox, account of the thoracic morphology of two species – has suggested that the presence of a ventral articulation between prosternum and fore coxae is a character peculiar to the Psocoptera and Mallophaga, but not all lice possess it. The meso- and metasterna are fused together but the nota of the pterothorax may be separate (some Amblycera) or united (Ischnocera). Postnotal sclerites are not differentiated. The thoracic segments are completely fused in *Haematomyzus*. The legs are rather similar throughout the order. The coxae are widely separated and the 1- or 2-segmented tarsi commonly bear a pair of claws, though in the Trichodectidae and Gyropidae, which infest mammals, the claws are single (von Kéler, 1952, 1955). There are no pulvilli, nor is an empodium developed.

The morphology of the abdomen and the genital appendages requires further investigation. The number of visible abdominal segments present in the adult varies from eight to ten. In *Lipeurus heterographus* the reduction to nine segments has been shown by Wilson (1936) to result from the fusion of the 1st and 2nd during embryonic development and of the 9th and 10th in postembryonic growth. The first and second abdominal segments are reduced or fused with adjacent segments. The male genitalia are represented by a median eversible aedeagus, often of complex structure and with parameres, but no other appendicular parts are present (Schmutz, 1955). The ovipositor may be absent or represented by one or two pairs of small gonapophyses.

Internal Anatomy – A general description of the internal organs is given by Snodgrass (1899). The *alimentary canal* (Haug, 1952; Fig. 311, A) is either an almost straight tube, or slightly convoluted, but always comparatively short. It has a well-developed crop, large mid gut, and short simple hind gut. A pair of large enteric caeca extend as outgrowths of the mid gut on either side of the crop. There are four Malpighian tubules and six prominent rectal papillae. Among the Amblycera, the crop is a simple expansion of the oesophagus; in the Ischnocera it is greatly developed, and is either connected with the gut by means of a narrow duct-like tube, as in *Trichodectes*, or assumes a more or less fusiform shape, and extends into the body-cavity to one side of the alimentary canal. In both the main suborders a pair of labial salivary glands, lying beneath the suboesophageal ganglion, opens by a median duct at the base of the labium. In the Ischnocera there is also a pair of thoracic salivary glands, each accompanied by a reservoir, which opens into the anterior part of the gut by an unpaired duct and also two supplementary glands with ducts opening separately into the front of the crop. In the Amblycera there is a pair of sausage-shaped or conical glands opening into the oesophagus in front of the crop and possibly corresponding to the supplementary glands of the Ischnocera. The *nervous system* (Mayer, 1954) is highly specialized: in *Docophoroides* the brain is laterally expanded in such a manner as to be U-shaped, the suboesophageal ganglion is exceptionally large, and is united with the thoracic chain by short thick connectives. The thoracic ganglia are three in number and connectives are wanting; there are no ganglia in the abdomen, the latter being innervated from the metathoracic ganglion. The *tracheal system* (Harrison, 1915) is disposed in two main trunks, opening to the exterior by seven pairs of spiracles (Fig. 309); of these, the first pair is prothoracic, and the remainder are abdominal and situated typically on segments 3 to 8 or, more rarely, on segments 2 to 7. In *Trimenopon* and *Gliricola* there are five pairs of abdominal spiracles on segments 3 to 7, while *Harrisoniella densa* is peculiar in possessing a pair of metathoracic spiracles. Details of spiracular structure are given by Webb (1946). The *heart* (Fulmek, 1906) is situated in the 7th and 8th or 8th segment of the abdomen. It is an extremely short chamber provided with two or three pairs of ostia, and is continued forwards as the aorta; the latter is swollen at its junction with the heart. The *female reproductive organs* consist of a pair of ovaries, each usually composed of five polytrophic ovarioles; in the Amblycera there is a tendency to reduction, and the ovarioles

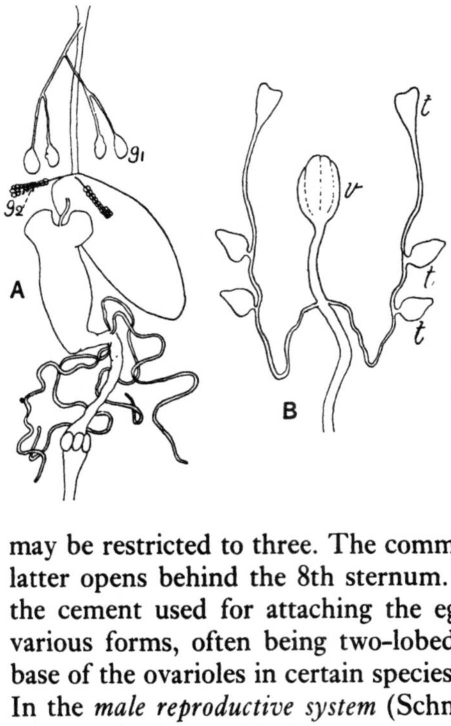

FIG. 311
A, digestive system of *Docophoroides brevis*
B, male reproductive organs of *Ricinus diffusus*
g_1, salivary glands; g_2, supplementary glands.
t, testis; *v*, vesicula seminalis. *After* Snodgrass, *Occas. Papers Calif. Acad. Sci.*, 1899.

may be restricted to three. The common oviduct leads into a vagina and the latter opens behind the 8th sternum. An unpaired accessory gland secretes the cement used for attaching the eggs and the spermatheca may assume various forms, often being two-lobed or double in the Amblycera. At the base of the ovarioles in certain species is an ampulla containing mycetocytes. In the *male reproductive system* (Schmutz, 1955; Fig. 311 B), the testes are composed of three (Amblycera) or two (Ischnocera) ovoid or pyriform follicles, which are quite separate from one another. Those of a side communicate with the corresponding vas deferens, and the two latter canals frequently discharge into a vesicula seminalis. This organ is compact and bilobed, often large, and is continuous distally with a tortuous ejaculatory duct.

Postembryonic Development – The eggs, of which each female usually produces less than 100, are cemented to the feathers or hair of the host, usually in certain favoured areas. They are small, elongate-oval and provided at the anterior pole with a cap, near the margin of which occur several micropyles (Balter, 1968). Posteriorly there is the so-called egg-stigma, consisting of a group of canals which partly or completely traverse the chorion and which assists in the attachment of the egg. Eclosion (Weber, 1939) occurs through internal pressure breaking off the cap around a preformed line of weakness and the rupture of the extra-embryonic membranes by a group of sharp cephalic hatching spines on the embryonic cuticle. The latter is then ruptured before the thorax and its appendages have emerged from the egg and the newly hatched nymph leaves it behind in the egg-shell. There are three nymphal instars which differ from the adults in their smaller size, lighter pigmentation and different chaetotaxy. Little is known of the duration of development but Martin (1934), who bred *Columbicola columbae* on pigeon-feathers in an incubator at 37° C, found the egg stage to last four days and each of the nymphal stages about seven days. For further information see Scott (1952). The number of generations per year under natural conditions is not known accurately, but there are clear indications of seasonal variations in the density of Mallophagan populations, several species having maxima in the winter and summer and minima in spring and autumn (von Kéler, 1969).

Affinities and Phylogeny – There are no fossil Mallophaga and opinions on the affinities and phylogeny of the biting lice are derived from morphological and taxonomic studies. The fact that the Mallophaga and the Psocoptera share such specializations as the sitophore sclerite and sclerotized filament in the hypopharynx, the pick-like lacinia (found in some Mallophaga) and the polytrophic ovarioles, all indicate that the biting lice arose from Psocopteran-like stock. It is also clear that the Mallophaga as defined in this chapter share many specialized features with the sucking lice (Siphunculata). These include the reduced antennae, eyes and spiracular arrangements, the more concentrated nervous system, the operculate eggs, with egg-stigma and micropylar apparatus, and the small number of nymphal instars (Königsmann, 1960). In some respects the Amblycera are markedly more primitive than a group comprising the Ischnocera, Rhynchophthirina and Siphunculata. Königsmann (*loc. cit.*), Clay (1970) and others therefore advocate that the phylogenetic relationships of these groups would be better expressed by a classification in which a single order (the Phthiraptera) is divided into four subordinal groups, the Amblycera, Ischnocera, Rhynchophthirina and Anoplura (= Siphunculata). See also von Kéler (1957).

Classification – As defined here there are about 2800 species of Mallophaga, mostly listed by Hopkins and Clay (1952). There are further bibliographies of the order by von Kéler (1939, 1960) and Eichler and Zlotorzycka (1969). The possibly paraphyletic nature and limits of the order are mentioned above. There has been some disagreement over the number and arrangement of families and the scheme indicated below follows Clay (1970) for the Amblycera and von Kéler (1969) for the Ischnocera.

Suborder 1. AMBLYCERA

Antennae 4- or 5-segmented, short and concealed in grooves at the side of the head; mandible with dorsal and ventral articulations; maxillary palp present, with 2 to 5 segments; labial palp 1-segmented or absent.

FAM. MENOPONIDAE. *Antennae 4- or 5-segmented, with 2 coeloconic sensilla on 4th or 4th and 5th segments; maxillary palps 4-segmented; labial palps present, with 5 distal setae; pro-, meso- and metanota separate; mid and hind legs each with 2 tarsal claws; 6 pairs of spiracles on 3rd to 8th abdominal segments; first abdominal tergum not fused with metanotum.* A large family, the genera of which have been keyed by Clay (1969). Among the main genera are *Actornithophilus* and *Austromenopon* on Charadriiformes, *Menacanthus* and *Myrsidea* on Galliformes and Passeriformes, and *Menopon* on Galliformes.

FAM. BOOPIDAE. *Generally as Menoponidae, but first abdominal tergum fused with metanotum and a seta, usually spiniform, on a protuberance at each side of mesonotum.* Almost exclusively on marsupials from Australia and Papua; *Heterodoxus spiniger* is more widely distributed, on dogs.

FAM. LAEMOBOTHRIIDAE. *As Menoponidae, but differing in chaetotaxy of the head and in having the meso- and metanota fused.* Contains large species (over 10 mm long) from raptorial birds, but other widely distinct hosts are also parasitized by *Laemobothrion*.

FAM. RICINIDAE. *No labial palps; clypeus and labrum enlarged; atypical head chaetotaxy; first abdominal tergum fused to metanotum. Ricinus* on song-birds, *Trochilocoetes* on humming-birds.

FAM. TRIMENOPONIDAE. *Four coeloconic sensilla opening into a cavity on the terminal segment of the 4-segmented antenna; maxillary palps 4- or 5-segmented; labial palp with four distal setae; five pairs of abdominal spiracles on lateral plates of 3rd to 7th segments.* On rodents and marsupials from S. America.

FAM. GYROPIDAE. *Mid and hind legs with a single tarsal claw; at least one pair of legs modified for clasping hairs of host.* On rodents in S. and C. America. Includes *Gyropus* and *Gliricola*, sometimes placed in separate families.

Suborder 2. ISCHNOCERA

Antennae 3- to 5-segmented, relatively long, not concealed; mandibles with anterior and posterior articulations; maxillary palps absent; labial palp 1-segmented.

FAM. DASYONYGIDAE. *Legs with one claw; antennae with 3 segments in male, 5 in female.* On hyraxes.

FAM. BOVICOLIDAE. *Legs with one claw; strong sexual dimorphism, the males short and broad, females more or less cylindrical.* On horses, camels, deer and Bovids, e.g. *Bovicola, Damalinia.*

FAM. TRICHODECTIDAE. *Both sexes broadly oval; legs short and thick, with tibiae broadened distally, with terminal process and one claw.* The main mammal-infesting family of Mallophaga, mostly on Carnivora but some also on rodents, hyraxes, lorises and Cebid monkeys, *Trichodectes, Stachiella, Felicola* and others. In a wider sense the Trichodectidae includes the previous two families.

FAM. PHILOPTERIDAE. *Legs with two claws; front margin of clypeus smoothly curved; last antennal segment cylindrical.* The major bird-infesting family of Mallophaga, including about 60 per cent of the European biting lice (Clay, 1951). Distributed over a wide variety of hosts and represented by such genera as *Anatoecus*, on ducks; *Goniodes* and its allies, mainly on galliform birds; *Degeeriella* on birds of prey; *Philopterus* and *Brueelia* on song-birds; *Columbicola* on doves; and *Ardeicola* on herons and storks.

FAM. TRICHOPHILOPTERIDAE. *Legs with two claws; clypeus with trapezoidal outline; last antennal segment clubbed; head with strong bristles.* Only *Trichophilopterus*, on lemurs.

Suborder 3. RHYNCHOPHTHIRINA

Head prolonged anteriorly into a rostrum; mandibles at apex of rostrum, working outwards; labium and maxillae vestigial; thoracic segments fused.

FAM. HAEMATOMYZIDAE. An aberrant family, containing only two species: *Haematomyzus elephantis*, found on the Indian and African elephants (Ferris, 1931; Weber, 1939, 1969), and *H. hopkinsi* on wart-hogs (Clay, 1963).

Literature on the Mallophaga

ASH, J. S. (1960), A study of the Mallophaga of birds with particular reference to their ecology, *Ibis*, **102**, 93–110.

BALTER, R. S. (1968), The microtopography of avian lice eggs, *Med. Biol. Illustr.*, 18, 166–179.

BEIER, M. (1936), Mallophaga, Federlinge oder Pelzfresser, In: Schulze, P. (ed.), *Biologie der Tiere Deutschlands*, **39**, 32 pp.

BUCKUP, L. (1959), Der Kopf von *Myrsidea cornicis* (DeGeer) (Mallophaga-Amblycera), *Zool. Jb.* (*Anat.*), 77, 241–288.

CLAY, T. (1949*a*), Some problems in the evolution of a group of ectoparasites, *Evolution*, **3**, 279–299.

—— (1949*b*), Piercing and sucking mouthparts in the biting lice (Mallophaga), *Nature*, **164**, 617.

—— (1951), An introduction to the classification of the avian Ischnocera (Mallophaga), Part 1, *Trans. R. ent. Soc. Lond.*, **102**, 171–194.

—— (1963), A new species of *Haematomyzus* Piaget (Phthiraptera, Insecta), *Proc. zool. Soc. Lond.*, **141**, 153–161.

—— (1969), A key to the genera of the Menoponidae (Amblycera: Mallophaga, Insecta), *Bull. Brit. Mus. nat. Hist.* (*Ent.*), **24**, 1–26.

—— (1970), The Amblycera (Phthiraptera: Insecta), *Bull. Brit. Mus. nat. Hist.* (*Ent.*), **25**, 73–98.

CLAY, T. AND MEINERTZHAGEN, R. (1943), The relationship between Mallophaga and Hippoboscid flies, *Parasitology*, **35**, 11–16.

COPE, O. B. (1940), The morphology of *Esthiopterum diomedeae* (Fabr.) (Mallophaga), *Microentomology*, **5**, 117–142.

—— (1941), The morphology of a species of the genus *Tetrophthalmus* (Mallophaga: Menoponidae), *Microentomology*, **6**, 71–92.

CRAUFURD-BENSON, H. J. (1941), The cattle-lice of Great Britain. Parts 1, 2. Biology, with special reference to *Haematopinus eurysternus*, *Parasitology*, **33**, 331–342; 343–358.

CRUTCHFIELD, C. M. AND HIXSON, H. (1943), Food habits of several species of poultry lice with special reference to blood consumption, *Florida Ent.*, **26**, 63–66.

EICHLER, W. (1963), Mallophaga, In: Bronn's *Klassen und Ordnungen des Tierreichs*, **5** (3), 7(b), 1–290.

EICHLER, W. AND ZLOTORZYCKA, J. (1969), Zeitgenössische Mallophagen-Literatur, *Angew. Parasit.*, **10**, 53–60, 44–124.

EWING, H. E. (1924), On the taxonomy, biology and distribution of the biting lice of the family Gyropidae, *Proc. U.S. nat. Mus.*, **63**, 1–42.

—— (1936), The taxonomy of the Mallophagan family Trichodectidae, with special reference to the New World fauna, *J. Parasitol.*, **22**, 233–246.

FERRIS, G. F. (1931), The louse of elephants, *Haematomyzus elephantis*, *Parasitology*, **23**, 112–127.

FULMEK, L. (1906), Beiträge zur Kenntnis des Herzens der Mallophagen, *Zool. Anz.*, **29**, 619–621.

HARRISON, L. (1915), The respiratory system of Mallophaga, *Parasitology*, **8**, 101–127.

HAUB, F. (1971), Der Kopf von *Ornithobius cygni* (Denny) (Mallophaga-Ischnocera), *Zool. Jb.* (*Anat.*), **88**, 450–504.

—— (1972), Das Cibarialsklerit der Mallophaga-Amblycera und der Mallophaga-Ischnocera (Kellogg) (Insecta), *Z. Morph. Tiere*, **73**, 249–261.

—— (1973), Das Cibarium der Mallophagen. Untersuchungen zur morphologischen Differenzierung, *Zool. Jb.* (*Anat.*), **90**, 483–525.

HAUG, G. (1952), Morphologische und histophysiologische Untersuchungen an den Verdauungsorganen der Mallophagen und Anopluren, *Zool. Jb.* (*Anat.*), **72**, 302–344.

HOPKINS, G. H. E. (1942), The Mallophaga as an aid to the classification of birds, *Ibis*, (14) **6**, 94–106.

—— (1949), The host-associations of the lice of mammals, *Proc. Zool. Soc. Lond.*, **119**, 387–605.

HOPKINS, G. H. E. AND CLAY, T. (1952), *A Check List of the Genera and Species of Mallophaga*. London, 362 pp. (For additions and corrections see *Ann. Mag. nat. Hist.*, (12), **6**, 434–448; (12), **8**, 177–190.)

KEIRANS, J. E. (1975), A review of the phoretic relationship between Mallophaga (Phthiraptera: Insecta) and Hippoboscidae (Diptera: Insecta), *J. med. Ent.*, **12**, 71–76.

KÉLER, S. VON (1938–39), Baustoffe zu einer Monographie der Mallophagen. I. Teil. Ueberfamilie der Trichodectoidea. II. Teil. Ueberfamilie der Nirmoidea (i), *Nova Acta Leop. Carol.*, **5**, 393–467; **8**, 1–254.

—— (1939), Uebersicht über die gesamte Literatur der Mallophagen, *Z. angew. Ent.*, **25**, 487–524.

—— (1944), Bestimmungstabelle der Ueberfamilie Trichodectoidea, *Stettin ent. Ztg.*, **105**, 167–191.

—— (1952), Über den feineren Bau des Tarsen bei *Pseudomenopon rowanae* v. Kéler, *Beitr. Ent.*, **2**, 573–582.

—— (1955), Einige Bemerkungen über den Bau der Tarsen von *Gyropus* und *Gliricola*, *Beitr. Ent.*, **5**, 293–308.

—— (1957), Über die Dezendenz und die Differenzierung der Mallophagen, *Z. Parasitenk.*, **18**, 5–160.

—— (1960), Bibliographie der Mallophagen, *Mitt. zool. Mus., Berlin*, **36**, 146–403.

—— (1963), 13. Ordnung: Läuslinge, Federlinge und Haarlinge. Mallophaga, *Tierwelt Mitteleuropas*, **4** (2), Hft. 7b, 31 pp.

—— (1969), 17. Ordnung Mallophaga (Federlinge und Haarlinge), In: Helmcke, J.-G., Starck, D. and Wermuth, H. (eds.), *Handbuch der Zoologie*, **4** (2), Lfg. 10, 1–72.

KÖNIGSMANN, E. (1960), Die Phylogenie der Parametabola, *Beitr. Ent.*, **10**, 705–744.

MARTIN, M. (1934), Life history and habits of the pigeon louse (*Columbicola columbae* (Linn.)), *Can. Ent.*, **66**, 1–16.

MATHYSSE, J. G. (1946), Cattle lice, their biology and control, *Cornell Univ. agric. Exp. Sta. Bull.*, **832**, 1–67.

MAYER, C. (1954), Vergleichende Untersuchungen am Skelett-Muskelsystem des Thorax der Mallophagen unter Berücksichtigung des Nervensystems, *Zool. Jb. (Anat.)*, **74**, 77–131.

NEUFFER, G. (1954), Die Mallophagenhaut und ihre Differenzierungen, *Zool. Jb. (Anat.)*, **73**, 450–519.

RIES, E. (1931), Die Symbiose der Läuse und Federlinge, *Z. Morph. Ökol. Tiere*, **20**, 233–367.

RISLER, H. (1951), Der Kopf von *Bovicola caprae* (Gurlt) (Mallophaga), *Zool. Jb. (Anat.)*, **71**, 325–374.

RISLER, H. AND GEISINGER, K. (1965), Die Mundwerkzeuge von *Gliricola gracilis* N. (Mallophaga: Amblycera), ein Beitrag zur Kopfmorphologie der Tierläuse (Phthiraptera), *Zool. Jb. (Anat.)*, **82**, 532–546.

SCHMUTZ, W. (1955), Zur Konstruktionsmorphologie des männlichen Geschlechtsapparates der Mallophagen, *Zool. Jb. (Anat.)*, **74**, 211–316.

SCOTT, M. T. (1952), Observations on the bionomics of the sheep body louse (*Damalinia ovis* L.), *Aust. J. agr. Res.*, **3**, 60–67.

SNODGRASS, R. E. (1899), The anatomy of the Mallophaga, *Occ. Pap. Calif. Acad. Sci.*, **6**, 145–229.

STÖWE, E. (1943), Der Kopf von *Trimenopon jenningsi* Kellogg und Paine. Eine morphologische Untersuchung unter besonderer Berücksichtigung des Nervensystems und der Drüsen, *Zool. Jb. (Anat.)*, **68**, 177–226.

SYMMONS, S. (1952), Comparative anatomy of the Mallophagan head, *Trans. zool. Soc. Lond.*, **27**, 349–436.

TIMMERMAN, G. (1957), Studien zu einer vergleichenden Parasitologie der Charadriiformes oder Regenpfeifervögel. Teil 1: Mallophaga, *Parasitol. Schriftenreihe, Jena*, **8**, 1–204.

—— (1965), Die Federlingsfauna der Sturmvögel und die Phylogenese des procellariiformen Vogelstammes, *Abh. Verh. Naturw. Ver. Hamburg*, (N.F.), 8 (Suppl.), 249 pp.

WEBB, J. E. (1946), Spiracle structure as a guide to the phylogenetic relationships of the Anoplura (biting and sucking lice), with notes on the affinities of the mammalian hosts, *Proc. zool. Soc. Lond.*, **116**, 49–119.

WEBER, H. (1939), Zur Eiablage und Entwicklung der Elefantenlaus *Haematomyzus elephantis* Piaget, *Biol. Zbl.*, **59**, 98–109; 397–409.

—— (1969), Die Elefantenlaus, *Haematomyzus elephantis* Piaget, 1869, Versuch einer konstruktionsmorphologischen Analyse, *Zoologica Stuttg.*, **41** (1), 1–154.

WILSON, F. H. (1936), The segmentation of the abdomen of *Lipeurus heterographus* Nitzsch. (Mallophaga), *J. Morph.*, **60**, 211–219.

WUNDRIG, G. (1936), Die Sehorgane der Mallophagen, nebst vergleichenden Untersuchungen an Liposceliden und Anopluren, *Zool. Jb.* (*Anat.*), **62**, 45–110.

ZLOTORZYCKA, J. (1965), Mallophaga parasitising Passeriformes and Pici. I–IV, *Acta Parasit. pol.*, **12**, 165–192; 239–282; 401–432; **13**, 41–70.

—— (1968), Parasitophyletische Probleme bei den Mallophagen von Passeres und Pici, *Angew. Parasit.*, **9**, 45–53, 97–119.

Order 18

SIPHUNCULATA (ANOPLURA; SUCKING LICE)

Apterous insects living as ectoparasites of mammals. Eyes reduced or absent. Ocelli absent. Antennae 3- to 5-segmented. Mouthparts highly modified for piercing and sucking, retracted within the head when not in use. Thoracic segments fused; tarsi 1-segmented, claws single. Thoracic spiracles dorsal. Cerci absent. Metamorphosis slight.

The insects included in this order are exclusively blood-sucking ectoparasites of mammals and about 300 species have been described. Of these, two species infest man and about a dozen occur on domestic animals; the remainder have been taken from several orders of mammals including monkeys, rabbits, mice, seals, elephants, etc. There is no doubt that a great number of species are still undescribed. As in the Mallophaga (q.v.) there is considerable host-specificity and closely related host species tend to be infested by similar Siphunculates, e.g. *Enderleinellus* is restricted to squirrels (Sciuridae) and *Pedicinus* to the Cynomorph monkeys. A comprehensive and critical review of Siphunculate host-relations is given by Hopkins (1949) while Webb (1949) has discussed phylogenetic aspects of the host-parasite association for the lice of Ferungulate mammals. Kellogg (1913) pointed out the close physiological relationships between certain of these parasites and the specific blood-characters of their hosts as determined by precipitin tests and it is possible that a lethal effect of abnormal host-blood on the essential symbiotic bacteria which occur in most Siphunculates is a factor helping to determine host-specificity.

The best known species of Siphunculata is *Pediculus humanus*, the common louse of man (Fig. 312). It infests people living under unhygienic conditions and who go for a number of days without change of clothing. This insect exists in at least two forms which have been regarded as separate species; they are *P. humanus capitis*, the head louse, and *P. humanus corporis*, the body louse. A detailed comparison of the two forms has been made by Busvine (1948) and earlier workers, whose results, however, do not coincide in all details. Morphologically, populations of the two forms differ significantly in total body size, head dimensions and antennal length, the body louse being larger with a larger head and longer antennae though there is appreciable overlapping in the frequency distributions of the measurements. Biologically they differ in that body lice live between the clothes and skin of the infested person and are more resistant to starvation than the form *capitis* which is confined to the hair of the head. The two forms breed together readily under experimental conditions, the fertile hybrids being mor-

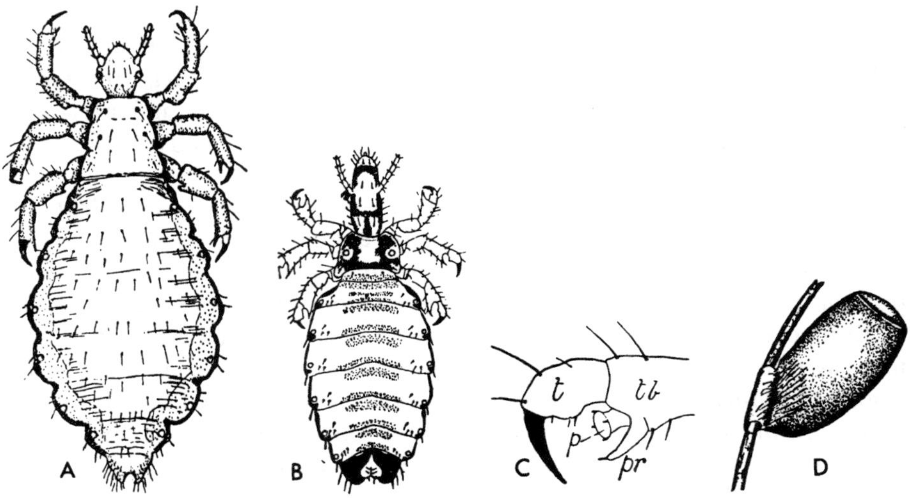

FIG. 312 A. Body louse (*Pediculus humanus*), female. B. Female hog louse (*Haematopinus suis*). C. Extremity of leg of *Haematopinus suis*. D. Egg attached to a bristle (*H. suis*)

t, tarsus; *tb*, tibia; *pr*, process of tibia; *p*, tibial pad.

phologically intermediate between the parents but in the F_2 and F_3 generations they become identical with the body strain. Earlier investigations suggested that the differences between the two forms were environmentally induced and that *capitis* reared on the body underwent a transformation to *corporis*. Busvine, however, found that head lice reared on the body for 43 generations did not change systematically in biometric characters.

The extensive literature on the biology and control of *Pediculus humanus* has been summarized by Buxton (1947). The species is of considerable medical importance since in addition to the relatively unimportant effects of irritation and inflammatory reaction caused by its feeding, it transmits four more serious diseases (see also Weyer, 1960). The most important of these is epidemic typhus, caused by the micro-organism *Rickettsia prowazeki* which is ingested by the louse when feeding on an infected person, multiplies in the gut and passes out in the faeces. Contact of the faeces (even after they have dried or when a louse is crushed) with a skin abrasion leads to infection, though it is likely that the disease is also acquired when dried, infective louse faeces are inhaled or come into contact with the conjunctiva. The disease is not transmitted by the feeding of infected lice. Another rickettsial disease transmitted by *Pediculus humanus* was the Trench Fever which appeared during the 1914–1918 War but afterwards died out completely. The mode of transmission probably resembled that of epidemic typhus. A third disease, murine typhus, related to the epidemic form, is normally flea-borne but there is evidence that it may sometimes become louse-borne. Finally, one form of relapsing fever (due to infection by *Borellia recurrentis*) is carried by the body louse, transmission occurring when a louse infected by feeding is crushed and the spirochaetes in its haemolymph enter through abraded skin.

The only other Siphunculate infesting man is *Pthirus pubis*, the Crab Louse, which is restricted mainly to the hair of the pubic and peri-anal regions but is more rarely found on other hairy parts of the body. So far as is known it does not transmit any disease.

Among other genera of the order one of the most prevalent is *Haematopinus* which is mostly parasitic upon ungulates: *H. suis* (Fig. 312) is the well-known hog louse which occurs on domestic and wild pigs in many parts of the world (Florence, 1921): *H. tuberculatus* is found on the buffalo in E. Europe and the Orient, and *H. eurysternus* occurs on domestic cattle and may, at times, prove a pest. Species of *Polyplax* find most of their hosts among the Muridae, and *P. spinulosus* transmits *Trypanosoma lewisi* from rat to rat. *Echinophthirius* and its allies exclusively infest marine mammals (seals, sea-lions and walruses), and the anomalous genus *Enderleinellus* occurs only on Sciuridae.

External Anatomy – The body of a louse is dorsoventrally flattened and only the abdomen is distinctly segmented. The head (Ramcke, 1965) is more or less conical and pointed and, in *Linognathus*, it is much attentuated and relatively little broadened behind the antennae. Though some of the sutures of the typical insect head are visible in the nymphs they are mostly obliterated in adult lice. There is no tentorium. The antennae are short and 3- to 5-segmented; in *Pediculus* and *Pthirus* they are 3-segmented in the first instar, but afterwards become 5-segmented, and in *Pedicinus* they are 3-segmented throughout life. The eyes are reduced to one ommatidium and sometimes absent but are relatively large in *Pediculus* (Wundrig, 1936; Webb, 1948).

The mouthparts (Fig. 313) are difficult to investigate owing to their minute size and delicate structure. Numerous, somewhat discordant, accounts have been given and these are summarized by Stojanovich (1945), whose description of the structure and homologies of the mouthparts in four species forms the basis of the following account (see also Snodgrass, 1944). The labrum forms the dorsal wall of a small snout-like proboscis which is armed internally with small teeth and which is everted to grip the host during feeding. Both cibarium and pharynx are well provided with dilator muscles and together form a powerful sucking pump. Opening off the ventral side of the cibarium is a well-developed pouch, the trophic sac, in which

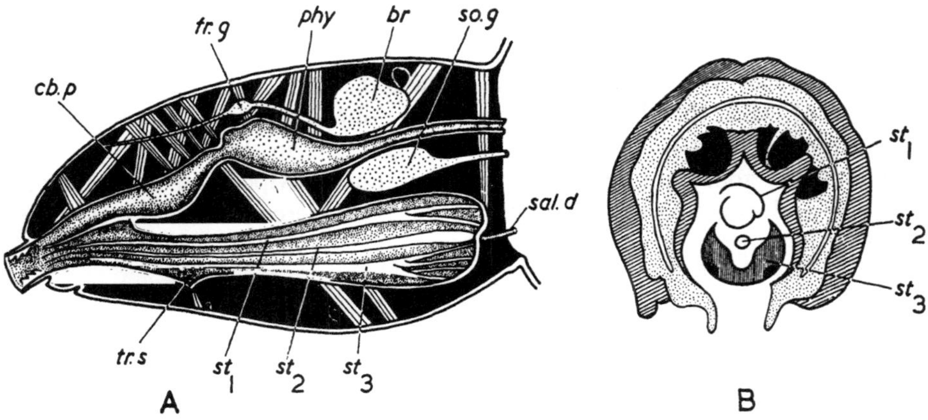

FIG. 313 A. Longitudinal section through head of Siphunculate louse (*after* Snodgrass, 1944). B. Transverse section through head and trophic sac of *Pediculus* (*from* Snodgrass, 1944, *after* Vogel, 1921)

br, brain; *cb.p*, cibarial sucking pump; *fr.g*, frontal ganglion; *phy*m pharynx; *sal.d*, salivary duct; *so.g*, suboesophageal ganglion; *st* 1–3, stylets; *tr.s*, trophic sac.

are accommodated three stylets, lying one above the other with the dorsal and ventral ones forked at their bases. These are the effective piercing organs. The dorsal stylet probably represents a highly modified hypopharynx, the middle stylet is regarded as a modification of the opening of the salivary duct and is pierced throughout its length by this duct while the ventral stylet is believed to be the greatly modified labium. Schölzel (1937), in his account of the embryological development of the mouthparts, showed that the mandibular rudiments disappear and it seems that a pair of small sclerotized structures situated lateroventral to the trophic sac are derived from the maxillary rudiments. It should, however, be mentioned that Fernando (1933) gives a somewhat different account of the development of the mouthparts, asserting that the dorsal stylet is formed of the fused maxillae and Ferris (1951) reaches a similar interpretation by considering the musculature and innervation of this stylet (see also Young, 1953; von Kéler, 1961). When the insect feeds its mode of action is probably as follows. The rostrum is everted and its denticles maintain a hold on the skin of the host. Special muscles come into play which draw the cibarium and pharynx forward, with the result that the cibarium and the opening of the trophic sac come into contact with the skin. The contraction of protractor muscles associated with the stylets brings the latter into action and they perforate the skin: at the same time saliva enters the puncture. By the periodic contraction of the cibarial and pharyngeal dilators blood is then sucked up from the wound (see von Kéler, 1966).

The *thorax* is relatively small and its segments are largely fused. The legs are strongly developed in accordance with a mode of life which requires appendages adapted for maintaining a firm hold on the host. The tarsi are single-segmented, and each ends in a powerful claw which works against a tibial process.

The *abdomen* is 9-segmented: the terga and sterna are, as a rule, thinly sclerotized, while the pleura are strongly developed and deeply pigmented. A copulatory organ is well developed in the male, and in the female there is a pair of short gonapophyses which are used during oviposition for grasping the hair and directing the alignment of the eggs. Cerci are wanting in both sexes.

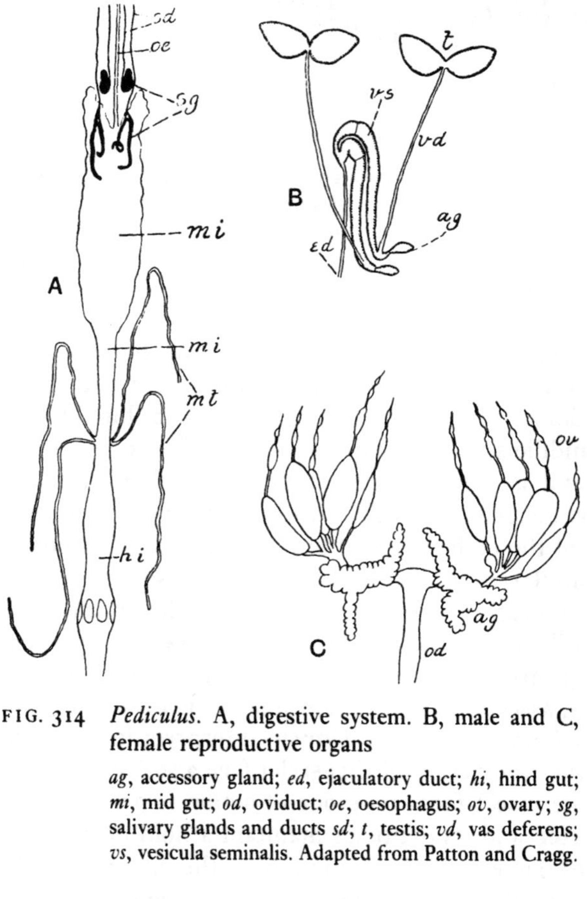

FIG. 314 *Pediculus*. A, digestive system. B, male and C, female reproductive organs

ag, accessory gland; *ed*, ejaculatory duct; *hi*, hind gut; *mi*, mid gut; *od*, oviduct; *oe*, oesophagus; *ov*, ovary; *sg*, salivary glands and ducts *sd*; *t*, testis; *vd*, vas deferens; *vs*, vesicula seminalis. Adapted from Patton and Cragg.

Internal Anatomy (Fig. 314) – Most of what is known concerning the internal anatomy relates to *Pediculus* and *Haematopinus*. The anterior portion of the fore gut has already been referred to, and the oesophagus passes directly to the mid gut, both crop and gizzard being undeveloped. The mid gut is a large chamber which narrows posteriorly, and occupies the greater portion of the abdominal cavity; in *Pediculus* a pair of large enteric caeca is present anteriorly. The hind gut presents no convolutions and receives four Malpighian tubules, and the rectum is provided with six sclerotized rectal papillae. There are two pairs of salivary glands in the thorax; one pair is elongate and tubular, the other compact and reniform; their ducts apparently combine to form the salivary canal already mentioned in relation to the mouthparts. A pair of glands, known as Pawlowsky's glands, open into the stylet sac and their secretion possibly lubricates the stylets. The *tracheal system* shows a general agreement with the simpler Mallophagan type (Harrison, 1915); there are usually seven pairs of spiracles, a mesothoracic pair being dorsal, and the abdominal spiracles opening on segments 3 to 8. The detailed structure of the spiracles has been described by Webb (1946; 1949) who considers that they provide characters of considerable phylogenetic importance. The *female reproductive organs* consist of five polytrophic ovarioles to each ovary and there is also a pair of accessory glands but a spermatheca is not always present. The *male reproductive organs* are composed of a pair of compact bilobed testes and the slender vasa deferentia either open into a pair of tubular vesiculae seminales (*Pediculus*) or discharge separately from the latter into the ejaculatory duct (*Pthirus*). In *Pediculus* copulation takes place at frequent intervals and this is probably related to the absence of a receptaculum in the female and presence of well-developed vesiculae seminales in the male; according to Bacot one male may fertilize 10 to 18 females. The *nervous system* is highly concentrated, the thoracic and abdominal ganglia being fused into a common mass (Hafner, 1971). The majority of Siphunculates possess symbiotic bacteria localized in mycetocytes which may occur separately in the gut-wall or grouped together in a special organ usually rather closely associated with the gut. In mature females the mycetocytes are found only in the ovarial ampullae, from which they pass into the eggs (Ries, 1931).

Postembryonic Development – Though *Pediculus humanus* often lays its eggs scattered and unattached, the other Siphunculates cement them to the hairs of the host in a manner similar to the Mallophaga. There is an egg-stigma at the posterior end which assists attachment and at the anterior pole is an approximately oval lid pierced by micropylar canals (5–24 in *Pediculus*) and often a group of micropylar chambers ('air cells'). *P. humanus* lays about 300 eggs at the rate of 8–12 daily; *Haematopinus eurysternus* not more than 24 (Craufurd-Benson, 1941). Eclosion (Weber, 1939) is very similar to the same process in the Mallophaga (q.v.). There are three nymphal instars, each usually having a distinctive chaetotaxy. The rate at which development occurs has been investigated for only a few species, e.g. under normal conditions *Haematopinus eurystenus* has an incubation period of 12 days and its nymphal development takes a further 12 days while for *Pediculus humanus* the corresponding figures are 8 and 16–19 days. Sexual maturity is attained after 1–3 days of adult life. The life-cycle of *Antarctophthirus* is closely linked to that of its host, the Weddell seal (Murray *et al.*, 1965).

Affinities and Phylogeny – The affinities of the Siphunculata have been mentioned briefly on p.664. There are no fossil lice but the anatomy of Recent species suggests that they are most closely related to the Ischnocera and Rhynchophthirina, with which they share the following specialized features (Königsmann, 1960): the similarly segmented antennae, the presence of an obturaculum, spiracular glands and symbionts; the testis of two follicles, and the eye composed of not more than one ommatidium. It seems not unlikely, therefore, that the Siphunculata arose from some Trichodectid-like stock living on early mammals. They have, however, undergone considerable further specialization of their own in the structure of the mouthparts, the loss of the tentorium and the fusion of the thoracic segments. The course of evolution within the Siphunculata is largely a matter for further study. Ferris (1951) has emphasized that Siphunculate evolution at about generic level is well correlated with the evolution of the hosts, but that there is a far less close relationship between the Siphunculate families and the hosts they infest.

Classification – Ferris (1919–35; 1951) has monographed the species of the world but the supra-generic classification of the order has not yet been satisfactorily settled. Webb (1946, 1949) has devised a classification based primarily on spiracular structure but it seems desirable that other characters should also be taken into account and the arrangement below is based on the work of Ferris (1951). It should be noted that the limits of the families recognized by this authority differ in important respects from those of earlier workers.

Twenty-five species occur on indigenous or domesticated British mammals, and in addition to Ferris's monographs the keys of von Kéler (1963), Jancke (1938) and Séguy (1944) may be used to identify the species of N.W. Europe. There is a summary of data on the occurrence and distribution of Siphunculata by Ludwig (1968).

FAM. ECHINOPHTHIRIIDAE. *Body densely clothed with thick setae, sometimes modified into scales; abdomen without sclerotized tergal, paratergal and sternal plates.* Confined to the Carnivora Pinnipedia, e.g. *Antarctophthirus*, *Echinophthirius*.

FAM. HAEMATOPINIDAE. *Setae normal; paratergal plates not projecting from body; abdominal cuticle wrinkled, sometimes with indistinct tergal and sternal plates.* Includes *Haematopinus*, with about a dozen species such as *H. suis* on pigs, *H. asini* on horses and zebras, and *H. eurysternus* on cattle.

FAM. HOPLOPLEURID E. *Paratergal plates projecting apically from body; tergal and sternal plates usually distinct.* The largest family, its members occurring mainly on rodents. *Polyplax* on Muridae and a few shrews (Insectivora); *Hoplopleura*, a large genus, on various rodents; *Haemodipsus*, with *H. ventricosus* on rabbits.

FAM. LINOGNATHIDAE. *Abdominal paratergal plates absent; abdomen membranous; 6 pairs of abdominal spiracles.* Restricted to Artiodactyla and Hyracoidea. *Linognathus*, with species on sheep, goats and cattle; *Solenopotes*.

FAM. NEOLINOGNATHIDAE. *Characters as given for Linognathidae but only one pair of spiracles, on eighth segment.* A peculiar group, probably related to the Hoplopleuridae. Two species on Macroscelidid insectivores.

FAM. PEDICULIDAE. *Characters as given for Haematopinidae but abdominal cuticle unwrinkled and membranous except for genital region.* On Primates, including man. *Pediculus, Pthirus.*

Literature on the Siphunculata

Ferris (1919–35; 1951) gives a valuable taxonomic bibliography and Hopkins (1949) provides numerous biological references and a host-parasite catalogue.

BUSVINE, J. R. (1948), The 'head' and 'body' races of *Pediculus humanus* L., *Parasitology*, 39, 1–16.

BUXTON, P. A. (1947), *The Louse, An account of the Lice which infest Man, their medical importance and control*, Arnold, London, 2nd edn, 164 pp.

CRAUFURD-BENSON, H. J. (1941), The cattle-lice of Great Britain. I, II. Biology, with special reference to *Haematopinus eurysternus*, *Parasitology*, 33, 331–342, 343–358.

EWING, H. E. (1932), The male genital armature in the order Anoplura, or sucking lice, *Ann. ent. Soc. Am.*, 25, 657–669.

FERNANDO, W. (1933), The development and homologies of the mouthparts of the head louse, *Q. Jl microsc. Sci.*, 76, 231–241.

FERRIS, G. F. (1919–35), Contributions towards a monograph of the sucking lice. I–VIII, *Stanford. Univ. Publ. Biol. Sci.*, 634 pp.

—— (1951), The Sucking Lice, *Mem. Pacific Coast ent. Soc.*, 1, 1–320.

FLORENCE, L. (1921), The hog louse, *Haematopinus suis:* its biology, anatomy and histology, *Cornell Univ. agric. Exp. Sta.*, *Memoir*, 51, 637–742.

FREUND, L. (1935), Läuse, Anoplura, In: Brohmer, Ehrmann and Ulmer, *Die Tierwelt Mitteleuropas*, 4: IX, 1–26.

HAFNER, P. (1971), Muskeln und Nerven des Abdomens besonders des männlichen Geschlechtsapparates von *Haematopinus suis* (Anoplura), *Zool. Jb.* (*Anat.*), 88, 421–449.

HARRISON, L. (1915), The respiratory system of Mallophaga, *Parasitology*, 8, 101–127.

HASE, A. (1931), Siphunculata; Anoplura; Aptera; Läuse, In: Schulze (ed.), *Biologie der Tiere Deutschlands*, 30, 58 pp.

HOPKINS, G. H. E. (1949), The host-associations of the lice of mammals, *Proc. zool. Soc. Lond.*, 119, 387–604.

JANCKE, O. (1938), Die Anopluren Deutschlands, In: *Tierwelt Deutschlands*, 35, 43–78.

JOHNSON, P. T. (1960), The Anoplura of African rodents and insectivores, U.S. Dept. Agric. Tech. Bull., 1211, 1–116.

KÉLER, S. VON (1961), Mandibelrudimente der Anopluren und ihre stammesgeschichtliche Bedeutung, *Beitr. Ent.*, 11, 930–942.

—— (1963), 14. Ordnung. Läuse, Anoplura. Nachträge und Berichtigungen, *Tierwelt Mitteleuropas*, 4 (2), Heft VIII, 14 pp.

—— (1966), Mandibelrudimente der Anopluren und ihre syngenische Bedeutung; III. Bau und Funktion des Stachels, *Z. Parasitenk.*, 27, 287–316.

KELLOGG, V. L. (1913), Ectoparasites of the monkeys, apes and man, *Science*, 38, 601–602.

KÖNIGSMANN, E. (1960), Die Phylogenie der Parametabola, *Beitr. Ent.*, **10**, 705–744.

KÜHN, H.-J. AND LUDWIG, H. W. (1967), Die Affenläuse der Gattung *Pedicinus*, *Z. zool. Syst. Evolutionsforsch.*, **5**, 144–297.

LUDWIG, H. W. (1968), Zahl, Vorkommen und Verbreitung der Anoplura, *Z. Parasitenk.*, **31**, 254–265.

MURRAY, M. D. AND NICHOLLS, D. G. (1965), Studies on the ectoparasites of seals and penguins I. The ecology of the louse *Lepidophthirus macrorhini* Enderlein on the southern elephant seal *Mirounga leonina* (L.), *Aust. J. Zool.*, **13**, 437–454.

MURRAY, M. D., SMITH, M. S. R. AND SOUCEK, Z. (1965), Studies on the ectoparasites of seals and penguins II. The ecology of the louse *Antarctophthirus ogmorhini* Enderlein on the Weddell seal, *Leptonychotes weddelli*, *Aust. J. Zool.*, **13**, 761–771.

RAMCKE, J. (1965), Der Kopf der Schweinelaus (*Haematopinus suis* L. Anoplura), *Zool. Jb.* (*Anat.*), **82**, 547–663.

RIES, E. (1931), Die Symbiose der Läuse und Federlinge, *Z. Morph. Ökol. Tiere*, **20**, 233–267.

SCHÖLZEL, G. (1937), Die Embryologie der Anopluren und Mallophagen, *Z. Parasitenk.*, **9**, 730–770.

SÉGUY, E. (1944), Insectes ectoparasites (Mallophaga, Anoploures, Siphonaptera), *Faune de France*, **43**, 684 pp.

SNODGRASS, R. E. (1944), The feeding apparatus of biting and sucking insects affecting man and animals, *Smithson. misc. Collns*, **104** (7), 113 pp.

STOJANOVICH, C. J. (1945), The head and mouthparts of the sucking lice (Insecta: Anoplura), *Microentomology*, **10**, 1–46.

VANZOLINI, P. E. AND GUIMARAES, L. R. (1955), Lice and the history of S. American land mammals, *Rev. brasil. Ent.*, **3**, 13–46.

WEBB, J. E. (1946), Spiracle structure as a guide to the phylogenetic relationships of the Anoplura (biting and sucking lice), with notes on the affinities of the mammalian hosts, *Proc. zool. Soc. Lond.*, **116**, 49–119.

—— (1948), Eyes in the Siphunculata, *Proc. zool. Soc. Lond.*, **118**, 575–577.

—— (1949), The evolution and host-relationships of the sucking lice of the Ferungulata, *Proc. zool. Soc. Lond.*, **119**, 133–188.

WEBER, H. (1929), Biologische Untersuchungen an der Schweinelaus (*Haematopinus suis*) unter besonderer Berücksichtigung der Sinnesphysiologie, *Z. vergl. Physiol.*, **9**, 564–612.

—— (1939), Zur Eiablage und Entwicklung der Elefantenlaus, *Haematomyzus elephantis* Piaget, *Biol. Zbl.*, **59**, 98–109, 397–409.

WEYER, F. (1960), Biological relationships between lice (Anoplura) and microbial agents, *Ann. Rev. Ent.*, **5**, 405–420.

WUNDRIG, G. (1936), Die Sehorgane der Mallophagen, nebst vergleichenden Untersuchungen an Liposceliden und Anopluren, *Zool. Jb.* (*Anat.*), **62**, 45–110.

YOUNG, J. H. (1953), Embryology of the mouthparts of Anoplura, *Microentomology*, **18**, 85–133.

Order 19

HEMIPTERA (RHYNCHOTA: PLANT BUGS, ETC.)

Two pairs of wings usually present; the anterior pair most often of harder consistency than the posterior pair, either uniformly so (Homoptera) or with the apical portion more membranous than the remainder (Heteroptera). Mouthparts piercing and suctorial, palps atrophied; the labium forming a dorsally grooved sheath in which lie two pairs of bristle-like stylets (modified mandibles and maxillae). Metamorphosis usually gradual, rarely complete.

The Hemiptera or true bugs are most easily recognized by the form of the mouthparts. These are adapted for piercing and sucking, and this habit is prevalent throughout life except in male Coccoidea, where the adult mouthparts are atrophied. The wings and other body structures vary greatly within the Hemiptera so that no further general morphological definition of the order can be given.

The Hemiptera cause a vast amount of direct and indirect damage to plants and for this reason there are few other orders of insects so inimical to man's welfare. Among the more destructive species are the Cotton Stainers (*Dysdercus* spp., Fig. 315), the Green Vegetable Bug *Nezara viridula*, the Chinch Bug *Blissus leucopterus*, various leaf-hoppers (Cicadellidae and related families), the white-flies (Aleyrodidae), the aphids or plant-lice (Aphidoidea) and the scale insects and mealy bugs (Coccoidea). Some Homoptera, notably among the Aphidoidea, are vectors of phytopathogenic viruses; the species *Myzus persicae* transmits over 50 such diseases, including mosaic and yellows of sugar-beet and leaf-roll of potatoes. Some Cicadellidae are also concerned in the transmission of aster yellows, 'curly top' of sugar-beet and 'streak' of maize.

Certain Heteroptera have developed a propensity for animal food, as in the predacious families Reduviidae, Nabidae and Anthocoridae, and in most Hydrocorisae. Both sexes of the Cimicidae, Polyctenidae, and some Reduviids such as *Triatoma* suck the blood of mammals or birds.

Hemiptera afford many instances of resemblance to insects of their own and other orders. Certain of the ant-like forms are very remarkable; thus the brachypterous form of the Coreid *Dulichius inflatus* closely resembles and associates with the ant *Polyrachis spiniger* and is furnished with pronotal and other spines rather similar to those of the ant. Another Coreid, *Alydus calcaratus*, is often found in England in company with *Formica rufa* and other ants, which its nymph closely resembles. Further cases of resemblance to insects of other orders are found in the Reduviidae.

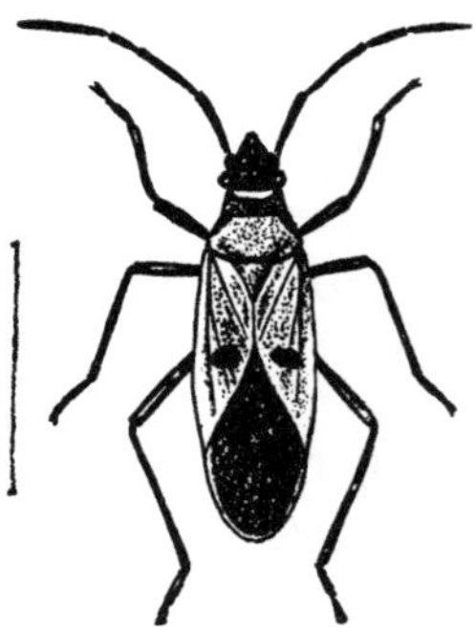

FIG. 315
Dysdercus cingulatus
After Distant in *Fauna of British India.*

Aquatic Hemiptera display many structural and functional adaptations to their environment, particularly with regard to locomotion and respiration (Torre-Bueno, 1916; Weber, 1930). In the surface dwellers (Amphibicorisae) the adaptations are less pronounced, the antennae free and unconcealed, and the legs not highly modified. These insects are clothed with a velvety pile to prevent wetting, and their respiratory mechanisms are simple. The Hydrocorisae, on the other hand, have the antennae concealed, the long antennae of above-water forms obstructing the freedom of motion of submerged insects. The legs are highly adapted for swimming and complex respiratory modifications occur.

External Anatomy

General accounts of the external anatomy of the order are given by Ekblom (1926–30), Weber (1930), Kramer (1950), Pesson (1951), Poisson (1951) and Jordan (1972).

The Head – The head (Evans, 1938, 1968; Spooner, 1938; Parsons, 1959*c*–64) is very variable both in form and in the inclination of its longitudinal axis, being porrect in most Heteroptera and usually deflexed in the Homoptera. In most cases the sclerites are compactly fused (Fig. 316), but a more or less distinct frons is found in some Cercopids, Cicadellids, Cicadids and Psyllids; in the two latter it bears a median ocellus. The clypeus is subdivided, the post-clypeus being a large sclerite, frequently much swollen in the Auchenorrhynchan Homoptera but less conspicuous in the Heteroptera where it extends well back on the dorsal surface of the head and its posterior limits are not recognizable externally. A small ante-clypeus and labrum are also present, the latter often narrow and acuminate (Stys, 1969). At the sides of the head are two pairs of more or less distinct sclerites

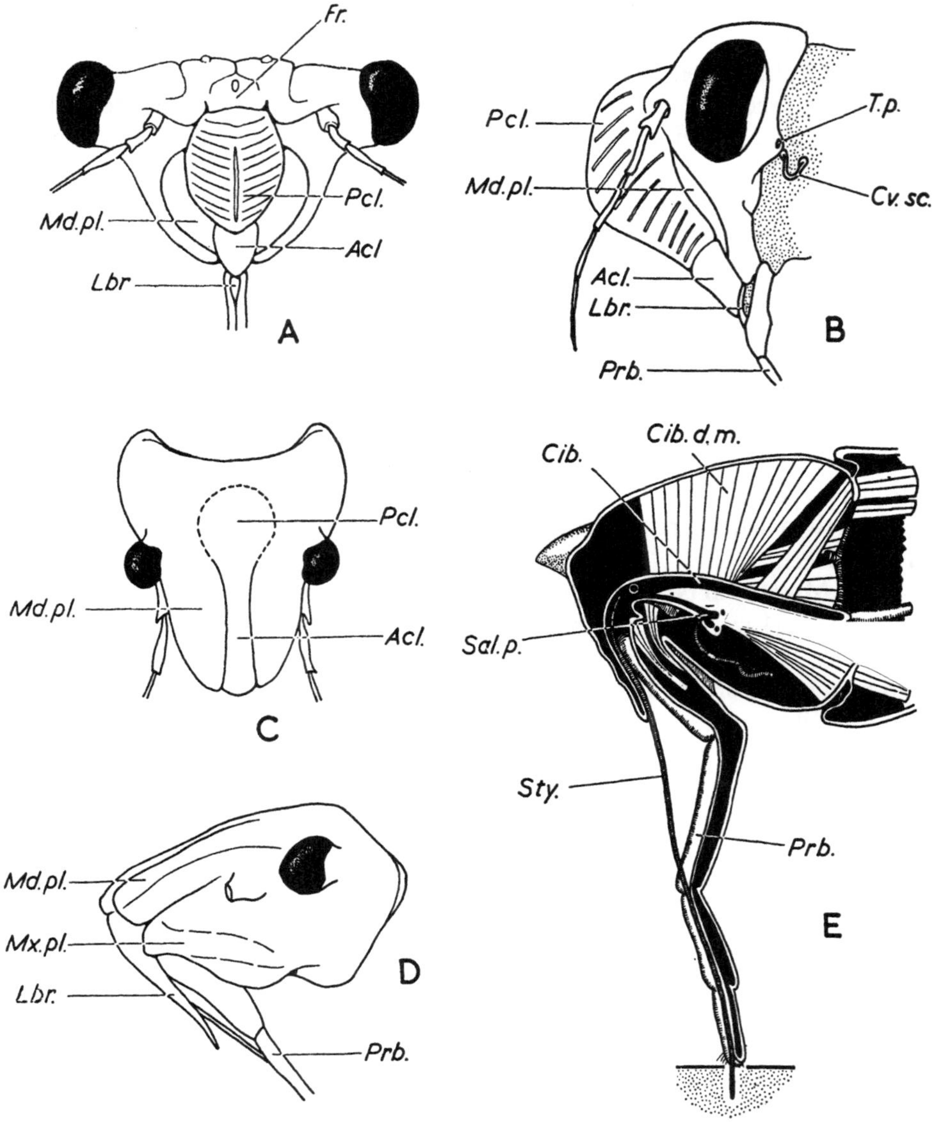

FIG. 316 Head and mouthparts of Hemiptera. A, B. Frontal and lateral views of *Magicicada septemdecim* (*after* Snodgrass, 1935). C, D. The same of a Pentatomid, *Euschistus variolarius* (*after* Snodgrass, 1935). E. Section of head of *Graphosoma italicum*, in feeding position (*after* Weber, 1930)

Acl, anteclypeus; *cib*, cibarium; *Cib.d.m*, cibarial dilator muscles; *Cv.sc*, cervical sclerite; *Fr*, frons; *Lbr*, labrum; *Md.pl*, mandibular plate; *Mx.pl*, maxillary plate; *Pcl*, postclypeus; *Prb*, proboscis (labium); *Sal.p*, salivary pump; *Sty*, stylets; *Tp*, tentorial pit.

associated internally with the bases of the mouthparts: the mandibular plates (sometimes known in the Homoptera as the lora) are probably parts of a modified hypopharynx (q.v.) while the maxillary plates, which are often fused with the genae, arise – at least in part – from the embryonic maxillary rudiments (see also Parsons, 1964; Evans, 1973). In all Heteroptera the maxillary plates meet and fuse ventrally to form the gula whereas this latter region remains small and membranous in the Homoptera. The tentorium is absent in Heteroptera and present, though sometimes reduced, in Homoptera. It may be mentioned that the frons and clypeus of many taxonomists are really the post- and anteclypeus respectively. Ocelli are usually present and frequently two in number (Heteroptera and most Auchenorrhyncha); three are present in Cicadidae, and many Sternorrhyncha. They are wanting in Pyrrhocoridae, Cimicidae, some Cicadellidae, most Hydrocorisae and in apterous species. In addition to compound eyes (Bedau, 1911; Kühn, 1926), ocular tubercles or supplementary eyes are present close to them in *Livia* and many aphids. The antennae have few segments, frequently only four or five; their maximum number is attained in the Sternorrhyncha, where 10 segments are found in Psyllids and 25 in the males of a few Coccids.

The Mouthparts (Figs. 316, 317) – These organs (e.g. Weber, 1928*b*; Pesson, 1944) are very alike in general structure in the different families, the similarity being correlated with the uniform nature of the feeding habits throughout the order. They are exclusively adapted for piercing and suction, the mandibles and maxillae being modified to form slender bristle-like stylets which rest in the grooved labium. The embryological studies of Newcomer (1948) and earlier workers have shown that the mandibles and maxillae develop quite normally from the first two pairs of postoral embryonic appendages. They subsequently become sunk to some extent within the head, and enclosed at their bases in pockets whose lining is continuous with the general integument. Both pairs of stylets are hollow seta-like structures, capable of limited protrusion and retraction by means of muscular action. They may extend back into the prothorax (e.g. Cranston and Sprague, 1961) and in many Homoptera and a few mycetophagous Heteroptera such as the Aradidae the stylets are extremely long, being looped or coiled upon themselves, and withdrawn into a pocket connected with the channel of the labium. This pocket takes various forms but in Coccoidea (Fig. 341) it is lined by thin cuticle, is situated between the central nervous system and the ventral body-wall and is known as the crumena.

The mandibular stylets form the anterior (outer) pair and, although usually free, may be closely interlocked with the maxillae as in *Lygus*; at their apices they are usually serrated. The posterior (inner) pair of stylets constitute part of the maxillae: the embryonic rudiments of the latter become two-segmented at an early stage, the basal segment thus formed gives rise to the maxillary plate, and the distal part to the maxillary stylet. Maxillary palps are absent. Each maxillary stylet tapers to a fine point and is grooved along its inner aspect; the groove is divided into two parallel channels by means of a longitudinal ridge which traverses the length of stylet. Seen in cross-section, the latter is shaped like a W, and the pair of stylets, by the approximation of their channels, form two extremely fine tubes. The dorsal one functions as the suction canal and communicates with the cibarial sucking pump: the ventral tube is the salivary canal through which the saliva is

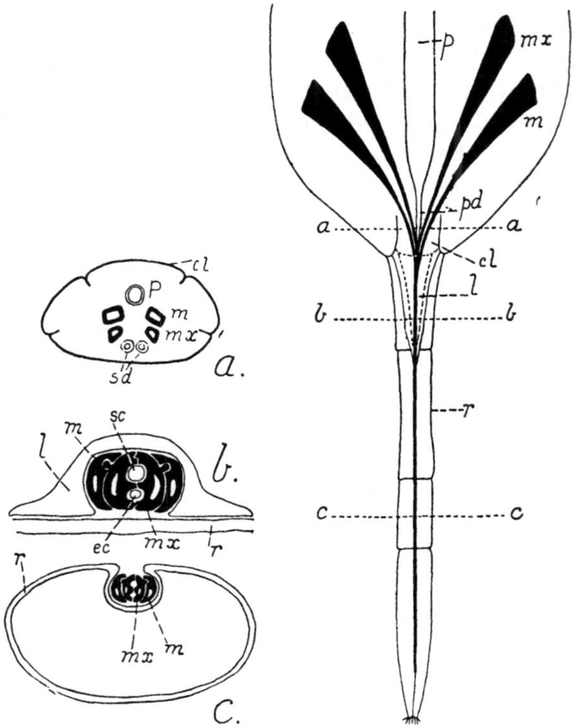

FIG. 317 Diagram of the mouthparts and adjacent region of the head of an Hemipterous insect

On the left are transverse sections across the regions bearing corresponding lettering; the magnifications of these sections are not uniform. *cl*, clypeus; *ec*, salivary canal; *l*, labrum; *m*, mandible; *mx*, maxilla; *p*, pharynx; *pd*, cibarial pump; *r*, rostrum; *sd*, salivary ducts; *sc*, suction canal.

discharged. Within the head the maxillary stylets diverge towards their bases, but externally they are closely interlocked, and appear as a single structure, as in *Anasa* (Tower, 1914), and *Psylla* (Grove, 1919); or the interlocking arrangement is wanting and they are simply apposed to one another (*Eriosoma*). In *Macrosteles* each stylet is traversed by a cavity containing up to five dendrites, presumably from sensilla that might be concerned in the selection of the host-plant and feeding site (Forbes and Raine, 1973). At the enlarged proximal ends of both pairs of stylets are ovoidal areas of tissue known as the *retort-shaped organs*, which secrete the new stylet that is developed at each nymphal moult. In many Hemiptera the bases of the stylets are attached to the head-capsule by mandibular and maxillary levers, sclerotized rods which extend outwards in a transverse direction and afford attachment to certain of the stylet muscles (Ekblom, 1926–30). The stylets themselves are enclosed in a sheath (rostrum) formed almost entirely by the labium which is dorsally grooved for their reception. At its base, however, the labial groove is wanting and in this region the sheath is roofed over by the labrum. Distally, the lips of the labial groove are approximated or fused to form a tube into which the stylets fit tightly. In the majority of Hemiptera the labium is either 4-segmented (Pentatomidae, Miridae, Lygaeidae, etc.) or, by reduction of the basal segment, apparently 3-segmented (most Reduviidae, Cicadidae, Psyllidae and Aleyrodidae); in Coccoidea and Corixidae it is always short and 1- or 2-segmented, the Corixid head and mouthparts being much specialized (Benwitz, 1956*a*; Jarial *et al.*, 1969). Its apex is provided with sensilla (e.g. Lo and Acton, 1969) and it performs no part in perforating the tissues of the host-plant. Labial palps are wanting, the so-called palps of Nepidae and Belostomatidae being small, secondary structures. The hypopharynx is highly specialized in Hemiptera. Part of it is visible as a small, well-sclerotized structure lying between the bases of the stylets where its median portion (sitophore) forms part of the floor of the cibarial sucking pump. Laterally, its walls are expanded and lamellate, becoming exposed externally as the mandibular plates. Such a condition is clearly seen in sections through the head of some Homoptera (Snodgrass, 1950), but in the Heteroptera the reduction of the lateral lamellae makes it less easy to see that the mandibular plates are actually part of the hypopharynx. Beneath the hypopharynx lies the salivarium which is modified into a powerful salivary pump, the walls of which are strongly sclerotized and into which runs the salivary duct (Popham, 1962; Parsons, 1963*b*).

The feeding mechanisms of Hemiptera have been studied by Weber (1928*b*), Pesson (1944), Kraus (1956) and others. While at rest the rostrum is flexed beneath the body, with its apex directed backwards. When the insect is about to feed, the rostrum is extended from its resting position and inclined downwards. In most Hemiptera the stylets are only slightly longer than the rostrum and some mechanism is needed to retract it when forcing the stylets into the plant. In Aphididae, for example, this is brought about by the proximal portion of the rostrum being telescoped into the body (Davidson). In *Lygus*, and other Heteroptera, the stylets are able to penetrate the tissues owing to the bending of the rostrum about its basal hinge (Awati, 1914). In Coccoidea and some Heteroptera the rostrum is very short, and the stylets extremely long; they are inserted into the plant, and afterwards withdrawn and looped within the body in the following way. Contraction of the protractor muscles of the stylets forces them into the plant for a short distance. They are then held, as in a clamp, by a specially modified muscular region of the labium while the contraction of the retractor muscles and relaxation of the protractors take up a little of the slack in the coiled or looped stylets, thus preventing their withdrawal. The protractors then begin a repetition of the cycle so that by degrees the stylets are forced deeper and deeper into the plant. It seems probable that all four stylets are not protracted simultaneously but that the mandibles are first pushed in one after the other, followed by the maxillae. The path of the stylets in the plant has been traced for many species (e.g. Flemion *et al.*, 1954; Pollard, 1973). Most of these insects feed upon the contents of phloem vessels and reach this tissue by a path which is mainly intercellular in aphides, but often passes through cortical cells in the leaf-hoppers. Once penetration has started, saliva is injected into the plant by the salivary pump and this reacts with the plant sap to form a tubular sheath around the stylets (Miles, 1972). It is also probable that enzymes in the saliva initiate extra-intestinal digestion, and that pectinases break down the middle lamella of the cell-walls to facilitate intercellular penetration. The withdrawal of sap into the alimentary canal of the feeding insect is usually said to occur through the activity of the cibarial sucking pump which is provided with powerful dilator muscles arising on the post-clypeus. The turgor pressure of the pierced plant cells also plays some role in the ascent of liquid through the feeding canal, especially in aphids (Mittler, 1957). Blood-sucking Reduviid bugs pierce the skin of their hosts by rapid alternating movements of the apically barbed mandibles, but further penetration is accomplished only by the maxillary stylets which probe deeply until they encounter and enter a capillary of suitable calibre, from which they then suck their blood meal (Lavoipierre *et al.*, 1959). *Cimex* differs from them in that the mandibles and maxillae penetrate deeply as a single fascicle (Dickerson and Lavoipierre, 1959).

The Thorax – The morphology of the Hemipteran thorax has not been extensively investigated, but a comparison of a number of genera has been made by Taylor (1918) while Larsén (1945*a*, *b*; 1950) has studied many Heteroptera (see also Weber, 1928, 1935; Parsons, 1963–71; Matsuda, 1962). Among Heteroptera the pronotum is tolerably uniform in its characters; it is always large, rarely marked off into separate sclerites, and forms the greater part of the thorax when viewed from above. The mesonotum frequently exhibits a fivefold division, thus presenting the maximum number of sclerites. Of these the most prominent is the scutellum, which is always well developed; in certain Pentatomoidea it extends posteriorly to cover the wings entirely, and imparts to the insect an apterous appearance. The metanotum is very variable; it may be well developed, as in *Anasa*, or reduced to a small region concealed beneath the mesoscutellum. It is never conspicuous, and is covered by the unexpanded wings. The sternites are, for the most part, fused with the respective pleura.

Among Homoptera there is more diversity of structure, though the Cicadidae may be regarded as fairly typical of the suborder (Kramer, 1950). The pronotum is almost always small and frequently collar-like, except in Membracidae where it may assume incredibly bizarre and grotesque forms and extends backwards over the abdomen. The mesothorax is the largest and most typical region, exhibiting the primary divisions into prescutum, scutum, scutellum and postnotum. In almost all Fulgorodea it bears well-developed tegulae, which are vestigial or absent in these species with reduced wings. The metanotum is usually well developed, and in Cicadellidae it is nearly as long as the mesonotum.

The Wings – Among Heteroptera there is a marked difference in the consistency of the two pairs of *wings*. The fore wings are termed hemelytra (*hemi-elytra*) and their proximal area is well sclerotized, resembling an elytron, with only the smaller distal portion remaining membranous. The hind wings are always membranous and are folded beneath the hemelytra at rest.

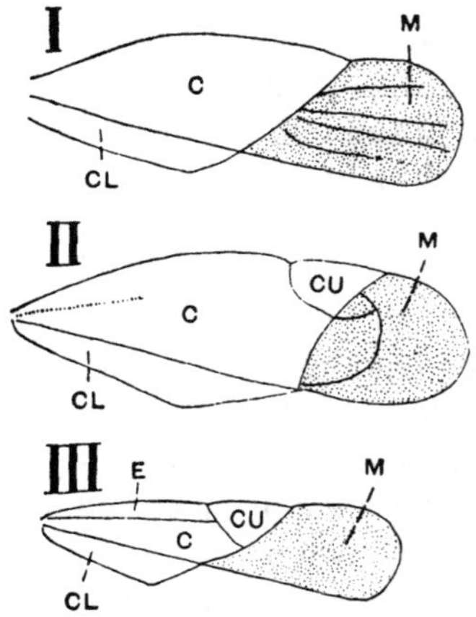

FIG. 318
Diagrams of the hemelytra of – I, a Lygaeid; II, a Mirid; III, an Anthocorid

C, corium; *CL*, clavus; *CU*, cuneus; *E*, embolium; *M*, membrane.

The hemelytra (Fig. 318) exhibit much diversity of structure and have been used taxonomically. The hardened basal portion is composed of two regions separated by a concave vein, Cu_2 – the *clavus* or narrower area next to the scutellum (when the wings are closed), and the *corium* or remaining broader portion. In the families Dipsocoridae and Anthocoridae a narrow strip of the corium, bordering on the costa, is demarcated from the remainder by R+M, and is known as the *embolium*. In the Miridae and Velocipedidae a triangular apical portion of the corium is separately differentiated to form the *cuneus*. Among Tingidae and some Gerroidea the differentiation into sclerotized and membranous regions is less distinct. In some cases the membranous area is much reduced or absent, but in the Enicocephalidae the hemelytra are entirely membranous. The two pairs of wings exhibit evident departures from a generalized condition. The venation is not used very extensively in taxonomy though it is proving of some value in the higher classification of certain groups (Hoke, 1926; China and Myers, 1929; Davis, 1957–66; Slater and Hurlbutt, 1957). The fore and hind wings are coupled in flight by a simple apparatus (Teodori, 1925–26) and in *Oncopeltus* it is known that the main forces propelling the insect are generated by the movements of the fore wing (Hewson, 1969).

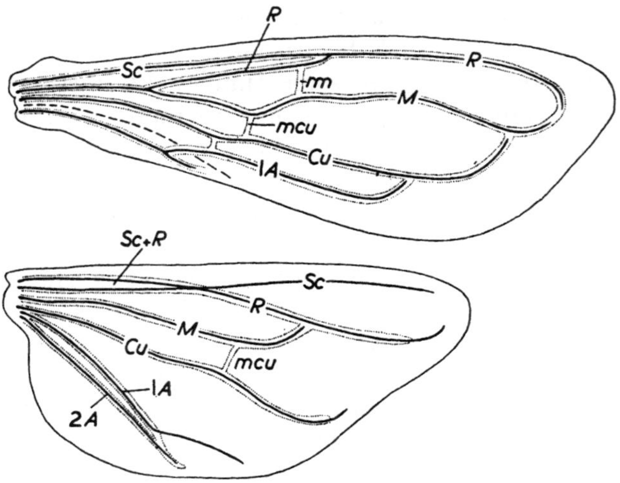

FIG. 319 Venation (dotted lines) and tracheation (full lines) of wings of *Triatoma* (*after* Usinger)

Among the Homoptera the fore wings are of uniform texture (Fig. 320) and are frequently of harder consistency than the hind pair. Apterous forms are the rule in female Coccoidea and Aphidoidea (sexuales), as well as in some parthenogenetic forms of the latter group; both apterous and alate males are sometimes present in the Aphidoidea and Coccoidea. Although there is great diversity of venation which is dealt with under the different families, the occurrence of fossil forms (Tillyard, 1926; Evans, 1956–64) and

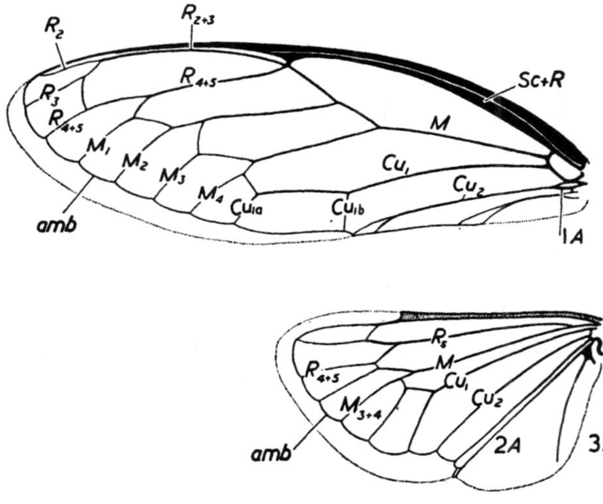

FIG. 320 Wing-venation of *Cicada orni* (*after* Silvestri, 1934, with Tillyard's nomenclature)

amb, ambient vein.

the nymphal tracheation have made it possible to determine the homologies of the wing-veins (Tanaka, 1926; Fennah, 1944; and see Leston, 1962).

Alary polymorphism occurs in some families, i.e. there are two well-marked types of individuals, apterous and macropterous, sometimes with intermediates or brachypterous forms. The phenomenon is evident, for example, in the Gerroidea (Larsén, 1931), Anthocoridae and Reduviidae among Heteroptera, and in certain Delphacidae and Cicadellidae among Homoptera. In *Perkinsiella* (Delphacidae) there is much specific variation in this respect: *P. saccharicida* has macropterous males and polymorphic females, while in *P. vitiensis* and *vastatrix* both sexes are dimorphic. In certain African Reduviidae (*Edocla*) the males are winged and the females apterous and physogastric; in other species of the genus both sexes are alike and physogastric. In *Paredocla*, there are both winged and apterous males; the latter resemble the females, which are also apterous, and all forms are found together (Jeannel, 1919). Many explanations have been advanced to account for alary polymorphism. It has been variously correlated with climate, season, mimicry, capability for leaping or rapid locomotion, and mode of life, whether arboreal or otherwise. Thus, in Africa, Jeannel states that most of the apterous and brachypterous genera of Reduviidae inhabit the hotter regions. In the European *Pyrrhocoris apterus* both pairs of wings may be either normally developed, or reduced to merely the sclerotized basal portions of the hemelytra, and the two forms vary very greatly both in local and seasonal occurrence. In the Aphidoidea, a highly polymorphic group, the development of wings is closely bound up with wider general problems of morph determination (p.721). For accounts of alary polymorphism and the related problems of flight polymorphism in aquatic Heteroptera, see, for example, Ekblom (1949, 1950), Larsén (1950), Young (1965) and Acton and Scudder (1969).

Legs – The legs of Hemiptera show appreciable variation in taxonomically important structures and also display many interesting adaptive modifications. The coxae articulate with the thorax by a rotary or a hinged joint in the Heteroptera while in the Fulgoroidea the hind coxae are fused with the metathorax. The number of tarsal segments never exceeds three; most Heteroptera have two or three segments, the Auchenorrhyncha have three and the Sternorrhyncha one or two. The pretarsal structures are used to define groups at various taxonomic levels (Dashman, 1953). In the Geocorisae paired claws are supported by an unguitractor plate of relatively constant form, produced into an empodium but without an arolium. Pulvilli (often miscalled arolia) are present, one under each claw and less well developed in the Cimicomorpha than the Pentatomomorpha (Goel and Schaefer, 1970; Goel, 1972). In the Auchenorrhyncha there is much detailed variation in the pretarsus (Fennah, 1945*a*) and in the Sternorrhyncha the form of the empodium is used to define the superfamilies. Jumping hind legs are particularly characteristic of many Auchenorrhyncha and Psyllidae, and raptorial fore legs occur in the Reduvioidea (Miller, 1942) and many

Hydrocorisae. The latter also have the middle and hind legs equipped for swimming under water while the Gerridae and Veliidae use theirs for sculling over its surface. Copulatory adaptations occur in the clasping male hind legs of the Gerrid *Rheumatobates* and in the fore tarsus of male Corixidae, with its row of paleal pegs (Popham, 1961). In both sexes of the Corixidae the fore tarsus is used to sweep fine particulate material towards the feeding apparatus and all three legs of this family contain well-developed glands which arise from the epidermis of the unguitractor apodeme and possibly produce a hydrofuge secretion (Benwitz, 1956*b*).

The Abdomen and External Genitalia – In its least modified condition, as in many Auchenorrhynchan Homoptera, 11 segments are present though the first two may be modified in connection with sound-producing organs (Cicadidae), the 8th and 9th undergo changes due to the development of external genitalia and the 10th and 11th are small annuli at the end of which the anus opens. Reduction of the number of well-defined segments is frequent. Thus, among the Homoptera, Crawford (1914) finds that the first three are suppressed or greatly reduced in Psyllidae while in Aphidoidea and female Coccoidea (Ferris, 1951) – where segmentation is often obscure – not more than 9 segments can be recognized. In the Heteroptera, the 10th and 11th segments are fused together and the first one or two may be reduced or absent (Scudder, 1963). Cerci are never found in the Hemiptera but the abdomen of most Pentatomomorphan Heteroptera bears *trichobothria*, groups of long sensory hairs whose arrangement is taxonomically important and whose electrophysiological properties suggest that they may be sound receptors (Drašlar, 1973).

The ovipositor is exhibited in a complete condition in many Auchenorrhynchan Homoptera (Snodgrass, 1933; Müller, 1942; Kramer, 1950) where it consists primitively of a pair of valves, each articulating with a small basal sclerite, on the 8th sternum, another similar pair on the 9th sternum and a third pair, softer, sheath-like and more dorsal in position, also developed from the 9th sternum. Their homologies seem to be similar to those of the more generalized Orthopteroid insects, but the gonangulum (p. 77) is fused with the ninth abdominal tergum. Some Heteroptera, particularly those which insert their eggs into plant tissues, retain a similar, well-developed ovipositor, but other Heteroptera have a smaller, flatter structure made up of only two pairs of valves (Scudder, 1959); it is of some limited taxonomic value (e.g. Slater, 1950). There is no ovipositor in the Aphidoidea and Coccoidea.

The comparative morphology of the male genitalia has been studied extensively though the homologies of the component parts are not fully established (Singh-Pruthi, 1925; Qadri, 1949; Marks, 1951; Fennah, 1945*b*; Snodgrass, 1957; Dupuis, 1955). The ninth abdominal segment forms a more or less compact pygophor and in many Auchenorrhynchan Homoptera this bears a pair of ventral sub-genital plates (perhaps homologous with the coxites and styles of the 9th segment) and a well-developed penis flanked by a pair of small parameres. A highly differentiated endophallus occurs in the Cicadidae (Orian, 1964) and the sclerotized structures of the aedeagus are especially important in the taxonomy of the Cicadellidae, many species of which can otherwise only be separated with difficulty (e.g. in the *Empoasca fabae* complex, cf. Ross and Moore, 1957). The paired periphallic structures have been lost in the more specialized Sternorrhyncha. In the Heteroptera the pygophor typically bears a pair of claspers (sometimes asymmetrical and probably equivalent to the parameres of the Auchenorrhyncha) as well as a median penis, whose endophallus is most highly differentiated in the Pentatomomorpha. The detailed structure and taxonomic value of the Heteropteran male genitalia have been well studied in certain groups (e.g. Ashlock, 1957; Baker, 1931; Bonhag and Wick, 1953; Davis, 1957–66; Kelton, 1959; Kullenberg, 1947; Piotrowsky, 1950 and Wagner, 1951).

Sound-producing Organs are of frequent occurrence among Heteroptera and five important types will be mentioned (see also Leston, 1957; Haskell, 1957; Leston and Pringle, 1964; Ashlock and Lattin, 1963).

(1) *The prosternal furrow* of many Reduviidae and Phymatidae. This furrow is cross-striated and stridulation is produced by the rugose apex of the rostrum working over it; it is well seen in *Reduvius personatus* and *Coranus subapterus*.

(2) *The strigose ventral areas* of certain Scutellerinae. These are found on each side of the median line of the apparent 4th and 5th abdominal sterna. On the inner side of the hind tibiae are wart-like tubercles, each bearing a subapical tooth. When the insect bends the tibia against the femur, and again extends it, the spinous tubercles pass across the strigose areas, thus enabling the insect to produce an audible sound by rapidly repeating the movements.

(3) *The pedal stridulating organs* of Corixidae. The males of some Corixidae produce sounds when a spinose area on the inside of each front femur is drawn over the edge of the clypeus (Mitis, 1935). The pala is not a stridulatory organ, nor has it been shown conclusively that the peculiar strigil of these insects is concerned with sound production.

(4) *Coxal stridulatory organs.* In *Ranatra* Torre-Bueno (1905) describes two opposing rasps, one on each coxa near the base with longitudinal striations, the other on the inner surface of the cephalic margin of the lateral plate of the coxal cavity. The latter plate is exceptionally thin and probably functions as a resonating organ.

(5) *The dorsal stridulatory organs* which are found in both sexes of *Tessaratoma papillosa* (Pentatomidae). The sound-producing organ consists of a striated surface or file situated one on each side of the dorsum of the abdomen close to the metathorax. On the under-surface of each wing, near the base, is a comb of strong teeth. The sclerite supporting the files is able to move backwards and forwards across the comb (Muir, 1907).

Among Homoptera, the sound-producing organs of the Cicadidae are complex structures peculiar to the family, and situated one on either side of the dorsal aspect of the base of the abdomen (p. 183). The remaining Auchenorrhyncha are usually regarded as being silent, but Ossiannilsson (1949) has shown that many leaf-hoppers possess the power of very quiet sound-production, the organs concerned being homologues of those found in the Cicadidae. See also Moore (1961).

Spiracles – Ten pairs of spiracles are normally present in the Heteroptera (Handlirsch, 1900). The first pair lies in the membrane between the pro- and mesothorax and is difficult to see. The second pair lies between the meso- and metathorax and the third pair between the metanotum and the first abdominal tergum, hidden by the wings. The 4th and following pairs lie on the ventral or, sometimes, the dorsal side of the successive abdominal segments. These general rules are subject to modification, especially in the Hydrocorisae (Parsons, 1970). Thus in *Nepa* there are 10 pairs of open spiracles in the nymph but fewer in the adult. The 4th, 5th and 9th pairs have atrophied and the 6th, 7th and 8th are highly modified sensory organs which perceive relative pressure differences (Baunacke, 1912; Oevermann, 1936; Thorpe and Crisp, 1947). The anatomy of the Heteropteran spiracles has been studied by Mammen (1912); those of the Geocorisae and Amphibicorisae possess an internal closing mechanism which has been lost in the Hydrocorisae.

In the Auchenorrhyncha there are usually 10 pairs of spiracles also, but among the Sternorrhyncha there is a wide range of variation. In the Aphidoidea there are usually 9 pairs, situated respectively on the pro- and metathorax and on the first 7 abdominal segments. Among Psyllidae Witlaczil (1885) finds 2 thoracic and 7 abdominal spiracles in the nymph of *Trioza*, while in the adult *Psylla mali* Awati (1915) states there are 2 thoracic and 3 abdominal pairs. In the Aleyrodidae the nymphs are closely applied to the leaf surface, and as the spiracles lie ventrally they are concealed. Air is conveyed to them by special breathing folds of the integument. Two pairs of thoracic spiracles are present, one pair between the anterior legs and the other between the posterior legs. Spiracles are also present behind the 2nd thoracic pair, apparently on the 1st abdominal segment, and a 4th pair exists alongside the vasiform orifice. Vestigial spiracles are apparently found in some genera on other abdominal segments. In the adult the distribution of the spiracles is very much the same as in the nymph. The respiratory system of both Psyllidae and Aleyrodidae, however, requires detailed investigation. In the nymphs and females of the Coccoidea there are commonly 2 pairs of spiracles on the ventral aspect of the thorax and abdominal spiracles are present in certain groups (Ferris, 1918). The primitive number of 2 thoracic and 8 abdominal pairs is found in *Xylococcus* (Oguma, 1919); in *Orthezia* (List, 1886) and *Monophlebus* there is one pair less on the abdomen. In *Icerya purchasi* the abdominal spiracles are reduced to 2 pairs but in certain other species of that genus there are at least 3 pairs.

Internal Anatomy

The Alimentary Canal – The entrance to the alimentary canal is the aperture of the cibarial sucking pump (Weber, 1928*b*; 1930; Snodgrass, 1935) which is situated at the base of the maxillary stylets. The walls of this tubular pump are strongly sclerotized and dilator muscles run from it to the postclypeus; in some aquatic Heteroptera it forms a triturating organ (Parsons, 1966; 1972). In the region of the epipharyngeal surface of the labrum there is, in most Hemiptera, a gustatory organ whose sensory cells communicate with the lumen of the cibarium through perforations in a sclerotized plate, which is a specialized development of the epipharyngeal membrane. The cibarial pump is followed by a small pharynx, the dilator muscles of which are attached to the frontal region of the head and the pharynx, in turn, communicates by a short oesophagus with the mid gut. The latter is modified considerably, but in different ways in the two sub-orders.

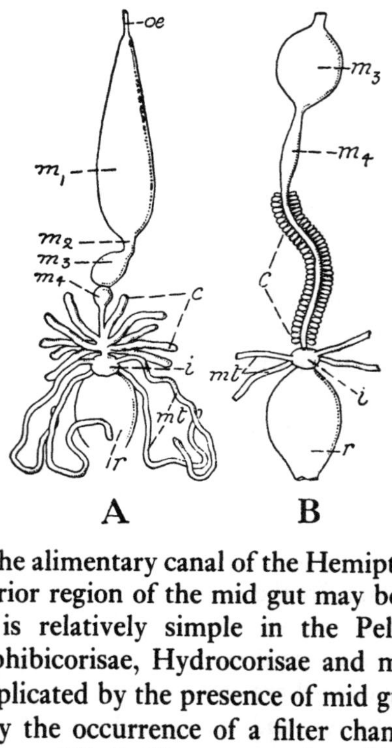

FIG. 321
Digestive system of Lygaeidae. A, *Blissus leucopterus*. B, *Oedancala dorsalis* (posterior portion only)

c, gastric caeca; *i*, ileum; m_1–m_4, chambers of mid gut; *mt*, Malpighian tubules; *oe*, oesophagus; *r*, rectum. *After* Glasgow (reduced).

The alimentary canal of the Hemiptera lacks a stomodaeal crop, though the anterior region of the mid gut may be dilated in a comparable manner. The gut is relatively simple in the Peloridiidae and in the Cimicomorpha, Amphibicorisae, Hydrocorisae and most Aphidoidea. Elsewhere it may be complicated by the presence of mid gut caeca, as in most Pentatomomorpha, or by the occurrence of a filter chamber in many Homoptera (Goodchild, 1963*a*, 1966). In Heteroptera the mid gut is normally differentiated into two, three or four segments: the most anterior, already mentioned, is sac-like and is usually followed by a tubular section, an ovoidal region, and a second, posterior tubular segment which bears the caeca in those species that possess them (Glasgow, 1914; Goodchild, 1952, 1963*b*; Marks, 1959; Parsons, 1959*a*; Miyamoto, 1961). The caeca vary considerably in number, size and form and usually contain various bacteria that are generally regarded as symbiotic (Steinhaus *et al.*, 1956; Buchner, 1965). These micro-organisms may be confined to the lumen of the caeca or occur intracellularly in their walls (Pierantoni, 1951), though similar bacteria may also be found in the general lumen of the mid gut in other species. Transmission to the progeny occurs through the female applying droplets of bacteria-containing excreta to the egg-mass or through the young nymphs feeding proctodaeally from the mother, as in the Cydnid *Brachypelta aterrima* (Schorr, 1957). In some Pentatomomorpha the caeca-bearing portion of the mid gut is separated anteriorly by a complete discontinuity of the gut in the adult or nymphal stages (Schneider, 1940; Goodchild, 1963*b*), while in the Pyrrhocoridae and some others (e.g. *Oncopeltus, Leptocorisa*) the caeca are vestigial or absent in one or both sexes. When present they may serve not only to harbour bacterial symbionts but perhaps also function as a site of water excretion (Goodchild, 1963*b*). The Pentatomid *Dalsira bohndorffi* is peculiar in that the edges of the anterior sac-like part of the mid gut enclose the enlarged ileum to form a kind of filter chamber (Goodchild, 1963*a*). The epithelium of the hind gut of many Heteroptera includes larger, more strongly basiphil cells which may be diffusely scattered, as in *Anasa tristis*, or restricted to pad-like anterior, dorsal or lateral regions of the rectum or, in Hydrocorisae, to parts of the ileum (Bahadur, 1963; Jarial and Scudder, 1970). In *Halosalda lateralis*, from estuarine habitats, the pads comprise two types of cells and may be concerned in salt regulation (Goodchild, 1969).

Among many Homoptera the oesophagus leads into a very capacious crop-like distension of the mid gut which occupies a large part of the abdominal cavity. The remainder of the mid gut is long and tubular and reflected on the first part in an ascending manner, with the result that its junction with the hind gut comes to lie very far forwards alongside the oesophagus (Fig. 322). Owing to this disposition, the insertions of the Malpighian tubules are likewise anteriorly situated, and these organs, together with the mid gut and the hind intestine, form a complex coil of tubes lying in the thorax and anterior abdomen. Many Cicadomorpha and Coccoidea have evolved from this arrangement a complex filter chamber, but such a structure is absent in the Fulgoroidea and most Aphidoidea (Forbes, 1964). In *Phalix titan* (Tettigometridae) the tubular mid gut is coiled within a membranous sheath composed of an inner thick epithelium and an outer peritoneal layer (Goodchild, 1963*a*). In *Typhlocyba* (Willis, 1949) the anterior and posterior portions of the mid gut are simply in loose contact, but in other Cicadelloids and Cercopoids the two closely apposed regions are enclosed in a peritoneal layer to form a filter chamber, from which also arise the Malpighian tubules (Licent, 1912; Weber, 1930, 1935; Gouranton, 1968*a*; Munk, 1967, 1968*a*). It has long been thought that this arrangement enables the excess liquid in the diet to pass directly into the hind gut without traversing the entire length of the mid gut. Experimental work with an artificially coloured food medium has shown that a flow of liquid does occur in this way and that it is accompanied by a slower movement through the whole of the mid gut (Munk, 1968*b*). In the Diaspididae the mid intestine is a closed sac (Fig. 323), unconnected with the hind gut. The food is entirely digested there and its components absorbed into the haemocoele, from which the well-developed Malpighian tubules presumably remove water and excretory products (Berlese, 1893–96; Pesson, 1942).

The prevalent number of Malpighian tubules in the Heteroptera is four, but in *Lethocerus* only two are present. In the Homoptera they are more variable: the Auchenorrhyncha have four but the Coccoidea usually have only two tubules of large diameter and with few convolutions (though *Icerya* has three and *Xylococcus* has four). Malpighian tubules are absent in the Aphidoidea. The proximal regions of the four tubules of nymphal

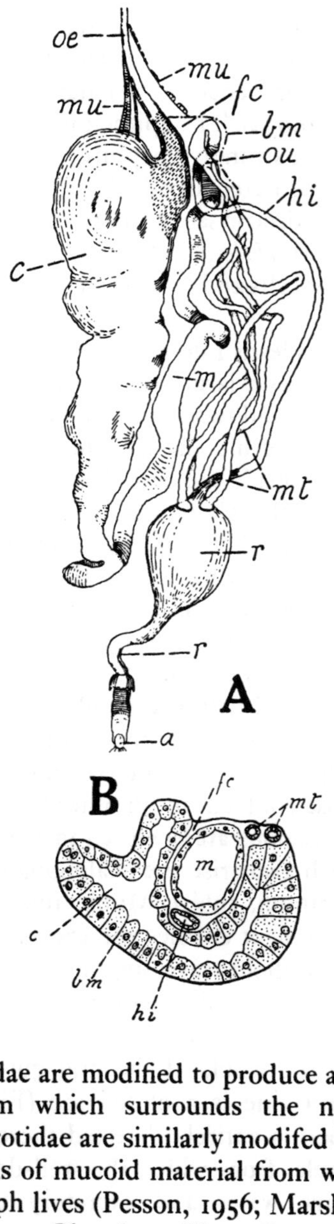

FIG. 322
A, Digestive system of a Membracid (*Tricentrus albomaculatus*). B, Transverse section taken just below line *ou* in

a, anus; *bm*, basement membrane; *c*, crop-like dilation of mid gut; *fc*, filter chamber; *hi*, hind intestine; *m*, mid intestine; *mt*, Malpighian tubules; *mu*, muscles; *oe*, oesophagus; *ou*, point of origin of Malpighian tubules; *r*, rectum. *After* Kershaw.

Cercopidae are modified to produce a surface-active substance that stabilizes the foam which surrounds the nymphs. The tubules of the related Machaerotidae are similarly modifed but also have another segment producing fibrils of mucoid material from which are constructed the case in which the nymph lives (Pesson, 1956; Marshall, 1964–65; Gouranton, 1968*b*).

Salivary Glands – The salivary (labial) glands of the Hemiptera have two main functions: the so-called accessory salivary glands act as excretory organs while the principal glands produce the characteristic salivary secretions (Miles, 1968, 1972). The accessory gland cells normally contain large intracellular vacuoles or canals or the whole gland is histologically similar to a Malpighian tubule. In the Heteroptera the general appearance of the gland (Fig. 324) depends more on the taxonomic position of the species than on its feeding habits: the accessory glands are tubular in the Pentatomomorpha and vesiculate in the Cimicomorpha (Southwood, 1955). In the Homoptera

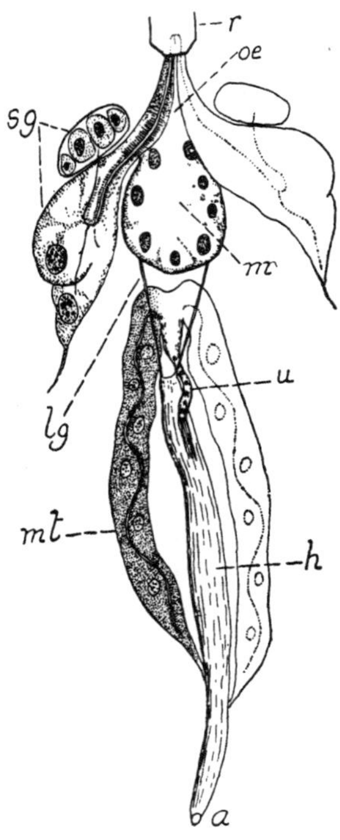

FIG. 323
Digestive system of *Lepidosaphes fulva* × 80

lg, ligament; *oe*, oesophagus; *r*, rostrum; *sg*, salivary glands; *u*, ureter. Other lettering as in Fig. 322. *After* Berlese, *Riv. di Pat. Veg.*, 5.

and the Pentatomomorpha, feeding occurs through the formation of a salivary sheath around the inserted stylets. Precursors of the sheath material are secreted mainly by the anterior and lateral lobes of the principal gland in the Heteroptera (Miles, 1967) while the posterior lobe secretes the hydrolytic enzymes of the saliva. A sheath is absent or greatly reduced in the Cimicomorpha, Hydrocorisae and Amphibicorisae. The glands of the Homoptera show considerable variation in the number and histological characteristics of the acini which are present. Only two acini occur in the Peloridiidae (Pendergrast, 1962), but up to 20 or so are found in some Cicadelloids, Cercopoids and Fulgoroids, the latter apparently lacking a distinct accessory gland (Sogawa, 1965; Balasubramanian and Davies, 1968). Among Cicadoids there are two clusters of tubular lobes in each gland, aphids have a single lobe composed of several kinds of cells, and the Coccoidea may have 8 separate lobes per gland, each with its own cell type (Moericke and Wohlfarth-Botterman, 1960; Pesson, 1944).

Scent Glands – Defensive scent glands, opening in the adult on the metepisternum or metasternum and in the nymphs on the dorsal side of the abdomen, are characteristic of most Heteroptera and do not occur in the other suborder. The adult glands (Carayon, 1971; Hepburn and Yonke, 1971; Schaefer, 1972) may open by a single median orifice or two closely adjacent pores (as in the Amphibicorisae and Hydrocorisae) or there may be separate metepisternal openings, each with its evaporating area of rugose cuticle, as in the Geocorisae. The paired glands may each have a reservoir of

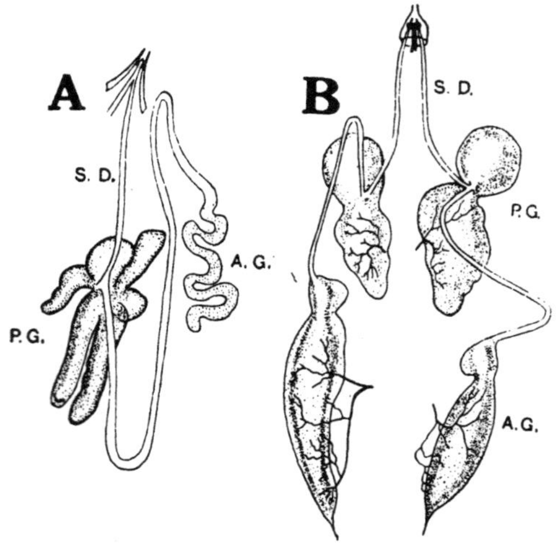

FIG. 324
Salivary glands of A, *Lygaeus apuans*: B, *Notonecta maculata*

SD, salivary duct; *AG*, accessory gland; *PG*, principal gland. *After* Bugnion, 1908 (reduced).

its own or they may share a common median reservoir. The gland cells have a complex 'end apparatus' (p.265) from which a fine ductule runs to the main collecting duct (Stein, 1966; Filshie and Waterhouse, 1968). The secretion contains a mixture of hydrocarbons and their derivatives, compounds such as *n*-hexanal and *n*-hexyl acetate predominating in the Coreoidea (Waterhouse and Gilby, 1965) while the scent of *Nezara viridula* includes *trans*-hex-2-enal, *trans*-dec-2-enal, *n*-tridecane, 4-keto-*trans*-hex-2-enal and other similar substances (Gilby and Waterhouse, 1965). The secretions are repellent and in sufficient quantities they can paralyse potential predators by entering the tracheal system, though the Heteroptera that produce the secretions are largely resistant to their effects (Remold, 1963; Gluud, 1968). In *Lethocerus* and some other species the glands occur only in males, suggesting that their secretions may then have an epigamic role (Carayon, 1971).

Wax Glands – These are prevalent in many Homoptera. They are usually unicellular (though multicellular wax glands are frequent in female Coccoidea) and may occur either singly or in groups (p. 265). They are well exhibited in the Oriental *Phromnia marginella*, where they are situated beneath series of sclerotized plates on the dorsum of the abdomen; each plate is studded with pores which are the apertures of the wax glands. In various Aphidoidea (*Pemphigus*, *Adelges*, *Eriosoma*, *Lachnus*, etc.) the plates are segmentally arranged in longitudinal series. The product of the glands is commonly a powdery secretion or dense flocculent threads.

The Nervous System (Pflugfelder, 1937) exhibits a very uniform and complete degree of concentration. The abdominal ganglia are to a large extent fused up with the thoracic centres, the last compound ganglion ending in a single or paired abdominal nerve which gives off lateral segmental branches. Various degrees of concentration are recognizable in the ventral ganglia.

(1) Three ventral ganglia present (*Lygaeus*, *Capsus*, *Notonecta*, *Aphrophora*, etc.) The suboesophageal and 1st thoracic ganglia are separate, while the abdominal ganglia are fused with those of the 2nd and 3rd thoracic segments to form a common centre.

(2) Two ventral ganglia present (Aphididae). The first is the suboesophageal ganglion, while the thoracic and abdominal ganglia are merged into a common centre. In the Nepidae the prothoracic and suboesophageal ganglia are apparently fused since the nerves supplying the first pair of legs issue from the latter centre.

(3) A single ganglionic centre formed by the coalescence of all the ventral ganglia (*Hydrometra* and Coccoidea).

For more detailed accounts of the nervous system of individual species see, among others, Ewen (1962, 1966), Graichen (1936), Guthrie (1961), Johansson (1957, 1958), Parsons (1960) and Wigglesworth (1959).

Dorsal Vessel – The heart and aorta have been studied comparatively in a selection of Heteroptera (Hinks, 1966). The Amphibicorisae and Geocorisae have a 2-chambered, posteriorly situated heart, whose aliform muscles are reduced, variable and do not form a dorsal diaphragm. In the Hydrocorisae the heart extends through much of the abdomen and its aliform muscles are better developed and lie in a diaphragm. Nephrocytes occur within the dorsal vessel and also adhere to the aliform muscles. The muscle fibres of the heart wall have an unusual helical arrangement in the few investigated Hydrocorisae; elsewhere they are circular. A well-developed heart is found in the Auchenorrhyncha and Psyllidae and may extend over 6 or 7 segments. In other Sternorrhyncha it is usually smaller and posterior in position and is apparently absent in some aphids (*Phylloxera*, *Eriosoma*), and some Coccoidea (Diaspididae).

Pulsatile Organs occur in various aquatic Heteroptera and some terrestrial species. They are present in each pair of legs and, owing to the opacity of the integument, are best observed in the nymphs. In the Hydrocorisae they are present at the base of the 1st tarsal segment of the fore legs, and at the base of the tibia in the other pairs (Brocher, 1909). In *Ranatra* each organ consists of a pulsatile membrane lying longitudinally in the cavity of the limb; it serves to ensure the circulation of the blood in the extremities. Pulsatile organs are also present in the tibiae of *Philaenus* and in aphids.

Reproductive System (Figs. 325, 326) – Each ovary has a variable number of acrotrophic ovarioles, usually containing one to four follicles. In the Heteroptera there are frequently seven ovarioles per ovary but four or five sometimes occur (Woodward 1950; Carayon, 1950*b*; Miyamoto, 1957). The form of the spermatheca varies between some of the major taxonomic groups of Heteroptera; in the Amphibicorisae it is connected anteriorly to the vagina by an additional fecundation canal, the Reduvioidea have 2 spermathecae, and the Cimicoidea have none though a distinct sclerotized portion of the vagina forms a bursa copulatrix in the Miridae, Nabidae and Velocipedidae (Pendergrast, 1957). Holmgren (1899) figures the reproductive system of certain Auchenorrhyncha and the number of ovarioles varies in the examples studied from three in *Eupteryx* to nine in *Philaenus*. In many Delphacidae the upper segment of the lateral oviduct is enlarged and glandular, secreting material that is coated over the surface of the eggs when they are laid (Strübing, 1956). Among Sternorrhyncha there are 8 or 9 very short ovarioles in *Psylla mali* (Awati, 1915), but in *P. alni* there are 40–50 (Witlaczil, 1885). In Coccoidea they are numerous, each consisting of a single follicle arising from a wide oviduct; in *Icerya* (Johnston, 1912), the oviducts are united anteriorly, forming a broad loop. In Aphidoidea the number of ovarioles varies in individuals of the same species, and different phases of the life-cycle. Thus, in *Viteus vitifolii* the apterous parthenogenetic forms have from one or two to thirty according to conditions (Foa, 1912), each with two follicles; in the alate females

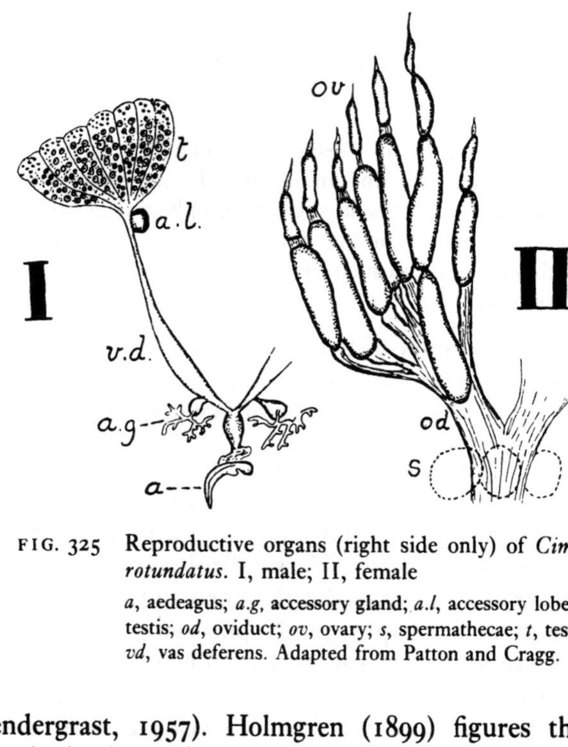

FIG. 325 Reproductive organs (right side only) of *Cimex rotundatus*. I, male; II, female

a, aedeagus; *a.g*, accessory gland; *a.l*, accessory lobe of testis; *od*, oviduct; *ov*, ovary; *s*, spermathecae; *t*, testis; *vd*, vas deferens. Adapted from Patton and Cragg.

there are usually two, and in the sexuales of this species and of *Eriosoma lanigerum* (Baker, 1915) there is a single unpaired unilocular ovariole. Accessory glands, two or three in number and either tubular or globose, are of widespread occurrence though they are wanting in the Diaspididae (Berlese, 1893–96).

The Cimicidae show unusual methods of haemocoelic fecundation, with associated modifications of the female reproductive system (Carayon, 1966). In *Cimex* the spermatozoa are transferred to the female when the male aedeagus penetrates the cuticle of a small, pouch-like *ectospermalege* (Ribaga's organ) lying at the right-hand posterior edge of the fourth abdominal sternum. The sperms are deposited in the underlying *mesospermalege*, a specialized tissue containing large numbers of amoebocytes which absorb the seminal secretions. The sperms then migrate peripherally,

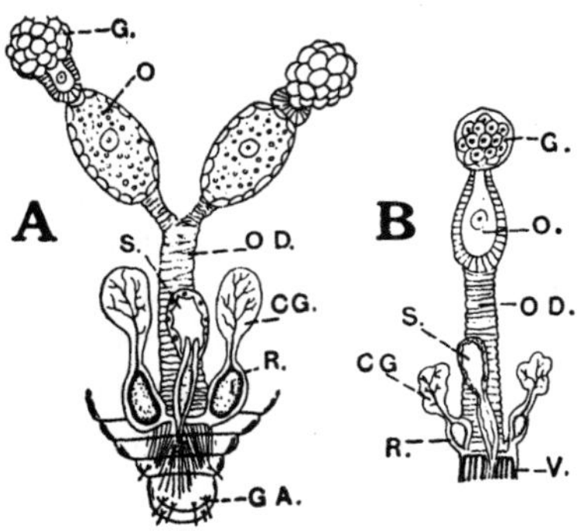

FIG. 326
Female reproductive organs of *Viteus vitifolii*
A, of winged agamic form: B, of sexual form. *CG*, colleterial gland and reservoir *R*; *G*, germarium; *GA*, genital aperture; *O*, ovum; *OD*, common oviduct; *S*, spermatheca; *V*, vagina. *After* Balbiani.

leaving the mesospermalege and traversing the haemocoele to enter the *sperm conceptacles* – paired enlargements of the wall of the common oviduct. These act in a manner analogous to spermathecae and from them the spermatozoa migrate in the walls of the lateral oviducts to reach the oocytes while these are still in the ovarioles. In *Primicimex* the sperms are injected directly into the haemocoele whereas in *Strictocimex* the mesospermalege is produced into a conducting strand that leads the sperms into the wall of the median oviduct. Other genera show various modifications of the same general mechanism and somewhat similar processes of fecundation occur in some Anthocoridae (e.g. *Xylocoris* and *Lyctocoris*) and in the Nabids *Prostemma* and *Alloeorhynchus* (Carayon, 1950, 1952).

The male reproductive system shows many variations. In the Heteroptera (Pendergrast, 1957) the number of testis follicles varies from 6–8 in most Geocorisae and from 4–7 in the Hydrocorisae, while the Gerroidea have only one or two follicles. In most Geocorisae the ejaculatory duct is enlarged anteriorly to form a bulbus ejaculatorius, but this is not found in the Gerroidea and the Hydrocorisae. The accessory glands are absent or vestigial in the Amphibicorisae, but long, tubular ectadenia occur in many Hydrocorisae and the Geocorisae have paired mesadenia and sometimes also ectadenia. Each testis in the Homoptera may consist of only a single follicle, as in the Coccoidea, but more often there is a small number (e.g. 4 or 5 in the Psyllidae) which may be free or united in a common sheath. Accessory glands are present and in the Psyllidae and Aleyrodidae part of the ejaculatory duct is modified to form a sperm pump (Schlee, 1969).

For further information on the reproductive system in Hemiptera see, among others, Bonhag (1955), Bonhag and Wick (1953), Davis (1955, 1956), Helms (1968), Leston (1961) and Wick and Bonhag (1955).

Mycetomes – Most Homoptera (some Typhlocybinae are a notable exception) harbour suposedly symbiotic bacteria, yeasts or other cell-inclusions which are most frequently confined to specialized cells, the mycetocytes. These cells may occur scattered in the gut-wall or fat-body (e.g. some Coccoidea) but more often are grouped into definite organs known as mycetomes. The latter vary greatly in size and position according to the insect and are sometimes very conspicuous structures described briefly on p. 257. The physiological relations between insect and micro-organism have not been adequately studied but in many cases it is known that symbionts in female insects migrate into the maturing egg and are thus transmitted to the offspring. A few experimental studies (e.g. on *Euscelis* by Schwemmler *et al.*, 1973) have shown that insects deprived of their symbionts develop abnormally or are less viable. It is, however, often impossible to culture the symbionts *in vitro* and they are sometimes regarded as degenerate or aberrant inclusions ('Blochmann bodies') which may be neither yeasts nor bacteria (Lanham, 1968, supported by the ultrastructural and histochemical data of Chang and Musgrave, 1969; see also Hinde, 1971). In the Heteroptera, where symbionts occur in the gastric caeca, mycetomes are rarely found, but they occur in *Ischnodemus* and a few others.

Immature Stages and Metamorphoses

Almost all Heteroptera are oviparous, but viviparity occurs in the Polyctenidae (Hagan, 1931) and apparently in the Lygaeid *Stilbocoris natalensis* (Carayon, 1961). In the Homoptera the Aphidoidea include viviparous parthenogenetic forms and the Coccoidea show various transitional stages from oviparity through ovoviviparity to viviparity. The eggs of the Heteroptera show considerable diversity of form and chorionic structure, and they vary in the mechanisms of respiration, fertilization and eclosion, often in ways that help to define major taxonomic groups (Southwood, 1956; Cobben, 1968). The chorion may be thin and solid, as in *Hebrus* and *Mesovelia*, but more usually there is an inner, porous, air-containing layer, differing in detail between the Geocorisae on the one hand and many Amphibicorisae and Hydrocorisae on the other. Respiratory channels (aeropyles) extend from the air-filled meshwork to the surface of the shell; they may be very numerous (nearly 200 in *Rhodnius*) and sometimes open on complicated chorionic processes (Hinton, 1962, 1969). In the Nepidae (Hinton, 1961) there are up to 26 long respiratory horns (Fig. 327), each comprising a central air-filled meshwork that connects through aeropyles with a peripheral plastron that extends over much of the horn. Micropyles, allowing access of spermatozoa for fertilization, may be represented by a single cephalic aperture in some Heteroptera or, more often, there are several and they are more widely distributed. They occur in both the Pentatomomorpha and Cimicomorpha (Miridae) and in some species they form tubes running down the centre of a stalked aeropyle. Among terrestrial Heteroptera the Cimicomorph egg is usually provided with a cap or operculum which is lifted by the young insect when it emerges; the Pentatomomorphan egg has no cap or an independently evolved cap-like structure, the pseudoperculum. Hatching spines or ridges are present on the embryonic cuticle of the head in both groups, but a median egg-burster is found only in the Pentatomomorpha. The Homopteran egg has been less fully studied

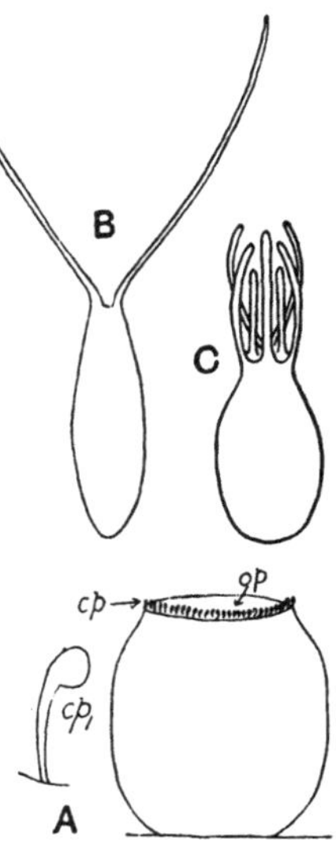

FIG. 327
Eggs of Heteroptera

A, a Pentatomid, *Euschistus* (*after* Heidemann). B, *Ranatra* and C, *Nepa* (*after* Schouteden). *op*, operculum; *cp*, chorionic processes; cp_1, one of the latter more enlarged.

(Cobben, 1965). Fulgoromorph eggs have a single dorsal or anterior micropyle, but Cicadids have three micropyles and Cicadellidae have from one to eight. Among Fulgoromorpha there may be an inner air-filled chorionic meshwork, and aeropylar horns with a plastron occur in the Tettigometridae and Tropiduchidae, though the Cicadomorpha lack these. They also differ in that eclosion occurs through a longitudinal split, whereas the Fulgoroids have an operculum. The Tettigometridae and Cixiidae lay their eggs in or on the soil, but the other Auchenorrhyncha embed them in plant tissues.

Postembryonic development is gradual in most Hemiptera, but colour changes are often very marked. The most sharply pronounced modifications are concentrated in the last ecdysis from the final immature instar to the imago. The external morphological changes during development involve the segments of the antennae and tarsi, the latter frequently not attaining their full number until the adult stage. The shape of the head and thoracic segments, more particularly the pronotum, undergo marked changes in different instars. Wing-rudiments are small or scarcely distinguishable in the 3rd instar but are evident in the 4th. Among Heteroptera the usual number of instars (including the adult) is 6 but *Dindymus sanguineus* is exceptional in passing through 9 instars. For keys to immature Heteroptera see Leston and Scudder (1956), Sweet and Slater (1961) and Decoursey (1971). In Homoptera the number of instars is subject to great variation; in *Psylla* and *Empoasca* there are 6, in aphids 5 except in the apterous Phylloxeridae, where there are 4; the latter number is also recorded in Aleyrodidae. The highest observed number is 7 (in *Magicicada*) and the lowest in Coccoidea where there are usually 3 instars in the females and 4 in the males. A holometabolous form of development has evolved independently in the Aleyrodidae and the male Coccoidea. In the latter the last immature instar is a quiescent pupa, but the Aleyrodidae lack a true pupal instar and accomplish their metamorphosis between the fourth larval instar and the adult.

Fossils and Phylogeny

Many Auchenorrhynchan wing-impressions are known from the Permian onwards, the early material falling into the families Archescytinidae, Prosbolidae and Scytinopteridae (Evans, 1956–64). The recent Peloridiidae are perhaps survivors of a similar but earlier Proto-Homopteran stock, from which a major evolutionary line probably radiated into the Fulgoroidea, though these are not known as fossils until the Triassic. A separate, second line led to the Cicadomorph Auchenorrhyncha. The origins of the Cercopoidea are not clear but among the Cicadoidea there is perhaps a relationship between the Tettigarctidae and the Permian Cicadoprosbolidae, while the Cicadelloidea may have arisen near two highly specialized Permo-triassic families, the Coleoscytidae and the Stenoviciidae. Among extant Sternorrhyncha, the Psyllidae are the most generalized and can be related to the Permian Protopsyllidae, themselves probably close to the Archescytinidae mentioned above. The structure of recent forms suggests, on the one hand, a relationship between the Psyllidae and the Aleyrodidae (Schlee, 1969) and, on the other hand, a connection between the Coccoidea and Aphidoidea (Theron, 1958). The known fossil aphids are largely from Tertiary amber and even the Mesozoic *Triassoaphis cubitus* already has a specialized venation, suggesting that the stock may have arisen in Permian or Carboniferous times (Heie, 1967). The Heteroptera probably arose from early Proto-Homopteran ancestors before the Permian; the details of their origin are not clear, though Schlee (1969) considers that the Peloridiidae and Heteroptera had a common ancestry. The Actinoscytidae (Upper Permian and Triassic), Ipsviciidae and Dunstaniidae (both Triassic) are among the earliest fossil Heteroptera and the suborder has numerous Jurassic and Cretaceous representatives. These do not, however, contribute much to theories of Heteropteran phylogeny, which depend largely on inferences from the morphology and taxonomy of recent families. Cobben (1968) considers that the terrestrial Heteroptera are widely polyphyletic and therefore rejects the concept of a single group of land-bugs, the Geocorisae. He believes that the Pentatomomorpha, the Cimicomorpha (in a strict sense), the Reduvioidea, the Saldidae and their allies, and some smaller superfamilies, are all major, independently evolved groups of terrestrial Heteroptera which, along with the monophyletic Hydrocorisae and the recent Amphibicorisae, all radiated from an extinct, ancestral Amphibicorisan stock. For further information on fossil Hemiptera and the possible phylogeny of the order, see the general references cited on pp. 748–781 and the catalogue of fossil Homoptera by Metcalf and Wade (1966).

Classification

Recent work has resulted in the recognition of an increasing number of Hemipteran families and no two authorities agree fully on a system of classification (e.g. Poisson, 1951; Pesson, 1951; Metcalf, 1951; Evans, 1966; China and Miller, 1959; Miller, 1971; Jordan, 1972; Cobben, 1968). Older, incomplete catalogues and guides to the order include those by Lethierry and Severin (1893–96), Bergroth (1908, 1913), Oshanin (1906–16) and van Duzee (1917). More recent works are cited below under the group concerned.

Suborder I. **HOMOPTERA**

A very diverse assemblage. Head more or less deflexed, gular region small and membranous or wanting. Wings usually sloping over the sides of the body, the fore pair generally of uniform consistency throughout; apterous forms frequent. Base of rostrum extending between anterior coxae. Pronotum small; trochantins usually large. Tarsi with 1–3 segments. Metamorphosis usually incomplete, sometimes complete in male, more rarely so in female.

The suborder is divided into three sections: the Coleorrhyncha, with only one small family, and two large divisions, the Auchenorrhyncha and Sternorrhyncha, each with four superfamilies and many families.

Series COLEORRHYNCHA

Small, flattened insects with a Tingid-like facies and a discontinuous distribution in the southern hemisphere. Short, 3-segmented antennae, concealed beneath head and without a terminal arista. Rostrum 4-segmented, its base partly ensheathed by propleura. Legs simple, tarsi 2-segmented. Male with relatively large pygophor, simple aedeagus and distinct parameres, but no subgenital plates.

FAM. PELORIDIIDAE. With the characters given above. This family includes about 20 species (China, 1962) from Chile, Patagonia, New Zealand, Tasmania, Australia and Lord Howe Island. They are mostly found in forests of the Southern Beech, *Nothofagus*, living in damp moss there and passing the dry season in leaf-mould on the forest floor, but the New Zealand *Oiphysa fuscata* is cavernicolous (Helmsing and China, 1937; Evans, 1941; Drake and Salmon, 1950). The taxonomic position of the Peloridiidae has been much debated; they have also been included in or near the Heteroptera (Jordan, 1972; Schlee, 1969) or regarded as very primitive Auchenorrhyncha (China, 1962). Their internal morphology certainly suggests a primitive condition; among other things, they lack a filter chamber in the gut and their central nervous system has four distinct ventral ganglia (Pendergrast, 1962). Many species lack hind wings, but the S. American *Peloridium hammoniorum* has brachypterous and macropterous forms in both sexes, the long-winged specimens being able to fly.

Series AUCHENORRHYNCHA

Antennae variable, generally short, with a terminal arista. Rostrum 3-segmented, arising plainly from the head and projecting backwards between the coxae. Tarsi 3-segmented. Active forms, often jumping and able to fly.

The subdivision of the Auchenorrhyncha presents many difficulties, but the main scheme adopted here is due to Evans (1951) who divides them into two primary groups, the Fulgoromorpha (containing a single superfamily, the Fulgoroidea), and the Cicadomorpha containing the superfamilies Cercopoidea, Cicadoidea and Cicadelloidea (=Jassoidea). The further division of each of these four superfamilies is again a matter of debate among specialists. The Auchenorrhyncha of the world have been catalogued by Metcalf (1927–47; 1954–71), who has also provided extensive bibliographies (Metcalf, 1945, 1960, 1962; Metcalf and Wade, 1963). There is a check-list of Palaearctic species by Nast (1972), a catalogue of Italian species by Servadei (1967) and various older regional monographs by Distant (1880–1900; 1902–18), Fowler (1894–1909) and Melichar (1896).

FIG. 328 *Hemiodoecus fidelis* (Peloridiidae) (*after* Poisson and Pesson)

In section Homoptères of *Traité de Zoologie*, 10 (1951), under direction of P.-P. Grassé, Masson & Cie. Éditeurs.

Superfamily **Fulgoroidea**

Head with postclypeus reduced, ocelli usually two and placed near the eyes; tegulae usually present; wings without ambient vein; 1A and 2A of fore wing joined apically to form a Y-vein; middle coxae elongate and placed wide apart; hind coxae immobile; antennal pedicel with numerous sensilla and a large sense-organ on basal segment of flagellum.

The Fulgoroidea include over 9000 species (indexed by Wade, 1960) and are arranged in 20 families, following the classification of Muir (1930). Metcalf (1938) and Fennah (1945*c*) deal with the taxonomy of many American species and Fennah (1956) has monographed those of Micronesia.

FAM. TETTIGOMETRIDAE. *Differ from other Fulgoroids in the segmented antennal flagellum and the ocelli lying inside the lateral carinae of the face; fore* *wings opaque, with reduced venation, more or less reticulate apically; ovipositor reduced or absent.* The 120 or so species are superficially similar to Cicadellids; they include some myrmecophilous species and are found chiefly in Europe but also occur in the Ethiopian region while one species is known from Peru. (See Metcalf, 1932; Fennah, 1952*b*.)

FAM. CIXIIDAE. *Second segment of hind tarsus not very small, with row of small spines apically, 6th–8th abdominal tergites without wax pores; hind tibia without large, moveable apical spur; claval vein running into commissure before apex; anal area of hind wing not reticulate.* This is one of the largest Fulgoroid families, about 1100 species having been described from all parts of the world (Muir, 1925). Very little is known of their biology, but the nymphs of *Oliarus felis* feed underground on the roots of grass in Australia and the same is probably true of the British species of *Cixius* (China, 1942).

FAM. DERBIDAE. *Characters of hind tarsus, hind tibia, abdomen and hind wing as given for Cixiidae; apical segment of labium about as long as wide.* Very delicate, usually long-winged insects of which over 750 species are known, some associated with palms and mostly tropical (Muir, 1918; Fennah, 1952*a*). None are British and few Palaearctic but *Melenia* occurs in the Mediterranean region.

Three small families, associated with the Cixiidae are as follows: the **ACHILIXIIDAE**, with about 12 species, are widely distributed (Fennah, 1947); the 50 or so species of **MEENOPLIDAE** are confined to the eastern hemisphere (Muir, 1925); the **KINNARIDAE**, with some 60 species (Fennah, 1942), include *Bytrois nemoralis*, which feeds on leaves of cocoa in Trinidad.

FAM. ACHILIDAE. *Also related to the Cixiidae, but with claval vein extending to apex of clavus; base of abdomen without the short appendages found in the Achilixiidae.* A moderately large family (about 240 species) of world-wide distribution, especially in the tropical zone (Fennah, 1950). The nymphs live under bark or in cavities in dead wood.

FAM. DELPHACIDAE (= Araeopidae). *Hind tibia with large, moveable, spur (calcar) at apex; median ocellus absent; ovipositor complete; males with large aedeagus and distinct parameres.* This family is the largest of the Fulgoroids with over 1300 species (Muir, 1915; Wagner, 1962; Le Quesne, 1960–69; Fennah, 1965). It is well represented in Great Britain where there are about 70 species, the biology of some being described by Hassan (1939), Rothschild (1964, 1966) and Waloff (1975). The Sugar-cane Leaf-hopper *Perkinsiella saccharicida* (Fig. 329) is very destructive in Queensland and was formerly so in the Hawaiian Islands before successful biological control (Muir, 1931; see also Williams, 1957); owing to the habit of oviposition in cane stalks this and other species are liable to transportation.

FAM. FULGORIDAE (Lantern-flies). *Tegmina, and usually also the hind wings, with reticulate system of supernumerary veins and cross-veins; ovipositor incomplete; aedeagus with a distinct theca surrounding the penis, the membrane between them usually bearing a complex armature.* A tropical family of about 600 species, including many brilliantly coloured insects, often of large size. In many genera the front of the head is greatly drawn out to form a huge hollow proboscis-like prolongation which was at one time believed to be luminous. Some species have the power of secreting quantities of a flocculent white wax which, in *Phenax*, streams behind as long filaments while the insect flies.

FAM. DICTYOPHARIDAE. *Allied to the Fulgoridae, but without reticulate venation in anal area of hind wing.* This family of over 500 species includes medium-sized Fulgoroids often with strongly modified heads bearing a distinct process. They are widely distributed and many species are confined to arid and semi-arid areas. *Retiala viridis* is a minor pest of coffee. See Melichar (1912) and Haupt (1929) for taxonomy.

FAM. ISSIDAE. *Second segment of hind tarsus with one apical spine on each side; tegmina reduced, often brachypterous; hind wings often absent.* A large family of about 1100 species, with several European representatives (Melichar, 1906; Doering, 1938–41; Fennah, 1954). Some of its members have a squat, beetle-like facies and *Augilia* and *Caliscelis* are unusual in possessing enlarged fore legs with foliaceous femora and tibiae.

FIG. 329 *Perkinsiella saccharicida*, male: magnified
After Kirkaldy, *Entom. Bull. Pt.* 9, *Hawaiian Sugar Planters' Assn.*

FAM. TROPIDUCHIDAE. *Allied to the Issidae; posterior angle of metanotum cut off by a groove or fine line.* Melichar (1951*b*) recognizes 140 species in this family, mostly from the Indo-Malaysian and Neotropical regions; a very few occur in southern Europe.

FAM. NOGODINIDAE. *Second segment of hind tarsus with one apical spine on each side; tegmina well-developed; venation with supernumerary veins and cross-veins; clavus without granules; clypeus usually with lateral carinae.* About 130 species have been described from Africa, America and the Indomalaysian region (Melichar, 1923).

FAM. FLATIDAE. *Characters as given for the Nogodinidae, but costal area of tegmen usually without cross-veins or, if these are present, then clavus granulate or clypeus without lateral carinae.* These beautiful moth-like species, often with delicately pigmented tegmina, are mainly tropical and sub-tropical; about 950 species having been reported from all the main zoogeographical regions. Both nymphs and adults frequently rest gregariously and the former are largely covered with long, curled, waxy filaments as in the Indian *Phromnia marginella*. The adults of some species occur in two conspicuously different colour forms and in an African species observed by Gregory, the insects were clustered on a stem with green individuals occupying the upper portion and red individuals situated just below them. In this attitude they were curiously like a red flowered spike with green unopened buds above. In other dimorphic species the colour forms have been observed intermixed; for a discussion of this subject see Imms (1914). There is a generic monograph by Melichar (1923), now partly out-dated. A related family, the **HYPOCHTHONELLIDAE**, is represented only by *Hypochthonella caeca* from S. Rhodesia (China and Fennah, 1952).

Five further families of Fulgoroidea, all allied to the Flatidae, may be mentioned briefly. The 80 or so species of **ACANALONIIDAE** (Melichar, 1923) occur mainly in the western hemisphere, with 6 species of *Philatis* known only from the Galapagos Islands. The **RICANIIDAE** include over 350 species, also moth-like in appearance; they are almost confined to the eastern hemisphere, especially the Ethiopian and Indomalaysian regions (Melichar, 1923). The **LOPHOPIDAE**, with about 120 species, are another mainly Old World group (Melichar, 1915*a*) and the **EURYBRACHIDAE** (170 species) are best represented in the Australian, Oriental and Ethiopian regions. *Eurybrachys tomentosa* is a pest of the sandal tree in S. India; its eggs are laid in a cluster covered with a white, flocculent secretion and it has 5 nymphal instars in 3 overlapping generations per year (Chatterjee, 1933). The **GENGIDAE** include only *Gengis panoplites* and *Microeurybrachys vitrifrons*, both from S. Africa (Fennah, 1949).

Superfamily **Cercopoidea** (Frog-hoppers, Cuckoo-spit insects)

Head with postclypeus greatly expanded, occupying most of the face and extending on to dorsal side; frons not distinct; two ocelli, rarely absent; tegmina opaque, tegulae absent; wings usually with ambient vein; 1A and 2A usually parallel or 2A absent; hind coxae mobile, short and conical; hind legs saltatorial, with tibiae elongate and bearing one or two large lateral spines and a double apical group of small spines. Male genitalia with subgenital plates.

Four families are recognized here, all sometimes treated as subdivisions of the Cercopidae (*sens. lat.*). There is an index to species by Wade (1963) and a bibliography by Metcalf (1960).

FAM. CERCOPIDAE. *Head narrower than pronotum, about as wide as anterior margin of scutellum; crown of head with disk usually convex or tectiform; ocelli on disk of crown, each at posterior end of a sulcus; eyes about as long as wide; pronotum hexagonal, its anterior margin straight or slightly arcuate.* This is the largest family with nearly 1400 species from all parts of the world and best represented in the Malaysian, Neotropical, Oriental, Ethiopian and Australian regions. The Palaearctic has only about 70 species, of which one, the conspicuous red and black *Cercopis vulnerata*, occurs in Britain. Cercopid nymphs are subterranean and, like those of the Aphrophoridae (see below), they surround themselves by froth. Species of *Tomaspis* feed on the leaves and roots of sugar cane and are injurious to this crop in the New World tropics, causing necrotic lesions on the leaves because of their toxic saliva ('frog-hopper blight'). For the taxonomy of the family see Lallemand (1949, 1961).

FAM. APHROPHORIDAE. *Eyes with horizontal diameter greater than vertical; pronotum with anterior margin strongly arcuate or sub-angulate, the antero-lateral margins usually short; scutellum shorter than pronotum, flat, without a spine; fore wing with clavus acute or obliquely truncate at apex.* This is a generally distributed but somewhat smaller group than the preceding family; about 300 of its 800 or so species are Palaearctic, with others predominating in the Ethiopian, Oriental and Malaysian regions. *Aphrophora, Philaenus* and *Neophilaenus* are well known in N.W. Europe. The nymphs of some genera are subterranean while others establish themselves on plants and become enveloped in a frothy substance commonly termed 'cuckoo spit'. It has been supposed that they are in this way protected from predacious insects and other Arthropods, but they are in fact sometimes seized from their spume by fossorial Hymenoptera and other enemies. Kirkaldy has observed that the froth protects their soft bodies from the sun, and when extracted from it and not allowed moisture they soon shrivel and die; probably there is truth in both explanations. In adaptation to this mode of life the nymphs have largely lost the power of leaping which is so characteristic of the adults, and are also nearly devoid of coloration. *Philaenus leucophthalmus* is the common 'cuckoo spit' insect of Europe and N. America; the biology of this and other species of the family are described by Osborn (1916), Speers (1941), Whittaker (1965, 1971) and Halkka *et al.* (1967). It has many colour forms and occurs on a wide range of wild and cultivated plants other than grasses, while *Neophilaenus lineatus* occurs almost entirely on the latter hosts. The production of the froth has been studied by Šulc (1911), Gahan (1918), Marshall (1964–65) and Gouranton (1968*b*). According to Gahan the tergites and pleurites of the 3rd to 9th abdominal segments, instead of ending as usual at the sides of the body, are curved beneath the abdomen as membranous extensions, which meet along the mid-ventral line. Between them and the true ventral surface of the abdomen there is thus formed a cavity into which the spiracles open. This chamber is closed anteriorly, but air can be admitted or expelled by means of a posterior V-shaped valve or slit. The frothing is the result of a fluid issuing from the anus, forming a film across this valve and becoming blown into bubbles by means of air expelled through it. The foam appears to be stabilized by a surface-active material secreted by cells of the proximal region of the Malpighian tubules of the nymphs. Granules in these cells coalesce to form material that is secreted into the lumen of the tubule and passes from there into the hind gut. In Madagascar the nymphs of *Ptyelus goudoti* discharge a clear liquid in such quantities as to resemble fine rain; Goudot has estimated that 70 individuals could emit a quart in 1¼ hours. For the taxonomy of the family see Lallemand (1912, 1949) and Doering (1930).

The **CLASTOPTERIDAE**, with some 85 species almost confined to the Americas, differ from the Aphrophoridae in the scutellum being longer than the pronotum and the clavus obtuse or obliquely truncate apically.

FAM. MACHAEROTIDAE. *Head narrower than pronotum, the face strongly inflated; scutellum acuminate apically or with a strong, curved spine; fore wing membranous apically, with one claval vein or with two which fuse distally; immature stages inhabiting fixed calcareous tubes.* This relatively small family of 100 or so species is mainly distributed in the Old World tropics (Maa, 1963). Their most distinctive biological feature is the construction by the nymph of curious serpuliform dwelling tubes about 1·5 cm long. The form of the tubes is characteristic of the species, they are attached to the branches of trees, and the nymphs live in them enclosed in froth similar to that of the other Cercopoids. The tubes consist of an organic framework of mucofibrils secreted by a distinct segment of the Malpighian tubules and subsequently mineralized, mainly with calcium carbonate derived from sphaerites of tricalcium phosphate secreted by the mid gut (Marshall, 1964–65; Marshall and Cheung, 1973). The life-histories and biology of some species are discussed by Lefroy (1909) and Evans (1940).

Superfamily **Cicadoidea** (Cicadas)

Head with distinct frons and three ocelli; tegmina typically transparent, tegulae absent; wings with ambient vein; 1A and 2A not forming a Y-vein; fore legs with enlarged femora; hind legs elongate, slender, not saltatorial; males with conspicuous stridulatory organs (tymbals) at base of abdomen, concealed by plate-like opercula developed from the metathoracic epimera; male genitalia without subgenital plates, aedeagus usually simple; ovipositor long, complete.

The cicadas are a moderately large, homogeneous group of about 4000 species, indexed by Wade (1964) and provided with a bibliography by Metcalf (1962). Although divided by him into two almost equally numerous families (Tibicinidae and Cicadidae *s.str.*) the vast majority of species are here united in a single family, the Cicadidae, with a second very small family, the Tettigarctidae.

FAM. CICADIDAE. *Females without tymbals; tympanal auditory organs present in both sexes; empodia absent.* Their large size and sound-producing powers make these insects familiar objects in the warmer regions of the world. Only about 100 species are Palaearctic and the sole British representative is *Cicadetta montana* which occurs in the New Forest and extends as far north as Finland on the European mainland. The capacity for sound-production (see p. 183) is limited to the males and varies very greatly in note and degree of intensity in different species. The sound has been variously compared to a knife-grinder, scissor-grinder and even a railway whistle. In the moist sub-Himalayan forest tracts of India the noise emitted by these insects is almost deafening, and extremely monotonous. Despite the number of described species, their biology has been little studied (Beamer, 1928; Myers, 1929). The nymphs so far known are subterranean, and the greatly enlarged and modified femora and tibiae of the fore legs are special adaptations to that mode of life. Among the best-known species are the 'periodical Cicadas' (*Magicicada septemdecim* and its allies) of the United States (Figs. 330, 331), which appear in great

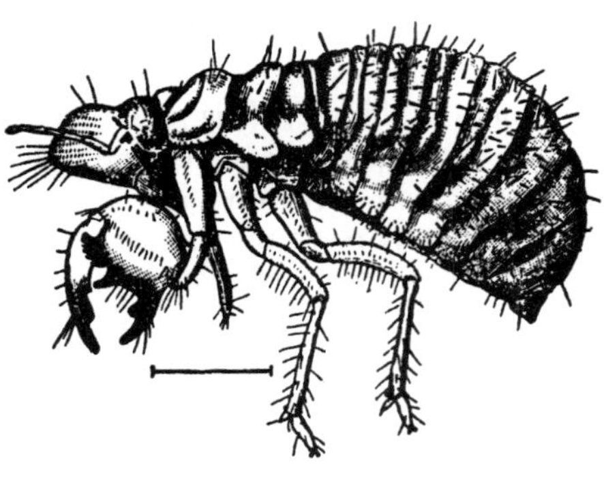

FIG. 330
Magicicada septemdecim nymph in 4th instar
After Marlatt, *U.S. Dept. Agr., Ent. Bull. 71.*

numbers after long intervals of time. Their periodical appearance is due to the nymphs requiring thirteen (in the south) or seventeen years (in the north) for their development, and the fact that the adults of one generation appear about the same time in vast numbers. A complex of sibling species is now known to be involved (Moore and Alexander, 1958; Alexander and Moore, 1963). Thus, among the 17-year forms, *M. septemdecim* and *M. cassini* are morphologically almost indistinguishable and occur together in the same area, but the females respond only to the distinctive calls of the conspecific males and cross-mating is thus prevented. The periodical cicadas have been intensively studied over very many years (e.g. Marlatt, 1907) and more than 20 distinct broods have been located in various parts of the country; a 17-year form has also been reared under field conditions from the egg. In

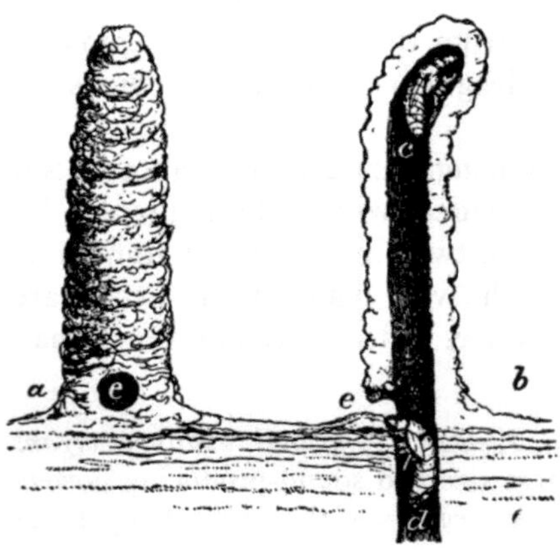

FIG. 331
Magicicada septemdecim, 'Earthen Chimneys'

a, front view; *b*, section; *c*, nymph in last instar awaiting time of change and at *d* ready for transformation; *e*, emergence hole. From Marlatt *after* Riley.

many districts several broods of different ages are known to co-exist, thus explaining the appearance of swarms of the insect several times during the 17-year cycle. The female deposits her eggs in slits which she makes in the twigs of trees, and the young emerge in about six weeks. They fall to the ground and there lead a subterranean life, 12–18 inches below the surface, sucking the juices from the finer roots of various trees. In May of the 17th year they regain the surface and, leaving their nymphal exuviae attached to tree-trunks, etc., emerge as perfect insects. When very abundant the nymphs practically honeycomb the soil; considering their size and numbers, the injury occasioned does not appear to be great but fruit growers sometimes experience a good deal of loss. Cicadas, incidentally, are unusual among Hemiptera in feeding on the xylem fluid; the surplus water they take in is rapidly shunted to the hind gut via the filter chamber and ionic resorption occurs in the ileum (Cheung and Marshall, 1973). Under certain circumstances the final-stage nymphs (often termed pupae) construct earthen cones or chimneys (Fig. 331), about 10 cms high, in which they live above ground for several weeks before emerging as adults. Several explanations have been offered for these structures and it may possibly be that in certain districts individuals prematurely reach the surface before they are prepared to become adults and construct cones as means of protection until they reach maturity. There appears to be a correlation between an unusually high local temperature and the occurrence of these cones, which are also said to be prevalent over burned areas. Taxonomic works on the Cicadidae include those of Distant (1889–92; 1906; 1912–14); Myers (1928) deals with their general morphology and Vasvary (1966) and Young (1975) with that of the sound-producing organs.

FAM. TETTIGARCTIDAE. *Pilose cicadas with complete tegminal venation; tymbals present in both sexes; tympanal auditory organs absent; empodia present. Tettigarcta tomentosa* from Tasmania and *T. crinita* from S.E. Australia are the only extant species, but the family is also represented by Mesozoic and Tertiary fossils from the N. hemisphere and is of some phylogenetic interest (Evans, 1941).

Superfamily **Cicadelloidea** (= Jassoidea) (Leafhoppers)

Postclypeus not greatly expanded on to dorsal side of head; frons not distinct; paired ocelli variously situated, on dorsal side of head, or on face, or between the two; tegmina transparent or opaque; tegulae absent; 1A and 2A not forming a Y-vein; hind coxae mobile, elongate; hind legs saltatorial, with tibiae ridged longitudinally and bearing longitudinal rows of spines on lateral margins.

Different authorities disagree on the taxonomic arrangement of the insects included here. The scheme below follows Evans (1951, 1966 and earlier papers) in recognizing two large, widely distributed families, the Membracidae and Cicadellidae (= Jassidae), and five smaller families with restricted distributions. Metcalf's *General Catalogue of the Homoptera* follows a different arrangement, as do several alternative systems (cf. Haupt, 1929; Oman, 1949; Wagner, 1951; Ross, 1957).

FAM. MEMBRACIDAE (Tree-hoppers). *Head modified, with crown almost vertical and face horizontal; pronotum enlarged and bearing a spine-like or variable (sometimes bizarre) process.* These insects (catalogued by Metcalf and Wade, 1965) may almost always be easily recognized by the pronotum, which is prolonged backwards into a prominent elevated hood or process, lying over the abdomen, and sometimes assuming the most grotesque forms (Fig. 332). The family, with over 2500 species, reaches the zenith of its development in the Neotropical region. The Palaearctic fauna includes only three genera, two of which, *Centrotus* and *Gargara*, are British. Biological studies on the family have been made by Branch (1914), Funkhouser (1917) and Couturier (1938). The eggs are usually deposited in small groups arranged in two nearly parallel slits cut in the twigs of trees and shrubs. The nymphal stages differ from the adults in the absence, or only partial development, of the pronotal process; the tergites are often furnished with elongate filaments or spinose projections. Certain genera (*Telamona*, *Thelia*, etc.) are affected by parasites which induce 'castration parasitaire', noticeable in the reduction or other modification of the external genitalia (Kornhauser, 1919). The life-history of *Vanduzea arquata*, a widely distributed N. American species, has been studied by Funkhouser (1915). It abounds on *Robinia* and appears to pass through two generations in the year. Both the nymphs and adults are commonly attended by ants, as is usual whenever Membracids are present in large numbers. The ants stroke the Membracids with their antennae, whereupon the latter insects exude a liquid from the retractile anal tube. The mutual relationships of the two groups of insects has attracted the attention of a number of observers (see Lamborn *et al.*, 1914). A few species have been noted to exhibit maternal solicitude; although usually leaping away at the first alarm, they refuse to move if disturbed while guarding their offspring. Among taxonomic papers see Goding (1926–39), Evans (1948), Funkhouser (1950) and Strümpel (1972).

Three small families, the **AETALIONIDAE** (about 50 species), the **NICOMIIDAE** (15 species) and the **BITURRITIIDAE** (6 species), are allied to the Membracidae.

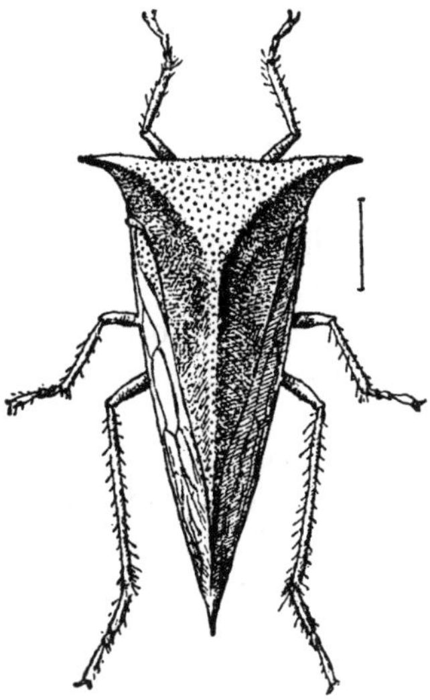

FIG. 332
Ceresa bubalus, enlarged
After Marlatt, *Ins. Life*, 7.

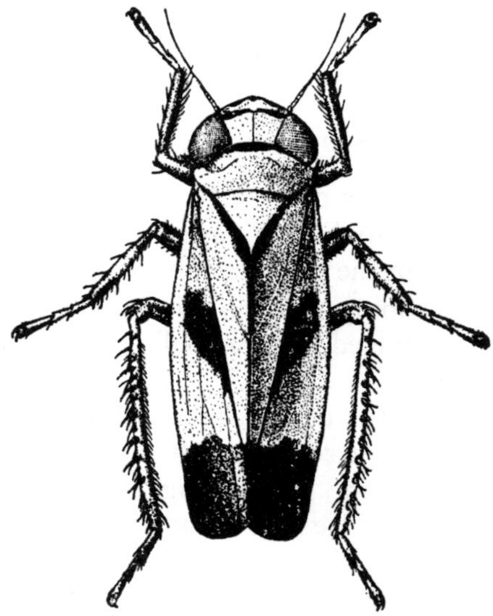

FIG. 333
Nephotettix apicalis × 12
After Misra, *Mem. Dep. Agric. India Entom. Ser.* 5.

FAM. EURYMELIDAE. *Confined to Australia, New Guinea and New Caledonia; ocelli on ventral side of head; tegmen with* M_{1+2} *extending to apex; hind tibia with one or more spines mounted on prominent bases.* Most of the 350 species feed on eucalypts, others on *Casuarina* or on the mistletoe *Loranthus* (itself on eucalypts). The nymphs, and sometimes also the adults, are gregarious and attended by ants. The females lay their eggs in batches in parallel slits in the twigs of the host-plant. Some species of *Pogonoscopus* live in ants' nests. For further information see Evans (1931, 1966).

FAM. CICADELLIDAE (= Jassidae). *Without the combination of characters given for the two above families of Cicadelloidea.* The members of this large group of some 8500 species have been divided into a number of families, but the lack of agreement between different authorities on their limits (see above) makes it preferable here to treat them as a single family. Except for the aphids they are probably the most abundant of all Homoptera and may readily be collected by sweeping grass, herbage and other foliage. They are slender, usually tapering posteriorly, and rest in a position ready for jumping. When disturbed they often leap several feet and readily take to flight. Their slender form and the appearance of the hind tibiae (Fig. 333) will enable most species to be distinguished from the Cercopidae at sight. The general biology of the family has been reviewed by DeLong (1971). Reproduction is usually bisexual and females lay up to 300 eggs, singly or in rows or clusters, beneath the epidermis of the host-plant, which they lacerate with the ovipositor. There are four to six (usually five) nymphal instars and though the arboreal species generally have only one generation per year, others may have up to five, depending on the locality; most of the common grassland species in Britain are bivoltine (Waloff, 1975). In temperate regions Cicadellids may overwinter in any of the stages, most often as the adult (e.g. *Empoasca*, *Erythroneura*) or the egg (*Typhlocyba*, *Idiocerus*). They are typically phloem feeders, but the Typhlocybinae feed on the mesophyll tissue of leaves; some species are polyphagous, others more restricted in their choice of hosts. Young adults disperse by flight (Johnson, 1969; Waloff, 1973) and a few species undertake considerable migrations, as shown in N. America by the sugar-beet leafhopper *Circulifer tenellus*, the potato leafhopper *Empoasca fabae* (a complex of sibling species; see Ross and Moore, 1957) and the six-spotted leafhopper *Macrosteles fasciifrons*. These and many other species are injurious to a variety of crops, especially rice (attacked by species of *Nephotettix*) and cotton (in which resistance to *Empoasca* species is correlated with hairiness of the leaves; see Parnell *et al.*, 1949). Injury to crops may be direct or through the transmission of phytopathogenic viruses or mycoplasma-like organisms (Nielson, 1968). Natural populations are subject to control by predators (some species of Sphecid wasps, for instance) and by parasitic Chalcids, Dryinids, Pipunculids and Strepsiptera, some of which are relatively non-specific (Waloff, 1973, 1975). Taxonomic works on the Nearctic species include those by Oman (1949), DeLong (1948), Young (1952) and Beirne (1956), and on the European species by Ribaut (1936; 1952) and Le Quesne (1960–69). Other more general taxonomic studies are listed above under Cicadelloidea.

A small group of about 30 species of Old World leafhoppers constitutes the family **HYLICIDAE**, the last of the seven Cicadelloid families recognized by Evans.

Series STERNORRHYNCHA (= Phytophthires)

Antenna usually well-developed, without a terminal arista, sometimes atrophied. Rostrum appearing to arise between the fore coxae or wanting. Tarsi 1- or 2-segmented. Many species with females and immature stages inactive or incapable of locomotion. These are the most specialized among the Hemiptera and fall into four well-defined and distinctive groups, the superfamilies Psylloidea, Aleyrodoidea, Aphidoidea and Coccoidea, of which the last two are each subdivided into several families.

Superfamily **Psylloidea**

Mouthparts present in both sexes; antennae usually 10-segmented; two pairs of wings, the front pair more strongly sclerotized than the hind; femora thickened; tarsi 2-segmented, with paired claws. Though sometimes subdivided into several families, the Psylloidea are here treated as a single family.

FAM. PSYLLIDAE (Chermidae: Jumping Plant Lice). The 1300 or so species of Psyllids are small insects about the size of aphids and bear a resemblance to minute cicadas. They are usually very active, their rapid movements being a combination of leaping and flying, but they are incapable of sustained flight. The act of leaping is performed with the aid of the hind legs which are larger and more muscular than the other pairs. The venation is simple and exhibits relatively few marked deviations among various genera. The most striking feature in the fore wing is the presence of a principal basal vein formed by the fusion of the stems of R, M and Cu (Fig. 334). In *Trioza* and its allies this compound vein divides distally into its three components while in *Psylla* and related genera it is bifurcate, dividing into R and $M+Cu_1$, the latter again dividing into M and Cu. In the hind wing the venation is extremely simple; R is represented by Rs only, M is unforked and Cu

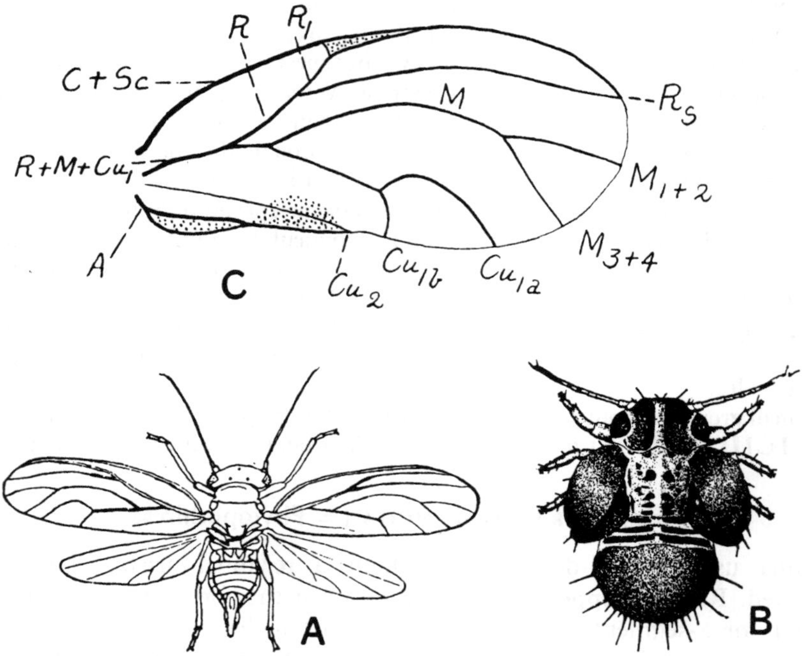

FIG. 334 A, *Psylla mali* (*after* Carpenter). B, *Psylla pyricola*, nymph in last instar (*after* Slingerland). C, *Psylla pyricola*, venation of fore wing. All enlarged

divided into Cu_1 and Cu_2 as in the fore wing. Cross-veins are absent from both wings and A is vestigial or wanting. A general account of the external anatomy of Psyllids is given by Crawford (1914) and of the venation by Patch (1909). The chief sources of information on the internal anatomy are papers by Witlaczil (1885) and Schlee (1969); an account of the morphology of *Pachypsylla* is given by Stough (1910) and of the head and thorax of *Psylla* by Weber (1929). There is a bibliography by Zacher (1916) and important taxonomic works by Aulmann (1913), Crawford (1914), Caldwell (1938), Tuthill (1943) and Heslop-Harrison (1949). The species of N.W. Europe may be identified from Haupt (1935) and the biology of the family is discussed by Lal (1934), Schaefer (1949) and Hodkinson (1974).

The life-history of the Apple Sucker *Psylla mali* (Speyer, W., 1929) may be regarded as fairly typical (Fig. 334). It passes the winter in the egg, the latter being laid about the beginning of September on the spurs of the food plant, around leaf-scars and in cracks on the new wood. The nymphs hatch in April and are flattened objects with whitish waxen threads projecting from the extremity of the abdomen. Five nymphal instars occur, and the different stages may be recognized by the increasing number of antennal segments; these organs are 2-segmented in the 1st instar and 7-segmented in the 5th. Wing-pads are evident in the 3rd instar and during later development they extend laterally in a prominent manner so as to make the insect appear nearly as broad as long (Fig. 334). The imago appears in early summer and the species is univoltine. The nymphs can be very injurious to apple in Britain, damaging the blossoms and stunting the shoots though the adults cause little injury. The Pear Sucker *P. pyricola* is very destructive in America and exhibits a different life-history. It is tri-voltine, hibernates as an imago and both nymphs and adults are injurious. The winter form of the imago differs from the summer type and was formerly regarded as a separate species. It is about one-third larger than the summer form and is much darker, particularly in the wing-veins. Certain species produce gall-like malformations on their food-plants; thus in Britain *Psylla buxi* causes the apical shoots of the box to become deformed into miniature cabbage-like growths, and *Livia juncorum* forms tassel-like galls on rushes. When Psyllids are abundant copious honey-dew is excreted by the nymphs on to the leaves and twigs. In *Psylla mali* a long waxen thread enclosing a central core of translucent liquid exudes through the anus and when the threads become broken up the fluid spreads over the leaves and twigs (Awati, 1915). In many of the Australian Spondyliaspinae the sedentary nymphs secrete a more or less elaborate test or 'lerp' under which they shelter. One of these is *Cardiaspina albitextura* which occurs on eucalyptus trees; its biology and population dynamics have been studied by Clark (1962, 1964).

Superfamily **Aleyrodoidea**

Mouthparts present in both sexes; antennae 7-segmented; two pairs of wings, of similar membranous consistency, opaque, whitish, clouded or mottled with spots or bands; legs long and slender; tarsi with two subequal segments and a pad-like empodium or spine between the claws. With a single, generally distributed family.

FAM. ALEYRODIDAE (**Aleurodidae**: White Flies). The 'white flies' are a much neglected group related to the Psyllidae; about 1100 have been described but the majority of the world's species are probably still unknown. For taxonomic accounts see Quaintance and Baker (1913–14; 1917), Sampson (1943; 1947), Zahradnik (1963) and Mound (1965; 1966) and among other biological papers, see Trehan (1940) and the bibliography by Trehan and Butani (1960; 1970). Both sexes are winged and are dusted with a characteristic mealy white powdery wax; all are small or minute with an average wing expanse of about 3 mm. *Trialeurodes vaporariorum* is the well known Greenhouse White Fly which is particularly injurious to tomato and cucumber, the insect infesting the lower surface of the leaves in all its stages (Fig. 335), (Weber, 1931; 1935; and, for a related pest, Butler, 1938*a*; 1938*b*). *Dialeurodes citri* is the Citrus White Fly which is destructive to Citrus in the southern United States (Morrill and Back, 1911). The most characteristic organ of the Aleyrodidae is the *vasiform orifice* which opens on the dorsal surface of the last abdominal segment. It is a conspicuous opening provided with an *operculum*, and situated within the orifice and beneath the operculum is a tongue- or strap-shaped organ known as the *lingula*. The latter is covered by the operculum in some species but projects beyond it in others. The anus opens within the orifice at the base of the lingula. Honey-dew is excreted in large quantities by all stages, particularly the larvae. It issues through the anus accumulating on the lingula, and this fact probably gave rise to the view that the latter organ secretes the honey-dew. The vasiform orifice is present both in the larval and adult stages and affords characters of taxonomic value.

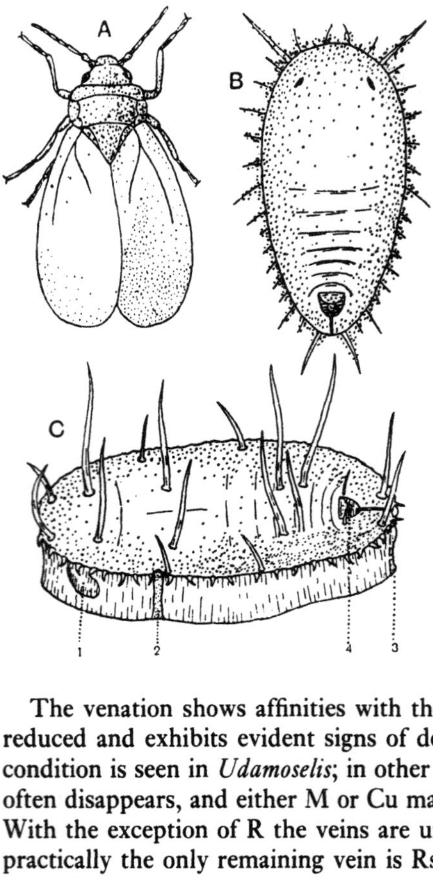

FIG. 335
Trialeurodes vaporariorum

A, imago × 50; B, larva in first instar × 150; C, puparium × 65. 1, adult eye; 2, thoracic breathing fold; 3, caudal breathing fold; 4, vasiform orifice. *After* Lloyd, *Ann. App. Biol.*, 9.

The venation shows affinities with the Psyllidae (e.g. *Trioza*); it is always much reduced and exhibits evident signs of degeneration (Fig. 335). The most primitive condition is seen in *Udamoselis*; in other genera C and Sc are more or less fused, R_1 often disappears, and either M or Cu may be present but are usually not coexistent. With the exception of R the veins are unbranched and in the most modified forms practically the only remaining vein is Rs.

Parthenogenesis is common in several species and probably occurs in many others, but the subject needs further investigation. Morrill and Back (1911) observe that virgin females of *D. citri* produce males. According to Schrader (1926) in *T. vaporariorum* there are two parthenogenetic races, one of which produces males and the other females; the fertilized females give rise to individuals of both sexes.

The eggs are very characteristic, with a pedicel which, in some cases, exceeds the length of the egg itself. According to Cary (1903), at the time of fertilization the lumen of the pedicel is filled with protoplasm. The spermatozoa move through it until they meet the female pronucleus which migrates until it comes to lie at the entrance to the pedicel. After fertilization the contents of the pedicel shrivel up. The eggs are attached to the leaves of the food-plant by this stalk, and are generally laid in a circle or arc, one or more rows deep. Four larval instars are present, the last forming a so-called pupal stage (Weber, 1934). The larvae are oval and greatly flattened and, after the first moult, the legs and antennae degenerate; during the 4th instar the imaginal organs become visible. It gradually becomes thicker and more opaque and is frequently adorned with conspicuous rods or filaments of wax. During the first part of this instar the insect feeds like the earlier larvae. Towards the end of the period it becomes inactive, remaining anchored to the leaf by its stylets, and the pharate adult with its appendages is then clearly visible within the outer case. The imago emerges by a T-shaped rupture of the dorsal wall of the last larval cuticle.

Superfamily **Aphidoidea**

Mouthparts present in both sexes; antennae variable, 1- to 6-segmented; polymorphic, but alate forms with two pairs of transparent membranous wings; legs long and slender; tarsi 2-segmented, the basal segment sometimes reduced, with paired claws; paired dorsal processes (cornicles) present on fifth abdominal segment. The subdivision of the group presents some difficulties. Two relatively small families, the Adelgidae and Phylloxeridae, form a distinct section in which only oviparous reproduction occurs. The remainder are now usually grouped into six families which, with some of their principal genera, are as follows: the Lachnidae include *Trama, Cinara, Stomaphis, Tuberolachnus, Lachnus* and others; the Chaitophoridae include genera like *Periphyllus* and *Chaitophorus* from deciduous trees and *Sipha* on grasses; the Callaphididae contain *Myzocallis, Drepanosiphum* and many more; the large family Aphididae (*s.str.*) is the dominant group (e.g. *Hyalopterus, Rhopalosiphum, Aphis, Dysaphis, Brachycaudus, Brevicoryne, Myzus, Capitophorus, Acyrtosiphon, Macrosiphum* etc.); the Thelaxidae is a relatively small group with *Anoecia, Mindarus, Hormaphis* and others; while the Pemphigidae include *Eriosoma, Schizoneura, Pemphigus* and *Forda*. The Aphidoidea, with over 3600 species, are especially characteristic of north temperate regions and have been intensively studied because of their great economic importance and the intrinsic biological interest of their complicated life-cycle and highly developed polymorphism. For the general treatment of these latter aspects it is convenient to retain the older usage and consider as a single family, the Aphididae (*s. lat.*), the six groups just mentioned.

FAM. APHIDIDAE (Aphids, greenfly, plant-lice). *Associated with woody or herbaceous plants; only sexual females oviparous, otherwise viviparous; cornicles and cauda usually present; last and penultimate antennal segment each with one primary rhinarium; R_s present in fore wing.* These familiar insects usually pass their life on the young shoots and foliage of plants but a few species live below ground on roots (e.g. *Trama*), some others occur on the branches of woody trees and shrubs (e.g. *Lachnus*) and a certain number are gall-formers (e.g. *Pemphigus, Hormaphis*). A few such as the woolly aphis, *Eriosoma lanigerum*, live both on the leaves or shoots and on the roots. The apterous generations of aphids, more especially when of flattened form (as in *Hormaphis*), are liable to confusion with the nymphs of other Sternorrhyncha, from which they may be distinguished by the following combination of characters: the 2-segmented tarsi with paired claws, the long several-jointed rostrum, the frequent presence of compound eyes and cornicles, and 9 pairs of lateral spiracles. Perhaps the most characteristic morphological feature of these insects are the cornicles or 'honey tubes' though these organs are greatly reduced in *Eriosoma* and some other genera. Réaumur believed their function to be excretory and later observers concluded wrongly that they secreted the sweet substance known as 'honey-dew'. It is now known that they are the secretory channels belonging to glands producing waxy secretions (mainly the triglycerides of hexanoic, hexadienoic and other acids). These are liberated when the aphid is attacked by a predator and act as alarm pheromones; they are perceived by the antennal receptors of other aphids, causing them to scatter, and the subject has attracted much recent study (Lindsay, 1969; Chen and Edwards, 1972; Wynn and Boudreaux, 1972; Callow *et al.*, 1973; Greenway and Griffiths, 1973; Nault *et al.*, 1973; Dixon and Stewart, 1975). Honey-dew is emitted through the anus (Broadbent, 1951). Many aphids also secrete a white waxy substance, either as a powder dusted over the surface of the body (e.g. *Hyalopterus*), or in flocculent threads (*Eriosoma*); in either case it is the product of epidermal glands.

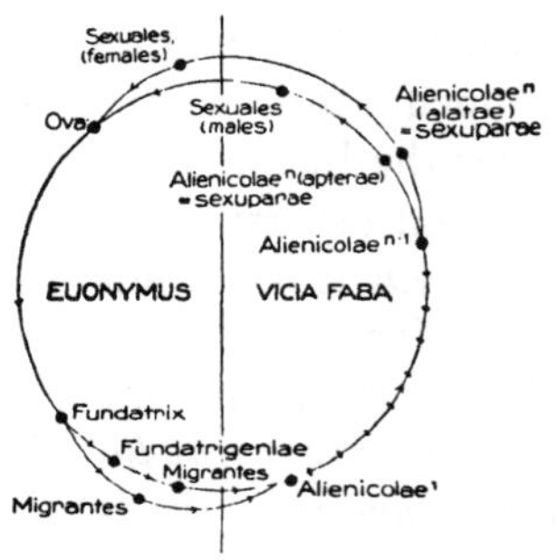

FIG. 336
Diagram of the life-cycle of *Aphis fabae* (based on observations by J. Davidson, at Rothamsted)

The venation of aphids has been studied in detail by Patch (1909); both the tracheae C and Sc are absent in all forms examined and the costal area of the adult wing is strengthened by a stout vein-like structure expanding distally into a stigma. This vein channel is interpreted as the fused main stems of all the principal veins.

Aphids are remarkable on account of their peculiar mode of development and the polymorphism exhibited in different generations of the same species (Kennedy and Stroyan, 1959; Ris Lambers, 1966). The associated phenomena concerning reproduction are – (1) parthenogenesis; (2) oviparity and viviparity; (3) the occurrence of generations in which the sexes are very unequally represented, males often being wanting and frequently rare. With regard to structure the phenomena are – (1) the production of totally different types of individual of the same sex either in the same or different generations; (2) the production of individuals with perfect and also atrophied mouthparts; (3) the production of individuals of the same sex but differing as to the gonads. Associated with habits are – (1) host alternation, involving migration to totally different plant hosts; (2) different modes of life of the same species on the same host; (3) different habits of individuals of the same generation.

In extreme cases almost all the above phenomena may occur associated with the annual cycle of an individual species. The most usual life-history of an aphid is as follows (Fig. 336). The winter is passed as eggs which are laid during the previous autumn by sexual females. With the advent of spring they hatch and give rise to apterous parthenogenetic viviparous females. The latter produce a new generation of similar forms among which a few winged females may occur. A variable number of generations of this kind are produced throughout the summer and winged viviparous females often become common. The latter are concerned with the migration and dispersal of the species and are produced in varying numbers in different generations (Johnson, 1954). At times these winged females appear in such swarms as to darken the sky and cover the vegetation. Those individuals which find plant hosts of the right species similarly reproduce on their own account. Towards the end of summer or in the autumn their progeny, and also those of the apterous forms which remained on the original plant, give rise to sexual males and females. These pair and the females are oviparous, their eggs overwintering on the food-plant, and the same cycle is repeated annually. In non-migratory aphids the whole life-cycle is spent on the same plant or on individuals of the same species. If any migration to other species of host does occur it is inconsiderable and an alternation of hosts is not essential to the life of the species. Among migratory forms well-known species are – *Pemphigus bursarius* which occurs on poplar and flies to the roots of various Compositae, returning to poplar in autumn; *Myzus persicae*, the primary (winter) host of which is peach and which occurs on a number of secondary host species (van Emden *et al.*, 1969); *Aphis fabae* which is found in autumn on the Spindle Tree (*Euonymus*), where it overwinters as the egg, and which, in May and June, flies to beans, sugar-beet, dock, poppies, etc., returning to the spindle tree in October (Jones, 1942).

The following types of individuals, arranged in sequence, are present in the life-cycle of migratory aphids (Figs. 336, 337). (1) The *Fundatrices*; usually apterous, viviparous, parthenogenetic females which emerge in spring from the overwintered eggs. The sense organs, legs and antennae are not so well developed as in succeeding apterous generations, the antennae, for example, being shorter and may comprise a smaller number of segments. The reduction of the parts is apparently correlated with increased reproductive capacity. The eyes are often smaller, or consist of fewer facets than in the succeeding generations, and there may be differences in the cornicles. In *Drepanosiphon platanoidis* and some others the fundatrices are exceptional in being winged. (2) *Fundatrigeniae*; apterous, parthenogenetic, viviparous females which are the progeny of the fundatrices and live on the primary host. (3) *Migrantes*; these usually develop in the second, third or later generations of fundatrigeniae and consist of winged parthenogenetic viviparous females. They develop on the primary host and subsequently fly to the secondary host. In *Drepanosiphon platanoidis* all the viviparous females are winged and consequently fundatrigeniae are wanting. (4) *Alienicolae*; parthenogenetic, viviparous females developing for the most part on the secondary host. They often differ markedly from the fundatrices and migrantes; many generations may be produced comprising both apterous and winged forms. (5) *Sexuparae*; parthenogenetic viviparous females which usually develop on the secondary host, the alate forms migrating to the primary host at the end of the summer. The sexuparae terminate the generations of alienicolae by giving rise to the sexuales. (6) *Sexuales*; these usually appear only once in the life cycle and consist of sexually reproducing males and females, the latter being oviparous. The females with rare exceptions (*Neophyllaphis, Tamalia*) are apterous, and distinguishable from the apterous viviparous generations of the same sex by the thickened tibiae of the hind legs, and the greater body length. The males are either winged or

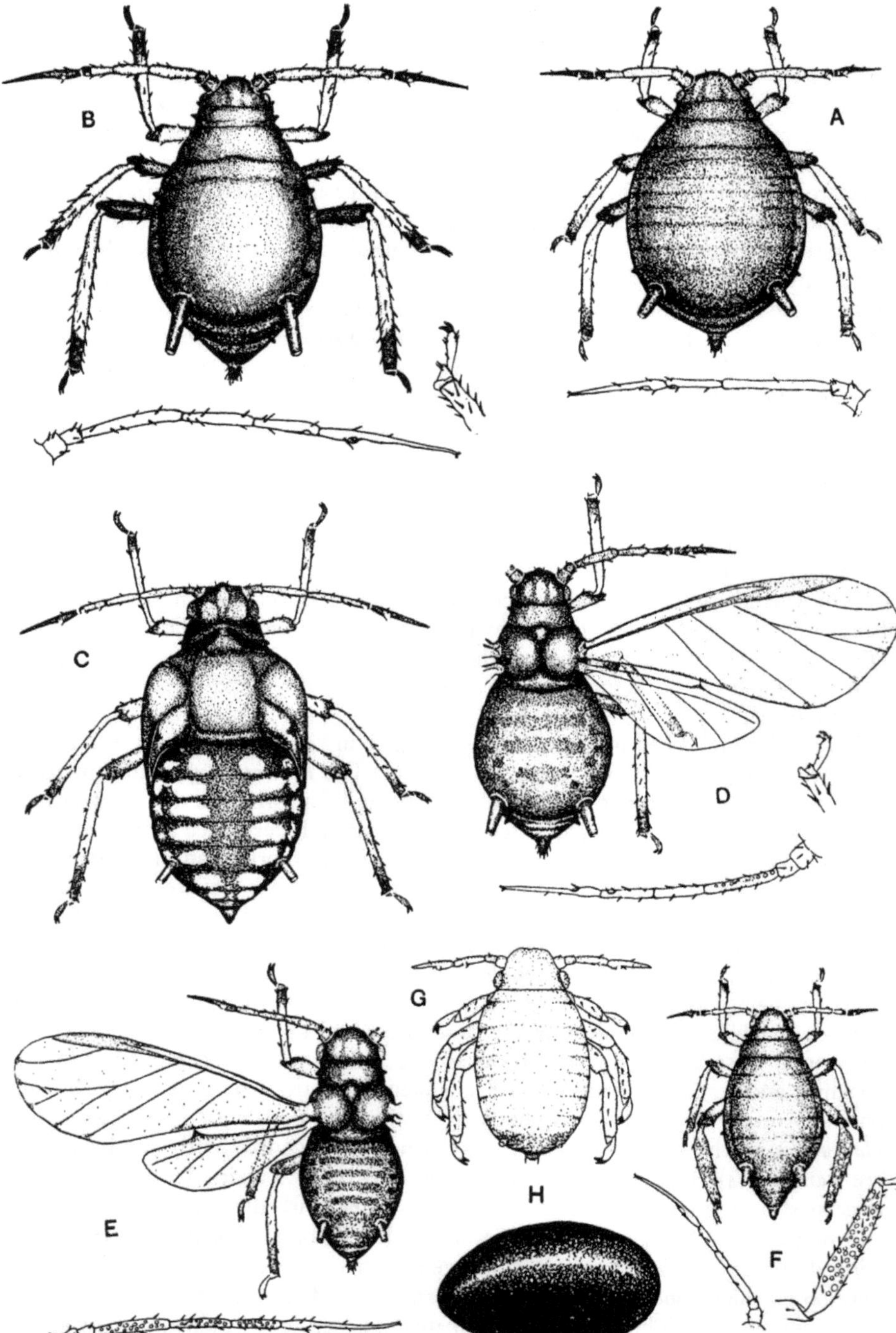

FIG. 337 *Aphis fabae*

A, fundatrix; B, apterous viviparous female; C, nymph of winged viviparous female; D, winged viviparous female; E, male; F, oviparous female; G, fundatrix, 1st instar; H, egg. The antennae are also shown under higher magnification, together with the tarsus in B and D and the hind tibia and tarsus in F. (From original drawings by J. Davidson.)

apterous, and in *Chaitophorus populeti* both types are produced. Intermediates between alate and apterous forms also occur. The sexuales exhibit various types of specialization among different genera, which reach their culmination in *Eriosoma* and its allies. In these instances both sexes are apterous, there are no functional mouthparts, the digestive system is degenerate, and the female lays only a single very large egg produced in a single ovary, the counterpart of the latter having atrophied (Baker, 1915). The eggs are laid on the primary host and in this stage the vast majority of species hibernate. Apterous parthenogenetic viviparous females may overwinter in several species, e.g. *Brevicoryne brassicae* and *Myzus persicae*.

With non-migratory species the terms migrantes and alienicolae are not applicable. In these cases the winged and wingless viviparous females are more conveniently referred to as fundatrigeniae alatae or apterae as the case may be, and either one or the other may give rise to the sexuparae.

A high degree of specialization is met in *Hormaphis* and the allied genera *Hamamelistes* and *Cerataphis*. They are gall-formers not only on the primary host, but often on the secondary one also; cornicles are vestigial or wanting and the sexuales are usually small and apterous. The alienicolae are flattened and scale-like, often with a marginal fringe of wax glands, which gives them a close resemblance to an Aleyrodid. The life-history may become greatly abbreviated and both the intermediate host and the aleyrodiform generations on it are eliminated, as in *Hormaphis hamamelidis*. According to Morgan and Shull (1910), in the vicinity of New York this species has only fundatrices, sexuparae and sexuales generations.

The physiological mechanisms underlying host-alternation and polymorphism are only partly elucidated, but these changes are probably, at least in some cases, the result of alterations in the physiology of the growing host-plant. Work by Kennedy and his colleagues (1950; 1951) on *Aphis fabae* and *Myzus persicae* shows that the aphids' readiness to feed and their rate of reproduction depend not only on the host species but also on the age and physiological condition of its leaves, actively growing and senescing leaves being preferred to mature or dying ones. The normal association of each seasonal form with its respective primary winter host and secondary summer host is therefore not obligatory but is due to the normal inaccessibility of the summer host to fundatrigeniae and to the normal unsuitability (due to maturity) of the winter host when winged alienicolae are about. Under the appropriate conditions, a reversal of normal relations was observed: fundatrigeniae of *A. fabae* colonized very young sugar-beet leaves and winged alienicolae started colonies on actively growing shoots of *Euonymus*.

So far as polymorphism is concerned, several factors operate directly or indirectly and interact in various ways. There are likely to be many different mechanisms and in only a few species have the processes been analysed experimentally. The mere fact that some species have apterous viviparae while in others they are alate, indicates that genetic control is involved and in *Aphis fabae* and *Myzus persicae*, for example, there are clonal differences in the readiness with which sexual forms are produced (De Fluiter, 1950; Blackman, 1971–74). Environmental influences may, however, also exert decisive effects. Thus crowding tends to promote the appearance of alate virginoparae, though the effect is produced differently in different species. In *Megoura viciae* apterous viviparae produce only apterous progeny when reared singly, but when crowded they give rise to alate offspring; the effect is prenatal and not due to nutritional factors (Lees, 1959–63, 1961, 1966). In *Brevicoryne* similar effects of crowding are due to contact between the parent and the larva whose differentiation is being controlled postembryonically, and in *Myzus persicae* they depend on direct contact between the larvae themselves (Bonnemaison, 1951). The production of sexual morphs is known in some species to be under photoperiodic control. In *Megoura viciae* exposure to short-day conditions induces the viviparous females to produce oviparae; there is a critical photoperiod of 14 hours 55 minutes at 15° C, shortening slightly as the temperature rises to a maximum of 23° C, above which the effect does not operate. Photoperiod is perceived directly by the mother (rather than through the host-plant) and this can occur while she is herself developing within the grandmother (Lees, 1961, 1966). Other species conform to a similar pattern, with differences in the length of the critical photoperiod and of the upper temperature limit, e.g. *Acyrthosiphon pisum*, in which the photoperiod also differs for the production of male and female progeny (Lamb and Pointing, 1972; and see also Blackman, 1975, for *Myzus persicae*). Suppression of sexuals can, however, occur in holocyclic species; the immediate offspring of fundatrices do not respond to ecological conditions that, in later generations, induce the formation of oviparae and males. Nor is it clear how the differentiation of sexual forms is controlled in subterranean species, where changes in illumination can have no direct effect. Mittler (1973) draws attention to the role of diet in affecting polymorphism. Aphids can perceive changes in the quality of their food through gustatory mechanisms and can respond by morphogenetic changes. The effects vary, however: *Myzus persicae* produces more apterae on a deficient artificial diet while *Phorodon humuli* may increase the production of alates when the host-plant grows under unfavourable conditions. The production of characteristic aestivating larval morphs of *Periphyllus* seems also to be determined through the host-plant (Essig and Abernathy, 1952). The mechanisms involved in these cases are no doubt complicated and are likely to be 'contingent' responses rather than direct nutritional effects. It is likely that the ultimate causes of the morphogenetic changes that underlie polymorphism are alterations in the endocrine balance during embryonic and postembryonic development, but the details of such control are not yet clear.

Aphids have remarkable powers of reproduction, as shown by the following data for *Dysaphis crataegi* (Baker and Turner, 1916). The fundatrix produces on an average 71 young. From 5 to 7 generations of spring forms occur and consist at first exclusively of fundatrigeniae, but migrantes appear in increasing numbers in each generation. The average number of young produced by the fundatrigeniae was 121 per female: the later generations were rather less prolific. The migrantes yielded on an average 18 young per female, the alienicolae 65, sexuparae 7, and the sexuales produced an average of 6 eggs per female. In practice aphid populations are subject to control by parasitic Hymenoptera and by predatory Aculeate Hymenoptera, Coccinellidae and the larvae of Syrphidae and various Neuroptera (e.g. Banks, 1955; Dunn, 1949; George, 1957). Vast numbers are destroyed by bad weather or failure to reach suitable host-plants.

The literature on the family is very extensive, but there is a bibliography by Smith (1972). Theobald's (1926–29) monograph on the British species has now been supplemented by the works of Ris Lambers (1933–34) and Stroyan (1950–64), while Ris Lambers (1938–53) has partially monographed the European species and Börner (1952) surveys the C. European forms. Other taxonomic works include Börner (1930, 1938), Oestlund (1942), Quednau (1954), Stroyan (1957–63), Bodenheimer and Swirski (1957), Heinze (1960–62), Eastop (1961, 1966, 1972), Pintera (1969) and Mackauer (1965). Among papers on the N. American fauna see Hottes and Frison (1931) and

Gillette and Palmer (1931–36). For lists of species and their host-plants see Davidson (1925), Patch (1938) and Averill (1945). The species of economic importance are reviewed by Börner and Heinze (1957) and there are discussions of various biological topics in Lowe (1973), van Emden (1972), Auclair (1963), Zwölfer (1957–58) and the reviews cited above. The role of aphids in the transmission of phytopathogenic viruses is dealt with by Broadbent (1953), Kennedy *et al.* (1963), Swenson (1968), Maramorosch (1969), Whitcomb and Davis (1970), Watson and Plumb (1972) and others.

FAM. PHYLLOXERIDAE. *On deciduous trees; both sexual and parthenogenetic females oviparous; wings held flat over abdomen at rest, fore wings without separate R_s, Cu and A arising on common stalk; ovipositor vestigial or absent; wax of apterous parthenogenetic female, if present, not flocculent.* This relatively small family includes such forms as *Phylloxera*, occurring on oaks, and *Viteus vitifolii*, the notorious Vine Phylloxera (Ordish, 1972). The life-history attains a high degree of complexity among Phylloxerids and in *Phylloxera quercus* Lichtenstein states that no less than twenty-one forms occur in the life-cycle. In *V. vitifolii* of the vine the life-history, in a summarized form, is as follows according to Grassi (1915; see also Maillet, 1957). The fundatrices are seldom met with on the European vine, and their fate on that plant has not been definitely settled. Grassi states that they usually perish, while those on the American vine produce leaf-galls; in no case do they develop on the roots as was formerly maintained. Given a suitable race of vine the fundatrices, therefore, are *gallicolae* or leaf-gall formers. They lay a large number of eggs and their progeny, or fundatrigeniae, are dimorphic when newly hatched. Grassi recognizes *neogallicolae-gallicolae*, or those which will become gallicolae, and *neogallicolae-radicicolae*, which pass to the roots and become *radicicolae*. The neogallicolae-gallicolae pass through several generations producing in each case both gallicolae and radicicolae. The former appear in greater numbers when the vine is in active growth and never develop on the roots. They may continue as radicicolae and hibernate as nymphs, or produce sexuparae. The latter are winged and fly to the aerial parts of the vine to lay their eggs, which are of two kinds – the larger being female-producing and the smaller giving rise to males. The sexuales are small and apterous; the females each lay a single large overwintering egg, on the bark of the trunk or branches; it hatches the following year into a fundatrix. The details of the life-cycle on the European vine in southern Europe have been debated, but it seems probable, from Grassi's account, that the radicicolae are the principal form on that host, and that gallicolae are seldom met. When, however, European vines are in contact with heavily galled American plants, it is stated that they are sometimes infested with neogallicolae-gallicolae derived from the latter.

FAM. ADELGIDAE (= Chermesidae). *On conifers; both sexual and parthenogenetic females oviparous; wings held roof-like over abdomen at rest; fore wings without separate R_s, Cu arising separately from A; sclerotized ovipositor present; apterous parthenogenetic females covered with flocculent wax.* This family includes such genera as *Adelges* and *Pineus* and all its members are confined to conifers. Two hosts are normally required for the complete life-cycle which extends over two years; the primary hosts are species of *Picea* (Spruce) and the secondary host may be *Larix*, *Pseudotsuga*, *Tsuga*, *Pinus* or *Abies*, according to the species of Adelgid concerned. In a few cases the life-cycle is confined to either the primary or the secondary host. The life-history (e.g. Cholodkovsky, 1907; Börner, 1909; Marchal, 1913; Chrystal, 1922;

Speyer, E. R., 1923; Cameron, 1936; Carter, 1971) is complex and has proved difficult to elucidate, especially as some species are difficult to distinguish anatomically. In a typical life-cycle, however, the following forms may be recognized (Marchal's terminology): (1) The apterous *sexuales* are produced on the primary host and the eggs which the females lay hatch in the autumn, giving rise to nymphs of (2) the apterous *fundatrices*. These nymphs overwinter and mature in the following spring when they lay parthenogenetic eggs from which hatch (3) the young *gallicolae*. The members of this stage immediately settle at the base of the spruce needles and produce a characteristic gall. Within the gall some of these young nymphs develop

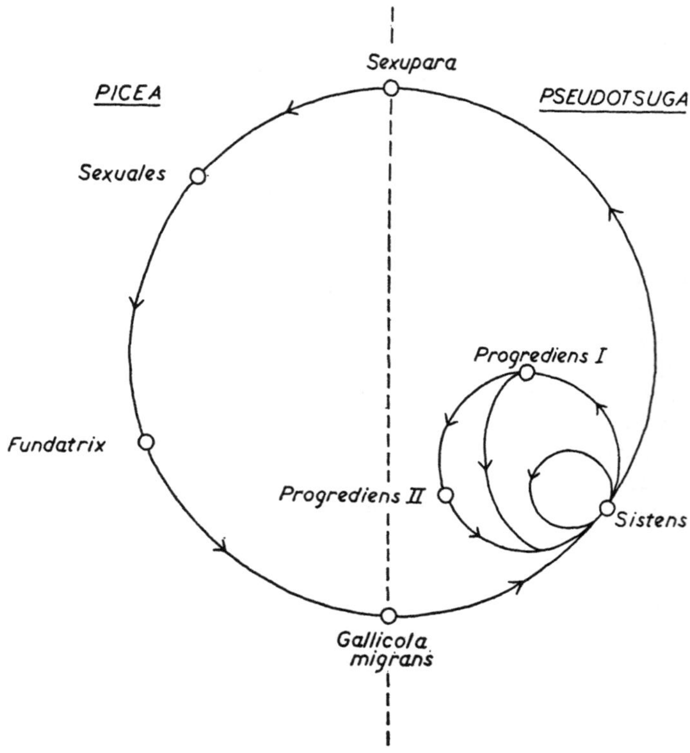

FIG. 338 Life-cycle of *Adelges cooleyi* in Scotland (based on Cameron, 1936)

into apterous *gallicolae nonmigrantes* which remain on the spruce and produce further fundatrices parthenogenetically while others develop into alate *gallicolae migrantes* which fly to the secondary host. Here they give rise parthenogenetically to (4) the apterous *exsules*. In *Pineus* the exsules are all alike, but in *Adelges* the first generation of exsules are known as *sistentes* and give rise parthenogenetically (after overwintering as a young nymph – the *neosistens*) to alate *sexuparae* and to apterous *progredientes*. The latter produce a number of generations, progrediens usually alternating with sistens; the sexuparae fly to the primary host where they lay parthenogenetic eggs from which develop the sexuales. The life-history of *Adelges cooleyi* is shown diagrammatically in Fig. 338. For the taxonomy of Adelgids, see especially Cholodkovsky (1895–96), Börner (1908), Annand (1928) and Steffan (1968).

Superfamily **Coccoidea** (Scale insects; mealy-bugs)

Sexually dimorphic. Females apterous, degenerate, larviform, scale-like, gall-like or covered with a waxy exudate; with functional mouthparts. Males more normal, usually with one pair of wings, mouthparts vestigial. Tarsi 1-segmented, with a single claw.

Modern workers regard this group as composed of several families which, in spite of great diversity in habits and structure, are all characterized by the more or less degenerate females (Fig. 340) which are apterous, obscurely segmented and may be scale-like, gall-like or with a waxy or powdery coating. Their legs and antennae are often atrophied, the tarsi, when present, are 1-segmented with a single claw and the rostrum is short. The early male instars usually resemble the females but adult males (Fig. 339) are more normal in appearance, recalling the Aleyrodidae in general

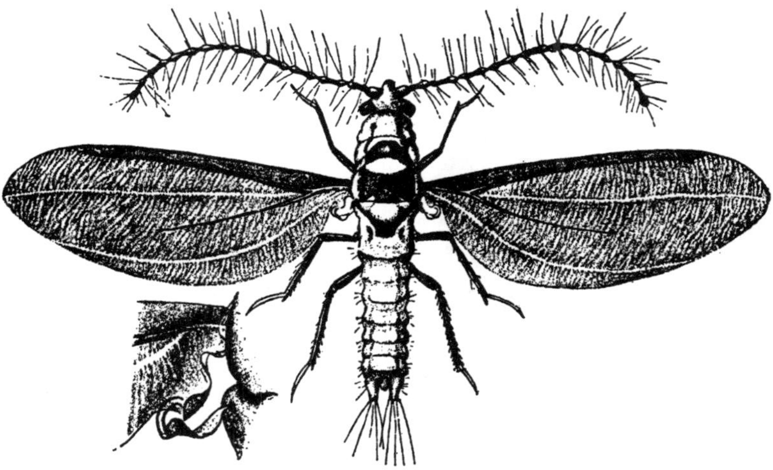

FIG. 339 *Icerya purchasi*, male, enlarged; on left, wing pocket and hooks more highly magnified
After Riley, *Ins. Life*, 1.

facies but with the hind wings reduced to haltere-like structures or completely apterous and with the mouthparts atrophied. The fore wing, when present, has a greatly reduced venation (Fig. 339; see Patch, 1909). In *Pseudococcus*, for example, R is unbranched and only a spur-like rudiment of Sc and a part of M are present. The hind wings in those species where they are developed are represented by a pair of halteres which develop from the metathoracic wing-buds (Witlaczil); each is furnished with one or more hooklets which engage in a basal pocket of the corresponding fore wing. The morphology of the males, previously neglected, is now well understood from the work of Theron (1958), Ghauri (1962), Giliomee (1966), Afifi (1968) and others (see also Boratynski and Davies, 1971). Other taxonomic works on the Coccoidea are mentioned below; its general biology is dealt with by Balachowsky (1939) and Bodenheimer (1935).

The form most usually encountered is the female and consequently such expressions as 'scale insect' or 'mealy bug' refer particularly to that sex. The host-plants are extremely numerous and probably include representatives of all orders of Phanerogams. It is therefore not surprising that the group includes important pests of cultivated plants, especially in tropical and subtropical areas and under glasshouse cultivation elsewhere. The San José Scale (*Quadraspidiotus perniciosus*) of deciduous fruits and the Red Scales of Citrus (*Aonidiella aurantii* and *Chrysomphalus ficus*) may serve as examples. The food-habits of the Coccoidea are variable, some species being monophagous (e.g. *Cryptococcus fagi* on *Fagus sylvaticus* and *Physokermes piceae* on *Abies excelsa*) while others feed on a wide variety of plants. *Lepidosaphes ulmi*, for example, is known to infest about 130 widely separate species and species of

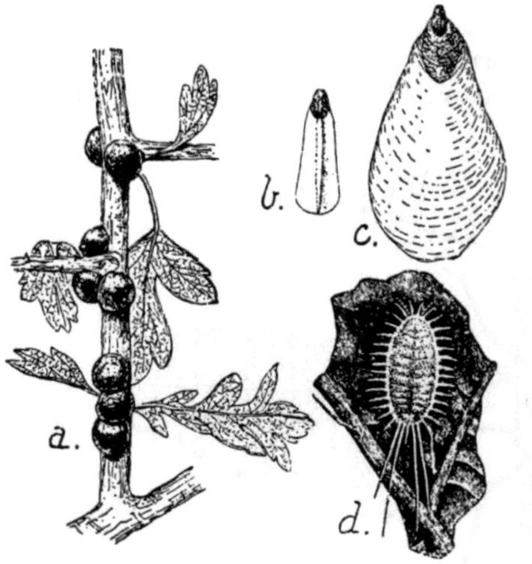

FIG. 340
A, *Parthenolecanium corni* on *Crataegus* (original); B, male and C, female scales of *Chionaspis salicis* (*after* Green); D, *Pseudococcus* female (*after* Comstock)

Pseudococcus occur on a variety of plants grown under glass. The facility with which the living insects can be transported over long distances on their hosts has resulted in many becoming almost cosmopolitan and even strict quarantine measures are not wholly successful in preventing their dispersal by man.

The life-history has been fully studied in relatively few cases. Reproduction may be bisexual or, in some cases, parthenogenetic and oviparous; ovoviviparous and viviparous forms are known, while *Icerya purchasi* is a functional hermaphrodite (p. 307). The eggs are protected in various ways, sometimes being enclosed in an ovisac of felted waxen threads (e.g. *Pseudococcus*) or, as in the Diaspididae, enclosed beneath the scale-like covering of the female, or between wax plates secreted from the end of the abdomen (*Orthezia*) or beneath the body of the female (*Parthenolecanium*). The first-stage nymphs ('crawlers') are provided with functional legs and their mobility ensures the dispersal of the species. Subsequent nymphal instars and the adult females are stationary, being attached by their mouthparts to the host, and though Pseudococcid nymphs always possess legs, one or more apodous instars occur in the life-cycles of other families. Females have one or two fewer instars than males, indicating that they are neotenic. The males, when winged, develop wing-pads in the last two instars which are commonly known as prepupal and pupal stages because of some resemblance to the corresponding stages of Endopterygote insects (Fig. 341, B). Mäkel (1942) gives a detailed account of the metamorphosis of a male coccid.

Like the aphids, many Coccoidea secrete honey-dew, which renders them attractive to ants and in some cases, e.g. *Pseudococcus kenyae*, tended by the ant *Pheidole punctulata*, the multiplication of the coccid colonies is greatly increased when ants are present (see Nixon, 1951, for a review of this subject).

The group is now considered to consist of over a dozen families, opinions differing on the exact number recognized and their precise limits. The taxonomy of the Coccoidea is still in an unsatisfactory state and is based almost entirely on characters provided by the adult female. Among older general taxonomic works are those of Lindinger (1912), Leonardi (1920) and MacGillivray (1921) while more recently

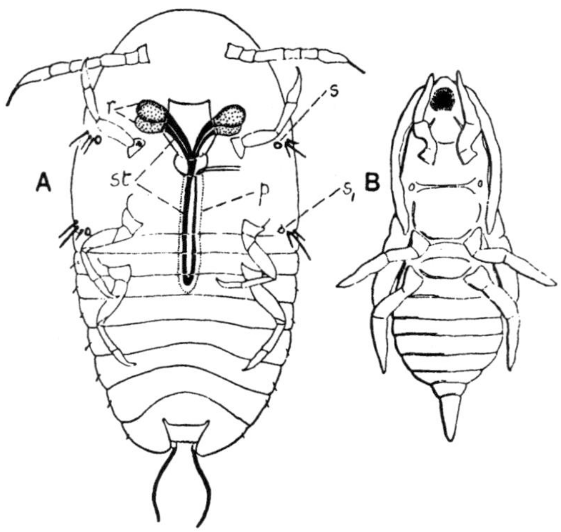

FIG. 341 A, larva of *Coccus hesperidum*, ventral view × 110

p, crumena; *r*, retort-shaped organs; *s*, s_1, spiracles; *st*, stylets.

B, male pupa of *Lepidosaphes fulva*, ventral view × 48

Adapted from Berlese, *Riv. di. Pat. Veg.*, 1893–95.

there have appeared accounts by Ferris (1937–55; 1957*b*), Borkhsenius (1958–63, etc.) and Balachowsky (1942; 1937–50) which maintain a much higher standard of description. Green (1927–28) has extended Newstead's (1901–03) earlier work on the British species and has monographed the species from Ceylon (Green, 1896–1922). Other taxonomic monographs are cited below: while for general biological accounts see Pflugfelder (1939) and Balachowsky (1932). There are bibliographies by Morrison and Renk (1957) and by Morrison and Morrison (1965); the latter have also provided an index of generic names (1966), Williams (1969) discusses family-group names and Ferris (1957*a*) has surveyed the history of work on the group.

FAM. MARGARODIDAE. – The females of this family possess rather distinctly segmented bodies, often covered with a waxy secretion. In *Monophlebus* this takes the form of a mealy coating while the bodies of the subterranean *Margarodes* are covered with pearl-like waxy scales and form the 'ground-pearls' which are collected and strung into necklaces in S. Africa and the Bahamas. *Margarodes* females are capable of undergoing prolonged quiescence ('encystment') accompanied by histolysis. The legs and antennae of Margarodids are usually well-developed and simple eyes are often found. The adult male is usually winged, has simple 10-segmented antennae and normally possesses compound eyes, *Steingelia* being an exception to the latter rule. *Icerya purchasi* (see Dingler, 1930) is injurious to citrus trees but has been controlled in California by the introduction of the predacious Coccinellid beetle *Rodolia cardinalis*. The family has been monographed by Morrison (1928) and Jakubski (1965).

FAM. ORTHEZIIDAE. A small group native to America and the Palaearctic region though *Orthezia insignis* is a widely distributed pest of many glasshouse crops. Females of this family show rather distinct segmentation of the body which is covered with white waxy plates. They have normal legs and antennae and simple eyes. The males have compound eyes. For a taxonomic treatment, see Morrison (1952).

FAM. LACCIFERIDAE (Lac Insects). This family is largely confined to the tropics and subtropics (Chamberlin, 1923–25). Its females are highly degenerate with vestigial antennae, no legs and an irregularly globular body enclosed in a dense resinous cell. The Indian Lac-insect, *Laccifer lacca*, is of great commercial importance since its secretion provides the stick-lac from which shellac is prepared. Its biology has been studied by Imms and Chatterjee (1915) who report its occurrence on many species of trees where it undergoes two generations annually and has dimorphic males – the first generation includes both alate and apterous forms, the second consists solely of wingless ones. *Gascardia*, of Madagascar, yields an inferior type of lac ('gum-lac') containing much wax.

FAM. PSEUDOCOCCIDAE. The familiar and distinctive 'mealy-bugs' belong here (Ferris, 1937–53; McKenzie, 1967; Afifi, 1968; Beardsley, 1960; Williams, 1962). They are typified by the large genus *Pseudococcus* with many species injurious to tropical and glasshouse crops: *P. njalensis* is a vector of a serious virus disease of cocoa ('swollen shoot'). The females are usually elongate-oval with distinct segmentation and generally covered with a mealy or filamentous waxy secretion which may be extended into lateral or terminal filaments. Legs are well-developed, antennae less so. *Ripersia* is subterranean.

FAM. ERIOCOCCIDAE. This family includes *Eriococcus* and its allies (formerly grouped in the Pseudococcidae) as well as *Kermes*, the species of which are confined to oaks where the adult female has a spheroidal, gall-like form with a strongly sclerotized integument. *Acanthococcus* (= *Eriococcus*) *devoniensis* causes galls on *Erica*. For taxonomy, see Ferris (1957*b*) and Hoy (1962, 1963).

FAM. DACTYLOPIIDAE. As restricted by Ferris (1937) this group includes only *Dactylopius*, with 9 species (De Lotto, 1974). *D. coccus* – a native of Mexico, living on various Cactaceae – produces the dyestuff cochineal which is extracted from the bodies of the females and used for colouring foodstuffs and cosmetics. The females are elongate-oval, convex and have small legs and antennae.

FAM. COCCIDAE (= **Lecaniidae**). This is one of the most important families and its members display considerable diversity of form. The segmentation of the female is obscure and the integument may be naked or covered with wax while the degree of development of antennae and legs varies considerably. Species of *Saissetia* are injurious to many cultivated plants (e.g. Smith, 1944) while *Ericerus pela* and *Ceroplastes ceriferus*, both Oriental, yield wax on a small commercial scale (see also Blumberg, 1935). For the males, see Giliomee (1966).

FAM. ACLERDIDAE. Contains only the subterranean *Aclerda*.

FAM. ASTEROLECANIIDAE. These scales are of variable form with reduced antennae and legs vestigial or absent. In the American *Cerococcus quercus*, the female is completely embedded in a mass of yellow wax, while in *Asterodiaspis minus* from the Holarctic region the females lie in pits on the twigs of oak and olive (see also Habib, 1943).

FAM. PHENACOLEACHIIDAE. Contains only one species (*P. zelandica*) on *Fagus*, *Cupressus* and *Podocarpus* in New Zealand.

FAM. STICTOCOCCIDAE. Includes only *Stictococcus*, the female of which is circular, flattend and with a wax-impregnated integument. The first instar shows unusually marked sexual dimorphism.

FAM. APIOMORPHIDAE. Sometimes known from the shape of the female as 'pegtop coccids', these are confined to Australia and New Zealand where they form galls on *Eucalyptus*.

FAM. CYLINDROCOCCIDAE. The members of this small family are restricted to the Southern hemisphere.

FAM. CONCHASPIDIDAE. These include a few tropical genera resembling the Diaspididae in the presence of a shield-like scale which covers the female, but differing from them in that the nymphal exuviae are not incorporated into the scale.

FAM. DIASPIDIDAE. A large family, the N. American species having been monographed by Ferris (1937–53) and McKenzie (1956). The adult females are specialized by the absence or vestigial nature of the antennae and the loss of legs and are covered by a hard waxy scale responsible for their common name of 'armoured scales'. Representatives of the family occur in all regions and include many injurious species such as the Red Scales of citrus (*Aonidiella aurantii* and *Chrysomphalus ficus*) and the San José Scale (*Quadraspidiotus perniciosus*) which attacks deciduous fruit trees. *Lepidosaphes ulmi*, the Mussel Scale, also belongs here. For a World catalogue see Borkhsenius (1966) and for the C. European species Schmutterer (1959). Beardsley and Gonzalez (1975) discuss the biology of the family.

Suborder 2. HETEROPTERA

Head porrect, gular region sclerotized. Wings folding flat over abdomen, the fore wings usually sclerotized basally and membranous apically. Base of rostrum not touching anterior coxae. Pronotum large, trochantins small. Tarsi commonly 3-segmented. Metamorphosis incomplete.

For the classification of this suborder, see the older papers by Reuter (1910, 1912), Schiödte (1870), Kirkaldy (1908), Ekblom (1929), China (1933) and Börner (1934) with more recent works by Leston *et al.* (1954), China and Miller (1959), Cobben (1968), Miller (1971) and Jordan (1972). Stys (1961), Kumar (1966–71; 1968), Manna (1958, 1962) and Popov (1971) provide further contributions on the higher classification of the group. The arrangement of families adopted here follows the threefold division into terrestrial, semi-aquatic and aquatic groups as represented by the series Geocorisae, Amphibicorisae and Hydrocorisae respectively (Leston *et al.*, 1954; but see also Cobben, 1968 and p. 702). Further morphological work is required before the exact composition, limits and relationships of all the families and superfamilies can be settled; the arrangement followed below relies much on Jordan (1972).

The very extensive taxonomic literature on the Heteroptera is partly accessible through the works cited on p. 748. In addition the following accounts of the fauna of restricted areas may be mentioned:

Europe: Fieber (1861), Stål (1870–76), Puton (1878–81), Reuter (1878–96), Stichel (1925–38; 1955–62), Hedicke (1935), Gulde and Jordan (1933–38), Kiritschenko (1951) and Wagner (1961; 1966–67). For the British fauna see Southwood and Leston (1959) and Macan (1965). **N. America**: van Duzee (1917), Britton *et al.* (1923), Parshley (1925), Blatchley (1926), Torre-Bueno (1939–46). **Central and S. America**: Distant (1880–1900), Champion (1897–1901). **Africa**: Stål (1864–66). **India**: Distant (1902–18).

General accounts of the Heteroptera are given by Jordan (1972) and Poisson (1951) while Weber (1929–35; 1930) and Miller (1971) deal with their biology; Butler (1923) and Southwood and Leston (1959) describe systematically the biology of the British species. Various aspects of the biology of aquatic Heteroptera are discussed by Hungerford (1919), Poisson (1924), Jordan (1928), Karny (1934), Jaczewski (1937) and Popov (1971).

Series 1. GEOCORISAE

Mainly terrestrial Heteroptera, a few littoral or semi-aquatic; body without ventral, hydrofuge pubescence; antennae longer than head; gula with rostral groove; legs not modified for swimming, the pretarsal claws apical; hemelytra with at least corium, clavus and membrane, or showing various degrees of brachyptery.

Section 1. CIMICOMORPHA

Antennae 4-segmented; hemelytra often with costal fracture and cuneus; hind wing with R and M joining distally; abdominal venter without trichobothria; eggs usually operculate, implanted in substrate by ovipositor of female; accessory salivary glands vesiculate; spermatheca absent or paired.

The section is divided into four superfamilies: Tingoidea, Reduvioidea, Cimicoidea and Dipsocoroidea.

Superfamily **Tingoidea**

Upper side almost always with lace-like reticulate sculpture over pronotum and hemelytra; antennae with short 1st and 2nd segments, the 3rd segment longest; rostrum 4-segmented, closely applied to gula at rest; pronotum usually extended posteriorly to cover scutellum and clavus.

FAM. TINGIDAE (Lace-bugs). *Scutellum concealed by pronotum and ocelli absent; body and hemelytra densely reticulate.* About 800 species are known, very many from the Mediterranean region (Drake and Davis, 1960; Drake and Ruhoff, 1960; 1965). They exhibit great variety of form, the prothorax often being produced into laminate outgrowths, or the whole body may be margined with closely set spines. In some genera there are crest-like modifications of the pronotum suggestive of the Membracidae. All species are plant feeders and sometimes occur in sufficient numbers to constitute minor pests. The eggs are frequently inserted upright in the plant tissue, and are invested with a brown viscid substance which hardens to form a cone-like elevation on the surface of the leaf. The immature stages are very different from the adults, the characteristic ornamentation of the latter not appearing until after the last moult. *Stephanitis pyri* attacks pear and apple in Europe, badly infested leaves dying, and *S. rhododendri* is a minor pest of rhododendrons (Johnson, 1936). Species of *Copium* are known to form galls on the flowers of *Teucrium* (Monod and Carayon, 1958) and *Teleonemia scrupulosa* has been used in the biological control of the weed *Lantana* (Harley and Kassulke, 1971). For the biology of some N. American species see Bailey (1951).

FAM. VIANAIDIDAE. *Small (c. 2 mm); without reticulation on hemelytra and pronotum; scutellum visible; hemelytra without division into corium and membrane; eyes rudimentary or absent. Anommatocoris* (=*Vianaida*) *coleoptrata* and two other species are associated with ants in S. and C. America. The group is sometimes treated as a subfamily of Tingidae. See Kormilev (1955) and Drake and Davis (1960).

Superfamily **Reduvioidea**

Predacious, with long head and large eyes; ocelli usually present; antennae usually geniculate after 1st segment, often with supernumerary segments; rostrum usually 3-segmented, powerful and curved away from ventral surface of head at rest; hemelytra with strong veins but no cuneus. The inclusion of this superfamily in the Cimicomorpha is disputed (cf. Cobben, 1968).

FAM. PHYMATIDAE. *Antennae with last (fourth) segment the largest, clavate or fusiform; mid and hind tarsi 2-segmented; fore legs raptorial, sometimes dilated, with chelate tibiae; abdomen usually with broad connexivium.* The 100 or so members of this mainly tropical family (Handlirsch, 1897; Evans, 1931; Tsing-Chao and Kuei-Shiu, 1956) are predacious, some species secreting themselves in flowers for the purpose of securing prey which may come within reach. In the oriental genus

Carcinocoris the whole body is margined with fine spines and the front tibia is articulated to the femur so as to form a pair of pincers. The prey of these insects consist of small adults of other orders and also Tenthredinid larvae. Balduf (1941) describes the biology of *Phymata pennsylvanica*.

FAM. REDUVIIDAE. *A very large family, exclusively predacious or blood-sucking; antennae often with intercalary segments, making up a total of as many as 40 segments; rostrum pointed, 3-segmented, capable of friction against stridulatory groove of prosternum; fore legs not raptorial but their tibiae often provided with a fossula spongiosa for adhesion; tarsi usually 3-segmented.* This extensive group shows a range of form that is hardly paralleled in any other family of insects. More than 3000 species are known, arranged in some 29 subfamilies, of which the Harpactorinae is the largest, with over 1000 species (Usinger, 1943; Carayon, Usinger and Wygodzinsky, 1958; China and Miller, 1959; and see Davis, 1958–66). The biology of many species is described by Readio (1927) and discussed generally by Miller (1971). The Reduviids are all predacious, usually living on the blood and body-contents of other insects and found on

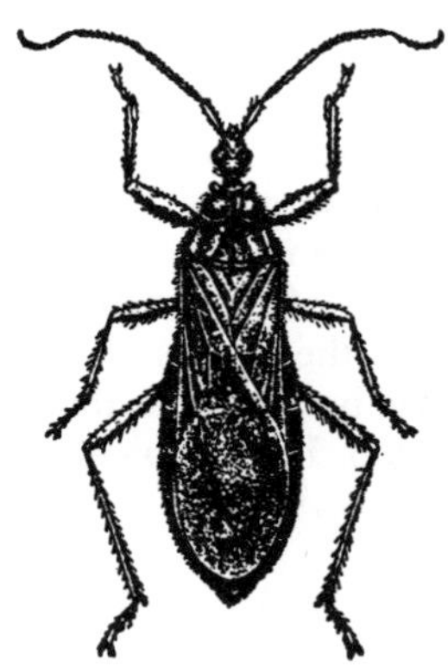

FIG. 342
Reduvius personatus
After Howard.

bushes, low herbage or the foliage of trees. The Triatominae, predominantly New World in distribution, suck the blood of mammals or birds (Abaloos and Wygodzinsky, 1951; Ryckman, 1962; Usinger *et al.*, 1966). *Rhodnius prolixus* and some polytypic species of *Triatoma* are the main vectors of *Schizotrypanum* (= *Trypanosoma*) *cruzi*, the causal organism of Chagas' disease, a fatal form of human trypanosomiasis in S. America (Usinger, 1944). *Triatoma rubrofasciata* also extends into Madagascar and South Asia; its nymphs are common in houses, where they are partially concealed by floor debris. There are numerous records of violent reactions to Triatomine bites but cases of severe illness lasting several days appear to be anaphylactic reactions on the part of hypersensitive individuals, others experiencing no ill effects. *Reduvius personatus* (Fig. 342) also frequents houses, normally preying upon *Cimex* and other insects (Immel, 1955); it is known to attack man, inflicting severe pain. Although uncommon in Britain it is widely distributed in Europe and has been introduced into North America. Certain members of the large genus *Acanthaspis* are also capable of inflicting painful punctures. Among the more exceptional members of the family is *Afrodecius* in which the 3rd segment of the rostrum is apposable to a process on the 2nd segment, suggesting an organ of prehension. The insect is African and resembles *Lycus* (Coleoptera) in form and coloration. *Rhaphidosoma* is apterous and greatly attenuated, resembling Phasmids.

Arilus cristatus is the Wheel Bug of North America, which frequents fruit-trees, preying on various soft-bodied larvae. *Harpactor costalis* preys on *Dysdercus cingulatus* in India, closely resembling it in coloration.

The **Elasmodemidae** are small, flat insects with a curved, transverse impression behind the eyes and fore legs that have thickened femora but are not raptorial. They are perhaps a subfamily of Reduviidae and are represented by S. American species of *Elasmodema* found in birds' nests and beneath bark (Wygodzinsky, 1944). The **Joppeicidae** contains only *Joppeicus paradoxus*, a predator of other insects on *Ficus* in Asia Minor (China, 1955). It is small and usually carries its rostrum in a forwardly directed position.

FAM. ENICOCEPHALIDAE (= **Henicocephalidae**). *Predacious Reduvioid insects with 4-segmented rostrum; prosternum without stridulatory groove; head constricted basally and behind eyes; hemelytra entirely membranous.* A small but widely distributed family of about 50 species, the members of which occasionally appear in swarms like midges; such swarms have been observed in South America, Tasmania and Ceylon. In the African *Aenictopechys alluaudi* the rostrum projects forwards and its apex is bifid. For taxonomy, see Usinger (1945), Jeannel (1942) and Villiers (1958).

Superfamily **Cimicoidea**

Antennae not geniculate, 2nd segment the longest; pronotum trapezoidal; hemelytra with costal fracture and cuneus in macropterous forms.

FAM. NABIDAE. *Thin antennae with 4 or 5 segments; rostrum 4-segmented; coxae rotatory; prosternal stridulatory groove absent; scent-gland openings present on metapleura; fore legs somewhat raptorial; tarsi 3-segmented; cuneus absent.* The Nabidae, partly monographed by Reuter and Poppius (1909), includes about 300 species from all regions, especially the tropics. Its members are all predacious and commonly encountered on herbage where they attack small, phytophagous insects and lay their eggs in plant stems (see, e.g. Mundinger, 1922). The males of some species have a stridulatory organ at the apex of the abdomen and according to Carayon (1950*a*) fertilization in *Prostemma* and *Alloeorrhynchus* involves the passage of spermatozoa into the haemocoele in a fashion recalling the remarkable mechanism known to exist in *Cimex* (p. 302). For this and other reasons they are therefore perhaps closer to the Cimicoid families than to the Reduvioid ones with which they were previously grouped. Harris (1928) describes the N. American species.

FAM. VELOCIPEDIDAE. *Medium-sized oval insects with prominent eyes; ocelli present; antennae long and thin; rostrum 3-segmented, the 2nd segment very long; metapleura with small scent-gland orifices; cuneus present.* This family, of uncertain affinities, includes only a few Oriental species of *Velocipeda* and *Scotomedes*. It is perhaps to be regarded as a subfamily of the Nabidae. The **Medocostidae** (Stys, 1967) with two W. African species are closely related to the Velocipedidae.

FAM. CIMICIDAE (**Acanthiidae**: Bed-bugs). *Oval flattened insects with very short hemelytra; rostrum lying in a ventral groove; ocelli absent; tarsi 3-jointed; parasites of mammals and birds.* A small but well-defined family of blood-sucking ectoparasites monographed by Usinger (1966). The bed-bugs (Fig. 343) belong to the genus *Cimex* and the two common species are *C. lectularius*, which is prevalent throughout Europe and N. America and is almost cosmopolitan, and *C. rotundatus* (= *hemipterus*) which abounds in southern Asia and also in Africa. They are particularly common in dirty houses, especially in large cities, and are nocturnal in habits, hiding by day in any convenient crevices about the walls, floors or furniture of rooms. Man is the host for both species, and the effect of their punctures varies very much with the individual person: with some people swelling and irritation may last for several days, with others the effects are slight. Pathologists have suspected the bed-bug of transmitting various diseases from infected to healthy persons, but definite confirmatory evidence has not been forthcoming. The eggs of bed-bugs are

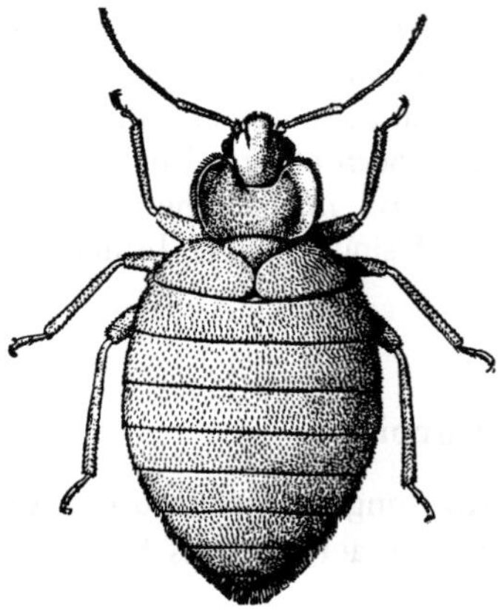

FIG. 343
Cimex lectularius, male, × 10. Europe, N. America, etc.
Reproduced by permission of the Trustees of the British Museum.

laid in crevices of wooden bedsteads and other objects; under favourable laboratory conditions they hatch in about eight days and the life-cycle is completed in seven weeks. Under normal conditions, however, the latter period may occupy six months or more. A fuller account of the structure and biology of these insects is given in papers by Puri (1924), Mellanby (1935) and Johnson (1942). Other members of the genus are parasites of birds and bats; *Oeciacus hirundinis* lives in martins' nests and *Haematosiphon inodorus*, from North and Central America, is a pest of poultry and has a very long rostrum.

FAM. POLYCTENIDAE. *Ectoparasites of bats and provided with ctenidia; rostrum 3-segmented; antennae and tarsi 4-segmented; eyes wanting; hemelytra short, of uniform consistency and devoid of a membrane.* Eighteen species are known (Ferris and Usinger, 1939; Usinger, 1946), all obligate parasites living deep in the fur of tropical bats belonging mainly to the genera *Molossus, Megaderma, Taphozous* and *Cynopterus*. The Polyctenids are characterized by the possession of one or more combs (ctenidia) of short flat spines – an armature which they share with *Platypsyllus*, the Nycteribiidae and Siphonaptera. They are viviparous, the embryos remaining in the ovarioles where they gradually mature. The young are born at an advanced stage but differ very considerably from the adults (Hagan, 1931) and two postnatal nymphal instars occur.

FAM. MICROPHYSIDAE. *Allied to the Anthocoridae but rostrum 4-segmented and tarsi 2-segmented; females usually brachypterous with swollen abdomen* *and no ocelli*. About 20 species are known, mostly Palaearctic. Some are myrmecophiles, others found among fallen leaves, moss, lichens, etc. They are predacious.

FAM. ANTHOCORIDAE. *Small, elongate-oval, flattened insects; ocelli present; rostrum 3-segmented; metathorax with scent-gland openings; cuneus and embolium present; tarsi 3-segmented.* This family of about 300 species (Reuter, 1885; Péricart, 1972; Carayon, 1972) is represented in all regions, its members being predacious on small arthropods and occurring mainly on foliage and flowers, under bark or among fallen leaves. *Orius insidiosus* is a major predator on larvae of the corn pest *Heliothis obsoleta* in the U.SA. (Barber, 1936). *Lyctocoris campestris* occurs in association with human habitations (granaries, thatched roofs, etc.) and it and some other species (e.g. *Anthocoris kingi*) are known to bite man occasionally. Some Anthocorids are myrmecophiles; some others live in birds' nests. For the immature stages and biology of some British species, see Sands (1957) and Hill (1961). Some species show haemocoelic fecundation and *Physopleurella pessoni* is viviparous (Carayon, 1956; 1957).

FAM. MIRIDAE (= Capsidae). *Medium or small, usually delicate insects; ocelli absent; rostrum 4-segmented; cuneus usually present, embolium indistinct; tarsi almost invariably 3-segmented.* A very large family of over 6000 species whose taxonomy presents many difficulties (Knight, 1941; Carvalho, 1952–60; Wagner and Weber, 1964; Leston, 1961). It constitutes the largest Palaearctic family of Heteroptera, with more than 1000 species from that region; over 180 occur in Britain and much information on the biology of the Swedish forms is given by Kullenberg (1944). Though the majority live on plant juices, some prey on small arthropods – e.g. *Blepharidopterus angulatus* is an important predator of the Fruit Tree Red Spider Mite, *Metatetranychus ulmi* (Collyer, 1952). Other Mirids do considerable damage to cultivated plants. Thus, *Plesiocoris rugicollis* (Fig. 180), the original host of which was *Salix*, now attacks apple, black- and red-currants (Petherbridge and Husain, 1918) while *Lygus pabulinus* (the Tarnished Plant-bug) is an almost cosmopolitan pest with many named varieties (Crosby and Leonard, 1914). *Helopeltis theivora*, formerly a serious pest of tea in S. E. Asia, can be controlled insecticidally (Cranham, 1966); *Sahlbergella singularis* and *Distantiella theobroma* are pests of cocoa in W. Africa (Entwistle, 1972).

FAM. ISOMETOPIDAE. *Closely related to the Miridae, from which they differ in presence of ocelli, Lygaeid-like antennae and, in some genera, the presence of enormous eyes.* There are over 60 species (Eyles, 1971–72) from all zoogeographical regions. They are associated with trees and low shrubs and some have enlarged hind femora and can jump. *Letaba bedfordi* is predacious on *Aonidiella aurantii*, the red scale of citrus, in S. Africa (Hesse, 1947).

Superfamily **Dipsocoroidea**

Small or very small species; antennae with two thick basal segments followed by two thin, long-haired segments; rostrum 3-segmented; fore wing with costal fracture but cuneus fused with membrane.

FAM. DIPSOCORIDAE (= Cryptostemmatidae, Ceratocombidae). *With the characters of the superfamily.* A widely distributed family of over 50 species, the members of which are mainly nocturnal and live in damp, concealed habitats among decaying leaves or under stones (McAtee and Malloch, 1925; Stys, 1970). Their affinities with the Cimicomorpha are disputed (China, 1955) and little is known of their biology. *Schizoptera* and related genera, sometimes placed in a family of their own, are included here; see Emsley (1969).

Section 2. PENTATOMOMORPHA

Antennae 4- or 5-segmented; hemelytra without costal fracture or cuneus; hind wing with R and M separate distally; abdominal venter usually with trichobothria; eggs not operculate and not implanted in substrate; accessory salivary glands long, tubular; spermatheca present, single. The section is divided into 6 superfamilies: Saldoidea, Aradoidea, Lygaeoidea, Pyrrhocoroidea, Coreoidea and Pentatomoidea.

Superfamily **Saldoidea**

Head broad, with prominent eyes, short broad labrum, and ocelli close together; antennae and rostrum long, 4-segmented; pronotum strongly narrowed anteriorly; hemelytra with corium, clavus and membrane, the latter bearing 4–5 cells but no other longitudinal veins.

FAM. SALDIDAE. (= **Acanthiidae**, Shore-bugs). *Eyes large and prominent; ocelli almost always present, not grouped on a pedunculate tubercle; rostrum long; tarsi 3-segmented.* The affinities of this family are disputed and it is sometimes placed in the Amphibicorisae (cf. Cobben, 1959; Gupta, 1963*a, b*). Most of the 150 or so described species frequent marshy places and the borders of streams, especially near the coast, and are predominantly Holarctic (Drake and Hoberlandt, 1952*b*; Cobben, 1959). They usually inhabit mud, moss, or salt-marsh plants and can fly and run rapidly. *Saldula orthochila* occurs in dry places while, at the other extreme, *S. pallipes* can withstand submergence by tides (Brown, 1948). So far as is known they are predacious or feed on insect remains. For the biology of *Salda littoralis*, see Jordan and Wendt (1938). *Aepophilus bonnairei*, formerly regarded as the sole representative of a separate family, has vestigial hemelytra and lives beneath stones below high-tide mark on the coasts of N.W. Europe (Lienhart, 1913).

Three small families are related to the Saldidae. The **Leptopodidae**, with ocelli placed on a pedunculate tubercle and a very short rostrum, occur mainly in the tropics and subtropics of the E. hemisphere (Drake and Hoberlandt, 1952*a*); *Leptopus marmoratus* is found under stones in C. Europe. The **Leotichiidae** include only two species of *Leotichius* from bat-inhabited limestone caves in S.E. Asia (China, 1933). The **Omaniidae** (Kellen, 1960; Cobben, 1970) comprise a few species of very small, beetle-like insects occurring on coral and limestone rocks in the littoral zone along the coasts of Arabia, Malaysia, Japan, Australia and some Pacific islands; they retreat into crevices when submerged by the tide.

Superfamily **Aradoidea**

Unusually flat insects, mainly subcorticolous or living in fungi; ocelli absent; stylets very long and coiled spirally in crumena when at rest; pronotal margin often toothed or granulate; tarsi 2-segmented and pretarsal claws without pulvilli; hemelytra, when present, always narrower than abdomen, which itself has broad connexivia.

FAM. ARADIDAE. *Head produced into two antennal tubercles; antennae projecting freely; hemelytra, when present, with clavus not meeting that of the opposite side behind the scutellum.* This family includes over 400 species from all regions (Usinger and Matsuda, 1959; Kumar, 1967). They usually occur beneath bark, in crevices on dead trees and among fungi. They are apparently mycetophagous but their biology is not well known (Tamanini, 1956). There are eight subfamilies, of which the Mezirinae (= Dysodiinae) have been treated as a separate family.

FAM. TERMITAPHIDIDAE. *Flat, blind, apterous, broadly-oval insects resembling woodlice; antennae 4-segmented, concealed beneath head; rostrum 4-segmented; tarsi 2-segmented.* Usinger (1942) recognizes nine species in this tropicopolitan family; all occur in termite galleries. The long stylets, coiled within the head when retracted, have suggested Aradoid affinities, but the resemblance may be the result of convergent adaptation to similar mycetophagous habits (China, 1931).

Superfamily **Lygaeoidea**

Medium to small, strongly sclerotized, usually more or less oval insects; antennae 4-segmented; rostrum 4-segmented, the first segment lying between the bucculae; ocelli present in winged forms; pretarsal claws with pulvilli; hemelytra with clavus longer than scutellum and at most five longitudinal veins (unbranched) in membrane. The subdivision of the Lygaeoidea and its relationships with the Coreoidea and Pyrrhocoroidea have been much discussed (e.g. Scudder, 1963; Schaefer, 1964; Kumar, 1968; Cobben, 1968).

FAM. LYGAEIDAE (= Myodochidae). *Small, dark or brightly-coloured forms; ocelli almost always present; antennae inserted well down on sides of head; thoracic gland openings present; membrane with 4 or 5 veins; coxae rotatory; tarsi 3-segmented.* A large family with over 2000 species from all regions (Slater, 1964); more than 70 are British (Fig. 344) and the family is well represented in N. America (Torre-Bueno, 1946). Most are plant feeders and usually occur in moss, surface rubbish, beneath stones or on low plants, but a few may be taken by sweeping vegetation. *Blissus leucopterus* is the American 'chinch bug', which is destructive to grasses and cereals; *Oxycarenus hyalipennis* is the Egyptian 'cotton stainer' and the cosmopolitan genus *Nysius* includes *N. vinitor*, which is destructive to fruit trees in Australia (see also Usinger, 1942). Among the predacious Lygaeids, *Geocoris punctipes* is a common enemy of the mite *Tetranychus telarius* on cotton in N. America (McGregor and McDonough, 1917). *Oncopeltus fasciatus*, the Milkweed

Bug, is widely used as a laboratory insect (Feir, 1974). Among other biological works on Lygaeidae see Servadei (1951), Tischler (1960) and Sweet (1964).

FAM. BERYTIDAE (= Neididae). *Elongate insects structurally resembling the Lygaeidae but with geniculate antennae and long, slender legs with apically clavate femora.* These are delicately formed insects, never very common, and sometimes known as 'stilt bugs'. In habits they are sluggish, frequenting undergrowth and meadows. Although they are probably universally distributed, their small size and fragility have caused them to be overlooked, and the tropical forms have been very little collected. Nine species occur in Britain. McAtee (1919) deals with the Nearctic species.

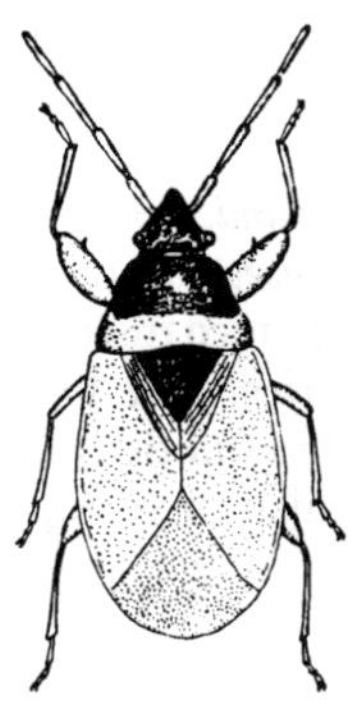

FIG. 344
Gastrodes grossipes (*Lygaeidae*) × 5. Britain

The 70 or more species of **Colobathristidae** with long legs and a basally constricted abdomen, are regarded by some as a subfamily of the Berytidae and by others as a separate though closely related family (see Stys, 1966). They occur mainly in the Indomalaysian and Neotropical regions, *Phenacantha saccharicida* being injurious to sugar-cane in Java.

FAM. PIESMATIDAE. *Small, elongate-oval insects; ocelli present in macropterous forms; rostrum 4-segmented; mandibular plates produced into a pair of horn-like structures at front of head; scutellum visible; fore wing reticulo-punctate; coxae rotatory; tarsi 2-segmented; pulvilli present.* Formerly grouped with the Tingidae, *Piesma* and its allies constitute a small family of mainly Palaearctic insects, all phytophagous and mostly associated with the Chenopodiaceae (Drake and Davis, 1958). The biology of *Piesma quadratum*, a pest of beet, has been described by Wille (1929).

FAM. THAUMASTOCORIDAE (= Thaumastotheriidae). *Small insects with broad head; ocelli present; rostrum short, 3-segmented; thoracic scent-glands absent; coxae rotatory, widely separated; tibiae usually with membranous apical appendage.* A small family with a relict distribution from Australia, India, Argentina and the Southern Antilles, these insects are traditionally placed near the Lygaeidae though Drake and Slater (1957) demonstrated their Cimicoid affinities. They are phytophagous, *Xylastodoris luteolus* feeding on the Royal Palm, *Oreodoxa regia* (Baranowski, 1958).

FAM. IDIOSTOLIDAE. *Ocelli present; prosternum, scutellum and thoracic sterna not punctured; tarsi 3-segmented; abdomen with many trichobothria, notably 4 on each side of the 7th sternum.* Two monotypic genera with a characteristic southern hemisphere distribution have been placed here – *Idiostolus* from Chile and *Trisecus* from Tasmania (Schaefer, 1966). They have also been regarded as forming a subfamily of the Lygaeidae, but may not even belong to the Pentatomomorpha (Cobben, 1968).

Superfamily Pyrrhocoroidea

Medium-sized, often brightly coloured, with conspicuous eyes but no ocelli; rostrum slender, 4-segmented; antennae 4-segmented; membrane with at most 5 longitudinal veins, which are unbranched.

FAM. PYRRHOCORIDAE. *With the characters of the superfamily; female with 7th abdominal sternum not divided.* A small family whose members exhibit strongly constrasting red and black coloration and include the well-known 'cotton stainers' (*Dysdercus*). The latter comprise many species, widely distributed in warm countries (Fig. 315). The name 'cotton stainer' is derived from their habit of piercing the bolls and thereby contaminating them with the fungus *Nematospora* which stains the fibres (Frazer, 1944; Pearson, 1958). *D. cingulatus* is a serious cotton pest in India and *D. sulphurellus* is prevalent in N. America. The widely distributed *Pyrrhocoris apterus* is the only British representative of the family, and is remarkable on account of its alary dimorphism. Kershaw and Kirkaldy (1908*a*) have followed the life-history of *Dindymus sanguineus* which is carnivorous, feeding on flies; the nymphs, however, apparently prefer termites.

FAM. LARGIDAE. *Very similar to the Pyrrhocoridae; females with 7th abdominal sternum divided.* Representatives of this family occur in New Guinea, Australia, the Philippines, China, India and Bolivia. It includes such genera as *Lohita, Largus* and *Iphita*. The Oriental *Lohita grandis* attains a length of over 5 cm and is sexually dimorphic, the male having the antennae and abdomen greatly elongated.

Superfamily Coreoidea

Medium to large insect, elongate or elongate-oval in form; antennae 4-segmented; clavus extending beyond scutellum; membrane with more than 5 longitudinal veins, which are often branched; legs strong, claws with pulvilli; more than 2000 species, often good fliers.

FAM. COREIDAE. *More or less oval rather than elongate in form; head much narrower than pronotum; juga not enclosing tylus anteriorly; scent-gland openings conspicuous; 5th abdominal tergum with margins sinuate but parallel, not narrower in mid-line.* The Coreidae are mostly brownish insects, occasionally red or yellow or even metallic green, often capable of producing a pungent scent and apparently all phytophagous. For their morphology and higher classification see Schaefer (1964–68). They are well represented in India, Africa and South America and include some pests of cultivated plants. *Anasa tristis*, the Squash Bug, attacks Cucurbitaceae in N. America (Beard, 1940) and species of *Amblypelta* and *Theraptus* are responsible for the premature fall of coconuts (Brown, 1959). *Leptoglossus* and other genera have unusual foliate expansions of the hind tibia, restricted to males but of unknown function.

FAM. ALYDIDAE. *Differ from the Coreidae (of which they are sometimes regarded as a subfamily) in the more elongate form; head as broad as pronotum; 4th antennal segment curved and longer than 3rd.* Representatives of this family occur in all regions, including such genera as *Protenor, Alydus, Coriscus* and *Leptocorisa. L. varicornis* is an Oriental pest of rice and millet and the group to which it belongs has been monographed by Ahmad (1965).

FAM. RHOPALIDAE (= Corizidae). *Also sometimes treated as a subfamily of the Coreidae, from which it differs in the adult scent-gland openings being slit-like or concealed; 5th abdominal tergum emarginate in front and behind so that it is shorter in the mid-line than laterally.* Members of the family occur in all regions, *Corizus, Rhopalus, Myrmus* and *Chorosoma* being among those found in Europe. Chopra (1967) deals with the higher classification of the family.

FAM. STENOCEPHALIDAE. *Large, more or less elongate insects; juga enclosing tylus anteriorly; with resemblances to both the Coreidae and the Lygaeidae.* This is almost entirely an Old World group, with only *Stenocephalus* and *Dicranocephalus* (see Lansbury, 1965–66).

FAM. HYOCEPHALIDAE. *Large, powerful insects, black or reddish-brown in colour; differ from other Coreoidea in having the membrane with only 4 veins, arising from 3 cells.* This is an Australian family, most of the species belonging to *Hyocephalus*. They feed on seeds and their morphology and relationships are discussed by Stys (1964*a*).

Superfamily **Pentatomoidea**

Strongly sclerotized body; antennae usually with 5 segments (less often with 4, rarely 3); pronotum usually 6-sided, mesoscutellum large, reaching at least to middle of abdomen and sometimes covering entire abdomen; legs with 2 or 3 tarsal segment; claws with pulvilli; membrane always with veins. The higher classification of the Pentatomoidea is difficult and debatable (Leston, 1958).

FAM. PLATASPIDIDAE (= Brachyplatidae, Coptosomatidae). *Strongly convex, broadly rounded, usually shiny insects; scutellum covering almost the entire abdomen; hemelytra long and bent beneath scutellum at rest; 2 tarsal segments.* The family includes some 500 species, mainly from the Ethiopian, Oriental, Australian and warmer parts of the Palaearctic region; *Coptosoma scutellatum* extends into parts of C. Europe, but is not British. Certain exotic genera (*Ceratocoris* and *Elapheozygum*) exhibit a remarkable sexual dimorphism, the males having the head greatly produced in front of the eyes, forming prominent horn-like projections. The related family **Lestoniidae** includes only *Lestonia haustorifera*, from New South Wales. It is a small, convex, tortoise-like insect with a flat ventral surface and explanate margins to the abdomen (McDonald, 1970).

FAM. CYDNIDAE (inc. **Thyreocoridae**). *Abdomen with 6 connexivia visible dorsally; scutellum normal or enlarged to cover hemelytra and abdomen; tibiae strongly spinose; antennae 5-segmented; tarsi 3-segmented.* This rather large family is represented in all regions by dark-coloured insects which live under stones and dead leaves or at the base of plants. Some are myrmecophilous. Taxonomic papers include those of Signoret (1881–84), McAtee and Malloch (1933) and Froeschner (1960). Southwood and Hine (1950) deal with the biology of *Sehirus bicolor* and Schorr (1957) with that of *Brachypelta aterrima.*

FAM. SCUTELLERIDAE. *Scutellum large, covering the entire abdomen; tarsi 3-segmented; membrane with veins; abdomen with 6 free sterna.* This is a widely distributed, largely tropical family, represented by such genera as *Odontoscelis, Eurygaster, Aphylum, Odontotarsus* and others. Some species are metallic or brightly coloured and range up to about 20 mm in length. *Tectocoris diophthalmus*, the Harlequin Bug of Queensland, attacks cotton while *Eurygaster integriceps* is an important pest (especially of wheat and barley) in the Middle East and areas near the Black Sea; the young adults of the latter species migrate in large numbers to their winter quarters (Zwölfer, 1930; Brown, 1965). *Aphylum*, with 2 Australian species, is sometimes placed in a family of its own, the **Aphylidae**, and is perhaps closer to the Plataspididae.

FAM. ACANTHOSOMATIDAE. *Antennae 5-segmented; tarsi 2-segmented; scutellum not reaching end of abdomen; 3rd abdominal sternum (2nd visible one) with a long, anteriorly directed spine.* The Acanthosomatidae, sometimes treated as a subfamily of the Pentatomidae, are predominantly phytophagous forms occurring in Africa, India, Java, the Philippines, Australia, Mexico and the Palaearctic region (Kumar, 1974). *Acanthosoma, Cyphostethus, Elasmostethus* and *Elasmucha* are found in Europe and *Elasmucha grisea* shows simple maternal care: the female protects the cluster of 20–50 eggs and the young nymphs with her body and wards off the attacks of parasitic Hymenoptera by vibrating her wings (Jordan, 1958*b*).

FAM. PENTATOMIDAE. *Antennae with 5 or, less often, 4 segments; tarsi 2- or 3-segmented; scutellum usually covering only about half the abdomen; hemelytra well-developed, membrane with 5–12 veins.* Even in the somewhat restricted sense employed here, this family is a large one, with well over 2500 species from all regions, especially the Neotropical, Ethiopian and Indo-Malaysian. The limits of the Pentatomidae and the mutual relationships of its subordinate groups are debatable, as may be seen from discussions by Leston (1958*a*, *b*) and Kumar (1965–71), the latter including much information on the morphology of the family. Some authorities define it so as to include the Scutelleridae and Acanthosomatidae, while others give separate family status to groups like the **Tessaratomidae, Dinidoridae** and **Urostylidae** (= **Urolabidae**), here regarded as subfamilies of the Pentatomidae.

The vast majority are vegetable feeders but members of the subfamily Asopinae are chiefly predacious, particularly upon Lepidopterous larvae (Dupuis, 1949). Nymphs of *Zicrona caerulea* are recorded by Kershaw and Kirkaldy (1908) in China to prey on larvae of *Haltica caerulea* while the adults attack the adults of that species. Others (e.g. *Picromerus bidens*) appear to live on either plant or animal tissue. In temperate regions the species appear to be mainly single-brooded, the nymphs occurring in spring or early summer and the adults later – many of the latter hibernate. The eggs (Esselbaugh, 1947) are usually barrel-shaped and deposited in compactly arranged masses. The nymphs are flattened and rounded in outline, their coloration is often striking and usually different from that of the adults; for observations on the biology of the family see Morrill (1910), Tischler (1938; 1939) and Esselbaugh (1949). The life-history of the oriental *Tessaratoma papillosa* has been followed by Kershaw (1907); the early stages are found on 'logan' and 'lichee' fruit trees which they apparently injure. Both sexes have the property of stridulation (p. 690) and are also able to eject an obnoxious fluid to a distance of 6–12 inches (Muir, 1907). In the later nymphs there are four pairs of dorsal abdominal odoriferous glands which atrophy in the adult, and are replaced by the usual ventral thoracic glands.

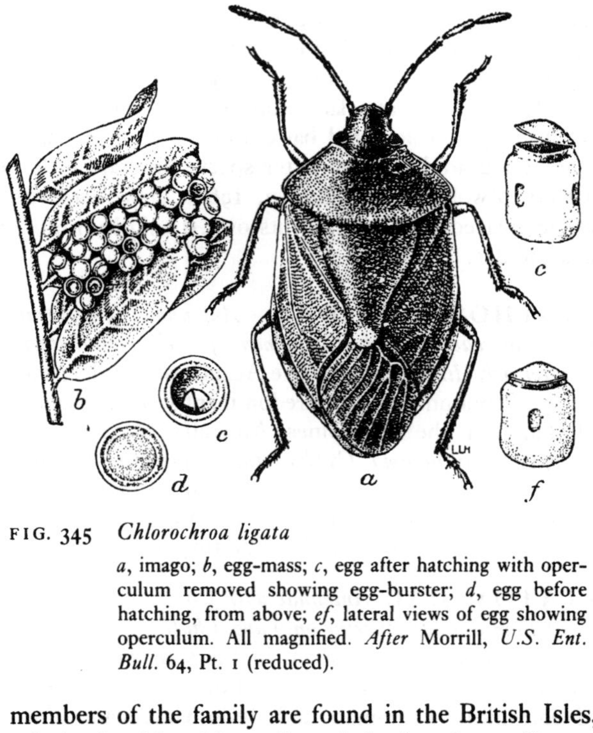

FIG. 345 *Chlorochroa ligata*

a, imago; *b*, egg-mass; *c*, egg after hatching with operculum removed showing egg-burster; *d*, egg before hatching, from above; *ef*, lateral views of egg showing operculum. All magnified. *After* Morrill, *U.S. Ent. Bull.* 64, Pt. 1 (reduced).

Only a few members of the family are found in the British Isles, occurring on various trees and shrubs. Notwithstanding their abundance, Pentatomids are not often major pests of crops. *Murgantia histrionica*, the Harlequin Cabbage Bug of the United States and C. America, is one of the best known and is especially partial to Cruciferae. *Chlorochroa ligata* is the 'Conchuela' of N. America and is injurious to various plants (Fig. 345). *Nezara viridula* is unusual in its almost world-wide distribution and attacks beans, tomatoes, lucerne, cotton and other crops. *Urostylis* and its allies are rather small, somewhat delicate, Coreid-like insects, usually green or greenish-brown in colour and found in Malaysia, Australasia and the eastern Palaearctic region. The largest members of the family are members of the Tessaratominae, some of which are brilliantly coloured. For a generic monograph of some subfamilies see Schouteden (1904–13); the catalogue of Kirkaldy (1909) gives the older bibliographical references, including those to habits and food-plants.

There are two small remaining families in the Pentatomoidea. The **Thaumastellidae** (Stys, 1964) includes only *Thaumastella aradoides* from Algeria and Iran, a Lygaeid-like insect that is placed in or near that family by some authorities. The **Phloeidae** (Leston, 1953) are large, flat insects with lamellate lateral extensions of the head and body. The antennae and tarsi are 3-segmented and the 3 Neotropical species occur on the bark of trees where their form and colour pattern help to conceal them.

Series 2. AMPHIBICORISAE

Aquatic Heteroptera, living on the water surface; antennae longer than head; underside of body with a hydrofuge pubescence; hemelytra never divided into clavus and corium, and membrane only rarely present; claws usually ante-apical. This section includes only one superfamily, the Gerroidea; its evolution and geographical distribution are discussed in relations to those of other aquatic Heteroptera by China (1954) and Hungerford (1958).

Superfamily **Gerroidea**

With the characters of the Amphibicorisae

FAM. HEBRIDAE (Naeogaeidae). *Very small, stoutly built forms; antennae usually 5-segmented; rostrum 4-segmented; clavus membranous and confluent with membrane.* A small family containing about 60 species of minute, semi-aquatic species found among *Sphagnum*, *Lemna*, etc. in marshes and other wet localities. *Hebrus* is widely distributed with two British species.

FAM. HYDROMETRIDAE. *Delicate rod-like insects with long head, 4- or 5-segmented antennae and 3-segmented rostrum; thoracic gland openings absent; legs thin and stilt-like; tarsi 3-segmented; claws apical.* There are about 70 species, mostly tropical, typified by *Hydrometra* (Fig. 346) which is found crawling slowly over the surface of stagnant water. There is a monograph by Hungerford and Evans (1934) and see also Sprague (1956).

FAM. MESOVELIIDAE. *Small forms with large head and eyes and long, delicate legs; antennae 4-segmented; rostrum 3-segmented; claws terminal.* This small family of over 20 species is widely distributed (Horvath, 1915, 1929; Gupta, 1963*a*, *b*). *Mesovelia* frequents the leaves of water-plants, in the stems of which its eggs are embedded (Hungerford, 1917). The New Guinea *Phrynovelia papua* is terrestrial, living among fallen leaves in forests.

FAM. VELIIDAE (Water Crickets). *Resemble Gerridae but stoutly built with 3-segmented rostrum; middle and hind legs not elongate.* The Veliidae include about 250 species from all regions, especially the Neotropical and Oriental. *Velia* occurs in streams and is often gregarious; macropterous forms are rare and the hemelytra, when present, are entirely membranous. *Rhagovelia* swims against the current of swift streams; the last tarsal segment of its middle pair of legs has a fan-like arrangement of hairs which spreads out and functions very much like the webbed feet of water-fowl. *Halovelia* and its allies are marine forms from the Indian and Pacific Oceans (Kenaga, 1941; Kellen, 1959). The classification of the Veliidae is discussed by China and Usinger (1949) and the biology of the C. American *Microvelia capitata* by Frick (1949).

FAM. GERRIDAE (Pond-skaters). *Moderately large, usually slender insects; rostrum 4-segmented; small median scent-gland opening on metathorax; hemelytra, when present, homogeneous; coxae rotatory; mid and hind legs usually elongate; tarsi 2-segmented, with ante-apical claws.* Like the other Amphibicorisan Heteroptera, the members of this family are semi-aquatic, living on the surface of still or running water where they feed mainly on dead insects or those floating there accidentally. The

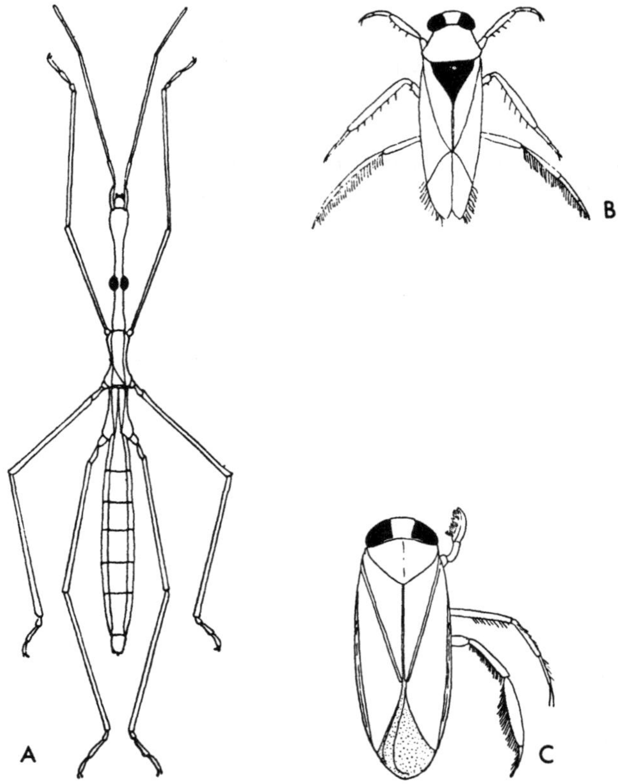

FIG. 346 A. *Hydrometra stagnorum* (× 8·5). B. *Notonecta glauca* (× 2). C. *Corixa* sp. (× 3).

body is covered ventrally with a dense, hydrofuge hair-pile and the wings are frequently reduced. The Gerridae include over 200 species from all regions, a number of genera being marine. *Gerris*, the common pond-skater, is cosmopolitan, laying its eggs in a group surrounded by a kind of mucilage and attached to submerged plants (Mitis, 1937; see also Sattler, 1957; Brinkhurst, 1960 and Darnhofer-Demar, 1969). *Halobates* and its allies are apterous and frequent the tropical and subtropical oceans, often occurring many hundreds of miles from land. They have been observed running over the surface of the sea in calm weather and feed upon dead, floating marine animals (Usinger, 1938; Barber, 1943; China, 1957; Herring, 1961). See Drake and Harris (1934) for an account of the Gerrinae of the W. hemisphere and Matsuda (1960) for the morphology and higher classification of the family.

Series HYDROCORISAE (= Cryptocerata)

Aquatic or, less often, semi-aquatic species; antennae shorter than the head and usually concealed beneath it; gula flat, without rostral groove; fore coxae free, hind coxae covered by episternum; pretarsus without pulvilli or 'pseudarolia'; hind legs used for swimming or, less often, for walking; egg without operculum. Three superfamilies are recognized: the Ochteroidea, Notonectoidea and Corixoidea.

Superfamily **Ochteroidea**

Semi-aquatic, riparian species; rostrum more or less elongate, 4-segmented; fore tarsi not spatulate; ocelli present.

FAM. OCHTERIDAE (Pelogonidae). *Small, oval insects with ocelli and large eyes; antennae 4-segmented; short but not concealed; rostrum long, 4-segmented; membrane with 2 series of cells; legs similar, tarsi with 1–3 segments.* About 20 species are known from all regions (Jaczewski, 1934; Schell, 1943). They are predacious, live at the margins of ponds or streams and can fly and jump actively.

FAM. GELASTOCORIDAE (Mononychidae: Toad-bugs). *Allied to the Ochteridae and of similar habits but with concealed, 3- or 4-segmented antennae, raptorial fore legs and short rostrum.* Over 70 species are known (Todd, 1955) and the biology of *Gelastocoris* is described by Hungerford (1922*a*).

Superfamily **Notonectoidea**

Aquatic species; rostrum more or less elongate, 3- or 4-segmented; antennae always concealed; fore tarsi not spatulate; ocelli absent.

FAM. NAUCORIDAE. *Small or medium, broadly oval, flattened insects; antennae 4-segmented, simple; rostrum strong, 3-segmented; membrane without veins; coxae hinged; fore legs usually strongly raptorial; hind legs fringed with swimming-hairs.* About 150 species, mostly tropical, make up this family (Usinger, 1941). They are mostly predacious insects frequenting both fresh and stagnant water. In the Oriental genus *Cheirochela* the fore legs are very powerful and chelate. They mostly haunt aquatic vegetation, among which they creep, coming to the surface to replenish their supply of air which is retained between the somewhat concave dorsum of the abdomen and the wings. *Aphelocheirus*, however, exhibits plastron respiration and is therefore independent of atmospheric oxygen (Thorpe and Crisp, 1947). The two British species belong respectively to the genera *Ilyocoris* and *Aphelocheirus*. Among biological papers may be mentioned Kramer (1935), Larsén (1927; 1931*b*), Usinger (1946*b*) and Ohm (1956).

FAM. BELOSTOMATIDAE (Giant Water Bugs). *Antennae 4-segmented; posterior legs adapted for swimming, the tibiae flattened and fringed with hairs; membrane reticulate; abdomen with 2 retractile apical appendages.* In this family are included the largest members of the Heteroptera and, in fact, of almost all insects,

Lethocerus grandis exceeding 11 cm in length. The 100 or so species are unrepresented in Britain, but prevalent in N. America, S. Africa and India. See De Carlo (1938), Cummings (1934) and Lauck and Menke (1961).

In habits they are very rapacious, feeding upon small fish, tadpoles, young frogs and insects. *Lethocerus* flies readily from one piece of water to another, is often attracted to lights and met with far away from water. The antennae in this genus are placed in ear-like pockets on the ventral surface of the head and not readily detected when in repose; the 2nd to 4th segments are provided with curious outgrowths whose significance is unknown. For the biology of various species see Rankin (1935), Lauck (1959) and Cullen (1969).

The life-history of *Zaitha flumineum* has been followed by Torre-Bueno (1906), and the average time taken from time of oviposition to full development is 50 days. Its favourite haunts are muddy-bottomed pools, where it lurks among the weeds. Both nymphs and adults obtain their air supply by piercing the surface film with the apex of the abdomen. The retractile appendages, when opposed, form a tube leading to the spiracles of the 6th abdominal segment. The dorsum of the abdomen in this family is somewhat concave, forming a reservoir under the wings which is ordinarily stored with air. In *Sphaerodema* and some other genera, the eggs are usually borne on the elytra of the males, being cemented to them by a waterproof secretion. According to Slater they are forcibly attached to the male by the female.

FAM. NEPIDAE (Water Scorpions). *Antennae 3-segmented; anterior legs strongly prehensile, posterior pair adapted for walking; tarsi 1-segmented, anterior pair clawless; abdomen with an apical breathing tube.* This family includes about 150 species, with only a few in the Holarctic and Australasian regions. The life-history of *Ranatra quadridentata* (N. America) has been followed by Torre-Bueno (1906) (see also Larsén, 1937). It occupies about 70 days from the time of oviposition to the adult stage, and the insect hibernates in the latter condition. The female is provided with a pointed toothed ovipositor, and the eggs in this genus are laid in notches cut in the petioles of water plants, each egg having a pair of apical filaments. Both nymphs and adults are capable of stridulation (Torre-Bueno, 1905). The respiratory tube is the most striking character of the family and consists of two elongate spine-like processes, each of which is grooved to form a half-canal. Torre-Bueno finds that *Ranatra* can move the two halves of the tube at will when beneath the water, and states that they are locked together by numerous hook-like bristles. When the insect is submerged the tube penetrates the surface film and air is conducted to a pair of spiracles placed at its base. Marshall and Severin found that the insect suffered no apparent ill-effects or inconvenience after the tube had been amputated. In the immature stages the tube is short and in newly hatched individuals it is absent.

Nepa differs in several important characters from *Ranatra*: the respiratory tube is short and among other features the eggs in (*N. cinerea*) are deposited in chains, adhering to one another by seven long filaments radiating from one extremity. Three pairs of 'false spiracles', situated respectively on the 3rd, 4th and 5th ventral abdominal segments, are present in both *Ranatra* and *Nepa*. They are sieve-like structures with the perforations occluded by a delicate membrane, and they are also provided with sensory setae. Baunacke (1912) made a detailed study of their histology and Thorpe and Crisp (1947) confirmed that they function by detecting relative differences in hydrostatic pressure between different spiracles. The general structure of *N. cinerea* has been studied by Hamilton (1931), but that of *Ranatra* needs further investigation.

FAM. NOTONECTIDAE (Backswimmers). *Body convex dorsally; head inserted into prothorax; rostrum 3- or 4-segmented, antennae 4-segmented; tarsi 2-segmented, anterior pair not flattened, posterior pair devoid of claws.* These insects differ from other aquatic Hemiptera in their habit of swimming on the back, which is shaped like the bottom of a boat. They are usually observed floating on the surface of the water with the long oar-like hind legs outstretched. They dive readily when alarmed, carrying a supply of air beneath the wings; they can also leap into the air and take to flight. Care is needed in handling them, as they can inflict painful punctures. *Notonecta* is almost universally distributed and markedly predacious, attacking small fish, tadpoles, etc., when kept in an aquarium. In *N. glauca* and its allies (Fig. 346) the female is provided with a piercing ovipositor with which it makes incisions in the stems of water plants, partially burying an egg in each notch; certain other species merely attach their eggs to the plants or other supports (Walton, 1936). The abdomen is keeled down the middle, with a longitudinal row of outwardly directed hairs on either side. These meet a corresponding series of similar hairs arising from near the pleura. In this way an air-filled channel is formed on either side of the abdomen, which enables the insect to respire when submerged. Larsén (1930) describes the biology of some species of *Notonecta*.

FAM. PLEIDAE. *Small, stout insects, strongly convex dorsally; rostrum and antennae 3-segmented; fore wing uniformly coriaceous, hind wing reduced; hind tibiae and tarsi without conspicuous swimming-hairs; tarsi 2- or 3-segmented.* Over 20 species have been described from all regions. The biology and external structure of the European *Plea atomaria* is described by Wefelscheid (1912).

FAM. HELOTREPHIDAE. *Small forms; head and thorax fused; antennae 1- or 2-segmented; rostrum 4-segmented; scutellum large; clavus and corium usually fused; hind wings usually absent or rudimentary.* The 15 or so species of this widely distributed family (Esaki and China, 1928; China, 1935; 1936) occur in still water with abundant vegetation (Usinger, 1937).

Superfamily **Corixoidea**

Aquatic species; body depressed; rostrum very short, 1- or 2-segmented; antennae 3- or 4-segmented, always concealed; ocelli present or absent; fore tarsus almost always spatulate and armed with a row of pegs in the male, where it forms the pala; hind tarsi also clawless.

FAM. CORIXIDAE. With over 200 species, this is the dominant family of aquatic Heteroptera; it includes some 33 British species, of which 15 belong to *Sigara*.

As a rule these insects remain at the bottom of the water, holding fast by the middle legs to various objects; now and again they ascend to the surface, swiftly propelled by the hind limbs. The abdomen is somewhat concave dorsally, forming a reservoir beneath the wings which serves to retain a supply of air. The feeding habits of the family are unusual in that its members are microphagous, consuming diatoms, the contents of algal cells, etc., (Sutton, 1951; Elliott and Elliott, 1967). *C. punctata* attaches its eggs to the stems and leaves of pond weeds by means of a glutinous substance. They are more or less onion-shaped, with one extremity prolonged into a blunt point. In some species the eggs are very numerous and closely grouped together; thus those of *Arctocorixa abdominalis* and *mercenaria* form very considerable masses which are used as food by the Mexican Indians. Bundles of reeds are placed in the water and collected at suitable intervals, and the eggs are detached by beating the reeds. The adult insects are also used as food in Mexico and Egypt. Hagemann (1910) has contributed some observations on the respiration of *Corixa* at successive stages in its life and on the structure of the spiracles. He also describes a tympanal organ in association with the 2nd pair, which may possibly serve for the perception of stridulatory sounds produced by the male.

The minute *Micronecta* lives in *Spongilla* and among water weeds and, like other members of the family, is able to stridulate, though this faculty is much less developed in the female than the male. The stridulatory organs of *Corixa* consist of the pedal organs mentioned on p. 690. There is no evidence that the abdominal strigil is used to produce sounds except in *Micronecta*. Among the numerous papers on this group, see Griffith (1945) and Peters (1962) on morphology, Young (1965) and Scudder (1971) on flight polymorphism, Hutchinson (1929) and Hungerford (1948) on taxonomy and Macan (1938; 1954) and Popham (1943–61) on biology.

Literature on the Hemiptera

ABALOOS, J. W. AND WYGODZINSKY, P. (1951), Las Triatominae Argentinas, *Univ. Nac. Tucumani Inst. Medicina Regional Publ.*, **601**, 179 pp.

ACTON, A. B. AND SCUDDER, G. G. E. (1969), The ultrastructure of the flight muscle polymorphism in *Cenocorixa bifida* (Hung.) (Heteroptera: Corixidae), *Z. Morph. Tiere*, **65**, 327–335.

AFIFI, S. A. (1968), Morphology and taxonomy of the adult males of the families Pseudococcidae and Eriococcidae (Homoptera: Coccoidea), *Bull. Br. Mus. nat. Hist.* (*Ent.*), Suppl. **13**, 210 pp.

AHMAD, I. (1965), The Leptocorisinae (Heteroptera: Alydidae) of the world, *Bull. Br. Mus. nat. Hist.* (*Ent.*), Suppl. **5**, 156 pp.

ALEXANDER, R. D. AND MOORE, T. E. (1962), The evolutionary relationships of 17-year and 13-year cicadas and three new species (Homoptera, Cicadidae, *Magicicada*), *Misc. Publ., Mus. Zool. Univ. Michigan*, **121**, 59 pp.

ANNAND, P. N. (1928), A contribution towards a monograph of the Adelginae (Phylloxeridae) of North America, *Stanford Univ. Publns biol. Sci.*, **6**, 1–146.

ARORA, G. L. AND SINGH, S. (1962), Morphology and musculature of the head and mouthparts of *Idiocerus atkinsoni* Leth. (Jassidae, Homoptera), *J. Morph.*, **110**, 131–140.

ASHLOCK, P. D. (1957), An investigation of the taxonomic value of the phallus in the Lygaeidae (Hemiptera-Heteroptera), *Ann. ent. Soc. Am.*, **50**, 407–426.

ASHLOCK, P. D. AND LATTIN, J. D. (1963), Stridulatory mechanisms in the Lygaeidae, with a new American genus of Orsillinae (Hemiptera, Heteroptera), *Ann. ent. Soc. Am.*, **56**, 693–703.

AUCLAIR, J. L. (1963), Aphid feeding and nutrition, *A. Rev. Ent.*, **8**, 439–490.

AULMANN, G. (1913), *Psyllidarum Catalogus*, Berlin: Junk, 82 pp.

AVERILL, A. W. (1945), Supplement to food-plant catalogue of the aphids of the world, including the Phylloxeridae, Index to genera and species of food-plants, *Bull. Maine agric. Exp. Stn*, **393–S**, 1–50.

AWATI, P. R. (1914), The mechanism of suction in the potato capsid bug, *Lygus pabulinus* Linn., *Proc. zool. Soc. Lond.*, **1914**, 685–733.

—— (1915), The apple sucker, with notes on the pear sucker, *Ann. appl. Biol.*, **1**, 247–272.

BAHADUR, J. (1963), Rectal pads in the Heteroptera, *Proc. R. ent. Soc. Lond.*, (A), 38, 59–69.

BAILEY, N. S. (1951), The Tingoidea of New England and their biology, *Entomologica am.*, **31**, 1–140.

BAKER, A. C. (1915), The woolly apple aphis, *U.S. Dept. Agric., Ent. Rep.*, **101**, 1–55.

BAKER, A. C. AND TURNER, F. W. (1916), Morphology and biology of the green apple aphis, *J. agric. Res.*, 5, 955–993.

BAKER, A. D. (1931), A study of the male genitalia of certain species of Pentatomidae, *Can. J. Res.*, 4, 148–179, 181–220.

BALACHOWSKY, A. (1932), Étude biologique des coccides du bassin occidentale de la méditerranée, *Encycl. Ent.*, **15**, 214 pp.

—— (1937–50), Les cochenilles de France, d'Europe, du Nord de l'Afrique et du bassin méditerranéen. I–V, *Actualités Sci. et Industr.*, **526**, 68 pp.; **564**, 59 pp.; 784, 111 pp.; **1054**, 154 pp.; **1087**, 163 pp.

—— (1942), Essai sur la classification des cochenilles, *Annls École nat. Agric. Grignon*, 3, 34–48.

BALASUBRAMANIAN, A. AND DAVIES, R. G. (1968), The histology of the labial glands of some Delphacidae (Hemiptera: Homoptera), *Trans. R. ent. Soc. Lond.*, **120**, 239–251.

BALDUF, W. V. (1941), Life history of *Phymata pennsylvanica americana* Melin (Phymatidae, Hemiptera), *Ann. ent. Soc. Am.*, 34, 204–214.

BANKS, C. J. (1955), An ecological study of Coccinellidae (Col.) associated with *Aphis fabae* Scop. on *Vicia fabae*, *Bull. ent. Res.*, 46, 561–587.

BARANOWSKI, R. M. (1958), Notes on the biology of the Royal Palm Bug *Xylastodoris luteolus* Barber (Hemiptera-Thaumastocoridae), *Ann. ent. Soc. Am.*, **51**, 547–551.

BARBER, G. W. (1936), *Orius insidiosus* (Say), an important natural enemy of the corn ear worm, *Tech. Bull. U.S. Dep. Agric.*, **504**, 1–24.

BARBER, H. G. (1943), Biological results of last cruise of Carnegie, VI. The *Halobates*, *Sci. Res. Cruise VII Carnegie, 1928–29, Biol.*, 4, 77–84.

BAUNACKE, W. (1912), Statisches Sinnesorgan bei den Nepiden, *Zool. Jb., Anat.*, 34, 179–346.

BEAMER, R. H. (1928), Studies on the biology of Kansas Cicadidae, *Kansas Univ. Sci. Bull.*, **18**, 155–263.

BEARD, R. L. (1940), The biology of *Anasa tristis* De Geer with particular reference to the Tachinid parasite, *Trichopoda pennipes* Fabr., *Bull. Connecticut agric. Exp. Sta.*, **440**, 597–679.

BEARDSLEY, J. W. (1960), A preliminary study of the males of some Hawaiian mealybugs (Homoptera: Pseudococcidae), *Proc. Hawaii. ent. Soc.*, **17**, 199–243.

BEARDSLEY, J. W. AND GONZALEZ, R. H. (1975), The biology and ecology of armored scales, *A. Rev. Ent.*, **20**, 47–73.

BEDAU, K. (1911), Die Fazettenaugen der Wasserwanzen, *Z. wiss. Zool.*, 97, 417–456.

BEIRNE, B. P. (1956), Leafhoppers (Homoptera: Cicadellidae) of Canada and Alaska, *Can. Ent.*, 88, Suppl. **2**, 180 pp.

BEKKER-MIGDISOVA, H. E. (1946), A contribution to the knowledge of the comparative morphology of Recent and Permian Homoptera, *Bull. Acad. Sci. U.R.S.S.*, 6, 741–766.

BENWITZ, G. (1956*a*), Der Kopf von *Corixa punctata* Ill. (*geoffroyi* Leach) (Hemiptera – Heteroptera), *Zool. Jb., Anat.*, **75**, 311–378.

—— (1956*b*), Die Beindrüsen der Corixiden, *Zool. Jb., Anat.*, **75**, 379–382.

BERGROTH, E. (1908), Enumeratio Pentatomidarum post catalogum bruxellensem descriptarum, *Mém. Soc. ent. Belg.*, **15**, 131–200.

—— (1913), Supplementum catalogi Heteropterorum bruxellensis, II. Coreidae, Pyrrhocoridae, Colobathristidae, Neididae, *Mém. Soc. ent. Belg.*, **22**, 125–183.

BERLESE, A. (1893–96), Le cocciniglie italiani viventi sugli agrumi. I–III, *Riv. Patol. veg., Padova*, **2**, 1–106; **3**, 107–201; **4** and **5**, 204–477.

BLACKMAN, R. L. (1971), Variation in the photoperiodic response within natural populations of *Myzus persicae* (Sulz.), *Bull. ent. Res.*, **60**, 533–546.

—— (1972), The inheritance of life cycle differences in *Myzus persicae* (Sulz.) (Hem., Aphididae), *Bull. ent. Res.*, **62**, 281–294.

—— (1974), Life cycle variation of *Myzus persicae* (Sulz.) (Hom., Aphididae) in different parts of the world, in relation to genotype and environment, *Bull. ent. Res.*, **63**, 595–607.

—— (1975), Photoperiodic determination of the male and female sexual morphs of *Myzus persicae, J. Insect Physiol.*, **21**, 435–453.

BLATCHLEY, W. S. (1926), *Heteroptera of eastern North America*, Nature Publ. Co., Indianapolis, 1116 pp.

BLUMBERG, B. (1935), The life cycle and seasonal history of *Ceroplastes rubens, Proc. R. Soc. Queensland*, **46**, 18–32.

BODENHEIMER, F. S. (1935), Studies on the zoogeography and ecology of Palaearctic Coccidae. I–III, *Eos*, **10**, 237–271.

BODENHEIMER, F. S. AND SWIRSKI, E. (1957), *Aphidoidea of the Middle East*, Weizmann Sci. Press, Jerusalem, 378 pp.

BONHAG, P. F. (1955), Histochemical studies of the ovarian nurse tissues and oocytes of the milkweed bug, *Oncopeltus fasciatus* (Dallas), I, II, *J. Morph.*, **96**, 381–440; **97**, 283–312.

BONHAG, P. F. AND WICK, J. R. (1953), The functional anatomy of the male and female reproductive systems of the milkweed bug, *Oncopeltus fasciatus* (Dallas), *J. Morph.*, **93**, 177–283.

BONNEMAISON, L. (1951), Contribution à l'étude des facteurs provoquant l'apparition des formes ailées et sexuées chez les Aphidinae, *Annls Épiphyt.*, **2**, 1–380.

BORATYNSKI, K. L. AND DAVIES, R. G. (1971), The taxonomic value of male Coccoidea (Homoptera) with an evaluation of some numerical techniques, *Biol. J. Linn. Soc.*, **3**, 57–102.

BORATYNSKI, K. L. AND WILLIAMS, D. J. (1964), A note on some British Coccoidea, with new additions to the British fauna, *Proc. R. ent. Soc. Lond.*, (B), **33**, 103–110.

BORKHSENIUS, N. S. (1958), On the evolution and phylogenetic interrelations of Coccoidea (Insecta, Homoptera), *Zool. Zh.*, **37**, 765–780.

—— (1960), Fn. U.S.S.R.: Homoptera 8. Suborder mealybugs and scales (Coccoidea), families Kermococcidae, Asterolecaniidae, Lecanodiaspididae, Aclerdidae, *Zool. Inst. Nauk S.S.S.R.* (*N.S.*), **77**, 1–283.

—— (1963), *Practical keys for identification of Coccoidea*, Moscow, Academy of Sciences, U.S.S.R., 311 pp.

—— (1965), Essay on the classification of the armoured scale insects (Homoptera, Coccoidea, Diaspididae), *Ent. Obozr.*, **44**, 372–376.

—— (1966), *A Catalogue of the Armoured Scale Insects* (*Diaspidoidea*) *of the World*, Acad. Sci. U.S.S.R., Moscow, 449 pp.

BÖRNER, C. (1908), Eine monographische Studie über die Chermiden, *Arb. biol. BundAnst. Land- u. Forstw.*, **6**, 75–320.

—— (1909), Zur Biologie und Systematik der Chermesiden, *Biol. Zbl.*, **29**, 118–125, 129–146.

—— (1930), Beiträge zu einem neuen System der Blattläuse, *Arch. Klass. Phylog. Ent.*, **1**, 115–194.

—— (1934), Ueber System und Stammesgeschichte der Schnabelkerfe, *Ent. Beih.*, **1**, 138–144.

—— (1938), Neuer Beitrag zur Systematik und Stammesgeschichte der Blattläuse, *Abh. naturw. Ver. Bremen*, **30**, 167–179.

—— (1952), Europae centralis Aphides. Die Blattläuse Mitteleuropas. Namen, Synonyme, Wirtspflanzen, Generationszyklen, *Schrift. Thüring. Land. Heilpflanz. Weimar*, **4** (3), 484 pp.

BÖRNER, C. AND HEINZE, K. (1957), Aphidina–Aphidoidea, In: Sorauer's *Handbuch der Pflanzenkrankheiten*, **5** (4), Lfg. 5, 577 pp.

BRANCH, H. E. (1914), Morphology and biology of the Membracidae of Kansas, *Kansas Univ. Sci. Bull.*, **8**, 73–115.

BRINKHURST, R. O. (1960), Studies on the functional morphology of *Gerris najas* DeGeer (Hem. Het. Gerridae), *Proc. zool. Soc. Lond.*, **133**, 531–559.

BRITTON, W. E. *et al.*, (1923), The Hemiptera or sucking insects of Connecticut, *Conn. N.H. Surv. Bull.*, **34**, 783 pp.

BROADBENT, L. (1951), Aphid excretion, *Proc. R. ent. Soc. Lond.*, (A), **26**, 97–103.

—— (1953), Aphids and virus diseases in potato crops, *Biol. Rev.*, **28**, 350–380.

BROCHER, F. (1909), Recherches sur la respiration des insectes aquatiques adultes. La Notonecte, étude biologique d'un insecte aquatique, avec un appendice sur la respiration des Naucores et des Corises, *Annls Biol. lacustr.*, **4**, 9–32.

BROWN, E. S. (1948), The ecology of Saldidae (Hemiptera-Heteroptera) inhabiting a salt-marsh, with observations on the evolution of aquatic habits in insects, *J. Anim. Ecol.*, **17**, 180–188.

—— (1959), Immature nutfall of coconuts in the Solomon Islands, I–III, *Bull. ent. Res.*, **50**, 97–133, 523–558, 559–566.

—— (1965), Notes on the migration and direction of flight of *Eurygaster* and *Aelia* species and their possible bearing on invasions of cereal crops, *J. Anim. Ecol.*, **34**, 93–107.

BUCHNER, P. (1954–67), Endosymbiose an Schildläusen. I–VIII. *Z. Morph. Ökol. Tiere*, **43**, 262–312, 397–424, 523–577; **45**, 379–410; **46**, 111–148, 481–528; **52**, 401–458; **56**, 275–362; **59**, 211–317.

—— (1965), *Endosymbiosis of Animals with Plant Micro-organisms*, Interscience, New York, 909 pp.

BUGNION, E. AND POPOFF, N. (1908), Le système nerveux et les organs sensoriels du Fulgore tacheté des Indes et de Ceylan (*Fulgora maculata*, Cicada), *J. Psych. Neurol.*, **13**, 326–354.

BUTLER, C. G. (1938*a*), On the ecology of *Aleurodes brassicae* Walk. (Hemiptera), *Trans. R. ent. Soc. Lond.*, **87**, 291–311.

—— (1938*b*), A further contribution to the ecology of *Aleurodes brassicae* Walk. (Hemiptera), *Proc. R. ent. Soc. Lond.*, (A), **13**, 161–172.

BUTLER, E. A. (1923), *A Biology of the British Hemiptera-Heteroptera*, Witherby, London, 682 pp.

CALDWELL, J. S. (1938), The jumping plant-lice of Ohio (Homoptera: Chermidae), *Bull. Ohio biol. Surv.*, **34**, 229–281.

CALLOW, R. K., GREENWAY, A. R. AND GRIFFITHS, D. C. (1973), Chemistry of the secretion from the cornicles of various species of aphids, *J. Insect Physiol.*, **19**, 737–748.

CAMERON, A. E. (1936), *Adelges cooleyi* Gillette (Hemiptera, Adelgidae) of the Douglas Fir in Britain: Completion of its life cycle, *Ann. appl. Biol.*, **23**, 585–605.

CARAYON, J. (1950*a*), Caractères anatomiques et position systématique des Hémiptères Nabidae (Note préliminaire), *Bull. Mus. Hist. nat., Paris*, **22**, 95–101.

—— (1950*b*), Nombre et disposition des ovarioles dans les ovaires des Hémiptères-Hétéroptères, *Bull. Mus. Hist. nat., Paris*, (2) **22**, 470–475.

—— (1952), Existence chez certains Hémiptères Anthocoridae d'un organe analogue à l'organe de Ribaga, *Bull. Mus. Hist. nat., Paris*, **24**, 89–97.

—— (1956), Trois espèces africaines de *Physopleurella* (Hemipt. Anthocoridae) dont l'une présente un nouveau cas de viviparité pseudoplacentaire, *Bull. Mus. Hist. nat., Paris*, (2) **28**, 102–110.

—— (1957), Introduction à l'étude des Anthocoridae omphalophores (Hemiptera-Heteroptera), *Annls Soc. ent. Fr.*, **126**, 159–197.

—— (1961), La viviparité chez les Hétéroptères, *Verh. II. int. Kongr. Ent.*, **1**, 711–714.

—— (1966), Traumatic insemination and the paragenital system, In: Usinger, R. L. (ed.), *Monograph of Cimicidae*, 81–166.

—— (1971), Notes et documents sur l'appareil odorant métathoracique des Hémiptères, *Annls Soc. ent. Fr.*, **7**, 737–770.

—— (1972), Caractères systématiques et classification des Anthocoridae (Hemiptera), *Annls Soc. ent. Fr.*, **8**, 309–349.

CARAYON, J., USINGER, R. L. AND WYGODZINSKY, P. (1958), Notes on the higher classification of the Reduviidae, with descriptions of a new tribe of the Phymatinae, *Revue Zool. Bot. afr.*, **57**, 256–281.

CARLO, J. A. DE (1938), Los Belostomidos americanos (Hemiptera), *Ann. Mus. argent. Ci. nat.*, **39**, 189–260.

CARTER, C. I. (1971), Conifer woolly aphids (Adelgidae) in Britain, *Forestry Commission Bull.*, **42**, 1–51.

CARVALHO, J. C. M. (1952), On the major classification of the Miridae (Heteroptera), *Ann. Acad. brasil. Cienc.*, **24**, 31–110.

—— (1955), Claves para os gêneros de Mirideos do mundo (Hemiptera), *Bol. Mus. Goeldi, Belém*, **11** (2), 1–151.

—— (1957–60), Catalogo dos Mirideos do mundo, *Arq. Mus. nac. Rio de Janeiro*, **44**, 1–158; **45**, 1–216; **47**, 1–161; **48**, 1–384; **51**, 1–194.

CARY, L. R. (1903), Plant house Aleyrodes, *Aleyrodes vaporariorum* Westw., *Maine agric. Exp. Sta. Bull.*, **96**, 20 pp.

CHAMBERLIN, J. C. (1923), A systematic monograph of the Tachardiinae or lac insects (Coccidae), *Bull. ent. Res.*, **14**, 147–212.

—— (1925), Supplement to a monograph of the Lacciferidae (Tachardiinae) or lac insects (Homopt., Coccidae), *Bull. ent. Res.*, **16**, 31–41.

CHAMPION, G. C. (1897–1901), Rhynchota Heteroptera, In: Godman & Salvin (eds.), *Biol. Centr. Amer.*, 2, 1–416.

CHANG, K. P. AND MUSGRAVE, A. J. (1969), Histochemistry and ultrastructure of the mycetome and its 'symbiotes' in the pear psylla, *Psylla pyricola* Foerster (Homoptera), *Tissue and Cell*, 1, 597–606.

CHATTERJEE, N. C. (1933), The life history and morphology of *Eurybrachys tomentosa* Fabr. Fulgoroidea (Homopt.), *Indian For. Rec.*, 18 (13), 26 pp.

CHEN, S. W. AND EDWARDS, J. S. (1972), Observations on the structure of secretory cells associated with aphid cornicles, *Z. Zellforsch.*, 130, 312–317.

CHEUNG, W. W. K. AND MARSHALL, A. T. (1973), Water and ion regulation in cicadas in relation to xylem feeding, *J. Insect Physiol.*, 19, 1801–1816.

CHINA, W. E. (1931), Morphological parallelism in the structure of the labium in the Hemipterous genera *Coptosomoides* gen. nov., and *Bozius* Dist. (family Plataspididae) in connection with mycetophagous habits, *Ann. Mag. nat. Hist.*, (10) 7, 281–286.

—— (1933), A new family of the Hemiptera-Heteroptera with notes on the phylogeny of the sub-order, *Ann. Mag. nat. Hist.*, (10) 12, 180–196.

—— (1935), New and little known Helotrephidae (Hemiptera, Helotrephidae), *Ann. Mag. nat. Hist.*, (10) 15, 593–614.

—— (1936), The first genus and species of Helotrephidae (Hemiptera) from the New World, *Ann. Mag. nat. Hist.*, (10) 17, 527–538.

—— (1942), A revision of the British species of *Cixius* Latr. (Homoptera), including the description of a new species from Scotland, *Trans. Soc. Brit. Ent.*, 8, 79–110.

—— (1954), The evolution of the waterbugs, *Bull. natn. Inst. Sci. India*, 7, 91–103.

—— (1955), A reconsideration of the systematic position of the family Joppeicidae Reuter (Hemiptera-Heteroptera), with notes on the phylogeny of the suborder, *Ann. Mag. nat. Hist.*, (12) 8, 353–370.

—— (1957), The marine Hemiptera of the Monte Bello islands, with descriptions of some allied species, *J. Linn. Soc. (Zool.)*, 43, 342–357.

—— (1962), South American Peloridiidae (Hemiptera-Homoptera: Coleorrhyncha), *Trans. R. ent. Soc. Lond.*, 114, 131–161.

CHINA, W. E. AND FENNAH, R. G. (1952), A remarkable new genus and species of Fulgoroidea (Homoptera) representing a new family, *Ann. Mag. nat. Hist.*, (12) 5, 189–199.

CHINA, W. E. AND MILLER, N. C. E. (1959), Check-list and keys to the families and subfamilies of the Hemiptera-Heteroptera, *Bull. Br. Mus. nat. Hist.* (Ent.), 8 (1), 45 pp.

CHINA, W. E. AND MYERS, J. G. (1929), A reconsideration of the classification of the Cimicoid families with the description of two new spider-web bugs, *Ann. Mag. nat. Hist.*, (10) 3, 97–125.

CHINA, W. E. AND USINGER, R. L. (1949), Classification of the Veliidae with a new genus from South Africa, *Ann. Mag. nat. Hist.*, (12) 2, 343–354.

CHOLODKOVSK, Y. N. (1895–96), Beiträge zu einer Monographie der Coniferen-Läuse, *Horae ent. Soc. Ross.*, 30, 1–102; 31, 1–61.

—— (1907), *Die Coniferen-Läuse* Chermes, *Feinde der Nadelhölzer*, Friedländer, Berlin, 44 pp.

CHOPRA, N. P. (1967), The higher classification of the family Rhopalidae, *Trans. R. ent. Soc. Lond.*, 119, 363–399.

CHRYSTAL, R. N. (1922), The Douglas Fir *Chermes*, *Bull. For. Comm., London*, **4**, 1–50.

CLARK, L. R. (1962), The general biology of *Cardiaspina albitextura* (Psyllidae) and its abundance in relation to weather and parasitism, *Aust. J. Zool.*, **10**, 537–586.

—— (1964), The population dynamics of *Cardiaspina albitextura* (Psyllidae), *Aust. J. Zool.*, **12**, 362–380.

COBBEN, R. H. (1959), Notes on the classification of Saldidae with the description of a new species from Spain, *Zoöl. Meded., Leiden*, **36**, 303–316.

—— (1965), Das aero-mikropylare System der Homoptereneier und Evolutionstrends bei Zikadeneiern (Hom. Auchenorrhyncha), *Zool. Beitr.*, **11**, 13–69.

—— (1968), Evolutionary trends in Heteroptera. Part 1. Eggs, architecture of the shell, gross embryology and eclosion, *Meded. Lab. Ent., Wageningen*, **151**, 1–475.

—— (1970), Morphology and taxonomy of intertidal dwarf-bugs (Heteroptera: Omaniidae fam. nov.), *Tijdschr. Ent.*, **113**, 61–90.

COLLYER, E. (1952), Biology of some predatory insects and mites associated with the Fruit Tree Red Spider mite (*Metatetranychus ulmi* (Koch)) in south-eastern England, I, *J. hort. Sci.*, **27**, 117–129.

COUTURIER, A. (1938), Remarques sur la biologie de *Ceresa bubalis* Fab., Membracide d'origine Americaine, *Revue Zool. agric. appl.*, **37**, 145–157.

CRANHAM, J. E. (1966), Tea pests and their control, *A. Rev. Ent.*, **11**, 491–514.

CRANSTON, F. P. AND SPRAGUE, I. B. (1961), A morphological study of the head capsule of *Gerris remigis* Say, *J. Morph.*, **108**, 287–298.

CRAWFORD, D. L. (1914), A monograph of the jumping plant lice or Psyllidae of the New World, *Bull. U.S. natn Mus.*, **85**, 1–182.

CROSBY, C. R. AND LEONARD, M. D. (1914), The tarnished plant bug (*Lygus pratensis* Linnaeus), *Bull. agric. Exp. Sta., New York*, **346**, 463–525.

CULLEN, M. J. (1969), The biology of giant water bugs (Hemiptera: Belostomatidae) in Trinidad, *Proc. R. ent. Soc. Lond.*, (A), **44**, 123–136.

CUMMINGS, C. (1934), The giant water bugs (Belostomatidae, Hemiptera), *Kansas Univ. Sci. Bull.*, **23**, 197–219.

DARNHOFER-DEMAR, B. (1969), Zur Funktionsmorphologie der Wasserläufer. I. Die Morphologie des Lokomotionsapparates von *Gerris lacustris* L. (Heteroptera: Gerridae), *Zool. Jb. (Anat.)*, **86**, 28–66.

DASHMAN, T. (1953), The unguitractor plate as a taxonomic tool in the Hemiptera, *Ann. ent. Soc. Am.*, **46**, 561–578.

DAVIDSON, J. (1925), *A List of British Aphides*, Longmans, Green & Co., London, 176 pp.

DAVIS, N. T. (1955), Morphology of the female organs of reproduction in the Miridae (Hemiptera), *Ann. ent. Soc. Am.*, **48**, 132–150.

—— (1956), The morphology and functional anatomy of the male and female reproductive systems of *Cimex lectularius* L. (Heteroptera, Cimicidae), *Ann. ent. Soc. Am.*, **49**, 466–493.

—— (1957–66), Contributions to the morphology and phylogeny of the Reduvioidea. I–III, *Ann. ent. Soc. Am.*, **50**, 432–443; **54**, 340–353; **59**, 911–924.

DECOURSEY, R. M. (1971), Keys to the families and subfamilies of the nymphs of North American Hemiptera–Heteroptera, *Proc. ent. Soc. Wash.*, **73**, 413–428.

DELONG, D. M. (1948), The leafhoppers or Cicadellidae of Illinois (Eurymelinae–Balcluthinae), *Bull. Ill. nat. Hist. Surv.*, **24**, 93–376.

—— (1971), The bionomics of leafhoppers, *A. Rev. Ent.*, **16**, 179–210.

DE LOTTO, G. (1974), On the status and identity of the cochineal insects (Homoptera: Coccoidea: Dactylopiidae), *J. ent. Soc. sth. Afr.*, **37**, 167–193.

DICKERSON, G. AND LAVOIPIERRE, M. M J. (1959), Studies on the methods of feeding of blood-sucking arthropods. II. The method of feeding adopted by the bed-bug (*Cimex lectularius*) when obtaining a blood meal from the mammalian host, *Ann. trop. Med. Parasit.*, **53**, 347–357.

DINGLER, M. (1930), Beiträge zur Biologie von *Icerya purchasi* Mask. (Coccidae, Monophlebinae), *Biol. Zbl.*, **50**, 32–49.

DISTANT, W. L. (1880–1900). In: Goodman and Salvin, *Biol. Centr.-Amer.*, Rhynchota Heteroptera, **1**, 1–462, Rhynchota Homoptera, 1–43.

—— (1889–92), *A Monograph of the Oriental Cicadidae*, Trustees Indian Mus., Calcutta, London, 158 pp.

—— (1906), *A Synonymic Catalogue of Homoptera Part I. Cicadidae*, Brit. Mus. (Nat. Hist.), London, 207 pp.

—— (1902–18), *Fauna of British India. Rhynchota*, Taylor & Francis, London, **1**, 438 pp.; **2**, 503 pp.; **3**, 503 pp.; **4**, 501 pp.; **5**, 362 pp.; **6**, 248 pp.; **7**, 210 pp.

—— (1912–14), Cicadidae, *Genera Insectorum*, **142**, 64 pp.; **158**, 38 pp.

DIXON, A. F. G. AND STEWART, W. A. (1975), Function of the siphunculi in aphids with particular reference to the sycamore aphid, *Drepanosiphum platanoidis*, *J. Zool., Lond.*, **175**, 279–289.

DOERING, K. C. (1930), Synopsis of the family Cercopidae (Homoptera) in N. America, *J. Kansas ent. Soc.*, **3**, 53–64, 81–102.

—— (1937–41), A contribution to the taxonomy of the subfamily Issinae in America north of Mexico (Fulgoridae, Homoptera), I–IV, *Kansas Univ. Sci. Bull.*, **24**, 421–467; **25**, 447–575; **26**, 83–167; **27**, 185–233.

DRAKE, C. J. AND DAVIS, N. T. (1958), The morphology and classification of the Piesmatidae (Hemiptera) with keys to the world genera and American species, *Ann. ent. Soc. Am.*, **51**, 567–581.

—— (1960), The morphology, phylogeny and higher classification of the family Tingidae, including the description of a new genus and species of the subfamily Vianaidinae (Hemiptera–Heteroptera), *Entomologica am.*, **39**, 1–100.

DRAKE, C. J. AND HARRIS, H. M. (1934), The Gerrinae of the Western Hemisphere (Hemiptera), *Ann. Carnegie Mus.*, **23**, 179–240.

DRAKE, C. J. AND HOBERLANDT, L. (1952*a*), Check list and distributional records of Leptopodidae (Hemiptera), *Acta ent. Mus. nat. Pragae*, **26** (373), 5 pp.

—— (1952*b*), Catalogue of genera and species of Saldidae (Hemiptera), *Acta ent. Mus. nat. Pragae*, **26** (376), 1–12.

DRAKE, C. J. AND RUHOFF, F. A. (1960), Lace-bug genera of the world, *Proc. U.S. natn. Mus.*, **112**, 1–105.

—— (1965), Lace-bugs of the world: a catalog (Hemiptera: Tingidae), *Bull. U.S. natn. Mus.*, **243**, 634 pp.

DRAKE, C. J. AND SALMON, J. T. (1950), A new genus and two new species of Peloridiidae (Homoptera) from New Zealand, *Zool. Publ. Victoria Univ. Coll.*, **6**, 1–7.

DRAKE, C. J. AND SLATER, J. A. (1957), The phylogeny and systematics of the family Thaumastocoridae (Hemiptera: Heteroptera), *Ann. ent. Soc. Am.*, **50**, 353–370.

DRAŠLAR, K. (1973), Functional properties of trichobothria in the bug *Pyrrhocoris apterus* (L.), *J. comp. Physiol.*, **84**, 175–184.

DUNN, J. A. (1949), The parasites and predators of potato aphids, *Bull. ent. Res.*, **40**, 97–122.

DUPUIS, C. (1949), Les Asopinae de la faune française (Hemiptera, Pentatomidae). Essai sommaire de synthèse morphologique, systématique et biologique, *Revue fr. ent.*, **16**, 233–250.

—— (1955), Les genitalia des Hémiptères Hétéroptères (Genitalia externes des deux sexes: Voies ectodermiques femelles), Revue de la morphologie. Lexique de la nomenclature, Index bibliographique analytique, *Mém. Mus. natn. Hist. nat., Paris*, (A) *Zool.*, **6** (4), 183–278.

EASTOP, V. F. (1958), *A Study of the Aphididae (Homoptera) of East Africa*, H.M.S.O., London, 126 pp.

—— (1961), *A Study of the Aphididae of West Africa*, London, British Museum (Natural History), 93 pp.

—— (1966), A taxonomic study of Australian Aphidoidea (Homoptera), *Aust. J. Zool.*, **14**, 399–592.

—— (1971), Keys for the identification of *Acyrthosiphon* (Hemiptera: Aphididae), *Bull. Br. Mus. nat. Hist.* (Ent.), **26**, 1–113.

—— (1972), A taxonomic review of the species of *Cinara* Curtis occurring in Britain (Hemiptera: Aphididae), *Bull. Br. Mus. nat. Hist.* (Ent.), **27**, 101–186.

EKBLOM, T. (1926–30), Morphological and biological studies of the Swedish families of Hemiptera Heteroptera, Pts I and II, *Zool. Bidr. Upps.*, **10**, 31–180; **12**, 113–150.

—— (1929), New contributions to the systematic classification of the Heteroptera, *Ent. Tidskr.*, **50**, 169–180.

—— (1949), Untersuchungen über den Flügelpolymorphismus bei *Gerris asper* Fieb., *Notul. Ent.*, **29** (2), 1–15.

— (1950), Über den Flügelpolymorphismus bei *Gerris odontogaster* Zett., *Notul. Ent.*, **30**, 4–49.

ELLIOTT, J. M. AND ELLIOTT, J. I. (1967), The structure and possible function of the buccopharyngeal teeth of *Sigara dorsalis* (Leach) (Hemiptera: Corixidae), *Proc. R. ent. Soc. Lond.*, (A), **42**, 83–86.

EMDEN, H. V. van (ed.) (1972), *Aphid Technology*, Academic Press, London and New York, 344 pp.

EMDEN, H. F. van, EASTOP, V. F., HUGHES, R. D. AND WAY, M. J. (1969), The ecology of *Myzus persicae*, *A. Rev. Ent.*, **14**, 197–270.

EMSLEY, M. G. (1969), The Schizopteridae (Hemiptera–Heteroptera) with the description of new species from Trinidad, *Mem. Am. ent. Soc.*, **25**, 1–154.

ENTWISTLE, P. F. (1972), *Pests of Cocoa*, Longman, London, 779 pp.

ESAKI, T. AND CHINA, W. E. (1928), A monograph of the Helotrephidae, subfamily Helotrephinae, *Eos*, **4**, 129–172.

ESSELBAUGH, C. O. (1947), A study of the eggs of the Pentatomidae (Hemiptera), *Ann. ent. Soc. Am.*, **39**, 667–691.

—— (1949), Notes on the bionomics of some mid-western Pentatomidae, *Entomologica am.*, **28**, 1–73.

ESSIG, E. O. AND ABERNATHY, F. (1952), *The Aphid Genus* Periphyllus *(family Aphididae): A Systematic, Biological and Ecological Study*, University of California Press, Berkeley, 166 pp.

EVANS, J. H. (1931), A preliminary revision of the ambush bugs of North America (Hemiptera, Phymatidae), *Ann. ent. Soc. Am.*, **24**, 711–736.

EVANS, J. W. (1938), The morphology of the head of Homoptera, *Pap. Proc. roy. Soc. Tasmania*, **1937**, 1–20.

—— (1939), The morphology of the thorax of the Peloridiidae (Homopt.), *Proc. R. ent. Soc. Lond.*, (B), 8, 143–150.

—— (1940), Tube-building Cercopids (Homoptera, Machaerotidae), *Trans. R. Soc. S. Aust.*, **64**, 70–75.

—— (1941), Concerning the Peloridiidae, *Aust. J. Sci.*, **4**, 95–97.

—— (1946–47), A natural classification of leaf-hoppers (Jassoidea, Homoptera). Parts I–III, *Trans. R. ent. Soc. Lond.*, **96**, 47–60; **97**, 39–54; **98**, 105–271.

—— (1948), Some observations on the classification of the Membracidae and on the ancestry, phylogeny and distribution of the Jassoidea, *Trans. R. ent. Soc. Lond.*, **99**, 497–515.

—— (1951), Some notes on the classification of leaf-hoppers (Jassoidea, Homoptera) with special reference to the Nearctic fauna, *Comment. biol. Soc. sci. fenn.*, **12** (3), 1–11.

—— (1956), Palaeozoic and Mesozoic Hemiptera, *Aust. J. Zool.*, **4**, 165–258.

—— (1957), Some aspects of the morphology and inter-relationships of extinct and recent Homoptera, *Trans. R. ent. Soc. Lond.*, **109**, 275–294.

—— (1963), The phylogeny of the Homoptera, *A. Rev. Ent.*, 8, 77–94.

—— (1964), The periods of origin and diversification of the superfamilies of the Homoptera–Auchenorhyncha (Insecta) as determined by a study of the wings of Palaeozoic and Mesozoic fossils, *Proc. Linn. Soc. Lond.*, **175**, 171–181.

—— (1966), The leafhoppers and froghoppers of Australia and New Zealand (Homoptera: Cicadelloidea and Cercopoidea), *Mem. Aust. Mus.*, **12**, 1–347.

—— (1968), Some relict New Guinea leafhoppers and their significance in relation to the comparative morphology of the head and prothorax of the Homoptera–Auchenorrhyncha (Homoptera: Cicadellidae: Ulopinae), *Pacif. Insects*, **10**, 215–229.

—— (1973), The maxillary plate of Homoptera–Auchenorrhyncha, *J. Ent.* (A), **48**, 43–47.

EWEN, A. B. (1962), The cephalic nervous system of *Adelphocoris lineolatus* (Goeze) (Hemiptera: Miridae), *Can. J. Zool.*, **40**, 1187–1193.

—— (1966), Histophysiology of the neurosecretory system and retrocerebral endocrine glands of the alfalfa plant bug, *Adelphocoris lineolatus* (Goeze) (Hemiptera, Miridae), *J. Morph.*, **111**, 255–274.

EYLES, A. C. (1971–72), List of Isometopidae (Heteroptera: Cimicoidea), *N.Z. Jl Sci. Technol.*, **14**, 940–944; **15**, 463–464.

FEIR, D. (1974), *Oncopeltus fasciatus*: a research animal, *A. Rev. Ent.*, **19**, 81–96.

FENNAH, R. G. (1942), New or little-known West Indian Kinnaridae (Homoptera, Fulgoroidea), *Proc. ent. Soc. Washington*, **44**, 99–110.

—— (1944), The morphology of the tegmina and wings in Fulgoroidea (Homoptera), *Proc. ent. Soc. Wash.*, **46**, 185–199.

—— (1945*a*), Characters of taxonomic importance in the pretarsus of Auchenorrhyncha (Homoptera), *Proc. ent. Soc. Wash.*, **47**, 120–128.

—— (1945*b*), The external male genitalia of Fulgoroidea (Homoptera), *Proc. ent. Soc. Wash.*, **47**, 217–229.

—— (1945*c*), The Fulgoroidea or lantern flies of Trinidad and adjacent parts of South America, *Proc. U.S. nat. Mus.*, **95**, 411–520.

—— (1947), A synopsis of the Achilixiidae of the New World (Homoptera, Fulgoroidea), *Ann. Mag. nat. Hist.*, (11), **3**, 183–191.

—— (1949), A new genus of Fulgoroidea (Homoptera) from South Africa, *Ann. Mag. nat. Hist.*, (12), **2**, 111–120.

—— (1950), A generic revision of Achilidae (Homoptera: Fulgoroidea), with descriptions of new species, *Bull. Brit. Mus.* (*nat. Hist.*) Entomol., **1** (1), 170 pp.

—— (1952*a*), On the generic classification of Derbidae (Fulgoroidea) with descriptions of new neotropical species, *Trans. R. ent. Soc. Lond.*, **103**, 109–170.

—— (1952*b*), On the classification of the Tettigometridae (Homoptera: Fulgoroidea), *Trans. R. ent. Soc. Lond.*, **103**, 239–255.

—— (1954), The higher classification of the family Issidae (Homoptera: Fulgoroidea) with descriptions of new species, *Trans. R. ent. Soc. Lond.*, **105**, 455–474.

—— (1956), Homoptera: Fulgoroidea, *Insects of Micronesia*, **6** (3), 72 pp.

—— (1965), Delphacidae from Australia and New Zealand (Homoptera: Fulgoroidea), *Bull. Br. Mus. nat. Hist.* (Ent.), **17**, 1–59.

FERRIS, G. F. (1918), A note on the occurrence of abdominal spiracles in the Coccidae, *Canad. Ent.*, **50**, 85–88.

—— (1937–55), *An Atlas of the Scale Insects of North America*, Stanford Univ. Press, Stanford, California, **1**, 1–136; **2**, 137–268; **3**, 269–384; **4**, 385–448; **5**, 1–278; **6**, 279–506; **7**, 1–233.

—— (1957*a*), A brief history of the study of the Coccoidea, *Microentomology*, **22**, 39–57.

—— (1957*b*), A review of the family Eriococcidae (Insecta, Coccoidea), *Microentomology*, **22**, 81–89.

FERRIS, G. F. AND USINGER, R. L. (1939), The family Polyctenidae (Hemiptera, Heteroptera), *Microentomology*, **4**, 1–50.

FIEBER, F. X. (1861), *Die europäischen Hemipteren, Halbflügler*, Gerold, Wien, 444 pp.

FILSHIE, B. K. AND WATERHOUSE, D. F. (1968), The fine structure of the lateral scent glands of the green vegetable bug, *Nezara viridula* (Hemiptera, Pentatomidae), *J. Microscopie*, **7**, 231–244.

FLEMION, F., LEDBETTER, M. C. AND KELLEY, E. S. (1954), Penetration and damage of plant tissues during feeding by the tarnished plant bug (*Lygus lineolaris*), *Contr. Boyce Thompson Inst. Pl. Res.*, **17**, 347–357.

FLUITER, H. J. de (1950), De invloed van de daglengte en temperatuur op het optreden van de geslachtsdieren bij *Aphis fabae* Scop., de zwarte bonenluis, *Tijdschr. PlZiekt.*, **50**, 265–285.

FOA, A. (1912), Riassunta teorico-pratica della biologia della Fillossera della vite, In: Grassi, B. *et al.*, *Contributo alla conoscenza delle Fillosserine ed in particolare della Fillossera della vite*, Rome, 456 pp.

FORBES, A. R. (1964), The morphology, histology and fine structure of the gut of the green peach aphid, *Myzus persicae*, *Mem. ent. Soc. Canada*, **36**, 74 pp.

FORBES, A. R. AND RAINE, J. (1973), The stylets of the six-spotted leafhopper, *Macrosteles fasciifrons* (Homoptera: Cicadellidae), *Can. Ent.*, **105**, 559–567.

FOWLER, W. W. (1894–1909), In: Godman and Salvin (eds), *Biol. Centr.-Amer.*, Rhynchota Homoptera, **1**, 77–147; **2**, 1–339.

FRAZER, H. L. (1944), Observations on the method of transmission of internal boll disease of cotton by the cotton stainer bug, *Ann. appl. Biol.*, **31**, 271–290.

FRICK, K. E. (1949), The biology of *Microvelia capitata* Guérin, 1857, in the Panama Canal Zone and its role as a predator on Anopheline larvae (Veliidae: Hemiptera), *Ann. ent. Soc. Am.*, **42**, 77–110.

FROESCHNER, R. C. (1960), Cydnidae of the Western Hemisphere, *Proc. U.S. natn. Mus.*, **111**, 337–680.

FUNKHOUSER, W. D. (1915), Life history of *Vanduzea arquata* Say (Membracidae), *Psyche*, **22**, 183–198.

—— (1917), Biology of the Membracidae of the Cayuga Lake basin, *Mem. Cornell Univ. agric. Exp. Stn*, **11**, 177–445.

—— (1950), Membracidae, *Genera Insect.*, 208, 383 pp.

GAHAN, C. J. (1918), Method of formation of cuckoo-spit by *Philaenus spumarius*, *Proc. ent. Soc. Lond.*, **1918**, cliii–clv.

GEORGE, K. S. (1957), Preliminary investigations on the biology and ecology of the parasites and predators of *Brevicoryne brassicae* (L.), *Bull. ent. Res.*, **48**, 619–629.

GHAURI, M. S. K. (1962), *The Morphology and Taxonomy of male Scale Insects*, (*Homoptera: Coccoidea*), Brit. Mus. (Nat. Hist.), London, 221 pp.

GILBY, A. R. AND WATERHOUSE, D. F. (1965), The composition of the scent of the green vegetable bug, *Nezara viridula*, *Proc. R. Soc.*, (B), **162**, 105–120.

GILIOMEE, J. H. (1966), Morphology and taxonomy of adult males of the family Coccidae, *Bull. Br. Mus. nat. Hist.* (Ent.), **19**, Suppl. 7, 168 pp.

GILLETTE, C. P. AND PALMER, M. A. (1931–36), The Aphidae of Colorado, Pts. I–III, *Ann. ent. Soc. Am.*, **24**, 827–934; **25**, 369–496; **27**, 133–255; **29**, 729–748.

GLASGOW, H. (1914), The gastric caeca and the caecal bacteria of the Heteroptera, *Biol. Bull.*, **26**, 101–170.

GLUUD, A. (1968), Zur Feinstruktur der Insektencuticula. Ein Beitrag zur Frage des Eigengiftschutzes der Wanzencuticula, *Zool. Jb.* (*Anat.*), **85**, 191–227.

GODING, F. W. (1926), Classification of the Membracidae of America, *Jl N.Y. ent. Soc.*, **34**, 295–317.

—— (1934, 1939), The Old World Membracidae, *Jl N.Y. ent. Soc.*, **42**, 451–480; **47**, 315–349.

GOEL, S. C. (1972), Notes on the structure of the unguitractor plate in Heteroptera (Hemiptera), *J. Ent.* (A), **46**, 167–173.

GOEL, S. C. AND SCHAEFER, C. W. (1970), The structure of the pulvillus and its taxonomic value in the land Heteroptera (Hem.), *Ann. ent. Soc. Am.*, **63**, 307–313.

GOODCHILD, A. J. P. (1952), A study of the digestive system of the West African cacao Capsid bugs, *Proc. zool. Soc. Lond.*, **122**, 543–572.

—— (1963*a*), Some new observations on the intestinal structures concerned with water disposal in sap-sucking Hemiptera, *Trans. R. ent. Soc. Lond.*, **115**, 217–237.

—— (1963*b*), Studies on the functional anatomy of the intestines of Heteroptera, *Proc. zool. Soc. Lond.*, **141**, 851–910.

—— (1966), Evolution of the alimentary canal in the Hemiptera, *Biol. Rev.*, **41**, 97–140.

—— (1969), The rectal glands of *Halosalda lateralis* (Fallén) (Hemiptera: Saldidae) and *Hydrometra stagnorum* (L.), (Hemiptera: Hydrometridae), *Proc. R. ent. Soc. Lond.*, (A), **44**, 62–69.

GOURANTON, J. (1968*a*), Ultrastructures en rapport avec un transit d'eaù. Étude de la 'chambre filtrante' de *Cicadella viridis* L., (Homoptera, Jassidae), *J. Microscopie*, **7**, 559–574.

—— (1968*b*), Sécrétion d'un mucus complexe par les tubes de Malpighi des larves de Cercopides (Homoptera), Role dans la formation de l'abri spumeux, *Annls Sci. nat. (Zool.)*, (Sér. 12) **10**, 117–126.

GRAICHEN, E. (1936), Das Zentralnervensystem mit Einschluss des sympathischen Nervensystems von *Nepa cinerea* L., *Zool. Jb., Anat.*, **61**, 195–238.

GRASSI, B. (1915), The present state of our knowledge of the vine *Phylloxera, Bull. Inst. Agric.*, **6**, 1269–1290.

GREEN, E. E. (1896–1922), *Coccidae of Ceylon*, London, Pts. I–V, 472 pp.

—— (1927–28), A brief review of the indigenous Coccidae of the British Islands, with emendations and additions, *Entomologist's Rec. J. Var.*, **40**, sep. pp. 1–14.

GREENWAY, A. R. AND GRIFFITHS, D. C. (1973), A comparison of triglycerides from aphids and their cornicle secretions, *J. Insect Physiol.*, **19**, 1649–1655.

GRIFFITH, M. E. (1945), The environment, life history and structure of the water boatman, *Rhamphocorixa acuminata* (Uhler) (Hemiptera, Corixidae), *Kansas Univ. Sci. Bull.*, **30** (2), 241–365.

GROVE, A. J. (1919), The anatomy of the head and mouthparts of *Psylla mali*, the apple sucker, with some remarks on the function of the labium, *Parasitology*, **11**, 456–488.

GULDE, J. AND JORDAN, K. H. C. (1933–38), *Die Wanzen Mitteleuropas. Hemiptera Heteroptera Mitteleuropas*, Frankfurt a. Main, Pts. **2–6**, 703 pp; Pt. **12**, 1–105; Literaturteil, 1–34.

GUPTA, A. P. (1963*a*), Comparative morphology of the Saldidae and Mesoveliidae, *Tijdschr. Ent.*, **106**, 169–196.

—— (1963*b*), A consideration of the systematic position of the Saldidae and Mesoveliidae (Hemiptera: Heteroptera), *Proc. ent. Soc. Wash.*, **65**, 31–38.

GUTHRIE, D. M. (1961), Anatomy of the nervous system in the genus *Gerris, Phil. Trans. R. Soc. Ser. B*, **244**, 65–102.

HABIB, A. (1943), The biology and bionomics of *Asterolecanium pustulatus* Ckll. (Hemiptera-Coccoidea), *Bull. Soc. Fouad Ier Ent.*, **27**, 87–111.

HAGAN, H. R. (1931), The embryogeny of the Polyctenid *Hesperoctenes fumarius* Westwood, with reference to viviparity in insects, *J. Morph.*, **51**, 1–118.

HAGEMANN, J. (1910), Beiträge zur Kenntnis von *Corixa, Zool. Jb., Anat.*, **30**, 373–426.

HALKKA, O., RAATIKAINEN, M., VAZARAINEN, A. AND HEINONEN, L. (1967), Ecology and ecological genetics of *Philaenus spumarius* (L.) (Homoptera), *Ann. zool. fenn.*, **4**, 1–18.

HAMILTON, M. A. (1931), The morphology of the water-scorpion, *Nepa cinerea* Linn. (Rhynchota, Heteroptera), *Proc. zool. Soc. Lond.*, **1931**, 1067–1136.

HANDLIRSCH, A. (1897), Monographie der Phymatiden, *Annln naturh. Hofmus. Wien*, **12**, 127–230.

—— (1900), Wieviele Stigmen haben die Rhynchoten? Ein morphologischer Beitrag, *Verh. zool.-bot. Ges. Wien*, **49**, 499–510.

HARLEY, K. L. S. AND KASSULKE, R. C. (1971), Tingidae for biological control of *Lantana camara* (Verbenaceae), *Entomophaga*, **16**, 389–410.

HARRIS, H. M. (1928), A monographic revision of the Hemipterous family Nabidae as it occurs in North America, *Entomologica am.*, (N.S.), **9**, 1–97.

HASKELL, P. T. (1957), Stridulation and its analysis in certain Geocorisae, *Proc. zool. Soc. Lond.*, **129**, 351–358.

HASSAN, A. I. (1939), The biology of some British Delphacidae (Homopt.) and their parasites with special reference to the Strepsiptera, *Trans. R. ent. Soc. Lond.*, **89**, 345–384.

HAUPT, H. (1929), Neueinteilung der Homoptera-Cicadina nach phylogenetisch zu wertenden Merkmalen, *Zool. Jb. Syst.*, **58**, 173–286.

—— (1935), Gleichflügler, Homoptera, In: *Die Tierwelt Mitteleuropas*, **4** (1), X, 115–X, 262.

HEDICKE, H. (1935), Ungleichflügler, Wanzen, Heteroptera, In: *Die Tierwelt Mitteleuropas*, **4** (1), X, 15–X, 113.

HEIE, O. E. (1967), Studies on fossil aphids (Homoptera: Aphidoidea), *Spolia zool. Mus. hauniensis*, **26**, 274 pp.

HEINZE, K. (1960–61), Systematik der mitteleuropäischen Myzinae mit besonderer Berücksichtigung der im Deutschen Entomologischen Institut befindlichen Sammlung Carl Börner, *Beitr. Ent.*, **10**, 744–842; **11**, 24–96.

—— (1962), Pflanzenschädliche Blattläuse aus den Familien Lachnidae, Adelgidae und Phylloxeridae, eine systematisch-faunistische Studie, *Dt. ent. Z.*, (N.F.), **9**, 143–227.

HELMS, T. J. (1968), Postembryonic reproductive-systems development in *Empoasca fabae*, *Ann. ent. Soc. Am.*, **61**, 316–332.

HELMSING, I. W. AND CHINA, W. E. (1937), On the biology and ecology of *Hemiodoecus veitchi* Hacker. (Hemiptera: Peloridiidae), *Ann. Mag. nat. Hist.*, (10) **19**, 473–48.

HEPBURN, H. R. AND YONKE, T. R. (1971), The metathoracic scent glands of coreoid Heteroptera, *J. Kansas ent. Soc.*, **44**, 187–210.

HERRING, J. L. (1961), The genus *Halobates*, *Pacific Insects*, **3**, 223–305.

HESLOP-HARRISON, G. (1949), Subfamily separation in the Homopterous Psyllidae, I., *Ann. Mag. nat. Hist.*, (12) **2**, 802–810.

HESSE, A. J. (1947), A remarkable new dimorphic Isometopid and two other new species of Hemiptera predaceous upon the red scale of Citrus, *J. ent. Soc. sth. Afr.*, **10**, 31–45.

HEWSON, R. J. (1969), Some observations on flight in *Oncopeltus fasciatus* (Hemiptera: Lygaeidae), *J. ent. Soc. Br. Columbia*, **66**, 45–49.

HILL, A. R. (1961), The biology of *Anthocoris sarothamni* Douglas & Scott in Scotland (Hemiptera: Anthocoridae), *Trans. R. ent. Soc. Lond.*, **113**, 41–54.

HINDE, R. (1971), The fine structure of the mycetome symbiotes of the aphids *Brevicoryne brassicae*, *Myzus persicae*, and *Macrosiphum rosae*, *J. Insect Physiol.*, **17**, 2035–2050.

HINKS, C. F. (1966), The dorsal vessel and associated structures in some Heteroptera, *Trans. R. ent. Soc. Lond.*, **118**, 375–392.

HINTON, H. E. (1961), The structure and function of the egg-shell in the Nepidae (Hemiptera), *J. Insect Physiol.*, **7**, 224–257.

—— (1962), The structure of the shell and respiratory system of the eggs of

Helopeltis and related genera (Hemiptera, Miridae), *Proc. zool. Soc. Lond.*, **139**, 483–488.
—— (1969), Respiratory systems of insect egg shells, *A. Rev. Ent.*, **14**, 343–368.
HODKINSON, I. D. (1974), The biology of the Psylloidea (Homoptera): a review, *Bull. ent. Res.*, **64**, 325–339.
HOKE, S. (1926), Preliminary paper on the wing-venation of the Hemiptera, *Ann. ent. Soc. Am.*, **19**, 13–34.
HOLMGREN, N. (1899), Beiträge zur Kenntnis der weiblichen Geschlechtsorgane der Cicadarien, *Zool. Jb., Syst.*, **12**, 403–410.
HORVATH, G. (1904), Monographia Colobathristinarum, *Ann. mus. nat. Hist. Hung.*, **2**, 117–172.
—— (1911), Revision des Leptopodides, *Ann. Mus. nat. Hist. Hung.*, **9**, 358–370.
—— (1911*a*), Nomenclature des familles des Hémiptères, *Ann. Mus. nat. Hist. Hung.*, **9**, 1–34.
—— (1915), Monographie des Mésoveliides, *Ann. Mus. nat. Hist. Hung.*, **13**, 535–556.
—— (1929), Mesoveliidae, In: Horvath and Parshley, *Gen. Cat. Hemipt.*, **2**, 1–15.
HORVATH, G. AND PARSHLEY, H. M. (1927–36), *General Catalogue of the Hemiptera,* Smith College, Northampton, Mass.
HOTTES, F. C. (1928), Concerning the structure, function and origin of the cornicles of the family Aphididae, *Proc. biol. Soc. Washington*, **41**, 71–84.
HOTTES, F. C. AND FRISON, T. H. (1931), The plant-lice or Aphididae of Illinois, *Ill. nat. Hist. Surv. Bull.*, **19**, 121–447.
HOY, J. M. (1962), Eriococcidae (Homoptera: Coccoidea) of New Zealand, *Bull. D.S.I.R. New Zealand*, **146**, 219 pp.
—— (1963), A catalogue of the Eriococcidae (Homoptera: Coccoidea) of the world, *Bull. D.S.I.R. New Zealand*, **150**, 260 pp.
HUNGERFORD, H. B. (1917), The life history of *Mesovelia mulsanti* White, *Psyche*, **24**, 73–84.
—— (1919), The biology and ecology of aquatic and semi-aquatic Hemiptera, *Univ. Kansas Sci. Bull.*, **11**, 3–265.
—— (1922*a*), The Nepidae of North America north of Mexico, *Univ. Kansas Sci.*, **14**, 425–469.
—— (1922*b*), The life history of the toad bug, *Gelastocoris oculatus* Fabr. (Gelastocoridae), *Univ. Kansas Sci. Bull.*, **14**, 145–171.
—— (1948), The Corixidae of the Western Hemisphere (including a monograph on *Trichocorixa* by R. I. Sailer), *Univ. Kansas Sci. Bull.*, **32**, 1–827.
—— (1958), Some interesting aspects of the world distribution and classification of aquatic and semi-aquatic Hemiptera, *Proc. 10th int. Congr. Ent.*, **1**, 337–348.
HUNGERFORD, H. B. AND EVANS, N. E. (1934), The Hydrometridae of the Hungarian National Museum and other studies in the family (Hemiptera), *Ann. Mus. nat. Hist. Hung.*, **28**, 31–112.
HUNGERFORD, H. B. AND MATSUDA, R. (1960), Keys to subfamilies, tribes, genera and subgenera of the Gerridae of the world, *Kansas Univ. Sci. Bull.*, **41**, 3–23.
HUTCHINSON, G. E. (1929), A revision of the Notonectidae and Corixidae of South Africa, *Ann. S. Afr. Mus.*, **25**, 359–474.
IMMEL, R. (1955), Zur Biologie und Physiologie von *Reduvius personatus* L., *Z. Morph. Ökol. Tiere*, **44**, 163–195.

IMMS, A. D. (1914), Observations on the Homopterous insect *Phromnia* (*Flata*) *marginella* Oliv. in the Himalayas, *Mem. Manchester Lit. phil. Soc.*, **58**, sep. pp., 12.

IMMS, A. D. AND CHATTERJEE, N. C. (1915), On the structure and biology of *Tachardia lacca* Kerr, with observations on certain insects predaceous or parasitic upon it, *Ind. For. Mem.*, **3**, 1–42.

JACZEWSKI, T. (1934), Notes on the Old World species of Ochteridae (Hemiptera), *Ann. Mag. nat. Hist.*, (10) **13**, 597–613.

—— (1937), Allgemeine Züge der geographischen Verbreitung der Wasserhemipteren, *Arch. Hydrobiol.* **31**, 565–591.

JAKUBSKI, A. W. (1965), *A Critical Revision of the Families Margarodidae and Termitococcidae*, British Museum (Nat. Hist.), London, 187 pp.

JANSSON, A. (1973), Stridulation and its significance in the genus *Cenocorixa* (Hemiptera, Corixidae), *Behaviour*, **46**, 1–36.

JARIAL, M. S. AND SCUDDER, G. G. E. (1970), The morphology and ultrastructure of the Malpighian tubules and hind gut in *Cenocorixa bifida* (Hung.) (Hemiptera, Corixidae), *Z. Morph. Tiere*, **68**, 296–299.

JARIAL, M. S., SCUDDER, G. G. E. AND TERAGUCHI, S. (1969), Observations on the labium of Corixidae (Hemiptera), *Can. J. Zool.*, **47**, 713–715.

JEANNEL, R. (1919), Insectes Hémiptères. iii. Henicocephalidae et Reduviidae, *Voy. Alluaud et Jeannel*, Paris, 113–313.

—— (1942), Les Hénicocéphalides. Monographie d'un groupe d'Hémiptères hématophages, *Annls Soc. ent. Fr.*, **110**, 273–386.

JOHANSSON, A. S. (1957), The nervous system of the milkweed bug, *Oncopeltus fasciatus* (Dallas) (Heteroptera, Lygaeidae), *Trans. Am. ent. Soc.*, **83**, 119–183.

—— (1958), Relation of nutrition to endocrine-reproductive functions in the milkweed bug, *Oncopeltus fasciatus* (Dallas), *Nytt Mag. Zool.*, **7**, 1–132.

JOHNSON, C. G. (1936), The biology of *Leptobyrsa rhododendri* Horváth (Hemiptera, Tingididae), the Rhododendron lace-bug, *Ann. appl. Biol.*, **23**, 342–368.

—— (1942), The ecology of the bed-bug, *Cimex lectularius*, L., in Britain, Report on research, 1935–40, *J. Hyg., Camb.*, **41**, 345–461.

—— (1954), Aphid migration in relation to weather, *Biol. Rev.*, **29**, 87–118.

—— (1969), *Migration and Dispersal of Insects by Flight*, Methuen, London, 763 pp.

JOHNSTON, C. E. (1912), The internal anatomy of *Icerya purchasi*, *Ann. ent. Soc. Am.*, **5**, 383–388.

JONES, M. G. (1942), The summer hosts of *Aphis fabae* Scop., *Bull. ent. Res.*, **33**, 161–165.

JORDAN, K. H. C. (1928), Zur Biologie der aquatilen Rhynchoten, *Isis Budissina*, **11**, 142–167.

—— (1958*a*), Lautäusserungen bei den Hemipteren-Familien der Cydnidae, Pentatomidae und Acanthosomatidae, *Zool. Anz.*, **161**, 130–144.

—— (1958*b*), Die Biologie von *Elasmucha grisea* (Heteroptera: Acanthosomidae), *Beitr. Ent.*, **8**, 385–397.

—— (1972), 20. Heteroptera (Wanzen), In: Helmcke, J.-G., Starck, D. and Wermuth, H. (eds.), *Handbuch der Zoologie*, **4** (2), Lfg. 16, 113 pp.

JORDAN, K. H. C. AND WENDT, A. (1938), Zur Biologie von *Salda littoralis* L. (Hem. Het.), *Stettin ent. Ztg.*, **99**, 273–292.

KARNY, H. H. (1934), *Biologie der Wasserinsekten, Ein Lehr- und Nachschlagbuch über die wichtigsten Ergebnisse der Hydro-Entomologie,* Wagner, Wien, 311 pp.

KELLEN, W. R. (1959), Notes on the biology of *Halovelia marianaram* Usinger in Samoa (Veliidae: Heteroptera), *Ann. ent. Soc. Am.*, **52**, 53–62.

—— (1960), A new species of *Omania* from Samoa with notes on its biology (Heteroptera: Saldidae), *Ann. ent. Soc. Am.*, **53**, 494–499.

KELTON, L. A. (1959), Male genitalia as taxonomic characters in the Miridae (Hemiptera), *Can. Ent.*, **91** (Suppl. 11), 72 pp.

KENAGA, E. E. (1941), The genus *Telmatometra* Bergroth (Hemiptera–Gerridae), *Kansas Univ. Sci. Bull.*, **27**, 169–183.

KENNEDY, J. S. AND BOOTH, C. O. (1951), Host alternation in *Aphis fabae* Scop. I. Feeding preferences and fecundity in relation to the age and kind of leaves, *Ann. appl. Biol.*, **38**, 25–64.

KENNEDY, J. S., DAY, M. F. AND EASTOP, V. F. (1962), *A Conspectus of Aphids as Vectors of Plant Viruses*, Commonwealth Institute of Entomology, London, 114 pp.

KENNEDY, J. S., IBBOTSON, A. AND BOOTH, C. O. (1950), The distribution of aphid infestation in relation to leaf age, I. *Myzus persicae* (Sulz.) and *Aphis fabae* Scop. on spindle trees and sugar beet plants, *Ann. appl. Biol.*, **37**, 651–679.

KENNEDY, J. S. AND STROYAN, H. L. G. (1959), Biology of aphids, *A. Rev. Ent.*, **4**, 139–160.

KERSHAW, J. C. W. (1907), Life history of *Tessaratoma papillosa* Thunb. with notes on the stridulating organ and stink-glands by Frederick Muir, *Trans. ent. Soc. Lond.*, **1907**, 253–258.

KERSHAW, J. C. W. AND KIRKALDY, G. W. (1908*a*), Biological notes on Oriental Hemiptera, *J. Bombay nat. Hist. Soc.*, **18**, 596–598.

—— (1908*b*), On the metamorphoses of two Hemiptera–Heteroptera from southern China, *Trans. ent. Soc. Lond.*, **1908**, 59–62.

KIRITSCHENKO, A. N. (1951), True Hemiptera of the European USSR, Key and bibliography (in Russian), *Opred. Faune SSSR, Moscow*, **42**, 1–423.

KIRKALDY, G. W. (1908), Some remarks on the phylogeny of the Hemiptera Heteroptera, *Can. Ent.*, **40**, 357–364.

—— (1909), *Catalogue of the Hemiptera (Heteroptera)*, Vol. I. Cimicidae [= Pentatomidae], Dames, Berlin, 392 pp.

KNIGHT, H. H. (1941), The plant bugs or Miridae of Illinois, *Bull. Ill. nat. Hist. Surv.*, **22**, 1–234.

KORMILEV, N. A. (1955), A new myrmecophil family of Hemiptera from the delta of Rio Paraná, Argentina, *Rev. ecuator. Ent. Parasit.*, **2**, 465–477.

KORNHAUSER, S. I. (1919), The sexual characteristics of the Membracid, *Thelia bimaculata* Fabr., *J. Morph.*, **32**, 531–636.

KRAMER, H. (1935), Beiträge zur Biologie von *Naucoris* mit besonderer Berücksichtigung der Atmung, *Arch. Hydrobiol.*, **28**, 523–534.

KRAMER, S. (1950), The morphology and phylogeny of Auchenorhynchous Homoptera (Insecta), *Illinois biol. Monogr.*, **20** (4), 111 pp.

KRAUS, C. (1957), Versuch einer morphologischen und neurophysiologischen Analyse des Stechaktes von *Rhodnius prolixus* Stål 1858, *Acta tropica*, **14**, 35–87.

KÜHN, O. (1926), Die Facettenaugen der Landwanzen und Zikaden, *Z. Morph. Ökol. Tiere*, **5**, 489–558.

KULLENBERG, B. (1944), Studien über die Biologie der Capsiden, *Zool. Bidr. Upps.*, **23**, 1–522.

—— (1947), Über Morphologie und Funktion des Kopulationsapparats der Capsiden und Nabiden, *Zool. Bidr. Upps.*, **24**, 217–418.

KUMAR, R. (1965–71), Morphology and relationships of the Pentatomoidea, I–V, *J. ent. Soc. Queensland*, **4**, 41–55; *Dt. ent. Z.*, **17**, 1–32 (with M. S. K. Ghauri); *Ann. ent. Soc. Am.*, **62**, 681–695; *Aust. J. Zool.*, **17**, 553–606; *Am. Midl. Nat.*, **85**, 63–73.

—— (1967), Morphology of the reproductive and alimentary systems of the Aradoidea (Hemiptera) with comments on relationships within the superfamily, *Ann. ent. Soc. Am.*, **60**, 17–25.

—— (1968), Aspects of the morphology and relationships of the superfamilies Lygaeoidea, Piesmatoidea and Pyrrhocoroidea (Hemiptera: Heteroptera), *Ent. mon. Mag.*, **103**, 251–261.

—— (1974), A revision of world Acanthosomatidae (Heteroptera: Pentatomoidea): keys to and descriptions of subfamilies, tribes and genera, with designation of types, *Aust. J. Zool.*, *Suppl.*, **34**, 1–60.

KUNKEL, H. AND KLOFT, W. (1974), Polymorphismus bei Blattläusen, In: Schmidt, G. H. (ed.), *Sozialpolymorphismus bei Insekten*, 152–201, Wissenschaftliche Verlagsgesellschaft, Stuttgart.

LAL, K. B. (1934), The biology of Scottish Psyllidae, *Trans. R. ent. Soc. Lond.*, **82**, 363–385.

LALLEMAND, V. (1912), Cercopidae, *Genera Insect.*, **143**, 167 pp.

—— (1949, 1961), Revision des Cercopinae (Hemiptera Homoptera). I, II, *Mém. Inst. r. Sci. nat. Belg.*, (2), **32**, 193 pp.; (2), **66**, 1–153 (with H. Synave).

LAMB, R. J. AND POINTING, P. J. (1972), Sexual morph determination in the aphid *Acyrthosiphon pisum*, *J. Insect Physiol.*, **18**, 2029–2042.

LAMBORN, W. A. *et al.* (1914), On the relationship between certain West-African insects, especially ants, Lycaenidae and Homoptera, *Trans. ent. Soc. Lond.*, **1913**, 436–524.

LANHAM, U. N. (1968), The Blochman bodies: hereditary intracellular symbionts in insects, *Biol. Rev.*, **43**, 269–286.

LANSBURY, I. (1965–66), A revision of the Stenocephalidae Dallas 1852 (Hemiptera–Heteroptera), *Ent. mon. Mag.*, **101**, 52–92; 145–160.

LARSÉN, O. (1927), Ueber die Entwicklung und Biologie von *Aphelocheirus aestivalis* Fabr., *Ent. Tidskr.*, **48**, 181–206.

—— (1930), Biologische Beobachtungen an schwedischen *Notonecta*-Arten, *Ent. Tidskr.*, **51**, 219–247.

—— (1931*a*), Beitrag zur Kenntnis des Pterygopolymorphismus bei den Wasserhemipteren, *Acta Univ. Lund* (*N.F.*), **27**, 30 pp.

—— (1931*b*), Beiträge zur Oekologie und Biologie von *Aphelocheirus aestivalis* Fabr., *Int. Rev. Hydrobiol.*, **26**, 1–19.

—— (1937), Zur Biologie von *Ranatra linearis*, *Opusc. ent.*, **1**, 112–119.

—— (1945*a*), Der Thorax der Heteropteren – Skelett und Muskulatur, *Acta Univ. Lund.* (*N.F.*), **41** (3), 96 pp.

—— (1945*b*), Das thorakale Skelettmuskelsystem der Heteropteren, Ein Beitrag zur vergleichenden Morphologie des Insektenthorax, *Acta Univ. Lund.* (*N.F.*), **41** (11), 83 pp.

—— (1950), Die Veränderungen im Bau der Heteropteren bei der Reduktion des Flugapparates, *Opusc. ent.*, **15**, 17–51.

LAUCK, D. R. (1959), The locomotion of *Lethocerus* (Hemiptera–Belostomatidae), *Ann. ent. Soc. Am.*, **52**, 93–99.

LAUCK, D. R. AND MENKE, A. S. (1961), The higher classification of the Belostomatidae (Hemiptera), *Ann. ent. Soc. Am.*, **54**, 644–657.

LAVOIPIERRE, M. M. J., DICKERSON, G. AND GORDON, R. M. (1959), Studies on the methods of feeding of blood-sucking arthropods, I: The manner in which triatomine bugs obtain their blood-meal, as observed in the tissues of the living rodent, with some remarks on the effects of the bite on human volunteers, *Ann. trop. Med. Parasit.*, **53**, 235–250.

LEES, A. D. (1959–63), The role of photoperiod and temperature in the determination of parthenogenetic and sexual forms in the aphid *Megoura viciae* Buckton, I–III, *J. Insect Physiol.*, **3**, 92–117; **4**, 154–175; **9**, 153–164.

—— (1961), Clonal polymorphism in aphids, *Symp. R. ent. Soc. Lond.*, **1**, 68–79.

—— (1966), The control of polymorphism in aphids, *Adv. Insect Physiol.*, **3**, 207–277.

LEFROY, H. M. (1909), *Indian Insect Life*, Thacker, Spink & Co., Calcutta and Simla, 786 pp.

LEONARDI, G. (1920), *Monografia delle cocciniglia italiane*, Della Torre, Portici, 555 pp.

LE QUESNE, W. J. (1960–69), Hemiptera Cicadomorpha; Hemiptera Fulgoromorpha, *R. ent. Soc. Handb. Ident. Brit. Insects*, **II** (2a), 1–64; **II** (2b), 65–148; **II** (3), 1–68.

LESTON, D. (1951), Alary dimorphism in *Nabis apterus* F. (Hem., Nabidae) and *Coranus subapterus* Deg., (Hem., Reduviidae), *Ent. mon. Mag.*, **87**, 242–244.

—— (1953), 'Phloeidae' Dallas: systematics and morphology, with remarks on the phylogeny of 'Pentatomoidea' Leach and upon the position of 'Serbana' Distant (Hemiptera), *Rev. brasil. Biol.*, **13**, 121–140.

—— (1957), The stridulatory mechanism in terrestrial species of Hemiptera Heteroptera, *Proc. zool. Soc. Lond.*, **128**, 369–386.

—— (1958*a*), Higher systematics of shield bugs (Hem. Pentatomidae), *Trans. 10th int. Congr. Ent.*, **1**, 325.

—— (1958*b*), Chromosome number and the systematics of Pentatomomorpha, *Trans. 10th int. Congr. Ent.*, **2**, 911–918.

—— (1961), Testis follicle number and the higher systematics of Miridae (Hemiptera–Heteroptera), *Proc. zool. Soc. Lond.*, **137**, 89–106.

—— (1962), Tracheal capture in ontogenetic and phylogenetic phases of insect wing development, *Proc. R. ent. Soc. Lond.*, (A), **37**, 135–144.

LESTON, D., PENDERGRAST, J. G. AND SOUTHWOOD, T. R. E. (1954), Classification of the terrestrial Heteroptera (Geocorisae), *Nature*, **174**, 91–92.

LESTON, D. AND PRINGLE, J. W. S. (1964), Acoustical behaviour of Hemiptera, In: Busnel, R.-G. (ed.), *Acoustic Behaviour of Animals*, Elsevier, Amsterdam, pp. 391–411.

LESTON, D. AND SCUDDER, G. G. E. (1956), A key to the larvae of the families of British Hemiptera–Heteroptera, *Entomologist*, **89**, 223–231.

LETHIERRY, L. AND SEVERIN, G. (1893–96), *Catalogue génerale des Hémiptères*, Bruxelles, **1**, 286 pp.; **2**, 277 pp.; **3**, 275 pp.

LICENT, P. (1912), Recherches d'anatomie et de physiologie sur la tube digestif des Homoptères supérieurs, *Cellule*, **28**, 1–161.

LIENHART, R. (1913), Habitat et géonomie d'*Aepophilus bonnairei* Signoret, *Annls Sci. nat.* (*Zool.*), **17**, 257–268.

LINDINGER, L. (1912), *Die Schildläuse* (*Cocciden*) *Europas, Nordafrikas und Vorderasiens, einschliesslich der Azoren, der Kanarien und Madeiras*, Stuttgart, 388 pp.

LINDSAY, K. L. (1969), Cornicles of the pea aphid, *Acyrthosiphon pisum* (Hemiptera, Homoptera, Aphididae); their structure and function. A light- and electron-microscope study, *Ann. ent. Soc. Am.*, **62**, 1015–1021.

LIST, J. H. (1885), *Orthezia cataphracta* Shaw, Eine Monographie, *Z. wiss. Zool.*, **45**, 1–86.

LO, S. E. AND ACTON, A. B. (1969), The ultrastructure of the rostral sensory organs of the water bug, *Cenocorixa bifida* (Hungerford) (Hemiptera), *Can. J. Zool.*, **47**, 717–722.

LOWE, A. D. (ed.) (1973), Perspectives in aphid biology, *Bull. ent. Soc. New Zealand*, **2**, 123 pp.

MAA, T. C. (1963), A review of the Machaerotidae (Hemiptera: Cercopoidea), *Pacific. Ins. Monogr.*, **5**, 1–166.

MACAN, T. T. (1938), The evolution of aquatic habitats with special reference to the distribution of Corixidae, *J. Anim. Ecol.*, **7**, 1–19.

—— (1954), A contribution to the study of the ecology of Corixidae (Hemipt.), *J. Anim. Ecol.*, **23**, 115–141.

—— (1965), A revised key to the British water bugs (Hemiptera-Heteroptera), 2nd edn., *Scient. Publs Freshwat. biol. Ass.*, **16**, 78 pp.

MACGILLIVRAY, A. D. (1921), *The Coccidae*, Scarab Publ. Co., Urbana, Illinois, 502 pp.

MACKAUER, M. (1965), Parasitological data as an aid in aphid classification, *Can. Ent.*, **97**, 1016–1024.

MAILLET, P. (1957), Contribution à l'étude de la biologie du Phylloxera de la vigne, *Annls Sci. nat.* (*Zool.*), **19**, 283–410.

MÄKEL, M. (1942), Metamorphose und Morphologie des *Pseudococcus*-Männchens mit besonderer Berücksichtigung des Skelettmuskelsystems, *Zool. Jb., Anat.*, **67**, 461–512.

MAMMEN, H. (1912), Über die Morphologie der Heteropteren- und Homopterenstigmen, *Zool. Jb., Anat.*, **34**, 121–178.

MANNA, G. K. (1958), Cytology and inter-relationships between various groups of Heteroptera, *Proc. 10th int. Congr. Ent.*, **2**, 919–934.

—— (1962), A further evaluation of the cytology and inter-relationships between various groups of Heteroptera, *Nucleus*, **5**, 7–28.

MARAMOROSCH, K. (ed.) (1969), *Viruses, Vectors and Vegetation*, Wiley, New York, 666 pp.

MARCHAL, P. (1913), Contribution à l'étude de la biologie des *Chermes*, *Annls Sci. nat.* (*Zool.*), **18** (11), 153–385.

MARKS, E. P. (1951), Comparative studies of the male genitalia of the Hemiptera (Homoptera-Heteroptera), *J. Kansas ent. Soc.*, **24**, 134–141.

—— (1959), A study of the midgut epithelium of five water bugs (Hemiptera: Cryptocerata), *J. Kansas ent. Soc.*, **32**, 77–83.

MARLATT, C. L. (1907), The periodical cicada, *Bull. U.S. Dept. Agric., Bur. Ent.*, (N.S.), **71**, 1–181.

MARSHALL, A. T. (1964–65), Spittle production and tube building by cercopoid nymphs (Homoptera), 1–3, *Q. Jl microsc. Sci.*, **105**, 257–262; 415–422; **106**, 37–44.

MARSHALL, A. T. AND CHEUNG, W. W. K. (1973), Calcification in insects: the dwelling-tube and midgut of Machaerotid larvae (Homoptera), *J. Insect Physiol.*, **19**, 963–972.

MATSUDA, R. (1960), Morphology, evolution and a classification of the Gerridae (Hemiptera-Heteroptera), *Kansas Univ. Sci. Bull.*, **41**, 25–632.

—— (1962), Morphology and evolution of the pleurosternal region of the pterothorax in Notonectidae and related families, *J. Kansas ent. Soc.*, **35**, 235–242.

MCATEE, W. L. (1919), Key to the Nearctic genera and species of Berytidae, *Jl N.Y. ent. Soc.*, **27**, 79–92.

MCATEE, W. L. AND MALLOCH, J. R. (1925), Revision of Cryptostemmatidae in the United States National Museum, *Proc. U.S. natn. Mus.*, **67**, 1–42.

—— (1933), Revision of the subfamily Thyreocorinae of the Pentatomidae (Hemiptera–Heteroptera), *Ann. Carnegie Mus. Pittsburg*, **21**, 191–412.

MCDONALD, F. J. D. (1970), The morphology of *Lestonia haustorifera* China (Heteroptera: Lestoniidae), *J. nat. Hist.*, **4**, 413–417.

MCGREGOR, E. A. AND MCDONOUGH, F. L. (1917), The red spider on cotton, *Bull. U.S. Dep. Agric.*, **416**, 1–72.

MCKENZIE, H. (1956), The armoured scale Insects of California, *Bull. Calif. Ins. Survey*, **5**, 209 pp.

—— (1967), *Mealy-bugs of California, with taxonomy, biology and control of North American species* (*Homoptera: Coccoidea: Pseudococcidae*), University of California Press, Berkeley, 525 pp.

MELICHAR, L. (1896), *Cicadinen* (*Hemiptera Homoptera*) *von Mitteleuropa*, Berlin, 364 pp.

—— (1906), Monographie der Issiden, *Abh. zool. bot. Ges., Wien*, **3** (4), 1–327.

—— (1912), Monographie der Dictyopharinen, *Abh. zool. bot. Ges., Wien*, **7** (1), 1–221.

—— (1915*a*), Monographie der Lophopinen, *Ann. Mus. nat. Hung.*, **13**, 337–385.

—— (1915*b*), Monographie der Tropiduchinen, *Verh. naturf. Ver. Brünn*, **53**, 82–225.

—— (1923), Acanaloniidae, Flatidae et Ricaniidae, *Genera Insect.*, Fasc. **182**, 185 pp.

MELLANBY, H. (1935), A comparison of the physiology of the two species of bed bugs which attack man, *Parasitology*, **27**, 111–122.

METCALF, Z. P. (1927–47), *General Catalogue of the Hemiptera*, Fasc. I–V: 581 pp. (1927); 24 pp. (1929); 144 pp. (1929); 1775 pp. (1932–47); 18 pp. (1947); (with contributions by W. D. Funkhouser *et al.*).

—— (1932), Fulgoroidea, Tettigometridae, In: China, W. E. and Parshley, H. M. (eds.), *General Catalogue of the Hemiptera*, **4** (1), 69 pp.

—— (1938), The Fulgorina of Barra Colorado and other parts of Panama, *Bull. Mus. comp. Zool. Harv.*, **82**, 275–423.

—— (1945), *A Bibliography of the Homoptera* (*Auchenorrhyncha*), Univ. N. Carolina, North Carolina, **1**, 886 pp.; **2**, 186 pp.

—— (1951), Phylogeny of the Homoptera Auchenorrhyncha, *Comment. biol.*, **12**, 1–14.

—— (1954–71), *General Catalogue of the Homoptera*, Fasc. IV–VIII; in parts, with contributions by V. Wade, and including indices and bibliographies.
—— (1960), A bibliography of the Cercopoidea (Homoptera: Auchenorrhyncha), *Pap. ent. Dept. N.C. agric. Exp. Sta., Raleigh*, **1135**, 1–262.
—— (1962), A bibliography of the Cicadoidea (Homoptera: Auchenorrhyncha), *N. Carolina St. Coll. Pap.*, **1373**, 229 pp.
METCALF, Z. P. AND WADE, V. (1963), *A bibliography of the Membracoidea and fossil Homoptera*, Raleigh, N. Carolina State University, 200 pp.
—— (1965), *General Catalogue of the Homoptera. Membracoidea*, U.S.D.A., Raleigh, N.C., 2 vols., 1–743; 745–1552.
—— (1966), A catalogue of the fossil Homoptera, *Gen. Cat. Homopt.*, Fasc. I (Suppl.), 245 pp.
MILES, P. W. (1967), The physiological division of labour in the salivary glands of *Oncopeltus fasciatus* (Dall.) (Heteroptera: Lygaeidae), *Aust. J. biol. Sci.*, **20**, 785–797.
—— (1968), Insect secretions in plants, *A. Rev. Phytopathol.*, **6**, 137–164.
—— (1972), The saliva of Hemiptera, *Adv. Insect Physiol.*, **9**, 183–255.
MILLER, N. C. E. (1942), On the structure of the legs in Reduviidae (Hemiptera-Heteroptera), *Proc. R. ent. Soc. Lond.*, (A), **17**, 49–58.
—— (1971), *Biology of the Heteroptera*, Classey, Hampton, Mddx., 2nd edn., 206 pp.
MITIS, H. von (1935), Zur Biologie der Corixiden. Stridulation, *Z. Morph. Ökol. Tiere*, **30**, 479–495.
—— (1937), Oekologie und Larvenentwicklung der mitteleuropäischen *Gerris*-Arten (Heteroptera), *Zool. Jb., Syst.*, **69**, 337–372.
MITTLER, T. E. (1957), Studies on the feeding and nutrition of *Tuberolachnus salignus* (Gmelin) (Homoptera, Aphididae), I. The uptake of phloem sap, *J. exp. Biol.*, **34**, 334–341.
—— (1973), Aphid polymorphism as affected by diet, In: Lowe, A. D. (ed.), *Perspectives in Aphid Biology*, pp. 65–75.
MIYAMOTO, S. (1957), List of ovariole numbers in Japanese Heteroptera, *Sieboldia*, **2**, 69–82 (additions and corrections, *ibid.*, **2**, 121–123).
—— (1961), Comparative morphology of alimentary organs of Heteroptera, with the phylogenetic consideration, *Sieboldia*, **2**, 197–259.
MOERICKE, V. AND WOHLFARTH-BOTTERMAN, K. E. (1960), Zur funktionellen Morphologie der Speicheldrüsen von Homopteren, I, IV, *Z. Zellforsch.*, **51**, 157–184; **53**, 25–49.
MONOD, T. AND CARAYON, J. (1958), Observations sur les *Copium* (Hemiptera-Tingidae) et leur action cécidogène sur les fleurs de *Teucrium* (Labiées), *Archs Zool. exp. gén.*, **95**, 1–31.
MOORE, T. E. (1961), Audiospectrographic analysis of sounds of Hemiptera and Homoptera, *Ann. ent. Soc. Am.*, **54**, 273–291.
MOORE, T. E. AND ALEXANDER, R. D. (1958), The periodical cicada complex (Homoptera: Cicadidae), *Proc. 10th int. Congr. Ent.*, **1**, 349–355.
MORGAN, T. H. AND SHULL, A. F. (1910), The life-cycle of *Hormaphis hammamelidis*, *Ann. ent. Soc. Am.*, **3**, 144–146.
MORRILL, A. W. (1910), Plant bugs injurious to cotton bolls, *Bull. U.S. Dept. Agric., Bur. Ent.*, **86**, 1–110.

MORRILL, A. W. AND BACK, E. A. (1911), White-flies injurious to citrus in Florida, *Bull. U.S. Dept. Agric., Bur. Ent.*, **92**, 1–109.

MORRISON, H. (1928), A classification of the higher groups and genera of the coccid family Margarodidae, *Tech. Bull. U.S. Dep. Agric.*, **52**, 1–240.

—— (1952), Classification of the Ortheziidae, *Tech. Bull. U.S. Dep. Agric.*, **1052**, 80 pp.

MORRISON, H. AND MORRISON, E. R. (1965), A selected bibliography of the Coccoidea, First supplement, *Misc. Publs U.S. Dep. Agric.*, **987**, 1–44.

—— (1966), An annotated list of generic names of the scale insects (Homoptera: Coccoidea), *Misc. Publs U.S. Dep. Agric.*, **1015**, 206 pp. (Suppl., 13 pp., by L. M. Russell, 1970).

MORRISON, H. AND RENK, A. V. (1957), A selected bibliography of the Coccoidea, *Misc. Publs U.S. Agric., Agric. Res. Serv.*, **734**, 222 pp.

MOUND, L. A. (1965), An introduction to the Aleyrodidae of Western Africa (Homoptera), *Bull. Br. Mus. nat. Hist.*, (Ent.), **17**, 115–160.

—— (1966), A revision of the British Aleyrodidae (Hemiptera: Homoptera), *Bull. Br. Mus. nat. Hist.*, (Ent.), **17** (9), 399–428.

MUIR, F. (1907), Notes on the stridulating organ and stink glands of *Tessaratoma papillosa*, *Trans. ent. Soc. Lond.*, **1907**, 256–258.

—— (1915), A contribution towards the taxonomy of the Delphacidae, *Can. Ent.*, **47**, 208–212; 261–270; 296–302; 317–320.

—— (1918), Notes on the Derbidae in the British Museum collection, *Ent. mon. Mag.*, **54**, 173–177; 202–207; 228–243.

—— (1925), On the genera of Cixiidae, Meenoplidae and Kinnaridae (Fulgoroidea, Homoptera), *Pan-Pacific Ent.*, **1**, 97–110; 156–163.

—— (1930), On the classification of the Fulgoroidea, *Ann. Mag. nat. Hist.*, (10), **6**, 461–478.

—— (1931), In: Williams, F. X. (ed.), *The Insects and other Invertebrates of Hawaiian Sugar Cane Fields*, Hawaiian Sugar Planters' Assoc., Hawaii, 400 pp.

MÜLLER, H. J. (1942), Ueber Bau und Funktion des Legeapparates der Zikaden (Homoptera, Cicadina), *Z. Morph. Ökol. Tiere*, **38**, 534–629.

MUNDINGER, F. G. (1922), The life history of two species of Nabidae (Hemip. Heterop.), *Nabis roseipennis*, Reut. and *Nabis rufusculus*, Reut., *N.Y. St. Coll. For., Tech. Publ.*, **16**, 149–167.

MUNK, R. (1967), Zur Morphologie und Histologie des Verdauungstraktes zweier Jassiden (Homoptera Auchenorrhyncha) unter besonderer Berücksichtigung der sog. Filterkammer, *Z. wiss. Zool.*, **175**, 405–424.

—— (1968*a*), Über den Feinbau der Filterkammer der Kleinzikade *Euscelidius variegatus* Kbm (Jassidae), *Z. Zellforsch.*, **85**, 210–224.

—— (1968*b*), Die Richtungen des Nahrungsflusses im Darmtrakt der Kleinzikade *Euscelidius variegatus* Kbm (Jassidae), *Z. vergl. Physiol.*, **58**, 422–428.

MYERS, J. G. (1928), The morphology of the Cicadidae (Homoptera), *Proc. zool. Soc. Lond.*, **1928**, 365–472.

—— (1929), *Insect Singers: a Natural History of the Cicadas*, Routledge, London, 304 pp.

NAST, J. (1972), *Palaearctic Auchenorrhyncha (Homoptera). An annotated check list*, Polish Sci. Publ., Warsaw, 550 pp.

NAULT, L. R., EDWARDS, L. J. AND STYER, W. E. (1973), Aphid alarm pheromones; secretion and reception, *Environmental Ent.*, **2**, 101–105.

NEWCOMER, W. S. (1948), Embryological development of the mouthparts and related structures of the milkweed bug, *Oncopeltus fasciatus* (Dallas), *J. Morph.*, **82**, 365–411.

NEWSTEAD, R. (1901–03), *Monograph of the Coccidae of the British Isles*, Ray Soc., London, **1**, 1–220; **2**, 1–270.

NIELSON, M. W. (1968), The leafhopper vectors of phytopathogenic viruses (Homoptera, Cicadellidae): taxonomy, biology and virus transmission, *Tech. Bull. U.S. Dep. Agric.*, **1382**, 386 pp.

NIXON, G. E. J. (1951), *The Association of Ants with Aphids and Coccids*, Commonwealth Inst. Ent., London, 36 pp.

OESTLUND, O. W. (1942), *Systema Aphididae. A guide to the phylogeny of the aphids or plant lice, Part I, The Lachnea*, Augustana Book Concern, Illinois, 78 pp.

OEVERMANN, H. (1936), Das statische Verhalten einiger Wasserwanzenarten, *Z. wiss. Zool.*, **147**, 595–628.

OGUMA, K. (1919), A new scale insect, *Xylococcus alni*, on alder, with special reference to its metamorphosis and anatomy, *J. Coll. Agric. imp. Univ. Hokkaido*, **8**, 77–109.

OHM, D. (1956), Beiträge zur Biologie der Wasserwanze *Aphelocheirus aestivalis* F., *Zool. Beitr.*, **2**, 359–386.

OMAN, P. W. (1949), The Nearctic leafhoppers (Homoptera: Cicadellidae), A generic classification and check list, *Mem. ent. Soc. Washington*, **3**, 253 pp.

ORDISH, G. (1972), *The Great Wine Blight*, London.

ORIAN, A. J. E. (1964), The morphology of the male genitalia of *Abricta ferruginosa* (Stål) (Homoptera: Cicadidae), *Proc. R. ent. Soc. Lond.*, (A), **39**, 1–4.

OSBORN, H. (1916), Studies of the life histories of the froghoppers of Maine, *Bull. Maine agric. Exp. Stn*, **254**, 265–288.

OSHANIN, B. (1906–10), Verzeichnis der paläarktischen Hemipteren mit besonderer Berücksichtigung ihrer Verteilung im russischen Reiche, *Ann. Mus. Zool. Acad. Sci. St. Petersburg*, 11–15. Bd. 1, Heteroptera, 1087 pp; Bd. 2, Homoptera, 492 pp.

—— (1912), *Katalog der paläarktischen Hemipteren (Heteroptera, Homoptera-Auchenorrhyncha und Psylloidea)*, Friedländer, Berlin, 187 pp.

—— (1916), Vadé mécum destiné à faciliter la détermination des Hémiptères, *Horae Soc. ent. Ross.*, **42**, 1–106.

OSSIANNILSSON, F. (1949), Insect drummers. A study on the morphology and function of the sound-producing organ of Swedish Homoptera-Auchenorrhyncha with notes on their sound production, *Opusc. ent. Suppl.*, **10**, 145 pp.

PARNELL, F. R., KING, H. E. AND RUSTON, D. F. (1949), Jassid resistance and hairiness of the cotton plant, *Bull. ent. Res.*, **39**, 539–575.

PARSHLEY, H. M. (1925), *Bibliography of North American Hemiptera-Heteroptera*, Smith College, Northampton, Mass., 252 pp.

PARSONS, M. C. (1959*a*), The mid gut of aquatic Hemiptera, *J. Morph.*, **104**, 479–525.

—— (1959*b*), The presence of a peritrophic membrane in some aquatic Heteroptera, *Psyche*, **64**, 117–122.

—— (1959*c*), Skeleton and musculature of the head of *Gelastocoris oculatus* (Fabricius) (Hemiptera-Heteroptera), *Bull. Mus. comp. Zool. Harv.*, **122**, 1–53.

—— (1960), The nervous system of *Gelastocoris oculatus* (Fabricius) (Hemiptera-Heteroptera), *Bull. Mus. comp. Zool. Harv.*, **123**, 131–199.

—— (1962), Skeleton and musculature of the head of *Saldula pallipes* (F.) (Heteroptera: Saldidae), *Trans. R. ent. Soc. Lond.*, **114**, 97–130.

—— (1963*a*), Thoracic skeleton and musculature of adult *Saldula pallipes* (F.) (Heteroptera: Saldidae), *Trans. R. ent. Soc. Lond.*, **115**, 1–37.

—— (1963*b*), The endoskeletal salivary pumping apparatus in representative Belostomatidae (Heteroptera), *Can. J. Zool.*, **41**, 1017–1024.

—— (1964), The origin and development of the Hemipteran cranium, *Can. J. Zool.*, **42**, 409–432.

—— (1966), Modifications of the food pumps of Hydrocorisae (Heteroptera), *Can. J. Zool.*, **44**, 585–620.

—— (1967), Modifications of the prothoracic pleuron in Hydrocorisae (Heteroptera), *Trans. R. ent. Soc. Lond.*, **119**, 215–234.

—— (1969), Skeletomusculature of the pterothorax and first abdominal segment in micropterous *Aphelocheirus aestivalis* F. (Heteroptera: Naucoridae), *Trans. R. ent. Soc. Lond.*, **121**, 1–39.

—— (1970), Respiratory significance of the thoracic and abdominal morphology of the three aquatic bugs *Ambrysus, Notonecta* and *Hesperocorixa* (Insecta: Heteroptera), *Z. Morph. Tiere*, **66**, 242–298.

—— (1971), The lateral thoracico-abdominal region in adults and fifth instar nymphs of an aquatic bug, *Notonecta undulata* Say (Insecta Heteroptera), *Z. Morph. Tiere*, **69**, 82–114.

—— (1972), Fine structure of the triturating devices in the food pump of *Notonecta* (Heteroptera: Notonectidae), *J. Morph.*, **138**, 141–167.

PATCH, E. M. (1909), Homologies of the wing veins of the Aphididae, Psyllidae, Aleyrodidae and Coccidae, *Ann. ent. Soc. Am.*, **2**, 101–135.

—— (1938), Food-plant catalogue of the aphids of the world, *Bull. Maine agric. Exp. Stn*, **393**, 1–431.

PEARSON, E. O. (1958), *The Insect Pests of Cotton in Tropical Africa*, London, Commonwealth Institute of Entomology, 355 pp.

PENDERGRAST, J. G. (1957), Studies on the reproductive organs of the Heteroptera with a consideration of their bearing on classification, *Trans. R. ent. Soc. Lond.*, **109**, 1–63.

—— (1962), The internal anatomy of the Peloridiidae (Homoptera: Coleorrhyncha), *Trans. R. ent. Soc. Lond.*, **114**, 49–65.

PÉRICART, J. (1972), Hémiptères Anthocoridae, Cimicidae et Microphysidae de l'ouest-paléarctique, *Faune de l'Europe et du Bassin méditerranéen*, **7**, 402 pp.

PESSON, P. (1942), Contribution à l'étude du tube digestif des Coccides, IV. Diaspinae, *Bull. Soc. zool. Fr.*, **66**, 230–238.

—— (1944), Contribution à l'étude morphologique et fonctionelle de la tête, de l'appareil buccale et du tube digestif des femelles de Coccides, *Monogr. Sta. Lab. Rech. agron. Paris*, **1944**, 266 pp.

—— (1951), Ordre des Homoptères. In: Grassé, *Traité de Zoologie*, **10** (2), 1390–1656.

—— (1956), Sécrétion d'une mucoproteine par les tubes de Malpighi des larves de Cercopides, son role dans la formation de l'abri spumeux, *Boll. Lab. Zool. gen. agr. Portici*, **33**, 341–349.

PETERS, W. (1962), The morphology of *situs inversus* in abdominal segmentation of *Krizousacorixa femorata*, Guérin (Heteroptera, Corixidae), *J. Morph.*, **110**, 141–156.

PETHERBRIDGE, F. R. AND HUSAIN, M. A. (1918), A study of the Capsid bugs found on apple trees, *Ann. appl. Biol.*, **4**, 179–205.

PFLUGFELDER, O. (1937), Vergleichend-anatomische, experimentelle und embryologische Untersuchungen über das Nervensystem und die Sinnesorgane der Rhynchoten, *Zoologica (Stuttgart)*, **34**, 102 pp.

—— (1939), Arthropoda, Insecta Coccina, In: Bronn, *Klassen und Ordnungen des Tierreichs*, **5** (3), Buch 8: 1–121.

PIERANTONI, U. (1951), La simbiosi in *Tropidothorax leucopterus* (Heteroptera, Lygaeidae), *Boll. Zool.*, **18**, 1–3.

PINTERA, A. (1969), Evaluation of some recent classification systems in aphidology (Homoptera), *Acta ent. bohemoslovaca*, **66**, 122–124.

PIOTROWSKI, F. (1950), Sur la morphologie de l'appareil copulateur des Hémiptères-Hétéroptères, avec considération speciale du groupe Pentatomoidaria Börner 1934, *Prace Kom. Biol. Wydz. mat.-przyr. poznan Tow. Przy. Nauk*, 38 pp.

POISSON, R. (1924), Contribution à l'étude des Hémiptères aquatiques, *Bull. biol. Fr. Belg.*, **58**, 49–305.

—— (1951), Ordre des Hétéroptères, In: Grassé, P. P., *Traité de Zoologie*, **10** (2), 1657–1803.

—— (1957), Hétéroptères aquatiques, *Faune de France*, **61**, 263 pp.

POLLARD, D. G. (1973), Plant penetration by feeding aphids (Hemiptera, Aphidoidea): a review, *Bull. ent. Res.*, **62**, 631–714.

POPHAM, E. J. (1943), Ecological studies of the commoner British Corixidae, *J. Anim. Ecol.*, **12**, 124–136.

—— (1947), Ecological studies of the mating habits of certain species of Corixidae and their significance, *Proc. zool. Soc. Lond.*, **116**, 692–706.

—— (1960), On the respiration of aquatic Hemiptera Heteroptera with special reference to the Corixidae, *Proc. zool. Soc. Lond.*, **135**, 209–242.

—— (1961), The function of the paleal pegs of Corixidae (Hemiptera Heteroptera), *Nature*, **190**, 742–743.

—— (1962), On the salivary pump of *Dysdercus intermedius* Dist. (Hemiptera, Heteroptera) and other bugs, *Proc. zool. Soc. Lond.*, **139**, 489–493.

POPOV, Y. A. (1971), Historical development of Hemiptera infraorder Nepomorpha (Heteroptera), *Trud. palaeont. Inst. Acad. Sci.*, USSR, **129**, 227 pp.

PURI, K. M. (1924), Studies on the anatomy of *Cimex lectularius, Parasitology*, **16**, 84–97; 269–278.

PUTON, A. (1878–81), *Synopsis des Hémiptères-Hétéroptères de France*, Deyrolle, Paris, 4 pts, 373 pp.

QADRI, M. A. H. (1949), On the morphology and postembryonic development of the male genitalia and their ducts in Hemiptera (Insecta), *J. zool. Soc. India*, **1**, 129–143.

QUAINTANCE, A. L. AND BAKER, A. C. (1913–14), Classification of the Aleyrodidae. I, II, *U.S. Dept. Agric. Bur. Ent. Tech. Ser.*, **27**, 1–93; 95–114.

—— (1917), A contribution to our knowledge of the white-flies of the subfamily Aleyrodinae (Aleyrodidae), *Proc. U.S. nat. Mus.*, **51**, 335–445.

QUEDNAU, W. (1954), Monographie der mitteleuropäischen Callaphididae (Zierläuse (Homoptera, Aphidina)) unter besonderer Berücksichtigung des ersten Jugendstadiums, I, *Mitt. biol. Zentanst., Berlin*, **78**, 1–71.

RANKIN, K. L. (1935), Life history of *Lethocerus americanus* (Leidy) (Hemiptera Belostomatidae), *Kansas Univ. Sci. Bull.*, **22**, 479–491.
READIO, P. A. (1927), Studies on the biology of the Reduviidae of America north of Mexico, *Univ. Kansas Sci. Bull.*, **28**, 1–291.
REMOLD, H. (1963), Über die biologische Bedeutung der Duftdrüsen bei den Landwanzen (Geocorisae), *Z. vergl. Physiol.*, **45**, 636–694.
REUTER, O. M. (1878–96), Hemiptera Gymnocerata Europae, *Acta Soc. Sci. fenn.*, **13**, 568 pp.; 179 pp.; **23**, 392 pp.
—— (1885), Monographia anthocoridarum orbis terrestris, *Acta Soc. Sci. fenn.*, **14**, 555–758.
—— (1891), Monographia ceratocombidarum orbis terrestris, *Acta Soc. Sci. fenn.*, **19**, 1–28.
—— (1910), Neue Beiträge zur Phylogenie und Systematik der Miriden nebst einleitenden Bemerkungen über die Phylogenie der Heteropterenfamilien, *Acta Soc. Sci. fenn.*, **37**, 2–172.
—— (1912), Bemerkungen über mein neues Heteropterensystem, *Ofv. Finska Vetensk-Societ. Forh.*, **54**, 1–62.
REUTER, O. M. AND POPPIUS, B. (1909), Monographia Nabidarum orbis terrestris. Pars prior, *Acta Soc. Sci. fenn.*, **37**, 1–62.
RIBAUT, H. (1936), Homoptères Auchénorhynches I. (Typhlocybidae), *Faune de France*, **31**, 1–228.
—— (1952). Ditto II. (Jassidae), *Faune de France*, **57**, 474 pp.
RIS LAMBERS, D. H. (1933–34), Notes on Theobald's 'The Plant lice or Aphididae of Great Britain'. I–III. *Stylops*, **2**, 169–176; **3**, 25–33.
—— (1938–53). Contributions to a monograph of the Aphididae of Europe, I–V, *Temminckia*, **3**, 1–43; **4**, 1–134; **7**, 173–319; **8**, 182–323; **9**, 1–176.
—— (1966), Polymorphism in Aphidoidea, *A. Rev. Ent.*, **11**, 47–78.
ROSS, H. H. (1957), Evolutionary developments in leafhoppers, *Syst. Zool.*, **6**, 87–97.
ROSS, H. H. AND MOORE, T. E. (1957), New species in the *Empoasca fabae* complex, *Ann. ent. Soc. Am.*, **50**, 118–122.
ROTHSCHILD, G. H. L. (1964), The biology of *Conomelus anceps* Germar (Homoptera: Delphacidae), *Trans. Soc. Br. Ent.*, **16**, 135–148.
—— (1966), A study of the natural population of *Conomelus anceps* (Germar) (Homoptera: Delphacidae) including observations on predation using the precipitin test, *J. Anim. Ecol.*, **35**, 413–434.
RYCKMAN, R. E. (1962), Biosystematics and hosts of *Triatoma protracta* complex in N. America. Hemip. Reduviidae, *Univ. Calif. Publs Ent.*, **27**, 148 pp.
SAMPSON, W. W. (1943), A generic synopsis of the Hemipterous superfamily Aleyrodoidea, *Entomologica am.*, **23**, 173–223.
—— (1947), Additions and corrections to 'A generic synopsis of the Aleyrodoidea', *Bull. Brooklyn ent. Soc.*, (N.S.), **42**, 45–50.
SANDER, K. (1956), Bau und Funktion des Sprungapparates von *Pyrilla perpusilla* Walker (Homoptera-Fulgoridae), *Zool. Jb., Anat.*, **75**, 383–388.
SANDS, W. A. (1957), The immature stages of some British Anthocoridae (Hemiptera), *Trans. R. ent. Soc. Lond.*, **109**, 295–310.
SATTLER, W. (1957), Beobachtungen zur Fortpflanzung von *Gerris najas* DeGeer (Heteroptera), *Z. Morph. Ökol. Tiere*, **45**, 411–428.
SCHAEFER, C. W. (1964–68), The morphology and higher classification of the Coreoidea (Hemiptera-Heteroptera), Parts I–IV, *Ann. ent. Soc. Am.*, **57**, 670–

684; *Misc. Publs ent. Soc. Am.*, 5, 1–76; *Occ. Pap. Univ. Conn.* (Biol. Sci. Ser.), 1, 153–199.
—— (1966), Some notes on Heteropteran trichobothria, *Mich. Ent.*, 1, 85–90.
—— (1972), Degree of metathoracic scent-gland development in the trichophorous Heteroptera (Hemiptera), *Ann. ent. Soc. Am.*, **65**, 810–821.
SCHAEFER, H. A. (1949), Biologische und ökologische Beobachtungen an Psylliden (Hemiptera), *Verh. naturf. Ges. Basel*, **60**, 25–41.
SCHALLER, F. (1952), Stridulation und Lautwahrnehmung von *Corixa striata*, *Z. vergl. Physiol.*, **33**, 476–486.
SCHELL, D. V. (1943), The Ochteridae (Hemiptera) of the Western Hemisphere, *J. Kansas ent. Soc.*, **16**, 29–47.
SCHIÖDTE, J. C. (1870), On some new and fundamental principles in the morphology and classification of Rhynchota, *Ann. Mag. nat. Hist.*, (4), **6**, 225–249.
SCHLEE, D. (1969), Phylogenetische Studien an Hemiptera, I–V, *Z. Morph. Tiere*, **64**, 95–138; 139–150; *Stuttg. Beitr. Naturk.*, **199**, 1–19; **210**, 1–27; **211**, 1–11.
SCHMUTTERER, H. (1959), Schildläuse oder Coccoidea. I. Deckelschildläuse oder Diaspididae, *Tierwelt Deutschlands*, **45**, 160 pp.
SCHNEIDER, G. (1940), Beiträge zur Kenntnis der symbiotischen Einrichtung der Heteropteren, *Z. Morph. Ökol. Tiere*, **36**, 595–644.
SCHORR, H. (1957), Zur Verhaltensbiologie und Symbiose von *Brachypelta aterrima* Först. (Cydnidae, Heteroptera), *Z. Morph. Ökol. Tiere*, **45**, 561–602.
SCHOUTEDEN, H. (1904–13), Pentatomidae, *Genera Insect.*, Fasc. **24**, 100 pp.; Fasc. **30**, 46 pp.; Fasc. **47**, 4 pp.; Fasc. **52**, 82 pp.; Fasc. **153**, 19 pp.
SCHRADER, F. (1926), Notes on the English and American races of the greenhouses whitefly (*Trialeurodes vaporariorum*), *Ann. appl. Biol.*, **13**, 189–196.
SCHWEMMLER, W. DUTHIOT, J.-L., KUHL, G. AND VAGO, C. (1973), Sprengung der Endosymbiose von *Euscelis plebejus* F. und Ernährung aposymbiontischer Tiere mit synthetischer Diät (Hemiptera, Cicadidae), *Z. Morph. Tiere*, **74**, 297–322.
SCUDDER, G. G. E. (1959), The female genitalia of the Heteroptera: morphology and bearing on classification, *Trans. R. ent. Soc. Lond.*, **111**, 405–467.
—— (1963), Adult abdominal characters in the lygaeoid-coreoid complex of the Heteroptera, and the classification of the group, *Can. J. Zool.*, **41**, 1–14.
—— (1971), The postembryonic development of the indirect flight muscles in *Cenocorixa bifida* (Hung.) (Hemiptera: Corixidae), *Can. J. Zool.*, **49**, 1387–1398.
SERVADEI, A. (1951), Note sull'*Heterogaster urticae* F. e sul genere *Heterogaster* Schill. (Hemiptera Heteroptera, Myodochidae), *Redia*, **36**, 171–220.
—— (1967), Rhynchota (Heteroptera, Homoptera Auchenorrhyncha), Catalogo topografico e sinonimico, *Fauna d'Italia*, **9**, 351 pp.
SINGH-PRUTHI, H. (1925), The morphology of the male genitalia in Rhynchota, *Trans. R. ent. Soc. Lond.*, **1925**, 127–254.
SIGNORET, V. (1881–84), Revision des groupes de Cydnides de la famille des Pentatomides, *Ann. Soc. ent. Fr.*, (6), **1**, 25–52, 193–218, 319–332, 423–436; **2**, 23–42, 145–168, 241–266, 465–484; **3**, 33–60, 207–220, 357–374, 516–534; **4**, 45–62, 117–128.
SLATER, J. A. (1950), An investigation of the female genitalia as taxonomic characters in the Miridae, *Iowa St. Coll. J. Sci.*, **25**, 1–81.
—— (1964), *A Catalogue of the Lygaeidae of the World*, Univ. Connecticut, Storrs, Conn., 2 vols., 1668 pp.

SLATER, J. A. AND HURLBUTT, H. W. (1957), A comparative study of the metathoracic wing in the family Lygaeidae, *Proc. ent. Soc. Wash.*, **59**, 67–79.

SMITH, C. F. (1972), Bibliography of the Aphididae of the world, *Tech. Bull. North Carolina agric. Exp. Stn*, **216**, 717 pp.

SMITH, R. H. (1944), Bionomics and control of the nigra scale, *Saissetia nigra*, *Hilgardia*, **16**, 225–288.

SNODGRASS, R. E. (1933), Morphology of the insect abdomen. Part II. The genital ducts and the ovipositor, *Smithson. misc. Coll.*, **89**, 148 pp.

—— (1935), *Principles of Insect Morphology*, McGraw-Hill, New York and London, 667 pp.

—— (1950), Comparative studies on the jaws of mandibulate arthropods, *Smithson. misc. Collns*, **116** (1), 85 pp.

—— (1957), A revised interpretation of the external reproductive organs of male insects, *Smithson. misc. Collns*, **135** (6), 60 pp.

SOGAWA, K. (1965), Studies on the salivary glands of rice plant leaf-hoppers. I. Morphology and histology, *Jap. J. appl. Ent. Zool.*, **9**, 275–290.

SOUTHWOOD, T. R. E. (1955), The morphology of the salivary glands of terrestrial Heteroptera (Geocorisae) and its bearing on classification, *Tijdschr. Ent.*, **98**, 77–84.

—— (1956), The structure of the eggs of the terrestrial Heteroptera and its relation to the classification of the group, *Trans. R. ent. Soc. Lond.*, **108**, 163–221.

SOUTHWOOD, T. R. E. AND HINE, D. J. (1950), Further notes on the biology of *Sehirus bicolor* (L.) (Hem., Cydnidae), *Ent. mon. Mag.*, **86**, 299–301.

SOUTHWOOD, T. R. E. AND LESTON, D. (1959), *Land and Water Bugs of the British Isles*, Warne, London, 436 pp.

SPEERS, C. F. (1941), The pine spittle bug (*Aphrophora parallela* Say), *Bull. N.Y. St. Coll. For., Tech. Publ.*, **54**, 65 pp.

SPEYER, E. R. (1923), Researches upon the larch *Chermes* (*Cnaphalodes strobilobius* Kalt.), and their bearing upon the evolution of the Chermesinae in general, *Phil. Trans. R. Soc.* (B), **212**, 111–146

SPEYER, W. (1929), Der Apfelblattsauger *Psylla mali* Schmidberger, *Monog. Pflanzenschutz*, **1**, 127 pp.

SPOONER, C. S. (1938), The phylogeny of the Hemiptera based on a study of the head-capsule, *Illinois biol. Monogr.*, **16** (3), 102 pp.

SPRAGUE, I. B. (1956), The biology and morphology of *Hydrometra martini* Kirkaldy, *Kansas Univ. Sci. Bull.*, **38**, 579–693.

STÅL, C. (1864–66), *Hemiptera Africana*, Holmiae, Offic. Norstedtiana, Stockholm, **1**, 256 pp.; **2**, 182 pp.; **3**, 200 pp.; **4**, 276 pp.

—— (1870–76), Enumeratio Hemipterorum, *K. svenska VetenskAkad. Handl.*, **9**, 1–232; **10**, 1–159; **11**, 1–163; **12**, 1–186; **14**, 1–162.

STEFFAN, A. W. (1968), Evolution und Systematik der Adelgidae (Homoptera: Aphidina). Eine Verwandtschaftsanalyse auf vorwiegend ethologischer, zytologischer und karyologischer Grundlage, *Zoologica* (*Stuttgart*), **40** (5), 1–139.

STEIN, G. (1966), Über den Feinbau der Duftdrüsen von Feuerwanzen (*Pyrrhocoris apterus* L., Geocorisae) I, II, *Z. Zellforsch.*, **74**, 271–290; **75**, 501–516.

STEINHAUS, E. A., BATEY, M. M. AND BOERKE, C. L. (1956), Bacterial symbiotes from the caeca of certain Heteroptera, *Hilgardia*, **24**, 495–518.

STICHEL, W. (1925–38), *Illustrierte Bestimmungstabellen der deutschen Wanzen (Hemiptera Heteroptera)*, Stichel-Verlag, Leipzig, 499 pp.

—— (1955–62), *Illustrierte Bestimmungstabellen der Wanzen. II. Europa.* Buchdruckerei Erich. Pröh, Berlin, **1**, 1–168; **2**, 170–907; **3**, 1–428; **4**, 1–838.

STOUGH, H. G. (1910), The hackberry Psylla *Pachypsylla cettidis-mammae* Riley, A study in comparative morphology, *Univ. Kansas Sci. Bull.*, **5**, 121–164.

STROYAN, H. L. G. (1950, 1955), Recent additions to the British aphid fauna. I, II, *Trans. R. ent. Soc. Lond.*, **101**, 89–124; **106**, 283–340.

—— (1957), Further additions to the British aphid fauna, *Trans. R. ent. Soc. Lond.*, **109**, 311–360.

—— (1957–1963), *A Revision of the British species of* Dysaphis. I, II, H.M.S.O., London, 59 pp.; 119 pp.

—— (1958), A contribution to the taxonomy of some British species of *Sappaphis* Matsumura 1918 (Homoptera, Aphidoidea), *J. Linn. Soc.*, **43**, 644–713.

—— (1964), Notes on hitherto unrecorded or overlooked British aphid species, *Trans. R. ent. Soc. Lond.*, **116**, 29–72.

STRÜBING, H. (1956), Über Beziehungen zwischen Ovidukt, Eiablage, und natürlicher Verwandtschaft einheimischer Delphaciden, *Zool. Beitr.*, **2**, 331–357.

STRÜMPEL, H. (1972), Beitrag zur Phylogenie der Membracidae Rafinesque, *Zool. Jb., Syst.*, **99**, 313–407.

STYS, P. (1961), Morphology of the abdomen and female ectodermal genitalia of the Trichophorous Heteroptera and bearing on their classification, *Verhandl. XI int. Kongr. Ent.*, **1**, 37–43.

—— (1946*a*), The morphology and relationship of the family Hyocephalidae (Heteroptera), *Acta zool. hung.*, **10**, 229–262.

—— (1964*b*), Thaumastellidae – a new family of Pentatomoid Heteroptera, *Čas. čsl. Spol. ent.*, **61**, 238–253.

—— (1966), Morphology of the wings, abdomen and genitalia of *Phaenacantha australiae* (Kirk) (Heteroptera, Colobathristidae) and notes on the phylogeny of the family, *Acta ent. bohemoslov.*, **63**, 266–280.

—— (1967), Medocostidae, A new family of Cimicomorphan Heteroptera based on a new genus and two new species from Tropical Africa, I. Descriptive part, *Acta ent. bohemoslov.*, **64**, 439–465.

—— (1969), On the morphology of the labrum in Heteroptera, *Acta ent. bohemoslov.*, **66**, 150–158.

—— (1970), On the morphology and classification of the family Dipsocoridae *s. lat.*, with particular reference to the genus *Hypsipteryx* Drake (Heteroptera), *Acta ent. bohemoslov.*, **67**, 21–46.

ŠULC, K. (1911), Ueber Respiration, Tracheensystem und Schaumproduktion der Schaumzikadenlarven (Aphrophorinae-Homoptera), *Z. wiss. Zool.*, **99**, 147–188.

SUTTON, M. F. (1951), On the food, feeding mechanism and alimentary canal of Corixidae (Hemiptera Heteroptera), *Proc. zool. Soc. Lond.*, **121**, 465–499.

SWEET, M. H. (1964), The biology and ecology of the Rhyparochrominae of New England (Heteroptera: Lygaeidae), Parts I & II, *Entomologia am.*, **43**, 1–124; **44**, 1–201.

SWEET, M. H. AND SLATER, J. A. (1961), A generic key to the nymphs of North American Lygaeidae (Hemiptera-Heteroptera), *Ann. ent. Soc. Am.*, **54**, 333–340.

SWENSON, K. G. (1968), Role of aphids in the ecology of plant viruses, *A. Rev. Phytopathol.*, **6**, 351–374.

TAMANINI, L. (1956), Ricerche zoologiche sui Monti Sibillini (Appennino umbro-marchiagiano). III, Caratteri morfologici e cenni biologici sull' *Aradus frigidus* Kiritschenko (Hemiptera: Heteroptera, Aradidae), *Mem. Mus. stor. nat., Verona*, **5**, 45–59.

TANAKA, T. (1926), Homologies of the wing veins of the Hemiptera, *Annotnes zool. jap.*, **11**, 33–53.

TAYLOR, L. H. (1918), The thoracic sclerites of Hemiptera and Heteroptera, *Ann. ent. Soc. Am.*, **11**, 225–249.

TEODORI, G. (1925–26), Sull'apparato di uncinamento fra elitre ed ali negli eterotteri, *Atti Accad. ven.-trent.*, (3) **15**, 49–54; **16**, 99–107.

THEOBALD, F. V. (1926–29), *The Plant-lice or Aphididae of Great Britain*, Headley, Ashford and London, **1**, 372 pp.; **2**, 411 pp.; **3**, 364 pp.

THERON, J. G. (1958), Comparative studies on the morphology of male scale insects (Hemiptera: Coccoidea), *Ann. Univ. Stellenbosch*, **34** (A) (1), 1–71.

THORPE, W. H. AND CRISP, D. J. (1947), Studies on plastron respiration, I–III, *J. exp. Biol.*, **24**, 227–269, 270–303, 310–382.

TILLYARD, R. J. (1926), Upper Permian insects of New South Wales, Part i. Introduction and the order Hemiptera, *Proc. Linn. Soc. N.S.W.*, **51**, 1–30.

TISCHLER, W. (1938–39), Zur Oekologie der wichtigsten in Deutschland an Getreide schädlichen Pentatomiden. I, II, *Z. Morph. Ökol. Tiere*, **34**, 317–366; **35**, 251–287.

—— (1960), Studien zur Bionomie und Ökologie der Schmalwanze *Ischnodemus sabuleti* Fall. (Hem. Lygaeidae), *Z. wiss. Zool.*, **163**, 168–209.

TODD, E. L. (1955), A taxonomic revision of the family Gelastocoridae (Hemiptera), *Kansas Univ. Sci. Bull.*, **37** (11), 277–475.

TORRE-BUENO, J. R. de la (1905), The tonal apparatus of *Ranatra quadridentata* Stål., *Can. Ent.*, **37**, 85–87.

—— (1906), Life-histories of North American water bugs. I. *Ranatra quadridentata* Stål., II. *Belostoma fluminea* Say, *Can. Ent.*, **38**, 242–252; 189–197.

—— (1916), Aquatic Hemiptera. A study in the relation of structure to environment, *Ann. ent. Soc. Am.*, **9**, 353–365.

—— (1939–46), A synopsis of the Hemiptera Heteroptera of America north of Mexico, *Ent. am.*, **19**, 141–304; **21**, 41–122; **26**, 1–141.

TOWER, D. G. (1914), The mechanism of the mouthparts of the squash bug, *Anasa tristis* Degeer, *Psyche*, **21**, 99–108.

TREHAN, K. N. (1940), Studies on the British whiteflies (Homoptera-Aleyrodidae), *Trans. R. ent. Soc. Lond.*, **90**, 575–616.

TREHAN, K. N. AND BUTANI, D. K. (1960, 1970), Bibliography of Aleyrodidae. I, II, *Beitr. Ent.*, **10**, 330–388; **20**, 317–335.

TSING-CHAO, M. AND KUEI-SHIU, L. (1956), A synopsis of the Old World Phymatidae, *Q. Jl Taiwan Mus.*, **9**, 109–156.

TUTHILL, L. D. (1943), The Psyllids of America north of Mexico (Psyllidae: Homoptera) (Subfamilies Psyllinae & Triozinae), *Iowa St. Coll. J. Sci.*, **17**, 443–667.

USINGER, R. L. (1937), Notes on the biology of *Hydrotrephes balnearius* (Helotrephidae, Hemiptera-Heteroptera), *Ent. mon. Mag.*, *73*, 179–180.

—— (1938), Biological notes on the pelagic water-striders (*Halobates*) of the

Hawaiian islands with descriptions of a new species from Waikiki (Gerridae, Hemiptera), *Proc. Hawaiian ent. Soc.*, **10**, 77–84.
—— (1941), Key to the subfamilies of Naucoridae with a generic synopsis of the new subfamily Ambrysinae (Hemiptera), *Ann. ent. Soc. Am.*, **34**, 5–16.
—— (1942), Revision of the Termitaphididae, *Pan-Pacific Ent.*, **18**, 155–159.
—— (1943), A revised classification of the Reduvoidea, with a new subfamily from South America (Hemiptera), *Ann. ent. Soc. Am.*, **36**, 602–618.
—— (1944), The Triatominae of North and Central America and the West Indies and their public health significance, *Publ. Hlth. Bull.*, **288**, 83 pp.
—— (1945), Classification of the Enicocephalidae, *Ann. ent. Soc. Am.*, **38**, 321–342.
—— (1946*a*), Polyctenidae, In: China & Parshley, *Gen. Cat. Hemipt.*, **5**, 18 pp.
—— (1946*b*), Notes and descriptions of *Ambrysus* Stål with an account of the life history of *Ambrysus mormon* Montd. (Hemiptera, Naucoridae), *Kansas Univ. Sci. Bull.*, **31**, 185–210.
—— (1942*b*), The genus *Nysius* and its allies in the Hawaiian Islands (Hemiptera, Lygaeidae, Orsillini), *Bull. Bishop. Mus. Honolulu*, **173**, 1–167.
—— (1966), *Monograph of Cimicidae (Hemiptera-Heteroptera)*, Thomas Say Foundation, Maryland, 585 pp.
USINGER, R. L. AND MATSUDA, R. (1959), *Classification of the Aradidae (Hemiptera-Heteroptera)*, British Museum (Nat. Hist.), London, 410 pp.
USINGER, R. L., WYGODZINSKY, P. AND RYCKMAN, R. E. (1966), The biosystematics of Triatominae, *A. Rev. Ent.*, **11**, 309–330.
VAN DUZEE, E. P. (1917), Catalogue of the Hemiptera of America north of Mexico, exclusive of the Aphidae, Coccidae and Aleurodidae, *Univ. Calif. Tech. Bull. Coll. agric. Exp. Stn*, **2**, 902 pp.
VASVARY, L. M. (1966), Musculature and nervous system of the thorax, of the sound mechanism, and of a typical pregenital abdominal segment of the male of the annual cicada, *Tibicen chloromera* (Walker) (Homoptera: Cicadidae), *Jl N.Y. ent. Soc.*, **74**, 2–55.
VILLIERS, A. (1958), Insectes Hémiptères Enicocephalidae, *Faune de Madagascar*, **7**, 5–78.
WADE, V. (1960), Species index. Fasc. IV. Fulgoroidea, *Gen. Cat. Hemipt.*, Fasc. IV, 78 pp.
—— (1963), General catalogue of the Homoptera. Species index. Fasc. VII. Cercopoidea, *N. Carol. St. Coll. Pap.*, 1598, 31 pp.
—— (1964), Species index. Fasc. VIII. Cicadoidea, *Gen. Cat. Hemipt.*, Fasc. VIII, 26 pp. (N. Carol. St. Coll. Pap., 1790.)
WAGNER, E. (1961), 1. Unterordnung: Ungleichflügler, Wanzen, Heteroptera (Hemiptera), In: Bröhmer, Ehrmann und Ulmer, *Tierwelt Mitteleuropas*, Bd. 4 (3), Heft Xa, 173 pp.
—— (1963), Untersuchungen über den taxonomischen Wert des Baues der Genitalien bei den Cydnidae, *Acta ent. Mus. Nat. Prag.*, **35**, 73–115.
—— (1966–67), Wanzen oder Heteroptera. I (Pentatomomorpha), II (Cimicomorpha), *Tierwelt Deutschlands*, **54**, 235 pp; **55**, 179 pp.
WAGNER, E. AND WEBER, H. H. (1964), Hétéroptères Miridae et Isometopidae, *Faune de France*, **67**, 589 pp.
WAGNER, W. (1951), Beitrag zur Phylogenie und Systematik der Cicadellidae (Jassidae) Nord- und Mitteleuropas, *Comment. biol., Soc. sci. fenn.*, **12** (2), 15–44.

—— (1962), Dynamische Taxionomie angewandt auf die Delphaciden Mitteleuropas, *Mitt. zool. StInst. Hamb.*, **60**, 111–180.

WALOFF, N. (1973), Dispersal by flight of leafhoppers (Auchenorrhyncha: Homoptera), *J. appl. Ecol.*, **10**, 705–730.

—— (1975), The parasitoids of the nymphal and adult stages of leafhoppers (Auchenorrhyncha: Homoptera) of acidic grassland, *Trans. R. ent. Soc. Lond.*, **126**, 637–686.

WALTON, G. A. (1936), Oviposition in the British species of *Notonecta* (Hemipt.), *Trans. Soc. Br. Ent.*, **3**, 49–57.

WATERHOUSE, D. F. AND GILBY, A. R. (1965), The adult scent glands and scent of nine bugs of the superfamily Coreoidea, *J. Insect Physiol.*, **10**, 977–987.

WATSON, J. A. AND PLUMB, R. T. (1972), Transmission of plant-pathogenic viruses by aphids, *A. Rev. Ent.*, **17**, 425–452.

WEAVER, C. R. AND KING, D. R. (1954), Meadow spittlebug, *Res. Bull. Ohio agric. Exp. Sta.*, **741**, 1–100.

WEBER, H. (1928*a*), Skelett, Muskulatur and Darm der schwarzen Blattlaus *Aphis fabae* Scop, *Zoologica, Stuttgart*, **28**, 120 pp.

—— (1928*b*), Zur vergleichenden Physiologie der Saugorgane der Hemipteren. Mit besonderer Berücksichtigung der Pflanzenläuse, *Z. vergl. Physiol.*, **8**, 145–186.

—— (1929), Kopf und Thorax von *Psylla mali* Schmidb. (Hemiptera-Homoptera), *Z. Morph. Ökol. Tiere*, **14**, 59–165.

—— (1929–35), Hemiptera. I–III, In: Schulze, *Biologie der Tiere Deutschlands*, **31**, 1–70, 71–208, 209–355.

—— (1930), *Biologie der Hemipteren. Eine Naturgeschichte der Schnabelkerfe*, Springer, Berlin, 543 pp.

—— (1931), Lebensweise und Umweltbeziehungen von *Trialeurodes vaporariorum* (Westwood) (Homoptera-Aleurodina), Erster Beitrag zu einer Monographie dieser Art; *Z. Morph. Ökol. Tiere*, **23**, 575–753.

—— (1934), Die postembryonale Entwicklung der Aleurodinen (Hemiptera-Homoptera), Ein Beitrag der Kenntnis der Metamorphose der Insekten, *Z. Morph. Ökol. Tiere*, **29**, 268–305.

—— (1935), Der Bau der Imago der Aleurodinen. Ein Beitrag zur vergleichenden Morphologie des Insektenkörpers, *Zoologica, Stuttgart*, **33** (89), 71 pp.

WEFELSCHEID, H. (1912), Über die Biologie und Anatomie von *Plea minutissima* Leach, *Zool. Jb., Syst.*, **32**, 389–474.

WHITCOMB, R. F. AND DAVIS, R. E. (1970), Mycoplasma and phytarboviruses as plant pathogens persistently transmitted by insects, *A. Rev. Ent.*, **15**, 405–464.

WHITTAKER, J. B. (1965), The distribution and population dynamics of *Neophilaenus lineatus* (L.) and *N. exclamationis* (Thun.) (Homoptera: Cercopidae) on Pennine moorland, *J. Anim. Ecol.*, **34**, 277–297.

—— (1971), Polymorphism of *Philaenus spumarius* (L.) (Homoptera: Cercopidae) in England, *J. Anim. Ecol.*, **37**, 99–111.

—— (1973), Density regulation in a population of *Philaenus spumarius* (L.) (Homoptera: Cercopidae), *J. Anim. Ecol.*, **42**, 163–172.

WICK, J. R. AND BONHAG, P. F. (1955), Postembryonic development of the ovaries of *Oncopeltus fasciatus* (Dallas), *J. Morph.*, **96**, 31–59.

WIGGLESWORTH, V. B. (1959), The histology of the nervous system of an insect, *Rhodnius prolixus* (Hemiptera), I, II, *Q. Jl microsc. Sci.*, **100**, 285–298, 299–313.

WILLE, J. (1929), Die Rübenblattwanze *Piesma quadrata* Fieb., *Monogr. Pflanzenschutz*, **2**, 116 pp.

WILLIAMS, D. J. (1962), The British Pseudococcidae, *Bull. Br. Mus. nat. Hist.*, (Ent.), **12**, 1–79.

—— (1969), The family-group names of the scale insects (Hemiptera: Coccoidea), *Bull. Br. Mus. nat. Hist.*, (Ent.), **23**, 315–341.

WILLIAMS, J. R. (1957), The sugarcane Delphacidae and their natural enemies in Mauritius, *Trans. R. ent. Soc. Lond.*, **109**, 65–110.

WILLIS, D. M. (1949), The anatomy and histology of the head, gut and associated structures of *Typhlocyba ulmi*, *Proc. zool. Soc. Lond.*, 118, 984–1001.

WITLACZIL, E. (1885), Die Anatomie der Psylliden, *Z. wiss. Zool.*, **43**, 569–638.

WOODWARD, T. E. (1950), Ovariole and testis follicle numbers in the Heteroptera, *Ent. mon. Mag.*, **86**, 82–84.

WYGODZINSKY, P. (1944), Contribuição ao conhecimento do genero *Elasmodema* Stål 1860 (Elasmodemidae, Reduvioidea, Hemiptera), *Rev. bras. Biol.*, **4**, 193–213.

WYNN, G. G. AND BOUDREAUX, H. B. (1972), Structure and function of aphid cornicles, *Ann. ent. Soc. Am.*, **65**, 157–166.

YOUNG, D. (1975), Chordotonal organs associated with the sound producing apparatus of cicadas (Insecta, Homoptera), *Z. Morph. Tiere*, **81**, 111–135.

YOUNG, D. A. (1952), A reclassification of Western Hemisphere Typhlocybinae (Homoptera, Cicadellidae), *Kansas Univ. Sci. Bull.*, **35**, 3–217.

YOUNG, E. C. (1965), The incidence of flight polymorphism in British Corixidae and description of the morphs, *J. Zool., Lond.*, **146**, 567–576.

ZACHER, F. (1916), Die Literatur über die Blattflöhe und die von ihnen verursachten Gallen, nebst einem Verzeichnis der Nährpflanzen und Nachträgen zum 'Psyllidarum Catalogus', *Zentbl. Bakt. ParasitKde*, **46**, 97–111.

ZAHRADNIK, J. (1963), Aleurodina (Mottenläuse), In: Bröhmer, Ehrmann and Ulmer, *Die Tierwelt Mitteleuropas*, Insekten I. Teil, Bd. 4 (3), Heft Xd: 19 pp.

ZWÖLFER, H. (1957–58), Zur Systematik, Biologie und Oekologie unterirdisch lebender Aphiden, *Z. angew. Ent.*, **40**, 182–221, 528–575; **42**, 129–171; **43**, 1–52.

ZWÖLFER, W. (1930), Beiträge zur Kenntnis der Schädlingsfauna Kleinasiens. I. Untersuchungen zur Epidemiologie der Getreidewanze *Eurygaster integriceps* Put., *Z. angew. Ent.*, **17**, 227–252.

Order 20

THYSANOPTERA (PHYSAPODA: THRIPS)

Small or minute slender-bodied insects with short 6- to 10-segmented antennae and asymmetrical piercing mouthparts with maxillary and labial palps. Prothorax well developed, free; tarsi 1- or 2-segmented, each with a terminal protrusible vesicle. Wings when present very narrow with greatly reduced venation and long marginal setae. Cerci absent. Metamorphosis accompanied by two or three inactive pupa-like instars.

These insects are commonly known as 'thrips'. The majority vary in length from 0·5–8 mm, the smaller forms being by far the most prevalent. They are mostly yellow, yellowish-brown or black and many are found among all kinds of growing vegetation, both on the flowers and about the foliage; others are subcortical or frequent moist decaying plant remains, particularly wood, leaf litter and fungi. Some species are predacious, or at least occasionally so, and suck the body-fluids of aphids, mites and other small insects. When disturbed some species crawl in a leisurely fashion, others run quickly or leap, and a large number are able to fly but do not readily do so. Many exhibit the habit of curving the apex of the abdomen upwards and in winged individuals this movement is generally preparatory to flight; the insect uses its legs to comb the long setae on the wings then flexes the abdomen to separate the wings. For general accounts of the order see Priesner (1968) and the detailed review of its biology, ecology and economic importance by Lewis (1973).

Most Thysanoptera feed by penetrating the living tissues of plants with their piercing mouthparts and imbibing the sap. It is therefore not surprising that they include a number of economically important species (Blunck, 1950; Bournier, 1970). Among the more widely distributed pests are *Thrips tabaci* on onions, *Scirtothrips aurantii* and *S. citri* on citrus in S. Africa and the United States respectively, species of *Caliothrips* and *Frankliniella* on cotton, *Limothrips cerealium* on cereals, and *Heliothrips haemorrhoidalis* as a pest of glasshouse crops in Britain. Most injurious Thysanoptera damage the leaves on which they feed, some affect the flowers or fruit and a smaller number induce the formation of galls. *Thrips tabaci* and species of

Frankliniella transmit spotted wilt virus, strains of which occur in tomato, tobacco, pineapples, lettuce and potatoes. On the other hand, several species play a role in the pollination of flowers, including those of crop plants, and one species, *Liothrips urichi*, was successfully introduced into Fiji to control the weed *Clidemia hirta* (Simmonds, 1933). Most winged Thysanoptera have a period of active dispersal by flight and in some abundant species such as *Limothrips cerealium* there are often mass flights, mainly associated with dry, sunny, settled weather (Lewis, 1964). Migration also occurs over longer distances through the agency of winds and of atmospheric convection and turbulence.

Sex determination in the Thysanoptera is by male haploidy and parthenogenesis occurs frequently. In several species, such as *Heliothrips haemorrhoidalis* and *Taeniothrips inconsequens*, there is thelyotoky and males are unknown or rare; in other cases arrhenotokous parthenogenesis takes place and sex-ratios may be very unequal, as in *Liothrips oleae* and *Taeniothrips simplex*. For further details see Pussard-Radulesco (1931), Bournier (1956) and Risler and Kempter (1961).

External Anatomy

Melis (1934*a*; 1936; 1939), Doeksen (1941) and Jones (1954) have given good general accounts of the external morphology of several species. The head (Risler, 1957; Mickoleit, 1963) is generally somewhat quadrangular in form with a pair of small but prominent compound eyes, the facets of which are relatively large, convex and rounded. Three ocelli are present on the vertex of winged forms but are absent from apterous ones. Nearly all the head sclerites are intimately fused, almost all traces of sulci being lost and the tentorium is greatly reduced. The antennae are 6- to 10-segmented, and are inserted close together in a very forward position. They bear setae and chemoreceptor sensilla (Slifer and Sekhon, 1974) whose distribution may be taxonomically important. The mouthparts are adapted for piercing and suction, certain of the organs being modified as stylets which are enclosed in a short cone or rostrum, projecting downwards from the ventral surface of the head (Fig. 348). Reyne (1927) has made a detailed study of the structure and development of the mouthparts and his interpretation is followed here. The mouth-cone is formed by the labrum and clypeus above, and the labium below, while the piercing organs are protruded through the short tubular base thus formed. Among the Terebrantia the mandibles of the two sides are totally unlike: the left organ is a strong sclerotized stylet while the right one is absent in all postembryonic stages. The maxillae consist of a pair of palp-bearing plates with associated stylets. The plates (which probably represent stipites) may be either symmetrical or unlike and they form the side walls of the mouth-cone. The palps are composed of 2–8 segments among different genera. Each stylet consists of a small basal piece articulating with the palp-bearing plate of its side, and a long piercing organ which is usually divided into a proximal and a distal element. The maxillary stylets are usually longer and more flexible in the Tubulifera; a few species show excessive elongation and convolution within the head capsule (Mound, 1970). The labium forms the trough-like floor of the mouth-cone and is divisible into a prementum and postmentum. The membranous apex of the prementum is more or less

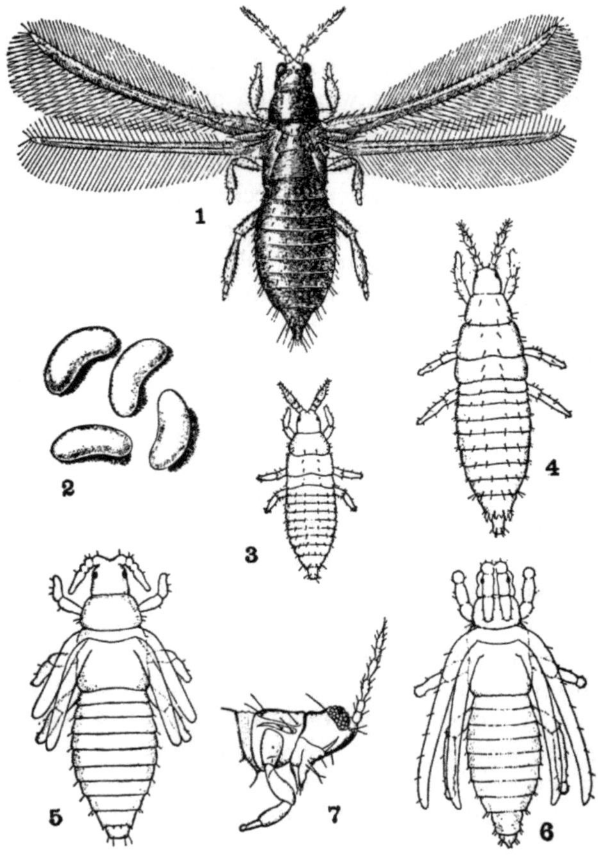

FIG. 347 The Pear Thrips (*Taeniothrips inconsequens*)

1, Imago; 2, eggs; 3, 1st instar nymph; 4, fully-grown nymph; 5, prepupa; 6, pupa; 7, lateral view of head of imago. Reduced from Foster and Jones, *U.S. Dept. Agric. Bull.*, 173.

bilobed and carries a pair of short labial pulps which are 1- to 4-segmented. A small hypopharynx is present. Among the Tubulifera certain differences in the mouthparts are noticeable. The mandible has come to articulate with the palp-bearing plate of its side and the two maxillary stylets have acquired separate, more posterior, articulations with the head-capsule. These are secondary differences, however, since the mouthparts of the Tubuliferan larvae are similar to those of the Terebrantia. The feeding mechanisms of

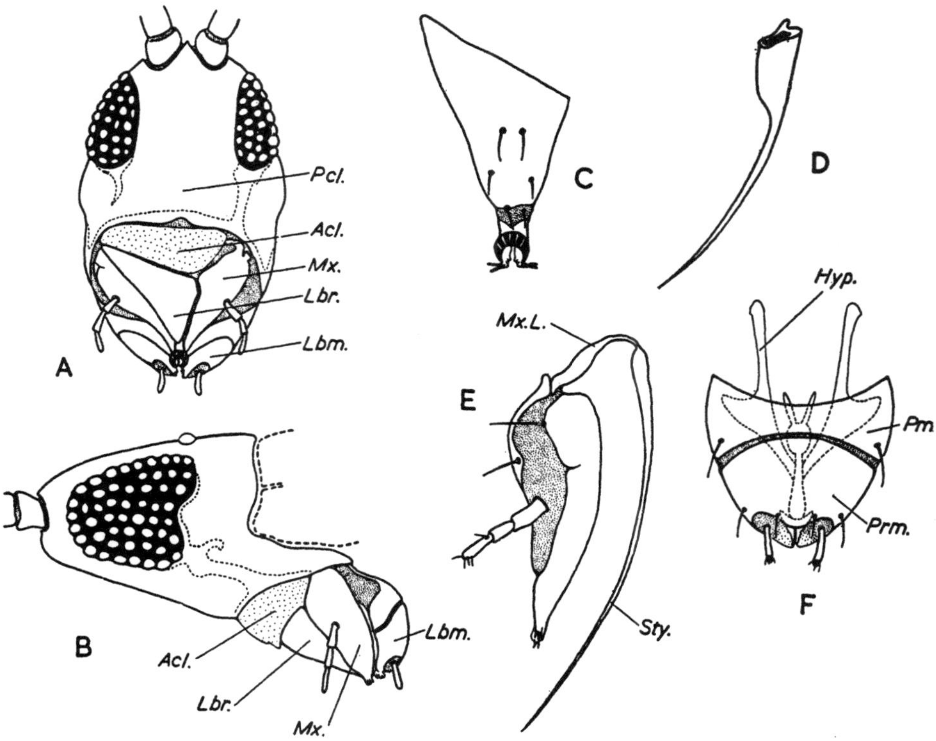

FIG. 348 Head and mouthparts of *Chirothrips hamatus* (*after* Jones, 1954). A. Frontal view of head. B. Lateral view of head. C. Labrum. D. Left mandible. E. Maxilla. F. Labium and hypopharynx

Acl, anteclypeus; *Hyp*, hypopharynx; *Lbm*, labium; *Lbr*, labrum; *Mx*, maxilla; *Mx.L*, maxillary lever; *Pcl*, postclypeus; *Pm*, postmentum; *Prm*, prementum. (N.B. *Pm* and *Prm* correspond to the submentum and mentum of other authors.)

Thysanoptera are not fully understood and may well differ somewhat between phytophagous forms which suck sap from the leaves and flowers of plants, predacious species that imbibe the body fluids of their prey and others that suck out the contents of fungal spores or pollen grains (Grinfel'd, 1959) or ingest intact spores. In general, thrips apply the mouth-cone to the surface of the plant or animal and the stylets are driven into the tissues. The exuding juices are then sucked up by a cibarial pump, either via the apex of the oral cone or through a channel formed by tongue-and-groove joints between the closely apposed maxillary stylets (Mound, 1971).

The prothorax is free and distinct, with a broad tergum, while the meso- and metathorax are compactly united (Mickoleit, 1961; Priesner, 1957). The legs are composed of the usual parts and only the tarsi present special features. These are 1- or 2-segmented, and bear a pair of weak claws. A remarkable protrusible vesicle is associated with the extremity of the tarsus, and it is to the presence of this organ that the other ordinal name Physapoda is due. When at rest, the vesicle (a modified arolium) is retracted and invisible, but when the insect is walking it is everted by blood pressure and enables the insect to walk upon almost any kind of surface (Heming, 1971, 1972). The wings are membranous, very narrow and strap-shaped; they have very few or no veins, and only rarely possess cross-veins. They are fringed with long setae and some species bear spines along the veins or their former courses. The wings of a side are coupled by several hooked spines near the base of the hind wing which engage a membranous fold on the anal area of the fore wing. The adults of many species exhibit striking variations in the degree of development of the wings (Bournier, 1961). A single species such as *Chirothrips manicatus* may have fully developed wings, reduced functionless wings, or be completely apterous. In other species both sexes may be winged or one winged and the other apterous; one or both sexes may be brachypterous or both may be wingless. When brachypterous forms occur among normally winged individuals the phenomenon is especially evident towards autumn.

The abdomen is long, tapering posteriorly and composed of 11 segments, though the first is reduced and the terminal ones may be modified in connection with the external genitalia. In the Tubulifera there is a long, cylindrical tenth segment and no ovipositor. In the Terebrantia the tenth segment is small and the ovipositor is a conspicuous serrated structure derived from paired appendages of the eighth and ninth abdominal segments. In males of both suborders there is an eversible aedeagus flanked by one or two pairs of parameres whose homologies are uncertain (De Gryse and Treherne, 1924); these male genitalia are sometimes taxonomically useful (e.g. Pitkin, 1972).

Internal Anatomy – Most of what is known concerning the internal structure of Thysanoptera is due to Jordan (1888), Uzel (1895), Klocke (1926), Müller (1927) and Sharga (1933). The digestive system is characterized by a cibarial sucking pump provided with radial muscles, a long oesophagus, a small crop, an extensive mid gut and four Malpighian tubules. The mid gut forms the largest portion of the alimentary canal and is divided into a capacious anterior chamber followed by a tubular coiled posterior region. The hind gut forms a straight passage to the anus and bears four rectal papillae. Two or three pairs of *salivary glands* are present and are located in the thorax and abdomen; their ducts unite to form a common canal opening into the front of the oesophagus. The *nervous system* is highly concentrated: the brain is well developed, the suboesophageal and prothoracic ganglia are fused while the meso- and metathoracic ganglia remain separate. A median nerve-cord passes backwards, but the ganglia have shifted forward and are concentrated into a single centre in the first abdominal segment. The *circulatory system* consists of a very short contractile heart in the 8th abdominal segment and a long aorta. In the *female reproductive* organs the ovaries each consist of four short panoistic ovarioles; a small pigmented spermatheca and a larger, sac-like accessory gland open into the vagina. The *male reproductive organs* consist of a pair of fusiform testes which communicate by rather short vasa deferentia with an ejaculatory duct. The latter is somewhat swollen proximally, forming an ampulla-like enlargement. At this point it receives the ducts of one or two pairs of relatively large accessory glands which considerably exceed the testes in size. The *tracheal system* is well developed and usually opens to the exterior by means of four pairs of spiracles. There is a pair on the mesothorax, another on the metathorax (small and easily overlooked in the Terebrantia) and a pair on the 1st and 8th abdominal segments.

Life-History and Metamorphoses (Fig. 347) – The eggs of the Terebrantia are more or less reniform, while those of the Tubulifera are commonly elongate-oval. In the first suborder the female cuts a slit with her saw-like ovipositor, laying the eggs singly in the tissues of the host plant. The Tubulifera lay their eggs externally, either singly or in groups, upon leaves, stems, under bark, etc. The newly hatched larvae resemble the adults in general facies and feeding habits but are less strongly sclerotized, without wing-pads and with fewer antennal segments (Priesner, 1926, 1958; Speyer and Parr, 1941). This stage is followed by a very similar second instar then, in the Terebrantia, by a prepupa and a pupa. The Tubulifera, however, have an additional pupal instar. The prepupa and pupa have conspicuous wing-pads but do not feed and their pretarsi lack the characteristic protrusible vesicles of the adults and first two larval stages (Heming, 1973). They differ in that the antennae of the pupa are bent back over the thorax. The prepupa and pupae are inactive but capable of slow movement when stimulated, and though *Heliothrips haemorrhoidalis* and a few others species pupate on the host-plant, most Thysanoptera do so among debris or under the ground in an earthen cell. The Thysanopteran life-cycle is thus especially interesting since the members of this Exopterygote group are clearly holometabolous and have presumably evolved the condition quite independently of the Endopterygote insects. The prepupa and pupa resemble the Endopterygote pupa in the differentiation of the reproductive system (Davies, 1961; Heming, 1970*a*, 1970*b*), the metamorphosis of the skeleto-muscular system (Davies, 1969), and the histolysis and replacement of the mid gut epithelium (Müller, 1927; Melis, 1934*b*). The number of generations passed through in the year differs from one species to another and from place to place. In cool temperate regions most species have one or two generations, with hibernation by adult females or, less often, by both sexes or by larvae or pupae (Lewis, 1973). In warm regions or under glasshouse conditions continuous breeding may occur with up to 12 generations annually and all stages present at any time.

Affinities – The Thysanoptera are an isolated group whose affinities are not well established. They are first known from the Jurassic but the fossils concerned provide no indication of the origin of the order. Morphologically, however, the thrips share with the remaining Hemipteroid insects such specializations as the absence of cerci, reduced tarsal segmentation, small number of Malpighian tubules and concentrated nervous system, while the stylet-like mandibles and maxillae perhaps suggest a common origin with the Hemiptera. Tertiary fossils (Priesner, 1929, 1968) include a relatively high proportion of Aeolothripoidea. This confirms the widely held view that the superfamily contains the most generalized Recent Thysanotera though there are also structural and behavioural reasons for regarding the Merothrippoidea as primitive (Mound and O'Neill, 1974). See also zur Strassen (1973).

Classification – About 5000 described species, from all major zoogeographical regions, have been catalogued by Jacot-Guillarmod (1970–74). The early monograph by Uzel (1895) has been supplemented by the generic keys of Priesner (1949) and a number of regional studies on the Thysanopteran faunas of North America (Watson, 1923; Cott, 1956; Bailey, 1957; Stannard, 1956, 1968), Europe (Priesner, 1928, 1964), Egypt (Priesner, 1960) and India (Ananthakrishnan, 1969). For the 180 or so British species see Morison (1949, 1957), Mound (1967, 1968) and Pitkin (1969). The following classification, based on Priesner (1968), distinguishes the two suborders Terebrantia and Tubulifera, containing five families and one family respectively. Two of these, the Thripidae and the Phlaeothripidae, are the major groups (including all the economically important species). They and the Aeolothripidae are each divided by Priesner into a few subfamilies, some of which are given family status in other classifications.

Suborder 1. TEREBRANTIA

Ovipositor saw-like; apex of abdomen conical in female, bluntly rounded in male; wings usually covered with microtrichia, fore wings with at least one longitudinal vein reaching to apex.

Superfamily **Aeolothripoidea**

Ovipositor curved dorsalwards; wings relatively broad, with several cross-veins; antennae 9-segmented.

FAM. AEOLOTHRIPIDAE. The best-known genus is *Aeolothrips*, with numerous species; they are predacious, the larvae pupate in a silken cocoon in the soil and the adults usually have transversely banded wings. *Melanthrips* is often associated with Cruciferae and *Ankothrips* with conifers.

Superfamily **Thripoidea**

Ovipositor curved downwards; wings more or less narrow, pointed; antennae with 6–10 segments, the last 1–3 forming a thin style.

FAM. HETEROTHRIPIDAE. *Antennae 9- or 10-segmented, with short, triangular sense-cones.* A mainly subtropical family, though *Heterothrips* is Nearctic; *H. azaleae* conveys the spores of the fungus causing azalea flower spot.

FAM. THRIPIDAE. *Antennae 6- to 9-segmented, with slender, simple or forked sense-cones.* This large and important family consists mainly of sap-feeding species arranged in over 160 genera and including many injurious forms. *Heliothrips haemorrhoidalis* is a polyphagous species damaging glasshouse crops and *Scirtothrips aurantii* is a pest of citrus in South Africa. Other economically important species include *Diarthrothrips coffeae* (on coffee), *Kakothrips robustus* (on peas in Europe), *Taeniothrips inconsequens* (on fruit trees in N. America), *T. simplex* (on gladioli), *Limothrips cerealium* (on cereals) and *Thrips tabaci*, mentioned above on p. 782. *Taeniothrips ericae* pollinates *Calluna* and is therefore indirectly responsible for successful sheep-rearing on the Faroes (Hagerup, 1950). The adults of several species, such as *Thrips fuscipennis, T. major, T. flavus, Taeniothrips atratus* and *T. vulgatissimus*, congregate in large numbers on many different flowers though their larvae feed on other plants.

FAM. UZELOTHRIPIDAE. Contains only the S. American *Uzelothrips scabrosus* with an unusually long terminal (7th) antennal segment.

Superfamily **Merothripoidea**

Ovipositor reduced or absent; pronotum usually with a longitudinal suture on each side; antennae 8- or 9-segmented.

FAM. MEROTHRIPIDAE. The 23 species are mostly placed in *Merothrips*, with a wide geographical distribution. They live in leaf litter or recently dead wood and many species include apterous and macropterous morphs, the latter sometimes bearing the stubs of their broken off wings.

Suborder 2. TUBULIFERA

Ovipositor absent; tenth abdominal segment usually tubular; wings without microtrichia, veins absent or with one vestigial vein.

FAM. PHLAEOTHRIPIDAE. With the characters of the suborder. This large family contains over 300 genera, with a variety of habits. *Haplothrips* includes predatory and phytophagous species, of which *H. tritici* is economically important, as is also the olive thrips *Liothrips oleae*. Others such as *Phlaeothrips* and *Cryptothrips* are associated with the bark of trees and several genera inhabit galls. *Megathrips* and its allies feed on fungal spores and occur in leaf litter or under bark, while *Urothrips* and related genera are found in the soil and among surface vegetation. The family includes the largest thrips, members of a few tropical genera such as *Phasmothrips* measuring up to 12 mm in length.

Literature on the Thysanoptera

ANANTHAKRISHNAN, T. N. (1969), Indian Thysanoptera, *C.S.I.R. Zool. Monogr.*, 1, 171 pp.

BAILEY, S. F. (1957), The thrips of California, Part 1. Suborder Terebrantia, *Bull. Calif. Insect Surv.*, 4 (5), 143–220.

BLUNCK, H. (1950), Thysanoptera, In: Sorauer, *Handbuch der Pflanzenkrankheiten*, Parey, Berlin, 5th edn, 4, 374–427.

BOURNIER, A. (1956), Contribution à l'étude de la parthenogenèse des Thysanoptères et de sa cytologie, *Arch. Zool. exp. gén.*, **93**, 221–317.

—— (1961), Remarques au sujet du brachyptérisme chez certains espèces de Thysanoptères, *Bull. Soc. ent. Fr.*, **66**, 188–191.

—— (1966), L'embryogenèse de *Caudothrips buffai* Karny, *Ann. Soc. ent. Fr.*, (N.S.), **11**, 415–435.

—— (1970), Principaux types de dégats de Thysanoptères sur les plantes cultivées, *Ann. Zool. Ecol. anim.*, **2**, 237–259.

CEDERHOLM, L. (1963), Ecological studies on Thysanoptera, *Opusc. ent. Suppl.*, **22**, 215 pp.

COTT, H. E. (1956), Systematics of the suborder Tubulifera (Thysanoptera) in California, *Univ. Calif. Publns Ent.*, **13**, 1–216.

DAVIES, R. G. (1961), The postembryonic development of the female reproductive system in *Limothrips cerealium, Proc. zool. Soc. Lond.*, **136**, 411–437.

—— (1969), The skeletal musculature and its metamorphosis in *Limothrips cerealium* Haliday (Thysanoptera: Thripidae), *Trans. R. ent. Soc.*, **121**, 167–233.

DE GRYSE, J. J. AND TREHERNE, R. C. (1924), The male genital armature of the Thysanoptera, *Can. Ent.*, **56**, 177–182.

DOEKSEN, J. (1941), Bijdrage tot de vergelijkende morphologie der Thysanoptera, *Meded. Landbouhoogesch., Wageningen*, **45** (5), 114 pp.

GRINFEL'D, E. K. (1959), The feeding of thrips (Thysanoptera) on pollen of flowers and the origin of asymmetry in their mouthparts, *Ent. Obozr.*, **38**, 798–804.

HAGERUP, O. (1950), Thrips pollination in *Calluna, Danske Vidensk. Selskab., Biol. Meded.*, **18**, 1–16.

HEMING, B. S. (1970*a*), Postembryonic development of the female reproductive system in *Frankliniella fusca* (Thripidae) and *Haplothrips verbasci* (Phlaeothripidae) (Thysanoptera), *Misc. Publ. ent. Soc. Am.*, **7** (2), 197–234.

—— (1970*b*), Postembryonic development of the male reproductive system in *Frankliniella fusca* (Thripidae) and *Haplothrips verbasci* (Phlaeothripidae) (Thysanoptera), *Misc. Publ. ent. Soc. Am.*, **7** (2), 235–272.

—— (1971), Functional morphology of the Thysanopteran pretarsus, *Canad. J. Zool.*, **49**, 91–108.

—— (1972), Functional morphology of the pretarsus in larval Thysanoptera, *Canad. J. Zool.*, **50**, 751–766.

—— (1973), Metamorphosis of the pretarsus in *Frankliniella fusca* (Hinds) (Thripidae) and *Haplothrips verbasci* (Osborn) (Phlaeothripidae) (Thysanoptera), *Canad. J. Zool.*, **51**, 1211–1234.

JACOT-GUILLARMOD, C. F. (1970–74), Catalogue of the Thysanoptera of the World. I–III, *Ann. Cape Prov. Mus. nat. Hist.*, **7**, 1–216, 217–515.

JONES, T. (1954), The external morphology of *Chirothrips hamatus* (Trybom) (Thysanoptera), *Trans. R. ent. Soc. Lond.*, **105**, 163–187.

JORDAN, K. (1888), Anatomie und Biologie der Physapoda, *Z. wiss. Zool.*, **47**, 541–620.

KLOCKE, F. (1926), Beiträge zur Anatomie und Histologie der Thysanoptera, *Z. wiss. Zool.*, **128**, 1–36.

LEWIS, T. (1964), The weather and mass flights of Thysanoptera, *Ann. appl. Biol.*, **53**, 165–170.

—— (1973), *Thrips, their Biology, Ecology and Economic Importance*, Academic Press, London and New York, 350 pp.

MELIS, A. (1934*a*) Tisanotteri italiani, Studio anatomico-morfologico e biologico del Liotripide dell'olivo (*Liothrips oleae* Costa), *Redia*, **21**, 1–183.
—— (1934*b*), Nuove osservazioni anatomo-istologiche sui diversi stati post-embrionali del *Liothrips oleae* Costa, *Redia*, **21**, 263–334.
—— (1936), Tisanotteri italiani, Genus *Taeniothrips, Redia*, **22**, 53–95.
—— (1939), Tisanotteri italiani, Genus *Haplothrips, Redia*, **25**, 37–86.
MICKOLEIT, E. (1963), Untersuchungen zur Kopfmorphologie der Thysanopteren, *Zool. Jb.* (*Anat.*), **81**, 101–150.
MICKOLEIT, G. (1961), Zur Thoraxmorphologie der Thysanoptera, *Zool. Jb.* (*Anat.*), **79**, 1–92.
MORISON, G. D. (1949), Thysanoptera of the London area, *London Naturalist Reprint*, **59**, 1–131.
—— (1957), A review of British glasshouse Thysanoptera, *Trans. R. ent. Soc. Lond.*, **109**, 467–534.
MOUND, L. A. (1967), The British species of the genus *Thrips*. Thysanoptera, *Ent. Gaz.*, **18**, 13–22.
—— (1968), A review of R. S. Bagnall's Thysanoptera collections, *Bull. Br. Mus. nat. Hist., Suppl.*, **11**, 181 pp.
—— (1970), Convoluted maxillary stylets and the systematics of some Phlaeothripine Thysanoptera from *Casuarina* trees in Australia, *Aust. J. Zool.*, **18**, 439–463.
—— (1971), The feeding apparatus of thrips, *Bull. ent. Res.*, **60**, 457–458.
MOUND, L. A. AND O'NEILL, K. (1974), Taxonomy of the Merothripidae, with ecological and phylogenetic considerations, *J. nat. Hist.*, **8**, 481–509.
MÜLLER, K. (1927), Beiträge zur Biologie, Anatomie, Histologie und inneren Metamorphose der Thrips-Larven, *Z. wiss. Zool.*, **130**, 215–303.
PITKIN, B. R. (1969), New records of Thysanoptera in the British Isles, *Ent. mon. Mag.*, **105**, 201–202.
—— (1972), A revision of the flower-living genus *Odontothrips* Amyot & Serville (Thysanopt., Thripidae), *Bull. Br. Mus. nat. Hist., Ent.*, **26**, 373–402.
PRIESNER, H. (1926), Die Jugendstadien der malayischen Thysanopteren, *Treubia*, **8**, (Suppl.), 1–264.
—— (1928), *Die Thysanopteren Europas*, Wagner, Vienna, 755 pp.
—— (1929), Bernstein-Thysanopteren, *Bernsteinforschungen*, **1**, 111–138.
—— (1949), Genera Thysanopterorum, Keys for the identification of the genera of the order Thysanoptera, *Bull. Soc. Fouad I. Ent., Cairo,* **33**, 31–157.
—— (1957), Zur vergleichenden Morphologie des Endothorax der Thysanopteren, *Zool. Anz.*, **159**, 159–167.
—— (1958), Geschlechtsunterschiede an den Larven der Thysanopteren, *Z. Wiener ent. Ges.*, **43**, 247–249.
—— (1960), A Monograph of the Thysanoptera of the Egyptian deserts, *Publ. Inst. Desert Egypte, Cairo*, **13**, 1–549.
—— (1964), Ordnung Thysanoptera (Fransenflügler, Thripse), *Bestimmungsbücher zur Bodenfauna Europas*, Lf. **2**, 242 pp.
—— (1968), Thysanoptera (Physapoda, Blasenfüsser), In: Kükenthal, *Handb. d. Zoologie*, (2nd edn, ed. Beier), **4**, (2), Lief. **5**, 32 pp.
PUSSARD-RADULESCO, E. (1931), Recherches biologiques et cytologiques sur quelques Thysanoptères, *Annls Epiphyt.*, **16**, 103–177.

REYNE, A. (1927), Untersuchungen über die Mundteile der Thysanopteren, *Zool. Jb., Anat.*, **49**, 391–500.

RISLER, H. (1957), Der Kopf von *Thrips physapus* L. (Thysanoptera, Terebrantia), *Zool. Jb., Anat.*, **76**, 251–302.

RISLER, H. AND KEMPTER, E. (1961), Die Haploidie der Männchen und die Endopolyploidie in einigen Geweben von *Haplothrips, Chromosoma*, **12**, 351–361.

SHARGA, U. S. (1933), On the internal anatomy of some Thysanoptera, *Trans. R. ent. Soc. Lond.*, **81**, 185–204.

SIMMONDS, H. W. (1933), The biological control of the weed *Clidemia hirta* D. Don., in Fiji, *Bull. ent. Res.*, **24**, 345–348.

SLIFER, E. H. AND SEKHON, S. S. (1974), Sense organs on the antennae of two species of thrips (Thysanoptera, Insecta), *J. Morph.*, **143**, 445–456.

SPEYER, E. R. AND PARR, W. J. (1941), The external structure of some thysanopterous larvae, *Trans. R. ent. Soc. Lond.*, **91**, 559–635.

STANNARD, L. J. (1957), The phylogeny and classification of the North American genera of the suborder Tubulifera (Thysanoptera), *Illinois biol. Monogr.*, **25**, 1–200.

—— (1968), The Thrips, or Thysanoptera, of Illinois, *Bull. Ill. nat. Hist. Surv.*, **29** (4), 552 pp.

STRASSEN, R. ZUR (1973), Insektenfossilien aus der unteren Kreide, 5. Fossile Fransenflügler aus mesozoischem Bernstein des Libanon, *Stuttg. Beitr. Naturk.*, A (Biol.), **256**, 1–51.

UZEL, H. (1895), *Monographie der Ordnung Thysanoptera*, The author, Königgrätz, 472 pp.

WATSON, J. R. (1923), Synopsis and catalogue of the Thysanoptera of North America, *Bull. Florida agric. exp. Sta.*, **168**, 100 pp.

Order 21

NEUROPTERA (ALDER FLIES, SNAKE FLIES, LACEWINGS, ANT LIONS, ETC.)

Small to rather large soft-bodied insects with usually elongate antennae. Mouthparts adapted for biting: ligula undivided or bilobed or often atrophied. Two pairs of very similar membranous wings, generally disposed in a roof-like manner over the abdomen when at rest. Venation primitive but with many accessory veins: costal veinlets numerous: Rs often pectinately branched. Abdomen without cerci. Larvae carnivorous, of a modified campodeiform type with biting or suctorial mouthparts: the aquatic forms usually with abdominal gills. Pupae exarate, decticous: wings with complete tracheation.

The heterogeneous group which formed the Neuroptera of Linnaeus is now divided into at least eight or nine well-defined orders, the original name being confined to the Megaloptera and Planipennia as enumerated below. The group thus restricted is still further dismembered by some authorities into two or three separate orders. Although it is evident that the Neuroptera exhibit at least three lines of evolution including marked divergence in their metamorphoses these several lines appear to be derivable from a common ancestral type. The species are rarely abundant in individuals, and most have weak powers of flight. They feed upon soft-bodied insects and liquid matter, such as honey-dew.

With the exception of the Coniopterygidae, the Neuroptera are separable from the Mecoptera by the venational features enumerated above. The mouthparts are well developed with biting mandibles, the maxillary palps are 5-segmented, the labial palps 3-segmented, and the ligula is reduced to the condition of a median and sometimes slightly bilobed process, or is totally atrophied. The wing-coupling apparatus is of the jugo-frenate type, though small, usually reduced and with distinct bristles; a frenulum, however, is present in Hemerobiidae. The tarsi have five segments and the abdomen ten. The morphology of the male genitalia (Killington, 1936; Tjeder *in* Tuxen

(1970); Acker, 1960; Adams, 1969). An aedeagus may be present (Hemerobiidae), absent (Osmylidae), or even present or absent within one family (Coniopterygidae). In the first-named family it is associated with sternite 10. Gonocoxites probably occur in most groups (though not in *Sialis*) and may be articulated to the 9th tergite or to a structure, the 'gonarcus', formed of the fused parameres. In both sexes, tergite 10 commonly bears a group of trichobothria on each side. In the female, the genital aperture is behind sternite 8 which is rather reduced and may bear a pair of gonapophyses. Sternite 9 is more or less divided and may function as an ovipositor and its lobes may, indeed, be the coxites of an appendage. This seems more probable in *Dilar* and *Raphidia* which have long ovipositors (Tjeder, 1937; but cf. Ferris and Pennebaker, 1939). The egg-pore and the copulatory pore are independent in the Megaloptera but not in the Planipennia (except the Mantispidae). The internal anatomy of the order has been very inadequately investigated. There are two pairs of thoracic and eight pairs of abdominal spiracles, and the ventral nerve-cord consists of three thoracic and generally seven abdominal ganglia. The digestive system is provided with a median dorsal food-reservoir, a peritrophic membrane is present, and the usual number of Malpighian tubes is eight: the ovaries consist of a variable number of usually polytrophic ovarioles.

The larvae exhibit great diversity of structure and mode of life, but are, in all cases, carnivorous; in a considerable proportion of the species they are aquatic. The latter forms usually carry segmentally arranged, and often jointed, abdominal processes.

The Neuroptera are divided in the present work into the suborders Megaloptera and Planipennia, which are treated separately below. The British species number 60, and for information concerning them see Killington (1936–37). Stitz (1927) deals with the C. European species and 4700 species of the order have been described.

Suborder I. MEGALOPTERA
(Alder Flies and Snake Flies)

Branches of veins without a marked tendency to bifurcate at the wing-margin or prothorax elongate and female with relatively long ovipositor. Larvae with biting mouthparts.

The Megaloptera fall into two superfamilies – the Sialoidea or 'alder flies' and Raphidioidea or 'snake flies'. Some authors treat these groups as orders but Achtelig (1967) shows that in at least the anatomy of the head *Corydalis* and *Chauliodes* are nearer to *Raphidia* than they are to *Sialis*. It seems better therefore to maintain one suborder the British species of which are described by Killington (1930).

The suborder is classified as follows:

1.(3). Prothorax quadrate. Exserted ovipositor wanting. Wings without a pterostigma. Larvae aquatic. Superfamily **Sialoidea**.

2. Wing-expanse 45–100 mm. Three ocelli present. Fourth tarsal segment simple. Larvae with 8 pairs of abdominal gills, a pair of terminal dorsal spiracles but no filament. **CORYDALIDAE.**
   Wing-expanse 20–40 mm. Ocelli absent. Fourth tarsal segment bilobed. Larva with 7 pairs of abdominal gills, no terminal dorsal spiracles but a long filament. **SIALIDAE.**

3(1). Prothorax elongate. Exserted ovipositor present. Wings with a pterostigma. Larva terrestrial. Superfamily **Raphidioidea. RAPHIDIIDAE.**

The SIALOIDEA are of special interest both on account of the large size and striking appearance assumed by certain of the species, and because the group includes the most generalized representatives of the Neuroptera. Like other primitive groups, the Sialoidea only include a small number of genera and species, but they have an almost world-wide although discontinuous distribution. They differ from other Neuroptera in the hind-wings being broad at their bases with the anal area folded fanwise when at rest (Fig. 349).

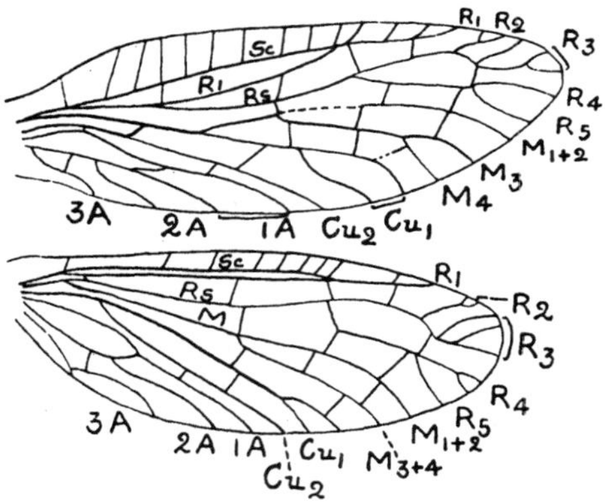

FIG. 349
Right wings of *Sialis lutaria*

The eggs are laid upon leaves, stones and other objects, usually not far from water. They are deposited regularly in compact masses: in *Sialis* each mass contains 200 to 500 eggs and in *Corydalis* the number amounts to two or three thousand. The eggs are cylindrical with rounded ends and dark brown in colour: at its free extremity each is provided with a conspicuous micropylar apparatus varying somewhat in form among different genera. The young larvae, after eclosion, make their way to the water: those of *Sialis* are found in the muddy bottoms of ponds, canals and slow-moving streams, while the larvae of *Corydalis* lurk under stones in rapidly flowing water. All Corydalid larvae seem to be able to live for long periods under stones in dried up streams; they can also breathe through their posterior spiracles if these are opened in the air. All the larvae of the Sialoidea are actively predacious, devouring other insect larvae, small worms, etc. The mouthparts resemble those of a Carabid larva, the mandibles being powerful and sharply toothed, while the maxillae exhibit the typical parts and the labium consists of a mentum, a dentate ligula and 3-segmented palpi. The antennae are prominent 4-segmented appendages, and the legs are well developed, ter-

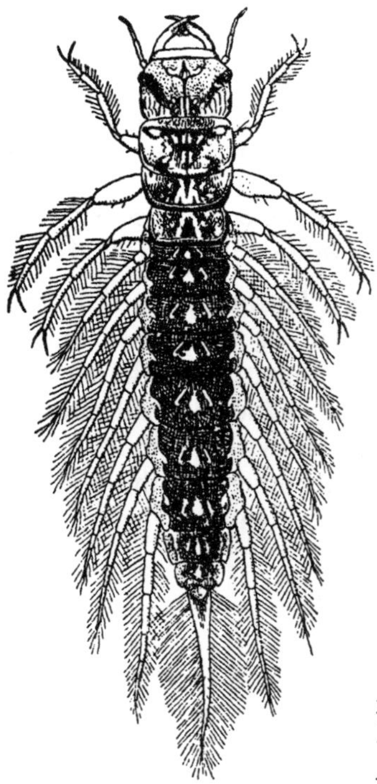

FIG. 350
Larva of *Sialis lutaria*, enlarged
After Lestage.

minating in paired claws. The larva of *Sialis* (Fig. 350) is provided with seven pairs of 5-segmented, lateral, segmentally arranged abdominal filaments or tracheal gills. Each of the latter is supplied by a tracheal branch and contains blood. On the 9th abdominal segment there is a terminal filament of a similar nature. In *Corydalis* and *Chauliodes* there are eight pairs of unjointed or imperfectly jointed filaments: in the former genus, and in *Neuromus*, each of the first seven abdominal segments also bears ventral, spongy tufts of accessory tracheal gills. The body in these three genera is terminated by a pair of hooked anal feet, without the gill-like filament of *Sialis*. There are eight pairs of small abdominal spiracles in *Sialis*, while in *Corydalis* and *Chauliodes* (Cuyler, 1958) thoracic spiracles are also present. Pupation occurs in the soil or in moss, etc., sometimes at a depth of several inches. The pupae are exarate and are able to work their way to the surface to allow the adults to emerge. In the common European *Sialis lutaria* the whole life-cycle occupies about a year. This species and *S. fuliginosa* are the only British members of the superfamily (Killington, 1930). *Corydalis* is North American and Asiatic and the male has enormously elongate, sickle-like mandibles and a wing-expanse ranging up to 150 mm. The larvae of *Neohermes* and *Chauliodes* are described by Smith, H. L. (1970). Some account of the anatomy of *Sialis* is given by Dufour (1841) and by Loew (1848). The skeleton and muscles of the head and thorax are described in *Chauliodes* by Kelsey (1954, 1957).

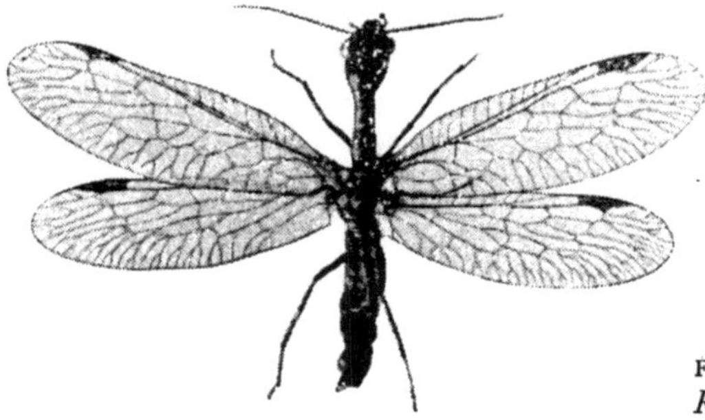

FIG. 351
Raphidia notata, Britain. × 2½

The RAPHIDIOIDEA are the most specialized Megaloptera and are entirely terrestrial in their habits. They occur in all continents except Australia and southern Africa. Most species are included in the genera *Raphidia* (Fig. 351) and *Inocellia*. The imagines are remarkable for the elongated prothorax which, together with the narrowed posterior region of the head, forms a kind of 'neck': unlike the Sialoidea, they possess an elongate setiform ovipositor. More than eighty species of the group are known, of which four, belonging to the genus *Raphidia*, have been recorded from Britain. They occur in wooded regions and are met with among rank herbage, on flowers or tree-trunks, etc. The eggs are inserted by means of the long ovipositor in slits in the bark: they are elongate-cylindrical with a

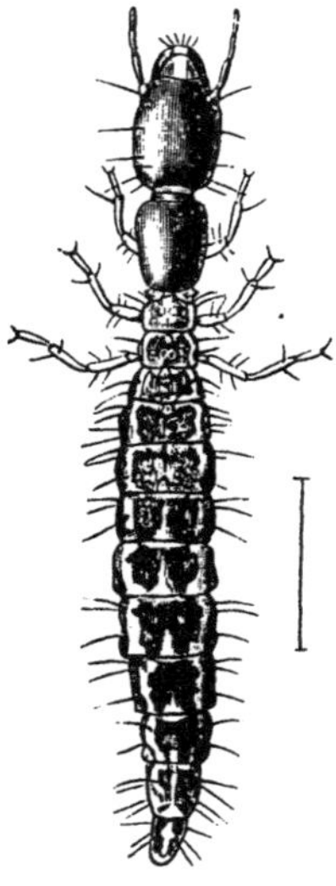

FIG. 352
Raphidia notata, larva
After Sharp, *Camb. Nat. Hist.*

small appendage at one extremity. The larvae occur under loose bark, particularly of conifers, and are very voracious, preying upon small soft-bodied insects which frequent similar situations. The larva of *Raphidia* (Fig. 352) is elongate and slender with a well sclerotized head and prothorax. The thoracic legs are long, the abdomen carries no processes or appendages, and the mouthparts resemble those of the imago. The pupa is more primitive than in any other of the Endopterygota and closely resembles the adult insect in its essential structural characters. Although first enclosed in a kind of cell the pupa emerges after a lapse of some time and, becoming active, crawls about until it finds a suitable place, where it remains until the eclosion of the imago. An interesting account of an American species, *Agulla astuta*, is given by Woglum and McGregor (1959).

Aspöck and Aspöck (1968, 1971 and other papers) have described the Palaearctic species of Raphidiidae and Carpenter (1936) the North American ones.

Suborder 2. PLANIPENNIA
(Lacewings, Ant Lions, etc.)

Branches of the veins usually conspicuously bifurcated at the margins of the wings. Larvae with suctorial mouthparts.

The Planipennia include the majority of the Neuroptera and an exceptional wealth of venational specialization occurs in the various families. Different as many of the families are in their imaginal characters, the group is well defined as a whole owing to the universal occurrence of suctorial piercing mouthparts in the larvae. Nearly all the Planipennia are terrestrial insects, a small number are more or less amphibious in their larval stages, and one or two genera have truly aquatic larvae.

The larvae of the Planipennia are universally predacious and are of considerable importance as destroyers of aphids and other injurious insects. The head is often large and very freely articulated with the prothorax. The mandibles and maxillae are long and exserted, being thereby adapted for seizing the prey (Meinert, 1889). The first-mentioned appendages are usually sickle-shaped and, in some families, armed with teeth. They are grooved along their ventral surfaces, and a lobe of the maxillae (perhaps the lacinia) which closely resembles them in size and shape, fits one into each groove: in this manner the two sets of appendages form a pair of imperfect suctorial tubes. The combined organs are deeply inserted into the prey and its juices are imbibed by means of the pumping action of the pharynx. At the base of each maxilla there is usually a pair of small sclerites – the cardo and stipes – but, as a rule, maxillary palpi are absent. The labium is greatly reduced, and its palps, although sometimes aborted, are very variable in different families. The antennae are filiform and often rather long. The prothorax is divided into three more or less distinct subsegments, but the meso- and metathorax are sometimes merged into the trunk and not sharply demarcated. The legs are long and slender and allow active movement; their tarsi are of one segment. The abdomen consists of ten segments and is devoid of cerci. The genitalia vary considerably but the homologies of the parts are not yet agreed (Tjeder, 1954; Archer, 1960; Adams, 1969). The larvae usually pass through three instars, except in *Ithone* where there are five: when about to pupate, they construct oval or spherical cocoons either of silk or of foreign particles bound together with that material. The pupae possess strong mandibles which are utilized in cutting through the cocoons to allow of the emergence of the imagines. The diet of the larvae consists solely of animal juices, and there is no through passage from the hind intestine to the anus. The Malpighian tubes are usually eight in number and, of these, six have acquired a secondary attachment by their distal extremities to the wall of the hind intestine. The tubes thus modified function as silk-producing organs in the last instar, the silken thread being emitted by means of an anal spinneret (Anthony, 1902). The respiratory system opens by nine pairs of spiracles, the 1st pair being prothoracic and the remainder abdominal in position.

The Planipennia fall into 16 families of which five are British and have been revised by Withycombe (1925), Killington (1936–37) and Fraser (1959). The families can be arranged, following Riek (1970), in five super-families.

Superfamily 1. **Coniopterygoidea**

Small insects with mealy wings; Rs with two branches, veins not bifurcating near margins. Larva with only two segments in antennae and labial palps.

FAM. CONIOPTERYGIDAE. This family includes about 240 species which are the smallest and most aberrant of the Neuroptera. They are extremely fragile insects bearing a general resemblance to aphids, with the body and wings covered with a whitish powdery exudation. The antennae are filiform and the segments vary between about 16 and 43 in number: the eyes are rather large and there are no ocelli. The mouthparts do not differ in any important features from those of other Planipennia. The venation is greatly simplified by reduction, and there are but few cross-veins (Fig. 353). The strongest claims these insects have to be regarded as Planipennia rests on the structural characters of their larvae. So far as known the eggs are laid upon various trees frequented by Aphididae, Coccidae or Acarina, and the resulting larvae prey upon those organisms (see Badgley, Fleschner and Hall, 1955). The larvae are more or less pyriform, tapering sharply towards the hinder extremity, and the legs are long and slender (Fig. 354). The antennae are

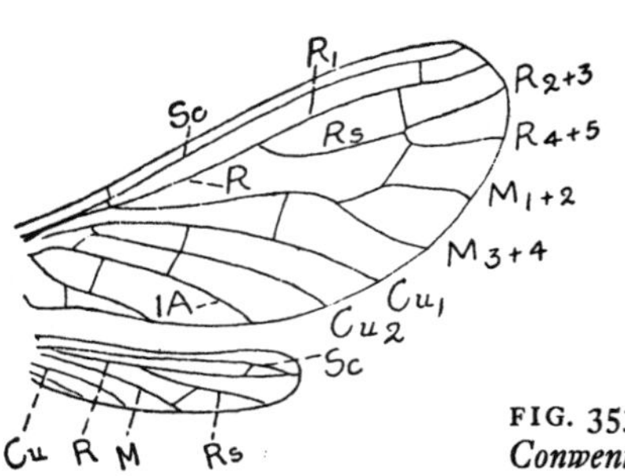

FIG. 353
Conwentzia psociformis, right wings × 12

few-segmented and fringed with rather long hairs: the mandibles and maxillae are short and stout piercing organs, and the labial palps are conspicuous clavate appendages projecting in front of the head. When about to pupate a cocoon is spun of silk emitted from the anus as in other Planipennia. According to Arrow (1917) the first generation of *Conwentzia psociformis* spins its cocoons on oak-leaves, while the second generation overwinters as larvae, which lie up in cocoons spun upon the trunk of that tree. The family has been monographed by Meinander (1972); although its members are not rare they need carefully looking for and, up to the present, only seven species have been found in Britain. Anatomically the larvae differ from other Planipennia in possessing only six Malpighian tubes and in the greatly concentrated abdominal nerve-cord.

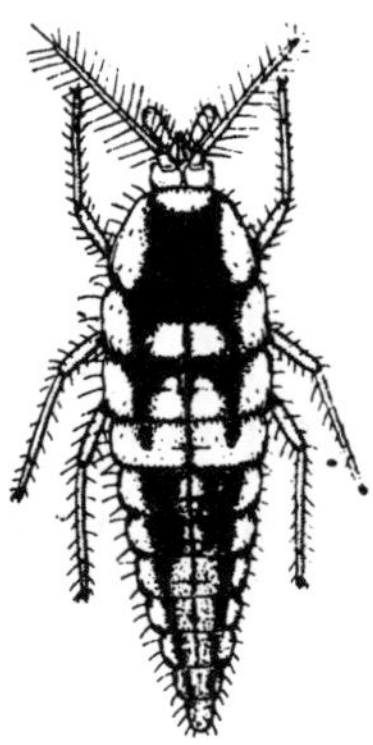

FIG. 354
Conwentzia psociformis, larva
After Withycombe, *Trans. Ent. Soc.*, 1922.

Superfamily 2. **Osmyloidea**

Costal veinlets numerous. Nygmata (defined spots with sensory bristles) occur between the two posterior branches of Rs and usually between Rs and M nearer the base of the wing.

FAM. ITHONIDAE. *Head small, ocelli absent. Venation much branched. Larva subterranean, resembling that of a scarabaeoid; maxillae stout. Rather large soft-bodied insects with broad wings of 40–70 mm expanse.* Three genera and about eight species are known: they frequent sandy localities in Australia and Tasmania (Tillyard, 1919*c*; 1922). Ithonidae are active runners taking refuge in dark crevices, etc., and when their wings are closed they bear a certain superficial resemblance to cockroaches. The eggs of *Ithone* are laid by means of a plough-like ovipositor in sand which adheres to them owing to a sticky secretion with which they are covered: the larva is soft, whitish, and blind, with small mandibles and stout maxillae: it normally preys upon scarabaeoid larvae.

FAM. OSMYLIDAE. *Ocelli present. Vein Sc joined to* R_1 *at apex, all branches of Rs arising from it after divergence from* R_1*, cross-veins numerous.* The Osmylidae are a considerable assemblage of beautiful insects, often with maculated wings, and *Osmylus fulvicephalus* is the largest British Neuropteran. This species occurs locally along the borders of clear streams where there is a dense growth of bushes, etc. Its larva (Fig. 355) lurks under stones or about moss, etc., either in or near the water. It is easily recognized by its long slender stylet-like mandibles and maxillae, which are only slightly curved upwards (see Wundt, 1961, for detailed anatomy of the head). Unlike the aquatic larva of *Sisyra* there are no gills and it breathes by means of thoracic and abdominal spiracles. According to Withycombe its natural food consists of Dipteran larvae. The family is found in most continents and larvae of several of the Australian genera, such as *Porismus*, live under bark. In Japan, *Spilosmylus flavicornis* lives in wet places and the larva feeds on small insects (Kawashima, 1957).

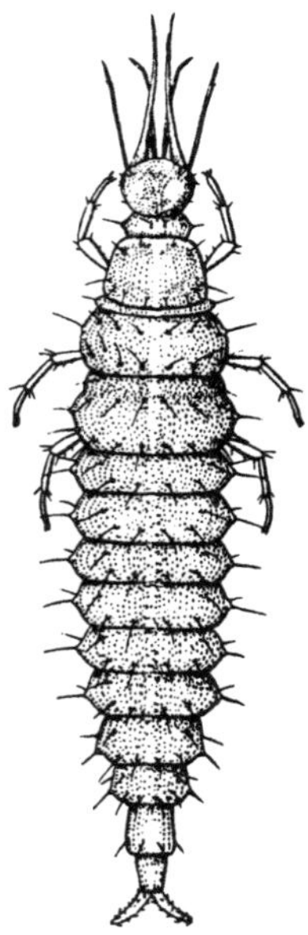

FIG. 355
Osmylus fulvicephalus, larva
After Withycombe, *Trans. Ent. Soc.*, 1922.

FAM. NEURORTHIDAE. *Ocelli absent. Venation as in the Osmylidae but far fewer cross-veins.* A very small but widely spread family with an aquatic larva (Zwick, 1967).

Superfamily 3. **Mantispoidea**

Veins Sc and R_1 connected towards apex by a cross-vein or pterostigma distinct; cell CuA not large and triangular. Trichosors (thickenings of wing-margin bearing several hairs) usually present, nygmata absent. Not more than 4 or rarely 5 cross-veins between veins R_1 and Rs.

FAM. DILARIDAE. *Small insects, Sc not connected to R_1, no pterostigma, Rs with only 3 main branches, only 1–2 cross-veins between R_1 and Rs. Male antennae pectinate. Female with a long ovipositor.* About 30 species in all but the Australian regions. The larvae (Gurney, 1947; MacLeod and Spiegler, 1961) live under the bark of dead trees, probably feeding on beetle eggs or larvae. The eggs are not stalked.

FAM. BEROTHIDAE. *Small to medium sized insects; antennae moniliform; Sc in the fore wing running into R_1 or into the costa and then joined to Sc by a cross-vein; outer margin of fore wing often emarginate, costa and sometimes other parts of females with peculiar seed-like scales.* The genus *Rachiberotha* has raptorial fore legs, rather like a Mantispid, but probably by convergence. Tjeder (1956), however, suggests that *Symphrasis*, usually placed in the Mantispidae, may really be closer to the Berothids. The **Polystoechotidae** are a small family (see Carpenter, 1940) with uncertain affinities, occurring in Western America (north and south). The venation is more of the Myrmeleontid type.

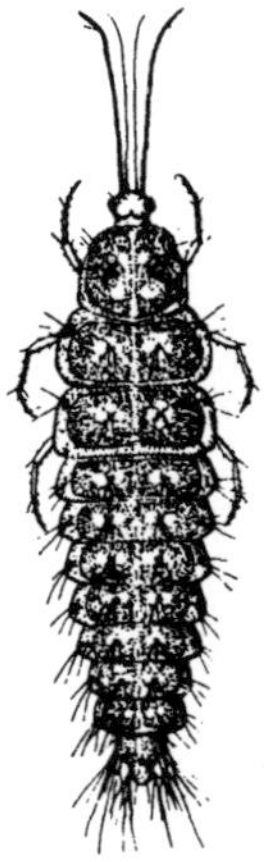

FIG. 356
Sisyra fuscata, larva
After Withycombe, *Trans. Ent. Soc.*, 1922.

FAM. SISYRIDAE. *Small, brownish, Hemerobiid-like insects, found near water. First antennal segment quadrate. Costal veinlets simple in fore wing* (Fig. 357). *Legs cursorial.* Larva (Fig. 356) living on freshwater sponges, with 7 pairs of ventral abdominal respiratory processes. The principal genera are *Sisyra* and *Climacia.* Brown (1952) has given an excellent account of the life-history of the latter and Anthony (1902) and Withycombe of *Sisyra.* The eggs are very small, resembling those of *Hemerobius*: they are laid in small clusters on leaves, piles and other objects standing in or overhanging water. The female covers each batch with a silken web as in the Psocoptera. The newly hatched larva swims through the water with *Cyclops*-like movements until it finds a sponge. The larva then clings to the surface of the sponge or descends into the open ostioles, piercing the sponge-tissue with its mouthparts. It is yellowish green or brownish, hairy, resembling that of a Chrysopid in general form (Fig. 356); it bears seven pairs of segmentally arranged, several-jointed abdominal gills, each supplied by tracheal branches. The antennae are long and setiform, while the mandibles and maxillae form a pair of almost equally elongate bristle-like stylets. Labial palpi are wanting and the legs are single-clawed. Pupation takes place above water in a finely-woven double cocoon. Three species of the genus occur in Britain, *S. fuscata* being common. Parfin and Gurney (1956) deal with the 17 species of the western hemisphere.

FAM. MANTISPIDAE. The members of this family are easily recognized by the elongate prothorax and the large raptorial anterior legs (but cf. Berothidae).

The latter appendages are formed very much the same as in the Mantidae (see p. 593) and fulfil similar functions. Each femur is armed with powerful spines and the tibia is adapted to fold closely on to it, the two parts forming a very effective prehensile organ for seizing the prey. The family occurs in most of the warm regions of the world and a few species occur in S. Europe. The life-history of *Mantispa styriaca* has been followed by Brauer (1869). The eggs are borne on long pedicels as in *Chrysopa* and the newly hatched larvae are elongate and campodeiform but are devoid of cerci. They pass into hibernation almost immediately and in the following

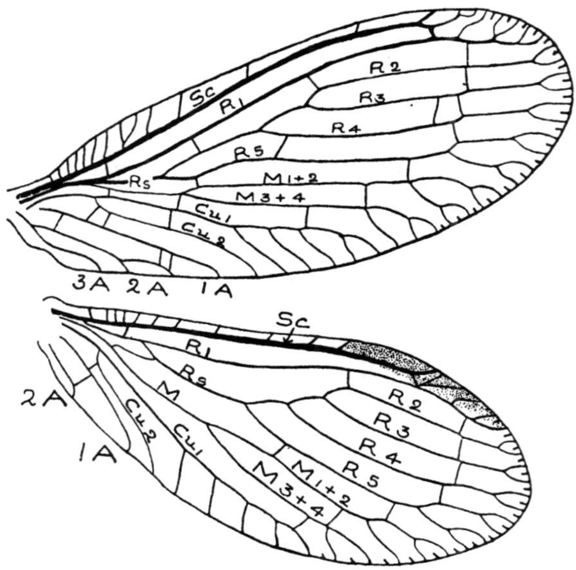

FIG. 357 Right wings of *Sisyra flavicornis*
After Comstock, *Wings of Insects.*

spring they seek out the egg-cocoons of the spider *Lycosa*. Only a single *Mantispa* larva enters each cocoon and it preys upon the young spiders, piercing them with the pointed mouthparts and imbibing their body-fluids. Feeding in this manner leads to an expansion of the larva which becomes so swollen as to resemble that of a miniature cockchafer. It subsequently undergoes ecdysis, and becomes transformed into an eruciform larva with a minute head, and small thoracic legs. It becomes mature a few days later, and spins a cocoon around itself, amidst the dried remains of its victims, within the original egg-bag of the spider. Pupation occurs within the last larval skin and the imago consequently has to pierce the latter and its own cocoon, and that of the spider, before it emerges into the open. The parent spider watches over her cocoon without hostility to the presence of the parasite. Some species are associated with Aculeate Hymenoptera, especially social wasps (Parfin, 1958). The larva of *Plega*, on the other hand, seems to act more like a predator on Scarabaeid and Noctuid larvae in the soil (Werner and Butler, 1965). Kuroko (1961) records that various Mantispids lay many thousands of eggs as happens in other parasites with such life-histories.

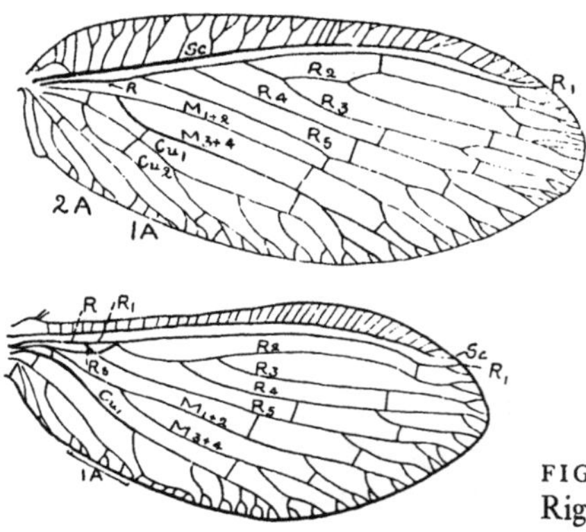

FIG. 358
Right wings of *Hemerobius*

Superfamily 4. **Hemerobioidea**

Veins Sc and R_1 connected by a cross-vein or else the antennae longer than the fore wing; vein CuA not bounding a large triangular area; Rs often arising as many stems or many cross-veins between R_1 and Rs. Trichosors usually present.

The families in this group include some of the commonest Neuroptera, with larvae predatory on aphids which they pierce with their curved jaws. The species are numerous, especially in temperate regions.

FAM. HEMEROBIIDAE. *Antennae moniliform. Fore wing* (Fig. 358) *with Rs arising as at least two branches, trichosors present. Larval body* (Fig. 359) *with short hairs, empodium not trumpet-shaped except sometimes in the first instar. Eggs* (Fig. 359) *not supported on stalks.* Hemerobiids are usually small or moderate-sized insects with obscure, brownish colours, the adults hiding by day and active by night. The wings

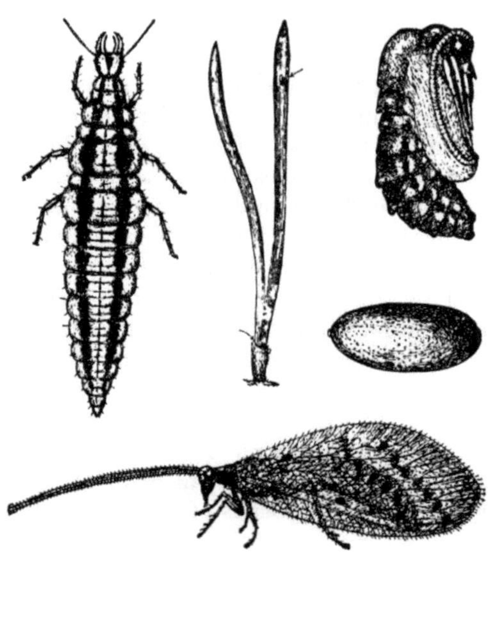

FIG. 359
Hemerobius stigma

are usually oval or, in *Micromus* and its allies, narrow; usually a wing-coupling apparatus including frenular bristles is present. The adults are predatory on the same small insects as the larvae. The eggs, unlike those of Chrysopidae, are devoid of pedicels and have a knob-like micropylar apparatus. The larvae (Fig. 359) are fusiform and smooth without tubercles of any kind, and the body hairs are simple. The mouthparts are rather stout and only slightly curved. A reduced pad-like empodium is present in the last two instars but a trumpet-like one in the first. They eat aphids and other Homoptera and mites. Laidlaw (1936) found that an adult of *Hemerobius stigma* could eat between 13 000 and 15 000 of the aphid *Adelges cooleyi* and the larva 3000. There are 29 British species, 50 in North America (Carpenter, 1940) and about 800 in the world (Tjeder, 1961).

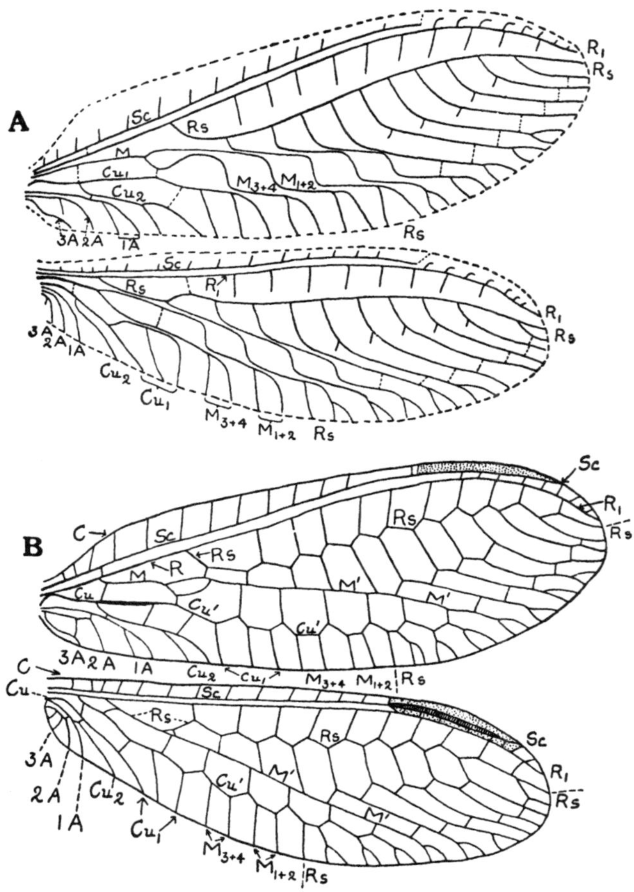

FIG. 360 *Chrysopa signata*. A, diagram of wing-tracheation. B, wing-venation

Cu′, pseudocubitus; M′, pseudomedia. *After* Tillyard, *Proc. Linn. Soc. N.S.W.*, 41.

FAM. CHRYSOPIDAE. *Antennae setiform, longer than fore wing. Fore wing* (Fig. 360) *with* Rs *arising as one stem, trichosors absent. Larval body* (Fig. 361) *often*

with long or hooked hairs on which the dead bodies of prey may be supported. Empodium trumpet-shaped. Eggs stalked. This family includes a large number of closely related species popularly known as 'green lacewings' or 'golden eyes'. Many have bright green bodies and appendages, with the wing-veins similarly coloured, and the eyes have a burnished metallic lustre. Certain of the species emit a disagreeable odour from a pair of prothoracic glands when handled, and have earned for the group the alternative name of 'stink flies'. The antennae of the Chrysopidae are filiform, and longer than they are in the Hemerobiidae, the segments being less distinctly demarcated. The venation (Tillyard, 1916*b*) is characterized (Fig. 360) by Rs arising from the main stem separately from R_1 which does not fuse distally with Sc and by the formation of a so-called pseudomedia and pseudocubitus. These are highly complex veins, the first being formed by the fusion of M_{1+2}, M_{3+4} and portions of the four

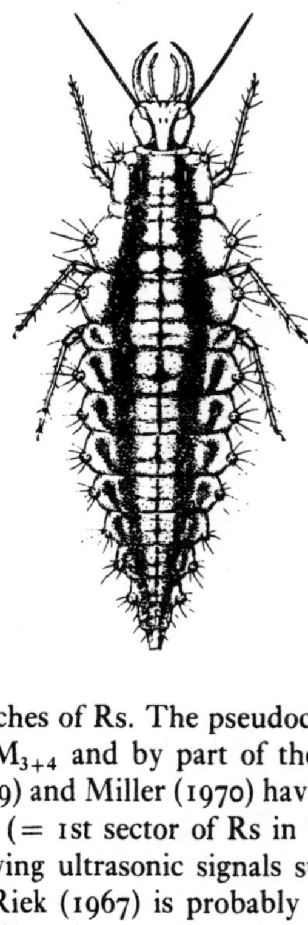

FIG. 361
Chrysopa carnea, larva in 3rd instar

After Withycombe, *Trans. Entom. Soc.*, 1922.

proximal branches of Rs. The pseudocubitus is formed by Cu, by the distal portion of M_{1+2} and M_{3+4} and by part of the three proximal branches of Rs. Miller and MacLeod (1969) and Miller (1970) have shown that a rounded swelling at the base of vein R + MA (= 1st sector of Rs in Tillyard's notation) surrounds an organ capable of perceiving ultrasonic signals such as the sounds made by bats. The organ described by Riek (1967) is probably not stridulatory but is homologous with the cenchri of sawflies (Arora, 1956) and serves to hold the wings in place when at rest. It is one of the features indicating a connection between the Neuroptera and the Hymenoptera.

Philippe (1972) gives a full account of the male and female reproductive organs. The eggs of the Chrysopidae are commonly laid in batches, and a small amount of secretory fluid accompanies each act of oviposition. A spot of this substance is applied to a leaf or other object and the abdomen is then uplifted, with the result that a viscous thread of the secretion is drawn out perpendicularly to the substratum. The thread rapidly hardens and is surmounted by an egg, the latter being thus supported upon a delicate pedicel. In *Chrysopa flava*, and certain other species, the pedicels of an egg-group are joined into a common bundle. Chrysopid larvae

(Fig. 361) resemble those of the Hemerobiidae in their general characters but differ according to Withycombe in the following points. They are shorter and broader with the jaws more slender and curved. The body is provided with setae arising from dorsolateral tubercles. The larvae are often concealed by the remains of their victims, which are retained in position by means of hooked hairs situated on the dorsal aspect of the abdomen. A trumpet-shaped empodium is present between the tarsal claws in all instars. *Chrysopa flava* differs from most other members of the family in its larva having no tubercles and in being more elongate. In coloration the larvae are exceedingly variable: the ground colour is generally white, yellowish or green, usually with darker markings of red, chocolate or black. They are familiar objects on aphid-infested vegetation and are commonly obscured by their coating of debris. Economically, they are of importance on account of the large numbers of soft-bodied insects which they consume: their prey consists principally of aphids, but Cicadellids, Psyllids, coccids, together with thrips and acari, are also attacked. According to Wildermuth ((1916) *Chrysopa californica* will destroy 300–400 aphids during its larval existence. Fourteen species of the family are British but 1350 are known, occurring in all regions except New Zealand (Tjeder, 1966).

FAM. PSYCHOPSIDAE. *Insects of moderate to large size. Antennae short, wings broad, with a vena triplica* (Fig. 362). *Larva with 10 distinct antennal segments, body bare of bristles, a trumpet-shaped empodium between the claws in all instars, living under bark. Eggs sessile, laid in large groups.* Although formerly regarded as a component part of the Hemerobiidae, this family is separable therefrom by its markedly different venational characters and the shortened antennae. The costal area of the wings is exceptionally deep and the three veins Sc, R_1 and Rs assume a parallel course as far as their terminal anastomosis (Fig. 362): they form, in this manner, a vena triplica which renders these insects easily recognizable. The biology of the Australian *Psychopsis elegans* has been followed by Tillyard (1919*b*) and the life-cycle

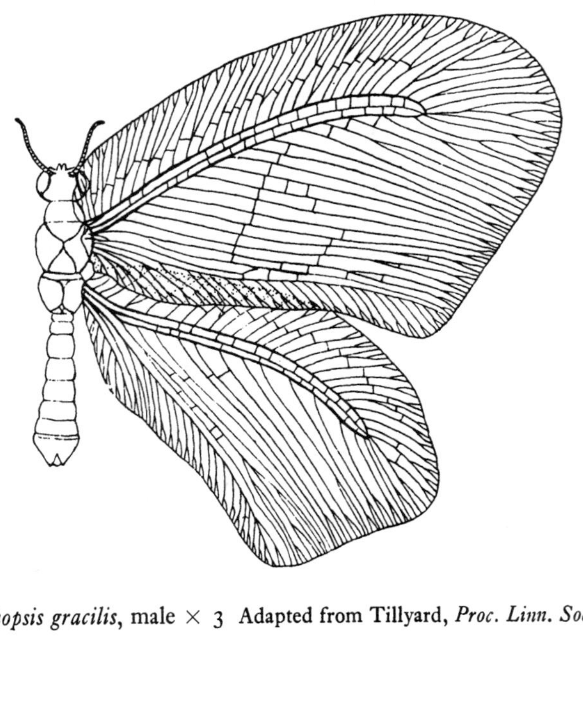

FIG. 362 *Psychopsis gracilis*, male × 3 Adapted from Tillyard, *Proc. Linn. Soc. N.S.W.*, 43.

occupies about a year. The eggs are laid in January or February upon the bark of trees, especially Eucalypti: they are oval and each is provided with a small micropylar projection. The larva is characterized by the great size of the mandibles, which are sickle-like and devoid of teeth: the head is also large and its broad base is closely connected with the prothorax without any visible 'neck'. In their habits these larvae are arboreal, living beneath bark: they probably only emerge from their hiding-places to seize the insects which come to feed upon the gum which exudes from the trees. There are three larval instars: about November they construct silken cocoons in crevices of the bark and the pupal stage lasts about three weeks. Psychopsidae are rare insects of nocturnal habits: 20 species (Kimmins, 1939) occur in Australia, S. Africa, Tibet, China and Burma.

Superfamily 5. **Myrmeleontoidea**

Antennae usually clubbed or thickened, often short. Wings long, veins Sc and R_1 clearly fused for a long distance at apex, not connected by cross-veins except in the Nymphidae; CuA bounding a large triangular cell; trichosors absent except in the Nymphidae. Larvae with gular plate covered by ventral extensions of the genae. Mandibles stout, curved and, except in some Nemopterids which have a very long thorax, with internal teeth.

FAM. NYMPHIDAE. A small group of insects confined to Australia and New Guinea which can be recognized by the presence of trichosors. The antennae are not thickened and they are transitional from the Psychopsidae to the Myrmeleontidae. *Nymphes* lays extraordinary horse-shoe shaped rings of stalked eggs. The larvae appear to live in rubbish. *Myiodactylus* has subcircular larvae with long lateral processes on the thorax and abdomen.

FAM. MYRMELEONTIDAE. (Ant lion flies.) *Wings long and narrow, without trichosors, fore wing with a long hypostigmatic cell* (Fig. 363) *below and beyond the fusion of Sc and* R_1. *Antennae short and thickened. Larva* (*ant lion*) *stout, bristles of body directed forwards; often living in a conical pit in which it traps its prey.* The adults are usually nocturnal, with long narrow wings and body. It is one of the largest families in the Neuroptera with more than 1200 species. During the day they hide among trees and bushes, only appearing on the wing towards dark. Myrmeleontidae are easily distinguished from other Neuroptera by their short knobbed antennae: their wings are usually marked with brown or black, and furnished with many accessory veins and cross-veins. They are closely related to the Ascalaphidae, but the latter insects have longer antennae and lack the elongate hypostigmal cell (Fig. 363). Although most abundant in tropical countries, species of *Myrmeleon* occur in Europe, one representative being found as far north as Finland, but the family is not found in the British Isles. The biology of *M. formicarius* was accurately observed by the early naturalist Réaumur. The ova are laid in sand and the newly emerged larvae excavate pits in the ground for catching their prey. The Myrmeleontid larva buries itself at the bottom of the pit, leaving only its large jaws protruding. An ant or other insect wandering over the edge of the pit usually dislodges the sand of the sloping sides and soon finds itself in difficulties. The ant lion jerks some of the sand by means of its head towards its victim and continues to do so until the latter is brought to the bottom of the pit. Here it is seized and not released until its juices are extracted. The larvae of this family (Meinert, 1889;

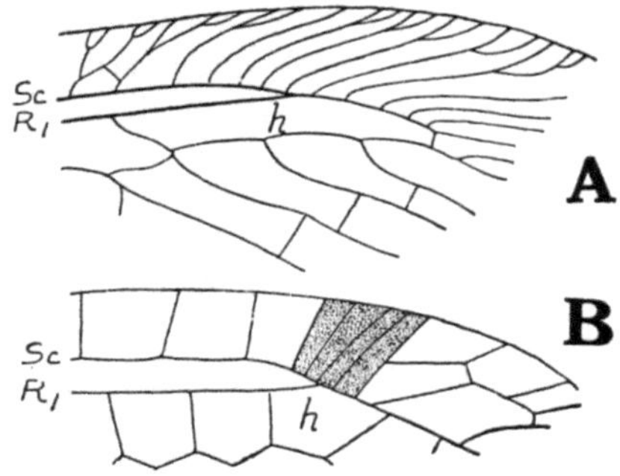

FIG. 363
Portion of fore wing of A, a Myrmeleontid; B, an Ascalaphid, showing hypostigmal cell *h*
Based on figures by Comstock.

FIG. 364
Myrmeleon, larva and pupa × 3. Switzerland
From enlarged photos by H. Main.

Redtenbacher, 1884; Gravely and Maulik, 1911) are flattened and ovoid with large heads, and long, protruding mandibles, armed with exceedingly sharp spiniform teeth (Fig. 364). The pit-forming habit is characteristic of *Myrmeleon* and several other genera, but the larva of *M. contractus* lives on the mud-covered trunks of trees in Bengal, and doubtless preys upon the ants which are constantly streaming up and down. Other larvae of this family hide away under stones and debris, or cover themselves with a coating of foreign substances, and thereby secure concealment. Some account of the anatomy of the imago is given by Dufour (1841) and Wheeler (1930) has published a striking review of the family. Markl (1954) classifies the family down to tribal level.

FAM. ASCALAPHIDAE. *Wings elongate but without a long hypostigmatic cell* (Fig. 363B). *Antennae long, distally sharply clubbed.* This family is closely related to the Myrmeleontidae and has a very similar distribution. Some of the species are active fliers, and are on the wing during daytime, hawking their prey after the manner of dragonflies: others, however, are nocturnal and very seldom seen. The eggs are deposited in rows upon grass stems, twigs, etc., and the batches are often fenced in below by circles of modified eggs (New, 1971) or repagula which possibly guard them from the attacks of predacious enemies. The larvae closely resemble ant lions and have similar dentate mandibles: they are often provided with lateral segmental processes fringed with modified setae (dolichasters). These processes are particularly well developed in *Pseudoptynx* and *Ulula*, while they are usually quite rudimentary in the Myrmeleontidae. The larvae do not construct pitfalls but live concealed on the ground among stones, leaves, etc., or more rarely on the bark of trees. The family has been monographed by Van der Weele (1908), who figures larvae of several genera: the life-history of *Ulula* is described by McClendon (1902). Several species are common in southern Europe and *Ascalaphus* (= *Libelloides*) *longicornis* occurs as far north as Paris.

FAM. STILBOPTERYGIDAE includes a few large species in Australia and S. America with venation much like that of Ascalaphidae but short, clubbed antennae like Myrmeleontidae.

FAM. NEMOPTERIDAE. A highly specialized family with enormously elongate, ribbon- or racket-like hind wings and with the head usually prolonged into a kind of rostrum. The antennae are long and not clubbed or thickened. They are striking and beautiful insects flying with a curious up-and-down motion after the manner of Ephemerids, with the long hind wings streaming in the air. The form of the latter is somewhat variable: in *Croce* they are filiform, and taper to a point, while in other cases they are sometimes expanded before their extremities (Fig. 365). The mid-rib, which lends support to these greatly attenuated organs, is formed, according to Comstock, by the closely approximated stems of R and M. The life-history of

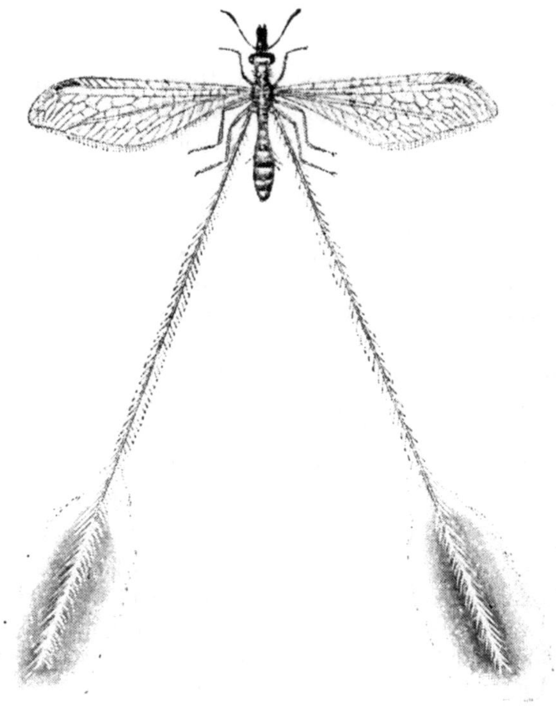

FIG. 365 *Croce filipennis* × 2·5. India

the Indian *Croce filipennis* (Fig. 366) occupies about a year (Imms, 1911). The imagines are crepuscular and frequent buildings. The eggs are laid among dust and refuse on floors, and the fully-grown larva has a large quadrate head and long, curved, finely dentated mandibles. The head is connected with the hind-body by a conspicuous 2-segmented 'neck'; the meso- and metathorax are imperfectly differentiated and merged into the abdomen. The larvae cover themselves with dust particles and are hard to detect: they prey upon Psocids and other small insects. The pupa is notable on account of its method of accommodating the long hind wings. These are many times coiled after the manner of watch-springs: they cross each other near their bases, so that the right wing lies on the left side and *vice versa*. The pupa is enclosed in a cocoon composed of sand and debris bound together by silk. In *Nina joppana* and *Pterocroce storeyi* the neck of the larva is so greatly attenuated that

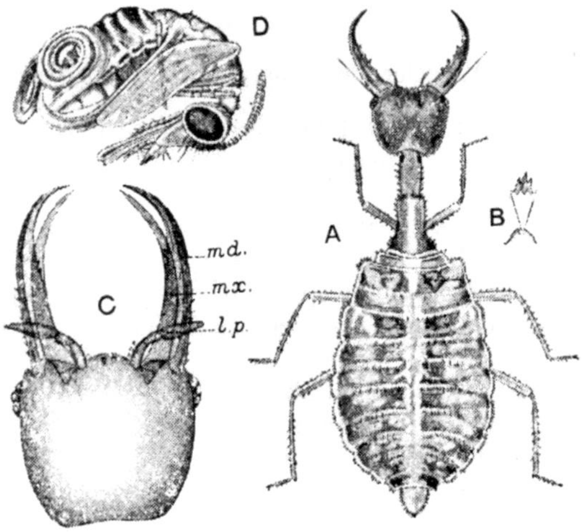

FIG. 366 *Croce filipennis*

A, larva in last instar. B, a dolichaster × 230. C, ventral aspect of head of larva × 50; *md*, mandible; *mx*, maxilla; *lp*, labial palp. D, pupa. *After* Imms, *Trans. Linn. Soc.*, 1911.

it equals in length the whole of the rest of the body (Fig. 367): these remarkable larvae have been found in caves in Egypt and Palestine (Withycombe, 1923). The family is widely distributed in S. Europe.

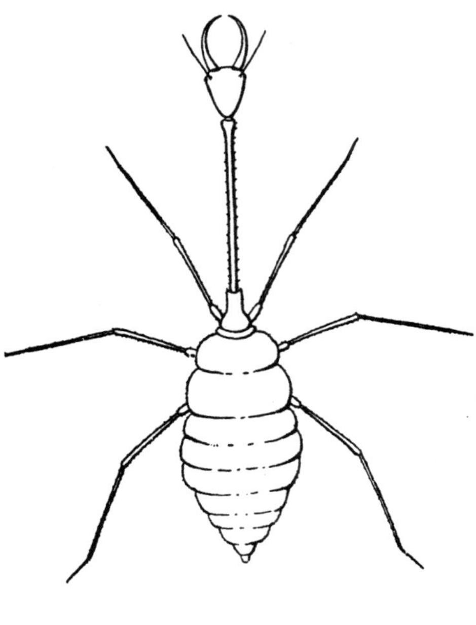

FIG. 367
Pterocroce storeyi, larva in last instar × circa 8
After Withycombe.

Literature on Neuroptera

Killington (1936–37) and Wheeler (1930) list many further papers on the order.

ACHTELIG, M. (1967), Über die Anatomie des Kopfes von *Raphidia flavipes* Stein und Verwandtschaftsbeziehungen der Raphidiidae zu Megaloptera, *Zool. Jb. (Anat.)*, **84**, 249–312.

ACKER, T. S. (1960), The comparative morphology of the male terminalia of Neuroptera (Insecta), *Microentomology*, **24**, 25–84.

ADAMS, P. A. (1969), A new genus and species of Osmylidae (Neur.) from Chile and Argentina, with a discussion of Planipennian genitalic homologies, *Postilla*, **141**, 1–11.

ANTHONY, M. H. (1902), The metamorphosis of *Sisyra*, *Am. Nat.*, **26**, 615–631.

ARORA, G. L. (1956), The relationship of the Symphyta (Hymenoptera) to other orders of insects on the basis of adult external morphology, *Res. Bull. Punjab Univ. (Zool.)*, **90**, 85–119.

ARROW, G. J. (1917), The life history of *Conwentzia psociformis* Curt., *Entomologist's mon. Mag.*, **53**, 254–257.

ASPÖCK, H. AND ASPÖCK, U. (1968), Vorläufige Mitteilung zur generischen Klassifizierung der Raphidiodea (Insecta, Neuroptera), *Ent. NachrBl.*, **15** (7–8), 53–64.

—— (1971), Raphidioptera (Kamelhalsfliegen), In: Kükenthal, W. (ed.), *Handbuch der Zoologie, IV Arthropoda, 2 Insecta*, **25**, 1–50.

BADGLEY, H. E., FLESCHNER, C. A. AND HALL, J. C. (1955), The biology of *Spiloconis picticornis* Banks (Neuroptera: Coniopterygidae), *Psyche, Cambridge*, **62**, 75–81.

BERLAND, L. AND GRASSÉ, P. P. (1951), Neuropteroidea, In: Grassé, P. P. (ed.), *Traité de Zoologie*, **10** fasc. 1, 1–69.

BRAUER, F. (1869), Beschreibung der Verwandlungsgeschichte der *Mantispa styriaca* Poda; und Betrachtungen über die sogenannte Hypermetamorphose Fabre's, *Verh. zool. bot. Ges. Wien*, **19**, 831–840.

BROWN, H. P. (1972), The life-history of *Climacia areolaris* (Hagen), a neuropterous 'parasite' of freshwater sponges, *Am. Midl. Nat.*, **47**, 130–160.

CARPENTER, F. M. (1936), Revision of the Nearctic Raphidioidea (recent and fossil), *Proc. Am. Acad. Arts and Sciences*, **71**, 89–157.

—— (1940), A revision of the Nearctic Hemerobiidae, Berothidae, Sisyridae, Polystoechotidae and Dilaridae (Neuroptera), *Proc. Am. Acad. Arts and Sciences*, **74**, 193–280.

CUYLER, R. D. (1958), The larvae of *Chauliodes* Latreille (Megaloptera: Corydalidae), *Ann. ent. Soc. Amer.*, **51**, 582–586.

DUFOUR, L. (1841), Recherches anatomiques et physiologiques sur les Orthoptères, les Hyménoptères et les Neuroptères, *Mém. Math. Savants étrangers*, **7**, 265–647.

—— (1848), Recherches anatomiques sur la larve à branchies extérieures du *Sialis lutarius*, *Ann. Sci. nat.*, (3), **9**, 91–99.

FERRIS, G. F. (1940), The morphology of *Plega signata* (Hagen) (Neuroptera, Mantispidae), *Microentomology*, **5**, 33–56.

FERRIS, G. F. AND PENNEBAKER, P. (1939), The morphology of *Agulla adnixa* (Hagen) (Neuroptera, Raphidiidae), *Microentomology*, **4**, 121–142.

FRASER, F. C. (1959), Mecoptera, Megaloptera, Neuroptera, *Hndbks. Ident. Brit. Ins.*, 1, parts 12–13, 40 pp.

GRAVELY, F. H. AND MAULIK, S. (1911), Notes on the development of some Indian Ascalaphidae and Myrmeleonidae, *Rec. Ind. Mus.*, 6, 101–110.

GURNEY, A. B. (1947), Notes on the Dilaridae and Berothidae with special reference to the immature stages of the nearctic genera (Neuroptera), *Psyche, Cambridge*, **54**, 145–169.

IMMS, A. D. (1911), On the life history of *Croce filipennis* Westw., *Trans. Linn. Soc. Lond.*, (2), 11, 151–160.

KAWASHIMA, K. (1957), Bionomics and earlier stages of some Japanese Neuroptera (1), *Spilosmylus flavicornis* (McLachlan) (Osmylidae), *Mushi*, **30**, 67–70.

KELSEY, L. P. (1954, 1957), The skeleto-muscular mechanism of the Dobson fly, *Corydalis cornutus*, Part I. Head and prothorax, Part II. Pterothorax, *Mem. Cornell agric. Exp. Sta.*, **334**, 1–51; **346**, 1–31.

KILLINGTON, F. J. (1930), A synopsis of the British Neuroptera, *Trans. ent. Soc. Hampshire and S. England*, **1929**, 1–36.

—— (1936–37), *A Monograph of the British Neuroptera*, 1 & 2, London, Ray Soc.

KIMMINS, D. E. (1939), A review of the genera of the Psychopsidae (Neuroptera), with a description of a new genus, *Ann. Mag. nat. Hist.*, (11), **4**, 144–153.

KUROKO, H. (1961), On the eggs and first instar larvae of two species of Mantispidae, *Esakia*, **3**, 25–32.

LAIDLAW, W. B. R. (1936), The Brown Lacewing flies (Hemerobiidae): their importance as controls of *Adelges cooleyi* Gillette, *Entomologist's mon. Mag.*, **71**, 164–174.

LOEW, H. (1848), Abbildungen und Bemerkungen zur Anatomie einiger Neuropterengattungen, *Linn. Ent.*, **3**, 345–386.

MacLEOD, E. G. AND SPIEGLER, P. E. (1961), Notes on the larval habitat and developmental peculiarities of *Nallachius americanus* (McLachlan) (Neuroptera: Dilaridae), *Proc. ent. Soc. Washington*, **63**, 281–286.

MARKL, W. (1954), Vergleichend-morphologische Studien zur Systematik und Klassifikation der Myrmeleoniden (Insecta, Neuroptera), *Verhandl. naturf. Ges. Basel*, **65**, 178–263.

MCCLENDON, J. F. (1902), The life history of *Ulula hyalina* Latreille, *Am. Nat.*, **36**, 421–429.

MEINANDER, M. (1972), A revision of the family Coniopterygidae (Planipennia), *Acta zool. Fenn.*, **36**, 1–357.

MEINERT, F. (1889), Contribution à l'anatomie des Fourmilions, *Overs. Dans. Selsk.*, **1889**, 43–66.

MILLER, L. A. (1970), Structure of the green lacewing tympanal organ (*Chrysopa carnea:* Neur. Chrysopidae), *J. Morphol.*, **131**, 359–382.

MILLER, L. A. AND MacLEOD, E. G. (1969), Ultrasonic sensitivity: a tympanal receptor in the Green Lacewing *Chrysopa carnea, Science*, **154**, 891–892.

NEW, T. R. (1971), Ovariolar dimorphism and repagula formation in some South American Ascalaphidae (Neuroptera), *J. Ent.*, (A), **46**, 73–77.

PARFIN, S. (1958), Notes on the bionomics of the Mantispidae (Neuroptera, Planipennia), *Ent. News*, **69**, 203–207.

PARFIN, S. AND GURNEY, A. B. (1956), The *Spongilla* flies, with special reference to those of the Western Hemisphere (Sisyridae, Neuroptera), *Proc. U.S. nat. Mus.*, **105**, 451–529.

PHILIPPE, R. (1972), Les appareils génitaux mâle et femelle de *Chrysopa perla* (Neuroptera), Étude anatomique, histologique et fonctionelle, *Ann. Soc. Ent. France*, (8), **3**, 693–705.

REDTENBACHER, J. (1884), Uebersicht der Myrmeleoniden-Larven, *Denk. Ak. Wien*, **48**, 335–368.

RIEK, E. F. (1967), Structures of unknown, possibly stridulatory, function on the wings and body of Neuroptera, with an appendix on other endopterygote orders, *Aust. J. Zool.*, **15**, 337–348.

—— (1970), Megaloptera. Neuroptera, In: *The insects of Australia*, *chap.* **28** and **29**, 405–471; 472–494, University Press, Melbroune.

SMITH, H. L. (1970), Biology and structure of the Dobson fly, *Neohermes californicus* (Walker) (Megaloptera, Corydalidae), *Pan-Pacific Ent.*, **46**, 142–150.

SMITH, R. C. (1922), The biology of the Chrysopidae, *Cornell Univ. agric. Exp. Sta. Mem.*, **58**, 1291–1372.

STITZ, H. (1927), Neuroptera, In: *Die Tierwelt Mitteleuropas, Insekten*, **3**, Lief. 1, 1–24.

TAUBER, C. A. AND TAUBER, M. J. (1968), *Lomamyia latipennis* (Neuroptera: Berothidae) life-history and larval descriptions, *Can. Ent.*, **100**, 623–629.

TILLYARD, R. J. (1915–19), Studies in Australian Neuroptera, i–viii, *Proc. Linn. Soc. N.S.W.*, **40**, 734–752 (1915); **41**, 41–70 (1916*a*); **41**, 221–248 (1916*b*); **41**, 269–332 (1916*c*); **43**, 116–122 (1918); **43**, 750–786 (1919*a*); **43**, 787–818 (1919*b*); **44**, 414–437 (1919*c*).

—— (1922), The life-history of the Australian moth-lacewing, *Ithone fusca* (order Neuroptera Planipennia), *Bull. ent. Res.*, **13**, 205–223.

TJEDER, B. (1937), A contribution to the phylogeny of the Dilaridae and the Raphidiidae (Neuroptera), *Opusc. Ent.*, **2**, 138–148.

—— (1954), Genitalic structures and terminology in the order Neuroptera, *Ent. Meddr*, **27**, 23–40.

—— (1956–57), Neuroptera Planipennia of southern Africa, **1–13**, In: Hanström, Brinck and Rudebeck (eds.), *South African Animal Life*, Swedish nat. Sci. Res. Council, Stockholm.

TUXEN, S. L. (ed.) (1970), *Taxonomist's glossary of genitalia in insects*, 2nd edition, Copenhagen, Munksgaard.

VAN DER WEELE, H. W. (1908), Ascalaphiden monographisch bearbeitet, In: *Cat. Coll. Selys.*, 8, 326 pp.

WERNER, F. G. AND BUTLER, G. D. JR. (1965), Some notes on the life history of *Plega banksi* (Neuroptera: Mantispidae), *Ann. ent. Soc. Amer.*, **58**, 66–68.

WHEELER, W. M. (1930), *Demons of the Dust*. Kegan Paul, Trench, Trubner & Co., London.

WILDERMUTH, V. L. (1916), California green lace-wing fly, *J. agric. Res.*, **6**, 515–525.

WITHYCOMBE, C. L. (1923), Systematic notes, on the Crocini (Nemopteridae) with descriptions of new genera and species, *Trans. ent. Soc. Lond.*, **1923**, 269–287.

—— (1925), Some aspects of the biology and morphology of the Neuroptera with special reference to the immature stages and their possible phylogenetic significance, *Trans. ent. Soc. Lond.*, **1924**, 303–411.

WOGLUM, R. S. AND MCGREGOR, E. A. (1959), Observations on the life-history and morphology of *Agulla astuta* (Banks) (Neuroptera: Raphidioidea Raphidiidae), *Ann. ent. Soc. Amer.*, **52**, 489–502.

WUNDT, W. (1961), Der Kopf der Larve von *Osmylus chrysops* L. (Neuroptera, Planipennia), *Zool. Jb.* (*Anat.*), **79**, 557–662.

ZWICK, P. (1967), Beschreibung der aquatischen Larve von *Neurorthus fallax* (Rambur) und Errichtung der neuen Planipennien-Familie Neurorthidae fam. nov, *Gewäss. Abwäss.*, **44–45**, 65–86.

Order 22

COLEOPTERA (BEETLES)

Minute to large insects whose fore wings, not used in flight, are modified into horny or leathery elytra which almost always meet to form a straight mid-dorsal suture: hind wings membranous, folded beneath the elytra, or often reduced or wanting. Mouthparts adapted for biting: ligula variably lobed. Prothorax large and mobile, mesothorax much reduced. Abdominal tergites often little sclerotized. Metamorphosis complete: larvae campodeiform or eruciform, seldom apodous, with mandibulate mouthparts: pupae adecticous and exarate, rarely obtect.

The Coleoptera number approximately 330 000 described species, and are consequently the largest order in the animal kingdom: about 3700 species inhabit the British Isles. Although they are the predominant insects of the present epoch beetles are not seen as frequently as members of other orders because of their more concealed habits. Their adaptability and the structural modifications which they exhibit have evidently contributed much to their dominance, for the adults of no other order of insects have invaded the land, air and water to the same extent. The habits of beetles, therefore, are extremely varied: they are more especially insects of the ground and either inhabit the soil itself, or the various decaying animal and vegetable substances. Dung, carrion, refuse of all kinds, humus, rotting wood and fungi all support large associations of Coleoptera. The members of twelve families are true aquatic insects while many other families have aquatic or semiaquatic representatives. The Chrysomeloids and most Curculionoids are usually met with in association with herbaceous plants, bushes and trees. Representatives of the most diverse families, whether they be aquatic or terrestrial, possess ample hind wings and readily take to flight. Several species are littoral and are daily submerged by the tides. A considerable number of beetles occur in close relation with man since they are found in wool, furs, hides, furniture, museum specimens and in dry stored foods and drugs. The great solidity of the integument in the majority of species has been an important factor in protecting them against enemies of various kinds. The various sclerites are fitted together with a precision that marks them out as truly marvellous pieces of natural mechanism.

Included in the order are some of the largest and also some of the most minute of living insects. Among the Scarabaeoids *Goliathus regius*, *Dynastes hercules* and *Megasoma elephas* attain a body-size not found outside the Coleoptera: *D. hercules* (including the cephalic horn) measures up to about 155 mm long and the Cerambycid *Macrodontia cervicornis* (including the mandibles) attains approximately the same dimension. On the other hand, among the Corylophidae and Ptiliidae are insects so minute that they may hardly reach a length of 0·5 mm.

The literature on Coleoptera has assumed enormous proportions. For a general introduction to the study of the order the student should consult the works of Fowler (1912), Jeannel and Paulian (1949) and Crowson (1960, 1968). For the British species the monograph by Fowler (1887–1913) is indispensable: the works of Reitter (1908–16) and of Freude, Harde and Lohse (1964–71) (not yet completed) will also prove valuable for purposes of identification while Horion (1940–65) gives their distribution and biology. The Palaearctic Coleoptera have been catalogued by Winkler (1924–32) and the species of the world by Schenkling and Junk (1910–40; revision in progress). Leng's catalogue of the North American species contains a very full bibliography of the systematic literature on the order up to 1920. The 3690 British species are listed by Kloet and Hincks (1945).

The Imago

EXTERNAL ANATOMY

The Head (Figs. 368, 371, 372) – The head is heavily sclerotized and there is probably no *epicranial sulcus*. A complete Y-shaped sulcus occurs, however, in some Hydrophilidae, but this may well be secondary (DuPorte, 1946; Snodgrass, 1947). In most of the Curculionoidea, and in a few isolated genera among other groups, the frons and vertex are prolonged anteriorly to

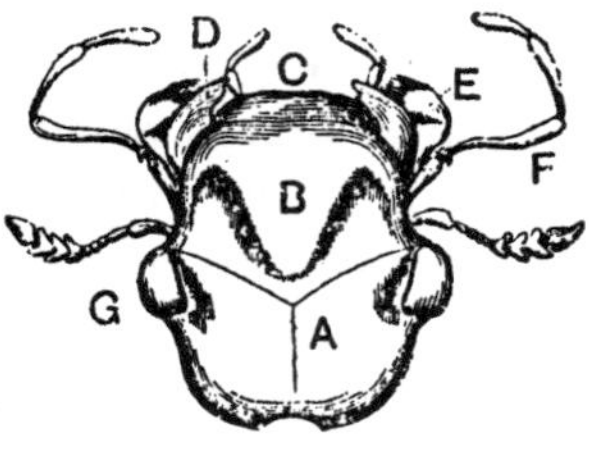

FIG. 368
Hydrophilus piceus, dorsal view of head

A, vertex; *B*, clypeus; *C*, labrum; *D*, mandible; *E*, maxilla and *F*, its palp; *G*, antenna. Adapted from Newport.

form a *rostrum* (Fig. 371). The latter bears the mouthparts at its apex and the antennae are also carried forward: as a rule the rostrum has a groove or *scrobe* on either side for the reception of the scape of the antenna. The *eyes* are very variable and may be totally absent. Eyeless Coleoptera are met with among cavernicolous species and in certain subterranean forms, including those living beneath boulders. Eyes are similarly wanting in *Platypsyllus* and *Leptinus*. In the males of many of the Lampyridae the eyes are very large and contiguous, or nearly so, above and beneath: in the females they are often very small. Occasionally the eyes are partially or almost completely divided by a corneous ridge as in *Trixagus* and *Dorcus*: or they may be completely separated into an upper and a lower eye on each side as in *Gyrinus* and *Amphiops*. *Ocelli* are rarely present but a pair of these organs are found in certain Staphylinidae and in *Pteroloma* (Silphidae), while most Dermestids have a single central ocellus. The *clypeus* is divisible into *anteclypeus* and *postclypeus*. The latter sclerite is fused with the frons and the dividing suture is wanting: the anteclypeus is often infolded and not visible from above (Stickney, 1923). Among Curculionoidea the reduced fronto-clypeal region is often termed the *epistoma*. The *labrum* is very variably developed but is present in nearly all the families: it may, however, be concealed beneath the clypeus, or fused with it, as in the majority of the weevils. The floor of the head, in the median line, is formed by the *gula* and the latter sclerite is marked off from the genae, on either side, by the *gular sulci*. Among most Curculionoidea, and a few other beetles (*Nicrophorus*, etc., Fig. 370), the gula is reduced or wanting and the genae meet in the mid-ventral line, and there is consequently only a single gular sulcus present.

The *antennae* exhibit a very wide range of variation and the usual number of segments is 11. They may, however, be 1-segmented, as in *Articerus*, or 2-segmented as in many Paussinae: on the other hand, they may consist of 27 segments or more in rare instances, and there are many transitions between these extremes.

The *mandibles* attain their extreme development in the males of many of the Lucanidae. In this family they often assume relatively enormous proportions and may be branched in an antler-like manner: in *Chiasognathus* their length exceeds that of the whole body (Fig. 369). In weevils of the genus *Curculio* they have a vertical movement, side by side, instead of being horizontal and opposed, owing to the dorsal position of their condyles. In the Curculionid subfamilies Brachyderinae and Otiorrhynchinae each mandible often bears a round or oval area with a raised margin. These structures are the *mandibular scars* which served as supports for the deciduous *provisional mandibles* (Marshall, 1916). The latter organs apparently enable the newly emerged imago to cut its way through the cocoon but are cast off soon after the insect has freed itself. In a few genera, however, they are permanently retained. In *Passalus cornutus*, certain Staphylinidae, Meloidae, and other beetles a movable inner lobe or *prostheca* is present (Blackwelder, 1934).

The *maxillae*, as a rule, are completely developed with the full number of elements present. In the terrestrial Adephaga and Dytiscidae the *galea* is generally 2-segmented and palpiform. The *lacinia* is frequently large and blade-like and may carry an articulated process, well shown in the Cicindelinae where it is claw-like. Specialization by reduction is frequent: thus a single maxillary lobe or *mala* is present, for example, in the Corylophidae and most of the Nitidulidae as well as among the Curculionoidea: in other members of the latter group the mala may be wanting. The *maxillary palpi* are generally 4-segmented, and more rarely 3-

segmented, while in *Aleochara* they are composed of five segments: in the Pselaphidae and Hydrophilidae these organs are very greatly developed.

FIG. 369
Chiasognathus grantii (Lucanidae), male and female
After Darwin, *Descent of Man.*

In the *labium* (Fig. 370) the *mentum* is large and well developed: the *submentum* is evident in some forms, including *Hydrophilus* and *Nicrophorus*, but is usually fused with the gula (Fig. 370, B) or no longer recognizable as an individual sclerite. The *prementum* is present but often folded under the mentum. The *ligula* is extremely variable: in some forms it is entire, in others it presents up to as many as five lobes or processes, apparently of a secondary nature. The *labial palps* are usually 3-segmented: more rarely, they are 2-segmented, while in certain Staphylinidae they are unjointed and bristle-like. The maxilla and labium of many genera are illustrated by Williams (1938).

The Thorax (Fig. 372) – The *prothorax* is the largest of the thoracic segments and is usually freely movable, the latter feature being a marked characteristic of the order. The *pronotum* is composed of a single sclerite and is entirely visible from above. The *pleuron* is frequently undivided into sclerites, and the *notopleural* sulcus between that region and the pronotum, on either side, is absent in the Polyphaga: in the latter case a single cuticular shield covers the whole of the dorsal and lateral regions. The pleurosternal sulci are distinct except in the Curculionoidea in which group the whole of the prothoracic sclerites are fused into an undivided annular band. The anterior *coxal cavities* are either entire, when they are closed behind by the

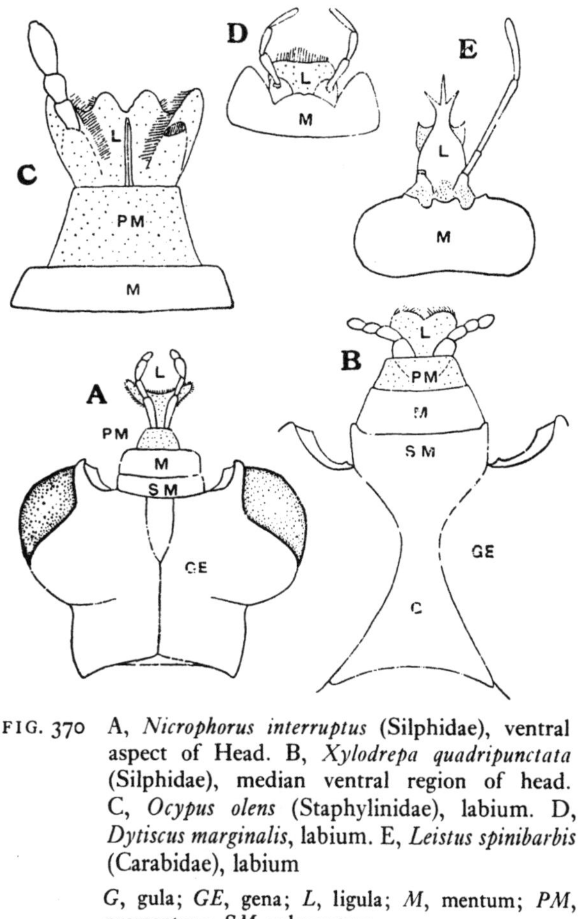

FIG. 370 A, *Nicrophorus interruptus* (Silphidae), ventral aspect of Head. B, *Xylodrepa quadripunctata* (Silphidae), median ventral region of head. C, *Ocypus olens* (Staphylinidae), labium. D, *Dytiscus marginalis*, labium. E, *Leistus spinibarbis* (Carabidae), labium

G, gula; *GE*, gena; *L*, ligula; *M*, mentum; *PM*, prementum; *SM*, submentum.

meeting of the prosternum and epimera, or by the meeting of the epimera alone: or they may be open, when the space is only bridged over by the membrane. The significance of the coxal cavities in the classification of the Adephaga is discussed by Bell (1967). The *meso-* and *metathorax* are fused together: the former segment is considerably reduced while the latter, on the contrary, is largely developed, except in species in which the wings are absent or non-functional. The tergum of both segments is divisible into *prescutum*, *scutum* and *scutellum* (Fig. 373). The latter sclerite is median in position and divides the scutum into two separated plates. The *metapost-*

notum is generally distinct but, according to Snodgrass, the corresponding sclerite of the mesothorax is wanting. With the exception of the *mesoscutellum* the entire dorsal surface of both segments is usually covered by the

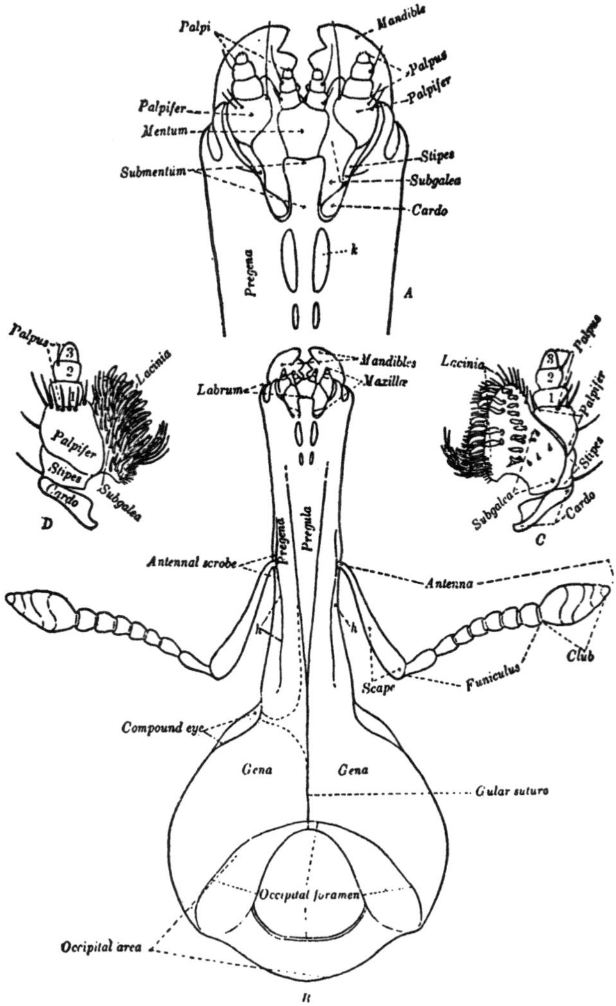

FIG. 371 *Pissodes strobi* (Curculionidae)

A, ventral aspect of head; B, apex of rostrum, ventral; C, interno-lateral and D, externo-lateral aspects of maxilla. Adapted from Hopkins, *U.S. Dept. Agric. Entom. Tech. Ser.* 20, pt. 1; subgalea + lacinia should be mala.

elytra. In the metathorax, a furca is invaginated from the sternum and the details of its structure are of some importance in classification (Crowson, 1938, 1944). The arms of the furca provide attachments for the leg muscles and are best developed in species which actively use their legs.

The *legs* are generally adapted for walking or running, but in many of the Scarabaeids and certain of the Carabidae, they are also modified for fossorial purposes. In the Dytiscidae the hind pair are flattened and used for swimming, while in the Gyrinidae both the middle and hind pairs are thus modified. In the Halticinae the hind femora are greatly enlarged for jumping. The legs of Coleoptera consist of the usual number of parts and the form and disposition of the coxae are of great importance in classification.

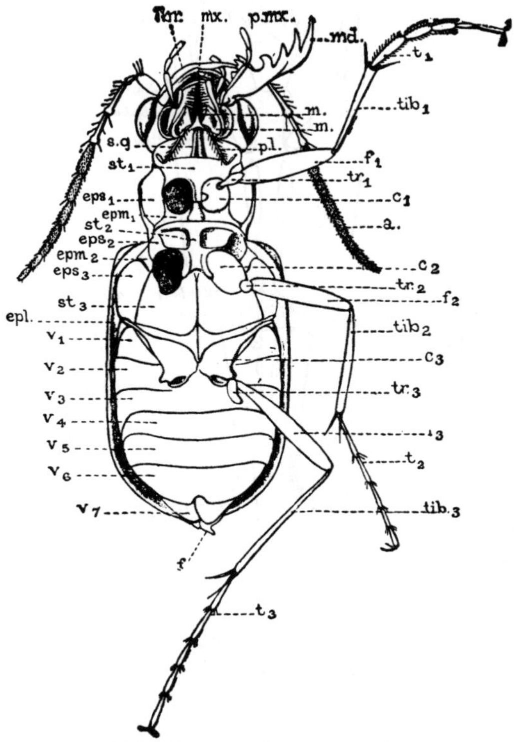

FIG. 372 *Cicindela*, ventral aspect: male

lbr, anterior margin of labrum; *md*, mandible; *mx*, maxilla and *p.m.x*, palp; *m*, *m*, mentum; *p.l*, labial palp; *s.g*, gular sutures; st_1–st_3, thoracic sterna; eps_1–eps_3, episterna; epm_1–epm_2, epimera; *epl*, epipleuron; v_1–v_3, visible abdominal segments; *f*, aedeagus; c_1–c_3, coxae; tr_1–tr_3, trochanters; f_1–f_3, femora; tib_1–tib_3, tibiae; t_1–t_3, tarsi. From Fowler (F.B.I.), *after* Ganglbauer.

The tarsal segments are extremely variable in number and afford valuable family and superfamily characters. The primitive 5-segmented condition is characteristic of the Adephaga, the Scarabaeids and many other Polyphaga. Among the Heteromera the fore and middle tarsi are 5-segmented, and the hind pair 4-segmented. In the Chrysomeloidea and Curculionoidea the 4th and 5th segments are anchylosed, the former being very small. In the Staphylinoidea the segments are very variable in number. Among many of the males of this group, and the Adephaga, one or more of the segments of the anterior tarsi, and sometimes of the middle pair also, are dilated and different from their fellows: this feature attains a high degree of specialization among the Dytiscidae.

The wings – The *elytra* are the highly modified mesothoracic wings and arise simultaneously with the hind wings: they develop in an exactly similar manner during the greater part of the larval life. In many Carabidae, Curculionidae and Ptinidae the hind wings are absent and the elytra are often firmly united so as to be immovable. In some Dytiscidae, Jackson (1952) finds considerable variation; some species are winged but have no flight muscles, in others there is a more complete reduction in some or all individuals. Many species are dimorphic with a winged and a more or less brachypterous form which is often capable of more rapid reproduction. In the Scolytinae the wing-muscles of the female may degenerate when she lays

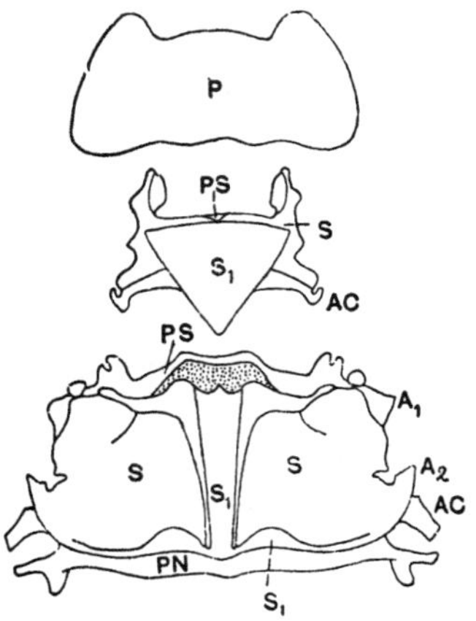

FIG. 373
Hydrophilus, dorsal aspect of thorax with the segments disarticulated

A_1, anterior wing process; A_2, posterior do.; *AC*, axillary cord; *P*, pronotum; *PN*, postnotum; *PS*, prescutum; *S*, scutum; S_1, scutellum. Partly after Snodgrass, *Proc. U.S. nat. Mus.*, 36.

eggs and then regenerate again before she flies to another oviposition-site (Chapman, 1957; Reid, 1958; Henson, 1961). In Coleoptera capable of flight the elytra are opened to form an angle with the body, and allow freedom of movement of the wings, but play no direct part in flight. The sides of the elytra are often reflexed to form the *epipleura* (Fig. 372) which conceal the pleura and are well exemplified in the Gyrinidae.

The hard texture of the elytra is due to the thickness of the lower layer of the cuticle, and also to the presence of pillars or trabeculae which connect the upper and lower elytral surfaces (Fig. 374). The cavity of the elytron is bounded by a thin epidermis and contains blood, nerves and tracheae, often together with numerous groups of gland cells: sometimes small lobules of fat-body are also evident. Comstock states that there is a very close similarity between the tracheation of the elytra and the hind wings, but in no case yet examined do the principal tracheae retain the primitive type of branching. The venation of the hind wings has been studied by Kempers (1909), Kühne (1915), Orchymont (1920), Forbes (1922), Suzuki (1969) and others. Three general types are recognizable (Figs. 375, 376).

(*a*) The Adephagid type. All the principal veins remain more or less

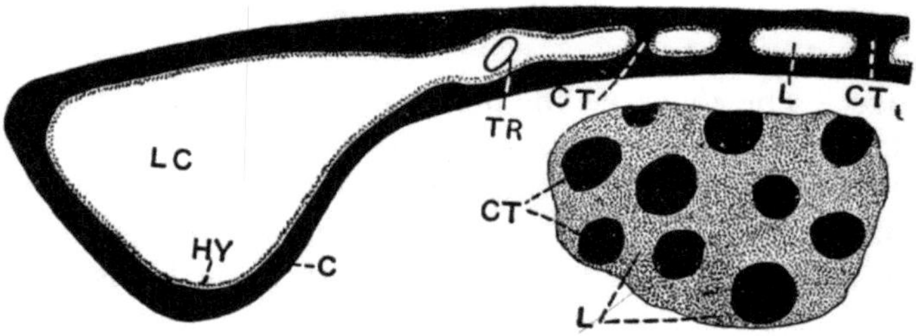

FIG. 374 Transverse section of a portion of an elytron of *Dytiscus* passing through the outer margin: below, a small area of an elytron seen in surface view (diagrammatic)

C, cuticle; *CT*, trabeculae; *HY*, epidermis; *L*, lacunae; *LC*, lateral blood channel; *TR*, trachea.

completely developed and are usually joined by a greater number of cross-veins than occur in other Coleoptera. M_1 is connected with M_2 by means of one or two transverse veins: when two are present an oblong cell, the *oblongum*, is formed which is very characteristic of the type. What appears to be the most generalized venation in the order is found in the Cupedidae.

(*b*) The Staphylinid type. Here the chief characteristic is exhibited in the disappearance of all the cross-veins, and the atrophy of the proximal portion of M, the remainder of that vein being isolated in the apical portion of the wing.

(*c*) The Cantharid type. In this type M and Cu often coalesce distally forming a very definite loop: at the point of junction a single vein (regarded as M) is continued to the wing-margin. R_2 frequently appears as a recurrent

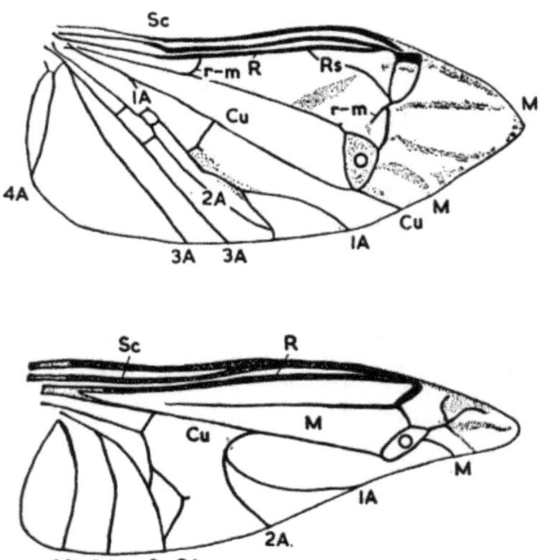

FIG. 375 Adephagid type of wing. Above, Cupedidae (*after* Forbes). Below, Carabidae (original)

O = oblongum.

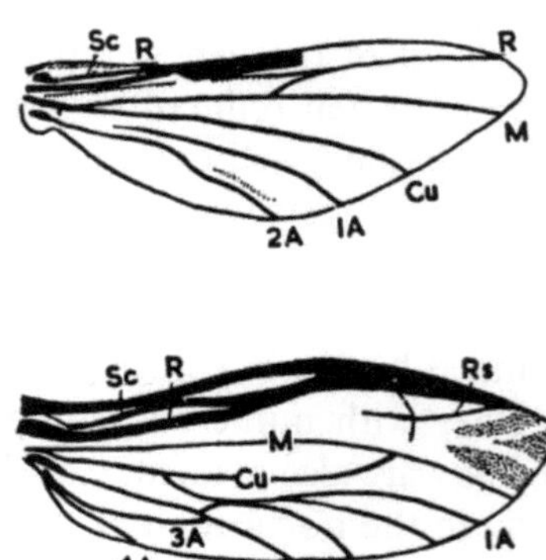

FIG. 376
Two types of wing in the Polyphaga. Above, Staphylinid type (*Ocypus*). Below, Cantharid type (*Cantharis*)

branch of the radius, and cross-veins are commonly present joining the cubital and anal veins. In some cases, the M loop is reduced to a mere hook, or may be absent (Passalidae and many Curculionoidea): when this type of modification occurs, and the cross-veins are atrophied, the Cantharid type is difficult to separate from the Staphylinid one.

The Abdomen – The number of segments comprising this region of the body is difficult to determine. As a rule the 1st tergum is membranous and one or more of the sterna from the 1st to the 3rd are aborted. The 1st sternite is never visible externally and is at most traceable in vestigial form at the back of the hind coxal cavities. Eight tergites are commonly visible externally, the 9th and 10th plates being invaginated. Five to seven sternites are visible externally and in this respect four types may be recognized:

(*a*) The Adephagid type. The hind coxae are immovably fused with the metathorax and completely divide the 1st visible abdominal sternite which is more or less fused with the next two, the sulci between them being partly obliterated (Fig. 372).

(*b*) The haplogastrous type. The 2nd abdominal segment exhibits a pleurite and a small lateral plate representing the sternite. Exceptionally, in some Staphylinidae, the 2nd sternite is complete (Fig. 377, B).

(*c*) The symphiogastrous type. The pleurite of the 2nd abdominal segment is fused to that of the 3rd and the sternite is membranous and nowhere visible externally (Fig. 377, C).

(*d*) The hologastrous type. In some Cantharoidea, the 2nd abdominal sternite is, perhaps secondarily, fully sclerotized and distinct from the 3rd (Fig. 377, A).

In many species the terminal abdominal segments of the female are retractile and tubular, thus functioning as an ovipositor, e.g. in the Cerambycidae. The 9th sternite is provided with paired structures to which the segmental appendages may contribute (Tanner, 1927). The male genitalia have been studied by Sharp and Muir (1912) and those of many species are also illustrated by Jeannel and Paulian (1949) and Jeannel (1955). The genitalia are withdrawn into the abdomen and concealed; they take the form of a tubular evagination; with certain associated sclerites, which arise between the 9th and 10th sternites. They have proved to have great value in classification.

Stridulating Organs – In one form or another these organs are present in the imagines of a large number of families and have been studied by Darwin (*Descent of Man*), Gahan (1900) and Dudich (1920–21). As the latter author remarks, wherever any part of the exoskeleton is subjected to the friction of an adjoining part by the movements of the insect there, in some species or another, these organs are likely to be found. Their position is not constant, even in different genera of the same family, and they are often similar in structure and location in genera belonging to widely different

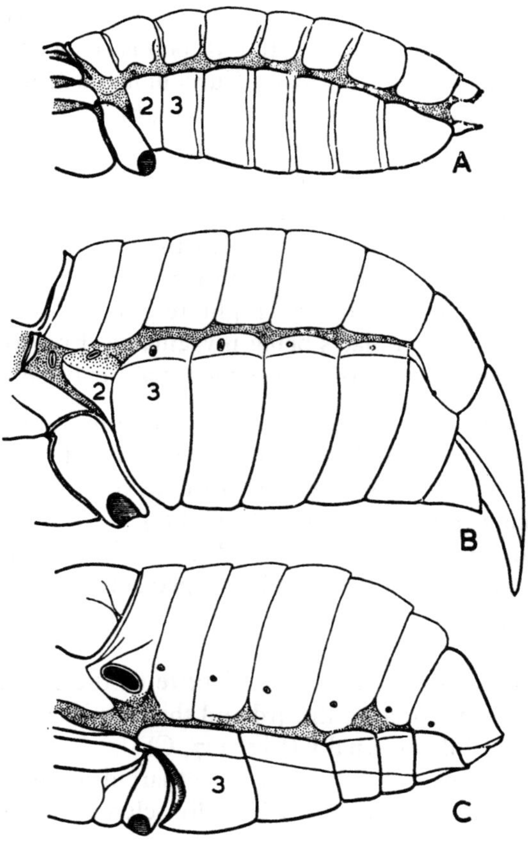

FIG. 377 Three types of abdomen in Coleoptera. A, Hologastran (*Rhagonycha limbata*, Cantharidae). B, Haplogastran (*Melolontha melolontha*, Scarabaeidae). C, Symphiogastran (*Hylobius abietis*, Curculionidae)

2, 3: 2nd and 3rd abdominal sterna.

families. They are most extensively developed in the Scarabaeoids where both the larvae and perfect insects are often capable of stridulation (see Arrow, 1904). Gahan divides these organs among Coleoptera into four groups according to where they are located, but it is only possible here to refer to one or two examples. In certain Nitidulidae and Endomychidae there is a file-like area on the crown of the head which is rasped by the anterior margin of the prothorax. In other cases (certain Tenebrionidae, Scolytinae, etc.) there is a file-like area on the underside of the head, sound being produced by friction with a projecting ridge on the prosternum. Stridulating organs are found on the mandibles and the maxillae in the larvae of various Scarabaeidae (Fig. 401). They are so arranged that a series of teeth on the maxillae rasp against some granulations on the ventral side of the mandibles, when the maxillae move forwards and backwards. Many of the Cerambycidae have stridulatory organs: in some cases the sound is produced by rubbing the hind margin of the prothorax over a striated area of the mesonotum: in others, it is produced by the friction of the hind femora against the edges of the elytra. The most remarkable stridulating organs (Fig. 401) are those met with in the larvae of the Lucanidae, Passalidae and of *Geotrupes* and its allies. They consist of a series of ridges or tubercles on the middle coxae, while the hind legs are modified in various ways as rasping organs. In certain of the Curculionidae there is a stridulating file on the underside of the elytra near their apices: the rasping is effected by a series of small tubercles situated on the dorsal side of the abdomen (Gibson, 1967; *Conotrachelus*). In some cases the file is present on the abdomen in the females, and on the elytra in the males, and the rasping organs are similarly reversed.

INTERNAL ANATOMY

The Digestive System – The digestive system of Coleoptera has been mainly studied by Dufour, whose results have been published in a series of papers (1824–40), and Bounoure (1919). Beauregard (1890) and Gupta (1965) have also studied the digestive organs in the Meloidae, Mingazzini (1889*a, b*) in the Scarabaeoids, Sedlaczek (1902) in the Scolytinae and Bordas (1903; 1904) in the Hydrophilidae and Silphidae. The mouth opens into the *pharynx* or widened commencement of the *oesophagus* and the latter region is a simple tube of variable length. At its hinder extremity the oesophagus expands to form the *crop* which is of very general occurrence although according to Beauregard it is absent in pollen-eating beetles such as *Zonitis*, *Sitaris* and *Mylabris*: it is large and capacious in *Carabus* (Fig. 378) and other genera. The oesophagus or crop, as the case may be, is followed by the *gizzard* which is usually a small chamber lined by sclerotized ridges or folds, or with spines or denticles whose arrangement may be of generic importance (Balfour-Browne, 1934, 1935; Thiel, 1936; Reichenbach-Klinke, 1953): it is present in many carnivorous and wood-boring Coleoptera, notably in the Curculionids (Aslam, 1961; Morimoto, 1962; Kissinger, 1963). The *mid intestine* is very variable in form, and is often of a complex nature. Its most characteristic feature is the presence of large numbers of small enteric caeca which often vary in character in different portions of the stomach. In the Carabidae and Dytiscidae the latter region is a simple slightly tortuous tube provided with numerous closely packed caeca, but the latter are usually wanting from its posterior portion. In *Meloe* the mid intestine is large and sac-like, occupying the greater part of the abdominal cavity. In the Scarabaeoidea (Fig. 379) it is very long and convoluted while in *Copris lunaris* it is thrown into a series of numerous coils after the manner of a watch-spring. In the Scolytinae the mid intestine is divisible into three regions: a sac-like anterior region, a narrow tubular middle portion and a wider posterior which is partially or completely invested with small caeca. The

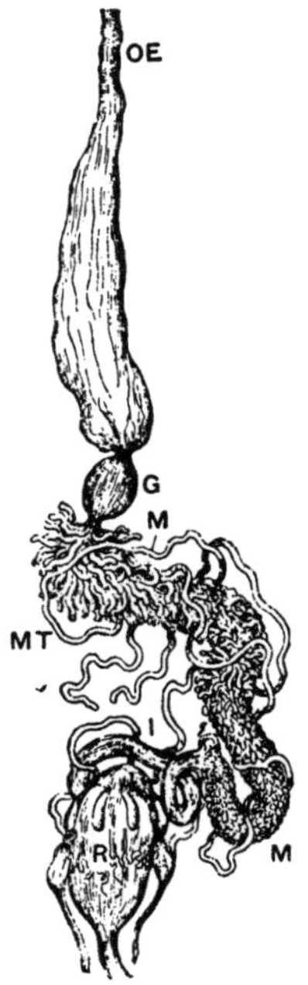

FIG. 378
Carabus monilis,
alimentary canal

OE, oesophagus; *G*, gizzard; *M*, mid gut; *I*, ileum; *R*, rectum; *MT*, Malpighian tubules. *After* Newport.

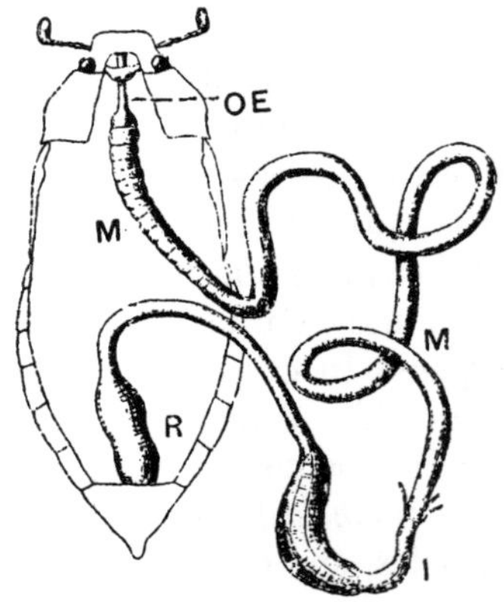

FIG. 379
Melolontha melolontha
alimentary canal

Lettering as in Fig. 378. Adapted from Bounoure.

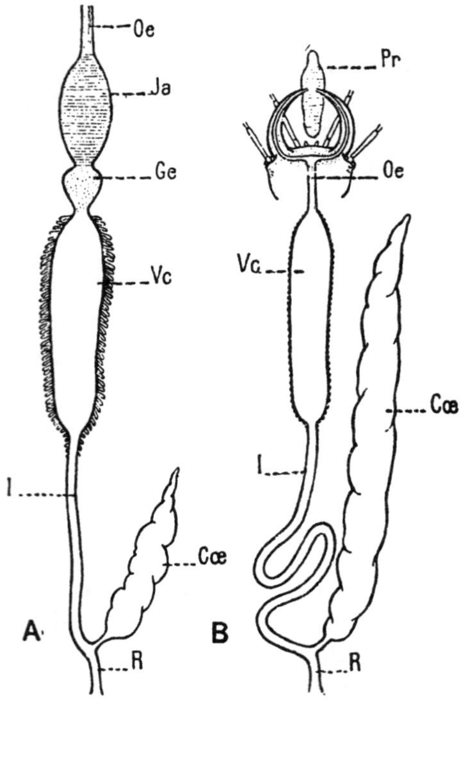

FIG. 380
Dytiscus marginalis; alimentary canal of A, imago; and B, larva

Oe, oesophagus; *Ja*, crop; *Ge*, gizzard; *Vc*, mid gut; *I*, hind gut; *Coe*, caecum; *R*, rectum; *Pr*, prey. *After* Portier.

hind intestine is always more or less convoluted: it is relatively short in the Carabidae, but long in *Dytiscus* and many other genera. In the Dytiscidae (Fig. 380) it gives off a conspicuous *rectal pouch*, an organ which is characteristic of that family (Bordas, 1906). Although a relatively small sac in *Ilybius* it attains enormous dimensions in *Dytiscus* and bears an apical tubular appendix: a posterior caecum is also present in *Silpha* and *Nicrophorus*. The *rectum*, when specially differentiated, is often a large chamber: *rectal papillae* are present in certain Passalidae and Silphidae, but as a rule they are wanting.

The *Malpighian tubules* are typically four or six in number and are of considerable importance in the classification of the families into major groups (Poll, 1932; Stammer, 1934). In the Lampyridae, where there are four of these vessels, the tubes of each pair unite distally, thus presenting the appearance of loops (Bugnion, 1920). A number of Coleoptera including *Donacia, Haltica, Cerambyx, Oedemera,* etc., exhibit the cryptonephric condition (Lison, 1937) in which the Malpighian tubules have two apparent terminations in the intestine owing to the fact that their distal extremities become applied to the walls of the colon or rectum, instead of remaining free as in most other insects. In no case, however, have any secondary openings into the hind intestine been discovered (Woods, 1916).

Gnathal glands have been little studied in Coleoptera and it is often stated that salivary glands are absent. Pradhan (1939) describes labial and maxillary glands in Coccinellids and Gupta (1937) records what appear to be labial glands in Tenebrionids. *Pygidial glands*, which are defensive in function, exist in many beetles and are very fully discussed by Berlese. They are paired organs secreting corrosive and pungent fluids which can sometimes be ejected to a distance of several inches. These glands open in close association with the anus and, among the Adephaga, they have been studied in detail by Dierckx (1899), Bordas (1899), Forsyth (1968, 1970), Moore and Wallbank (1968), Schildknecht, Winkler and Maschwitz, 1968, and Aneshansley and Eisner (1969). In *Pterostichus melanarius* (= *vulgaris*), for example, each gland consists of spherical acini composed of gland cells: each acinus opens by a separate duct into the common canal of its side.

Among the Carabidae many chemical compounds such as salicylaldehyde, *n*-alkane, formic acid, *m*-cresol and *iso*-butyric acid have been identified in the secretions, different components tending to be characteristic of each tribe. However, the Paussinae + Ozaeninae have evolved, independently of the Brachynini, an explosive, enzyme-catalysed reaction between quinols and hydrogen peroxide to form quinones and gaseous oxygen. This may be ejected at 100° C at frequent intervals.

In the staphylinid genera *Staphylinus*, *Stenus*, etc., and also in *Blaps* eversible foetid anal glands are present. Forsyth (1970) records that Dytiscidae and Hygrobiidae also have thoracic defensive glands.

The Nervous System – The most important differences in the nervous system are exhibited in the ventral cord (Brandt, 1879). As a rule the commissures retain their double nature, a feature which is well exhibited in the thorax of most beetles. The most generalized type of nervous system is found in the Cantharidae where, in addition to the supra- and infra-oesophageal centres, there are three thoracic ganglia and seven or eight abdominal ganglia. The latter number is maintained in *Dictyopterus* and seven abdominal ganglia are found in *Cantharis* and *Lampyris*. Reduction in the number of abdominal ganglia, unaccompanied by a similar specialization of the thoracic centres, may be traced through a number of genera. Thus, in *Cicindela* and *Tenebrio* there are six ganglia in the abdomen; in *Silpha*, *Mordella* and *Creophilus* there are five: in *Donacia*, *Meloe* and *Callidium* there are four: in *Cassida* there are three and in *Chrysolina* and *Coccinella 7-punctata* there are two. Among the Scarabaeoids (*Geotrupes*, *Aphodius*, etc.) the abdominal ganglia are merged into the metathoracic ganglion to form a common centre. In a number of other Coleoptera the meso- and metathoracic ganglia are closely united or merged together owing to the disappearance of the connectives between them. This feature is characteristic of many other Scarabaeoids (*Melolontha*, *Passalus*, *Lachnosterna*, *Phyllopertha*, *Cetonia*) and the centre thus formed also includes the fused ganglia of the abdominal chain. In the Curculionidae there are usually two separate abdominal centres, in *Gyrinus* one, and in *Nicrophorus* five. The maximum specialization is found in *Serica brunnea* and *Amphimallon solstitialis*. In the former all the thoracic and abdominal ganglia unite to form a single complex: in the latter species Brandt states that coalescence has proceeded still further, the infra-oesophageal ganglion being also involved in the fusion.

The Circulatory System – The structure of the dorsal vessel has only been investigated in a few examples. The heart is divided into a variable number of chambers and is continued as the aorta through the thorax into the head where it becomes branched at its apex. In *Melolontha* Straus-Dürckheim found nine chambers with eight pairs of ostia. In *Lucanus* Newport described seven chambers and a similar number of pairs of alary muscles.

The Respiratory System – The tracheal system attains its highest degree of differentiation among the actively flying members of the Scarabaeoidea, particularly in *Geotrupes* and *Melolontha*. Its trunks are greatly ramified and in many species there is an elaborate system of air-sacs. The latter structures do not attain a great size, their chief characteristic being the large numbers present. In *Melolontha* they occur throughout the body, even penetrating into the recesses of the head (Straus-Dürckheim). In *Lucanus* (male) the large massive head and mandibles are filled with air-sacs, especially the mandibles. Newport states that they are developed in rows from long tracheae which penetrate the jaws, and the latter apparently unwieldy structures are thus rendered extremely light.

As a rule ten pairs of spiracles are present: the first is situated between the pro- and mesothorax and the remaining pairs are metathoracic and abdominal in position. Among the Scarabaeids (Ritcher, 1969) and certain

Curculionids and other Coleoptera, the eighth pair of abdominal spiracles is either absent or vestigial and non-functional. In the Scolytinae the number of functional abdominal spiracles varies from five to seven. Special modifications in aquatic species are mentioned under the families concerned.

The Reproductive System – The *male reproductive organs* have been investigated by Dufour (1825), Escherich (1894), Bordas (1900), Williams (1945) and others. They consist of the testes, the vasa deferentia, one or

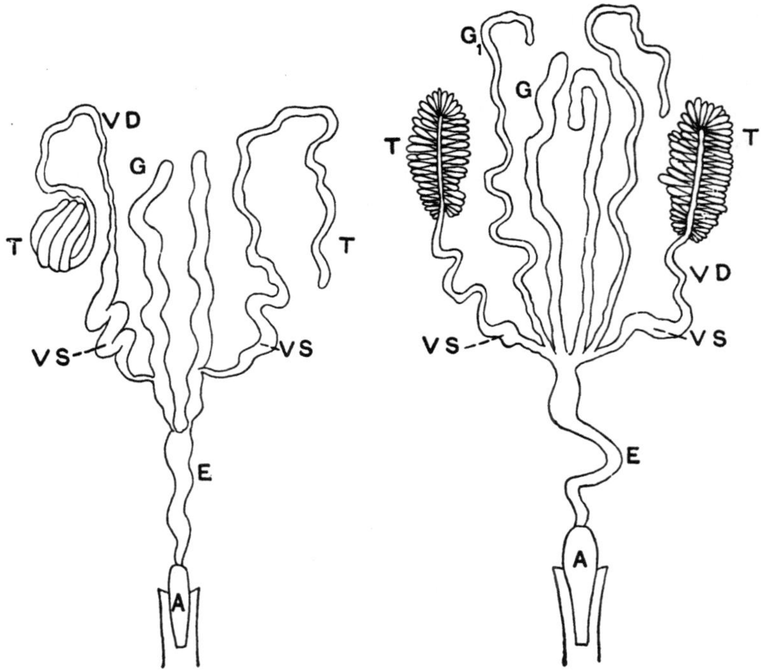

FIG. 381 Male reproductive organs of: left, Adephaga; right, Polyphaga. The right testis in the Adephaga is represented uncoiled

A, aedeagus; *E*, ejaculatory duct; *G*, accessory gland (ectadenes); G_1, accessory gland (mesadenes); *T*, testis; *VD*, vas deferens; *VS*, vesicula seminalis. Adapted from Bordas.

more pairs of accessory glands and a median ejaculatory duct. Vesiculae seminales are often present as dilatations of the vasa deferentia. Two general types of reproductive organs are recognized by Bordas and are based upon characters afforded by the testes (Fig. 381). In the Adephaga, these organs are simple and tubular and more or less closely coiled, each being enclosed in a membrane. In the second type the testes are compound and divided into a number of separate follicles. The latter may be rounded capsules, each communicating with the vas deferens by means of a separate duct, as in the Chrysomelids, Curculionids and Scarabaeids. Or, the testicular follicles may be composed of aggregations of small rounded or oval sessile sacs which open directly into the vas deferens (most other Polyphaga).

The accessory glands exhibit many differences with regard to their position, number and mode of origin. Escherich (1894) has divided them into ectadenia and mesadenia: the former are believed to arise as ectodermal invaginations of the ejaculatory duct, while the latter are stated to be of mesodermal origin, since they are formed as outgrowths of the vasa deferentia. Definite ontogenetic evidence is needed, however, to substantiate these conclusions.

The *female reproductive organs* (Stein, 1847; Williams, 1945) may likewise be divided into two types, according to whether the ovarioles are polytrophic or acrotrophic in character. The former type is characteristic of the Adephaga and the latter type is found, so far as known, throughout the Polyphaga. The ovarioles vary greatly in number: thus in *Ips typographus, Hylobius abietis* and *Sitona lineatus* there are two ovarioles to each ovary: in *Ocypus olens* there are 6–7, in certain Elateridae four, in *Dorcus* and *Saperda carcharias* twelve, in *Byrrhus pilula* there are about twenty, and in the Meloidae they are extremely short and much more numerous. In some Coleoptera (*Dytiscus*) a colleterial gland is present in association with each oviduct. A *spermatheca* is generally present and opens, by a slender and often exceedingly long duct, either into the vagina or the bursa copulatrix. An accessory gland, of variable character, is generally found in connection with the spermatheca. In many Coleoptera a second passage or 'canal of fecundation' leads from the spermatheca or its duct and opens into the vagina near the point of union of the two oviducts (Fig. 382). This canal is believed to allow of the direct passage of the spermatozoa from the spermatheca to the eggs. A *bursa copulatrix* is present as a diverticulum of the wall of the vagina. It is believed that the spermatozoa are received into this sac during copulation and subsequently make their way into the spermatheca. A *spermatophore* is found in some species but not in others (Khalifa, 1949). The process of fecundation in Coleoptera, however, is very little understood and the significance of the frequently great length of the spermathecal duct is unknown.

Metamorphoses

The Egg – The eggs of Coleoptera are usually ovoid in form and rarely exhibit any marked diversity of form of structure as is seen, for example, in the Hemiptera and Lepidoptera. In *Ocypus* they are unusually large and few in number, while in the Meloidae they are small and the number laid by a single female may run into several thousand. Certain cave-dwelling beetles (some Anisotomidae-Bathysciinae, some Carabidae-Trechinae) lay a single large egg from which a larva emerges and pupates without feeding or moulting; this appears to be an adaptation to a precarious food-supply (Deleurance-Glaçon, 1963*a* and *b*). Many Coccinellidae lay their eggs in batches on leaves, the Hydrophilidae enclose them in cocoons, while among Cassidinae they are protected by highly specialized oothecae (cf. Gressitt and Kimoto, 1963: 985). In the Curculionidae they are frequently deposited in deep holes drilled by the rostrum of these beetles in the food-plant. In the Scolytinae the females have the habit of entering into the trunk or plant within which the eggs are laid. In several genera – *Nicrophorus* (Pukowski, 1933), *Platystethus* (Hinton, 1944), some Scarabaeinae (*Copris*) (Halffter and Matthews, 1966), and some Scolytinae – the females show maternal care, protecting and sometimes feeding the young larvae (Lengerken, 1954).

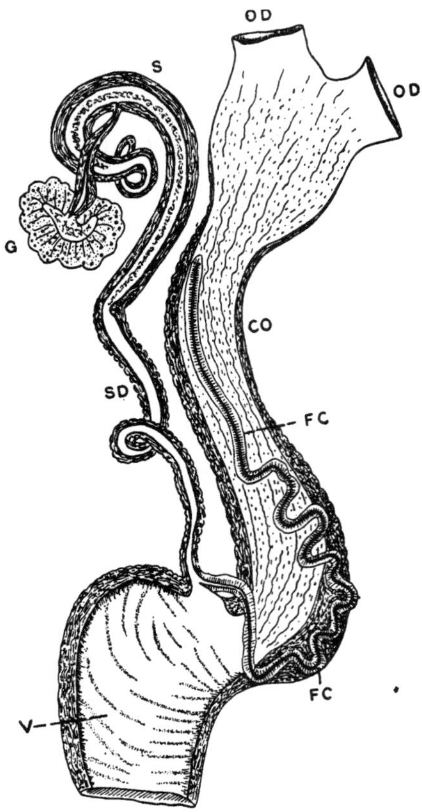

FIG. 382
Oodes helopioides (Carabidae), proximal portion of female reproductive organs

OD, paired oviduct; *CO*, common oviduct; *S*, spermatheca with *SD* duct and *G* gland; *FC*, fecundation canal; *V*, vagina. *After* Stein.

The Larva – In Coleopterous larvae the head is well developed, the mouthparts are adapted for biting and do not differ in their essential features from those of the adults (Böving, 1936; Anderson, 1936). Such larvae never possess abdominal feet, but they are generally provided with thoracic legs: cerci (or urogomphi) may be present or absent. The tracheal system is generally peripneustic with usually nine pairs of spiracles: the first pair is located, as a rule, between the pro- and mesothorax, and the remaining pairs are situated on the first eight abdominal segments. There is, in many cases, a marked similarity among larvae of the same family. This is well exhibited for example in the Carabidae, Buprestidae and Curculionidae. On the other hand, the larval differences found among the Chrysomelidae are scarcely paralleled in any other family of insects. Some of the most remarkable forms occur in the aquatic families Haliplidae, Gyrinidae and Hydrophilidae with their special adaptations to life in the water. Among terrestrial larvae, those of the Dermestidae, with their dense clothing of tufted hairs, are totally different in appearance from all other Coleoptera.

The primitive campodeiform larva (Fig. 383) is characteristic of the Adephaga, many of the Staphylinoidea, and of the first instar in the Meloidae and Rhipiphoridae. Among other of the Staphylinoidea and the vast majority of the Cucujoidea, the larvae are more highly modified and, although they incline to the campodeiform type, they are transitional between the latter and the eruciform type (Fig. 384). Among the Chrysomeloidea, Curculionoidea and Scarabaeoidea the eruciform larva is prevalent. The extreme apodous type is characteristic of the great majority

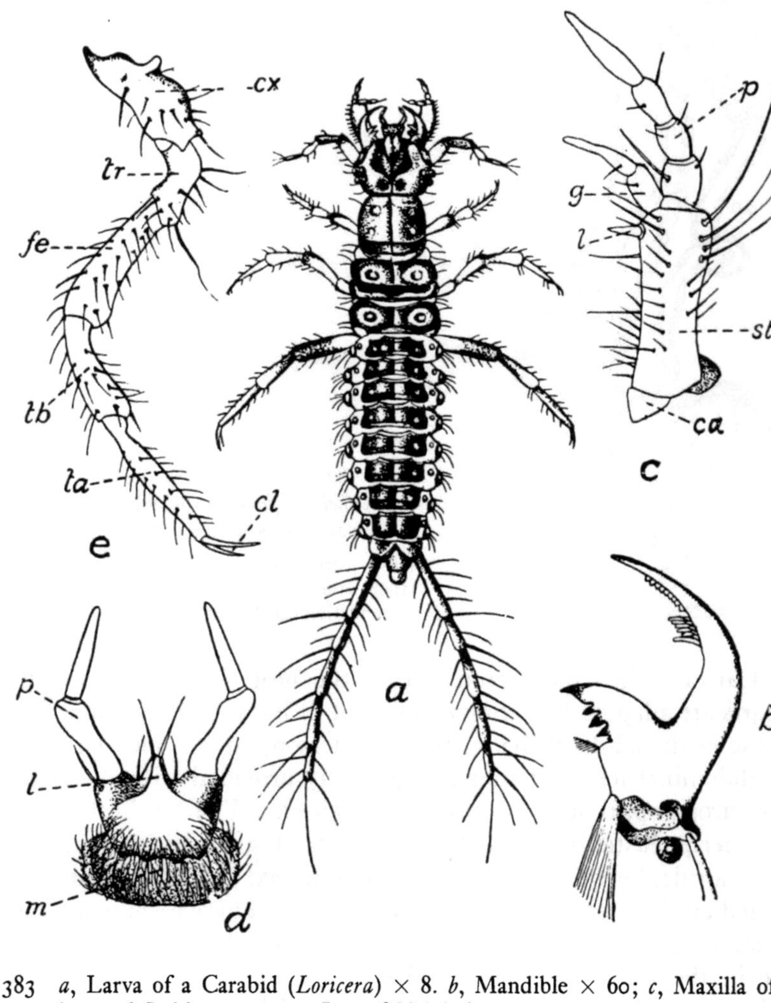

FIG. 383 *a*, Larva of a Carabid (*Loricera*) × 8. *b*, Mandible × 60; *c*, Maxilla of *Nebria* larva; *d*, Labium × 32; *e*, Leg of *Nebria* larva × 24

From Carpenter *after* Schiödte.

of the Curculionoidea. It is also met with in certain of the Cerambycidae and Buprestidae, in the dung-feeding larva of *Cercyon*, and in the Eucnemidae, while an apodous stage occurs in the ontogeny of members of the Meloidae and Bruchidae. It is a comparatively easy matter, therefore, to arrange a graduated series of larval Coleoptera. At the head of such a series is the active, armoured campodeiform type, with well-developed antennae and mouthparts, completely formed legs with tarsi and paired claws, and movable jointed cerci: larvae of this nature are seen in some Carabidae. At the other extreme are the soft apodous maggots of the Curculionidae, with their vestigial antennae, reduced mouthparts and no cerci. The mode of life is the

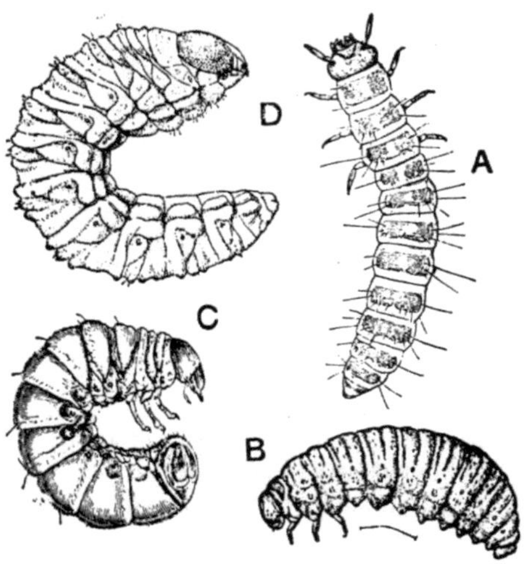

FIG. 384
Coleopterous larvae

A, campodeiform (Cucujidae); *after* Chittenden, *U.S. ent. Bull.*, 4 n.s. B, eruciform (Chrysomelidae); *after* Chittenden, *U.S. Dept. Agric. Year Book*, 1896. C, scarabaeiform (Scarabaeidae), *after* Riley. D, eruciform and apodous (Curculionidae), *after* Chittenden, *U.S. ent. Bull.*, 23 n.s.

primary modifying factor in the development of larval types and, in the absence of an exposed or predatory life, structural changes sooner or later supervene and attain their culminating point in the degenerate internal-feeding larvae that live surrounded by an abundance of nutriment. Hypermetamorphosis is known to occur in a few Coleoptera. It is well exemplified in the Meloidae whose first instar is a campodeiform larva, and in the later development modified campodeiform, eruciform and apodous stages may be passed through in the ontogeny of an individual species (Fig. 188). Hypermetamorphosis similarly prevails in the Rhipiphoridae, Micromalthidae, in *Lebia scapularis* and in the parasitic Staphylinids of the genus *Aleochara*.

The head bears a variable number of ocelli: thus there may be six on either side as in most Carabidae and Hydrophilidae, four in the Cicindelinae, or they may be reduced to a single one, and even this may degenerate into a mere pigment spot. In many larvae which are internal feeders ocelli are totally wanting. Antennae are well developed in campodeiform larvae, and are very long in those of the Helodidae: almost every stage in reduction may be traced until they are represented by papilla-like vestiges as in the Curculionidae. The mandibles are large and exserted in predacious forms, and in the Dytiscidae they are specially modified for suctorial purposes. The proximal inner edge of the mandible often has a *mola*, a thickened, sometimes crenulate or plicate area which serves for grinding, especially in vegetarian species. In larvae which feed internally in wood and other plant tissues, they are short and stout. Structures resembling superlinguae are comparatively well developed in the Dascillidae, and vestigial structures of a similar nature occur in the Scarabaeidae (Carpenter and MacDowell, 1912): rudiments have also been found by Mangan in the Dytiscidae. The maxillae (Figs. 383, 385) are always well developed: their palpi are variable, being long in *Gyrinus* and *Stenus*, while in eruciform larvae they are often reduced to the condition of 2-segmented papillae. In the majority of Coleopterous larvae there is a single lobe or mala which is often composed of two segments.

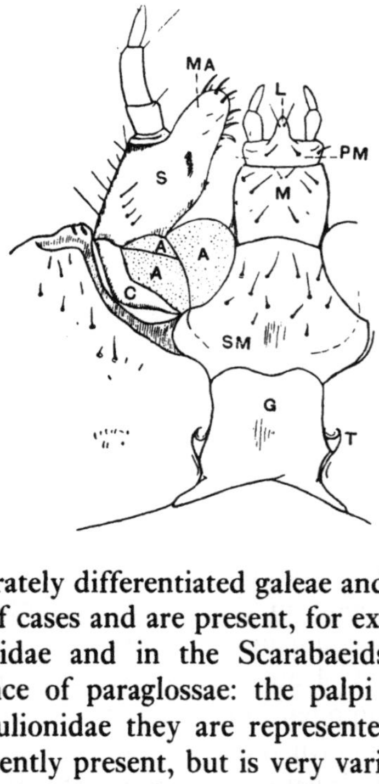

FIG. 385
Gula, labium and right maxilla of a Tenebrionid larva (*Embaphion*)

AA, articulating areas; *C*, cardo; *G*, gula; *L*, ligula; *M*, mentum; *MA*, mala; *PM*, prementum; *S*, stipes; *SM*, submentum; *T*, tentorial pit. Adapted from Böving, *J. agric. Res.*, 22, 1921.

Separately differentiated galeae and laciniae are evident, however, in a number of cases and are present, for example, in *Agriotes*, the Byrrhidae, certain Silphidae and in the Scarabaeids. The labium is characterized by the absence of paraglossae: the palpi are commonly 2-segmented but in the Curculionidae they are represented by unjointed tubercles. The glossa is frequently present, but is very variable, and in many genera it is not separately distinguishable. In *Silpha* the ligula is represented by a pair of rounded lobes which are perhaps to be regarded as being those of a divided glossa. The legs exhibit different degrees of development: among the Adephaga they are undoubtedly primitive and are characterized by the presence of a distinct tarsus and, sometimes, paired claws. According to Jeannel and Paulian (1949) Carabid larvae usually have one claw but some of the larger ones (e.g. *Carabus*) have a second claw which is really a modified bristle. These features are lost in the Polyphaga, where the tarsus is not separately differentiated, and the claws are single. Exceptions are extremely few, but in the first instar of the Micromalthidae and Meloidae a tarsus is present and the claws are paired. The abdomen is 10-segmented and, among the Carabidae and Staphylinidae, the anal segment is often tubular and functions as a pseudopod. Cerci are well developed jointed appendages in many campodeiform larvae: in other cases they may be fixed and unjointed (*urogomphi*). The morphology of the rigid horny anal processes of many larvae is not understood: they have the appearance of being non-appendicular outgrowths of the body-wall, but when their development is studied they may prove, in some cases, to be highly modified cerci.

The respiratory system is subject to comparatively few modifications. The position of the first pair of spiracles is somewhat variable: although commonly intersegmental, they may as in *Cantharis* (*Telephorus*) be located on the mesothorax. Well-developed metathoracic spiracles occur in the Lycidae but in other families they are absent or vestigial. The most striking variations occur in aquatic larvae (Hinton, 1947): *Peltodytes* and *Gyrinus* are apneustic, and respire by means of filamentous processes of the body-wall, while certain of the Hydrophilidae are metapneustic.

Information on the internal anatomy of Coleopterous larvae is fragmentary and very scattered. The head and its muscles have been studied in *Cantharis* by Lagoutte (1966). The alimentary canal has been investigated by Portier (1911) in the Dytiscidae and Hydrophilidae, by Payne (1916) in *Cantharis*, by Woods (1916; 1918) in *Haltica*, and by Mingazzini (1889) in the Scarabaeids. In the latter group and also in *Cantharis* and *Calosoma* it pursues a straight course from the mouth to the anus, the hind intestine in these instances being short (Figs. 380, 386). In the Dytiscidae and Scolytinae the gut is convoluted owing to the increase in length of the hind intestine. A well-developed crop is present, for example, in *Sitophilus* but in *Cantharis, Haltica* and *Dendroctonus* it is represented by a small distal enlargement of the oesophagus. A gizzard is present in the latter genus, while both crop and gizzard are wanting in the Dytiscidae and Hydrophilidae. The mid intestine is very variable, but always forms a large portion of the gut, and frequently is differentiated into several distinct regions. In *Cantharis* it is a large simple sac, but in many larvae it is coiled and tubular, as for example, in the Dytiscidae, Hydrophilidae, and also in *Haltica* and *Dentroctonus*. Differentiation into separate regions is evidenced by change of calibre, by the histological structure, and the presence or absence of enteric caeca. In *Oryctes*, and other Scarabaeids, the latter structures are very large and are restricted to three annular bands (Fig. 387): in *Sitophilus* (= *Calandra*) they are represented by numerous papilla-like outgrowths. An extensive caecum is sometimes present in relation with the hind intestine. In *Dytiscus* it occupies a considerable part of the body cavity and a large caecum is also present in many Scarabaeids. In the Coprinae the larvae have a characteristic dorsal hump which serves for the accommodation of this organ. As in most adults, labial glands are wanting. The Malpighian tubes, as a rule, are similar in number and character to those of the imagines. The nervous system generally consists of three thoracic and seven or eight abdominal ganglia. In

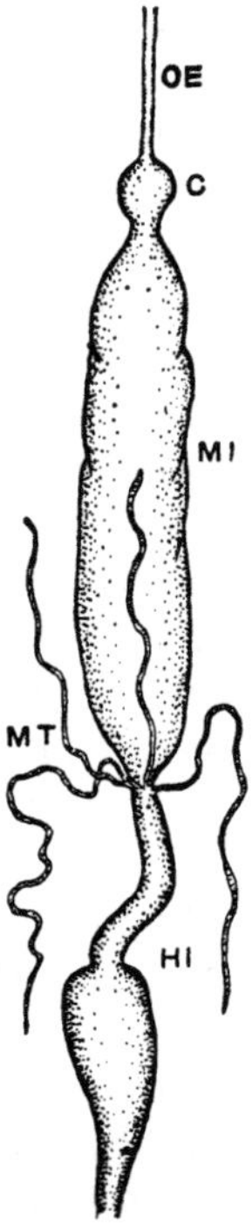

FIG. 386
Cantharis, alimentary canal of larva

OE, oesophagus; *C*, crop; *MI*, mid gut; *MT*, Malpighian tubules, *HI*, hind gut.

FIG. 387
Oryctes nasicornis (Scarabaeidae), alimentary canal of larva

OE, oesophagus; $C_1C_2C_3$, enteric caeca; *MI*, mid gut; *S*, sac of hind gut; *R*, rectum. *After* Mingazzini.

Coccinella 7-punctata the abdominal ganglia are concentrated in the anterior segments of the hind body, the intervening commissures being very much abbreviated. In *Melolontha*, and other Scarabaeids, the whole of the ventral ganglia are concentrated in the thorax.

The heart has been observed by Payne in *Cantharis*. It is of an extremely narrow calibre and apparently exhibits no division into chambers: nine pairs of alary muscles are present. Segmental glands have been described in a few cases: a pair is present on each of the thoracic and abdominal segments in *Ocypus* (Georgevitsch, 1898), *Chrysomela* (Berlese) and *Cantharis* (Payne).

The Pupa – The pupae in this order are adecticous, of the exarate type, pale-coloured, and are invested by a thin, soft cuticle. In some of the Staphylinidae they are obtect (Hinton, 1946*a*: 318), being covered by an exudation that solders the appendages down to the body and forms a hardened coat. In the Coccinellidae the pupae likewise have a hardened integument and are, moreover, often conspicuously coloured. A large number of Coleoptera pupate in earthen cells below ground: many others pupate within the food plant. A cocoon is frequently present, but the nature and origin of the substance by means of which it is produced needs investigation. In certain of the Curculionidae the cocoon is formed by a product of the Malpighian tubules, while among several of the Scarabaeids it is described as being formed from the contents of the posterior caecum. Many of the Cerambycidae construct pupal cells largely impregnated with carbonate of lime. The naked exposed pupae of the Coccinellidae are often protected by the persistent remains of the last larval skin (cf. Hinton, 1946*b*).

Literature on the Metamorphoses – The most important publications on the life-histories of these insects are those of Schiödte (1862–81), which are written in Latin and have excellent illustrations, and of Böving and Craighead (1931). The complete literature on the transformations of European Coleoptera, up to 1894, has been collated and arranged by Rupertsberger, while Beutenmüller (1891) has catalogued the references to those of the American species. A series of papers by van Emden (1939–49) provides keys for the identification of the British species and gives many references to the literature. Bertrand (1972) deals with all known aquatic larvae.

Classification of Coleoptera

Most recent classifications of the beetles derive from that of Ganglbauer which has been expounded in more detail in relation to previous systems by Gahan (1911), Fowler (1912) and Leng (1920). The suborder Archostemmata has been more recently added (Kolbe, 1908; Böving and Craighead, 1931). Further valuable discussions of the subject have been published by Peyerimhoff (1933) and Meixner (1935). A series of papers by Crowson (1950–54, revised 1955, reprinted with additions up to Dec. 1967 in 1968) has been much used in the account which follows. In 1955 he erected a fourth suborder, the Myxophaga, for several small families. Arnett (1968) provides keys to the beetles of the U.S.A. down to generic level.

SUBORDERS

1. Hind coxae immovably fixed to the metasternum, completely dividing the 1st visible abdominal sternite which is more or less fused with the 2nd and 3rd. Notopleural sulcus present in prothorax. Wings usually with 2 *m-cu* cross-veins defining an oblongum. 4 simple Malpighian tubules, testes tubular, coiled inside a membranous sheath, only 1 pair of male accessory glands, ovarioles polytrophic. Larva normally with a tarsus and 1 or 2 claws developed in the legs, mandibles without a molar area. Hypopharynx without a sclerome, labrum fused to head-capsule. **Adephaga** (p. 840).

– Hind coxae rarely fused to the metasternum, if so not dividing first visible abdominal sternite. Notopleural sulcus rarely present in prothorax. Wings usually with base of vein R missing, rarely complete, oblongum rarely distinct. Larva if with distinct claw-bearing tarsus, mandibles with a mola, labrum free and hypopharynx with a sclerome. **2**

2. Wings with Rs complete to base, oblongum usually distinct, apex of wing spirally rolled in repose. Metepisternum partly bounding mid coxal cavities. Notopleural sulcus usually present in prothorax. Antennae filiform or serrate. Larva with a claw-bearing tarsus, at least in first instar. Hypopharynx with characteristic darkened sclerome. **Archostemata** (p. 840).

– Wings with base of Rs absent, oblongum rarely present, apex of wing not spirally rolled. Metepisternum rarely reaching mid coxal cavities. Notopleural sulcus rarely distinct in prothorax, if so antennae clubbed. Larva never with a claw-bearing tarsus; hypopharynx without a darkened sclerome. 3

3. Notopleural sulcus usually distinct in prothorax. Wings with oblongum more or less distinct, sometimes rolled apically in repose. Maxilla without a distinct galea. Size very small. **Myxophaga** (p. 840).

– Notopleural sulcus never distinct in prothorax. Wings never with a distinct oblongum. Testes follicular. Ovarioles acrotrophic. **Polyphaga** (p. 846).

Suborder ARCHOSTEMATA

This includes the families **Cupedidae** and **Micromalthidae**. The first family has 6 genera and 25 species (Atkins, 1963) found in both hemispheres, including Australia. Mesozoic fossils are not uncommon in British rocks and are elsewhere found from the Upper Permian onwards. The structure of the adult shows several affinities with the Adephaga but its internal anatomy is unknown apart from some brief notes by Atkins (1958). The larva, however, is a specialized wood-feeder (Snyder, 1913; Fukuda, 1938). The body widens somewhat posteriorly and terminates in a stout anal spine; there are no cerci and the legs are short and have one claw which sometimes forks. The Micromalthidae has one species, *Micromalthus debilis* of the U.S.A., since introduced into Hawaii and S. Africa. The larva develops in timber, especially in mines, and undergoes the most complex developmental cycle of any beetle (Barber, 1913; Pringle, 1938; Scott, 1941), including four larval forms (cf. p. 304).

Suborder MYXOPHAGA

This group includes one superfamily, the **Sphaerioidea**, with four small families, the **Sphaeriidae**, the **Hydroscaphidae**, the **Torridincolidae** and the **Lepiceridae**; the **Calyptomeridae**, originally placed here by Crowson, are now excluded. All the species seem to be found in wet places by streams and the larvae of the first three (Hinton, 1967*b*) all have spiracular gills on the first eight abdominal segments and in the Hydroscaphidae one also on the thorax. The first two familes are somewhat widespread while the Torridincolidae (Steffan, 1964) includes one species from Rhodesia and the Lepiceridae two from C. America. The Torridincolidae have the only beetle larva in which there is a plastron to supply the gill with oxygen.

Suborder ADEPHAGA

This includes a single superfamily, the **Caraboidea**. In many respects, such as the larval structure, the group retains a primitive character. In other ways, however, for instance in the structure of the hind legs, the Adephaga are probably specialized for an active predatory life which was not pursued by their ancestors. A few only of the Carabidae have become secondarily vegetarian (Johnson and Cameron, 1969).

Two groups which are now included as subfamilies of the Carabidae have normally been given family rank, viz. **Cicindelidae** and **Paussidae**. Their separation, however, was mainly based on their characteristic appearance and a few special modifications. Bell and Bell (1962) would also include the Rhysodidae as a tribe allied to the Scaritini.

FAM. RHYSODIDAE. The members of this small family (Arrow, 1942) are readily distinguished from other Adephaga by their stout and conspicuously moniliform antennae (Fig. 388); the metasternum has no groove in front of the hind

FIG. 388
Rhysodes boysi

coxae and the hind wing lacks the oblongum. There are more than 125 species, mostly tropical, but with a few extending into southern Europe. All stages live in rotten wood and the larvae (Grandi, 1974) are of a reduced type without cerci.

FAM. CARABIDAE. This important family comprises about 25 000 described species (Fig. 389) and is distributed throughout the world. In temperate regions its members are almost entirely ground beetles occurring in the soil, under stones, in moss and rotting wood, under bark, etc. The elytra in many species are firmly soldered together and the wings are often atrophied. In the tropics there are numerous arboreal genera, with well-developed wings and considerable powers of flight. In many genera the legs are slender, and adapted for running; in others (*Clivina*, *Dyschirius*, etc.) they are shorter, and are used for digging. Although a considerable number of the species is metallic or otherwise brightly coloured, the majority have the sombre dark coloration of ground insects. Many Carabidae, in their general configuration, bear a resemblance to the Tenebrionidae, but may be easily separated upon tarsal characters. Although both the larvae and adults are essentially carnivorous a few have been recorded as devouring cereals and the seeds of plants, the habit being noted in species of *Ophonus*, *Zabrus*, *Omophron* and *Amara*. *Ophonus rufipes* sometimes causes damage to strawberries. *Calosoma* largely preys upon Lepidopterous larvae and *C. sycophanta* has been imported in large numbers from Europe into N. America, in order that its predacious habit may be utilized in destroying the larvae of the gipsy and brown-tail moths (Burgess, 1911).

The Cicindelinae or Tiger beetles are very actively predacious and are characterized by the markedly prominent eyes, the large and acutely toothed mandibles, and by the lacinia usually terminating in an articulated hook. The legs are long or very long, and there are generally six ventral abdominal segments visible in the female and seven in the male (Fig. 372). The subfamily comprises about 2000 species (cf. Horn, 1938), the majority being denizens of tropical and subtropical lands. About half its members belong to the genus *Cicindela* and to the latter are assigned the four British representatives of the family. Tiger beetles are often brightly coloured, although they seldom appear conspicuous in their natural surroundings. Their movements are very active, they run with extreme rapidity and many quickly take to the wing. Although their flights are of short duration, their darting movements render it extremely difficult to follow their course with the eye.

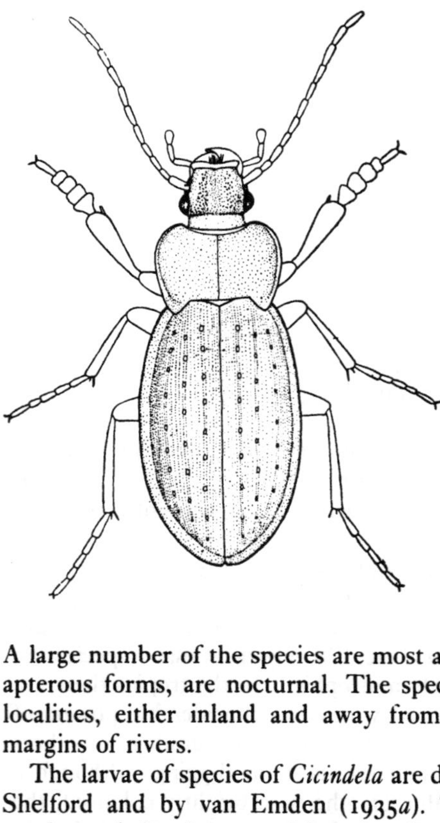

FIG. 389
Carabus nemoralis × $\frac{5}{2}$. Europe

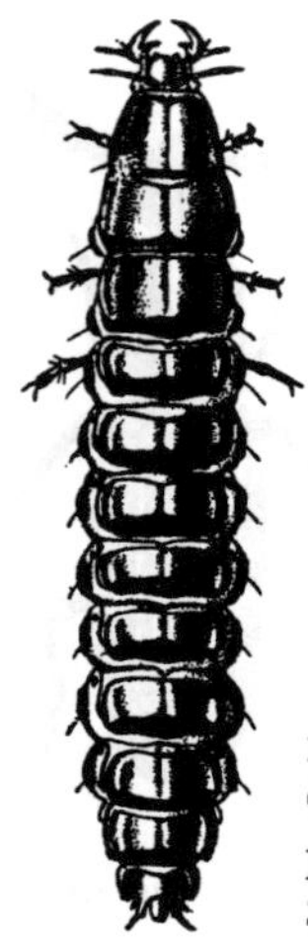

FIG. 390
Carabus cancellatus larva × $\frac{7}{4}$
From Fowler (F.B.I.), *after* Schiödte.

A large number of the species are most active in hot sunshine but others, including apterous forms, are nocturnal. The species of *Cicindela* chiefly affect open sandy localities, either inland and away from water, or on the seashore or along the margins of rivers.

The larvae of species of *Cicindela* are described and figured by Schiödte, by V. E. Shelford and by van Emden (1935*a*). They are characterized by the head and prothorax being larger and broader than the rest of the body. The mandibles are large and there are six ocelli (two vestigial) on each side. The legs are rather long and slender, the tarsi bear paired claws and there are no anal cerci. The most characteristic organ consists of a pair of hooks arising from a swollen base on the dorsal side of the 5th abdominal segment. These larvae are ground dwellers, living in burrows which may extend for a foot or more in the earth. The broadened head and prothorax occupy the entrance to the burrow, and its curiously bent body enables the larva to maintain a firm contact with the sides of its abode. This is mainly achieved by the dorsal hooks already mentioned, and the legs also assist in this respect. The food consists of other insects that may wander near the mouth of the burrow and, when the prey is sufficiently near, the larva suddenly throws back its head, seizes the victim with its long sharp jaws, and draws it within the retreat where it is devoured. According to V. E. Shelford the larva of *Cicindela purpurea* requires twelve or thirteen months for its growth and during that time it passes through three ecdyses. The larva of *Neocollyris* has been described by R. Shelford (1907) and by Docters van Leeuwen (1910). It is of the typical Cicindelid form but there is only a single pair of ocelli on each side of the head. In the place of the pair of dorsal abdominal hooks there is a series of three smaller hooks on either side of the same segment. This larva bores into the shoots of tea and coffee plants and, according to Docters van Leeuwen, that of *Tricondyla* is very similar in structure and habits.

Species of *Anophthalmus*, and other genera, are devoid of eyes and live in caves or beneath stones deeply embedded in the ground. The tribe Brachynini, as already noted (p. 829), contains some species which eject an explosive volatile liquid. The Paussinae (Fig. 391), often treated as a family but connected to more ordinary forms by the Ozaeninae, are known by having fewer than six visible abdominal sternites and by the antennae which are clubbed and often with few segments. In

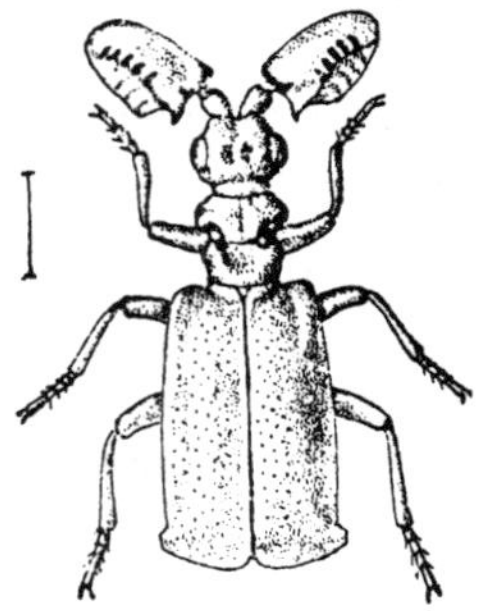

FIG. 391
Paussus testaceus, Tenasserim
After Fowler (F.B.I.).

Protopaussus they are of the simple 11-segmented Carabid type; in *Cerapterus* and other genera they are 10-segmented and exceedingly broad and compressed; in *Pleuropterus* most of the segments are soldered together and in *Paussus* and many other genera they are apparently 2-segmented. The 2nd segment is greatly enlarged to form a club, which assumes the most bizarre shapes, and is probably developed as the result of the consolidation of an originally multi-articulate flagellum. According to Wasmann (1910) the antennal development is correlated with the growth of a glandular exudatory tissue which produces an aromatic secretion attractive to ants. This tissue is found not only in the enlarged antennal segments but also beneath the body-wall of the head, prothorax and apex of the abdomen. Its positions are indicated by the presence of tufts of yellow hairs or groups of cuticular pores which facilitate the diffusion of its secretion. The latter is eagerly licked by the ants off the bodies of their Paussid inquilines, who make a return for the hospitality they receive (cf. p. 1243). The metamorphoses of the subfamily have received very little attention: the larva of *Paussus* is of a modified carabid type (Böving, 1907; van Emden, 1936). There are about 400 species, all tropical or subtropical, and perhaps all associating with ants.

Carabid larvae (Fig. 390) are very active, linear or elongate in form, with ten abdominal segments, and the legs are terminated by one or two claws. The head carries a pair of sharp caliper-like mandibles and there are six ocelli on either side. The 9th abdominal segment carries a pair of cerci of variable length and the 10th segment is tubular in form, and generally provided with a pair of protrusible vesicles (Kemner, 1912–13). In addition to the writings of Schiödte and Xambeu, a number of Carabid larvae are described by Böving (1910; 1911), Dimmock and Knab (1904), Kemner (1912; 1913), and van Emden (1942) provides a key to the genera.

The Carabidae are divided into many subfamilies, the largest being the Pterostichinae with over 5000 species. The Carabinae comprise many of the larger and more striking forms (Fig. 389), and the Thyreopterinae includes the arboreal Malayan genus *Mormolyce*, in which the lateral borders of the elytra are produced into broad leaf-like expansions. The Pseudomorphinae are likewise an aberrant group, and have the head grooved on either side for the reception of the antennae.

The detailed classification of the Carabidae is still partly under discussion. Ali (1967) shows that some of the proposed subfamilies do not agree with features in the internal anatomy and he considers that the tribes are better founded than the larger groups. Some features in the modern treatment of the group appear in the works of Britton (1970) and Lindroth (1969).

FAM. AMPHIZOIDAE. A very small family (Edwards, 1950) consisting of four species which are indigenous to N. America and one to Tibet. The front coxae are spherical and their cavities open behind. The legs are not adapted to swimming. They frequent cold, rapid streams where they cling to stones and timber, but are not capable of swimming. They feed on drowned insects. The larva of *Amphizoa* is described by Hubbard (1892): it is likewise aquatic, the side margins of the segments are extended into lamellate prolongations and the larva bears a close resemblance to that of a Silphid. Six ocelli are present on either side, the tarsal claws are paired, and there are eight abdominal segments terminated by a pair of short spine-like cerci. The only pair of functional spiracles is terminal, the remaining pairs being obsolete.

FAM. HYGROBIIDAE. Like the Amphizoidae this is a very small family with a remarkably discontinuous geographical range, its single genus *Hygrobia* occurring in Britain and S. Europe, central Asia and Australia. The species are aquatic but, unlike those of *Amphizoa*, the legs are adapted for swimming. They also have conical fore coxae the cavities of which are closed behind. *Hygrobia hermanni* is capable of loud stridulation which is produced by rubbing the apex of the abdomen against a file on the inner aspect of the elytra. The larva of this species is figured by Schiödte; the spiracles are minute and functionless, and it respires by means of a series of ventral branchiae. The body has a club-shaped appearance owing to the greatly enlarged head and prothorax and the narrow linear abdomen. The latter is terminated by two very long cerci and a median process of very similar proportions. For an account of the biology of this species see Balfour-Browne (1922).

FAM. HALIPLIDAE. A family of about 200 small aquatic beetles of very wide geographical range. The hind coxae are produced into large plates which cover the first two or three sternites. They and their larvae feed upon algae and *Chara* in both running and standing water, where they are found among aquatic vegetation or under stones. Three genera and rather more than a dozen species are British. Their larvae are very peculiar and quite distinct from those of any other family of Coleoptera. The whole body is invested by segmentally arranged groups of fleshy process, which are long and thread-like in *Peltodytes*, and shorter in *Haliplus*. In the latter genus there are eight pairs of abdominal spiracles, but in *Peltodytes* spiracles are wanting and the processes of the body-wall function as tracheal gills (Bertrand,

1928). There is much information on their biology in a paper by Seeger (1971*a* and *b*). He shows that the setigerous tubercles on the tergites and sternites are a new type of very small tracheal gill, only 25–55 μm in diameter.

FAM. NOTERIDAE. *Distinguished from the Dytiscidae by their more convex upper surface without visible scutellum; hind coxae small and produced into plates which cover the trochanter.* The larva lacks the constriction behind the head seen in the Dytiscidae and although apparently predacious, the mandibles are grooved on the inner side, not pierced by a tube. There are about 150 widely spread species.

FAM. DYTISCIDAE (True Water Beetles). Although this family occurs all over the world it is more especially characteristic of the Palaearctic region: about 4000 species are known, over 100 being British (Balfour-Browne, 1940–50). Its members frequent both running and standing water, one or two species inhabit thermal springs, while others occur in brackish or more or less salt water. The remarkable eyeless genus *Siettitia* has been found in subterranean waters in France. Compared with the Noteridae, they are less convex above, the scutellum is usually visible and the hind coxae are larger and not produced into plates which cover the trochanters. The structure and classification of the family form the subject of a comprehensive memoir by Sharp (1882) and this authority points out that, although the Dytiscidae are aquatic in their larval and imaginal instars, they are to be regarded as modified terrestrial Adephaga. In this connection it may be noted that (1) in their general structure and venation they resemble the Carabidae, the main differences being in the form of the metasternum, the hind coxae and natatorial legs; (2) they drown more quickly than many land beetles do, the imagines can exist perfectly well on land, and are capable of prolonged flight; (3) the pupae, so far as is known, are terrestrial. These insects may be readily distinguished from the Hydrophilidae, which they resemble in general shape, by their filiform antennae: Dytiscidae are, furthermore, exclusively carnivorous both as larvae and adults. The hind legs function as swimming organs, and are greatly flattened, widely separated and fringed with long hairs. In the males of certain genera the first three segments of the fore tarsi are dilated to form highly efficient adhesive pads which are provided beneath with cup-like suckers. The latter are moistened with a glutinous secretion and, according to Blunck (1912), this product indirectly aids adhesion after the manner of grease in an air-pump and, directly, by increasing the adhesive force. The male, by the aid of these sucker-pads, is enabled to retain hold of the female for many hours continuously. The best known member of the family is *Dytiscus marginalis*, a species which has been more fully studied from every aspect than any other example of the Coleoptera. The eggs of this insect are laid singly, each in an incision made by the ovipositor in the stem of a water-plant. The larva is extremely voracious and preys upon various aquatic animals including molluscs, worms, insects, tadpoles and even small fishes. The victim is pierced by the long sickle-shaped mandibles which, as Meinert and others have shown, are traversed internally by a fine, almost closed in, groove. The latter communicates at the base of the mandible with a transverse conduit which, along with its fellow of the opposite side, opens into the cibarium. A secretion of the mid gut is injected through the channelled mandibles into the prey and digestion of the tissues of the latter takes place externally. By means of the pumping action exerted by the pharynx the liquefied food is imbibed through the mandibular canals and thence into the gut. For details concerning the structure of the mouthparts and the physiological questions involved vide Portier (1911). In the imago, on the other hand, the mandibles are masticatory and digestion takes place wholly internally. The larva swims with the aid of its legs which are fringed with hairs and are efficient oars: it is also capable of making sudden movements by throwing its body into serpent-like curves. The last two abdominal segments and the small pair of terminal lobes are fringed with hairs, which enable the larva to hang head downwards, suspended from the surface film. In this position it is able to take in air by the caudal pair of spiracles: the remaining seven pairs of the latter organs are rudimentary and closed. When fully fed, the larva makes its way to the moist earth near the water, and there constructs a cell in which pupation takes place. In the adult beetle the last two pairs of spiracles are markedly larger than those preceding. When the insect comes to the surface to breathe the caudal extremity rises above the water, thus placing the enlarged spiracles in communication with the atmosphere. A supply of air, furthermore, is retained beneath the elytra and clings to the felted hairs covering the abdominal terga. This is utilized during submergence and is renewed when the beetle comes to the surface, the elytra being slightly elevated to allow of the free entry of air beneath them.

The literature on *Dytiscus* is very extensive: for further details concerning its structure and biology reference should be made to the great monographic work edited by Korschelt (1923). The larvae of Dytiscidae and of the three preceding families are described by Bertrand (1928).

FAM. GYRINIDAE (Whirligig Beetles). Included in this family are about 700 species which are surface swimmers. They are mostly gregarious and sometimes occur in large congregations. Individuals are seen constantly darting in graceful curves around one another with an agility that renders their movements difficult to follow with the eye. The various species are very uniform in appearance, being ovoid or elliptical, more or less flattened, and of a steely-black or bronze lustre. The antennae are very different from the prevalent Adephagid type, being extremely short and stout, auriculate basally, and inserted beneath the front. The eyes are divided into upper and lower organs, and it has been suggested that the former are adapted for aerial vision and the latter for use beneath the water. The fore legs are long and prehensile: in the male the tarsi are often dilated and provided with suckers. The hind legs are broad, greatly flattened, and highly adapted for swimming, while the middle pair are similarly modified, but in a lesser degree. Larsén describes the muscles which are involved in their complex movements. *Gyrinus* is chiefly carnivorous and its eggs are laid end to end in rows upon submerged water plants. The larva (Fig. 392) is elongate with deeply constricted segments, the mandibles are pointed and perforated by a sucking canal, and the legs long with paired claws. Each of the first eight abdominal segments bears a pair of plumose tracheal gills, and two pairs of similar organs are carried on the 9th segment. Pupation takes place in a cocoon which is attached to water plants. *Orectochilus*, the only other common British genus, is mainly nocturnal in habits. Ochs (1969) summarizes much information on their habits and distribution.

Suborder POLYPHAGA

This suborder includes the majority of beetles and their grouping into superfamilies is a matter of great difficulty. There is a number of exceptional

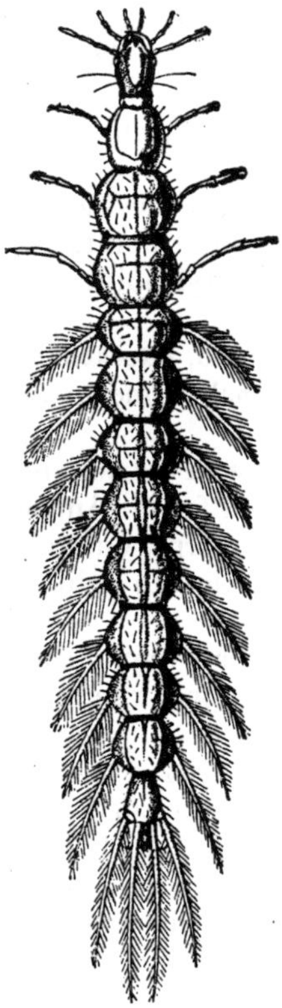

FIG. 392
Gyrinus marinus, Larva × 6
From Fowler (F.B.I.), *after* Schiödte.

FIG. 393
Cupes latreillei
After Lameere.

genera which are difficult to fit into the system or which still need further study. The following key, based on that of Crowson (1968), must be used with care.

Superfamilies of the Polyphaga

1. Pleural sclerites of the true 2nd abdominal sternite distinct from those of 3rd, sternite itself represented by small lateral plates (Fig. 337, B), rarely fully developed when the elytra are much shorter than the abdomen. Tarsi usually with 5 segments, fore tibia often spinose or toothed. Antennae filiform or clubbed, club often made up of last 5 segments (**Haplogastra**). **2**

– Pleural sclerite of the true 2nd abdominal sternite almost always fused to that of 3rd, sternite itself usually entirely membranous (Fig. 377, C), more rarely fully sclerotized. Antennal club, if developed, not usually made up of the last 5 segments (**Symphiogastra**). **5**

2. Antennae usually 10-segmented, with 3–7 apical segments produced on one side to form a lamellate club. Species usually stout. **Scarabaeoidea** (p. 857).

– Antennae never with a club of this type. Species usually less stout. Wing-venation of Staphylinid type (Fig. 376). **3**

3. Maxillary palpus nearly always longer than the antenna of which the first 3–5 segments are glabrous, the next segment cup-like, and the last ones form a strong

pubescent club. Head usually with a Y-shaped impressed line on the front. Wings usually with Cantharid venation (Fig. 376). Habits mostly aquatic. **Hydrophiloidea** (p. 850)

– Antennae not so constructed and not shorter than the maxillary palpi. Head without a Y-shaped impressed line. Habits rarely aquatic. **4**

4. Antennae with last 3 segments rarely forming a compact club; if they do, 1st segment not elongate. Exoskeleton rarely very hard and shining, elytra truncate and usually leaving more than 2 abdominal segments exposed. **Staphylinoidea** (p. 852)

– Antenna geniculate with last 3 segments forming a compact club. Exoskeleton hard, black, shining, elytra truncate and leaving 1 or 2 abdominal segments exposed. **Histeroidea** (p. 851)

5. Hind coxa almost always with a vertical posterior surface and with the postero-ventral edge produced into a plate which partly covers the retracted femur. Antennae filiform, serrate or pectinate, very rarely the last 1–4 segments sharply differentiated from the rest. Tarsi nearly always 5-segmented, rarely with the 4th reduced and enclosed in the lobes of the 3rd. **6**

– Hind coxa rarely with a vertical posterior surface and the postero-ventral edge produced into plates; if so, the antennae has the last 3 segments very long or forming a club. **14**

6. Mid coxae widely separated, hind coxae close together with broad femoral plates, fore coxae transverse with exposed trochantins. Tarsi with 5 segments, none lobed below, last not longer than the others together. Head with no distinct fronto-clypeal sulcus. **Byrrhoidea** (p. 863)

– If the mid coxae are widely separated, the hind coxae are also widely separated or the tarsi have 1 or more segments lobed beneath, or the head has a distinct fronto-clypeal sulcus. **7**

7. Fore coxa more or less projecting, hind coxa with a femoral plate. **8**

– If the fore coxa projects, the hind coxa has incomplete femoral plates and there are more than five visible abdominal sternites. **9**

8. Hind wing with 5 anal veins. Metasternal transverse sulcus complete. Hind coxa with posterior face oblique and only slightly excavated. **Dascilloidea** (p. 862)

– Hind wing with not more than 4 anal veins. Metasternal transverse sulcus not reaching sides. Hind coxa with posterior face vertical, strongly excavated. **Eucinetoidea** (p. 863)

9. Fronto-clypeal sulcus distinct. Tarsi with segments rarely lobed beneath, if so the hind margin of the pronotum is crenulate; last segment of tarsus often as long as the others together. **Dryopoidea** (p. 864)

– Fronto-clypeal sulcus not distinctly developed. Last segment of tarsi not as long as rest together. Hind margin of pronotum not crenulate. **10**

10. Fore coxa projecting with large exposed trochantin, hind coxa with large femoral plate. Empodium large and plurisetose. Antennae pectinate or flabellate. First visible abdominal sternite nearly always in same plane as metasternum. **Rhipiceroidea** (p. 866)

– If fore coxa projects the femoral plates of hind coxa are incomplete or absent, empodium absent and small and bisetose. **11**

11. Metasternum with well-marked transverse sulcus. Suture between first two visible abdominal sternites more or less obliterated. Abdominal tergites all well sclerotized. Prothorax normally immovable on mesothorax, prosternal intercoxal

process received into a deep mesosternal cavity. Antennae short, serrate. Fore coxa small, rounded. Tarsi with ventral adhesive lobes at least on segments 2–4. **Buprestoidea** (p. 866)

– Metasternum without transverse sulcus. First two visible abdominal sternites as clearly separated as 2nd and 3rd. Abdominal tergites weakly sclerotized. Prothorax more or less movable on the mesothorax. Tarsi rarely with more than one segment with adhesive lobes beneath. **12**

12. Front coxae transverse or slightly projecting, trochantins never connate to sternum or hypomeron, usually exposed. Hind angles of prothorax never acute. Prosternal process well-developed, received into a mesosternal pit. First abdominal sternite with a well-marked intercoxal keel. **Artemotopoidea** (p. 866)

– Front coxae rounded and with trochantinal apodeme connate to sternum or hypomeron, *or* prosternal process incomplete or absent, *or* front coxae strongly projecting; hind angles of prothorax more or less acute, *or* first abdominal sternite without an intercoxal keel. **13**

13. Hind coxa almost always with complete femoral plates, fore coxa nearly always rounded, trochantins hidden. Prothorax nearly always with acute hind angles, prosternal sternites, 1st in same plane as metasternum. **Elateroidea** (p. 868)

– Hind coxa with femoral plates very narrow, incomplete, or absent, fore coxa large, projecting. Prothorax usually with obtuse hind angles, prosternal intercoxal process not or scarcely received into mesosternum. 6 or 7 visible abdominal sternites, 1st not in same plane as metasternum. **Cantharoidea** (p. 870)

14. Fore coxa usually somewhat projecting, hind coxa often with a more or less distinct femoral plate. Tarsi of 5 segments, 1st sometimes very small. Antennae nearly always with last 3 segments differentiated from rest. 5 visible abdominal sternites. **15**

– If fore coxa is projecting, tarsi are usually heteromerous or apparently with 4 segments, hind coxa without femoral plate. Ocelli absent. **16**

15. Prothorax not hood-like. Tarsi never with 1st segment very small, trochanters normal, their junction with femora very oblique. 1 or 2 ocelli often present. Antennae not filiform, rarely serrate, last 3 segments not greatly elongate. **Dermestoidea** (p. 873)

– Prothorax nearly always produced over the head like a hood. Tarsi with 1st segment very small or trochanters elongate and joined to femur at a transverse sulcus. Ocelli absent. **Bostrychoidea** (p. 874)

16. Tarsi 5-segmented but 4th segment very small and concealed by the lobes of the 3rd, 1st to 3rd with adhesive lobes beneath. Transverse sulcus of metasternum usually distinct laterally. **16**

– If tarsi are 5-segmented with the 4th segment very small, the antennae have a marked 3-segmented club, the head is not rostrate and the gular sulci are distinct. Transverse suture of metathorax not distinct laterally. **18**

17. Head not rostrate, or if slightly so, gular sulci distinct and separate. Antenna without a 3-segmented club, not received into a groove. **Chrysomeloidea** (p. 892)

– Head more or less produced into a rostrum, gular sulci nearly always confluent. Antenna usually geniculate and clubbed, 1st segment retractable into a groove (scrobe). **Curculionoidea** (p. 897)

18. Tarsi 5-segmented. Fore coxa projecting, or transverse and empodium conspicuous and bisetose. Abdomen with 5 or 6 visible sternites. **Cleroidea** (p. 876)

– If tarsi are all 5-segmented, fore coxa is rounded or transverse. Empodium small or absent. **19**

19. Tarsi 5-segmented, filiform. All coxae more or less projecting. Antennae short, more or less serrate. Abdomen with 6 or 7 visible sternites. **Lymexyloidea** (p. 878)

– If the fore coxa projects, the tarsi are heteromerous. Antennae filiform or clubbed, rarely serrate. Abdomen normally with 5 visible sternites, if with 6 or 7, tarsi nearly always heteromerous. **Cucujoidea** (p. 878)

Superfamily 1. **Hydrophiloidea**

The adults of this group are nearly always known by the elongate maxillary palps which seem to have taken on the tactile functions of the antennae as the latter became involved in respiration. They have six simple Malpighian tubules and the larval maxilla has separate galea and lacinia though the latter may be reduced. Although the majority of species are aquatic, they are less fully adapted for such a life than the Dytiscidae. The group is divided into five families of which the first is much the largest.

FAM. HYDROPHILIDAE. A large family comprising about 2000 species which are especially numerous in the tropics. The adults live upon decomposing vegetable matter and, in many cases, the larvae have a similar habit but those of *Hydrophilus* and its allies are predacious. A large number of the species have elongate maxillary palpi (Fig. 368) and, on this account, the family has often been termed the Palpicornia: this character, however, is not always very evident. The long palpi perform the functions of antennae, the latter being used in respiration by the submerged insect. Although a large number of the Hydrophilidae are truly aquatic, the family name is inappropriate as a considerable number are land insects. The latter are met with in damp or marshy places or among vegetable refuse, while *Cercyon* and *Sphaeridium* are common in dung. Species of *Helophorus* sometimes cause damage to root-crops. One of the best known members of the family is *Hydrophilus piceus* which is almost the largest British Coleopteron. It is less perfectly adapted for swimming than *Dytiscus* and does not have the agility that characterizes predacious insects. Much has been written on this species, especially with reference to its peculiar mode of respiration. A dorsal air-reservoir is present beneath the elytra and there are ventral hairy tracts which also serve to retain an air-film. On either side of the thorax and abdomen there is a longitudinal tract of delicate pubescence bounded above by the overhanging edges of the prothorax and elytra. The spiracles open into these linear tracts, and the latter also communicate with the dorsal air-reservoir. When the insect rises to renew its air supply the body is slightly inclined to one side so as to bring the angle between the head and prothorax, on one side of the body, to the surface. The hairy antennal club plays an important part in breaking the surface film, and facilitating the entry of air into the cleft already mentioned, and its passage into the lateral tracts. The complete details of the respiratory process are too lengthy for discussion here but are referred to in the works of of Miall (1912), Portier (1911) and Hrbáček (1950).

The eggs of *Hydrophilus, Hydrochara, Hydrobius* and other genera are enclosed in cocoons of a remarkable construction (vide Portier): the latter are usually attached to grass or floating objects, but *Helochares* fasten them to their own bodies. The larvae of the family do not admit of any general description on account of their great diversity of form and structure: those of a number of forms have been studied by d'Orchymont (1913), Böving and Henriksen (1938), and the African ones by Bertrand (1962). Several of the aquatic genera, including *Hydrophilus*, are metapneustic and the spiracles are placed on the last body segment in a kind of atrium. *Hydrochara* and *Berosus* have long fringed gill-like structures on the first seven abdominal segments: in *Helophorus aquaticus* the larva is strongly sclerotized, the thoracic terga are entire, and each of the first eight abdominal segments is protected by four transverse plates. The larvae of *Cercyon* and *Sphaeridium* are degenerate and grub-like with the legs atrophied or vestigial. In the majority of the larvae of this family cerci are present and sometimes elongate.

FAM. HYDRAENIDAE. There are about 300 members of this widely spread family which in some respects resembles the Staphylinoidea. They have six or seven visible abdominal sternites and the antennae nearly always with a five-segmented pubescent club, preceded by a cupule. Larval abdomen usually ending in a pair of hooks. The species live in water, stagnant or flowing, or in the film on damp rocks, feeding on filamentous algae. British genera are *Hydraena* and *Ochthebius*.

The family **Spercheidae** includes one widely distributed small genus. Both adult and larva walk on the underside of the water surface-film. The adult has the fore coxal cavities closed behind; antenna with not more than three segments before the pubescent cupule. Tarsi with a large, plurisetose empodium. The eggs are fastened to their own bodies.

FAM. GEORYSSIDAE. This small family has only recently been placed in the present group. It has most of the characters of the Hydrophiloidea but the maxillary palpi are short, the head is deflexed and the first two abdominal sternites are fused. The small beetles are found on sand or mud on the edge of water; the larva is not certainly known.

Superfamily 2. Histeroidea

This group consists of one large family and two very small ones. The latter are little known and their larvae have not been described. The Histeridae have six Malpighian tubules and the predatory larva has falcate mandibles and a maxilla with no lacinia but with a small galea borne on an elongate palpiger.

FAM. SPHAERITIDAE. The single genus *Sphaerites* contains three species. *S. glabratus* is found in coniferous woods in Scotland, probably associating with fungus. The discovery of the larva would probably make the systematic position of the group more certain. In the adult the fore tibia is not externally dentate and the antennae are not geniculate.

FAM. SYNTELIIDAE. The genus *Syntelia* (Fig. 394) has a few species in Mexico and the Orient where it has been found at sap exuding from wounded trees. The adult has dentate fore tibia and geniculate antennae like the Histeridae but the coxae are all approximated and front ones projecting.

FAM. HISTERIDAE. The Histeridae are a large family (ca. 2500 species, Wenzel, 1944) of compact hard, shining beetles with geniculate and strongly clubbed antennae. The elytra are truncated behind leaving the two apical segments exposed. For the most part they are black or brown insects, but in some cases the elytra are marked with red, and a few species are metallic. When alarmed they simulate death and closely retract the antennae and legs beneath the body. *Hister* (Fig. 395) frequents dung and carrion: *Hololepta* and *Platysoma* live beneath bark and are greatly flattened: others are cylindrical and live in the burrows of wood-boring insects. Several genera are found in ants' nests and others in those of termites. The larvae have a soft and often much wrinkled integument, very short legs

FIG. 394
Syntelia indica
After Fowler (F.B.I.).

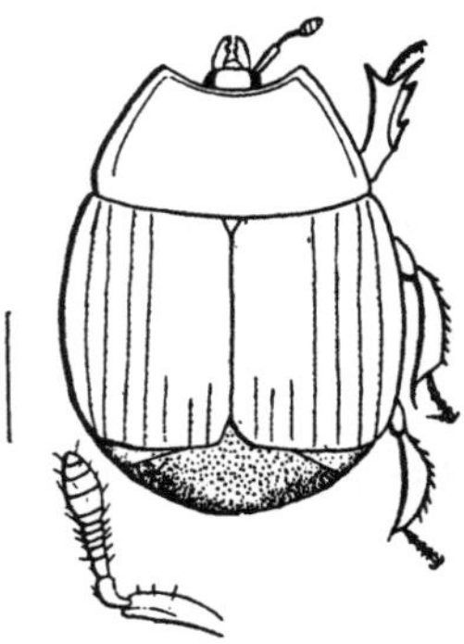

FIG. 395
Hister unicolor. Europe

and no ocelli or labrum. The mandibles and palpi are prominent, while the broad 9th abdominal segment bears short 2-segmented cerci. So far as known they are carnivorous; the larva of *Saprinus virescens* preys upon that of *Phaedon, Hister pustulosus* attacks Noctuid larvae, while the dung-feeding and lignicolous forms probably prey upon Dipteran and other larvae. The genus *Niponius* is chiefly characterized by the very large head and slender tarsi: its larva frequents the burrows of Scolytinae and probably preys upon the immature stages of the latter. *Niponius* (Gardner, 1935) occurs in Japan, the Himalayas and Borneo: for its larva, see Gardner (1930).

Superfamily 3. **Staphylinoidea**

In this very large group, the elytra are usually short, leaving at least some abdominal tergites exposed; the venation is of a special type (Fig. 376) in which veins M and Cu are not connected, and there are four Malpighian tubules. In the larva the galea and lacinia of the maxilla are more or less fused and the palpiger is not elongate. In habits, the species are mould-eaters or predatory, very rarely phytophagous (some Silphidae). Very few of them are found in dry situations.

FAM. LIMULODIDAE. Adult with a plate covering half the length of the femur. Eyes absent, the antennae with a 3-segmented club and retractile into grooves beneath the head. Five genera and about 33 species of myrmecophilous beetles are placed in this family (Seevers and Dybas, 1943). They are mostly American, with one genus in Australia and one in Africa.

FAM. PTILIIDAE. The 300 members of this family are very minute: the Neotropical *Nanosella fungi* is stated to be the smallest known Coleopteran and measures 0·25 mm long, while the maximum size in any species is only about 2 mm. The elytra are variable in length and the wings very narrow, with a marginal fringe of exceptionally long setae. These insects occur in all kinds of decaying vegetable matter, in fungi and under bark. The larvae (Perris, 1876) have 3-segmented antennae, peripneustic respiration and articulated urogomphi. About 60 British species are now recognized (cf. Matthews, 1872). According to Dybas (1966) some species of *Eurygyne* are parthenogenetic.

FAM. LEPTINIDAE. *The eyes are absent, fore coxae small, rounded; head behind forming a keel which overlaps the pronotum. Leptinus testaceus* is not very rare in Britain on small rodents or in their burrows. It is also sometimes found elsewhere, e.g. in nests of *Bombus* and the larva has been described by Reid (1942). *Silphopsyllus* is found in Russia on the Desman and is more strictly confined to its host, apparently by its adaptation to a certain temperature. *Platypsyllus castoris*, found on the beaver in Europe and N. America, is much more modified in adult structure, though the larva (Bugnion and du Buysson, 1924) is much less aberrant. The head of the adult is provided with a comb-like row of spines near the hind margin, the mandibles are vestigial, and the elytra are short, leaving six abdominal segments exposed. The wings are absent. *Leptinillus* also lives on the American beaver. The eggs are laid in the earth of lodges but the larvae live on the skin products of the host (Parks and Barnes, 1955; Wood, 1964).

FAM. ANISOTOMIDAE. Formerly placed in the Silphidae, these beetles may be distinguished by the form of the antennae which have a 5-segmented club. According to Crowson, some of club-segments contain peculiar sensory vesicles, best developed in genera like *Liodes* which breed in subterranean fungi. The Coloninae with 100 species and the Anisotominae with about 400 species mostly breed in subterranean Ascomycetes and the adults are nocturnal. *Scotocryptus* lives in the nests of *Melipona*.

The Catopinae which some authors place in a separate family with 800 species have varied habits. Some are mere scavengers with a tendency to haunt the nests of mammals, more rarely of insects. A large group of species, especially in the Bathysciinae, are cave dwellers, often highly modified like *Leptodirus* with its narrow head and thorax, globose elytra and long appendages. Such species are also often blind with long tactile setae. The biology of these beetles and their larvae is described by Deleurance-Glaçon (1963*a*).

FAM. SILPHIDAE. The members of this family (Fig. 396) are mostly rather large and are commonest in the Holarctic region. There are about 200 species. The fore coxal cavities are open behind, and the coxa has an exposed trochantin. The elytra are often somewhat shortened and there are 6 visible abdominal sternites. Nearly all the species feed on carrion and *Silpha* comprises the roving carrion-beetles (Fig. 396). The larvae of some species wander in search of decomposing vegetable matter, those of *Phosphuga atrata* are predacious upon snails and that of

Xylodrepa 4-punctata upon Lepidopteran larvae. On the other hand, the species of *Aclypea* are vegetarian and often damage beet and other root-crops. *Nicrophorus* includes the well-known burying beetles which bury the bodies of small vertebrates by excavating beneath them. The eggs are laid in a gallery leading from the buried corpse and the larvae (Fig. 397) in their earlier instars are fed by the female (Pukowski, 1933).

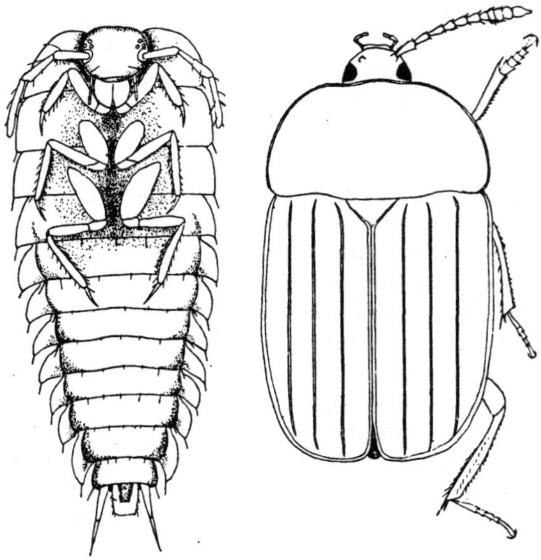

FIG. 396 *Silpha tristis*, larva and imago

FIG. 397 *Nicrophorus vespillo*, larva × 3

From Fowler (F.B.I.), *after* Schiödte.

FAM. SCAPHIDIIDAE. The members of this family are fungivorous or occur in rotting wood both as larvae and adults. They are small, oval, convex and very shining insects with filiform or slightly clavate antennae. The fore coxa is hidden, the hind coxae more or less widely separated and the first visible abdominal sternite as long as the next three together. Their affinities have been much disputed, but they are probably allied to the Staphylinidae. Only about 200 species are known, and the few British representatives belong to the genera *Scaphium*, *Scaphidium* and *Scaphisoma*. The larva of the last-mentioned genus is described by Perris: it is of a modified campodeiform type with elongate hairs along the sides, rather long antennae, and greatly reduced cerci.

FAM. SCYDMAENIDAE. The elytra in this family cover the abdomen and the maxillary palps have the penultimate segment large and the last one small. They are almost all very small insects: they are very widely distributed and more than 1200 species are known, about two dozen being British. They mostly occur in moss, under bark, etc., or in ants' nests, often in company with Pselaphidae. Although related to the latter family, their 5-segmented tarsi, and longer elytra, afford a ready means of separation. They are more closely allied to the Silphidae and chiefly differ from the latter in the coarser eye-facets and the separated hind coxae. The larva of *Scydmaenus tarsatus* is figured by Meinert (1888): it is flattened and onisciform in general shape with laterally expanded margins to the segments. Scarcely anything appears to be known of the biology of the family.

FAM. PSELAPHIDAE. *Abdomen not flexible though partly exposed; body often with deep foveae. Tarsi normally 3-segmented. Maxillary palps with last segment very large. Claws unequal.* A large family (ca. 5000 species) of very small reddish or yellow beetles bearing a resemblance to ants. Although worldwide in distribution it attains its greatest development in the tropics. The species mostly live in ants' nests; they present great diversity of form, the antennae and maxillary palpi being especially remarkable. The Pselaphinae usually have 11-segmented antennae and greatly developed maxillary palpi, notably in the males of certain genera. The members of this subfamily are less highly modified than the Clavigerinae; some are known to be myrmecophilous, while others occur under bark, among moss, or in caves. The Clavigerinae are sometimes regarded as a separate family, and are true symphiles. The antennae are composed of one to six segments and rival those of the Paussinae in their specialization: the maxillary palpi are greatly reduced or rudimentary and are evidently no longer needed in species which are fed by their hosts. At the base of the abdomen there is an extensive hollow which is surrounded by tufts of golden-yellow hair diffusing a substance that the ants are fond of. The European *Claviger testaceus* is well known and lives in the nests of *Lasius*: the ants feed it with regurgitated food and individuals have been kept under observation by Janet for over four years. The chief authority on the Pselaphidae is Raffray and some of the more remarkable forms are figured in his monograph (1908): rather more than 30 species are British (Pearce, 1957). The larvae (Besuchet, 1956) are of a general Staphylinid type with small, fixed urogomphi which may be absent. The antennae have three segments. They pupate without any special protection and are carnivorous in habit.

FAM. STAPHYLINIDAE (Rove Beetles: Fig. 398). The most obvious feature of this family is seen in the often very short elytra, hence the older name of Brachelytra for the group. Notwithstanding the small size of these organs, they conceal large well-developed wings, which are complexly folded away beneath them. On the other hand the unfolding of the wings can take place with great rapidity, thus allowing the insect to resort to almost instantaneous flight. In a few genera (*Olophrum*, *Lathrimaeum*, etc.) the elytra are larger than usual, leaving only the apex of the abdomen uncovered. The head is very variable in form and size and frequently differs in the sexes: the antennae are 10- or 11-segmented and either filiform or more or less clubbed. The eyes are very variable in development though rarely wanting and, in a few cases, a single ocellus or a pair of these organs is also present. The number of segments to the tarsi is inconstant and the latter are sometimes heteromerous (see Blackwelder, 1936). The abdomen is frequently terminated by a pair of styliform appendages, and certain species exhibit the curious habit of curling the distal portion of the hind body over the back in a threatening manner. The Staphylinidae include 27 000 species of which over 800 inhabit the British Isles. The majority of species are small and inconspicuous, but a few are brightly coloured and the largest British species, *Ocypus olens* (Fig. 398), attains the exceptional length of 28 mm. Members of the family abound where there is decaying organic matter, including dung and dead animals, and many are predacious. *Stenus* with its protrusible labium

(Schmitz, 1943; Weinreich, 1968) attacks such insects as Collembola. More than 300 species are known to be myrmecophilous (see p. 1242): thus *Myrmedonia* includes species preying upon dead or disabled ants, while other genera live as tolerated guests of Doryline ants (Seevers, 1965) and exhibit a remarkable mimetic resemblance to the latter. *Dinarda* is a synoekete in the nests of certain species of *Formica* and the Aleocharine genera *Lomechusa* and *Atemeles* are highly evolved symphiles which are assiduously tended by ants. *Atemeles* spends the summer with *Formica* and the winter with *Myrmica* (Hölldobler, 1970). Numerous termitophilous genera have been brought to light by Silvestri, Trägårdh and Seevers. Certain of these are viviparous, and *Corotoca*, *Spirachtha*, *Termitomimus* and other genera are physogastric, the abdomen assuming bizarre forms (Seevers, 1957; Kistner, 1970 and many other papers).

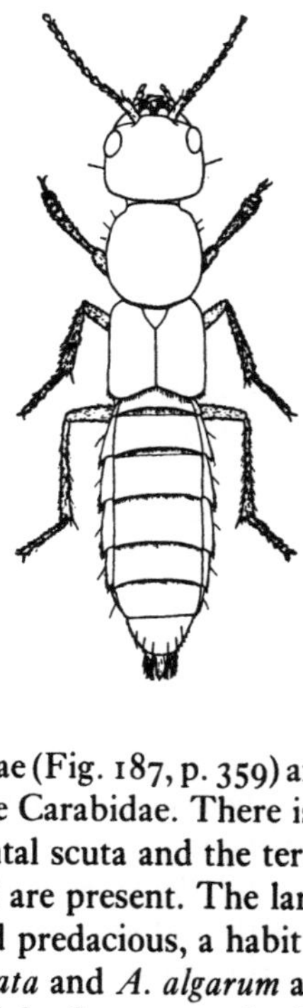

FIG. 398
Ocypus olens, male × 2.
Europe

Staphylinid larvae (Fig. 187, p. 359) are typically campodeiform and often closely resemble those of the Carabidae. There is no distinct labrum, the body is protected by sclerotized segmental scuta and the terminal segment is tubular. The legs have only one claw and cerci are present. The larvae of certain species are definitely known to be carnivorous and predacious, a habit which is apparently very general. The larvae of *Aleochara bilineata* and *A. algarum* are pupal parasites of cyclorrhaphous Diptera. The life-history of the former species has been followed by Wadsworth (1915). The newly hatched larva is campodeiform and gnaws its way into the puparium of its host. It subsequently undergoes hypermetamorphosis, becoming eruciform, with obvious degeneration in adaptation to a parasitic life. The degenerate eruciform type of larva also occurs in *Lomechusa* as an adaptation to myrmecophilous habits. The larvae of *Syntomium* and *Micropeplus* are aberrant, being short and broad and markedly onisciform. A considerable number of Staphylinoid larvae has been described by Paulian (1941) and Kasule (1966) gives a key to the subfamilies of British larvae. For a generic synopsis of the family see Eichelbaum (1909); see also Moore (1964) for the American subfamilies and Coffait (1972, work still in progress) for the west Palaearctic fauna.

Superfamily 4. **Scarabaeoidea**

This is one of the most distinct sections of the Coleoptera and the species, besides the characters given in the key, have the following features in common: four Malpighian tubules, dentate fore tibia with one apical spur, 8th abdominal tergite forming an exposed pygidium, spiracle non-functional, larva without tergal plates or cerci and almost always with cribriform spiracles. The species are primarily fossorial, and the burrowing habit persists to a greater or less degree in the majority of the species. In form they are compact and very stoutly built; they are endowed with remarkable muscular powers but they walk without much agility, and in an ungainly fashion. Nearly all species, however, are active fliers: apterous forms are relatively few and, although most frequent in the female, they may occur in both sexes. In some members of the group the colours are bright and striking, and the head and thorax are often ornamented with remarkable cuticular outgrowths, producing some of the most bizarre forms in the insect world. Sexual dimorphism is a very characteristic phenomenon, the differences affect almost every part of the body and, in many cases, the males and females of a species are so unlike that they have been relegated to different genera (Fig. 369) (Arrow, 1944). Scarabaeoids are also remarkable for the variety of their stridulating organs, not only in the imagines, but more particularly among the larvae: the sound produced is usually very highly pitched, and often inaudible to the human ear, if the insect be held more than a few inches distant. The eggs are large and few in number: they are noteworthy from the fact that they have been observed to change their form and size considerably during growth after deposition. During the larval stage which usually lasts two years, these insects feed upon dead vegetable or animal matter, roots, or dung and occur in the ground, in the decaying parts of trees, or in debris, etc. The larvae (Fig. 401) are described by Schiödte, Perris and others. They are easily recognized and exhibit great similarity. They are broad and fleshy, whitish or greyish white and the body is often curved in the form of a letter C; the legs are well developed, but are rarely used for locomotion. The majority of species lie upon the back or side and are surrounded by sufficient food to render active movement unnecessary. The head is large and downwardly inclined and strongly sclerotized; the three thoracic segments are short, bringing the legs closely together, and the last two to four abdominal segments have a somewhat inflated appearance being much larger than those preceding. Eyes are seldom present, but the antennae are well developed and 2- to 5-segmented. The mandibles are powerful and exposed, and the maxillae terminate either in one or two lobes. The prothorax and first eight abdominal segments each bear a pair of cribriform spiracles. A general account of the group is given by Arrow (1910) or Jeannel and Paulian (1949).

FAM. PASSALIDAE. The members of this family are somewhat flattened, parallel-sided black or dark brown insects. The elytra (Fig. 399A) cover the abdomen and are deeply longitudinally striated, and the mandibles are not specially developed in the male. There are five visible sternites; the antenna is not geniculate and the club segments are not closely apposed. The mandibles have a movable dorsal tooth and the prementum is deeply emarginate. About 500 species have been described and they inhabit decaying wood in the moist warm forests of the world. None are European, and only a single species occurs in America north of Mexico. Ohaus has claimed that the two parent beetles are accompanied by several larvae which they tend throughout life until maturity is attained, but this observation has not been confirmed. The

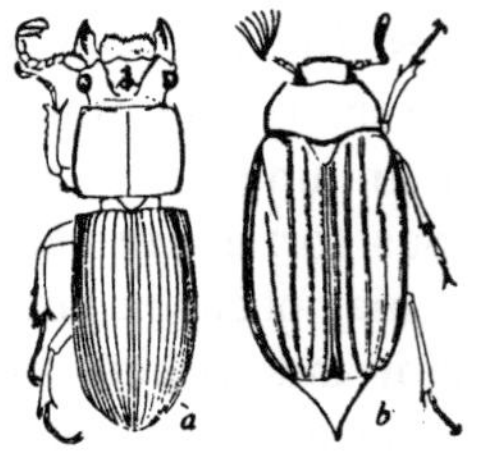

FIG. 399
a, *Passalus interruptus*. Paraguay; *b*, *Melolontha melolontha* (Scarabaeidae). Britain

adults disintegrate the wood and chew it into a condition suitable for consumption by their progeny. The larvae (Gravely, 1916) are more elongate and less markedly crescentic than those of most Scarabaeids (Fig. 401B). They are, furthermore, active and have the first two pairs of legs relatively long: the third pair are greatly modified, each leg being reduced to a very short coxa and a more elongate trochanter. The latter is adapted to form an organ which works across a striated area on the mesocoxa, thus producing a squeaking noise. Stridulation is effected in the adults by friction between the wings and the upper surface of the abdomen.

FAM. LUCANIDAE (Stag Beetles). In these insects the abdomen is covered by the elytra but the latter are almost always devoid of longitudinal striae. There are five visible abdominal sternites. The antennae are geniculate with the club loose. The prementum is entire. Stag beetles are familiar on account of the great development of the mandibles in the males which in some cases attain a length equal to that of the rest of the body (Fig. 400). The significance of these enormous mandibles is not clear: notwithstanding their formidable appearance in *Lucanus cervus*, for example,

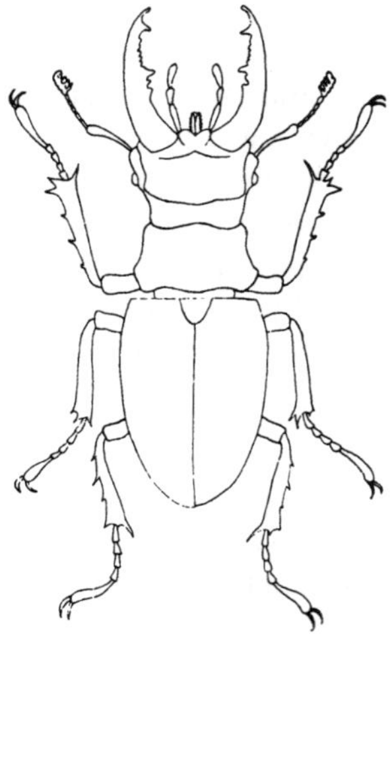

FIG. 400
Lucanus cervus, male, natural size. Europe

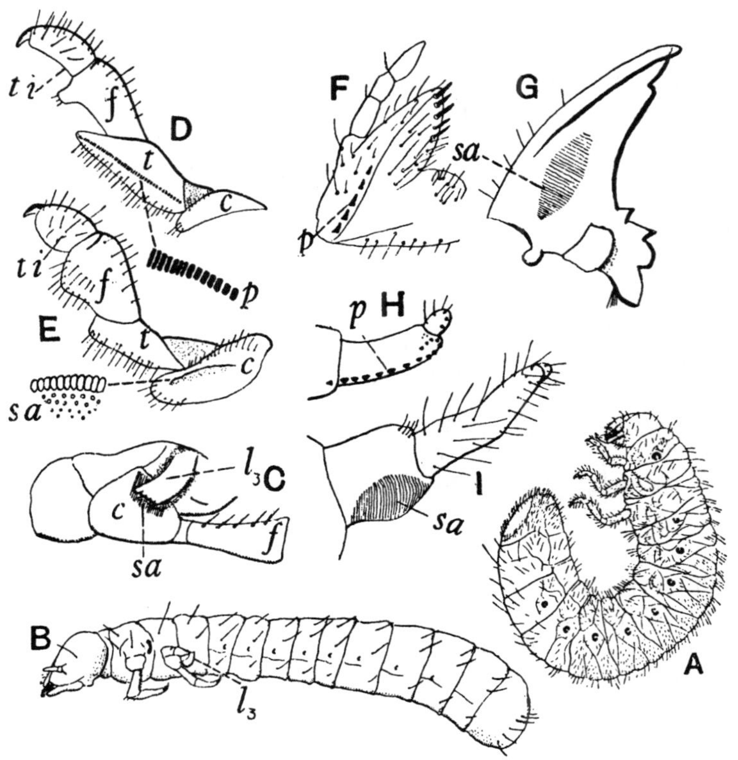

FIG. 401 Scarabaeoid larvae and stridulatory organs

A, larva of *Anomala polita* (Scarabaeidae: *after* Gardner). B, larva of *Passalus* with C enlarged detail of part of 2nd leg and reduced 3rd leg, D, *Lucanus cervus*, 3rd leg and E, 2nd leg. F, *Phyllopertha horticola*. left maxilla and G right mandible. H, *Geotrupes* 3rd leg and I 2nd leg. *c*, coxa; *f*, femur; l_3, reduced 3rd leg; *p*, plectrum; *sa*, stridulatory area; *t*, trochanter; *ti*, tibia.

they are not as strong, or as capable of inflicting as severe a bite as the short stout mandibles of the female. The male insects are usually much larger than those of the other sex and they exhibit great variation in size among individuals of the same species. These variations are coupled with striking differences in the development of the head and mandibles and it is often possible to distinguish large (teleodont), small (priodont) and intermediate (mesodont) forms. In other cases there are no intermediates known between the extremes and species like *Odontolabis sinensis* consequently exhibit what has been termed high and low dimorphism. Lucanid larvae (van Emden, 1935*b*) inhabit the rotting wood of trees or their roots. They possess well-developed antennae and legs, the maxillae are single-lobed and they differ from many Scarabaeid larvae in that the segments are not raised into three folds. The larva of *Lucanus cervus* stridulates by rubbing certain hard ridges on the third pair of legs over a rugose area at the base of the second pair: the third pair, however, is not specially modified or reduced in size as in the Passalidae and *Geotrupes*. The duration of larval existence in this family does not appear to have been definitely ascertained: in *L. cervus* it lasts about four years, while certain other species are stated to require six years to complete their development. Pupation (Fig. 402) takes place in a cell formed of gnawed wood fragments. The Holarctic genus *Sinodendron* is of an aberrant character, the species being completely cylindrical and instead of the mandibles differing in the male, the latter sex carries a cephalic horn. The thorax is very truncated in front and the antennae are short and non-geniculate. The larva occurs in rotting wood of ash, etc.: it is more slender than the usual type and gradually narrowed posteriorly. About 750 species of Lucanidae are known but only three genera, each with a single species, occur in the British Isles. For a criticism of the classification of the family see Holloway (1960).

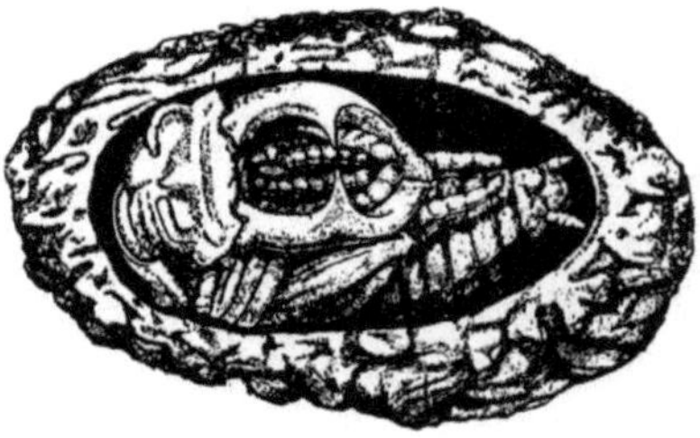

FIG. 402
Lucanus cervus, male pupa in its cell
After Roesel.

FAM. TROGIDAE. *There are five visible abdominal sternites. The segments of the antennal club are closely co-adapted. The surface is dull, scaly or setose.* The 170 described species are widely distributed and most fall within the genus *Trox* of which there are two British species. The species are found in dry places and feed on carrion or dung. The larva is without stridulatory files and the claws are unusually long and acute.

FAM. ACANTHOCERIDAE. *There are five visible abdominal sternites. The segments of the antennal club are closely co-adapted.* The surface is smooth and shining and the insect can roll up into a ball. This includes about 120 tropical species, mostly associated with rotten wood.

FAM. GEOTRUPIDAE. *There are six visible abdominal sternites, elytra covering the whole abdomen where the spiracles are all in the pleural membrane. There are 11 antennal segments.* The 'dor' beetles are large convex insects mostly of coprophilous habits. *Lethrus*, however, is apterous and the larval burrow is stored with green plant material, so that they may be injurious to crops. *Geotrupes* with six British species is a very familiar insect. It is the 'shard-borne beetle' of Shakespeare and the beetle which 'wheels his droning flight' of Gray's Elegy. *Geotrupes* constructs burrows several centimetres deep in the earth, below a patch of dung, and portions of the latter are carried down to serve as food for the larvae. Each burrow or a number of cells leading out of it is filled at its blind end with a plug of dung in which a single egg is deposited (Teichert, 1955–56; Howden, 1955). The larvae stridulate very much after the manner of Passalids, but the hind legs are less reduced and the positions of the file and rasping organ are reversed. The adults stridulate by rubbing together a file on the hind coxae and the sharp edge of the coxal cavity. About 300 species have been described.

FAM. SCARABAEIDAE (Chafers, etc.). A very large family of more or less convex insects, with the mandibles not specially developed in the males, and with the elytra not usually completely covering the abdomen. The abdomen has six visible sternites in which some of the spiracles lie. The antennae have 8–10 segments. The significance of the spiracles in their classification is discussed by Ritcher (1969). Over 17000 species are known and about 90 occur in the British Isles. A classification of this extensive group, with a table of subfamilies, is given by Arrow. A key to a number of the larvae is provided by Perris (1931) and by Böving and Craighead (1937). Ritcher (1958) has reviewed their biology.

The Cetoniinae are typically represented in England by the 'rose chafer' – *Cetonia aurata*. They are exceedingly brilliantly coloured, mostly diurnal insects, especially found in the tropics, and number about 2600 species. Their mouthparts are adapted for dealing with soft or liquid food and the labrum is membranous and concealed; the mandibles, with few exceptions, are thin and incapable of biting, and the maxillae are invested with long hairs. The larvae are generally found among roots, in decaying wood, accumulations of dead leaves and other plant refuse. The life-histories of *Cetonia, Oxythyrea* and *Potosia* have been followed by Fabre (1903). The larvae of *P. cuprea*, and other species, inhabit the nests of *Formica* where they have been found consuming the woody material of which these habitations are composed. The Cremastochilini are exceptional in being mostly sombre-coloured nocturnal insects, living as larvae and adults in the nests of ants and termites.

The 1400 species of Dynastinae include some of the largest and most striking of all Coleoptera. The majority of these species are black and, being nocturnal or crepuscular in habits, they are not very often seen at large. They are chiefly remarkable on account of the extreme development of sexual dimorphism which is exemplified in the presence of large horny processes in the males. On the head there is usually a slender, recurved, and sometimes toothed or bifurcated frontal horn: on the prothorax there are commonly one or more processes which often arise from the margins of a dorsal cavity. In a few cases, e.g. *Oryctes rhinoceros*, both sexes are horned. Many species possess stridulating organs consisting of a file-like area on the penultimate tergum which is rasped by the apices of the elytra. The Dynastinae are almost all tropical, and more especially Neotropical. Very little is known of their biology, but their larvae have been found in tree-trunks and in compost while the adults sometimes feed on palm-leaves. Several species are injurious, their larvae attacking the roots of sugarcane and rice. *O. rhinoceros* is a great pest of coconut plantations, destroying the tissue at the leaf-bases and providing for the onset of decay. Banks (1906) states that it will also develop freely in vegetable refuse and in soil. *O. nasicornis* is often found in decomposing bark refuse of tanneries in S. Europe.

The largest subfamily, the Melolonthinae (9000 species), include the 'cockchafers' and the common European *Melolontha melolontha* (Fig. 379) formed the subject of the classical anatomical memoir by Straus-Dürckheim. They differ from the two preceding groups in the presence of an evident sclerotized labrum. The larvae feed among decaying vegetable matter or among the roots of plants and are, in some cases, exceedingly injurious. In the case of *M. melolontha* the eggs are laid in several batches of fifteen or more during early summer and are deposited at a depth 6 to 8 inches in the ground. The larvae hatch after an interval of about three weeks, and the insect remains in this stage for three years in England, and for a longer or shorter period in other countries according to climatic conditions. During the cold months the larvae descend into the ground but, for the rest of the year, they come nearer the surface and devour the roots of corn, grass, etc., sometimes causing great injury. At the end of the third summer, they form oval pupal cells at a depth of two feet or more in the soil. The adults emerge about October but do not leave the ground until about the following May, when they are common about oak and other trees upon whose foliage they feed.

The Rutelinae are likewise an extensive subfamily (2500 species) and many of the species are brightly coloured. In general facies they resemble the Melolonthinae but are usually separable on account of the mobile claws which are of unequal size. They are represented in Britain by *Phyllopertha horticola*, whose adults often devour the leaves and blossoms of roses and fruit trees, and *Anomala aenea*. Species of *Lachnosterna* are very destructive, as larvae, in N. America.

The Aphodiinae are more or less oblong convex species of small size with concealed labrum and mandibles. They are often found abundantly in dung (Madle, 1934; Schmidt, 1935) but other species feed on roots. The extensive genus *Aphodius* is represented in Britain by about 40 species and the subfamily has 1200.

The Scarabaeinae with 2000 species are round or oval and often very convex beetles living almost entirely in dung. Their mandibles are membranous and incapable of biting. Much of our knowledge of their biology is due to Fabre and more recently Halffter and Matthews (1966), and the curious dung-rolling habit of the sacred *Scarabaeus* of the Egyptians has attracted attention from very early times. Similar habits are met with among the allies of this insect living in S. Europe, Asia and Africa. The ball is composed of dung which, in this form, is transported to a suitable retreat as food for the beetle itself. The mass of dung which contains the egg is pyriform and it is constructed in a separate underground chamber of material brought there for the purpose. In *Copris* this chamber is very large and is the combined work of the male and female. It contains 2–7 pyriform cells of dung, each containing a single egg, and the 'nest' is guarded and tended by the female. In some Indian species of *Cartharsius* and *Heliocopris* the egg balls are very large and coated with clay. When first discovered they were thought to be ancient stone cannon balls, and Lefroy mentions one being found 8 feet below ground. In certain species, including *Copris hispanus*, the female, instead of dying after oviposition, tends her brood to maturity and then produces a second generation, but the number of eggs laid in each case does not appear to exceed four. Several genera are myrmecophilous and have the usual secretory glands and hair tufts indicative of symphiles. In north America *Phanaeus vindex* has some importance as a vector of the spirulid Nematodes of pigs (Fincher, Davis and Stewart, 1971). The subfamily is represented in Britain by *Copris lunaris* and several species of *Onthophagus*. The larvae of British Scarabaeids have been tabulated by van Emden (1941) and Ritcher (1966) may also be consulted.

There are two or three much smaller subfamilies besides the large ones mentioned above. Two of these, the Glaphyrinae and the Hybosorinae, can also be treated as families.

Superfamily 5. **Dascilloidea**

The principal characters of this group are, in the adult, the filiform or slightly serrate antennae; the more or less conically projecting fore coxae whose cavities are completely open behind; the complete transverse metasternal sulcus; the hind coxae with the posterior surface oblique and not much excavated; the side margins of the pronotum complete and the hind margin never crenulate; five anal veins; six simple Malpighian tubules. In the larva, the labrum is free, the mandible has a distinct mola and a ventral tubercle, maxillary galea and lacinia distinct, legs well developed, tergites more or less corneous, cerci absent. In appearance, it resembles a scarabaeid and its cribriform spiracles have no closing apparatus. There is one family.

FAM. DASCILLIDAE. About 50 species, mostly of moderate size, are usually found on flowers in the adult stage. The larva (Gahan, 1908) feeds on grass-roots, sometimes to an injurious extent, and resembles a Scarabaeid. The ocelli are absent and its spiracles are cribriform.

Superfamily 6. **Eucinetoidea**

This small group has recently been separated by Crowson from the previous one. The metasternal sulcus does not extend to the margins; the hind coxa has its posterior face vertical and deeply excavated and there are not more than 4 anal veins. The larva has annular spiracles with a closing apparatus.

FAM. CLAMBIDAE. About 60 very small species with a broad head, capable of being reflexed against the underside of the prothorax; antennae with 8–10 segments; all legs with 4 tarsal segments. Crowson found some larvae resembling those of *Eucinetes* on wet sticks but others are thought to be myrmecophilous.

FAM. EUCINETIDAE. About 25 species, mostly in the genus *Eucinetes*, are widely distributed and found in rotten wood. The adults are long, narrow and boat-shaped. The postgena is not keeled; hind coxae very large; all tarsi 5-segmented, simple.

FAM. HELODIDAE. These small insects have a thin and very loosely articulated exoskeleton. There are five tarsal segments of which the fourth is bilobed. The hind coxae are clearly narrower than their metasternum and the postgena is not keeled. They are found near water in which the larvae live. The latter are provided with rectal tracheal gills (Beier, 1949; Bernet-Kempers, 1944; Bertrand, 1974) and are unique in having multiarticulate antennae. *Scirtes* is notable for its power of jumping (Lombardi, 1928). There are about 360 species of which 13 are British.

Superfamily 7. **Byrrhoidea**

With the elimination of several dubious elements, this group is reduced to the single well-defined family, the Byrrhidae. In the adult, the head is deflexed with no visible clypeus, the fore coxa is large with an exposed trochantin, the broad prosternal intercoxal process is received into an emargination of the flat mesosternum, and there are six simple Malpighian tubules. The larvae are relatively stout, sometimes capable of rolling up, and without cerci or gills.

FAM. BYRRHIDAE. *The antennae are gradually thickened to the apex and the mentum is small and hidden.* There are about 270 species which mostly occur on the ground beneath stones, at roots of grasses or in moss. Their most striking feature is the power these beetles have of withdrawing their appendages in close contact with the body and remaining motionless: in this attitude they are hard to detect and often closely resemble their surroundings. The best known British species of the family is *Byrrhus pilula*, which is often found on paths in spring. Its life-history is in need of investigation and, according to Chapuis and Candèze, the larva is cylindrical and fleshy and may be recognized by the large size and breadth of the prothorax and the last two abdominal segments. The head is short and broad, the antennae very short and there is a pair of ocelli on either side. The pronotum is markedly sclerotized and sculptured, and the last abdominal segment carries a pair of retractile locomotory processes. The larvae occur beneath turf or moss and are about 18 mm long.

FAM. NOSODENDRIDAE. The antennae have a sharply three-segmented club and the mentum is large. The 30 species of the family are widely distributed and all stages are found under bark. The larva has been figured by Böving and Craighead.

Superfamily 8. **Dryopoidea**

The beetles of this group (Hinton, 1939), are mostly subaquatic but except in some Elmids which exhibit plastron respiration (p. 225) they show few special adaptations. The larvae, however, often have rectal or abdominal tracheal gills and a mobile operculum by which the 10th abdominal segment can be tightly enclosed within the 9th. The adult may have four or six Malpighian tubules which are cryptonephric in the Dryopidae. The detailed classification of the European forms has been discussed by Steffan (1961).

FAM. PSEPHENIDAE. *Front coxae projecting; margins of pronotum usually crenulate; maxillary palps inserted more laterally than usual, second segment as long as next two together.* This family with a few species in India and America is interesting on account of its remarkable larvae (Fig. 403) (Böving, 1926). The latter occur in swift rivers and in waterfalls, *Psephenus* being especially abundant in the rapids of Niagara, while larvae of this and other genera are also plentiful in the Himalayan rivers. They are flattened, rounded or ovoid and almost scale-like in form: the margins of the body are greatly expanded and consequently the appendages are not visible from above. There is a great variety of respiratory devices. They cling with great tenacity to stones, etc., and the whole body appears to act in a suckerlike fashion rendering these larvae difficult to remove. Respiration takes place either by means of abdominal gills, or by the aid of a retractile tuft of anal filaments which is only visible in the undisturbed living larva. The larvae are apneustic in *Psephenoides* and the pupae are submerged and soldered down to the stones upon which the larval life was passed: they closely resemble the larvae when viewed from above, but are armed with tufts of long filamentous gills (Hinton, 1947). The British species, *Eubria palustris*, is now referred to this family of which Hinton (1955) gives an interesting account. *Eubria* larva has a grill-like cribriform spiracle.

The next three families have the front coxae projecting, the antennae long, and the maxillary palps normally exserted with the second segment not very long.

FIG. 403 A Psephenid larva, Himalayas. A, dorsal; B, ventral. × 15

FAM. EURYPOGONIDAE. *Pronotal hind margin not crenulate, head without an occipital keel.* A few species in N. Asia and in N. America constitute this family which is placed here primarily because of the structure of their larvae.

FAM. PTILODACTYLIDAE. *Pronotal hind margin crenulate, head usually with an occipital keel, antennae not retractile into grooves.* A little-known family of about 170 tropical species. Some are Elaterid-like, and the males have enlarged eyes and flabellate antennae. The larvae are either aquatic or live in rotten wood.

FAM. CHELONARIIDAE. *Pronotal hind margin crenulate, head usually with an occipital keel, legs and antennae retractile into grooves.* While the adult beetles of this family resemble Byrrhids, the larvae resemble those of Dryopids or Elmids. There are about 40 species in the tropics of Asia and especially of America.

FAM. HETEROCERIDAE. *Front coxae not projecting, antennae short and thick, tarsi 4-segmented.* These small beetles are densely pubescent and live in galleries which they excavate in the mud bordering pools and streams. Their larvae inhabit the same situations and may be recognized by the prominent mandibles, the very broad thoracic segments and the much narrower abdomen: the whole body is strongly setose. The family is very widely distributed and about 150 species are known, several being indigenous to the British Isles.

FAM. LIMNICHIDAE. *Antennae more or less filiform. Front coxae not projecting, mid coxae widely separated but hind coxae contiguous.* The adult beetles resemble Byrrhidae except in the presence of a fronto-clypeal sulcus, but Hinton (1939) has shown that there is a close relationship to the next family. There are about 70 species, principally Palaearctic with one British.

The next two families have the fore coxae not projecting and the mid and hind coxae both similarly separated.

FAM. DRYOPIDAE. *Antennae short with a 6-segmented, pectinate club.* As now defined (Hinton, 1955) this family has about 250 species and is widely distributed. They show all transitions from a terrestrial to a purely aquatic life. Such British forms as

Dryops occupy an intermediate position. Their larvae live in damp earth beneath stones or feed on waterlogged wood and somewhat resemble those of Elateridae. Bertrand (1962) has described a number of African larvae.

FAM. ELMIDAE. *Antennae slender without a pectinate club of 6 segments.* Three hundred species of this widely distributed family are known and all stages may be found together in the weeds in running water. As in the previous family all stages in adaptation to aquatic respiration are found. The larvae are typically onisciform and cling to weeds or stones (cf. Hinton, 1940).

Superfamily 9. Rhipiceroidea

The adults of this group are known by their strongly flabellate antennae and by the nose-like projection which replaces a free labrum. There is one family, the **Rhipiceridae** (Sandalidae) with membranous processes beneath the first four tarsal segments. There are about 65 species, found in most regions, and one N. American *Sandalus* is a parasite of immature Cicadas.

Superfamily 10. Artematopoidea

This recently defined group (Crowson, 1973) includes three families (Brachypsectridae since withdrawn) which seem to be derived from the same stock that gave rise to the next two superfamilies, so that their characters are rather diverse.

FAM. CALLIRHIPIDAE. *Tarsi without membranous lobes below. Elytra without a hook-like process beneath the apex. Antennae pectinate or flabellate.* There are rather more than 100 species, found in most regions. The only known larvae live in rotten wood and resemble those of Elaterids.

FAM. ARTEMATOPIDAE. *At least tarsal segments 3 and 4 with membranous lobes. Apex of elytra with a tongue-like process beneath apex. Antennae not flabellate, rarely pectinate.* A rather small number of widely spread species of the biology of which almost nothing is known.

Superfamily 11. Buprestoidea

The single family **Buprestidae** has often been placed near the Elaterids, but among other differences the adults have a scutellary stria on the elytra, a well-marked transverse sulcus on the metasternum and six cryptonephric Malpighian tubules, while the larva has a free labrum.

It is an essentially tropical family comprising over 11 500 species; relatively few are European and only five genera with twelve species occur in Britain. They are among the most brilliantly coloured of all insects and some species, owing to the splendour of their metallic lustre, are used in embroidery and in jewellery (Fig. 404). They are typically inhabitants of hot moist forests and are exceedingly active on the wing, often taking flight at the least alarm. The larvae (Xambeu, 1892–93) are distinct from those of other Coleoptera, and characterized by the great expansion of the prothorax and the slender hind body which imparts to them a clubbed appearance (Fig. 405). The head is small and almost entirely withdrawn into the thorax, the antennae extremely short, and there are no ocelli. The legs are vestigial or absent, the abdominal segments are nine in number and there are no anal processes

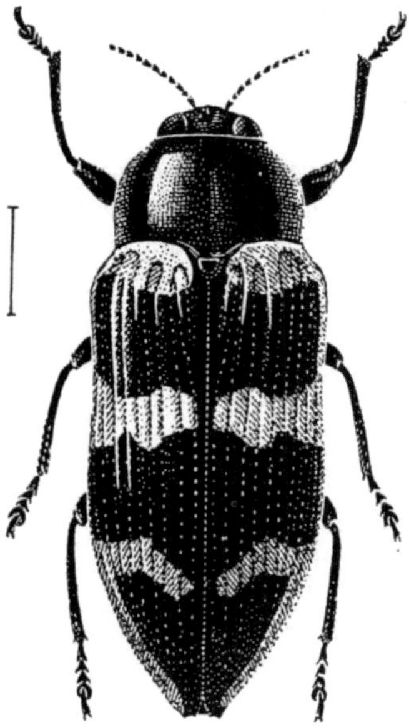

FIG. 404
Stigmadera interstitialis (Buprestidae). Australia
After Carter.

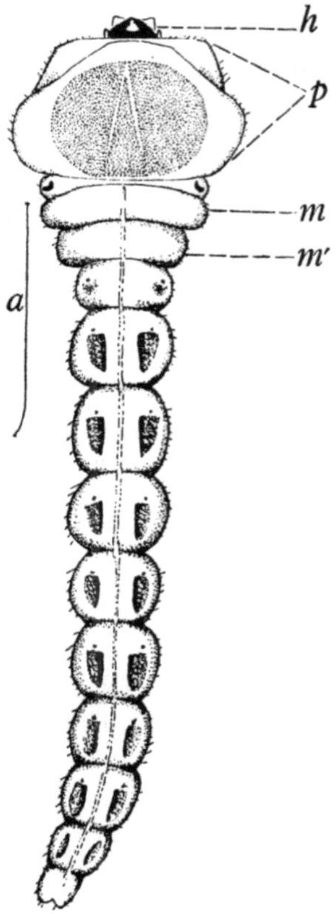

FIG. 405
Chrysobothris (Buprestidae), larva
a, actual length; *h*, head; *p*, prothorax; *m*, mesothorax; *m'*, metathorax.

except sometimes in *Agrilus*. There are nine pairs of spiracles, the first pair being situated between the pro- and mesothorax or on the latter segment. The larvae mostly gnaw rather broad flattened galleries in or beneath the bark of trees or in roots: some are found in the stems of herbaceous plants and a few mine leaves or make galls. Some of the genera of this family are exceedingly large: *Agrilus* comprises nearly 700 species while *Sphenoptera* and *Chrysobothris* each include about 300 and the Australian *Stigmodera* nearly 400. For further information on the family reference should be made to the work of Kerremans (1906–14) and Théry (1942).

Superfamily 12. Elateroidea

In this superfamily, the adults have small fore coxae with concealed trochantins, hind angles of pronotum acutely projecting, no transverse metasternal suture, hind coxae contiguous, and freely ending (normally four) Malpighian tubules. The larvae have no free labrum, no channel along the mandibles and no median epicranial sulcus. The majority of species fall in the familiar family, the Elateridae. All except the Ceraphytidae have femoral plates developed on the hind coxae.

FAM. CEBRIONIDAE. *Labrum free; at least one abdominal sternite free; head with a transverse ridge above the antennal sockets to which the clypeus does not extend; mandibles bent, very prominent.* About 200 of these rather large beetles are found, mostly in regions with a Mediterranean climate. The larva is not unlike a large 'wireworm' but with a large gula and an eversible ventral cervical membrane. It burrows in the soil. The female is often wingless and does not leave the subterranean burrow where it is sought out by the active male.

FAM. ELATERIDAE. *Labrum free; at least one free abdominal sternite; head with a transverse ridge above the antennal sockets or clypeus extending in front of them; tibiae with two spurs.* A large family of some 7000 species (Fig. 406) found in all the

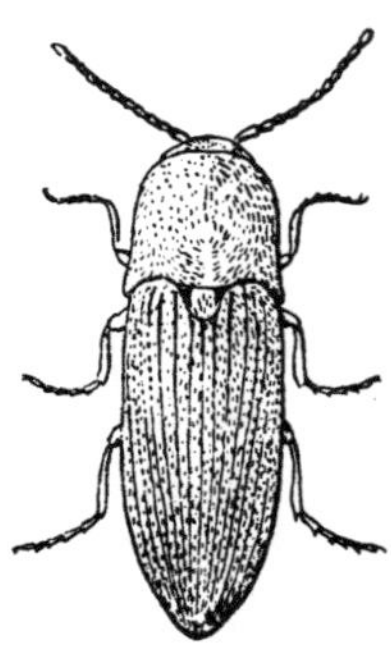

FIG. 406
Agriotes obscurus
(Elateridae), Europe
(Reproduced by permission of the Ministry of Agriculture.)

main regions. The adults possess the power of leaping when lying on their back. The mechanism of this act is not entirely clear, but the existence of the saltatory power is connected with the mobility of the articulation between the pro- and mesothorax. As a preliminary, the apex of the prosternal process catches against the edge of the mesosternal cavity. When, however, the process slips over the catch it is driven with considerable force into the mesosternal cavity accompanied by a clicking sound. The force imparted by this jerking movement causes the insect to pivot on its elytra and to spring into the air. The elasticity of its skeleton also seems to assist in bringing about the leap (Evans, 1972, 1973) (Fig. 407C).

The classification into subfamilies is difficult (Crowson, 1961). The Elaterinae have the antennae inserted near the eyes. They are mostly sombre coloured elongate insects, but a few are red or have metallic colours. The most remarkable species are the 'fire-flies' (*Pyrophorus*) which are mainly Neotropical. *P. noctilucus* emits an exceptionally bright light from a rounded yellow area on either side of the thorax and, when on the wing, an additional source of light is revealed at the base of the ventral surface of the abdomen. The eggs and larvae are also luminous. In the young larva the photogenic organ is situated at the junction of the head and thorax: in older larvae there are numerous small lateral organs in addition. The photogenic organs are very similar in structure to those of the Lampyrinae and are dealt with on p.185. The larvae of the Elaterinae are elongate and cylindrical and very tough-skinned (Fig. 408). The head is corneous and flattened, the antennae very short and 3-segmented, eyes are present, the labrum is not defined, and the trunk-segments are very alike. The whole body is usually reddish-brown or yellow, owing to the strong sclerotization of all the segments, and the legs are short. The prothorax is the largest, and the 9th segment is specially differentiated and exceedingly variable, thus affording important generic and specific characters. It is often corneous and margined with teeth and may terminate in single or paired processes which, in their

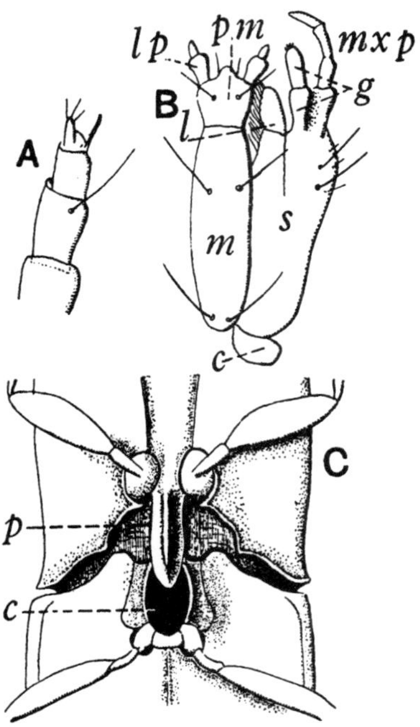

FIG. 407
Elateridae

A, right antenna. B, labium and left maxilla of *Agriotes* larva. C, base of pro- and meso-sternum of *Athous* (adult) showing process *p* and cavity *c* of leaping apparatus; *g*, galea; *l*, lacinia; *lp*, labial palp; *m*, mentum; *mxp*, maxillary palp; *pm*, prementum; *s*, stipes.

turn, may be simple, bifid or denticulate. In *Agriotes* this segment is relatively simple and bears a pair of dark-coloured pits possibly sensory in function. The larva of *Cardiophorus* is very different from the prevailing Elaterid type, being extremely long and vermiform, owing to the great development of the intersegments of the abdomen. The 9th segment of the latter region bears a pair of recurved hooks and a terminal fascicule of setae. Larvae of certain genera are exceedingly injurious to agriculture and are known as 'wireworms': under this category are species of *Agriotes* (Fig. 408), *Limonius, Athous* and others. The 'wireworm' group of larvae are root-feeders and are extremely destructive to pastures, cereals, root-crops, etc.: no effectual method of control has yet been devised. Other larvae are lignicolous and xylophagous (*Melanotus*, etc.), or carnivorous (e.g. *Athous rhombeus*). Exact observations regarding the length of the larval stages are greatly needed: in the case of *Agriotes obscurus*, which is probably the commonest English wireworm, Rymer Roberts considers that the early estimate of five years is approximately correct. This species constructs an earthen pupal cell, and the pupal instar only occupies about

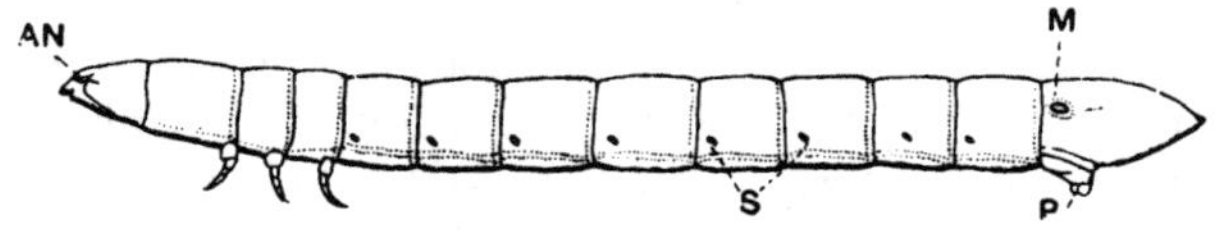

FIG. 408 Larva of *Agriotes obscurus*. × 5
AN, antenna; *M*, margin of pit; *P*, anal pseudopod; *S*, spiracles.

three weeks. A large number of Elaterid larvae have been described by Beling (1883), Henriksen (1911), Hyslop (1917), Guéniat (1935) and van Emden (1945): for the metamorphosis of *Agriotes obscurus* see Ford (1917) and Rymer Roberts (1919–22).

FAM. TRIXAGIDAE (Throscidae). There are rather more than 200 of these small Elaterid-like beetles of which six are British. They differ from the Elateridae chiefly in not having two tibial spurs. The larvae live in decaying vegetable matter.

FAM. CEROPHYTIDAE. In the adult the labrum is concealed and the hind coxa lacks femoral plates. About a dozen species found in Europe and America. All stages are found in hollow trees.

FAM. EUCNEMIDAE. *Labrum concealed; all five abdominal sternites connate; mandibles stout, prominent.* Small Elaterid-like beetles, commonest in warm climates and attached to rotten wood. Of the 1000 known species, three are British, the commonest being *Melasis buprestoides*. The larval head is greatly reduced, the legs are absent, and there are horny posterior processes.

FAM. PEROTHOPIDAE. Three North American species of *Perothops* resemble Eucnemids but the fifth abdominal sternite is free, the mandibles are slender and the tarsal claws pectinate. The early stages are unknown (unless the larva in Böving and Craighead (1931; Pl. 81, E-G) belongs here).

Superfamily 13. **Cantharoidea**

It is not easy to find diagnostic characters for all the adults of this group. The integument is usually soft and the parts loosely articulated; there are usually four Malpighian tubules which are never cryptonephric. The larval cuticle is usually soft and pubescent and the mandibles are internally channelled or perforated in connection with their predatory mode of life.

FAM. DRILIDAE. *Prosternum long in front of fore coxae; antennal sockets lateral; female usually apterous.* About 80 species, mostly European, of which *Drilus flavescens* is British. The larva preys on snails (Crawshay, 1903). Barker (1969) describes the biology of the W. African, *Selasia unicolor*, which feeds on the snail *Limicolaria*. The female is at least ten times as large as the male.

FAM. PHENGODIDAE. *Prosternum short in front of the fore coxae; antennal sockets more or less widely separated under the sides of the front, 12-segmented, segment 3 short; male abdomen not entirely covered, female apterous.* An American family of about 50 species in which luminous organs producing light of more than one colour may be present in both adult and larva. The eggs and pupae may also be luminous to some extent. It is probable that all species feed on Millipedes (Tiemann, 1967).

FAM. LAMPYRIDAE. *Prosternum short in front of front coxae; antennal sockets approximated or facing dorsally; trochanters short, hind coxae with femoral plates in male; usually sexually dimorphic, luminous organs at least in one sex.* There are about 1700 species of 'glow-worms' and 'fire-flies' and two are British. They differ from the Lycidae in having the middle coxae contiguous, whereas, in the latter group, they are spaced apart. They are nocturnal insects, and most of the members are provided with luminous organs which emit a more or less bright light usually, although not invariably, strongest in the female. These organs are borne on certain

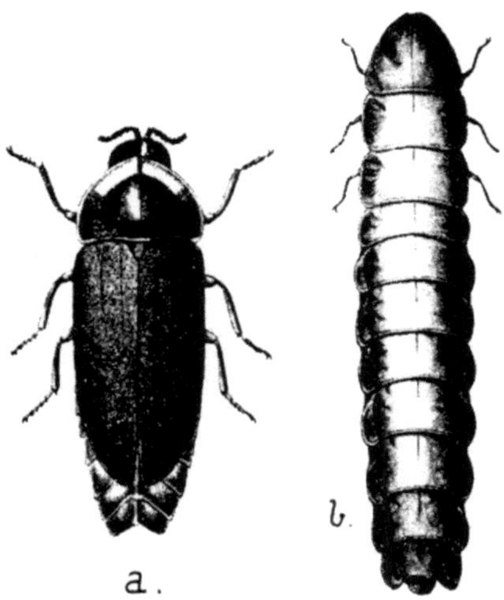

FIG. 409
Lamprophorus tardus (Lampyridae), India. *a*, male × 2; *b*, female, nat. size

of the hind segments of the abdomen – often the 6th and 7th in the male, and the latter segment in the female. The eggs, larvae and pupae are also luminous to a variable degree. There is an extensive literature on the structure and physiology of the photogenic organs (see p. 185): as a rule the light is pale yellowish green. The function of this luminescence is difficult to conceive with reference to the immature stages but, in the adults, it serves in most cases to bring the sexes together. In many forms the male is winged with greatly developed eyes, and the female devoid of both elytra and wings (see Fig. 409), being larviform with the eyes small. This dimorphism is well exhibited in the common European 'glow-worm' *Lampyris noctiluca*. The biology of this insect has been studied by Newport (1837), Fabre and others, and its larva by Vogel (1915). The adult insect takes little or no food, but the larvae (Fig. 410) are carnivorous, feeding upon snails and slugs, which they seize with their sharp sickle-like mandibles. The latter are each traversed by a fine canal through which a dark-coloured secretion is injected into the tissues of the prey. As there are no salivary glands, the secretion is apparently produced by a pair of acinose glands near the anterior end of the mid gut which probably open at the base of the mandibles. The secretion has the property of breaking down the tissues of the mollusc, and digestion is largely external, the larva imbibing its prepared meal by means of the pumping action of its cibarium. Unlike the *Dytiscus* larva, the food appears to be taken in through the mouth which is guarded by a mass of hairs precluding the entry of anything excepting small particles. The larvae of *Photinus* and *Photuris* have been studied by Williams (1916) and their luminous organs are situated on the 8th abdominal segment. They persist in the pupa but are replaced in the imago by structures located on the 6th and 7th segments. In the Mediterranean 'fire-flies' (*Luciola*) both sexes are winged. They are gregarious, but the females are less so than the males and are rarely seen. Lloyd (1971) has reviewed communication by means of light-signals. Quite elaborate specific codes have been evolved with

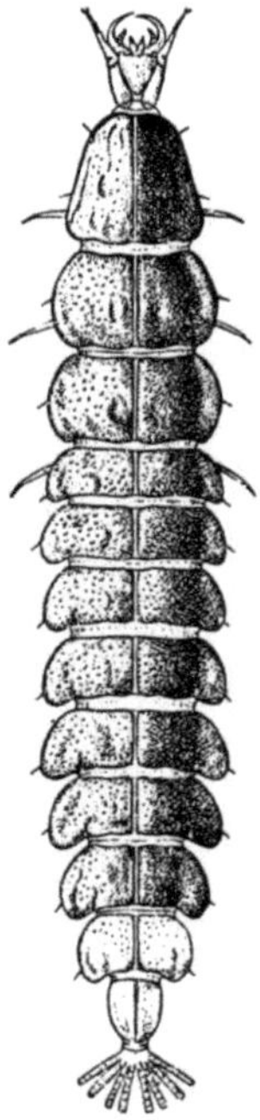

FIG. 410
Luciola lusitanica (Lampyridae), larva × 8. France
After Bugnion.

differences in the number, quality and frequency of light-pulses and in the interval before the female replies. Buck and Buck (1972) find that in *Photuris greeni*, the male gives two flashes, 1·1–1·7 s apart and the female replies after 0·8–1·1 s. In some species of *Photuris* the females mimic the flashes of other species and eat the males of those species which are attracted.

FAM. CANTHARIDAE. *Luminous organs absent; trochanters short, very obliquely joined to femora; fourth tarsal segment with a bilobed ventral lobe, antennal sockets more or less approximated, facing dorsally.* This large family includes 3500 species (41 British) which are active and probably always predacious both in the larval and adult stage (cf. Jackson and Crowson, 1969; Janssen, 1963). Some of the best known members are species of *Cantharis* and *Rhagonycha*, often called 'soldier beetles' which frequent flowers and herbage. Their somewhat flattened larvae are found in the soil or among moss, etc. They have a velvety appearance due to a covering of fine hairs (Payne, 1916). The head is flat, the antennae short, and there is a single ocellus behind each: the anal segment has a ventral pseudopod but there are no cerci.

FAM. LYCIDAE. *Luminous organs absent; trochanters long, transversely joined to the femora; fourth tarsal segment with an entire ventral lobe; antennal sockets closely approximated, facing antero-dorsally.* The adults of this large family (3000 species) are diurnal and found on leaves and flowers or under bark: the 4 British members are rare and local. The group is principally tropical and the species are often brightly coloured and conspicuous and are distasteful to birds, etc. *Duliticola* of the E. Indies has a normal male but an apterous female – the so-called 'trilobite' larva (Mjöberg, 1925). A number of species enter into large mimetic assemblages of which they are probably key members (e.g. Marshall and Poulton, 1902; Darlington, 1938; Linsley, Eisner and Klots, 1961). Eisner and Kafatos (1962) show that in Arizona the males of *Lycus loripes* produce a substance which attracts other males and leads to aggregations; the females are not attracted. This is thought to increase the aposematic effect.

Three or four other small families are placed in this superfamily. The **Brachypsectridae** with the prosternum produced into an anterior lobe and the first pair of coxae transverse and not projecting, has a few species in Asia and in California where one species lives under bark and feeds on spiders. The **Karumiidae** (Arnett, 1964) have an unmodified tenth abdominal tergite which does not form part of the genitalia. Six species in the Near East and in C. and S. America seem to live with termites. The **Homalisidae** resembling the Lycids have a few species in the Mediterranean region. The **Telegeusidae**, resembling the Phengodids, have three species in North America.

Superfamily 14. Dermestoidea

The four families grouped here by Crowson (1959) have often been much more widely separated in the system. In the adult there seem to be usually six cryptonephric Malpighian tubules and in the larva some of the tergites are setose or spinose. Some of the adults have dorsal ocelli which very rarely occur in the order.

FAM. DERMESTIDAE. *Hind coxal cavities not closed behind; metasternum of normal length, without a transverse sulcus. Head often with a median ocellus.* A family of about 700 species of small or moderate sized beetles usually invested with fine hair or with scales. They mostly inhabit furs, hides, wool and other integumentary substances as well as bacon, cheese, etc., and are exceedingly destructive as larvae. Some, from their habits, have become almost cosmopolitan and 16 species occur in Britain. Out in the field many act as scavengers in removing offensive animal matter. The adults of *Anthrenus* have been found in natural history specimens and also on flowers: its larvae are extremely destructive, and are the enemy of the collector. *Tiresias* occurs under loose bark among cobwebs, probably feeding upon the insect remains present. *Thaumaglossa* is predatory on the eggs of Mantids. The strange genus *Thylodrias*, with very aberrant adult structure and a larviform female, has a normal Dermestid-type larva which feeds principally on dead insects. *Dermestes* includes many species, some of which occur in dead animals and others are more frequently met with in dwellings, museums, etc., where they attack hides, furs, bacon, etc. The larvae of this family (Rees, 1943; Hinton, 1945) have their upper surface covered with a complex clothing of hairs of various lengths. The hairs are often aggregated into terminal or lateral tufts which, in some cases at least, can be raised at will or even rapidly vibrated: the function of this investment is perhaps protective. When a larva is about to pupate the integument splits down the back and remains as a pupal covering.

FAM. THORICTIDAE. There are about 70 species in this family, many of which are myrmecophiles. They are commonest in the Mediterranean region but *Thorictodes* has been widely spread in stored foods. Beal (1961) demotes this family to a tribe of the Dermestidae from which they differ in their elongate trochanters and widely separated, unexcavated hind coxae; they never have an ocellus.

FAM. DERODONTIDAE. *Fore coxal cavities closed behind, hind coxae excavated, metasternum not long, narrowed in front; two dorsal ocelli or ocellus-like tubercles.* Ten Holarctic species whose larvae live in slime-fungus or, in *Laricobius*, are predatory on Adelgids.

FAM. SAROTHRIIDAE. *Fore coxal cavities closed behind, hind coxae not excavated, metasternum very long, parallel-sided; no ocelli or tubercles.* Four little-known species in the Indo-Australian region.

Superfamily 15. **Bostrychoidea**

The families united to form this group contain species often associated with timber and sometimes very destructive. The adult integument is rather hard and the pronotum largely developed and hood-like; there are six cryptonephric Malpighian tubules and, as in the preceding group, their ends are said to be attached at one side of the gut. The larva has a soft body without sclerotized dorsal plates and without setae and it maintains a C-like posture.

FAM. ANOBIIDAE. *Hind coxal cavities excavated; antennal insertions separated by more than length of first segment.* The 1100 species which make up this family are often very destructive to wood, less often to other kinds of stored products. Their larvae (Böving, 1954) resemble those of the Scarabaeids in their crescentic form; they are likewise fleshy and the terminal abdominal segments are generally larger than those preceding. The antennae are very short and there are no anal processes or cerci. *Anobium punctatum* (*striatum*) and *Xestobium rufovillosum* (*tessellatum*) are very destructive to furniture, rafters and flooring, their larvae boring into the solid wood (Fig. 411): the small round exit holes made by the adult beetles are very familiar objects. The name of 'death-watch' is often applied to both these species but belongs more properly to the latter insect. The tapping noise is a sexual call and is heard most often in April to May when pairing takes place. The beetle jerks its body forward several times in rapid succession, each time striking the lower part of the front of the head against the surface upon which it is standing (Gahan). *Lasioderma serricorne* and *Stegobium paniceum* are cosmopolitan: they injure a great variety of stored materials, etc. The former species attacks cigarettes, cigars, drugs, ginger, etc., and the latter is destructive to biscuits, flour, bread, many drugs including opium and aconite, together with a wide range of other substances.

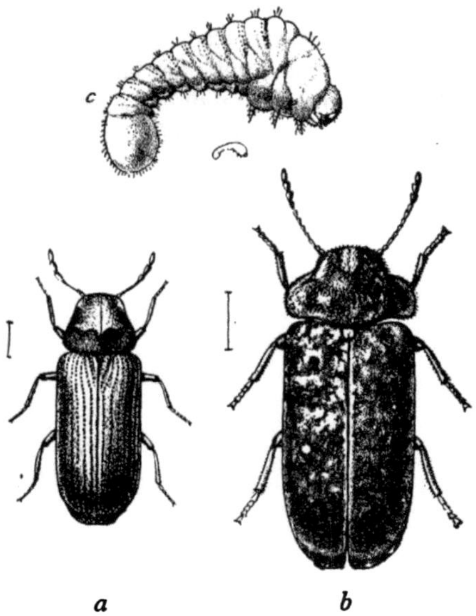

FIG. 411
a, *Anobium punctatum*; *c*, its larva; *b*, *Xestobium rufovillosum*
After Gahan (reproduced by permission of the Trustees of the British Museum).

FAM. PTINIDAE. *Hind coxae more or less separated; antennal sockets on front, separated by less than length of first segment; trochanters distally truncate.* No member of this family seems to be a wood-borer, but they are associated with stored foods, dead insects, excrement, or dry vegetable matter. The larvae (Hall and Howe, 1953) are superficially similar to those of Anobiids but the abdominal tergites lack rows of spinules and the anterior spiracle is in the prothorax and not intersegmental. There are about 700 species of which 20 have been recorded in Britain, some as introductions. A series of papers on their biology is summarized by Howe (1959). The Australian **Ectrephidae** (5 species) and the neotropical **Gnostidae** (2 species) are probably specialized, myrmecophilous Ptinidae.

FAM. BOSTRYCHIDAE. *Trochanters distally obliquely joined to femora; antennae with less than eleven segments, the last three forming a club; hind coxae contiguous, without a femoral plate; fore coxae projecting, cavities open behind; pronotum hood-like.* The members of this family make cylindrical burrows in felled timber or dried wood, and occasionally attack unhealthy standing trees. They exhibit a great variety of sculpture while the body is often truncated posteriorly and armed with small projections. Species of *Sinoxylon* and *Dinoderus* are very destructive to felled trees and bamboo in India. Their larvae resemble those of the Anobiidae and are similarly curved posteriorly, but the head is greatly reduced and the thorax more enlarged: the larva of *Bostrychus capucinus* is figured by Perris, and Anderson (1939) has tabulated those of many genera. The family is worldwide (430 spp.) but represented in Britain only by a few rare species.

FAM. LYCTIDAE (Powder post beetles). *Trochanters distally oblique to the femora. Antennae with 11 segments and a club of 2 segments. Hind coxae separated, without femoral plate; fore coxae rounded, their cavities closed behind. Pronotum flattened.* This family is closely allied to the one preceding. The larvae also are very similar and Lameere was of opinion that this similarity outweighed any adult differences. They are small elongate insects, found both in freshly cut and old timber, palings and furniture; only the wood of broad-leaved trees appears to be attacked. *Lyctus* larvae are often mistaken for those of the Anobiidae but may be easily distinguished by the more retracted head and large 8th abdominal spiracle. The Lyctidae have 70 species and several occur in the British Isles (Fig. 412).

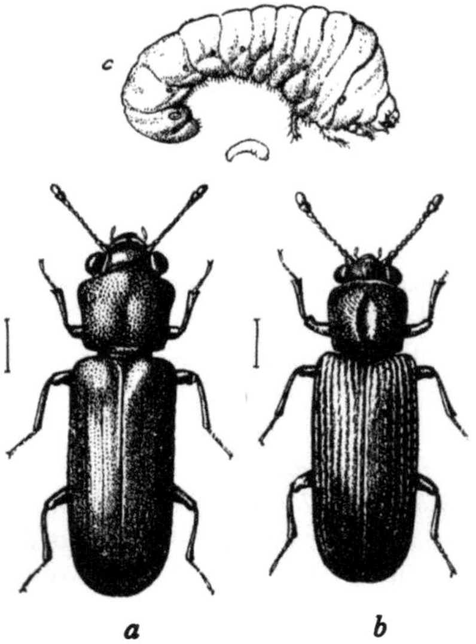

FIG. 412
a, *Lyctus brunneus*; *c*, its larva; *b*, *Lyctus linearis*

After Gahan (reproduced by permission of the British Museum).

Superfamily 16. **Cleroidea**

The adults of this group have five tarsal segments, no femoral plate on the hind coxa, the prosternal intercoxal process is not received by the mesosternum, and there are six cryptonephric Malpighian tubules whose ends are attached regularly round the gut. The larval mouthparts are more or less protruded and there are two horny urogomphi. Their habits are nearly always predatory, at least in the larva. Crowson's views on the group are given in his papers (1964, 1970).

FAM. TROGOSSITIDAE. *Antennae clubbed. Fore coxae transverse; first tarsal segment much shorter than second. Gular sulci are least partly confluent.* Although there are over 600 species in this family they are mostly tropical, only three genera and as many species being British. They vary greatly in form, some being elongate and cylindrical, others almost hemispherical, but Crowson (1970) separates several of the more aberrant genera into distinct small families (Lophocateridae, Peltidae, etc.). Several genera inhabit decaying trees, and prey upon the larvae of other lignicolous insects, while others occur in fungi. The cosmopolitan 'Cadelle' *Tenebroides mauritanicus* (Fig. 413) is found in flour, grain and many other stored products: it is often injurious but the damage it causes is to some extent counterbalanced by its also being predacious. Its whitish cylindrical larva is furnished with long setae along the sides and the thoracic terga are protected by sclerotized shields: the last abdominal segment is brown-black and bears two strong spines. The larva of *Nemosoma* is described by Erichson (1848): that of *Thymalus* by Chapuis and Candèze: and the larva of *Temnochila* by Perris. *Peltis* and its allies which are widely spread in America and the Indo-Australian region differ from the Trogossitidae in being broad and rounded with fore coxal cavities open behind. Crowson (1964, 1966) puts them in the Peltidae.

The family **Chaetosomatidae** with three New Zealand species is similar to the Trogossitidae but the antennae are filiform and the abdomen has 6 not 5 visible abdominal sternites. The body is setose.

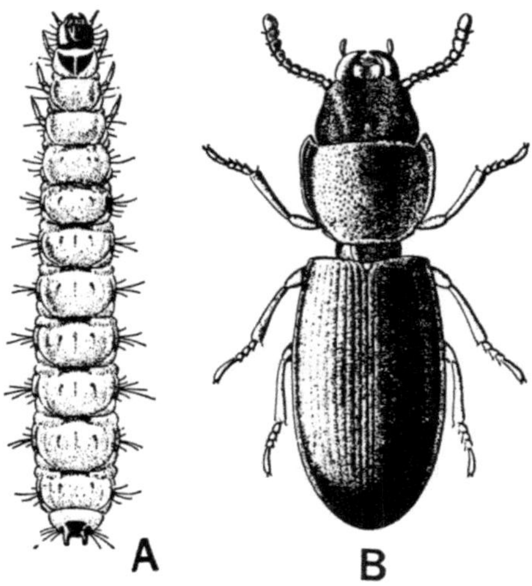

FIG. 413
Tenebroides mauritanicus (Trogossitidae). A, larva; B, imago. Enlarged
After Fletcher and Ghosh, *Proc. 3rd ent. Meeting, Pusa*, 1919.

FAM. CLERIDAE. *Fore coxae projecting; gular sutures separate but close to one another; tarsi with fewer than 5 segments, some lobed beneath; labial palps with last segment triangular.* An extensive family (nearly 3400 species), mainly tropical, many of whose members are of graceful form and beautiful coloration. They are mostly found on plants or tree-trunks, but a few (*Necrobia, Corynetes*) occur in carcasses and skins. In the larval stage they are mostly predacious and feed upon wood- and bark-boring Coleoptera. In Texas, both sexes of the Clerid *Thanasimus dubius* are attracted to the sex-pheromone, frontalin, released by its prey, *Dendroctonus* (Scolytinae) (Vité and Williamson, 1970). *Necrobia* is probably saprophagous and also feeds upon Dipterous larvae of the same habit; the nearly cosmopolitan *N. rufipes* is destructive to stored hams. *Corynetes* sometimes preys upon *Anobium*, and *Trichodes* is known to infest the nests of *Apis, Chalicodoma* and other bees. The general appearance of Clerid larvae can be gathered from a valuable paper by Böving and Champlain (1920). They are frequently bright red, brown, pink or otherwise vividly coloured, and are more or less elongate and cylindrical, or slightly flattened. The pronotum is strongly sclerotized but the remaining segments are usually fleshy except the 9th, which carries a hard shield bearing two corneous processes, and the abdomen often has ambulatory swellings.

FAM. MELYRIDAE. *Fore coxa projecting; gular sulci widely separated; tarsi 5-segmented, filiform; antennae filiform.* This large family of nearly 4000 species of which about 20 are British shows considerable diversity in structure and needs thorough revision. Many of them superficially resemble Cantharidae but this is probably due to convergence (Jackson and Crowson, 1969). Their habits generally resemble those of Clerids. *Malachius* frequents flowers and is characterized by the presence of lateral protrusible vesicles at the sides of the thorax and abdomen (cf. Schmidt, 1949).

FAM. PHLOIOPHILIDAE. The British species *Phloiophilus edwardsii* is the only one certainly attributable to this family. It has been found very locally under bark but the larva, which might throw much light on its affinities, has not yet been described. The adult resembles the Melyridae but the antennae have a loose, 3-segmented club and the body is not setose.

The family **Phytosecidae** with a few species, mainly in Australia and New Zealand, seems to fall in the Cleroidea but its position is rather obscure.

Superfamily 17. **Lymexyloidea**

This group is restricted by Crowson to the single family, the **Lymexylidae** with about 40 species.

They are elongate insects with soft integument. The maxillary palpi are flabellate in the male. The larva has an enlarged prothorax, well-developed legs and the 9th abdominal tergite though often greatly modified does not bear paired processes. There are said to be six cryptonephric Malpighian tubules. There are about 50 species of world-wide distribution. They are capable of boring into hard wood, doing at times considerable damage by drilling cylindrical holes and also through associated fungus infections. The curious larvae of the British *Lymexylon* and *Hylecoetus* are figured by Westwood (1839) and that of *Melittomma* by Gahan (1908). *M. insulare* has done much damage to coconut-palms in the Seychelles (Brown, E. S., 1954). *Atractocerus* has rudimentary elytra but ample wings, and its long flexible abdomen gives its species the appearance of Staphylinids. In tropical regions it is sometimes attracted to light.

Superfamily 18. **Cucujoidea**

Different authorities vary greatly in their treatment of the vast number of species assembled here. In the system proposed by Crowson the group is divided into two not very clear-cut sections, corresponding roughly with the older groupings – Clavicornia and Heteromera. The further division of these into families is a matter of so much difficulty, subject to such differences of opinion that the student is advised to become thoroughly acquainted with the structure and appearance of the more easily recognizable families in the first instance, and gradually to identify the remainder with the aid of a reference collection.

Section **Clavicornia**

Tarsi never 5–5–4 in both sexes. Front coxae never projecting. If the trochanters are obliquely attached to less than the full width of the femur, then the tarsi are

5–5–5, 3–3–3 or apparently 3–3–3. Antennae usually clubbed. All visible abdominal sternites nearly always movably articulated to one another. Sometimes less than seven pairs of abdominal spiracles. Larva usually with a mandibular prostheca and rarely with a distinct median epicranial suture (cf. Verhoeff, 1923).

FAM. NITIDULIDAE. *Tarsi 5–5–5, rarely 4–4–4. Fore and mid coxae very transverse, trochantins exposed. Abdomen usually with the last 1 or 2 tergites not covered by the elytra, with only 6 (rarely 5) pairs of functional spiracles.* A large family of about 2200 species which are extremely variable in form, structure and habits (Fig. 414). A large number inhabit flowers and, in some cases, are restricted to particular species of the latter: others are found in fungi or in decaying animal matter. The larvae have been studied by Böving and Rozen (1962). Those of certain species of *Meligethes* are sometimes injurious to cultivated Cruciferae while those of *Cybocephalus* attack scale-insects. The few tropical American species of *Smicrips* are very similar to the Nitidulidae though Crowson places them in a separate family.

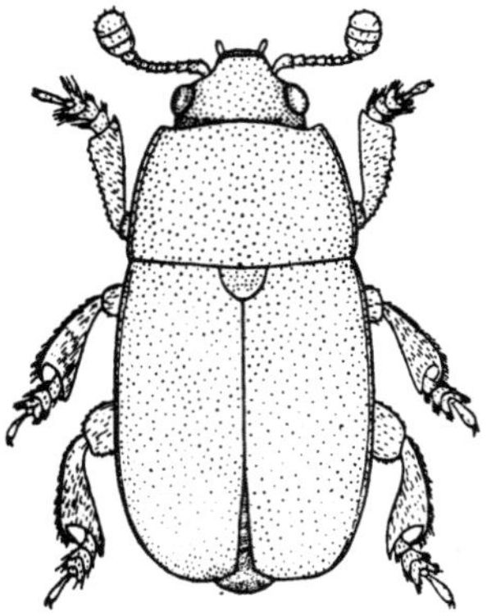

FIG. 414
Meligethes aeneus
(Nitidulidae). Europe. × 22

FAM. RHIZOPHAGIDAE. *Antennal club of 1 or 2 segments (10 and 11 fused; 9 sometimes enlarged). Form usually parallel-sided, with elytra truncate and 1 abdominal tergite uncovered. Tarsi 5–5–5, or 5–5–4 in male, or 4–4–4 in both sexes. Fore coxal cavities closed behind.* A small family (200 species) allied to and sometimes included in the preceding one. The adults and larvae are usually found under bark, often associating with Scolytinae on which they are predacious. *Monotoma* with nine British species is found in decaying vegetable matter and is sometimes placed in a separate family.

FAM. SPHINDIDAE. *Tarsi 5–5–5 in ♀, 5–5–4 in ♂, 1st segment the smallest, empodium small. Fore coxae transverse, trochantins exposed, cavities closed behind.* About 30 small species (two British) which feed on Mycetozoa. Crowson (1952) has erected a new family the **Protocucujidae** with one Chilean and one Australian species which fall nearest to the Sphindidae but have the penultimate tarsal segment the smallest and a large bisetose empodium. The same author places the few and little-known species of *Hypocoprus* in a family with somewhat similar affinities.

FAM. CUCUJIDAE. *Aedeagus inverted, with the parameres ventral. If mid coxal cavities are closed outwardly by the sterna, the 1st tarsal segment is much shorter than the 2nd. Antennae often filiform.* A family of moderate size (500 species) with about 15 British species which are mostly found under bark or in borings of other beetles. Their habits seem most often to be predatory but in *Laemophloeus* both predacious and grain-eating species occur. The larva (see Perris) has no epicranial sulcus, annular spiracles and a pair of urogomphi.

Two other groups are hardly more than subfamilies of the preceding. The **Passandridae** (98 species) have the gular sulci confluent (very unusual within the group), tarsi 5–5–5, usually small with no segment lobed beneath. They are mostly found in warmer climates and *Laemotmetus ferrugineus* is sometimes introduced into Britain with cargoes of grain, etc. The few larvae which are known are predatory on wood-boring beetles.

The **Silvanidae** (400 species) are like the Cucujidae but the third tarsal segment is lobed beneath, the fourth shorter than the first, tarsi 5–5–5 in both sexes; antennae clubbed. These beetles occur mostly on plants or among plant materials and several are found in warehouses. Some species of *Silvanus* are found under bark, often associating with some particular species of Scolytine or other beetle. Wheeler (1921) discovered in British Guiana two species (*Coccidotrophus socialis* and *Eunausibius wheeleri*) which live, along with their early stages, in the hollow leaf-petioles of *Tachigalia paniculata*. They are accompanied by a mealy bug (*Pseudococcus bromeliae*) whose honey-dew is solicited by the beetles and their larvae.

Three interesting small families with vaguely Cucujoid affinities have recently been described by Sen Gupta and Crowson (1969*a*). They are found in Australia (**Boganiidae**) or Australia, N. Zealand and Chile (**Caviognathidae, Phloeostichidae**). All, except one group of the Boganiidae, have mandibular *mycangia* – cavities for the probable transference of fungal spores, such as are also found in the Cucujidae and Sphindidae. The few larvae known seem to associate with fungi and they each have 3 or 4 monotypic genera.

FAM. HELOTIDAE. *Mid coxal cavities not outwardly closed by sterna, fore coxal cavities closed behind, trochantins hidden, all coxae widely separated. Elytra usually with metallic depressions, abdomen covered.* About 80 species are found in the warmer parts of Asia. The larva is described by Rymer Roberts (1958) and is said to be phytophagous. The **Propalticidae** with about a dozen species, mostly in the warmer regions of the old world, is another small family, perhaps closest to the next one. The adults are capable of jumping and are said to occur under bark.

FAM. CRYPTOPHAGIDAE. *Mid coxal cavities closed outwardly by sterna. Elytral epipleura indistinct or developed only on anterior half. Tarsi usually 5–5–4 in the male. Elytra without scutellary striole, surface pubescent.* About 800 species are now placed in this family whose definition has gradually been refined. The numerous species of *Cryptophagus* are essentially fungus-feeders, but some of them are found in nests of bees or wasps. *Antherophagus* may be found on flowers but the larva is found in nests of *Bombus*. The adults occasionally cling to the legs of the bees when in flight. *Telmatophilus* lives in all stages on the heads of *Typha* and other water plants. Some of the larvae are illustrated by Hinton (1945).

FAM. BIPHYLLIDAE. *Mid and hind trochanters of the heteromerous type* (*p.* 884). *Tarsi 5–5–5, claws simple. Fore trochantin hidden.* There are about 200 principally tropical species. Of the two found in Britain, *Biphyllus lunatus* is associated with the fungus *Daldinia concentrica* and *Diplocoelus fagi* is found under bark.

FAM. BYTURIDAE. *Mid and hind trochanters of the heteromerous type. Tarsi 5–5–5, claws toothed. Fore trochantin exposed.* This family includes a few genera rather closely similar to the British *Byturus. B. tomentosus* infests raspberry and allied plants and the adults may do great damage to the blossoms and the larvae to the fruit (Massee, 1954). The larva according to Böving and Craighead shows, like the adult, some heteromerous affinities, but the tarsal formula, in particular, is in disagreement.

FAM. LANGURIIDAE. *Fore coxal cavities open behind. Tarsi never 5–5–4, claws simple. Pronotum with more or less distinct posterior impressions. Form rather narrow and parallel-sided, elytral epipleura complete.* About 400 species, mostly found in Asia and in N. America (cf. Villiers, 1943), but one or two have been introduced into Europe in stored grain, etc. The adults can stridulate by rubbing the anterior margin of the pronotum against a file on the vertex. In some species, the female mandibles are strongly asymmetrical. The species seem to be phytophagous and the adults are found on leaves or flowers.

The **Lamingtoniidae** (Sen Gupta and Crowson, 1969*b*) includes one Australian monotypic genus with affinities to the Languriids.

FAM. EROTYLIDAE. *Fore coxal cavities closed behind. Tarsi with 5 segments, 4th small, claws simple. Pronotum without posterior impressions. Last segment of maxillary palpi broad. Form more or less broadly oval, elytral epipleura complete.* There are nearly 1600 species (seven British) of this family of which about half are found in S. America. The American species have been revised by Boyle (1956). As far as is known, the early stages (Rymer Roberts, 1938) are always spent in the fruiting bodies of fungi.

FAM. PHALACRIDAE. *Tarsal claws toothed or appendiculate. Fore coxal cavities open behind. Tarsi 5–5–5, 2nd and 3rd bilobed, 4th small. Small, convex, shining insects. Larvae with paired urogomphi.* The members of this family are easily recognized when once known as they are all of very similar appearance. There are more than 500 species (17 British) widely distributed in all the main regions. The larvae of the genus *Phalacrus* develop on smuts on grasses, etc. (Friederichs, 1908). The species of *Olibrus* are found on flowers, especially those of Compositae. The larva develops in the flower and when full-grown bores down the stem to pupate in the ground. According to Heeger (1857) there may be six generations in the year.

FAM. CERYLONIDAE. *Tarsi 4–4–4, segments 2 and 3 similar and not lobed beneath. Mid coxal cavities broad, closed externally by sterna. Palps with cylindrical segments.* This is a small group (nearly 200 species) proposed by Crowson to contain species which have usually been placed in the Colydiidae. Their larvae are tabulated by Sen Gupta and Crowson (1973). The species of *Cerylon* are usually found under bark and *Murmidius* is found in vegetable detritus, often in warehouses: its strange oval larva, fringed with spinulose setae, is figured by Böving and Craighead.

FAM. CORYLOPHIDAE. *Very small beetles, head more or less concealed beneath pronotum. Tarsi 4–4–4, segment 3 much smaller than 2 which is more or less lobed beneath. Last segment of maxillary palpi not enlarged. Larval mandibles usually not falcate nor body setose.* These minute beetles are mostly found in rotting vegetation or decayed wood. *Sericoderus* has been reared on mould but some species associate with coccids. The pupa is obtect as in the next family. There are about 300 species of which 11 are British.

FAM. COCCINELLIDAE. *Beetles of moderate size, convex, head partly concealed by pronotum. Tarsi 4–4–4, 3rd segment concealed in the deeply bilobed second. Last segment of maxillary palpi securiform. Larval mandibles normally falcate and body usually setose.* This very important family comprises nearly 5000 species, for the most part brightly coloured and spotted. The greater number of the species are carnivorous and predacious, feeding during the larval and adult stages upon aphids, coccids and

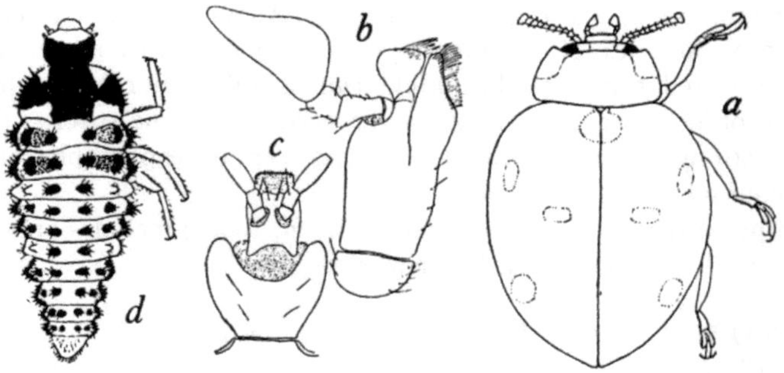

FIG. 415 Coccinellidae

a, *Coccinella septempunctata*; *b*, maxilla; *c*, labium; *d*, a coccinellid larva. All magnified.

occasionally on other soft-bodied insects. They are, therefore, of very great importance in reducing the numbers of injurious species. A comparatively small group are phytophagous but they rarely cause serious damage. Structurally, the carnivorous forms (Fig. 415) (Coccinellinae) are characterized by the mandibles having simple or bifid apices and each jaw being armed with a basal tooth. Even in this group, however, *Psyllobora 22-punctata* feeds as a larva on mildews and others live in dung. The herbivorous species (Epilachninae) lack the basal tooth and the apex of the mandible is multidentate. The Tetrabrachinae form a third subfamily: very little is known about their habits but the mandibles are of the carnivorous type. They are distinguished by the tarsi being evidently 4-segmented.

When disturbed many members of the family discharge a bitter, amber-coloured fluid. It is usually emitted through pores situated around the tibio-femoral articulations, but in *Epilachna* the pores have a much wider distribution. According to McIndoo (1916) the exuded liquid is a secretory product of epidermal gland cells: other writers have regarded it as the blood of the insect. Porta (1903) found that the secretion had a poisonous effect upon vertebrates but had no influence upon insects. It is regarded as defensive in function.

Several members of the family, notably the common species of *Adalia*, are remarkable for their wide range of colour variation. Another peculiarity is the markedly gregarious habits of certain species during hibernation and aestivation: at times these insects have been found in 'masses', see Hagen (1962) and Hodek (1973). One of the best known members of the family is *Rodolia cardinalis* which has been imported from Australia into California for purposes of controlling *Icerya purchasi* – a serious enemy of Citrus cultivation in the latter country. The beetle proved so effective a controlling agent that it has since been imported into all countries where the coccid has become injurious.

Coccinellid larvae (Böving, 1917) are soft-bodied and variously coloured: they are often of a leaden or other dark hue spotted with yellow or white. There are three ocelli on either side, the mandibles are sickle-shaped with molar bases (except in the Epilachninae) and the legs are long and slender. The terga are usually provided with segmental tubercles and spines and the abdomen tapers distally, but never bears the urogomphi so characteristic of other families. In some genera (*Hyperaspis*, *Scymnus* and *Platynaspis*) the spines are wanting and the whole body is covered with a white flocculent secretion. In *Chilocorus* (Fig. 416) the body is protected by long integumental processes.

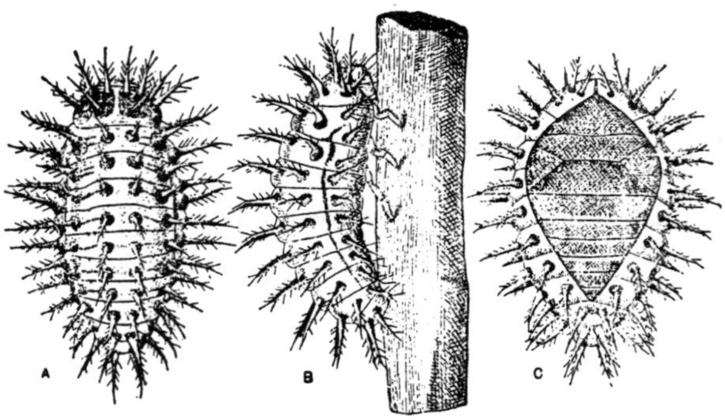

FIG. 416 Larva of *Chilocorus* (Coccinellidae). A, dorsal; B, lateral; C, pupa
After Silvestri.

The usual number of instars appears to be four, and the complete development of *Adalia bipunctata* in England was found to occupy about 34 days in captivity (Hawkes, 1920), an average of 20 days being spent as a larva. In California, Clausen (1915) found the average developmental period was 26 days. The eggs of Coccinellids are yellow, and disposed in batches, with their long axes perpendicular to the surface of the leaves upon which they are laid. Palmer in America found that the number laid by *Coccinella 9-notata* varied from 435 to 1047: in *A. bipunctata* Hawkes states that the average number lies between 140 and 148 with 418 as the maximum. The number of aphids daily consumed by the larva of this species is stated by Clausen to be 14 while in *Coccinella californica* it is about 20. During the entire larval period he found that the number consumed varied between 216 and 475 for different species: the adults are usually even more voracious. *Hyperaspis binotata* is a coccid feeder and according to Simanton (1915) it will destroy 90 adults and 3000 nymphs during its period of larval existence. When about to pupate, Coccinellid larvae usually suspend themselves by the caudal extremity which is attached by means of a secretion to plants, palings and other objects. The pupae are usually conspicuously coloured and are either surrounded by the larval exuviae, or the latter are pushed back to the anal extremity.

The larvae of the Epilachninae (Kapur, 1950) are invested with long branched processes of the body-wall. Members of this subfamily are often destructive to the foliage of potatoes, Cucurbitaceae, etc., especially in N. America. The only British member of the group is *Subcoccinella 24-punctata* whose larva gnaws the parenchyma of clover and other plants.

For further information on the biology of the family see especially Hodek (1973) who includes a key by Klausnitzer to many of the larvae. Donisthorpe (1919; 1920) has followed the complete life-history of *Coccinella distincta* – a species found in association with ants. Kapur (1970) has discussed the phylogeny of the group.

FAM. ENDOMYCHIDAE. *Species of moderate size, convex. Tarsi 4–4–4, 3rd segment concealed in the deeply bilobed 2nd. Last segment of maxillary palps not enlarged. Larva without branched setae on body, mandibles not falcate.* A family of more than 600 species chiefly met with among fungi on timber in tropical forests. Many have brilliant colours and are variable in form and size. Among the few British species the black and red *Endomychus coccinens* and the small *Mycetaea hirta* are the best known. The latter occurs in dung, vegetable refuse and often in warehouses. The larvae may be ovate with the tergites expanded to conceal the pleura; the 9th segment occasionally bears urogomphi.

FAM. DISCOLOMIDAE (Notiophygidae). *Tarsi 3–3–3. All coxae similar and strongly transverse, their lateral parts concealed by the sterna.* A small tropical family with about 170 species (John, 1954), many of them wingless, found on the bark of tree-trunks. The larvae (van Emden, 1932) may be found with the adults. They are flattened, oval creatures.

FAM. LATHRIDIIDAE. *Tarsi 3–3–3, coxae not strongly transverse, with the outer part concealed by the sterna. Fore coxal cavities closed behind. Trochanters not elongate.* There are about 600 species in this widely spread family and apparently all of them feed in all stages on fungi, especially moulds, and on Mycetozoa. The numerous species found indoors, in cellars and warehouses, are probably attracted to fungal growth. Crowson (1952) proposes a family, the **Merophysiidae**, to include a few genera such as *Holoparamecus* and *Merophysia* which have usually been considered Lathridiids. In these adults, the fore coxal cavity is open behind and the trochanters are elongate. The larva of *Holoparamecus* has urogomphi which are not present in Lathridiids.

The family **Aculagnathidae** with one Australian species living with ants has peculiar, apparently suctorial mouthparts but according to Besuchet (1972) might well be placed in the next family.

Section **Heteromera**

Tarsi 5–5–4 in both sexes, or 4–4–4 (very rarely 3–4–4 in males, or 3–3–3). Fore coxae usually projecting; if not, trochanters obliquely attached to less than the full width of the femur (heteromerous type) and the first three visible abdominal sternites are fused. Abdomen with 7 pairs of spiracles. Larva rarely with a mandibular prostheca and often with a median epicranial sulcus. Abdullah (1973) gives a key to larvae at family level.

FAM. COLYDIIDAE. *Tarsi 4–4–4 (very rarely 3–3–3). Mesepimera not reaching middle coxal cavities. Fore coxae not at all protuberant. Antennal sockets usually beneath margins of front. Larval mandibles symmetrical.* A family of more than 1400 species, found especially in the tropics, but well developed in New Zealand. There is great diversity of form and the definition of the family is still a matter of opinion. According to Crowson (1953), *Monoedus* (= *Adimerus*) clearly belongs here though some authors have placed it in a separate family. For *Meryx*, with a few Australasian species, Crowson proposes a new family in which the mesepimera reach the mid coxal cavities and the antennae are 11-segmented with a weak 3-segmented club. Some Colydiids are found deep underground, associated with buried wood or roots. Others, e.g. *Bothrideres*, are parasitic in the larval stage on wood-boring beetles or bees. Probably most of the species are in some way predatory.

FAM. CIIDAE. *Head deflexed, more or less covered by the pronotum. Tarsi 4–4–4, 4th segment small. Fore coxae somewhat transverse, their cavities open behind and internally. Larvae with paired urogomphi.* The correct position of this family has been disputed; their superficial resemblance to the Bostrychidae is evident and the larvae show some resemblances to the Cleroidea. The reduced number of antennal segments in the adult (8–10) and the 3-segmented club suggest rather a position among the Clavicornia but it is now agreed that they are Heteromera. The 250 species (26 British) are found in all the principal regions and they are all associated with woody fungi or with fungus-impregnated wood. Many of them seem to breed gregariously and Lawrence (1967) finds that *C. fuscipes* is partly parthenogenetic (thelytokous).

The family **Prostomidae** usually placed as a group of the Cucujidae seems really to have affinities in the present region (Crowson, 1968: 212).

FAM. MYCETOPHAGIDAE. *Tarsi 4–4–4 in ♀, 3–4–4 in ♂. Mesepimera reaching mid coxal cavities. Fore coxae somewhat projecting. Antennal sockets not beneath sides of front, antennal club usually with 4–5 segments. Larval mandibles asymmetrical.* A small family of about 200 species, mostly associating with fungi but *Berginus* develops on flowers. Some of the larvae are described by Hinton (1945): spine-like urogomphi may be present or absent.

FAM. PTEROGENIIDAE. *Tarsi 5–5–4, no segment lobed. Fore coxae ovate, without concealed extensions, their cavities open behind. No antennal club.* Crowson (1953) proposes this family for two Indo-Malayan genera of very uncertain affinities.

FAM. NILIONIDAE. *First 3 visible abdominal sternites connate. Trochanters not of the heteromerous type. Form broad and rounded, elytral epipleura very broad.* Insects somewhat resembling large Coccinellids in shape and associated with fungi on trees. There are about 40 species in tropical countries. *Nilio* is American but if the forms placed in the Leiochroini are also included, the family occurs also in the Old World.

FAM. TENEBRIONIDAE. *First 3 visible abdominal sternites connate: Trochanters of heteromerous type. Fore coxae not projecting. Claws simple.* One of the largest families of Coleoptera comprising more than 15 000 species which exhibit an extraordinarily wide range of superficial dissimilarity: the larvae on the other hand are strikingly uniform in character. Many are ground beetles, usually black in colour, and often bear a superficial resemblance to the Carabidae. These forms are very often apterous, or have vestigial wings, and the elytra are frequently immovable. Many of the wood-feeding species have ample wings. The species of *Blaps* often occur in cellars and outbuildings. *Tenebrio molitor* and *T. obscurus* are nearly cosmopolitan: they are found in all stages in meal, flour and stored goods, their larvae being known as 'meal worms' (Fig. 417). *Tribolium* has very similar habits and *T. ferrugineum* and *T. confusum* are likewise widely spread, through commerce, in granaries and stores (Fig. 418). An account of the biology of *Tenebrio* is given by Cotton and St. George (1929). A very large number of species is found in desert habitats. Species of other genera live in dung, in dead animal matter, in fungi, under

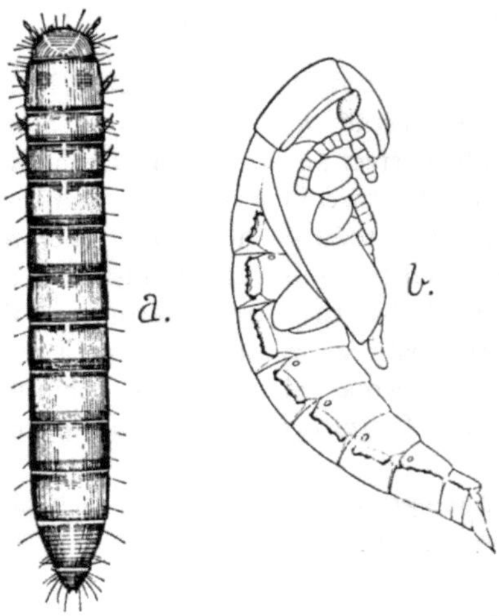

FIG. 417
Tenebrio obscurus, larva and pupa: enlarged

FIG. 418
Tribolium ferrugineum (Tenebrionidae), cosmopolitan. × 14

bark, etc. The larvae of this family bear a tolerably close resemblance to those of the Elateridae, but the labrum is plainly visible and the terminal segment of the abdomen rarely attains the complexity found in that family. For an account of Tenebrionid larvae see Korschefsky (1943) and Marcuzzi and Rampazzo (1950). The subdivisions of the family and its relations to the closest forms which are now excluded is discussed by Doyen (1972). In his system, the Nilionidae, Lagriidae and Alleculidae are treated as subfamilies of the Tenebrionidae.

The **Zopheridae**, a small and principally American group, were first separated from the Tenebrionidae by Böving and Craighead on the basis of larval structure, but the adult also differs in having movable abdominal sternites. The genus *Synchroa* (formerly Melandryidae) has been put in a new family near to the Zopheridae. The **Boridae** is another very small group usually placed in the Tenebrionidae but the fore coxal cavities are completely open. It has a wide distribution. The **Perimylopidae** with 4 genera and 8 subantarctic species (Watt, 1970) also has more or less open fore coxal cavities.

FAM. LAGRIIDAE. *First 3 visible abdominal sternites connate. Trochanters of heteromerous type. Front coxae somewhat projecting, prosternal intercoxal process narrow.* There are nearly 400 species in this family which is widely distributed (except

New Zealand). The only British representative is *Lagria hirta*: the larva of this species is somewhat broader and more active than those of the Tenebrionidae and the segments are furnished with lateral tufts of hairs (Schiödte). The head is very short and the last body-segment is bifid at the apex. The pupa is remarkable on account of the long broad clavate processes which project from most of the abdominal segments.

FAM. ALLECULIDAE (Cistelidae). *Tarsal claws pectinate. Antennae filiform or pectinate. Fore coxae not projecting, prosternal intercoxal process broad.* The members of this family are easily recognized by the structure of their claws. There are about 1100 species, found in all parts of the world, usually on flowers or leaves. The larvae are found in rotten wood or humus and that of *Gonodera* (*Cistela*) is figured by Westwood (1839). Binaghi (1949) describes the biology of *Omophlus lepturoides* which damages potatoes.

FAM. MONOMMIDAE. *Fore coxae small, rounded, separated by a broad prosternal process, their cavities widely open behind. Propleuron grooved to receive antenna.* About 200 species found in most of the warmer parts of the world and breeding in rotten wood. *Monomma indicum* lives in decaying stems of the paw-paw. Freude (1958) revises the African species and summarizes his work on the world-fauna.

FAM. ELACATIDAE (= Othniidae). *Fore coxae small, rounded to projecting, prosternal process between them usually broad, their cavities closed behind. Mesepisterna not nearly meeting in front of mesosternum.* Under this name Crowson unites several groups (*Othnius*, *Eurystethus* = *Aegialites*, and others) which have in the past been placed in separate families or scattered through other groups. *Othnius* is widespread in warm countries and lives in decaying vegetable matter. The larva has been figured by Fukuda (1962). The species of *Eurystethus* are found on the coasts of western North America. Crowson has also erected a family **Inopeplidae** for *Inopeplus* which has usually been placed in the Cucujidae though it appears to be more allied to the Salpingidae.

FAM. SALPINGIDAE. *Fore coxae rounded, separated by a broad prosternal process, cavities internally closed. Mid coxal cavities closed by sterna. Tarsal segments simple. Antennae clubbed.* A family of rather small size but of wide distribution. The British genera *Salpingus* and *Lissodema* superficially resemble Carabids, whereas *Rhinosimus* has a well-marked rostrum. The British species are found under bark but very little is known of their habits. The larval spiracles are biforous and there are complex urogomphi (see Perris). Near the Salpingidae Crowson places the several genera forming his family **Cononotidae**: the adults have filiform antennae and the last segment of the maxillary palpi is securiform.

FAM. PYTHIDAE. *Mid coxal cavities not closed externally by the sterna. Fore coxal cavities completely open both externally and internally. Antennae short or clubbed.* As now dismembered, this family is relatively small though widely distributed in Eurasia and America. One species of *Pytho* is British. The larva has cribriform spiracles and broad, almost lamelliform, urogomphi. The **Mycteridae** include a number of genera which have often been placed in the Pythidae though the mid coxal cavities are closed by the sterna. They are perhaps closer to the Salpingidae but the penultimate tarsal segment is lobed. *Mycterus* superficially resembles a weevil. It is found on flowers and is reputedly British. A probable larva was found in dead pine bark in Spain (Crowson, 1964*a*).

FAM. HEMIPEPLIDAE. *Tarsi with last 2 segments lobed beneath. Mid coxal cavities closed by sterna. Fore coxae oblique, not projecting.* This family, which has been segregated from a position in the Cucujidae, has a small number of species in the warmer regions. The adults and larvae live together under the leaf-bases of palms.

FAM. TRICTENOTOMIDAE. *Large, Prionid-like species. Mid coxal cavities closed by the sterna, fore coxae strongly transverse. Antennae long, last 3 segments differentiated.* About a dozen species are known from the forests of the Oriental region. The larva has been figured by Gahan (1908).

FAM. PYROCHROIDAE (Cardinal Beetles). *Antennae long, filiform to pectinate. Tarsi with penultimate segment lobed beneath. Mid coxal cavities not closed outwardly by sterna, fore coxal cavities completely open both internally and externally.* A family of rather more than 100 species occurring principally in the north temperate region. They are of rather large size and often partly scarlet in colour. The adults are often found under bark where the larval life is spent but they also occur on leaves and flowers. The larva of *Pyrochroa* is figured by Schiödte (Fig. 419*a*) and those of the three British species are distinguished by van Emden (1943).

FAM. MELANDRYIDAE (Serropalpidae). *Pronotum with well-marked side borders. Head not sharply constricted at neck. Tibial spurs simple or serrate, in former condition, antennae often with 4-segmented club.* About 600 species have been placed in this family which is principally found in woodland areas in temperate regions. The species are associated with decaying wood or with woody fungi: *Osphya*, however, is found in flowers. The species of *Orchesia* can jump actively with the aid of the tibial spurs. Some authorities recognize a family **Tetratomidae** for those genera in which these spurs are simple. The constitution of these groups is discussed by Crowson (1966*a*) and Viedma (1966) describes their larvae.

FAM. SCRAPTIIDAE. *Pronotum with more or less distinct side borders, about as wide as elytra at shoulders. Tibial spurs pubescent. Tarsi with 3rd and 4th segments more or less lobed, claws simple.* About 350 species of this family are distributed over most of the principal regions (Franciscolo, 1964). The large genus *Anaspis* which has usually been referred to the next family is mainly Holarctic. The species of Scraptiidae are mostly associated with rotten wood or fungi, but the adults of *Anaspis* are abundant in flowers. The larva of *Scraptia* is principally characterized by the great length and spoon-shaped form of the last abdominal segment (Böving and Craighead).

FAM. MORDELLIDAE. *Tibial spurs pubescent. Pronotum with more or less distinct side borders, about as wide as elytra at shoulders. Tarsi with penultimate segment not or slightly lobed, claws serrate and with bristle-like lobe beneath.* Most species of this family, of which there are about 650, can be recognized by having the 7th abdominal segment produced into a more or less long spine. The larva of *Tomoxia* (Schiödte) develops in decaying wood but that of *Mordellistena* is phytophagous and burrows in the stems of various plants. The adults of many species are found on Umbelliferae but they are rarely common in Britain. The genera of the family have been revised by Ermisch (1950).

FAM. RHIPIPHORIDAE. *Antennae strongly flabellate, at least in ♂. Tarsal claws serrate but not appendiculate, penultimate tarsal segment simple. Pronotum posteriorly about as wide as elytra at shoulders.* This family with more than 250 species is noted for its specialized parasitic habits. The European *Pelecotoma*, whose larva is said to be predatory on those of other beetles in rotten wood, has probably the simplest type of life history. The Rhipiphorini are parasites of the larvae of aculeate Hymenoptera and exhibit hypermetamorphosis. *Metoecus paradoxus* is a parasite in nests of *Vespula*, particularly *V. vulgaris*. According to Chapman (1870; 1891) the

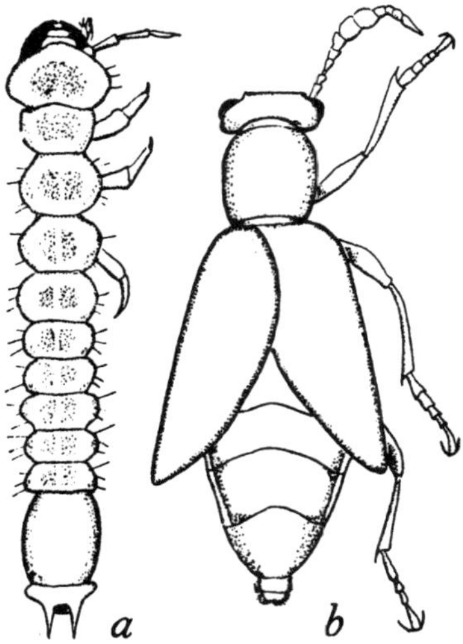

FIG. 419
Heteromera
a, *Pyrochroa*, larva; *b*, *Meloe proscarabaeus*, Imago.

newly hatched larva is black and campodeiform, resembling that of *Meloe*, but how it enters the nest of its host does not appear to have been ascertained, since the eggs are laid in old wood. Two hypotheses have been put forward – either the eggs are transferred along with wood fragments by the wasps to their nests, or the young larvae attach themselves to the wasps while the latter are gathering wood. Within the nest the larva becomes an endoparasite of that of *Vespula*: it subsequently becomes an ectoparasite, and gradually devours the whole of its host. Pupation takes place in the cell of the attacked individual, and the adult beetles are found up to the number of twenty or more in a single nest: more rarely they occur on flowers, etc., outside the nest. Among other genera *Macrosiagon* has a very similar life-history, but utilizes *Odynerus* as its host (Grandi, 1937). *Rhipidius* is an endoparasite of Blattids; the female is apterous and larviform and does not leave the body of its host (Stamm, 1936). The classification of the group is discussed by Selander (1957) and a number of genera is illustrated by Manfrini de Brewer (1963).

FAM. MELOIDAE (Oil Beetles, Blister Beetles). *Head strongly deflexed, neck narrow. Tarsal claws appendiculate and usually serrate. Adult soft-bodied with long* *legs, fore coxal cavities open behind.* This is one of the most interesting of all groups of Coleoptera on account of the remarkable life-histories of its members and the general occurrence of hypermetamorphosis. About 2000 species have been described and they are very widely distributed: *Meloe* (Fig. 419*b*), *Sitaris* and *Lytta* occur in Britain. The classification of this family is not yet settled. Larval characters do not seem always to agree with those of the adults (MacSwain, 1956; Selander, 1964; Kaszab, 1963, 1969). The female beetles lay a very large number of eggs (often 2000 to 10 000) which is explicable on the grounds that the subsequent life-history is extremely precarious, and very large numbers of larvae perish in the first instar. Oviposition takes place in the soil or on the surface of the ground, and the resulting larvae prey upon the eggs of Orthoptera and aculeate Hymenoptera. In their first instar they are minute, active, hard-skinned, campodeiform larvae known as triungulins. At this stage they are principally engaged in seeking out their hosts: having discovered the latter, they subsequently undergo ecdysis and change either into soft-bodied, short-limbed eruciform larvae or, more rarely, into a modification of the campodeiform type known as the caraboid stage. The next succeeding instars differ from the preceding, and the second, or later larva, passes into a resting period when the insect assumes the pseudo-pupal or 'coarctate' condition. The latter is followed by a further larval instar which is succeeded by the pupa.

The biology of *Sitaris* has been investigated by Fabre (1857) and Mayet (1875). The eggs of *S. muralis* are deposited near the nests of *Anthophora* about August. The newly hatched triungulins remain lethargic and hibernate until spring when they become more active. A certain number succeed in attaching themselves to the hairy bodies of the male bees, which appear earlier than the females. When opportunity allows, they pass to the female bees and so get carried to the nests of the latter. *Anthophora* constructs cells in the ground, in each of which there is a supply of nectar and a single egg. When the bee deposits an egg on the nectar, a triungulin slips off her body, alights on the egg, and becomes imprisoned in the sealed-up cell. It consumes the contents of the egg, and changes into a fleshy ovoid eruciform larva with vestigial legs. In this instar it feeds upon the nectar and pollen stored by its host, and subsequently changes into the so-called pseudo-pupal condition within the larval skin. After about one month, a certain number of individuals pass through the subsequent instars and appear as beetles the same year. More usually, they winter in the pseudo-pupal condition and, in spring, assume a second eruciform stage, which differs comparatively little from the earlier one. No food is taken during this period, and the larvae soon change into ordinary coleopterous pupae from which emerge the adult beetles.

Riley (1878) has studied the biology of *Epicauta vittata* in N. America. This insect deposits its eggs in parts of the ground frequented by the locust *Melanoplus* (= *Caloptenus*). Triungulins emerge in due course, and explore the soil until they discover the egg-capsules of the Orthopteron. Having found the latter, a single triungulin eats its way in and commences to devour the contained eggs. After a few days ecdysis takes place, and the larva passes into the caraboid or second instar. After about a week, ecdysis again occurs, and the larva becomes curved in shape. From its general body-form this instar is known as the scarabaeoid stage. The succeeding instar is very similar and, when fully grown, the larva deserts the egg-capsule, and changes nearby into the pseudo-pupal stage in which it hibernates. In spring it undergoes further changes, and in the sixth instar it is only slightly different from the scarabaeoid stages. From this condition it passes into the pupa and subsequently into the imago (see Fig. 188, p. 361).

The life-history of *Meloe* has been partially followed by Newport (1845–53) and is apparently very similar to that of *Sitaris*. Its triungulins do not appear to exercise much discrimination, and although their hosts are *Anthophora* and *Andrena*, they have often been found attached to other bees and also hairy Coleoptera and Diptera. Large numbers consequently perish through selecting the wrong host, while still greater numbers probably never discover a host at all. The second instar corresponds with Riley's caraboid stage, although it more closely resembles the scarabaeoid larvae in general form. In this condition it feeds upon the stored nectar and pollen, and afterwards transforms into a legless pseudo-pupa. This form moults and the final larval instar is a thick-bodied apodous grub. Space excludes references to the biology of other members of the family, and the student should consult the work of Beauregard (1890) for further information, also numerous more recent papers by Cros.

The 'Spanish fly', *Lytta* (*Cantharis*) *vesicatoria*, of southern Europe is rarely found in England. It yields the pharmaceutical product cantharidin ($C_{10}H_{12}O_4$) which is prepared from the dried insects. The elytra are alone used in pharmacy and contain more of the active principle than the soft parts collectively. Species of *Mylabris* are known to yield a larger amount of cantharidin than *Lytta* and are also used commercially. Gerber, Church and Rempel (1971) give a detailed account of the reproductive system of *Lytta*.

The family **Cephaloidae** with about 11 species in E. Asia and N. America is very like the Meloidae in the adult stage but the head is less deflexed and the neck broader; the larva is much more like a Mordellid and has no hypermetamorphosis.

FAM. OEDEMERIDAE. *Claws simple or toothed, tarsi with 1 or 2 of the penultimate segments lobed. Head little deflexed, without a marked neck. Eyes emarginate.* The 1500 species of this family are widely distributed but especially abundant in temperate regions. The larvae are soft-bodied and develop in timber. *Nacerdes* whose adult greatly resembles one of the Cantharidae breeds in water-logged timber, especially in maritime situations. Rozen has discussed the classification of the group and illustrated many larvae.

FAM. ANTHICIDAE. *Head more or less strongly deflexed and usually constricted into a narrow neck. All visible abdominal sternites connate. Tarsi with penultimate segment lobed beneath, antepenultimate simple.* This large family has about 1700 species, but some authorities place some of them in a separate family, the **Pedilidae**. The classification is discussed by Abdullah (1969). The species are mostly found in vegetable refuse, often in damp situations. Some of them inhabit the burrows of *Bledius* (Staphylinidae) and a number are halophilous. The larvae are found with the adults but their food is not known. Several species of *Anthicus* and *Notoxus* are attracted to *Meloe* beetles (Abdullah, 1964) and have been found hanging by their mandibles to the abdominal segments. Sometimes they have been attracted in large numbers to the dead bodies of *Meloe* put out to dry by coleopterists.

FAM. ADERIDAE (**Xylophilidae**). *First 2 visible abdominal sternites connate. Tarsi with penultimate segment small, antepenultimate lobed beneath. Head deflexed and often constricted into a neck. Trochanters not of heteromerous type.* These small beetles are widely distributed with about 700 species. Three of the large genus *Aderus* (*Xylophila*) are British. The beetles are usually found in dead wood or in vegetable refuse but very little is known of their habits (see Baguena Corella, 1948).

The family **Petriidae** with a few species in C. Asia requires further elucidation. Several important structural details have not yet been recorded.

Superfamily 19. **Chrysomeloidea**

This vast assemblage of phytophagous or xylophagous insects is rather well defined. There is every gradation from the short squat Chrysomelid type to the elongate Cerambycids with long antennae. There are six cryptonephric Malpighian tubules. In the larva, the thoracic legs are more or less developed, the mandible lacks a mola, and the antennae have three segments.

FAM. CERAMBYCIDAE (Longicornia). *Antennae usually at least two-thirds as long as body, capable of being flexed backwards, arising on strong tubercles. All tibiae with two spurs. Claws nearly always simple.* The longicorn beetles number about 20 000 species mostly of elongate form and attractive coloration (Fig. 420). Some of the members, such as *Macrotoma heros* and *Titanus giganteus*, are among the largest of insects. The family exists throughout the world wherever there is woody vegetation, and

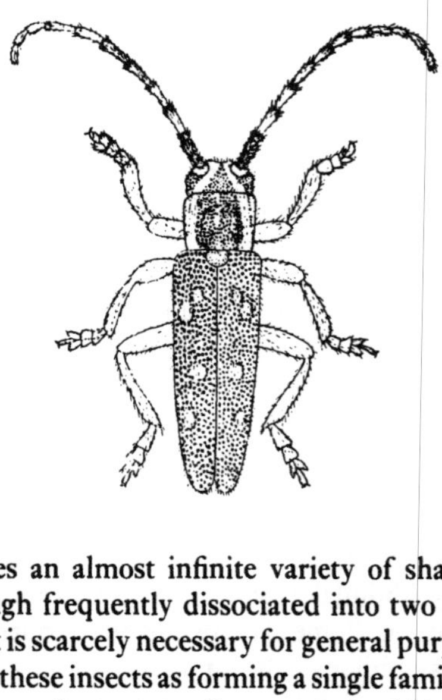

FIG. 420
Saperda populnea (Cerambycidae), Europe and N. America. × 2

includes an almost infinite variety of shape and ornamentation among its species. Although frequently dissociated into two or three divisions, each of separate family rank, it is scarcely necessary for general purposes to do otherwise than follow Sharp and regard these insects as forming a single family. The number of genera and species found in Europe is very small compared with those of the tropics. Only a few outlying representatives of this extensive group occur in Britain, and they comprise about 65 species, several being exceedingly rare or doubtfully indigenous. Linsley (1961–64) has published part of an extensive monograph of the N. American species. A number of forms are well known for their cryptic coloration, while others exhibit a close mimetic resemblance to insects of other families and also of other orders. One of the most striking cases of cryptic coloration is afforded by the African *Petrognatha gigas* whose whole upper surface resembles dead velvety moss and its irregular antennae are very like dried tendrils or twigs. The common British species *Clytus arietis* bears a resemblance to a

Vespid: it, furthermore, runs actively and exhibits antennal movements highly suggestive of those of a wasp. An interesting digression on these subjects will be found in the work of Fowler (1912) where a number of instances is enumerated. Many Cerambycidae possess the faculty of stridulating: in some cases the sound is caused by the hind margin of the prothorax working against a specialized striated area at the base of the scutellum: in others sound is produced by the friction of the hind femora against the edges of the elytra. In the Hawaiian *Plagithmysus* both types of organs are present in the same insect.

The larvae of the Cerambycidae bore for the most part into the wood of trees, but a few are confined to the roots or pith of herbaceous plants. Most species affect dead or decaying trees, some selecting moist and others dry wood. Certain species bore into the bark or into the sap or heart-wood of living trees and a few, such as *Saperda*, live in stems. The pupal habits are likewise varied, this instar occurring in the wood, between the latter and the bark, or in the bark. The pupa lies in the final larval burrow or in a special gallery leading therefrom and, in either case, a closed chamber is formed by the entrance being plugged with frass or fibrous chips. Many species adopt further measures for sealing up the pupal chamber (Beeson, 1919). In these cases a large amount of calcium carbonate is produced by the Malpighian tubes, this substance being mixed with gummy or silky matter and utilized for constructing an operculum which completely closes the pupal cell. In other cases the whole of the latter may be lined by an eggshell-like coating of the same substances. Cells which are closed or lined in this manner are protected from various enemies and are also probably enabled to maintain the requisite moisture-content. On account of their concealed mode of life, the larvae (Fig. 421) are soft and fleshy and of a whitish or yellowish colour: they are, furthermore, often finely pubescent. The form of the larvae is largely correlated with their habits, the bark-boring species being more or less flattened while those living in wood or stems tend to become cylindrical. The head is invaginated into the prothorax and is usually small and transverse, but in the Lamiinae it is longer than broad. The prothorax is large and is broader than the remaining trunk-segments. The 9th abdominal segment is often longer than those preceding and somewhat vesicular: in the Aseminae it bears a pair of spines. Thoracic legs are generally present, but are usually so much reduced as to be non-functional: in most of the Lamiinae they are wanting. Locomotion takes place by the aid of dorsal and ventral segmentally arranged abdominal swellings which, in some genera, bear cuticular asperities. In many larvae a variable number of the anterior abdominal segments bear small asteriform structures known as pleural discs which are the points of attachment of chorodotonal organs (see p. 127).

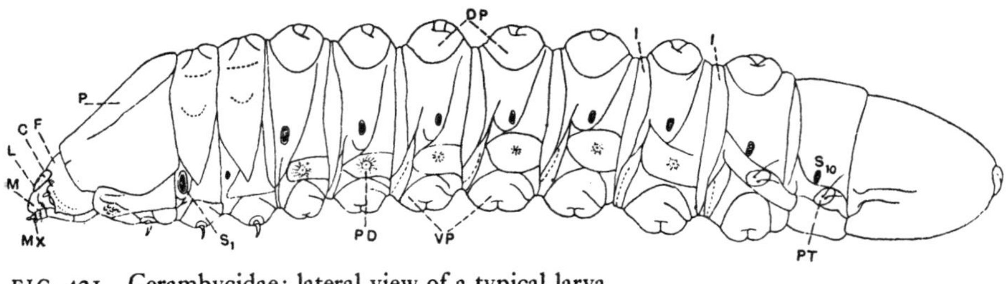

FIG. 421 Cerambycidae: lateral view of a typical larva

C, clypeus; *DP*, dorsal pseudopods; *F*, frons; *I*, intersegmental region; *L*, labrum; *M*, mandible; *MX*, maxilla; *P*, pronotum; *PD*, pleural disc; *PT*, pleural tubercle; S_1, 1st thoracic spiracle; S_{10}, 8th abdominal spiracle; *VP*, ventral pseudopods. Adapted from Craighead, *U.S. Dept. Agric. Office of Sec. Report*, 107.

The writings of Perris and Schiödte include descriptions of a number of larvae belonging to this family, while van Emden (1939) and Duffy (1953) have tabulated the British species and Craighead (1923) some of the Canadian ones. Duffy (1957, 1960, 1963) has monographed the larvae of African, Neotropical and Australasian species. Among life-history studies those of Ritchie (1920) on *Saperda* (Fig. 420) and Crawshay (1907) on *Tetropium* may be mentioned. In *Saperda carcharias* the life-cycle occupies about four years in Scotland, the first winter being passed in the egg stage. In *Tetropium gabrieli*, on the other hand, the life-history is of one year's duration and hibernation occurs in the larval stage. The shorter cycle is more usual, but the relative supply of moisture and the nutrient qualities of the food tend to increase or diminish the normal time by months or even years. Several instances are recorded in which wood, made into furniture many years, has been found to contain larvae which finally emerged as imagines (Craighead).

FAM. BRUCHIDAE (Lariidae). *Antennae short, not capable of being flexed backwards; if inserted on tubercles then usually less than two tibial spurs on one or more tibia. Claws often split or appendiculate. Ligula absent. Eyes deeply emarginate. Vertex not grooved. Pronotum with side margins. Elytra striate but no scutellar striole. Hind femur thickened and often toothed. First visible abdominal sternite as long as next three together.* The position of this family has often been discussed, many authors having regarded it as being closely related to the Anthribidae (p. 897) through the genus *Bruchela* (*Urodon*). On the other hand, the Bruchidae are also in many respects allied to the Sagrinae in the Chrysomelidae. Over 1200 species are known and their larvae mostly live in seeds of Leguminosae, causing great injury to peas, beans, lentils, etc.: those of certain other species attack coconuts and palm-nuts. On account of this habit they are very often carried from one country to another in cargoes of seeds. About a dozen species have been found in the British Isles, several being direct introductions from other lands (Herford, 1935).

The eggs of Bruchidae are usually laid on the young seed pods, as for example in *Bruchus pisorum*, and the larvae mine their way through until they reach the seed. In *Pachymerus chinensis* and *Acanthoscelides obtectus* the eggs are laid either upon the pods or the seeds, while *Bruchus pruinosus* lays them on the seed. In *B. pisorum* of the pea only a single larva enters a seed and dried peas are unattacked. *A. obtectus* readily attacks dried beans and in suitable climates six generations may occur in the year in the same batch of seed. The larvae in this family are eruciform and grub-like with thick bodies, becoming curved in the later development when they resemble those of the Curculionidae. The head is small and often narrower than the prothorax, with short stout mandibles. The first instar differs from those that follow in possessing legs and prominent spinous pronotal processes. In *B. pisorum* and *B. fabae* Riley states that the legs are slender and 3-segmented, but atrophy once the boring life within the seed is assumed. The retention of these appendages in certain species requires further investigation. Thus, in *A. obtectus* they are similarly present in the first instar, but most writers state that they subsequently atrophy. Razzauti, however, finds that they persist throughout life in the form of papilla-like vestiges. Owing probably to the nutritious nature of the endosperm upon which it feeds, a single larva usually devours only a small amount of nutriment, but where many occur in a single seed, as in *A. obtectus*, destruction is more complete. Pupation takes place as a rule within the seed.

Utida (1972) has shown that in *Callosobruchus maculatus* two forms exist, an active flying one of paler colour and a darker, inactive one which does not fly. Though partly genetically controlled, the active form is largely induced by crowding, high temperature, food of low water-content and unusual photoperiods.

FAM. CHRYSOMELIDAE. This family competes very closely with the Curculionidae as regards number of species and over 20 000 have been described. They are extremely closely allied to the Cerambycidae, and there appear to be no definite and constant structural differences separating the two families. As a rule, the Chrysomelidae are very different in general appearance: their antennae are only of moderate length, and the eyes do not embrace their points of insertion: the upper surface of the body is generally bare and shining, frequently with metallic coloration. They lack some or all of the special characters noted for the Bruchidae.

Crowson recognizes eleven subfamilies, but many authors would separate the Halticinae from the Galerucinae and the Cassidinae from the Hispinae. The Sagrinae (Crowson, 1946) are large brilliantly coloured tropical insects with strongly thickened hind femora. According to Sharp the larva of *Sagra splendida* lives in swellings on the stems of *Dioscorea*. The Donaciinae are elongate and usually metallic insects common in temperate climates. They are aquatic in the pre-imaginal stages and in *Haemonia* the adults also live beneath the water. The metamorphoses of *Donacia* have been investigated by MacGillivray (1903), Böving (1910*a*) and others. The larvae feed submerged at the roots or in the stems of water plants. They are elongate, subcylindrical whitish creatures with short, hooked thoracic legs. The abdomen is terminated by a pair of spinous processes, the structure and functions of which have been much discussed (Houlihan, 1969). They enable the insect to perforate the plant tissues and insert its caudal extremity into the air spaces for purposes of respiration. When feeding, they gnaw holes in the plants and, by means of their specially modified mouthparts, they extract the sap which is pumped into the digestive system by the aid of the pharynx. The pupae are enclosed in tough cocoons attached to the roots of the host plants. The Criocerinae are represented in Britain by a few species of *Lema*, *Lilioceris* and *Crioceris*. Their larvae are short, thick, fleshy grubs which feed externally on the leaves of plants. Some have the habit of concealing themselves with coverings of excrement while other, and often closely allied species, do not possess this trait. The asparagus beetle (*Crioceris asparagi*) is familiar to growers of that vegetable, and *Lema melanopa* is occasionally injurious to growing cereals.

The Chrysomelinae include the greater number of the species of the family and 40 are British (Fig. 422). Their larvae live exposed on plants and are short and convex, frequently with leathery pigmented integument: those of *Orina* are well figured by Chapman (1903) and of *Leptinotarsa* by Tower (1906). The latter genus includes the well known Colorado potato beetle (*L. decemlineata*). The larvae and imagines of *Phaedon* are destructive to Cruciferae, particularly mustard, the former feeding in companies on the leaves. Kimoto (1962 and other papers) has described the larvae of many Japanese species. According to Champion and Chapman (1901) certain species of *Orina* are viviparous and this same method of reproduction is recorded by Williams (1914) in *Phytodecta viminalis*; see summary in Hagan (1951).

The Clytrinae are characterized by the peculiar structure of the abdomen which appears to be correlated with the formation of a case which envelops the egg. The larvae are also enclosed in cases which are composed, at least partially, of excrement. Owing to their concealed life they are usually devoid of pigment and resemble small

Scarabaeid larvae in general form. The larval cases of *Cryptocephalus* and *Clytra* are described by Weise, Fabre and others: in the former genus they are carried almost erect and the larvae move with a jerky action. The life-history of *Clytra quadripunctata* has been studied by Donisthorpe (1902): its larvae live in nests of *Formica rufa* and their cases are composed of a mixture of earth and excrement.

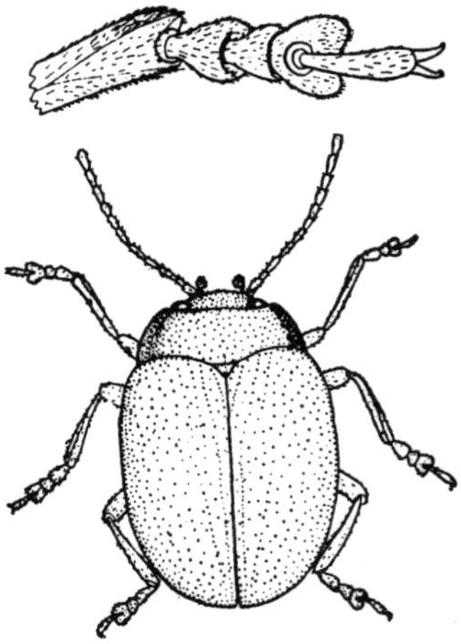

FIG. 422
Chrysolina staphylea (Chrysomelidae). Europe and N. America. × 4. Above – tarsus more highly magnified

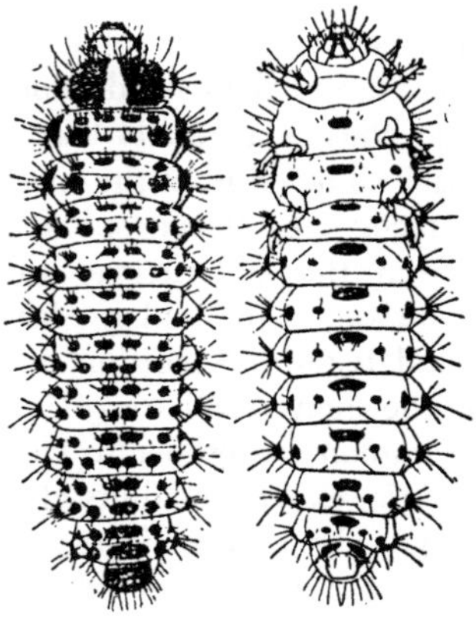

FIG. 423
Galerucella luteola (Chrysomelidae), larva, dorsal and ventral aspects. Enlarged
After Silvestri.

The Galerucinae (Fig. 423) and Halticinae comprise a number of highly injurious species. Their larval habits are extremely varied: many feed openly on the parenchyma of leaves, others live in roots, and a considerable number are leaf-miners. The Turnip Flea Beetle, *Phyllotreta nemorum*, and other members of the genus, are exceedingly destructive to Cruciferae, especially to the turnip. This genus and other members of the Halticinae have greatly developed leaping powers which reside in the swollen hind femora.

The Hispinae and Cassidinae are closely allied. The species of *Hispa* are usually covered with long stout upright spines: they are mainly tropical and their subfamily is unrepresented in the British fauna. Their larvae so far as known are leaf-miners. The Cassidinae include the 'tortoise' beetles, and have the lateral margins of the body greatly expanded which gives these insects a flattened shield-like appearance. Many are notable for their extremely brilliant coloration which fades very quickly after death. Their metamorphoses are of a remarkable character (Muir and Sharp, 1904): in certain species the eggs are enclosed in an ootheca often of complex structure, in others the ootheca is very small and imperfect and a layer of excrement is laid over it. The larvae are short and oval, somewhat flattened and spiny, often assuming bizarre forms: they usually cover their bodies with excrement which is supported and attached by a forked caudal process. The cast skins also form part of this adventitious covering, and the excrement may either form a solid pad, attached to the exuviae, or assume the condition of long filaments. Gressitt and Kimoto (1963) describe the larvae and oothecae of a number of Chinese species.

Superfamily 20. Curculionoidea

In terms of the number of species this is one of the largest of all the groups of the Coleoptera and can in many respects be regarded as the most highly evolved. They are allied to the Chrysomeloidea as shown by the resemblances between the Anthribidae and the Bruchidae. The antennae, however, are nearly always clubbed or the last segments are to some extent differentiated. The head is characteristically produced into a rostrum but this is hardly developed in such forms as the Scolytinae. The gular sulci are commonly confluent, but they are separate in the Belidae and obsolete in the Anthribidae. In the larva, the hypopharyngeal bracon, a transverse bar on the inner side of the labium, is a characteristic structure which is absent in the Chrysomeloidea. The thoracic legs are nearly always absent and when developed, as in the Anthribidae, have no more than two segments. Only the head and occasionally the pronotum are at all well sclerotized, and urogomphi are absent. In nearly every species the larvae are phytophagous.

The number of families into which the group should be divided is a matter still not completely settled. In the tabulation below, which follows Crowson, certain families (e.g. Belidae, Apionidae) are recognized which most authors have treated as subdivisions of the Curculionidae, whereas the Scolytinae and Platypodinae are usually given family rank.

FAM. NEMONYCHIDAE (Rhinomaceridae). *Maxillary palps normal, flexible. Labrum distinct and separate, gular sutures more or less obsolete. Pronotum more or less bordered at sides. First four abdominal sternites more or less connate.* This small group (18 species) has usually been included in the Curculionidae or the Attelabidae though the larva shows marked affinities to the Anthribidae: in both groups the larval mandible has a molar area which is otherwise lacking in the superfamily. The single British species of *Cimberis* (*Rhinomacer*) lays its eggs on the male blossoms of the pine.

FAM. ANTHRIBIDAE. *Like the previous family but gular sulci distinct. Pronotum without side borders. Five free abdominal sternites.* The remaining families of weevils have rigid maxillary palps and the labrum not fully free. This large family with nearly 2400 species is mainly tropical and is particularly numerous in the Indo-Malayan region. Its species are chiefly met with in old wood, dead branches and in fungi, but *Brachytarsus* is predacious on scale-insects. The anomalous genus *Bruchela* (*Urodon*) has been placed in the Bruchidae but it seems to have the essential characters of the present superfamily. The larva develops in the seed-capsules of *Reseda*. A few Anthribids such as *Xenocerus* have very elongate antennae and closely resemble Cerambycidae. The Anthribid larvae have been tabulated by Anderson (1947).

FAM. BELIDAE. *The gular sulci short but separate; antennae filiform and the tarsi apparently 4-segmented.* This small group (170 species) has not generally been given family status though the adult characters seem to be distinctive. The species are found in Australia, New Zealand and S. America and are attached to a variety of plants.

The **Oxycorynidae**, recognized as a family by Kolbe and more recently by Crowson, has 20 species and contains the genera *Oxycorynus* of S. America developing in the flowers of *Prosopanche* (Hydnoraceae) and *Metroxena* of the E. Indies living in palm fruits. Kuschel (1959) discusses the relationships of the Nemonychidae, Belidae and Oxycorynidae, especially their S. American species.

FAM. AGLYCYDERIDAE (PROTERHINIDAE). The members of this family, which has about 120 species, are known from their allies by having apparently 3-segmented tarsi, otherwise only known in some Apionidae, and the antennae are filiform. The species of *Proterhinus* are mostly found in the Pacific region, especially in Hawaii, but *Aglycyderes* which is sometimes placed in a separate family has a single species in the Canary Is. The larvae of *Proterhinus* (Anderson, 1941) may live in dead branches, under bark, in stems, or as leaf-miners.

FAM. ATTELABIDAE. *Maxillary palps with 4 segments. Labial palps inserted in the mentum ventrally, not in deep pits. Antennae clubbed, not geniculate.* The 300 weevils placed in this family are responsible for some of the familiar 'leaf-rolls' on various species of trees. These are constructed by the female after making certain cuts in the leaf-blade and an egg is laid in the centre of the roll. This is seen in *Attelabus, Apoderus* and *Rhynchites* and Buck (1952) and Daanje (1964) have given accounts of the behaviour of *Deporaus betulae*. In species of *Caenorhinus*, such as *C. germanicus* which attacks strawberry and raspberry, a leaf or shoot is girdled so that withered material is provided for the development of the larva. In that stage, the vertex is overlapped by the prothorax, the frontal sulci reach the articulations of the mandible and the abdominal segments each have two dorsal folds. *Allocorynus* of tropical America which breeds in the male cones of the Cycad, *Zamia*, probably belongs here. There are 20 British species of the family.

FAM. BRENTHIDAE (Brentidae). *Antennae not geniculate, scarcely clubbed. Labial palps minute, in deep pits. Trochanters normal.* A group of narrow elongate beetles numbering about 1300 species which are almost confined to wooded tropical countries, with one species in the Mediterranean region. The size of the individuals of a species is often subject to a great range of variation and the males are usually much larger than the females. In many species the two sexes are structurally very different (Fig. 424). In such

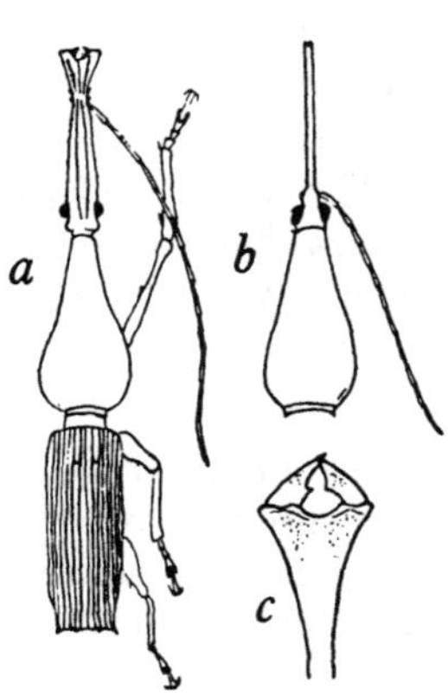

FIG. 424
a, *Ectocemus cinnamonii*, male, Borneo; *b*, head and thorax of female; *c*, *Eutrachelus*, Sumatra, extremity of head of male (Brenthidae)

instances the rostrum of the male is broad and rudimentary and the mandibles are large and are said to be used in seizing and carrying about the female before pairing. The females have minute jaws, but the rostrum is very slender, often equalling or exceeding the body in length. The early stages, so far as is known, are passed in wood, and the rostrum of the female is used for boring holes in which the eggs are laid. The larvae have the frontal sulci reaching the articulation of the mandibles, the abdominal segments each have 3–4 dorsal folds, and thoracic legs (otherwise only found in the Nemonychidae and most Anthribidae) are usually present. The African species have been revised by Damoiseau (1969).

FAM. APIONIDAE. *Antennae clavate, rarely geniculate, trochanters elongate. Ventral surface of mentum without projecting setae.* The familiar genus *Apion* is easily recognized by its non-geniculate antennae, but *Nanophyes* which has usually been located elsewhere in the Curculionidae has antennae of the common weevil type though it resembles *Apion* in its long trochanters. Another subfamily, the Eurhynchinae, includes the well-known Sweet Potato Weevil, *Cylas formicarius*: the antennal club is considerably less compact than in *Apion* and the trochanters are short. Most of these weevils do not have the strongly sclerotized proventriculus which is characteristic of most Curculionidae. In the larva, the abdominal segments have only two dorsal folds and the frontal sulci extend to the mandibular articulations. The genus *Apion*, of which there are 79 British species, develops in seeds, stems or roots of plants: several species are injurious to leguminous crops. There are more than 1000 species in the family of which Kissinger (1968) gives a useful review.

FAM. CURCULIONIDAE (Weevils, including the bark beetles, Scolytinae and Platypodinae). *Antennae nearly always geniculate; trochanters very elongate. Ventral surface of mentum with a projecting seta or tuft of bristles.* Even after removing the several groups which have here been treated as families, the weevils are still an immense group with more than 60 000 species. Their arrangement in a logical system of subfamilies and tribes is a matter of great difficulty. In one section which comes nearest to the Apionidae the larva has only two dorsal tergal abdominal folds and the adult is without a differentiated proventriculus: this group includes such British genera as *Rhynchaenus, Stenopelmus* and *Mecinus*. The other larger section includes species in which there are three or four larval tergal folds and the adult has a sclerotized proventriculus with eight valves. Two large subdivisions can be recognized (van Emden, 1938; 1952) in this section: the Adelognatha (Otiorrhynchinae, Brachyderinae, etc.) in which there are provisional mandibles (p. 818), the rostrum is short, the mentum is expanded so as largely to conceal the maxillae, the larval antennae do not project and the larva lives in the soil, feeding on the roots of plants. In the Phanerognatha, there are no provisional mandibles, the maxillae are not hidden by the mentum, the larva has conically projecting antennae, and very rarely lives in soil, feeding upon roots.

Kissinger (1964) gives a useful survey of the classification of the family in which he includes the Apionidae. Aslam (1961) examines how far the characters of the internal organs support the traditional groupings. He finds that the Adelognatha and Phanerognatha are well-defined, the former having a long tubular, posteriorly sclerotized vagina and a relatively unconcentrated central nervous system. He also finds little to support the separation of the Apionidae as a family.

The vast majority of weevils can be recognized by the pronounced rostrum, the geniculate clubbed antennae and the reduced rigid palps. The function of the rostrum in the female is often that of a boring instrument, a hole being drilled by it for placing the eggs: in some species the eggs are inserted far into holes previously made by the ovipositor, but whether the rostrum plays any part in this act or not is uncertain: in a number of species it is not used for either of these purposes and its function, like that of the male rostrum, is not known. In many instances this organ exhibits sexual differences, being better developed in the female than in the male. This dimorphism is well exhibited among British weevils in *Curculio*, and in the S. African *Antliarrhinus* which oviposits in cones of cycads. The rostrum in the female is about three times the whole length of the body and six times the length of the corresponding organ in the male. Unlike other Coleoptera, an exceptionally large proportion of the species are clothed with scales, but very little attention has been devoted to their form and structure. As a rule weevils have a sombre coloration, but most of the common British species of *Phyllobius* and *Polydrosus* have bright green scales and the Papuan *Eupholus* is sky-blue and the brilliancy of this colour rivals even that of the Lycaenidae. The diamond beetles (*Cyphus* and *Entimus*) of Brazil are probably the most resplendent of all Coleoptera. In some weevils the colour is produced by a fine powdery exudation which is readily abraded and, in a few cases at least, it is stated to be renewed during the life of the insect. Gressitt (1966) finds that the Leptopine weevil, *Gymnopholus*, in New Guinea often has a number of cryptogams (fungi, algae, etc.) growing on its rough dorsal surface. Various small animals such as mites live among the epiphytes. Similar but less marked symbiosis occurs in a Colydiid and weevils of several other subfamilies.

The larvae of Curculionidae are apodous and exhibit great similarity of form. The vast majority are internal or subterranean feeders; *Phytonomus*, however, has larvae which feed externally on leaves. No part of plants, from the roots to the seed, is entirely free from the attacks of one or more species of weevil. A few genera are aquatic, their larvae inhabiting the submerged parts of water plants. The imagines of such genera as *Bagous, Eubrychius* and *Litodactylus* are likewise aquatic and the two first mentioned swim by means of the hind legs. In some of the genera whose larvae feed openly, the larvae maintain their position by means of a viscous secretion, as in *Phytonomus* and *Cionus*. The larvae of *Hylobius* and of some species of *Pissodes* burrow in the wood of the Coniferae, either in the larger roots or in the trunk. Many *Otiorrhynchus, Sitona*, etc., affect roots. *Sitophilus* is entirely a seed feeder, and species of *Gymnetron, Ceuthorrhynchus*, etc., form either stem or root galls. *Rhynchaenus* mines leaves. When about to pupate, certain species construct cocoons from a product of the Malpighian tubules which is worked up by means of the larval mouthparts (Knab, 1915). It exudes from the anus and forms the reticulate cocoons of *Phytonomus*, the parchment-like capsules of *Cionus*, the chalky nodular cocoons of certain species of *Larinus* and the felted cocoons of *Rhynchaenus*. In the two first mentioned genera the material is the same as that which enveloped the larvae (see above). Labial spinnerets occur in some weevil larvae, and it is likely that the latter may spin cocoons in the strict sense of that operation or contribute material from stomodeal glands to the cocoon-forming substance. The bulk of Curculionid larvae pupate in the soil or in the substance of the food-plant.

A number of species in the subfamilies Otiorrhynchinae and Brachyderinae are parthenogenetic and lack males. Such species are nearly always polyploid, usually triploid (Suomalainen, 1969).

In a family of the size of the Curculionidae it is not remarkable that a number of the species are highly injurious, either as larvae or as imagines also. The Granary Weevil (*Sitophilus* (= *Calandra*) *granarius*) deposits its eggs in the grains of wheat or barley and has become widely distributed through commerce. The cosmopolitan Rice Weevil (*S. oryzae*) affects a variety of food substances including rice and other cereals. *Anthonomus grandis* is the Mexican Cotton Boll Weevil – the most serious enemy of the cotton crop in America, where it is estimated to destroy an equivalent of 400 000 bales annually. It is a comparatively recent introduction, having entered Texas about 1892 from tropical America. The eggs are laid in cavities made in the flower buds which usually fail to develop. Under suitable conditions the whole life-history only occupies two to three weeks. *Anthonomus pomorum* is the Apple-blossom Weevil, locally destructive in many parts of England. It is univoltine and the eggs are laid in the unopened blossom buds. The larvae feed upon the inner parts of the flower and on the receptacle: growth of the flower ceases, the petals dying and forming a kind of brown cap, hence the name of 'capped blossom' for this affection of the tree. *Rhynchophorus ferrugineus* is the Palm Weevil, which infests the toddy and coconut palms. The eggs are laid in the soft tissue at the bases of the leaf-sheaths, in wounds, or in cuts made by the toddy drawer. The larvae tunnel the stems in all directions and pupate in fibrous cocoons. *Hylobius abietis* is extremely injurious to young conifers: the weevils gnaw the bark and cambial layer, thus reducing or stopping the flow of sap. The larvae, on the other hand, are not injurious and mostly live below ground in the roots of trees that have been felled. Certain species of *Sitona* are pests of leguminous crops, and the larvae of *Ceuthorrhynchus pleurostigma* form conspicuous galls on the roots of cabbages and swedes. Scherf (1964) illustrates many Curculionid larvae and gives many details of their food and habits.

The subfamilies Scolytinae and Platypodinae have usually been given family rank; the cylindrical shape (Fig. 425) of their more typical members is characteristic of wood-borers and recurs in the Anobiidae and Bostrychidae. The majority of species bore into the bark and between the latter and the wood: others may attack the roots, solid wood or twigs, while still other species attack shrubs and a small number select herbaceous plants. A few species bore into the fruit or seed of palms, etc.,

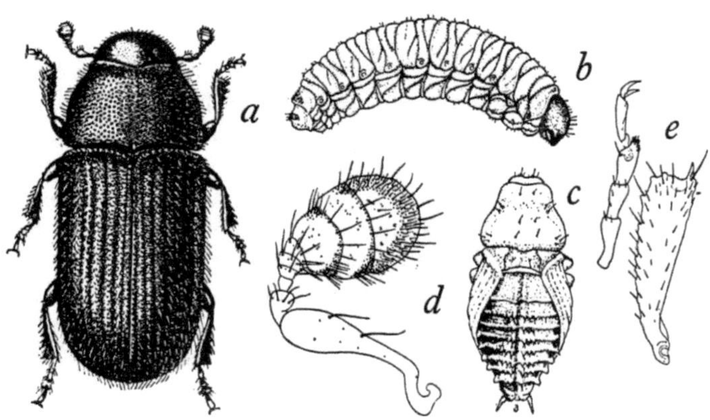

FIG. 425 Scolytinae

a, *Dendroctonus*; *b*, larva; *c*, pupa; *d*, *Gnathotrichus*, antenna; *e*, hind tibia and tarsus. (*a*, *b*, *c after* Hopkins: *d*, *e after* Schedl.)

young fir cones or the wood of casks and barrels. The larvae and adults feed upon the starches, sugars, etc., found in the host plants, or upon fungi which grow in the brood galleries. Owing to such habits these insects are of great interest to the forester. The method of attack is first to construct an entrance tunnel through the bark which, in the wood-boring forms, is carried deeply into the tree: in the bark-feeding species it does not reach further than the surface of the wood. From the inner end of the entrance tunnel two or more egg-tunnels are cut vertically, transversely or radially between the bark and wood (Fig. 426). With many species a nuptial chamber is excavated at the end of the entrance tunnel and, in such cases, the egg-tunnels originate from it. In most species this chamber is probably constructed by the male. The eggs are laid in niches along the walls of the egg-tunnels and the larvae excavate slender mines or larval burrows usually at right angles to them. The larval burrows are generally filled with excrement and their calibre increases as the larvae grow. The form and arrangement of the egg-galleries and larval burrows exhibit various features characteristic of each species or group of

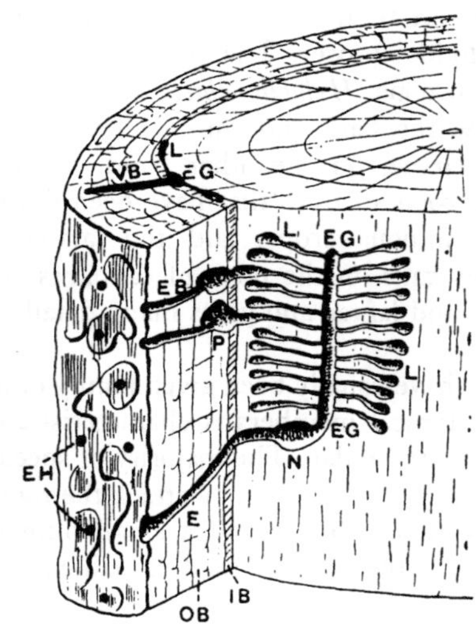

FIG. 426
Schematic figure illustrating the tunnellings of a bark beetle (Scolytinae) in a branch of a conifer

E, entrance burrow; *EB*, exit burrow; *EG*, egg-gallery; *EH*, exit holes; *IB*, inner bark; *L*, larval galleries; *N*, nuptial chamber; *OB*, outer bark; *P*, pupal cell; *VB*, ventilating burrow.

species and consequently these excavations are of particular taxonomic value. The extremities of the larval burrows are widened to form the pupal cells, and the adult beetles finally construct exit burrows leading from the pupal cells to the exterior. 'Ventilating burrows' are also often constructed: they are located in the roof of an egg gallery and extend to or near the exterior of the tree. Although perhaps serving for ventilation in some cases, they appear to serve more usually for the storage of boring dust, or as an opening through which this material may be ejected.

The social habits and relations of the sexes in this family are of a remarkable nature. As Hopkins remarks, there is a wide range of variation from simple or unorganized and intensive polygamy to specialized or organized polygamy, and a gradual reduction in the number of females associated with a single male from one male and sixty or more females (*Xyleborus*), to one male and two females (*Ips*) and finally to specialized monogamy (*Scolytus*). With many species copulation takes place on the bark of old trees or after alighting on the new host tree. Monogamous species often pair in the entrance tunnel and polygamous species in or near the nuptial chamber. In *Xyleborus ferrugineus*, however, the virgin female can produce a progeny of haploid males, incapable of flight. If she then pairs with some of her sons, the offspring are diploid winged females and haploid males. If the unfertilized female flies off the process may be repeated (Peleg and Norris, 1973).

Often trees are subject to mass-attack by many beetles simultaneously. This is now known to be due, at least in *Ips* and *Dendroctonus*, to complex but fully identified substances (pheromones) released by the females (McNew, 1970) or by both sexes (Wood, 1973).

The ambrosia beetles (Platypodinae) penetrate the wood and their larvae are nourished by certain fungi which develop upon the walls of the burrows. A carefully prepared bed or layer of chips and excreta is provided by the female beetle, upon which the fungus develops – ambrosia being the name applied to this fungus-food. The mycelium spreads to the various galleries staining them dark brown or black owing to the action of the fungus upon the wood. Certain species of fungi appear to be associated with individual species of beetles. Unless eaten off regularly, the fungus develops and spreads rapidly and during wet weather it may block up the galleries and kill the occupants. The transportation of the reproductive bodies of the fungus from one tree to another has received diverse explanations and takes place, either fortuitously or intentionally, through the agency of the beetles themselves. In the case of *Xyleborus* it has been stated that the conidia are either voided in the excreta or carried in the crops of the female beetles, and regurgitated when a fungus-bed is being prepared. In other cases it has been found that the brushes on the front of the head in the female of certain species retain the conidia among their hairs, and facilitate transportation. In *Diapus furtivus* Beeson (1917) has observed the same method of conveyance. He also states that groups of large prothoracic pores are found in many Platypodinae and are each filled with a globule of fatty secretion to which the spores readily adhere. He has observed the latter germinating *in situ* but they speedily become separated from the insect once the latter is established in its tunnel. Among the best known genera of ambrosia beetles are *Xyleborus, Trypodendron, Crossotarsus, Diapus* and *Platypus*. In many genera special cavities (mycangia) have been developed in which fungus spores or particles are carried (Farris, 1965). In *Xyloterinus*, Abrahamson and Norris (1966) find a thoracic cavity in the female which carries the 'ambrosia' fungus and a smaller oral cavity which carries yeasts in both sexes.

Scolytine larvae are apodous and usually closely resemble those of other Curculionidae. The literature on the family is very extensive; among the more important contributions are those of Hopkins (1909; 1915), which are accompanied by a full bibliography, and the writings of Hagedorn, Swaine, Nüsslin, Fuchs, Schedl, etc.: works on forest entomology should also be consulted. Kleine (1939) has collected the literature up to 1938: see also Chamberlain (1939) and Balachowsky (1949).

For the Curculionidae as a whole, the classification has been discussed by Marshall (1916) and the French species have been treated by Hoffmann (1950, 1954). From the morphological standpoint, Hopkins' study of the larval and imaginal structure of *Pissodes* (1911) is valuable. Important life-history studies are those by Trägårdh (1910) on *Rhynchaenus* (*Orchestes*), Jackson (1920–22) on *Sitona* and Günthart (1949) on *Ceuthorrhynchus*. Morimoto (1962) tries to relate adult structure, both external and internal, and the characters of the larvae to the phylogeny and classification of the Japanese weevils.

Literature on Coleoptera

ABDULLAH, M. (1964), *Protomeloe argentinensis*, a new genus and species of Meloidae, *Ann. Mag. nat. Hist.*, (13) **7**, 247–254.

—— (1969), The natural classification of the family Anthicidae with some ecological and ethological observations, *Dt. ent. Z.*, **16**, 323–366.

—— (1973), Larvae of the families of Coleoptera III. Heteromera, Cucujoidea: a key to the world families including their distinguishing characters, *J. nat. Hist.*, **7**, 535–544.

ABRAHAMSON, L. P. AND NORRIS, D. M. (1966), Symbiotic relationships between microbes and ambrosia beetles. I. The organs of microbial transport and perpetuation of *Xyloterus politus*, *Ann. ent. Soc. Am.*, **59**, 877–880.

ALI, H. A. (1967), The higher classification of the Carabidae and the significance of internal characters (Coleoptera), *Bull. ent. Soc. Egypte*, **51**, 211–231.

ANDERSON, W. H. (1936), A comparative study of the labium of Coleopterous larvae, *Smithson. misc. Coll.*, **95** no. 13, 29 pp.

—— (1939), A key to the larval Bostrychidae in the United States National Museum (Coleoptera), *J. Washington Acad. Sci.*, **29**, 382–391.

—— (1941), On some larvae of the genus *Proterhinus* (Coleoptera: Aglycyderidae), *Proc. Hawaiian ent. Soc.*, **11**, 25–35.

—— (1947), Larvae of some genera of Anthribidae (Coleoptera), *Ann. ent. Soc. Am.*, **40**, 489–577.

ANESHANSLEY, D. J. AND EISNER, T. (1969), Explosive secretory discharge of bombardier beetles (*Brachinus*: Col., Carabidae), *Science*, **165**, 61–63.

ARNETT, R. H. Jr. (1964), Notes on the Karumiidae (Coleoptera), *Coleopts. Bull.*, **18**, 65–68.

—— (1968), *The Beetles of the United States, A Manual for Identification*, (Revised), Amer. ent. Inst., Ann Arbor.

ARROW, G. J. (1904), Sound production in the Lamellicorn beetles, *Trans. ent. Soc. Lond.*, **1904**, 709–750.

—— (1910), *The fauna of British India, including Ceylon and Burma. Coleoptera Lamellicornia* (*Cetoniinae and Dynastinae*), Taylor & Francis, London.

—— (1942), The beetle family Rhyssodidae, with some new species and a key to those at present known, *Proc. R. ent. Soc. Lond.*, (B) **11**, 171–183.

—— (1944), Polymorphism in giant beetles, *Proc. zool. Soc. Lond.*, **113** (1943), 113–116.

ASLAM, N. A. (1961), An assessment of some internal characters in the higher classification of the Curculionidae *s.l.*, *Trans. R. ent. Soc. Lond.*, **113**, 417–489.

ATKINS, M. D. (1958), On the phylogeny and biogeography of the family Cupedidae (Coleoptera), *Canad. Ent.*, **90**, 532–537.

—— (1963), The Cupedidae of the world, *Canad. Ent.*, **95**, 140–162.

BAGUENA CORELLA, L. (1948), *Estudio sobre los Aderidae* (*Coléopteros Heterómeros*), *Xylophilidae sive Euglenidae*, Inst. Estud. Afr., Madrid.

BALACHOWSKY, A. (1949), Coléoptères Scolytides, *Faune de France*, **50**, [2+] 320 pp., P. Lechevalier, Paris.

BALFOUR-BROWNE, F. (1922), The life history of the water-beetle *Pelobius tardus* Herbst, *Proc. zool. Soc. Lond.*, 1922, 79–97.

—— (1934), The proventriculus in the Dytiscidae (Col.) as a taxonomic character, *Stylops*, **3**, 241–244.

—— (1935), The proventriculus in the Dytiscidae (Col.) as a taxonomic character. Second note, *Stylops*, **4**, 191.

—— (1940–50), *British Water Beetles*, 2 vols., Ray Soc., London.

BANKS, C. S. (1906), The principal insects attacking the Coconut palm, *Philippine J. Sci.*, **1**, 143–167; 211–218.

BARBER, H. S. (1913), The remarkable life-history of a new family (Micromalthidae) of beetles, *Proc. biol. Soc. Washington*, **26**, 185–190.

BARKER, J. F. (1969), Notes on the life-cycle and behaviour of the drilid beetle *Selasia unicolor* (Guérin), *Proc. R. ent. Soc. Lond.*, (A) **44**, 169–172.

BEAUREGARD, H. (1890), *Les insectes vésicants*, Anc. Libr. Germer. Ballière et Cie., Paris.

BEESON, C. F. C. (1917), The life-history of *Diapus furtivus* Sampson, *Indian For. Rec.*, **6**, 1–29.

—— (1919), The construction of calcareous opercula by Longicorn larvae of the group Cerambycini (Coleoptera, Cerambycidae), *Indian For. Bull.*, **38**, 1–10.

BEAL, R. S. Jr, (1961), Coleoptera, Dermestidae, *Insects of Micronesia*, **16**, 107–131.

BEIER, M. (1927), Vergleichende Untersuchungen über das Centralnervensystem der Coleopterenlarven, *Z. wiss. Zool.*, **130**, 174–250.

—— (1949), Körperbau und Lebensweise der Larve von *Helodes hausmanni* Gredler (Col. Helodidae), *Eos*, **25**, 49–100.

BELING, T. (1883), Beitrag zur Metamorphose der Käferfamilie der Elateriden, *Dt. ent. Z.*, **27**, 129–144; 257–304.

BELL, R. T. (1967), Coxal cavities and the classification of the Adephaga (Coleoptera), *Ann. ent. Soc. Am.*, **60**, 101–107.

BELL, R. T. AND J. R. (1962), The taxonomic position of the Rhysodidae (Coleoptera), *Coleopt. Bull.*, **16**, 99–106.

BERNET-KEMPERS, K. W. J. (1944), De larven der Helodidae (Cyphonidae), *Tijdschr. Ent.*, **86**, (1943), 85–91.

BERTRAND, H. (1928), Les larves et nymphes des Dytiscides, Hygrobiides, Haliplides, *Encycl. Entom.*, **10**, Paris.

—— (1962), Contribution à l'étude des premiers états des Coléoptères aquatiques de la région éthiopienne, (2) Famille Dryopidae. (3) Famille Psephenoididae. (4) Famille Hydrophilidae *s. lat.* (Palpicornia auct.), *Bull. Inst. franç. Afr. noire*, **24**, 710–777; 778–793; 1065–1114.

—— (1964), Contribution à l'étude des premiers états des Coléoptères aquatiques de la région éthiopienne (6e note). Famille Helodidae, *Bull. Inst. franç. Afr. noire*, (A) **26**, 513–579.

—— (1972), *Larves et nymphes des Coléoptères du globe*, Imprimerie Paillart, Abbeville, 804 pp.

BESUCHET, C. (1956), Larves et nymphes de Psélaphides (Coléoptères), *Rev. suisse Zool.*, **63**, 697–705.

—— (1972), Les Coléoptères Aculagnathides, *Rev. suisse Zool.*, **79**, 99–145.

BEUTENMÜLLER, W. (1891), Bibliographical catalogue of the described transformations of North American Coleoptera, *J. N.Y. micr. Soc.*, **7**, 1–52.

BINAGHI, G. (1949), Sull'*Omophlus* (*Odontomorphus*) *lepturoides* F. (Col. Alleculidae); quale notevale e poco noto parassita dei tuberi di patata, *Mem. Soc. ent. ital.*, **28**, 33–60.

BLACKWELDER, R. E. (1934), The prostheca or mandibular appendage, *Pan-Pacific Ent.*, **10**, 111–113.

BLACKWELDER, R. E. (1936), Morphology of the Coleopterous family Staphylinidae, *Smithson. misc. Collns.*, **94** no. 13, 102 pp.

BLUNCK, H. (1912), Beitrag zur Kenntnis der Morphologie und Physiologie der Haftscheiben von *Dytiscus marginalis* L., *Z. wiss. Zool.*, **100**, 459–492.

BORDAS, L. (1899), Les glandes défensives ou glandes anales des Coléoptères, *Ann. Fac. Sci. Marseilles*, **9**, 205–249.

—— (1900), Recherches sur les organes reproducteurs mâles des Coléoptères, *Ann. Sci. nat. Zool.*, **11**, 283–448.

—— (1903), L'appareil digestif des Silphidae, *C.R. Acad. Sci. Paris*, **137**, 344–346.

—— (1904), Anatomie et structure histologique du tube digestif de l'*Hydrophilus piceus* E. et de l'*Hydrous caraboides* L., *C.R. Soc. biol. Paris*, **56**, 1099–1100.

—— (1906), L'ampoule rectale des Dytiscides, *C.R. Soc. biol. Paris*, **61**, 503–505.

BOUNOURE, L. (1919), *Aliments, Chitine et Tube Digestif chez les Coléoptères*, A. Hermann et Fils, Paris.

BÖVING, A. G. (1907), Om Paussiderne og Larven til *Paussus kannegieteri* Wasm., *Vid. Medd. naturh. For.*, **9**, 109–136.

—— (1910–11), Nye Bidrag til Carabernes Udvicklingshistorie, i–ii, *Ent. Meddr*, **3**, 319–376; **4**, 129–180.

—— (1910*a*), Natural history of the larvae of Donaciinae, *Internat. Rev. Hydrobiol.*, **3** Suppl., 108 pp.

—— (1914), Notes on the larva of *Hydroscapha* and some other aquatic larvae from Arizona, *Proc. ent. Soc. Washington*, **16**, 169–174.

—— (1917), A generic synopsis of the Coccinellid larvae, etc., *Proc. U.S. natn. Mus.*, **51**, 621–650.

—— (1926), The immature stages of *Psephenoides gahani* Champ. (Coleoptera, Dryopidae), *Trans. ent. Soc. Lond.*, **1926**, 381–388.

—— (1936), Description of the larva *Plectris aliena* Chapin and explanation of new terms applied to the epipharynx and raster, *Proc. ent. Soc. Washington*, **38**, 169–185.

—— (1954), Mature larvae of the beetle family Anobiidae, *Biol. Meddr*, **22**, no. **2**, 1–299.

BÖVING, A. G. AND CHAMPLAIN, A. B. (1920), Larvae of North American beetles of the family Cleridae, *Proc. U.S. natn. Mus.*, **57**, 575–649.

BÖVING, A. G. AND CRAIGHEAD, F. C. (1931), Larvae of Coleoptera, *Entomologica Amer.*, **11**, 1–351.

BÖVING, A. G. AND HENRIKSEN, K. L. (1938), The developmental stages of the Danish Hydrophilidae (Ins. Coleoptera), *Vid. Meddr naturh. For.*, **102**, 27–162.

BÖVING, A. G. AND ROZEN, J. G. Jr. (1962), Anatomical and systematic study of the mature larvae of the Nitidulidae (Coleoptera), *Ent. Meddr*, **31**, 265–299.

BOYLE, W. W. (1956), A revision of the Erotylidae of America north of Mexico (Coleoptera), *Bull. Amer. Mus. nat. Hist.*, **110**, 61–172.

BRANDT, E. (1879), Vergleichend-anatomische Untersuchungen über das Nervensystem der Käfer (Coleoptera), *Hor. Soc. ent. Ross.*, **15**, 51–67.

BRITTON, E. B. (1970), Coleoptera (Beetles), In: *The insects of Australia*, Mackerras, I. M. (ed.), Chapter 30, Melbourne.

BROWN, E. S. (1954), The biology of the coconut pest *Melittomma insulare* (Col., Lymexylonidae), and its control in the Seychelles, *Bull. ent. Res.*, **45**, 1–66.

BUCK, H. (1952), Untersuchungen und Beobachtungen über den Lebensablauf und das Verhalten des Trichterwicklers *Deporaus betulae* L., *Zool. Jb. Allg. Zool.*, **63**, 154–236.

BUCK, J. AND BUCK, E. (1972), Photic signalling in the firefly *Photinus greeni, Biol. Bull.*, **142**, 195–205.

BUGNION, E. (1920), Les anses malpighiennes des Lampyrides, *Bull. Soc. zool. France*, **45**, 133–144.

BUGNION, E. AND BUYSSON, H. du (1924), Le *Platypsyllus castoris* Ritz., *Ann. Sci. nat. Zool.*, (10) **7**, 83–130.

BURGESS, A. F. (1911), *Calosoma sycophanta*: its life history, behavior and successful colonisation in New England, *U.S. Dept. Agric. Bur. Ent. Bull.*, **101**, 94 pp.

CARPENTER, G. H. AND MacDOWELL, M. C. (1912), The mouthparts of some beetle larvae, etc., *Quart. J. micr. Sci.*, **57**, 373–396.

CEREZKE, H. F. (1964), The morphology and functions of the reproductive systems of *Dendroctonus monticolae* Hopk. (Coleoptera: Scolytidae), *Can. Ent.*, **96**, 477–500.

CHAMBERLAIN, W. J. (1939), *The Bark and Timber Beetles of North America north of Mexico. The taxonomy, biology and control of 575 species belonging to 72 genera of the superfamily Scolytoidea.* Corvallis vi + 513 pp.

CHAMPION, G. C. AND CHAPMAN, T. A. (1901), Observations on *Orina*, a genus of viviparous and ovo-viviparous beetles, *Trans. ent. Soc. Lond.*, **1901**, 1–17.

CHAPMAN, J. A. (1957), Flight muscle change during adult life in the Scolytidae, *Can. Dept. Agric., For. Biol. Div., Bimonthly Prog. Rep.*, 13 (1).

CHAPMAN, T. A. (1870), On the parasitism of *Rhipiphorus paradoxus* and some facts towards a life history of *Rhipiphorus paradoxus, Ann. Mag. nat. Hist.*, (4), **5**, 191–198; **6**, 314–326.

—— (1891), On the oviposition of *Metoecus* (*Rhipiphorus*) *paradoxus, Entomologist's mon. Mag.*, (2), **2**, 18–19.

—— (1903), A contribution to the life history of *Orina* (*Chrysochloa*) *tristis* Fabr., *Trans. ent. Soc. Lond.*, **1903**, 245–261.

CLAUSEN, C. P. (1915), A comparative study of a series of aphid-feeding Coccinellidae, *J. econ. Ent.*, **8**, 487–491.

COFFAIT, H. (1972), Coléoptères Staphylinidae de la région paléarctique occidentale. I. Généralités, sous-familles Xantholiniae et Leptotyphlinae, *Nouvelle Rev. Ent.*, Suppl. **2**, **2**, 1–651.

COTTON, R. T. AND ST. GEORGE, R. A. (1929), The meal worms, *Tech. Bull. U.S. Dept. Agric.*, **95**, 37 pp.

CRAIGHEAD, F. C. (1923), North American Cerambycid larvae, etc., *Dept. Agric. Canada, Bull.* n.s. **27**, 238 pp.

CRAWSHAY, L. R. (1903), On the life history of *Drilus flavescens* Rossi, *Trans. ent. Soc. Lond.*, **1903**, 39–52.

—— (1907), The life history of *Tetropium gabrieli* Ws. = *T. fuscum* Sharp = *T. crawshayi* Sharp, etc., *Trans. ent. Soc. Lond.*, **1907**, 183–212.

CROWSON, R. A. (1938), The metendosternite in Coleoptera: a comparative study, *Trans. R. ent. Soc. Lond.*, **87**, 397–416.

—— (1944), Further studies on the metendosternite in Coleoptera, *Trans. R. ent. Soc. Lond.*, **94**, 273–310.

—— (1946), A revision of the genera of the Chrysomelid group Sagrinae (Coleoptera), *Trans. R. ent. Soc. Lond.*, **97**, 75–115.

—— (1950–54), The classification of the families of British Coleoptera, *Entomologist's mon. Mag.*, **86**, **87**, **88**, **89**, **90**, *passim*. Revised and issued as a book, London, 1955.

CROWSON, R. A. (1959), Studies on the Dermestoidea (Coleoptera), with special reference to the New Zealand fauna, *Trans. R. ent. Soc. Lond.*, **111**, 81–94.

—— (1960), The phylogeny of the Coleoptera, *Annual Rev. Ent.*, **5**, 111–134.

—— (1961), On some interesting new characters of classificatory importance in adults of Elateridae (Coleoptera), *Entomologist's mon. Mag.*, **96**, 158–161.

—— (1962), Observations on the beetle family Cupedidae, with descriptions of two new fossil forms and a key to the recent genera, *Ann. Mag. nat. Hist.*, (13) **5**, 147–157.

—— (1964*a*), Observations on the relationships of the genera *Circaeus* Yablok. and *Mycterus* Clairv., with a description of the presumed larva of *Mycterus* (Col., Heteromera), *Eos*, **40**, 99–107.

—— (1964*b*), A review of the classification of Cleroidea (Coleoptera), with descriptions of two new genera of Peltidae and of several new larval types, *Trans. R. ent. Soc. Lond.*, **116**, 275–327.

—— (1966*a*), Observations on the constitution and subfamilies of the family Melandryidae (Coleoptera), *Eos*, **41** (1965), 507–573.

—— (1966*b*), Further observations on Peltidae (Coleoptera, Cleroidea), with descriptions of a new subfamily and of four new genera, *Proc. R. ent. Soc. Lond.*, (B) **35**, 119–127.

—— (1968), The natural classification of the families of Coleotera. (Reprint 1967.) E. W. Classey Ltd., Hampton, Middlesex, 214 pp.

—— (1970), Further observations on the Cleroidea (Col.), *Proc. R. ent. Soc. Lond.*, (B), **39**, 1–20.

—— (1973), On a new superfamily Artematopoidea of polyphagan beetles, with the definition of two new fossil genera from the Baltic Amber, *J. nat. Hist.*, **7**, 227–238.

DAANJE, A. (1964), Ueber die Ethologie und Blattrolltechnik von *Deporaus betulae* L. und ein Vergleich mit den anderen blattrollenden Rhynchitiden und Attelabinen (Coleoptera, Attelabinae), *Verh. K. ned. Akad. Wet. Afd. Natuurk.*, **56**, 1–215.

DAMOISEAU, R. (1969), A monograph of the Brentidae (Coleoptera) from the continent of Africa, *Ann. Mus. Roy. Afr. centr.* (*Zool.*), (1967), **160**, 1–507.

DARLINGTON, P. J. JR. (1938), Experiments on mimicry in Cuba, with suggestions for future study, *Trans. R. ent. Soc. Lond.*, **87**, 681–695.

—— (1950), Paussid beetles, *Trans. Amer. ent. Soc.*, **76**, 47–142.

DELEURANCE-GLAÇON, S. (1963*a*), Recherches sur les Coléoptères troglobies de la sous-famille des Bathysciinae, *Ann. Sci. nat., Zool., Paris*, (12), **5**, 1–172.

—— (1963*b*), Contribution à l'étude des Coléoptères cavernicoles des Trechinae, *Ann. Spéléol.*, **18**, 227–265.

DIERCKX, F. (1899; 1901), Étude comparée des glandes pygidiennes chez les Carabides et les Dytiscides, etc., *La Cellule*, **16**, 61–176; **18**, 255–310.

DIMMOCK, G. AND KNAB, F. (1904), Early stages of Carabidae, *Bull. Springfield Mus.*, **1**, 1–55.

DOCTERS VAN LEEUWEN, W. (1910), Ueber die Lebensweise und die Entwicklung einiger holzbohrenden Cicindeliden-Larven, *Tijdschr. Ent.*, **53**, 18–40.

DONISTHORPE, H. ST. J. K. (1902), The life history of *Clythra quadripunctata* L., *Trans. ent. Soc. Lond.*, **1902**, 11–24.

—— (1919–20), The myrmecophilous lady-bird, *Coccinella distincta* Fald., its life-history and association with ants, *Ent. Rec.*, **31**, 214–222; **32**, 1–3.

DOYEN, J. T. (1972), Familial and subfamilial classification of the Tenebrionoidea (Coleoptera) and a revised generic classification of the Coniontini (Tentyrinae), *Quaest. Entomol.*, **8**, 357–376.

DUDICH, E. (1920–21), Ueber den Stridulationsapparat einiger Käfer, *Ent. Blatt.*, **16**, 146–161; **17**, 136–140; 145–155.

DUFFY, E. A. J. (1953), *A Monograph of the Immature Stages of British and Imported Timber Beetles (Cerambycidae)*, Brit. Mus. (Nat. Hist.), London.

—— (1957), *A Monograph of the Immature Stages of African Timber Beetles (Cerambycidae)*, Brit. Mus. (Nat. Hist.), London, viii + 338 pp.

—— (1960), *A Monograph of the Immature Stages of Neotropical Timber Beetles (Cerambycidae)*, Brit. Mus. (Nat. Hist.), London, vii + 327 pp.

—— (1963), *A Monograph of the Immature Stages of Australian Timber Beetles (Cerambycidae)*, Brit. Mus. (Nat. Hist.), London, viii + 235 pp.

DUFOUR, L. (1824–40), Recherches anatomiques sur les Carabiques et sur plusieurs autres Coléoptères, etc., *Ann. Sci. nat. Paris*, (1), **2**, **3**, **4**, **5**, **6**, **8**, **14**; (2) **1**, **3**, **13**, *passim*.

DUPORTE, E. M. (1946), Observations on the morphology of the face in insects, *J. Morph.*, **79**, 371–418.

DYBAS, H. S. (1966), Evidence for parthenogenesis in the featherwing beetles, with a taxonomic review of a new genus and eight new species (Coleoptera: Ptiliidae), *Fieldiana, Zoology*, **51**, 11–52.

EDWARDS, J. G. (1950), Amphizoidae (Coleoptera) of the world, *Wasmann J. Biol.*, **8**, 303–332.

EICHELBAUM, F. (1909), Katalog der Staphyliniden-Gattungen nebst Angabe ihrer Literatur ... geographischen Verbreitung und ihrer bekannten Larvenzustände, *Mem. ent. Soc. Belge*, **17**, 71–250.

EISNER, T. AND KAFATOS, F. C. (1962), Defense mechanisms of Arthropods X. A pheromone-promoting aggregation in an aposematic distasteful insect, *Psyche, Cambridge*, **69**, 53–61.

EMDEN, F. I. van (1932), Die Larven von *Discoloma casideum* Reitt. (Col. Colyd.) und *Skwarraia paradoxa* Lac. (Col. Chrysomel.), *Zool. Anz.*, **101**, 1–17.

—— (1935*a*), Die Larven der Cicindelidae. I. Einleitendes und alocosternale Phyle, *Tijdschr. Ent.*, **78**, 134–183.

—— (1935*b*), Die Gattungsunterschiede der Hirschkäferlarven, ein Beitrag zum natürlichen System der Familie (Col. Lucan), *Stett. ent. Ztg.*, **96**, 178–200.

—— (1936), Eine interessante zwischen Carabidae und Paussidae vermittelnde Käferlarve, *Arb. Phys. angew. Ent.*, **3**, 250–256.

—— (1938), On the taxonomy of Rhynchophora larvae (Coleoptera), *Trans. R. ent. Soc. Lond.*, **87**, 1–37.

—— (1939–49), Larvae of British Beetles, I, II, III, IV, V, VI, and suppl. VII, *Entomologist's mon. Mag.*, **75**, **76**, **77**, **78**, **79**, **81**, **83**, **84**, **85**, *passim*.

—— (1942), A key to the genera of larval Carabidae (Col.), *Trans. R. ent. Soc. Lond.*, **92**, 1–99.

—— (1952), On the taxonomy of Rhynchophora larvae: Adelognatha and Alophinae (Insecta, Coleoptera), *Proc. zool. Soc. Lond.*, **122**, 651–795.

ERICHSON, W. F. (1848), *Naturgeschichte der Insekten Deutschlands*, Verlag der Nicolaischer Buchhandlung, Berlin (Vol. 3).

ERMISCH, K. (1950), Die Gattungen der Mordelliden der Welt, *Ent. Bl.*, **45–46** (1949–50), 34–92.

ESCHERICH, K. (1894), Anatomische Studien über das männliche Genitalsystem der Coleopteren, *Z. wiss. Zool.*, **57**, 620–641.

EVANS, M. E. G. (1972), The jump of the click beetle (Coleoptera: Elateridae) – a preliminary study, *J. Zool.*, **167**, 319–336.

—— (1973), The jump of the click beetle (Coleoptera: Elateridae) – energetics and mechanics, *J. Zool.*, **169**, 181–194.

FABRE, J. H. (1857), Mémoire sur l'hypermétamorphose et les moeurs des méloides, *Ann. Sci. nat. Paris*, (4), **7**, 299–365.

—— (1903), *Souvenirs entomologiques. Huitième série. I. Les Cétoines*, 5th edn, Libr. Ch. Delagrave, Paris.

FARRIS, S. H. (1965), Repositories of symbiotic fungus in the ambrosia beetle *Monarthrum scutellaris* Lec. (Coleoptera: Scolytidae), *Proc. ent. Soc. Brit. Columbia*, **62**, 30–33.

FINCHER, G. F., DAVIS, R. AND STEWART, T. B. (1971), Flight activity of coprophagous beetles on a swine pasture, *Ann. ent. Soc. Am.*, **64**, 855–860

FORBES, W. T. M. (1922), The wing-venation of the Coleoptera, *Ann. ent. Soc. Am.*, **15**, 328–345.

FORD, G. H. (1917), Observations on the larval and pupal stages of *Agriotes obscurus* L., *Ann. appl. Biol.*, **3**, 97–115.

FORSYTH, D. J. (1968), The structure of the defence glands in the Dytiscidae and Gyrinidae (Coleoptera), *Trans. R. ent. Soc. Lond.*, **120**, 159–182.

—— (1970), The structure of the defence glands of the Cicindelidae, Amphizoidae and Hygrobiidae (Col.), *J. Zool.*, **160**, 51–69.

FOWLER, W. W. (1887–1913), *The Coleoptera of the British Islands*, Lovell Reeve & Co., London.

—— (1912), *Coleoptera. General introduction and Cicindelidae and Paussidae. Fauna of British India including Ceylon and Burma*, Taylor and Francis, London.

FRANCISCOLO, M. E. (1964), Nota preliminare sulla filogenia degli Scraptiidae (Coleoptera, Heteromera), *Atti Accad. naz. ital. Ent. Bologna Rc.*, **11**, 175–181.

FREUDE, H. (1958), Die Monommidae der afrikanischen Region (Coleoptera). (IV. Teil der Monommiden der Welt mit Zusammenfassung der Ergebnisse), *Ann. Mus. Congo Belge (8). Sci. Zool.*, **61**, 7–115.

FREUDE, H., HARDE, K. W. AND LOHSE, G. A. (1964–76), *Die Käfer Mitteleuropas*, Vols. 1–5, 7–9, Goecke & Evers, Krefeld.

FRIEDERICHS, K. (1908), Ueber *Phalacrus corruscus* als Feind der Brandpilze des Getreides und seine Entwicklung an brändigen Ähren, *Arb. biol. Anst. Berlin*, **6**, 38–52.

FUKUDA, A. (1938–39), Description of the larva and pupa of *Cupes clathratus*, *Trans. nat. Hist. Soc. Formosa Taihoku*, **28–29**, 390–393, 75–82.

—— (1962), Description of the larva of *Elacatis kraatzi* Reitter (Elacatidae, Coleoptera), *Kontyu*, **30**, 17–20.

GAHAN, C. J. (1900), Stridulating organs in Coleoptera, *Trans. ent. Soc. Lond.*, **1900**, 433–452.

—— (1908), On the larva of *Trictenotoma childreni* Gray, *Melitomma insulare* Fairmaire and *Dascillus cervinus* Linn., *Trans. ent. Soc. Lond.*, **1908**, 275–282.

—— (1911), On some recent attempts to classify the Coleoptera in accordance with their phylogeny, *Entomologist*, **44**, 121–125; 165–169; 214–219; 245–248; 259–262; 312–314; 348–351; 392–396.

GARDNER, J. C. M. (1930), The early stages of *Niponius andrewesi* Lew. (Col. Hist.), *Bull. ent. Res.*, **21**, 15–17.

—— (1935), Fam. Histeridae, subfam. Niponiinae, *Genera Insect.*, **202**, 6 pp. Desnet-Verteneuil, Brussels.

GEORGEVITSCH, J. (1898), Die Segmentaldrüsen von *Ocypus*, *Zool. Anz.*, **21**, 256–261.

GERBER, G. H., CHURCH, N. S. AND REMPEL, J. G. (1971), The anatomy and physiology of the reproductive systems of *Lytta nuttalli* Say (Col. Meloidae). I. The internal genitalia, *Can. J. Zool.*, **49**, 523–533.

GIBSON, L. P. (1967), Stridulatory mechanisms and sound production in *Gonotrachelus* (Coleoptera: Curculionidae), *Ann. Ent. Soc. Am.*, **60**, 43–54.

GRANDI, G. (1925), Contributo alla conoscenza biologica e morfologica di alcuni Lamellicorni filliofagi, *Bol. Lab. Zool. Portici*, **18**, 159–224.

—— (1937), L'ipermetabolia dei Ripiforidi, *Mem. Accad. Sci. Ist. Bologna* (9), **4**, 13 pp.

—— (1974), Comparative morphology and ethology of insects with a specialized diet, *Rhysodes germari* Ganglb., *Boll. Ist. Entomol. Univ. Bologna*, **30**, 32–47.

GRAVELEY, F. H. (1916), Some lignicolous beetle larvae from India and Borneo, *Rec. Ind. Mus.*, **12**, 137–175.

GRESSITT, J. L. (1966), Epizoic symbiosis: the Papuan weevil genus *Gymnopholus* (Leptopiinae) symbiotic with cryptogamic plants, oribatid mites, rotifers and nematodes, *Pacific Insects*, 8, 221–280.

GRESSITT, J. L. AND KIMOTO, S. (1963), The Chrysomelidae (Coleopt.) of China and Korea. Part II, *Pacific Ins. Monogr.*, **1A**, 1–299; (1961), **1B**, 301–1026; Suppl. **5**, 921–932.

GUÉNIAT, E. (1935), Contribution à l'étude du développement et de la morphologie de quelques Elatérides (Coléoptères), *Mitt. schweiz. ent. Ges.*, **16**, 167–298.

GÜNTHART, E. (1949), Beiträge zur Lebensweise und Bekämpfung von *Ceuthorrhynchus quadridens* Panz. und *Ceuthorrhynchus napi* Gyll. mit Beobachtungen an weiteren Kohl- und Rapsschädlingen, *Mitt. schweiz. ent. Ges.*, **22**, 441–591.

GUPTA, A. P. (1965), The digestive and reproductive systems of the Meloidae (Coleoptera) and their significance in the classification of the family, *Ann. Ent. Soc. Amer.*, **58**, 442–474.

GUPTA, R. L. (1937), On the salivary glands in the order Coleoptera. Part I. The salivary glands in the family Tenebrionidae, *Proc. nat. Acad. Sci. India*, **7**, 181–192.

HAGAN, H. R. (1951), *Embryology of Viviparous Insects*, The Ronald Press Co., New York.

HAGEN, K. S. (1962), Biology and ecology of predacious Coccinellidae, *Ann. Rev. Ent.*, **7**, 289–326.

HALFFTER, G. AND MATTHEWS, E. G. (1966), The natural history of dung beetles of the subfamily Scarabaeinae (Coleoptera, Scarabaeidae), *Folia ent. mex.*, **12–14**, 1–312.

HALL, D. W. AND HOWE, R. W. (1953), A revised key to the larvae of the Ptinidae associated with stored products, *Bull. ent. Res.*, **44**, 85–96.

HAWKES, O. A. M. (1920), Observations on the life-history, biology, and genetics of the lady-bird beetle, *Adalia bipunctata* (Mulsant), *Proc. zool. Soc. Lond.*, **1920**, 475–490.

HEEGER, E. (1857), Beiträge zur Naturgeschichte der Insekten, *S.B. Akad. Wiss. Wien*, **24**, 315–334.

HENRIKSEN, K. L. (1911), Oversigt over de danske Elateridelarven, *Ent. Meddr*, **4**, 225–252.

HENSON, W. R. (1961), Laboratory studies on the adult behaviour of *Conophthorus coniperda* (Schwarz) (Coleoptera: Scolytidae). I. Seasonal changes in the internal anatomy of the adult, *Ann. ent. Soc. Am.*, **54**, 698–701.

HERFORD, G. M. (1935), A key to the members of the family Bruchidae (Col.) of economic importance in Europe, *Trans. Soc. Brit. Ent.*, **2**, 1–32.

HINTON, H. E. (1939), An inquiry into the natural classification of the Dryopoidea, based partly on a study of their internal anatomy (Col.), *Trans. R. ent. Soc. Lond.*, **89**, 133–184.

—— (1940), A monographic revision of the Mexican water-beetles of the family Elmidae, *Novit. zool.*, **42**, 217–396.

—— (1944), Some general remarks on sub-social beetles, with notes on the biology of the Staphylinid, *Platystethus arenarius* (Fourcroy), *Proc. R. ent. Soc. Lond.* (A), **19**, 115–128.

—— (1945), *A Monograph of the Beetles associated with Stored Products*, Vol. I, Brit. Mus. (Nat. Hist.), London.

—— (1946*a*), A new classification of insect pupae, *Proc. zool. Soc. Lond.*, **116**, 282–328.

—— (1946*b*), The 'gin-traps' of some beetle pupae; a protective device which appears to be unknown, *Trans. R. ent. Soc. Lond.*, **97**, 473–496.

—— (1947), The gills of some aquatic beetle pupae (Coleoptera, Psephenidae), *Proc. R. ent. Soc. Lond.*, (A), **22**, 52–60.

—— (1955), On the respiratory adaptations, biology and taxonomy of the Psephenidae, with notes on some related families (Coleoptera), *Proc. zool. Soc. Lond.*, **125**, 543–568.

—— (1967*a*), Structure and ecdysial process of the larval spiracles of the Scarabaeoidea, with special reference to those of *Lepidoderma*, *Austr. J. Zool.*, **15**, 947–953.

—— (1967*b*), On the spiracles of the larvae of the suborder Myxophaga (Coleoptera), *Austr. J. Zool.*, **15**, 955–959.

HODEK, I. (1973), *Biology of Coccinellidae, with keys for identification of larvae by co-authors G. I. Savöiskaya & B. Klausnitzer*, Prague, Czechoslovak Acad. Sci., The Hague, Dr W. Junk N.V.

HOFFMANN, A. (1950, 1954), Coléoptères Curculionides, *Faune de France*, **52**, 1–486; **59**, [iii] + 487–1208.

HÖLLDOBLER, B. (1970), Zur Physiologie der Gast-Wirt-Beziehungen (Myrmecophilie) bei Ameisen. II. Das Gastverhältnis der imaginalen *Atemeles pubicollis* Bris. (Col., Staphylinidae) zu *Myrmica* und *Formica* (Hym., Formicidae), *Z. vergl. Physiol.*, **66**, 215–250.

HOLLOWAY, B. A. (1960), Taxonomy and phylogeny in the Lucanidae (Insecta: Coleoptera), *Rec. Dominion Mus.*, Wellington, **3**, 321–365.

HOPKINS, A. D. (1909), The genus *Dendroctonus*, *U.S. Bur. Ent. Tech. Ser.*, **17**, xiii + 164.

—— (1911), Contributions towards a monograph of the Barkweevils of the genus *Pissodes*, *U.S. Dept. Agric. Tech. Ser.*, **20**, 1–68.

—— (1915), Preliminary classification of the superfamily Scolytoidea, *U.S. Dept. Agric. Ent. Tech.*, **17**, 165–232.

HORION, A. (1940–65), *Faunistik der deutschen Käfer*, Parts 1–10, Düsseldorf, Frankfurt a/M, München, Überlingen-Bodensee.

HORN, W. (1938), 2000 Zeichnungen von Cicindelinae, *Ent. Beih.*, **5**, 1–71.

HOULIHAN, D. F. (1969), Respiratory physiology of the larva of *Donacia simplex*, a root-piercing beetle, *J. Insect Physiol.*, **15**, 1517–1536.

HOWDEN, H. F. (1955), Biology and taxonomy of North American beetles of the subfamily Geotrupinae with revisions of the genera *Bolbocerosoma*, *Eucanthus*, *Geotrupes* and *Peltogeotrupes* (Coleoptera), *Proc. U.S. natn. Mus.*, **104**, 151–319.

HOWE, R. W. (1959), Studies in beetles of the family Ptinidae. XVII. Conclusions and additional remarks, *Bull. ent. Res.*, **50**, 287–326.

HRBÁČEK, J. (1950), On the morphology and function of the antennae of the Central European Hydrophilidae (Coleoptera), *Trans. R. ent. Soc. Lond.*, **101**, 239–256.

HUBBARD, H. G. (1892), Description of the larva of *Amphizoa lecontei*, *Proc. ent. Soc. Washington*, **2**, 341–346.

HYSLOP, J. A. (1917), The phylogeny of the Elateridae based on their larval characters, *Ann. ent. Soc. Amer.*, **10**, 241–263.

JACKSON, D. J. (1920–22), Bionomics of weevils of the genus *Sitones* injurious to leguminous crops in Britain, *Ann. appl. Biol.*, **7**, 269–298; **9**, 93–115.

—— (1928), The inheritance of long and short wings in the weevil *Sitona hispidula* with a discussion of wing reduction among beetles, *Trans. Roy. phys. Soc. Edinburgh*, **55**, 665–735.

—— (1952), Observations on the capacity for flight of water beetles, *Proc. R. ent. Soc. Lond.*, (A), **27**, 57–70.

JACKSON, G. J. AND CROWSON, R. A. (1969), A comparative anatomical study of the digestive systems of *Malachius viridis* F. (Col., Melyridae) and *Rhagonycha usta* Gemm. (Col., Cantharidae), with observations on their diet and taxonomy, *Entomologist's mon. Mag.*, **105**, 93–98.

JANSSEN, W. (1963), Untersuchungen zur Morphologie, Biologie und Oekologie von *Cantharis* L. und *Rhagonycha* Eschsch. (Cantharidae, Col.), *Z. wiss. Zool.*, **169**, 115–202.

JEANNEL, R. (1955), L'édéage. Initiation aux recherches sur la systématique du Coléoptères, *Publ. Mus. Hist. nat. Paris*, no. **16**, 155 pp.

JEANNEL, R. AND PAULIAN, R. (1949), Coléoptères, In: Grassé, P. P. (ed.), *Traité de Zoologie*, **9**, 771–1077. Masson et Cie., Paris.

JOHN, H. (1954), Familiendiagnose der Notiophygidae (= Discolomidae, Col.), etc., *Ent. Bl.*, **50**, 9–75.

JOHNSON, N. E. AND CAMERON, R. S. (1969), Phytophagous ground beetles (Coleoptera: Carabidae), *Ann. ent. Soc. Am.*, **62**, 909–914.

KAPUR, A. P. (1950), The biology and external morphology of the larvae of Epilachninae (Coleoptera, Coccinellidae), *Bull. ent. Res.*, **41**, 161–208.

—— (1970), *Phylogeny of Ladybeetles*, Fifty-seventh Indian Science Congress, Kharagpur, 1970. Sect. Zoology and Entomology.

KASULE, F. K. (1966), The subfamilies of the larvae of Staphylinidae (Coleoptera) with keys to the larvae of the British genera of Steninae and Proteininae, *Trans. R. ent. Soc. Lond.*, **119**, 261–283.

KASZAB, Z. (1963), Merkmale der Adaptation, Spezialisation, Konvergenz, Korrelation und Progression bei den Meloiden (Coleoptera), *Acta Zool. hung.*, **9**, 135–175.

KASZAB, Z. (1969), The system of the Meloidae (Col.), *Mem. Soc. ent. Ital.*, **48**, 241–248.

KEMNER, A. (1912–13), Beiträge zur Kenntnis einiger schwedischen Koleopterenlarven, *Ark. Zool.*, **7** no. 31, 31 pp.; **8** no. 13, 1–13, 15–23.

—— (1925–26), Zur Kenntnis der Staphylinidenlarven, *Ent. Tidskr.*, **46**, 61–77; **47**, 133–170.

KEMPERS, K. J. W. (1899–1909), Het Adersystem der kevervleugels, *Tijdschr. Ent.*, **42**, 180–208; **43**, 172–199; **44**, 13–39; **45**, 53–71; **51**, ix–xvi; **52**, 272–283.

KERREMANS, C. (1906–14), *Monographie des Buprestides*, J. Janssens, Brussels. 7 vols.

KHALIFA, A. (1949), Spermatophore production in Trichoptera and some other insects, *Trans. R. ent. Soc. Lond.*, **100**, 449–479.

KIMOTO, S. (1962), A phylogenetic consideration of Chrysomelinae based on immature stages of Japanese species (Coleoptera), *J. Fac. Agric. Kyushu Univ.*, **12**, 105–116 and following papers.

KISSINGER, D. G. (1963), The proventricular armature of Curculionidae (Coleoptera), *Ann. ent. Soc. Amer.*, **56**, 769–771.

—— (1964), Curculionidae of America north of Mexico. A key to the genera. *South Lancaster, Mass. Taxonomic Publns*, 143 pp.

—— (1968), Curculionidae subfamily Apioninae of North and Central America with reviews of Apioninae and world subgenera of *Apion* Herbst. (Coleoptera), *Taxonomic Publications, South Lancaster*, Mass., vii + 559.

KISTNER, D. H. (1970), Revision of the old world species of the termitophilous tribe Corotocini (Col. Staphylinidae). III. The genera *Idiogaster*, *Paracorotoca* and *Fulleroxenus*, with notes on the relationships of their species, postimaginal growth and larvae, *J. ent. Soc. Sthn. Africa*, **33**, 157–192.

KLEINE, R. (1939), Die Gesamtliteratur der Borkenkäfer (Ipidae und Platypodidae) bis einschliesslich 1938, *Stett. ent. Ztg.*, **100**, 1–184.

KLOET, G. S. AND HINCKS, W. D. (1945), *A Check List of British Insects*, The authors, Stockport.

KNAB, F. (1915), The secretions employed by rhynchophorous larvae in cocoon-making, *Proc. ent. Soc. Washington*, **17**, 154–158.

KOLBE, H. J. (1908), Mein System der Coleopteren, *Z. wiss. InsektBiol.*, **4**, 116–123; 153–162; 219–226; 246–251; 286–294; 389–400.

KORSCHEFSKY, R. (1943), Bestimmungstabelle der bekanntesten deutschen Tenebrioniden- und Alleculidenlarven, *Arb. physiol. angew. Ent.*, **10**, 58–68.

KORSCHELT, E. (1923–24), *Der Gelbrand* Dytiscus marginalis L., Verlag Wilhelm Engelmann, Leipzig, 2 vols.

KREMER, J. (1917–19), Beiträge zur Histologie der Coleopteren mit besonderer Berücksichtigung des Flügeldeckengewebes, etc, *Zool. Jb., Anat.*, **40**, 105–154; **41**, 175–272.

KÜHNE, O. (1915), Der Tracheenverlauf im Flügel der Koleopteren-Nymphe, *Z. wiss. Zool.*, **112**, 602–718.

KUSCHEL, G. (1959), Nemonychidae, Belidae y Oxycorynidae de la fauna chilena, con algunas consideraciones biogeograficas, *Invest. nes. zool. chil.*, **5**, 229–271.

LAGOUTTE, S. (1966), Squelette et musculature céphaliques de la larve de *Cantharis rustica* Fall. (Col. Cantharidae), *Trav. Lab. Zool. Stn. Aquic. Grimaldi Dijon*, **70**, 1–29.

LAMEERE, A. (1900–03), Notes pour la classification des Coléoptères, *Ann. Soc. ent. Belg.*, **44**, 355–377; **47**, 155–165.

LARSÉN, O. (1966), On the morphology and function of the locomotor organs of the Gyrinidae and other Coleoptera, *Opusc. Ent. Suppl.*, **30**, 242 pp.

LAWRENCE, J. F. (1967), Biology of the parthenogenetic fungus beetle *Cis fuscipes* Mellié (Coleoptera: Ciidae), *Breviora*, **258**, 1–14.

—— (1971), Revision of the North American Ciidae (Col.), *Bull. Mus. comp. Zool.*, **142**, 419–522.

LECONTE, J. L. AND HORN, G. H. (1883), Classification of the Coleoptera of North America, *Smithson. misc. Coll.*, **26** no. 507, 567.

LENG, C. W. (1920), *Catalogue of the Coleoptera of America north of Mexico*, J. D. Sherman jr., Mount Vernon, N.Y.

LENGERKEN, H. von (1924–27), Coleoptera, *Biologie der Tiere Deutschlands*, **40**, 346 pp., Gebrüder Borntraeger, Berlin.

—— (1954), *Die Brutfürsorge- und Brutpflegeinstinkte der Käfer*, Akademische Verlaggesellschaft M. B. H., Leipzig. 2nd edn.

LINDROTH, C. H. (1961–69), The ground-beetles (Carabidae, excl. Cicindelinae) of Canada and Alaska, *Opusc. Ent. Suppl.*, **20, 24, 29, 33, 34, 35**: 1192.

LINSLEY, E. G. (1961–1972), The Cerambycidae of North America, *Univ. Calif. Publ. Ent.*, **18, 19, 20, 21, 22, 69**, 880 pp.

LINSLEY, E. G., EISNER, T. AND KLOTS, A. B. (1961), Mimetic assemblages of sibling species of Lycid beetles, *Evolution*, **15**, 15–29.

LISON, L. (1937), Étude histophysiologiques sur le tube de Malpighi des insectes. III, *Arch. Biol.*, **48**, 321–360; 489–512.

LLOYD, J. E. (1971), Bioluminescent communication in insects, *A. Rev. Ent.*, **16**, 97–122.

LOMBARDI, D. (1928), Contributo alla conoscenza della *Scirtes hemisphaericus* L. (Coleoptera-Helodidae), *Boll. Lab. Ist. sup. Ent. Bologna*, **1**, 236–258.

MACGILLIVRAY, A. D. (1903), Aquatic Chrysomelidae and a table of families of Coleopterous larvae, *N.Y. State Mus. Bull.*, **68**, 288–327.

MACSWAIN, J. W. (1956), A classification of the first instar larvae of the Meloidae (Coleoptera), *Calif. Publ. Ent.*, **12**, 1–182.

MADLE, H. (1934), Zur Kenntnis der Morphologie, Oekologie und Physiologie von *Aphodius rufipes* Linn. und einiger verwandten Arten, *Zool. Jb., Anat.*, **58**, 303–396.

MANFRINI DE BREWER, M. (1963), Contribucion al conocimento de las Ripiphoridae Argentinas (Coleoptera), *Op. lilloana*, **11**, 6–106.

MARCUZZI, G. AND RAMPAZZO, L. (1950), Contributo alla conoscenza delle forme larvali dei Tenebrionidi (Col. Heteromera), *Eos*, **36**, 63–117.

MARSHALL, G. A. K. (1916), *The fauna of British India. Coleoptera, Rhynchophora, Curculionidae*, Taylor & Francis, London.

MARSHALL, G. A. K. AND POULTON, E. B. (1902), Five years observations and experiments (1896–1901) on the bionomics of South African insects, etc., *Trans. ent. Soc. Lond.*, **1902**, 287–584.

MASSEE, A. M. (1954), *The Pests of Fruits and Hops*, 3rd edn., Crosby Lockwood & Son, Ltd., London.

MATTHEWS, A. (1872), *Trichopterygia illustrata et descripta*, E. W. Janson, London (supplement by Mason, P. B., 1900).

MAYET, V. (1875), Mémoire sur les moeurs et les métamorphoses d'une nouvelle espèce de Coléoptère de la famille des vésicants, le *Sitaris colletis*, *Ann. Soc. ent. France*, (5), **5**, 65–92.

MCINDOO, N. E. (1916), The reflex 'bleeding' of the Coccinellid beetle, *Epilachna borealis*, *Ann. ent. Soc. Amer.*, **9**, 201–223.

MCNEW, G. I. (1970), The Boyce Thompson Institute programme in forest entomology that led to the discovery of pheromones in barkbeetles (Col., Scolytidae), *Contr. Boyce Thompson Inst.*, **24**, 25–262.

MEINERT, F. (1888), *Scydmaenus*-larven, *Ent. Meddr*, **1**, 144–50.

MEIXNER, J. (1935), Coleopteroidea, In: Kükenthal, W. (ed.), *Handbuch der Zoologie*, **4**, Insecta 2, Lief. 3–5, Walter de Gruyter & Co., Berlin.

MIALL, L. C. (1912), *The Natural History of Aquatic Insects*, Macmillan & Co., London.

MINGAZZINI, P. (1889*a*), Ricerche sul canale digerente delle larva dei Lamellicorni fitofagi, *Mitth. Zool. Sta. Neapel*, **9**, 1–112.

—— (1889*b*), Ricerche sul tubo digerente dei Lamellicorni fitofagi, *Bull. soc. nat. Nap.*, **3**, 24–30.

MJÖBERG, E. (1925), The mystery of the so-called 'Trilobite larva' or 'Perty's larvae' definitely solved, *Psyche*, Cambridge, **32**, 119–157.

MOORE, B. P. AND WALLBANK, B. E. (1968), Chemical composition of the defensive secretion in Carabid beetles and its importance as a taxonomic character, *Proc. R. ent. Soc. Lond.*, (B) **37**, 62–72.

MOORE, I. (1964), A new key to the subfamilies of the Nearctic Staphylinidae and notes on their classification, *Coleopt. Bull.*, **18**, 83–91.

MORIMOTO, K. (1962), Comparative morphology, phylogeny and systematics of the superfamily Curculionoidea of Japan, I, *J. Fac. Kyushu Univ., Fukuoka*, **11**, 331–373.

MUIR, F. AND SHARP, D. (1904), On the egg-cases and early stages of some Cassididae, *Trans. ent. Soc. Lond.*, **1904**, 1–24.

NEWPORT, G. (1837), On the natural history of *Lampyris noctiluca*, *J. Proc. Linn. Soc., Lond.*, **1**, 40–71.

—— (1845–53), On the natural history, anatomy and development of the oil beetle (*Meloë*), etc., *Proc. Linn. Soc. Lond.*, **1**; *Trans. Linn. Soc. Lond.*, **20**, **21**, *passim*.

OCHS, G. (1969), Zur Ethökologie der Taumelkäfer (Col., Gyrinidae), *Arch. Hydrobiol. (Suppl.)*, **35**, 373–410.

ORCHYMONT, A. d' (1913), Contribution à l'étude des larves hydrophilides, *Ann. Biol. lacustre*, **6**, 173–214.

—— (1920), La nervation alaire des Coléoptères, *Ann. Soc. ent. France*, **89**, 1–50.

PARKS, J. T. AND BARNES, J. W. (1955), Notes on the family Leptinidae including a new record of *Leptinillus validus* Horn in North America (Coleoptera), *Ann. ent. Soc. Am.*, **48**, 417–421.

PAULIAN, R. (1941), Les premiers états des Staphylinoidea (Coleoptera). Étude de morphologie comparée, *Mém. Mus. Hist. nat. Paris*, (n.s.) **15**, 361 pp.

PAYNE, O. G. M. (1916), On the life-history and structure of *Telephorus lituratus* Fallén, *J. zool. Res.*, **1**, 1–32.

PEARCE, E. J. (1957), Coleoptera (Pselaphidae), *Hndbks. Ident. Brit. Ins.*, **4**, part 9, 32 pp., R. ent. Soc. Lond.

PELEG, B. AND NORRIS, D. M. (1973), Haploid versus diploid *Xyleborus ferrugineus* (Col. Scolytidae), *Ann. ent. Soc. Am.*, **66**, 180–183.

PERRIS, E. (1876–77), Larves de coléoptères, *Ann. Soc. Linn. Lyon*, **22**, 259–418; **23**, 1–430.

PEYERIMHOFF, P. (1933), Les larves des Coléoptères d'après A. G. Böving et F. C. Craighead et les grands criteriums de l'ordre, *Ann. Soc. ent. France*, **102**, 359–412.

POLL, M. (1932), Contribution à l'étude des tubes de Malpighi des Coléoptères, *Rec. Inst. Zool. Torley-Rousseau*, **4**, 47–80.

PORTA, A. (1903), La funzione epatica negli insetti, *Anat. Anz.*, **22**, 447–448; **24**, 97–111.

PORTIER, P. (1911), Recherches physiologiques sur les insectes aquatiques, *Arch. Zool. exp.*, (5), **8**, 89–379.

PRADHAN, S. (1939), Glands in the head capsule of Coccinellid beetles with a discussion on some aspects of gnathal glands, *J. Morph.*, **61**, 47–66.

PRINGLE, J. A. (1938), A contribution to our knowledge of *Micromalthus debilis* (Coleoptera), *Trans. R. ent. Soc. Lond.*, **87**, 271–286.

PRIORE, R. (1966), Anatomia ed istologia della *Rodolia cardinalis* Muls., *Boll. Lab. Ent. agr. Filippo Silvestri*, **24**, 247–316.

PUKOWSKI, E. (1933), Oekologische Untersuchungen an *Necrophorus* F., *Z. Morph. Oekol. Tiere*, **27**, 518–586.

RAFFRAY, A. (1908), Pselaphidae, *Genera Insectorum*, **64**, Desnet-Verteneuil, Brussels.

REES, B. E. (1943), Classification of the Dermestidae (larder, hide, and carpet beetles) based on larval characters, etc., *Misc. Publ. U.S. Dept. Agric.*, **511**, 18 pp.

REICHENBACH-KLINKE, H. H. (1953), Die Entwicklung des Proventrikels der Coleopteren mit besonderer Berücksichtigung der carnivoren Arten der Unterordnung Polyphaga (Coleoptera), *Ent. Blatt.*, **49**, 2–17.

REID, J. A. (1942), A note on *Leptinus testaceus* Müller (Coleoptera: Leptinidae), *Proc. R. ent. Soc. Lond.* (A) **17**, 35–37.

REID, W. R. (1958), Internal changes in the female Mountain Pine Beetle, *Dendroctonus monticolae* Hopk., associated with egg laying and flight, *Can. Ent.*, **90**, 464–468.

REITTER, E. (1908–16), *Fauna Germanica, Die Käfer*, K. G. Lutz Verlag, Stuttgart, 5 vols.

—— (1909), Coleoptera, In: von Brauer, *Süsswasserfauna Deutschlands*, Verlag Gustav Fischer, Jena.

RILEY, C. V. (1878), *United States entomological Commission. First annual report.* 477 + 294 pp.

RITCHER, P. O. (1958), Biology of Scarabaeidae, *A. Rev. Ent.*, **3**, 311–334.

—— (1966), White grubs and their allies. A study of North American scarabaeoid larvae, *Oreg. St. Monogr. Stud. Ent.*, **4**, 219 pp.

—— (1969), Spiracles of adult Scarabaeoidea (Coleoptera) and their phylogenetic significance. 1. The abdominal spiracles. 2. Thoracic spiracles and adjacent sclerites, *Ann. ent. Soc. Am.*, **62**, 869–880; 1388–1398.

RITCHIE, W. (1920), The structure, bionomics and economic importance of *Saperda carcharias* Linn., the large poplar longicorn, *Ann. appl. Biol.*, **7**, 299–343.

RITTERSHAUS, K. (1927), Studien zur Morphologie und Biologie von *Phyllopertha horticola* L. und *Anomala aenea* Geer (Coleopt.), *Z. Morph. Oekol. Tiere*, **8**, 271–408.

ROZEN, J. G. Jr. (1960), Phylogenetic systematic study of larval Oedemeridae (Coleoptera), *Misc. Publ. ent. Soc. Amer.*, **1**, 35–38.

RUPERTSBERGER, M. (1880), *Biologie der Käfer Europas*, The author, Linz.
—— (1894), *Die biologische Literatur über die Käfer Europas von 1880 an*, The author, Linz.
RYMER ROBERTS, A. W. (1919–22), On the life-history of 'wireworms' of the genus *Agriotes* Esch., etc., *Ann. appl. Biol.*, **6**, 116–134; **8**, 193–215; **9**, 306–324.
—— (1958), On the taxonomy of the Erotylidae (Coleoptera), with special reference to the morphological characters of the larvae. II, *Trans. R. ent. Soc. Lond.*, **110**, 245–285.
SCHENKLING, G. AND JUNK, W. (1910–40), *Coleopterorum catalogus*, **1–70**, W. Junk Verlag, Berlin. A second edition is still in print.
SCHERF, H. (1964), Die Entwicklungsstadien der mitteleuropäischen Curculioniden (Morphologie, Bionomie, Oekologie.), *Abh. senckenb. naturforsch. Ges.*, **506**, 1–335.
SCHILDER, M. (1949), Zahl und Verbreitung der Coleoptera, *Biol. Zbl.*, **68**, 385–397.
SCHILDKNECHT, H., WINKLER, H. AND MASCHWITZ, U. (1968), Vergleichend chemische Untersuchungen der Inhaltsstoffe der Pygidialwehrblasen von Carabiden, *Z. Naturf.*, **23** B, 637–644.
SCHIÖDTE, J. C. (1861–83), De metamorphosi eleutheratorum observationes, *Nat. Tidsskr.*, (3), **1–13**, *passim*.
SCHMIDT, G. (1935), Beiträge zur Biologie der Aphodiinae (Coleoptera, Scarabaeidae), *Stett. ent. Ztg.*, **96**, 293–350.
SCHMIDT, H. (1949), Biologische und morphologische Untersuchungen an Malachiiden. (Col. Malacodermata), *Ent. Bl.*, **41–44** (1945–48), 167–177.
SCHMITZ, G. (1943), Le labium et les structures buccopharyngiennes du genre *Stenus* Latreille, *La Cellule*, **49**, 291–334.
SCOTT, A. (1941), Reversal of sex-production in *Micromalthus*, *Biol. Bull.*, **81**, 420–431.
SEDLACZEK, W. (1902), Ueber den Darmkanal der Scolytiden, *CentrBl. ges. Forstwesen*, **28**, 23 pp.
SEEGER, W. (1971*a*), Autökologische Laboratoriumsuntersuchungen an Halipiden mit zoogeographischen Anmerkungen, (Haliplidae, Coleoptera). *Arch. Hydrobiol.*, **68**, 528–574.
—— (1971*b*), Morphologie, Bionomie und Ethologie von Halipliden, unter besonderer Berücksichtigung funktionsmorphologischer Gesichtspunkte (Haliplidae, Coleoptera), *Arch. Hydrobiol.*, **68**, 400–435.
SEEVERS, C. H. (1957), A monograph on the termitophilous Staphylinidae (Coleoptera), *Fieldiana*, **40**, 1–334.
—— (1965), The systematics, evolution and zoogeography of Staphylinid beetles associated with army ants (Coleoptera, Staphylinidae), *Fieldiana*, **47**, 137–351.
SEEVERS, C. H. AND DYBAS, H. S. (1943), A synopsis of the Limulodidae (Coleoptera), etc., *Ann. ent. Soc. Am.*, **36**, 546–586.
SELANDER, R. B. (1957), The systematic position of the genus *Nephrites* and the phylogenetic relationships of the higher groups of Rhipiphoridae (Coleoptera), *Ann. ent. Soc. Am.*, **50**, 88–103.
—— (1964), Sexual behavior in blister beetles (Coleoptera, Meloidae). I. The genus *Pyrota*, *Can. Ent.*, **96**, 1037–1082.
SEN GUPTA, T. AND CROWSON, R. A. (1969*a*), Further observations on the

family Boganiidae, with definition of two new families Cavognathidae and Phloeostichidae (Co.), *J. nat. Hist.*, 3, 571–590.

—— (1969*b*), On a new family of Clavicornia (Col.) and a new genus of Languriidae, *Proc. R. ent. Soc. Lond.*, (B), 38, 125–131.

—— (1973), A review of the classification of Cerylonidae (Coleoptera, Clavicornia), *Trans. R. ent. Soc. Lond.*, **124**, 365–446.

SHARP, D. (1882), On aquatic carnivorous Coleoptera or Dytiscidae, *Trans. Roy. Dublin Soc.*, (2), **2**, 179–1003.

SHARP, D. AND MUIR, F. (1912), The comparative anatomy of the male genital tube in Coleoptera, *Trans. ent. Soc. Lond.*, **1912**, 477–642.

SHELFORD, R. (1907), The larva of *Collyris emarginatus* Dej., *Trans. ent. Soc. Lond.*, **1907**, 83–90.

SHELFORD, V. E. (1909), Life-histories and larval habits of the Tiger beetles (Cicindelidae), *J. Linn. Soc. Zool.*, **30**, 157–184.

SIMANTON, F. L. (1915), *Hyperaspis binotata*, a predatory enemy of the Terrapin scale, *J. agric. Res.*, **6**, 197–203.

SNODGRASS, R. E. (1947), The insect cranium and the 'epicranial suture'. *Smithson. misc. Coll.*, **107** no. 7, 52 pp.

SNYDER, T. E. (1913), Record of the rearing of *Cupes concolor* Westw., *Proc. ent. Soc. Washington*, **15**, 30–31.

STAMM, R. H. (1936), A new find of *Rhipidius pectinicornis*. Thbg. (*Symbius blattarum* Sund.) (Col. Rhipiphor.), *Ent. Meddr*, **19**, 286–297.

STAMMER, H. J. (1934), Bau und Bedeutung der malpighischen Gefässe der Coleopteren, *Z. Morph. Oekol. Tiere*, **29**, 196–217.

STEFFAN, A. W. (1961), Vergleichend-mikromorphologische Genitaluntersuchungen zur Klärung der phylogenetischen Verwandtschaftsverhältnisse mitteleuropäischer Dryopoidea (Coleoptera), *Zool. Jb.* (*Syst.*), 88, 255–356.

—— (1964), Torridincolidae, coleopterorum nova familia e regione ethiopica, *Ent. Z.*, **74**, 193–200.

STEIN, F. (1847), *Ueber die Geschlechtsorgane und den Bau des Hinterleibes bei den weiblichen Käfern*, Diencker & Humblot, Berlin.

STICKNEY, F. S. (1923), The head-capsule of Coleoptera, *Illinois biol. Monogr.*, 8,1–104.

SUOMALAINEN, E. (1969), Evolution in parthenogenetic Curculionidae, In: Dobzansky, T., Hecht, M. K. and Steere, W. C. (eds.), *Evolutionary Biology*, 3, 261–296, Meredith Corp., New York.

SUZUKI, K. (1969), Comparative morphology and evolution of the hind-wings of the family Chrysomelidae (Coleoptera). I. Homology and nomenclature of wing-venation in relation to allied families, *Kontyû*, **37**, 32–40.

TANNER, V. M. (1927), A preliminary study of the genitalia of female Coleoptera, *Trans. Amer. ent. Soc.*, **53**, 5–50.

TASSELL, E. R. van (1965), An audiospectrographic study of stridulation as an isolating mechanism in the genus *Berosus* (Coleoptera: Hydrophilidae), *Ann. ent. Soc. Am.*, **58**, 408–413.

TEICHERT, M. (1955–56), Biologie und Brutfürsorgemassnahmen von *Geotrupes mutator* Marsh. und *Geotrupes stercorarius* L. (Col. Scarab.). Nahrungsspeicherung von *Geotrupes vernalis* L. und *Geotrupes stercorosus* Scriba (Coleopt. Scarab.), *Wiss. Zs. Univ. Halle Math. Nat.*, **52**, 187–218; **54**, 669–672.

THÉRY, A. (1942), Coléoptères Buprestides, *Faune de France*, **41**, Paul Lechevalier, Paris.

THIEL, H. (1936), Vergleichende Untersuchungen an den Vormägen von Käfern, *Z. wiss. Zool.*, **147**, 395–432.

TIEMANN, D. L. (1967), Observations on the natural history of the western banded glowworm *Zarhipis integripennis* (Le Conte) (Coleoptera: Phengodidae), *Proc. Calif. Acad. Sci.*, **35**, 235–264.

TOWER, W. L. (1906), An investigation of evolution in Chrysomelid beetles of the genus *Leptinotarsa, Carneg. Inst. Publ.*, **48**, 320 pp.

TRÄGÅRDH, I. (1910), Contribution towards the metamorphosis and biology of *Orchestes*, etc., *Arch. Zool.*, **6**, no. 7, 25 pp.

UTIDA, S. (1972), Density-dependent polymorphism in the adult of *Callosobruchus maculatus* (Col., Bruchidae), *J. Stored Prod. Res.*, **8**, 111–126.

VERHOEFF, K. W. (1893), Vergleichende Untersuchungen über die Abdominalsegmente und die Copulationorgane der männlichen Coleoptera, *D. ent. Z.*, **1893**, 113–170.

—— (1894), Vergleichende Morphologie des Abdomens der männlichen und weiblichen Lampyriden, etc., *Arch. Naturg.*, **60**, 129–210.

—— (1923), Beiträge zur Kenntnis der Coleopteren-Larven mit besonderer Berücksichtigung der Clavicornia, *Arch. Naturg.*, (A), **89**, Heft 1, 1–109.

VIEDMA, M. G. DE (1966), Contribución al conocimento de las larvas de Melandryidae en Europa (Coleoptera), *Eos*, **41** (1965), 483–506.

VILLIERS, A. (1943), Étude morphologique et biologique des Languriidae, *Publ. Mus. Hist. nat. Paris*, **6**, 1–98.

VITÉ, J. P. AND WILLIAMSON, D. L. (1970), *Thanasimus dubius*: prey-perception (Col., Cleridae), *J. Insect Physiol.*, **16**, 233–239.

VOGEL, B. (1915), Beiträge zur Kenntnis des Baues und der Lebensweise der Larve von *Lampyris noctiluca, Z. wiss. Zool.*, **112**, 291–432.

WADSWORTH, J. T. (1915), On the life-history of *Aleochara bilineata* Gyll., a staphylinid parasite of *Chortophila brassicae* Bouché, *J. econ. Biol.*, **10**, 1–27.

WASMANN, E. (1910), Zur Kenntnis der Gattung *Pleuropterus* und anderer Paussiden, *Ann. Soc. ent. Belg.*, **54**, 392–402.

WATT, J. C. (1970), Coleoptera Perimylopidae of South Georgia, *Pacific Ins. Monogr.*, **23**, 243–253.

WEINREICH, E. (1968), Ueber den Klebfangapparat der Imagines von *Stenus* Latr., (Coleopt. Staphylinidae) mit einem Beitrag zur Kenntnis der Jugendstadien dieser Gattung, *Z. Morph. Oekol. Tiere*, **62**, 162–210.

WENZEL, A. L. (1944), On the classification of the Histerid beetles, *Publ. Field Mus. Chicago*, **28** (Zool. ser.), 51–151.

WESENBERG-LUND, C. (1913), Biologische Studien über Dytisciden, *Internat. Rev. Hydrobiol.* (5) **1**, 1–129.

WESTWOOD, J. O. (1839), *Introduction to the Modern Classification of Insects*, Vol. I, Longman, Brown, Orme, Green and Longmans, London.

WHEELER, W. M. (1921), A study of some social beetles in British Guiana and of their relations to the ant-plant, *Tachigalia, Zoologica, N.Y.*, **3**, 35–134.

WILLIAMS, C. B. (1914), *Phytodecta viminalis*, a viviparous British beetle, *Entomologist*, **47**, 249–250.

WILLIAMS, F. X. (1916), Photogenic organs and embryology of Lampyrids, *J. Morph.*, **28**, 145–186.

—— (1917), Notes on the life-history of some North American Lampyridae, *J. N.Y. ent. Soc.*, **25**, 11–33.

WILLIAMS, I. W. (1938), The comparative morphology of the mouthparts of the order Coleoptera treated from the standpoint of phylogeny, *J. N.Y. ent. Soc.*, **46**, 245–288.

WILLIAMS, J. L. (1945), The anatomy of the internal genitalia of some Coleoptera, *Proc. ent. Soc. Washington*, **47**, 73–87.

WINKLER, A. (1924–32), *Catalogus Coleopterorum regionis palaearcticae*, The author, Vienna.

WOOD, D. M. (1964), Studies on the beetles *Leptinillus validus* (Horn) and *Platypsyllus castoris* Ritsema (Coleoptera: Leptinidae) from Beaver, *Proc. ent. Soc. Ontario*, **95**, 33–63.

—— (1973), Selection and colonization of Ponderosa pine by bark beetles. Insect/Plant relationships, *Sympos. R. ent. Soc. Lond.*, **6**, 101–117.

WOODS, W. C. (1916), The Malpighian vessels of *Haltica bimarginata* Say. (Coleoptera), *Ann. ent. Soc. Am.*, **9**, 391–406.

—— (1918), The alimentary canal of the larva of *Altica bimarginata* Say, *Ann. ent. Soc. Am.*, **11**, 283–313.

XAMBEU, V. (1892–93), Moeurs et métamorphoses d'insectes, *Rev. d'Ent.*, **11** and **12**, *passim*.

Order 23

STREPSIPTERA (STYLOPIDS)

Small, endoparasitic insects: males free-living, mouthparts of a degenerate biting type, antennae conspicuous and flabellate, fore wings reduced to small club-like structures, metathorax greatly developed, hind wings large and fan-shaped, trochanters absent. Pupae exarate, adecticous. Female normally remaining in host, enclosed in a puparium, sexual openings segmental, 3 to 5 in number: a few females leave the hosts and have a larviform structure with terminal gonopore; larval development hypermetamorphic.

The order Strepsiptera comprises about 370 species, listed by Kinzelbach (1971*b*), of very anomalous insects whose larvae are endoparasitic. The majority of females remain all their lives in a puparium which protrudes slightly from the body of the host. In common parlance the adults are termed 'stylops' and an insect harbouring these parasites is said to be 'stylopized'. Their hosts consist principally of members of the Homoptera Auchenorrhyncha and superfamilies Vespoidea, Sphecoidea and Apoidea among the Hymenoptera. More recently there have also been records of parasitized Diptera (Philip, 1950, *Chrysops*, Tabanidae; Platystomatidae, Riek, 1970). Among the most extensively parasitized hosts are species of some of the common genera of the Homopterous Delphacidae and the genera *Polistes*, *Halictus* and *Andrena* among Hymenoptera. The last-mentioned genus is more often attacked than any other and it includes a very long list of parasitized species. The majority of writers have included the Strepsiptera among the Coleoptera, placing them near the heteromerous families Meloidae and Rhipiphoridae, mainly on account of similarities in the larvae and metamorphosis. The resemblances to these families with specialized life-histories may well be due to convergent evolution and probably do not indicate close relationship. The characters of the Strepsiptera are so peculiar that it seems best to treat them as a separate order. They may perhaps have been derived from the same general stock as the Coleoptera but such a view is largely speculative (Pierce, 1909; 1936, 1964; Bohart, 1941; Jeannel, 1945; Crowson, 1955). Kinzelbach (1971*a*) considers the relationships of the order are quite uncertain and that the resemblances to the Coleoptera may be convergent.

About 370 species are known, the majority from the Holarctic region, though the order is also represented in all zoogeographical regions including

New Zealand (Gourlay, 1953): a synopsis of British species of *Stylops* and *Halictoxenos* is given by Perkins (1918). The classification at the generic and specific level has been much confused by the too easy assumption that each parasite is confined to a single host species, which is certainly untrue (cf. Hassan, 1939). All members of the order are small or minute, the males commonly measuring about 1·5–4·0 mm in length. In colour they are either black, or some shade of brown, and the protruding female puparium is usually yellowish-brown.

External Anatomy (Fig. 427)

In the males the integument is very thin and in many parts transparent. The head is transverse, the compound eyes with a rather larva-like structure

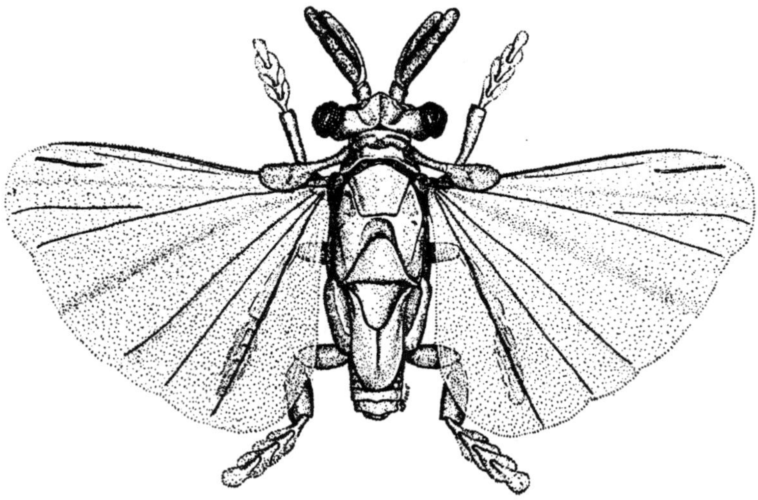

FIG. 427 *Stylops shannoni*, N. America: male, enlarged
After Pierce, *Proc. U.S. Nat. Mus.*, 54.

(Strohm, 1910; Wachmann, 1972) are very protuberant, and there are no ocelli, at least externally. The antennae are 4- to 7-segmented, but are variable in form and of peculiar structure: the 3rd segment is flabellate, giving the antennae a bifurcate appearance, and the succeeding segments may also be similarly produced, the antennae then appearing as if branched. The surface of the segments is studded with complex sensory organs. The mouthparts exhibit modification and great reduction from the normal biting type. Only vestiges of parts corresponding with the labrum and labium are recognizable; the mandibles are usually narrow and sickle-like or, more rarely, they are almost bristle-like, while a pair of 2-segmented palps alone represent the maxillae. In the thorax the first two segments are greatly reduced, but the metathorax is very large, occupying at least half the length of the body. The legs are chiefly used for clinging to the female during copulation: the tarsi are ordinarily 2- to 4-segmented, without claws (except *Mengenilla* and Mengeidae), and usually each segment is provided with a ventral adhesive pad. The anterior wings are represented by small club-like processes which function like the halteres of Diptera (Ulrich, 1930), but the hind wings are relatively large and fan-shaped, with radiating veins. The venation is degenerate; in the most generalized forms eight simple longitudinal veins are recognizable but their homologies in relation to the pupal wing-tracheation have not been determined, and there are no cross-veins. The abdomen is 10-segmented: an aedeagus is located on the 9th sternum: cerci are absent.

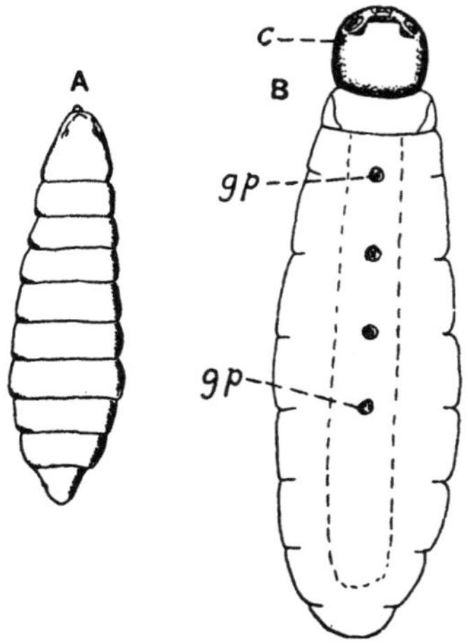

FIG. 428
Xenos vesparum (Stylopidae). A, fully-grown male larva. B, adult female

c, cephalothorax; *gp*, genital pores. Adapted from von Siebold.

The female (Fig. 428) is highly modified through degeneration in accordance with a permanently endoparasitic life. She is larviform, apodous, and enclosed in the persistent larval cuticle. The head and thorax are adnate, forming a cephalothorax which is separated by a constriction from the long sac-like abdomen. Antennae and eyes are wanting and the mouthparts are vestigial: mandibles are always present. The thorax is separated ventrally from the head by the aperture of the brood canal, which is a passage between the body of the female and the last larval cuticle, leading from the genital apertures to the exterior. In the Mengenillidae, however, the female is free-living and the head and thorax are distinct. The eyes, antennae and legs are functional and copulation and emergence of the first-stage larvae takes place through a single posterior orifice.

Internal Anatomy

Our knowledge of the internal anatomy of Strepsiptera is mainly due to Nassonow whose work forms the basis of the account given by Pierce (1909).

The alimentary canal is an unconvoluted tube of simple structure. In the male it exhibits three well-marked regions – the fore, middle and hind intestine – but there is no communication between the two latter parts of the gut: in the adult female the hind intestine has atrophied, the posterior end of the stomach being in contact with the integument of the last abdominal segment, there being no anal opening. The Malpighian tubes are absent. The nervous system is highly concentrated in both sexes: in the male the brain assumes much larger proportions than in the female, owing to the presence of the antennary and visual centres. The para-oesophageal connectives pass to a common ganglionic mass formed by the union of all the ventral ganglia up to, and including, the ganglia of the 2nd or 3rd abdominal segment: a median abdominal nerve-cord terminates in a nervous centre formed by the coalescence of the posterior ganglia. The tracheal system opens to the exterior by one or two pairs of thoracic spiracles and, in the male and in female Mengeoids, up to eight pairs of abdominal spiracles. The reproductive system is very similar in the larvae of both sexes and consists of a pair of tubes lying one on either side of the gut. In the adult male, these organs maintain their paired structure, and communicate with the exterior by means of a common duct. In the female, the reproductive organs are stated to disintegrate, and the egg-masses are scattered through the body space. Cuticular invaginations, which develop into funnel-like tubes, function as genital ducts. The number of these apertures normally varies from two to five, but there are two rows of 3–8 pores in *Deinelenchus* and three transverse rows of 12–14 pores in *Stichotrema*: they are segmentally disposed on the median ventral region of the 2nd and following abdominal segments. According to Noskiewicz and Poluszyński (1935) polyembryony occurs in *Halictoxenos*.

Biology and Host Relations

The biology of these insects has been mainly studied with reference to species parasitizing Hymenoptera. The most complete study of the life-history of any species is the account given by Nassonow (1892) (in Russian) of *Xenos vesparum*. Male Strepsiptera are free-living, and usually only survive a few hours after emerging from their hosts. The females, on the other hand, remain permanently endoparasitic, and only the cephalothorax is visible externally, where it protrudes through the body-wall of the wasp or bee. The males are by no means rare insects, but their small size and brief life cause them to elude the observation of most entomologists. They emerge from their hosts early in the morning, very soon after the latter have taken to the wing. Pairing takes place by the male alighting on the host, and inserting the aedeagus into the aperture of the brood canal of the female (Perkins); in some cases it appears probable that the eggs are able to develop parthenogenetically. The larvae hatch within the body of the female and issue in large numbers (sometimes several thousand) through the genital canals previously alluded to. They pass into the space (or brood pouch) between the ventral surface of the parent and the persistent larval cuticle, ultimately emerging through the aperture of the brood canal. They then remain upon the body of the host until opportunity is afforded for escape. At this stage the young larvae bear a resemblance to the triungulins of *Meloe*: they are very minute, active creatures with a group of ocelli and well-formed legs (though without trochanters), and pairs of long caudal setae (Fig. 429).

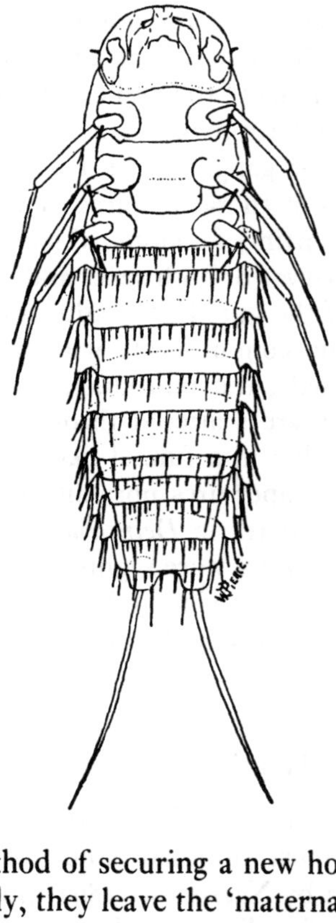

FIG. 429
Stylops californica, triungulin, ventral view, enlarged
After Pierce, *loc. cit.*

Their method of securing a new host has rarely been directly observed but, presumably, they leave the 'maternal' hosts when the latter are on flowers, in the nest, or in other situations. If liberated on to flowers, they probably attach themselves to other host individuals that come along, and become transported thereby to the nests. Within the latter, they seek out the larvae, and speedily burrow through the body-wall and become endoparasitic. In *Corioxenos*, however, the triungulins jump by means of their caudal bristles on to *Antestia* (Pentatomidae) to which they are apparently attracted by colour (Kirkpatrick, 1937). Having entered the host, the stylopid larvae undergo ecdysis, assuming an apodous maggot-like form in the second instar (Fig. 428). Their subsequent history has been followed by Nassonow in the case of *Xenos* and Lauterbach (1954) in *Stylops*. Nutrition appears to take place by the filtration of the host's blood through the delicate cuticle of the parasite. The parasite does not penetrate the organs of the host, but occupies the body space between them, pushing them out of position. At the 7th instar the parasitic larva works its way outwards, so as to protrude from the abdomen of the host and, at this stage, the wasp or bee has assumed the pupal condition. In the case of *Stylops* protrusion usually takes place through the intersegmental membrane between the 4th and 5th abdominal segments. The male parasite now undergoes pupation and the pupa is enclosed in the exuviae of the two preceding instars. The rounded tuberculated apex of the puparium, thus formed, is the only region visible externally and the winged insect emerges by pushing open an operculum. The female parasite is recognizable by the flattened disc-like cephalothorax, the large white grub-like after-body remaining with the abdomen of the host.

Both sexes of the host are liable to parasitization but, in most cases, the largest number of attacked examples are females. As a rule, male parasites are the commoner, but both sexes may occur in the same individual host. The latter often nourishes several parasites: Pierce (1909) mentions the exceptional number of 31 larvae in a single host and states that the largest number of male puparia found exserted from the body of a host is 15. The effects of stylopization on the hosts have been studied by Pérez (1886), Smith and Hamm (1914), Perkins (1918) and others with reference to *Andrena* and by Salt (1931) in Vespoids; the whole subject is also discussed by Pierce and by Wheeler. In the first-mentioned genus, in some species stylopized examples often exhibit a shorter and more globular abdomen with increased pilosity, the head is usually smaller than in normal specimens, while the punctation of the body becomes finer, but different individuals do not necessarily react similarly to the presence of the parasite. These changes are common to both sexes and affect the specific characters. Much confusion has consequently arisen through the founding of new species on stylopized individuals. The following changes affect the secondary sexual characters. (1) Parasitized females have the pollen-collecting apparatus so diminished that the hind legs resemble those of the males. (2) When the clypeus or frons in the males is normally marked with a greater amount of yellow than in the females, stylopization may result in the females acquiring the yellow coloration of the males, and individuals of the latter sex having the light colour very markedly diminished. (3) The sting is curtailed in size and the copulatory apparatus of the male suffers reduction. Certain other minor changes may also occur and Pérez concludes that, in the case of *Andrena*, secondary sexual modifications induced by stylopization are inversions of development, and that parasitized examples are not merely diminished individuals, but that the female acquires certain characters belonging to the male and the male develops certain of those which pertain to the female. In *Andrena vaga*, however, there is only parasitic castration, especially by the female parasite (Brandenburg, 1953). Useful accounts of the relationships between *Andrena* and *Stylops* are Linsley and MacSwain (1957) and Borchert (1963).

Before maturity the parasites live on the fat-body and blood-tissues of the hosts. As already mentioned, they do not directly attack the other organs but the latter undergo partial atrophy through inadequate nutrition. The gonads become more or less reduced in size, and the oocytes degenerate in their follicles. There is no evidence that the females are ever fertile, but the males are known to be capable of producing spermatozoa, and parasitized examples of both sexes of *Andrena* have been taken *in copula*.

The effects of parasitism by *Elenchus* on Delphacidae are discussed by Hassan (1939) who shows that they are approximately proportional to the time which the parasite spends within the host. This varies with the season and is greatest in those hosts in which *Elenchus* overwinters. Here, as in other Strepsiptera, the effect of a male parasite is greater than that of a female.

According to Ogloblin (1939) and Luna de Carvalho (1959), the family Myrmecolacidae has an extraordinary life-history. The genus *Myrmecolax* was long known from a single male bred from an ant in Ceylon about 100 years ago. Pierce (1909) described one more Mexican species from four males. Ogloblin pointed out that the family Stichotrematidae was known only from females; one species was a parasite of a Tettigoniid in New Guinea and he himself found several more as parasites of Gryllids and Mantids in Brazil. In the same region he found several Myrmecolacids, all males, attacking different genera of ants. The exuviae of first-stage larvae in the ants are of the same type as that stage in *Stichotrema*, and Ogloblin suggests that in the Myrmecolacids the males develop in ants, the females in orthopteroids. Bohart (1951) described ten species of males from a collection of about 300 caught at light in the Philippine Islands. The Mengenillidae, which Kinzelbach puts in a separate suborder, are the only Strepsiptera in which the adult female is free living. The early stages are now known to develop in Lepismatids (Carpentier, 1939; Silvestri, 1942). The puparia are found under stones and both sexes exhibit many generalized features.

Classification – The Strepsiptera may be divided into 9 families arranged in 5 superfamilies (Kinzelbach, 1971*b*). Seventeen species of three families are known in Britain.

The **Mengenilloidea** have free-living females with a single posterior genital opening and in the male a strong vein MA_1. **Mengeoidea,** known only from a single fossil male in the Baltic Amber, lack this vein but like the previous superfamily has five tarsal segments. The **Corioxenoidea** seem to be a connecting link between these primitive forms and the higher families. The few recorded hosts are Hemiptera. The **Halictophagoidea** have the stipes fused to the head-capsule and only 3 tarsal segments. They are mostly parasites of Auchenorrhyncha but *Tridactylophagus* lives in an Orthopteroid. The **Bohartilloidea** are known only from a male caught in Honduras. The antennae have antennal segments 5–6 flabellate, the mandibles are short, the base of the maxilla swollen and there is no vein MA_1. The **Stylopoidea** include the **ELENCHIDAE** (parasites mostly of Delphacids), **MYRMECOCOLACIDAE** (with ♂ attacking ants and ♀ Orthopteroids) and the **STYLOPIDAE** (parasites of Aculeate Hymenoptera). They all lack flabellae on antennal segments 5–6 (if they have so many segments), the mandibles are very long, the base of the maxilla is not swollen and there are at least traces of vein MA_1. The British species fall into *Halictophagus*, *Elenchus*, *Stylops* (in *Andrena*) and *Halictoxenos* (in *Halictus*).

Literature on the Strepsiptera

BAUMERT, D. (1958), Mehrjährige Zuchten einheimischer Strepsipteren und Homopteren. 1. Hälfte. Larven und Puppen von *Elenchus tenuicornis* Kirby. 2. Hälfte. Imagines, Lebenszyklus und Artbestimmung von *Elenchus tenuicornis* Kirby (ferner Zusammenfassung und Literatur zur Gesamtabteilung), *Zool. Beitr.*, (N.S.), 3, 365–421.

BOHART, R. M. (1941), A revision of the Strepsiptera with special reference to the species of North America, *Univ. Calif. Publ. Ent.*, 7, 91–160.

—— (1951), The Myrmecolacidae of the Philippines (Strepsiptera), *Wasmann J. Biol.*, 9, 83–103.

BORCHERT, H. M. (1963), Vergleichend morphologische Untersuchungen an Berliner *Stylops* (Strepsipt.) zwecks Entscheidung der beiden Spezifitätsfragen. 1. Gibt es an unseren Frühjahres-Andrenen (Hymenopt, Apidae) mehrere *Stylops*-Arten und 2. Gibt es Wirtsspezifitäten?, *Zool. Beitr.*, (N.S.), 8, 331–445.

BRANDENBURG, J. (1953), Der Parasitismus der Gattung *Stylops* and der Sandbiene *Andrena vaga* Pz., *Z. Parasitenk.*, 15, 457–475.

CARPENTIER, F. (1939), Sur le parasitisme de la deuxième forme larvaire d'*Eoxenos laboulbenei* Peyer., *Bull. Ann. Soc. ent. Belg.*, 79, 451–468.

CROWSON, R. A. (1955), *The Natural Classification of the families of Coleoptera*, Nathaniel Lloyd & Co., London.

GOURLAY, E. S. (1953), The Strepsiptera, an insect order new to New Zealand, *N.Z. Entomologist*, 1, 3–8.

HASSAN, A. I. (1939), The biology of some British Delphacidae (Homopt.) and their parasites with special reference to the Strepsiptera, *Trans. R. ent. Soc. Lond.*, 89, 345–384.

HOFENEDER, K. AND FULMEK, L. (1942–43), Verzeichnis der Strepsiptera und ihre Wirte, *Arb. phys. angew. Ent.*, 9, 179–185, 249–283; 10, 32–58, 139–169, 196, 230.

JEANNEL, R. (1945), Sur la position systématique des Strepsiptères, *Rev. franç. Ent.*, 11, 111–118.

KINZELBACH, R. K. (1971*a*), Strepsiptera (Fächerflügler), In: Kükenthal, W. (ed.), *Handbuch der Zoologie*, IV. Arthropoda. 2. Insecta, Walter de Gruyter & Co., Berlin.

—— (1971*b*), Morphologische Befunde an Fächerflüglern und ihre phylogenetische Bedeutung (Insecta: Strepsiptera), *Zoologica Stuttg.*, 119, Pt. I, 1–128; Pt. 2, 129–256.

KIRKPATRICK, T. W. (1937), Colour vision in the triungulin larva of a Strepsipteron (*Corioxenos antestiae* Blair), *Proc. R. ent. Soc. Lond.*, (A), 12, 40–44.

LAUTERBACH, G. (1954), Begattung und Larvengeburt bei den Strepsipteren. Zugleich ein Beitrag zur Anatomie der *Stylops*-Weibchen, *Z. Parasitenk.*, 16, 255–297.

LINSLEY, E. G. AND MacSWAIN, J. W. (1957), Observations on the habits of *Stylops pacifica* Bohart (Coleoptera: Stylopidae), *Publ. Univ. Calif. Ent.*, 11, 395–430.

LUNA DE CARVALHO, E. (1959), Segunda contribuição para o estudo dos Estrepsipteros angolenses. (Insecta, Strepsiptera), *Publ. cult. Comp. Diam. Angola, Lisbon,* 41, 125–154.

NASSONOW, N. V.* (1892), On the metamorphosis of the Strepsiptera, *Warsaw Univ. News*, 1892, 1–36. (In Russian.)

—— (1893), On the morphology of *Stylops melittae*, *Warsaw Univ. News*, 1893, 1–30. (In Russian.)

NOSKIEWCZ, J. AND POLUSZYŃSKI, G. (1935), Embryologische Untersuchungen an Strepsipteren. II. Teil. Polyembryonie, *Zool. polon. Lwow*, 1, 53–94.

OGLOBLIN, A. A. (1939), The Strepsipterous parasites of ants, *Verh. VII Kongr. Ent.*, 2, 1277–1284.

PARKER, H. L. AND SMITH, H. D. (1933), Additional notes on the Strepsipteron *Eoxenos laboulbenei* Peyerimhoff, *Ann. ent. Soc. Am.*, 26, 217–233.

—— (1934), Further notes on *Eoxenos laboulbenei* Peyerimhoff with a description of the male, *Ann. ent. Soc. Am.*, 27, 468–479.

PÉREZ, J. (1886), Des effets du parasitisme des *Stylops* sur les apiaires du genre *Andrena, Act. Soc. Linn. Bordeaux*, 40, 21–60.

PERKINS, R. C. L. (1905), Leaf-hoppers and their natural enemies. Pt. iii. Stylopidae, *Bull. Hawaiian Sugar Planters' Assoc. Exp. Sta.*, 1, 90–111.

—— (1918*a*), Synopsis of British Strepsiptera of the genera *Stylops* and *Halictoxenus, Entomologist's mon. Mag.*, 54, 67–72.

—— (1918*b*), Further notes on *Stylops* and stylopized bees, *Entomologist's mon. Mag.*, 54, 115–128.

—— (1918*c*), On the assembling and pairing of *Stylops, Entomologist's mon. Mag.*, 54, 129–131.

PHILIP, C. B. (1950), New North American Tabanidae (Diptera). Parts I–III, *Ann. ent. Soc. Am.*, 42 (1949), 451–460.

PIERCE, W. D. (1909), A monographic revision of the twisted-winged insects comprising the order Strepsiptera Kirby, *Bull. U.S. nat. Mus.*, 66, xii + 232 pp.

—— (1936), The position of the Strepsiptera in the classification of insects, *Ent. News*, 42, 257–263.

—— (1964), The Strepsiptera are a true order, unrelated to the Coleoptera, *Ann. ent. Soc. Am.*, 57, 603–605.

RIEK, E. F. (1970), Strepsiptera, In: Mackerras, I. (ed.), *The Insects of Australia*, Chapter 31, C.S.I.R.O., Divn. Entomology, Melbourne.

SALT, G. (1927), The effects of stylopization on aculeate Hymenoptera, *J. exp. Zool.*, 48, 223–330.

—— (1931), A further study of the effects of stylopization on wasps, *J. exp. Zool.*, 59, 133–164.

SILVESTRI, F. (1942), Nuove osservazioni sulla *Mengenilla parvula* Silv. (Insecta Strepsiptera), *Acta pontif. Acad. Sci. Roma*, 6, 95–96.

SMITH, G. AND HAMM, A. H. (1914), Studies in the experimental analysis of sex. Pt. ii. On *Stylops* and stylopization, *Quart. J. micr. Sci.*, 60, 436–461.

* Nassonow's work was translated into German by Sipiagin and edited by Hofeneder in 1910. Untersuchungen zur Naturgeschichte der Strepsipteren von N. V. Nassonow. *Ber. Naturw.-med. Ver. Innsbruck*, 33, 206 pp.

STROHM, K. (1910), Die zusammengesetzten Augen der Männchen von *Xenos rossii*, *Zool. Anz.*, **36**, 156–159.

ULRICH, W. (1927), Strepsiptera, *Biologie d. Tiere Deutschlands*, **41**, 1–103, Gebrüder Borntraeger, Berlin.

—— (1930), Die Strepsipteren-Männchen als Insekten mit Halteren an Stelle der Vorderflügel, *Z. Morph. Oekol. Tiere*, **17**, 552–624.

—— (1943), Die Mengeiden (Mengenillini) und die Phylogenie des Strepsipteren, *Z. Parasitenk.*, **13**, 62–101.

WACHMANN, E. (1972), Zum Feinbau des Komplexauges von *Stylops* spec. (Insecta, Strepsiptera), *Z. Zellforsch.*, **123**, 411–424.

WHEELER, W. M. (1910), The effects of parasitic and other kinds of castration in insects, *J. exp. Zool.*, **8**, 377–438.

Order 24

MECOPTERA (SCORPION FLIES)

Slender, moderate or small-sized insects with elongate, filiform antennae, head usually produced into a vertically deflected rostrum, with biting mouthparts: ligula wanting. Legs long and slender. Wings similar and membranous, carried longitudinally and horizontally in repose: venation primitive, Rs dichotomously branched, Cu_1 *simple. Abdomen elongate with short cerci, male genitalia prominent. Larvae eruciform with biting mouthparts and three pairs of thoracic legs: abdominal feet present or absent. Pupae exarate, decticous: wings with reduced tracheation.*

This small order comprises less than 400 species, the greater number of which belong to the genera *Panorpa* and *Bittacus*. The majority of the members of the group are easily recognized by the beak-like prolongation of the front of the head, and their often maculated wings. The 'scorpion flies' (*sen. str.*) belong to the Panorpidae, which include many species widely spread over the northern hemisphere (Fig. 430). Their vernacular name is due to the fact that the males carry the terminal segment of the abdomen upwardly curved, somewhat after the manner of scorpions. The Bittacidae are very slender *Tipula*-like insects with prehensile tarsi: they are found in most parts of the world excepting the northern portion of the Holarctic region. The Boreidae are characterized by their vestigial wings and occur in Europe and N. America. The order is represented in the British Isles by three species of *Panorpa* and a single species of *Boreus* (Hobby and Killington, 1934).

The Mecoptera are essentially terrestrial insects undergoing their transformations in the soil: an exception is found in *Nannochorista*, of which the larva is aquatic. Both their larvae and imagines may be carnivorous, but the extent to which the Panorpidae prey upon living uninjured insects or other animals is doubtful. Brauer and Felt have reared larvae of *Panorpa* upon fragments of meat, but Miyake found wounded or dead insects more acceptable. *Panorpodes* and *Brachypanorpa* seem to be phytophagous (Byers, 1965). The adults are mostly found in shaded situations where there is a growth of

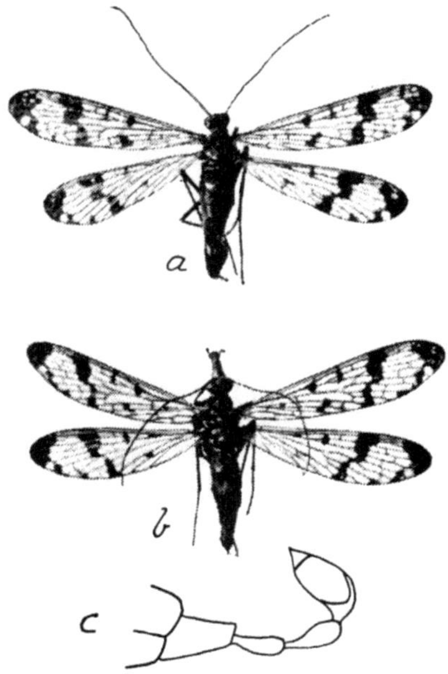

FIG. 430
Panorpa communis. A, male; B, female (from Photos by W. J. Lucas); C, apex of abdomen of male
After MacLachlan.

rank herbage. *Bittacus* rests suspended from grasses or twigs by its fore legs, and preys upon small Diptera, seizing them by means of its raptorial tarsi. *Boreus* lives among moss or beneath stones in autumn and early winter, appearing occasionally on the surface of snow; it feeds upon vegetable matter.

External Anatomy (Fig. 431)

The anterior region of the head (Heddergott, 1938; Hepburn, 1969*b*; Mickoleit, 1971*a*) is usually prolonged into a rostrum which is formed by the elongation of parts of the head-capsule together with the clypeus, labrum and maxillae. The compound eyes are well developed and there are usually three ocelli. The antennae are more or less filiform and many-jointed, there being about 40–50 segments in *Panorpa*, and about 16–20 in *Bittacus*. The mandibles are slender and elongate: they are only dentate at their apices, each bearing from 1 to 3 sharp teeth. The maxillae are complete: their palps are 5-segmented, and the galeae and laciniae are hairy lobes of somewhat complex structure (Miyake, 1912). The labium consists of an elongate submentum, not always clearly differentiated from the short mentum: the prementum exhibits traces of a bilobed structure, but the ligula has disappeared. The labial palps are 1- to 3-segmented; in some cases they are in the form of fleshy lobes in which, according to Crampton, traces of pseudotracheae may be present, resembling those found in the labium of Diptera. The mouthparts of *Nannochorista* are considerably specialized (Tillyard, 1917; Imms, 1944; Hepburn, 1969*b*). The labrum forms a sharply projecting process, the mandibles are vestigial, and the labial palpi (paraglossae of

Tillyard) are partially fused at their bases. This genus, which is accorded separate family rank by Tillyard, exhibits a tendency towards the development of suctorial mouthparts and foreshadows the condition found in the lower Diptera.

The prothorax is very small except in *Notiothauma*, its largest region being the notum, which is divided by transverse lines into four areas. Both the meso- and metathorax are well developed. The legs are generally adapted

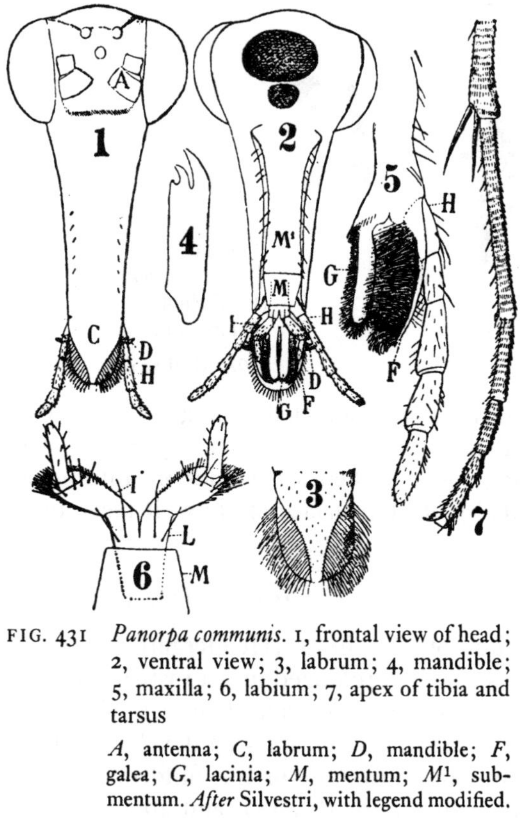

FIG. 431 *Panorpa communis.* 1, frontal view of head; 2, ventral view; 3, labrum; 4, mandible; 5, maxilla; 6, labium; 7, apex of tibia and tarsus

A, antenna; *C*, labrum; *D*, mandible; *F*, galea; *G*, lacinia; *M*, mentum; *M*[1], submentum. *After* Silvestri, with legend modified.

for walking, the claws are usually paired and in *Panorpa* they are strongly pectinated. In *Bittacus* the claws are single, and the 4th and 5th tarsal segments are provided with fine teeth along their inner margins: the 5th segment is capable of closing on to the 4th after the manner of the blade of a pocket-knife. The two pairs of wings are similar in form and nearly equal in size: in many species they are conspicuously spotted or banded. These organs are totally absent in the Californian *Apterobittacus* and the Tasmanian *Apteropanorpa*: in the males of the Boreidae (Fig. 432) they are represented by two pairs of slender bristle-like vestiges, and in the females there is a single pair of scale-like lobes on the mesothorax. In the Nannochoristidae and Choristidae there is a definite wing-coupling apparatus with a well-developed frenulum (see p. 55). Microtrichia are generally

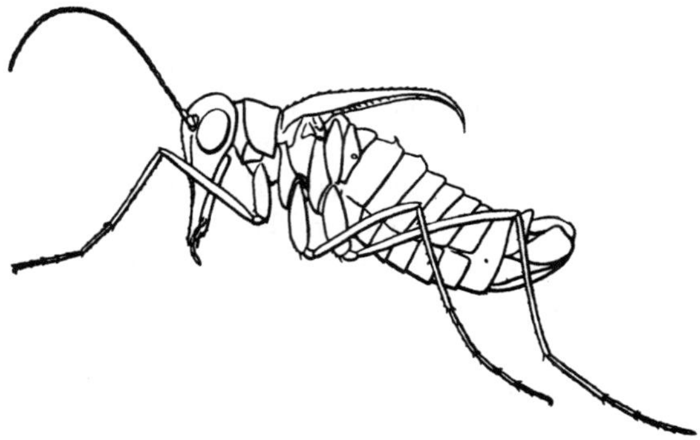

FIG. 432
Boreus hyemalis, male × 15
After Withycombe.

present, and macrotrichia occur on the veins and their branches, but not on the cross-veins; the latter type of seta is also often present on the wing-membrane. The venation is extremely archaic, the principal veins and their primary branches (excepting those of Cu_1) frequently being present (Fig. 433). The wing tracheae, on the other hand, are highly specialized by reduction. The primary dichotomies of the veins usually occur fairly close to the bases of the wings, and cross-veins are numerous, but without definite arrangement. In their venational features the two pairs of wings are also very alike, the principal difference being the basal fusion of Cu_2 and 1A in the hind wing. A marked deviation from the primitive type is exhibited in *Nannochorista* in which R_{2+3} is unforked and M + Cu_1 are fused for about half their length.

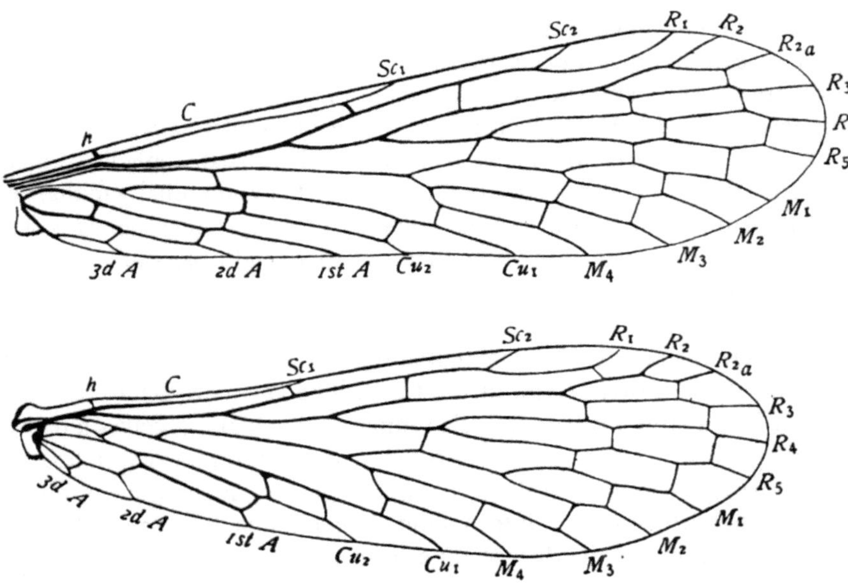

FIG. 433 Wings of *Panorpa*
After Comstock, *Wings of Insects.*

The abdomen has 10 distinct segments but there appears to be an 11th segment in the female at least of *Panorpa* (Ferris and Rees, 1939). In the male of that genus (Grell, 1942) the hind margin of the 9th sternum is prolonged into a deeply cleft process, the two arms of which are styliform. The 9th tergum is prolonged into a subquadrate plate. Between the dorsal and ventral processes thus formed there is a pair of laterally inserted 2-segmented claspers. A sperm-pump, probably homologous with that of the Diptera, occurs in all Mecoptera except the Boreidae (Mickoleit). The 10th segment is very inconspicuous, and bears a pair of short 1-segmented cerci. Between the base of the claspers is the longitudinally cleft aedeagus. In the female the 7th to 10th segments are cylindrical, and each is telescoped into the preceding segment: at the apex of the abdomen is a pair of 2-segmented cerci.

Internal Anatomy

The internal anatomy (Miyake, 1912; Grell, 1938; Potter, 1938*b*) has been investigated in several genera. In *Panorpa* the *alimentary canal* is a tolerably straight tube, the only convolution present occurring in the hind intestine. The oesophagus is curiously dilated at two points along its course to form what appears to be a kind of muscular pumping-apparatus. A short distance further back there is an elliptical chamber which is regarded as the proventriculus: the latter is provided with longitudinal and circular muscles, and its inner lining is beset with numerous long setae. These *acanthae* much resemble those of the Siphonaptera (Hepburn, 1969*a*; Richards, 1965). They are used to retain solid matter while the juices are pressed out for the purpose of external digestion of the food. The mid intestine is an elongate tube of large calibre, and the commencement of the hind intestine is marked by the insertions of 6 Malpighian tubes. A pair of tubular salivary glands is also present. The *nervous system* consists of the usual cephalic centres (Bierbrodt, 1942), 3 thoracic and 6–8 (♂) or 5–7 (♀) abdominal ganglia: the first of the latter is located in the metathorax. The *respiratory system* is well developed: there are two pairs of thoracic and six to eight pairs of abdominal spiracles. The *reproductive system* in the male consists of a pair of testes, each composed of 3–4 follicles arranged side by side around a longitudinal axis: the vasa efferentia are densely convoluted (except *Boreus*), forming a kind of epididymis at the posterior end of the testis. The two vasa deferentia open separately into a large median vesicula seminalis which also receives a pair of accessory glands. Each ovary consists of 7–19 polytrophic ovarioles, the number varying according to the species. The two oviducts unite to form a common canal which opens into a kind of genital pouch: the latter also receives the opening of the duct leading from a small pyriform spermatheca and that of the duct of a pair of colleterial glands. The genital pouch communicates with the exterior on the 9th abdominal segment but in *Boreus* there is a true vagina.

Life-History and Metamorphosis

The eggs of several species have been obtained by confining the adults in vessels containing damp soil. In the European and American species of *Panorpa* they are laid in small batches in crevices in the soil: in the Japanese *P. klugi* Miyake mentions nearly 100 eggs being deposited in a group. In form they are ovoid in *Panorpa* and more or less cuboidal in *Bittacus*. The life-history of *Panorpa* was first observed by Brauer (1863); larval development has been described in *Panorpa communis* by Rottmar (1966), in *P. nuptialis* by Byers (1963) while the most complete account is that of Miyake (1912) which refers to *P. klugi* (Fig. 434). The first stage larva is yellowish-grey with the head testaceous. It is eruciform and bears a close resemblance to a caterpillar. The head is rather large with prominent 3-segmented antennae and it bears a group of about 20–28 simple eyes on either side. The

FIG. 434 *Panorpa*, larva in last instar × 5
Adapted from Miyake.

mandibles are sharply toothed, and the maxillae are divided in lobes apparently corresponding with a galea and lacinia: the maxillary palpi are 4-segmented. The labium is small and its palpi 3-segmented. The thorax bears 3 pairs of legs, each composed of 4 segments: the abdomen is 10-segmented and the first 8 somites each carry a pair of abdominal feet. A median dorsal sclerotized shield is present on all the body segments. The first 9 abdominal shields each carry a pair of annulated processes, the last two pairs being considerably the larger: the 10th segment bears a single median process of a similar character together with a curious retractile lobed vesicle on its ventral side. Nine pairs of spiracles are present: they are located on the prothorax and first 8 abdominal segments. After the first ecdysis the annulated processes practically disappear except those of the last three segments. Boese (1973) describes and gives a key to separate 13 N. American fourth instar larvae. The number of ecdyses that occur has not been observed. Felt (1895), from head-measurements, recognized seven stages in *Panorpa rufescens*. Pupation takes place in an earthen cavity below ground: the pupa is of the usual exarate type and is capable of movement when disturbed: according to Miyake it works its way to the surface prior to the emergence of the imago. The European species probably pass through a single generation in the year. The larva of *Boreus* is strongly curved: the thoracic legs are well developed but there are no abdominal feet. It lives among moss and, when about to pupate, constructs a vertical tube leading near to the surface. The internal anatomy of the larvae of *Panorpa* and *Boreus* is described by Potter (1938), Grell (1938) and Bierbrodt (1942).

The relationship of the Mecoptera to other orders has been considered in detail by Mickoleit in a number of papers in which he has studied their structure, especially their musculature (1965, 1966, 1967, 1968, 1971*b*). They are still regarded as on the whole nearest to the Diptera though more different than Imms supposed.

Classification – The classification of the Mecoptera has been reviewed by Byers (1965), applying numerical methods to 88 characters. It is agreed that the living forms fall into two suborders, the **Protomecoptera** and the **Eumecoptera.** The former are mostly fossil but include the Chilean *Notiothauma reedi* (with ocelli), the American *Merope* and the W. Australian *Austromerope* (without ocelli). The **Notiothaumidae** and the **Meropeidae** have very full venation with more than four branches to Rs and many randomly placed cross-veins. The male genitalia are not swollen.

The **Eumecoptera** include the **Bittacidae** with their elongate body and legs and single tarsal claw, opposable to the tarsus, and a number of families more or less close to the **Panorpidae,** differing especially in their venation. Byers (1965) recognizes 5 families, viz. besides the cosmopolitan Panorpidae, the Australian **Choristidae** (Riek, 1973), **Nannochoristidae** and **Apteropanorpidae** and the American **Panorpodidae.** The Holarctic **Boreidae** are the subject of disagreement. Most authors regard them merely as another family but Hinton (1958), following Crampton, makes them a separate order, the **Neomecoptera.** The adult, for instance, has panoistic not polytrophic ovarioles; lacks cerci and the six rectal glands; Mickoleit adds that they lack a sperm-pump. In the larva, there is no epistomal suture, the cardo is distinct from the basistipes and has a tentorial adductor muscle, a postmentum is developed and there are no prolegs. A good case can be made out for treating the group as a suborder. Cooper (1974) gives a good account of its biology.

Literature on Mecoptera

APPLEGARTH, A. G. (1939), Larva of *Apterobittacus apterus* Maclachlan (Mecoptera: Panorpidae), *Microentomology*, 4, 109–120.

BIERBRODT, E. (1942), Der Larvenkopf von *Panorpa communis* L. und seine Verwandlung, mit besonderer Berücksichtigung des Gehirns und der Augen, *Zool. Jb., Anat.*, 68, 49–136.

BOESE, A. E. (1973), Descriptions of larvae and key to fourth instars of North American *Panorpa* (Mecoptera: Panorpidae), *Univ. Kansas Sci. Bull.*, 5, 163–186.

BRAUER, F. (1863), Beitrag zur Kenntnis der Panorpiden-Larven, *Verh. zool. bot. Ges. Wien*, 13, 307–324.

BYERS, G. W. (1963), The life history of *Panorpa nuptialis* (Mecoptera: Panorpidae), *Ann. ent. Soc. Am.*, 56, 142–149.
—— (1965), Families and genera of Mecoptera, *Proc. 12th internat. Congr. Ent. London*, 1964, 123.
CARPENTER, F. M. (1931), Revision of the Nearctic Mecoptera, *Bull. Mus. comp. Zool.*, 72, 205–277.
—— (1935), New Nearctic Mecoptera, with notes on other species, *Psyche*, Cambridge, 42, 105–22.
COOPER, K. W. (1974), Sexual biology, chromosomes, development, life histories and parasites of *Boreus*, especially of *B. notoperates*, a southern Californian *Boreus II.* (Mecoptera: Boreidae), *Psyche*, Cambridge, 81, 84–120.
CURRIE, G. A. (1932), Some notes on the biology and morphology of the immature stages of *Harpobittacus tillyardi* (Order Mecoptera), *Proc. Linn. Soc. N.S.W.*, 57, 116–122.
FELT, E. P. (1895), The Scorpion-flies, *Rep. N.Y. Ent.*, 10 (1894), 473–480.
FERRIS, G. F. AND REES, B. E. (1939), Morphology of *Panorpa nuptialis* Gerstaecker (Mecoptera: Panorpidae), *Microentomology*, 4, 79–108.
GRELL, K. G. (1938), Der Darmtraktus von *Panorpa communis* L. und seine Anhänge bei Larve und Imago, *Zool. Jb., Anat.*, 64, 1–86.
—— (1942), Der Genitalapparat von *Panorpa communis* L., *Zool. Jb., Anat.*, 67, 513–588.
HANDLIRSCH, A. AND BEIER, M. (1936), Panorpatae oder Mecoptera, In: Kükenthal, W. (ed.), *Handbuch der Zoologie*, 4, Heft 2, Insecta 2, Lief. 6, 1467–1490, Walter de Gruyter & Co., Berlin.
HEDDERGOTT, H. (1938), Kopf und Vorderdarm von *Panorpa communis* L., *Zool. Jb., Anat.*, 65, 229–294.
HEPBURN, H. R. (1969*a*), The proventriculus of Mecoptera, *J. Georgia ent. Soc.*, 4, 159–167.
—— (1969*b*), The skeleto-muscular system of the Mecoptera, *Univ. Kansas Sci. Bull.*, 48, 721–765.
HINTON, H. E. (1958), The phylogeny of the Panorpoid orders, *A. Rev. Ent.*, 3, 181–206.
HOBBY, B. M. AND KILLINGTON, F. J. (1934), The feeding habits of British Mecoptera; with a synopsis of the British species, *Trans. Soc. Brit. Ent.*, 1, 39–49.
IMMS, A. D. (1944), On the constitution of the maxillae and labium in Mecoptera and Diptera, *Quart. J. micr. Sci.*, 85, 73–96.
ISSIKI, S. (1933), Morphological studies on the Panorpidae of Japan, etc., *Jap. J. Zool.*, 4, 315–416.
MERCIER, L. (1915), Caractère sexuel secondaire chez les Panorpes. Le rôle des glandes salivaires des mâles, *Arch. Zool. Paris*, 55 (notes et revues), 1–5.
MICKOLEIT, G. (1965), Ueber die morphologische Deutung des caudalen Sternocoxalmuskels im Pterothorax der Neuropteroidea, *Zool. Jb., Anat.*, 82, 521–531.
—— (1966), Zur Kenntnis einer neuen Spezialhomologie (Synapomorphie) der Panorpoidea, *Zool. Jb., Anat.*, 83, 483–496.
—— (1967), Das Thoraxskelett von *Merope tuber* Newman (Protomecoptera), *Zool. Jb., Anat.*, 84, 313–342.
—— (1968), Zur Thoraxmuskulatur der Bittacidae, *Zool. Jb., Anat.*, 85, 386–410.

MICKOLEIT, G. (1971*a*), Das Exoskelett von *Notiothauma reedi* MacLachlan, ein Beitrag zur Morphologie und Phylogenie der Mecoptera (Insecta), *Z. Morph. Ökol. Tiere*, **69**, 318–362.

—— (1971*b*), Zur phylogenetischen und funktionellen Bedeutung der sogenannten Notalorgane der Mecoptera (Insecta, Mecoptera), *Z. Morph. Okol. Tiere*, **69**, 1–8.

MIYAKE, T. (1912), The life history of *Panorpa klugi* M'Clachlan, *Tokyo J. Coll. Agric.*, **4**, 117–139.

POTTER, E. M. (1938*a*), The internal anatomy of the larvae of *Panorpa* and *Boreus* (Mecoptera), *Proc. R. ent. Soc. Lond.*, (A), **13**, 117–130.

—— (1938*b*), The internal anatomy of the order Mecoptera, *Trans. R. ent. soc. Lond.*, **87**, 467–501.

RICHARDS, A. G. (1965), The proventriculus of adult Mecoptera and Siphonaptera, *Ent. News*, **76**, 253–256.

RIEK, E. F. (1972), Mecoptera, In: Mackerras, I. (ed.), *The insects of Australia*, Chapter 32, C.S.I.R.O., Divn Entomology, Melbourne.

—— (1973), A revision of the Australian Scorpion flies of the family Choristidae (Mecoptera), *J. Aust. ent. Soc.*, **12**, 103–112.

ROTTMAR, B. (1966), Ueber Züchtung, Diapause und postembryonale Entwicklung von *Panorpa communis* L., *Zool. Jb., Anat.*, **83**, 497–570.

SETTY, L. R. (1931), The biology of *Bittacus stigmaterus* Say (Mecoptera, Bittacusidae), *Ann. ent. Soc. Am.*, **34**, 467–484.

STEINER, P. (1930), Studien an *Panorpa communis* L. i. Zur Biologie. ii. Zur Morphologie und Postembryonalen Entwicklung des Kopfskeletts, *Z. Morph. Ökol. Tiere*, **17**, 1–67.

STITZ, H. (1908), Zur Kenntnis des Genitalapparats der Panorpaten,. *Zool. Jb., Anat.*, **26**, 537–564.

TILLYARD, R. J. (1917), Studies in Australian Mecoptera. No. 1. The new family, Nannochoristidae, with descriptions of new genera and four new species, etc., *Proc. Linn. Soc. N.S.W.*, **42**, 284–301.

—— (1918), Do., No. 2. The wing-venation of *Chorista australis* Klug. *Proc. Linn. Soc. N.S.W.*, **43**, 395–408.

—— (1919), The Panorpoid complex. Part 3. The wing-venation, *Proc. Linn. Soc. N.S.W.*, **44**, 533–718.

—— (1935), Evolution of scorpion-flies and their derivatives (order Mecoptera), *Ann. ent. Soc. Am.*, **28**, 1–45.

WITHYCOMBE, C. L. (1922), On the life-history of *Boreus hyemalis* L., *Trans. ent. Soc. Lond.*, **1921**, 312–318.

Order 25

SIPHONAPTERA (FLEAS)

Small, apterous, laterally compressed insects whose adults are ectoparasites of warm-blooded animals. Eyes absent, 2 ocelli usually present: antennae short and stout, reposing in grooves: mouthparts modified for piercing and sucking, maxillary and labial palpi present. Thoracic segments free: coxae very large, tarsi 5-segmented. Larvae elongate, eruciform and apodous. Pupae exarate, adecticous, enclosed in cocoons.

The Siphonaptera (Aphaniptera) or fleas may be readily distinguished from other apterous parasitic insecta since they are strongly compressed laterally instead of being dorso-ventrally flattened. They constitute a sharply defined order of insects without close connection with any other group. Supposed traces of mesothoracic wings are found in some pupae (Sharif, 1935) but Poenicke (1969) denies that they are homologous and they do not occur in the Pulicidae. Some resemblances can be seen to the Diptera and Mecoptera (Potter, 1938) and in their metamorphoses they possess certain features in common with the Dipteran Nematocera.

Fleas are blood-sucking ectoparasites of mammals and birds. They are negatively phototactic and respond to warmth. Humphries (1968) discusses host-finding in *Ceratophyllus gallinae* which seems to respond especially to light and gravity. The shadow cast by a potential host attracts the flea. When a host dies the fleas leave as soon as the body cools and seek fresh hosts which are not always of the same species. Many kinds are apparently confined to one species of animal while others infest a range of hosts. The relation between different species of fleas and their hosts, however, is not a very close one, and in the absence of the elective species of the latter, many will feed readily on the blood of other animals. Their chief method of progression is walking but they can also progress by leaping. According to Mitzmain the maximum vertical height attained by the leap of *Pulex irritans* is 19·7 cm, while the horizontal range may extend to 33 cm.

Nearly 1400 species of the order have been described of which 53 are known to occur in the British Isles (see Smit, 1957). During the present century the work of the Indian Plague Commission, and of many independent observers, has resulted in a great increase of knowledge relating to these insects. Many species have proved capable of transmitting bubonic plague (Brumpt, 1949). In India the species mostly implicated is the rat flea,

Xenopsylla cheopis (Fig. 435). The rat is particularly susceptible to this disease, and the flea itself becomes infected with the plague bacillus by feeding upon an infected animal. When the latter dies, the fleas desert the body, and many find their way to man, particularly when the human population lives under crowded conditions in rat-infested quarters. It is impossible here to detail the evidence as to the manner in which the rat flea transmits the disease to man as the problem is a complicated one. It may be pointed out, however, that while feeding, the contents of the gut of the flea, which

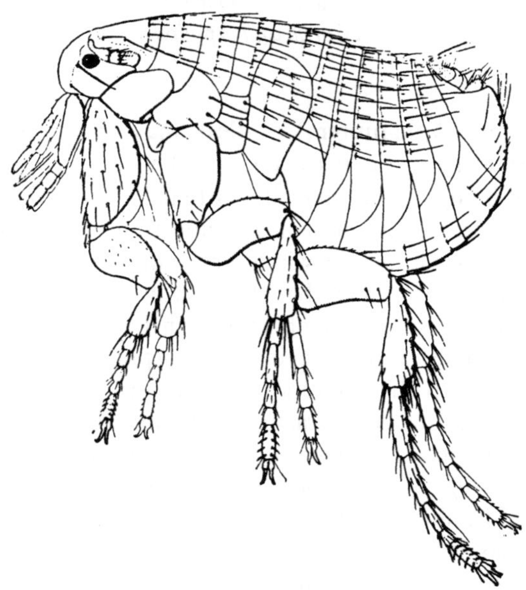

FIG. 435 *Xenopsylla cheopis* (Pulicidae), male. × 20
After Waterston. Reproduced by permission of the Trustees of the British Museum.

contain the pathogenic bacteria, are voided from time to time. This excretory matter may be introduced beneath the skin by scratching. Bacot and Martin have demonstrated that plague-infested fleas also convey the disease by a method comparable to inoculation. When placed on a host they suck vigorously but owing to the fact that, in a certain number of the insects, the digestive canal is blocked by a dense mass of rapidly multiplying plague bacteria, the blood that is imbibed fails to enter the stomach and is regurgitated into the puncture. Since this blood is now contaminated with bacteria derived from the previous host, the disease is thus transmitted to the new host. This seems to be the common method of transmission. Rats are not the only animals attacked by plague, squirrels and other rodents being also liable to the disease; consequently fleas of any species, which attack both ground rodents and man, in lands where plague is prevalent, are to be regarded as possible agents in the transmission of the malady.

One of the most familiar of these insects is the cosmopolitan human flea *Pulex irritans*. Although man used to be its favourite host it is found now mainly on pig-farms. The extensive genus *Xenopsylla* includes the plague flea *par excellence* (*X. cheopis*, Fig. 435) which has been previously alluded to: it is almost tropicopolitan and is a scarce vagrant in the British Isles. *Ctenocephalides* (Fig. 436) includes the dog and cat fleas (*C. canis* and *C. felis*) both of which occur on dogs and cats. The rabbit flea (*Spilopsyllus*

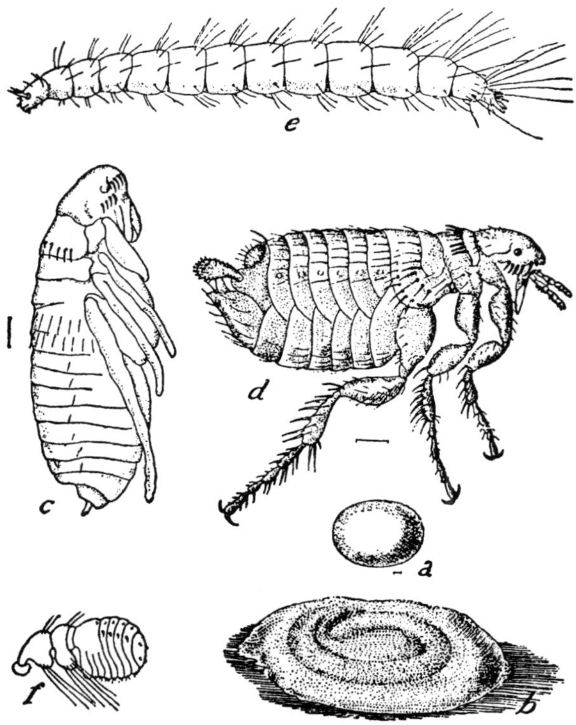

FIG. 436 *Ctenocephalides canis* (Pulicidae)

a, egg; *b*, larva in cocoon; *c*, pupa; *d*, imago; *f*, antenna of imago; *e*, *Ceratophyllus fasciatus*, larva. From Bishopp, *U.S. Dept. Agric. ent. Bull.* 248 (all except *e*, *after* Howard).

cuniculi) commonly affects the ears of rabbits, and sometimes becomes transferred to cats while the latter are hunting those animals. The genus *Ceratophyllus* includes a number of species affecting birds: *C. gallinae* is frequent in hen-houses and in the nests of many wild birds. *Leptopsylla segnis* is harboured by the mouse and species of *Nycteridopsylla* and *Ischnopsyllus* are essentially bat parasites. In addition to the foregoing species, which occur in Great Britain, mention needs to be made of the well-known 'jigger' or 'chigoe' (*Tunga penetrans*) of the tropics whose females remain attached to the skin in one position for the greater part of their existence. The modifications of external structure associated with this habit are so marked that the abdomen becomes distended to the size of a small pea, the insect bearing a close resemblance to a tick. It has a large number of hosts and its attacks are usually confined to the feet: in man it particularly affects the toes. Instead of remaining at the surface of its host, the fertilized female burrows into the flesh, until it may become completely embedded, though most of the eggs reach the exterior before hatching.

External Anatomy

The body in the Siphonaptera (Snodgrass, 1946) is strongly compressed, and well sclerotized, with the evident advantage of enabling these insects readily to work their way among the hair or feathers of the host. There is usually a prominent armature of spines and bristles which are sharply inclined backwards, thus aiding forward progression, and the claws of the feet are strong in conformity with the necessity for grasping. The *head* (Wenk, 1953) is very closely attached to the thorax with a small cervical sclerite at each side. Situated on the middle line of the frons there is, in many species, a tubercle which has been thought to help in opening the cocoon but this is very doubtful. The two *ocelli* (cf. p.146; also Wachmann, 1972) are laterally displaced, and may be deeply pigmented but, in a number of species, they are vestigial or absent. The lateroventral border of the head often carries a row of powerful spines forming the *genal comb* which is present on either side: these organs are frequently referred to as ctenidia and are well seen, for example, in *Ctenocephalides*. The *antennae* are lodged in antennal grooves and are short and stout with three evident segments. The terminal portion is pectinated and exhibits a number of annular divisions, which vary in completeness of development in different genera, and sometimes in different sexes. A strong interantennal groove often connects the two sockets. The *mouthparts* (Fig. 437) are adapted for piercing and sucking, and the most important organs are the *laciniae* of the maxilla. These structures are rather broad blades which are serrated along the distal two-thirds of their length. Proximally, the inner surfaces of the laciniae are in contact with the short hypopharynx and, where the latter organ terminates, they are closely apposed to the epipharynx above. Each lacinia is distally grooved along its inner aspect from the point where the hypopharynx ceases, and they form together a channel through which the saliva is ejected. Basally, the lacinia articulates with the maxillary lobe (stipes) by means of a small rod-like sclerite which imparts to it considerable freedom of movement. The *labrum* is small and hidden beneath the clypeus but the *epipharynx* is a long slender organ which is ventrally grooved, and closely approximated to the laciniae, the combined organs thus forming an afferent channel through which blood is sucked up. The *hypopharynx* is a small sclerite which is concave ventrally and incurved at the margins: within the area thus defined the salivary pump and its operating muscles are lodged. Anteriorly, the hypopharynx is prolonged into a small process, which is perforated by the salivary duct, and extends for a short distance between the epipharynx and the laciniae. The *maxillary lobes* are mainly derived from the stipes: each consists of a single lobe, on either side of the mouth, and a 4-segmented palpus: they are not cutting organs and apparently do not enter the puncture made by the laciniae. The labium is formed of a *postmentum* more or less fused to the head and an articulated *prementum* which carries distally a pair of labial palps: these usually have 5 segments but may have 1 or 3. The stylets are held in the grooved prementum by the appressed palps. In feeding, the laciniae rock on the sclerite articulating with the stipes, so that they have an up and down movement like a pneumatic drill; the epipharynx penetrates with them in a passive manner. The muscles of the salivary pump inject saliva into the perforation thus formed and the cibarial and pharyngeal pumps draw up the blood.

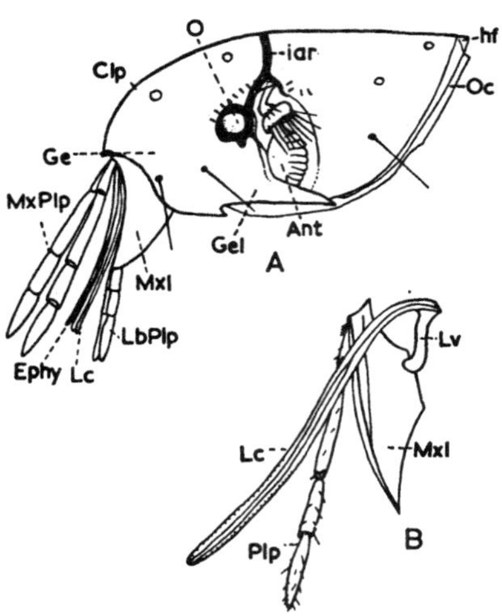

FIG. 437
A. Head of *Pulex irritans*, female. B. Right maxilla of *Anomiopsyllus* sp. (Both redrawn from Snodgrass, 1946)

Ant, antenna; *Clp*, clypeus; *Ephy*, epipharynx; *Gel*, subantennal lobe of gena; *hf*, head flange; *iar*, interantennal ridge; *LbPlp*, labial palpus; *Lc*, lacinia or maxillary stylet; *Lv*, lever of lacinia; *Mxl*, palpus-bearing lobe of maxilla; *MxPlp*, maxillary palpus; *O*, ocellus; *Oc*, occiput; *Plp*, palpus.

The *thorax* is composed of three quite distinct segments which admit of a certain amount of movement. In many species the hind margin of the pronotum carries a row of stout spines forming the *pronotal comb*. The terga are simple, broad, arched plates and the metathorax is characterized by its greatly developed epimera which overlie the base of the abdomen. The legs are adapted for clinging and leaping with large flattened coxae, short stout femora and elongate 5-segmented tarsi (Jacobson, 1940). The jump makes use of the energy stored in the pleural arch partly in a patch of resilin. The energy is released by a click-mechanism (Rothschild *et al.*, 1975).

The *abdomen* is composed of 10 segments, the first of which has the sternum wanting, and the last three segments are modified for sexual purposes (Günther, 1961; Rothschild and Hopkins, 1971). The 9th segment in the male is reduced in the tergal region. The 9th sternum is V-shaped, one branch of the V projecting beyond the large 8th sternum. To it are articulated a pair of two-segmented claspers. The *aedeagus* which projects between the claspers has an extremely complex structure. The combined 9th and 10th segments consist of a dorsal sensory plate or *sensilium* (= pygidium) and a more posterior *proctiger*, provided with dorsal and ventral plates. In the female the terminal segments are less modified than in the male. The 8th segment is large and the 9th bears a sensilium, as in the male and the proctiger consists of a small dorsal plate bearing a conical setiferous process known as the *stylet* and the corresponding sternum is represented by a small ventral plate.

Internal Anatomy

The mouth is surrounded by the ring-like base of the epipharynx and forms the definitive opening into the alimentary canal. The pharynx is an elongate chamber with strongly sclerotized dorsal and ventral walls. It is followed by a long oesophagus of very small calibre, which leads into a somewhat conical organ termed the proventriculus. The inner walls of the latter are beset with a series of long backwardly directed sclerotized rods or acanthae (Richards, 1965; Richards and Richards, 1969). It has been claimed that the function of this arrangement is to break up the ingested corpuscles by a rhythmic movement of the spines. The stomach, when fully distended, occupies a large part of the abdominal cavity, and, near its junction with the hind intestine, are found the insertions of the four Malpighian tubes. Near the termination of the hind intestine there are six rectal papillae resembling those found among Diptera. The *salivary glands* consist of a pair of ovoid sacs on either side: their ducts eventually combine to form a common canal which enters the salivary pump beneath the hypopharynx. The *nervous system* is exceptionally primitive in that the ventral nerve-cord consists of three thoracic and seven abdominal ganglia in the female and one more abdominal ganglion in the male (Minchin, 1915): these centres are very much approximated owing to the great reduction in length of the intervening connectives. The *male reproductive organs* consist of a pair of fusiform testes whose contents pass down extremely fine vasa deferentia: the latter unite to form a single passage opening into a small vesicula seminalis. The ejaculatory duct is associated with the copulatory organ. The *female reproductive organs* are composed of a pair of ovaries, each formed of from four to eight panoistic ovarioles. Attached to the vagina are one or two sclerotized spermathecae whose shape and size differ among various species. The *respiratory system* is well developed and communicates with the exterior by ten pairs of spiracles: two pairs of the latter are located on the thorax, the remainder being abdominal in position. In a few sedentary species some of the spiracles may lose their function.

Biology and Metamorphosis (Fig. 436)

The eggs of these insects are ovoid and white or cream in colour: unlike those of many ectoparasites they are not glued to the hair or feathers of the host. When deposited on the body of the latter they readily fall off and are normally found in the haunts or sleeping-places of the animal parasitized. In houses fleas breed in the cracks of floors, under matting or beneath carpets and almost always in dirty dwellings. Rat fleas often breed in granaries, barns, etc., particularly in those where there is an accumulation of floor litter. The dried excrement, feathers straw, etc., which accumulate in chicken houses also afford a favourable environment. The incubation period varies on an average from three to ten days, according to temperature, and the young larva ruptures the chorion by the aid of a hatching-spine on the dorsal side of the head. The larvae (Bacot and Ridewood, 1914; Sikes, 1930; Sharif, 1937*b*) are active, whitish, vermiform objects usually measuring about 4 mm in length when fully grown. They are non-parasitic and feed upon particles of organic matter found in the host's lair, or among the dust and dirt which collects on the ground in the vicinity. In some species, however, blood which has passed through the body of adult fleas appears to be a necessary part of their nutriment. Larval Siphonaptera possess a well-developed head but are devoid of both eyes and legs: in their general characters they resemble the larvae of certain Nematocera. The antennae are single-segmented but rather prominent, the mandibles are very definitely toothed and the maxillae assume a curious brush-like form with small 2-segmented palps: each labial palp is composed of a short basal segment surmounted by stout setae. The trunk consists of three thoracic and ten abdominal somites, each of which is armed with a band of outstanding bristles, while the 10th bears a pair of anal struts. Spiracles are present on the pro- and metathorax and first eight abdominal segments. After undergoing two ecdyses, the larva spins a cocoon which is concealed by the fine particles of debris adherent to its outer surface. The adults remain quiescent for a variable period before emerging from the cocoons, and they often issue in large numbers in response to slight mechanical stimuli. The vibrations set up by persons walking about a disused room, for example, have been explained as being the cause of the emergence of an abundance of fleas within a very short time. When newly emerged, the adults can remain alive for a considerable period without food, but they take the first opportunity of reaching their particular host. As a general rule the female needs to imbibe the blood of the normal host before becoming capable of laying fertile eggs.

The condition of the host may greatly influence reproduction. Female rabbit fleas normally feed and become sexually mature on pregnant does. At parturition, they transfer to the young and then pair and lay eggs which they will not do on the doe. They can also mature on the young and will then pair without transfer to a new host. These physiological and behavioural functions are mainly controlled by the sex hormones of the host (Rothschild and Ford, 1973). There is a somewhat similar effect on the maturation and copulation of the male, the effects being greatest on the young and least on the buck rabbit; frequency of copulation is also increased by transfer to a new host (Rothschild, Ford and Hughes, 1970).

The period occupied by the complete developmental cycle varies in different species and in different countries. Thus *Pulex irritans* in Europe requires from 4 to 6 weeks, while *Xenopsylla cheopis* in India passes through a complete generation in about 3 weeks: on the Pacific coast the life-cycle of the latter species occupies, according to Mitzmain, 9 to 11 weeks.

Classification – The latest classification is that of Rothschild and Hopkins (1953–66). They recognize two superfamilies. The **Pulicoidea** lack an outer external ridge on the mid coxa and an apical tooth outside the hind tibia, while the sensilium has 8 or 14 pits. The **Ceratophylloidea** usually possess the coxal ridge and hind tibial tooth and have 16, less often 14, pits on the sensilium. They recognize 17 families of widespread **Pulicidae** in which the hind coxa has strong internal bristles and the sensilium has 14 pits on each side. The other family, the **Tungidae,** has no coxal bristles and 8 pits on each side. *Tunga* is the Chigoe flea of which the females bore into the skin but the other members of the family live more normally.

The bulk of the fleas fall into the **Ceratophylloidea**; among them are the bat fleas, **Ischnopsyllidae,** which have a preoral comb of two stout bristles on each side. The **Hystrichopsyllidae** include the very large mole flea, *Hystrichopsylla talpae*, in which the metanotum lacks marginal spinelets. The families in which these spinelets are present include the **Ceratophyllidae** which have no genal comb and in which the uppermost of the three ocular setae is in front of the eye, and the **Leptopsyllidae** which may have a genal comb and in which the upper seta of the ocular row is above the eye. They both include some parasites of birds and some of mammals.

Riek (1970) records two fossil fleas from the Lower Cretaceous of Gyppsland, Australia. One rather resembles *Echidnophaga* (Pulicidae) but the other is quite unusual, with long antennae.

Literature on Siphonaptera

BACOT, A. W. AND RIDEWOOD, W. G. (1914), Observations on the larvae of fleas, *Parasitology*, 7, 157–175.

BRUMPT, E. (1949), *Précis de Parasitologie*, vol. II, Paris.

COSTA LIMA, A. da AND HATHAWAY, C. R. (1946), Pulgas. Bibliografia, catalogo o animais por elas sugadas, *Monogr. Inst. Osw. Cruz.*, **4**, 317 pp.

EWING, H. E. AND FOX, I. (1943), The fleas of North America, *Misc. Publ. U.S. Dep. Agric.*, **500**, 142 pp.

GÜNTHER, K. K. (1961), Funktionell-anatomische Untersuchung des männlichen Kopulationsapparates der Flöhe unter besonderer Berücksichtigung seiner postembryonalen Entwicklung (Siphonaptera), *Dtsch. ent. Z.*, N.F. **8**, 258–349.

HOLLAND, G. P. (1949), The Siphonaptera of Canada, *Tech. Bull. Dep. Agric. Canada*, **70**, 306 pp.

HUMPHRIES, D. A. (1968), The host-finding behaviour of the hen flea, *Ceratophyllus gallinae* (Schrank) (Siphonaptera), *Parasitology*, 58, 403–414.

JACOBSON, H. (1940), Über die Sprungmuskulatur des Uferschwalbenflohes *Ceratophyllus styx* Roth., *Z. Morph. Ökol. Tiere*, 37, 144–154.

JORDAN, H. E. K. (1947), On some phylogenetic problems within the order of Siphonaptera (= Suctoria), *Tijdschr. Ent.*, 88 (1946), 79–93.

MINCHIN, E. A. (1915), Some details in the anatomy of the rat-flea, *Ceratophyllus fasciatus* Bosc., *J. Queckett micr. Club*, **12**, 441–464.

MITZMAIN, M. B. (1910), General observations on the bionomics of the rodent and human fleas, *U.S. Pub. Health Bull.*, 38, 34 pp.

PERFILJEW, P. P. (1926), Zur Anatomie der Flohlarven, *Z. Morph. Ökol. Tiere*, 7, 102–126.

POENICKE, H. W. (1969), Ueber die postlarvale Entwicklung von Flöhen (Insecta, Siphonaptera), unter besonderer Berücksichtigung der sogenannten 'Flügelanlagen', *Z. Morph. Ökol. Tiere*, **65**, 143–186.

POTTER, E. M. (1938), The internal anatomy of the order Mecoptera, *Trans. R. ent. Soc. Lond.*, 87, 467–502.

RICHARDS, A. G. (1965), See Mecoptera (p. 940).

RICHARDS, P. A. AND RICHARDS, A. G. (1969), Acanthae: a new type of cuticular process in the proventriculus of Mecoptera and Siphonaptera, *Zool. Jb., Anat.*, **86**, 158–176.

RIEK, E. F. (1970), Lower Cretaceous flea (Siphonaptera), *Nature*, **227**, 746–747.

ROTHSCHILD, M. *et al.* (1975), Jumping mechanism of *Xenopsylla cheopis*, I, II, III, *Phil. Trans. R. Soc., Lond.*, **271B**, 457–515.

ROTHSCHILD, M. AND HOPKINS, G. H. E. (1953–71), *An illustrated catalogue of the Collection of Fleas, etc.* Parts 1–5, Brit. Mus. (Nat. Hist.), London.

ROTHSCHILD, M. AND FORD, B. (1972), Breeding cycle of the flea *Cediopsylla simplex* is controlled by breeding cycle of host, *Science*, 178, 625–626.

—— (1973), Factors influencing the breeding of the rabbit flea (*Spilopsylla cuniculi*): spring-time accelerator and a kairomone in nestling rabbit urine with notes on *Cediopsylla simplex*, another 'hormone bound' species, *J. Zool., Lond.*, **170**, 87–137.

ROTHSCHILD, M., FORD, B. AND HUGHES, M. (1970), Maturation of the male rabbit flea (*Spilopsylla cuniculi*) and the oriental rat-flea (*Xenopsylla cheopis*): some effects of mammalian hormones on development and impregnation, *Trans. zool. Soc. Lond.*, **32**, 105–188.

SHARIF, M. (1935), On the presence of wing buds in the pupa of Aphaniptera, *Parasitology*, **27**, 461–464.

—— (1937*a*), On the life-history and biology of the rat-flea, *Nosopsyllus fasciatus* (Bosc.), *Parasitology*, **29**, 225–238.

—— (1937*b*), On the internal anatomy of the larva of the rat-flea, *Nosopsyllus fasciatus* (Bosc.), *Philos. Trans. Lond.*, (B), **227**, 466–538.

SIKES, E. K. (1930), Larvae of *Ceratophyllus wickhami* and other species of fleas, *Parasitology*, **22**, 242–259.

SMIT, F. G. A. (1957), Siphonaptera, *Hndbks. Ident. Brit. Ins., R. ent. Soc. Lond.*, **1**, pt. 16, 94 pp.

SNODGRASS, R. E. (1946), The skeletal anatomy of fleas (Siphonaptera), *Smithson. misc. Coll.*, **104**, 89 pp.

WACHMANN, E. (1972), Das Auge des Hühnerflohes *Ceratophyllus gallinae* (Schrank) (Insecta, Siphonaptera), *Z. Morph. Ökol. Tiere*, 73, 315–324.
WEIDNER, H. (1937), Beiträge zur Kenntnis der Biologie des Fledermausflohes *Ischnopsyllus hexactonus* Kol., *Z. Parasitenk.*, 9, 543–548.
WENK, P. (1953), Der Kopf von *Ctenocephalus canis* (Curt.) (Aphaniptera), *Zool. Jb., Anat.*, 73, 103–164.

Order 26

DIPTERA (TWO-WINGED OR TRUE FLIES)

Insects with a single pair of membranous wings, the hind pair modified into halteres. Mouthparts suctorial, usually forming a proboscis and sometimes adapted for piercing: mandibles rarely present: labium usually distally expanded into a pair of fleshy lobes. Prothorax and metathorax small and fused with the large mesothorax: tarsi commonly 5-segmented. Metamorphosis complete, larvae eruciform and apodous, frequently with the head reduced and retracted: tracheal system variable, most often amphipneustic, pupa either free or enclosed in the hardened larval cuticle or puparium, adecticous, primitively obtect but in higher forms exarate: wing-tracheation reduced.

The Diptera are one of the largest orders of insects including over 85 000 described species, and approximately 5200 species are known from the British Isles. Structurally Diptera are among the most highly specialized members of their class. The imagines of most species are diurnal and the majority are either flower-lovers, which feed upon nectar, etc., or frequently decaying organic matter of various kinds. Although these two habits predominate, a considerable number of flies are predacious and live on various insects. In addition to the foregoing, there are other Diptera which have acquired blood-sucking habits, and besides man many other vertebrates may be resorted to by one or other species. Excluding the Muscidae and the so-called Pupipara, this habit is largely confined to the female. The blood-sucking forms include almost the whole of the Culicidae, besides the Simuliidae, Phlebotominae, Tabanidae and the Glossinidae and allied forms, also certain members of the Ceratopogonidae and Muscidae. In virtue of this propensity the order has acquired great medical significance. The pathogenic organisms of some of the most virulent diseases such as malaria, sleeping sickness, elephantiasis and yellow fever are transmitted to man through the intermediary of blood-sucking Diptera.

External Anatomy

The work of Crampton (1942) forms a useful introduction to the external anatomy of the Diptera.

The Head (Fig. 438) is remarkable on account of its mobility and is usually of relatively large size. An extensive portion of its area is occupied by the *compound eyes* which, as a rule, are considerably larger in the male than the female. When the eyes of the two sides are contiguous they are stated to be *holoptic*, and when markedly separated *dichoptic*; very occasionally the holoptic condition is found in the female as well as the male. In some species the upper facets are larger and more conspicuous than the lower, a peculiarity rarely seen in the female. It assumes its most extreme development in the Bibionidae where the two areas of different facets are sharply defined (Fig. 84). Between or slightly behind the eyes are the *ocelli*: the latter are usually three in number and are generally arranged in the form of a triangle: in some

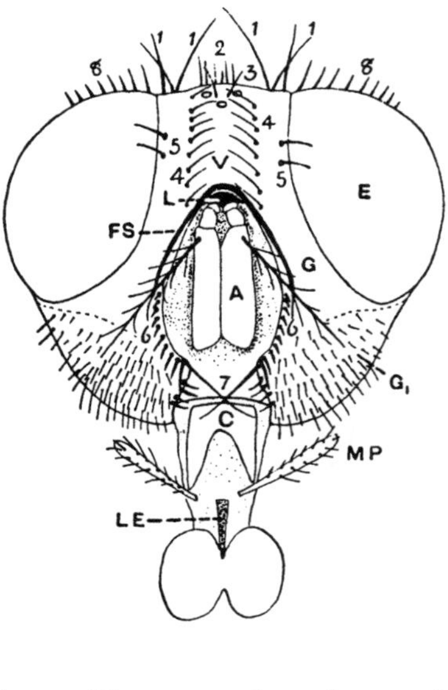

FIG. 438
Frontal view of head of a Cyclorrhaphous fly (Schizophora)

A, antenna; *C*, clypeus; *E*, eye; *FS*, frontal suture; *G*, gena and lower portion G_1 or jowl; *L*, lunule; *LE*, labrum; *MP*, maxillary palp; *V*, vertex. The numerals refer to the chaetotaxy (p. 961).

families ocelli are wanting. A complete Y-shaped epicranial sulcus has been described in *Mycetophila* but this is a secondary structure. The terminology of the regions of the head in general use is confusing owing to the multiplicity of names which have been employed: many do not admit of wide application and are often devoid of morphological value. In a Muscoid fly the 'front'* is regarded as the region between the eyes, and is limited by a line drawn through the bases of the antennae and by the upper margin of the head. In holoptic flies the space between the eyes and the basal line of the antennae is the *frontal triangle*: the triangular region bearing the ocelli and often bounded by grooves or depressions is the *ocellar triangle*. The region enclosed by the frontal suture is the *face* which is demarcated laterally by the *facial ridges* (facialia or vibrissal ridges) and distally by the epistoma. At the lower extremities of the facial ridges are two prominences or *vibrissal angles*

* In most Diptera almost the whole of the anterior surface of the head appears to be formed by the vertex: the true frons is either of very limited extent or merged with the clypeus.

carrying the vibrissae. The antennae are frequently lodged in *antennal grooves* or foveae which may be separated by a median *facial carina*. The *genae* (parafacials or cheeks) comprise the region lying between the face and the anterior margin of the eye on either side, while the *jowls* are the lower portions of the genae below the eyes. The upward continuations of the genae, along the inner border of the eyes, are known as the *orbits* or *parafrontals*. The *epistoma* is the distal border of the face and, in front of it, is a sclerite which is here regarded as the *clypeus* (or frontoclypeus). In many Nematocera the frontoclypeus is a well-defined region, but in some Brachycera and all Cyclorrhapha the clypeus (tormae of Peterson) appears to be separated off as a distinct sclerite. The latter is frequently a crescentic or semilunar plate, lying in the membrane of the rostrum, and forming the anterior or dorsal wall of the fulcrum.

The ptilinum or frontal sac is a characteristic cephalic organ of Cyclorrhapha and its presence is indicated externally by the arched *frontal* or *ptilinal suture*. The latter lies transversely above the antennae and extends downwards on each side of them, thus presenting a U-shaped form. The suture is of the nature of an extremely narrow slit, along the margins of which the wall of the head is invaginated to form a membranous sac or ptilinum, and the walls of the latter are seen to consist of the same layers as the integument. The outer surface of the ptilinum is roughened owing to the presence of minute scales or spines of various forms (Strickland, 1953). When viewed in sections taken through the head, the ptilinum is seen lying in the cavity of the latter in front of the brain. Attached to its inner surface in certain positions, are groups of slender muscle-fibres which apparently aid in retracting the organ (Laing, 1935). The function of the ptilinum is to thrust off the anterior end of the puparium at the time when the contained imago is ready to emerge and to force the fly through soil, etc. (Fraenkel, 1936). This is accomplished by the sac being exserted and distended in front of the head, under pressure from within (Fig. 439). When fully protruded it is in the form of a bladder, which presses upon the wall of the puparium until the latter ruptures. After the emergence of the fly, the ptilinum is

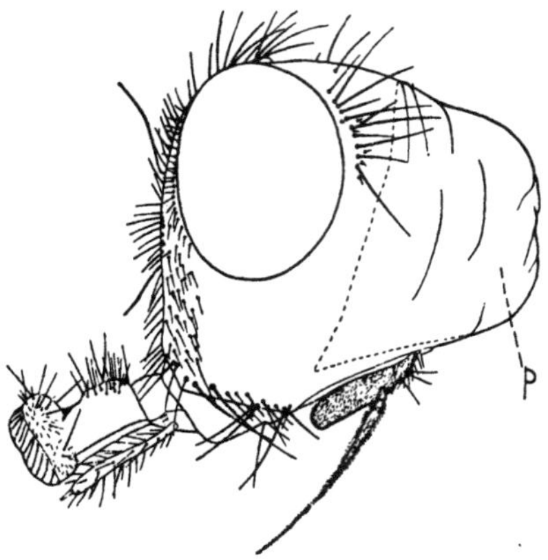

FIG. 439
Head of *Calliphora* (seen immediately after emergence from pupa) with ptilinum *p* inflated

withdrawn into the head-cavity and is no longer functional. The only outward manifestation of its existence is seen in the presence of the frontal suture. In the Aschiza the frontal suture is vestigial or absent and there is no ptilinum. Just above the bases of the antennae in the Cyclorrhapha is a small crescentic sclerite known as the *frontal lunule*: in the Schizophora it is separated by the frontal suture from the part of the head immediately above.

The antennae (Fig. 440) furnish some of the most important characters in the classification of Diptera. They are seen in the least modified condition among the Nematocera, where the flagellum consists typically of a variable

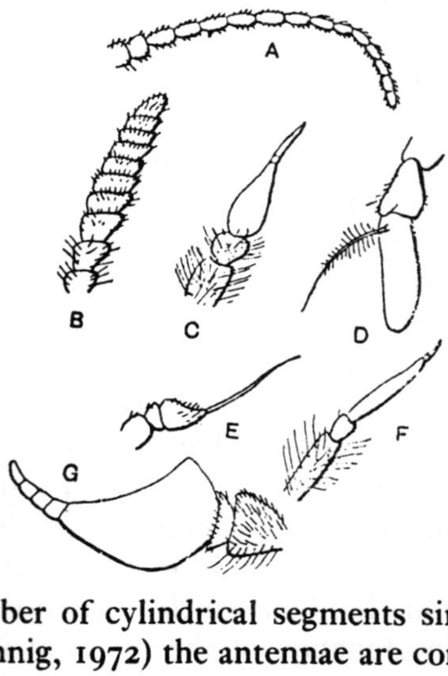

FIG. 440
Antennae. A, Mycetophilid. B, *Bibio*. C, *Empis*. D, *Sarcophaga*. E, *Rhagio*. F, *Bombylius*. G, *Tabanus*

number of cylindrical segments similar to one another. In the Brachycera (Hennig, 1972) the antennae are composed, as a rule, of a smaller number of dissimilar elements. They consist of 2 or 3 evident basal segments carrying a terminal appendage, which corresponds to the greater part of the flagellum in Nematocera. This appendage may be distinctly annulated or jointed, or very much attentuated when it is known as a *style*. If it is still more slender and bristle-like it is termed an *arista*, which is a characteristic feature of the Cyclorrhapha. Morphologically, there is no clearly marked distinction between a style and an arista: the former, however, is always terminal while the latter is usually dorsal and rarely terminal. In the Cyclorrhapha the antennae similarly consists of three basal segments of which the third is the largest and most complex and carries the arista. The various forms of the arista are of classificatory value, and they may be either bare, plumose, or pectinate.

The mouthparts of Diptera exhibit a wide range of structure in adaptation to diverse habits, and there are some differences of opinion in interpreting the morphology of certain of the component parts. The generally accepted homologies as presented by Dimmock (1881) are confirmed by Kellogg (1899) who, from a study of the larval head in Nematocera, observed that the developing imaginal mouthparts are found in unmistakable correspondence or homologous relations with the larval counterparts. A similar conclusion was arrived at by Miall in his study of the head of *Chironomus*. There is,

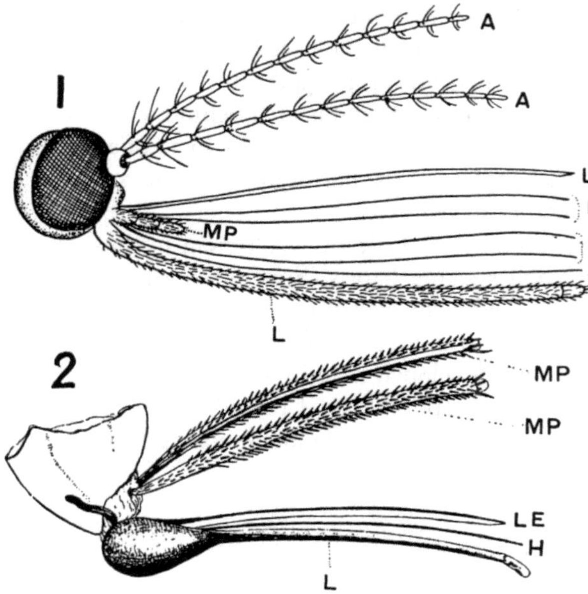

FIG. 441 Mouthparts of 1, *Culex*; 2, *Glossina palpalis*

A, antenna; *H*, hypopharynx; *L*, labium; *LE*, labrum; *M*, mandibles; M_1, maxillae and *MP*, palpi; (2 after Stephens and Newstead).

however, some diversity of opinion with reference to the homologies of the maxillae and their palpi (see Imms, 1944; Hoyt, 1952).

Although the most generalized type of mouthparts in Diptera is far removed from the orthopterous condition, the following features can be recognized. (1) The *labrum* which is dorsally well sclerotized but ventrally more membranous, this surface sometimes being termed the epipharynx. (2) *Mandibles* are absent or reduced except in many of the blood-sucking forms. (3) The *maxillae* are very rarely if ever complete: the basal sclerites may be separate and distinct, or either the cardo or stipes may be wanting. A single maxillary lobe or lacinia (Imms, 1944) is generally evident among Orthorrhapha. It is very variable in development and may be almost filiform (*Exoprosopa*), rod-like (*Sciara, Trichocera*), or totally wanting (*Tipula, Dolichopus*). The maxillary palpi are important for classificatory purposes: they may consist of four complete segments, but in the more highly specialized forms they are reduced to single-segmented organs. (4) The *labium* forms the proboscis which is usually expanded distally to form a pair of prominent fleshy lobes or *labella*. Crampton (1921) brought forward evidence which suggests that the latter organs are the reduced and modified labial palps. In most Nematocera the labella are free, but in the higher Diptera coalescence takes place to a greater or lesser degree. With the beginning of coalescence fine trachea-like food channels or *pseudotracheae* become evident: they attain their most complete development in the Calyptratae where the fusion reaches its maximum. In the majority of Diptera a posterior sclerotized plate is present near the base of the labella and is probably the counterpart of the prementum (Fig. 443), the rest of the mentum being represented by the median membranous area behind. (5) The *hypopharynx* is probably universally present and is either in the form of a lanceolate organ or a greatly attenuated stylet. It is perforated by the salivary duct and is frequently considerably developed.

The mouthparts attain their fullest development in those Nematocera and Brachycera with blood-sucking habits (Snodgrass, 1943; Downes, 1958). In these forms the trophi, with the exception of the palps and labium, are either stylet-like or blade-like, and adapted for piercing. The females, moreover, have exceptionally well-developed mandibles. In the males the latter organs are usually atrophied, except in the Tabanidae, the Simuliidae and the Ceratopogonidae. The labrum in these blood-sucking forms is grooved or U-

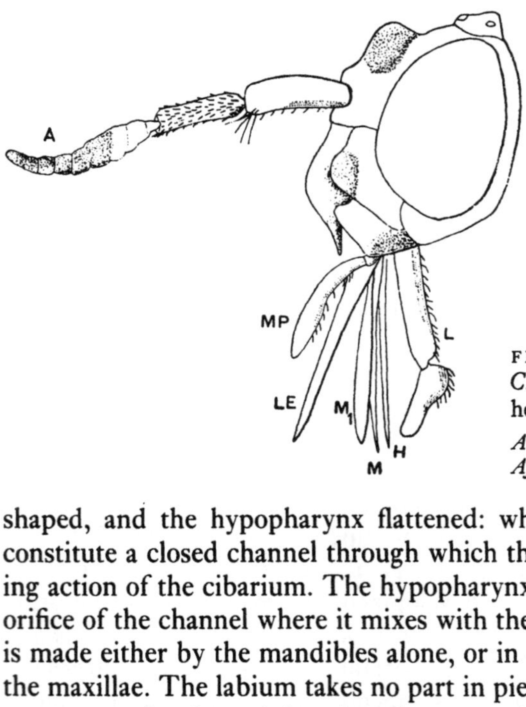

FIG. 442
Chrysops (Tabanidae), lateral view of head × 15

A, antenna; other lettering as in Fig. 441. *After* Surcouf and Gonzalez-Rincones.

shaped, and the hypopharynx flattened: when apposed, the two elements constitute a closed channel through which the blood is drawn by the pumping action of the cibarium. The hypopharynx conveys the saliva to the distal orifice of the channel where it mixes with the blood. The wound on the host is made either by the mandibles alone, or in conjunction with the laciniae of the maxillae. The labium takes no part in piercing: it is grooved dorsally and serves as a sheath retaining the other appendages when at rest (Fig. 441). In the Tabanidae (Fig. 442) both mandibles and maxillae are flattened and blade-like, minutely serrated distally: the labrum is shaped like a double-edged sword, and overlies a similar but more slender hypopharynx. In addition to functioning as a sheath for the other mouthparts, the labium in Tabanids is also an organ for imbibing liquid matter from moist surfaces, which is absorbed by the pseudotracheae present on the labella. In the Culicidae (Fig. 441) specialization has been carried a step further, all the mouthparts are more elongated and the piercing organs are modified into extremely fine needle-like stylets. The labella have many sensory hairs on their distal margins and are mainly tactile in function: the method of feeding in this family is dealt with on pp. 985–6. In the predacious Brachycera (Asilidae and Empididae) the labium is hardened and horny with the labella small, and usually with poorly developed pseudotracheae. The laciniae are rigid and blade-like, being seemingly adapted for perforating the prey: both the labrum and hypopharynx are large and strong.

In most Cyclorrhapha all the mouth-organs contribute to the formation of the proboscis. Its morphology is difficult to appreciate owing to the modification which has resulted through the reduction of the maxillae, and the increased development of membranous areas, in order to impart to the organ the maximum flexibility. The anatomy of the proboscis has been most fully studied in *Calliphora* (Fig. 443). In this genus (Graham-Smith, 1930) it consists of a proximal and somewhat cone-shaped basal portion or *rostrum*,

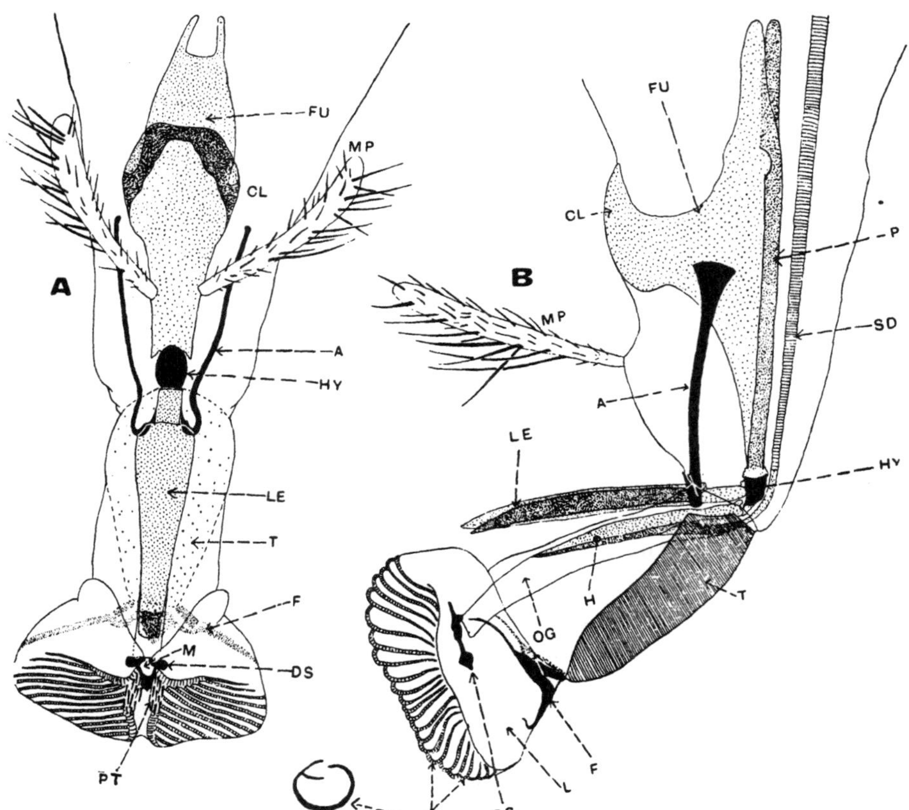

FIG. 443 Proboscis of *Calliphora*. A, frontal; B, lateral view

A, apodeme (stipes); *CL*, clypeus; *DS*, discal sclerite; *F*, furca; *FU*, fulcrum; *H*, hypopharynx; *HY*, theca; *L*, labellum; *LE*, labrum; *M*, mouth; *MP*, maxillary palpi; *OG*, oral groove; *P*, course of pharynx; *PS*, pseudotracheae; *PT*, prestomal teeth; *SD*, salivary duct; *T*, prementum.

and a distal portion or *haustellum*. Morphologically the rostrum belongs to the head and carries anteriorly the *maxillary palps*. Situated within this region is a complex cuticular framework known as the *fulcrum*. According to Snodgrass (1943; 1953), the fulcrum is derived from the inflected margins of the clypeus to which its anterior portion is hinged. The proximal portion of the fulcrum is quadrangular in section, and the distal portion U-shaped, the anterior or roof-like portion being wanting in this region. Between the lower end of the fulcrum and the base of the labrum is a small U-shaped *theca* which lies on the pharyngeal wall, and serves to keep the lumen of the cibarium distended. The haustellum carries the labrum and the hypopharynx on its anterior (or dorsal) face, and these organs are situated in a furrow formed by its projecting membranous sides. The haustellum is continuous with the apex of the rostrum and, on its posterior aspect, it is strengthened by the *prementum*. The latter articulates distally with a short rod or *furca*, and arising therefrom are two divergent arms which form the principal skeleton of the oral lobes. The membrane investing the oral or distal surface of the labella contains a series of food channels or *pseudotracheae* which pass from its outer edges to the inner margins. These channels are kept open by a framework consisting of a series of incomplete sclerotized rings which impart to them an appearance resembling tracheae. Each ring is bifurcate at one end and single at the other – the single and bifurcate extremities alternating. The pseudotracheae open on the external surface of the oral lobes by means of the space which lie between the forked extremities of the rings: inwardly they communicate with the oral aperture. The latter is situated in a small oral pit between the labella: the sides of the depression are bordered by a row of prestomal teeth which are greatly developed in *Ochromyia* and blood-sucking Muscids. The proboscis is adapted for sucking up liquids, and none but the smallest solid particles are able to enter the food channels. Under certain conditions the labella may be strongly reflected which allows of the protrusion of the *prestomal teeth* and their use as rasping organs. When this occurs food can be sucked directly through the oral aperture which admits particles of larger size (e.g. pollen grains in Syrphidae) than could traverse the pseudotracheae. When the proboscis is protruded, the rostrum is extended by means of the distension of the lateral air-sacs at its base, and probably of certain of the cephalic air-sacs also. The haustellum, on the other hand, is brought into use by means of the contraction of its extensor muscles and, finally, the labella are extended and rendered turgid by means of blood-pressure. The retraction of the proboscis is brought about by the contraction of its numerous muscles.

In the blood-sucking Muscidae and the Hippoboscidae and their allies the proboscis itself has become modified to form the principal organ of penetration. It differs from that of most Cyclorrhapha in its horny consistency and swollen bulbous base: owing to the elongation of the haustellum the proboscis can no longer be concealed when retracted. In *Stomoxys* the labella are small oval lobes, devoid of pseudotracheae, and have their outer membrane provided with plate-like teeth adapted for cutting. The labrum and hypopharynx are shorter than the proboscis and, consequently, do not perform any part in the making of the wound: they have, furthermore, thin and flaccid distal extremities. In *Glossina* (Fig. 441) the proboscis is embraced by the elongate palps when at rest, and specialization has proceeded still further. The labella are even less evident, and the slender labrum lies throughout in close contact with the labial groove and, for this reason, has lost much of its rigidity. In *Hippobosca*, *Olfersia* and their allies the basal portion of the proboscis is sunk within the head, the distal part of the organ alone being visible. It bears no labella but the cutting teeth exhibit a bilateral arrangement. The labrum is much stouter than in the preceding genera and, instead of lying within the labial groove, it forms the roof of the latter. The hypopharynx in *Hippobosca* is a slender flattened organ containing the salivary duct between its two layers: at its upper end the dorsal lamina fuses with the labrum and the ventral lamina merges into the lining of the labial groove.

The principal general papers regarding the mouthparts of Diptera are those of Dimmock (1881), Kellogg (1899), Peterson (1916), Frey (1921), Snodgrass (1943), Imms (1944) and Hoyt (1952): for *Glossina* and their allies the reader is referred to papers by Jobling (1926–33).

The Tentorium is characterized by three pairs of arms and a reduced body: the primitive invaginations persist to a greater or less degree in most Diptera as intracranial tunnels. As a rule the most prominent invaginations are those of the anterior arms (well seen in *Chironomus* and *Anopheles*) which are situated some distance below the antennae, and are often located within the arms of the V-shaped suture. The invaginations of the dorsal arms lie just below the bases of the antennae, but as a rule they are wanting: those of the posterior arms are situated near the ventrolateral margins of the occiput (Peterson, 1916).

The Thorax (Fig. 444) is characterized by the great development of its second segment which carries the wings, and the correlated reduction of the segments in front and behind it (Young, 1921). The two latter regions are little more than anterior and posterior bands, whose active function is the support of the fore and hind legs. The sclerites are well exhibited for preliminary study in Tipulidae (Mickoleit, 1962), but among Cyclorrhapha real difficulty will be experienced owing to the specialization which has occurred.

The *pronotum* in Tipulids is represented by a band-like scutum and scutellum, but is still more reduced in the higher Diptera. The *mesonotum* forms the greater part of the dorsal aspect of the thorax and is clearly divisible into prescutum, scutum and scutellum: the postnotum of this segment is well developed in Tipulidae, Culicidae and other Nematocera, but is subdivided among the higher Diptera. The boundary between the prescutum and scutum is known as the *transverse suture* and, although complete and V-shaped in Tipulids, it is generally incomplete in the middle line in

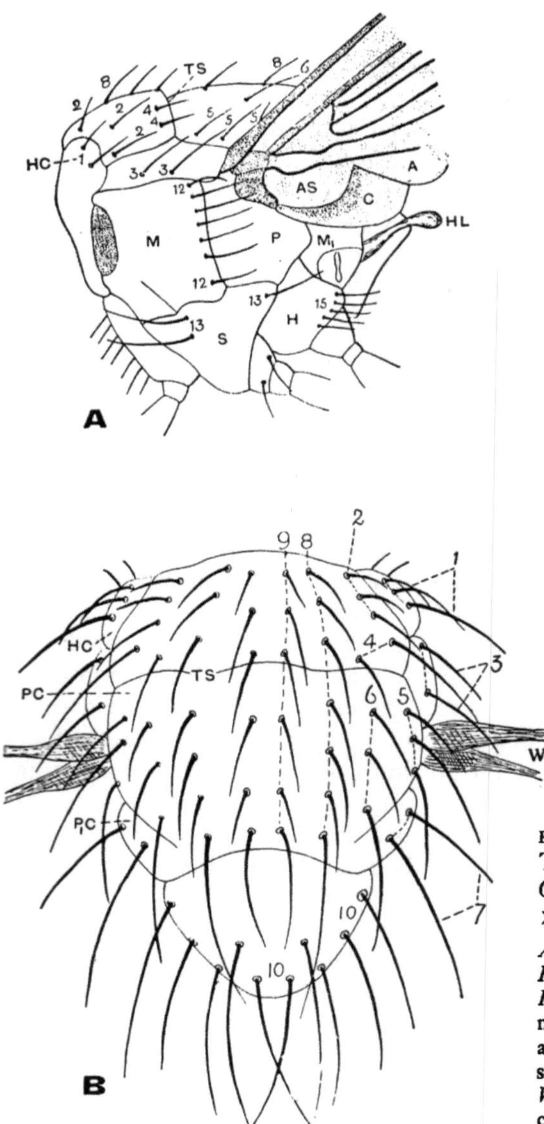

FIG. 444
Thorax of A, *Lucilia caesar* × 11; B, *Compsilura concinnata*, dorsal view × 11

A, alula; *AS*, antisquama; *C*, calyptra; *H*, hypopleuron; *HC*, humeral callus; *HL*, haltere; *M*, mesopleuron; M_1, metapleuron; *P*, pteropleuron; *PC*, prealar callus; P_1C, postalar callus; *S*, sternopleuron; *TS*, transverse suture; *W*, wing-base. The numerals refer to the chaetotaxy, see pp. 961–2. Adapted from Surcouf and Gonzalez-Rincones.

other Diptera. On the mesonotum are certain more or less prominent lateral swellings on either side, which are recognized by systematists, and known as calli or callosities. The *prealar callus* is situated just above the root of the wing, while the *humeral callus* forms the anterodorsal angle of the postpronotum, and the *postalar callus* constitutes the prominent posterodorsal angle of the scutum. The *metanotum* is always reduced and band-like, and is continuous laterally with the epimera.

The pleura do not present any serious difficulties among typical

Nematocera and Brachycera, and both episterna and epimera can be recognized in each segment. Among Cyclorrhapha, however, the interpretation of the pleurites is in a far from satisfactory condition, and the extensive use of chaetotaxy for classificatory purposes demands the definition of these plates with precision. The terminology of Osten-Sacken, although of limited application, has much to recommend it for the somewhat paradoxical reason that it has no strict morphological value. In cases where homologies are uncertain, a purely conventional terminology presents more chances of fixity, and can coexist with the growth of a more scientific system, based upon increasing knowledge of comparative morphology. In Osten-Sacken's nomenclature the pleural regions are identified in relation to certain well-defined sutures. (1) The *notopleural suture*, running from the humeral callus to the wing-base, thus separating the mesonotum from the pleuron: (2) the *sternopleural suture*, running below the notopleural suture and separating the mesopleura from the sternopleura: (3) the *mesopleural suture*, passing downwards from the wing-base to the middle coxa. The *mesopleuron* (= dorsal part of mesepisternum) is the area in front of the root of the wing between the noto- and sternopleural sutures: the *pteropleuron* (= dorsal part of mesepimeron) lies below the root of the wing and behind suture 3: the *sternopleuron* (= ventral part of mesepisternum) is situated below suture 2 and above the anterior coxa: the *metapleuron* (= lateral part of mesopostnotum) lies behind the pteropleura and to the outside of the metanotum: the *hypopleuron* (= meropleurite, composed of ventral part of the mesepimeron and the meron) is the region above the middle and posterior coxae and below the metapleuron (See Fig. 444). For further information on the thorax of Diptera see Osten-Sacken (1884), Edwards (1925*a*), Crampton *et al.* (1942) and Speight (1969).

Chaetotaxy – The study of the arrangement of the *macrochaetae* or differentiated bristles of flies is termed by Osten-Sacken *chaetotaxy*. His important paper (1884) emphasized the value of these structures for classificatory purposes, and their application has been greatly extended by more recent writers, notably Girschner. A knowledge of chaetotaxy is essential for the systematic study of Diptera and the following are the most important of the macrochaetae (Figs. 438, 444).

A. CEPHALIC BRISTLES – 1. *Vertical:* inner and outer pairs situated close to and rather behind the upper inner corner of the eye. 2. *Postvertical:* just behind the ocelli. 3. *Ocellar:* one pair in the ocellar triangle. 4. *Interfrontal:* a double row in front of the ocelli, external to the frontal suture, often descending to the base of the antennae. 5. *Orbital:* one or more on each side of the front near the orbit, behind 4, and immediately below 1. 6. *Facial:* a series above 7, on either side of the face external to the antennae. 7. *Vibrissae:* stout, placed close to the sides of the epistoma. 8. *Postorbital:* a row nearly parallel with the posterior margin of the eye.

B. THORACIC BRISTLES – 1. *Humeral:* one or more on the humeral callus. 2. *Posthumeral:* near the inner edge of the humeral callus. 3. *Notopleural:* one pair between the humeral callus and the base of the wing. 4. *Presutural:* one or more immediately in front of the transverse suture on either side. 5. *Supra-alar:* between

3 and 7, above the root of the wing. 6. *Intra-alar:* several between 5 and 8. 7. *Postalar:* behind 5, on postalar callus. 8. *Dorsocentral:* a row on either side of 9, on the inner part of the mesoscutum. 9. *Acrostichal:* a row along each side of median line. 10. *Scutellar:* along the margin of the scutellum.

C. LATERAL THORACIC BRISTLES – 11. *Propleural:* immediately above coxae of fore legs. 12. *Mesopleural:* on the mesopleura. 13. *Sternopleural:* on the sternopleuron. 14. *Metapleural:* on the metapleura. 15. *Hypopleural:* on the hypopleura.

D. ABDOMINAL BRISTLES – 1. *Marginal:* inserted dorsally on the margins of the segments (Tachinidae). 2. *Discal:* one or more pairs near the middle of the segments. 3. *Lateral:* one or more near the lateral margins of the segments.

The Legs do not call for any detailed mention and, except in a few abnormal forms, the tarsi are 5-segmented. In many Acalyptratae a differentiated bristle is present on the outer border of the tibiae, a short distance below the apex, and quite distinct from the tibial spurs. It is known as the *preapical bristle* and considerable importance has been ascribed to it for classificatory purposes. For the same reason the pads of the feet are noteworthy: thus, pulvilli may be wanting or vestigial in many Nematocera, or may be replaced by a single pad-like arolium (Scatopsinae). In the Stratiomyidae, Tabanidae, etc., both the pulvilli and the arolium are pad-like, while among the Asilidae there is a stiff and bristle-like empodium. Two pad-like pulvilli are the rule among Cyclorrhapha.

Wings are usually present but are wanting or vestigial in a certain number of forms. Apterous or sub-apterous species are principally found in maritime and insular genera (Clunioninae, Ephydridae, etc.), parasites (Hippoboscidae and their allies), and among species inhabiting ants' and termites' nests (Phoridae, *Termitomastus*). Occasional apterous species, not associated with the above modes of life, occur in various families, notably *Chionea, Epidapus* (female) and certain Sphaeroceridae.

The venation of the more generalized members of the order shows a tolerably close approximation to the hypothetical primitive type, the chief differences being the atrophy of Cu_2 and the vestigial condition of 2A and 3A. Neither accessory nor intercalary veins are developed, and only the chief cross-veins are present. A very primitive dipteran wing is seen in the Tanyderid *Protoplasa* which exhibits all four branches of Rs and M, while there is no tendency towards the apical coalescence of adjacent veins. It has been pointed out by Comstock that in all Nematocera, in which Rs is 3-branched, R_2 and R_3 remain distinct: while in those Brachycera that have Rs 3-branched (Fig. 446) R_4 and R_5 are separate. Among certain other of the Brachycera Rs is 2-branched only, and this condition obtains among the Cyclorrhapha. According to Tillyard Cu_1 of Comstock is in reality M_4, while its basal section is Comstock's *m-cu* cross-vein. The lettering of the venational figures is in accordance with this interpretation.

On the posterior margin of the wing, near the base, there is frequently a free lobe or *alula*, and on the inner side of the latter there are often one or two additional lobes or *squamae*. When two squamae are present, the one nearest the alula is known as the *antisquama*, the squama being the lobe nearest the thorax (Fig. 444). In the Calyptratae the squama is large, usually covering the haltere, and is often referred to as the *calyptron* (or calypter). All three lobes are well seen in *Musca* and *Calliphora*.

With the exception of a few apterous forms (e.g. *Melophagus, Braula*, etc.), *halteres* (balancers) are universally present among Diptera. They develop from the dorsal metathoracic wing-buds, and are consequently the highly modified counterparts of the posterior wings. Further evidence for their origin is seen in certain mutations described by Morgan in *Drosophila* in one of which the halteres are replaced by hind wings with clearly recognizable venation (Fig. 445). Each haltere consists of a dilated basal portion or

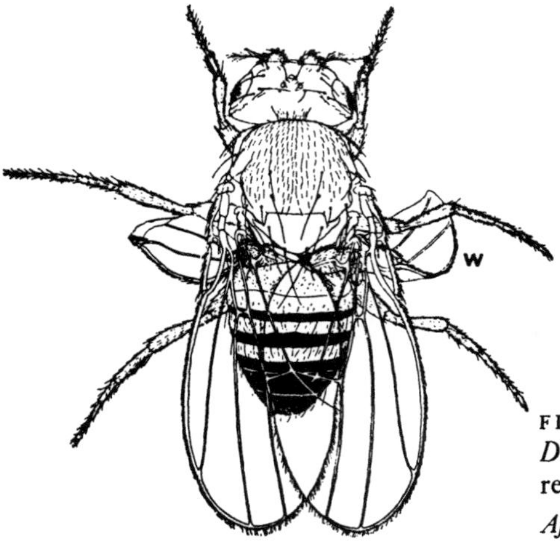

FIG. 445
Drosophila melanogaster: Mutant with halteres replaced by hind wings *W*
After T. H. Morgan, *Publ. Carnegie Inst.*, 327, 1923.

scabellum, which supports a delicate *pedicel* or stalk, surmounted by a knob-like extremity or *capitellum*. The scabellum articulates freely with the metathorax, and is moved by four muscles arising from its proximal border (Lowne): the halteres are, therefore, freely movable and are capable of vibration. It is in the scabellum that the principal sensory structures of the haltere are located. These consist in *Calliphora* of three groups of minute so-called chordotonal organs invested by a thin integument, and three highly sculptured elevations of the cuticle containing larger and more complex structures – the two *scapal organs* (scalae of Lowne) and the *basal organ* (cupola of Lowne). Both scapal and basal organs exhibit thin transparent areas, each of which overlies a minute vesicle enclosing a central refractive spot. The cavity of the haltere contains blood and a fine tracheal branch. The nerve supplying this appendage is the largest in the thorax. Binet (1894) has demonstrated that the majority of its fibres arise from the brain, and traverse the thoracic ganglia on their course to the metathoracic centre; from there they pass onwards to the scabellum, and are distributed to the several sense organs (Lowne, 1890; Weinland, 1890). Pringle (1948) has shown that the halteres, which during flight are rapidly oscillated, function as gyroscopic organs and allow the fly to control any tendency to roll or yaw. Experiments conducted with certain species show that, if the capitellum and part of the pedicel of a haltere be amputated, flight becomes clumsy and difficult: if both halteres are treated alike the power of flight is almost entirely lost, and insects so mutilated can only fly a few centimetres, and usually fall vertically when thrown into the air.

In the **Abdomen** the first segment is usually much reduced. Of the segments that follow the 2nd–11th are present in *Tipula*, but among the Cyclorrhapha the number is difficult to ascertain, and rarely more than 4 or

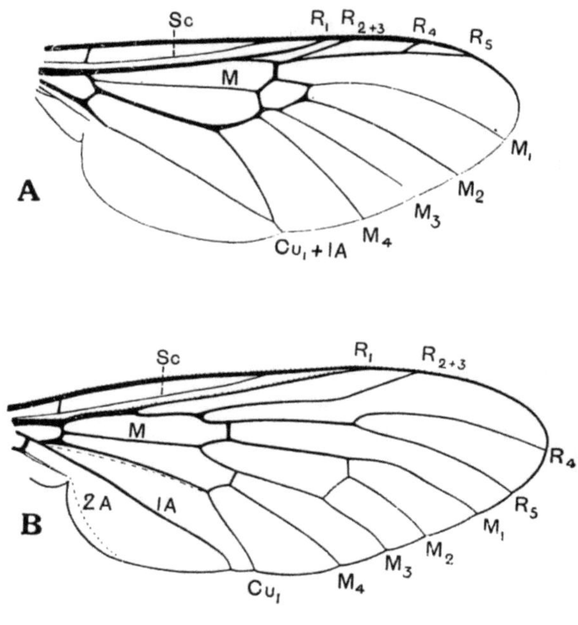

FIG. 446 Venation of Brachycera. A, *Geosargus* (Stratiomyiidae); B, *Rhagio*

5 are evident without dissection. In *Dacus* Miyake (1919) finds 11 segments present in both sexes, the 1st segment being represented by its reduced sternum. In the female *Musca* the visible segments are the 2nd to the 5th, while the 6th to 10th segments form the rectractile functional ovipositor or *oviscapt*. The latter organ is formed in this manner in the majority of Diptera, but in the Tipulidae a valvular ovipositor may be present (Snodgrass, 1903). In the male, the 9th and 10th segments are often curved ventrally and with the genitalia form the *hypopygium*. In a number of families, the hypopygium is twisted through 180°, so that the anus may become ventral to the genital aperture (*hypopygium inversum*). Segments 6–8 may in such cases become much modified and asymmetrical; the structures found in the Cyclorrhapha are still very difficult to interpret (Lamb, 1922; Feuerborn, 1922; Crampton, 1944; Hardy, 1944; Salzer, 1968; Fletcher, 1970; Griffiths, 1972). They have a *hypopygium circumversum* in which after

inversion the anus has apparently moved back into a position dorsal to the genitalia though it will be found that the ductus ejaculatorius is looped round the hind gut. A circumverse hypopygium is also found in some Dolichophodidae (Bährmann, 1966). The genitalia include *claspers*, made up of a movable coxite and style, and an aedeagus with accessory structures lying between sternites 8 and 9 (Edwards, 1920).

There are two *spiracles* in the thorax and may be eight in the abdomen. Their position in the latter and the number which remain functional is modified in higher groups. Their regulatory mechanism is discussed by Hassan (1944).

Internal Anatomy

The Alimentary Canal is generally only slightly convoluted among Nematocera, but is more coiled among Brachycera. In the Cyclorrhapha it exhibits greater complexity, its length being much increased mainly owing to the greater extension of the mid intestine (Singh and Judd, 1966) (Fig. 98).

The cibarium forms the sucking apparatus by means of which the food is drawn up through the proboscis and passed into the oesophagus. The original circular lumen in these parts becomes modified, and the sclerotized lining is developed as two or three hardened plates. The latter afford a basis for the attachment of dilator muscles, and are capable of being drawn apart by their contractions. In this manner the lumen is increased, and the food sucked up through the siphon formed by the mouthparts. In the Brachycera, the anterior part of the pharynx is also furnished with muscles and forms a second pump. In Culicidae the cibarium is provided with dilator muscles but the principal pumping apparatus is pharyngeal. The blood is first pumped into the cibarium and from thence it passes into the pharynx, a valve situated between these two regions precluding a return flow.

The *oesophagus* passes through the neck into the thorax where it divides. One branch enters the proventriculus and the other is continued backwards as the slender duct of the food reservoir. The *proventriculus* is the homologue of the gizzard and has a well-marked musculature: it never contains denticles and a valve is usually present. The proventriculus is wanting in *Phlebotomus*, *Simulium* and *Culicoides*, is elongate and tubular in *Tabanus* and in Cyclorrhapha it is much reduced and disc-like, mainly consisting of its valvular portion. The *food-reservoir* (or crop) is the most characteristic feature of the digestive canal. It is situated in the anterior region of the abdomen and is, morphologically, a diverticulum of the oesophagus. Although present in most families it is wanting in certain Asilidae, Oestridae and in *Hippobosca* and *Melophagus*. In *Musca* it is a bilobed sac with very thin walls composed of a single layer of small flattened cells, external to which is a network of muscle fibres; internally it is lined by a delicate cuticle. The usual position of the food reservoir and its duct is ventral, but in *Tabanus* these parts are dorsal. In the Culicidae, instead of a single sac, three oesophageal diverticula are present, of which two are dorsolateral, while a third and larger sac is ventral. The function of the food reservoir is that of a storage chamber into which the nutriment is passed as it is sucked up: its contents then become gradually emptied into the mid gut. The time the food remains in the reservoir varies greatly: thus in *Musca* it may not be emptied for several days, while in *Tabanus* it is usually empty and possibly its contents are quickly regurgitated into the mid gut (Patton and Cragg). As a rule, after a meal the reservoir is distended with food, as has been demonstrated by allowing flies to feed upon a coloured liquid.

The *mid intestine* in Nematocera is a pyriform or fusiform sac: in the Culicidae its anterior region, or cardia, is elongate and tubular, and leads into a dilated chamber or stomach. Among Cyclorrhapha the mid gut is no longer dilated but is tubular throughout, and thrown into numerous convolutions. It is divisible into an anterior region – the *ventriculus* or *chyle stomach*, followed by a narrower and much longer *proximal intestine*. The *Malpighian tubules* are generally four in number: in most Cyclorrhapha they arise in pairs from a common duct on either side. *Psychoda* and the Culicidae are exceptional in possessing five Malpighian tubules: in *Culicoides* there are only two (Bugnion).

The *hind intestine* is divisible into the *distal intestine* and *rectum* which are often separated by a *rectal valve* (Singh and Judd, 1966; Hori, 1967). The former, in many Diptera, is naturally separable into a narrow coiled *ileum* and a wider region or *colon*. The rectum is a pyriform or rounded chamber provided with a variable number of papillae which may be either two (*Chironomus*), four (*Musca*, *Calliphora*, etc.) or six (*Anopheles* and *Tabanus*).

The *salivary glands* are usually elongate and tubular but exhibit great variation in length. In the Culicidae they are situated in the thorax and each gland is trilobed: a layer of secretory cells surrounds the cavity of each lobe, and the smaller central lobe (formerly known as the poison gland) differs somewhat in histological features. The salivary gland cells often contain giant chromosomes which have provided much information for taxonomic and evolutionary studies. The common salivary duct passes to the base of the hypopharynx, where it expels the secretion down the salivary groove to the apex of that organ. In the Tabanidae the glands extend into the anterior part of the abdomen, while in *Musca* they are considerably longer than the total length of the body.

Labial Glands are frequently present on the proboscis at the bases of the labella. In *Musca* they are spherical aggregations of gland cells. According to Hewitt (1914) the ducts are intracellular, each arising from a vacuole. They pass outwards from the gland to form a number of larger ducts which unite and open into the oral pit by means of a pair of median pores. The secretion of the labial glands serves to moisten the surface of the labella.

The Nervous System (Brandt, 1879; Künckel d'Herculais, 1879) presents many modifications, almost every transition being found between the Nematocera, with 3 thoracic and 7 abdominal ganglia, and the

Calyptratae in which all of the ganglia of the ventral chain are fused into a single thoracic mass (Fig. 66). There is, furthermore, a marked relation between the degree of concentration of the nervous system and specialization in other directions. A graduated series illustrating the progressive concentration of the nervous system may be exemplified as follows.

1. Two or three thoracic centres and always six abdominal centres: 1st abdominal ganglion united with the metathoracic and the 7th and 8th abdominal ganglia fused (most Nematocera also Asilidae, Empididae, Bombyliidae and *Xylophagus*). 2. Three thoracic and five abdominal centres (Scenopinidae). 3. Two thoracic and four abdominal centres (Therevidae). 4. Two thoracic and no abdominal centres (Dolichopodidae). 5. One thoracic and five abdominal centres (Tabanidae, Stratiomyidae). 6. One thoracic and two abdominal centres (Syrphidae). 7. One thoracic and one abdominal centre (Conopidae and most Acalyptratae). 8. A single thoracic centre (Calyptratae).

In the Nematocera, and also the Rhagionidae and Asilidae, the nervous system of the imagines exhibits only a slightly greater concentration than in their larvae. Stratiomyidae, Syrphidae, Conopidae and certain Acalyptratae exhibit decentralization in the imago compared with the larva. In the Calyptratae the concentration of the larval nervous system is persistent in the imago. In *Musca* and other Calyptratae the nervous system exhibits the highest stage of concentration. The brain and infra-oesophageal ganglion are closely united to form a compact mass perforated by a foramen for the oesophagus. The thoracic and abdominal ganglia are intimately fused to form a common ganglionic mass situated in the thorax. Posteriorly, the nervous system is prolonged as a median abdominal nerve giving off lateral segmental nerves, two pairs in the thorax, and the remainder in the abdomen (Fig. 66D).

In the **Female Reproductive System** there is a variable number of polytrophic ovarioles (Fig. 447). The latter are fewest in number in larviparous species: thus in *Glossina*, *Musca bezzii* and *Termitoxenia*, a single ovariole is present on each side, while in *Melophagus* and *Hippobosca* there are two. The majority of Diptera, however, are oviparous and the ovarioles are much more numerous, their number varying from about 5 to over 100. In *Chironomus* the morphology of the ovaries is peculiar: each consists of a central axis radiating out from which is a large number of short ovarioles, the whole being enclosed in a delicate membrane (Miall and Hammond).

Spermathecae are universally present: they are usually conspicuous dark brown or black globular sacs, lined with thick cuticle. There may be a single spermatheca present (*Anopheles*, *Simulium*), or two (*Mansonia*, *Phlebotomus*, *Dacus*), or three (*Culex*, *Aedes aegypti*, the Tabanids and most Calyptratae). A pair of tubular accessory glands is usually present opening into the dorsal region of the vagina. Small and unpaired in *Anopheles*, they are large in Tabanids, elongate and filiform in *Musca*, *Hypoderma* and most other Calyptratae. Their normal function apparently is to secrete a viscid substance which enables the eggs to adhere to one another or to the substratum

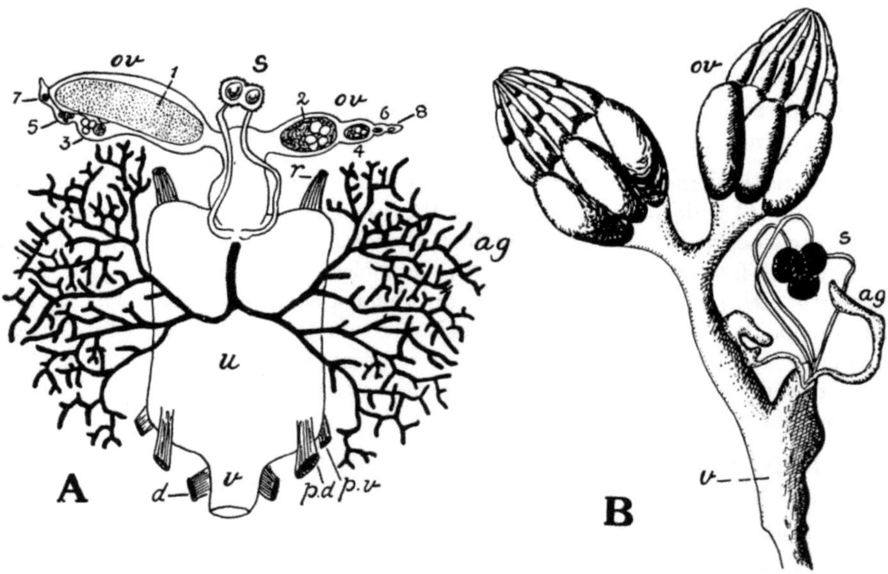

FIG. 447 Female reproductive organs. A, *Glossina palpalis*, adapted from Roubaud, 1908; B, *Limnophora* (*Melanochelia*) *riparia* (Muscidae), *after* Keilin, 1917

ag, accessory glands; *d*, dilator muscle of vagina, *v*; *pd*, *pv*, dorsal and ventral protractor muscles of uterus, *u*; *r*, retractors of same; *ov*, ovary; *s*, spermatheca. The numerals in A refer to the relative ages of the oocytes.

upon which they are laid. In *Glossina* and its allies they secrete a milky fluid which serves to nourish the intra-uterine larva.

Viviparity is not infrequent among Calyptratae and is of general occurrence in *Glossina* and its allies: among other Diptera it is exceedingly rare but occurs in *Chironomus stercorarius*. Viviparous Diptera may be divided into two main groups as follows (Keilin, 1916):

GROUP I – Species whose larvae hatch from the eggs in the uterus of the parent but exhibit no special adaptations to an intra-uterine life. Included in this group are numerous Tachinidae which produce a large number of minute eggs and the larvae emerge within the uterus which is greatly elongated for their reception. In some species the larvae hatch outside the parent, immediately after the eggs have been laid. Larviparity is characteristic of the Sarcophaginae, but in this group the eggs are larger and fewer: usually 40 to 80 are produced at a time and the larvae are deposited in their first instar. In a number of other species (*Theria muscaria* = *Helicobosca distinguenda*, *Mesembrina meridiana*, *Hylemyia strigosa* = *strenua*, *Musca larvipara*, *Dasyphora pratorum*, etc.) a single very large egg is produced at a time and the parental uterus is enlarged to form an incubatory pouch. In these instances the larva is retained for a variable time within the parent before deposition and the extreme condition is afforded by *Dasyphora pratorum* in which it has attained its 3rd instar at the time of extrusion.

GROUP II includes *Glossina* and allied families. The larva lies in the uterus of the parent and is nourished by the product of special nutritive glands. The secretion is discharged at the apex of a papilla and absorbed directly through the mouth of the larva. The following special adaptations to an intra-uterine life are exhibited. The buccal apparatus is reduced to a single basal sclerite: the mid gut is a closed sac which does not communicate with the hind intestine and, moreover, is greatly elongated to form a food-reservoir; there are no salivary glands. The hind intestine is greatly shortened and forms a receptacle for the accumulation of waste products excreted by the Malpighian tubes. The larvae when deposited are mature and shortly afterwards pupate.

In the **Male Reproductive System** (Keuchenius, 1913) the *testes* are, as a rule, ovoid or pyriform and frequently pigmented. The vasa deferentia are generally short and become confluent distally to form a common ejaculatory duct. In association with the latter, in many Diptera, is a muscular *ejaculatory sac* probably concerned with regulating the discharge of the seminal fluid. Paired accessory glands are often present.

In *Chironomus*, *Phlebotomus* and *Tabanus* the first portion of the common genital canal is enlarged and functions as a *vesicula seminalis* from which a narrow ejaculatory duct leads to the aedeagus; in these genera accessory glands are wanting. In *Culex* each vas deferens enlarges distally to form a vesicula seminalis, and two pyriform accessory glands open into a very short ejaculatory duct. In *Musca* there are no accessory glands and the ejaculatory duct is a long winding canal: *Calliphora* (Graham-Smith, 1939) (Fig. 147) resembles *Musca* but differs in the presence of accessory glands. In *Dacus* the latter consist of about 16 blind tubules (Miyake, 1919), while in *Hypoderma* there is a small unpaired globular gland. In the Hippoboscidae and allied forms the genital organs attain their greatest complexity, and the testes are in the form of compactly coiled tubules resembling balls of thread. The reproductive organs of *Glossina* resemble those of Hippoboscids rather than of Muscidae, the testes being similar densely coiled tubes. The ejaculatory sac is an organ of variable structure: in *Musca* it contains a sclerotized, phylliform *ejaculatory apodeme* which aids in propelling the seminal fluid along the genital canal during copulation (Hewitt). In *Dacus* the ejaculatory sac is very large, while in *Phlebotomus* its place appears to be taken by an organ termed by Grassi the 'pompetta' – a piston-like chamber provided with a movable rod: since the opening of the ductus ejaculatorius is near the lower end of this chamber, the latter is believed to regulate the seminal flow after the manner of a pump.

The Heart has been very little investigated: in *Musca* (Hewitt) and *Calliphora* (Lowne) it is divided into four large chambers, corresponding to the visible abdominal segments, and a small anterior chamber: each chamber in *Musca* has a pair of dorsolateral ostia at its posterior end. Anteriorly the heart is prolonged as a tube of narrow calibre.

The most important feature of the **Tracheal System** is the great development of air-sacs, particularly among Cyclorrhapha. In *Musca* and *Volucella* the air-sacs occupy more space than any other organs, and the haemocoele is consequently much reduced. The largest and most prominent of the air-sacs are the abdominal: numerous sacs are also present in the thorax and head (Hewitt, 1914; Künckel d'Herculais, 1879).

Literature – General works on the anatomy of adult Diptera are extremely few: a good deal of information will be found in the writings of Dufour (1851) and the textbook of Patton and Cragg (1913). For the detailed structure of individual types see Miall and Hammond (1892) for *Chironomus*; Christophers (1960) for *Aedes aegypti*; Nuttall and Shipley (1901–03) for *Anopheles*; Künckel d'Herculais (1875) for *Volucella*; Hewitt (1914) for *Musca*; Lowne (1890) for *Calliphora*; Tulloch (1906) for *Stomoxys*; Cragg (1912) for *Haematopota*; and Roubaud (1909) for *Glossina*; Owsley (1946) for Asilidae.

Eggs

Dipteran eggs often have a simple external structure, particularly if they develop in permanently wet habitats. A few species (e.g. some *Tipula*) undergo a short diapause before development. The eggs of a number of genera, especially those which lay in such substances as cow dung or fermenting matter, have one or more tubular flattened projections which form a plastron. These project outside of the substrate and enable the insect to breathe with great efficiency when covered by a film of water (Hinton, 1960*a*, 1963). Some eggs, notably those of some Culicids, are organized into groups, called egg-rafts in this family.

Larvae

No other order of insects presents so great a diversity of larval habits as the Diptera. Only four families have the great majority of their species phytophagous in the larval state, i.e. Cecidomyidae, Tephritidae, Agromyzidae and Chloropidae, while the Mycetophilidae and Platypezidae are fungivorous. The saprophagous habit is largely in evidence among the Anthomyidae. Other notable scavengers are the Bibionidae, Sepsidae, Phoridae, Heleomyzidae and Scatophaginae. True parasitism, either internal or external, obtains in the Tachinidae (s.l.), Oestridae, Pipunculidae, Conopidae, Bombyliidae, Acroceridae, Nemestrinidae and in a few Acalyptrates. Next to the parasitic Hymenoptera, the Diptera constitute the most important natural controlling agency over the increase of other insects. The predacious habit occurs in many families, particularly among the Brachycera, and in numerous members of the Syrphidae and Muscidae. With the exception of many Sciomyzidae and Ephydridae, the truly aquatic larvae belong mostly to the Nematocera and to the Stratiomyidae and Tabanidae among the Brachycera.

In their larval instars many Diptera affect the operations of man or his person. The four phytophagous families enumerated above include some of the most serious pests of agriculture and fruit-growing. The larvae of the pear and wheat midges, of the Mediterranean fruit fly, the frit and gout flies are cases in point. Among the Muscidae, the larvae of the cabbage root fly and onion fly do much damage to those vegetables.

The science of parasitology is concerned with many Dipteran larvae which directly affect the bodies of man and domestic animals. Under the term *myiasis* are included all affections produced by the larvae among vertebrates, and more particularly mammals. The species concerned are either parasitic or saprophagous, and it is frequently possible to distinguish between primary myiasis, which is induced by true parasites, and secondary myiasis which is brought about by saprophagous larvae. The latter only follows on diseased conditions or wounds, and usually where there is microbial infection.

From the clinical standpoint myiasis in man may be grouped as follows:

1. CUTANEOUS MYIASIS: the larvae primarily concerned are those of *Dermatobia*, *Cordylobia* and *Bengalia*. Species of *Hypoderma* and *Gasterophilus* also occasionally induce myiasis. 2. MYIASIS OF THE CRANIAL CAVITIES (orbital, nasal and auditory): caused by larvae of *Oestrus*, *Rhinoestrus*, *Gasterophilus* and *Dermatobia*. When of a secondary nature it is commonly due to larvae of *Sarcophaga*, *Musca*, or *Cochliomyia*: auditory myiasis appears to be always of a secondary nature and follows some purulent affection of the ear. 3. MYASIS OF THE DIGESTIVE CANAL: larvae of at least 18 genera have occurred in the alimentary tract, but probably many pass through without causing recognizable symptoms.

In almost all cases of human myiasis the occurrence of Dipterous larvae is occasional and their presence is a departure from their normal host or mode of life.

Dipterous *larvae* (Fig. 448) are devoid of true legs, locomotion often either taking place by means of pseudopods, or by the aid of groups of shagreen-like spinules, frequently located on swellings of the body-wall. The greatest number of undoubted segments present is twelve, e.g. three thoracic and nine abdominal. Departures from this generalized condition are not infrequent: thus in some larvae the number is less than twelve, atrophy or fusion of one or more of the somites having taken place. In larvae possessing more than twelve apparent segments (Anisopodidae and Therevidae) two explanations have been offered. Either certain segments have undergone secondary division, or the intersegments have become greatly enlarged so as to assume the appearance of true segments. Keilin (1915) has shown that six groups of sensory organs are present in all Dipteran larvae, and are in direct relation with the imaginal leg-buds, thus occupying the positions of the ancestral thoracic limbs.

The number of families in which a well-developed *head* is present is small; it is fully formed in the Culicidae, which are described as being 'eucephalous', as well as in most other Nematocerous larvae (except Cecidomyidae and Tipulidae). At the opposite extreme is the so-called 'acephalous' condition present in the Cyclorrhapha, where the head is vestigial (Fig. 448). Many Dipteran larvae (e.g. Brachycera) are in a 'hemicephalous'

FIG. 448 A typical Cyclorrhaphous larva (*Hylemyia*)
h, head; *a.s*, *p.s*, anterior and posterior spiracles.

or intermediate condition, a reduced head or 'jaw-capsule' being present. This type of head is incomplete posteriorly and can be withdrawn into the thorax. A similar condition is found in the Tipulidae, the skin of the neck being attached to the middle region of the head with the result that the latter is incomplete posteriorly, and permanently imbedded within the body.

The *antennae* are variously formed, very rarely prominent, and are composed of 1 to 6 segments. They are best developed in active larvae which need to seek out their food (Nematocera). In the Mycetophilidae, many Brachycera and most Cyclorrhapha they are reduced to the condition of small papillae.

The *mouthparts* are variable in character in different groups and are exhibited in their least modified condition in certain families of Nematocera. Thus, in *Bibio* (Fig. 449) there is a definite labrum, mandibles are well developed and move in the horizontal plane and the maxillae are represented by a single lobe or mala and an evident palpus on either side. The labium is in the form of a median plate with a strongly sclerotized hypopharynx lying above it on the pharyngeal aspect: labial palpi are wanting. Among the Brachycera the same parts, although variously modified, are more or less evident but the mandibles work in the vertical plane. In the Cyclorrhapha the typical mouthparts have undergone atrophy in correlation with the reduction of the head: the maxillae and labium are scarcely recognizable other than by the papillae representing their palps. In this group of Diptera there is a very characteristic framework of articulated sclerites, the whole being known as the *cephalopharyngeal skeleton* (Fig. 450) (Muirhead-Thomson, 1937). This structure is a secondary development and is composed in the mature larva of the following principal sclerites. The most anterior are the *mouth-hooks* or *mandibular sclerites* which articulate basally

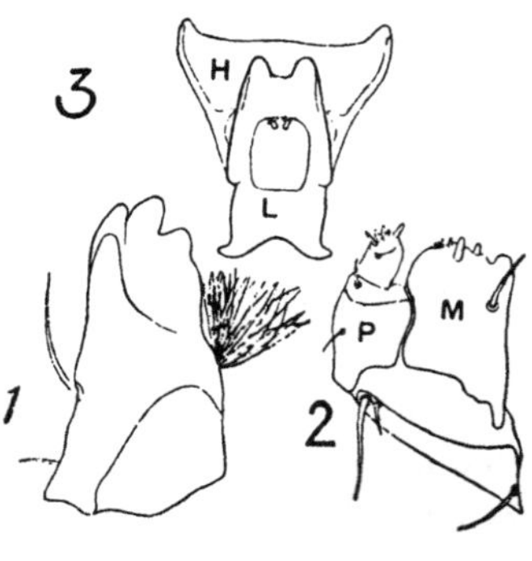

FIG. 449
Mouthparts of larvae of *Bibio marci*
1, mandible. 2, maxilla. *P*, stipes; *M*, mala. 3, *L*, labium; *H*, hypopharynx. *After* Morris.

with the hypostomal or *intermediate sclerite.* The latter is H-shaped, its halves being joined by a transverse bar: the hypostomal sclerites receive the opening of the salivary duct. Behind this sclerite is the much larger *basal* or *pharyngeal sclerite.* The latter is formed of two lateral, vertical lamellae which unite ventrally forming a trough in which is lodged the pharynx. In many species a cuticular arc (*dentate sclerite*) unites the bases of the mandibular sclerites: various other small accessory sclerites are frequently present, notably in carnivorous species.

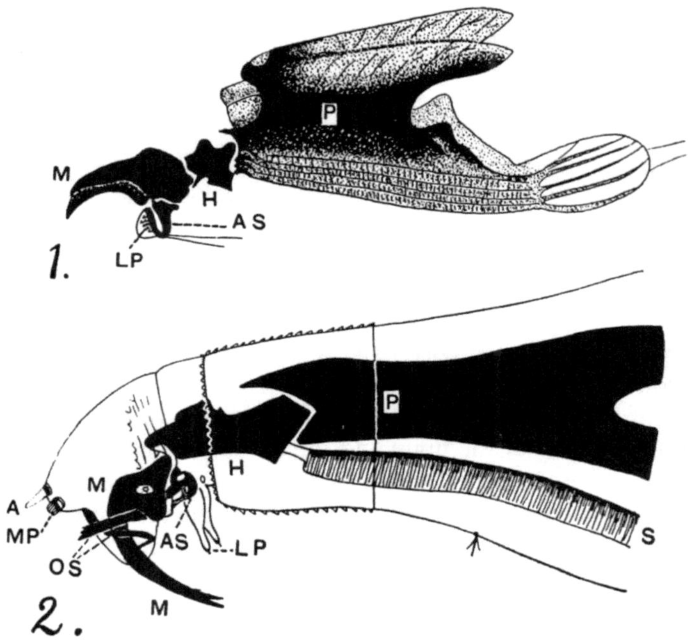

FIG. 450 Cephalopharyngeal skeleton of larvae of 1, *Musca domestica* and 2, *Limnophora* (= *Melanochelia*) *riparia*

A, antenna; *AS*, dentate sclerite; *H*, hypostomal sclerite; *LP*, labial palp; *M*, mandibular sclerites; *MP*, maxillary palp; *OS*, accessory oral sclerites; *P*, pharyngeal sclerite; *S*, salivary duct. Adapted from Keilin, *Parasitology*, 9, 1917.

Keilin has shown that in saprophagous larvae the floor of the pharyngeal sclerite is beset with longitudinal ridges which project into the cavity of the pharynx: larvae feeding on living animal or vegetable tissues are devoid of pharyngeal ridges or, if the latter be present (as in *Pegomyia*) they are reduced (Fig. 451) (cf. Hennig, 1935). Furthermore, in phytophagous larvae the mandibular sclerites are usually toothed, and in carnivorous larvae they are sharply pointed: in the parasitic forms the buccal armature undergoes marked reduction.

The profound changes which have led to the reduction of the head and the atrophy of the normal biting mouthparts in the larvae of the

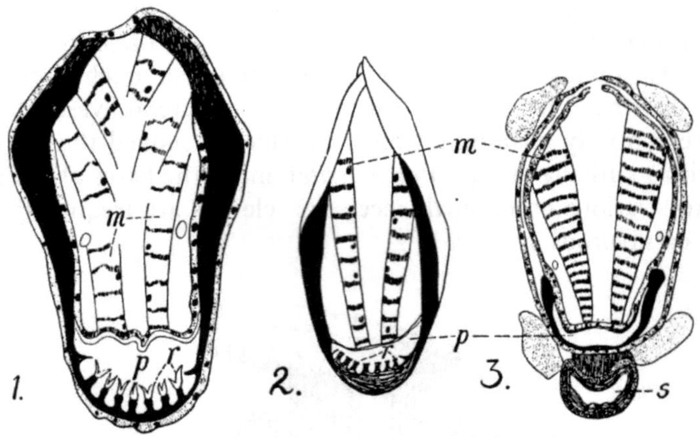

FIG. 451 Transverse sections of the pharynx of Dipterous larvae. 1, *Muscina assimilis* (Muscidae) (saprophagous). 2, *Pegomyia nigritarsis* (Anthomyiidae) (phytophagous). 3, *Syrphus ribesii* (carnivorous)

m, dilator muscles; *p*, cavity of pharynx; *r*, ridges; *s*, salivary duct. *After* Keilin, 1915.

Cyclorrhapha appear to be correlated with the two series of factors: firstly, degeneration consequent upon a life passed in the immediate proximity of an abundance of food, and also a change in the manner of feeding; secondly, to the backward shifting of the brain and the development of the imaginal head within the larval metathorax. For a general discussion of the head and its modifications among Dipteran larvae, and the structure of the mouthparts, the student is referred to the writings of Holmgren (1904), Becker (1910), de Meijere (1916), Keilin (1915), Keilin and Tate (1940), Anthon (1943), Cook (1949) and Hennig (1948–52).

Although Dipteran larvae are apodous in the true sense of the term pseudopods are present in numerous genera. According to Hinton (1955), leg-like structures have evolved on at least 27 independent occasions in the Diptera. Thus in *Chironomus*, *Thaumalea* and *Simulium* a pair is present on the prothoracic and anal segments. In *Dicranota* five pairs are evident on segments 6 to 10: in *Eristalis* there are 7 pairs while in *Atherix* and *Clinocera* there are 8. Circlets of pseudopods are present in the abdominal region in *Laphria* and the Tabanidae.

The *tracheal system* (Keilin, 1944; Whitten, 1955, 1960*a* and *b*) presents features of great systematic value and the most prevalent type is the amphipneustic one (Fig. 452). The primitive or peripneustic condition is almost entirely confined to Nematocera: the maximum number of pairs of spiracles present is 10 (*Bibio*) while 9 pairs occur in Scatopsinae, Cecidomyidae and a few others. Indications of a former peripneustic condition are seen in other forms in the presence of solid stigmatic cords leading from spiracular scars to adjacent tracheae. Such scars show where the trachea was pulled out of the body at the previous moult. The larvae of Cyclorrhapha, when newly

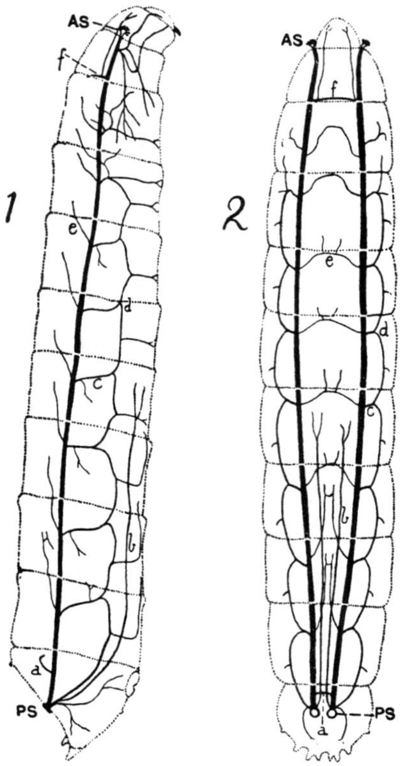

FIG. 452
Amphipneustic tracheal system of 3rd instar larva of *Hylemyia* (Anthomyoidae)
1, lateral; 2, dorsal view. *AS*, anterior spiracle; *PS*, posterior do.

hatched, are metapneustic, becoming amphipneustic in the 2nd and 3rd instars. Among the Aschiza, the prothoracic spiracles in some genera are apparently non-functional, the tracheal system being physiologically metapneustic. The propneustic condition is extremely rare but is stated to obtain in *Polylepta*. Apneustic larvae are found in *Chaoborus*, in Chironomidae and also in *Ceroplatus* and *Atherix*.

In peri- and amphipneustic larvae the 1st pair of spiracles is carried on the prothorax: in the former condition the 2nd pair is borne on the metathorax or 1st abdominal segment, never on the mesothorax. The segmental position of the posterior pair of spiracles is variable: it is frequently on the last segment whether it be the 11th, as in many Brachycera, or the 12th segment, as in *Dicranota*, *Ptychoptera* and numerous Cyclorrhapha. In the Culicidae and certain Brachycera the spiracles are situated on the penultimate segment, and in the Therevidae and Scenopinidae on the ante-penultimate segment. In certain metapneustic larvae (Culicidae, *Dixa, Psychoda*, some Tipulidae, etc.) the two main tracheal trunks give off a plexus of fine tracheal branches in the neighbourhood of the spiracles, and pass to the walls of the posterior region of the heart. These branches are very thin-walled, and it appears probable that the blood is brought into close contact with the oxygen contained therein, and in this way they function as a kind of lung (Imms, 1907).

Accessory respiratory organs in the form of gills are found among aquatic larvae. In certain Chironomids two pairs of 'blood-gills' are situated on the 11th segment, and a similar number of smaller blood-gills are present around the anus. In Culicidae (Wigglesworth, 1933), these structures have been shown to extract chloride ions from the water. Tracheal gills are much more frequent than blood-gills: they may be ventral as in the two pairs on the last segment of *Dicranota*, segmental as in *Phalacrocera*, caudal as in Culicidae, or rectal as in *Simulium* and *Eristalis*. The caudal retractile processes of *Pedicia* and other aquatic Tipulid larvae are probably of a similar nature.

The *alimentary canal* in larvae of the Nematocera, and certain Brachycera, is a short tube but little convoluted. In most Cyclorrhapha (Guyénot, 1907) it is greatly lengthened and complexly coiled upon itself. The oesophagus is prolonged backwards into the mid gut to form an oesophageal valve, or cardia, of varying complexity. In many Cyclorrhapha a food reservoir is present as in the adult. The usual number of Malpighian tubules is four, which may arise separately from the hind gut, as in *Tabanus* and *Stomoxys*. In *Musca* and most other Cyclorrhapha they united basally in pairs, each pair communicating with the hind gut by means of a short duct. In the Culicidae, *Psychoda*, *Ptychoptera* and the Blephariceridae five Malpighian tubules are present.

Salivary glands are found in all Dipteran larvae and generally take the form of hollow vesicles lined with a single layer of cells. *Mandibular glands* occur in *Sciara* and extend almost the whole length of the body (Keilin, 1913). Metamerically arranged *epidermal glands* are found in Tipulid larvae (Fig. 132), and small *peristigmatic glands* are found in association with the spiracles of many Dipterous larvae.

The *heart* consists of a series of eight chambers in *Anopheles* and other Nematocera. In species of *Chironomus* it is formed of a single enlarged chamber situated in the 11th segment and provided with two pairs of ostia. In *Musca* it comprises three chambers situated in the terminal segments, while in *Dicranota* no distinct chambers are evident (Miall). In all cases the heart is prolonged through the thorax as the aorta, which terminates in the head near the brain. A short distance behind the latter is a glandular structure, generally forming a ring (Weismann's ring) around the aorta (Thomsen, 1951; cf. p. 278).

The *nervous system* in Nematocerous larvae (Brandt, 1882; Brauer, 1884) consists of the usual supra- and infra-oesophageal ganglia and, as a rule, 3 thoracic and 8 abdominal ganglia. Among the Brachycera this generalized condition of 11 postcephalic ganglia is present in the Rhagionidae, Asilidae, Therevidae and Dolichopodidae. The Tabanidae are intermediate between these families and the Cyclorrhapha, reduction and concentration resulting in only 1 thoracic and 5 abdominal ganglia being present. In *Stratiomyia* all the ganglia are fused into a single ovoid ganglionic mass and a similar conditions is the rule throughout the Cyclorrhapha (Fig. 128). The position of the brain varies among Nematocera; although usually present in the head as in *Culex*, in *Tipula* and *Ptychoptera* it is situated partially in the head and prothorax, while in *Dicranota, Psychoda* and certain Chironomidae it lies wholly in the prothorax. In *Calliphora* and other Cyclorrhapha it is situated in the metathorax.

The Pupa

Most Nematocera have 4 larval instars (6 in Simuliidae); in the Brachycera the number varies as a rule between 5 and 8; in the Cyclorrhapha there are only 3 instars. Pupation takes place by one of two methods. In the Nematocera and Brachycera the skin is normally cast at pupation but in the Stratiomyidae and some Cecidomyidae the exuviae persist and enclose the pupa. In the Cyclorrhapha the pupa is enclosed in the larval skin which hardens forming an outer shell or puparium. The puparium is ovoid or barrel-shaped and quite immobile (Fig. 192). The actual process is quite complicated. The mature larva ceases feeding, evacuates the gut and becomes the 'postfeeding larva' (Fraenkel and Bhaskaran, 1973). The process of 'pupariation' is relatively rapid but inside the puparium there is at first a 'prepupa'. After the larval-pupal apolysis, the pupa appears but is at first 'cryptocephalic' with the head retracted. Later the head is suddenly everted and the ordinary exarate pupa appears. The pupa-adult apolysis occurs before emergence, so that for some time there is a pharate adult which sheds both larval and pupal cuticles on emergence.

A number of pupae or puparia (Simuliidae, Blephariceridae, Deuterophlebiidae, some Chironomidae, Tipulidae, Empididae, Dolichopodidae and Canacidae) have spiracular gills (Hinton, 1962, 1967). Such a gill is a modification of the body-wall near a spiracle, sometimes also of the spiracle, to allow the pharate adult to obtain oxygen. It seems that in all cases a plastron is involved and the gill can work in the air or under water. A pupal shelter is formed in several Nematocerous families and among the Brachycera a cocoon is present in certain Dolichopodidae. Among Cyclorrhapha it is very rare but is found in a few genera of Muscidae. In the lower Diptera prothoracic and 7 pairs of abdominal spiracles are usually evident; aquatic Nematocera, however, are propneustic. In the puparia of the Cyclorrhapha remains of the larval spiracles can be seen. In *Musca* communication with the air is maintained by means of a pair of pupal spiracles in the form of small spine-like projections between the 5th and 6th segments of the puparium (Hewitt). Similar, though more prominent, respiratory structures are prevalent in other Cyclorrhapha.

Literature on the Larval and Pupal Stages – Malloch (1917) and Verrall (1909) give much information about Dipteran larvae and the first author keys to the early stages of Nematocera and Brachycera. A very complete tabulation and summary has been published by Hennig (1948–52). Aquatic Dipterous larvae have been much studied notably by Meinert

(1886), Miall (1895), Grünberg (1910), Williams (1939), Johannsen (1937), Johannsen and Thomsen (1937), Lenz (1941) and Thienemann (1944). Other studies of Nematocera are those of Mayer (1934) (Ceratopogonidae), Madwar (1937) (Mycetophilidae), Sellke (1936) (Tipulidae), and Keilin and Tate (1940) (Trichoceridae, Anisopodidae). For larvae of Cyclorrhapha the works of Banks (1912), Keilin (1915; 1917) may be referred to and for particular families – Heiss (1938) and Bhatia (1939) (Syrphidae), Muirhead-Thomson (1937) (Anthomyidae), and Butt (1937), Phillips (1946) and Keilin and Tate (1943) (Tephritidae). Hayes (1938) has published a list of published keys for identification of larvae and Séguy (1950) a review of the biology of the order.

Classification of Diptera

Introductions to the classification of the order will be found in Verrall (1901, 1909) and Oldroyd *et al.* (1949 onwards) for the British species and in Hendel (1928, 1936–37). Lindner (1924–74) may be consulted for the Palaearctic fauna; for the N. American forms consult Curran (1934) and Cole (1969). Up to date catalogues exist of the N. American species (Stone *et al.*, 1965) and are appearing or in preparation for S. America (Papavero, 1966 onwards), Oriental region (Delfinado and Hardy, 1973–74) and Ethiopian region (in preparation at the British Museum). Kertész's World catalogue (1902–10) is incomplete.

Diptera were first divided into the Nematocera with long, many-segmented antennae and Brachycera with apparently 3-segmented ones. Much later Brauer pointed out that the more primitive forms (Nematocera and some Brachycera) emerged from the pupa by a straight or longitudinal slit (= Orthorrhapha) while the higher Brachycera emerge by pushing a circular cap off a puparium (Cyclorrhapha). The most convenient arrangement seems to be to recognize three suborders, the Nematocera, the Brachycera and the Cyclorrhapha which show a progressive but far from uniform increase in specialization and of which the first and third and probably the second are monophyletic.

Suborder I. NEMATOCERA

Larvae usually with a well-developed exserted head and horizontally biting mandibles; pupa obtect, free except in some Cecidomyidae. Antennae of imago many-segmented (except in Nymphomyidae), usually longer than the head and thorax, the majority of the segments usually alike, not forming an arista. Palpi usually 4- or 5-segmented, pendulous. Pleural suture of mesothorax straight (except in the Psychodidae). Discal cell generally absent, cubital cell when present widely open.

Suborder II. BRACHYCERA

Larvae with an incomplete head, usually retractile, and with vertically biting mandibles; pupa obtect, free except in the Stratiomyidae. Antennae of imago shorter than the thorax, very variable, generally 3-segmented with the last elongate; arista or style when present terminal. Palpi porrect, 1- or 2-segmented. Pleural suture twice bent (except in the Acroceridae). Discal cell almost always present, cubital cell contracted before the wing-margin or closed.

Suborder III. CYCLORRHAPHA

Larvae with a vestigial head: pupation in a puparium, pupa exarate. Antennae of imago 3-segmented (4 in Platypezidae) with an arista usually dorsal in position. Palpi 1-segmented. Discal cell almost always present. Pleural suture twice bent. Cubital cell contracted or closed. Head with a frontal lunule and usually a ptilinum (absent in the Aschiza).

Suborder I. NEMATOCERA

A certain number of members of this section exhibit exceptional morphological characters. Thus, among the Culicidae the palpi are stiff and projecting, not pendulous as in other families. When the antennae are short, and apparently only annulated (Simuliidae and Bibionidae), the widened cubital cell and pendulous palpi indicate their affinities with this suborder. The anomalous Nymphomyidae with Brachycera-like antennae have characteristic venation and long fringes to the wings. When the palpi are 1- or 2-segmented (certain Cecidomyidae) the antennal and venational characters remove all doubts.

Superfamily **Tipuloidea**

FAM. TIPULIDAE (Daddy-long-legs or Crane Flies). *Antennae long 6- to many-segmented, ocelli wanting. Legs always long and deciduous, mesonotum with V-shaped transverse suture, discal cell present. Ovipositor valvular, horny. Larvae metapneustic, anal extremity with fleshy retractile processes.* The Tipulidae include some of the largest species of Nematocera. The number of antennal segments is extremely variable and they are occasionally pectinate or serrate in the male, but not plumose. The front of the head is prolonged forwards to a greater or lesser degree, and in a few genera an elongate proboscis is present: throughout the family ocelli are absent. The thoracic musculature is described by Mickoleit (1962). The V-shaped mesonotal sulcus is characteristic but it is wanting in the apterous genus *Chionea*. An appreciable number of species which live in cold places or which, like *Chionea*, emerge in winter, have reduced wings (Byers, 1969*a*).

The larvae are hemicephalous, the head being deeply embedded in the prothorax and incomplete posteriorly. The antennae are well developed, the labium is large and toothed anteriorly, and there is usually a large and heavily sclerotized hypopharynx. The body is elongate-cylindrical, either with or without pseudopods, 11- or 12-segmented, and usually ashy grey or brownish in colour. Frequently the first 6 abdominal segments are subdivided and, as a rule, the anal segment is truncated, and bears the spiracles. Around the latter is a series of fleshy retractile processes; in aquatic species these processes are often fringed with hairs and protrusible blood-gills are present. In the terrestrial forms the hair fringes and gills are usually greatly reduced. The pupae are very elongate, and the thoracic respiratory horns are either slender or plate-like.

The larvae of *Tipula* (British species, Chiswell, 1956) may be taken as representative of the family, and live among grass, roots, etc., decaying vegetation, or are aquatic (Bodenheimer, 1924; Sellke, 1936). The larva of *Dicranota* lives in the beds of ponds and streams where it preys upon the worm *Tubifex*. It is characterized by paired retractile pseudopods on segments 6–10, and on the 12th segment there are 3 pairs of outgrowths of the nature of gills (Miall, 1893). The larva of *Ctenophora* and its anatomy has been studied by Anthon (1908) and that of *Holorusia* by Kellogg (1901) and the west Palaearctic Tipulinae by Theowald (1957): descriptions of the larvae and pupae of many other genera are given by Malloch (1917), Byers (1961), Alexander (1920) and Brodo (1967).

The larvae of the Cylindrotominae are very remarkable: they are green in colour, and aquatic or terrestrial, feeding upon mosses or Angiosperms. The body is provided either with filaments or leaf-like outgrowths: Miall and Shelford (1897) have made a detailed study of the larva of *Phalacrocera* which is aquatic, feeding upon mosses, and the whole body is invested with numerous elongate filamentous processes. In *Cylindrotoma* (Cameron, 1918) the larva is terrestrial with lateral plate-like outgrowths: it feeds openly like a caterpillar on various phanerogamic plants, and the pupa is attached to the food-plant by means of the partially cast exuviae. Among the Limnobiinae (Eberhard Lindner, 1959) the larvae of some *Limnobia* are fungivorous, while those of *Dicranomyia* are mostly aquatic or semi-aquatic. A single species of the latter genus is exceptional in having a leaf-mining larva in the Hawaiian Islands. The larva of *Erioptera squalida* lives in the water and can pierce the roots of *Glyceria maxima* with its posterior spiracular processes; the pupa obtains its oxygen from the roots (Houlihan, 1969).

FAM. TRICHOCERIDAE (Winter Gnats). *Small, Tipulid-like flies, but legs not deciduous and ocelli present.* The adult flies commonly dance in swarms during the winter, or more rarely during the summer at high altitudes. The eucephalous, amphipneustic larvae are much more like those of the Anisopodidae than those of the Tipulidae, which the adults resemble. They develop in humus, often in groups. The larva is described by Keilin and Tate (1940).

FAM. PTYCHOPTERIDAE. *Moderate-sized flies resembling Tipulids, but legs not deciduous and a characteristic vena spuria between R and M.* The adult flies of this small family are usually found at the edges of streams or ditches. The larvae (Miall, 1895*a*) are found in damp situations, usually in muddy water where they feed upon the vegetable matter contained therein. They are long and slender with well-developed pseudopods armed with spinules: the spiracles are minute, and are borne at the apex of an extremely slender tube formed by the greatly prolonged terminal segments of the body. The pupal respiratory organs are unequal in length, one being many times longer than the other.

Superfamily Psychodoidea

FAM. TANYDERIDAE. *Flies of rather small size, Rs and M both 4-branched. Mandibles present in a reduced state* (Fig. 453). The flies of this family are regarded as the most primitive of all living Diptera (Williams, 1933). The larva which is aquatic somewhat resembles those of *Ptychoptera* (Alexander, 1930). There are about 30 species; one in the U.S.A., a few oriental and African, and a number in New Zealand (Alexander, 1927).

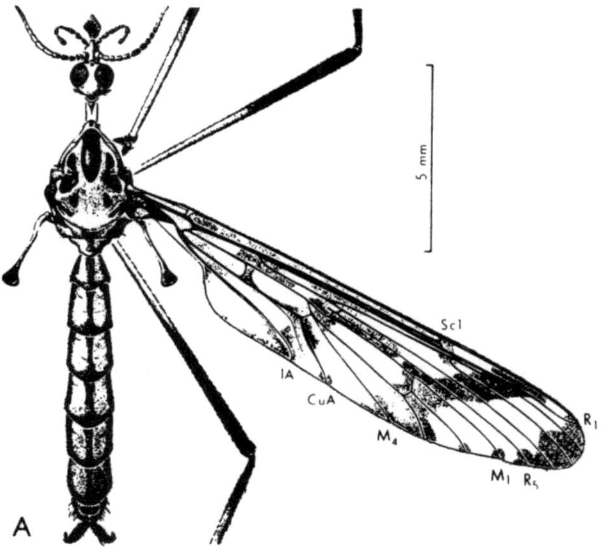

FIG. 453 *Eutanyderus oreonympha* (Tanyderidae).
From Colless, D. H. Diptera *in* Insects of Australia, 1970 fig. 16a.

FAM. PSYCHODIDAE (Moth-flies, Sand-flies). *Minute moth-like flies, the legs, body and wings clothed with long coarse hairs, often admixed with scales. No ocelli; wings with Rs 4-branched and no obvious cross-veins. Larvae usually aquatic or saprophagous, of variable structure, generally amphipneustic.* These fragile insects are to be found in close proximity to the larval habitat and are commonly met with in dark or shaded damp situations; some are frequent on windows and are often attracted to a light at night. Females of the genus *Phlebotomus* (Fig. 454) and *Lutzomyia* in the subfamily Phlebotominae (often now called a family) feed by sucking the blood of vertebrates. In addition to man, other mammals, birds and reptiles are used as hosts; and in the case of *P. minutus* lizards and geckos are probably the principal animals preyed upon. The well-known 'Pappataci' or 'three-day' fever was proved in 1908 by Doerr, in Herzegovina, to be carried by *P. papatasi*; possibly other species of the genus also function as carriers. In other regions, species are carriers of kala-azar and forms of leishmaniasis. Townsend has brought forward evidence indicating that the disease known as verruga in Peru is transmitted by a species of *Phlebotomus*. The eggs in this genus are elongate and dark brown; the larvae have mostly been found in damp, dark places such as crevices in rocks and stone walls, in drains, unclean cellars, moist earth, etc. Their minute size, however, renders them extremely difficult to discover and further information is greatly needed with reference to their habitat. When fully grown the larva is 5–8 mm long, and provided with elongate caudal bristles which may be almost as long as the body. Decaying organic matter appears to be their chief food. The pupa is found in similar situations and usually carries the larval exuvia at its anal extremity. For further information on this genus reference should be made to papers by Grassi (1907), Perfil'ev (1937), Raynal (1934) and Lewis (1971, 1974).

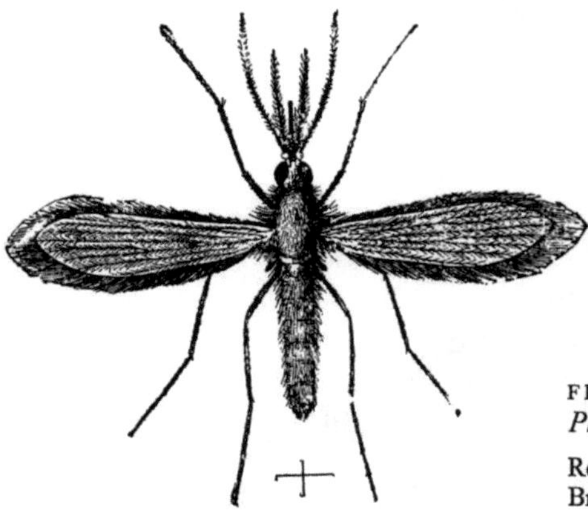

FIG. 454
Phlebotomus sp., female, Sudan. × 12

Reproduced by permission of the Trustees of the British Museum.

The Psychodinae (for British species see Tonnoir, 1940) have very doubtfully been accused of the blood-sucking habit which, at any rate, is extremely rare in this subfamily. Many of the adults have remarkable secondary sexual characters (Feuerborn, 1922*b*). Their larvae possess a well-developed head and 12 trunk segments, the first and last carrying a pair of spiracles (Dell, 1905; Malloch, 1917 and Miall, 1895*a*). The last segment is drawn out and provided with 4 fleshy outgrowths clothed with elongate hairs. By this means it forms a kind of siphon surrounding the posterior spiracles, and is protruded to the surface of the water. The thoracic and 1st abdominal segments are transversely divided, the remaining abdominal segments being triannulate. Dorsally, the larva carries a series of strongly sclerotized plates bearing sensory hairs. These plates are present on each segment (*Pericoma*) or only on the posterior segments (*Psychoda*). The larvae are aquatic, or live in fluid organic matter of various kinds, including sewage filter-beds. Those of *Ulomyia* and *Maruina* live in cascades. In the latter genus they are provided with ventral sucker discs as in Blephariceridlarvae, and moreover are metapneustic (Müller, 1895). The larvae of *Trichomyia* lives in decaying wood (Keilin, 1914). It is narrowly cylindrical, smooth, and devoid of dorsal plates and setae; the segments are undivided, and the tracheal system is amphipneustic with no respiratory siphon. In the remarkable genus *Termitomastus* found in the nests of Neotropical termites, the wings are reduced to strap-like rudiments: two other termitophilous genera, *Termitodipteron* and *Termitadelphos*, occur in the nests of *Eutermes* in Peru.

The American species have been revised by Quate (1955) who also deals with the larvae. Satchell and Tonnoir (1953) have described the Australian species. Jung (1956) deals with the European fauna, including the larva.

FAM. NYMPHOMYIIDAE. *Small flies with reduced venation, with long fringes to the wings, mouthparts atrophied, antennae with 3 stout segments and a short style, third segment sometimes annulated, eyes meeting beneath head but separated above, 2 ocelli present. Larvae aquatic.* These extraordinary flies were discovered by Tokunaga in 1932 in a fast-flowing stream, in Japan. He has since published several studies of this species which seems to be a highly specialized Nematoceran, retaining, however, some archaic features (Tokunaga, 1935; 1936). The larva has not been described, but the pupa (Tokunaga, 1935*a*) shows some affinity to the Psychodidae (Hennig, 1950). Ide (1965) and Cutten and Kevan (1970) have described two more genera, each with a single species, from the E. Himalayas and Canada. The second paper also describes the larva which has 7 abdominal and 1 anal pair of pseudopods bearing claw-like pectinate spines. It does not at all resemble a Psychodid larva.

Superfamily **Culicoidea**

FAM. DIXIDAE. *Insects almost devoid of hairs and scales, antennae elongate, about 16-segmented, filiform apically. Venation as in Culicidae; proboscis somewhat projecting, not adapted for piercing; ocelli absent. Larvae metapneustic and aquatic, usually assuming a* U*-shaped attitude.* A small family comprising the genera *Dixa* and *Neodixa* with more than 100 widely distributed species. These insects have been variously included in the Culicidae and Tipulidae. They are readily separable from most of the latter by the absence of the discal cell and the V-shaped thoracic suture: the filiform non-plumose antennae are totally different from those organs in the Culicidae, and find their parallel in *Trichocera*. They closely resemble the Culicidae, however, in their venation but differ therefrom in the absence of scales from the wings. The larva of *Dixa* frequents shady, weedy pools or streams and might be mistaken for that of *Anopheles*. It is eucephalous with 12 trunk segments, the 4th and 5th each bearing a pair of ventral pseudopods armed with curved spinules: segments 5–10 in certain species carry a dorsal shield fringed by setae. The pupa closely resembles that of the Culicidae.

FAM. CULICIDAE (Mosquitoes). *Very slender flies, generally with an elongate piercing proboscis and no ocelli: the palpi stiff and not pendulous. Legs long, antennae densely plumose in the males, pilose in the females. Wings fringed with scales along the posterior margin and the veins. Larvae and pupae aquatic and very active: the former metapneustic, with an enlarged thoracic mass.* The remarkable discoveries in their life-histories, and the part played by the adults as disease carriers, has given an enormous stimulus to the study of mosquitoes. More than 1600 species have been described and at least 36 species occur in Britain. Culicidae are almost world-wide in distribution, but the tropics are much richer in genera and species than northern latitudes. In arctic regions they are extremely abundant during the short summer, though few in species. In these parts they often occur far from the haunts of man and frequently in regions uninhabited by quadrupeds. In Lapland their numbers even exceed those of most tropical regions. For a full account of the biology of the family the student is referred to the standard treatise by Howard, Dyar and Knab (1912) and an interesting account by Gillett (1971). For the anatomy of mosquitoes see Nuttall and Shipley (1901–03) and Christophers (1960). For the larval anatomy of *Anopheles* see Imms (1907–08) and of *Culex* see Raschke (1887). The mouthparts of a mosquito (Fig. 441) have already been described (pp. 956–7): in the Chaoborinae, often now treated as a family, they are very short, concealed, and not adapted for piercing. Except in a few genera the whole body, legs and wings are in part, or entirely, clothed with scales.

The eggs (Hinton, 1968*a* and *b*) are deposited on or near the surface of the water, and the number laid by a single individual varies from 40–100 (*Anopheles maculipennis*) up to 300 or more (*Culex pipiens*). They may occur singly, as in *Anopheles* or *Aedes*, or collectively to form a compact mass or egg-raft as in *Culex* (Bates, 1949) and other genera. The eggs vary in shape in different genera: those of *Anopheles* are boat-shaped with a conspicuous float on either side: in *Culex* they are fusiform, in *Toxorhynchites* somewhat club-shaped, while those of *Aedes* are ovoid

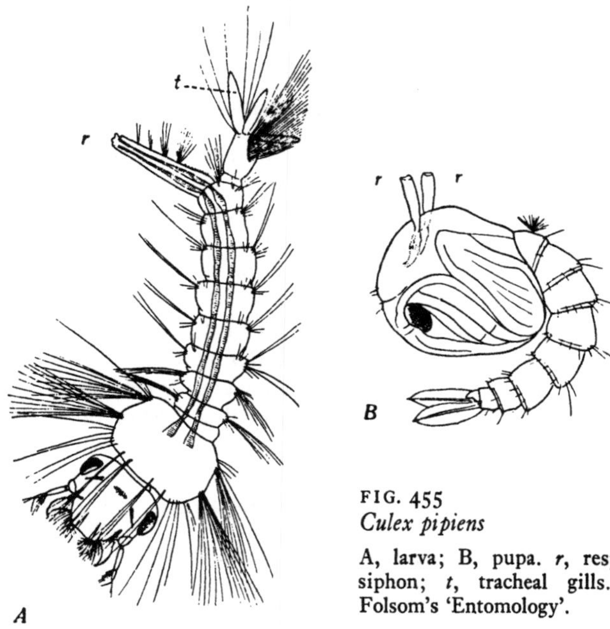

FIG. 455
Culex pipiens

A, larva; B, pupa. *r*, respiratory siphon; *t*, tracheal gills. From Folsom's 'Entomology'.

and surrounded by a series of small air-chambers which aid in floating.

Mosquito larvae (Fig. 455) have a well-developed mobile head: the eyes vary according to the age of the larva and, as a rule, both the primitive larval eyes and the developing compound organs of the imago are present. A pair of dense tufts of long hair, or feeding brushes, are present over the mouth on either side of the head. By means of the movement of these brushes a current is set in motion which wafts microscopic food-particles towards the mouth. The thoracic segments are fused to form a single broad, rounded region. Nine abdominal segments are present, and the anal somite is surrounded at its apex by four tracheal gills. These organs are small in surface feeders such as *Anopheles*, but larger in *Aedes* which is a bottom feeder. The respiratory system is metapneustic, and opens on the dorsal surface of the 8th segment. The spiracles are placed either on a quadrilateral area raised slightly above the preceding segment (Anophelinae), or more usually at the apex of a respiratory siphon. The larvae of *Chaoborus* (Peus, 1934) is a highly specialized type, being almost completely transparent and apneustic. It is provided with a pair of pigmented air-sacs in the thorax and a second pair in the 7th abdominal segment: these structures act as organs of flotation, respiration being cutaneous.

When at rest, and during feeding, the larvae of Anophelines float horizontally just beneath the surface-film with the palmate hairs and spiracular area in contact with the latter. In the Culicines the larvae bring the apex of the siphon in contact with surface and hang head downwards, inclined at an angle with the surface film.

In their feeding habits, mosquito larvae may be phytophagous or carnivorous. As a rule they feed upon minute algae and other particles contained in the water. Certain forms, however, are carnivorous. These may be readily recognized either by the mouth-brushes being replaced by stout spines, which serve to seize the prey, or by the prehensile antennae (Chaoborinae). The organisms most frequently preyed upon are other mosquito larvae.

As a rule, mosquito larvae are only able to exist in small numbers in permanent waters on account of the presence of predators, such as fish and insect larvae. Their habitat is extremely varied, thus *Anopheles lutzi* breeds in the cups of epiphytic and pitcher plants. The larvae of *A. vagus* frequent shallow rain-filled pools such as abound in India during the monsoon. *A. sundaicus* occurs in pools flooded by the sea at high tides: larvae of *Taeniorhynchus* live at the roots of aquatic plants in swamps, inserting their modified siphons into the tissues, and thus deriving their supply of oxygen. *A. fluviatilis* frequents sub-Himalayan hill streams; *A. multicolor* lives in the waters of Saharan oases containing 40 gm of chlorides per litre and *A. stephensi* abounds in Bombay, living in the waters of wells and cisterns. The pupae are very active, and respire by means of a pair of breathing trumpets communicating with the anterior spiracles. They float at the top of the water with their trumpets attached to the surface film.

When at rest Anophelines can usually be distinguished from other mosquitoes by the fact that they settle with the proboscis and the long axis of the body in one straight line, while in the Culicines the abdomen is usually parallel with, or inclined towards, the surface upon which the insect rests (Fig. 456). The length of the life-cycle of mosquitoes is primarily dependent upon temperature: thus, that of *A. aegypti* is normally 15–20 days, but may be as short as 11 days. The adult refuses to feed below 23° C and is quite inactive at 20° C.

Economically mosquitoes are of the utmost significance owing to their functioning as the intermediary hosts of malaria, yellow fever, filariasis, dengue and numerous viruses. Increased knowledge of these insects, and the diseases transmitted by them, has rendered vast areas of tropical countries no longer a menace to the life of the European. The experimental researches of Ross on the malaria *Plasmodium* have conclusively proved that this parasite passes through two periods of multiplication during its life-cycle: the first is one of asexual reproduction (schizogony) and occurs mainly in the blood of man. The second or sexual cycle (sporogony) takes place in the mosquito, and commences with the entry of blood containing suitable forms of the parasite into the stomach of the insect. After fertilization the zygote bores into the gut-wall where it becomes encysted. The cyst increases enormously in size, and eventually ruptures, liberating great numbers of sporozoites into the haemocoelic cavity of the insect. Those sporozoites, which bore their way into the salivary glands, are then able to be transmitted to another human being through the punctures of the mosquito, and there continue their development.

When the female mosquito feeds the tip of the labium is first brought against the skin, and then the pointed mandibles and maxillae are inserted. The labrum is also

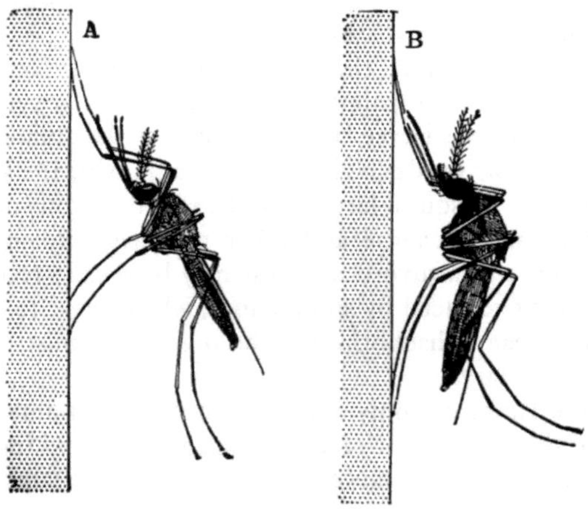

FIG. 456 Resting attitudes of A, *Anopheles*, B, *Culex*
Reproduced by permission of the Trustees of the British Museum.

inserted into the puncture along with the hypopharynx. The labium is then doubled back in the form of a loop as the mouthparts become more deeply inserted. According to Macgregor (1931) two methods of feeding occur. In the method just described the ingested food is drawn continuously into the stomach. In the second method, termed 'discontinuous feeding', the mouthparts are not disengaged from the labium and the tip of the proboscis is merely dipped from time to time below the surface of the fluid to be imbibed. The ingested food then passes first to the oesophageal diverticula which act as food reservoirs. These organs also function as air separators in which 'air locks', between discontinuous sections of fluid, are removed before such fluid passes into the stomach. Control is exercised over ingested foods to the extent that blood is allowed to pass to the stomach and sugary solutions into the diverticula. It appears that in nearly all mosquitoes the entry and digestion of blood in the stomach is a necessary condition for ovulation.

A number of Anopheline mosquitoes are now known to be carriers of the plasmodia of one or more forms of malaria and, as the habits of these species greatly vary, a knowledge of their bionomics is of the highest importance from the standpoint of public health. Among the more important carriers are *Anopheles maculipennis* group (Europe), *A. culicifacies* (India), *A. minimus* (Assam to China and the Philippines), *A. gambiae* and *A. funestus* (Africa), *A. darlingi* and *A. albimanus* (Central and S. America). Malaria has now been eliminated from the U.S.A. An important recent development is the recognition within many of the older species of 'micro-species' with substantial differences in their biology and efficacy as vectors (Bates, 1949).

Aedes aegypti (Christophers, 1960) is one of the commonest mosquitoes of the tropics and subtropics of the world, and occurs largely along the coasts and the courses of the larger rivers. In 1881 Finlay observed the incidence of this mosquito and yellow fever in Cuba, and succeeded in transmitting the disease through the agency of its punctures. In 1899 an American commission sent to Cuba definitely proved that yellow fever is carried by *A. aegypti*. As the result of anti-*Aedes* measures in the Panama Canal zone, which was at one time a notorious region for this disease, urban yellow fever has become non-existent there. The eggs of this mosquito are laid upon any accumulation of stagnant water however small, old tins, broken bottles, holes in trees often being utilized but not ground pools. The adult is most easily identified by the lyre-shaped white mark on the thorax: it is essentially a domestic species rarely found away from towns and villages.

Culex pipiens fatigans is an almost tropicopolitan mosquito of great economic significance. Along with other species it is a carrier of *Wuchereria bancrofti* which produces elephantiasis. *Aedes aegypti* and related species are able to transmit the virus of dengue from one human being to another. Several species of *Aedes* are intermediary hosts of *Dirofilaria immitis* of the dog and several species of Culicidae transmit the plasmodium of avian malaria from one bird to another.

The literature on the Culicidae is now extremely large. Stone *et al.* (1959, with supplements) provide a world catalogue of the species. Perhaps the best general work on the family is that by Howard, Dyar & Knab (1912) which deals with the species of North and Central America. The general biology of the group is discussed by Hackett (1937) and Bates (1949) and the behaviour of the adults by Muirhead-Thomson (1951). Boyd (1949) deals especially with the relation of mosquitoes to malaria. Among the more important regional taxonomic studies are those of Hopkins (1952) and Edwards (1941) (Ethiopian), Edwards (1921) (Palaearctic), Natvig (1948) (Scandinavian), Marshall (1938) (British), Matheson (1944) (N. American), Lane (1939 and revised 1953) has catalogued the large and peculiar S. American fauna, Belkin (1962) (S. Pacific). Mattingly (1974) gives a key to the world genera (all stages). The work of Russell *et al.* (1946) includes keys to the Anophelini of the world.

FAM. BLEPHARICERIDAE. *Elongate flies with long legs: eyes in both sexes often holoptic, and usually bisected into areas of different-sized ommatidia: ocelli present. Thorax with transverse suture: wings with a complex network of permanent folds. Mouthparts in female adapted for lacerating. Larvae aquatic, onisciform, with ventral suckers.* A small family of very wide but discontinuous geographical range. It is confined to hilly or mountainous districts and is unrepresented in the British Isles. The adults frequent the borders of streams; they are weak fliers, and are less often met with than the larvae. The females are predacious (Pryor, 1948), preying upon small Diptera, and the males probably feed upon nectar. The wings possess a fine network, or 'secondary venation', of creases or folds in the membrane, which have not been obliterated after emergence from the pupa. The larvae inhabit swiftly running hill streams where they fix themselves by means of their ventral suckers (Rietschel, 1961) to rocks and stones, usually in places where the current is swiftest. The head (Anthon and Lyneborg, 1967), thorax and first two abdominal segments are fused together, and the remaining segments are deeply incised laterally. A longitudinal row of median ventral suckers, usually six in number, is their most characteristic feature: to the outside of each sucker is a group of digitate processes which are regarded by Kellogg as being tracheal gills. The tracheal system (Whitten, 1963) is peripneustic: the spiracles are minute and situated ventrally, but in all probability are closed. The pupae are broad, and flattened beneath, adhering tenaciously to rocks, etc.; the respiratory horns (Hinton, 1962) are lamellate, and the legs extend almost to the apex of the abdomen. For information about the family the works of Mannheim (1935), of Tonnoir (1924; 1930) (larvae) and Edwards (1929) (adults) should be consulted.

FAM. DEUTEROPHLEBIIDAE. *Antennae filiform, very elongate. Wings with a network of creases: ocelli, mouthparts and true venation absent.* This small family consists of a single genus, *Deuterophlebia*, which is perhaps allied to the Blephariceridae and occurs in N. America, Japan and the mountains of C. Asia. Pennak (1945) has revised the group and summarized its biology. The larva has seven pairs of large segmental outgrowths bearing suckers and posteriorly what have been described as anal blood-gills.

FAM. SIMULIIDAE (Blackflies). *Small stoutly built flies with short legs and elongate mandibles. Wings broad, anterior veins thickened, the others faint. Antennae 11-segmented, scarcely longer than the head: ocelli wanting, the males holoptic. Larvae in running water attached to rocks, etc. by the anal extremity, spiracles closed.* A small family of world-wide distribution and including the familiar 'buffalo gnats' of America (Fig. 457). In the males the eyes have the upper facets markedly larger than

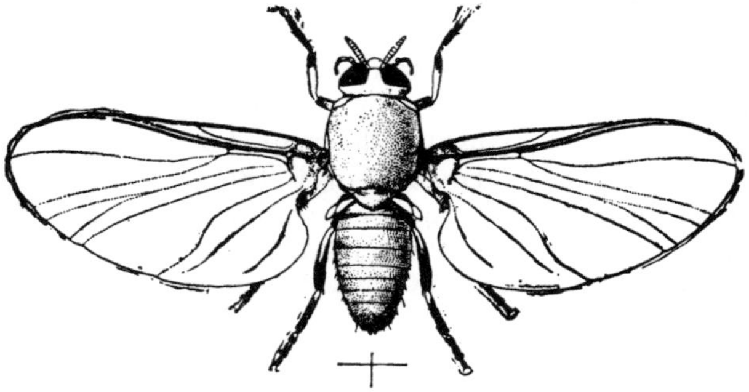

FIG. 457 *Simulium venustum*, female. × 12. N. America
Reproduced by permission of the Trustees of the British Museum.

the lower, and the 1st tarsal segment is usually much dilated. Both sexes possess elongated piercing mandibles which are broader in the female than in the male (cf. Nicholson, 1945; Chance, 1970). The females of certain species are active blood-suckers and in some countries are veritable scourges. *S. indicum*, the 'potu' fly, is a troublesome pest in parts of the Himalayas, and *S. columbaschense* is at times a great scourge of man and domestic animals, particularly in regions bordering on the Danube. It often appears in enormous swarms and the flies attack the orifices of the body entering the ears, nostrils, margins of the eyes, etc., in great numbers, and their punctures produce an inflammatory fever often resulting in death. *S. damnosum* bites man in C. Africa and carries the worm, *Onchocerca*, the cause of a filariasis. Certain other species are pests in many parts of N. America; *S. meridionale* causes the death of numerous turkeys and chickens in Virginia, and *S. pecuarum* is the common American buffalo gnat. One Canadian species is specifically and apparently chemically attracted to the loon, *Gavia immer* (Lowther and Wood, 1964; Fallis and Smith, 1964). In a few instances species of this genus have been observed to suck blood of other insects.

Adult Simuliidae occur in the neighbourhood of streams and rivers and their reproductive biology is described by Davies and Petersen (1956); the eggs are laid either on herbage or stones, above or beneath the surface of the water. Britten (1915) has observed the female of *S. maculatum* submerged to a depth of 1 foot during oviposition: the eggs are laid on vegetation and are coated with a gelatinous secretion.

The larvae are invariably aquatic and require swiftly flowing water for their environment, and for this reason they are often found congregated in the vicinity of rapids and waterfalls, etc. Information concerning the metamorphoses of the family is given by Johannsen (1903–05), Puri (1925) and Smart (1944). In the larvae the head is complete and is characterized by the large maxillae and the prominent mouth brushes. On the ventral aspect of the thoracic region there is a foot-like protuberance provided with hooklets: it functions as a kind of sucker and is formed by the fusion of a pair of pseudopods. On the anal segment there is a second sucker, armed with concentric series of stout hooklets and, associated with the anus, is a group of rectal gills which vary in number in different species. Nine pairs of minute spiracles are present from the mesothorax to the 7th abdominal segment; but respiration is performed by means of the rectal gills, which contain blood and are supplied with tracheoles (Taylor, 1902). The larvae fix themselves to their substratum by means of the anal sucker and, as a means of locomotion, they loop the body after the manner of Geometrid caterpillars, bringing the anal extremity forwards beside the anterior sucker. Some species attach themselves to the bodies of fresh-water prawns or of Ephemerid nymphs and may gain some sustenance from the body to which they are attached (Fontaine, 1964; Williams, 1968; Disney, 1971).

Before pupation the larva forms a pocket-like cocoon open above; the pupal respiratory organs are composed of long tube-like filaments, which protrude from the cocoon and obtain oxygen from the moving water. For the British species of *Simulium* and their larvae reference should be made to Smart (1944); the classification of the family as a whole is also discussed by Smart (1945). It has recently been shown (e.g. Dunbar, 1958) that there are sometimes sibling species, distinguished mainly by their chromosome banding. The African species have been revised by Freeman and de Meillon (1953) and Crosskey (1969).

FAM. THAUMALEIDAE. *Antennae of 2 apparent segments terminated by a 10- or 11-segmented style-like appendage: palpi longer than antennae: eyes holoptic in both sexes. Larvae aquatic, amphipneustic, Chironomid-like.* A small family readily distinguishable from all other Nematocera by the structure of the antennae. The adults are small sluggish insects and in Britain *Thaumalea testacea* may be swept from grass and other herbage bordering hill-streams. The larva of this species has been described by Saunders (1923), and in general appearance it resembles that of a Chironomid. Prothoracic and anal pseudopods are present, together with paired dorsal anal blood-gills, and spiracles are evident on the first and penultimate segments. The pupa is almost entirely covered with small warts which even extend on to the short respiratory horns. Its anal segment is provided with a pair of slender upwardly directed processes and two elongate setae. For a revision of the family, see Lindner (1930).

FAM. CERATOPOGONIDAE (Biting Midges). *Small or very small, gnat-like flies: antennae plumose in male, pilose in female. Head not concealed by the thorax: ocelli absent: mouthparts adapted for piercing: fore legs not elongate. Larvae without prothoracic pseudopods: aquatic or terrestrial.* The small flies of this family are all in some way predatory in the adult stage. They may suck vertebrate blood (*Culicoides*), or blood of large insects such as moths, caterpillars or dragonflies (*Forcipomyia*, etc.)

or catch insects smaller than themselves (*Palpomyia*, etc.). More than 500 species are known, so far principally from the Holarctic region, though they are found in all the continents. The family falls into two groups – those with aquatic, vermiform larvae whose imagines are more or less bare-winged (*Culicoides*, *Bezzia*) and those with terrestrial larvae found in sap, under bark or in decaying organic matter. The latter include *Forcipomyia* and *Dasyhelea* whose imagines have hairy wings. The distinctions, however, are not absolute as some larvae of *Culicoides* are terrestrial and certain of those of *Forcipomyia* are aquatic. For the British species, see Edwards (1926*b*). Macfie (1940) has tabulated the world genera and de Meillon (1937) and Lee (1948) have dealt with S. African and Australian species, respectively. The larvae have been described by Saunders (1924–25), Johannsen and Thomsen (1937) and Kettle and Lawson (1952).

FAM. CHIRONOMIDAE (Midges). *Delicate gnat-like flies: antennae conspicuously plumose in the males, pilose in the female. Head small, often concealed by the thorax: ocelli absent. Mouthparts poorly developed but mandibles present in one genus* (Downes and Colless, 1967). *Fore legs elongate. Anterior wing-veins more strongly marked than posterior. Larvae apneustic: aquatic.* These insects bear a general resemblance to Culicidae but may be distinguished by the wings being unscaled. The adults occur in great numbers in the vicinity of lakes, ponds and streams: many appear on the wing just before sunset, and exhibit a characteristic gregarious habit of 'dancing' in the air in swarms. During these evolutions the number of females present does not appear to be large and, when pairing is accomplished, the mated couple leave the swarm. About 2000 species have been described: in Britain nearly 400 species are listed but more await discovery. The antennae are 6- to 15-segmented and the mouthparts are poorly developed. In *Chironomus* no food is taken during adult life and the digestive canal is consequently shrunken and empty.

The eggs of Chironomidae are laid in a mass, enveloped by transparent mucilage secreted by the accessory gland of the female: these egg-masses or ribbons vary in shape and number, and arrangement of the eggs therein, in different species. The larvae usually inhabit slow streams and ponds, or even puddles or water troughs. A few species can live at great depths, having been obtained from the bottom of Lake Geneva and Lake Superior. Several species occur in the sea, both in shore pools and at a depth of 15–20 fathoms: vast numbers frequent the salt lakes adjoining the Suez Canal. A typical kind of Chironomid larvae, such as that of *C. dorsalis*, has a well-developed head and 12 trunk segments, with a pair of pseudopods on the prothorax and last abdominal segment: in other forms pseudopods are present on the prothorax only or, more rarely, are absent. Two pairs of elongate blood-gills may be present on the 11th segment, and two pairs of papilla-like anal gills are placed around the anus. In *C. dorsalis* the tracheal system is greatly reduced and limited to the thorax, where there are two pairs of closed spiracles. A number of species are red, owing to the presence of haemoglobin dissolved in the blood-plasma, and are commonly known as 'blood-worms'. It was pointed out many years ago by Lankester that haemoglobin occurs among invertebrates when increased facilities for oxygenation are required, as by burrowing forms and those which lurk in the mud of stagnant pools. Surface-haunting Chironomid larvae are generally green. The larvae usually live in tubes either free, or attached to stones, etc., and composed of mud particles or of vegetable fragments, sticks, particles of green leaves, algae, etc. A species of *Polypedilum* which lives in shallow rock-pools in Africa is capable of surviving almost complete desiccation; in this condition it can survive long periods at the temperature of liquid helium or short ones at over 100°C (Hinton 1960*b*). The pupae may be active (*Tanypus*), float at the surface of the water, or remain at the bottom of the water: in the latter case they rest in the old larval tube which is often provided with an operculum. The pupal respiratory organs either consist of a pair of much-branched filaments, or of simple tubes: they are rarely absent. The literature on Chironomid larvae is now very extensive, owing to the numerous investigations of limnologists. References may be made to the papers by Johannsen (1937) and Brundin (1947). The Clunioninae include certain remarkable maritime genera whose larvae live among algae, and the adults are apterous. Among them are *Belgica* from Patagonia, *Halirytus*, Kerguelen I., and the European *Clunio*, the males of which are winged. For an account of marine Chironomids see Edwards (1926). Hashimoto (1965), dealing with the Japanese *Clunio takahashii*, shows that there is one type of female in the genus with head and legs very reduced. Such females are incapable of walking and are fertilized in the larval case.

Parthenogenesis is known to occur in a few Chironomids and results in the production of females only. The first observations were made by Grimm on *Tanytarsus* in 1870 and have been confirmed by Zavrel. Both the pupae and newly-emerged imagines are parthenogenetic. *Corynoneura celeripes* and *Chironomus clavaticornis* also lay parthenogenetic eggs (Edwards, 1919). Paedogenesis has been recorded in the larva of *Tanytarsus dissimilis* by Johannsen in America (1910) but see p. 305. For the classification of the family see Edwards (1929*b*), Coe in Oldroyd *et al.* (1950) and Goetghebuer and Lenz in Lindner (1936–50). The paper by Brundin (1966) on the transantarctic distribution of midges is of great interest and gives an introduction to the Andean fauna. Saether (1969) describes some of the N. American species.

Superfamily **Anisopodoidea**

FAM. ANISOPODIDAE (RHYPHIDAE). *Discal cell present: eyes in male often holoptic, ocelli evident. Antennae 16-segmented, about as long as thorax. Larvae amphipneustic, saprophagous.* A small family of gnat-like flies represented in all zoogeographical regions. In the presence of a discal cell *Anisopus* (Fig. 34) differs from other Nematocera excepting the Tipulidae, but is separable from the latter on account of the absence of the V-shaped mesonotal suture and the presence of ocelli. The male genitalia are inverse. The whitish larva of this genus is well known (Keilin and Tate, 1940) and lives in decaying vegetable matter and manure. It is about 10 mm long, elongate-cylindrical, and devoid of pseudopods. The thoracic segments are longer than broad, and those of the abdomen are separated by intercalary rings, giving the appearance of an increased number of segments. Spiracles are present on the prothoracic and last abdominal segments. *Mycetobia* as far as adult characters go would be placed in the Mycetophilidae, but Edwards (1916) and Keilin showed its true affinities by a study of the larva which occurs in sap and in fungi growing on trees.

Superfamily **Bibionoidea**

FAM. BIBIONIDAE. *Antennae 8- to 16-segmented, usually shorter than the thorax, the segments usually bead-like and closely apposed* (Fig. 440). *Wings large, anterior veins usually more strongly marked than posterior* (Fig. 458). *Eyes in males*

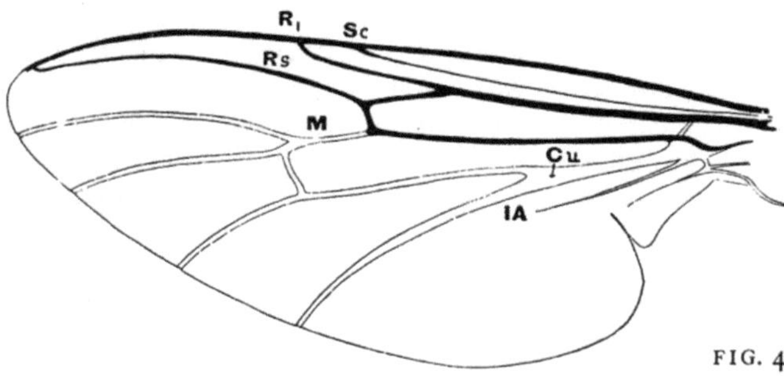

FIG. 458 *Bibio marci*, venation

usually holoptic or approximated, ocelli present. Larvae terrestrial, peripneustic. This family is somewhat variable in adult structure and is split into five by Hendel; it seems better, however, to treat them together, at least until the metamorphoses of some of the aberrant types are known. Typical Bibionidae are robust flies, often pubescent, with shorter legs and wings than most other Nematocera. In the males the eyes often occupy nearly the whole of the head and the upper facets are much larger than the lower, the two series being sharply differentiated. Certain species exhibit colour dimorphism, the females often being reddish-brown, while the males are entirely black.

The species of the subfamily Bibioninae frequent meadows, grassy hillsides or decaying vegetation and often appear in great numbers. Their larvae feed gregariously at the roots of grasses, cereals, hops and in leaf mould. Those of *Bibio* (Morris, 1921–22, Perrandin, 1961) are often gregarious and, structurally, they are the most primitive of all Dipterous larvae. They are 12-segmented with a large exserted head, well-developed mouthparts and are devoid of pseudopods (Fig. 459). Each segment is provided with a band of short fleshy processes, the latter attaining their greatest length on the 11th and 12th segments: the first segment is transversely divided and carries two bands of these processes. Open spiracles are present on each segment except the 2nd and 11th, the hindmost pair being considerably larger than its fellows. Pupation occurs in an earthen cell below ground.

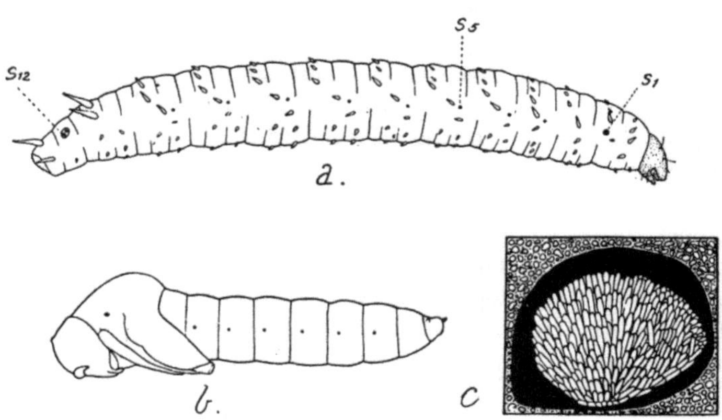

FIG. 459 *Bibio marci*

a, larva: s_1, s_5, s_{12}, spiracles of their respective segments; *b*, pupa; *c*, egg-mass in subterranean chamber. *After* Morris, *Bull. entom. Res.*, 1921.

The Scatopsinae are now often treated as a family. The adults lack the tibial spurs of Bibionids. Their larvae live in decaying organic matter and pupation takes place in the persistent larval skin. In *Scatopse* (Lyall, 1929), the larva is provided with longitudinal and transverse bands of hairs: nine pairs of spiracles are present on the 1st segment and on the 4th to 11th segments respectively, the hindmost pair being carried on stout cuticular pedicels. The 11th segment bears two posteriorly directed processes fringed with long hairs. For the British Bibionidae, see Edwards (1925*b*).

FAM. HYPEROSCELIDAE. A few genera resembling the Scatopsinae but vein Rs forked and M_{1+2} fused with Rs to a considerable distance beyond its base (Hardy and Nagatomi, 1960). The larvae which live in rotten wood are very peculiar with no head-capsule or mouthparts (Mamaev and Krivoscheina, 1969). The species are rare but found all over the world outside the tropics.

FAM. PACHYNEURIDAE. *Antennae with 16 or 17 elongate segments, in all almost as long as the thorax. Wings well-developed, veins of almost uniform thickness. Eyes holoptic or separate, ocelli present. Larvae not known.* A few rare species in the Holarctic region and the adjacent parts of the Oriental region. The family **Axymyiidae** (Mamaev and Krivoscheina, 1966) includes two or three species which have been regarded as allied to the Bibionidae or Anisopodidae. The adult rather resembles the latter group but the male genitalia do not seem to be inverse and the wing has no discal cell. The larvae and pupae are specialized, the former with a long, terminal, tube-like extension, the latter with a strong head-shield and prothoracic horns. In the family **Cramptonomyiidae** are placed three monotypic, Holarctic genera which seem even more close to the Anisopodidae.

Superfamily **Mycetophiloidea**

FAM. MYCETOPHILIDAE (Fungus Gnats). *Small flies provided with ocelli, antennae long, usually lacking whorls of hair in the male; coxae elongate, tibiae spurred. Larvae smooth and vermiform with a small dark head, 8 pairs of spiracles, and living gregariously in fungi or decaying vegetable matter.* The fungus gnats are delicate flies of a small or medium size, often bearing a resemblance to gnats or midges, and exceedingly numerous in individuals and species. Upwards of 2000 species are known, and the geographical range of the family is very wide. In coloration fungus gnats are seldom striking – blacks, browns and yellowish hues predominating. The body is elongate and compressed, with the thorax more or less arched, and sometimes markedly so. The antennae (Fig. 440) are almost always long and filiform, and composed of 12–17 segments, 16 being a common number. The tibiae are slender and armed with apical spurs, and the tarsal claws are toothed or pectinate. Sexual dimorphism occurs in a few species of the *Sciarinae*. In *Sciara semialata*, for example, the male possesses greatly reduced wings while the female is normal. In *Pnyxia scabiei* the female is destitute of both halteres and wings, while the male exists in two forms – one with reduced and the other with normal wings.

The larvae of a number of species have been described and the valuable paper by Osten-Sacken (1862) should be consulted together with more recent work by Keilin (1919), Madwar (1937) and Plassmann (1972). They are soft and whitish, with a small black or brown strongly sclerotized head, and 12 body segments. They are elongate and vermiform in shape, and generally sufficiently transparent to reveal much of their inner anatomy. The cuticle is smooth and devoid of hairs or setae; on the ventral surface there are often transverse swellings which, in many cases, are furnished with minute spines, aiding in locomotion. The antennae are always very short and frequently almost absent; they are better developed in *Bolitophila* than in most other genera. Situated below the antennae an oval pellucid spot is often present (*Bolitophila, Mycetophila, Leia, Epidapus,* etc.) which is probably of the nature of an ocellus. The respiratory system is peripneustic with 8 pairs of spiracles. The latter are found on the prothoracic and first 7 abdominal segments, the prothoracic pair being the largest. Exceptional genera include *Ditomyia* and *Symmerus* with 9 pairs of spiracles (Keilin, 1919); *Platyura* is devoid of spiracles and provided with protrusible anal gills and *Speolepta* is propneustic.

The imagines are found in a variety of situations, most commonly in damp or dark places, where there is fungoid growth, or decaying vegetation. Cellars, sheds, manure heaps and damp secluded parts of woods furnish many species. One of the characteristic features of these flies is their power of leaping, the hind legs being adapted for the purpose; many species simulate death when disturbed. The popular name of fungus gnat is derived from the fact that the larvae feed upon fungi more often than any other substance. A number of species however are found in rotting wood and other decaying organic matter, including leaf-mould and manure. The larvae are markedly gregarious, and many species construct a loose slimy web on their pabulum. Some of these, such as *Platyura* (Mansbridge, 1933), are predatory, feeding on small insects and worms which are killed by oxalic acid, secreted on the web. Most larvae, however, feed on fungi or other vegetable matter, often burrowing in their food and lining the tunnels with a slimy secretion, as in *Exechia*; *Phronia* (Steenberg, 1943) lives in a case made of particles of excrement, shaped into the form of an *Ancylus*-shell.

Larvae (Perrandin, 1961; *Lycoria* = *Sciara*) of the Sciarinae (often treated as a separate family) have been found in decaying apples, pears, turnips, potatoes, etc., and sometimes attack seedlings. In certain species of *Sciara* they exhibit the curious habit of travelling in vast numbers, so closely together as to almost constitute a single mass. This phenomenon is not infrequent at certain seasons in woods in Germany, Sweden, Russia and also in N. America. The migratory columns are elongate in form, and have been termed 'snake worms' in the United States, on account of their snake-like movements and appearance, which are said to resemble a thin grey reptile. They progress as a single mass with the larvae several deep over each other, and the movement is stated to be at the rate of about an inch a minute. In Europe they have been termed the 'army worm', but in America this expression is more properly applied to certain Noctuid caterpillars. Lintner mentions the stream of larvae as often being 4–5 m long, 5–7 cm broad, and perhaps 1 cm in thickness. In the United States one species has been reared and identified as *Sciara fraterna*; the common European 'army worm' is *Sciara militaris*. No satisfactory explanation has yet been advanced to account for the assemblage of these hordes of footless larvae.

Several species exhibit luminosity, which appears to have been first observed in the larvae and pupae of *Ceroplatus sessiodes* by Wahlberg in 1838. A particularly brilliant light was observed in a New Zealand species (*Arachnocampa luminosa*), by Hudson who remarks that the light emitted from a single larva kept in a caterpillar cage may be seen streaming out of the ventilators at a distance of several feet. Wheeler and Williams (1915) describe it as being emitted from the distal portion of the Malpighian tubules. The pupa and female imago are also strongly luminous but, according to Norris (1894), the male does not exhibit this property.

Some Mycetophilids spin true cocoons for pupation while others construct a fragile case of earthy material: the pupa in *Leia* is simply suspended by means of loose threads. In *Epicypta* the larval skin is adapted to form a shell in which to pupate but the pupa itself is free. The eggs are laid singly or in small groups, occasionally in strings, on whatever substance serves as food for the larvae. Many species pass through several generations in the course of a year, and as a general rule larval and pupal life is of short duration although certain species hibernate as pupae. According to Johannsen the time occupied from the egg to the adult may not exceed two weeks in midsummer. Edible mushrooms are frequently attacked by larvae of *Sciara, Exechia* and *Mycetophila*. They completely riddle the plants and may ruin a whole mushroom bed. Not infrequently they are introduced into the mushroom cellars through the agency of the manure used in the beds. According to Hopkins there are forms of potato scab and rot which are not due to fungoid disease, but are the direct result of the attacks of species of *Sciara* and *Epidapus*. *Sciara tritici* damages roots and stems of young wheat plants, and Johannsen remarks that there is no lack of evidence that Sciarinae damage the roots of cucumbers, grass and potted plants.

The most important works on the family are those of Johannsen (1909) and, for Palaearctic species, those by Landrock (1926–27) and Lengersdorf (1928–30) in Lindner's *Die Fliegen*. Edwards (1925*a*) describes most of the British species.

FAM. PERISSOMATIDAE. A very few small flies with curious venation and each eye separated into two complete parts. They breed in fungi and are known from Australia and Chile (Colless and McAlpine, 1970).

FAM. CECIDOMYIDAE (Gall Midges). *Minute delicate flies with long moniliform antennae adorned with conspicuous whorls of hair; ocelli present or absent. Wings with few longitudinal veins, for the most part unbranched, and with no obvious cross-veins. Coxae not elongate, tibiae devoid of spurs. Larvae peripneustic with a reduced head and usually a sternal spatula.* The Cecidomyidae include a large number of fragile and often very minute insects. The antennal characters, and the greatly simplified venation, enable these midges to be easily recognized. Among the best known species is the Hessian Fly (*Mayetiola destructor*) whose larvae are often destructive to wheat (Enock, 1891); from Europe it has been introduced into N. America and New Zealand. The Pear Midge (*Contarinia pyrivora*) is one of the most serious pests of that fruit in Europe; its larvae feed gregariously in the young fruitlets, which become deformed and subsequently decay.

A general monograph on the family is that of Kieffer (1900*a*) while the plant galls are described and catalogued by Houard. A good deal of information is also given in the numerous reports of Felt and more recently by Harris (1966) and most of the British galls are dealt with in the works of Connold and Swanton; for the Cecidomyids affecting cultivated plants, see Barnes (1946–49).

Larval Cecidomyidae exhibit great diversity of habits and may be classified as follows.

I. Zoophagous species of which very few are true parasites: Kieffer instances *Endaphis perfidus* which parasites *Drepanosiphon platanoidis*. A considerable number are predacious, preying mainly upon Homoptera, but others attack Acari; a few (species of *Lestodiplosis*) attack Dipteran larvae and pupae, including those of other Cecidomyidae.

II. Saprophagous species. Kieffer records species which live among the excrement of Tipulids and Lepidopterous larvae, and a few are found among decaying vegetable matter.

III. Phytophagous species which may be divided into (*a*). Those which live on or within various parts of plants without producing any gall formation. A number of species live on the spikelets of Gramineae, others in the flowers of Compositae, in fruit, or among fungi. (*b*) Gallicolous forms: a few live in galls formed by Coleoptera, Trypetidae and other Cecidomyidae. (*c*) Cecidogenous or true gall-making species. The vast majority of the family come under this category and all parts of the plant may be affected. Felt (1911*a*) computed that 438 species, included in 44 genera, of American Cecidomyidae affected 177 plant genera comprised in 66 families. Of these 146 species formed bud-galls, 44 fruit-galls, 218 leaf-galls, 130 stem-galls, and 4 formed root-galls. The Compositae, Salicaceae and Gramineae are the most frequently selected both in America and Europe. One of the most generalized of true gall-makers is *Rhabdophaga* which is partial to *Salix*, producing simple deformities such as bud and subcuticular galls. *Dasyneura* also forms comparatively simple leaf- and bud-galls on various plants.

The larval structure is dealt with at length by Kieffer, and the larvae are usually rather short and somewhat narrowed at both extremities. They vary in colour, being frequently white, yellow, orange or bright red, and occasionally brown. The head is very small and incompletely differentiated; pigment spots are present but there are no eyes. Thirteen trunk segments are evident, the first being intercalated between the head and prothorax. There are nine pairs of spiracles situated on the prothorax and first 8 abdominal segments. According to Kieffer the larva of *Rhinomyia perplexa* is exceptional in possessing 10 pairs, the additional pair being situated on the anal segment. The most characteristic structure associated with Cecidomyid larvae is the sternal spatula or so-called 'breast bone' (Fig. 460), which is situated mid-ventrally on the thorax. It is an elongate sclerite either toothed, pointed, or bilobed anteriorly: in some genera it is wanting. The function of this organ has been variously interpreted as an organ of perforation used for abrading plant tissues, as a locomotory organ, or for changing the position of the larva in its cocoon or case. Many larvae possess the power of leaping (*Contarinia*, etc.) and, according to Giard, in performing this act the anal crochets lock in the extremity of the spatula. The larva is thus curved into a loop, perpendicular to the surface upon which it is resting. By means of a sudden release of the tension it may be projected a distance of several centimetres.

Two methods of pupation occur in Cecidomyidae. In the usual method the pupa is enclosed in a cocoon which may be either single or double: in *Mayetiola* and *Chortomyia* the outer layer is a puparium, formed by the persistent larval skin.

The adults usually bear *circumfili* on the antennae: they are best developed in the males and are curious looped filaments or tortuous threads. In *Contarinia* and its allies each loop is fused basally with its fellows, thus producing an apparent whorl around each segment (Fig. 460, C). Their function is obscure, but it is presumably sensory. For the occurrence of paedogenesis in this family, see p. 305. There has been much recent work on this topic, partly because some of the species attack commercial mushrooms (Möhn, 1960; Nikolei, 1961; Whyatt, 1961, 1967; Kaiser, 1972).

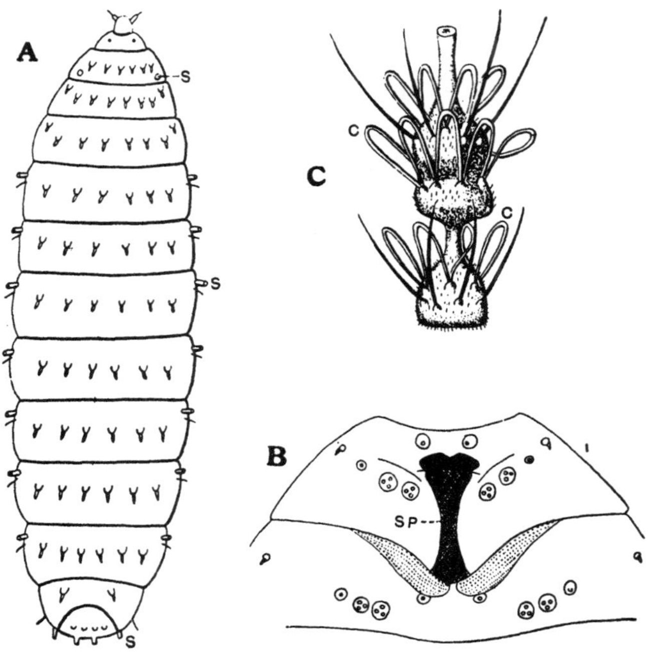

FIG. 460 A, larva of *Contarinia pyrivora*, dorsal view; *s*, spiracles. B, ventral aspect of 1st thoracic segment (1) showing sternal spatula *SP*. C, antennal segment of *Xylodiplosis praecox* male showing circumfili *c*

Adapted from Kieffer, *Ann. Soc. ent. Fr.*, 1900.

An extraordinary wingless fly with reduced, narrow thorax, physogastrous abdomen and eyes with a single ommatidium is placed in a new genus *Baeonotus* by Byers (1969*b*). It falls near the Cecidomyiidae though he makes a new family for it. The single known female specimen was found in forest litter in Virginia.

Suborder II. BRACHYCERA

The Brachycera include sixteen families whose antennal characters are so varied that the student will probably recognize many of its members more readily by means of the venation and the short porrect palps. This applies particularly in the case of *Xylophagus, Rhachicerus* and *Coenomyia* where the 3rd antennal segment is annulated to such an extent as to resemble the flagellum of some Nematocera and, furthermore, the style is wanting. It is also noteworthy that the discal cell is absent in the Dolichopodidae and certain Empididae. For a discussion of the affinities of the Brachycera and their larvae the student is referred to the introductory pages of Verrall's work (1909) and of Hennig (1954). The arrangement of the families is still a matter for discussion (Steyskal, 1953; Hennig, 1967). With the exception of the Stratiomyidae, the head in Brachycerous larvae (Fig. 461) is usually retractile within the thorax. The tracheal system is typically amphipneustic, and rarely peripneustic or metapneustic. If we except Stratiomyidae, the pupa is free and not enclosed in the larval skin: very rarely a cocoon is present (Dolichopodidae and *Drapetis*). The pupae may be recognized by their thorny appearance, spines being present on the antennal sheaths and other regions of the head and thorax. The abdominal segments are also usually provided with girdles of spines and the terminal somite is armed with pointed processes. The prothoracic respiratory organs are usually sessile.

Superfamily **Tabanoidea**

FAM. RHAGIONIDAE. *Bristleless flies; with the 3rd antennal segment usually non-annulated with a terminal style* (Fig. 440). *Some or all the tibiae spurred, squamae practically absent. Wing-veins well defined, not concentrated anteriorly. Pulvilli and arolium pad-like.* The Rhagionidae include rather elongate flies of sombre coloration, usually thinly pilose or almost bare (Fig. 462). Over 400 species have been described, of which about 19 are British. Normally they are predacious upon other insects, but it is recorded that the female of *Symphoromyia* in America and *Spaniopsis* in Tasmania are bloodsuckers (cf. Ross, 1940).

The metamorphoses of many Rhagionidae are known (Beling, 1875; 1882; Greene, 1926); the larvae (Fig. 461 C) are cylindrical, with a small exserted head succeeded by 11 trunk segments, which may or may not bear pseudopods. The

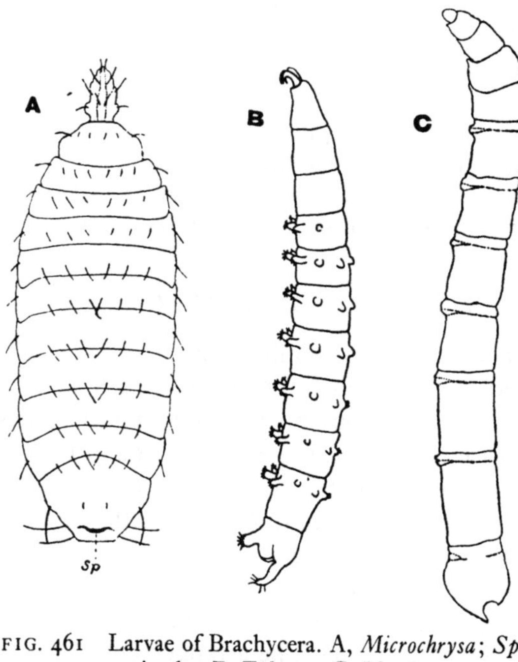

FIG. 461 Larvae of Brachycera. A, *Microchrysa*; *Sp*, spiracles. B, *Tabanus*. C, *Rhagio*

A and C *after* Cameron, *J. econ. Biol.*, 8; B *after* King.

abdominal segments often possess transverse denticulate, ventral swellings which aid in locomotion. The last segment is modified and marked by longitudinal folds or grooves, or provided with hairy processes. Rhagionid larvae are carnivorous, preying upon other insects or their larvae: according to Marchal the larva of *R. tringaria* lives upon small Oligochaetes. Their usual habitat is in the earth or in leaf mould: *Atherix*, however, is aquatic. The larva of *Atherix* has been described by Malloch (1917) and Nagatomi (1961): the head is minute and each abdominal segment bears a pair of pseudopods capped by spines. The sides of the body are fringed with numerous filamentous processes which have been regarded as gills, no spiracles having been detected. The anal segment carries a pair of prominent, hairy, backwardly directed processes. The females of *Atherix* deposit their eggs in masses on dry twigs, etc., overhanging water, into which the larvae fall upon hatching out.

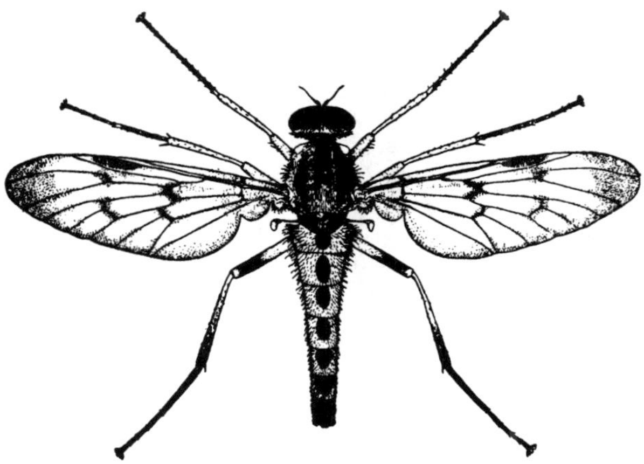

FIG. 462 *Rhagio scolopacea*, male. × 3. Britain
After Verrall.

Many individuals lay their eggs on the same cluster, and afterwards die on the spot, often in numbers. As their dead bodies adhere together, large incrustrations are thus formed. In Oregon the Indians at one time collected these masses of eggs and flies for food (Aldrich). Females of *Vermileo* and *Lampromyia* (Hemmingsen, 1963) lay their eggs in sand, and the larvae construct conical pitfalls for the capture of their prey, after the manner of 'ant lions'. The 5th segment of the larva bears a ventral mobile pseudopod which assists in seizing and holding the prey. The 10th and 11th segments each carry a transverse row of long hooklets which serve as organs for boring and fixation (cf. Wheeler, 1930).

FAM. XYLOPHAGIDAE (= ERINNIDAE). *Bristleless flies with 3rd antennal segment annulated, squamae small, at least mid tibia with spurs, vein C encompassing the wing, arolium padlike.* The members of this somewhat heterogeneous primitive family are difficult to place in the system. The adult has a more or less elongate form and the larva has an almost complete head and lives in decaying wood (Verrall, 1909; Perris, 1870; Malloch, 1917). *Xylophagus* has two British species and *Coenomyia*, often put in a separate family, occurs in southern Europe.

The gigantic *Pantophthalmus*, also often put in a separate family, and its allies are found in S. America where the larva has been studied by Thorpe (1934). (See also, Rapp and Snow, 1945.)

FAM. TABANIDAE (Horse Flies and Clegs). *Bristleless flies of stout build with the 3rd antennal segment annulated but devoid of a style* (Fig. 440). *Eyes very large, laterally extended; proboscis projecting, adapted for piercing in the female* (Fig. 442). *Squamae large, pulvilli and arolium pad-like.* An extensive family of moderate to large sized flies (Fig. 463), including about 2000 species, which are distributed over the whole world: 28 species are British. They are more or less flattened insects and, as a rule, mottled brown, tawny or grey in colour; *Chrysops*, however, has more con-

FIG. 463 *Tabanus maculicornis*, female. × 3. Britain
After Verrall.

spicuous hues. During life the eyes are iridescent, exhibiting brilliant shades of green often marked with bands or spots of brown or dark purple. In British Tabanids the proboscis is always rather short, but almost every transition may be found among the various genera of the family culminating in species of *Pangonia*, where it may be more than twice the length of the body.

Horse flies are active on warm sunny days, and the females are well-known blood-suckers, whereas the males mostly subsist upon honeydew and on the juices of flowers. In the absence of blood, the females will also imbibe these same substances (Hine). Many species are swift fliers, and those of *Tabanus* are particularly trouble-some to horses and cattle. The piercing action of the proboscis is often painful, but is seldom accompanied by inflammation. Experimental evidence indicates that the disease of horses known as surra is transmitted by the punctures of *T. striatus*, and other species, and that these flies play an important part in spreading the infection. According to Leiper *Chrysops dimidiata* is a vector of *Loa loa* which is responsible for the affliction known as Calabar swellings among the natives of West Africa.

Pangonia may often be found hovering over flowers on the borders of forests; species have been observed to attack both man and cattle in various parts of the world. Their method of attack varies considerably in the experience of different observers. The labium is not adapted for piercing, the latter operation probably being performed by the other trophi, the proboscis only being used for sucking up the blood. In some species this is performed on the wing, in others it is stated to take place after the insect has alighted (Tetley, 1918). The species of *Haematopota* or 'clegs' are voracious blood-suckers and especially frequent damp meadows. They are notable for their quietness of approach, and often the pricking sensation of their punctures is the first intimation of their presence. *H. pluvialis* is the most abundant British Tabanid, and is particularly troublesome to man. According to Portchinsky, in parts of Russia, these flies are so numerous and offensive that agricultural operations have to be carried out at night. By covering the pools frequented by Tabanids with a thin layer of petroleum he succeeded in destroying large numbers of these troublesome insects, which were killed by the oil adhering to their bodies.

The eggs of Tabanidae are spindle-shaped and white, brown, or black; they are deposited in compact masses on the leaves and stems of plants, growing in water or marshy places. The larvae (Fig. 461 B) (Stammer, 1924; Teskey, 1969) are 12-segmented with a relatively small retractile head, well-developed antennae and strong mouth-hooks. The trunk is cylindrical, tapering at both extremities, and usually longitudinally striated; there is a circle of prominent fleshy pseudopods around each of the first abdominal segments. They are metapneustic with the spiracles placed closely together in a vertical fissure at the anal end of the body. Near the hind extremity of the larva of *Tabanus* is a pyriform sac, narrowing posteriorly into a fine tube which opens at the surface between the last two segments. Within the sac is a series of capsules, each containing a pair of minute black pyriform bodies which are attached to the walls by means of delicate pedicels. The whole structure is known as *Graber's organ*, and can readily be seen through the integument of the living larva. It is well supplied by nerves and is presumably sensory in function. The larvae of *Tabanus* and *Chrysops* are closely alike but according to Malloch (1917) in *Chrysops* the thoracic segments are either smooth, or less markedly striated than the abdominal, and the apical antennal segment is much longer than the one preceding. In *Tabanus* the striation is uniformly well developed over the body, and the terminal antennal segment is shorter than the preceding one. The larva of *Goniops* differs from the usual Tabanid form in that the hindmost segments are stouter than those preceding thereby imparting to the body a pyriform or club-shaped appearance (McAtee, 1911). The larva of *Haematopota* resembles that of *Tabanus* but according to Perris (1870) and Lundbeck (1907) it is amphipneustic.

Tabanid larvae have been found in a variety of moist situations – in damp soil bordering ponds and streams, under stones in similar places, in mud, wet rotting logs, etc. They are carnivorous, devouring small earthworms, crustacea and insect larvae. The pupae are markedly elongate and cylindrical. They are characterized by the thoracic spiracles being connected subcutaneously with a large cavity on either side of the median line, near to the anterior margin of the thorax. Each abdominal segment carries 1–2 dorsal bands of closely contiguous setae and a weaker series ventrally. The terminal segment is armed with six stout pointed projections.

The literature on the family is considerable and for a general account, including the British species, consult Verrall (1909), Lundbeck (1907), and the work of Edwards *et al.* (1939). Chvála and Lyneborg (1972) have monographed the European species and Mackerras (1954) reviews the modern classification. Important studies of foreign species are those of Efflatoun (1930), Olsoufieff (1937) and Oldroyd (1952). For the morphology and biology see Bromley (1926) on *Tabanus*, Cragg (1912) on *Haematopota*, the textbook of Patton and Cragg (1913) and Marchand (1920).

The **PELECORHYNCHIDAE** including the single primitive genus,

Pelecorhynchus of S. Chile and E. Australia, used to be placed in the Tabanidae but has reduced mandibles and is not a blood-sucker. It seems to be nearer to *Coenomyia* but R_4 and R_5 are divergent and the scutellum does not have two spikes.

FAM. STRATIOMYIDAE. *Bristleless flies with the 3rd antennal segment annulated, squamae small, tibiae rarely with spurs; scutellum often conspicuously developed, sometimes with spines or projections, wing-veins crowded near costa and more strongly pigmented than those behind, C not encompassing the wing. Pulvilli and arolium pad-like.* The Stratiomyidae (Fig. 464) are small to rather large flies, more or less flattened and usually with white, yellow or green markings: in the Geosarginae, however, the prevailing colour is metallic. About 1400 species are known and of these about 50 are British. They are not usually strong fliers and occur on umbelliferous and other flowers and herbage, especially in damp situations. The Solvinae (= Xylomyidae) are the most aberrant members both as regards their venation and in the presence of tibial spurs (also seen in some Beridinae); they are annectent between this family and the Xylophagidae.

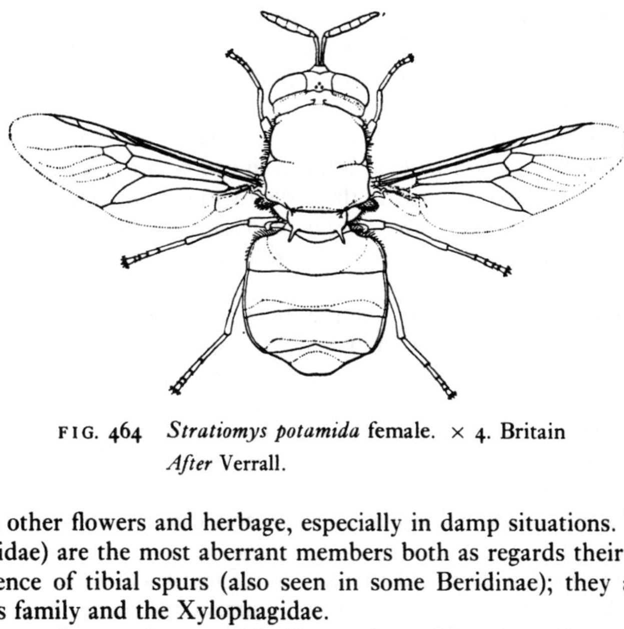

FIG. 464 *Stratiomys potamida* female. × 4. Britain
After Verrall.

The metamorphoses of a considerable number of Stratiomyidae are known: the eggs are laid on plants near the edge of water, or even on the surface of the water, also in dung and in the soil. The larvae are carnivorous or saprophagous, and either terrestrial or aquatic, the terrestrial larvae being largely scavengers: those of the Solvinae occur in rotting wood. Stratiomyid larvae exhibit considerable diversity of form: all have a peculiar thick leathery skin impregnated with calcareous matter. The head is small and exserted, and there are 11 trunk segments, none of which bear pseudopods (Fig. 461A). Although often described as being peripneustic, it is doubtful whether they are functionally so: they appear to be physiologically metapneustic or in some cases amphipneustic. The lateral spiracles, with the exception of the prothoracic pair, are minute and difficult to detect. In *Stratiomys* 9 pairs are present; they are situated on the 1st and 3rd thoracic segments, and on each of the first 7 abdominal segments. Although stated by Brauer to be peripneustic, Miall remarks that the lateral spiracles are closed. The terminal or posterior spiracles are always open and are situated in a horizontal fissure, fringed with hairs in the aquatic forms. The larvae of *Stratiomys* and *Odontomyia* are greatly elongate and taper towards the anal extremity. In the former genus the last segment is much drawn out and tubular: these larvae live in water or mud, and hang from the surface film by means of the tail coronet of feathery hairs which is spread out in an asteriform manner. When the larva descends, the hairs are drawn inwards and enclose a large air bubble: as the latter becomes used up the larva returns to the surface. The larva of *Solva* resembles that of *Geosargus*, being broad with parallel sides, and the usual leathery skin is impregnated with calcareous matter. It is regarded as being amphipneustic with prothoracic and terminal spiracles: according to Lundbeck non-functional lateral abdominal spiracles are also present. Stratiomyid pupae differ from other Brachycera in being enclosed within the larval skin. Tables of the larval characters for each subfamily are given by Verrall (1909): further information will be found in the works of Lundbeck (1907), Irwin-Smith (1921–23) and McFadden (1967).

Superfamily **Asiloidea**

FAM. THEREVIDAE. *More or less elongate densely pubescent flies with slender non-prehensile legs. 3rd antennal segment with an apical (sometimes jointed) style.* R_1 *usually long, cell* M_3 *present. Empodium absent, or represented by a weak bristle.* A

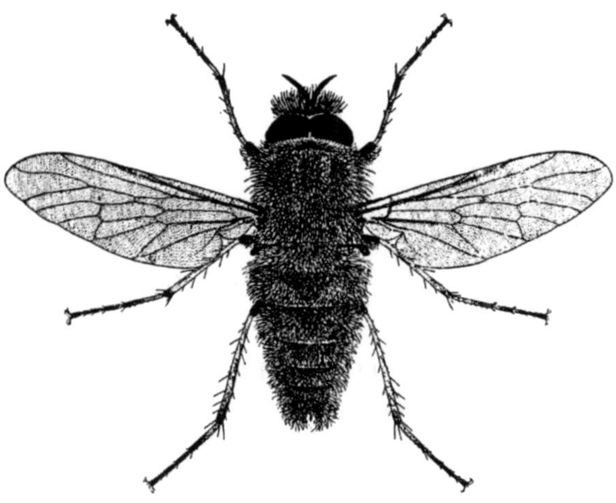

FIG. 465 *Thereva nobilitata*, male. × 4. Britain
After Verrall.

small family including nearly 500 described species (Fig. 465) of which ten are British (Collin, 1948). They exhibit a resemblance to some Asilidae, but the weaker legs and the non-protuberant eyes enable them to be readily separated. In habits these flies are commonly stated to be predacious, but very few direct observations appear to have been made. The proboscis is rather prominent, and provided with fleshy labella, instead of the horny apex as in the Asilidae. The larvae (Fig. 466) of several species are known to be predacious upon those of other insects, including wireworms, etc. They live in the soil, among leaf-mould, in fungi, decaying wood, etc., and exhibit quick serpent-like movements. They are smooth and vermiform, bearing an extremely close resemblance to the larvae of *Scenopinus* (see below). The larva of *Thereva* (Malloch, 1915) has a small though distinct head followed by 20 segment-like divisions. The labrum is hook-like, and the mandibles also exhibit a hooked form: small antennal papillae are present but no eyes. A pair of prominent lateroventral bristles are found on each thoracic segment, and three pairs of bristles on the 10th abdominal segment: the tracheal system is amphipneustic. At the anal extremity are two small styliform processes. The pupa has thorn-like, projecting antennae and a long curved spine at the base of each wing.

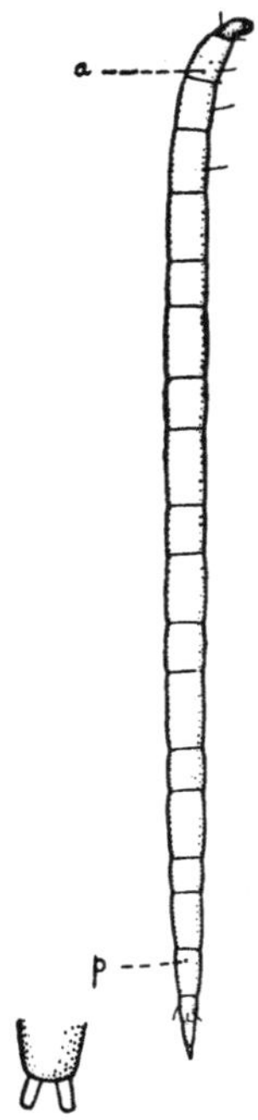

FIG. 466
Larva of *Thereva* and anal segment more enlarged

a, anterior and *p*, posterior spiracles.

FAM. SCENOPINIDAE (OMPHRALIDAE). *Narrow oblong flies devoid of true pubescence or bristles. First two antennal segments short, the third elongate and devoid of a style or arista. Vein* M_{1+2} *terminating before the apex of the wing.* R_1 *short. Pulvilli small. Empodium bristle-like.* This family includes rather small dark coloured flies and more than 200 species are known (Kelsey, 1969). The adults are occasionally found on windows, or about stables and outbuildings. The larva of *Scenopinus* (see Perris, 1870) resembles that of *Thereva*. It is amphipneustic, white and vermiform, with serpent-like movements. The head is brown and well-developed, and is followed by 20 apparent segments. Most of the abdominal somites are subdivided by a strongly marked constriction, thus giving the appearance of an increased number of segments: the terminal segment bears two small styles. At one time it was believed that these larvae fed upon neglected carpets, horse-rugs, etc.: there is little doubt, however, that they are predacious upon the larvae of *Tinea pellionella* and of other insects. Scenopinid larvae have also been found in *Polyporus*, in branches of trees, and other situations.

FAM. MYDIDAE. *Very large flies devoid of bristles and obvious pubescence. Antennae terminating in a jointed and usually clubbed style. Venation complex;* R_1 *very* *long receiving several succeeding veins before its apex, R_4, R_5 and M_{1+2} bent forwards towards the apex of the wing. Pulvilli moderately large, no empodium.* A family of mostly exotic forms with a few moderate-sized species occurring in southern Europe. It includes the largest known Diptera, and the adults are stated to be predacious, but only very scanty observations are available. The larvae have been found in decaying wood and, in some cases, are known to be predacious upon Coleopterous larvae. *Mydas heros* lays its eggs in the nests of *Atta sexdens* (leaf-cutting ant) and its larvae very probably prey on those of *Coelosis* spp. (Dynastinae) (Papavero and Wilcox, in: Papavero, 1968). For the general structure of the family see Bequaert (1961).

FAM. APIOCERIDAE. *Rather large elongate bristly flies: antennae with or without a short style, palps spatulate. Venation rather similar to Mydidae, M_1 terminating before wing apex. Two pulvilli are present and the empodium is wanting or bristle-like.* A very small family comprising about 100 species, none of which is European but 70 are Australian (Paramonov, 1953*a*). They are thickly pilose insects with a non-piercing proboscis. The larva of the Australian *Apiocera maritima* has been described by English (1947); it is very similar to the same stage in an Asilid.

FAM. NEMESTRINIDAE. *Rather large bristleless flies with many of the veins running parallel with the hind margin of the wing: Sc and R_1 very long. 3rd antennal segment simple with a terminal style, pulvilli and arolium pad-like but often minute.* A family of about 200 species (Fig. 467), none of which occur in the British Isles, but 57 are Palaearctic and 14 are European. The 56 Australian species are revised by Paramonov (1953*b*). They are for the most part inhabitants of hot and arid regions where there is a minimum of rainfall. They mainly frequent flowers, hovering over them while imbibing the nectar. The proboscis is very variable and often long, or very long; in *Nemestrina longirostris* it is about four times the whole length of the insect.

The larvae of this family have a very small retractile head and 12 trunk segments:

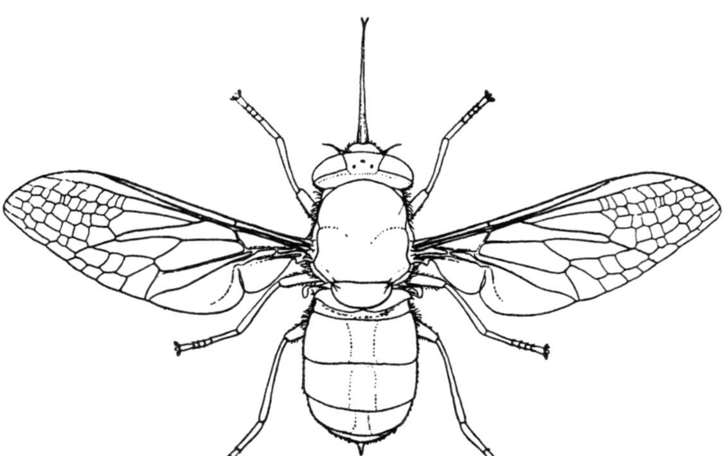

FIG. 467 *Nemestrina perezi*, female. × 3·25
After Verrall.

the tracheal system is amphipneustic, with the posterior spiracles spaced apart in a transverse fissure. The life-history of *Hirmoneura* has been partially observed by Brauer (1883) and Handlirsch (1882). Its habits resemble to some extent those of the Bombyliidae and hypermetamorphosis also occurs. The young larva is slender and provided with a pair of pseudopods on the 6th and 12th segments, which are not present at a later stage. It appears that this species is parasitic upon *Amphimallon solstitialis*, and probably upon other Coleoptera. The eggs are deposited in clusters within the burrows of Coleoptera (other than *Amphimallon*) from whence the newly-hatched larvae issue in large numbers. They are stated to place themselves in an erect position by means of their terminal hooklets, and are blown away by the wind. Their subsequent history is unknown, but it is believed that they attach themselves to the body of the female *Amphimallon*, and are thus carried to the place in the earth where the latter lays her eggs. In *Neorhynchocephalus* which parasitizes grasshoppers, the young larva first punctures a host trachea but soon pierces the external cuticle and plugs the hole with the plate bearing its metapneustic spiracles. The posterior part of the larva then gradually grows into a long tube so that the anchorage does not prevent the larva feeding throughout the host's interior (Prescott, 1961).

FAM. ACROCERIDAE (CYRTIDAE). *Bristleless flies with the head very small and almost entirely composed of the eyes which are holoptic in both sexes. Thorax humped, squamae exceedingly large: abdomen greatly inflated and globular. Pulvilli and arolium pad-like.* A small family of medium-sized flies including about 200 species which are readily distinguishable from all other Brachycera. Although occurring in all parts of the world Acroceridae are local and uncommon: two genera, *Oncodes* and *Acrocera* (Fig. 468) are found in the southern portion of England (Verrall, 1909). (See also Sabrosky, 1948 for the N. American species). So far as known, their larvae are parasitic upon spiders living in the egg-cocoons, or attached to the abdomen of their host. The eggs are black, and have been noted on dead twigs and on *Equisetum*: those of *Oncodes* are laid in masses (Maskell, 1888). The life-history of *Pterodontia* has been partially followed by King (1916). The newly-hatched larva bears a striking resemblance to the triungulin of *Stylops*: it is strongly sclerotized, and armed dorsally and ventrally with segmental bands of powerful spines and pectinate scales. At the caudal extremity of the 8th abdominal segment is a sucker, which is flanked by a long anal seta on either side. In addition to a looping leech-like movement, the larva is able to leap by standing erect upon its sucker, with the caudal setae bent beneath

FIG. 468 *Acrocera globulus*, male. × 7·5. Britain
After Verrall.

the body: a sudden straightening of the setae lifts the larva 5 or 6 mm in the air. The insect bores its way into the host by penetrating the articular membranes of the legs, and lives endoparasitically: according to King there are no spiracles. The older larvae in this family are short and stout, and apparently amphipneustic: König (1894) and Maskell state that the younger larvae, of presumably *Oncodes*, are metapneustic; they similarly possess leaping powers. According to Brauer (1869), the larva of *Astomella* lives within the abdomen of the spider, with its hind spiracles penetrating the lung-books of the latter. The pupae in this family are devoid of spines or bristles, and differ from those of other Diptera in the great size of the thorax, which exceeds the abdomen in length.

FAM. ASILIDAE (Robber Flies). *Usually elongate bristly flies with a horny proboscis adapted for piercing, and the palps never spatulate. Vein R_1 very long, M_1 terminating some distance beyond the apex of the wing. Legs powerful and prehensile: pulvilli large, empodium bristle-like.* The Asilidae are moderate to very large-sized flies (Fig.470), always bristly, and in *Laphria* also densely hairy. They constitute the largest family of Brachycera, numbering at least 4000 species: in the British Isles 27 species are recognized. Hull (1962) has described and made keys for the genera of the world and listed their species. The adults are predacious in habit, their powerful legs being adapted for grasping the prey. The proboscis is firm and horny, directed downwards or obliquely forward. A prominent tuft of hairs, forming a 'mouth-beard', and the protuberant eyes are characteristic of the family. The conspicuous male genitalia and the corneous ovipositor are also well marked features.

The prey of Asilidae is extremely varied and information on the subject has been collected by Whitfield (1925), Hobby (1931*a*) and others. It appears that the females are far more commonly found with prey than the males; it is remarkable as Poulton adds, that the stings of Aculeates, the distasteful properties of the Danainae, Acraeinae and of the odoriferous *Lagria*, the hard cuticle of Coleoptera and the aggressive powers of the Odonata are alike insufficient protection against these voracious insects. Whether Asilids inject any poison into their victims or not has yet to be ascertained. It has been recorded that the captured insect collapses very rapidly after being perforated by the proboscis, which suggests that some toxic secretion may be present.

Asilid larvae live in soil, sand, wood, or in leaf-mould, and are either predacious or scavengers. They are cylindrical with a small, dark coloured, pointed head and are amphipneustic, the spiracles being situated on the prothoracic and penultimate segments. The mouthparts comprise a hook-shaped labrum, knife-like mandibles and large broad maxillae with 2-segmented palps. Small papilla-like antennae are present but no eyes. The anterior abdominal segments are provided either with ventral intersegmental areas, or circlets of pseudopods (*Laphria*) resembling those of Tabanid larvae. Ten or eleven segments are present, the higher number depending upon whether a short and indistinct segment-like swelling at the anal extremity is regarded as a true somite or not. The pupae are remarkably spined about the head: the abdominal segments have a dorsal girdle of spines, a ventral girdle of bristles, and the apex of the abdomen also bears spinous projections. The larva of *Laphria* has been found beneath bark and in the burrows of Longicorn larvae living in *Pinus*: it has been figured by Perris (1870) and later by Sharp (see Verrall, 1909). For the metamorphoses of other genera see Melin (1923) and Lundbeck.

FAM. BOMBYLIIDAE (Bee Flies). *Often densely pubescent with elongate slender legs, and often a long projecting proboscis. 3rd antennal segment simple, style small or vestigial, and not more than 2-segmented* (Fig. 440F). *Cell M_3 absent. Pulvilli sometimes and an empodium always rudimentary.* Most of the flies of this family are moderate or rather large in size, often bearing bristles and dense pubescence (Fig. 469). The proboscis is usually very long and projecting forwards, but is sometimes short with broad labella. The wings are often darkly marbled and, when at rest, they remain half opened or outspread. The females sometimes have terminal combs of spines on the last abdominal segments by means of which they lay their eggs in sand (Mühlenberg, 1968). Although only 12 species frequent Britain, considerably over 2000 are known: cf. Hesse (1938) for the large S. African fauna.

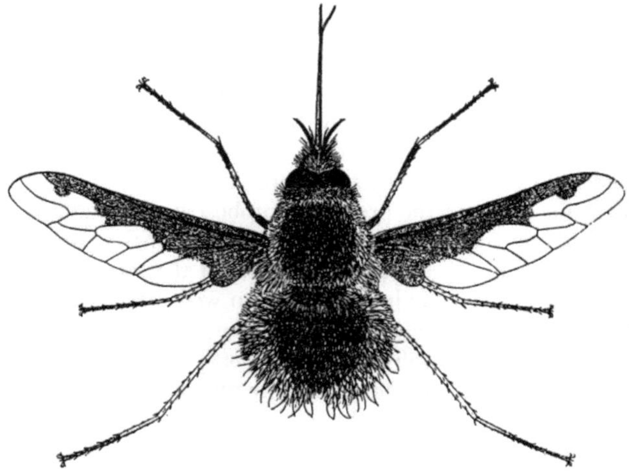

FIG. 469 *Bombylius major*, male. × 3. Britain *After* Verrall.

The larvae are parasites and when young, they are elongate and slender, with a very small head, and 12 trunk segments (Merle, 1972). They are metapneustic; each thoracic segment may bear a pair of long setae, and a further pair is carried at the anal extremity. They undergo hypermetamorphosis and, when fully grown, are cylindrical or somewhat flattened, with a small retractile head and no eyes: the spiracles are found on the prothoracic and penultimate segments. The pupae are very characteristically spined on the head, with bands of hooklets across the dorsal side of the abdomen.

The larvae of *Argyramoeba* are parasites on those of solitary bees and fossorial wasps. The life-history of *A. trifasciata* has been observed by Fabre. The eggs are deposited on the ground, near the nest of the host (*Chalicodoma muraria*), and it appears that the young larva has to make its way into the cell of the bee. The pupa is armed with cephalic spines for the purpose of piercing the masonry enclosing its host. *Argyramoeba anthrax* (*sinuata*) has been bred from nests of *Anthophora*, *Chalicodoma* and *Osmia* and an account of its life-history is given by Verhoeff (1891).

Several species of *Anthrax* are parasitic upon Noctuid larvae or pupae, Aculeate

Hymenoptera and also upon the eggs of Orthoptera. Other members of the genus are hyperparasites attacking Hymenopterous or Dipterous parasites of Lepidoptera.

The larvae of *Bombylius* are parasitic upon solitary bees (*Andrena, Halictus, Colletes*, etc.): those of *B. minor* have been studied by Nielsen (1903) who states that the young larva is very like that of *Argyramoeba* in form. At this stage it feeds upon the pollen store in the cell of *Colletes*, but when it attains a length of 2 mm it attacks its host larva: it subsequently moults, becoming maggot-like and amphipneustic. The life-history of *B. major* has been observed by Chapman (1878): the eggs were deposited on a sloping bank while the fly was on the wing, and descriptions of the larva and pupa agree in the main with those of *B. minor*.

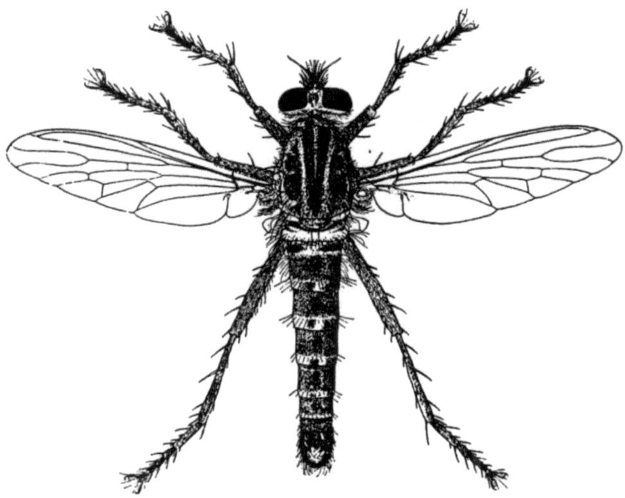

FIG. 470 *Philonicus albiceps*, male (Asilidae). × 3. Britain
After Verrall.

Larvae of *Systoechus* live as parasites in the egg-cases of the locusts *Oedipoda* and *Stauronotus*: the larva and pupa of *S. oreas* are described and figured by Riley *et al.* (1880). According to Künckel d'Herculais (1905) the larva of *Systropus* parasitizes larvae of the Lepidopteran *Limacodes. Spogostylum* is parasitic upon *Xylocopa* and other bees, and is also recorded from two genera of the Coleoptera: for the life-history of *S. anale*, a parasite of Cicindeline larvae, see Shelford (1913).

Superfamily **Empidoidea**

Dyte (1967) distinguishes between the larvae of the two families in the group and indicates some of their characters and Ubrich (1971) compares their thoracic skeleton and musculature.

FAM. EMPIDIDAE. *Bristly flies with a horny proboscis adapted for piercing; the style or arista (if present) almost always terminal. Cells M and 1st M_2 separate, cell Cu generally short. Empodium linear–membranous, or setiform.* A family of medium to very small sized flies of grey, yellowish, or dark coloration, very rarely metallic. About 3000 species are known, and in Britain there are over 300 representatives.

The proboscis is of variable length, and is generally rigid and downwardly projecting. The legs often display sexual characters, the male exhibiting special structural features such as thickened femora, tibiae or tarsi (Fig. 471). Empididae may be distinguished from the Asilidae by the absence of the face-beard and the much shorter cubital cell. Their species are predacious upon smaller insects and, according to Poulton (1906), they prey most frequently upon Diptera; in this feature they are sharply contrasted with the Asilidae. Their mouthparts have been described by Krystoph (1961) and their genitalia by Ubrich (1972).

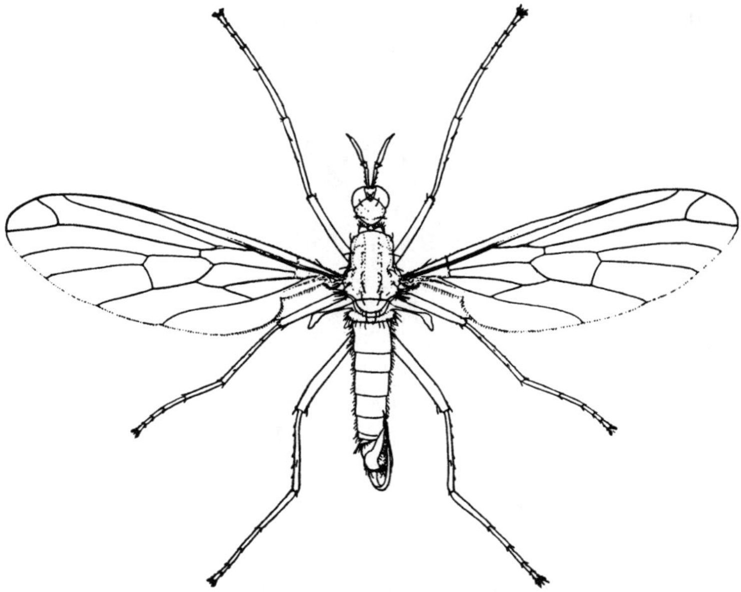

FIG. 471 *Empis trigramma*, male. × 6. Britain
After Verrall.

Species of *Empis*, *Hilara* and *Rhamphomyia* may often be observed 'dancing' or swarming in the air after the manner of Chironomids – a behaviour which is concerned with the meeting of the sexes. Either one or both sexes may perform these aerial evolutions and, in many species of the above genera, the males catch the prey, and kill but do not devour it. On meeting a female the latter receives the prey and feeds upon it during coitus: when copulation is accomplished the female drops the prey. The true significance of this remarkable habit is not understood (Hamm, 1909). An American species, *Empis aerobatica*, makes a curious frothy balloon, enclosing a small prey, which is probably transferred to the female during copulation; it is often released after the latter function is accomplished. Species of *Hilara* envelop their prey in a slight web before offering it to the female: the web is constructed by the male from a secretion of glands opening on the fore tarsi (Eltringham and Hamm, 1928). *H. sartor* constructs a more extensive web than other species, and a whole literature has grown up around the subject of the origin and significance of this structure (Wheeler, 1924).

Larval Empididae are cylindrical, more or less spindle-shaped, with a very small retractile head and 11 trunk segments. They are amphipneustic, and most of the abdominal segments are provided with transverse ventral swellings, or more strongly developed pseudopods. The anal segment is somewhat rounded, and provided with a small terminal protuberance or spine, above which lie the posterior spiracles. Empidid larvae live in soil or among leaves and humus, in decaying wood, among moss, etc.: a few, such as *Hemerodromia*, are aquatic. Only scanty observations have been made with regard to their feeding habits but, in a few cases, they have been found to be carnivorous. K. G. V. Smith (1969*a*) in describing the Southern African species gives a useful introduction to the classification of the family. Collin (1961) deals with the British species. Knutson and Flint (1971) in Chile found the pupa of *Neoplasta* (Hemerodromiinae) inside the cocoon of *Cailloma* sp. (Trich., Rhyacophilidae) together with the dead caddis pupa. Another fly-pupa of the same subfamily was found in the cocoon of *Mortoniella*.

The larvae of several genera are briefly described by Beling (1882; 1888); the metamorphosis of *Hemerodromia* is dealt with by Brocher (1909) and that of *Roederoides* by Needham and Betten (1901). Kieffer (1900*b*) has described the larva and pupa of *Empis meridionalis* and Brauer (1884) figures the larva of *Hilara lurida*.

FAM. DOLICHOPODIDAE. *Small bristly usually metallic green or blue-green flies with a dorsal or terminal arista, and a short fleshy proboscis. Cells M and 1st M_2 confluent, cell Cu very short. Two pulvilli and a linear or narrowly lobiform empodium.* A large family comprising more than 2000 species of which at least 250 are British (Fig. 472). The various species occur among grass and low herbage, generally in damp places; several genera frequent the sea-shore. The work of Parent (1938) is useful for the West European fauna and Robinson (1970) outlines the classification of the American species. The family comes closest of any of the Brachycera to the Cyclorrapha but it is probably only by convergence that some genera have a circumverse hypopygium (Bährmann, 1966).

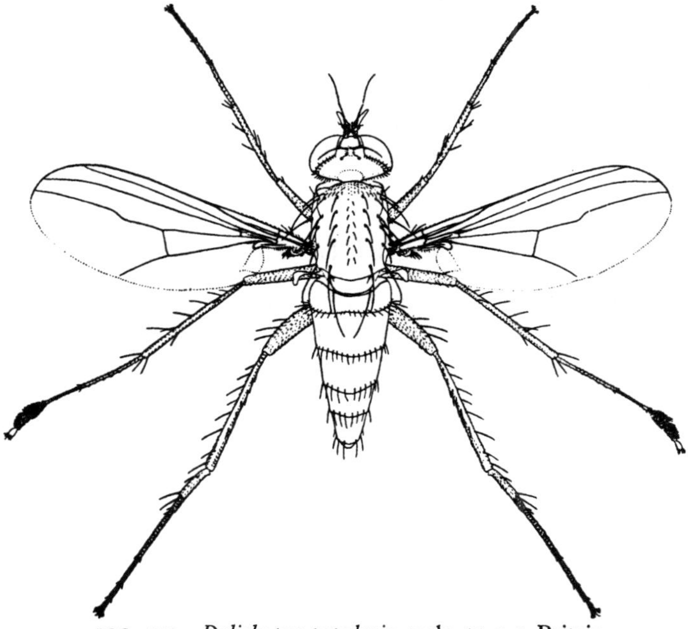

FIG. 472 *Dolichopus popularis*, male. × 7·5. Britain *After* Verrall.

The venation of the family is very similar to that of the Ephydridae, which certain species also resemble in their behaviour and habitat. The secondary sexual characters in the males attain a remarkable degree of development and may affect almost any of the outer parts of the body.

In adult life the Dolichopodidae are predacious upon minute soft-bodied insects, etc., which they envelop by means of the labella while extracting the juices. According to Becher (1882) some amount of mastication of the prey takes place on account of the mobility of the labrum during feeding. In *Melanderia* (Snodgrass, 1922) remarkable horny processes of the labella aid in grasping the prey. The adult mouthparts in general have been studied by Cregan (1941). Many species occur on flowers and undoubtedly feed upon nectar.

The larvae of this family have been found in a variety of situations and live beneath the ground, in rotten wood, among humus, etc., while others are aquatic. In *Aphrosylus* the larva lives among cast-up weed on the sea-shore, while those of several species of *Medetera* prey upon the larvae and pupae of wood-boring Coleoptera. Most of the larvae of this family are probably carnivorous. They are elongate and cylindrical, 12-segmented, with a small retractile head, and most of the abdominal segments bear pseudopods armed with locomotory spinules. The last segment is obliquely truncated, often slightly swollen, and carries four short protuberances. The tracheal system is amphipneustic, and both pairs of spiracles are small; exceptions are met with in *Medetera* (peripneustic) and *Argyra* (metapneustic).

The pupae are, as a rule, short and stout with a pair of elongate thoracic respiratory horns. Lundbeck states that the larva of *Dolichopus* forms an earthen pupal cell, lining its interior with a secretion forming a dense film-like layer. At one extremity the latter is wanting over a smaller area through which the pupal horns protrude. As the cocoon is apparently impenetrable to air Lundbeck thus explains the significance of the long pupal horns, so characteristic of the family. In other cases the cocoon is constructed of wood fragments, etc., and is lined by silken material. Although the metamorphoses of a number of species have been described the life-history has rarely been followed in any detail. Marchand (1918) has described the larva and pupa of *Argyra*, Perris (1870) those of *Medeterus*, and *Thrypticus* has been studied by Johannsen and Crosby (1913) and also by Lübben (1908).

Suborder III. CYCLORRHAPHA

The Cyclorrhapha are divided into two sections as given below. Their relationships are discussed by Griffiths (1972).

Section A. ASCHIZA

Frontal suture absent: lunule usually indistinct or absent: ptilinum absent. Cell Cu elongated (except in Phoridae) and extending more than half-way to wing-margin.

Section B. SCHIZOPHORA

Frontal suture and lunule distinct: ptilinum always present. Cell Cu short or vestigial (except in Conopidae).

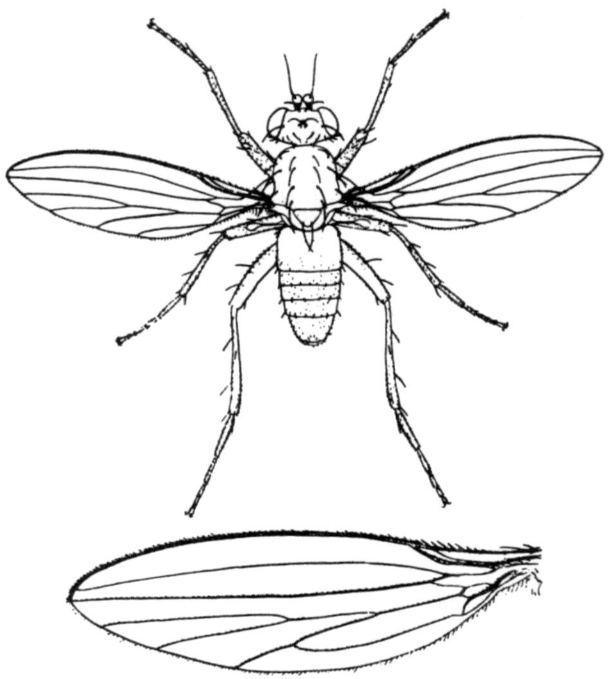

FIG. 473 *Lonchoptera latea* ♂, *Lonchoptera tristis* ♀ wing.
From Verrall, G. H. British Flies, 5. 1909 figs. 51, 52

Section A. ASCHIZA

Superfamily **Lonchopteroidea**

FAM. LONCHOPTERIDAE. *3rd antennal segment rounded or globular with a long terminal or subdorsal arista. Wings pointed at the apex and with no obvious cross-veins. Sc free, Rs with its two branches closely approximated at the wing-apex. Empodium wanting.* This family (Fig. 473) includes a few small, slender, bristly flies, usually pale coloured, and often found along the borders of shady streams. The larvae of *Lonchoptera* have been found among leaves and other vegetable matter. According to de Meijere (1900) they are amphipneustic, much flattened and with long anterior and posterior setae. The head is vestigial but the capsule is better developed than in other Schizophora; there are only 10 apparent trunk segments, of which the last appears to be of a composite nature, and bears a pair of widely separated spiracles (Fig. 474).

Superfamily **Phoroidea**

FAM. SCIADOCERIDAE. *First two antennal segments more or less clearly visible, arista dorsal. Sc free at base only, forks of Rs and M normal or basal part of M missing to beyond fork.* A small group which includes one Australian and one Patagonian species and two or three fossils, the oldest from the Upper Cretaceous

(Hennig, 1964; McAlpine and Martin, 1966). The larva (Fuller, 1934) resembles that of the Phoridae.

FAM. IRONOMYIIDAE. *Arista terminal. Sc free at base and apex, stigma present, forks of Rs and M normal, latter at distal end of discal cell.* There is a single Tasmanian species whose position has been uncertain (McAlpine, D. K., 1967). A second species has recently been found in the Canadian amber (McAlpine, J. F., 1973).

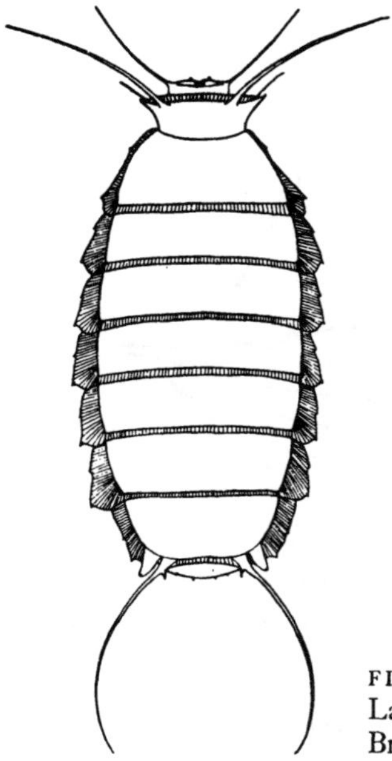

FIG. 474
Larva of *Lonchoptera*; magnified. Britain

FAM. PHORIDAE. *Antennae apparently consisting of one large segment, which conceals the others, and bearing a long apical or subdorsal arista. Wings often vestigial or absent. Anterior veins very heavily developed, and joining costa along the proximal half of the latter; Sc free at base: remaining veins weak and abnormally distributed.* A family (Schmitz, 1929) of small or minute greyish-black or yellowish flies (Fig. 475): they are active runners and present a curious humped appearance. Their habits are varied but the adults are frequently met with among decaying vegetation, while others occur in the nests of ants and termites. The wings exhibit a wide range of variation as regards degree of development, especially among the females; certain apterous and micropterous genera are only known from that sex. In *Ecitomyia* (female) for example, the wings are narrow and strap-like, and in *Puliciphora* (female) they are totally wanting.

The affinities of the Phoridae have been much discussed (Brues, 1907). They seem to be connected through the Sciadoceridae to the Platypezidae (Hendel, 1936–37). The larvae of Phoridae (Keilin, 1911) live in decaying vegetable matter and dead animals, especially *Helix* (Schmitz, 1917): others are myrmecophilous and some are parasites. They resemble those of other Cyclorrhapha in their general morphology and consist of a reduced head, 3 thoracic and 8 abdominal segments.

Furthermore, they agree with this suborder in being metapneustic in the 1st instar and amphipneustic subsequently. Each segment bears metamerically arranged bands of papillae. Pupation takes place in the larval skin, and the pupa carries a pair of elongate processes on the 2nd abdominal segment, which appear as the anterior respiratory horns on the puparium.

One of the most remarkable of all Phoridae is *Termitoxenia* (often placed in a separate family, Schmitz, 1928; 1939). It inhabits termite nests and has the wings reduced to minute vestiges; there are two types of individuals – stenogastric and physogastric. Wasmann (1900) claimed that it had lost the larval and pupal stages and that it was hermaphrodite, the same individual becoming successively male and

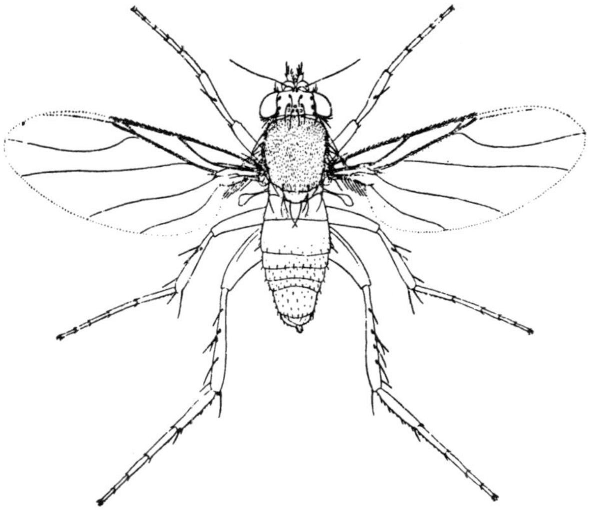

FIG. 475 *Aneurina caliginosa* (= *Phora urbana*), male (Phoridae). × 10. Britain
After Verrall.

female. It appears, however, from the observations of Mergelsberg (1935) and Wasmann (1940) that while most, perhaps all, the species have a normal metamorphosis, the adults are true hermaphrodites, an ovary and active sperm having been found even in a specimen extracted from the puparium. The method of reproduction is not known.

In *Puliciphora*, de Meijere (1913) has also found stenogastric and physogastric individuals, along with larval and pupal stages. On the other hand, according to Schmitz (1917), *Wandolleckia*, which lives on the bodies of the W. African snail *Achatina*, is ametabolous, and has both types of individuals present: it is presumed to be a protandric hermaphrodite. Further investigation of these remarkable genera is still needed. As Keilin (1919) remarks, the only way to prove that protandric hermaphroditism exists is the discovery of spermatogenesis in stenogastric forms. *Thaumatoxenia* is probably the most highly modified of all Phoridae (Trägårdh, 1908: Börner, 1908): originally regarded as belonging to the Hemiptera, further study has shown that it is a Phorid. The world species have been catalogued by Borgmeier (1968) and he has also revised the N. American species (1963–66).

FAM. PLATYPEZIDAE. *Small thinly pilose flies with the hind tarsi remarkably dilated. 3rd antennal segment elongate and often pyriform, arista terminal. Sc free, cell R_5 open, base of fork of M_{1+2} sometimes obliterated.* A small family including about 100 species of which more than a score are British (Verrall, 1901). They are usually to be met with dancing in the air in companies or running over herbage. Their most striking feature is afforded by the hind tarsi whose basal three or four segments are dilated, flattened, or ornamented in a curious manner, and very different in the male from the female. In the male (Kessel, 1968) the genitalia are not fully circumverted until after the male emerges from the pupa.

An account of metamorphosis in *Callimyia* is given by de Meijere (1900*b*); Willard (1914) also figures the larva and pupa of *Platypeza agarici*. The larvae are broad and flattened with the sides bordered by long setae: in *Callimyia* the whole margin is deeply incised, each incision being strongly serrated. The trunk comprises 10 or 11 segments, the head and first segment being wholly ventral. The tracheal system is amphipneustic, with the anterior spiracles placed beneath the body; the posterior pair is inconspicuous and rather widely separated. So far as known the larvae live in Agaricini. Flies of the genus *Microsania* are attracted to wood-smoke but possibly breed in fungi which develop after fires. The genera of the family have been revised by Kessel and Maggioncalda (1968).

Superfamily **Syrphoidea**

FAM. PIPUNCULIDAE. *Thinly pilose or almost bare flies with a very large subhemispherical mobile head formed almost entirely of the eyes. Antennae with a usually long dorsal arista. Wings much longer than the abdomen, cell R_5 open; tibiae devoid of spurs or with very weak ones. Ovipositor horny, exserted.* A very distinct family of small dark flies, many of which pertain to the genus *Pipunculus*. They have a markedly hovering habit, and are usually to be taken on flowers, or by sweeping miscellaneous herbage. Their most striking feature is the great size and mobility of the head; the 3rd antennal segment is of peculiar shape, being sometimes prolonged into a curious beak-like process. For general information on the family the reader is referred to the works of Perkins (1905) and Coe (1966); Hardy (1943) has reviewed the N. American species and Aczél (1943) has surveyed the literature.

The larvae are endoparasites of Homoptera Auchenorrhyncha. They are narrowed anteriorly, and capable of a good deal of extension and retraction: segmentation is obscure but apparently 10 or 11 somites are present. The anterior spiracles are small, and situated a short distance behind the mouth; the posterior pair is dark coloured, approximated, and placed some distance in front of the anal extremity. The puparium is provided with a pair of anterior spiracular tubercles, while the posterior spiracles are very much as in the larva. Dehiscence of the puparium usually occurs by the detachment of the dorsal plate through which the spiracular horns project. The head of the larval parasite is directed towards that of the host, and the fully grown parasite fills the greater part of the abdomen of the latter. In certain cases it has been found that 'castration parasitaire' results, and the abdomen of the female host is stated to undergo structural modification (Giard, 1889; Keilin and Thompson, 1915), but further research is greatly needed. When the Pipunculid larva quits its host, it usually escapes at the junction of the metathorax and abdomen, either above or below, the segments being ruptured at that point. It falls to the ground and buries itself beneath the soil or among rubbish, etc.

FAM. SYRPHIDAE (Hover Flies). *Moderate to large sized flies with brightly coloured markings, almost always bristleless. Arista, with few exceptions, dorsal. Certain of the veins forming a kind of secondary margin parallel with the outer wing-margin: cell R_5 closed; vena spuria present between R and M.* The Syrphidae are one of the largest and most sharply defined families of Diptera. They are usually very brightly coloured flies and may be striped, spotted or banded with yellow on a blue, black, or metallic ground-colour. The black and yellow coloration often imparts to them a superficial resemblance to wasps: other species are densely hairy and resemble bumble bees. Nearly all members of this family are attracted to flowers and may frequently be observed poised in the air, their wings vibrating with extreme rapidity, hence the name of 'hover flies'. The vena spuria (Fig. 476) is one of their most characteristic features and is rarely found in other Diptera. It is a vein-like thickening of the wing-membrane and may be distinguished from true veins in being fainter, and terminating without association with other veins.

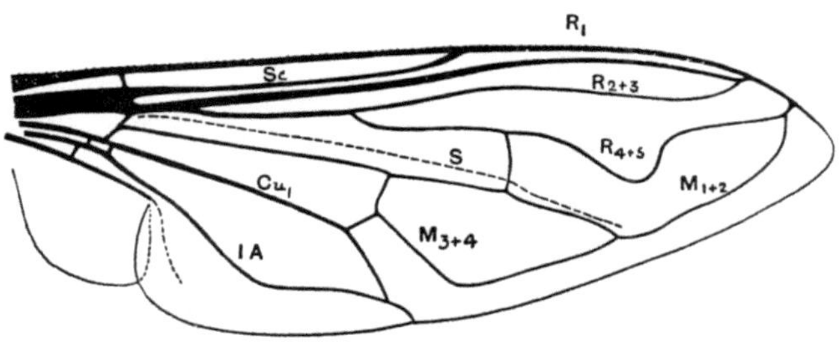

FIG. 476 Venation of *Eristalis* (Syrphidae)
s, vena spuria.

The larval habits of Syrphidae are extremely varied. They may be: (*a*) *Phytophagous*, feeding externally upon plants (*Mesogramma polita*) or internally in bulbs (*Merodon equestris, Eumerus strigatus*), or within stems or in fungi (*Cheilosia*). (*b*) *Carnivorous*, living predaciously upon aphids and the nymphs of other Homoptera (species of *Pipiza, Paragus, Melanostoma, Baccha, Syrphus,* etc.); a few species, such as *Dasysyrphus tricinctus*, eat mainly caterpillars and sawfly larvae (Schneider, 1969). (*c*) *Saprophagous*, living in decaying organic material, dung, liquid mud, or dirty water (species of *Eristalis, Helophilus, Sericomyia, Syritta, Tropidia,* etc.); in the sap and wet, rotting wood of diseased parts of trees (*Xylota, Mallota, Myathropa, Myolepta, Ceria,* etc.); or as scavengers in the nests of ants and termites (*Microdon*) or of Aculeate Hymenoptera (*Volucella*). Verrall remarks that probably all the European species of *Volucella* are scavengers in the nests of large Aculeates, feeding upon diseased larvae or pupae and, towards the end of the season, sometimes on living larvae and pupae. *Volucella bombylans* occurs in the nests of *Bombus* while the species *zonaria, pellucens* and *inanis* are found in the nests of social wasps.

The eggs of the aphidophagus Syrphidae are easily recognized and Chandler (1968) has described those of some British species. Morphologically, Syrphid larvae (Fig. 477) may be recognized by the following characters. The head is greatly reduced and carries a pair of short fleshy, sensory processes. The cuticle is tough or leathery, and segmentation is obscure owing to the transverse corrugation of the body, but apparently 11 somites are present. The tracheal system is amphipneustic with the anterior spiracles on the 2nd apparent segment: the posterior pair is situated on two tubes of very variable length, which are fused together down the median line.

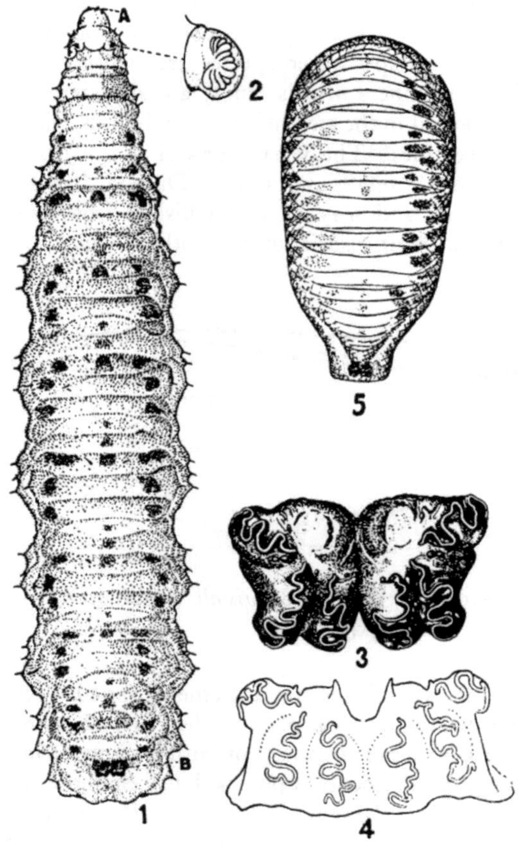

FIG. 477 *Metasyrphus nitens*

1, larva × 7·5; A, antenna; B, posterior respiratory organ. 2, anterior spiracle more enlarged. 3, end view of posterior respiratory organ × 75. 4, posteroventral view of same × 75. 5, puparium, dorsal view × 7·5. Adapted from Metcalf, *Bull.* 253 *Maine agric. exp. Sta.*

Three principal types of Syrphid larvae may be recognized. 1. The aphidivorous type with the ventral aspect flattened, the body much attenuated anteriorly, and the posterior respiratory tubes very short. The body is frequently marked with green or brown, and the general appearance is rather slug-like; all have a marked capacity for changing their shape (*Syrphus*, *Melanostoma*, etc.). 2. The short-tailed filth-inhabiting type with the body cylindrical and not attenuated anteriorly, and the respiratory tubes short. Three pairs of lateral fleshy protuberances are present on the 11th segment, and groups of branched plumose hairs are placed around the hind spiracles (*Syritta*, *Tropidia*, etc.). 3. The rat-tailed type with the body terminating in a long flexible respiratory process which, in some species, is capable of being extended several times the length of the body (*Eristalis*, *Helophilus*, etc.) (Doležil, 1972).

In addition to the above three types there are several anomalous larvae. That of *Microdon* is broadly ovoid in outline and flattened ventrally, and is bordered by a row of marginal spines. The dorsal surface is convex and there is no evident segmentation or anterior spiracles. In general appearance it is slug-like and, when first described, it was regarded as a new genus of land Mollusca. This curious larva has been studied in detail by Cerfontaine (1907) and others. In the boring larva of *Merodon equestris* the body is cylindrical and much contracted, with rounded extremities: it comes nearest to type 2, but there are no fleshy protuberances on the 11th segment. In *Volucella bombylans* the larva is rather broad and fleshy, tapering anteriorly. The body is provided with numerous small lateral spinous outgrowths and larger terminal processes on the last segment (Künckel d'Herculais, 1875).

Prior to pupation Syrphid larvae come to rest in some suitable place on, or near, their habitat. In many species the caudal segments are cemented to a leaf, twig or other support with a secretion apparently derived from the hind gut; in other cases the larva buries itself in the soil or other medium. The puparium, as a rule, is considerably inflated dorsally and laterally: spiracles are present on the puparium in the region of the 4th or 5th larval segment and may be either sessile (*Melanostoma*) or elevated upon conspicuous horns (*Volucella*, *Eristalis*). Eclosion of the imago usually takes place by means of a dorsal rupture of the puparium. A good deal of information concerning the structure and biology of Syrphid larvae will be found in the writings of Metcalf (1916–17) and Hartley (1961, 1963); Bhatia (1939) and Scott (1939) deal with the aphidophagous larvae. The internal anatomy of both the larva and imago is dealt with by Künckel d'Herculais (1875) in his monograph on *Volucella*. The larva of *Eristalis* is described by Miall (1895) and its tracheal system has been studied in detail by Wahl (1899).

Economically, the predacious larvae of this family are notable in being important enemies of aphids, coccids and other Homoptera. The capacity of Syrphid larvae for the rapid destruction of aphids is remarkable and Metcalf states that a larva of *Metasyrphus nitens*, which had not been fasting previously, caught and destroyed 21 examples of the large aphid *Pterocomma flocculosa* in 20 minutes. The entire insect is never devoured, but only the soft and readily assimilated body-contents are sucked out. Notwithstanding the great size of the family, and its varied larval habits, very few Syrphids are in any sense injurious to man. The 'Corn-feeding Syrphid Fly', *Mesogramma polita*, occurs in several states of N. America where its attacks are occasionally considered serious: the larvae feed upon the pollen grains and the saccharine cells in the axils of the leaves. Larvae of *Merodon* and *Eumerus* attack and destroy bulbs of *Narcissus*, *Amaryllis*, etc., and may occur separately or in association. They are well known pests in Europe and have been introduced, along with their host plants, into N. America and other parts. Larvae of a few species of the family, more particularly those of *Eristalis*, have been found as accidental parasites in the human body causing myiasis of the intestine.

While Verrall (1901) and Coe (1953) describe the British species, Vockeroth (1969) gives a revised classification of the world Syrphini.

Section B. SCHIZOPHORA

The classification of this group is a matter of great difficulty owing to the number of genera annectent between the families. The number of these recognized by some modern authors is very large. In the recent past it has been usual to divide up these flies into at least three major groups – the Calyptratae with the calyptrae enlarged and a distinct external groove on the second antennal segment (*Musca*-like flies), the Acalyptratae with the calyptrae usually small and usually no groove on the second antennal segment, and the Pupipara (a group purely of convenience) in which the wings are often more or less reduced, reproduction is by direct production of puparia (the rest of the development taking place within the mother) and the species are external parasites of vertebrates.

It is now more usual to break up and regroup these families in such a way that the old major groups are no longer useful. The number of families recognized varies from 47 (Colless and McAlpine, 1970) to 66 (Griffiths, 1972) though in Griffiths' arrangement a number of families which others separate are joined and many new ones recognized. The classification of the group is also discussed by Hennig (1958, 1969, 1971) whose scheme differs considerably from that of Griffiths (see also Roback, 1951). Unfortunately, the position of some of the families is still doubtful. In the present volume, only a selection of the families is described and a modified version of Griffiths' scheme is used.

Superfamily **Lonchaeoidea**

Antenna with second segment cleft. Not more than one proclinate orbital bristle present. Frons with dense setulae. Costa broken at end of Sc. Male 6th abdominal sternite symmetrical, 8th tergite present. Female 7th tergite and sternite fused, 8th segment elongate.

FAM. LONCHAEIDAE. *Sc incomplete. Arista present. Postvertical bristles not divergent. Vibrissae absent or present and male holoptic. Female with an oviscapt.* In the genus *Lonchaea* the male eyes are more or less approximated and the colours are dark. The larvae are scavengers in rotting vegetation or dung, or feed upon living plants, or live under bark and feed on the frass of bark beetles. Some of the phytophagous species make galls on grasses and *L. aristella* does much damage to figs in the Mediterranean region. The larva of *Lonchaea* has been described by Cameron (1913) and Silvestri (1917*b*). There are between 200 and 300 species (Morge, 1963).

FAM. CRYPTOCHETIDAE. *At once known by the absence of an arista. There is also only a single pair of abdominal spiracles.* The larvae are endoparasites of coccids. In *Cryptochetum iceryae*, which parasitizes *Icerya*, there are four larval instars (see Thorpe, 1941). The 1st-instar larva is an embryo-like sac devoid of tracheae and mouthparts, with the digestive canal closed: caudally it bears a pair of finger-like processes. In the successive instars the caudal processes increase in length and become filamentous until, in the last stage, they are much longer than the whole body: these organs appear to be mainly respiratory in function. There is a small number of species, mainly in tropical countries. In the past, they have been associated with the Agromyzidae but this is certainly incorrect.

Superfamily **Lauxanioidea**

Antenna with second segment not cleft. At least two superior orbital bristles present. Postvertical bristles convergent. Costa unbroken, anal vein not reaching the wing-margin. Male 6th and 7th abdominal sternites symmetrical, 7th tergite well-developed, dorsal, 8th tergite lost, sternum reduced or absent. Female 7th tergite and sternite separate, 8th segment short.

FAM. LAUXANIIDAE (Sapromyzidae). *Vibrissae absent. Wing with Sc complete. Tibiae with preapical bristles. Male accessory glands repeatedly branched.* As many as 1200 species (47 British) of this family are known. The larvae are mostly scavengers, feeding on dead leaves; they have numerous transverse rows of tooth-like projections. That of *Lauxania aenea* attacks the leaves of clover (Marchal, 1897*b*).

The **CELYPHIDAE** is a small family of flies, mostly oriental in distribution, known by the greatly enlarged scutellum which covers the wings when at rest. Metamorphoses unknown.

FAM. CHAMAEMYIIDAE (Ochthiphilidae). *Vibrissae absent. Postvertical bristles sometimes absent. Wing with Sc complete, vein R_1 not bristly above. Female with four spermathecae.* These minute, grey, sometimes black-spotted flies appear to be predatory on coccids and aphids in their early stages. This habit is well-known in the genus *Leucopis* and in *Chamaemyia* of which the larva is said to inhabit galls the same habit may be suspected. In both sexes of *Leucopis* there is a file at the end of the 2nd abdominal tergite and a scraper on the hind femur (Schirnhaus, 1971).

The **BRAULIDAE** of which the adults superficially resemble some of the Hippoboscids are possibly allied to the preceding family (Imms, 1942). The adults of *Braula coeca* (Fig. 478) are usually found clinging to queen or worker honeybees, especially the former. The eggs are laid on the walls or beneath the cappings of honey cells, Skaife (1921) having first shown that the genus is oviparous. The larvae

FIG. 478 *Braula coeca.* × 24. *After* Carpenter.

make tunnels in the wax, boring from one cell to another but seem to do little damage. The pupa is unique among the Cyclorrhapha in that it is enclosed in the unmodified last larval cuticle which does not become brown or barrel-shaped. The affinities of the family have been much disputed; the adult characters seem to be largely due to adaptive convergence, but the larva shows some resemblance to that of *Leucopis*.

The curious S. American fly, *Eurychoromyia*, also seems to fall near the Chamaemyiidae in that it also has four spermothecae. It has a very broad, bare head with small eyes (McAlpine, J. F., 1968).

Superfamily **Drosophiloidea**

Antenna with 2nd segment cleft. At least two superior orbital bristles, none proclinate. Postvertical bristles convergent or absent. Costa broken at end of Sc. Male 6th and 7th abdominal sternites symmetrical, 7th tergite well-developed, dorsal, 8th tergite present, large, sternite also large. Female with 7th tergite and sternite separate, 8th segment short.

FAM. DROSOPHILIDAE. *Vibrissae present. Arista dorsally pectinate. Postvertical bristles convergent. Costa broken at end of* R_1 *and sometimes at end of humeral vein, Sc present but reduced,* Cu_2 *and 1A present.* The flies belonging to this family usually have light red eyes, and are commonly taken by sweeping herbage. Others are prevalent about flowing sap, decaying fruit, cider presses, wine vats, vinegar factories, etc., where they are attracted by certain by-products of fermentation. A few (*Scaptomyza*) are leaf-miners and others, e.g. *Acletoxenus* which feeds on Aleyrodids, have predatory larvae; at least two Caribbean species of *Drosophila* live with land-crabs, the larvae developing either beneath the 3rd maxilliped or in the gill-chamber (Carson, 1967; Carson and Wheeler, 1968). The eggs of *Drosophila* are often spindle-shaped, bearing elongate processes at one extremity: as the eggs are frequently submerged in fluid media, with the filaments at the surface, it has been suggested that the latter are concerned with respiration. The larva is 11-segmented with each segment surrounded by a girdle of minute hook-like spines (*D. funebris*), or the body may be uniformly invested with these structures (*D. fenestrarum*). Three pairs of conical lateral outgrowths are borne on the anal segment, together with a longer median retractile process, carrying the posterior spiracles. The pupae are fusiform with the anterior dorsal surface flattened to form an ovoid plate which is forced upwards to allow of the eclosion of the imago. Arising from this plate is a pair of stalked, digitate or plumose spiracular processes. Traces of the anal processes of the larvae are also evident upon the puparium. Details of the early stages are given by Sturtevant (1921) and Mayer (1935). The genus *Drosophila* is now very large with over 1000 species and there are more than 400 species in the Hawaiian Islands where there has been an explosive evolution (Hardy, D. E., 1965).

The species are genetically isolated from one another by a variety of methods, some cytological but most behavioural involving response to visual, tactile, acoustic and chemical stimuli (Spieth, 1952; Bennet-Clarke and Ewing, 1968). Throckmorton (1962) has considered their phylogeny, using a wide range of characters for his comparisons. It is interesting that it does not usually seem to result from a series of regular bifurcations such as has sometimes been maintained is universal. Two small groups, the **Camillidae** and the **Curtonotidae** are often separated from the

Drosophilidae; their metamorphoses are unknown. In the former, the female abdomen is almost entirely membranous; in the latter, the vein Sc is complete, the 6th and 7th tergites of the male are reduced and the 8th sternite is lost. In all three families the third antennal segment sends out a process which projects into the second.

FAM. EPHYDRIDAE. *Vibrissae usually present. Arista sometimes bare. Postvertical bristles absent or replaced by divergent postocellar bristles. Costa broken at humeral vein and at end of* R_1, *Sc reduced.* The flies of this family are black or darkly coloured, inhabiting marshy places, damp meadows, etc. Jones (1906) has described the life-history of *Ephydra millbrae* which is aquatic. The eggs are partially clothed with hairs and are attached to floating vegetation, etc. The amphipneustic larvae are densely covered with short pubescence, with the anterior spiracles 7-branched, and a pair of respiratory tubes emerge from a terminal anal siphon. Eight pairs of conspicuous pseudopods, armed with hooks, are present on the abdomen. The puparium is provided with an elongate siphon whose apex rests at the surface of the water. Larvae of *Notiphila* occur in the stems of water plants, while those of *Hydrellia modesta* are found in the leaves of *Potamogeton* and are metapneustic throughout life (Keilin, 1915). The pupa of *Notiphila* (Varley, 1937; Houlihan, 1969) obtains oxygen from the roots of water-plants which are pierced by the posterior spiracular processes. Other larvae occur in salt or alkaline waters, particularly those of *Ephydra hians* and *E. californica* which often appear in such vast numbers as to have been used by the N. American and Mexican Indians as food. A few species are parasitic, e.g. *Trimerina* on eggs of spiders. An account of the anatomy of the larva of *E. riparia* is given by Trägårdh (1903) and the metamorphosis of *Teichomyza fusca* by Vogler (1900) (cf. also Johannsen, 1935; Hennig, 1943*b*). The larva of *Psilopa petrolei* (Thorpe, 1930) is a biological curiosity since it lives in pools of crude petroleum in California. Morphologically it differs little from many other Ephydrids and its adaptation to its mode of life appears to be physiological. The food of this larva consists of insects trapped in the oil and possibly of metabolites of the bacteria which occur in that medium.

The genus *Diastata* with about 20 Holarctic species and metamorphoses unknown, has often been put in a separate family but Griffiths unites them with the Ephydrids though he leaves *Campichoeta* to form a a new small family.

Superfamily **Nothyboidea**

Antenna with second segment not cleft. At least two superior orbital bristles present, none proclinate. Postvertical bristles if present divergent. Male 6th abdominal sternite symmetrical, 7th sternite and 8th tergite lost. Female with 7th tergite and sternite separate, 8th segment short.

FAM. NOTHYBIDAE. *Ocellar and postvertical bristles absent. Postscutellum conical, larger than scutellum. Base of vein M interrupted, axillary lobe and alula wanting, anal vein very close to wing-margin.* Fewer than 10 oriental species; biology not known (Aczél, 1955).

FAM. PSILIDAE. *Vibrissae absent. Ocellar triangle large. Costa more or less interrupted well before end of* R_1 *where vestigial Sc would end, veins 1A and* Cu_2

developed. Sternopleural and humeral bristles not developed. About 100 flies with a mainly Holarctic distribution are included (cf. Collin, 1944). The larvae are phytophagous and *Psila rosae* is the well-known carrot fly whose larvae cause much damage by eating into the tap root of the carrot and other roots (Körting, 1940); see also several papers by Petherbridge, Wright and others, mostly in *Ann. appl. Biol.*, 1942–47.

FAM. PERISCELIDIDAE. *Costa complete, Sc vestigial, R_1 reaching middle of wing, C ending at R_5 or at M_1 in Somatia.* A small family which in the past was placed near the Drosophilidae. The adults are attracted to sap flows in which the larvae perhaps develop.

The small family **Teratomyzidae**, with a few species in Australia, New Zealand and Chile, is placed here by Griffiths.

The remaining families of Diptera are placed by Griffiths in his superfamily Muscoidea. This is, however, so large and diverse that we have raised eight of his lower groups to superfamily level. The basic characters of the whole group are: second antennal segment not cleft except in the Muscoidea *s.s.* At least two superior orbital bristles present, none proclinate, postvertical bristles divergent. Costa often broken. Male with 6th abdominal sternite more developed on left side, 7th sternite asymmetrical, extending dorsally on left side, 7th tergite a lateral vestige, sternite large. Female with 7th abdominal tergite and sternite separate, 8th short.

Superfamily **Tanypezoidea**

Antenna with second segment not cleft. Inferior orbital bristles absent, two superior orbitals and verticals, ocellar bristles normal. No prestomal teeth, hyoid sclerite absent, pseudotracheae opening direct into mouth. Costa broken at end of Sc which runs to wing margin, cross-vein *im* present. Abdomen with spiracles 2–5 in the membrane, 7 lost. Female with 3 spermathecae.

FAM. TANYPEZIDAE. *Vibrissae absent. Tibiae with preapical bristles. Legs and abdomen elongate, mid and hind tarsus as long as their tibiae. Male holoptic.* A few holarctic and S. American species: the genus *Strongylophthalmomyia* (Australian and Oriental) is also now placed here though Sc is incomplete. Foote (1970) has described the larva of *Tanypeza* which he bred in the laboratory on damp vegetable refuse. He suggests that it resembles the larva of *Micropeza*.

Griffiths also places here the family **Heteromyzidae** which includes a few species which have been put in the Heleomyzidae. The larva of *Tephrochlamys*, a saprophage, is described by Lobanov (1970).

Superfamily **Muscoidea**
(Calyptratae of Griffiths)

Second antennal segment cleft. Inferior orbital bristles present, also at least two superior orbitals, two verticals except perhaps in the Hippoboscidae.

Prestomal teeth and hyoid sclerite present, pseudotracheae opening into two main channels that run into the mouth. Costa broken at end of Sc which is distinct to wing margin. Vestige of 7th abdominal tergite lost.

FAM. SCATOPHAGIDAE (CORDILURIDAE). *Body more or less strongly bristly. Eyes widely separated in male. Usually only one sternopleural bristle and no hypopleural bristles. Vein M_1 straight, costal spine reduced or absent. Abdomen with 6 pregenital segments.* The species are mostly phytophagous and include leaf- and stem-borers, such as *Cordilura* and *Norellia*. *Hydromyza* feeds on water plants in the larval stage (Hickman, 1935). A few larvae seem to be predacious and others are saprophagous, such as those of the well-known dung-fly, *Scatophaga stercoraria*, whose life-history has been described by Cotterell (1920). The adults are probably all predatory on small insects (Hobby, 1931*b*).

FAM. ANTHOMYIIDAE. *Vein Cu_1 + 1A reaching the wing margin though often faint distally. Lower calypter not longer than the upper one* (Fig. 479A). In the present, more restricted sense the family includes a large number of species whose larvae (Fig. 448) are mainly phytophagous; *Fucellia* breeds in seaweed on the beach. Smith (1948) discusses the members of the family which are of agricultural importance such as the larvae of *Erioischia brassicae*. This species is extremely destructive to vegetables of the *Brassica* tribe and also affects wild Cruciferae. It destroys the roots of these plants and the eggs are deposited round the stem near soil level. *Hylemyia coarctata* is the Wheat Bulb Fly, which is a serious pest in many parts of Europe: it is exceptional in laying its eggs on bare soil and not necessarily in proximity to its host plant. The larvae of *Pegomyia* are leaf-miners and those of *P. hyoscyami* are destructive to beet and mangolds. The work of Karl (1928) is useful in the identification of the species.

FAM. MUSCIDAE. *Vein Cu_1 + 1A not reaching the margin of the wing. Lower calypter nearly always longer than the upper one. Prosternum sclerotized.* Most of the species of Muscidae are small to rather large flies, many of which bear a general resemblance to the house-fly. Although the family includes the haematophagous genera *Stomoxys*, *Lyperosia* and *Haematobia*, in which both sexes suck the blood of man and other mammals by means of piercing mouthparts, the vast majority of its members are innocuous in this respect. The larvae usually inhabit decaying organic matter, more especially of vegetable origin. Of these latter the majority are saprophagous, while the rest are carnivorous preying upon other Dipterous larvae, small Oligochaeta, etc., which inhabit the same medium.

Musca domestica may be taken as a typical representative of the family. It is, as a rule, most abundant during the hottest months of the year and in Europe and N. America attains its greatest numbers from July to September. According to Roubaud, the insect usually does not hibernate but continues reproduction during winter in warm rooms and stables, but further research is much needed with particular reference to various climatic conditions. The eggs are cylindrical-oval, 1 mm long, with two curved rib-like thickenings along the dorsal side: they are laid in masses of 100–150 and the usual number deposited by a single fly in a life of about 2½ months is probably 600–1000. Dunn (1922), however, states that in Panama a single female may deposit 2387 eggs during 31 days after emergence. The chief breeding places are accumulations of horse manure or stable refuse, but human and other excrement is often selected, and also most kinds of fermenting animal and vegetable substances, particularly the contents of ash bins, etc. At a temperature of 25–35° C the larvae hatch in 8–12 hours. The first instar larva is 2 mm long, metapneustic, and each posterior spiracle opens by a pair of small, oblique, slit-like apertures. This stadium lasts 24–36 hours under favourable conditions. The second instar larva is amphipneustic with larger posterior spiracles and, at a temperature of 25–35° C, the stadium lasts about 24 hours. The third instar is also amphipneustic and measures about 12 mm long when fully grown. The anterior spiracles have 6–8 processes, and each

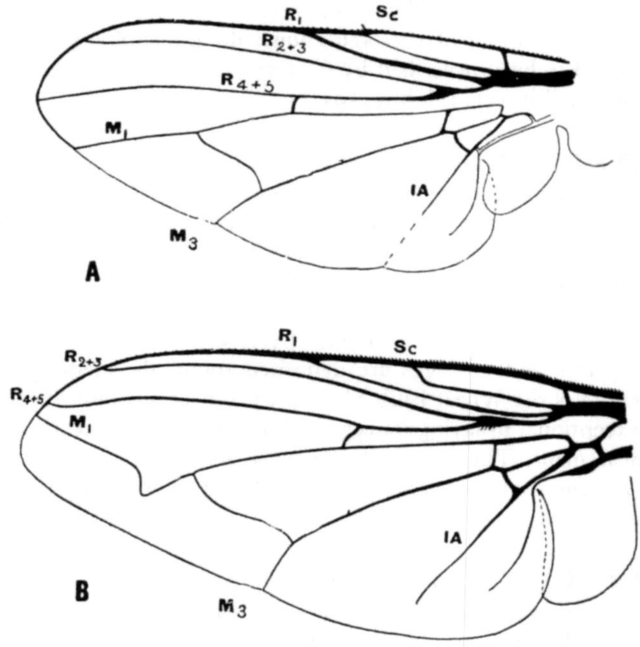

FIG. 479 Venation of A, *Hylemyia strigosa* (Anthomyiidae) and B, *Calliphora erythrocephala*

posterior spiracle is a D-shaped ring surrounding three sinuous slits. Incubated at 35° C this stadium lasts 3–4 days, and the pupal stadium averages 4–5 days. The developmental cycle, from the egg to the eclosion of the imago, varies in different parts of the world with the temperature and other factors. According to Herms it varies from an average of 44·8 days at 16° C to an average of 10·4 days at 30° C. Roubaud states that, in a heap of actively fermenting manure in warm weather, the life-cycle may only require six days. The house-fly has an important bearing upon the welfare of man as a carrier of the germs of summer diarrhoea, typhoid and possibly other diseases: there is also evidence pointing to the probability of its acting as a carrier of the eggs of certain species of intestinal worms. Preventive and remedial measures are numerous, and of these the most important is the elimination of the larval breeding places, or the rendering of the latter fly-proof or unsuitable for the insect. The destruction of adult flies must also form part of any system of eradication: this may be brought about by use of fly traps baited with attractive chemotactic substances, by means of adhesive fly-papers, or by spraying walls with

DDT. The literature on this species has assumed extensive proportions: a general account of the insect and its relation to man is given by Hewitt (1914) and West (1951). Observations on its breeding habits are given by Newstead (1908) and Roubaud (1922) and the reproductive organs are described by Rivosecchi (1958), while the relation of the insect to medical science and sanitation is dealt with by Graham-Smith (1913). Additional observations on this insect and other common flies will be found in papers by the latter author (1916 and 1919).

The biting house-fly *Stomoxys calcitrans* breeds principally in horse manure and stable refuse, but the larva may also be found in grass mowings, sewage beds, etc. The life-history has been studied by Newstead (1906), Bishopp (1913), Mitzmain (1913) and others. The duration of the life-cycle depends upon temperature, humidity and the nature of the food supply, while the minimum period for complete development is 23–32 days. According to Newstead two important conditions are necessary for development – an absence of light and an abundance of moisture. The ova are laid in batches of 60–70 and about 600 is the greatest number deposited by one female (during 65 days). At an average day temperature of 22° C, and 18·3° C by night, the larvae hatch out in 2–3 days. When fully grown they measure 11–12 mm long, and differ from the larvae of the house-fly in that the hind spiracles are rounded and smaller, with the three apertures on each plate only slightly curved instead of being sinuous. In England, during August, the larval period lasts 2 to 3 weeks and 9 to 13 days are spent as a pupa: in the tropics the pupal stage may only last 4 days. The larvae of the Muscinae are described by Schumann (1963).

A number of forms, Lispinae, Eginiinae, Coenosiinae and Phaoniinae, which were previously placed in the Anthomyiidae are now included here (Hennig, 1965). Lispinae and Coenosiinae are predacious on small insects. The Phaoniinae include many rather large flies, often found in woodland habitats and laying their eggs in a variety of decaying materials. Their larvae in many cases (*Limnophora* = *Melanochelia*, *Phaonia*, *Mydaea*, *Hydrotaea*) are carnivorous, feeding especially on other Dipterous larvae (Keilin, 1917). The larvae of the African *Passeromyia* are blood-sucking parasites of birds (Rodhain and Bequaert, 1916). The larvae of certain species of *Mydaea* are also subcutaneous avian parasites (Nielsen, 1911).

The **Fanniidae** have also in the past been associated with these flies but as shown by Hennig (1965) they differ in several characters – the female orbits are wide and lack proclinate superior orbital bristles; vein Sc is bent so that it is almost at once well separated from R_1; vein 2A is bent and Cu_1 + 1A shortened. The larvae which live in dung and other decaying organic matter are broad and somewhat flattened with paired segmental outgrowths (Lyneborg, 1970). The British species of this and the previous family are revised by D'Assis Fonseca (1968) while Chilcott (1960) deals with the Nearctic adults and larvae.

FAM. GASTEROPHILIDAE. *Hairy rather than bristly flies. Mouthparts reduced, functionless. Wing with cross-vein r–m near base of wing, veins M_1 and M_3 straight and with Cu_1 + 1A distally weak or not reaching margin.* The structure of the adult flies is to some extent anomalous and many authors treat them as forming no more than a subdivision of the Oestridae. The larval stages are found in the alimentary canal of mammals – *Gasterophilus* (Equidae), *Gyrostigma* (Rhinoceros), *Cobboldia* (Elephants).

The horse bot-flies lay their eggs on the hair. In *G. intestinalis* they are found on various regions, preferably the fore legs. The young larvae hatch upon the application of moisture and friction supplied by the licking of the horse; they are ingested, and attach themselves to the walls of the stomach. *G. nasalis* oviposits on the hairs beneath the jaws, and to some extent on the shoulders, etc. The larvae attach themselves to the pharynx, stomach and duodenum. *G. haemorrhoidalis* lays its eggs singly on the hairs around the lips: the larvae attach themselves to the stomach-wall, eventually migrating to the rectum, where they become re-attached. Before leaving the host they again become attached close to the vent and protrude therefrom. In all three species the larvae are ultimately voided through the anus and pupate in the ground.

The most complete account of the metamorphoses and habits of *Gasterophilus* is that of Dinulescu (1932) while Zumpt and Paterson (1953) deal especially with the African species.

FAM. OESTRIDAE (Warble or Bot Flies). *Oral fossa very small. Bristles reduced or absent, or replaced by woolly hairs. Vein M_1 bent forwards. Postscutellum distinct.* A comparatively small family of stoutly built more or less pilose flies, often bee-like in appearance. The antennae are short and partially sunken in facial grooves and the venation almost always is of the Tachinid type. There are no sternopleural setae while the hypopleural setae are represented by a group of hairs. The oviscapt is extensile and often long but not adapted for piercing: the eggs are laid on the body hairs of the hosts and are provided with special clasping flanges. The larvae are endoparasites of mammals but, with few exceptions, the life-histories are imperfectly known owing to difficulties attending observations of the cycle in the living animal. While the Oestridae are usually regarded as a separate family, they have definite affinities with the Tachinidae with which they are associated by some authorities.

Oestridae are more frequently met with as larvae than as adults, and a number of species have been described from the larval phase only: only six species occur in Britain. Parasitization of the mammalian hosts occurs in the nasal and pharyngeal cavities, and beneath the skin. As a rule each species parasitizes a single species of host, and each genus or group of allied species attacks allied hosts. The larvae, when fully grown, are broadly cylindrical or somewhat barrel-shaped, narrowing relatively little at the extremities, and never tapering anteriorly in a manner comparable with other cyclorrhaphous larvae. Twelve segments are present with the first two much reduced and annular. The body-wall is very tough with lateral swellings and groups of spinules. As a rule Oestrid larvae are amphipneustic, the anterior spiracle lying in a deep pit. Carpenter and Pollard (1918) have detected the presence of 6 pairs of vestigial lateral abdominal spiracles in *Hypoderma* and *Oedemagena*. Mouth-hooks are present in all 1st-stage Oestrid larvae but subsequently they may become reduced or vestigial.

The larvae feed upon the serous and other exudations into the tissues of their hosts, which fluids are usually either altered or increased owing to irritation induced by the presence of the parasites. When mature the larvae leave their hosts and pupate in the ground or among surface litter.

Hypoderma (sometimes placed in a separate family) includes the well known 'warble flies', *H. bovis* and *H. lineatum*. The adults are active from May to August and the eggs are mostly laid on the hairs of the flanks, legs and feet of cattle. According to Hadwen *H. lineatum* lays 1–14 eggs on a single hair, usually between the point of the hock and the ischium, and on the inside of the legs. *H. bovis* lays its eggs singly on the hairs, chiefly about the legs. In both species they hatch in 4 to 5 days, and the larvae bore their way beneath the skin, and migrate for several months through the body, until they reach the wall of the gullet. Here they are found from late summer until winter: from December onwards they start to appear beneath the skin of the back. Later, the skin is pierced and the posterior spiracles then communicate with the exterior. From February until May or later the fully grown larvae are found in the swellings or 'warbles' on either side of the spine of the host-animal. Ultimately each larva works its way out and falls to the ground where it pupates. The pupal instar lasts about 5 to 6 weeks. Squeezing out the larvae is the best remedy at present available as no efficient preventive methods have been devised (Min. Agric., 1926). The injuries caused by the perforation of the hide, and the deterioration of the flesh, and reduction in the milk occasioned by the presence of these larvae, entail great losses to the trades concerned. Further research is needed to ascertain the course followed by the young larvae during their migration from the skin to the gullet. Not infrequently they are found in the spinal canal having apparently deviated from their normal path after leaving the gullet. Most of what is known of their biology is contained in the papers of Hadwen (1912; 1916), Carpenter and his co-workers (1908 onwards), Cameron (1932), Eichler (1941) and Dinulescu (1961).

The Sheep Nostril Fly (*Oestrus ovis*) is usually larviparous, depositing its larvae in the nostrils of sheep. The young larvae migrate into the frontal sinuses of the head where they attach themselves to the mucous membrane. When mature they release their hold and leave the animal. The presence of these parasites causes nasal discharge in the sheep and often obstruction of the air passages. The Human Warble-fly (*Dermatobia hominis*) is widely distributed in N. and S. America. The females seize mosquitoes, particularly *Psorophora*, and attach the eggs to these vectors (Bates, 1943): more rarely Muscidae are utilized for this purpose. When the mosquito, or other carrying insect, settles on man the warmth evidently induces eclosion of the larvae which bore their way beneath the skin and cause warble-like swellings. In addition to man most kinds of domestic animals function as hosts. This genus, together with *Cuterebra* (parasites of rodents and marsupials) are sometimes placed in a family **Cuterebridae** because of the elongate scutellum and reduced postscutellum (postnotum).

FAM. CALLIPHORIDAE. *Pteropleural and hypopleural bristles present. Postscutellum well developed. Second abdominal sternite entirely overlapped by the tergite.* This family includes a very large number of species whose larvae may be saprophagous, or flesh-feeders or parasites of various Arthropods, the parasitic habit being less developed in this family than in the Tachinidae. The British species are dealt with in the works cited under the Tachinidae and the works of Hall (1948) and Senior-White *et al.* (1940) may be consulted for N. American and Oriental species.

The Sarcophaginae are characterized by the arista being plumose up to, or slightly beyond, the middle and bare distally: macrochaetae are usually only present on the distal part of the abdomen, the disk being rarely bristly, and the eyes are but little approximated in the male. The subfamily includes comparatively few genera but numerous species, often very much alike. For the most part they are uniformly coloured flies, with a grey longitudinally striped thorax, and marbled abdomen. The larvae (Thompson, 1921) are of the Muscid type and taper anteriorly with the posterior extremity rounded. Transverse bands of denticles differentiate the segments, and the posterior spiracles are situated in a deep stigmal pit bearing, as a rule, three straight subparallel slits. The larvae occur in decaying animal or vegetable matter or are parasites of insects and other animals (Aldrich, 1915). Their hosts include Orthoptera, Lepidopterous larvae, adult Coleoptera, scorpions, earthworms, etc.; snails are also not infrequently utilized as hosts. According to Pantel (1910) the parasitic larvae lie free within the body of their insect hosts, and do not acquire any organic connection with the latter as in Tachinids. Species of *Sarcophaga* (or flesh flies) are larviparous, with large eggs, and the uterus is greatly enlarged to form an incubatory pouch: on an average a female will deposit 40–80 larvae in their 1st instar. Although mainly living in decaying flesh the habits of this genus are extremely varied. Several species parasitize grasshoppers (Kelly, 1914), their larvae boring beneath the body-wall of the host soon after deposition. Others have been found beneath the skin of tortoises, in the stomachs of frogs, in snails (Keilin, 1919) or causing nasal myiasis in man: *S. haematodes*, however, is coprophagous. Fabre observed that the carrion fly *S. carnaria* will deposit its larvae from a height of 26 inches, and that the ordinary wire meat cover affords imperfect protection, since the larvae can fall through the mesh unless the latter is very fine. *Wohlfartia magnifica* is abundant in Russia, causing great suffering to domestic animals owing to even the smallest wound becoming infected with its larvae: in man it often causes myiasis of the ear, nose, eyes, etc. In *Theria muscaria*, a parasite of snails, the larviparous method of reproduction reaches a high degree of specialization: the female produces a single enormous egg which give rise to a correspondingly large-sized larva (Keilin, 1916). The species of *Miltogramma*, *Metopia* and their allies live, as larvae, in the nests of solitary bees and wasps which burrow in the ground. The female fly deposits young larvae in the nests of the bees, or on the prey of the wasps, and the larvae devour the food of their hosts.

The Calliphorinae (Fig. 479 B) are very often metallic green or blue flies and are distinguished by the weak development of the macrochaetae, which are usually absent from the dorsal surface of the abdomen: the arista is markedly plumose, usually for nearly its whole length. Many of the species are of importance in medical and veterinary science and in the typical genus *Calliphora*, which includes the well-known 'blue-bottles' or 'blow-flies', the larvae occur in carrion, flesh, etc. *Lucilia* includes the 'green-bottle flies'. The almost cosmopolitan *L. caesar* breeds in carrion and excrement while *L. sericata*, the 'sheep maggot fly', lays its eggs on the wool of sheep: its larvae bore into the flesh, causing death when present in large numbers. In Australia the sheep blow-fly problem of cutaneous myiasis is one of great importance. The species *Lucilia cuprina* is of primary significance in this connection and passes its larval development on the living sheep. *L. sericata* and species of *Calliphora* may also play a part in initiating the attack. They are followed by secondary flies, including *Chrysomyia rufifacies*, and other forms, which take advantage of the diseased conditions thus set up. Tertiary flies may also participate during the end-stages of the attack on the sheep. There is thus involved a complex biological association of larvae which entails great losses among the flocks (Tillyard and Seddon, 1933). Among other Calliphorinae producing myiasis is *Auchmeromyia luteola*, whose larva is the Congo 'floor maggot', frequenting the floors of native huts, and is an ectoparasite sucking human blood (Roubaud, 1913). *Protocalliphora azurea* is an ectoparasite in the nests of swallows, larks, sparrows and other birds: its larvae suck the blood of nestlings, attaching themselves to the skin by means of a suctorial disk on the 1st segment (Coutant, 1915; Roubaud, 1915). The larvae of *Cochliomyia macellaria* (facultative parasite), *C. hominvorax* (obligate parasite), the 'screw-worm flies' of N. and S. America, and of *Cordylobia anthropophaga*, the 'tumbu fly' of Africa, cause cutaneous myiasis in man and other mammals.

The 'cluster fly' *Pollenia rudis* is a parasite of earthworms of the genus *Allolobophora* (Keilin, 1915). The eggs are laid in the earth in September, and the young larva probably makes its way through the genital aperture into the vesicula seminalis of its host, where it remains during the winter. At the beginning of May it awakens and enters the body-cavity, if it has not already done so earlier. For a period of one to four days it migrates forwards and, during the last part of the journey, its spiracular extremity is directed towards the prostomium of the worm. Arriving at the latter region, it wears through the body-wall by means of the denticles around the anal segment, and the spiracles are thus placed in communication with the exterior. Six to ten days after perforating the prostomium the larva moults and, growing considerably, eats its way into the pharynx of the worm. After a further period of nine days it passes into the 3rd instar, and gradually eats its way backwards until only the hinder segments of the host remain: pupation subsequently takes place, and the imago appears in 35 to 45 days. A very similar host relationship occurs in the genus *Onesia* (Fuller, 1933).

The Rhinophorinae include a small number of species which are mostly parasitic on terrestrial Isopods (Bedding, 1973).

The Dexiinae include a number of genera whose species are parasitic in beetles but the Neotropical *Calodexia* lays in Orthopteroids which are disturbed by army ants.

FAM. TACHINIDAE. *Pteropleural and hypopleural bristles present. Postscutellum little developed and not convex. Second abdominal sternite with its sides visible, lying above those of tergites.* This is a very large assemblage of flies whose classification presents great difficulty (Fig. 480); Mesnil (1939, and in Lindner, 1944–52) is in the process of publishing a new system. The British species have been tabulated

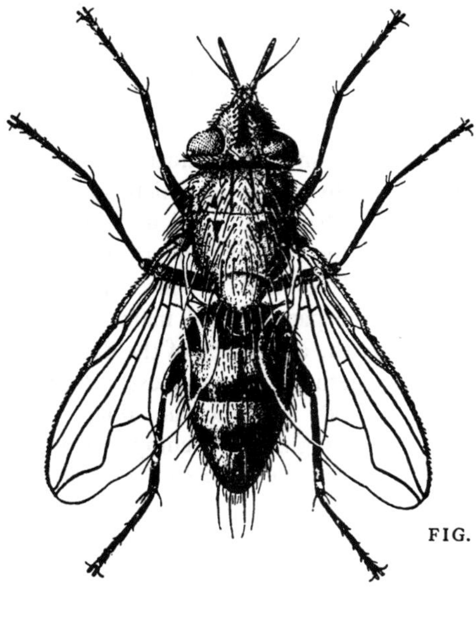

FIG. 480 *Hyperectina cinerea* (Tachinidae), female
From *Bull.* 1429 *U.S. Dept. of Agric.*

by Wainwright (1928; 1932), Day (1948) and van Emden (1954). The larvae are uniformly parasites of other insects, or rarely of other arthropods. They are conspicuously bristly flies, thinly or not at all pilose and with the abdomen usually clothed with marginal, lateral and discal setae: the arista is most often bare (Fig. 480). Their habits are very much alike and they are mostly found among vegetation, particularly on flowers. Biologically they are of great interest and importance and Pantel (1910) has divided the species into a number of groups according to their manner of placing the eggs. Briefly, it may be said that many species have ovoid eggs, flattened below, and cement them to the skin of the host; the resulting larvae speedily bore their way internally (*Gymnosoma*, *Thrixion*, *Winthemia*, *Eutachina*, etc.). Others are virtually viviparous since the larvae hatch immediately and are deposited on the bodies of the future hosts (*Exorista, Voria, Plagia,* etc.). Numerous species lay abundant small darkly coloured eggs on plants: these eggs are swallowed with the food and hatch within the bodies of the hosts (*Sturmia*, *Rhacodineura*, *Zenillia*, *Gonia* (= *Salmacia*), etc.). A considerable number of Tachinids lay their eggs in situations frequented by the hosts: these hatch almost immediately into migratory and often armoured larvae which bore their way into the first suitable host (*Digonochaeta*, *Echinomyia*, etc.). There are again others which pierce the host with a special spine-like oviscapt and deposit their eggs, or larvae, internally (*Ocyptera*, *Alophora*, *Compsilura*).

Tachinids select as their hosts larval and adult Coleoptera, Orthoptera and Hemiptera, but most often parasitize Lepidopterous larvae and to a lesser degree those of Hymenoptera: in a few instances they are known to select larval Diptera. Their larvae are usually broadly cylindrical, tapering but little towards the anterior extremity and with rather indistinct segmentation. The anterior spiracles are small but the posterior pair is conspicuous and often darkly coloured owing to sclerotization. Most of what is known of the biology of the family is due to the researches of Pantel (1898–1910), Nielsen (1909), Thompson (1923 onwards), Baer (1920) and Clausen (1940).

Within their hosts the larval life of Tachinids presents many variations but, in some stage of existence, they respire free air either by means of a perforation in the body-wall of their host, or by means of a secondary connection with the tracheal system of the latter. In either case, the larva is enclosed in a sheath ('gaine de fixation' of Pantel, funnel or siphon of other observers) which may be either primary or secondary.

(1) The PRIMARY SHEATH: this is always cutaneous in origin, and is formed as an ingrowth from the lips of the original perforation by means of which the larva enters the host. This perforation persists as an air-hole ('soupirail' of Pantel) and the larva hangs, head downwards, with its spiracles respiring free air through the aperture. The sheath consists of an inner sclerotized layer and an outer layer of epidermis; it grows around and closely embraces the parasite and maintains the latter in position. The sheath may be complicated by the adherence of the degenerating surrounding tissues which are often soldered together by the profuse secretion of the epidermis. In this manner muscle fibres, fat-body and tracheae, along with dead phagocytes may become involved, the whole forming a dense, compact sheath surrounding the parasite. This type of sheath occurs in *Echinomyia fera*, *Winthemia 4-pustulata*, etc.

(2) The SECONDARY SHEATH: this may be either cutaneous or tracheal in origin according to the position of the air-hole. In species in which this type of sheath

obtains, the parasite lives for a while free in the body-cavity of its host as in *Thrixion* or within some particular organ (nervous system, muscles, etc.) as in *Plagia trepida* and *Sturmia sericariae* (see more especially Pantel, 1909; Sasaki, 1886). Sooner or later, owing to the respiratory needs of increasing growth, it seeks communication with the air. By means of the anal extremity the larva gradually bores its way either through the integument, or into a tracheal trunk, and thus forms a secondary air-hole. Whichever situation is chosen, a sheath grows round the larva either by means of an ingrowth of the integument (*Thrixion*) or as an outgrowth from the wall of a trachea or of an air-sac (*Blepharidea*, *Bucentes*, *Gymnosoma*). In either case the parasite becomes enclosed as in the primary sheath. Whichever way it is formed, the sheath is a pathological reaction of the host against irritation and microbic infection induced by the presence of the parasite. In *Compsilura* and *Sturmia* the parasite acquires a direct connection with a spiracle of the host, and the sheath under these circumstances is little more than a collar-like rim around the caudal end of the parasite.

The mode of life of the parasite within its host varies not only among different Tachinids, but also during the life of an individual species. Thus in *Thrixion*, for example, the larva devours only the blood and fat-body and forsakes the host while the latter is alive. Furthermore, it does not void excretory matter until it leaves its host. In similar cases, in which the first diet of the larva consists of the blood plasma of the host, the surrounding sheath is closed, absorption taking place according to Pantel by means of 'physiological filtration'; at a later stage the buccal armature pierces the sheath and the larva then commences to devour the fat-body. The greater number of Tachinids rupture the surrounding sheath in the 3rd instar and, becoming free in the body-cavity of the host, they commence to devour the vital organs of the latter. In certain other Tachinids a still more complex mode of life is followed: thus in *Sturmia sericariae* (Sasaki, 1886), which parasitizes the silkworm, the eggs are deposited on mulberry leaves and are swallowed along with the leaf-tissue by the host. The eggs hatch in the gut of the latter, and the young larvae bore their way through the wall of the digestive canal, and penetrate into the ganglia of the nervous system. At a later stage they forsake the latter, and acquire connections with the spiracles of their host. Other species similarly live an intra-organic life within the nervous system, muscles, etc., of the hosts, during part of their existence (Pantel, 1909; 1910). Such species have remarkably small eggs adapted for being swallowed by their host. Pupation in Tachinids takes place as a rule in the soil: in some species, however, such as *Carcellia gnava*, which is a parasite of *Malacosoma neustria*, the pupal stage occurs within the pupa of the host.

The subfamily Phasiinae includes a number of genera which attack Hemiptera and much information on their biology has been published by Dupuis (1963).

The remaining four families are all viviparous, the female depositing mature larvae or puparia and the last three families used for that reason to be put in a suborder, the Pupipara.

FAM. GLOSSINIDAE. *Arista plumose with feathered hairs. Proboscis needle-like, porrect, at rest entirely sheathed by palpi. Prosternum membranous.* The species of *Glossina* (Fig. 472) or Tsetse flies (Swynnerton, 1936; Buxton, 1955; Mulligan, 1970; Potts, W. H., 1974) are now well known to be the carriers of the pathogenic agents

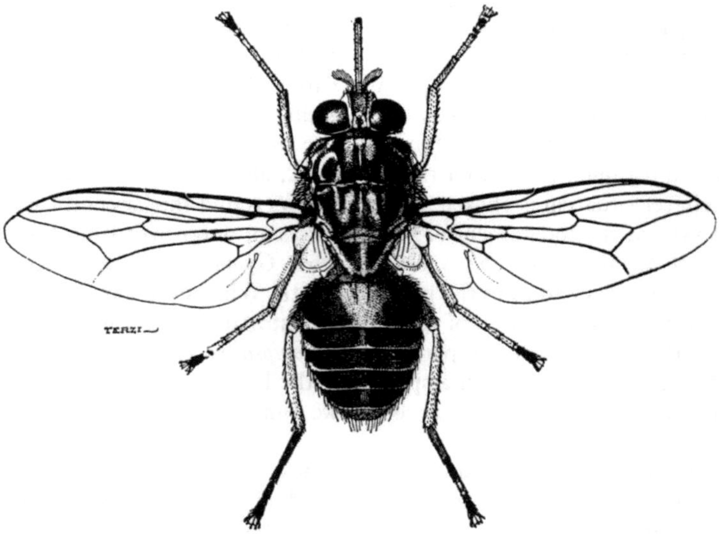

FIG. 481 *Glossina palpalis*, female. × 5

Reproduced by permission of the Trustees of the British Museum.

of certain virulent diseases in Africa. Thus *Glossina palpalis* (Fig. 481) transmits *Trypanosoma gambiense*, the causal agent of sleeping sickness, from man to man by means of its piercing mouthparts. In a similar manner *G. morsitans* transmits *Trypanosoma rhodesiense* which is responsible for the more local or Rhodesian form of that disease. *Glossina morsitans* is also the chief carrier of the trypanosomes which cause among domestic animals the disease known as nagana. Although widely distributed through tropical and sub-tropical Africa, species of *Glossina* do not occur continuously throughout that area, but are largely restricted to patches of forest and bush where there is warmth, damp and shade, such tracts being known as 'fly belts'. The larvae in this genus are nourished within the uterus of the parent and, when mature, are deposited singly and at intervals in a shady situation on the ground. When newly born, the larva is yellowish-white, with a black posterior extremity bearing a pair of polypneustic spiracular lobes. It speedily burrows or otherwise conceals itself and pupates, the imago appearing about a month later.

The next three families while like *Glossina* pupiparous are in the adults much more modified for life as ectoparasites of their vertebrate hosts, their wings often being reduced. Schlein (1970) compares some of the characters of the three families. The winged forms do not fly any considerable distance, and all species are adepts at clinging to their hosts and working their way among the hairs or feathers. The claws are highly developed, and toothed or spined for the purpose. The abdomen is indistinctly segmented and, like the rest of the body, tough and leathery. All species are blood-sucking ectoparasites of mammals and birds, but do not utilize man as a normal host. The larvae are retained within the uterus of the parents, where they are nourished by the secretion of the greatly developed accessory glands. When fully mature they are deposited on the ground, or in the abodes of their hosts, and almost immediately change to pupae. A list of the Palaearctic species arranged according to their hosts is given by Bezzi (in Becker, 1905). For the structure of the proboscis see Jobling (1926–28), and for the general biology and morphology of the families consult Massonat (1909) and Falcoz (1926). The phylogeny and distribution is discussed by Speiser (1908) who has also monographed the world's genera.

FIG. 482
Hippobosca rufipes. × 2.
S. Africa

Reproduced by permission of the Trustees of the British Museum.

FAM. HIPPOBOSCIDAE. *Head sunk into an emargination of the thorax. Palps neither leaf-like nor forwardly projecting, forming a sheath to the proboscis. Eyes round or oval, ocelli present or absent. Antennae inserted into a depression, 1-segmented, with or without a terminal bristle or long hairs. Legs short and stout, claws strong and often toothed. Wings present or absent.* These insects (Fig. 482) are dorsoventrally flattened and of a tough leathery consistency, both features being correlated with an ectoparasitic life. The family includes such well-known insects as the 'forest fly' *Hippobosca equina* which affects horses and cattle, and the sheep 'ked' *Melophagus ovinus.* Among other British species *Ornithomyia avicularia* is a parasite of many wild birds and *Lipoptena cervi* is found on deer. All these species have a very extensive geographical range and *O. avicularia* has been carried by birds almost all over the world. The degree of development of the wings differs greatly in various members of the family. In *Hippobosca* and *Ornithomyia* they are fully formed: in *Allobosca* they are vestigial while in *Melophagus* (Fig. 483) wings and halteres are absent. Both sexes of *Lipoptena cervi* are winged but upon discovering a host they

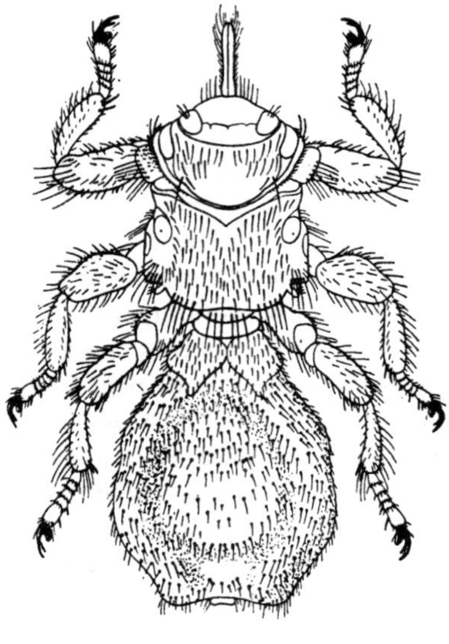

FIG. 483
Melophagus ovinus, magnified

soon cast their wings near the bases. The palps in this family are rigid organs projecting forwardly downwards and forming a partial sheath to the proboscis. The latter is curved and slender, protrusible, and hidden from view when retracted. Both sexes are equally active blood-suckers, but their punctures are seldom painful. The females produce at intervals single larvae which are whitish, or pale yellow, with a black cap at the posterior end which involves the spiracles. They are immobile with little or no traces of segmentation, and very soon transform into puparia. For the genera of the family consult Bequaert (1940) and an account of the structure and biology of the members of the family is given by Bequaert (1953–57) and Maa (1963): the British species are discussed and figured by Edwards *et al.* (1939).

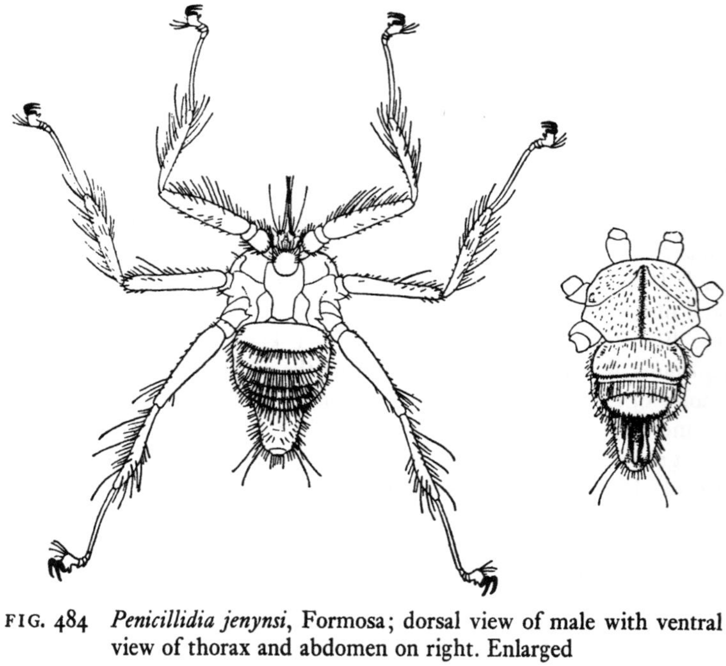

FIG. 484 *Penicillidia jenynsi*, Formosa; dorsal view of male with ventral view of thorax and abdomen on right. Enlarged

FAM. NYCTERIBIIDAE. *Head folded back at rest in a groove on dorsum of thorax, eyes when present vestigial; antennae 2-segmented, terminated by a dendriform arista. A fan-shaped evertible comb of spines (ctenidium) inserted in a hollow at the anterior end of the thorax; legs elongate, wings absent.* A small family of highly modified and completely apterous insects (Fig. 484) parasitic upon bats. They are more particularly characteristic of the Old World, the countries bordering on the Indian Ocean being especially rich in species; *Nycteribia* is British. Frequently, a single species of Nycteribiid may utilize several species of host which may be of different genera or, more rarely, of different families. Conversely a species of bat may support several species of Nycteribiids: thus, at least 9 species of the latter have been recorded from *Miniopterus schreibersi* in various countries. Information on the general structure and classification of the family is given by Speiser (1901) and Theodor (1967): the papers of Kolenati (1862) and Scott (1917) should also be consulted and that of Ryberg (1939) for their biology. Rodhain and Bequaert (1916*b*) have published a detailed account of the behaviour of *Cyclopodia greeffi*. The larvae are deposited in a less advanced stage of development than those of certain Hippoboscidae and the spiracles are posterodorsal in position. The puparia were found adhering to the perches and parts of the cages in which the hosts were confined. According to Müggenberg (1892) a ptilinum is wanting in this family.

FAM. STREBLIDAE. *Head not flexed on dorsum or thorax; eyes when present small, no ocelli; antennae in pits, 2-segmented. Palps leaf-like, projecting in front of the head but not sheathing the proboscis. Hind coxae enlarged, claws not distinctly toothed. Wings well-developed, vestigial or absent.* A small family widely distributed throughout the tropics and warm regions of the world. In habits they are exclusively parasitic upon bats, and much of what is known concerning the family is included in a paper by Speiser (1900). *Ascodipteron* (Queensland, E. Indies to Africa) is one of the most remarkable of Pupipara and is unique on account of the degeneration undergone by the female. Both sexes are winged but exhibit marked differences in the structure of the proboscis: in the female the labellar teeth are very large and blade-like, on the other hand, in the male, the teeth are exceedingly small. On reaching its host (*Miniopterus*) the female makes a way beneath the skin near the base of the ear or sometimes elsewhere, and casts both legs and wings. In this situation she develops into a greatly enlarged, flask-shaped sac, with the hinder extremity communicating with the exterior (Muir, 1912). Typical genera of the family include *Nycteribosca* which extends into Europe, *Raymondia* and *Strebla* (see Jobling, 1936, 1938, 1939, 1949, 1951). Theodor (1968) has revised the African species and Zeve and Howell (1962–63) give a full account of the external anatomy of the genus *Trichobius*. The various species deposit larvae which are either attached to the substrate or dropped to the ground.

Superfamily **Micropezoidea**

Second antennal segment not cleft. Inferior orbital bristles absent, two superior orbital bristles present. Abdominal spiracles 2–5 in the membrane. Male with 6th abdominal tergite as long as the 5th, 7th sternite asymmetrical, extending dorsally on the left, 7th tergite reduced to a small lateral ventral vestige, 8th tergite present as a vestige. Cerci not linked to lower telomere.

FAM. MICROPEZIDAE. *Costa unbroken, Sc complete. Vibrissae absent. Arista dorsal. Tibiae with preapical bristles. Legs long and stilt-like, front legs shorter.* A few hundred species, especially common in the Neotropical region (cf. McAlpine, D. K., 1975); 8 species are British (Collin, 1945). The adults often sit upon dead wood and a few larvae have been bred from this habitat and from roots (Berg, 1947; Steyskal, 1964; Wallace, 1969).

FAM. NERIIDAE. *Like the last family but arista terminal or nearly, fore legs almost as long as the others.* About fifty species in the damp tropics and subtropics (Aczél, 1961). The larvae live in decaying vegetable matter.

Associated with these families are two small groups, the **Cypselosomatidae** and the **Pseudopomyzidae** (McAlpine, D. K., 1966).

Superfamily **Diopsoidea**

Second antennal segment not cleft. No inferior orbital bristles, one or no vertical bristle, postverticals reduced or absent. Costa unbroken, Sc and anal vein continued to margin, cross-vein *im* present. Abdominal spiracles 2–5 in the membrane, male with 6th tergite as long as 5th, 7th sternite forming a complete ventral band, 7th tergite lost, 8th sternite large, tergite lost, 7th left spiracle in sternite.

FAM. DIOPSIDAE. *Eyes usually borne on long stalks. Vibrissae absent, head bristles generally reduced. Anterior femora thickened with spines beneath. Aedeagus short.* While the long stalk which supports the eye is characteristic, it re-occurs in other families (Otitidae, Drosophilidae, etc.) and is absent in *Centrioncus*. The early stages are saprophagous or phytophagous (Descamps, 1957; Tan, 1967) and the adults are sometimes extremely abundant by streams in E. Africa. Wickler and Seibt (1972) describe their behaviour which includes sexual combats. The 150 species are mostly found in Africa and the Orient but extend to New Guinea and N. America (Shillito, 1971; Steyskal, 1972).

FAM. SYRINGOGASTRIDAE. *Eyes not stalked. Ocellar bristles present. Hind femora thickened with spines. Alula and squamae greatly reduced. Aedeagus long and coiled.* There are about 8 Neotropical species in the genus *Syringogaster* (Prado, 1969).

Superfamily **Sciomyzoidea**

Second antennal segment normally without cleft. Vibrissae usually weak or absent. Postvertical bristles divergent. Costa unbroken, Sc complete. Male 6th tergite at most half as long as 5th, 6th sternite more developed on left side, 7th sternite asymmetrical, extending dorsally on left or not defined.

FAM. COELOPIDAE (Kelp flies). *Antennae sometimes received into scrobal hollows, peristomium bristly. Postvertical bristles convergent. Legs bristly or woolly, preapical bristles present. Vein 1A extends to margin.* These flies breed in seaweed at the high tide mark where the adults often occur in great numbers.

FAM. DRYOMYZIDAE. *Clypeus well exposed. Postvertical bristles divergent. Tibiae with preapical bristles.* Includes a few species which associate with excrement or decaying fungi but *Helcomyza* (*Actora*) which has a precoxal bridge in the prosternum breeds on the sea-shore and all stages can withstand immersion (Joseph, 1880).

FAM. SCIOMYZIDAE. *Antennae not received into a scrobal hollow, peristomium not bristly. Postvertical bristles divergent, parallel or absent. Clypeus hidden. Tibiae with preapical bristles.* Steyskal (1965) has considered the classification but Griffiths would make two of the subfamilies, Phaeomyiinae and Heliosciomyzinae, into separate families. As far as is known, all Sciomyzidae are predatory on land and freshwater snails and because of the interest in their possible role in limiting schistosomiasis they have been much studied in the last 20 years. More than 500 species are now known and Rozkošný (1966) gives a key to the larvae. Stephenson and Knutson (1966) record the distribution of the British species. A few species like those of *Pteromicra* or *Sciomyza* lay their eggs on the shells of land snails and are more or less specific to one snail species (Berg, 1961; Knutson and Berg, 1967). Some *Tetanocera* exclusively attack slugs (Trelka and Foote, 1970) while some *Renocera* and *Knutsonia* attack bivalves (Sphaeriidae) (Foote and Knutson, 1970). Other genera such as *Sepedon* may live in one *Succinea* for the first part of their life and then pass on rapidly through a succession of varied prey (Knutson *et al.*, 1976*b*; Knutson, 1970). The genera *Pherbiella* and *Colobaea* pupate behind calcareous septa built across the interior of the shell with material derived from the Malpighian tubules (Knutson *et al.*, 1967*a*).

FAM. ROPALOMERIDAE. *Eyes swollen, vertex depressed between them. Hind tibiae compressed and curved. Otherwise rather like next family.* A few American species, mostly Neotropical, which appear to live in damp places and to associate with the sap of trees.

FAM. SEPSIDAE. *Palps reduced. Vibrissae present or absent. Vein R_1 bare above. Tibiae without preapical bristles, only front femora with stout bristles or spines beneath and then chiefly in the males.* About 200 species of these small flies with a somewhat ant-like build are known. They usually wave their wings which have a black spot at the tip. The larvae are saprophagous and have been found in dung. The adults are often very abundant on the larval food and on flowers.

Two other small groups have been given family status. The **Megamerinidae** with three genera and a few Palaearctic and Oriental species have the hind femora thickened with internal spines. *Cremifannia* which perhaps mainly because of its habits has been placed in the Chamaemyiidae is placed in a separate family in the present group by Griffiths.

Superfamily **Anthomyzoidea**

Second antennal segment with no cleft. Ocellar bristles developed; inferior orbitals present, at least two superior orbitals, two verticals, postverticals usually divergent. Costa usually unbroken, cross-vein *im* present. Abdominal spiracles 2–7 in the membrane. Male with 6th abdominal tergite as long as 5th, 6th sternite more developed on the left side, or reduced or lost; seventh sternite asymmetrical, extending dorsally on left, or lost or fused to 8th sternite, 7th tergite a latero-ventral vestige or lost; 8th tergite lost. Female with seventh spiracle on 7th tergite, three spermathecae.

FAM. HELEOMYZIDAE. *Costa more or less interrupted at point where Sc ends. Postvertical bristles convergent, vibrissae present. Costa usually with spines. Aedeagus often elongate.* There are about 300 species of small or moderate-sized flies in this family of which 63 are British. The larvae are scavengers in fungi, excrement, etc.; a number of species has been bred from the nest of birds and mammals and some occur in caves. A few larvae have been described by Lobanov (1970).

FAM. ANTHOMYZIDAE. *Costa more or less interrupted before the end of R_1, Sc vestigial. Vein 1A and cell Cu_2 developed. Vibrissae present. Fore femur with a* *spine-like bristle beneath.* The family **Opomyzidae** is very similar but has no true vibrissae and no spine on the fore femur. These families include a few species of small flies whose early stages are spent in the stems or leaves of grasses (Hennig, 1952). The British species have been tabulated by Collin (1944*b*; 1945*b*). The early stages of *Opomyza* and *Geomyza* have been described by Thomas (1933, 1934, 1938).

FAM. ASTEIIDAE. *Small flies with characteristic long, narrow wings; costa unbroken, Sc vestigial, r-m near wing-base, no median or anal cells.* The larva of *Liomyza* lives in dried plant stems and that of *Asteia* in compost. F. X. Williams found one larva in the water of a sheathing leaf base.

FAM. SPHAEROCERIDAE (BORBORIDAE). *Costa more or less interrupted where the incomplete Sc would run out. Vibrissae present. First segment of hind tarsi compressed and broadened.* In these small, usually black-bodied, flies, the individual species often have a very wide distribution. They mostly breed in decaying plant materials or excrement. Some species of *Ceroptera* hang on to the backs of *Scarabaeus* and other dung-beetles and probably lay their eggs in the burrows. There are a number of short-winged or apterous species, such as *Anatalanta* of Kerguelen Is. and a series of genera found in the tree-heath zone of the E. African mountains. The British species were tabulated by Richards (1930) and a number of puparia were figured by Okely (1974). Griffiths places several other small families in this group. The **Rhinotoridae** include small Neotropical and Australian flies which D. K. McAlpine regards as a tribe of the Heleomyzidae. An Australian species has been bred from a longicorn burrow in a fig-tree. The **Borboropsidae** with one European species has also usually been placed in the Heleomyzidae. The small Holarctic flies of the family **Trixoscelidae** are usually regarded as a subfamily of the Heleomyzidae; their metamorphoses are unknown. The **Chyromyidae** are more distinct. There is a membranous central facial area and the palps are very short. The British species are yellow flies, often seen on windows; one has been bred from birds' nests. The **Aulacigastridae** have been placed near the Drosophilidae. The larva breeds in tree-sap (Robinson, 1953).

Superfamily **Agromyzoidea**

Second antennal segment not cleft. Inferior orbital bristle present, normally at least two superior orbitals, ocellar and vertical bristles present, postverticals divergent. Costa broken at R_1 or Sc, cross-vein *im* usually present. Abdominal spiracles 2–5 in the membrane, tergite 6 as long as 5, 6th sternite asymmetrical, reduced or lost, 7th sternite asymmetrical or lost or fused with 8th. Female with three spermathecae; usually no oviscapt.

This is a large assemblage of families some of which are perhaps not yet well established.

FAM. CLUSIIDAE. *Costa interrupted where the complete Sc runs out, cross-veins r–m and im close together, postvertical bristles divergent, vibrissae present.* A small but widely distributed family whose larvae live in rotting wood. That of *Clusiodes albimana* has been described by Perris.

FAM. AGROMYZIDAE. *Costa interrupted where Sc or R_1 runs out. Vibrissae present, postvertical bristles divergent, anterior orbitals sometimes directed inwards. Female with an oviscapt.* A rather large family of small flies whose larvae are leaf-miners, stem-borers or more rarely gall-makers. Their range of food-plants varies greatly; thus many species of *Phytomyza* are attached to one or a few allied species of plant while *P. horticola* and *Liriomyza pusilla* utilize species of many natural orders. *Melanagromyza aeneiventris* tunnels into the stems and roots of Compositae, while *A. pruinosa* mines the cambium of birch and hazel. *Phytomyza syngenesiae* does considerable damage to Chrysanthemums (cf. Griffiths, 1967 for the taxonomy). The majority of species of the family pass through several generations in the year – five or more in the case of *Liriomyza pusilla*. The larvae are cylindrical, tapering somewhat anteriorly, and more or less truncated posteriorly. The mouth-parts are conspicuous on account of their dark colour and strong sclerotization: on the ventral surface of the anal segment is a small sucker-like organ. The posterior spiracles are situated at the apices of backwardly projecting processes of variable length, usually contiguous and porrect. The puparia are broadly fusiform with the segments well defined: both the anterior and posterior spiracles are prominent and projecting. Pupation either occurs in the larval mine or in the soil. For information on the metamorphoses of various species reference should be made to the papers of Hendel in Lindner (1934–36) and Frick (1952); Phillips (1914) for *Agromyza parvicornis*; Malloch (1915) for *A. pruni*; Webster and Parks (1913) for *L. pusilla*; Miall and Taylor (1907) and Smulyan (1914) for *Phytomyza*; and Barnes (1933) for the cambium miner *Dizygomyza* (= *Phytobia*) *cambii*.

Spencer (1973) has recently revised the 300 British species and has published a series of papers, such as Spencer (1969), covering the faunas of much of the world while Hering (1957) tabulated the European species. Schirnhaus (1971) has described a stridulating organ in either both sexes or in the male of some Agromyzids; a file at the end of the 2nd abdominal tergite and a scraper on the hind femur. A similar organ occurs in some *Leucopis* (Chamaemyiidae).

The **Fergusoninidae** are Australian and make galls on *Eucalyptus* but their relationship to the Agromyzidae is very problematical.

FAM. MORMOTOMYIIDAE. A single hairy species with reduced wings lives on bat-guano in caves in Kenya. Its position is doubtful but it is apparently not related to the Scatophagidae as suggested by van Emden (1950).

FAM. ODINIIDAE. *Costa broken at R_1, Sc evanescent, cross-veins r–m and im rather close together, basal, median and anal cells well defined. Male sixth abdominal tergite shortened, seventh and eighth lost.* A few small species which associate with the sap-fluxes of trees.

FAM. TETHINIDAE. *Male with 6th abdominal sternite, 7th tergite and sternite and 8th tergite lost. Postvertical bristles convergent. Sc close to R_1 and evanescent distally. Labella elongate. Female with two spermathecae.* A few small grey flies which live by the sea but not usually in the wettest habitats.

FAM. MILICHIIDAE. *Costa interrupted at humeral cross-vein and at end of R_1, Sc reduced. Postvertical bristles convergent. Cell Cu_1 and vein 1A absent. No* *sclerotized spermathecae.* Small flies which seem to be fundamentally saprophagous but in several genera have developed peculiar habits. In *Desmometopa* the adults are often seen sucking the prey of larger predators (Reduviids, spiders); others, such as *Phyllomyza*, live with ants. *Leptometopa latipes* has been bred from excrement and Hennig (1956) has described its larva. *Meoneura* (found on carrion or in birds' nests) and *Carnus* are sometimes placed in a family, **Carnidae.** *Carnus hemapterus* is exclusively found in birds' nests and the wings break off by the time the adults are mature; the flies feed on the blood of the hosts (Eichler, in Hennig 1952) but the larval habits are not known.

FAM. CHLOROPIDAE. *Costa interrupted at R_1 or sometimes more proximally, Sc hardly visible except as a fold, veins 1A and base of M_4 not developed, cell Cu_2 absent. Ocellar triangle very large. Postvertical bristles convergent. Male 6th and 7th abdominal tergites lost, aedeagus reduced. Female with no sclerotized spermathecae.* Small, bare, often light-coloured flies, plentiful among miscellaneous herbage about roadsides, meadows, etc. The larvae are essentially phytophagous, although a few species are predacious such as those of *Thaumatomyia* which feed on root-aphids (Parker, 1918). The larva of *Oscinella frit* (Balachowsky and Mesnil, 1935–36) is very narrow in proportion to its length. The anterior spiracles are 6-lobed, and the posterior pair open at the apices of short tubular projections at the hind extremity of the body: each spiracle has three circular openings separated by sclerotized ridges. In *Oscinosoma* the larva is less elongate and more musciform with ambulatory swellings on the abdominal segments: the spiracles closely resemble those of *Oscinella* (see Silvestri, 1917*a*). The anatomy of the larva of *Platycephala* was studied by Wandolleck (1899). The 'frit-fly' *Oscinella frit* (Fig. 485) is a pest of

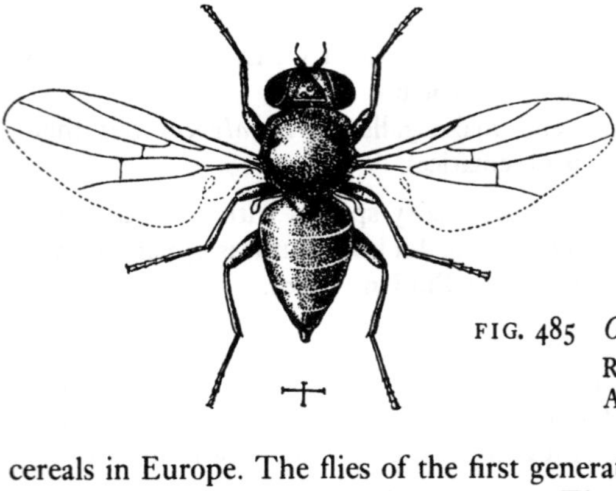

FIG. 485 *Oscinella frit* (Chloropidae)
Reproduced by permission of the Ministry of Agriculture.

cereals in Europe. The flies of the first generation oviposit in May on the leaves or stems of spring oats and various grasses. The larvae migrate to the shoots causing the death of the central leaves. Flies of the second generation oviposit during July on the ears of oats, and the larvae feed on the spikelets and young grain. Oviposition in the third generation occurs during September on winter cereals and various grasses. Winter is passed in the larval condition at the bases of the shoots which they ultimately destroy. The 'gout fly' *Chlorops taeniopus* (Fig. 486) lays its eggs during June on the leaves of spring barley or occasionally on couch grass. The larvae migrate into the shoots which become thickened and the leaves are distorted. If the barley ear is about to be formed the larva eats a groove down one side of it and the internode. The ear fails to grow away from the ensheathing leaf. The flies of the second generation oviposit from the middle of August until the middle of October mainly on couch grass, but sometimes on self-sown or winter cereals. The shoots become greatly thickened, the leaves distorted, and no ear is formed (Frew, 1923).

In *Lipara lucens* whose larvae live in galls on *Phragmites* stems, Mook and Bruggemann (1968) find that both sexes vibrate the reed by rocking their bodies; females answer the males and this leads to sexual encounters. The four European species of *Lipara* each produces a different specific vibration to which each female responds (Chvála *et al.*, 1974). Certain species of *Siphunculina* and *Hippelates* probably transmit conjunctivitis and other eye diseases in the Orient, California, etc. Such flies have spinous pseudotracheae which appear to make incisions in the conjunctiva and so aid the entry of pathogenic organisms carried on their bodies (Graham-Smith, 1930*b*).

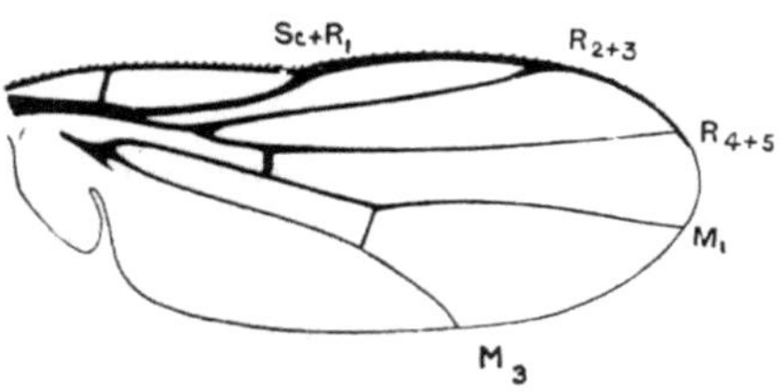

FIG. 486 Venation of *Chlorops taeniopus*

Acartophthalmus (eyes with fine hairs; costa broken near humeral cross-vein, Sc complete, base of M_4 and vein 1A complete, cell Cu_2 present; male 6th abdominal tergite lost, 7th and 8th sternites fused; female with long slender cerci) may be placed in a separate family here rather that within the Clusiidae. There are two Holarctic species which associate with carrion and decaying fungi.

FAM. CONOPIDAE. *Ptilinal groove sometimes shortened. Proboscis very long and often conspicuously jointed. Palps reduced. Cell Cu_2 long and reaching or nearly reaching the wing margin. Male 6th, 7th and 8th abdominal tergites lost, aedeagus sometimes long and coiled. Female with oviscapt often oblique to long axis (except* Stylogaster) *and covered by 7th tergite.* It was considered in the past that the Conopids were intermediate between the Aschiza and the Schizophora because in some genera the frontal suture is very short but this seems to be secondary and the ptilinum is always well-developed. There are also no grounds for regarding them as intermediate between the Calyptrates and Acalyptrates.

A family of more or less elongate, moderate-sized flies, thinly pilose or almost bare, and frequently bearing a striking resemblance to solitary wasps or other Hymenoptera. Most species visit flowers; they are slow fliers and less than 20 kinds occur in the British Isles. The larvae are often endoparasites of adult bees and wasps or of Orthoptera (de Meijere, 1904; 1912*b*). The species of *Conops* parasitize *Bombus, Odynerus, Sphex, Vespa,* etc.: *Physocephala* parasitizes *Apis, Xylocopa* and *Bombus*, while *Myopa* is known to attack *Andrena, Bombus* and *Vespa* (Séguy, 1928). The flies of the very distinct genus *Stylogaster* hover over the swarms of driver ants, *Anomma* and *Eciton*, but lay their eggs on cockroaches and various Muscid and Calliphorid flies which the ants disturb. However, none of them has as yet been bred (Rettenmeyer, 1961; Smith, 1967, 1969*b*; Stuckenberg, 1963).

The eggs of the Conopidae are elongate-oval with a group of hooks, filaments, or other outgrowths at the micropylar end. In certain cases the eggs are stated to be deposited on the bodies of the hosts during flight. The larvae (Smith, 1966) are generally found in the region of the anterior abdominal segments of the host, and are attached by their hinder extremity to a large trachea or air-sac: the exact relation between the Conopid larva and the tracheal system of the host is in need of further investigation. The mouthparts are greatly reduced and the larvae are mainly haemophagous. In general shape the larvae are ovoid or pyriform with considerable powers of changing their form. Their most conspicuous features are the large convex plates of the posterior spiracles. The latter are complex structures, and distributed over the surface of each spiracle is a series of small sieve-like areas.

FAM. RICHARDIIDAE. *Costa interrupted well before R_1 which is thickened where it ends and then forms a short convex bow, Sc complete. Tibiae with preapical bristles. Male with spiracles 6 and 7 present on left or on both sides. Female with an oviscapt.* There are about 100 Neotropical species whose metamorphoses are unknown.

FAM. PIOPHILIDAE. *Costa interrupted well before R_1, Sc complete. Vibrissae present, postvertical bristles not convergent. Male with 6th and 7th spiracles lost on both sides. Female without an oviscapt.* Another group of small flies with larvae of saprophagous habit. *Piophila casei* is the well-known 'Cheese Skipper' whose larvae may do much damage to cheese and other fatty foods. In the last larval instar it is capable of jumping by attaching its anal end to the substratum, bending itself into a circle with its mouthparts engaged near the anus, and then suddenly releasing its hold (Wille, 1922). The non-domestic species mostly seem to breed in carrion; the metamorphoses of the group are reviewed by Hennig in Lindner (1943).

The **Thyreophoridae** is a very small family of flies, very rare in the adult stage, and associated with old, dried-out carcases. In the adult the antennae lie in deep grooves.

The **Neottiophilidae** with R_1 bristly above and 1A reaching the wing-margin are moderate-sized flies of yellowish colour. The larva lives in the nests of various birds, sucking the blood of nestlings (Tate, 1954).

FAM. TEPHRITIDAE (TRYPETIDAE). *Costa more or less interrupted before end of R_1, Sc complete or sending a perpendicular fold to R_1, wings usually spotted or banded* (Fig. 487). *Inner orbits completely sclerotized, nearly always with inwardly directed anterior orbital bristles, postverticals not convergent. Male with a long, flexible, coiled aedeagus. Female with an oviscapt.* The 'fruit-flies' form an easily recognizable and natural family of almost cosmopolitan distribution. More than 1500 species are known of which 72 are British. The wings as a rule are conspicuously marbled and a horny, flattened oviscapt is very characteristic: in *Toxotrypana* it exceeds the rest of the insect in length. The standard systematic work on the family is that of Loew (1862);

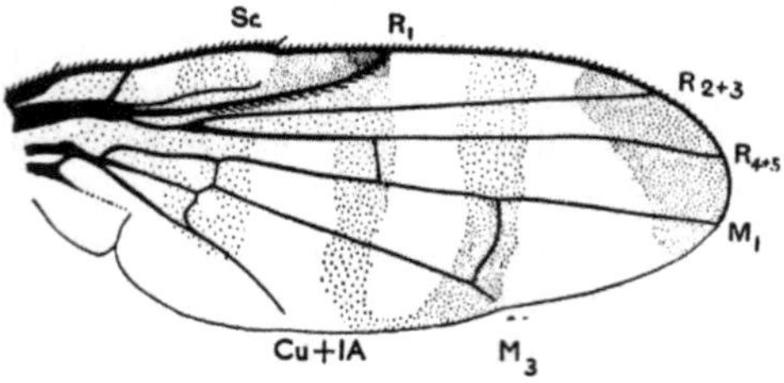

FIG. 487 Venation of *Trypeta tussilaginis*

more recent papers are those of Bezzi (1913), Hendel in Lindner (1927), and Munro (1947), while Collin (1947) should be consulted for British genera.

The larvae are phytophagous and those of several species are well known (Phillips, 1946). When fully grown they are rounded or barrel-shaped: a pair of rounded anal tubercles are present, and the posterior spiracles each contain three simple slits. The prothoracic pair are many-lobed, from about 14 to 38 processes being present. The larvae may be grouped under four headings with reference to their habits. (1) Living in fruits, preferably of the fleshy type: *Dacus, Ceratitis,* etc. (2) Living in the flower heads of Compositae: *Tephritis*, some *Urophora*. (3) Leaf- or stem-miners, *Philophylla, Euribia*, etc. (4) Gall formers on various parts of plants, some *Urophora. Ceratitis capitata* is the well-known Mediterranean fruit-fly (Quaintance, 1912) whose larva attacks almost all commercial and other succulent fruits, and becomes extremely injurious wherever established. The eggs are deposited inside the fruit, and the whole life-cycle occupies about 30 to 40 days, pupation taking place in the ground. This species occurs throughout the tropics and warmer regions, including the Mediterranean countries of Europe. *Philophylla* (= *Acidia*) *heraclei* (see Smith and Gardner, 1922) is the celery fly whose larvae cause considerable damage by mining the leaves of that plant and the parsnip: it also affects certain wild Umbelliferae. The life-history of *Urophora solstitialis* (correctly *jaceana*) is described by Wadsworth (1914) and that of *Dacus cucurbitae* by Back and Pemberton (1917). Observations on the anatomy of the larva and adult of *Dacus tsuneonis* are given by Miyake (1919).

FAM. OTITIDAE. *Costa is usually unbroken, Sc meets the costa at an acute angle and is not abruptly elbowed distally, cell Cu_2 is pointed distally. In the male there are 5 evident abdominal segments and an elongate rolled up aedeagus: in the female there are 6 abdominal segments with a flattened corneous oviscapt.* A large family many of whose members have mottled wings and consequently resemble Tephritidae. The flies are commonly met with wherever there is abundant vegetation, and nearly twenty species are British. The larva of *Platystoma lugubre* (= *umbrarum*) is described by Perris (1855) and that of *Euxesta nitidiventris* by Brues (1902); Allen and Foote (1967) and Valley, Novak and Foote (1969) describe larvae of further genera. The latter species lives beneath the bark of dead trees and is 11-segmented with the head nearly as long as the first two thoracic segments. The larva of *E. notata* attacks oranges, apples, onions, cotton bolls, etc., and has also been found in human excrement.

The Otitidae are today often split up into about seven families. The most distinct of these groups is the **Pyrgotidae** in which the adults usually lack ocelli. The larvae of these are parasitic in Scarabaeid beetles such as *Phyllophaga* and *Popillia*. The adult female is nocturnal and lays her egg through the soft cuticle of the abdominal tergites when the beetle is in flight (Clausen, 1940). The **Tachiniscidae** are large, bristly flies, like bumblebees and there is one S. African and one Neotropical genus. The **Phytalmiidae** with one oriental genus of elongate, long-legged flies, often with a neck like a giraffe. In the **Ulidiidae** which includes such well-known British genera as *Seioptera* and *Physiphora* (= *Chrysomyza*), the larvae are saprophagous or less often phytophagous as in some species of *Euxesta*. The **Pterocallidae** have an enlarged costal cell, cell Cu_2 pointed distally, cell R_5 long, narrow, parallel-sided, oviscapt sometimes very long. *Myennis octopunctata* breeds in melons and the species are mainly Holarctic and Neotropical. The **Platystomatidae** (*Platystoma* mentioned above) lack the pointed Cu_2 cell, cross-veins *r-m* and *im* are approximated and the antennae sunk in deep depressions. The species are mainly found in the warmer parts of the Old World. Some species attack bulbs.

The **Pallopteridae** which have often been placed in the Lonchaeidae seem better put here. The larvae live in the seed-heads of composites.

Literature on Diptera

ACZÉL, M. (1943), Sammelreferat der bionomisch-ökologischen Literatur über Dorylaiden, etc., *Dt. ent. Z.*, **1943**, 1–27.

—— (1955), Nothybidae, a new family of Diptera, *Treubia*, **23**, 1–18.

—— (1961), A revision of the American Neriidae (Diptera, Acalyptrata), *Stud. Ent.* (N.S.), **4**, 257–346.

ALDRICH, J. M. (1915), The economic relations of the Sarcophagidae, *J. econ. Ent.*, **8**, 242–246.

ALEXANDER, C. P. (1919–20), The crane-flies of New York. I & II, *Cornell. agric. Exp. Sta., Mem.*, **25**, 763–993; **38**, 695–1133.

—— (1927), Diptera. Fam. Tanyderidae, *Gen. Insect.*, Fasc. **189**, 1–13.

—— (1930), Observations on the Dipterous family Tanyderidae, *Proc. Linn. Soc. N.S.W.*, **5**, 221–230.

ALLEN, E. J. AND FOOTE, B. A. (1967), Biology and immature stages of three species of Otitidae (Diptera) which have saprophagous larvae, *Ann. ent. Soc. Am.*, **60**, 826–836.

ANTHON, H. (1943), Zum Kopfbau der primitivsten bisher bekannten Dipterenlarve: *Olbiogaster* sp. (Rhyphidae), *Ent. Meddr*, **23**, 303–320.

ANTHON, H. AND LYNEBORG, L. (1967), The cuticular morphology of the larval head-capsule in Blepharoceridae (Diptera), *Spolia Zool. Mus. haun.*, **27**, 11–54.

ANTHON, S. I. (1908), The larva of *Ctenophora angustipennis* Loew, *J. Morph.*, **19**, 541–560.

BACK, E. A. AND PEMBERTON, C. E. (1917), The Melon fly in Hawaii, *U.S. Dept. Agric. Bull.*, **491**, 64 pp.

BAER, W. (1920), Die Tachinen als Schmarotzer der schädlichen Insekten, *Z. angew. Ent.*, **6** and **7**, 185–246; 97–163, 349–423.

BÄHRMANN, R. (1966), Das Hypopygium von *Dolichopus* Latreille unter besonderer Berücksichtigung der Muskulatur und der Torsion (Diptera: Dolichopodidae), *Beitr. Ent.*, **16**, 61–72.

BALACHOWSKY, A. AND MESNIL, L. (1935–36), *Les Insectes Nuisibles aux Plantes Cultivées*, Etab. Busson, Paris.

BANKS, N. (1912), The structure of certain Dipterous larvae, with particular reference to those in human foods, *U.S. Bur. Ent. Tech. Ser.*, **22**, 1–44.

BARNES, H. F. (1933), A cambium-miner of basket-willow (Agromyzidae) and its inquiline gall-midge (Cecidomyidae), *Ann. appl. Biol.*, **20**, 498–519.

—— (1946–49), *Gall Midges of Economic Importance*, Parts I–VI, Crosby, Lockwood & Son, Ltd., London.

BATES, M. (1943), Mosquitoes as vectors of *Dermatobia* in eastern Columbia, *Ann. ent. Soc. Am.*, **36**, 21–24.

—— (1949), *The Natural History of Mosquitoes*, Macmillan Co., New York.

BECHER, E. (1882), Zur Kenntnis der Kopfbildung der Dipteren, *Wien. ent. Ztg.*, **1**, 49–54.

BECKER, R. (1910), Zur Kenntnis der Mundteile und des Kopfes der Dipterenlarven, *Zool. Jb., Anat.*, **29**, 281–314.

BECKER, T. *et al.* (1903–17), *Katalog der paläarktischen Dipteren*, 4 vols, G. Wissonlényi in Hodmezövásárhely, Budapest.

BEDDING, R. A. (1973), The immature stages of Rhinophorinae (Diptera: Calliphoridae) that parasitise British woodlice, *Trans. R. ent. Soc. Lond.*, **125**, 27–44.

BELING, K. W. T. (1875), Beitrag zur Metamorphose der zweiflügeligen Insecten, *Arch. Naturg.*, **41**, 31–57.

—— (1882), Beitrag zur Metamorphose der zweiflügeligen Insecten aus den Familien Tabanidae, Leptidae, Asilidae, Empidae, Dolichopodidae und Syrphidae, *Arch. Natur.*, **48**, 186–240.

—— (1888), Beitrag zur Metamorphose einiger zweiflügeligen Insecten aus den Familien Tabanidae, Empidae und Syrphidae, *Verh. k.k. zool. bot. Ges. Wien*, **38**, 1–4.

BELKIN, J. (1962), *The mosquitoes of the South Pacific (Diptera, Culicidae)*, Vol. I, Univ. Calif. Press, Los Angeles and Berkeley.

BENNET-CLARKE, H. AND EWING, A. W. (1968), The wing mechanism involved in courtship of *Drosophila*, *J. exp. Biol.*, **49**, 117–128.

BEQUAERT, J. (1940), Moscas parásitas pupíparas de Colombia y Panamá, *Rvta. Colomb. Cienc.*, **3**, 414–418.

—— (1953; 1955; 1957), The Hippoboscidae or louse-flies (Diptera) of mammals and birds, *Entomologica am.*, N.S. **32–33**, 442 pp.; **35**, 183 pp.; **36**, 194 pp.

BEQUAERT, M. (1961), Contribution à la connaissance morphologique et la classification des Mydaidae (Diptera), *Bull. Inst. Sci. nat. Belg.*, **37**, no. 19, 1–18.

BERG, C. O. (1947), Biology and metamorphosis of some Solomon Islands Diptera. Part 1. Micropezidae and Neriidae, *Occ. Pap. Mus. Zool. Univ. Michigan*, **503**, 1–14.

—— (1961), Biology of snail-killing Sciomyzidae (Diptera) of North America and Europe, *Verh. XI internat. Kongr. Ent. Wien*, **1960**, **1961**, 197–202.

BEZZI, M. (1913), Indian Trypaneids (Fruit-Flies) in the collection of the Indian Museum, *Mem. Indian Mus.*, **3**, 53–168.

BHATIA, M. L. (1939), Biology, morphology and anatomy of aphidophagous Syrphid larvae, *Parasitology*, **31**, 78–129.

BINET, A. (1894), Contribution á l'étude du système nerveux sous-intestinal des insectes, *J. Anat. Physiol.*, **30**, 449–580.

BISHOPP, F. C. (1913), The Stable Fly (*Stomoxys calcitrans* L.) an important livestock pest, *J. econ. Ent.*, **6**, 112–126.

BODENHEIMER, F. (1924), Beiträge zur Kenntnis der Kohlschnake (*Tipula oleracea* L.), *Arch. Naturg.*, (A), **90**, Heft. 2, 61–108.

BORGMEIER, T. (1963–66), Revision of the North American Phorid flies, *Stud. Ent.*, **6**, 1–256; **7**, 257–416; **8**, 1–160.

—— (1968), A catalogue of the Phoridae of the world (Diptera, Phoridae), *Stud. ent.*, **11**, 1–367.

BÖRNER, C. (1908), *Braula* und *Thaumatoxena*, *Zool. Anz.*, **32**, 537–549.

BOYD, M. F. (1949), *Malariology*, 2 vols., W. B. Saunders & Co., Philadelphia and London.

BRANDT, E. (1879), Vergleichend-anatomische Untersuchungen über das Nervensystem der Zweiflügler (Diptera), *Hor. Soc. ent. Ross.*, **15**, 84–101.
—— (1882), Recherches sur le système nerveux des larves des insectes Diptères, *C.R. Acad. Sci. Paris*, **94**, 982–985.
BRAUER, O. F. (1869), Beitrag zur Biologie der Acroceriden, *Verh. k.k. zool. bot. Ges. Wien*, **19**, 737–740.
—— (1883), Ergänzende Bemerkungen zu A. Handlirsch's Mittheilungen über *Hirmoneura obscura*, *Wien. ent. Ztg.*, **2**, 25–26 (cf. also pp. 114 and 208).
—— (1884), Die Zweiflügler des kaiserlichen Museums zu Wien. III, *Denkschr. Akad. Wiss. Wien*, **47**, 1–100.
BRITTEN, H. (1915), A note on the oviposition of *Simulium maculatum* Mg., *Entomologist's mon. Mag.*, **51**, 170–71.
BROCHER, F. (1909), Métamorphoses de *l'Hemerodromia precatoria* Fall., *Ann. Biol. lacustre*, **4**, 44–45.
BRODO, F. (1967), A review of the subfamily Cylindrotominae in North America (Diptera: Tipulidae), *Kansas Univ. Sci. Bull.*, **47**, 71–115.
BROMLEY, S. W. (1962), The external anatomy of the black horse-fly *Tabanus atratus* Fab. (Diptera: Tabanidae), *Ann. ent. Soc. Am.*, **19**, 440–460.
BRUES, C. T. (1902), Notes on the larvae of some Texan Diptera, *Psyche, Cambridge* **9**, 351–355.
—— (1907), The systematic affinities of the Dipterous family Phoridae, *Biol. Bull.*, **12**, 349–359.
BRUNDIN, L. (1947), Zur Kenntnis der schwedischen Chironomiden, *Ark. Zool.*, **39A**, No. 3, 1–95.
—— (1966), Transantarctic relationships and their significance, as evidenced by Chironomid midges. With a monograph of the subfamilies Podonominae and Aphroteniinae and the austral Heptagyiae, *K. svensk. Vetensk. Akad. Handl.*, (4) **11**, 1–472.
BUTT, F. H. (1937), The posterior stigmatic apparatus of Trypetid larvae, *Ann. ent. Soc. Am.*, **30**, 487–491.
BUXTON, P. A. (1955), *The natural history of Tse-tse flies*, School of Hygiene and Tropical Medicine, London, Mem. 10, 816 pp.
BYERS, G. W. (1961), The Crane-fly genus *Dolichopeza* in North America, *Kansas Univ. Sci. Bull.*, **42**, 665–924.
—— (1969*a*), Evolution of wing reduction in Crane flies (Diptera: Tipulidae), *Evolution*, **23**, 346–354.
—— (1969*b*), A new family of Nematocerous Diptera, *J. Kansas ent. Soc.*, **42**, 366–371.
CAMERON, A. E. (1913), On the life-history of *Lonchaea chorea* Fabricius, *Trans. ent. Soc. Lond.*, **1913**, 314–322.
—— (1918), Life-history of the leaf-eating crane-fly, *Cylindrotoma splendens*, Doane, *Ann. ent. Soc. Am.*, **11**, 67–89.
—— (1932), The nasal bot fly, *Cephenomyia auribarbis* Meigen (*Diptera*, Tachinidae) of the Red deer, *Cervus elephas* L., *Parasitology*, **24**, 185–195.
CARPENTER, G. H. *et al.* (1920), The warble flies: fifth report on experiments and observations as to life-history and treatment, *J. Dept. Agric. tech. Instruct. Ireland*, **20**, 452–459 (and *ibid.*, 1908 and later).
CARPENTER, G. H. AND POLLARD, F. J. S. (1918), The presence of lateral spiracles in the larva of *Hypoderma*, *Proc. R. Irish Acad.*, **34** (B), 73–84.

CARSON, H. L. (1967), The association between *Drosophila carcinophila* Wheeler and its host, the land crab *Gecarcinus ruricola* (L.), *Am. midl. Nat.*, **78**, 324–343.

CARSON, H. L. AND WHEELER, M. R. (1968), *Drosophila endobranchia*, a new drosophilid associated with land crabs in the West Indies, *Ann. ent. Soc. Am.*, **61**, 675–678.

CERFONTAINE, P. (1907), Observations sur la larve d'un Diptère, du genre *Microdon*, *Arch. biol.*, **23**, 367–410.

CHANCE, M. M. (1970), The functional morphology of the mouthparts of blackfly larvae (Dipt., Simuliidae), *Quaestiones Ent.*, **6**, 245–284.

CHANDLER, A. E. J. (1968), A preliminary key to the eggs of some of the commoner aphidophagous Syrphidae (Diptera) occurring in Britain, *Trans. R. ent. Soc. Lond.*, **120**, 199–217.

CHAPMAN, T. A. (1878), On the economy, etc. of *Bombylius*, *Entomologist's mon. Mag.*, **14**, 196–208.

CHILCOTT, J. G. T. (1960), A revision of the Nearctic species of Fanniinae (Diptera: Muscidae), *Can. Ent.*, Suppl. **4**, 295 pp.

CHISWELL, J. R. (1956), A taxonomic account of the last instar larvae of some British Tipulinae (Diptera: Tipulidae), *Trans. R. ent. Lond.*, **108**, 409–484.

CHRISTOPHERS, S. R. (1901), The anatomy and histology of the adult female mosquito, *Rep. Malaria Comm. R. Soc. Lond.*, **4**, 20 pp.

—— (1960), *Aedes aegypti (L.) the Yellow Fever mosquito, its life history, bionomics and structure*, Cambridge Univ. Press, London.

CHVÁLA, M., DOSKOCIL, J., MOOK, J. H. AND POKORNÝ, V. (1974), The genus *Lipara* Meigen (Diptera, Chloropidae), systematics, morphology, behaviour, and ecology, *Tijdschr. Ent.*, **117**, 1–25.

CHVÁLA, M. AND LYNEBORG, L. (1972), *The horse Flies of Europe (Diptera, Tabanidae)*, Copenhagen, Ent. Soc. Copenhagen.

CLAUSEN, C. P. (1940), *Entomophagous Insects*, McGraw-Hill Book Co., London.

COE, R. (1950), Tipulidae, Chironomidae, In: Coe, R. L., Freeman, P. and Mattingly, P. F. (eds.), Diptera 2, *Handbks. Ident. Brit. Ins.*, **9** (2), R. ent. Soc. Lond.

—— (1953), Diptera: Syrphidae, *Handbks. Ident. Brit. Inst.*, **10** (1).

—— (1966), Diptera: Pipunculidae, *Hndbks. Ident. Brit. Ins.* **10** (2c).

COLLESS, D. H. AND MCALPINE, D. K. (1970), Diptera, In: *The Insects of Australia*, Chapter 34, Divn. Entomology, C.S.I.R.O., Melbourne.

COLE, F. R. (1969), *The Flies of Western North America*, Univ. Calif. Press, Berkeley, 693 p.

COLLIN, J. E. (1944*a*), The British species of Psilidae (Diptera), *Entomologist's mon. Mag.*, **80**, 214–224.

—— (1944*b*), The British species of Anthomyzidae (Diptera), *Entomologist's mon. Mag.*, **80**, 265–272.

—— (1945*a*), British Micropezidae, *Ent. Rec.*, **57**, 115–119.

—— (1945*b*), The British species of Opomyzidae (Diptera), *Ent. Rec.*, **57**, 13–16.

—— (1947), The British genera of Trypetidae (Diptera), with notes on a few species, *Ent. Rec.*, **58**, 36 and suppl. 1–14.

—— (1948), British Therevidae, *Proc. R. phys. Soc. Edinburgh*, **23**, 95–102.

—— (1961), *British Flies*, **6**, pts 1–3, Cambridge Univ. Press, London.

COOK, E. F. (1949), The evolution of the head in the larvae of the Diptera, *Microentomology*, **14**, 1–57.

COTTERELL, G. S. (1920), The life-history and habits of the yellow dung-fly (*Scatophaga stercoraria*); a possible blow-fly check, *Proc. zool. Soc. Lond.*, **1920**, 629–647.

COUTANT, A. F. (1915), The habits, life-history, and structure of a blood-sucking Muscid larva (*Protocalliphora azurea*), *J. Parasit.*, **1**, 135–150.

CRAGG, F. W. (1912), The structure of *Haematopota pluvialis* (Meigen), *Sci. Mem. Indian Med.*, **55**, 1–33.

CRAMPTON, G. C. (1921), The sclerites of the head, and the mouthparts of certain immature and adult insects, *Ann. ent. Soc. Am.*, **14**, 65–103.

—— (1925), Phylogenetic study of the thoracic sclerites of the non-tipuloid Nematocerous Diptera, *Ann. ent. Soc. Am.*, **18**, 49–69.

—— (1944), A comparative morphological study of the terminalia of male Calyptrate Cyclorrhaphous Diptera and their Acalyptrate relatives, *Bull. Brooklyn ent. Soc.*, **39**, 1–31.

—— *et al.* (1942), Guide to the insects of Connecticut. Part VI. The Diptera or true flies of Connecticut, First fascicle., *Bull. Conn. geol. nat. Hist. Surv.*, **64**, pp. 509. Reprinted 1966.

CREGAN, H. B. (1941), Generic relationships of the Dolichopodidae (Diptera) based on a study of the mouthparts, *Illinois biol. Monogr.*, **18**, 1–48.

CROSSKEY, R. W. (1969), A re-classification of the Simuliidae (Diptera) of Africa and its islands, *Bull. Brit. Mus. nat. Hist. (Ent.) Suppl.*, **14**, 1–195.

CURRAN, C. H. (1934), *The families and genera of North American Diptera*, The author, New York.

CUTTEN, F. E. A. AND KEVAN, D. K. MCE. (1970), The Nymphomyiidae (Diptera), with special reference to *Palaeodipteron walkeri* Ide and its larva in Quebec, and a description of a new genus and species, *Can. J. Zool.*, **48**, 1–24.

D'ASSIS FONSECA, E. C. M. (1968), Muscidae, *Hndbks. Ident. Brit. Ins.*, Diptera: Cyclorrhapha: Calyptrata, **10** (4b), R. ent. Soc. Lond.

DAVIES, D. M. AND PETERSEN, B. V. (1956), Observations on the mating, feeding, ovarian development, and oviposition of adult black flies (Simuliidae, Diptera), *Can. J. Zool.*, **34**, 615–655.

DAY, C. D. (1948), *British Tachinid flies. Tachinidae* (*Larvaevoridae and Calliphoridae*), Arbroath (reprinted from North Western Naturalist).

DELFINADO, M. D. AND HARDY, D. E. (1973–74), *A Catalog of the Diptera of the oriental regions*, Parts I–III, Univ. Hawaii Press, Honolulu.

DELL, J. A. (1905), The structure and life history of *Psychoda sexpunctata* Curtis, *Trans. ent. Soc. Lond.*, **1905**, 293–311.

DESCAMPS, M. (1957), Recherches morphologiques et biologiques sur les Diopsidae du Nord-Cameroun, *Bull. Sci. Min. Colon F. Outre Mer*, **7**, 1–154.

DIMMOCK, G. (1881), *The Anatomy of the Mouthparts and the Sucking Apparatus of some Diptera*, A. Williams & Co., Boston, U.S.A.

DINULESCU, G. (1932), Recherches sur la biologie des Gastrophiles. Anatomie, physiologie, cycle évolutif, *Ann. Sci. nat. Paris*, (10), **15**, 1–183.

—— (1961), Diptera. Fam. Oestridae. *Fauna Repub. pop. rom. Insecta*, **11**, fasc. 4, 165 pp.

DISNEY, R. H. L. (1971), Associations between blackflies (Dipt. Simuliidae) and prawns (Crustacea, Atyidae), with a discussion of the phoretic habit in simuliids, *J. Anim. Ecol.*, **40**, 83–92.

DOLEŽIL, Z. (1972), Developmental stages of the tribe Eristalini (Diptera, Syrphidae), *Acta ent. Bohemoslov.*, **69**, 339–350.

DOWNES, J. A. (1958), The feeding habits of biting flies and their significance in classification, *A. Rev. Ent.*, **3**, 249–266.

DOWNES, J. A. AND COLLESS, D. H. (1967), Mouthparts of the biting and blood-sucking type in Tanyderidae and Chironomidae (Diptera), *Nature*, **214**, 1355–1356.

DUFOUR, L. (1851), Recherches anatomiques et physiologiques sur les diptères, accompagnées de considérations relatives à l'histoire naturelle de ces insectes, *Mém. prés. à l'Acad. Sci. Paris, Sci. math. phys.*, **11**, 171–360.

DUNBAR, R. W. (1958), The salivary gland chromosomes of two sibling species of black flies included in *Eusimulium aureum* Fries, *Can. J. Zool.*, **36**, 23–44.

DUNN, L. H. (1922), Observations on the oviposition of the house-fly, *Musca domestica* L. in Panama, *Bull. ent. Res.*, **13**, 301–305.

DUPUIS, C. (1963), Essai monographique sur les Phasiinae (Diptères Tachinaires parasites d'Héteroptères), *Mém. Mus. Hist. nat. Paris* (*n.s.*) *Zool.*, **26**, 1–461.

DYTE, C. E. (1967), Some distinctions between the larvae and pupae of the Empididae and Dolichopodidae (Diptera), *Proc. R. ent. Soc. Lond.*, (A), **42**, 119–127.

EDWARDS, F. W. (1916), On the systematic position of the genus *Mycetobia* Mg. (Diptera Nematocera), *Ann. Mag. nat. Hist.*, (8), **17**, 108–115.

—— (1919), Some parthenogenetic Chironomidae, *Ann. Mag. nat. Hist.*, (9), **3**, 222–228.

—— (1920), The nomenclature of the parts of the male hypopygium of Diptera Nematocera, with special reference to mosquitoes, *Ann. trop. Med. Parasit.*, **14**, 28–40.

—— (1921), A revision of the mosquitoes of the Palaearctic region, *Bull. ent. Res.*, **12**, 263–351.

—— (1925*a*), British fungus-gnats (Diptera, Mycetophilidae). With revised generic classification of the family, *Trans. ent. Soc. Lond.*, **1924**, 505–662.

—— (1925*b*), A synopsis of British Bibionidae and Scatopsidae (Diptera), *Ann. appl. Biol.*, **12**, 263–275.

—— (1926*a*), On marine Chironomidae (Diptera); with descriptions of a new genus and four new species from Samoa, *Proc. zool. Soc. Lond.*, **1926**, 779–806.

—— (1926*b*), On the British biting midges (Diptera, Ceratopogonidae), *Trans. ent. Soc. Lond.*, **74**, 389–426.

—— (1929*a*), Blepharoceridae, *Diptera of Patagonia and South Chile based mainly on material in the British Museum* (*Natural History*), Part **2**, fasc. **2**, 33–75.

—— (1929*b*), British non-biting midges (Diptera, Chironomidae), *Trans. ent. Soc. Lond.*, **77**, 279–430.

—— (1941), *Mosquitoes of the Ethiopian Region, III. Culicine adults and pupae*, Brit. Mus. (Nat. Hist.), London.

EDWARDS, F. W., OLDROYD, H. AND SMART, J. (1939), *British Blood-sucking Flies*, Brit. Mus. (Nat. Hist.), London.

EFFLATOUN, H. C. (1930), A monograph of Egyptian Diptera, Part IV. Family Tabanidae, *Mém. Soc. ent. Egypte*, **4**, 1–114.

EICHLER, W. (1941), Morphologische und biologische Merkmale mitteleuropäischer Dasselfliegen und ihrer Larven, *Z. Parasitenk.*, **12**, 95–106.

ELTRINGHAM, H. AND HAMM, A. H. (1928), On the production of silk by the species of the genus *Hilara* Meig. (Diptera), *Proc. Roy. Soc.*, (B), **102**, 327–338.

VAN EMDEN, F. I. (1950), *Mormotomyia hirsuta* Austen (Diptera) and its systematic position, *Proc. R. ent. Soc. Lond.*, (B), **19**, 121–128.

—— (1954), Tachinidae and Calliphoridae, *Hndbks. Ident. Brit. Ins.*, Diptera: Cyclorrapha: Calyptrata (1A), R. ent. Soc. Lond.

—— (1965), *The fauna of India and adjacent countries*, Diptera 7, Muscidae part 1, Publications of Government of India, Calcutta.

ENGLISH, K. M. J. (1947), Notes on the morphology and biology of *Apiocera maritima* Hardy (Diptera, Apioceridae), *Proc. Linn. Soc. N.S.W.*, **71** (1946), 296–302.

ENOCK, F. (1891), The life-history of the Hessian fly, *Cecidomyia destructor*, Say, *Trans. ent. Soc. Lond.*, **1891**, 329–366.

FABRE, J. H. (1921), *Souvenirs Entomologiques*, **3**, P. Lechevalier, Paris.

FALCOZ, L. (1926), Diptères Pupipares, *Faune de France*, **14**, P. Lechevalier, Paris.

FALLIS, A. M. AND SMITH, S. M. (1964), Ether extracts from birds and CO_2 as attractants for some ornithophilic Simuliids, *Can. J. Zool.*, **42**, 723–730.

FELT, E. P. (1911*a*), Hosts and galls of American gall midges, *J. econ. Ent.*, **4**, 451–475.

—— (1911*b*), Summary of food habits of American gall midges, *Ann. ent. Soc. Am.*, **4**, 55–62.

FEUERBORN, H. J. (1922*a*), Das Hypopygium 'inversum' und 'circumversum' der Dipteren, *Zool. Anz.*, **55**, 189–213.

—— (1922*b*), Der sexuelle Reizapparat der Psychodiden, etc., *Arch. Naturg.*, **88A**, Heft 4, 1–137.

FLETCHER, L. W. (1970), Abdominal and genitalic homologies in the screw-worm, *Cochliomyia hominivorax* (Dip. Calliphoridae), established by a genetic marker, *Ann. ent. Soc. Am.*, **63**, 490–495.

FONTAINE, J. (1964), Commensalisme et parasitisme chez les larves d'Ephéméroptères, *Bull. mens. Soc. linn. Lyon*, N.S. **33**, 163–174.

FOOTE, B. A. (1970), The larva of *Tanypeza longimana* (Dipt., Tanypezidae), *Ann. ent. Soc. Am.*, **63**, 235–238.

FOOTE, B. A. AND KNUTSON, L. V. (1970), Clam-killing fly larvae (Dipt., Sciomyzidae), Nature, **226**, 466.

FRAENKEL, G. (1936), Observations and experiments on the blow-fly (*Calliphora erythrocephala*) during the first day after emergence, *Proc. zool. Soc. Lond.*, **1935**, 893–904.

FRAENKEL, G. AND BHASKARAN, G. (1973), Pupariation and pupation in Cyclorrhaphous flies (Diptera): terminology and interpretation, *Ann. ent. Soc. Am.*, **61**, 418–422.

FREEMAN, P. AND DE MEILLON, B. (1953), *Simuliidae of the Ethiopian region*, Brit. Mus. (Nat. Hist.), London.

FREW, J. G. H. (1923), On the larval anatomy of the gout-fly of barley (*Chlorops taeniopus*), etc., *Proc. zool. Soc. Lond.*, **1923**, 703–821.

FREY, H. (1921), Studien über den Bau des Mundes der niederen Diptera Schizophora, etc., *Acta Soc. Fauna Fl. fenn.*, **48**, 245 pp.

FRICK, K. B. (1952), A generic revision of the family Agromyzidae (Diptera) with a catalogue of new world species, *Univ. California Publ. Ent.*, **8**, 339–452.

FULLER, M. E. (1933), The life history of *Onesia accepta* Malloch (Diptera, Calliphoridae), *Parasitology*, **25**, 342–352.
—— (1934), The early stages of *Sciadocera rufomaculata* White (Dipt. Phoridae), *Proc. Linn. Soc. N.S.W.*, **59**, 9–15.
GIARD, A. (1889), Sur la castration parasitaire des *Typhlocyba* par une larve d'Hyménoptère *l'Aphelopus melaleucus*, Dalm. et par une larve de Diptère (*Ateloneura spuria*, Meig.), *C.R. Acad. Paris*, **109**, 708–710.
GILLETT, J. D. (1971), *Mosquitos*, Weidenfeld & Nicolson, London.
GOETGHEBUER, M. AND LENZ, F. (1936–50), In: Lindner, E. (ed.), *Die Fliegen*, **13**, Tendipedidae, E. Schweitzerbart'sche Verlagsbuchhandlung, Stuttgart.
GRAHAM-SMITH, G. S. (1913), *Flies in Relation to Disease (non-bloodsucking flies)*, Cambridge University Press, Cambridge.
—— (1916), Observations on the habits and parasites of common flies, *Parasitology*, **3**, 440–544.
—— (1919), Further observations on the habits and parasites of common flies, *Parasitology*, **11**, 347–384.
—— (1930*a*), Further observations on the anatomy and function of the proboscis of the blow-fly, *Calliphora erythrocephala* L., *Parasitology*, **22**, 47–115.
—— (1930*b*), The Oscinidae (Diptera) as vectors of conjunctivitis, and the anatomy of their mouthparts, *Parasitology*, **22**, 457–467.
—— (1939), The generative organs of the blow-fly, *Calliphora erythrocephala* L. with special reference to musculature and movements, *Parasitology*, **30** (1938), 441–476.
GRASSI, B. (1907), Richerche sui flebotomi, *Mem. Soc. ital. Sci.*, (3) **14**, 353–394.
GREENE, C. T. (1926), Descriptions of larvae and pupae of two-winged flies belonging to the family Leptidae, *Proc. U.S. nat. Mus.*, **79** Part. 2, 1–20.
GRIFFITHS, G. C. D. (1967), Revision of the *Phytomyza syngenesiae* group (Diptera, Agromyzidae), including the species hitherto known as '*Phytomyza atricornis* Meigen', *Stuttg. Beitr. Ent.*, **177**, 28 pp.
—— (1972), *The phylogenetic classification of Diptera Cyclorrapha, with special reference to the structure of the male postabdomen*, Dr W. Junk N.V., The Hague.
GRÜNBERG, K. (1910), Diptera, Teil 1. In: Brauer, A. (ed.), *Die Süsswasserfauna Deutschlands*, Verlag Gustav Fischer, Jena.
GUYÉNOT, E. (1907), L'appareil digestif et la digestion de quelques larves de mouches, *Bull. sci. Fr. et Belg.*, **41**, 353–370.
HACKETT, L. W. (1937), *Malaria in Europe: an ecological study*, Clarendon Press, Oxford.
HADWEN, S. (1912), Warble flies: The economic aspect and a contribution on the biology, *Dept. Agric. Canada, Health of Animals Br.*, Bull. **16**, 18 pp.
—— (1916), A further contribution to the biology of *Hypoderma lineatum, Dept. Agric. Canada, Health of Animals Br.*, Sci. Bull. **21**, 10 pp.
HALL, D. G. (1948), *The Blowflies of North America*, Thomas Say Foundation, no. 4, Lafayette, Indiana.
HAMM, A. H. (1909), Observations on *Empis opaca* F. and further observations on the Empinae, *Entomologist's mon. Mag.*, **45**, 132–134.
HANDLIRSCH, A. (1882), Die Metamorphose und Lebensweise von *Hirmoneura obscura*, etc., *Wien ent. Ztg.*, **1**, 224–228.
HARDY, D. E. (1943), A revision of the Nearctic Dorilaidae (Pipunculidae), *Kansas Univ. Sci. Bull.*, **28**, 1–231.

HARDY, D. E. (1965), Family Drosophilidae, *Insects of Hawaii*, 12, Honolulu.

HARDY, D. E. AND NAGATOMI, A. (1960), An unusual new Nematocera from Japan and a new family name, *Pacific Ins.*, 2, 263–267.

HARDY, G. H. (1944), The copulation and terminal segments of Diptera, *Proc. R. ent. Soc. Lond.*, (A), 19, 52–65.

HARRIS, K. M. (1966), Gall midge genera of economic importance (Diptera: Cecidomyiidae) Part 1. Introduction and subfamily Cecidomyiinae; supertribe Cecidomyiidi, *Trans. R. ent. Soc. Lond.*, 118, 313–358.

HARTLEY, J. C. (1961), A taxonomic account of the larvae of some British Syrphidae, *Proc. zool. Soc. Lond.*, 136, 505–573.

—— (1963), The cephalopharyngeal apparatus of Syrphid larvae and its relationship to other Diptera, *Proc. zool. Soc. Lond.*, 141, 261–280.

HASHIMOTO, H. (1965), Discovery of *Clunio takehashii* from Japan, *Japan. J. Zool.*, 15, 13–29.

HASSAN, A. A. G. (1944), The structure and mechanism of the spiracular regulatory apparatus in adult Diptera and certain other groups of insects, *Trans. R. ent. Soc. Lond.*, 94, 103–153.

HAYES, W. P. (1938), A bibliography of keys for the identification of immature insects. Part 1. Diptera, *Ent. News.*, 49, 246–251.

HEISS, E. M. (1938), A classification of the larvae and puparia of the Syrphidae of Illinois, exclusive of aquatic forms, *Illinois biol. Monogr.*, 16, 142 pp.

HEMMINGSEN, A. M. (1963), The ant-lion-like sand trap of the larva of *Lampromyia canariensis* Macquart (Diptera Leptidae-Rhagionidae, Vermileoninae), *Vidensk. Meddr dansk. naturh. Foren*, 125, 237–267.

HENDEL, F. (1927), Trypetidae, In: Lindner, E. (ed.), *Die Fliegen*, 49, 16–19, E. Schweitzerbart'sche Verlagsbuchhandlung, Stuttgart.

—— (1928), Diptera, allgemeiner Teil, In: Dahl, F. (ed.), *Tierwelt Deutschlands*, Verlag Gustav Fischer, Jena.

—— (1936–37), Diptera, In: Kükenthal, W. (ed.), *Handbuch der Zoologie*, Bd. 4, Insecta 2, Lief 8–11, Walter de Gruyter & Co., Berlin.

HENNIG, W. (1935), Der Filterapparat im Pharynx der Cyclorrhaphenlarven und die biologische Deutung der Madenform, *Zool. Anz.*, 111, 131–139.

—— (1943*a*), Piophilidae, In: Lindner, E. (ed.), *Die Fliegen*, 40, E. Schweitzerbart'sche Verlagsbuchhandlung, Stuttgart.

—— (1943*b*), Uebersicht über die bisher bekannten Metamorphosestudien der Ephydriden, etc., *Arb. morph. taxon. Ent.*, 10, 105–138.

—— (1948–52), *Die Larvenformen der Dipteren*, 1–3, Akademie Verlag, Berlin.

—— (1954), Flügelgeäder und System der Dipteren unter Berücksichtigung der aus dem Mesozoikum beschriebenen Fossilien, *Beitr. Ent.*, 4, 245–388.

—— (1956), Beitrag zur Kenntnis der Milichiiden-Larven (Diptera: Milichiidae), *Beitr. Ent.*, 6, 138–145.

—— (1958), Die Familien der Diptera Schizophora und ihre phylogenetischen Verwandtschaftsbeziehungen, *Beitr. Ent.*, 8, 505–688.

—— (1964), Die Dipteren-Familie Sciadoceridae in Baltischen Bernstein (Diptera: Cyclorrhapha Aschiza), *Stuttg. Beitr. Naturkd.*, 127, 1–10.

—— (1965), Vorarbeiten zu einem phylogenetischen System der Muscidae (Diptera: Cyclorrhapha), *Stuttg. Beitr. Naturkd.*, 141, 100 pp.

—— (1967), Die sogenannten 'niederen Brachycera' in Baltischen Bernstein (Diptera: Fam. Xylophagidae, Xylomyidae, Rhagionidae, Tabanidae), *Stuttg. Beitr. Naturkd.*, 174, 1–51.

—— (1969), Neue Uebersicht über die aus dem Baltischen Bernstein bekannte Acalyptratae (Dipt., Cyclorrhapha), *Stuttg. Beitr. Naturkd.*, **209**, 1–42.

—— (1971), Neue Untersuchungen über die Familien der Diptera Schizophora (Diptera: Cyclorrhapha), *Stuttg. Beitr. Naturkd.*, **226**, 1–76.

—— (1972), Eine neue Art der Rhagionidengattung *Litoleptis* aus Chile, mit Bemerkungen über Fühlerbildung und Verwandtschaftsbeziehungen einiger Brachycerenfamilien (Diptera: Brachycera), *Stuttg. Beitr. Naturkd.*, **242**, 1–18.

HERING, E. M. (1957), *Bestimmungstabellen der Blattminen von Europa einschliesslich des Mittelmeerbeckens und der Kanarischen Inseln.* Dr. W. Junk N. V., 's Gravenhage.

HESSE, A. J. (1938), A revision of the Bombyliidae of Southern Africa. Part I, *Ann. S. Afr. Mus.*, **34**, 1–1053.

HEWITT, C. G. (1914), *The House-fly*, Cambridge University Press, Cambridge.

HICKMAN, C. P. (1935), External features of the larva of *Hydromyza confluens, Proc. Acad. Sci. Indiana*, **44**, 212–216.

HINTON, H. E. (1955), On the structure, function, and distribution of the prolegs of the Panorpoidea, with a criticism of the Berlese-Imms theory, *Trans. R. ent. Soc. Lond.*, **106**, 455–540.

—— (1960*a*), The structure and function of the respiratory horns of the eggs of some flies, *Phil. Trans. R. Soc. Lond.*, **243**, 45–73.

—— (1960*b*), Cryptobiosis in the larvae of *Polypedilum vanderplanki* Hint. (Chironomidae), *J. Ins. Physiol.*, **5**, 286–300.

—— (1962), The structure and function of the spiracular gills of *Deuterophlebia* (Deuterophlebiidae) in relation to those of other Diptera, *Proc. zool. Soc. Lond.*, **138**, 111–122.

—— (1963), The respiratory system of the egg-shell of the blowfly, *Calliphora erythrocephala* Meig., as seen with the electron microscope, *J. Ins. Physiol.*, **9**, 121–129.

—— (1967), Structure of the plastron in *Lipsothrix* and the polyphyletic origin of the plastron in Tipulidae, *Proc. R. ent. Soc. Lond.*, (A) **42**, 35–38.

—— (1968*a*), Structure and protective devices of the egg of the mosquito *Culex pipiens, J. Ins. Physiol.*, **14**, 145–161.

—— (1968*b*), Observations on the biology and taxonomy of the eggs of *Anopheles* mosquitoes, *Bull. ent. Res.*, **57**, 495–508.

HOBBY, B. M. (1931*a*), The British species of Asilidae (Diptera) and their prey, *Trans. ent. Soc. S. England*, **6** (1930), 1–42.

—— (1931*a*), The prey of dung-flies (Diptera), Cordyluridae, *Proc. ent. Soc. Lond.*, **6**, 47–49.

HOLMGREN, N. (1904), Zur Morphologie des Insektenkopfes. I, *Z. wiss. Zool.*, **76**, 439–477.

HOPKINS, G. E. (1952), *Mosquitoes of the Ethiopian Region.* I, *Larval bionomics of mosquitoes and taxonomy of Culicine larvae.* 2nd edn. (Additions by P. F. Mattingly.) London (B.M.).

HORI, K. (1967), Comparative anatomy of the internal organs of the Calyptrate Muscids. Consideration of the phylogeny of the Calyptratae, *Sci. Rep. Kanazawa Univ.*, **12**, 215–254.

HOULIHAN, D. F. (1969), The structure and behaviour of *Notiphila riparia* and *Erioptera squalida* (Dipt.), *J. Zool.*, **159**, 249–267.

HOWARD, L. O., DYAR, H. G. AND KNAB, F. (1912), *The Mosquitoes of North and Central America and the West Indies*, Carnegie Inst., Washington.

HOYT, C. P. (1952), The evolution of the mouthparts of adult Diptera, *Microentomology*, **17**, 61–125.

HULL, F. M. (1962), Robber flies of the world. The genera of the family Asilidae, *Bull. U.S. natn. Mus.*, **224**, parts 1 and 2, 907 pp.

IDE, F. P. (1965), A fly of the archaic family Nymphomyiidae (Diptera) from North America, *Can. Ent.*, **97**, 497–507.

IMMS, A. D. (1907), On the larval and pupal stages of *Anopheles maculipennis* Meigen, *J. Hyg.*, **7**, 291–318.

—— (1908), On the larval and pupal stages of *Anopheles maculipennis* Meigen, Part iii. *Parasitology*, **1**, 103–133.

—— (1942), On *Braula coeca* Nitzsch and its affinities, *Parasitology*, **34**, 88–100.

—— (1944), On the constitution of the maxillae and labium in Mecoptera and Diptera, *Quart. J. micr. Sci.*, N.S., **85**, 73–96.

IRWIN-SMITH, V. (1921–23), Studies in life-histories of Australian Diptera Brachycera. Part 1. Stratiomyidae. Nos. 1–4, *Proc. Linn. Soc. N.S.W.*, **45** (1920), 505–530; **46**, 252–255; 425–432; **48**, 49–81.

JOBLING, B. (1926), A comparative study of the head and mouthparts in the Hippoboscidae (Diptera Pupipara), *Parasitology*, **18**, 319–349.

—— (1928), The structure of the head and mouthparts in the Nycteribiidae (Diptera Pupipara), *Parasitology*, **20**, 254–272.

—— (1933), Revision of the structure of head, mouthparts and salivary glands of *Glossina palpalis* Rob.-Desv, *Parasitology*, **85**, 73–96.

—— (1936), A revision of the subfamilies of the Streblidae and the genera of the subfamily Streblinae, etc., *Parasitology*, **28**, 355–380.

—— (1938), A revision of the species of the genus *Trichobius* (Diptera Acalypterae, Streblidae), *Parasitology*, **30**, 358–387.

—— (1939), On the African Streblidae (Diptera Acalypterae) including the morphology of the genus *Ascodipteron* Adens and a description of a new species, *Parasitology*, **31**, 147–165.

—— (1949). Host-parasite relationship between the American Streblidae and the bats, with a new key to the American genera, etc., *Parasitology*, **39**, 315–329.

—— (1951), A record of the Streblidae from the Philippines and other Pacific Islands, etc., *Trans. R. ent. Soc. Lond.*, **102**, 211–246.

JOHANNSEN, O. A. (1903–05), Aquatic Nematocerous Diptera, *New York State Mus., Bull.*, **68**, 328–441; **86**, 76–327.

—— (1909), Diptera. Fam. Mycetophilidae *in* Wytsman, P. *Gen. Insect.*, **93**, Disnet & Verteneuil, Brussels.

—— (1910), Paedogenesis in *Tanytarsus, Science*, (N.S.), **32**, 768.

—— (1935), Aquatic Diptera. Pt. II. Orthorrhapha Brachycera and Cyclorrhapha, *Mem. Cornell agric. Exp. Sta.*, **177**, 62 pp.

—— (1937), Aquatic Diptera. Pt. III, Chironomidae. *Mem. Cornell agric. Exp. Sta.*, **205**, 84 pp.

JOHANNSEN, O. A. AND CROSBY, C. R. (1913), The life-history of *Thrypticus muhlenbergiae* sp. nov., *Psyche*, Cambridge, **20**, 164–166.

JOHANNSEN, O. A. AND THOMSEN, L. C. (1937), Aquatic Diptera. Parts IV and V, *Mem. Cornell agric. Exp. Sta.*, **210**, 80 pp.

JONES, B. J. (1906), Catalogue of the Ephydridae with bibliography and description of new species, *Univ. Calif. Publn. Entom.*, 1, 153–198.

JOSEPH, G. (1880), Anatomische und biologische Bemerkungen über *Actora aestuum* Meigen, *Zool. Anz.*, 3, 250–252.

JUNG, H. F. (1956), Beiträge zur Biologie, Morphologie und Systematik der europäischen Psychodiden (Diptera), *Dt. ent. Z.*, N.F., 3, 97–257.

KAISER, P. (1972), Ueber hormonale Regelung der Pädogenese bei den viviparen Gallmücken (Diptera: Cecidomyiidae), *Ent. Mitt. (Hamburg)*, **4** no. 77, 259–262.

KARL, O. (1928), Zweiflügler oder Diptera. III. Muscidae, In: Dahl, F., *Die Tierwelt Deutschlands*, Verlag Gustav Fischer, Jena.

KEILIN, D. (1911), Recherches sur la morphologie larvaire des Diptères du genre *Phora*, *Bull. sci. Fr. et Belg.*, **45**, 27–88.

—— (1913), Sur diverses glandes des larves des Diptères, *Arch. Zool. gén. et exp.*, **52**, 1–8 (notes et revues).

—— (1914), Sur la biologie d'un Psychodide à larve xylophage, *Trichomyia urbica*, *C.R. Soc. biol. Paris*, **76**, 434–437.

—— (1915), Recherches sur les larves de Diptères Cyclorhaphes, *Bull. sci. Fr. et Belg.*, **49**, 15–198.

—— (1916), Sur la viviparité chez les Diptères et les larves de Diptères vivipares, *Arch. Zool. gén. et exp.*, **55**, 393–415.

—— (1917), Recherches sur les Anthomyides à larves carnivores, *Parasitology*, **9**, 325–450.

—— (1918), On the structure of the larvae and the systematic position of the genera *Mycetobia* Mg., *Ditomyia* Winn., and *Symmerus* Walk. (Diptera Nematocera), *Ann. Mag. nat. Hist.*, (9), **3**, 33–42.

—— (1919), On the life-history and larval anatomy of *Melinda cognata* Meigen (Diptera Calliphorinae) parasitic in the snail *Helicella* (*Heliomanes*) *virgata* Da Costa, with an account of other Diptera living on molluscs, *Parasitology*, **11**, 430–455.

—— (1944), Respiratory systems and respiratory adaptations in larvae and pupae of Diptera, *Parasitology*, **36**, 1–66.

KEILIN, D. AND TATE, P. (1940), The early stages of the families Trichoceridae and Anisopodidae (= Rhyphidae) (Diptera: Nematocera), *Trans. R. ent. Soc. Lond.*, **90**, 39–62.

—— (1943), The larval stages of the Celery-fly (*Acidia heraclei* L.) and of the Braconid *Adelura apii* (Curtis); with notes upon an associated parasitic yeast-like fungus, *Parasitology*, **35**, 27–36.

KEILIN, D. AND THOMPSON, W. R. (1915), Sur le cycle évolutif des Pipunculides, parasites intracoelomiques des Typhlocybes, *C.R. Acad. Sci. Paris*, **78**, 9–12.

KELLOGG, V. L. (1899), The mouthparts of the Nematocerous Diptera, *Psyche*, Cambridge, **8**, 303, 327, 346, 355, 363.

—— (1901), The anatomy of the larva of the giant crane-fly (*Holorusia rubiginosa*), *Psyche*, Cambridge, **9**, 207–213.

KELLY, E. O. G. (1914), A new sarcophagid parasite of grasshoppers, *J. agric. Res.*, **2**, 435–446.

KELSEY, L. P. (1969), A revision of the Scenopinidae (Diptera) of the world, *Bull. U.S. natn. Mus.*, **277**, 1–336.

KERTÉSZ, K. (1902–10), *Catalogus Dipterorum*, Wilhelm Engelmann, Budapest.

KESSEL, E. L. (1968), Circumversion and mating positions in Platypezidae – an expanded and emended account (Diptera), *Wasmann J. Biol.*, **26**, 243–253.

KESSEL, E. L. AND MAGGIONCALDA, E. A. (1968), A revision of the genera of Platypezidae, with descriptions of five new genera and considerations of phylogeny, circumversion, and hypopygia (Diptera), *Wasmann J. Biol.*, **26**, 33–106.

KETTLE, D. S. AND LAWSON, J. W. H. (1952), The early stages of British biting midges *Culicoides* Latreille (Diptera: Ceratopogonidae) and allied genera, *Bull. ent. Res.*, **43**, 421–467.

KEUCHENIUS, P. E. (1913), The structure of the internal genitalia of some male Diptera, *Z. wiss. Zool.*, **105**, 501–536.

KIEFFER, J. J. (1900*a*), Monographie des Cécidomyides d'Europe et d'Algérie, *Ann. Soc. ent. France*, **69**, 181–384.

—— (1900*b*), Beiträge zur Biologie und Morphologie der Dipterenlarven, *Ill. Zs. Ent.*, **5**, 131–133.

KING, J. L. (1916), Observations on the life-history of *Pterodontia flavipes* Gray, *Ann. ent. Soc. Am.*, **9**, 309–321.

KNUTSON, L. V. (1970), Biology of snail-killing flies in Sweden (Dipt. Sciomyzidae), *Entomol. Scand.*, **1**, 307–314.

KNUTSON, L. V., BERG, C. O., EDWARDS, L. J., BRATT, A. D. AND FOOTE, B. A. (1967), Calcareous septa formed in snail-shells by larvae of snail-killing flies, *Science*, **156**, 522–523.

KNUTSON, L. V., NEFF, S. E. AND BERG, C. O. (1967), Biology of snail-killing flies from Africa and southern Spain. (Sciomyzidae: *Sepedon*), *Parasitology*, **57**, 487–505.

KNUTSON, L. V. AND BERG, C. O. (1967), Biology and immature stages of malacophagous Diptera of the genus *Knutsonia* Verbeke (Sciomyzidae), *Bull. Inst. r. Soc. Sci. nat. Belg.*, **43** (7), 1–60.

KNUTSON, L. V. AND FLINT, O. S. (1971), Pupae of Empididae in pupal cocoons of Rhyacophilidae and Glossosomatidae (Dipt. and Trich.), *Proc. ent. Soc. Washington*, **73**, 314–320.

KOLENATI, F. A. (1862), Beiträge zur Kenntnis der Phthyriomyiarien, *Hor. Soc. ent. Ross.*, **2**, 1–109.

KÖNIG, A. (1894), Ueber die Larve von *Oncodes, Verh. k.k. zool. bot. Ges. Wien*, **44**, 163–166.

KÖRTING, A. (1940), Zur Biologie und Bekämpfung der Möhrenfliege (*Psila rosae* F.) in Mitteldeutschland, *Arb. phys. angew. Ent.*, **7**, 209–232, 269–285.

KRIVOSCHEINA, N. P. AND MAMAEV, B. M. (1970), The family Cramptonomyiidae (Dipt. Nematocera) new for the U.S.S.R., its morphology, ecology and phylogenetic relationships, *Ent. Obozr.*, **49**, 886–898 (in Russian).

KRYSTOPH, H. (1961), Vergleichend-morphologische Untersuchungen an den Mundteilen bei Empididen (Diptera), *Beitr. Ent.*, **11**, 824–872.

KÜNCKEL D'HERCULAIS, J. (1875), *Recherches sur l'organisation et le développement des Volucelles*, G. Masson, Paris.

—— (1879), Recherches morphologiques et zoologiques sur le système nerveux des Insectes diptères, *C.R. Acad. Sci. Paris*, **89**, 491–494.

—— (1905), Les Lépidoptères Limacodides et leurs Diptères parasites, Bombyliides du genre *Systropus, Bull. sci. Fr. Belg.*, **39**, 141–151.

LAING, J. (1935), On the ptilinum of the blow-fly (*Calliphora erythrocephala*), *Quart. J. micr. Sci.*, **87**, 497–521.

LAMB, C. G. (1922), The geometry of insect pairing, *Proc. Roy. Soc.*, (B), **93**, 1–11.
LANDROCK, K. (1926–27), Fungivoridae, In: Lindner, E. (ed.), *Die Fliegen der paläarktischen Region*, **8**, E. Schweitzerbart'sche Verlagsbuchhandlung, Stuttgart.
LANE, J. (1939), Catalogo dos mosquitos neotropicos, *Bol. Biol. Ser. monogr*, **1**, 218 pp., Univ. São Paulo.
—— (1953), *Neotropical Culicidae*, I & II, Univ. S. Paulo.
LAWSON, J. W. H. (1951), The anatomy and morphology of the early stages of *Culicoides nubeculosus* Meigen (Diptera: Ceratopogonidae = Heleidae), *Trans. R. ent. Soc. Lond.*, **102**, 511–574.
LEE, D. J. (1948), Australian Ceratopogonidae (Diptera, Nematocera). Parts I–V, *Proc. Linn. Soc. N.S.W.*, **72** (1947), 313–356; **73**, 57–70.
LENGERSDORF, F. (1928–30), Lycoriidae, In: Lindner, E. (ed.), *Die Fliegen der paläarktischen Region*, E. Schweitzerbart'sche Verlagsbuchhandlung, Stuttgart.
LENZ, F. (1941), Die Jugendstadien der sectio Chironomariae (Tendipedini) connectentes (Subf. Chironominae = Tendipedinae). Zusammenfassung und Revision, *Arch. Hydrobiol.*, **38**, 1–69.
LEWIS, D. J. (1971), Phlebotomid sandflies, *Bull. Wld. Health Org.*, **44**, 535–551.
—— (1974), The biology of Phlebotomidae in relation to leishmaniasis, *A. Rev. Ent.*, **19**, 363–384.
LINDNER, EBERHARD (1959), Beiträge zur Kenntnis der Larven der Limoniidae (Diptera), *Z. Morph. Ökol. Tiere*, **48**, 209–319.
LINDNER, ERWIN (1924–74), *Die Fliegen der paläarktischen Region*, Stuttgart (incomplete).
—— (1930), Thaumaleidae, In: Lindner, E. (ed.), *Die Fliegen der paläarktischen Region*, E. Schweitzerbart'sche Verlagsbuchhandlung, Stuttgart.
LOEW, F. (1862), *Die europäischen Bohrfliegen* (Trypetidae), Staatsdruckerei, Wien.
LOBANOV, A. M. (1970), The morphology of the mature larvae of Helomyzidae (Diptera), *Zool. Zh.*, **49**, 1671–1675 (Russian, English summary).
LOWNE, B. T. (1890–95), *The anatomy, morphology, and development of the blow-fly* (Calliphora erythrocephala), 2 vols, R. H. Porter for the author, London.
LOWTHER, J. K. AND WOOD, D. M. (1964), Specificity of a black fly, *Simulium euryadminculum* Davies, towards its host, the common loon, *Can. Ent.*, **96**, 911–913.
LÜBBEN, H. (1908), *Thrypticus smaragdinus* Gerst. und seine Lebensgeschichte, *Zool. Jb., Syst.*, **26**, 319–332.
LUNDBECK, W. (1907–27), *Diptera Danica*, **1–7**, Copenhagen.
LYALL, E. M. (1929), The larva and pupa of *Scatopse fuscipes* Mg. and a comparison of the known species of Scatopsid larvae, *Ann. appl. Biol.*, **16**, 630–638.
LYNEBORG, L. (1970), Taxonomy of European *Fannia* larvae (Diptera: Fanniidae), *Stuttg. Beitr. Naturkd.*, **215**, 1–28.
MAA, T. C. (1963), Genera and species of Hippoboscidae (Diptera): types, synonymy, habitats and natural groupings, *Pacific Ins. Monog.*, **6**, 1–186.
MACFIE, J. W. S. (1940), The genera of Ceratopogonidae, *Ann. trop. Med. Parasit*, **34**, 13–30.
MACGREGOR, M. E. (1931), The nutrition of adult mosquitoes: preliminary contribution, *Trans. R. Soc. Trop. Med. Hyg.*, **24**, 465–472.
MACKERRAS, I. M. (1954), The classification and distribution of Tabanidae (Diptera). I. General review, *Austr. J. Zool.*, **2**, 431–454.

MADWAR, S. (1937), Biology and morphology of the immature stages of Mycetophilidae (Diptera, Nematocera), *Philos. Trans. Lond.*, (B), **227**, 1–110.

MALLOCH, J. M. (1915), Some additional records of Chironomidae for Illinois and other notes on other Illinois Diptera, *Illinois State Lab. nat. Hist. Bull.*, **11**, 305–364.

—— (1917), A preliminary classification of the Diptera exclusive of Pupipara, based upon larval and pupal characters, etc. Part 1., *Illinois State Lab. nat. Hist. Bull.*, **12**, 161–409.

MAMAEV, B. M. AND KRIVOSCHEINA, N. P. (1966), New data on the taxonomy and biology of Diptera of the family Axymyiidae, *Ent. Obozr.*, **45**, 168–180 (Russian).

—— (1969), New data on the morphology and ecology of Hyperoscelidae (Dipt. Nematocera), *Ent. Obozr.*, **48**, 933–942 (Russian).

MANNHEIM, B. T. (1935), *Beiträge zur Biologie und Morphologie der Blepharoceriden (Dipt.)*, Robert Noske, Leipzig.

MANSBRIDGE, G. H. (1933), On the biology of some Ceratoplatinae and Macrocerinae (Diptera, Mycetophilidae), *Trans. R. ent. Soc. Lond.*, **81**, 75–92.

MARCHAL, P. (1897*a*), Les Cécidomyies des céréales et leurs parasites, *Ann. Soc. ent. France*, **66**, 1–105.

—— (1897*b*), Note sur la biologie de *Lauxania aenea* Fall., Diptère nuisible au trèfle, *Bull. Soc. ent. France*, **1897**, 216–217.

MARCHAND, W. (1918), The larval stages of *Argyra albicans* Lw. (Diptera, Dolichopodidae). *Ent. News*, **29**, 216–220.

—— (1920), The early stages of Tabanidae (Horse flies), *Rockefeller Inst. Med. Res. Monogr.*, **13**, 203 pp.

MARSHALL, J. F. (1938), *The British Mosquitoes*, Brit. Mus. (Nat. Hist.), London.

MASKELL, W. M. (1888), On *Henops brunneus*, Hutton, *Trans. N.Z. Inst.*, **20**, 106–108.

MASSONAT, E. (1909), Contribution à l'étude des Pupipares, *Ann. Univ. Lyon.*, N.S. **128**, 356 pp. (Thèse).

MATHESON, R. (1944), *Handbook of the Mosquitoes of North America*, 2nd edn, Ithaca.

MATTINGLY, P. F. (1974), *Culicidae* (*Mosquitoes*), In: Smith, K. G. V. (ed.), *Insects and other arthropods of medical importance*, Chapter 3, Brit. Mus. (Nat. Hist.), London.

MAYER, K. (1934), Die Metamorphose der Ceratopogonidae (Dipt.), *Arch. Naturg.*, n.f. **3**, 205–288.

—— (1935), Die Metamorphose einiger Drosophiliden aus Niederländisch-Indien, *Arch. Hydrobiol.*, suppl. **13**, 462–473.

MCALPINE, D. K. (1966), Description and biology of an Australian species of Cypselosomatidae (Diptera), with a discussion of family relationships, *Austr. J. Zool.*, **14**, 673–685.

—— (1975), The subfamily classification of the Micropezidae and the genera of Eurybatinae (Diptera: Schizophora), *J. Ent. (B)*, (1974) **43**, 231–245.

MCALPINE, J. F. (1967), A detailed study of Ironomyiidae (Diptera: Phoroidea), *Can. Ent.*, **99**, 224–236.

—— (1968), Taxonomic notes on *Eurychoromyia mallea* (Diptera: Eurychoromyiidae), *Can. Ent.*, **100**, 819–823.

—— (1970), First records of Calyptrate flies in the Mesozoic era (Dipt., Calliphoridae), *Can. Ent.*, **102**, 342–346.

—— (1973), A fossil ironomyiid fly from Canadian amber (Diptera: Ironomyiidae), *Can. Ent.*, **105**, 105–111.

MCALPINE, J. F. AND MARTIN, J. E. H. (1966), Systematics of Sciadoceridae and relatives with descriptions of two new genera and species from Canadian Amber and erection of family Ironomyiidae (Diptera: Phoroidea), *Can. Ent.*, **98**, 527–544.

MCATEE, W. L. (1911), Facts in the life history of *Goniops chrysocoma, Proc. ent. Soc. Washington,* **18**, 21–29.

MCFADDEN, M. W. (1967), Soldier fly larvae in America north of Mexico, *Proc. U.S. nat. Mus.*, **121**, 1–72.

DE MEIJERE, J. H. C. (1895), Ueber zusammengesetzte Stigmen bei Dipterenlarven, etc, *Tijdschr. Ent.*, 38, 65–100.

—— (1900*a*), Ueber die Larve von *Lonchoptera*, etc., *Zool. Jb.*, *Syst.*, **14**, 87–132.

—— (1900*b*), Ueber die Metamorphose von *Callomyia amoena* Meig., *Tijdschr. Ent.*, **43**, 223–231.

—— (1904), Beiträge zur Kenntnis der Biologie und der systematischen Verwandtschaft der Conopiden, *Tijdschr. Ent.*, **46**, 144–225.

—— (1912*a*), Ueber die Metamorphose von *Puliciphora* und über neue Arten der Gattungen *Puliciphora* Dahl und *Chonocephalus* Wand., *Zool. Jb.*, Suppl. **15**, 141–154.

—— (1912*b*), Beiträge zur Kenntnis der Conopiden, *Tijdschr. Ent.*, **55**, 184–207.

—— (1916), Beiträge zur Kenntnis der Dipteren-Larven und Puppen, *Zool. Jb.*, *Syst.*, **40**, 177–322.

DE MEILLON, B. (1937), Studies on insects of medical importance from Southern Africa and adjacent territories. Part 4, *Pub. S. Afr. Inst. med. Res.*, **40**, 305–411.

MEINERT, F. (1886), Die eucephale Mygelarves, *Dansk. Selsk. vid. Skr.*, **3**, 373–493 (French résumé); 305–411.

MELIN, D. (1923), Contributions to the knowledge of the biology, etc., of the Swedish Asilids, *Zool. Bidrag. Uppsala,* **8**, 1–317.

MERGELSBERG, O. (1935), Ueber die postimaginale Entwicklung (Physogastrie) und den Hermaphroditismus bei afrikanischen Termitoxenien (Dipt.), *Zool. Jb.*, *Anat.*, **60**, 345–398.

MERLE, P. DU (1972), Morphologie de la larve planidium d'un Diptère Bombyliidae, *Villa brunnea, Ann. Soc. ent. France*, N. S. **8** (4), 915–950.

MESNIL, L. (1939), Essai sur les Tachinaires (Larvaevoridae), *Monogr. Sta. Lab. rech. agron. Paris*: 67 pp.

—— (1944–52), Larvaevorinae (Tachininae), In: Lindner, E. (ed.), *Die Fliegen*, **64**, E. Schweitzerhart'sche Verlagsbuchhandlung, Stuttgart.

METCALF, C. L. (1916–17), Syrphidae of Maine and second report, *Maine agric. expt. Sta. Bull.*, **253**, **263**, 193–264; 153–176.

MIALL, L. C. (1893), *Dicranota*; a carnivorous Tipulid larva, *Trans. ent. Soc. Lond.*, **1893**, 235–253.

—— (1895*a*), *The Natural History of Aquatic Insects*, Macmillan & Co., London.

—— (1895*b*), The life-history of *Pericoma canescens* (Psychodidae), etc., *Trans. ent. Soc. Lond.*, **1895**, 141–153.

MIALL, L. C. AND HAMMOND, A. R. (1892), The development of the head of the imago of *Chironomus, Trans. Linn. Soc. Zool.*, **5**, 265–279.

MIALL, L. C. AND SHELFORD, R. (1897), The structure and life-history of *Phalacrocera replicata,* etc., *Trans. ent. Soc. Lond.*, **1897**, 343–366.

MIALL, L. C. AND TAYLOR, T. H. (1907), The structure and life-history of the Holly-fly, *Trans. ent. Soc. Lond.*, **1907**, 259–283.

MICKOLEIT, G. (1962), Die Thoraxmuskulatur von *Tipula vernalis* Meigen. Ein Beitrag zur vergleichenden Anatomie des Dipterenthorax, *Zool. Jb.*, *Anat.*, **80**, 213–244.

MINISTRY OF AGRICULTURE AND FISHERIES (1926), *Report of the departmental committee on Warble Fly pest*, 48 pp, London.

MITZMAIN, M. B. (1913), The bionomics of *Stomoxys calcitrans*, Linnaeus, a preliminary account, *Philippine J. Sci.*, **8B**, 29–48.

MIYAKE, T. (1919), Studies on the fruit-flies of Japan. 1: Japanese Orange-fly, *Bull. imp. exp. Sta. Japan*, **2**, 85–165.

MÖHN, E. (1960), Studien an paedogenetischen Gallmückarten (Diptera, Itonidae). I Teil., *Stuttg. Beitr. Naturk.*, **31**, 1–11.

MOOK, J. H. AND BRUGGEMANN, C. G. (1968), Acoustical communication by *Lipara lucens* (Diptera, Chloropidae), *Entomologia exp. appl.*, **11**, 397–402.

MORGE, G. (1963, 1967), Die Lonchaeidae und Pallopteridae Österreichs und angrenzender Gebiete, *Naturkndl. Jb.*, *Stadt Linz*, **1963**, 123–312; **1967**, 141–212; 242–295.

MORRIS, H. M. (1921–22), The larval and pupal stages of the Bibionidae, *Bull. ent. Res.*, **12**, 221–232; **13**, 189–195.

MÜGGENBERG, F. H. (1892), Der Rüssel der Diptera Pupipara, *Arch. Naturg.*, **58**, 287–332.

MÜHLENBERG, M. (1968), Zur Morphologie der letzten Abdominalsegmente bei weiblichen Wollschwebern (Diptera, Bombyliidae), *Zool. Anz.*, **181**, 277–279.

MUIR, F. (1912), Two new species of *Ascodipteron, Bull. Mus. comp. Zool.*, **54**, 349–366.

MUIRHEAD-THOMSON, R. C. (1937), Observations on the biology and larvae of the Anthomyidae, *Parasitology*, **29**, 273–358.

—— (1951), *Mosquito Behaviour in relation to Malaria Transmission and Control in the Tropics*, Edward Arnold and Co., London.

MÜLLER, F. (1895), Contribution towards the history of a new form of larvae of Psychodidae (Diptera) from Brazil, *Trans. ent. Soc. Lond.*, **1895**, 479–481.

MULLIGAN, H. W. (ed.) (1970), *The African Trypanosomiases*, G. Allen & Unwin, London.

MUNRO, H. K. (1947), African Trypetidae (Diptera), *Mem. ent. Soc. S. Africa*, **1**, 284 pp.

NAGATOMI, A. (1961), Studies in the aquatic snipe flies of Japan, Part III. Descriptions of the larvae, Part IV, Descriptions of the pupae (Diptera, Rhagionidae), *Mushi*, **35**, 11–27; 29–38.

NATVIG, L. R. (1948), Contributions to the knowledge of the Danish and Fennoscandian mosquitoes, Culicini, *Norsk. ent. Tidskr.*, suppl. **1**, 567 pp.

NEEDHAM, J. G. AND BETTEN, C. (1901), Aquatic insects in the Adirondacks, *Bull. N.Y. State Mus.*, **47**, 383–612.

NEWSTEAD, R. (1906), On the life-history of *Stomoxys calcitrans*, Linn. *J. econ. Biol.*, **1**, 157–166.

—— (1908), On the habits, life-cycle and breeding places of the common house-fly (*Musca domestica* Linn.), *Ann. trop. Med.*, **1**, 507–520.

NICHOLSON, H. P. (1945), The morphology of the mouthparts of the non-biting black fly, *Eusimulium dacotense* D. and S. as compared with those of the biting species, *Simulium venustum* Say (Diptera: Simuliidae), *Ann. ent. Soc. Amer.*, **38**, 281–297.

NIELSEN, J. C. (1903), Ueber die Entwicklung von *Bombylius pumilus* Meig., eine Fliege, welche bei *Colletes daviesana* Smith schmarotzt, *Zool. Jb., Syst.*, **18**, 647–658.

—— (1909), Jagttagelser over entoparasitiske Muscidelarver hos Arthropoder, *Ent. Meddr*, **4**, 1–126. (English summary.)

—— (1911), *Mydaea anomala* Jaenn., a parasite of South-American birds, *Vid. Meddr nat. Foren Kobenhavn*, **1911**, 195–208.

NIKOLEI, E. (1961), Vergleichende Untersuchungen zur Fortpflanzung heterogener Gallmücken unter experimentellen Bedingungen, *Z. Morph. Ökol. Tiere*, **50**, 281–329.

NORRIS, A. (1894), Observations on the New Zealand Glow-worm, *Bolitophila luminosa, Entomologist's mon. Mag.*, **30**, 202.

NUTTALL, G. F. AND SHIPLEY, A. E. (1901–03), The structure and biology of *Anopheles* (*Anopheles maculipennis*), *J. Hyg.*, **1**, **3**, 45–77; 166–215.

OKELY, E. F. (1974), Description of the puparia of twenty-three British species of Sphaeroceridae (Diptera, Acalyptratae), *Trans. R. ent. Soc. Lond.*, **126**, 41–56.

OLDROYD, H. (1952, 1957), *The horse-flies* (*Diptera: Tabanidae*) *of the Ethiopian region*, I–III, Brit. Mus. (Nat. Hist.), London.

OLDROYD, H. *et. al.* (1949 onwards), *Handbooks for the identification of British Insects*, **9**, London, R. ent. Soc.

OLSOUFIEFF, N. G. (1937), Faune de l'U.R.S.S. Insectes Dipteres. Fam. Tabanidae, (Russian with German descr.) Leningrad, *Inst. Zool. Acad. Sci. U.R.S.S.*, **7**, No. 2, 433 pp.

OSTEN-SACKEN, C. R. (1862), Characters of the larvae of Mycetophilidae, *Proc. ent. Soc. Philadelphia*, **1**, 151–172.

—— (1884), An essay on comparative chaetotaxy, etc., *Trans. ent. Soc. Lond.*, **1884**, 497–517.

OWSLEY, W. B. (1946), The comparative morphology of internal structures of the Asilidae (Diptera), *Ann. ent. Soc. Am.*, **38**, 33–68.

PANTEL, J. (1898), Essai monographique sur les caractères extérieur, la biologie et l'anatomie d'une larve parasite du groupe des Tachinaires, *La Cellule*, **15**, 1–290.

—— (1909), Notes de neuropathologie comparée; ganglions de larves des insectes parasités par des larves d'insectes, *Le Neuraxe*, **10**, 269–297.

—— (1910), Recherches sur les Diptères à larves entomobies, *La Cellule*, **26**, 25–216.

PAPAVERO, N. (ed.) (1966–), *A catalogue of the Diptera of the Americas south of the United States*, Dept. Zoologia, Univ. S. Paulo.

PAPAVERO, N. AND WILCOX, J. (1968), Family 34, Mydidae. In: Papavero, N. (1966).

PARAMONOV, S. J. (1953*a*), A review of the Australian Apioceridae (Diptera), *Austr. J. Zool.*, **1**, 449–537.

—— (1953*b*), A review of Australian Nemostrinidae (Diptera), *Austr. J. Zool.*, **1**, 242–290.

PARENT, O. (1938), Diptères Dolichopodidae, *Faune de France*, **35**, Paul Lechevalier, Paris.

PARKER, J. R. (1918), The life history and habits of *Chloropisca glabra* Meig., a predaceous oscinid, *J. econ. Ent.*, 11, 368–380.

PATTON, W. S. AND CRAGG, F. W. (1913), *A Textbook of Medical Entomology*, Christian Literature Society for India, London.

PENNAK, R. W. (1945), Notes on mountain midges (Deuterophlebiidae) with a description of the immature stages of a new species from Colorado, *Amer. Mus. Novit.*, **1276**, 10 pp.

PERFIL'EV, P. P. (1937), Faune de l'U.R.S.S. Insectes Diptères. III. No. 2. Psychodidae (Phlebotominae), Moscow, *Inst. Zool. Acad. Sci., U.R.S.S.*, 144 pp.

PERKINS, R. C. L. (1905), Leaf-hoppers and their natural enemies, Part IV. Pipunculidae, *Hawaiian Sugar Planters' Assoc. Div. Entom., Bull.* 1, Part 4, 123–157. (Suppl., *ibid.*, Bull. 1, Part 10 (1906)).

PERRANDIN, J. (1961), Recherches sur l'anatomie céphalique des larves Bibionides et de Lycoriides (Dipt. Nematocera), *Trav. Lab. Zool. Dijon*, **41**, 1–47.

PERRIS, E. (1855), *Histoire des Métamorphoses de quelques Insectes*, Dessain, Liège.

—— (1870), Histoire des insectes du Pin maritime. Diptères, *Ann. Soc. ent. France*, (4), **10**, 135–232.

PETERSON, A. (1916), The head-capsule and mouth-parts of Diptera, *Univ. Illinois biol. Monogr.*, **3**, 1–62.

PEUS, F. (1934), Zur Kenntnis der Larven und Puppen der Chaoborinae (Corethrinae auct.), *Arch. Hydrob.*, **27**, 641–668.

PHILLIPS, V. T. (1946), The biology and identification of Trypetid larvae (Diptera: Trypetidae), *Mem. Amer. ent. Soc.*, **12**, 161 pp.

PHILLIPS, W. J. (1914), Corn-leaf blotch miner, *J. agric. Res.*, **2**, 15–31.

PLASSMANN, E. (1972), Morphologisch-taxonomische Untersuchungen an Fungivoridenlarven (Diptera), *Dt. ent. Z.*, **19**, 73–99.

POTTS, W. H. (1974), Glossinidae (Tse-tse flies), In: Smith, K. G. V. (ed.), *Insects and other Arthropods of medical importance*, Brit. Mus. (Nat. Hist.), London.

POULTON, E. B. (1906), Predaceous insects and their prey, *Trans. ent. Soc. Lond.*, **1906**, 323–409.

PRADO, A. P. DO. (1969), Syringogastridae, una nova familia de Dipteros Acalyptratae, com a descrição de seis espécies novas do gênero *Syringogaster* Cresson, *Stud. Ent.*, **12**, 1–32.

PRATT, H. S. (1893), Beiträge zur Kenntnis der Pupiparen. Die Larve von *Melophagus ovinus*, *Arch. Naturg.*, **59**, 151–200.

PRESCOTT, H. W. (1961), Respiratory pore construction in the host by the Nemestrinid parasite *Neorhynchocephalus sackeni* (Diptera), with notes on respiratory tube characters, *Ann. ent. Soc. Am.*, **54**, 557–566.

PRINGLE, J. W. S. (1948), The gyroscopic mechanism of the halteres of Diptera, *Philos. Trans.*, **233B**, 347–384.

PRYOR, M. G. M. (1948), Mouthparts and feeding habits of Blepharoceridae (Diptera), *Proc. R. ent. Soc. Lond.*, (A), **23**, 67–70.

PURI, M. (1925), On the life history and structure of the early stages of Simuliidae (Diptera, Nematocera), *Parasitology*, **17**, 295–369.

QUAINTANCE, A. L. (1912), The Mediterranean fruit-fly, *U.S. Dept. Agric. Ent. Circ.*, **160**, 25 pp.

QUATE, L. W. (1955), A revision of the Psychodidae (Diptera) in America north of Mexico, *Univ. Calif. Publ. Ent.*, **10**, 103–273.

RAPP, W. F. AND SNOW, W. E. (1945), Catalogue of Pantophthalmidae of the world, *Rev. Ent.*, **16**, 252–254.
RASCHKE, W. (1887), Die Larve von *Culex nemorosus, Arch. Naturg.*, **53** (2), 133–163.
RAYNAL, J. (1934), Contribution à l'étude des Phlébotomes d'Indo-Chine: généralités, *Arch. Inst. Pasteur Indochine*, **19**, 337–369.
RETTENMEYER, C. W. (1961), Observations on the biology and taxonomy of flies found over swarming raids of army ants (Diptera: Tachinidae, Conopidae), *Univ. Kansas Sci. Bull.*, **42**, 993–1066.
RICHARDS, O. W. (1930), The British species of Sphaeroceridae (Borboridae, Diptera), *Proc. zool. Soc. Lond.*, **1930**, 261–345.
RIETSCHEL, P. (1961), Funktion und Entwicklung der Haftorgane der Blepharoceridenlarven, *Z. Morph. Ökol. Tiere*, **50**, 239–265.
RILEY, C. V. *et. al.* (1880), *Second report of the United States Entomological Commission for the years 1878–79 relating to the Rocky Mountain Locust, etc.*, Washington. Cf. also Riley, C. V. (1880), *Ann. Ent.*, **3**, 279–283.
RIVOSECCHI, L. (1958), Gli organi della riproduzione in *Musca domestica, Rc. Ist. sup. Sanità*, **21**, 458–488.
ROBACK, S. S. (1951), A classification of the Muscoid Calyptrate Diptera, *Ann. ent. Soc. Am.*, **44**, 327–361.
ROBINSON, H. (1970), The subfamilies of the family Dolichopodidae in North and South America (Dipt.), *Pap. Avulsos Zool.*, **53**, 23–62.
ROBINSON, I. (1953), The postembryonic stages in the life cycle of *Aulacigaster leucopeza* (Meigen) (Diptera Cyclorrhapha: Aulacigasteridae), *Proc. R. ent. Soc. Lond.*, (A), **28**, 77–84.
RODHAIN, J. AND BEQUAERT, J. (1916*a*), Matériaux pour une étude monographique des Diptères parasites de l'Afrique, *Bull. sci. Fr. Belg.*, **50**, 53–165.
—— (1916*b*), Observations sur la biologie de *Cyclopodia greeffi* Karsch Nycteribiide parasite d'une chauve-souris congolaise, *Bull. Soc. zool.*, **40** (1915), 248–262.
ROSS, H. H. (1940), The Rocky Mountain 'Black-fly' *Symphoromyia atripes* (Diptera: Rhagionidae), *Ann. ent. Soc. Am.*, **33**, 254–257.
ROUBAUD, E. (1909), La *Glossina palpalis*, etc., Thèse. Paris.
—— (1913), Recherches sur les Auchméromyies Calliphorines à larves succeuses de sang de l'Afrique tropicale, *Bull. sci. Fr. et Belg.*, **47**, 105–202.
—— (1915), Hématophagie larvaire et affinités parasitaires, d'une mouche calliphorine, *Phormia sordida*, parasite des jeunes oiseaux, *Bull. Soc. Path. exot.*, **8**, 77–79.
—— (1922), Recherches sur la fécondité et la longévité de la mouche domestique, *Ann. Inst. Pasteur*, **36**, 765–783.
ROZKOŠNÝ, R. (1966), Zur Morphologie und Biologie der Metamorphosestadien mitteleuropäischer Sciomyziden (Diptera), *Acta sc. nat. Acad. Sci. Bohemoslav.*, **1**, 117–160.
RUSSELL, P. F., WEST, L. S. AND MANWELL, R. D. (1946), *Practical Malariology*, W. B. Saunders & Co., London.
RYBERG, O. (1939), Beiträge zur Kenntnis der Fortpflanzungsbiologie und Metamorphose der Fledermausfliegen Nycteribiidae (Diptera Pupipara), *Verh. 7. int. Kongr. Ent.*, **2**, 1285–1299.
SABROSKY, C. W. (1948), A further contribution to the classification of the North

American spider parasites of the family Acroceratidae (Diptera), *Amer. Midland Nat.*, **39**, 382–430.

SAETHER, O. A. (1969), Some Nearctic Podonominae, Diamesinae, and Orthocladiinae (Diptera, Chironomidae), *Bull. Fish. Res. Bd. Can.*, **170**, 1–154.

SALZER, R. (1968), Konstruktionsanatomische Untersuchung des männlichen Postabdomens von *Calliphora erythrocephala* Meigen (Insecta: Diptera), *Z. Morph. Ökol. Tiere*, **63**, 155–238.

SASAKI, C. (1886), On the life-history of *Ugimya sericaria*, Rondani, *J. Coll. Sci. Univ. Japan*, **1**, 1–46.

SATCHELL, G. H. AND TONNOIR, A. L. (1953), The Australian Psychodidae (Diptera), *Austr. J. Zool.*, **1**, 357–448.

SAUNDERS, L. G. (1923), On the larva, pupa and systematic position of *Orphnephila testacea*, Meig. (Diptera, Nematocera), *Ann. Mag. nat. Hist.*, (9), **11**, 631–640.

—— (1924), On the life history and the anatomy of the early stages of *Forcipomyia*, (Diptera, Nemat., Ceratopogoninae), *Parasitology*, **16**, 164–213.

—— (1925), On the life history, morphology and systematic position of *Apelma* and *Thyridomyia* n.g. (Diptera, Nemat., Ceratopogoninae), *Parasitology*, **17**, 252–277.

SCHIRNHAUS, M. von (1971), Unbekannte Stridulationsorgane bei Dipteren und ihre Bedeutung für Taxonomie und Phylogenetik der Agromyziden (Diptera: Agromyzidae et Chamaemyiidae), *Beitr. Ent.*, **21**, 551–579.

SCHLEIN, Y. (1970), A comparative study of the thoracic skeleton and musculature of the Pupipara and the Glossinidae (Dipt.), *Parasitology*, **60**, 327–373.

SCHMITZ, H. (1917), Biologische Beziehungen zwischen Dipteren und Schnecken, *Biol. CentrBl.*, **37**, 24–43.

—— (1928), *Revision der Phoriden*, Ferd. Dummlers Verlag, Berlin.

—— (1939), Beiträge zu einer Monographie der Termitoxeniidae (Diptera), I–V, *Broteria*, **34**, 22–40, 55–70, 132–146, 147–162; **35**, 53–63, 133–148.

—— (1940), Zum Ausbau der Systematik der Termitoxeniidae (Dipt.), *6. Congr. int. Ent. Madrid*, **1935**, 9–15.

SCHNEIDER, F. (1969), Bionomics and physiology of aphidophagous Syrphidae (Dipt.), *A. Rev. Ent.*, **14**, 103–124.

SCHUMANN, H. (1963), Zur Larvalsystematik der Muscinae nebst Beschreibung einiger Musciden- and Anthomyiidenlarven, *Dt. ent. Z.*, (N.F.), **10**, 134–163.

SCOTT, E. I. (1939), An account of the developmental stages of some Aphidophagous Syrphidae (Dipt.) and their parasites (Hymenopt.), *Ann. appl. Biol.*, **26**, 509–532.

SCOTT, H. (1917), Notes on Nycteribiidae, with descriptions of two new genera, *Parasitology*, **9**, 593–610.

SÉGUY, E. (1928), Études sur les mouches parasites, I, *Encycl ent.* (A), **9**, 240 pp.

—— (1950), La biologie des Diptères, *Encycl. ent.*, **26**, 609 pp.

SELLKE, K. (1936), Biologische und morphologische Studien an schädlichen Wiesenschnaken (Tipulidae, Dipt.), *Z. wiss. Zool.*, **148**, 465–555.

SENIOR-WHITE, R., AUBERTIN, D. AND SWART, F. (1940), *The Fauna of British India, including the remainder of the Oriental region. Diptera, VI. Family Calliphoridae,* Taylor & Francis, London.

SHELFORD, V. E. (1913), The life-history of the bee-fly (*Spogostylum anale* Say), etc., *Ann. ent. Soc. Am.*, **6**, 213–225.

SHILLITO, J. F. (1971), The genera of the Diopsidae, (*Zool.*) *J. Linn. Soc. Lond.*, **50**, 287–295.

SILVESTRI, F. (1917*a*), Descrizione di una specie di *Oscinosoma* (Diptera: Chloropidae) osservato in fruttescenze di caprifico, *Bull. Lab. Zool. gen. agrar. Portici*, **12**, 147–154.

—— (1917*b*), Sulla *Lonchaea aristella* Beck. (Diptera: Lonchaeidae) dannosa alle inflorescenze e fruttescenze del Caprifico e del Fico, *Bull. Lab. Zool. gen. agrar. Portici*, **12**, 123–146.

SINGH, S. B. AND JUDD, W. W. (1966), A comparative study of the alimentary canal of adult calyptrate Diptera, *Proc. ent. Soc. Ontario*, **96** (1965), 29–80.

SKAIFE, S. H. (1921), On *Braula coeca*, Nitzsch, a dipterous parasite of the honey bee, *Trans. R. Soc. S. Africa*, **10**, 41–48.

SMART, J. (1944), The British Simuliidae with keys to the species in the adult, pupal and larval stages, *Sci. Publ. Freshwater biol. Ass. Brit. Emp.*, **9**, 57 pp.

—— (1945), The classification of the Simuliidae (Diptera), *Trans. R. ent. Soc. Lond.*, **95**, 463–528.

SMITH, K. G. V. (1966), The larva of *Thecophora occidentalis*, with comments upon the biology of Conopidae (Diptera), *J. Zool.*, **149**, 263–276.

—— (1967), The biology and taxonomy of the genus *Stylogaster* Macquart, 1835 (Diptera: Conopidae, Stylogasterinae) in the Ethiopian and Malagasy regions, *Trans. R. ent. Soc. Lond.*, **119**, 47–69.

—— (1969*a*), The Empididae of Southern Africa (Dipt.), *Ann. Natal Mus.*, **19**, 1–342.

—— (1969*b*), Further data on the oviposition by the genus *Stylogaster* Macquart (Dipt., Conopidae, Stylogasterinae), *Proc. R. ent. Soc. Lond.* (A), **44**, 35–37.

SMITH, K. M. (1948), *A Textbook of Agricultural Entomology*, 2nd edn., Cambridge.

SMITH, K. M. AND GARDNER, J. C. M. (1922), *Insect Pests of the Horticulturalist: their nature and control*, London. Vol. 1.

SMULYAN, M. T. (1914), The Marguerite fly (*Phytomyza chrysanthemi* Kowarz), *Mass. agric. Exp. Sta. Bull.*, **157**, 52 pp.

SNODGRASS, R. E. (1903), The terminal abdominal segments of female Tipulidae, *J. New York ent. Soc.*, **11**, 177–183.

—— (1922), Mandible substitutes in the Dolichopodidae, *Proc. ent. Soc. Washington*, **24**, 148–152.

—— (1943), The feeding apparatus of biting and disease-carrying flies, *Smithson. misc. Coll.*, **1**, 51 pp.

—— (1953), The metamorphosis of a fly's head, *Smithson. misc. Coll.*, **122**, 25 pp.

SPEIGHT, M. V. D. (1969), The prothoracic morphology of acalypterates (Diptera) and its use in systematics, *Trans. R. ent. Soc. Lond.*, **121**, 325–421.

SPEISER, P. (1900–1), Ueber die Strebliden, etc., *Arch. Naturg.*, **66**, **67**, 31–70, 11–77.

—— (1908), Die geographische Verbreitung der Diptera Pupipara und ihre Phylogenie, *Z. wiss. InsBiol.*, **4**, 241–246, 301–305, 420–427, 437–447.

SPENCER, K. A. (1969), The Agromyzidae of Canada and Alaska, *Mem. ent. Soc. Can.*, **64**, 311 pp.

—— (1972), Diptera. Agromyzidae, *Hndbk. Ident. Brit. Ins.*, **10**, pt. **5**, *R. ent. Soc., Lond.*

—— (1973), *Agromyzidae (Diptera) of economic importance*, Series entomologica, Junk, W. B. V. (ed.), **9**, The Hague.

SPIETH, H. T. (1952), Mating behaviour within the genus *Drosophila* (Diptera), *Bull. Amer. Mus. nat. Hist.*, **99**, 395–474.

STAMMER, H. J. (1924), Die Larven der Tabaniden, *Z. Morph. Ökol. Tiere*, **1**, 121–170.

STEENBERG, C. M. (1943), Études sur les larves du genre *Phronia* (Fungivoridae, Nematocera), *Ent. Meddr*, **23**, 337–351.

STEPHENSON, J. W. AND KNUTSON, L. V. (1970), The distribution of the snail-killing flies (Dipt., Sciomyzidae) in the British Isles, *Entomologist's mon. Mag.*, **106**, 16–21.

STEYSKAL, G. C. (1953), A suggested classification of the lower Brachycerous Diptera, *Ann. ent. Soc. Am.*, **46**, 237–242.

—— (1964), Larvae of Micropezidae (Diptera) including two new species that bore in ginger roots, *Ann. ent. Soc. Am.*, **57**, 292–296.

—— (1965), The subfamilies of the Sciomyzidae of the world (Diptera: Acalyptratae), *Ann. ent. Soc. Am.*, **58**, 593–594.

—— (1972), A catalogue of species and key to the genera of the family Diopsidae (Diptera: Acalyptratae), *Stuttg. Beitr. Naturkd.*, **234**, 1–20.

STONE, A., KNIGHT, K. L. AND STARKE, H. (1959), *A synoptic catalog of the Mosquitoes of the world*, Thomas Say Foundation, **6**, 358 pp., Washington, D.C.

STONE, A. *et al.* (1965), *A catalog of the Diptera of America north of Mexico*, U.S. Dept. Agric., Agricultural Res. Serv., Washington D.C.

STRICKLAND, E. H. (1953), The ptilinal armature of flies (Diptera, Schizophora), *Can. J. Zool.*, **31**, 263–299.

STUCKENBERG, B. R. (1963), A study on the biology of the genus *Stylogaster*, with descriptions of a new species from Madagascar (Diptera: Conopidae), *Rev. Zool. Bot. Afr.*, **68**, 251–275.

STURTEVANT, A. H. (1921), The North American species of *Drosophila, Carnegie Inst. Washington, Publn.* **301**, 150 pp.

SWYNNERTON, C. F. M. (1936), The Tsetse flies of East Africa, *Trans R. ent. Soc. Lond.*, **84**, 579 pp.

TAN, K. B. (1967), The life-histories and behaviour of some Malayan stalkeyed flies (Diptera: Diopsidae), *Malay Nat. J.*, **20** (1–2), 31–38.

TATE, P. (1954), Notes upon the biology and morphology of the immature stages of *Neottiophilum praeustum* (Meigen, 1826) (Diptera: Neottiophilidae), *Parasitology*, **44**, 111–119.

TAYLOR, T. H. (1902), On the tracheal system of *Simulium, Trans. ent. Soc. Lond.*, **1902**, 701–716.

TESKEY, H. J. (1969), Larvae and pupae of some eastern North American Tachinidae (Dipt.), *Mem. ent. Soc. Can.*, **63**, 1–147.

TETLEY, H. (1918), The structure of the mouthparts of *Pangonia longirostris* in relation to the probable feeding habits of the species, *Bull. ent. Res.*, **8**, 253–267.

THEODOR, O. (1967), *An illustrated catalogue of the Rothschild collection of Nycteribiidae (Diptera) in the British Museum (Natural History)*, Brit. Mus. (Nat. Hist.), London.

—— (1968), A revision of the Streblidae (Diptera) of the Ethiopian region, *Trans. R. ent. Soc. Lond.*, **120**, 313–373.

THEOWALD, BR. (1957), Die Entwicklungstadien der Tipuliden (Diptera, Nematocera), inbesondere der West-paläarktischen Arten, *Tijdschr. Ent.*, **100**, 195–308.

THIENEMANN, A. (1944), Bestimmungstabellen für die bis jetzt bekannten Larven und Puppen der Orthocladiinen (Diptera, Chironomidae), *Arch. Hydrobiol.*, **39**, 551–664.

THOMAS, I. (1933, 1934, 1938), On the bionomics and structure of some Dipterous larvae infecting cereals and grasses, I, II, III, *Ann. appl. Biol.*, **20**, 707–721; **21**, 519–529; **25**, 181–196.

THOMPSON, W. R. (1921), Contribution à la connaissance des formes larvaires des Sarcophagides, *Bull. Soc. ent. France*, **1921**, 27–31, 219–222.

—— (1923), Recherches sur la biologie des Diptères parasites, *Bull. biol. Fr. Belg.*, **57**, 174–237.

THOMSEN, M. (1951), Weismann's ring and related organs in larvae of Diptera, *K. Dansk. Vid. Selsk. biol. Skr.*, **6**, No. 5, 32 pp.

THORPE, W. H. (1930), The biology of the Petroleum fly (*Psilopa petrolei* Coq.), *Trans. ent. Soc. Lond.*, **78**, 331–334, (cf. also, 1932, *Nature*, **130**, 437).

—— (1934), Observations on the structure, biology and systematic position of *Pantophthalmus tabaninus* Thunb. (Dipt. Pantophthalmidae), *Trans. R. ent. Soc. Lond.*, **82**, 5–22.

—— (1941), The biology of *Cryptochaetum* (Diptera) and *Eupelmus* (Hymenoptera) parasites of *Aspidoproctus* (Coccidae) in East Africa, *Parasitology*, **33**, 149–168.

THROCKMORTON, L. H. (1962), The problem of phylogeny in the genus *Drosophila*. Studies in Genetics. II. Research Reports on *Drosophila* genetics, taxonomy and evolution, *Univ. Texas, Publ.*, **6205**, 207–343.

TILLYARD, R. J. AND SEDDON, H. R. (1933), The sheep blow-fly problem in Australia, *Rep. Joint Blowfly Cttee., C.S.I.R. and N.S.W. Dept. Agric.*, **1**, 136 pp.

TOKUNAGA, M. (1935*a*), A morphological study of a Nymphomyid fly, *Philipp. J. Sci.*, **56**, 127–214.

—— (1936), The central nervous, tracheal and digestive systems of a Nymphomyid fly, *Philipp. J. Sci.*, **59**, 189–216.

—— (1935*b*), On the pupae of the Nymphomyid fly (Diptera), *Mushi*, **8**, 44–52.

TONNOIR, A. L. (1924), Les Blepharoceridae de la Tasmanie, *Ann. Biol. lacustre*, **13**, 5–69.

—— (1930), Notes on Indian Blepharocerid larvae and pupae, etc., *Rec. Ind. Mus.*, **32**, 161–214.

—— (1940), A synopsis of the British Psychodidae (Dipt.) with descriptions of new species, *Trans. Soc. Brit. Ent.*, **7**, 21–64.

TOWNSEND, C. H. T. (1934–44), *Manual of Myiology*, 15 parts, C. Townsend & Filhos, São Paulo.

TRÄGÅRDH, I. (1903), Beiträge zur Kenntnis der Dipterenlarven, *Ark. Zool.*, **1**, 1–42.

—— (1908), Contributions to the knowledge of *Thaumatoxena* Bredd. & Börn, *Ark. Zool.*, **4**, No. 10, 1–12.

TRELKA, D. G. AND FOOTE, B. A. (1970), Biology of slug-killing *Tetanocera* (Dipt. Sciomyzidae), *Ann. ent. Soc. Am.*, **63**, 877–895.

TULLOCH, P. H. G. (1906), The internal anatomy of *Stomoxys*, *Proc. Roy. Soc.* (B), **77**, 523–531.

UBRICH, H. (1971), Zur Skelett- und Muskelanatomie des Thorax der Dolichopodiden und Empididen (Diptera), *Veröff. Zool. Staatssamml. München*, **15**, 1–44.

UBRICH, H. (1972), Zur Anatomie des Empididen-Hypopygiums (Diptera), *Veröff. Zool. Staatssamml. München*, **16**, 1–28.

VALLEY, K., NOVAK, J. A. AND FOOTE, B. (1969), Biology and immature stages of *Eumetopiella rufipes, Ann. ent. Soc. Am.*, **62**, 227–234.

VARLEY, G. C. (1937), Aquatic insect larvae which obtain oxygen from the roots of plants, *Proc. R. ent. Soc. Lond.*, (A), **12**, 55–60.

VERHOEFF, C. (1891), Biologische Aphorismen über einige Hymenopteren, Dipteren und Coleopteren, *Verh. naturf. Ver. Rheinl. Westf.*, **48**, 1–80.

VERRALL, G. H. (1901), *British Flies*, 8, *Syrphidae*, Gurney & Jackson, London.

—— (1909), *British Flies*, 5, *Stratiomyidae*, etc., Gurney & Jackson, London.

VOCKEROTH, J. R. (1969), A revision of the genera of the Syrphini (Diptera: Syrphidae), *Mem. ent. Soc. Can.*, **62**, 176 pp.

VOGLER, C. H. (1900), Beiträge zur Metamorphose der *Teichomyza fusca, Ill. Z. Ent.*, **5**, 1–4, 17–20, 33–36.

WADSWORTH, J. T. (1914), Some observations on the life-history and bionomics of the knapweed gall-fly *Urophora solstitialis* Linn., *Ann. appl. Biol.*, **1**, 142–169.

WAHL, B. (1899), Ueber das Tracheensystem und die Imaginalscheiben der Larve von *Eristalis tenax* L., *Arb. zool. Inst. Wien*, **12**, 48–98.

WAINWRIGHT, C. J. (1928), The British Tachinidae (Diptera), *Trans. ent. Soc. Lond.*, **76**, 139–254.

—— (1932), The British Tachinidae (Diptera). First supplement, *Trans. ent. Soc. Lond.*, **80**, 405–424.

WALLACE, J. B. (1969), The mature larva and pupa of *Calobatina geometroides* (Cresson) (Dipt., Micropezidae), *Ent. News*, **80**, 317–321.

WANDOLLECK, B. (1899), Zur Anatomie der cyclorraphen Dipterenlarven. Anatomie der Larve von *Platycephala planifrons* (F), *Abh. zool. Mus. Dresden*, Fortschr., **7**, 39 pp.

WASMANN, E. (1900), *Termitoxenia*, ein neues flügelloses, physogastres Dipterengenus aus Termitennestern, *Z. wiss. Zool.*, **67**, 599–617.

WEBSTER, F. M. AND PARKS, T. H. (1913), The serpentine leafminer, *J. agric. Res.*, **1**, 59–87.

WEINLAND, J. (1890), Ueber die Schwinger (Halteren) der Dipteren, *Z. wiss. Zool.*, **51**, 55–166.

WELCH, P. S. (1914), Observations on the life history and habits of *Hydromyza confluens* Loew, *Ann. ent. Soc. Am.*, **7**, 135–147.

WEST, L. S. (1951), *The Housefly, its Natural History, Medical Importance, and Control*, Constable & Co. Ltd., London.

WHEELER, W. M. (1924), The courtship of the Calobatas, etc., *J. Heredity*, **15**, 485–495.

—— (1930), *Demons of the Dust*, Kegan Paul, French, Trubner & Co., Ltd, New York.

WHEELER, W. M. AND WILLIAMS, F. X. (1915), The luminous organ of the New Zealand glow-worm, *Psyche*, Cambridge, **22**, 36–43.

WHITFIELD, F. G. S. (1925), The relation between the feeding-habits and the structure of the mouthparts of the Asilidae (Diptera), *Proc. zool. Soc. Lond.*, **1925**, 599–630.

WHITTEN, J. M. (1955), A comparative morphological study of the tracheal system in larval Diptera, Pt. I, *Quart. J. micr. Sci.*, **96**, 257–278.

—— (1960*a*), The tracheal pattern in selected Diptera Nematocera, *J. Morphol.*, **107**, 233–257.
—— (1960*b*), The tracheal system as a systematic character in larval Diptera, *Syst. Zool.*, 8 (1959), 130–139.
—— (1963), The tracheal pattern and body segmentation in the Blepharocerid larva, *Proc. R. ent. Soc. Lond.* (A), **38**, 39–44.
WHYATT, I. J. (1961), Pupal paedogenesis in the Cecidomyiidae (Diptera). I., *Proc. R. ent. Soc. Lond.* (A), **36**, 133–143.
—— (1967), Pupal paedogenesis in the Cecidomyiidae (Diptera). 3. A reclassification of the Heteropezini, *Trans. R. ent. Soc. Lond.*, **119**, 71–98.
WICKLER, W. AND SEIBT, U. (1972), Zur Ethologie afrikanischer Stielaugenfliegen (Diptera, Acalyptrata), *Z. Tierpsychol.*, **31**, 113–130.
WIGGLESWORTH, V. B. (1933), The function of the anal gills of the mosquito larva, etc., *J. exp. Biol.*, **10**, 1–37.
WILLARD, E. (1914), Two new species of *Platypeza* found at Stanford University, *Psyche*, Cambridge, **21**, 166–168.
WILLE, J. (1922), Biologische und physiologische Beobachtungen und Versuche an der Käsefliegenlarve (*Piophila casei* L.), *Zool. Jb., Allg. Zool.*, **39**, 301–320.
WILLIAMS, F. X. (1939), Biological studies in Hawaiian water-loving insects. Part III. Diptera or flies, *Proc. Hawaiian ent. Soc.*, **10**, 281–315.
WILLIAMS, I. W. (1933), The external morphology of the primitive Tanyderid Dipteron *Protoplasa fitchii* O.S. with notes on the other Tanyderidae, *J. New York ent. Soc.*, **41**, 1–34.
WILLIAMS, T. R. (1968), The taxonomy of the East African river-crabs and their association with the *Simulium neavei* complex, *Trans. R. Soc. trop. Med. Hyg.*, **62**, 29–34.
YOUNG, B. P. (1921), Attachment of the abdomen to the thorax in Diptera, *Mem. Cornell agric. Exp. Sta.*, **44**, 255–282.
ZEVE, V. H. AND HOWELL, D. E. (1962–63), The comparative external morphology of *Trichobius corynorhini*, *T. major* and *T. sphaeronotus* (Diptera, Streblidae), *Ann. ent. Soc. Am.*, **55**, 685–794; **56**, 2–17, 127–138.
ZUMPT, F. AND PATERSON, H. E. (1953), Studies on the family Gasterophilidae, with keys to the adults and maggots, *J. ent. Soc. S. Afr.*, **16**, 59–72.

Order 27

LEPIDOPTERA (BUTTERFLIES AND MOTHS)

Insects with 2 pairs of membranous wings; cross-veins few in number. The body, wings and appendages clothed with broad scales. Mandibles almost always vestigial or absent, and the principal mouthparts generally represented by a suctorial proboscis formed by the maxillae. Larvae eruciform, peripneustic, frequently with 8 pairs of limbs. Pupae usually adecticous and more or less obtect, and generally enclosed in a cocoon or an earthen cell; a few primitive forms decticous and exarate. Wing tracheation complete.

Lepidoptera are the most familiar and easily recognizable of all insects, and it is in this order that coloration has reached the highest degree of specialization. These insects have always been popular objects for study, and more than 100 000 species have been described. Staudinger and Rebel (1901) enumerated over 9500 Palaearctic species which are represented by about 2400 in the British Isles.

On the whole the imagines exhibit a remarkable constancy as regards their fundamental structure, and this uniformity has led to great difficulties in evolving a division of the order into major groups for classificatory purposes. On the other hand, the more superficial or adaptive characters exhibit almost endless variation in the larvae. As might be anticipated from this structural similarity, the habits of these insects are remarkably uniform. The imagines live entirely upon the juices of flowers, over-ripe fruit, honey-dew and other liquid substances: in a considerable number of species the mouthparts have atrophied. The larvae possess masticatory mouthparts and differ from those of other orders in feeding, with but few exceptions, entirely upon phanerogamic plants.

Economically Lepidoptera are of a great importance in the larval stage. The majority of injurious species devour the foliage and shoots of trees and crops; a smaller number bore into the stems or attack underground parts, and several species are injurious to timber; others attack manufactured goods such as carpets, clothing and their like, while a few are extremely destructive to stored products, including grain, flour, etc. Several predacious species are enemies of *Laccifer lacca* and injurious to lac cultivation, and one or two species live in beehives, destroying and fouling the combs. The Saturniidae and *Bombyx mori* (Rolet, 1913), on the other hand, confer a direct benefit upon man from the fact that they yield silk of commercial value.

Among the more recent general works on the order are those of Seitz (1906 *et seq.*) on the larger Lepidoptera of the world, and Hering (1933) on the European forms. Snodgrass (1961) also provides a very good short introduction. The world's species are listed in the catalogue edited by Wagner (1911, etc.) and those of the Palaearctic region by Staudinger and Rebel (1901). The leading treatises on the British species are those of Meyrick (1928), Barrett (1893–1907) and Tutt (1899–1909). The work of the last-mentioned author contains a great deal of biological information but was not completed. Vorbrodt and Müller-Rutz (1911–14) have dealt with the Swiss species. Works on the butterflies (that is superfamilies Hesperioidea and Papilionoidea) are particularly numerous: the Swedish and Italian species have been dealt with by Nordström, Wahlgren, and Tullgren (1935–41) and by Verity (1940–50), respectively; those of North America by Edwards (1868–97) and Clark (1932), both works also containing much general information. Among numerous other works, the volumes by Godman and Salvin (1879–1901) on Central America, by Corbet and Pendlebury (1956), Talbot (1939, 1947), and by Woodhouse and Henry (1942) on the Oriental region, and by Common and Waterhouse (1972) on the butterflies of Australia, are important.

Head-capsules of Lepidopterous larvae have been found in Cretaceous amber (Mackay, 1970) and the evolution of the order probably paralleled that of the flowering plants.

The Imago

EXTERNAL ANATOMY

The Head (Fig. 488) – The greater part of the head is formed by the *fronto-clypeus* and large globular *compound eyes*. The *ocelli* are two in number and lie close behind the eyes: they are seldom conspicuous, and generally much concealed by scales or often apparently absent though they may be concealed within the head (Eaton, 1971; Dickens and Eaton, 1973). In some families a pair of sensory organs known as the *chaetosemata* (Jordan, 1923; Eltringham, 1925*c*) are also present. The anterior region of the head (Short, 1951) is occupied by the large *fronto-clypeus* which is frequently delimited dorsally by a *transfrontal sulcus* (Duporte, 1956). The labrum is narrow and pointed in *Micropterix* and its allies but forms a short transverse plate in other Lepidoptera. It is provided with a small pointed median projection which is usually regarded as an extension of the *epipharynx* and often also with lateral ones known as *pilifers*. Between the fronto-clypeus and the eyes are the narrow *genae* and, when mandibular rudiments are present, they either articulate or fuse with the latter sclerites. The *antennae* (Jordan, 1898) are composed of an indefinite number of segments and vary greatly in length

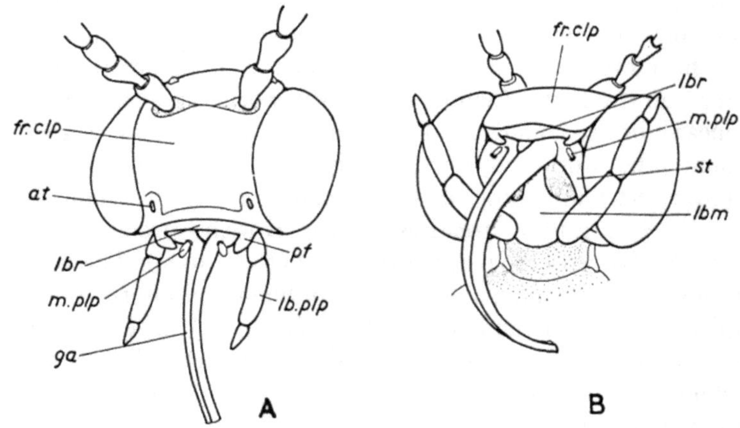

FIG. 488 Head and mouthparts of Lepidoptera. A. Frontal view (*after* Snodgrass, 1935). B. Ventral view (based on Weber, 1933)

a.t, anterior tentorial pit; *fr.clp*, fronto-clypeus; *ga*, galea; *lbm*, labium; *lbr*, labrum; *lb.plp*, labial palp; *m.plp*, maxillary palp; *pf*, pilifer; *st*, stipes.

and structure. In the male they frequently show an increased development as compared with the female which is particularly well exhibited in the Saturniidae. They are generally scaled dorsally and very often ventrally also: in some cases scales are absent as in the Saturniidae and many butterflies.

Mouthparts – In the majority of Lepidoptera, mandibles are totally wanting and the maxillae are highly modified to form a suctorial proboscis. The latter is composed of the two greatly elongated galeae, each being channelled along its inner face, and the two are held together by means of hooks and interlocking spines. In this manner the combined grooves form a tube through which liquid food is imbibed. The laciniae are either entirely atrophied or, according to Berlese, rudiments thereof may be embodied in the base of the proboscis. When fully developed, the maxillary palpi are 5- or 6-segmented and usually more or less folded, as in the Tineidae; in the great majority of Lepidoptera they are either much reduced or wanting, their functions presumably being assumed by the labial palpi. Among Noctuidae they are 2- to 3-segmented; in the Sphingidae, Papilionoidea and most Geometridae they are single-segmented (Walter, 1884). The labium is reduced to a small plate on the ventral aspect of the mouth: its palps are normally 3-segmented and vary greatly in size, shape and scaling. A hypopharynx is present on the floor of the mouth and in *Danaus* it is provided with gustatory sensilla.

When not in use the proboscis is spirally coiled and stowed away beneath the thorax: it presents an extraordinary variation in length, attaining its maximum in the Sphingidae. In *Danaus*, according to Burgess (1880), each half of the proboscis is seen to be composed of an immense number of sclerotized rings, which are incomplete since they are absent from its inner or grooved aspect. These rings are separated by intervening bands of membrane which admit of the spiral coiling of the organ. Each ring is made up of a row of quadrangular plates which are provided with spine-like processes directed towards the proboscis channel, hence the plates are somewhat nail-like in form. Scattered over the surface of the proboscis, and more especially at the apex, are small circular plates each bearing a minute central papilla, which are perhaps tactile in function (Börner, 1939). According to Breitenbach they are often developed into denticulate spines which enable the proboscis to lacerate the tissues of fruit and imbibe their juices: this condition is particularly well exhibited in *Aletia xylina*. The interior of each half of the proboscis is hollow and occupied throughout its length by a nerve and a trachea, but the bulk of its cavity accommodates two sets of muscles which diagonally cross it.

According to Eastham and Eassa (1955) the proboscis is extended by blood-pressure, supported by a stipital valve (Pradhan and Aren, 1941), rendering the elastic organ dorsally convex. Some contraction of the intrinsic muscles then leads to extension. Retraction and coiling are due to natural elasticity. The evolution of these peculiar mouthparts is discussed by Kristensen (1968*a*) and the musculature and metamorphosis of Lepidopterous head by Eassa (1963*a* and *b*).

In some Lepidoptera (*Orgyia*, *Zeuzera*, etc.) the proboscis is reduced and non-functional, the two galeae remaining separate; in many others the galeae are represented by two minute papillae (*Hepialus*) or entirely atrophied. In the reduced or atrophied condition it is evident that no food can be imbibed and the mouth may be wanting also, as in the Saturniidae (cf. Naumann, 1937, and Gohrbrandt, 1940).

The mouthparts are exhibited in their most primitive form in *Sabatinca* (Tillyard, 1923*b*) where they are clearly of the mandibulate type. The mandibles are functional dentate organs, with evident ginglymus and condyle, and movable by means of well-developed abductor and adductor muscles. In the maxillae cardo and stipes are evident, the galea is short and 2-segmented, the lacinia blade-like, and the palpi are long and 5-segmented. In the labium, however, there is no ligula and lobes formerly regarded as paraglossae are in reality processes of the palpi (Tillyard): the basal sclerites are represented by a single mental plate. The hypopharynx in *Micropterix ammanella* is laterally provided with small accessory pieces which are regarded by Busck and Böving as the superlinguae. In *Eriocrania* the mandibles are non-dentate and in *Mnemonica* they are unsclerotized with the ginglymus and condyle rudimentary: proof that these are true mandibles is afforded by the fact that they lie within those of the pupa. In both the above genera the laciniae are lost, and the 2-segmented galeae are greatly elongated. The terminal segment of the galea of either side is opposed to that of its fellow, thus exhibiting the first step in the formation of the Lepidopterous proboscis. Vestigial mandibles are also present in *Hepialus*, in various Tineoids including *Argyresthia*, *Tinea*, *Tineola* and *Hyponomeuta*, and in Sphingidae (Walter, 1885; Packard, 1895; Petersen, 1900; Short, 1951). Some authors have mistaken the pilifers for the mandibles, but Kellogg shows that both occur together in *Protoparce*.

The maxillae of *Tegeticula* are exceptional in exhibiting sexual dimorphism: in the male they are normal but the galeae are quite separate, and in the female there is an elongate inner lobe often known as the maxillary tentacle. The two latter organs are adapted for holding a large mass of pollen beneath the head: their morphology is doubtful and it has been suggested that they are the greatly produced palpifers.

In the **Thorax** (Fig. 489) (Weber, 1924, 1928) the prothorax is evident in the lower forms but compressed and reduced in all the higher families where it assumes the form of a collar. It frequently carries a pair of small lateral processes or *patagia* which are peculiar to the higher Lepidoptera and appear as thin, lobe-like, erectile expansions, well-developed in many Noctuidae (e.g. *Agrotis*). The mesothorax is the largest and most prominent segment of the three; its tergum consists of a narrow band-like prescutum, a very large, longitudinally divided scutum and a well-developed more or less rhomboidal scutellum. *Tegulae* are particularly well developed and very characteristic of

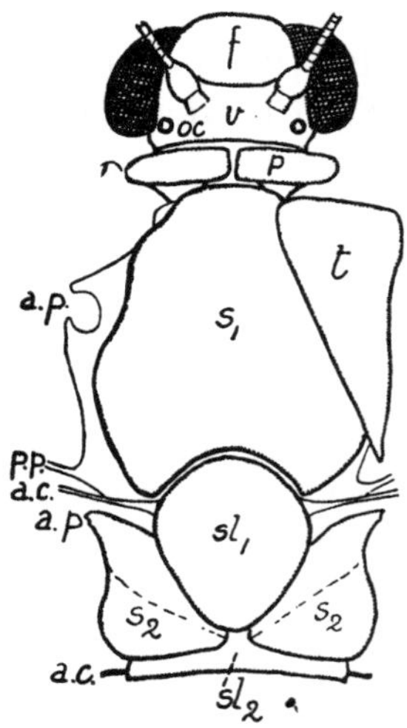

FIG. 489
Dorsal view of head and thorax of *Nactua pronuba* (left tegula removed)

a.c, axillary cord; *a.p*, anterior wing process; *oc*, ocellus; *p*, patagium; *pp*, posterior wing process; s_1, mesoscutum; sl_1, mesoscutellum; s_2, metascutum; sl_2, postnotum; *t*, tegula.

the order; each is carried on a special tegular plate of the notum supported by means of a tegular arm arising from the base of the pleural wing process (Snodgrass). The metathorax is relatively small as compared with the previous segment; Snodgrass finds that in *Phassus* (Hepialidae), however, it is larger and more like the mesothorax than is usual among the higher insects. In most other Lepidoptera it is very much shortened antero-posteriorly and greatly reduced. A postnotum is present in both the meso- and metathorax but largely concealed. Common (1969) describes a roughened area on each side of the metanotum which engages with a roughened area beneath the fore wing. It is found in all the more primitive superfamilies and some of the higher ones. They doubtless correspond with the *cenchri* (p. 1201).

With regard to the **Legs** a meron is present in relation with the meso- and metathoracic coxae and, as a rule, the coxae have but little mobility upon the pleuron, the principal movement of the base of the leg being in the articulation between the coxa and trochanter (Snodgrass). The anterior legs exhibit special features in certain families of butterflies and are reduced and modified so as to become useless for walking, either in the male only (Riodinidae) or in both sexes (Nymphalidae). The anterior tibiae are comparatively short in most Lepidoptera and in certain families they are provided on the inner surface with a peculiar lamellate spur ('epiphysis') which is regarded by Haase as the vestige of an organ formerly developed for cleaning the antennae. The mid and hind tibiae usually have one and two pairs of spurs, respectively. Frequently in the male the posterior tibiae (more rarely the middle pair) are provided with an expansible tuft of hair which is located in a groove and functions as a scent-producing organ. The tarsi are normally 5-segmented, the first segment being much the longest and in the males of certain Lycaenidae it is conspicuously swollen. In the Pieridae the claws are exceptional in being cleft or bifid, and among Lycaenidae either one or both claws are wanting in the male. In some degenerate females of the Psychidae the legs have atrophied.

Wings – The most characteristic feature is the covering of overlapping scales which are, morphologically, flattened and highly modified macrotrichia. Transitional stages between the latter and short broad scales are readily observable and the identity of the two types of structures is clearly established. Thus, in *Prototheora* Tillyard mentions that macrotrichia remain in an unmodified condition on the veins. The scales on the wing-membrane lying closest to the veins are linear and narrow, becoming shorter and broader the further they are away from a vein. Microtrichia or *aculei* are present on the wing-membrane except in most Ditrysia.

The innervation of the wings has been studied by Vogel (1911) who finds that each wing is supplied by three nerve branches whose fibres are ultimately distributed to the various sensory organs present. Vogel (1912) recognizes four types of the latter, each organ having a sensory cell at its base. Possibly tactile are sensory scales and setae, while certain papillae suggest on account of their structure an orientating function. At the bases of the wings are chordotonal organs and, in some cases, a well-developed 'tympanal organ' is associated with them (see pp. 135, 1082). The scales of Lepidoptera are secreted by evaginated and greatly enlarged ectodermal cells – the formative cells of Semper (Fig. 490). Their structure and development have been studied in considerable detail, more especially by Mayer (1896) and Reichelt (1925). Each scale is provided with a short pedicel which fits into a minute socket in the wing-membrane. In the more primitive forms they are irregularly scattered but in the butterflies, for example, a regular arrangement is very noticeable. On its exposed or outer surface, each scale is ornamented with longitudinal ridges or striae, often with transverse trabeculae between them. These ridges are in the form of longitudinal thickenings of the outer scale-wall, and their presence imparts rigidity very much after the manner of the corrugations of a sheet of roofing iron. In many cases these striae are extremely fine, and Kellogg (1894) found that in a species of *Morpho* they are placed from 0·0007 to 0·00072 mm apart, or at the rate of 35 000 to the inch,

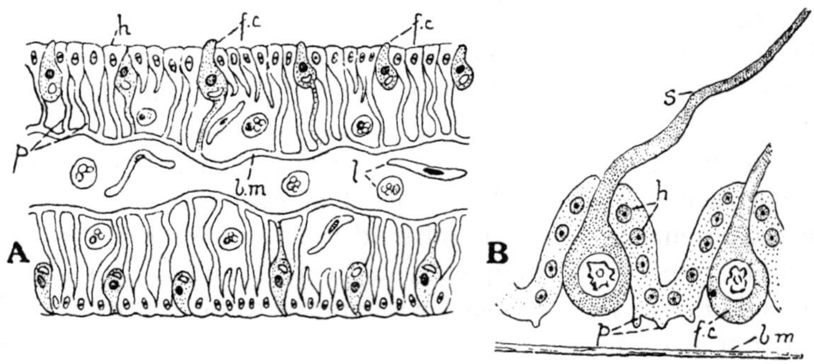

FIG. 490 A, portion of a young pupal wing of *Nymphalis antiopa* in longitudinal section. B, the same of *Danaus plexippus*, about 8 or 9 days before emergence

h, epidermis; *f.c*, formative cell of scale; *l*, leucocytes; *bm*, basement membrane; *p*, processes of epidermal cells; *s*, developing scale. *After* Mayer, *Bull. Mus. Harvard*, 1896.

and are responsible for producing beautiful iridescent colours (see p. 19). Seen in microtome sections scales are greatly flattened hollow sacs (Fig. 492) strengthened by minute transverse bars. Although they may only contain air, in the majority of cases a layer of pigment is enclosed between the two walls. In surface view they exhibit a wide range of variation of both form and sculpturing. In the males of various Lepidoptera groups of more specialized scales or *androconia* (plumules) occur on the upper surface of the wings and likewise assume very varied shapes (Fig. 491). They are found either scat-

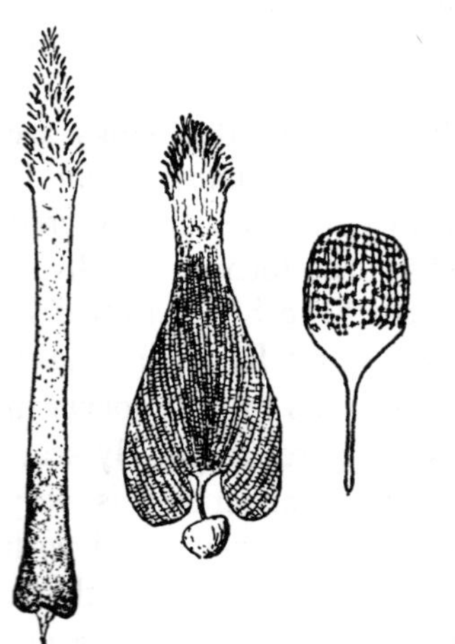

FIG. 491
Androconia of male butterflies

From Comstock *after* Kellogg.

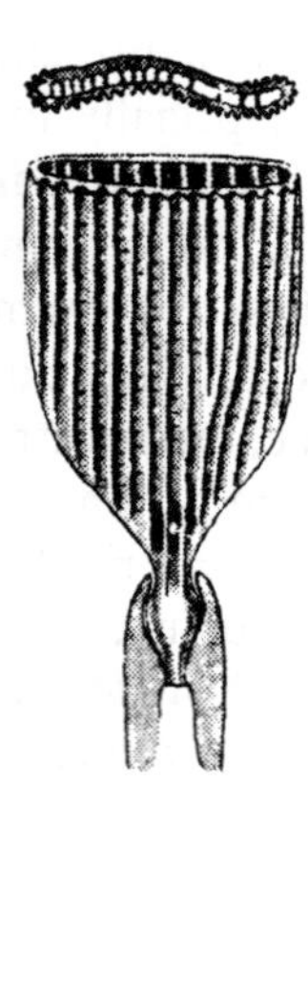

FIG. 492
Upper (i.e. exposed) portion of a scale of *Danaus plexippus* with the distal portion cut away to show the cross bars: above is seen a scale in transverse section.

After Mayer, *loc. cit.*, 1896.

tered over portions of the wings, or in limited areas such as the 'brand' or discal patch of *Thymelicus*, the discal patch of certain Lycaenids, as well as on folds of the wings and other situations. Physiologically they are scent scales which serve as the outlets of odoriferous glands (Thomas, 1893); they are often fringed distally, with each tip of the fringe finely divided, thus probably ensuring the ready diffusion of the odour so characteristic of many Lepidoptera. Among the Danaine butterflies (Nymphalidae) a glandular scent patch is present on each hind wing and the odoriferous secretion is exuded at the surface of the wing by means of cuticular 'cups'. These latter are provided with a covering membrane pierced in the centre by a minute pore. Each cup is protected by a small scale differing from normal wing-scales in size and shape (Eltringham, 1915). In *Amauris niavius* the insect has been observed to brush the odoriferous area with the anal tuft of hairs which thus acquires some of the characteristic odour. Included in the anal tuft are numerous delicate filaments having the property of breaking up transversely into minute particles thus forming a kind of dust which assists in the diffusion of the scent. It is noteworthy that Dixey has shown that in certain Pieridae an alcoholic extract may be made from the wings and it possesses the same odour as the species concerned.

With regard to the venation (Figs. 493, 494, 495) wherever specialization is evident it has been the result of the atrophy or coalescence of veins and not by addition. Throughout the order the principal cross-veins are few in number and vein M_4 is distally fused with Cu_{1a}. The researches of Tillyard

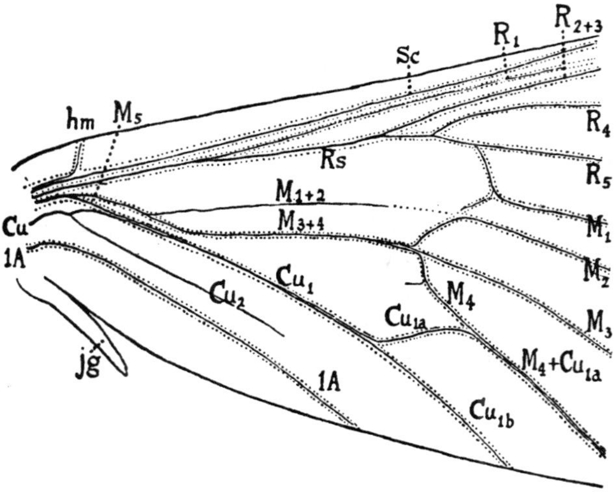

FIG. 493 Basal part of fore wing of *Charagia* (Hepialidae) showing venation (double dotted lines) and tracheation × 4

(Most authors do not recognize M_5 and regard M_4 as obsolete)

After Tillyard, *Proc. Linn. Soc. N.S.W.*, 44, pt. 3.

(1919*a*) provide strong evidence indicating that 1A of Comstock is in reality Cu_2, a conclusion which has been adopted in the present work. One of the most characteristic features of the Lepidopterous wing is the *trigamma* or 3-pronged fork, whose prongs are represented by M_3, Cu_{1a} and Cu_{1b} and whose base completes the closure of cell M or its regional equivalent. Among the Cossidae the stem R_{4+5} (chorda of Turner, 1918) divides the cell R into the basal cell 1st R, and an apical cell 2nd R (areole of Turner). In the vast majority of Lepidoptera, however, the stem R_{4+5} has atrophied and also the main stem M. This condition has resulted in the formation of a single enormous discal cell on account of cells R + M + 1st M_2 thus becoming confluent.

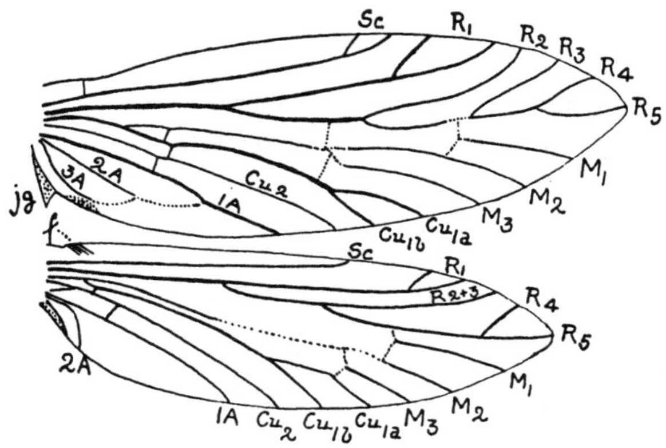

FIG. 494 Venation of Eriocraniidae (*Mnemonica subpurpurella*)

jg, fibula; *f*, costal spines. Adapted from Tillyard, *Proc. Linn. Soc. N.S.W.*, 44, pt. 1.

The most primitive type of venation is found in the family Micropterigidae where that of both pairs of wings is closely alike. Most of the archaic features are exhibited in *Mnemonica* (Fig. 494) in which Sc and R_1 are separate in both pairs of wings, and bifurcated in the fore wings; Rs is 3-branched in the hind wings and the three branches of Cu are complete. Some authors label them CuA_1, CuA_2 and CuP. In the family Hepialidae (Fig. 493) both Sc and R_1, although almost always distinct, are typically unbranched, and there is a considerable reduction or partial atrophy of Cu_2 in one or both pairs of wings.

Among the Ditrysia there is a marked divergence in the venation of the two pairs of wings (Fig. 495) but no annectent type has yet been discovered unless in the Gracillariidae (Busck, 1914). The most ancient type of venation is found among the Cossidae (Turner) which, however, exhibits the characteristic specializations of the hind wing, viz. – the fusion of Sc and R_1, the reduction of Rs, and the coalescence of 1A and 2A. As we ascend the Lepidopterous series the vein Cu_2 disappears from both pairs of wings.

The *wing-coupling apparatus* attains a high degree of specialization among various Lepidoptera (Griffith, 1898; Tillyard, 1918; Braun, 1917; 1919; 1924; Philpott, 1924; 1925). In the Micropterigidae, Eriocraniidae and females of some other Monotrysia, the base of the fore wing is produced into a small lobe or *fibula*. This rests on the *upper* surface of the hind wing and partially engages with the small *costal spines*. In the Hepialidae, the *jugum* replaces and is only an enlargement of the fibula. It also rests on the upper surface of the hind wing but there are no costal spines. A reduced or vestigial jugum occurs in the families of the Microlepidoptera (Sharplin, 1963). In the males of the Nepticuloidea and Incurvarioidea, the fibula and costal

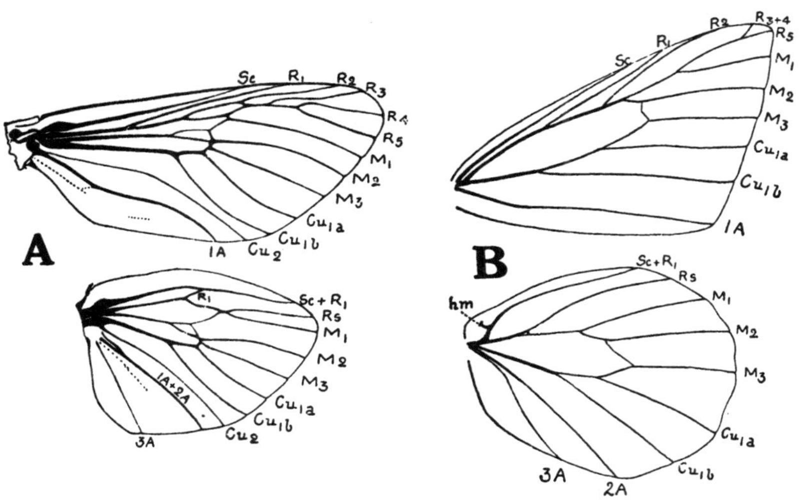

FIG. 495 Venation of Ditrysia. A, *Prionoxystus robiniae* (Cossidae). *After* Comstock, lettering modified. B, *Pieris brassicae* (Pieridae). Original

spines are reduced or absent but there is a true *frenulum* or spine composed of a fused group of bristles arising from a humeral sclerite proximal to the origin of the spines. Among the Ditrysia there are two main types of wing-coupling apparatus, viz. the *frenate* and the *amplexiform*. In the *frenate type* a sexual difference is very noticeable; thus in the male the frenulum consists of a single stout bristle which, however, can be clearly seen to be composed of several setae fused together; in the female the bristles remain separate and vary from 2 to 9 in number. In strongly flying males the frenulum is often large and powerful, while among species in which the females are weak fliers or fly but little the frenulum is correspondingly reduced. In both sexes it arises from a small swelling at the humeral angle of the hind wing, and passes beneath the fore wing where its apex is retained in position by a locking mechanism or *retinaculum*, and in this manner the wings are held together. The retinaculum (Braun, 1919; 1924) varies somewhat in structure, especially among the Tineoidea. Typically, in the female, it is little more than a group of somewhat stiffened hairs or scales arising near Cu_1. In the male, there is usually a strong hook from the underside of Sc but the cubital retinaculum is often present as well (e.g. *Synemon*, Castniidae). In many Sesiidae, in which both sexes are swift fliers, the females exhibit the male type of frenulum and possess the hook-like retinaculum. In the *amplexiform type* the frenulum is lost, and the two wings of a side are maintained together owing to their overlapping to a very considerable degree. This condition is met with for example in the Saturniidae, Lasiocampidae and in all the families of the butterflies. The humeral lobe of the hind wing is enlarged and often strengthened by the development of one or more short humeral veins, and projects far beneath the fore wing. In the Castniidae both the frenulum and humeral lobe are well developed, and from such a condition as this it is evident that the amplexiform type may have been derived through the loss of the frenulum. The cause which necessitated the change is obscure but may perhaps be correlated with a change in the manner of flight. Intermediates between the above two types of wing-coupling apparatus are to be met with; thus in *Bombyx mori*, the frenulum is vestigial and the humeral lobe well developed; this same condition is found among other frenulum-losers such as the Lacosomidae.

In the females of certain Geometridae and Psychidae, also of *Orgyia*, etc., wings are either totally wanting, or reduced to small non-functional vestiges. This flightless condition evidently confines the females to a great extent to their larval food-plants and it is noteworthy that the latter are almost always very common and generally distributed species. The fact that the flightless females of the Geometridae, etc., belong to forms which occur during the colder months of the year has often been commented upon. This peculiarity has been explained as being an adaptation to prevent their leaving the food-plant and perishing owing to inclement weather. Some other explanation, however, needs to be formulated to account for the flightless condition of such eminently summer insects as *Orgyia* and the Psychidae.

The Abdomen consists of ten segments; the 1st segment is reduced and its sternum wanting or wholly membranous, the 7th and 8th are sometimes slightly modified in relation to the genitalia and the 9th and 10th segments are greatly modified in the latter respect. On either side of the metathorax or the base of the abdomen in many Lepidoptera there is a complex organ, the *tympanum*. This structure is well seen in the Geometridae and appears as a bladder-like vesicle closely associated with the 1st abdominal spiracle of its side and certain of the neighbouring tracheal air-sacs. It is innervated from the last thoracic ganglion and from its general structure (Eggers, 1919; 1928; Richards, 1933) and from more recent experimental work (Roeder and Treat, 1957, 1962; Roeder, 1964, 1966) it is known to perceive the high-pitched sounds emitted by bats whose attacks the moths are able to avoid. The position of the organ and the structure of the associated sclerites vary considerably and are of great taxonomic importance (Kiriakoff, 1953). The tympanum occurs only in the larger nocturnal Lepidoptera but not in some of the swifter fliers such as the Sphingidae nor

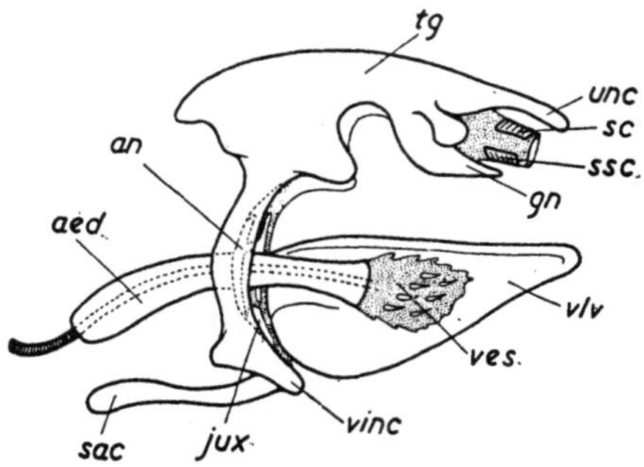

FIG. 496
Diagrammatic lateral view of male genitalia of Lepidoptera, left valve removed (based on Viette, 1948)

aed, aedeagus; *an*, anellus; *gn*, gnathos; *jux*, juxta; *sac*, saccus; *sc*, scaphium; *ssc*, subscaphium; *tg*, tegumen; *unc*, uncus; *ves*, vesica; *vinc*, vinculum; *vlv*, valve.

in more primitive families such as Hepialidae or Cossidae (but see p. 1116). Some Sphingidae, however, have a different acoustic receptor organ formed by the second palpal segment and the pilifer (Roeder and Treat, 1970; Roeder, 1972). Some Arctiidae, such as the Neotropical *Melese laodamia*, can generate ultrasonic signals from a modified region of the metakatepisternum which may perhaps confuse the echo-location of bats (Blest, Collett and Pye, 1963). In some Nymphalids (Swihart, 1967), a small sac at the base of the hind wing and less importantly fore wing can detect sounds.

The nomenclature of the male genitalia has become much involved, but summaries have been published by Viette (1948) and Klots *in* Tuxen (1970). The 9th segment or *tegumen* (Figs. 496, 497) is a narrow ring encircling the apex of the body and its sternal region or *vinculum* is usually invaginated to form a median *saccus* which extends into the preceding segment. A pair of *claspers* or valves (harpes of Pierce) are hinged to the *vinculum* and form the most prominent organs of the external genitalia. The *harpes* are spine-like structures often present in the inner aspect of the claspers. Attached to the hind margin of the 9th tergum is a median process or *uncus* which is usually

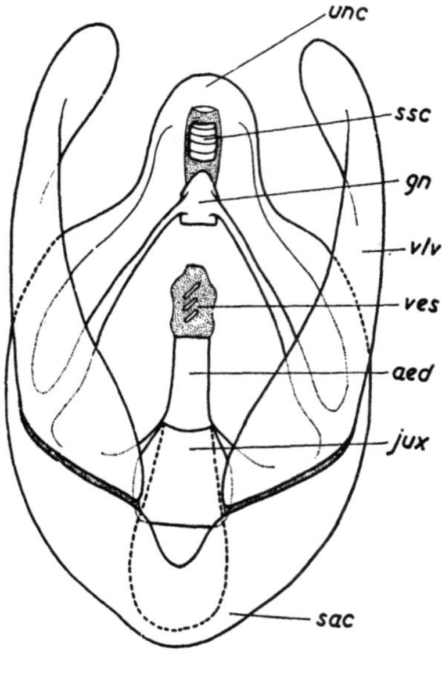

FIG. 497
Diagrammatic ventral view of male genitalia of Lepidoptera
Lettering as in Fig. 496.

hook-like or bifid, and in many Lepidoptera there is a median ventral sclerite or *gnathos* lying a short distance below it. The uncus and the gnathos are usually regarded as the tergum and sternum of the 10th segment, but according to Zander (1903) they are secondary processes, the segment itself remaining membranous. The anus opens just beneath the uncus and between that sclerite and the gnathos. The *aedeagus* is situated below the gnathos and is enclosed in a sheath and at the point where the latter joins the body there is a sclerotized support or *juxta* (penis-funnel or ringwall). For further information on the male genitalia reference should be made to the writings of Zander (1903), Pierce (1909–43) and Mehta (1933). In the female the terminal segments of the abdomen are sometimes attenuated and telescoped, thus functioning as a retractile ovipositor. An exserted sclerotized oviscapt is present in most Monotrysia and in some Ditrysia.

The great majority of female Lepidoptera produce pheromones, generally from glandular cells in the intersegmental ectoderm near the genital orifice. Many of these substances have now been chemically identified (Jacobson, 1973). Less often the males produce sex-pheromones from glands in a variety of situations (e.g. Papilionoidea, Danainae), often associated with brush-like organs (Birch, 1972).

INTERNAL ANATOMY

The Digestive System – The cavity of the proboscis communicates with the pharynx and we owe to Schmitt (1938) an account of the structure of the latter organ. It is an ovoid chamber provided with powerful muscular walls and issuing from between the fibres of the latter are five muscles, which pass outwards to be attached to the head-capsule. When the latter muscles contract the pharyngeal cavity enlarges and a partial vacuum is created; this becomes filled by an ascent of fluid through the proboscis. The walls of the pharynx then contract, thereby forcing the food backwards into the oesophagus, and the closure of a pharyngeal valve precludes the return flow down the proboscis. The oesophagus is a long tube of very narrow calibre (Mortimer, 1965) and, in the more primitive forms, expands distally into a well-developed crop (Monotrysia, Cossidae, Psychidae, many Tineoids, *Attacus, Apocheima*). In other species the crop takes the form of a lateral dilatation connected with the oesophagus by means of a wide-mouthed channel (*Adela*, some Tineoids, Zygaenidae, certain Saturniidae, *Ematurga*, etc.). In the majority of Lepidoptera the crop forms a large food reservoir connected with the fore intestine by a short narrow duct. The stomach is a straight tube of relatively small capacity, and the hind intestine consists of a narrow coiled ileum, a distended chamber or colon, and a short muscular rectum. Le Grice (1968) points out that the S. African, fruit-sucking Noctuid, *Serrodes*, whose habits expose it to desiccation, has a rectal sphincter and many rectal pads which are absent or less numerous in the case of species of more sheltered habitats. Salivary glands take the form of a long coiled filamentous tube on either side, the silk glands of the larva degenerating in the pupa and being no longer evident. The Malpighian tubes are six in number, three of a side opening by a common duct into the commencement of the ileum. Exceptions are found in certain Tineina (*Tinea pellionella*, *Tineola biselliella* and *Monopis rusticella*) which possess only a single pair, and in *Galleria mellonella* there are similarly two vessels but each is irregularly ramified (Cholodkovsky, 1887; Bordas, 1920; Dauberschmidt, 1933).

The Nervous System (see Newport, 1834; Brandt, 1879; Petersen, 1900; Buxton, 1917; Bretschneider, 1921; 1924) exhibits a certain amount of concentration with regard to the ganglia of the ventral nerve-cord. The most primitive condition is found in *Hepialus* in which there are three thoracic and five abdominal ganglia. In the Micropterigidae and also *Tinea pellionella, Cossus, Sesia, Zygaena, Phalera* and *Ematurga* the 4th and 5th abdominal ganglia are fused into a large common centre. The majority of Lepidoptera, however, are characterized by two thoracic and four abdominal ganglia; those of the meso- and metathorax are fused and the abdominal ganglia lie in the 2nd to 6th segments. Nüesch (1957) illustrates the detailed anatomy of the thorax in *Telea*. The Psychidae are primitive but variable: thus Petersen records three thoracic and six abdominal ganglia in the female *Canephora unicolor*, while in *Psyche intermedia* and other species there are four abdominal ganglia in both sexes.

The Dorsal Vessel has been very little investigated: Newport states that in most Lepidoptera there are eight pairs of lateral ostia, and in *Danaus* Burgess states that slight constrictions divide the heart into a number of segments corresponding to those of the abdomen. In *Protoparce*, as Brocher has pointed out, the aorta makes a sharp loop in the thorax and at the apex of the bend it is connected with a pulsatile organ. Recently Hessel (1966, 1969) has examined a considerable number of Ditrysia. He found a pulsatile thoracic chamber in most families though not in the Papilionidae.

The Male Reproductive Organs (Fig. 498A) have been studied by Cholodkovsky (1884) in many species, and also by Stitz (1900), Petersen (1900) and Ruckes (1919). Typically each testis consists of four follicles exhibiting varying degrees of coalescence while among the higher Lepidoptera the two organs are intimately fused into a single median gonad. *Nemophora* (*Nemotois*) is exceptional in that each gonad consists of twenty follicles. Two principal types of reproductive system are distinguishable. (1) The testes are paired and each is enclosed in a separate scrotum. In *Hepialus* the follicles are separate and the gonad presents a digitate appearance: this condition is evidently the most primitive found in the order. In other cases the follicles are compressed together and surrounded by a common scrotum. This type is met with in the Micropterigidae, certain Saturniidae, *Bombyx mori, Maculinea arion, Parnassius* and a few others. (2) The testes are fused and enclosed in a common scrotum: in some cases the paired nature of the gonad is still evident while in others the fusion is complete. This type (no. 2) is the prevalent one, and usually the follicles are spirally wound around the longitudinal axis of the gonad. The organs in *Platysamia cecropia* have been studied by Ruckes and in *Bombyx mori* by Verson. The testes lie in a dorso-lateral position, close to the alimentary canal and just beneath the 5th and 6th abdominal terga. The vasa deferentia are narrow tubes which enlarge proximally to form the vesiculae seminales. Each receives a long filamentous accessory gland but, according to Ruckes, the structure of the latter is not markedly glandular, its walls being provided with longitudinal muscles fibres and it appears probable that the gland serves, along with the vesiculae seminales, as a receptacle for storing the spermatozoa. The vesiculae seminales unite to form a common ductus ejaculatorius which terminates in a bulbus ejaculatorius at the base of the aedeagus.

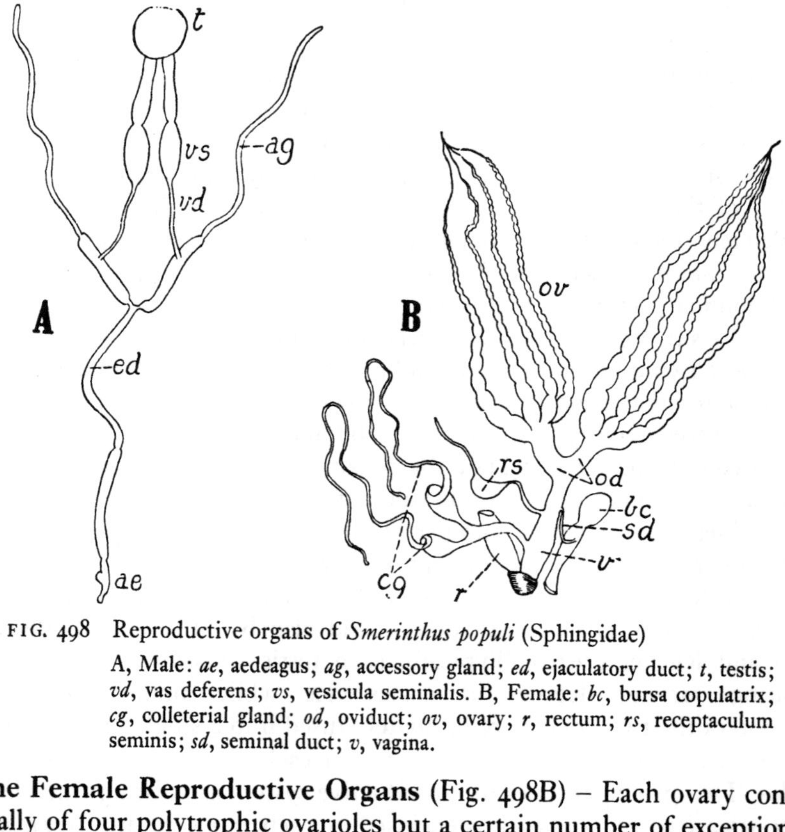

FIG. 498 Reproductive organs of *Smerinthus populi* (Sphingidae)

A, Male: *ae*, aedeagus; *ag*, accessory gland; *ed*, ejaculatory duct; *t*, testis; *vd*, vas deferens; *vs*, vesicula seminalis. B, Female: *bc*, bursa copulatrix; *cg*, colleterial gland; *od*, oviduct; *ov*, ovary; *r*, rectum; *rs*, receptaculum seminis; *sd*, seminal duct; *v*, vagina.

The Female Reproductive Organs (Fig. 498B) – Each ovary consists typically of four polytrophic ovarioles but a certain number of exceptions to this rule are known among the lower members of the order. Thus, there are six ovarioles to each ovary in *Cochliotheca helix*, 10 to 12 in *Adela*, 14 in *Conopia scoliaeformis*, and 12 to 20 in *Nemophora* (= *Nemotois*). The structure of the female genital apparatus is of fundamental importance in the classification of the order (Petersen, 1900, 1904; Busck, 1932; Williams, 1941, 1947; Oiticica, 1948; Bourgogne, 1949; Stekolnikov, 1967; Mutuura, 1972). There may be some differences of opinion owing to the difficulty of assigning sclerites at the posterior end of the abdomen to their correct segment. In the Zeugloptera, there is a single cloacal opening on segment X.

In the Dacnonypha and some Monotrysia there is a single cloacal opening on segment IX or possibly in some forms on VIII. In the Hepialidae and, apparently, the Neopseustidae, there is a separate anal opening (anus) and two adjacent genital pores (copulatory, oviporal) on segment IX or VIII and IX. Finally, in the Ditrysia there are two quite distinct apertures, copulatory on VIII and oviporal on IX, with the anus behind the ovipore. It seems probable that the Hepialid type may eventually be made the basis of a fifth suborder but the structures of some of these forms still seem to be imperfectly understood; Mutuura's valuable account suffers from not indicating the position of the anus. In the Zeugloptera and some Monotrysia, the rectum and the common oviduct join near the aperture so that the *cloacal duct* is short; in other Monotrysia and in the Dacnonypha there is a long duct and the terminal part of the abdomen forms an oviscapt. Where there are two genital pores, a special sperm duct (*ductus seminalis*) or, in some Hepialids, a surface furrow, connects the ductus bursae to the oviduct. In the more primitive condition, as in some Psychidae, this duct is very short and broad; in higher forms it is often very long and narrow. The *receptaculum seminis* in which sperm is stored, joins the common oviduct near the point where the ductus seminalis enters it (Fig. 499). It seems that no living form retains a primitive arrangement, which was probably a single genital aperture on sternite VIII. The backward extension of the common oviduct beyond this may have first occurred as a groove, as is suggested by the ontogeny of these structures (Hatchett-Jackson, 1890; Dodson, 1937). The condition in *Hepialus* is difficult to interpret but appears to be specialized, at least as regards the position of the two apertures.

A pair of ramified or filiform colleterial glands open into bladder-like ducts which communicate with the common oviduct just behind the aperture of the receptaculum seminis. In many species (*B. mori*, etc.) an accessory gland is also present in relation with the latter structure, and the whole organ then resembles a colleterial gland in general appearance and has often been referred to as such. During copulation, a spermatophore is deposited in the bursa copulatrix (Norris, 1932). The muscles in the wall of that chamber probably force the spermatozoa into the ductus seminalis. They subsequently enter the oviduct and then pass up the duct leading into the receptaculum seminis, and are stored in the latter organ until the eggs enter the oviduct for fertilization. For detailed information on the female genital system reference should be made to the works of Eidmann (1929), Petersen (1900; 1904) and Stitz (1901).

The Tracheal System communicates with the exterior by means of usually nine pairs of spiracles, two being thoracic and the remainder abdominal in position: the pair on the 8th segment of the abdomen, although present in the larva, is aborted.

The general **Literature** dealing with the morphology of adult Lepidoptera is relatively small. The principal anatomical treatise is that of Petersen (1900) and a good deal of information on the external structure will

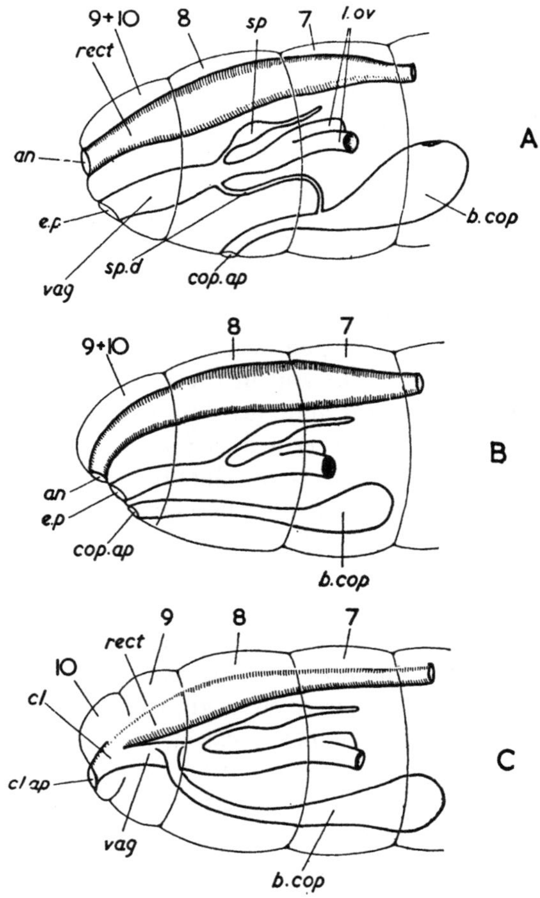

FIG. 499 Relationships of female reproductive ducts and rectum in Lepidoptera. A. Ditrysian type. B. Hepialid type. C. Micropterigid type

an, anus; *b.cop*, bursa copulatrix; *cl*, cloaca; *cl.ap*, cloacal aperture; *cop.ap*, copulatory aperture; *e.p*, egg pore; *l.ov*, lateral oviduct; *rect*, rectum; *sp*, receptaculum seminis; *sp.d*, sperm duct; *vag*, vagina; 7–10, 7th–10th abdominal segments.

be found in Rothschild and Jordan's monograph (1903). The most completely investigated species is *Bombyx mori* whose anatomy has been studied by many workers including Blanc, Verson, Tichomirov and others: for the general structure of *Danaus*, see Burgess (1880), while Brandt has dealt with that of *Hepialus* (1880) and the Sesiidae (1890), and Nigmann (1908) with *Acentropus*, Della Beffa (1938) with Gracillariidae, Madden (1944) with a Sphingid, Freeman (1947) with a Tortricid and Matthes (1948) with the Psychidae. Brock (1971) describes the thoracic structure of many Ditrysia and Bharadwaj, Chandran and Chadwick (1974) summarize much information on cervical and thoracic musculature.

The Egg

The eggs of Lepidoptera (see Tutt, 1899) are roughly divisible into two forms: (1) ovoid or flattened, with the long axis horizontal: in this type the shell is usually only ornamented with rough pittings and rarely with longitudinal ribs; (2) upright and either fusiform, spherical or hemispherical, with the axes either equal, or the vertical axis the longest. The ornamentation is usually more complex and often exhibits a cell-like structure divided by longitudinal ribs.

The *micropyle* is usually placed in a slight depression at one extremity of the horizontal axis of an ovoid type of egg, and at the summit in the upright form. It consists of a number of minute radiating microscopic canals by means of which the spermatozoa gain access into the interior of the egg.

The average number of eggs laid by many species is high, sometimes exceeding 1000 (*Noctua fimbriata, Zeuzera pyrina*), and they are deposited in a great variety of ways and positions. Certain Hepialids, and also *Cerapteryx graminis*, drop their eggs at random among the herbage on which the larvae feed. Others, such as *Malacosoma neustria* and *Alsophila aescularia*, deposit them in orderly necklace-like rings around the twigs of their respective food-plants. Certain Geometridae lay them in imbricate groups, while the Incurvariids are provided with a complex cutting apparatus with which they excise pockets in a leaf. The duration of the egg stage is subject to great variation: in *Sterrha virgularia* (= *Idaea seriata*) it may be as short as two days, but for species which hatch out during the year of deposition 10–30 days may be taken as the usual developmental period. A number of species hibernate in this stage, which is then often of longer duration than the combined larval, pupal and imaginal periods.

The Larva

Lepidopterous larvae have a well-developed head; 3 thoracic and 10 evident abdominal segments. Nine pairs of spiracles, borne respectively on the prothoracic and first 8 abdominal segments, are present. In the *head* (Figs. 500, 501) the median epicranial sulcus is well-developed and the frons is usually represented by a pair of narrow oblique plates termed the *adfrontals*

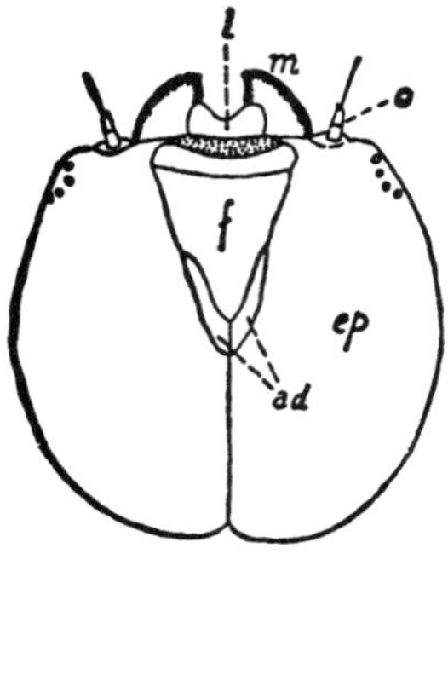

FIG. 500
Macrothylacia rubi, frontal view of head of fully-grown larva (Lasiocampidae)

a, antenna; *ad*, adfrontal sclerites or frons; *f*, clypeus; *ep*, epicranial plate; *l*, labrum; *m*, mandible.

(Short, 1951; but cf. Hinton, 1947). Both clypeus and labrum are evident and the typical number of ocelli is 6 which are situated just behind, and a little above the bases of the short 3-segmented antennae (Dethier, 1941). The mandibles are powerful and adapted for mastication; in sap-feeding larvae, however, they are concerned with the laceration of tissues and may even be wanting (*Phyllocnistis*). The maxilla consists of a cardo and stipes;

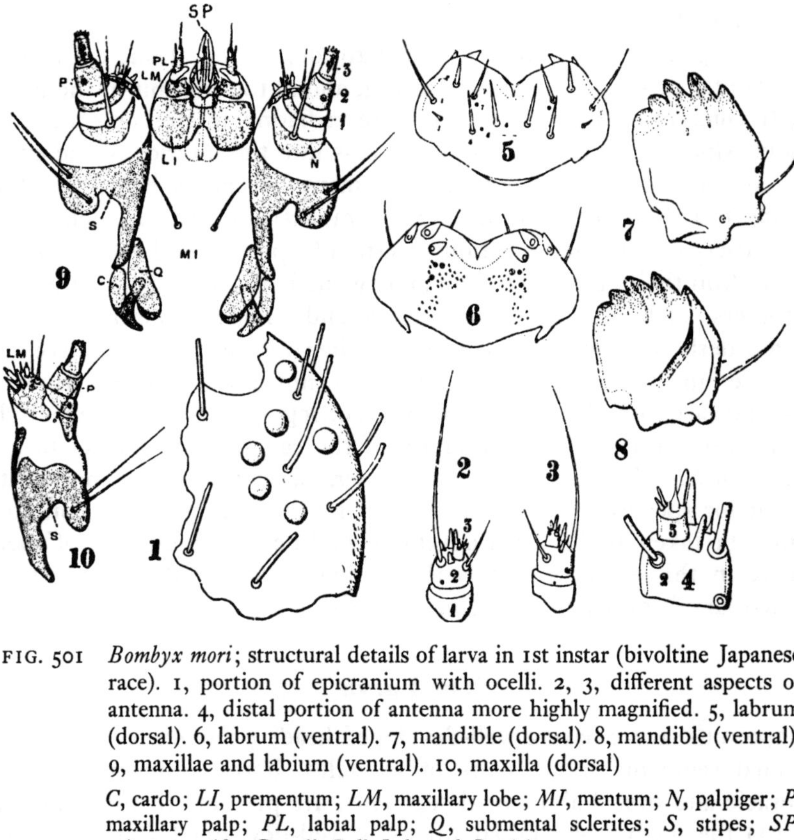

FIG. 501 *Bombyx mori*; structural details of larva in 1st instar (bivoltine Japanese race). 1, portion of epicranium with ocelli. 2, 3, different aspects of antenna. 4, distal portion of antenna more highly magnified. 5, labrum (dorsal). 6, labrum (ventral). 7, mandible (dorsal). 8, mandible (ventral). 9, maxillae and labium (ventral). 10, maxilla (dorsal)

C, cardo; *LI*, prementum; *LM*, maxillary lobe; *MI*, mentum; *N*, palpiger; *P*, maxillary palp; *PL*, labial palp; *Q*, submental sclerites; *S*, stipes; *SP*, spinneret. *After* Grandi, *Boll. Lab. zool. Portici*, 1922.

there is usually a single maxillary lobe and the palpi are 2- or 3-segmented organs. The ventral region of the head, between the proximal portions of the maxillae, is occupied by the labium. The mentum is relatively very large and lightly sclerotized; the submentum is usually divided into a pair of triangular sclerites. Distally, the prementum carries a median process or *spinneret*. The labial palpi usually each consist of a principal cylindrical, and a minute apical segment. On the oral surface of the labium is a median pad or hypopharynx. Paired lobes, which have been interpreted as superlinguae, overlie the sides of the hypopharynx and have been recognized in *Mnemonica* by Busck and Böving (1914), and by De Gryse (1915) and Heinrich (1918) in other Lepidopterous larvae. In some leaf-mining larvae, the mouthparts are much modified (Jayewickreme, 1940).

The *thorax* carries a pair of legs on each segment; these are 5-segmented and the terminal segment or tarsus is provided with a single curved claw. The *abdomen* commonly bears five pairs of so-called 'prolegs' which are present on segments 3 to 6 and on 10: the first 4 pairs may be termed the abdominal feet and remaining pair the claspers. A typical abdominal leg is a fleshy, more or less conical, retractile projection whose apex or *planta* is rounded or flat. The latter is provided with a series of hooks or crochets which aid the larvae in locomotion, and to the centre of the planta is attached a muscle by means of which it can be completely inverted. The arrangement of the crochets is diverse, and the variations present afford important classificatory characters (Fig. 502).

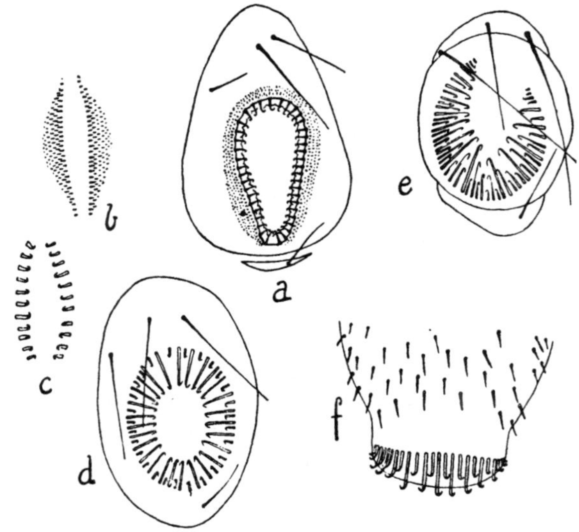

FIG. 502 Arrangement of crochets

a, multiserial circle; *b*, transverse multiserial bands; *c*, transverse uniserial bands; *d*, biordinal uniserial circle; *e*, penellipse; *f*, biordinal mesoseries. Adapted from Fracker, 1915.

In the detailed studies of Fracker (1915) the following terminology is adopted with reference to the arrangement of the crochets. In the most generalized forms the planta bears a complete circle of well-developed hooks, surrounded by several circles of smaller ones. This arrangement is a *multiserial circle* and is found in the Hepialidae, *Yponomeuta*, etc. When the crochets are absent from the mesial and lateral parts of the circle, as in *Adela*, two transverse *multiserial bands* are formed. When the outer circles of smaller crochets disappear we get a *uniserial circle*. The latter occasionally has crochets of uniform length (*uniordinal*), but more usually they are of two lengths alternating (*biordinal*). When a portion of a uniserial circle is wanting, and the remainder is more than a semicircle in extent, we get a *penellipse* as in the Psychidae; the gap, moreover, is variable in position. Frequently more than half the circle may be absent, and a *mesoseries* results, as in nearly all the higher Lepidoptera excepting the Hesperiidae.

Departure from the usual number of abdominal limbs are the rule in certain families. Thus in the Geometridae they are generally present only on the 6th and 10th segments. In the early instars of many Noctuidae the abdominal feet on the 3rd and 4th segments are rudimentary, and the method of progression resembles that of Geometrid larvae; the limbs of those segments generally attain their full development in a later instar. In the Plusiinae and several other subfamilies, however, they are permanently absent and the looping habit is maintained throughout life. Larvae of the Micropterigidae are exceptional in possessing eight pairs of abdominal limbs. At the opposite extreme are certain leaf-mining larvae, including those of *Phyllocnistis* and *Eriocrania*, which are totally apodous.

The armature of the body consists of simple hairs or setae, tubercles of various types, and *verrucae*: the latter are somewhat elevated portions of the cuticle bearing tufts of setae. More rarely the body-wall is produced into spinous processes or *scoli* as in the Saturniidae, or into a median dorsal horn as in the Sphingidae: other modifications are dealt with under the respective families. The setae are arranged in a definite manner, and have been extensively studied by Dyar (1895), Fracker (1915), Gerasimov (1935), Hinton (1946), Mutuura (1956), Hasenfüss (1963, 1969) and Wasserthal (1970). According to Hinton, the setae are of two types, viz. microscopic, probably *proprioceptor* setae and long *tactile* setae which are more numerous. The setae, especially in the first instar, can be homologized in the various families, but the detailed arrangement provides important taxonomic characters. There is also some evidence that setae, attached basally to sensory cells, may mediate the perception of low frequency sounds (Klots, 1969). Before the anus there is sometimes an *anal comb* (Frost, 1919).

Repugnatorial glands are a common feature and there is an extensive literature on the subject. In the Papilionidae there occur very characteristic organs known as *osmeteria*. An osmeterium consists of a bifurcate protrusible sac which is thrust out through a slit in the 1st thoracic segment. It exhales a distinct odour varying according to the species and in some cases is extremely disagreeable. In many larvae, including those of the Nymphalidae, certain Noctuidae and Notodontidae a ventral defensive gland is present in the form of an internal sac opening on to the prothoracic sternum, and is capable of discharging a jet of spray. In the Lymantriidae a pair of eversible glands is present on the dorsum of the 6th and 7th abdominal segments. In many Lycaenidae also there is a dorsal gland on the 7th abdominal segment, its presence being indicated by a transverse slit through which a minute globular vesicle may be protruded. In the Megalopygidae there are lateral abdominal glands permanently everted, and metamerically arranged (Packard). Many larvae obtain protection through the possession of *urticat-*

ing hairs which bristle with minute lateral points. Whether their irritating properties are due to mechanical action alone, or to the presence of a poisonous secretion, has not been satisfactorily ascertained. These urticating hairs are known to most entomologists who have handled larvae pertaining to the Lymantriidae, Lasiocampidae or Arctiidae. Such structures evidently produce marked irritation if they come into contact with the epithelial lining of the digestive tracts of an insectivorous bird or mammal. *Glandular hairs* are present in some larvae and take the form of hollow, smooth setae. Being filled with a poisonous secretion and extremely liable to fracture, they are capable of causing great irritation and smarting when a larva bearing such setae is handled. In certain Megalopygidae these setae are developed into spines and, according to Packard, the secretion is formed in specialized epidermal cells situated at their bases.

A very large number of larvae obtain protection by other means which may be grouped under three chief headings: (1) Concealment. This is evident in case-bearers such as *Coleophora*, the Psychidae, etc., while in *Stigmella*, *Phyllonorycter* and other Tineoids, the larvae are leaf-miners, and in numerous Tortricidae they are leaf-rollers. Others construct silken galleries or spin together adjacent leaves as in *Gelechia*, *Cynthia* and *Drepana*; in certain Lymantriidae, and species of *Yponomeuta*, the larvae live gregariously in dense silken webs. (2) Protective resemblance. This extensive subject has received a good deal of attention from Poulton and other observers. Protection is attained owing to the remarkable resemblance which many larvae exhibit to portions of their food-plant, or other objects in their immediate environment. Perhaps the most striking instances are afforded by Geometrid larvae which bear such a close resemblance to twigs as to render detection often a matter of very great difficulty. The fully-grown larva of *Stauropus fagi* resembles a withered and irregularly curled-up leaf of its food-plant (*Fagus*). Tutt (1899) states that the larva of *Smerinthus ocellatus* bears a remarkable resemblance to a curled apple leaf, its lateral stripes giving an idea of light and shadow on the supposed leaf. The larva of *Anarta myrtilli* with its intricate green pattern is hardly discernible while resting on a twig of heather. A very long list of such instances of protective resemblance might be drawn up, and the phenomenon has probably been induced in the first instance by the presence of chlorophyll in the food-plants, derivatives of which are utilized in the larval coloration. In certain cases experiments of Poulton tend to show that larval coloration may be due to 'phytoscopic', rather than phytophagic influences. In other words, it is the superficial colour of a leaf, for example, rather than its pigmentary substance, that functions as a stimulus in producing differences of coloration under varying environmental conditions. Larvae of *Catocala*, when subjected to green surroundings, become bluish-green, and in a darkly-coloured environment become bluish-grey. Similarly it has been found that those of *Opisthograptis luteolata* and other Geometridae tend to exhibit responses of a like nature. We are unacquainted with the mechanism that produces this result, but it is suggested by Poulton that the reflection of light, from the immediate environment of a susceptible larva, produces a nervous response resulting in a physiological change in the accumulation of pigment within the epiderm. In addition to the writings of this authority an admirable discussion of the subject is given by Tutt (1899). (3) Warning coloration. This is evident in striking colours or patterns which readily catch the eye and their possessors usually feed openly and are distasteful to insectivorous vertebrates.

It has already been mentioned that Lepidopterous larvae feed almost entirely upon Phanerogamic plants. There is probably not a single family of the latter that is not resorted to by one or more species of these insects. In N. America Scudder states that 52 families are represented in the foodplants of butterflies alone. Exceptions to the habit of feeding upon Phanerogamic plants do occur, but they are not numerous; references thereto will be found in the sections devoted to the Noctuidae and Tineoidea.

The number of ecdyses passed through varies greatly in different species and, in some instances, even within the limits of a single species. Edwards (1880) finds that four moults is the usual number in N. American butterflies, with an additional moult in hibernating larvae. Buckler records nine moults in *Nola aerugula* (= *centonalis*), while in *Acronycta* five is the usual number; Gosse (1879) finds the same in *Attacus atlas*, and Soule records a similar number in other Lepidoptera. Species of *Smerinthus* undergo three or four moults, *Sphinx ligustri* six, and three occur in *Callosamia promethea*. *Arctia caja*, on the other hand, may moult seven times – four before hibernation and three after; the number, however, varies between five and eight (Tutt). In a few cases a sexual difference has been noted, the female larva undergoing one more moult than the male, as in *Orgyia*. Chapman observes (1887) that, in *O. antiqua*, larvae which moult three times produce males, those which moult five times produce females, and those which moult four times give rise to imagines of both sexes.

The Internal Anatomy of Lepidopterous larvae is relatively simple. The *digestive canal* is a straight or almost straight tube, from the mouth to the anus (Fig. 503). The oesophagus is short and frequently enlarged posteriorly (in the mesothorax). The stomach is a tube of wide calibre, extending to the hind margin of the 6th abdominal segment or to the middle of the 7th segment, and is lined by a peritrophic membrane. It is provided with conspicuous muscle bands and, in *Protoparce* for example, its walls are transversely constricted by means of the circular fibres and further divided by six bands of longitudinal muscles. Enteric caeca are rare, but in some species small diverticula are present near the anterior end of the stomach. The hind intestine is always extremely short and devoid of convolutions: in some cases it is divisible into three more or less globular chambers separated by constrictions and probably corresponding to the ileum, colon and rectum. In other larvae two dilatations (colon and rectum) only are present. while in further examples the hind gut consists of a single large chamber (Bordas, 1911). With very few exceptions, six Malpighian tubules are present (Poll,

1939), and they open, on either side, by means of a common duct into a small excretory chamber communicating with the hind intestine. The common duct bifurcates and one branch subdivides, thus giving rise to three tubules to a side. The *silk glands* are the most conspicuous appendages of the

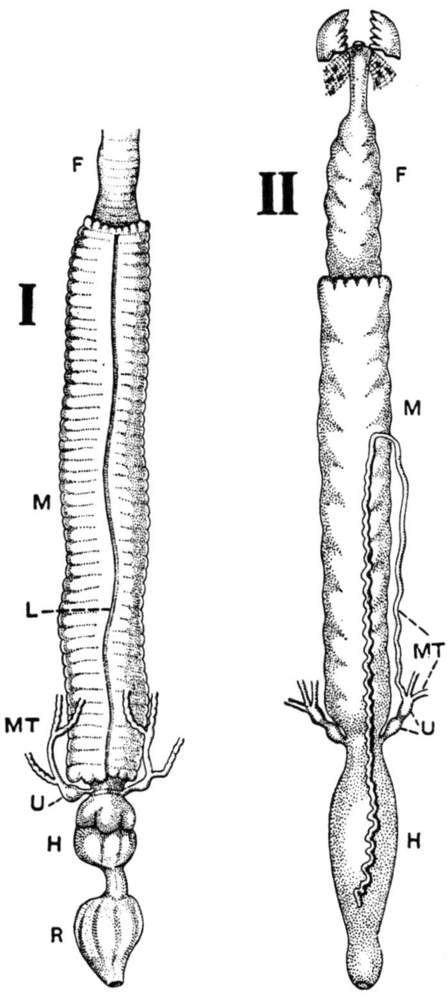

FIG. 503
Alimentary canal of larva of I, *Acherontia atropos* (Sphingidae); II, *Phragmatobia fuliginosa* (Arctiidae)

F, fore gut; *H*, hind gut; *L*, dorsal longitudinal muscle band; *M*, mid gut; *MT*, Malpighian tubules; *R*, rectum; *U*, excretory chamber. *After* Bordas, 1911.

digestive system (Fig. 504). Morphologically they are labial glands homologous with the true salivary glands of other insects. Each gland is in the form of an elongate cylindrical tube of exceedingly variable length, and it lies partly at the side of and partly beneath the digestive canal. These glands are longest in the Saturniidae and Bombycidae: thus in *Telea polyphemus* they measure about seven times the length of the body and are complexly folded, while in *Bombyx mori* they are four times the body length, and folded so as to envelop the hinder region of the gut. Anteriorly, each gland is prolonged to form a duct, and the two latter converge and unite to open at the apex of a median cylindrical organ known as the spinneret. The morphology of this structure has not been satisfactorily ascertained, but it appears to be the highly modified ligula. It will be recalled that the labial glands of insects normally open on the hypopharynx, but in Lepidopterous larvae their aperture has been carried beyond that organ on to the anterior margin of the labium. Histologically, silk glands consist of a single layer of extraordinarily large secretory cells disposed around a central cavity. The cells have large characteristically branched nuclei, and are limited exteriorly by a peritoneal membrane: internally the gland cavity is lined by cuticle,

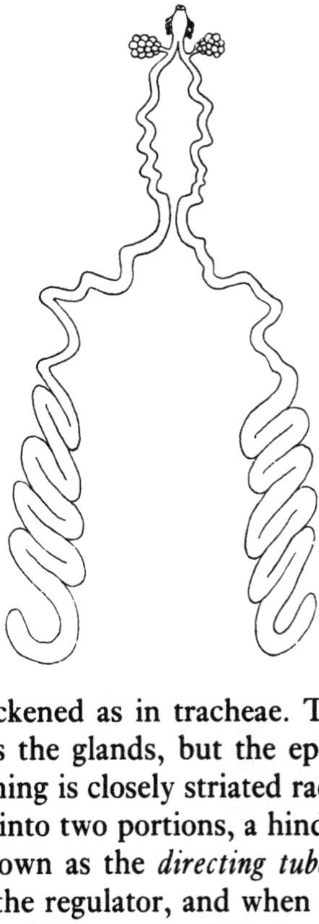

FIG. 504
Spinning glands with small acinose accessory glands of the larva of *Saturnia pyri*
After Bordas, 1910.

spirally thickened as in tracheae. The silk ducts possess the same essential histology as the glands, but the epithelial cells are more flattened, and the cuticular lining is closely striated radially (Fig. 139). The spinning apparatus is divisible into two portions, a hinder part, or *silk-regulator*, and an anterior division known as the *directing tube*. When the regulator muscles contract, silk enters the regulator, and when the tip of the tube touches anything, the silk adheres. Movements of the head then draw it out. The entire spinning apparatus lies within the spinneret, and the thread as it issues from the aperture of the latter is in the form of a double ribbon-like band. Associated with the silk glands in most species is a pair of *accessory glands*, often termed Filippi's glands, though they were recognized by Lyonnet so long ago as 1762 (Bordas). They are paired organs, often voluminous, and each opens by a separate duct into the silk duct of its side. In *Arctia caja* and *Cydia pomonella* they are rudimentary, and reduced to a group of follicles surrounding the silk duct. Among the Sphingidae they are also rudimentary or entirely absent. The function of these glands is to secrete a substance of a liquid or viscid nature which enables the two threads to adhere within the spinneret and, at the same time, facilitates the process of hardening. *Mandibular glands* (Fig. 136) are present in almost all Lepidopterous larvae, and are situated in the thorax one on either side of the fore intestine. They communicate with the buccal cavity by means of a pore placed on the inner side of the base of each mandible. As a rule they are tubular and often of considerable length, but in *Papilio alexenor* and *Stauropus fagi* they are short and sac-like. Histologically they consist of the same layers as the silk glands and their nuclei are lobed or irregular in form. Functionally they are salivary glands and, in some cases, according to Bordas, they may exercise a defensive rôle also. In *Ephestia kuehniella*, the secretion prevents larvae from approaching one another too closely and also indicates to its Ichneumonid parasite, *Venturia canescens*, where to probe with its ovipositor (Corbet, 1971).

The *nervous system* is subject to but little variation. In addition to the usual cephalic ganglia the central nervous system consists of three thoracic and seven or eight abdominal ganglia. The connectives between the meso- and metathoracic ganglia are, typically, double and widely separated, but those uniting the remaining ventral ganglia appear as single cords. As a rule, the 7th and 8th abdominal ganglia are intimately united owing to the elimination of the connective between them (cf. Swaine, 1920–21). In *Sphida* the number of paired nerves arising from the terminal ganglion suggests that three or more nerve-centres have undergone coalescence (DuPorte): in *Cossus* the 7th and 8th abdominal ganglia are separate and united by a short connective (Brandt). The *dorsal vessel* extends from the 8th abdominal segment into the 1st segment, or the commencement of the metathorax, and from there it is continued as the *aorta* into the head. According to Newport there are nine chambers separated by eight pairs of lateral ostia. The *reproductive organs* take the form of a pair of small ovoid bodies situated in the 5th abdominal segment and in close relation with the dorsal vessel on either side. They are present in the newly hatched larvae and undergo a certain amount of differentiation during later instars. The ovaries are slightly larger than the testes and may also be recognized histologically by the rudiments of ovarioles.

The Literature on Lepidopterous larvae is very extensive: larvae of the British species are illustrated by Buckler (1887–99), while for the European species reference should be made to the work of Hofmann (1893) and Blaschke (1914). For a general account of the external structure of the larvae of the order the works of Tutt (1899), Forbes (1910), Werner (1958), Peterson (1959), Hasenfüss (1960) and Mackay (1972) are useful: for the butterflies see Scudder (1889). For the larval characteristics of the different families and diagnostic keys, see Dyar (1894), Forbes (1910) and Fracker (1915). The internal anatomy has been mainly studied in isolated species, notably in *Cossus* by Lyonnet (1762), *Bombyx mori* by Blanc (1889) and others, and *Protoparce* by Peterson (1912). The digestive system and

Malpighian tubules have been extensively studied by Bordas (1910, 1911); and many investigators, more especially Helm (1876), Gilson (1890) and Lesperon (1937), have devoted attention to the silk and other glands. Wailly (1896–97) discusses commercial silk-production. The nervous system has been studied by Newport (1832), Brandt (1879), Cattie (1881) and DuPorte (1915).

The Pupa

The change from the larva to the pupa usually first becomes evident by cessation of feeding. In many cases the larvae desert the food-plant and wander in search of a suitable site in which to undergo the transformation. The contents of the digestive canal are voided and the larval skin loses much of its characteristic colour, becoming darker and wrinkled. The body becomes contracted and distended, the epidermis secretes a fresh layer beneath the old cuticle, and ecdysis is greatly aided by the secretion of the exuvial glands which gradually loosens the two layers. When the latter process is complete, dehiscence of the larval skin takes place along the middle of the dorsal aspect of the thorax, and the exuvia is gradually slipped off from behind, thus liberating the pupa. In the majority of species pupation takes place in a cocoon of some description, which is constructed by the larva. It may be composed of silk as in Bombycidae, Saturniidae, Lasiocampidae, etc.; or of leaves drawn together by a silken meshwork, or of a mixture of silk and various foreign particles. In other cases, as in *Dicranura* and *Cerura*, the cocoon is formed of gnawed fragments of wood agglutinated together by means of a fluid secretion which quickly hardens. Also, in the construction of the earthen cells of many Noctuidae the soil particles are cemented together by a fluid secretion, and no silk appears to be utilized. Among the butterflies the pupa is very frequently naked and protectively coloured, and suspended by the caudal extremity which is hooked on to a small pad of silk: the latter, and the silken girdle which is often present, represent the last vestige of a cocoon. The usual division of the body into head, thorax and abdomen is easily recognized in the pupa and the general external structure has been studied by Poulton (1890–91), Packard (1895), Chapman (1893–96), Mosher (1916) and others (see Figs. 191 and 505).

The Head – The *vertex* forms the dorsal area of the head behind the epicranial suture while the region anterior to the latter is the *fronto-clypeus*. In a few generalized forms, however, the frons and clypeus are separately demarcated. Invaginations of the anterior arms of the *tentorium* are evident as small pores or slit-like openings associated with the lateral margins of the clypeus. The *labrum* is usually very distinct but a clypeo-labral suture appears seldom to be developed: in many families the labrum bears lateral projections or *pilifers* and according to Mosher they are notably conspicuous in the Pyralidae and Papilionoidea. Definite *genae* are rarely evident except among a few of the lower forms. The *eyes* are always prominent and are divided into smooth and sculptured portions, the former being regarded as the true pupal eye. The *antennae* exhibit less marked sexual differences than in the imago and, in *Saturnia pavonia* for example, the pupal differences are extremely small in the two sexes, notwithstanding their divergence in the imago. *Mandibles* are only functional in the Zeugloptera and the Dacnonypha: in *Eriocrania* they are very large and are used by the pupa to cut its way through the cocoon. In other families they are only represented by small elevated areas. The *maxillae* are exceedingly variable, and attain their greatest development in certain Sphingidae where their great length is

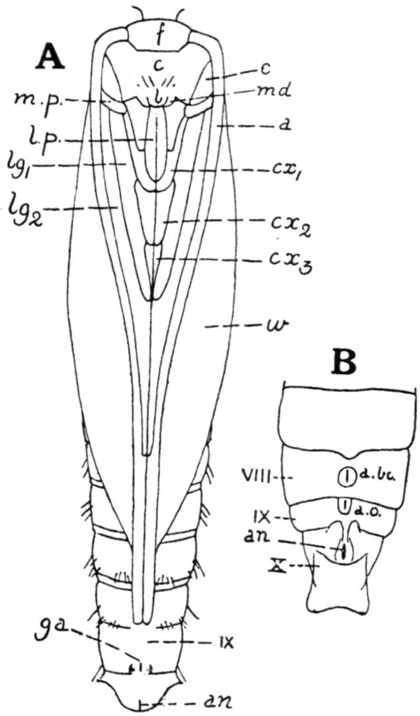

FIG. 505
A, *Tinea pellionella*, male pupa, ventral aspect (adapted from Mosher, 1916). B, *Pieris brassicae*, terminal segments of female pupa, ventral aspect

a, antenna; *a.bc*, aperture of bursa copulatrix; *an*, anus; *a.o*, aperture of oviduct; *c*, clypeus; cx_1–cx_3, coxae; *e*, eye; *f*, frons; *ga*, male genital aperture; *l*, labrum; *l.p*, labial palp; $lg_1 lg_2$, legs; *md*, mandible; *m.p*, maxillary palp; *w*, wing; VIII–X, 8th to 10th abdominal segments.

accommodated by their becoming looped to form the familiar 'jug-handle' appendage. *Maxillary palps* are wanting in certain groups, notably in the Cossidae, Hepialidae and in butterflies. *Labial palps* are visible in many pupae but, in others, they are almost entirely concealed by the maxillae.

The Thorax – The three segments are distinct on the dorsum but ventrally they are concealed by the appendages. The anterior pair of wings almost entirely conceals the posterior pair, except for a narrow strip along the dorsal margin of the latter. Among the apterous or subapterous females of certain genera the pupal wings are likewise less developed than in the male. In *Erannis defoliaria* and *Lycia zonaria* the sexual divergence is but little marked in the pupa, although the female imagines are almost apterous. In such forms as *Orgyia*, and the Psychidae, the degeneration appears to be sufficiently ancient to have allowed a corresponding reduction of the wings of the female pupae. The thoracic *spiracles* consist of a single pair placed between the pro- and mesothorax, towards the dorsal aspect.

The Abdomen – Ten abdominal segments are present and a certain number are always fixed and immovable. The greatest number of free segments is found in the more generalized forms; thus in *Mnemonica* all the segments are movable excepting the last three (Mosher). In the Hepialidae and Psychidae the 1st segment is fixed and segments 2 to 7 are free in the male and 2 to 6 in the female; in the Cossidae the first two abdominal segments are fixed and consequently the movable segments are 3 to 7 in the male and 3 to 6 in the female; in the Noctuidae, Geometridae, Sphingidae, etc., the only free segments are the 4th, 5th and 6th in both sexes, while among certain of the butterflies all the segments are immovable. *Spiracles* are present on the first eight segments: the first pair is usually covered by the wings and the last pair is vestigial. In male pupae the genital aperture is situated on the 9th sternum and in the female there is either a single common aperture on the 8th sternum (10th sternum in *Micropterix* according to Chapman) or, more usually, two apertures which are associated with the 8th and 9th sterna. These openings in some cases become confluent and represent those of the bursa copulatrix and oviduct respectively. The anus is carried on the caudal margin of the 10th segment, and this somite is produced to form the *cremaster*, which is the homologue of the suranal plate of the larva. It is an organ of attachment and exhibits many modifications: it may take the form of a pointed spine or of hooklets, and the latter may be grouped together, or scattered irregularly over the surface of the anal segment. In many of the more generalized families the cremaster is absent, while among the butterflies, with their suspended pupae, it is particularly well developed. One of its functions is to anchor the pupal case when the adult pulls itself out of it.

Internal Structure – The internal anatomy differs in important features both from that of the larva and imago but more closely approaches the latter. The digestive system has undergone extensive modifications as compared with that of the larva; the oesophagus is long and narrow and the stomach greatly reduced in size. The food-reservoir is undeveloped and the hind intestine less convoluted than in the imago. The larval silk glands have atrophied, and the salivary glands of the imago replace them. The changes undergone by the nervous system have been studied in great detail by Newport, and briefly it may be said that it undergoes a gradual process of concentration during about the first 60 hours of pupal life. By that time its whole arrangement is very nearly as it exists in the imago.

Types of Pupae and Method of Emergence from the Cocoon – Chapman (1893*a*) divides Lepidopterous pupae into two main groups, the Incompletae and Obtectae. The *pupae incompletae* have the appendages often partially free and more than three of the abdominal segments are mobile. As pointed out by Hinton (1946), a more fundamental contrast is between the *decticous* (p. 366) *exarate* pupa of the Zeugloptera and Dacnonypha and the *adecticous obtect* or *incomplete* pupa of the remaining Lepidoptera. Dehiscence is accompanied by the freeing of segments and the appendages previously fixed, and the pupae exhibit considerable power of motion, usually emerging from the cocoon to allow of the escape of the imago. They are provided with a varied armature of hooks, processes and spines to facilitate the process. Many species also work their way to the surface of the ground or to the entrance of the larval gallery in the case of those whose larvae are internal feeders. Most pupae incompletae possess some kind of hard process adapted for tearing open the cocoon. This *cocoon cutter*, as it may be termed, is well seen in *Phyllonorycter hamadryadella* and according to Packard there are rough knobs or slight projections answering the same purpose in the Hepialidae, *Megalopyge*, *Zeuzera* and in *Datana*. The *pupae obtectae* represent a more highly specialized type: they are smooth and rounded and the only free segments in both sexes are the 4th, 5th and 6th. Dehiscence takes place by an irregular fracture, the pupa rarely emerges from the cocoon, and a cremaster is generally present. This pupa is prevalent in all the higher Lepidoptera, and exhibits a hard exterior, the appendages being all soldered down to form a smooth surface. The areas which are hidden are covered by a delicate pellicle and there is no separation of the appendages during emergence. Certain species (*Saturnia pavonia*, *Deilephila elpenor*) have retained the habit of pupal emergence, but in other forms the presence of the cremaster and the reduced mobility of the abdominal segments usually preclude it. Many different methods have been adopted to allow of the freeing of the imago. These may consist of weak places in the cocoon, a particular arrangement of the silk to allow of easy egress (*Saturnia pavonia*), a softening fluid applied by the emerging insect (certain Saturniids, *Dicranura*), provisional imaginal spines (Attacine moths), etc.

Classification

The familiar division of the Lepidoptera into Rhopalocera (butterflies) and Heterocera (moths) has little to recommend it other than long usage. Again, the old divisions of Macro- and Microlepidoptera were founded mainly upon the size criterion. The adoption of these two groups led to the inclusion of certain families among the Macrolepidoptera, whereas their true affinities lay with the division which comprised the 'micros' in a literal sense. As generally understood, the 'Microlepidoptera' are the Zeugloptera, Dacnonypha, Monotrysia and the sections Tortricoidea, Pyralidoidea and Tineoidea of the Ditrysia in the present work. Comstock's classification was founded upon the venation and wing-coupling apparatus. He recognized two suborders, the Jugatae and Frenatae – the former possessing a jugum and the latter a frenulum. The presence of a frenulum, however, is too variable even within the limits of a single family to have very much classificatory value. In 1895 Packard laid stress upon Walter's researches on the mouthparts and separated the order into the Lepidoptera Laciniata (or Protolepidoptera) and the Lepidoptera Haustellata, the main feature being the presence of biting mouthparts in the former suborder (which includes *Micropterix*) and their absence in all other Lepidoptera. The Haustellata he further divided into the Palaeolepidoptera (which includes the Eriocraniidae) and the Neolepidoptera. The latter he divided into two sections corresponding in the main to the Pupae Incompletae and Pupae Obtectae of Chapman. In the same year Meyrick brought out a classification based upon the venation in conjunction with other features and in 1895 Hampson published a revision of his earlier scheme (1892) also founded upon the venation. In addition to the above-mentioned systems, the eggs have been examined by Chapman (1896) and Tutt (1899), while classifications based upon larval characters have been advanced by Dyar (1894), Forbes (1910), Fracker (1915) and others: Mosher (1916) has re-examined the pupa from the same standpoint. The recent tendency (Busck, 1932; Börner, 1939; Hinton, 1946; Bourgogne, 1951) to base the primary divisions on the female genital system seems to be well-founded. Four suborders, the Zeugloptera, Dacnonypha, Monotrysia and Ditrysia are recognized in the present work but it might be better to make a fifth, removing the Hepialid-like moths from the Monotrysia. There is still much discussion of the relationships between the members of the lower suborders and the following works may be consulted: Common (1970), Hinton (1946), Kristensen (1967, 1968*a*, *b*, *c*), Mutuura (1972) and Stekolnikov (1967).

There is no agreed classification of the many groups of the Ditrysia whose phylogeny is discussed by Brock (1971). The arrangement adopted is a conservative one. The generic classification is increasingly based on features of the genitalia.

Key to the suborders

1. Adult with functional mandibles, maxilla with lacinia developed, galea not haustellate. Larva with a small transverse post-clypeus and with 8 pairs of abdominal legs, each terminating in a single hook.

Female bursa copulatrix opening into common oviduct which joins the rectum to form a short cloaca with its aperture behind sternite IX; sternites VIII and IX without apodemes. Male sternite IX fused with tergite, in ventral view square, without anterolateral apodemes. Wings with aculei, venation of fore and hind pair very alike, Rs 4-branched, in hind wing R_1 not running into Sc. Fore wing with fibula, hind wing with some costal spines but no frenulum. Pupa decticous, exarate. Suborder **ZEUGLOPTERA**.

– Adult without functional though sometimes with vestigial mandibles, maxilla without lacinia, galea more or less haustellate unless the mouthparts are very reduced. Larva with a triangular post-clypeus and not more than 7 pairs of abdominal legs. 2

2. Female with 1 or 2 genital openings behind sternite IX. Wings more or less distinctly aculeate (except Heliozelidae). Male sternite IX without a saccus. 3

– Female with the opening of the bursa copulatrix on sternite VIII and that of the egg-pore on sternite IX. Wings not or, rarely, a little aculeate, venation of fore and hind wings different, in latter R_1 running into Sc and Rs a single vein (except in a few Gracillariidae). Fore wing without jugum or fibula, hind wing with frenulum or else the coupling is amplexiform. Male sternite IX usually U- or V-shaped, often produced into a large saccus. Larva with not more than 5 pairs of crochet-bearing abdominal legs. Pupa obtect, without functional mandibles. Suborder **DITRYSIA.**

3. Venation of fore and hind wings much alike, hind wing with Rs 3- or 4-branched, R_1 separate from Sc, frenulum not developed. 4

– Venation reduced, different in fore and hind wings, hind pair with Rs unbranched and R_1 coincident with Sc. Male with rudimentary fibula on fore wing and strong frenulum on hind; female with stronger fibula but hind wing with no frenulum but with a more distal group of costal spines. Pupa obtect, mandibles rudimentary. Female with bursa copulatrix opening into the common oviduct which joins the rectum to open behind sternite IX. Suborder **MONOTRYSIA.** Superfamilies **Nepticuloidea, Incurvarioidea.**

4. Female with bursa copulatrix opening into the common oviduct which joins the rectum to form a long cloacal duct with a single aperture behind sternite IX; sternites VIII and IX with long apodemes. Male sternite IX nearly square with 2 short anterolateral apodemes. Fore wing with fibula, hind wing with some costal spines. Larva apodous. Pupa exarate, with hypertrophied mandibles. Suborder **DACNONYPHA.**

– Female bursa copulatrix, common oviduct and rectum all opening separately behind sternite IX; sternites VIII and IX without apodemes. Male sternite IX a small widely U-shaped sclerite. Fore wing with a strong humeral veinlet and a long jugum, hind wing without costal spines, aculei not numerous. Larva subterranean, with 5 pairs of crochet-bearing abdominal legs. Pupa obtect, with rudimentary mandibles. Suborder **MONOTRYSIA.** Superfamily **Hepialoidea.**

Suborder ZEUGLOPTERA

Adult with functional mandibles, lacinia developed, galea not haustellate.

FAM. MICROPTERIGIDAE. *Mouthparts well developed, ocelli present, tibial spurs present, wing-coupling apparatus fibulate, without a frenulum.* This family is of great importance from the standpoint of phylogeny as it includes the most primitive of all Lepidoptera. The family is widely spread but with only a few dozen species (5 in Britain; Heath, 1958). They are small diurnal moths with a wing-expanse sometimes less than 7 mm, and rarely exceeding 15 mm. The fore wings are ovate-lanceolate with metallic colouring. Like many ancient groups, they enjoy an extremely wide distribution, but the family has probably yet to be identified in many parts of the world. The well-known British genus is *Micropterix* (*Eriocephala*), and *M. calthella* is a common insect during late spring in many parts of the British Isles. The New Zealand genus *Sabatinca* exhibits the most primitive venation, which is almost identical with that of the Trichopteran *Rhyacophila*. Functional mandibles and laciniae are present and the galeae are free (Fig. 506), there being no proboscis (Hannemann, 1956). As in all members of the family the ligula is atrophied and the labial palps are 3-segmented organs. These insects are pollen feeders and use their maxillae for the purpose. The male genitalia are also unusual, especially in their musculature (Hannemann, 1957). The larva of *Micropterix* is usually stated to feed on moss but apparently mainly eats dead leaves and is characterized by the presence of eight pairs of abdominal limbs (see Chapman, 1894; Martynova, 1950; Lorenz,

1961). These appendages closely resemble the thoracic legs in being jointed and each is terminated by a claw. The mouthparts exhibit both lacinia and galea and the five ocelli are grouped into a sort of compound eye. The body bears eight rows of metamerically arranged globose processes. The larva of *Sabatinca* (Tillyard, 1923*a*) lives among liverworts and has a similar number of reduced abdominal limbs: the pupa is characterized by the possession of functional mandibles. According to Yasuda (1962), the larvae of *Neomicropteryx* also associates with a liverwort in Japan.

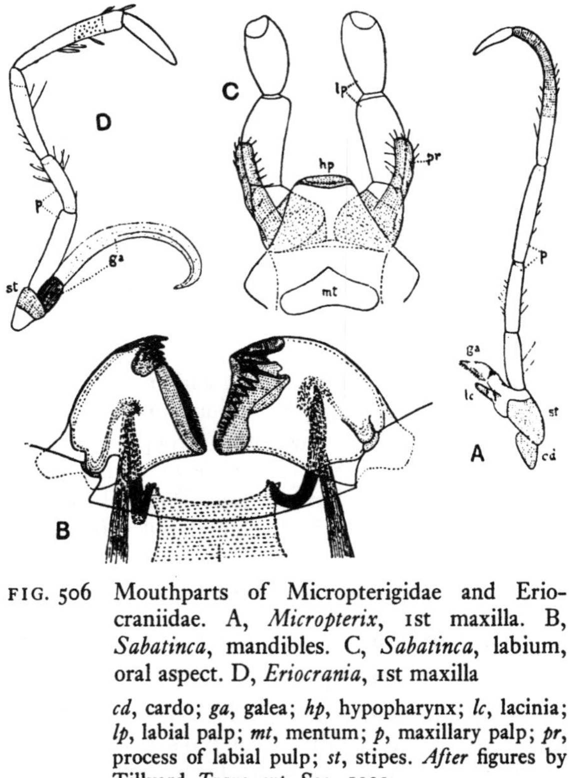

FIG. 506 Mouthparts of Micropterigidae and Eriocraniidae. A, *Micropterix*, 1st maxilla. B, *Sabatinca*, mandibles. C, *Sabatinca*, labium, oral aspect. D, *Eriocrania*, 1st maxilla

cd, cardo; *ga*, galea; *hp*, hypopharynx; *lc*, lacinia; *lp*, labial palp; *mt*, mentum; *p*, maxillary palp; *pr*, process of labial pulp; *st*, stipes. *After* figures by Tillyard, *Trans. ent. Soc.*, 1923.

Much difference of opinion has been expressed with regard to the systematic position of the Micropterigidae, and their affinities are fully discussed by Tillyard (1919*a*). Both the latter observer and Meyrick regard these insects as being true Lepidoptera, Comstock considers that they are terrestrial Trichoptera while Chapman (1917), followed by Hinton (1946), put the genus *Micropterix* into an independent order – the Zeugloptera. Tillyard enumerates four salient differences between this family and the Trichoptera, viz. M_4 is not present as a separate vein in the fore wing whereas it exists in archaic Trichoptera; the pupal wing-tracheation is complete whereas in Trichoptera it is reduced to two tracheae only; the characteristic Trichopterous wing-spot is absent: and broad scales with numerous striae are present, whereas scales only appear in a few isolated and highly specialized Trichoptera, and then only of narrow primitive form with few striae. Most authors, however, have laid insufficient stress on the primitive mouthparts of both adult and larva and on the peculiar, probably specialized, female genitalia.

Suborder **DACNONYPHA**

FAM. ERIOCRANIIDAE. The adult moths (Kristensen, 1968*b*) have lost long. Wings aculeate, venation of the fore and hind wings very similar. Pupal mandibles hypertrophied, larva apodous.

Superfamily **Eriocranioidea**

FAM. ERIOCRANIIDAE. The adult moths (Kristensen, 1968*b*) have lost the laciniae and the galeae are adapted to form a short proboscis. Mandibles are frequently erroneously stated to be absent: they are reduced though distinct, and are non-denticulate (Walter, 1885). They are visible within the bases of the pupal mandibles and possess strongly developed abductor and adductor muscles identical with those of the pupa (Busck and Böving). Ocelli are present. The larvae of *Eriocrania* are apodous leaf-miners in birch, hazel, oak and chestnut: the head is very small and partly hidden by the large prothorax and the usual number of spiracles are present. Pupation takes place in a tough cocoon of silk and earthen particles and the pupa closely resembles that of a Trichopteran. It is of the typical exarate type, with the appendages free and the abdominal segments movable. The most conspicuous organs are the long curved serrated mandibles which are used to rupture the cocoon and aid the pupa in making its way to the surface of the soil. For the metamorphosis and detailed structure of all stages of *Mnemonica*, see Busck and Böving (1914); and for the pupa of *Eriocrania*, see Chapman (1893*b*).

The **Mnesarchaeidae** are much more specialized and are represented by the New Zealand genus *Mnesarchaea*. Ocelli and mandibles are wanting, the maxillary palps are 3-segmented only, laciniae are absent and the galeae form a rudimentary proboscis used as a sucking-organ. According to Mutuura (1972) the female has copulatory apertures as in the Hepialids. Their metamorphoses are unknown. The **Neopseustidae** are rare and little known but Davis (1975) recognizes 3 genera and seven species, mostly from the Indo-Chinese region but one from Chile. Davis would place them in a separate superfamily. The maxillary palps have five segments and the ocelli are absent (Kristensen, 1967, 1968*c*; Mutuura, 1971). The **Agathiphagidae** with ocelli absent have the mandibles much more like those of the Zeugloptera but the mid tibia has two pairs of spurs. The larva feeds on *Agathis* (Dumbleton, 1952; Kristensen, 1967). The **Lophocoronidae** (Common, 1973) have only four maxillary palp segments. There are three Australian species but the female and the larva are unknown.

Suborder **MONOTRYSIA**

Female with two genital openings and an anus on or behind sternite 9 or with a cloacal opening on the same segment. *Either* fore and hind wings with similar venation and a jugum present *or* very small moths with reduced venation in the hind wing and a frenulum, at least in the male. Pupa adecticous, larva often with crochet-bearing legs but sometimes apodous.

Superfamily **Hepialoidea**

Venation of fore and hind wings similar. Female with two genital openings on segment IX.

FAM. HEPIALIDAE (Swift Moths). *Antennae very short, mouthparts vestigial. Wing-coupling apparatus of jugate type, the jugal lobe elongate and resting upon the hind wing. Tibial spurs absent.* A family comprising about 300 species which are widely distributed but best represented in Australia (Tindale, 1932–42) and New Zealand (Dumbleton, 1966). It is a peculiarly isolated group and although primitive in many features of the external and internal anatomy it is specialized in certain others. The significance of the peculiar female genitalia (p. 1087) is uncertain. The species are extremely rapid fliers and vary greatly in size: some are relatively gigantic, attaining a wing-expanse of about 180 mm. Although the five British representatives are sombre-coloured insects certain of the great Australian and S. African forms (*Charagia, Leto*) are magnificently decorated with green and rose or adorned with metallic markings. The European species are crepuscular, or fly before dusk, and in two cases at least the mating habits are exceptional in that the female seeks the male. In *Hepialus humuli* the male is commonly white and is readily sought out by the female: in *H. hectus* the female discovers the male by means of an odour diffused by the latter. The larvae are subterranean, feeding upon roots, or are internal wood feeders. Those of several European species are described by Fracker (1915), and Quail (1900) and Evans (1941) have contributed observations on the metamorphoses of certain Australian forms. They are elongate, devoid of colour pattern and both tufted and secondary setae are wanting. The crochets are disposed in a complete multiserial circle. The pupae are unusually elongate and active and are armed with spines, toothed ridges and cutting plates on the abdominal segments, which are special adaptations for making their way to the surface. The 2nd to 6th abdominal segments are free in the female, and the 7th also in the male (see Packard, 1895).

The **Prototheoridae**, with seven S. African and Australian species, is related to the Hepialidae but has somewhat more primitive venation. The Australian and Oriental **Palaeosetidae** are also related to the Hepialidae. They have reduced maxillae.

Superfamily **Nepticuloidea**

Wing-venation reduced, especially in hind wing. Fore wing usually with no closed discal cell. Male with frenulum. Female with short cloacal duct and fleshy ovipositor. First antennal segment expanded into an 'eyecap'.

FAM. NEPTICULIDAE. *Nepticula* and its allies include the smallest of the Lepidoptera, *N. microtheriella* having a wing-expanse of only 3 to 4 mm. The wings are clothed with aculei and are narrowly lanceolate, with a peculiar venation unlike that of all other Lepidoptera (Fig. 507C). A jugum is present on the fore wing in the female together with a row of hooked spines on the hind wing: in the male a true frenulum is present (Braun). The larvae are mostly leaf-miners and are devoid of jointed legs or crochets: two pairs of leg-like swellings are present on the thorax and similar structures on the 2nd to 7th abdominal segments. There are 67 British species, and it is nearly world-wide in range.

The family **Opostegidae** has the wing-venation (Fig. 507B) extremely reduced with no cell and the larva is an apodous miner in stems or bark. The group is widespread and there are four British species.

Superfamily **Incurvarioidea**

Wing-venation reduced in hind wing. Fore wing with a closed discal cell. Male with frenulum. Female with long cloacal duct and sclerotized ovipositor. First antennal segment not expanded.

FAM. HELIOZELIDAE. These are the only Monotrysia in which the scales of the head are depressed, as in many of the higher Tineoids. There are about 100 widely distributed (not in N. Zealand) species. The larvae are apodous leaf-miners.

FAM. INCURVARIIDAE. In *Adela* and *Nemophora* the moths are often metallic and fly in sunshine. In the males, the antennae are longer than in the females and often many times longer than the insect: the eyes in the males are often greatly enlarged and approximated dorsally. The larva in later instars lives in a case and feeds on vegetable refuse. *Incurvaria* and its allies do not have long male antennae and the haustellum is reduced. The larva is a leaf-miner or lives in a case on vegetable refuse (Saalas, 1935–36).

FAM. PRODOXIDAE. In this family there is an intimate relationship between the moths and species of Yucca. The female *Tegeticula yuccasella* is associated with *Y. filamentosa* and by the aid of her specially modified mouthparts collects the pollen and applies it to the pistil in which she has deposited an egg. In this way development of the fruit, upon which the larva feeds, is ensured. In *Prodoxus* the above relationship does not exist and the larva feeds in the stems or on the seed-pod (Powell and Mackie, 1966). The family has ben revised by Davis (1967). The **Tischeriidae** have somewhat less complete venation and the larva, which is a miner, sometimes has five pairs of abdominal legs. The species are widespread and moderately numerous.

Suborder **DITRYSIA**

Female with a copulatory pore on sternite VIII and an egg-pore on sternite IX. Fore wing without jugum or fibula; hind wing often with frenulum, its venation reduced. Pupa adecticous, obtect. Larvae normally with crochet-bearing abdominal prolegs.

This large group includes more than 97 per cent of the species in the order. There is a wide range in size and structures and its detailed classification is still under discussion.

Superfamily **Tineoidea** (Figs. 507, 509)

Maxillary palpi often fully developed: labial palpi with 3rd segment usually slender and pointed. Cu_2 generally present in both wings but often reduced in fore wing. Hind wing with $Sc + R_1$ free, less often joined to cell by a bar. M_1 and Rs separate, sometimes approximated or stalked: or, venation degenerate in many small species: or, wings divided into plumes. Aculei sparse and restricted or usually absent.

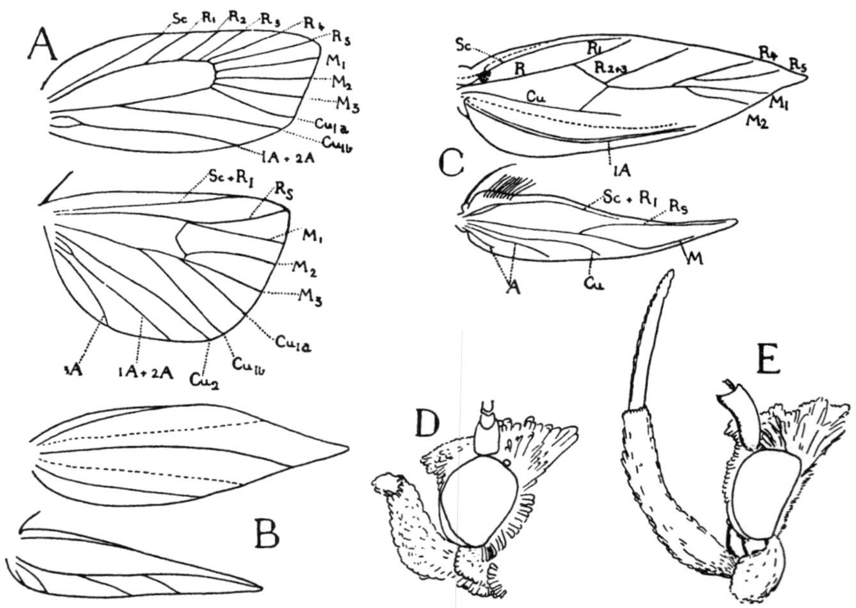

FIG. 507 Wings of A, *Acleris* (Tortricidae), B, *Opostega* (Opostegidae) and C, *Obrussa* (Nepticulidae). D, side view of head showing labial palp of *Cydia pomonella* (Tortricidae): E, the same of *Cryptolechia tentoriferella* (Oecophoridae). Adapted from Meyrick (A, B), Braun (C) and Forbes (D, E)

If broadly drawn, this is a very large group including nearly half the Microlepidoptera. Common, however, would divide it into four, excluding the Yponmeutoidea, Gelechoidea and the Copromorphoidea. Only a selection of the families are dealt with here.

Tineoidea (in the restricted sense)

Head sometimes with rough scales. Antenna sometimes with 'eyecap'. Haustellum not scaly, maxillary palpi with 1–5 segments. Vein M often present in the discal cell.

FAM. TINEIDAE. A world-wide family of over 2400 species characterized by the head being usually rough-haired, the proboscis short or absent and the maxillary palpi often long. The labial palpi are usually porrected and the posterior tibiae hairy. The wings have all the usual veins present and separate, the hind pair being narrow. *Tinea* is universally distributed; its larvae show diverse habits and sometimes live in portable cases. Those of most European species feed upon various dry animal or plant material. Thus, the larva of *T. vastella* feeds upon dried fruit, horns of antelopes, and other dried matter: those of *T. pellionella, Tineola bisselliella* and *Trichophaga tapetzella* are 'clothes moths' – household pests attacking clothing, carpets, furs, feathers, etc. A revision of the European species, including their early stages and foods has been published by Petersen (1969).

FAM. PSYCHIDAE (Bag-worm Moths). A family with 800 species and an extremely wide distribution: about 150 species occur in the Palaearctic region but very few are British. The family has evolved along totally different lines in the two sexes, the males being highly specialized and swift fliers, while the females include the most degenerate of all Lepidoptera (Fig. 508). In the former sex the wings are thinly clothed with hairs and imperfect scales, and are almost devoid of markings. The labial palpi are very short, the antennae are strongly bipectinated, and the frenulum exceptionally large. The proboscis is absent and the tibiae have spurs. In the male vein Cu_2 is present in the fore wing. The females are always apterous, but

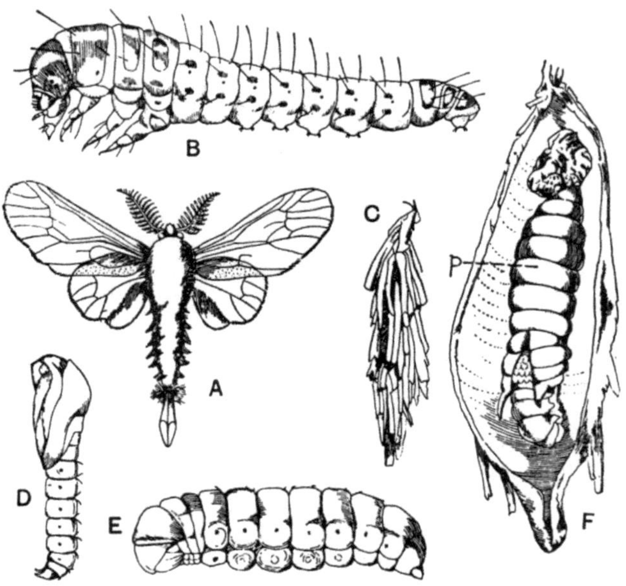

FIG. 508 *Thyridopteryx ephemeraeformis* (Psychidae)
A, male imago; B, larva; C, larval 'bag'; D, male pupa; E, female pupa; F, female imago within 'bag'; *p*, pupa case. All enlarged. Adapted from Howard and Chittenden, *U.S. Bur. Entom.*, *Circ.* 97.

exhibit various degrees of degeneration: in extreme forms the antennae, mouthparts and legs are totally wanting. The larvae inhabit cases which exhibit great variety of shape and of materials used in their construction: they carry their cases with them as they move about their food-plants. These habitations are formed of silk covered with fragments of leaves, twigs, grass and other objects. In *Cochliotheca* they are wholly constructed of silk and are extremely close copies of *Helix*-like shells. Pupation takes place within the larval case, and the pupae are provided with a row of sharp spinules on the abdominal segments. There is much diversity of structure in the female pupae: thus according to Heylaerts wings are present in *Psyche* (= *Fumea*), while in *Thyridopteryx* and *Oiketicus* there are no traces either of these organs or of antennae, maxillae, or eyes and only slight vestiges of legs are present (Mosher). The imago of this sex is little more than an egg-sac and spends her whole life within the larval habitation. Copulation takes place by the male alighting on the case and inserting his protrusible abdomen between the wall of the former and the ventral surface of the female. *Psyche* is exceptional in that the female emerges from the case prior to copulation. Parthenogenesis is known to occur in *Cochliotheca crenulella* var. *helix* and in *Solenobia* and *Luffia* (Seiler, 1923, etc.). Heylaerts (1881) has monographed the European species and gives much general information on the family: for the habits and structure of *Acanthopsyche opacella* (= *atra*), see Chapman (1900); and Matthes (1948) for *Amicta*. The larvae of a few species are pests of fruit or other trees in N. America, S. Africa, etc. The smaller species of Psychidae are, in the male, quite like Tineids but the larger ones in some respects resemble the Zygaenoidea. Davis (1964) has monographed the American species (ca. 74). Dierl (1970) separates out a few African and Oriental species, previously placed in the genus *Melusina*, as members of a new family, **Compsoctenidae.** They differ especially in the male and female genitalia. The larvae are stem-borers or live in tubes in the ground (cf. *Melusina energa* of Ceylon, Fryer).

FAM. GRACILLARIIDAE. A cosmopolitan family of about 1000 small species with narrow, long-fringed wings. They are often recognizable by their habit of resting with the fore part of the body upraised by the rather widely separated two anterior pairs of legs. The head is smooth-scaled except for a posterior fringe; antenna without an 'eyecap'. According to Busck (1914), some species retain more branches of Rs in the hind wing than do any other members of the suborder. The larvae are leaf-miners: when young they are very much flattened with blade-like mandibles and vestigial maxillae and labium. At this stage they lacerate the cells and suck the exuding sap: later they usually undergo hypermetamorphosis, acquiring normal mouthparts and devour the parenchyma. The mines in *Phyllonorycter* (= *Lithocolletis*) are small blotches, of which one surface is silk-lined and caused to contract, thus producing a hollow chamber: the contracting surface may be either on the upper or lower leaf surface, but is constant for a species (Meyrick). The two larger genera, *Phyllonorycter* and *Gracillaria*, together include over 450 species.

FAM. PHYLLOCNISTIDAE. *Phyllocnistis* with over 50 species includes some of the smallest and most delicate of all moths, and is remarkable in that the larvae are apodous. The antenna has an 'eyecap' and the hind tibia has a dorsal row of stout bristles.

Yponomeutoidea

FAM. SESIIDAE (AEGERIIDAE: Clearwings). This family is distinguishable by Sc + R_1 in the hind wings being concealed by a fold of the costa. Their most striking character, however, is the absence of scales from the greater part of both pairs of wings: the antennae are often dilated or knobbed and the abdomen is terminated by a conspicuous fan-like tuft of scales. The fore wings are extremely narrow owing to the great reduction of the anal area and in most species the bristles of the frenulum in the female are consolidated as in the male. The family is characteristic of the northern hemisphere, and the species are diurnal, flying rapidly during warm sunshine. Many resemble wasps, bees, ichneumons, etc., in appearance, which is largely due to their clear wings, slender bodies and often bright colours. They are in many ways an aberrant group, especially as regards the internal anatomy (see Brandt, 1890). The larvae feed in the wood of trees and bushes or in the root-stocks of plants. They are colourless with greatly reduced setae; the abdominal feet bear two transverse bands of uniordinal crochets, and a single row on the anal claspers.

Among other characters Fracker states the spiracles of the 8th segment are much larger and higher up than on other abdominal segments. Pupation takes place in the larval gallery and the pupae are provided with various forms of cutting plates for working their way to the surface: these are mostly situated on the head which is heavily sclerotized. There are two rows of spines on most of the abdominal segments which extend around to the ventral surface, and a definite cremaster is wanting. Owing to their internal feeding habit several species have attracted the notice of economic entomologists, particularly the European and American Currant Borer (*Sesia tipuliformis* = *Synanthedon salmachus*) and the Peach Tree Borer (*Sanninoidea exitiosa*) of the latter continent (Gossard and King, 1918). The West African *Ceritrypetes* is exceptional in preying on *Ceroplastes* (Bradley, 1956). Over 100 species of the family are Palaearctic and no less than 90 belong to the genus *Sesia*: 15 species have been found in the British Isles but several are rare and local. The N. American species have been revised by Engelhardt (1946).

FAM. GLYPHIPTERIGIDAE. A large family (about 900 spp., 12 British), especially abundant in the southern hemisphere. The ocelli are prominent. The hind wing venation is not much reduced and the hind tibia is not bristly. *Glyphipterix* includes metallic-winged moths which fly in the sunshine and whose larvae feed chiefly on grasses and sedges. In *Chloreutis* the moths resemble Tortrices in form and their larvae often form webs among seeds or leaves.

The **Douglasiidae** with a few small species resemble the Glyphipterigids but the venation of the hind wing is reduced and the discal cell is open.

FAM. HELIODINIDAE. A moderate-sized cosmopolitan family, whose members when at rest have the trait of displaying the hind legs either upraised or applied to the back or sides of the body: the posterior legs have the tibiae and tarsi furnished with whorls of bristles. Out of about 400 known species of the family one is British.

FAM. YPONOMEUTIDAE. A family of about 800 species whose tropical representatives are often brightly coloured and of relatively large size. The ocelli are

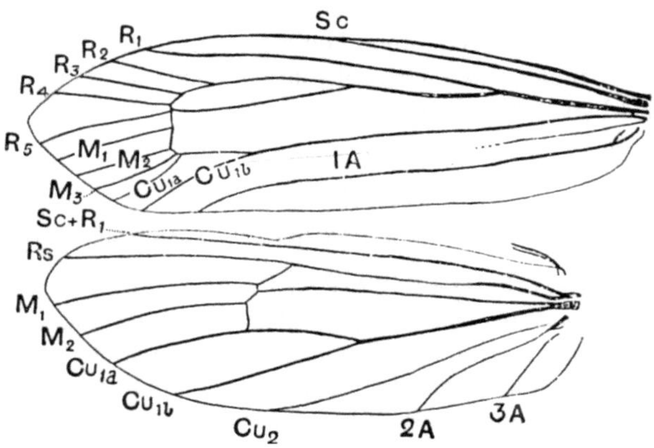

FIG. 509 *Yponomeuta euonymella*, left wings. × 15

usually small or absent and the maxillary palps well-developed and projecting. The small ermine moths (*Yponomeuta*, Fig. 509) are very widely distributed and their larvae live gregariously on shrubs and fruit trees. The large genus *Argyresthia* is well represented in Europe and N. America, the larvae living in shoots, leaf-buds, fruit, etc. The European Yponomeutidae are revised by Friese (1960) who lists the food-plants. The subfamily **Plutellinae** includes about 200 species, occurring in most regions. The larvae feed in a slight web in leaves or occasionally mine leaves or stems: those of the cosmopolitan genus *Plutella* feed on the leaves of Cruciferae. The Diamond-back Moth (*Plutella maculipennis*: Fig. 510) is very destructive to vegetables and, owing to its ability to flourish in about all climates, it has become one of

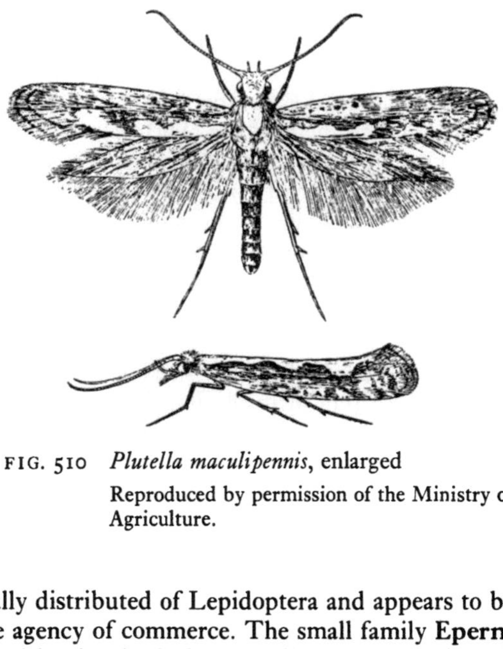

FIG. 510 *Plutella maculipennis*, enlarged
Reproduced by permission of the Ministry of Agriculture.

the most universally distributed of Lepidoptera and appears to be still extending its range through the agency of commerce. The small family **Epermeniidae** is related to the **Yponomeutidae** but both the hind tibia and tarsus have stiff bristles. There are somewhat less than 100 species, found on most continents. The larvae are at first miners but later live under a web, especially on Umbelliferae. The European species have been revised by Gaedike (1966).

Gelechioidea

Haustellum more or less densely scaled at base

FAM. COLEOPHORIDAE (EUPISTIDAE). In this small family the only extensive genus is *Coleophora*, which is represented by over 400 Holarctic species: about 90 species occur in N. America and 98 species in Britain. They are narrow-winged insects, usually recognizable by the antennae being held in a porrect position in repose. The maxillary palps are rudimentary. The hind wings are narrower than their fringe with vein Rs close to Sc + R_1. The fore wing has R_1 arising from the discal cell well before its upper angle. With regard to the larval habits, Meyrick remarks that they are leaf-mining when very young, afterwards inhabiting a portable case. The latter is attached to a leaf or seed-vessel and the larva bores into the interior. In the case of leaves a pale blotch is usually produced, with a distinctive round hole in one membrane (see Barasch, 1934).

FAM. ELACHISTIDAE. A restricted family whose most important genus is *Elachista*, with over 200 species. The hind wings are narrow but vein Rs is central, remote from Sc + R_1. In the fore wing vein R_2 arises from the discal cell well before its upper angle. The larvae mine leaves, especially of grasses or allied orders of plants (see Braun, 1948).

FAM. SCYTHRIDAE. Fore wing with vein R_2 arising near upper angle of the discal cell, R_5 ends behind the costa. A cosmopolitan family, especially developed in S. Europe and S. Africa. The larvae live in a web or in silken tubes among stems or flowers of various plants.

FAM. STATHMOPODIDAE. Hind wing very narrow. Hind legs raised in repose, hind tibiae or tarsus with whorls of bristles. A family of moderate size, most species being placed in *Stathmopoda*, some of whose larvae are predacious on Coccids while others bore into leaves or fruit.

FAM. OECOPHORIDAE. *Antennae usually with a basal pecten. Hind wings usually broad with Rs and M separate and parallel* (Fig. 507E). A family comprising at least 3000 species, many of which are Australian: 25 genera and 78 species are British and there are 117 N. American species (Clarke, 1941). The larvae feed among spun leaves or seeds, in decaying wood, etc. In the large genus *Depressaria* and its allies the larvae affect more especially Compositae and Umbelliferae: the common *Agonopterix heracliana* spins together the flower heads and seeds of the parsnip and other plants in Europe and N. America. *Blastobasis* and its allies are often regarded as a separate family: their larvae feed on dry refuse, seeds, etc., or live as parasites of scale-insects.

FAM. ETHMIIDAE. *Hind wings not very narrow, vein Cu_2 clearly present, M_2 arising nearer M_1 than M_3.* The wings are often boldly marked. The larvae are partial to Boraginaceae, living under a slight web on the leaves.

FAM. COSMOPTERYGIDAE. *Antennae with a slight basal pecten. Fore wings lanceolate or linear: hind wings as in Gelechiidae.* A neglected but widely distributed group of about 1200 species of small narrow-winged moths, of which 35 are British. The classification of the N. American species is discussed by Hodges (1962). The larvae have varied habits, usually fixed for a genus: many are leaf-miners, some feed in shoots or seeds, others among dry refuse or attack scale-insects. The species of *Cosmopyteryx* are elegantly marked with black, orange and gold and the larvae usually form blotch mines in leaves; nearly 300 species occur in Hawaii.

FAM. XYLORYCTIDAE. *Antennal pecten absent. Hind wings with Rs and M_1 basally approximated or stalked. Maxillary palps usually with four segments.* This family includes some of the largest Tineoidea, and is especially well represented in Australia with a smaller number of species in S. America, India, etc. Some of the finest species belong to the genera *Cryptophasa*, *Maroga*, *Uzucha* and *Xylorycta*: they are often conspicuously coloured and attain a wing-expanse up to about 75 mm. The larvae are concealed in shelters or coverings, or tunnel in wood, carrying in leaves for food.

FAM. GELECHIIDAE. *Antennae rarely with basal pecten. Fore wings trapezoidal. Hind wings with Rs and* M_1 *stalked or approximated at base, posterior margin usually sinuate.* This family includes nearly 400 genera and about 4000 species of small moths represented nearly all over the world. A few forms have the hind wings elongate-ovate and resemble Oecophoridae, but may be separated from them by Rs and M_1 being basally approximated or stalked. The larvae usually feed among spun leaves or shoots, sometimes in seedheads or roots, but are seldom leaf-miners or case-bearers (Meyrick). One of the best known species is the nearly cosmopolitan Angoumois Grain Moth (*Sitotroga cerealella*), whose larvae are destructive to wheat, maize, etc. The Pink Boll-worm (*Platyedra gossypiella*) is the widest spread and one of the most destructive of all cotton pests, few cotton regions being free from it. The Potato Tuber Moth (*Phthorimaea operculella*) is a widespread pest of stored potatoes, more rarely affecting the field crop, and *Holcocera pulverea* is an important enemy of lac in India, its larvae being predacious on the latter insect. Over 130 species of the family are British.

FAM. STENOMIDAE. *Head smooth scaled, ocelli absent. Maxillary palpi with four segments. Hind tibia with long scale-hairs. Fore wing with* R_4 *and* R_5 *usually separate,* R_5 *to apex. Hind wing broader than fore wing,* $Sc + R_1$ *approaching Rs or else Rs curved forwards.* Especially S. American but also in Indo-Australia and Madagascar (Duckworth, 1973). Larvae living between leaves or mining in leaves or stems.

Copromorphoidea (Alucitoidea)

This group is sometimes placed outside the Tineoids, nearer to the Pyraloidea. Haustellum not scaled.

FAM. COPROMORPHIDAE. *Head smooth scaled. Hind wing with all three branches of M present. Fore wing usually with raised scale-tufts.* A few Australian and S. American species. The larva of one Australian species attacks shoots and stems of figs.

FAM. ORNEODIDAE (ALUCITIDAE: Many-plume Moths). A small isolated family characterized by both pairs of wings being cleft into six or more narrow plume-like divisions, densely fringed with hairs along both margins (Fig. 511). They are related to the Pyralids and Tineids, but exhibit no close affinity with the Pterophoridae. With the exception of *Orneodes hexadactyla*, which is European, the various species have a restricted range: the former insect is common in Britain, where it is the sole representative of the family. The larvae burrow into shoots, flower-stalks and buds giving rise to galls, and the known food-plants include *Lonicera*, *Scabiosa* and *Stachys*. They are hirsute, cylindrical and rather stout; the crochets are uniordinal, arranged in a complete circle. The pupae are very different from those of the Pterophoridae and have affinities with Tineids and Pyralids. A

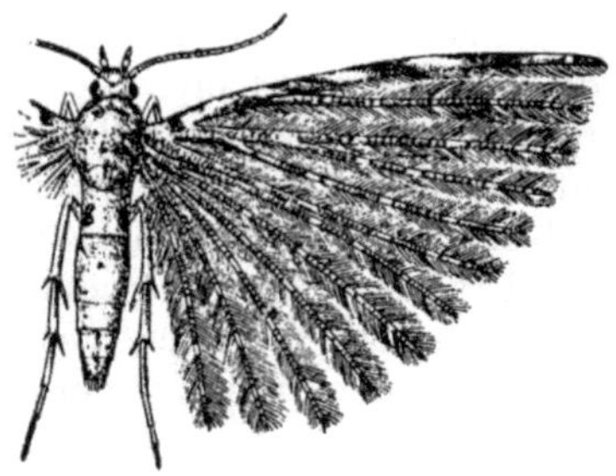

FIG. 511
Orneodes pygmaea, enlarged, Ceylon

After Fletcher.

cocoon is formed on the surface of the ground and consists of loose silk or of fine earthen particles. Most of what is known concerning the family will be found in papers by Chapman (1896*a*), Hofmann (1898), Fletcher (1910) and by Meyrick (1910).

FAM. CARPOSINIDAE. *One or two branches of M absent in the hind wing. Fore wing with raised scale-tufts. Hind tibia with long hair-like scales.* A moderate number of species, especially in Australia and Hawaii. The larvae seem to burrow in bark. Swatschek (1958) figures those of two species.

Superfamily **Cossoidea** (Figs. 495, 512)

Both pairs of wings with Cu_2 *present and M furcate within the cell. Areole present in fore wing. Haustellum absent.*

FAM. COSSIDAE (Goat Moths, Carpenter Moths). Insects of moderately large or exceedingly large size, the females of *Duomitus leuconotus* attaining a wing-expanse of 180 mm. The family is generally distributed and, according to Turner (1918), it retains the most ancient form of venation among Ditrysia. The antennae are

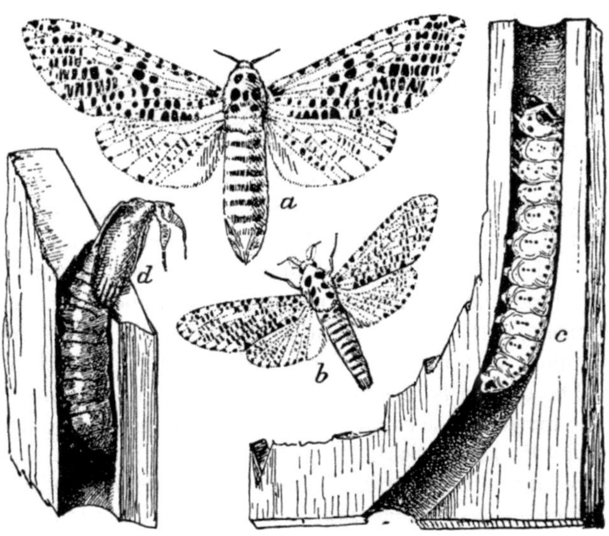

FIG. 512 *Zeuzera pyrina* (Cossidae)

a, female; *b*, male; *c*, larva; *d*, pupal case. Actual size. *After* Howard and Chittenden. *U.S. Dept. Agric. Circ.* 109 (reduced).

frequently bipectinate in both sexes, rarely simple: in other cases they are bipectinate in the male for a portion of their length and filiform distally. The frenulum is sometimes short and apparently non-functional, more often it is well developed: in the female it may consist of as many as nine bristles (Hampson). These moths are nocturnal fliers and lay their eggs on the bark of trees, or in the tunnels from which they have emerged. The larvae are internal feeders boring large galleries in the wood of forest, shade and fruit trees or in the pith of reeds, etc., often causing serious injury. The head is closely united to the enlarged prothorax, and the mandibles are very large. Only primary setae are present and the full number of limbs is retained, the crochets being usually either bi- or tri-ordinal, arranged in a complete circle. In certain species the larvae attain a very large size and in *Cossus cossus* (*ligniperda*) and *Prionoxystus robiniae* they live for at least two years. The pupae lack maxillary palpi: the 3rd to 6th abdominal segments are movable in the female and the 7th also in the male. The dorsum of the segments is armed with a toothed ridge along each margin and a cocoon of silk and gnawed wood is usually constructed. *Cossus* is one of the most primitive genera and is universally distributed. *Xyleutes* includes numerous species found in all regions, particularly Australia, *Zeuzera* includes the Leopard Moth (*Z. pyrina*) of which the larva burrows in the branches of fruit-trees (Fig. 512) and *Z. coffeae*, known as the 'White Borer' of coffee. According to Common the genus *Dudgeonea* has a rudimentary tympanum at the base of its abdomen.

Superfamily **Zygaenoidea**

Maxillary palpi absent or vestigial. Proboscis usually atrophied. Tympanal organs wanting. Posterior tibial spurs very short with middle spurs often absent. Both wings with Cu_2 present (rarely wanting in fore or hind wing only) and M almost always present within the cell. Hind wing with $Sc + R_1$ remote from Rs beyond the cell.

FAM. MEGALOPYGIDAE (LAGOIDAE). *Hind wing with vein $Sc + R_1$ coincident with cell to middle or beyond it.* An essentially American family with only few Palaearctic species which occur in Africa. Their affinities apparently lie nearest to the Cochlidiidae, particularly with regard to larval characters. According to Dyar their larvae possess two series of abdominal feet. The normal ones occur on segments 3 to 6 and on 10, and are provided with crochets; the secondary feet lie on segments 2 to 7 and are of the nature of sucker-discs. Mosher states that in *Lagoa* the pupa has the head and thoracic segments free, and abdominal segments 1 to 6 are free in the female, with segment 7 also in the male. The whole pupal covering is thin and membranous with the appendages entirely free from each other and from the body-wall. The cocoon is furnished with a circular operculum to allow of the emergence of the imago. An account of the metamorphoses and anatomy of *Lagoa crispata* is given by Packard (1894).

The **Heterogynidae** are an extremely small family represented by the southern European genus *Heterogynis*. The larvae are not case-bearers, and the females resemble those of the Psychidae in being vermiform and degenerate. They are stated to remain in the cocoons and lay their eggs there.

The **Chrysopolomidae** are similarly a very small family comprising only two genera and about 24 species which inhabit parts of Africa. The vein $Sc + R_1$ in the hind wing anastomoses with the cell and the palps are present.

FAM. METARBELIDAE (TERAGRIIDAE). A small tropical family of Ethiopian and Oriental range whose larvae, so far as is known, are wood-borers: *Indarbela quadrinota* bores in the branches of guava (Srivastava, 1962). The moths are nocturnal and closely resemble the Cossidae, but have a more reduced venation. In the hind wing vein $Sc + R_1$ is free of the cell and vein Cu_2 is absent; the proboscis and chaetosemata are absent.

FAM. LIMACODIDAE (COCHLIDIIDAE, HETEROGENIDAE, EUCLEIDAE). A small family allied to the Zygaenidae and Megalopygidae and including less than 40 Palaearctic species: *Heterogenea* and *Cochlidion* are British. Their larvae are commonly known as 'slug caterpillars', which have thick, short fleshy bodies, a small retractile head and minute thoracic legs. Segmentation is indistinct and there are no abdominal feet, but according to Chapman (1894) secondary sucker-discs are present on the first eight abdominal segments. A valuable series of papers on the structure of these anomalous larvae has been contributed by Dyar and Morton (1895–99). Those of different genera have very little in common beyond the features enumerated: many are smooth and glabrous while others are provided with a conspicuous armature of spine-bearing scoli which, in the case of *Empretia stimulea*, are said to be poisonous. The pupae strongly resemble those of the preceding family and are enclosed in a hardened oval or round cocoon. The latter is provided with an operculum which is constructed by the larva and allows of the free escape of the imago (Fig. 513).

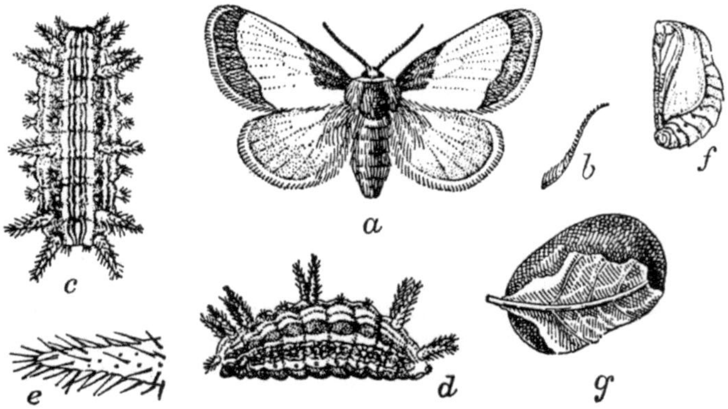

FIG. 513 *Euclea indetermina*, N. America (Limacodidae)

a, female imago; *b*, antenna of male; *c*, *d*, larva; *e*, scolus, much enlarged; *f*, pupa; *g*, cocoon. *After* Chittenden, *U.S. Dept. Agric. Ent. Bull.* 124.

FAM. ZYGAENIDAE. The members of this family closely resemble the Ctenuchidae but are readily separable therefrom by the presence of Cu_2 in the hind wings (Fig. 514B). Many are very brilliantly coloured and there is considerable diversity of structure. They are diurnal in habit, with a slow heavy flight, and are inclined to be very locally distributed. The larvae (Fig. 514B) possess the full number of limbs, and, so far as is known, they are short and cylindrical with numerous verrucae from which arise short hairs; they live exposed on herbaceous plants. The pupae are enclosed in tough elongate membranous cocoons above ground; owing to their great capacity for movement, they are enabled to work their way out prior to the emergence of the imagines.

The subfamily Zygaeninae is characteristic of the Palaearctic region where it is represented by 12 genera and over 100 species; 2 genera and 10 species inhabit the British Isles. *Zygaena* includes the 'Burnets' which have the antennae distally enlarged and *Ino* includes the brilliant metallic green 'Foresters'. The Chalcosiinae are far the largest group and are essentially tropical, only two species entering the Palaearctic region. Many species are butterfly-like with slender bodies and broad large wings; in *Elcysma* and *Histia* the hind wings are tailed. The Phaudinae are a small and very aberrant subfamily in which the mouthparts are wanting. In *Himantopterus* the hind wings are filiform as in the Nemopteridae (p. 810) and there is no frenulum: the genus is placed by some authorities in a family of its own – the **Himantopteridae** (Thymaridae).

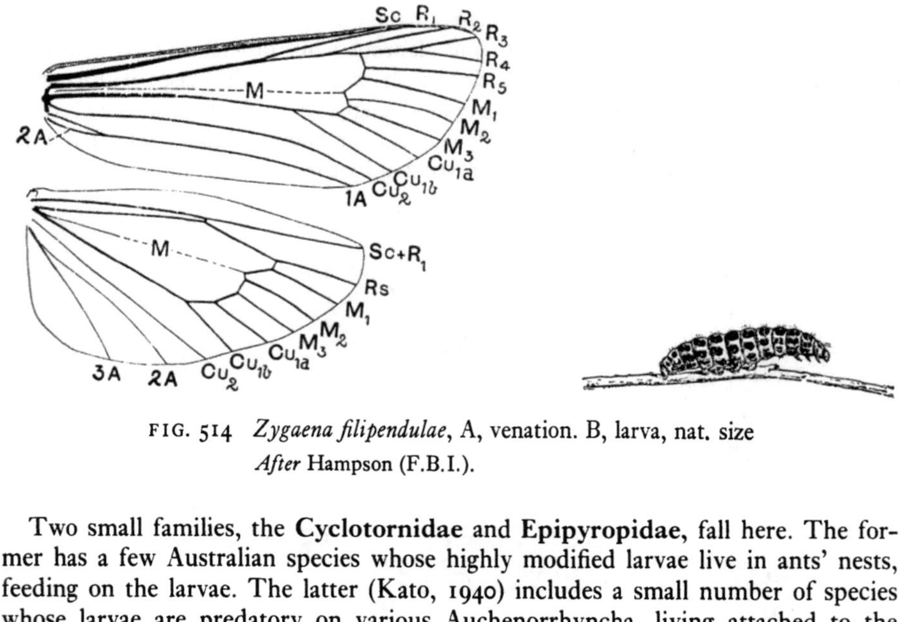

FIG. 514 *Zygaena filipendulae*, A, venation. B, larva, nat. size
After Hampson (F.B.I.).

Two small families, the **Cyclotornidae** and **Epipyropidae**, fall here. The former has a few Australian species whose highly modified larvae live in ants' nests, feeding on the larvae. The latter (Kato, 1940) includes a small number of species whose larvae are predatory on various Auchenorrhyncha, living attached to the victim.

Superfamily **Castnioidea**

Maxillary palpi present. Proboscis present or absent. Chaetosema absent. Antennae clavate. Tympanal organs wanting. Tibial spurs present. Fore wing with Cu_2 present and M developed within the cell. Hind wing with Cu_2 present or absent, $Sc + R_1$ remote from Rs beyond the cell.

FAM. CASTNIIDAE. Included in this family are about 160 brightly coloured day-flying moths often bearing a resemblance either to Nymphaline butterflies or 'Skippers'; they are confined to tropical America and the Indo-Malayan and Australian regions. Their metamorphoses have been very little studied. The eggs are

Related to the Thyrididae are the small Indo-Australian families **Tineodidae** and **Oxychirotidae**. In the latter family *Cenoloba* has each wing divided into two plumes. Its larva, according to Common, feeds on the fallen seeds of the Mangrove, *Avicennia*. The Tineodidae have relatively narrow wings and long legs.

FAM. HYBLAEIDAE. *Hind wing with vein Cu_2 fully developed and vein $Sc + R_1$ not approximated to Rs distally. Tympanal organs absent. Haustellum naked.* These moths superficially resemble Noctuids but besides lacking tympana, they have prominent maxillary palps and the larva is Pyraloid. The group includes a few Indo-Australian species.

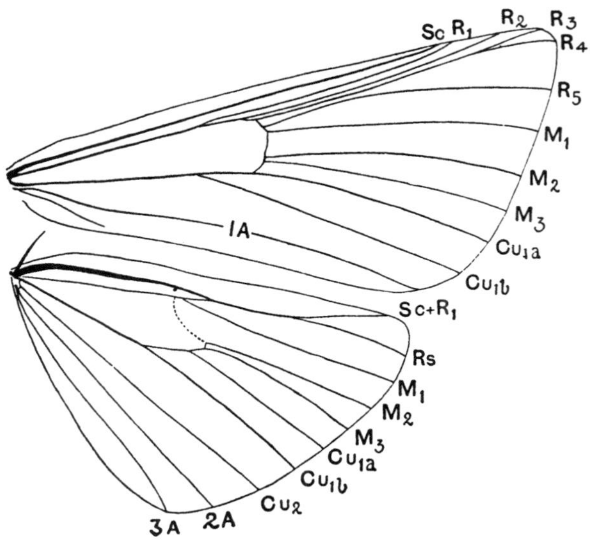

FIG. 515 *Pleuroptya ruralis* (Pyraustinae) venation

FAM. PYRALIDAE. *Hind wing with vein Cu_2 fully developed but vein $Sc + R_1$ approximated to, or fused with, Rs distally* (Fig. 515). *Haustellum usually scaly. Tympana present at the base of the abdomen.* This group has often been split into five families, the Phycitidae, Crambidae, Galleriidae, Pyralidae and Pyraustidae but these are now more often treated as subfamilies. The early stages of many of the European genera are described by Schwarz (1964) and the larvae in more detail by Hasenfüss (1960).

The Galleriinae are a small but widely distributed group whose larvae feed on a variety of dried substances, including the combs of beehives and of wasps' nests, dried fruits, and in a few cases in roots, beneath bark, etc. Pupation takes place in a peculiarly tough cocoon. The best known species is the Wax Moth, *Galleria mellonella*, which has become artifically spread among hives in many parts of the world, including Australia. The biology and method of nutrition of this species have been studied by Metalnikov (1908). Its larva and that of the Lesser bee moth, *Achroia*, have an enzyme for digesting beeswax. The Crambinae include the Grass Moths, of which the old genus *Crambus* (now split up, BleszińSki, 1963) included about 400 species (25 British). They are mostly small insects with narrow elongate fore wings and porrected labial palpi. They are extremely abundant in grassland and rest by day in an upright position with the wings folded on the stems. Their larvae usually feed in silken galleries on grasses, reeds and allied plants, or on moss. Among other forms *Diatraea saccharalis* is the American Sugar-cane Borer and species of *Chilotraea* are similar pests in India.

The Phycitinae are a very large group with elongate fore wings which lack vein R_5: the hind wings have, on the dorsal side, a well-defined pecten of hairs on the lower margin of the cell near the base. These insects are exceptional in that the frenulum is simple in both sexes. Secondary sexual characters are well seen in the swollen basal antennal segment of the males, and the same sex is often provided with a conspicuous row, or tuft, of hairs or scales on the fore wings. The larvae vary greatly in habits and usually live in silken tubes by day, coming out to feed at night. Nearly 50 species of the family are British, and over 800 are found in the Palaearctic region. Heinrich (1956) has revised the American species. *Ephestia* includes the Mediterranean Flour Moth (*E. kuehniella*) whose larvae are great pests in flour mills; those of other species attack dried fruits, biscuits and other commodities, *E. cautella* being the nearly cosmopolitan Fig Moth. The Indian-meal Moth (*Plodia interpunctella*) is even more widely distributed and attacks maize, figs and seeds of various kinds. *Laetilia coccidivora* is remarkable on account of its predacious larva which lives upon various Scale Insects in N. America. A detailed study of the metamorphosis and larval and pupal structure of a Phycitine is given in Beeson's paper (1910) on the Oriental Toon Moth (*Hypsipyla robusta*), which is a shoot-borer. Larvae of various other genera live in rolled or spun leaves, others affect flower-heads, and many live on the bark of trees. The Anerastiinae, which have a vestigial proboscis, are sometimes regarded as a separate subfamily.

The Pyralinae are fairly numerous in the tropics, but scarce elsewhere, and absent from New Zealand. The larvae feed, as a rule, upon dry or decaying vegetable substances. Those of the cosmopolitan *Pyralis farinalis* form silken galleries among corn and flour debris; species of *Aglossa* mainly live among hay and chaff refuse, while *Synaphe angustalis* is found in damp moss.

The Schoenobiinae have the proboscis vestigial and their larvae live among aquatic plants. The anomalous genus *Acentropus* is the most truly aquatic of all Lepidoptera, and its structure and biology has been studied in detail by Berg (1941). The young larva tunnels in the petioles of *Potamogeton* and other water plants; it subsequently constructs a tube of portions of leaves spun together, but open at the two extremities. A cocoon is spun in a rather similar leaf-shelter, the pupa being almost completely submerged. Respiration in the larva appears to be cutaneous at first and it is only in the later stages that the tracheae become filled with air. The females are dimorphic: the rare long-winged forms are aerial while those with reduced wings live entirely in the water, using their fringed mid and hind legs for swimming (cf. Kokociński, 1969).

The Nymphulinae are of special interest for the reason that some species are also aquatic. In the genus *Nymphula*, the larvae are usually leaf-miners at first and live throughout life below the surface of the water. Their biology has been frequently studied, notably by Miall, Müller (1892) and Welch (1916). Two definite larval types occur; those without gills when mature (*N. nympheata*), and those in which such organs are present (*Paraponyx statiotata*, etc.). The life-history of *N. maculalis* has been studied by Welch, who states that tracheal gills are wanting in the first instar but increase numerically after each moult. The pupa is enclosed in a silken cocoon on the submerged surface of a leaf, and the imago is not affected by contact with water during emergence. The method of respiration in this genus requires further study: during early life it is cutaneous and spiracles, if present, are closed. In *N. nympheata* and also in *Cataclysta lemnata* respiration subsequently takes place by open spiracles. In other species it is performed by means of tracheal gills: non-functional spiracles co-exist with the latter in *N. stratiotata*, but apparently not in *N. maculalis*.

The Scopariinae are a small group characterized by a raised tuft of scales in the cell of the fore wing. The large genus *Scoparia* mostly inhabits temperate regions and is extensively developed in New Zealand. The larvae feed on moss and lichen, among which they form silken galleries. The Pyraustinae are the largest group with 42 British species, and differ from the Scopariinae in the absence of raised scales from the fore wings. They are common in nearly all parts of the world and are exceedingly abundant in the tropics. Their larvae usually feed in a slight web among spun-up leaves, or in stems, fruits or roots. The most notorious species is the European Corn Borer (*Ostrinia nubilalis*), which is an introduced pest of corn (maize) in Ontario and the eastern U.S.: although abundant on the continent of Europe, it is rare in Britain.

Superfamily **Pterophoroidea**

Ocelli, chaetosemata and tympanal organs absent. Haustellum naked, maxillary palps reduced. Wings normally split into plumes.

FAM. PTEROPHORIDAE (Plume Moths). These insects are readily distinguishable by their deeply fissured wings; the anterior pair is longitudinally cleft into two, or more rarely three or four divisions, and the hind pair into three. There are no maxillary palps, and all the species are extremely lightly built with very elongate fore wings, and unusually long and slender legs armed with prominent tibial spurs. The species are nowhere numerous and 37 inhabit the British Isles. *Agdistis*, and two other genera, are exceptional in possessing undivided wings. The larvae mostly feed exposed on flowers and leaves but sometimes internally in stems or seed vessels, the Compositae being more frequently selected than any other order of plants. They are long and cylindrical with numerous secondary setae. The abdominal feet are long and stem-like with uniordinal crochets. The pupae (Chapman, 1896*a*) are attached by the cremaster and occur above ground, sometimes in a slight cocoon. The body is roughened with short spines or with small groups of longer barbed spines arising from small elevations. Unlike the Pyralidae, there are no maxillary palps, and the deep furrow between the 9th and 10th abdominal terga is likewise absent. Among British species one of the commonest is *Pterophorus pentadactyla* whose larva feeds upon *Convolvulus*: the larva of *Agdistis staticis* selects *Statice limonium* and that of *Buckleria paludum* feeds upon the leaf-tentacles of *Drosera* (Chapman, 1906). (See also Lange, 1939.) Yano (1963) describes the larvae and male genitalia of the 57 Japanese species.

Superfamily **Papilionoidea** (Fig. 495B)

Antennae slender with an abrupt club. Labial palps moderately long, more or less rough-haired, terminal segment rather pointed. Maxillary palps obsolete. Fore wings with Cu_2 *absent,* M_2 *arising from or above middle of transverse vein. Hind wings without frenulum:* Cu_2 *absent;* $Sc + R_1$ *arising out of cell near base, thence strongly curved and diverging.*

This and the next superfamily which some authors unite together include those insects commonly known as butterflies and are frequently regarded as constituting a group (Rhopalocera) of equal systematic value to the whole of the remainder of the Lepidoptera or moths (Heterocera). There is, however, no scientific justification for according to these insects any higher rank than that of two allied superfamilies. They are characterized by the antennae being clubbed or dilated, the absence of a frenulum and by the humeral lobe of the hind wing being greatly developed. In other Lepidoptera the antennae are not clubbed or dilated except in infrequent cases, and in such instances a frenulum is present. *Euschemon* has been regarded as a moth and either given separate family rank or placed in the Castniidae but although the male has a frenulum it is probably the most archaic of all butterflies and a member of the Hesperiidae. The Papilionoidea are a tolerably natural group, but there

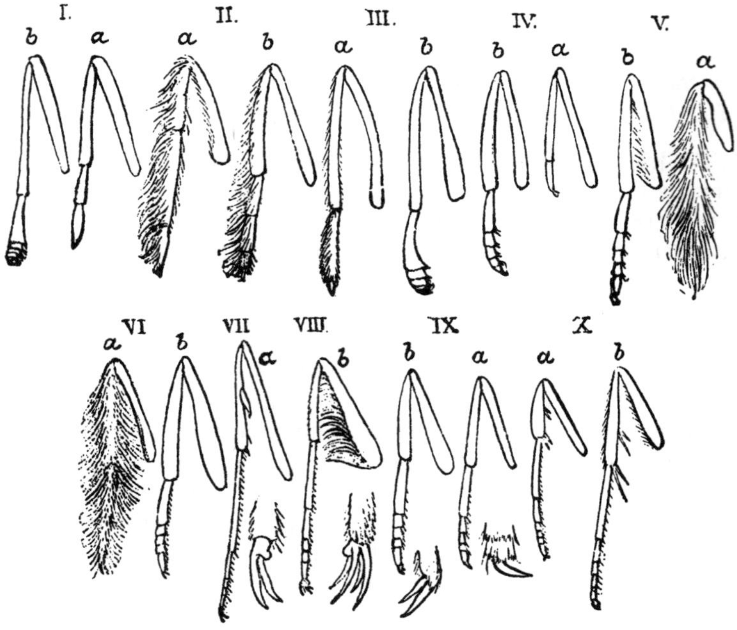

FIG. 516 Fore legs of Papilionoidea

a, male; *b*, female. I, Danainae; II, Satyrinae; III, Nymphalinae; IV, Acraeinae; V, Libytheinae; VI, Riodinidae; VII, Papilionidae; VIII, Pieridae; IX, Lycaenidae; X, Hesperiidae. *After* Bingham (F.B.I.).

is no general consensus of opinion as to their phylogeny. Both Hampson and Meyrick regard them as being derived from the Pyraline family Thyrididae while other authorities derive them from the Castniidae.

The families are largely separated by the nature and degree of reduction of the tarsi (Fig. 516) which in some groups are used in chemoreception (Fox, 1966). Their phylogeny is discussed by Kristensen (1971). Lewis (1974) illustrates in colour about 5000 of the species.

FAM. NYMPHALIDAE. *Anterior legs of both male and female useless for walking.* The dominant family of the butterflies and one of the largest of all Lepidoptera, including about 5000 described species. The fore legs in both sexes are reduced in size, usually folded on the thorax, and functionally impotent: the tibiae are short and clothed with long hairs, hence the name of 'brush-footed' butterflies.

The Danainae (Euploeiinae, Limnadidae) have the antennal club often but little pronounced, and the whole antenna is devoid of scales: the fore feet in the female terminate in a corrugate knob. The larvae are smooth and cylindrical, with two to four pairs of fleshy processes, at least on the mesothorax, and often on one or more of the abdominal segments. They are all very strikingly marked with black and yellow, red, or green. The imagines have developed what must be, to our senses at any rate, an acrid disagreeable odour and taste, accompanied with a leathery consistency of body which evidently protects them from insectivorous enemies. In the majority of forms secondary sexual characters in the form of androconia, tufts of hairs, etc., having peculiar odours, are prominent (Bingham). In *Danaus gilippus* a ketone produced by the male hair-pencils is perceived by short, thin-walled sensilla on the female antennae and excites her to pair (Myers and Brower, 1969). The subfamily occurs in all warmer regions and well-known genera are *Danaus, Euploea* (Fig. 517) and *Amauris*.

FIG. 517
Euploea godarti (Nymphalidae), male. India × $\frac{1}{1}$
After Bingham (F.B.I.).

The Ithomiinae differ from the Danainae in that the female has a true though somewhat shortened fore tarsus. The antennae are devoid of scales and the wings are elongate, often in great part translucent, and thinly scaled. The subfamily is Neotropical, and many species exhibit colour resemblances to the Heliconiinae or to the Pieridae.

The Satyrinae are a world-wide group which includes the common 'Meadow-browns', 'Heaths', 'Graylings' and 'Marbled Whites'. They are easily recognizable by certain of the veins at the base of the fore wings being greatly swollen, and by the strongly adpressed palps. They are small to medium-sized butterflies, frequently some shade of brown or tawny in colour, with a variable number of eye-like or annular spots. Their powers of flight are not greatly developed, and they are largely shade-loving insects, cryptically coloured on the underside. The larvae feed mostly upon Gramineae: they are fusiform and green, yellowish, or brown marked with longitudinal lines. In appearance they bear a resemblance to Noctuid larvae. The head is often bilobed or horned, the prothorax constricted, and the body is clothed with small papillae bearing short secondary setae. The segments are divided into annulets, and the suranal plate is bifurcate, bearing a pair of short backwardly directed processes. The pupae are similar in general form to those of the Nymphalinae but are devoid of tubercles, and have few prominent ridges. They are generally suspended by the cremaster, and there is no median silken belt: a few are subterranean and, in some cases, they construct a slight cocoon or cell. *Erebia* is characteristic of the Alps of Europe, but also occurs on the mountains of Asia, S. Africa and N. America: two species inhabit N. Britain. In *Melanitis* the bases of the veins of the fore wings are normal and not swollen. *M. ismene* extends across the southern half of Africa through the Oriental region to Australia: it has both wet and dry season forms and numerous local races. The Neotropical genera *Cithaerias* and *Haetera* have delicate transparent wings, with the scales almost wanting.

The Morphinae are exclusively tropical, and have the discal cell in the hind wings open; there is also a cradle-like depression along their inner margins for the reception of the abdomen. The species of *Morpho* are large, and have an extensive wing-expanse in proportion to the size of the body. They are brilliant metallic blue insects peculiar to the forests of tropical America. The eastern representatives of the group do not equal their S. American allies either in size or brilliancy.

The Brassolinae are likewise Neotropical, and are very large insects with the discal cell of the hind wings closed. They are deeply and richly coloured, and the under surface is marked with eye-spots and intricate lines. *Caligo* is one of the most familiar genera.

The Acraeinae are essentially African insects and the majority belong to the extensive genus *Acraea*: a few are Oriental and S. American. The wings are elongate and sparsely scaled, or more or less diaphanous. These insects appear to be largely immune from insectivorous enemies in all stages, and the imagines readily exude a nauseous fluid. The females in certain species develop an abdominal pouch very much as in *Parnassius* (see p. 1130).

The Heliconiinae form one of the most characteristic groups of Neotropical butterflies and are peculiar to that region. The fore wings are about twice as long as broad; the fore tarsus in the male is elongate and single-segmented, and 4-segmented in the female. They are medium-sized insects, many of which are stated to be protected owing to possessing nauseous or evil-smelling properties. They are closely related to the Nymphalinae in all their stages, but the imagines are readily distinguished by the closed discal cell. Gilbert (1972) has shown that species of *Heliconius* eat pollen and can digest it to support prolonged oviposition.

The Nymphalinae constitute the largest of the subfamilies: the discal cell in both pairs of wings is very often open or closed only by an imperfect veinlet. The palps are large and usually broad anteriorly. The fore tarsi in the male are unjointed and in the female four or five segments are present. In Britain, as in most parts of the world, they constitute the dominant group of butterflies and include the 'Fritillaries'

(*Argynnis, Brenthis* and *Melitaea*); the 'Tortoiseshells' and 'Peacock' (*Nymphalis*); the 'Purple Emperor' (*Apatura*) and other familiar insects. The larvae are almost always cylindrical and armed with numerous scoli. In *Apatura* and *Charaxes* they are smooth with tentacle-like processes on the head, and a pair of posteriorly directed anal processes. Müller (1886) gives a very complete account of the metamorphoses of many Brazilian species, and discusses the significance of colour pattern and its relation to the scoli. The pupae are very characteristic and are often armed with prominent tubercles on the surface of the body: there are usually seven rows on the abdomen, and there is a pointed projection on either side of the head in many species. The pupa is suspended head downwards by the cremaster, unsupported by a median girdle. Among the more notable species may be mentioned the central European *Araschnia levana* which produces two annual generations so dissimilar that they were formerly regarded as the two species, *A. levana* and *A. prorsa*. The seasonal forms of S. African *Precis* are equally striking (Marshall, 1902). *Cynthia cardui*, the 'Painted Lady', is probably the widest distributed of all Lepidoptera. *Apatura* occurs over the northern hemisphere, and is represented in Britain by *A. iris*, which is local in oak woods of the southern counties. The Indo-Malayan *Kallima* includes the 'leaf butterflies', remarkable on account of the extraordinarily perfect resemblance to leaves which is exhibited by the under surface of the closed wings. *Charaxes* includes large butterflies very widely distributed through the eastern hemisphere to Australia, and the hind wings are produced at veins M_3 and Cu_{1b} into long slender tails.

FAM. RIODINIDAE (NEMEOBIIDAE). *Anterior legs reduced in male, functional in female.* An extensive family comprising over 1000 species which are characteristic of the Neotropical region. A few species are found in the United States and approximately 100 occur in the eastern hemisphere. For the most part they are small butterflies, with short broad fore wings, and the fore legs in the male are imperfect and brush-like, with one-segmented tarsi devoid of claws: in the female the fore legs are functionally perfect but distinctly smaller than the remaining pairs. The vast majority of the species belong to the subfamily Nemeobiinae which has a single European representative, *Hamearis* (= *Nemeobius*) *lucina*. The latter insect extends its range into Britain where it is local but not rare. The Libytheinae may be easily recognized by the very long and closely approximated porrect palps. The widely distributed genus *Libythea* includes a single Palaearctic species *L. celtis*, which occurs in central Europe. The affinities of this subfamily have given rise to much discussion, and certain authorities relegate it to the Nymphalidae while others regard it as forming a separate family. The Nemeobiinae on the other hand are more nearly related to the Lycaenidae. The larvae of the Libytheinae bear considerable resemblance to those of Pieridae: each segment is divided into annulets, and numerous secondary setae are present. The pupae are short and smooth and suspended perpendicularly. Larvae of the Nemeobiinae exhibit marked diversity of form: in some cases they are onisciform, attenuated at the extremities, and covered with a variable development of secondary setae. Both larvae and pupae resemble those of the Lycaenidae rather than any other family.

FAM. LYCAENIDAE (Blues, Coppers, Hair-streaks). *Anterior tarsi of male more or less abbreviated, one or both claws absent.* A family of small to moderate-sized butterflies well represented in most regions. Over 280 species are Palaearctic, and 18 have been recognized as British, though several are either no longer met with or are casual and extremely rare. The predominant colour of the upper surface of the wings is metallic blue or coppery, dark brown, or orange; on the under side coloration is more sombre, with dark-centred eye-spots or delicate streaking. The antennae are ringed with white and a rim of white scales surrounds each eye; the hind wings are frequently provided with delicate tail-like prolongations. The legs are all functional and used for walking but, in the males, the anterior tarsi are more or less abbreviated, or with one or both claws wanting. The sexes frequently exhibit great differences in coloration; thus in *Lysandra corydon*, the male is pale shining blue and the female iridescent brown. The great majority of the larvae are onisciform, tapering towards the extremities, and with broad projecting sides concealing the limbs (Fig. 518A). This type of body-form resembles that of *Zygaena* more than of any

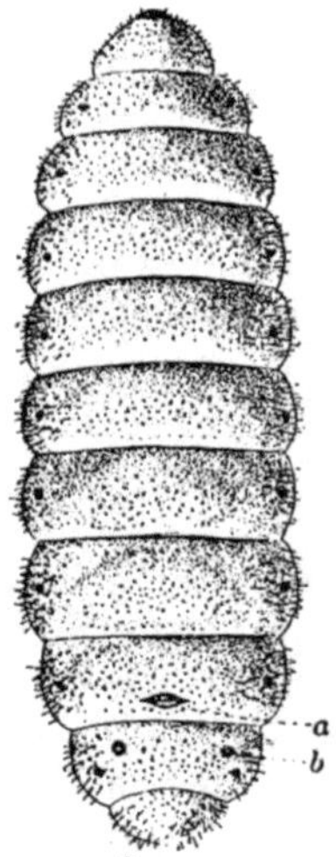

FIG. 518A
A Lycaenid larva

a, aperture of gland; *b*, one of the pairs of extensile organs. *After* Wheeler, 'Ants'.

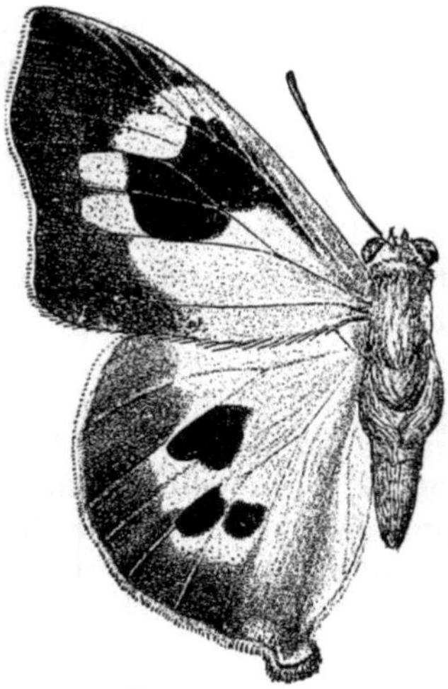

FIG. 518B
Liphyra brassolis (Lycaenidae) $\times \frac{1}{1}$

After Bingham (F.B.I.).

other Lepidoptera. Secondary setae are usually numerous, but some larvae are smooth or dorsally corrugated; many are clothed with a short pile, others are armed with bristle-bearing verrucae and a few are hairy. The pupa is relatively short and stout, anteriorly rounded, and with little or no freedom of motion in the abdominal segments, which fit together to form a smooth surface. Generally it is attached at the anal extremity and secured by a central girth of silk: there are, however, a number of exceptions and in some cases the pupa is subterranean. For an account of the metamorphosis of several species of the family the student is referred to a series of papers by Chapman (1911–20) and Malicky (1969). The larvae in some cases are known to be carnivorous: that of *Gerydus chinensis* feeds upon aphids in China (Kershaw, 1905). *Maculinea arion* (Europe) is phytophagous up to the last instar when it enters nests of *Myrmica* and becomes carnivorous preying upon the ant larvae (Chapman, 1916). The larva of the American *Feniseca tarquinius* is wholly carnivorous feeding upon woolly aphids (*Eriosoma*, etc.), while that of *Spalgis epius* is recorded by Green as preying upon coccids. Larvae of other species are frequently sought after by ants, who use their antennae to stroke them and induce them to yield drops of fluid secretion. The latter is apparently the product of a dorsal gland situated on the 7th abdominal segment (see Hinton, 1951). The Indo-Australian *Liphyra brassolis* (Fig. 518B) is the most remarkable member of the family, being totally unlike other forms in any of its stages (Chapman, 1902; 1903). Its larva is flattened, and has a very hard smooth sclerotized covering, devoid of evident segmentation: the jaws are sharply toothed and adapted for tearing and piercing rather than mastication. This curious larva is found associated with *Oecophylla smaragdina* and is believed to prey upon the brood of the latter, its hard covering serving as a protection against the ants. Pupation takes place in the larval skin: the pupa shrinks away from the cuticle and is loosely enclosed in the puparium thus formed. The newly emerged imago is covered with a number of loosely attached scales which may serve as protection against the ants, as they certainly cause the latter trouble when enveloped by them (Dodd, 1902). *Euliphyra mirifica* similarly frequents nests of the same ant in W. Africa, and its greatly modified larva has been described by Eltringham (1913).

FAM. PIERIDAE (Whites, etc.). *Legs normal. Hind wing with two anal veins.* Included in this family are some of the very commonest of all butterflies; they are mostly of medium size and usually either white, yellow or orange marked with black. The six legs are well-developed and similar in both sexes, and the claws of the feet are bifid or toothed. Several taxonomists have united this family with the Papilionidae, to form a single group, but the distinctness of the characters in the two cases does not appear to warrant this procedure. The larvae are rather elongate with the segments divided into annulets, and the body bears numerous secondary setae varying in size: the crochets are bi- or triordinal arranged in a mesoseries. The larvae are further characterized by the absence of osmeteria, fleshy filaments and cephalic or anal horns.

The pupae are suspended in an upright position attached by the caudal extremity and a central band of silk: they may be readily distinguished by the single median cephalic projection or spine, and the hind wings are not visible ventrally (Mosher). *Pieris* includes the common White or Cabbage butterflies whose larvae, in several species, are extremely destructive to cruciferous vegetables in Europe and N. America. In this respect *Pieris rapae* is probably the most injurious of all butterflies. Larvae of other members of the family feed chiefly on plants belonging to the Leguminosae and Capparidaceae. *Euchloe* and *Synchloe* include the 'Orange Tips', *Colias* the 'Clouded Yellows' and *Gonepteryx* the 'Brimstones' or 'Sulphurs': all are characteristic of the northern hemisphere. Certain species of Pieridae have the habit of migrating in large numbers, which has attracted the notice of travellers in many parts of the world (Williams, 1930). No certain explanation for these flights has been put forward: clouds of butterflies chiefly of *Appias* and *Catopsilia* may stream past the observer for hours at a time, all going in one direction (Bingham). Baker (1969) has suggested that in the spring (in the Old World) butterflies fly on a northerly gradient, using a sun-compass reaction. The reaction may be reversed in the autumn. He suggests that at lower temperatures, larger larvae and adults which lay more eggs are produced. In the autumn, however, it is advantageous to lay the eggs at higher temperatures so that they reach the overwintering stage before development ceases.

FAM. PAPILIONIDAE (Swallow-tails). *Legs normal. Hind wings with one anal vein.* An extensive family of pre-eminently tropical butterflies including some of the most magnificent of all insects. About 600 species are known; less than 70 of these are Palaearctic, and about 30 range into America north of Mexico. In the British Isles the sole representative is *Papilio machaon* which is local and now restricted to certain fenny districts in East Anglia. The wings of these insects are extraordinarily variable in shape and, in the majority of species, the hind pair is provided with conspicuous tail-like prolongations which are marginal extensions in the region of vein M_3. The prevailing ground colour is generally black, strikingly marked with shades of yellow, red, green or blue. The larvae are smooth or provided with a series of fleshy dorsal tubercles or sometimes with a raised prominence on the 4th segment. Except in *Parnassius*, in which secondary setae and verrucae are evident, the body is practically devoid of setae. An osmeterium is situated on the prothorax (see p. 1092) and when retracted its presence is revealed by a dorsal groove through which it is everted. The pupae are variable in form: the head bears two lateral cephalic projections and the hind wings are visible ventrally. Suspension takes place at the caudal extremity in an upright position, and the pupa is

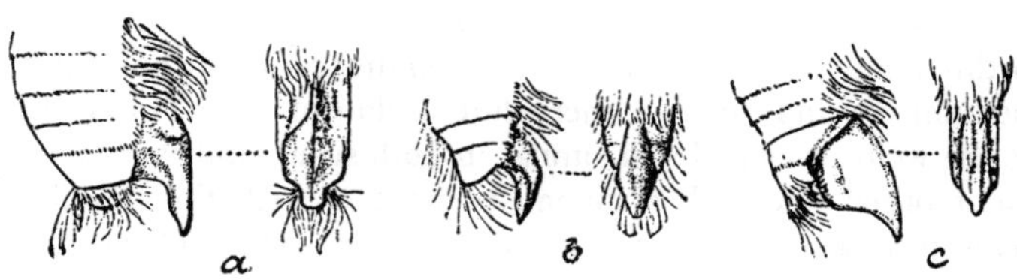

FIG. 519 Anal pouches (ventral and lateral aspects) of three species of *Parnassius* (Papilionidae)
After Bingham (F.B.I.).

further secured by a median silken girdle. In *Thais* there is a cephalic as well as an anal attachment and *Parnassius* is exceptional in that the pupa is not suspended but occurs in a slight silken web among leaves. The imagines of many species of the family have the sexes extraordinarily different both in form and colour, and often in habits also. In numerous instances the females are polymorphic while, in other cases, this peculiarity extends to both sexes. As examples may be mentioned the Oriental *Papilio memnon* which has three distinct forms in each sex and two of these in the female are tailless. The North American *Iphiclides ajax* has three distinct seasonal forms, viz. those appearing in early spring, late spring and summer. The African *Papilio dardanus* is represented by different races or subspecies in various regions of that continent, and each of these possesses from one to five different forms of the female which, for the most part, are close mimics of certain Danaine butterflies. The greater number of the species of the family are included in the genus *Papilio*: those of the *Ornithoptera* group comprise the finest of all butterflies and they form the subject of a sumptuous monograph by Rippon (1890–1910). In the Oriental genus *Leptocircus* the fore wings have a transparent scaleless band, and the tails are exceedingly long. *Parnassius* occurs in the mountains of the Holarctic region chiefly in Central Asia. Both pairs of wings are diaphanous, with few scales, and the tails are wanting. During copulation the females receive from the male a corneous anal pouch exhibiting specific variations in form (Fig. 519).

Superfamily **Hesperioidea**

Antennae dilated apically to form a gradual club which often ends in a hook, bases remote. Labial palps more or less rough-haired, maxillary palps wanting. Fore wing with Cu_2 wanting, M_2 arising from or below middle of transverse vein, none of the veins stalked and Rs with four branches. Hind wing with no frenulum except in male of Euschemon, *Cu_2 absent, $Sc + R_1$ arising out of cell near base, then rapidly diverging.*

FAM. HESPERIIDAE (Skippers). These insects derive their popular name from their erratic darting flight which is different from the more sustained aerial evolutions of other butterflies. They form an extremely large family, generally distributed, but not ranging into New Zealand (see Mabille and Boullett, 1909; also Mabille, 1903). The antennae are relatively widely separated at their bases, and their apices are generally prolonged beyond the club to form a small recurved point. The abdomen is stout, the wings are proportionately less ample than in most butterflies, and the venation of a markedly distinct type. As a general rule the larvae are moderately stout and taper towards both extremities; secondary setae are small, or absent dorsally, and the crochets are tri-ordinal arranged in a circle. In the Hesperiinae the head is large and attached to a strongly constricted 'collar' while in the Megathyminae it is small and partially retractile (Fracker). They frequently live concealed, drawing together leaves by means of silk, or inhabit webs or galleries: those of the Megathyminae are borers. The pupa is devoid of angular points or projections and is usually enclosed in a slight cocoon among leaves: in other cases it is exposed and attached by the caudal extremity, and also by means of a median band of silk. The eggs are spherical or oval, flattened beneath, smooth or reticulated, and sometimes ribbed (Meyrick). The vast majority of the species belong to the Hesperiinae and eight are indigenous to Britain. The Megathyminae include the Giant Skippers which have the apex of the antennae devoid of a recurved point, and the wing-veins are peculiarly specialized and greatly strengthened in the male. The group is mainly a tropical one and unrepresented in the Palaearctic region. The Euschemoninae were once regarded as a family of moths since the males possess a frenulum. *Euschemon* is the most archaic of all butterflies and according to Tillyard (1919*a*) its larvae and pupae exhibit definite Hesperiid characters.

Superfamily **Geometroidea**

Maxillary palps vestigial or atrophied. Tympanal organs in abdomen. Cu_2 absent from both wings: fore wings almost always with M_2 not nearer M_3 than M_1 at base and with $1A + 2A$ forming a basal fork.

FAM. DREPANIDAE (Hook Tips). *Tympanal organs dorsal except in many males. $Sc + R_1$ in hind wing remote from Rs. Fore wing with M_2 and M_3 approximated at base.* A rather small family mainly developed in the Indo-Malayan portion of the Oriental region. Its members exhibit considerable diversity of structure and, as a rule, have the apex of the fore wing falcate. The eggs are rounded-oval with the surface finely punctured. The larvae are somewhat slender without the claspers on segment 13, and the anal extremity is prolonged into a slender projection which is raised in repose; certain of the other segments are often humped. The pupa is enclosed in a cocoon, usually among leaves above ground. *Drepana* is the chief genus with 9 Palaearctic species. *Cilix* has the fore wings nonfalcate, and there are five species whose ranges extend from Britain to China.

FAM. THYATIRIDAE (CYMATOPHORIDAE). *Tympanal organs dorsal. Hind wing with* $Sc + R_1$ *close to or joined to Rs beyond the cell. Fore wing with* M_2 *not nearer to* M_3 *than to* M_4 *at base.* A relatively small family resembling the Noctuidae and mainly restricted to the northern hemisphere. Thirteen genera are Palaearctic, four being represented by common species in the British Isles, the most familiar being the 'Buff Arches' (*Habrosyne derasa*: Fig. 520) and the 'Peach Blossom' (*Thyatira batis*). The larvae are cylindrical and bear no secondary setae: the abdominal feet carry biordinal crochets arranged in a curved mesoseries, and the claspers are reduced in size. Pupation occurs in a rather slight cocoon among leaves.

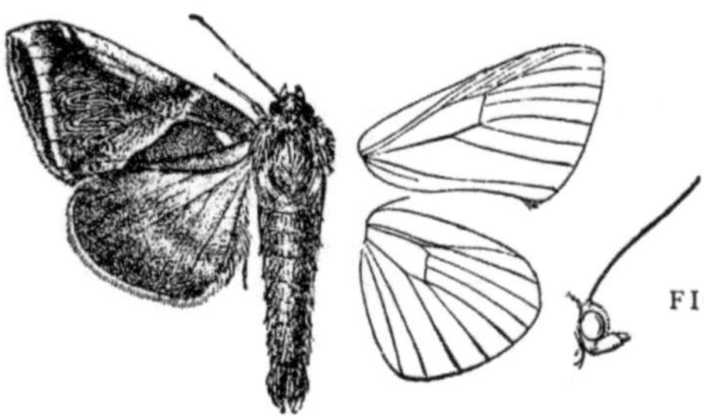

FIG. 520 *Habrosyne derasa* (Cymatophoridae), male, Palaearctic region. × $\frac{1}{1}$ *After* Hampson (F.B.I.).

FAM. GEOMETRIDAE (Carpets, Waves, Pugs, etc.). *Tympanal organs ventral. Hind wing with* $Sc + R_1$ *remote from Rs beyond the cell, furcate at extreme base. Fore wing with* R_{4+5} *connected with* R_{2+3}. A very large family consisting of some 12 000 species which are almost always of slender build with relatively large wings (Fig. 521). Their flight is never strong and, when at rest, the wings are often laid horizontally. Most of the adults have a cryptic colour-pattern, the method of whose evolution is illustrated in the industrial melanism of *Boarmia* (Kettlewell, 1973). Both frenulum and proboscis are generally present, but in a few cases either the one or the other may be wanting. In some genera the females have greatly degenerate wings or are completely apterous as in *Alsophila, Operophtera, Erannis* and *Apocheima* and its allies. The larvae are elongate and usually very slender: as a rule abdominal legs are only developed on the 6th and 10th segments and progression takes place by drawing the posterior somites close to those of the thorax, the body thus forming a loop. The whole body is then extended in the direction desired and the looping action repeated. In some instances abdominal legs appear on segments other than those normally carrying them. Thus in *Colotois pennaria* a pair is present on the 5th segment but disappears with the fourth moult while in *Alsophila aescularia* they are developed on the same segment and persist throughout the larval period. In *Archiearis notha* Sharp states that rudimentary abdominal feet are present on the 3rd to 5th segments in the newly hatched larva, but attain greater development when the latter is fully grown. It is evident that in this species the larva is much more a Noctuid than a Geometrid in its morphology. The vast majority of the larvae of the family bear an exceedingly close resemblance to twigs, or the thicker veins of leaves, and can only be detected with difficulty when at rest. McGuffin

(1958) gives a key to the subfamilies of larvae and deals at length with those of the Larentiinae. In the pupa there are no maxillary palps, the first two pairs of legs are longer than in most other Lepidoptera, and there is often a deep dorsal furrow between the 9th and 10th abdominal segments. A slight cocoon is spun between leaves or the pupa is subterranean. The family is divided into six subfamilies by Hampson while Meyrick (1928) regards each division as constituting a separate family under a different name. Over 3000 species occur in the Palaearctic region and, of these, about 270 are British. In *Boarmia*, and its allies, a fovea is present on the underside of the fore wing at the base of the anal region; it is generally confined to the male, is often hyaline and sometimes glandular. This structure is present in many species, and it is suggested by Meyrick that it may be a scent-producing organ. The posterior tibiae, also in the male, are often enlarged and contain an

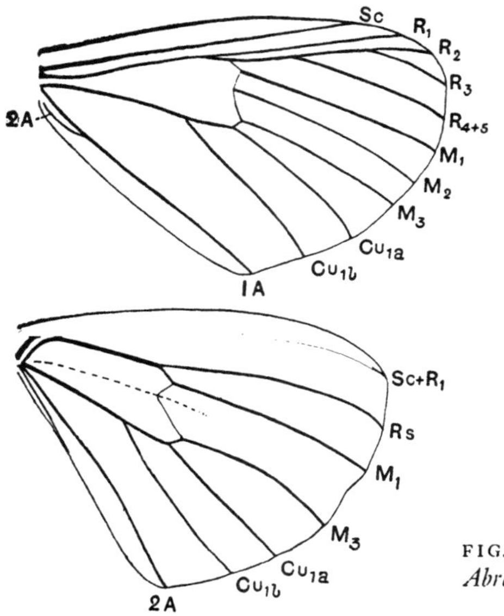

FIG. 521
Abraxas grossulariata (Geometridae), venation

expansible tuft of hairs. *Archiearis, Alsophila* and a few other genera are regarded by Meyrick as being the most primitive of all Geometers and constitute his family Monocteniadae. *Archiearis* has also been referred to a family of its own, while other authorities have regarded it as a Noctuid. The larvae of *Palaecrita* are known as Canker Worms and are pests of fruit and shade trees in N. America. Those of the Winter Moth (*Operophtera brumata*) and of species of *Erannis* are well-known defoliators of similar trees in Europe.

FAM. URANIIDAE. *Tympanal organs ventral in female, at sides of tergite 2 in male. Frenulum absent. Venation as in next family.* A very widely distributed but exclusively tropical family occurring in both the old and new worlds. They are often large slender-bodied moths, many of which are diurnal in habit. *Chrysiridia, Nyctalaemon* (Fig. 522) and *Urania* include exquisitely coloured insects resembling Papilionid butterflies: others bear a likeness to Geometrid moths. The larvae exhibit great diversity of structure but have the full number of abdominal limbs (Hampson, 1895; Gosse, 1881): in two genera they are known to feed on Euphorbiaceae. Those of *Nyctalaemon* and *Epicopeia* are figured by Hampson: in *E. polydora* (Himalayas) the body is invested with a thick covering of long white cottony filaments. In *Chrysiridia ripheus* there is an armature of black spatulate processes (Eltringham, 1924). The pupae are enclosed in loosely woven, silken cocoons. The Asiatic genus *Epicopeia* has a vestigial frenulum and is often relegated to a separate family – the Epicopeiidae – which has been monographed by Janet and Wytsmann (1903).

The small family **Sematuridae** with a few, mostly Neotropical species is sometimes considered allied to the Uraniidae, although it has no tympanal organs.

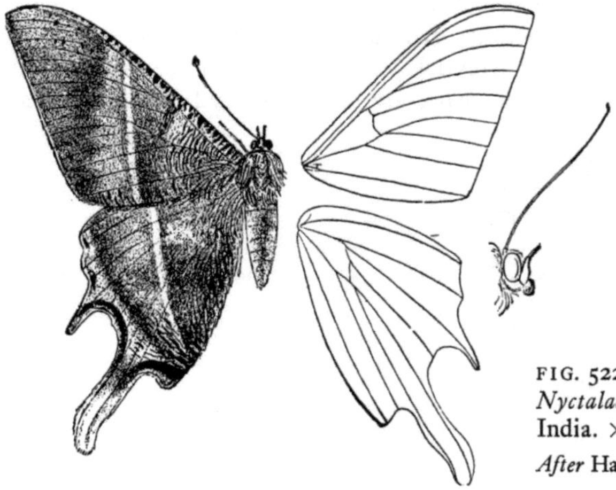

FIG. 522
Nyctalaemon patroclus (Uraniidae), male, India. × ½
After Hampson (F.B.I.).

FAM. EPIPLEMIDAE. *Tympanal organs ventral. Hind wing with Sc + R_1 remote from Rs beyond the cell. Fore wing with R_{4+5} remote from R_{2+3}, usually stalked with M_1, frenulum present.* A group of about 550 inconspicuous species, only doubtfully separate from the Uraniidae. They occur on all continents but are best developed in Papua and the adjacent islands. They commonly rest during the day with the fore wings rolled up in a peculiar manner while the hind pair is applied to the sides of the body: in this attitude they resemble spiders.

The small family **AXIIDAE** with tympanal organs in the seventh abdominal segment includes about 10 Mediterranean species.

Superfamily **Calliduloidea**

Maxillary palps and tympanal organs absent. Frenulum present though small. Proboscis present. Antennae simple or pectinate. Chaetosema present. Cu_2 absent from both wings. Fore wing with M_2 basally approximated to M_3.

The **Pterothysanidae** include slender moths with large wing-expanse bearing a resemblance to Geometridae. In the hind wing Sc + R_1 is remote from Rs beyond the cell. They are chiefly African, but the genus *Pterothysanus* inhabits the eastern Orient and is easily recognized by the long double hair-fringe which adorns the inner margin of the hind wings. The related family **Callidulidae** are day-flying moths, bearing a close resemblance to certain Thecline or other butterflies. In the hind wing Sc + R_1 is approximated to Rs beyond the cell. The family is essentially Oriental and does not occur in Europe.

Superfamily **Bombycoidea**

Maxillary palps and tympanal organs absent. Frenulum almost always atrophied or vestigial: proboscis rarely developed. Chaetosema absent. Antennae pectinated, especially in male. Cu_2 *absent from both wings: hind wing with* $Sc + R_1$ *usually diverging from cell and Rs, or only connected with cell by a cross-vein* (R_1).

The Bombycoidea or 'frenulum-losers' are chiefly distinguished by the loss or the absence of characters. Their main feature is the atrophy of the frenulum and, correlated with it, the basal widening of the humeral area of the hind wing. A frenulum occurs in some Bombycidae, but elsewhere only vestiges occasionally persist in the superfamily.

FAM. ENDROMIDIDAE. *Hind wing with* $Sc + R_1$ *connected to cell by a cross-vein,* M_2 *arising behind middle of cell, nearer to* Cu_{1a} *than to* M_1*. Proboscis absent.* The family includes only one species *Endromis versicolor*, which is a rather large day-flying moth, widely distributed in N. and C. Europe but extremely local in Britain. It frequents the vicinity of woods, its larvae feeding on birch and other trees. Seitz also includes with it the anomalous species *Mirina christophi* of Amurland.

FAM. LASIOCAMPIDAE (Eggars, Lappet-moths). *Hind wing with costal area greatly widened basally and supported by usually two or more stout humeral veins from subcostal cell between bases of Sc and* R_1. Usually moderate to large sized densely-scaled moths, with stout bodies, and the humeral lobe of the hind wings prominent. The proboscis is atrophied, there are no ocelli, and the antennae are bipectinated in both sexes. These insects are widely distributed with more than 1000 species but absent from New Zealand and are most abundant in the tropics. The eggs are smooth and oval, and the larvae stout with a more or less dense clothing of secondary hairs which obscure the primitive setae. They are often provided with lateral downwardly directed hair-flanges, and hairy subdorsal tufts or dorsal humps on the anterior segments. The full number of abdominal limbs is present, and the crochets are biordinal, arranged in a mesoseries. The pupae resemble those of the Bombycidae but differ in the presence of an epicranial sulcus and in the labial palps being unconcealed. The body is provided with numerous setae and there is no cremaster. A dense, rather firm, oval cocoon of hair and silk is commonly present and met with above ground. *Lasiocampa* is a small genus of large moths common in Europe and its species are usually swift fliers in sunshine: most members of the family, however, are nocturnal. *Malacosoma neustria* (Fig. 523) is the Lackey Moth of Europe, whose larvae live gregariously in webs during their earlier stages, and are very destructive to the foliage of fruit trees. The larvae of *M. americana* have a similar habit and are commonly known as 'tent-caterpillars', their webs measuring 2 feet or more in length.

FAM. ANTHELIDAE. *Frenulum present in male. Fore wing with one or two areoles (areas enclosed between anastomosing long veins), Sc in hind wing without short costal branches.* A family with nearly 100 species in Australia and New Guinea. Larvae with verrucae and dense setae. They feed on *Acacia, Eucalyptus*, etc., but sometimes transfer to the introduced *Pinus radiata*.

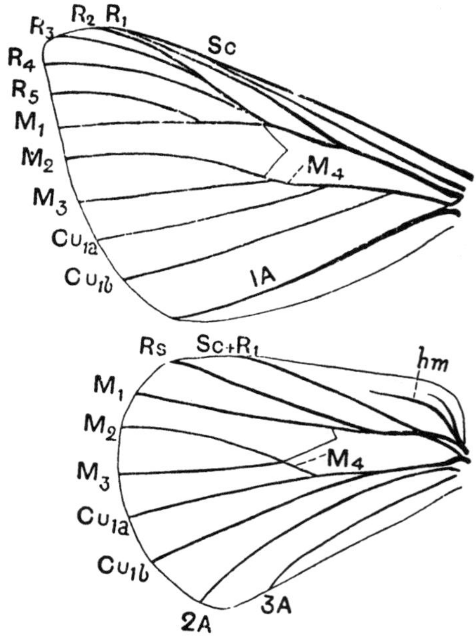

FIG. 523
Malacosoma neustria (Lasiocampidae), venation

FAM. LACOSOMIDAE (PEROPHORIDAE, MIMALLONIDAE). *Hind wing with $Sc + R_1$ diverging from cell base, M_2 arising at or beyond middle of cell, nearer M_1 than Cu_{1a}, frenulum vestigial, Cu_2 present as rudiments, two anal veins.* These insects are moderate sized rather stout-bodied moths found in N. and S. America. Their affinities are doubtful and they have been placed as exceptional members of the Drepanidae or Psychidae. They are remarkable on account of the larval habit of making suspended protective cases of the leaves of the food-plants. In some instances the case is only constructed by the mature larva, the latter living previously under a web, and in at least one species the larva constructs a covering of its own excrement. There is a considerable literature on these larvae and further information is given by Sharp (1899: 378). The life-history of *Lacosoma chiridota* is described by Dyar (1900).

FAM. BOMBYCIDAE (including **EUPTEROTIDAE**). *Hind wing with $Sc + R_1$ connected with cell by a cross-vein, fore wing with M_1 free or shortly stalked on Rs. Proboscis absent.* In this group a frenulum is usually present in *Eupterote* (Fig. 524) and its allies and absent in *Bombyx* and related genera. None is British and the family is mainly Ethiopian and Oriental. The antennae are markedly pectinate in both sexes and the larvae are of two types. In *Eupterote*, etc., they are tufted with long hair and secondary setae are always numerous, but distinct verrucae are wanting. In *Bombyx* and its allies the larvae are glabrous and elongate, usually with a mediodorsal horn on the 8th abdominal segment: they form dense silken cocoons. The larva of *Bombyx mori* is the well-known silkworm, an inhabitant of China which has been introduced into many parts of the world for commercial purposes. It is now entirely domesticated and is not known in the wild state. A number of local races exist, and these have been regarded by Hutton, Cotes and others as distinct species. They differ chiefly in the number of annual broods which are largely

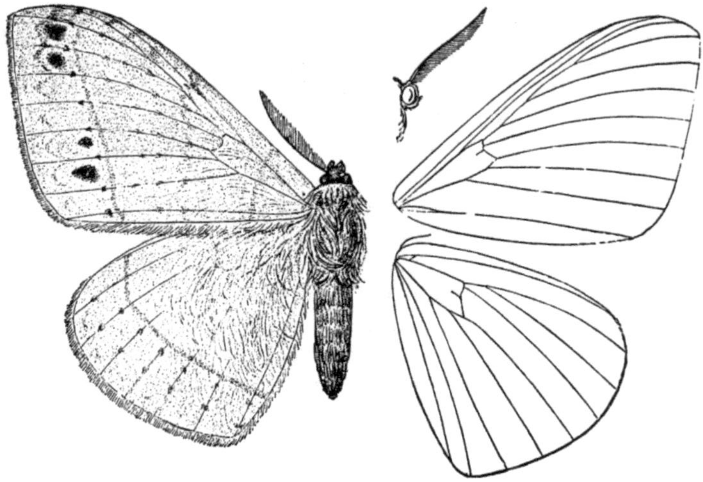

FIG. 524 *Eupterote fabia* (Bombycidae), male, India. × 1
After Hampson (F.B.I.).

dependent upon climate. The natural food in all cases is the leaves of mulberry, and the silk produced is white or yellow.

FAM. LEMONIIDAE. *Resembling the last family but fore wing with M_1 and Rs on a common stalk.* This family includes *Lemonia* with ten Palaearctic species. In the adults, the claw of the fore leg is much enlarged.

FAM. SATURNIIDAE (ATTACIDAE). *Hind wing with vein $Sc + R_1$ diverging from cell-base, M_2 arising at or in front of middle of cell, nearer M_1 than Cu_{1a}, Cu_2 and frenulum completely absent, very rarely more than one anal vein. Tibiae without spurs.* In this family (which includes the Hemileucidae) the antennae are prominently bipectinate in both sexes, the rami being longest in the males; the labial

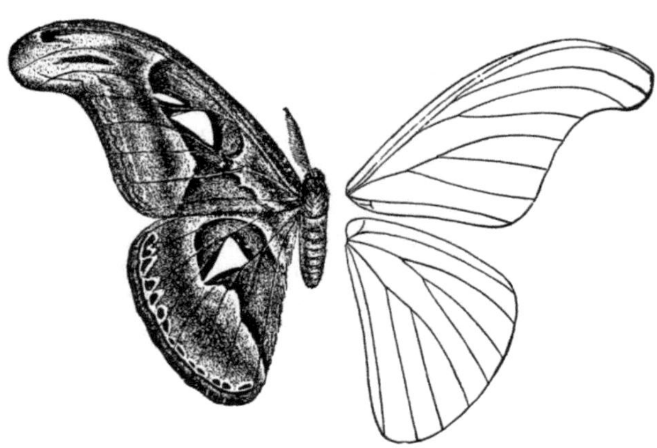

FIG. 525 *Attacus atlas* (Saturniidae), male, India × $\frac{1}{3}$
After Hampson (F.B.I.).

palps are minute and there is no frenulum. It includes a number of large, or very large, tropical insects with but few representatives in temperate regions; almost all are characterized by a transparent eye-spot near the centre of each wing. The only British species is the Emperor moth *Saturnia pavonia: S. pyri* is the largest European Lepidopterous insect. *Attacus* ranges from Mexico and S. America to Africa, and throughout the Oriental region to Japan. *A. atlas* (Fig. 525) and *A. edwardsi* are among the largest moths in the world, the females having a wing-expanse of about 25 cm. Saturniid larvae are very highly specialized (Fig. 526); they are stout and smooth and differ from most other families in possessing scoli or at least rudiments thereof. The position and number of the scoli vary very greatly in different genera, and for a detailed study of their arrangement reference should be made to Fracker's paper (1915); in *Saturnia* they are subequal in size on all the segments. The pupae have the antennae broadly pectinate in both sexes, with the

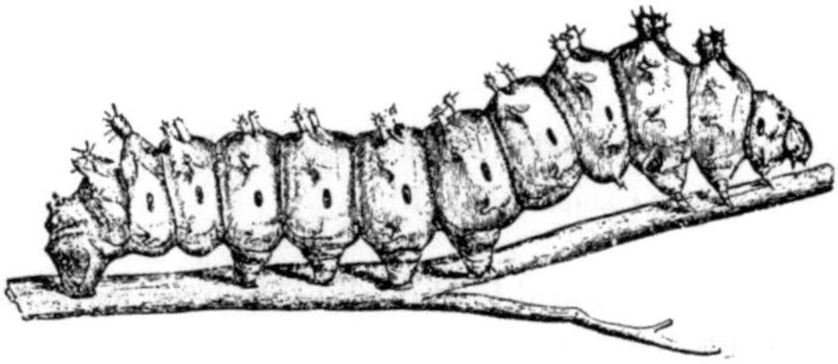

FIG. 526 *Platysamia cecropia* (Saturniidae), larva
After Riley.

axis of the flagellum very prominent. The maxillae are always short, not more than $\frac{1}{3}$ the length of the wings, and the cremaster, if present, is very short. A dense firm cocoon is always formed and is very characteristic of the family: several species yield silk of commercial value. *Antheraea yamamai* is the Japanese oak silkworm which is reared on a large scale in that country, and was introduced into Europe in 1861. *A. pernyi*, the Chinese oak silkworm, yields Shantung silk which is pale buff in colour and largely exported. *A. paphia* and *assama* are polyphagous. See Lefroy and Ghosh (1912), Packard (1914) and Michener (1952).

FAM. RATARDIDAE. A small group found in the Oriental region. The median vein is preserved within the cell of the fore wing.

FAM. CITHARONIIDAE (CERATOCAMPIDAE). *Hind wing with vein $Sc + R_1$ diverging from cell-base, M_2 arising at or in front of middle of cell, nearer to M_1 than to Cu_{1a}, two anal veins. Tibiae with spurs.* Large or medium sized moths with stout hairy bodies, and powerful wings. The antennae are bipectinate for about half their length only, and both proboscis and tibial spines are present. The family is a small one, unknown in Europe, but well represented in N. America. The larvae are thinly hairy, and are armed with unbranched scoli on the 1st to 6th abdominal segments, and a large mediodorsal scolus on the 8th segment. The pupae are roughened with spines on the thorax and abdomen, the metathorax is provided with oblong lateral tubercles, and the cremaster is bifurcate. Transformation occurs in the ground, no cocoon being formed. For a monograph of the family see Packard (1905; 1914): keys to the larva are given by Fracker (1915), and the pupae have been studied by Mosher (1914).

FAM. BRAHMAEIDAE. *Hind wing with vein $Sc + R_1$ basally approximated to Rs beyond the cell. Proboscis present.* A very small group of tropical moths related to the Saturniidae, but readily distinguishable by the presence of a proboscis, and the large upturned labial palpi. They are large, sombre-coloured insects with very complex wing-patterns, and the antennae are bipectinated in both sexes. *Brahmaea* occurs in Africa and through southern Palaearctic Asia to China: the life-history is described by Packard (1914).

FAM. CARTHAEIDAE. *$Sc + R_1$ approaching cell near base and not connected by a thin cross-vein. Proboscis present, labial palpi conspicuous. Frenulum present.* One large species in Western Australia whose larvae feed on *Dryandra* (Proteaceae) (Common, 1966).

Two other small families described by Jordan (1924) are S. American and their members have $Sc + R_1$ similarly connected to the cell and have a short proboscis but they lack a frenulum; these are the **Oxytenidae** and the **Cercophanidae**.

Superfamily **Sphingoidea**

Antennae gradually thickened into a club with the apex pointed and usually hooked. Proboscis and frenulum almost always strongly developed. Cu_2 absent from both wings. Fore wing with M_1 arising from stem of R_{3-5} or basally approximated to it. Hind wing with $Sc + R_1$ connected with cell by a cross-vein (R_1) and approximated to Rs beyond the cell. Tympanal organs absent.

A somewhat isolated group with a single family whose affinities lie towards the Notodontidae.

FAM. SPHINGIDAE (Hawk Moths). An important family of moderate-sized to very large moths, including at least 1000 species, which are distributed over almost the whole world. It is essentially a tropical group which is represented in the British Isles by 8 genera and 17 species. *Hyles* (= *Celerio*) *lineata* is cosmopolitan and others such as *Acherontia atropos, Daphnis nerii* and *Agrius* (= *Herse*) *convolvuli* (Fig. 527) have a very wide geographical range. The imagines are easily recognizable by the elongate fore wings and their very oblique outer margin. The antennae are thickened towards or beyond the middle and are pointed at the apices which are nearly always hooked: in the male the antennae are ciliated with partial whorls. The proboscis may be developed to a length which is not attained by any other Lepidoptera, but it is very variable. In *Cocytius* (tropical America) it measures 25 cm long while the opposite extreme is found in *Polyptychus* in which it is reduced to a pair of tubercles. The frenulum and retinaculum are present in all generalized forms, but in some instances they are reduced or vestigial. In the Humming Bird Moths (*Macroglossum*) and the Bee Hawk Moths (*Hemaris*) the apex of the abdomen is provided with an expansile, truncated tuft of hairs. In the latter genus the disc of the wings is transparent, the fugitive scales present on newly-emerged specimens being very quickly lost. Sphingidae have an exceptionally powerful flight and hover over flowers as they feed on the wing: most are crepuscular and nocturnal but a few (*Macroglossum, Hemaris*, etc.) are diurnal.

The larvae are smooth, or with a granulated skin, but the latter feature is often only present in the first instar. The 8th abdominal segment almost always bears an obliquely projecting dorsal horn – relatively longer in the first than the later instars. The pupa occurs free in a cell in the ground, or in a very loose cocoon on the surface, between leaves, etc. The 5th and 6th abdominal segments are free and there is always a cremaster. Various methods of accommodating the proboscis are noticeable in the pupa and in some genera this organ projects from the body in a conspicuous manner so as to resemble the handle of a pitcher. Harris (1972) gives an interesting account of the relation between larval food and adult taxonomy.

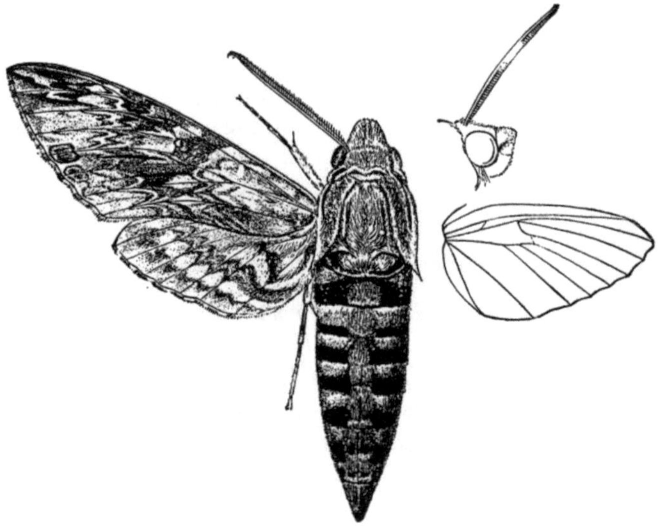

FIG. 527 *Agrius convolvuli* (Sphingidae), male and venation of hind wing. × $\frac{1}{1}$
After Hampson (F.B.I.).

The Death's Head moths (*Acherontia*) are remarkable in several respects: the imagines have been noted to enter beehives to rob them of honey, and they possess the faculty of sound production. The note emitted is a shrill chirping sound and many hypotheses have been advanced to account for it. The literature thereon is fully discussed by Tutt (1899); the sound was originally attributed to friction but it seems probable that the real cause is the forcing of air from the pharynx through the base of the proboscis (see p. 185). The imago is occasionally audible before emergence from the pupa, but the larva emits a different type of sound. Most observers agree that when irritated it produces a series of rapidly repeated 'cracking' notes resembling those emitted during the discharge of successive electric sparks, and the sounds are made by the mandibles when sharply brought together. The principal works on the family are those of Rothschild and Jordan (1903; 1907) and Carcasson (1968) deals with the Ethiopian (especially E. African) species. For a general study of the larvae, consult Forbes (1911), while the larval colour changes and their significance are discussed by Piepers (1897); for the life-history of *H. convolvuli*, see Poulton (1888).

Superfamily Noctuoidea

Maxillary palpi minute. Tympanal organs present in metathorax. Cu_2 *absent from both wings; fore wing usually with* M_2 *basally approximated to* M_3 *and with 1A + 2A not forming a definite basal fork.*

FAM. NOTODONTIDAE (Prominents, etc.). *Fore wing with* M_2 *parallel to* M_3 *or approximated to* M_1. Insects with moderately stout bodies and rather elongate fore wings: they are generally distributed but absent from New Zealand and poorly represented in Australia. The imagines are exclusively nocturnal, and are often attracted to a light, otherwise the various species are usually only obtained as larvae. A large number of the larvae of this family are well figured by Packard (1896); they mostly feed exposed on trees and shrubs, seldom affecting herbaceous plants. According to Fracker all exhibit secondary setae on the abdominal limbs and, in some genera (*Phalera*, etc.), these setae are present on the body also. The anal claspers are frequently modified into slender processes which are erected when in repose: the latter habit is also exhibited by *Notodonta* which has the claspers unmodified. The pupa only exhibits a small proximal portion of the labial palpi, maxillary palpi are absent, and the maxillae do not reach the caudal margin of the wings: the abdomen is punctate and a cremaster usually present (Mosher). *Notodonta* is characteristic of the temperate regions of the northern hemisphere and in this genus, *Ptilodon* (= *Lophopterix*) and others there is a tuft of projecting scales on the middle of the hind margin of the fore wings. *Stauropus* is Indo-Malayan with a single European species *S. fagi*: its larva is very remarkable on account of the great length of the 2nd and 3rd pairs of thoracic legs. The anal extremity is inflated and claspers are replaced by two slender processes. In repose both extremities are abruptly erected, and in the curious attitude thus presented the larva, when irritated, has been regarded by Müller as resembling a spider; when at rest it was compared by Birchall to a twig with unopened buds, and by other observers to a dead and crumpled leaf. The larva of *Dicranura vinula* is a very striking and familiar object: it is provided with a pair of roughened tubercles on the prothorax, and a prominent fleshy protuberance on the metathorax. The anal claspers are modified into a pair of long slender processes containing extensible filaments, and the histology and mechanism of these organs have been investigated by Poulton (1887). This larva, and also those of other members of the family, is provided with a ventral prothoracic gland (Latter, 1897) having the power of ejecting an irritating fluid. The latter in the case of *D. vinula* has been found to consist of formic acid. The pupa in this species, and in those of *Cerura*, is enclosed in a hard wood-like cocoon on the bark of trees. The escape of the imago is facilitated by the cocoon being thinner anteriorly and the labrum of the imago bears two sharply pointed processes used for scraping the inner surface of the cocoon, in order to break a way through. At the same time, a secretion of potassium hydroxide is produced from the mouth in order to soften the cocoon. The eyes, and median portion of the head of the pupa, persist as a shield protecting those same parts in the imago until emergence is effected (Latter, 1892; 1895).

FAM. DIOPTIDAE. *First anal vein usually indicated in the hind wing. Fore wing with* M_3 *and* Cu_1 *stalked together.* Quite closely allied to the previous family but much more slender moths, wings often brightly coloured with transparent windows. A few hundred American, mostly Neotropical moths. The Californian *Phryganidia* occasionally defoliates oak-trees.

FAM. CTENUCHIDAE (AMATIDAE, SYNTOMIDAE). *Fore wing with $Sc + R_1$ fused with Rs, M_2 basally approximated to M_3.* This family comprises about 2000 species and is most abundant in the tropics; no representatives are indigenous to the British Isles and *Syntomis phegea* is the commonest of the few European forms. They are small to medium-sized moths (Fig. 528), usually inactive and largely diurnal in habit. The proboscis is generally well developed, the labial palpi are small and porrect and the retinaculum bar-shaped. Although often included among the Zygaenidae, they appear to be nearest related to the Arctiidae. Many are brilliantly coloured, and a number of species bear a striking resemblance to Aculeata, Tenthredinidae and other insects (Kaye, 1913). The resemblance is heightened by the frequently basally constricted abdomen and the general shape and coloration; in many cases the wings have transparent areas devoid of scales. In the Neotropical genus *Trichura* the males of certain species are provided with a long filamentous appendage arising from the terminal abdominal segment. This structure attains a length equal to that of the whole body of the insect, but its significance appears to be unexplained. The larvae are short, and armed with verrucae bearing numerous setae and they closely resemble those of the Arctiidae. They feed on grasses or (Comstock and Henne, 1967) lichens (*Parmelia*). Pupation takes place in a cocoon of silk and felted hairs: according to Mosher the pupa of *Ctenucha* is indistinguishable from that of an Arctiid.

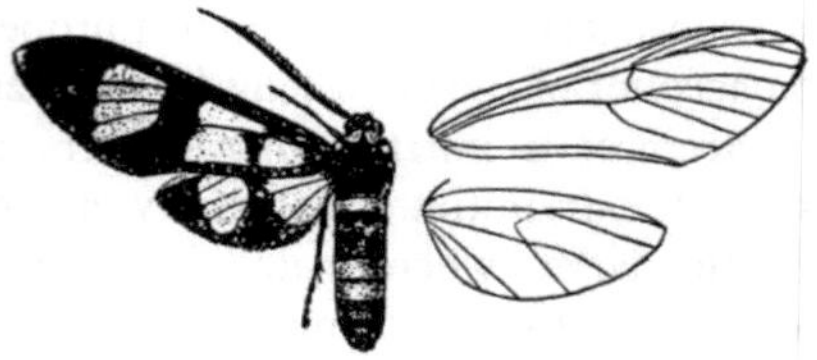

FIG. 528
Euchromia polymena (Ctenuchidae), male, India. × $\frac{1}{1}$
After Hampson (F.B.I.).

FAM. ARCTIIDAE (LITHOSIIDAE: Tiger Moths, etc.). *Fore wing with M_2 basally approximated to M_3, $Sc + R_1$ separate from Rs. Hind wing with $Sc + R_1$ anastomosing with cell to or to beyond middle.* An assemblage of usually stout-bodied moths, often with moderately broad wings, which are frequently conspicuously spotted, banded or otherwise marked with bright colours. Most species are nocturnal in habit and are attracted to a light. The family is tolerably well represented in nearly all zoogeographical regions, but attains its greatest development in the tropics. Over 3500 species are known and, of these, 40 inhabit the British Isles. According to Meyrick *Callimorpha* (= *Panaxia*) is the most ancestral form, but it is placed by Hampson in the Hypsidae. In its general affinities the family comes nearest to the Noctuidae. It is noteworthy that species of several genera are known to be capable of sound-production, but the mechanism thereof has not been adequately studied.

The Arctiinae comprise the 'Tiger' and 'Ermine' moths with their allies. They are brightly coloured insects with extremely diverse patterns, and individual species exhibit an extraordinarily wide range of variation with respect to the latter. The extensive genus *Arctia* includes the common 'Tiger' Moth (*A. caja*) which extends through the northern Palaearctic region to Japan. *Utetheisa pulchella*, although casual in Britain, occurs through the greater part of the Old World, with very similar species in Australia. The larvae are clothed with dense long hairs which they utilize along with silk to construct their cocoons; those of the Palaearctic species hibernate and feed principally upon low herbaceous plants. The Lithosiinae include those moths which are popularly termed 'Footmen'; they are diurnal or crepuscular in habit and, in typical genera, the fore wings are long and very narrow. The larvae are sparsely hairy, and commonly feed upon lichens growing about tree trunks and in other situations.

FAM. AGARISTIDAE (PHALAENOIDIDAE). *Hind wing with $Sc + R_1$ separated from Rs, anastomosing with cell near base. Antennae with shaft more or less dilated distally.* A small family absent from Europe and including over 60 genera embracing about 300 species. They are largely tropical, only two Palaearctic species being listed by Staudinger and Rebel; others occur in N. America and Australia. In general facies and vivid coloration they resemble the Arctiinae and many are diurnal in habit (Fig. 529). They are very similar to the Noctuidae in

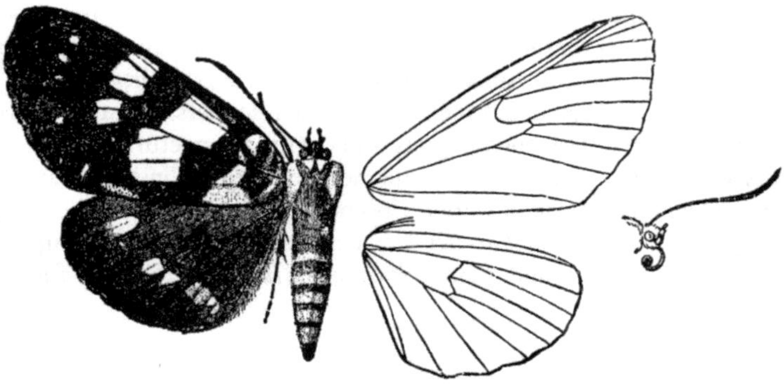

FIG. 529 *Eusemia adulatrix* (Agaristidae), female, India. × $\frac{1}{1}$
After Hampson (F.B.I.).

structure, and also in larval features, but their type of coloration and antennal characters serve to distinguish them. According to Hampson (1892) in *Aegocera tripartita* a portion of the membrane of the anterior wing is dilated and ribbed; a clicking sound is produced during flight probably by friction on the greatly enlarged mid-tarsal spines.

FAM. NOCTUIDAE (AGROTIDAE). *Hind wing with $Sc + R_1$ separate from Rs, connected with the cell by a bar. Antennae with the shaft not dilated* (Figs. 530, 531). This family includes a larger number of described species than any other group of Lepidoptera; about 1800 are Palaearctic (cf., Kozhantshikov, 1937) and approximately 3500 are known from N. America. They are eminently nocturnal insects attracted to a light and to the collector's sugar mixture, while *Plusia* and its allies frequent flowers at dusk. The family exhibits a monotonous similarity of structure particularly with regard to the venation and labial palpi; the maxillary palpi are vestigial. A frenulum is always present and the proboscis very rarely atrophied. In some species, the proboscis is stout enough to pierce fruit or even the skin of vertebrates. Allied species with a more flexible proboscis feed at wounds or take in liquid from the eye (Bänziger, 1968, 1969). The colour of the fore wings is nearly always cryptic and sombre, thus assimilating the insect to its surroundings

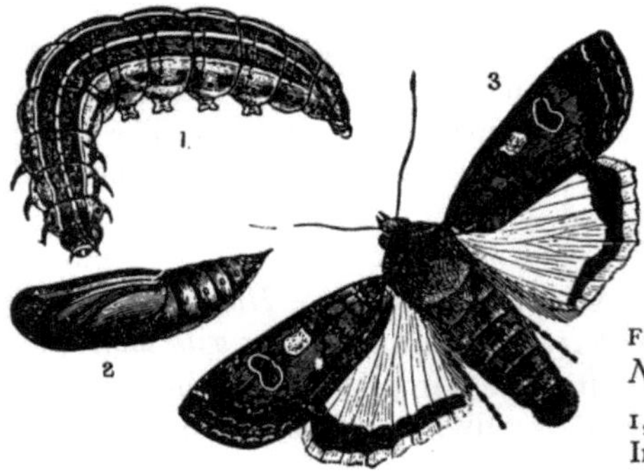

FIG. 530
Noctua pronuba (Noctuidae), Europe
1, larva; 2, pupa; 3, imago × $\frac{1}{1}$. *After* Curtis, 'Farm Insects'.

(Figs. 530, 531). Being protected in this manner it passes the day resting with folded wings on tree-trunks, etc., to a large extent concealed from its enemies. In the larvae of the majority of species primary setae only are present, and the crochets are generally in a uniordinal mesoseries. There are usually four pairs of abdominal feet, but among the Catocalinae, Plusiinae and Hypeninae the 1st pair, or the 1st and 2nd pairs, are more or less aborted and the larvae are semiloopers. Most of the larva feed upon foliage; they are often polyphagous and many are nocturnal while a few are stem-borers and live concealed. Among the more exceptional instances are the larvae of *Eublemma* which are predacious upon Coccoidea, those of *E. amabilis* being one of the most important enemies of *Laccifer lacca*. Species of *Cryphia* feed upon lichens; *Nonagria* lives in stems of marsh plants; *Parascotia fuliginaria* utilizes fungi growing

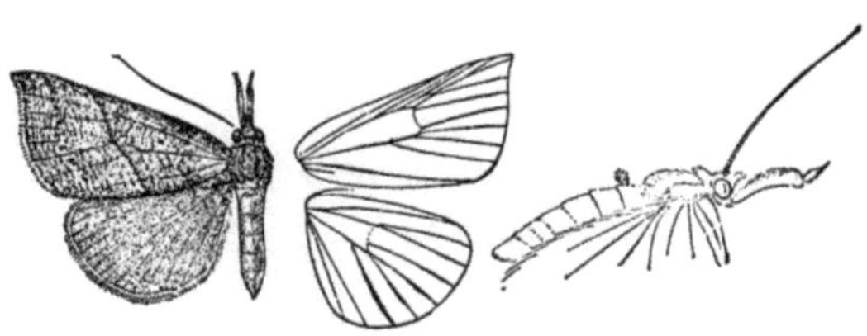

FIG. 531 *Hypena proboscidalis* (Hypeninae), male, Palaearctic region. × $\frac{1}{1}$
After Hampson (F.B.I.).

on rotting wood; and larvae of *Hadena* often select the seed capsules of Caryophyllaceae. The larvae of the American species have been tabulated by Crumb (1956; see also Ripley, 1925) and of the European ones by Beck (1960). According to Mosher the pupae, with few exceptions, are characterized by the presence of labial palpi and of maxillae which extend to the caudal margin of the wings. Numerous genera have the prothoracic epimera exposed, and those lacking labial palps possess setae arranged around the scars of the larval verrucae as in Arctiidae. They differ from the latter, however, in that the cremaster bears hooked setae. Pupation takes place as a rule in an earthen cell below ground, and the pupal cuticle is retained within the cell by the cremaster: in *Plusia* and its allies a cocoon is usually present and is spun between leaves, etc. The eggs of Noctuidae are spherical and generally ribbed and reticulated.

Certain Noctuid larvae (*Agrotis*, etc.) are known as 'Cut-worms'; they are more or less abundant every year and in N. America rank among the worst of insect pests (see Gibson, 1915*b*). The larva of *Leucania unipuncta* is the notorious and almost cosmopolitan 'Army Worm' so called from its habit of appearing in enormous numbers; as food becomes exhausted these larvae assume a gregarious marching habit seeking fresh fields. It is particularly injurious to cereals in the United States and Canada, and for a full account of its habits see Gibson (1915*a*). Some of these larvae exhibit a 'phase' variation in colour analogous to that seen in locusts (Faure, 1943). The larvae of the Antler Moth *Cerapteryx graminis* are periodically exceedingly destructive to upland pastures in N.W. Europe; a severe outbreak in Britain took place in 1917. Among other destructive species is *Aletia argillacea* whose larva is the well-known Cotton Worm of N. America; that of *Heliothis armigera* is the Boll-worm which is injurious to cotton bolls and the fruit of other economic plants on that same continent. *Diataraxia oleracea* is a serious pest in tomato houses in England (Lloyd, 1920).

The Chloeophorinae are a very small group which is sometimes regarded as a separate family (Cymbidae) or placed in the Arctiidae: they are frequently green insects found among the herbage of trees and shrubs. The larvae are never prominently hairy and the cocoon is boat-shaped. In *Pseudoips* (*Halias*) the larva is smooth and feeds in the open while in *Earias* it is hirsute and lives among rolled leaves, etc.; that of *E. insulana* is the destructive Egyptian Cotton Boll-worm, widely distributed in the tropics.

The small family **Nolidae** with less than a hundred species resembles the Noctuidae but they are smaller than most of that family and have scale-tufts on the wings. In the hind wing Sc + R_1 fuses with Rs from its base to near the middle of the wing, as in a few Noctuidae.

FAM. LYMANTRIIDAE (LIPARIDAE, OCNERIIDAE: Tussock Moths). *Hind wing with* $Sc + R_1$ *separate from Rs. Proboscis atrophied.* The Lymantriidae are mostly moderate-sized insects, rarely brilliantly coloured, and the antennae of the males are very prominently bipectinate to the apex. The family is hardly distinguishable from the Noctuidae on any venational feature: as a rule the bipectinate male antennae, and the absence of ocelli, afford more easily recognizable characters. The caudal extremity of the female is often provided with a large tuft of anal hairs which are deposited as a covering for the egg-masses. The larvae are hairy, generally densely so, often with thick compact dorsal tufts on certain segments (Fig. 532). Osmeteria are frequently present on the 6th and 7th abdominal segments. Larvae of the common European Gold Tail, *Euproctis similis* (= *chrysorrhoea*) are provided with urticating hairs composed of barbed spicules. It appears uncertain whether their irritating properties are mechanical only, or are partly due to a poisonous secretion bathing these spicules. Eltringham (1913) has shown that the female collects the spicules, which are present on the cocoon, by brushing the latter with the anal tuft, and subsequently distributes them over the egg-mass. The pupae are enclosed in cocoons above ground, and are characterized by the presence of very evident setae arranged around the scars of the larval verrucae. The best known member of this family is *Lymantria dispar*, the common Gipsy Moth of Europe, which was introduced into N. America about 1868 along with the Brown Tail (*Euproctis phaeorrhoea*). These species have now become serious pests of shade and foliage trees on that continent. *Lymantria monacha* is the Nun Moth whose larvae are often a serious pest in the forests of Germany. In *Orgyia* wings are vestigial or absent in the female (Fig. 532).

The **Hypsidae** differ from the preceding family in the presence of a well-developed proboscis. The larvae are thickly covered with long hairs and construct a slight pupal cocoon. *Hypsa* occurs in Africa, throughout the Orient and allied genera in tropical Australia. The **Thaumetopoeidae** are a very small Palaearctic family of few species, perhaps better treated as a subfamily of the Notodontidae. The larvae are tufted with long hair and secondary setae are always numerous, but distinct verrucae are wanting. Larvae of *Thaumetopoea* are known as processionary caterpil-

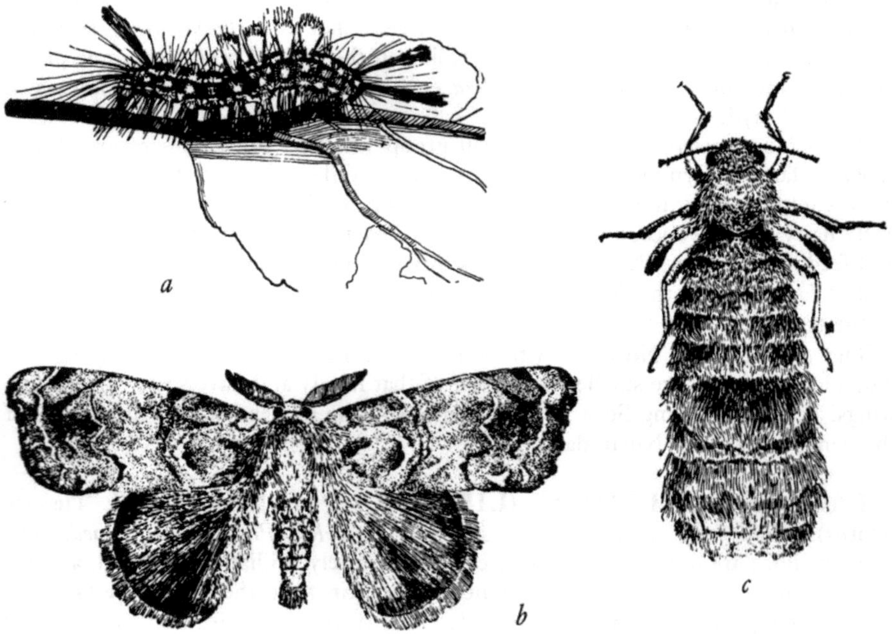

FIG. 532 *Hemerocampo vetusta* (Lymantriidae) *a*, larva; *b*, male; *c*, female
From Essig, E. O. Insects and mites of western North America. 1958. figs. 692, 693, 694

lars and exhibit gregarious habits. *T. processionea* is the well-known European processionary moth. Its larvae march in columns, each being headed by a leader, the column gradually becoming broader behind. It is believed that the individuals guide themselves and maintain their positions by means of threads spun by the leaders of each of the files. Brindley (1910) has observed these columns in the case of *T. pinivora* and conducted a series of experiments. He concludes, however, that the threads secreted by individuals on the march are of very slight importance either in forming the procession or maintaining its integrity. The larvae endeavour to maintain head and tail contact with the members of their file and this appears to be of primary significance in forming the procession.

Literature on Lepidoptera

BAKER, R. R. (1969), The evolution of the migratory habit in butterflies, *J. Anim. Ecol.*, **38**, 703–746.

BÄNZIGER, H. (1968), Preliminary observations on a skin-piercing bloodsucking moth (*Calyptra eustrigata* (Hmps.)) (Lep., Noctuidae) in Malaya, *Bull. ent. Res.*, **59**, 159–163.

—— (1969), Erste Beobachtungen über früchtestechende Noctuiden in Europa, *Mitt. schweiz. ent. Ges.*, **42**, 1–10.

BARASCH, A. (1934), Natürliche Gruppierung der mitteleuropäischen Coleophoriden (Lep.). etc., *Dt. ent. Z.*, **1934**, 1–116.

BARRETT, C. G. (1893–1907), *The Lepidoptera of the British Isles*, **1–11**, L. Reeve & Co., London.

BECK, H. (1960), *Die Larvalsystematik der Eulen (Noctuidae) aus dem Zoologischen Institut der Universität Erlangen*, Abh. Larvalsyst. Ins. Nr. 4, Berlin, 406 pp.

BEESON, C. F. C. (1910), The life history of the Toon Shoot and Fruit Borer, *Hypsipyla robusta*, Moore, etc., *Indian For. Rec.*, **7**, 71 pp.

BERG, K. (1941), Contributions to the biology of the aquatic moth *Acentropus niveus* (Oliv.), *Vidensk. Meddr dansk natur. Foren.*, **105**, 57–139.

BHARADWAJ, R. K., CHANDRAN, R. S. AND CHADWICK, L. E. (1974), The cervical and thoracic musculature of Lepidoptera. I, II & III, *J. nat. Hist.*, 8, 291–300; 311–331; 333–343.

BIRCH, M. C. (1972), Male abdominal brush-organs in British Noctuid moths and their value as a taxonomic character. I, II, *Entomologist*, **105**, 185–205; 233–244.

BLANC, L. (1889), La tête du *Bombyx mori* à l'état larvaire, anatomie et physiologie, Extrait de *Trav. Lab. études de la Soie*, Lyon.

BLASCHKE, P. (1914), *Die Raupen Europas mit ihren Futterpflanzen*, The author, Annaberg.

BLEST, A. D., COLLETT, T. S. AND PYE, J. D. (1963), The generation of ultrasonic signals by a New World Arctiid moth, *Proc. R. ent. Soc. Lond.* (B), **158**, 196–207.

BLESZYŃSKI, H. (1963), Studies on the Crambidae (Lepidoptera), Pt 40, A review of the genera of the family Crambidae with data on their synonymy and types, *Acta zool. Cracoviensia*, 8, 91–132.

BORDAS, L. (1910), Les glandes céphaliques des chenilles de Lépidoptères, *Ann. Sci. nat. Zool.*, (9), **10**, 125–198.

—— (1911), L'appareil digestif et les tubes de Malpighi des larves des Lépidoptères, *Ann. Sci. nat. Zool.*, (10), **14**, 191–274.

—— (1920), Étude anatomique et histologique de l'appareil digestif des Lépidoptères adultes, *Ann. Sci nat. Zool.*, (10), **3**, 175–250.

BÖRNER, C. (1939), Die Grundlagen meines Lepidopterensystems, *Verh. VII Int. Kongr. Ent. 1938*, **2**, 1372–1424.

BOURGOGNE, J. (1949), Un type nouveau d'appareil génital femelle chez les Lépidoptères, *Ann. Soc. ent. France*, **105**, 69–79.

—— (1951), Lépidoptères, In: Grassé, P. P., *Traité de Zoologie*, **10**, fasc. 1. Masson et Cie., Paris.

BRADLEY, J. D. (1956), A new clearwing moth from West Africa predaceous on scale-insects (Lep., Aegeriidae), *Entomologist*, **89**, 203–205.

BRADLEY, J. D. AND TREMEWAN, W. G. (1973), *British Tortricoid moths. Conchylidae and Tortricinae*, Ray Soc., London.

BRANDT, E. (1879), Vergleichend-anatomische Untersuchungen über das Nervensystem der Lepidopteren, *Hor. Soc. ent. Ross.*, **15**, 68–83.

—— (1880), Ueber die Anatomie der *Hepialus humuli*, *Zool. Anz.*, **3**, 186–187.

—— (1890), On the anatomy of *Sesia tipuliformis* and *Trochilium apiforme*, L. (Transl.), *Ann. Mag. nat. Hist.*, (6), **6**, 185–190; from *Hor. Soc. ent. Ross.*, 1889: 332.

BRAUN, A. F. (1917), Observations on the pupal wings of *Nepticula* with comparative notes on other genera, *Ann. ent. Soc. Am.*, **10**, 233–239.

—— (1919), Wing structure of Lepidoptera and the phylogenetic and taxonomic value of certain persistent Trichopterous characters, *Ann. ent. Soc. Am.*, **12**, 349–366.

—— (1924), The frenulum and its retinaculum in the Lepidoptera, *Ann. ent. Soc. Am.*, **17**, 234–256.

—— (1948), The Elachistidae of North America (Microlepidoptera), *Mem. Amer. ent. Soc.*, **13**, 110.

BRETSCHNEIDER, F. (1921), Ueber das Gehirn des Wolfsmilchschwärmers (*Deilephila euphoriae*), *Z. Naturw.*, **57**, 423–462.

—— (1924), Ueber das Gehirn eines Bärenspinners (*Callimorpha dominula*, die Jungfer), *Z. Naturw.*, **60**, 147–173.

BRINDLEY, H. H. (1910), Further notes on the procession of *Cnethocampa pinivora*, *Proc. Cambr. philos. Soc.*, **15**, 575–580.

BROCK, J. P. (1971), A contribution towards an understanding of the morphology and phylogeny of the Ditrysian Lepidoptera, *J. nat. Hist.*, **5**, 29–102.

BUCKLER, W. (1887–99), *The Larvae of British Butterflies and Moths*, Ray Soc., London, 9 vols.

BURGESS, E. (1880), Contributions to the anatomy of the milkweed butterfly, *Danais archippus*, Fabr., *Anniv. Mem.*, *Boston. Soc. nat. Hist.*, 16 pp.

BUSCK, A. (1914), On the classification of the Microlepidoptera, *Proc. ent. Soc. Washington*, **16**, 46–54.

—— (1932), On the female genitalia of the Microlepidoptera and their importance in the classification and determination of these moths, *Bull. Brooklyn ent. Soc.*, **26** (1931), 199–211.

BUSCK, A. AND BÖVING, A. (1914), On *Mnemonica auricyanea*, Walsingham, *Proc. ent. Soc. Washington*, **16**, 151–163.

BUXTON, P. A. (1917), On the protocerebrum of *Micropteryx*, *Trans. ent. Soc. Lond.*, **1917**, 112–153.

CARCASSON, R. H. (1968), Revised catalogue of the African Sphingidae (Lepidoptera) with descriptions of the East African species, *J. E. African nat. hist. Soc. natn. Mus.*, **26**, no. 3, 1–148.

CARPENTER, G. D. H. AND FORD, E. B. (1933), *Mimicry*, Methuen & Co., London.

CATTIE, J. T. (1881), Beiträge zur Kenntnis der Chorda supraspinalis der Lepidoptera und des zentralen, peripherischen und sympathischen Nervensystems der Raupen, *Z. wiss. Zool.*, **35**, 304–320.

CHAPMAN, T. A. (1887), On the moulting of the larva of *Orgyia antiqua*, *Entomologist's mon. Mag.*, **23**, 224–227.

—— (1893*a*), On some neglected points in the structure of the pupae of Heterocerous Lepidoptera, *Trans. ent. Soc. Lond.*, **1893**, 97–119.

—— (1893*b*), On a Lepidopterous pupa (*Micropteryx purpurella*) with functionally active mandibles, *Trans. ent. Soc. Lond.*, 255–265.
—— (1894), Some notes on the Micro-Lepidoptera whose larvae are external feeders, and chiefly on the early stages of *Eriocephala calthella* (Zygaenidae, Limacodidae, Eriocephalidae), *Trans. ent. Soc. Lond.*, **1894**, 335–350.
—— (1895), Notes on pupae, *Ent. Rec.*, **6**, 286–288.
—— (1896*a*), Notes on pupae, *Orneodes*, *Epermenia*, *Chrysocorys* and *Pterophorus*, *Trans. ent. Soc. Lond.*, **1896**, 129–147.
—— (1896*b*), On the phylogeny and evolution of the Lepidoptera from the pupal and oval standpoint, *Trans. ent. Soc. Lond.*, 567–587.
—— (1900), Note on the habits and structure of *Acanthopsyche opacella* H.-Sch., *Trans. ent. Soc. Lond.*, **1900**, 403–410.
—— (1902), The classification of *Gracilaria* and allied genera, *Entomologist*, **35**, 81–88, 138–142, 159–164.
—— (1903), The larva of *Liphyra brassolis*, Westw., *Entomologist*, **36**, 36.
—— (1906), Observations on the life history of *Trichoptilus paludum* Zell., *Trans. ent. Soc. Lond.*, **1906**, 133–154.
—— (1911–20), Papers on the biology of the Lycaenidae, *Trans. ent. Soc. Lond.*, **1911**, 148–159; **1912**, 393–406; **1912**, 409–411; **1914**, 285–308; **1914**, 309–313; (1915) **1914**, 469–481; (1915*a*) **1914**, 482–484; (1916) **1915**, 291–312; (1916*a*) **1915**, 397–410; (1916*b*) **1915**, 411–423; (1916*c*) **1915**, 424-427; **1916**, 156–180; (1917) **1916**, 315–321; (1920) **1919**, 443–449; (1920*a*) **1919**, 450–465.
—— (1917), *Micropteryx* entitled to ordinal rank: Order Zeugloptera, *Trans. ent. Soc. Lond.*, **1916**, 310–314.
CHOLODKOVSKY, N. (1880), Ueber die Hoden der Schmetterlinge, *Zool. Anz.*, **3**, 115–117.
—— (1884), Ueber die Hoden der Lepidopteren, *Zool. Anz.*, **7**, 564–568.
—— (1885), Ueber den Geschlechtsapparat von *Nematois metallicus*, Pod., *Z. wiss. Zool.*, **42**, 559–568.
—— (1887), Sur la morphologie de l'appareil urinaire des Lépidoptères, *Arch. biol.*, **6**, 497–514.
CLARK, A. H. (1932), The butterflies of the district of Columbia and vicinity, *Smithson. Inst. U.S. Nat. Mus.*, *Bull.* **137**, 337 pp.
CLARKE, J. E. G. (1941), Revision of the North American moths of the family Oecophoridae, with descriptions of new genera and species, *Proc. U.S. nat. Mus.*, **90**, 33–286.
COMMON, I. B. F. (1966), A new family of Bombycoidea (Lepidoptera) based on *Carthaea saturnioides* Walker from Western Australia, *J. ent. Soc. Queensland*, **5**, 29–36.
—— (1969), A wing-locking or stridulatory device in Lepidoptera, *J. Aust. ent. Soc.*, **8**, 121–125.
—— (1970), Lepidoptera, *The Insects of Australia*, Chapter 36, C.S.I.R.O., Divn. Entomology, Melbourne.
—— (1973), A new family of Dacnonypha (Lepidoptera) based on three new species from Southern Australia with notes on the Agathiphagidae, *J. Aust. ent. Soc.*, **12**, 11–23.
COMMON, I. B. F. AND WATERHOUSE, D. F. (1972), *Butterflies of Australia*, Angus & Robertson, Sydney.

COMSTOCK, J. A. AND HENNE, C. (1967), Early stages of *Lycomorpha regulus* Grinnell, with notes on the imago (Lepidoptera: Amatidae), *J. Res. Lepid.*, **6**, 275–280.

CORBET, A. S. AND PENDLEBURY, H. M. (1956). *The butterflies of the Malay Peninsula*, 2nd edn (Corbet, A. S. and Riley, N. D., eds), Oliver & Boyd, London.

CORBET, S. A. (1971), Mandibular-gland secretion of larvae of the flour moth, *Anagasta kuehniella*, contains an epideictic pheromone and elicits oviposition movements in an hymenopterous parasite *Venturia canescens*, *Nature*, **232**, 481–484.

CRUMB, S. E. (1956), The larvae of the Phalaenidae, *Tech. Bull. U.S. Dep. Agric.*, no. **1135**, 1–356.

DAUBERSCHMIDT, K. (1933), Vergleichende Morphologie des Lepidopterendarmes und seiner Anhänge, *Z. angew. Ent.*, **20**, 204–267.

DAVIS, D. R. (1964), Bagworm moths of the Western Hemisphere (Lepidoptera: Psychidae), *Bull. U.S. natn. Mus.*, no. **244**, 1–233.

—— (1967), A revision of the moths of the subfamily Prodoxinae (Lepidoptera: Incurvariidae), *Bull. U.S. natn. Mus.*, no. **255**, 1–170.

—— (1975), Systematics and zoogeography of the family Neopseustidae with the proposal of a new superfamily (Lepidoptera: Neopseustoidea), *Smithson. Contrib. Zool.*, **210**, 1–45.

DE GRYSE, J. J. (1915), Some modifications of the hypopharynx in Lepidopterous larvae, *Proc. ent. Soc. Washington*, **17**, 173–178.

DELLA BEFFA, G. (1938), I. Microlepidotteri minatori delle foglie dei Pioppi, *Boll. Lab. sperim. Osserv. fitopat.*, **14** (1937), 1–18.

DETHIER, V. L. (1941), The antennae of Lepidopterous larvae, *Bull. Mus. comp. Zool.*, **87**, 455–528.

DICKENS, J. C. AND EATON, J. L. (1973), External ocelli in Lepidoptera previously considered to be inocellate, *Nature*, **242**, 205–206.

DIERL, W. (1970), Compsoctenidae: ein neues Taxon von Familienstatus (Lepidoptera), *Veröff. Zool. Staats. Samml. Münch.*, **14**, 1–41.

DODD, F. P. (1902), Contribution to the life-history of *Liphyra brassolis*, Westw., *Entomologist*, **35**, 153–156, 184–188.

—— (1912), Some remarkable ant-friend lepidoptera of Queensland, with supplement by E. Meyrick, *Trans. ent. Soc. Lond.*, **1911**, 577–590.

DODSON, M. (1937), Development of the female genital ducts in *Zygaena* (Lepidoptera), *Proc. R. ent. Soc. Lond.*, (A), **12**, 61–68.

DUCKWORTH, W. D. (1973), The Old World Stenomidae: a preliminary survey of the fauna, notes on relationships and a revision of the genus *Eriogenes* (Lepidoptera: Gelechioidea), *Smithson. Contrib. Zool.*, **147**, 21 pp.

DUMBLETON, L. J. (1952), A new genus of seed-infesting micropterygid moths, *Pacif. Sci.*, **6**, 17–29.

—— (1966), Genitalia, classification and zoogeography of the New Zealand Hepialidae (Lepidoptera), *N.Z. J. Sci.*, **9**, 920–981.

DUPORTE, E. M. (1915), On the nervous system of the larva of *Sphida obliqua* Walker, *Trans. R. Soc. Canada*, **8** sec. 4, 225–253.

—— (1956), The median facial sclerite in larval and adult Lepidoptera, *Proc. R. ent. Soc. Lond.*, (A), **31**, 109–116.

DYAR, H. G. (1894), A classification of Lepidopterous larvae, *Ann. N.Y. Acad. Sci.*, **8**, 194–232.
—— (1900), Notes on the larval cases of Lacosomidae (Perophoridae) and life-history of *Lacosoma chiridota* Grt, *Ann. N.Y. Acad. Sci.*, **8**, 177–180.
—— (1907), The life-histories of the New York slug caterpillar, *Ann. N.Y. Acad. Sci.*, **15**, 219–226.
DYAR, H. G. AND MORTON, E. L. (1895), The life-histories of the New York slug caterpillars, *J. N.Y. ent. Soc.*, **3–7**, **3**, 145–157; **4**, 1–9; **4**, 167–190; **5**, 1, 5, 10, 51, 61, 167; **6**, 1–9, 94–98, 151–158, 241–246; **7**, 61–68, 234–253.
EASSA, Y. E. E. (1963*a*), The musculature of the head appendages and the cephalic stomodaeum of *Pieris brassicae* (Lepidoptera: Pieridae), *Ann. ent. Soc. Am.*, **56**, 500–510.
—— (1963*b*), Metamorphosis of the cranial capsule and its appendages in the Cabbage butterfly, *Pieris brassicae*, *Ann. ent. Soc. Am.*, **56**, 510–521.
EASTHAM, L. E. S. AND EASSA, Y. E. E. (1955), The feeding mechanism of the butterfly *Pieris brassicae*, *Phil. Trans. R. Soc. Lond.*, (B), **239**, 1–43.
EATON, J. L. (1971), Insect photoreceptor: an internal ocellus is present in Sphinx moths, *Science*, **173**, 822–823.
EDWARDS, W. H. (1868–97), *The Butterflies of North America*, Amer. ent. Soc., Philadelphia, 3 vols.
—— (1880), On the number of molts of butterflies, etc., *Psyche*, Cambridge, **3**, 159–161, 171–174.
EGGERS, F. (1919), Das thoracale bitympanale Organ einer Gruppe der Lepidoptera Heterocera, *Zool. Jb.*, *Anat.*, **41**, 273–376.
—— (1928), Die stiftführenden Sinnesorgane, *Zool. Bausteine*, **2** Hft. 1, 354 pp.
EIDMANN, H. (1929–31), Morphologische und physiologische Untersuchungen am weiblichen Genitalapparat der Lepidopteren, *Z. angew. Ent.*, **15**, 1–66; **18**, 57–112.
ELTRINGHAM, H. (1913), On the scent apparatus in the male of *Amauris niavius* Linn., *Trans. ent. Soc. Lond.*, **1913**, 309–406.
—— (1915), Further observations on the structure of the scent organs in certain male Danaine butterflies, *Trans. ent. Soc. Lond.*, **1915**, 152–176.
—— (1924), On the early stages of *Chrysiridia ripheus* Dru., *Trans. ent. Soc. Lond.*, **1923**, 439–442.
—— (1925*a*), On the source of the sphragidal fluid in *Parnassius apollo* (Lepidoptera), *Trans. ent. Soc. Lond.*, **1925**, 11–14.
—— (1925*b*), On the abdominal glands in *Heliconius* (Lepidoptera), *Trans. ent. Soc. Lond.*, **1925**, 269–275.
—— (1925*c*), On a new organ in certain Lepidoptera, *Trans. ent. Soc. Lond.*, **1925**, 7–9.
ENGELHARDT, G. P. (1946), The North American clear-wing moths of the family Aegeriidae, *Bull. U.S. nat. Mus.*, No. **190**, 222 pp.
EVANS, J. W. (1941), Tasmanian grass-grubs, *Bull. Dep. Agric. Tasmania*, (N.S.), No. **22**, 23 pp.
FAURE, J. C. (1943), The phases of the lesser Army worm, *Laphygma exigua* (Hübn.), *Farming in S. Africa*, **18**, 69–78.
—— (1943*a*), Phase variation in the Army worm, *Laphygma exempta* (Walk.), *Union S. Africa Dept. Agric. & For. Sci. Bull.*, **234**, 17 pp.

FLETCHER, T. B. (1910), The plume-moths of Ceylon. Part ii. The Orneodidae, *Spolia zeylan.*, **6**, 150–169.
—— (1920), Life-histories of Indian Micro-Lepidoptera i–ix, *Mem. Dep. Agric. India Ent. Ser.*, **6**, 1–217.
—— (1934), The same (2nd Ser.), *Imp. Council agric. Res., Sci. Monog.* **4** (1933), 85 pp.
FORBES, W. T. M. (1910), A structural study of some caterpillars, *Ann. ent. Soc. Am.*, **3**, 94–132.
—— (1911), A structural study of the caterpillars. 2. The Sphingidae, *Ann. ent. Soc. Am.*, **4**, 261–279.
—— (1924), Lepidoptera of New York and neighbouring states, *Cornell Univ. agric. Exp. Sta. Mem.*, **68** (1923), 729 pp.
FOX, R. M. (1966), Forelegs of butterflies. 1. Introduction, Chemoreception, *J. Res. Lepid.*, **5**, 1–12.
FRACKER, S. B. (1915), The classification of Lepidopterous larvae, *Illin. biol. Monogr.*, **2**, 1–169.
FREEMAN, T. N. (1947), The external anatomy of the Spruce budworm *Choristoneura fumiferana* (Clem.) (Lepidoptera, Tortricidae), *Can. Ent.*, **79**, 21–31.
FRIESE, G. (1960), Revision der paläarktischen Yponomeutidae unter besonderer Berücksichtigung der Genitalien (Lepidoptera), *Beitr. Ent.*, **10**, 1–131.
FROST, S. W. (1919), The function of the anal comb of certain Lepidopterous larvae, *J. econ. Ent.*, **12**, 446–447.
GAEDIKE, R. (1966), Die Genitalien der europäischen Epermeniidae (Lepidoptera: Epermeniidae), *Beitr. Ent.*, **16**, 633–692.
GENTHE, K. W. (1897), Die Mundwerkzeuge der Mikrolepidopteren, *Zool. Jb., Syst.*, **10**, 373–471.
GERASIMOV, A. M. (1935), Zur Frage der Homodynamie der Borsten von Schmetterlingsraupen, *Zool. Anz.*, **112**, 177–194.
GIBSON, A. (1915*a*), The Army-Worm. *Cirphis* (*Leucania*) *unipuncta* Haw., *Dep. Agric. Canada Ent. Bull.*, **9**, 34 pp.
—— (1915*b*), Cutworms and their control, *Dep. Agric. Canada Ent. Bull.*, **10**, 31 pp.
GILBERT, L. E. (1972), Pollen feeding and reproductive biology of *Heliconius* butterflies, *Proc. natn. Acad. Sci. U.S.A.*, **69**, 1403–1407.
GILMER, P. M. (1925), A comparative study of the poison apparatus of certain lepidopterous larvae, *Ann. ent. Soc. Amer.*, **18**, 203–239.
GILSON, G. (1890), Recherches sur les cellules sécrétantes. La soie et les appareils séricigènes, *La Cellule*, **6**, 117–182.
GODMAN, F. D. AND SALVIN, O. (1879–1901), *Biologia centrali Americani, Lepidoptera Rhopalocera*, 1–3, London.
GOHRBRANDT, I. (1940), Die Reduktion des Saugrüssels bei den Noctuiden und die korrelativen Beziehungen zur Ausbildung der Flügel und der Antennen, *Z. wiss. Zool.*, **152**, 571–597.
GOSSARD, H. A. AND KING, J. L. (1918), The Peach tree borer, *Ohio agric. Exp. Sta. Bull.*, **329**, 87 pp.
GOSSE, P. H. (1879), *The Great Atlas Moth of Asia*, London.
—— (1881), *Urania sloanus* at home, *Entomologist*, **14**, 241–245.

GRIFFITH, G. C. (1898), On the frenulum of the Lepidoptera, *Trans. ent. Soc. Lond.*, **1898**, 121–132.

HAMPSON, SIR G. F. (1892), On stridulation in certain Lepidoptera, and on the distortion of the hind wings in males of certain Ommatophorinae, *Proc. zool. Soc. Lond.*, **1892**, 188–189.

—— (1892–96), *The Moths of India*, London, 4 vols.

—— (1895–1920), *Catalogue of the Lepidoptera Phalaenae in the British Museum*, 13 + 2 supp. vols., London.

—— (1897), On the classification of the Thyrididae – a family of the Lepidoptera Phalaenae, *Proc. zool. Soc. Lond.*, **1897**, 603–633.

HAMPSON, SIR G. F. AND DURRANT, J. H. (1918), Some small families of the Lepidoptera, etc., *Novit. Zool.*, **25**, 366–394.

HANNEMANN, H. J. (1956), Die Kopfmuskulatur von *Micropteryx calthella* (L.) (Lep.). Morphologie und Funktion, *Zool. Jb.* (*Anat.*), **75**, 177–206.

—— (1957), Die männliche Terminalia von *Micropteryx calthella* (L.). (Lep. Micropterygidae), *Dt. ent. Z.*, (N.F.), **4**, 209–222.

HARRIS, P. (1972), Food-plant groups of the Semanophorinae (Lep., Sphingidae): a possible taxonomic tool, *Can. Ent.*, **104**, 71–80.

HASENFÜSS, I. (1960), Die Larvalsystematik der Zünsler (Pyralidae) aus dem Zoologischen Institut der Universität Erlangen, *Abh. Larvalsyst. Ins.*, **5**, 263 pp.

—— (1963), Eine vergleichend-morphologische Analyse der regulären Borstenmuster der Lepidopterenlarven, etc., *Z. Morph. Ökol. Tiere*, **52**, 197–364.

—— (1969), Zur Homologie der Borstenmusterelemente der Larvenkopfkapsel einiger monotrysicher Lepidoptera, *Beitr. Ent.*, **19**, 289–301.

HATCHETT-JACKSON, W. (1890), Studies in the morphology of the Lepidoptera, *Trans. Linn. Soc. Lond.*, (2), **5**, 145–186.

HEATH, J. (1958), The British Eriocraniidae and Micropterygidae, *Proc. & Trans. S. London ent. nat. Hist. Soc.*, **1957**, 115–125.

HEINRICH, C. (1918), On the Lepidopterous genus *Opostega* and its larval affinities, *Proc. ent. Soc. Washington*, **20**, 27–34.

—— (1956), American moths of the subfamily Phycitinae, *Bull. U.S. natn. Mus.*, **207**, 581 pp.

HELM, F. E. (1876), Ueber die Spinndrüsen der Lepidopteren, *Z. wiss. Zool.*, **26**, 434.

HERING, M. (1926), *Biologie der Schmetterlinge*, Julius Springer, Berlin.

—— (1933), Die Schmetterlinge, In: Brohmer, P., Ehrmann, P. and Ulmer, G., *Die Tierwelt Mitteleuropas*, Ergänzungsband 1 (1932), Verlag Quelle & Mayer, Leipzig.

HESSEL, J. H. (1966), A preliminary comparative anatomical study of the mesothoracic aorta of the Lepidoptera, *Ann. ent. Soc. Am.*, **59**, 1217–1227.

—— (1969), The comparative morphology of the dorsal vessel and accessory structures of the Lepidoptera and its phylogenetic implications, *Ann. ent. Soc. Am.*, **62**, 355–370.

HEYLAERTS, F. J. M. (1881), Essai d'une monographie des Psychides, etc., *Ann. Soc. ent. Belg.*, **25**, 29–73.

HINTON, H. E. (1946), On the homology and nomenclature of the setae of lepidopterous larvae, with some notes on the phylogeny of the Lepidoptera, *Trans. R. ent. Soc. Lond.*, **97**, 1–37.

HINTON, H. E. (1947), The dorsal cranial area of caterpillars, *Ann. Mag. nat. Hist.*, (11), **14**, 843–852.
—— (1951), Myrmecophilous Lycaenidae and other Lepidoptera – a summary. *Proc. Trans. S. London Ent. nat. Hist. Soc.*, **1949–50**, 111–175.
HODGES, R. W. (1962), A revision of the Cosmopterigidae of America north of Mexico, with definition of the Momphidae and Walshiidae (Lepidoptera: Gelechioidea), *Ent. Am.* (*N.S.*), **42**, 1–171.
HOFMANN, E. (1893), *Die Raupen der Gross-Schmetterlinge Europas*, C. Hoffmannscher Verlagsbuchhandlung, Stuttgart.
HOFMANN, O. (1898), Die Orneodiden (Alucitiden) des paläarktischen Gebietes, *Dt. ent. Z. Iris*, **11**, 329–359.
JACOBSON, M. (1973), *Insect sex pheromones*, Academic Press, London and New York.
JANET, A. AND WYTSMANN, P. (1903), Lepidoptera Heterocera Fam. Epicopiidae, *Gen. Ins.*, **16**, 5 pp., Desnet-Verteneuil, Brussels.
JAYEWICKREME, S. H. (1940), A comparative study of the larval morphology of leaf-mining Lepidoptera in Britain, *Trans. R. ent. Soc. Lond.*, **90**, 63–105.
JORDAN, K. (1898), Contributions to the morphology of Lepidoptera, *Novit. Zool.*, **5**, 374–415.
—— (1923), On a sensory organ found on the head of many Lepidoptera, *Novit. Zool.*, **30**, 155–258.
—— (1924), On the Saturnoidean families Oxytenidae and Cercophanidae, *Novit. Zool.*, **31**, 135–193.
KATO, M. (1940), A monograph of the Epipyropidae (Lepidoptera), (Japanese, English summ.) *Ent. World*, **8**, 67–94.
KAYE, W. J. (1913), A few observations on mimicry, *Trans. ent. Soc. Lond.*, **1913**, 1–10.
KELLOGG, V. L. (1893), The sclerites of the head of *Danais archippus*, Fab., *Kansas Univ. Quart.*, **2**, 51–57.
—— (1894), The taxonomic value of the scales of Lepidoptera, *Kansas Univ. Quart.*, **3**, 45–89.
—— (1895), The mouthparts of the Lepidoptera, *Am. Nat.*, **29**, 546–556.
KERSHAW, J. C. W. (1905), The life history of *Gerydus chinensis* Felder., *Trans. ent. Soc. Lond.*, **1905**, 1–4.
KETTLEWELL, B. (1973), *The Evolution of Melanism*, Clarendon Press, Oxford.
KIRIAKOFF, S. A. (1953), De gehoororganen en de systematiek der Lepidoptera, *Ent. Ber.*, **14**, 245–250.
KLOTS, A. B. (1969), Audition by *Cerura* larvae (Lepidoptera: Notodontidae), *J. N.Y. ent. Soc.*, **77**, 10–11.
KOKOCIŃSKI, W. (1969), Appareil respiratoire des formes adultes du papillon aquatique *Acentropus niveus* Oliv. (Lepidoptera, Pyralidae), *Zoologica poloniae*, **19**, 559–588.
KOZHANTSHIKOV, I. (1937), Faune de l'U.R.S.S., *Insectes Lépidoptères*, xiii, no. 3. (Noctuidae subf. Agrotinae; full English summary.) Izdatel'stro 'Nauka', Moscow.
KRISTENSEN, N. P. (1967), Erection of a new family in the Lepidopterous suborder Dacnonypha, *Ent. Meddr*, **35**, 341–345.
—— (1968*a*), The morphological and functional evolution of the mouthparts in adult Lepidoptera, *Opusc. Ent.*, **33**, 69–72.

—— (1968*b*), The anatomy of the head and the alimentary canal of adult Eriocraniidae (Lep., Dacnonypha), *Ent. Meddr*, **36**, 239–315.
—— (1968*c*), The skeletal anatomy of the heads of adult Mnesarchaeidae and Neopseustidae (Lep., Dacnonypha), *Ent. Meddr*, **36**, 137–151.
—— (1971), Dagsommerfuglenes storsystematik en oversigt over nyere undersøgelser, *Ent. Meddr*, **39**, 201–233.
LANGE, W. H. (1939), Early stages of California Plume moths no. 1, *Bull. S. Calif. Acad. Sci.*, **38**, 20–26.
LATTER, O. H. (1892), The secretion of potassium hydroxide by *Dicranura vinula* (imago) and the emergence of the imago from the cocoon, *Trans. ent. Soc. Lond.*, **1892**, 287–292.
—— (1895), Further notes on the secretion of potassium hydroxide by *Dicranura vinula* (imago) and similar phenomena in other Lepidoptera, *Trans. ent. Soc. Lond.*, **1895**, 399–412.
—— (1897), The prothoracic gland of *Dicranura vinula*, and other notes, *Trans. ent. Soc. Lond.*, **1897**, 113–125.
LEFROY, H. M. AND GHOSH, C. C. (1912), Eri silk, *Dept. Agric. India Mem.*, **4**, 130 pp.
LE GRICE, D. S. (1968), Some observations on the morphology and histology of the alimentary canal of the fruit-sucking moth *Serrodes inara* Cram. (Lepidoptera: Noctuidae), *S. Afr. J. agric. Sci.*, **11**, 789–796.
LESPERON, L. (1937), Recherches cytologiques et expérimentales sur la soie et sur certaines méchanismes excréteurs, chez insectes, *Arch. Zool. exp. gén.*, **79**, 1–156.
LEWIS, H. L. (1974), *Butterflies of the World*, Harrap, London.
LLOYD, Ll. (1920), The habits of the glasshouse tomato moth *Hadena* (*Polia*) *oleracea* and its control, *Ann. appl. Biol.*, **7**, 66–102.
LORENZ, R. E. (1961), Biologie und Morphologie von *Micropterix calthella* L., (Lep. Micropterigidae), *Dt. ent. Z.*, (N.F.) **8**, 1–23.
LYONNET, P. (1762), *Traité anatómique de la chenille qui ronge le bois du saule*, The author, La Haye.
MABILLE, P. (1903), Lepidoptera, Rhopalocera, Fam. Hesperiidae, *Gen. Ins.*, **17**, 210 pp., Desnet-Verteneuil, Brussels.
MABILLE, P. AND BOULLETT, E. (1909), Essai de revision de la famille des Hespérides, *Ann. Sci. nat.*, (9) **16**, 167–207.
MacKAY, M. R. (1959), Larvae of North American Olethreutidae (Lepidoptera), *Can. Ent.*, **91** Suppl. 10, 3–338.
—— (1962*a*), Larvae of the North American Tortricinae (Lepidoptera: Tortricidae), *Can. Ent.*, Suppl. **28**, 1–182.
—— (1962*b*), Additional larvae of the North American Olethreutinae (1) (Lepidoptera: Tortricidae), *Can. Ent.*, **94**, 626–643.
—— (1970), Lepidoptera in Cretaceous amber, *Science*, **167**, 379–380.
—— (1972), Larval sketches of some Microlepidoptera, chiefly North American, *Mem. ent. Soc. Can.*, 88, 1–83.
MADDEN, A. H. (1944), The external morphology of the adult tobacco horn-worm (Lepidoptera, Sphingidae), *Ann. ent. Soc. Am.*, **37**, 145–160.
MALICKY, H. (1969), Uebersicht über Präimaginalstadien, Bionomie und Oekologie der Mitteleuropäischen Lycaenidae (Lepidoptera), *Mitt. ent. Ges. Basel*, **19**, 25–91.

MARLATT, C. L. (1905), The Giant Sugar-cane borer (*Castnia licus* Fab.), *Bull. U.S. Bur. Ent.*, **54**, 71–75.
MARSHALL, G. A. K. (1902), Five years' observations and experiments (1896–1901) on the bionomics of South African insects, etc., *Trans. ent. Soc. Lond.*, **1902**, 287–584.
MARTYNOVA, YE. F. (1950), The structure of *Micropteryx* caterpillars (Lepidoptera, Micropterygidae), *Entom. obozr.*, **31**, 142–150.
MATTHES, E. (1948), *Amicta febretta.* Ein Beitrag zur Morphologie und Biologie der Psychiden (Lepidoptera), *Mem. Estud. Mus. zool. Coimbra*, **184**, 779 pp.
MAYER, A. G. (1896), The development of the wing-scales and their pigment in butterflies and moths, *Bull. Mus. comp. Zool.*, **29**, 209–236.
—— (1897), On the color and color-patterns of moths and butterflies, *Proc. Boston. Soc. nat. Hist.*, **27**, 243–330.
MCGUFFIN, W. C. (1958), Larvae of the Nearctic Larentiinae (Lepidoptera: Geometridae), *Can. Ent.*, **90** Suppl. 8, 3–104.
MEHTA, D. R. (1933), On the development of the male genitalia and the efferent genital ducts in Lepidoptera, *Quart. J. micr. Sci.*, (N.S.), **76**, 35–61.
METALNIKOV, S. (1908), Recherches expérimentales sur les chenilles de *Galleria mellonella, Arch. Zool. exp. gén.*, (4), **8**, 489–588.
MEYRICK, E. (1910), Lepidoptera Heterocera. Fam. Orneodidae, *Gen. Ins.*, fasc. **108**, 4 pp., Desnet-Verteneuil, Brussels.
—— (1912), Lepidoptera Heterocera. Fam. Micropterygidae, *Gen. Ins.*, fasc. **132**, 9 pp., Desnet-Verteneuil, Brussels.
—— (1928), *Revised Handbook of British Lepidoptera,* Macmillan & Co., London.
MIALL, L. C. (1903), *The Natural History of Aquatic Insects*, Macmillan & Co., London.
MICHENER, C. D. (1952), The Saturniidae (Lepidoptera) of the Western Hemisphere, *Bull. Amer. Mus. nat. Hist.*, **98**, art. 5, 341–501.
MORTIMER, T. J. (1965), The alimentary canals of some adult Lepidoptera and Trichoptera, *Trans. R. ent. Soc. Lond.*, **117**, 67–93.
MOSHER, E. (1914), The classification of the pupae of the Ceratocampidae and Hemileucidae, *Ann. ent. Soc. Am.*, **7**, 277–300.
—— (1916), A classification of the Lepidoptera based on characters of the pupa, *Bull. Illin. Lab. nat. Hist.*, **12**, 15–159.
MÜLLER, G. W. (1892), Beobachtungen an im Wasser lebenden Schmetterlingsraupen, *Zool. Jb., Syst.*, **6**, 617–630.
MÜLLER, W. (1886), Südamerikanische Nymphalidenraupen, *Zool. Jb.*, **1**, 417–678.
MUTUURA, A. (1956), On the homology of the body areas in the thorax and abdomen and new system of the setae on Lepidopterous larvae, *Bull. Univ. Osaka Prefect.*, Ser. B **6**, 93–122.
—— (1971), A new genus of homoneurous moth and the description of a new species. (Lep., Dacnonypha, Neopseustidae), *Can. Ent.*, **103**, 1129–1136.
—— (1972), Morphology of the female terminalia in Lepidoptera and its taxonomic significance, *Can. Ent.*, **104**, 1055–1071.
MYERS, J. AND BROWER, L. P. (1969), A behavioral analysis of the courtship pheromone receptors of the queen butterfly, *Danaus gilippus berenice, J. Insect. Physiol.*, **15**, 2117–2130.
NAUMANN, F. (1937), Zur Reduktion des Saugrüssels bei Lepidopteren und deren Beziehung zur Flügelreduktion, *Zool. Jb., Syst.*, **70**, 381–420.

NEWPORT, G. (1832), On the nervous system of the *Sphinx ligustri*, Linn. and on the changes which it undergoes during a part of the metamorphosis of the insect, *Phil. Trans. Roy. Soc. Lond.*, **122**, 383–398.
NIGMANN, M. (1908), Anatomie und Biologie von *Acentropus niveus* Oliv., *Zool. Jb., Syst.*, **26**, 489–560.
NORDSTRÖM, F., WAHLGREN, E. AND TULLGREN, A. (1935–41), *Svenska Fjarillar,* Nordisk Familjeboks Förlags Aktiebolag, Stockholm.
NORRIS, M. J. (1932), The structure and operation of the reproductive organs of the genera *Ephestia* and *Plodia* (Lepidoptera, Phycitidae), *Proc. zool. Soc. Lond.*, **1932**, 595–611.
NÜESCH, H. (1957), Die Morphologie des Thorax von *Telea polyphemus* Cr. (Lepid.). II. Nervensystem, *Zool. Jb., Anat.*, **75**, 615–642.
OBRAZTSOV, N. S. (1954–68), Die Gattungen der paläarktischen Tortricidae, *Tijdschr. Ent.*, **97**, **98**, **99**, **100**, **101**, **102**, **103**, **104**, **110**; summary and index, **11**, 1–48.
OITICICA, J. (1948), Sôbre a genitalia das fêmeas de Hepialidae (Lepidoptera), *Summ. Brasil. Biol.*, 1 fasc. 16, 384–403.
PACKARD, A. S. (1894), A study of the transformations and anatomy of *Lagoa crispata*, a Bombycine moth, *Proc. phil. Soc.*, **32**, 275–292.
—— (1895), On a new classification of the Lepidoptera, *Am. Nat.*, **29**, 636–647, 788–809.
—— (1896–1914), Monograph of the Bombycine moths of America north of Mexico, Parts 1–3, *Mem. nat. Acad. Sci.*, **7** (1896), 390 pp., **9** (1905), 272 pp.; **12** (1914), 516 pp.
PETERSEN, G. (1969), Beiträge zur Insekten-Fauna der D.D.R. (Lepidoptera, Tineidae), *Beitr. Ent.*, **19**, 311–388, English summary.
PETERSEN, W. (1900), Beiträge zur Morphologie der Lepidopteren, *Mem. Acad. Sci. St. Petersb.*, (8), **9**, 144 pp.
—— (1904), Die Morphologie der Generationsorgane der Schmetterlinge, etc., *Mem. Acad. Sci. St. Petersb.*, (8), **16**, 84 pp.
PETERSON, A. (1912), Anatomy of the tomato-worm larva, *Protoparce carolina, Ann. ent. Soc. Am.*, **5**, 246–269.
—— (1959), *Larvae of insects. An introduction to the Nearctic species,* 3rd edn, Columbus, Ohio.
PHILPOTT, A. (1924), The wing-coupling apparatus in *Sabatinca* and other primitive genera of Lepidoptera, *Rep. Austral. Assoc. Adv. Sci.*, **16**, 414–419.
—— (1925), On the wing-coupling apparatus of the Hepialidae, *Trans. ent. Soc. Lond.*, **1925**, 331–340.
—— (1926), The venation of the Hepialidae, *Trans. ent. Soc. Lond.*, **1926**, 531–535.
—— (1927), Notes on the female genitalia in the Micropterygoidea, *Trans. ent. Soc. Lond.*, **75**, 319–323.
PIEPERS, M. C. (1897), Ueber die Farbe und den Polymorphismus der Sphingidenraupen, *Tijdschr. Ent.*, **40**, 27–105.
PIERCE, F. N. AND OTHERS (1909–43), *The Genitalia of the British Noctuidae*, etc., 7 vols., The authors, Liverpool and Oundle.
POLL, M. (1939), Contribution à l'étude de l'appareil urinaire des chenilles de Lépidoptères, *Ann. Soc. Zool. Belg.*, **69** (1938), 9–52.
POULTON, E. B. (1884), Notes upon, or suggested by, the colours, markings, and

protective attitudes of certain Lepidopterous larvae and pupae, etc., *Trans. ent. Soc. Lond.*, **1884**, 27–60.

POULTON, E. B. (1885), The essential nature of the colouring of phytophagous larvae (and their pupae), etc., *Proc. Roy. Soc.*, 38, 269–315.

—— (1885*a*), Further notes upon the markings and attitudes of lepidopterous larvae, etc., *Trans. ent. Soc. Lond.*, **1885**, 281–329.

—— (1886–88), Notes upon lepidopterous larvae, etc., *Trans. ent. Soc. Lond.*, **1886**, 137–179; **1887**, 281–321; **1888**, 516–606.

—— (1887*a*), An inquiry into the cause and extent of a special colour-relation between certain exposed lepidopterous pupae and the surfaces which immediately surround them, *Philos. Trans.*, 178 B, 311–441.

—— (1890), The external morphology of the lepidopterous pupa, etc., *Trans. Linn. Soc. Zool.*, (2), 5, 187–212, 245–263.

—— (1892), Further experiments upon the colour-relation between certain lepidopterous larvae, pupae, cocoons and imagines and their surroundings, *Trans. ent. Soc. Lond.*, **1892**, 293–487.

—— (1894), The experimental proof that the colours of certain lepidopterous larvae are largely due to modified plant pigments derived from food, *Proc. Roy. Soc.*, 54, 417–430.

POWELL, J. A. AND MACKIE, R. A. (1966), Biological inter-relationships of moths and *Yucca whipplei* (Lepidoptera: Gelechiidae, Blastobasidae, Prodoxidae), *Univ. Calif. Publs. Ent.*, 42, 1–46.

PRADHAN, S. AND AREN, N. S. (1941), Anatomy and musculature of the mouthparts of *Scirpophaga nivella* (Pyralidae), with a discussion on the coiling and uncoiling mechanisms of the proboscis in Lepidoptera, *Indian J. Ent.*, 3, 179–195.

QUAIL, A. (1900), Life histories in the Hepialid group of Lepidoptera, etc., *Trans. ent. Soc. Lond.*, **1900**, 411–432.

REICHELT, M. (1925), Schuppenentwicklung und Pigmentbildung auf den Flügeln von *Lymantria dispar*, etc., *Z. Morph. Ökol. Tiere*, 3, 477–525.

RICHARDS, A. G. (1933), Comparative skeletal morphology of the Noctuid tympanum, *Entomologica Amer.*, 13 (1932), 1–43.

RIPLEY, L. B. (1925), The external morphology and postembryology of Noctuid larvae, *Illin. Biol. Monogr.*, 8 (1924), 243–344.

RIPPON, R. H. F. (1890–1910), *Icones Ornithopterorum: a monograph of the Papilionine tribe* Troides *Hübner, etc.*, The author, London.

ROEDER, K. D. (1964), Aspects of the Noctuid tympanic nerve response having significance in the avoidance of bats, *J. Insect Physiol.*, 10, 529–546.

—— (1966), Acoustic sensitivity of the Noctuid tympanic organ and its range for the cries of bats, *J. Insect Physiol.*, 12, 843–859.

—— (1972), Acoustic and mechanical sensitivity of the distal lobe of the pilifer in choerocampine hawk moths, *J. Insect Physiol.*, 18, 1249–1264.

ROEDER, K. D. AND TREAT, A. E. (1957), Ultrasonic reception by the tympanic organ of Noctuid moths, *J. exp. Zool.*, 134, 127–157.

—— (1962), The detection and evasion of bats by moths, *Ann. Rep. Smithsonian Inst.* (1961), 455–464.

—— (1970), An acoustic sense in some hawkmoths (Lep., Sphingidae, Choerocampinae), *J. Insect Physiol.*, 16, 1069–1086.

ROLET, A. (1913), *Les Vers à Soie*, Doin, Paris.

ROTHSCHILD, W. AND JORDAN, K. (1903), *A Revision of the Lepidopterous Family Sphingidae*, Tring, The authors, London.
—— (1907), Lepidoptera Heterocera. Fam. Sphingidae, *Gen. Ins.*, **57**, 157 pp.
RUCKES, H. (1919), Notes on the male genital system in certain Lepidoptera, *Ann. ent. Soc. Am.*, **12**, 192–209.
SAALAS, U. (1935–36), *Incurvaria pectinea* Hw. (Lep., Incurvariidae), *Ann. ent. fenn.*, **1**, 113–137; **2**, 1–16.
SCHMITT, J. B. (1938), The feeding mechanism of adult Lepidoptera, *Smithsonian misc. Coll.*, **97**, no. 4, 28 pp.
SCHWARZ, R. (1964), Beitrag zur Morphologie, Biologie und Oekologie der mitteleuropäischen Crambidae (Lepidoptera), *Z. ArbGem. öst. Ent.*, **16**, 46–67.
SCUDDER, S. H. (1888–89), *The Butterflies of the Eastern United States and Canada*, etc., 3 vols., The author, Cambridge, Mass.
SEILER, J. (1923), Die Parthenogenese der Psychiden, *Zs. indukt. Abstammungslehre*, **30**, 286; **31**, 1–99.
SEITZ, A. (1906 *et seq.*), *Macro-lepidoptera of the World*, Fritz Lehmann Verlag, Stuttgart.
SHARP, D. (1899), *The Cambridge Natural History*, Macmillan & Co., London.
SHARPLIN, J. (1963–64), Wing base structure in Lepidoptera. I, II, III, *Can. Ent.*, **95**, 1024–1050, 1121–1145; **96**, 943–949.
SHORT, J. R. T. (1951), Some aspects of the morphology of the insect head as seen in the Lepidoptera, *Proc. R. ent. Soc. Lond.*, (A), **26**, 77–88.
SILVESTRI, F. (1924), Contribuzioni alla conoscenza dei Tortricidi delle Querce I, II, *Boll. Lab. Portici*, **17**, 41–107.
SNODGRASS, R. E. (1961), The caterpillar and the butterfly, *Smithsonian misc. Coll.*, **143**, no. 6, 1–51.
SOULE, C. G. (1895), Uncertainty of the duration of any stage in the life history of moths, *Psyche*, Cambridge, **7**, 191.
SRIVASTAVA, A. S. (1962), A preliminary study of the life-history and control of *Indarbela quadrinotata* Wlk. (Metarbelidae: Lepidoptera), *Proc. natn. Acad. Sci. India*, **328**, 265–270.
STAUDINGER, O. AND REBEL, H. (1901), *Catalog der Lepidopteren des paläarktischen Faunengebietes*, Friedländer & Sohn, Berlin, 3rd edn.
STEKOLNIKOV, A. A. (1967), Functional morphology of the copulative organs in archaic moths and the general directions in the evolution of genitalia in Lepidoptera, *Ent. Obozr.*, **46**, 670–689. Russian: translated in *Ent. Rev. Washington*, **46**, 400.
STITZ, H. (1900–01), Der Genitalapparat der Mikrolepidopteren. 1 & 2, *Zool. Jb., Anat.*, **14**, 135–176; **15**, 385–434.
SWAINE, J. M. (1920–21), The nervous system of the larva of *Sthenopis thule* Strecker, *Can. Ent.*, **52**, 275–283; **53**, 29–34.
SWATSCHEK, B. (1958), Die Larvalsystematik der Wickler (Tortricidae und Carposinidae), *Abh. Larvalsyst. Ins.* **3**, 269 pp.
SWIHART, S. L. (1967), Hearing in butterflies (Nymphalidae: *Heliconius; Ageronia*), *J. Insect Physiol.*, **13**, 469–476.
TALBOT, G. (1939), *The Fauna of British India including Ceylon and Burma, Butterflies*. I and II, Taylor & Francis, London.
THOMAS, M. B. (1893), The androconia of Lepidoptera, *Am. Nat.*, **27**, 1018–1020.
TILLYARD, R. J. (1918), The Panorpoid complex 1. The wing coupling apparatus

with special reference to the Lepidoptera, *Proc. Linn. Soc. N.S.W.*, **43**, 286–319.

TILLYARD, B. J. (1919*a*), On the morphology and systematic position of the family Micropterygidae (sens. lat.), *Proc. Linn. Soc. N.S.W.*, **44**, 95–136.

—— (1919*a*). The Panorpoid complex. 3. The wing-venation. *Proc. Linn. Soc. N.S.W.*, **44**, 533–718.

—— (1923*a*), On the larva and pupa of the genus *Sabatinca* (Order Lepidoptera, Family Micropterygidae), *Trans. ent. Soc. Lond.*, **1922**, 437–453.

—— (1923*b*), On the mouthparts of the Micropterygoidea (Lep.), *Trans. ent. Soc. Lond.*, **1923**, 181–206.

TINDALE, N. B. (1928), Preliminary note on the life history of *Synemon* (Lepidoptera, fam. Castniidae), *Rec. S. Aust. Mus.*, **4**, 143–144.

—— (1932–45), Revision of the Australian Ghost moths (Lepidoptera Homoneura, Family Hepialidae). Parts 1–5, *Rec. S. Austr. Mus.*, **4**, 497–534; **5**, 14–43; **5**, 275–332; **7**, 15–46; **7**, 151–168.

TURNER, A. J. (1918), Observations on the Lepidopteran family Cossidae and on the classification of the Lepidoptera, *Trans. ent Soc., Lond.*, **1918**, 155–190.

—— (1947), A review of the phylogeny and classification of the Lepidoptera, *Proc. Linn. Soc. N.S.W.*, **71** (1946), 303–338.

TUTT, J. W. (1899–1909), *A Natural History of the British Lepidoptera*, 9 vols, (incomplete), Swan Sonnenschein & Co., London.

TUXEN, S. L. (ed.) (1970), *Taxonomist's glossary of genitalia in insects*, 2nd edn, Munksgaard, Copenhagen (A. B. Klots, Lepidoptera, 115–130).

VERITY, R. (1940–47), *Le Farfalle diurne d'Italia,* vols. 1–4 (incomplete), Casa Ed. Marzocco, S.A., Florence.

VIETTE, P. (1948), Morphologie des genitalia mâles des Lépidoptères, *Rev. franç. Ent.*, **15**, 141–161.

VOGEL, R. (1911), Ueber die Innervierung der Schmetterlingsflügel, etc., *Z. wiss. Zool.*, **98**, 68–134.

—— (1912), Ueber die Chordotonalorgane in der Wurzel der Schmetterlingsflügel, *Z. wiss. Zool.*, **100**, 210–244.

VORBRODT, K. AND MÜLLER-RUTZ, J. (1911–14), *Die Schmetterlinge der Schweiz*, 2 vols, Druck und Verlag J. Wyss, Bern.

WAGNER, H. (ed. and others) (1911–36), *Lepidopterorum Catalogus*, W. Junk, Berlin.

WAILLY, A. (1896–97), Silk-producing Lepidoptera, *Entomologist*, **29**, 157–159, 208–210, 235–239, 274–279, 352–356; **30**, 39–44.

WALTER, A. (1884), Palpus maxillaris lepidopterorum, *Jenaische Z. Naturw.*, (11), **19**, 121–173.

—— (1885), Zur Morphologie der Schmetterlingsmundtheile, *Jenaische Z. Naturw.*, **18**, 19–27.

WASSERTHAL, L. (1970), Generalisierende und metrische Analyse des primären Borstenmusters der Pterophoriden-Raupen (Lepidoptera), Eilarven als Objekte systematischer Untersuchungen, *Z. Morph. Ökol. Tiere*, **68**, 177–254.

WEBER, H. (1924), Das Thorakalskelett der Lepidopteren, *Z. ges. Anat.*, Abt. 1, **73**, 277–331.

—— (1928), Die Gliederung der Sternopleuralregion des Lepidopterenthorax, *Z. wiss. Zool.*, **131**, 181–254.

WELCH, P. S. (1916), Contribution to the biology of certain aquatic Lepidoptera, *Ann. ent. Soc. Am.*, **9**, 159–187.

WERNER, K. (1958), Die Larvalsystematik einiger Kleinschmetterlingsfamilien (Hyponomeutidae, Ortholeliidae, Acrolepidae, Tineidae, Incurvariidae und Adelidae), *Abh. Larvalsyst. Ins.*, **2**, 145 pp., 212 pp.

WHALLEY, P. E. S. (1971), The Thyrididae (Lepidoptera) of Africa and its islands: a taxonomic and zoogeographic study, *Bull. Brit. Mus.* (*Nat. Hist.*) *Ent.*, suppl. 17, 1–198.

WILLIAMS, C. B. (1930), *The Migration of Butterflies*, Oliver & Boyd, London.

WILLIAMS, J. L. (1941), The internal genitalia of Yucca moths, and their connection with the alimentary canal, *J. Morph.*, **69**, 217–222.

—— (1947), The comparative anatomy of the internal genitalia of some Tineoidea (Lepidoptera, Gracillariidae-Tischeriidae), *Proc. R. ent. Soc. Lond.*, (A), **22**, 8–17.

WOODHOUSE, L. G. O. AND HENRY, G. M. R. (1942), *The Butterfly Fauna of Ceylon*, Govt. Record Office, Colombo; Dubin & Co, London.

YANO, K. (1963), Taxonomic and biological studies of Pterophoridae of Japan (Lepidoptera), *Pacific Ins.*, **5**, 65–209.

YASUDA, T. (1962), On the larva and pupa of *Neomicropteryx nipponensis* Issiki, with its biological notes (Lepidoptera, Micropterygidae), *Kontyû*, **30**, 130–136.

ZANDER, E. (1903), Beiträge zur Morphologie der männlichen Geschlechtsanhänge der Lepidopteren, *Z. wiss. Zool.*, **74**, 557–615.

Order 28

TRICHOPTERA (CADDIS FLIES)

Small to moderate-sized moth-like insects with setaceous antennae. Mandibles vestigial or absent: maxillae single-lobed with elongate palpi: labium with a small median glossa or large hypopharyngeal haustellum and well-developed palps. Wings membranous, more or less densely hairy, and held roof-like over the back in repose. Fore wings elongate, hind wings broader with a folding anal area: venation generalized: cross-veins few. Tarsi 5-segmented. Larvae aquatic, more or less eruciform and usually living in cases: body terminated by hooked caudal appendages. Pupae exarate, decticous: wing tracheation reduced.

The Trichoptera are weakly flying insects of moth-like appearance found in the vicinity of water (Figs. 533, 534). They are unfamiliar to the general student, whose acquaintance with the order is usually restricted to the case-bearing larvae which frequent ponds and streams. The imagines are mostly obscurely coloured, being generally some shade of brown, often with darker markings. They are not often seen on the wing unless disturbed, and they rest on herbage, trees, or stones: their flight is of short and uncertain duration. The Leptoceridae, however, fly over the water in the daytime. Many species are nocturnal: some are attracted to a light, others to the moth-collector's sugar, and a few visit flowers. They have seldom been observed in the act of feeding but the mouthparts are adapted for licking fluid nourishment. In their general affinities they are allied to the Lepidoptera and are

FIG. 533
Hydroptila angustella. × 5
After McLachlan.

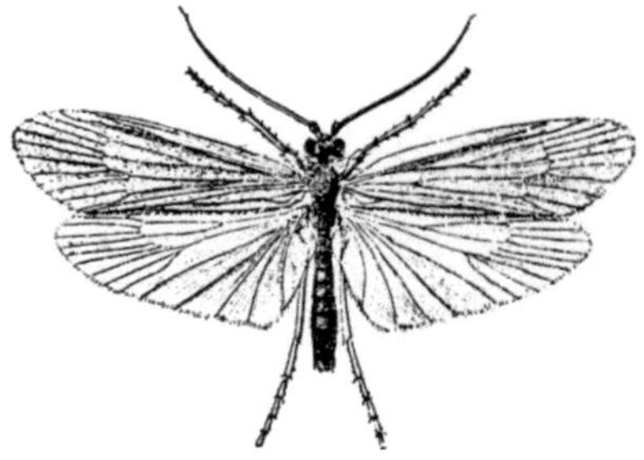

FIG. 534
Halesus guttatipennis (Limnephilidae) × circa 2
After McLachlan.

only separable from the latter upon comparatively slight characters. In the Trichoptera, however, a thyridium is generally present on each wing, M_4 is separate from Cu_{1a} in the fore wing and broad scales are rarely found and never numerous. About 5000 species of the order are known and, of these, 188 inhabit the British Isles. The principal work on the European forms is that of McLachlan (1874–84), and Mosely (1939) has monographed the British species. The best modern introductions to the order are by Ulmer (1909) and Betten (1934).

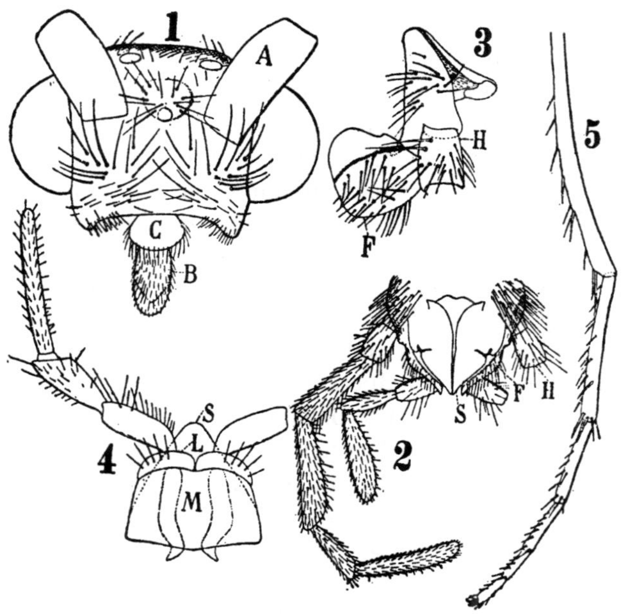

FIG. 535 *Limnephilus*. 1, head, frontal view. 2, maxillae and labium (inner aspect). 3, maxilla. 4, labium. 5, tarsus and apex of tibia of 3rd leg

A, base of antennae; *B*, labrum; *C*, clypeus; *F*, galea; *H*, base of maxillary palp; *L*, prementum; *M*, mentum; *S*, hypopharynx. *After* Silvestri.

Anatomy (Fig. 535)

The antennae are multi-articulate and setaceous, frequently several times the length of the wings: when in repose they are held closely porrected in front of the head. The compound eyes are usually small, but occasionally they occupy nearly the whole of the head in the male. Ocelli are either three in number or wanting. The structure of the mouthparts has been described by Crichton (1957). The clypeus is narrow and transverse, while the labrum is generally somewhat elongated. The mandibles are atrophied, or vestigial, in many genera such as *Phryganea*, *Limnephilus*, *Anabolia*, etc. (Lucas, 1893), but are better developed in certain others. The maxillae are small and closely associated with the labium: they are ordinarily provided with a single lobe or mala, the palpi are elongated and 5-, rarely 6-segmented in the females, but in the males the segments are more variable. The labium consists of a well-developed mentum, sometimes a small glossa, and usually 3-segmented palpi. There is a prominent hypopharynx which bears the aperture of the salivary glands and is often developed into an haustellum with pseudotrachea-like channels. In the Australian *Plectrotarsus* the labrum and labium are greatly elongated, forming a kind of rostrum, and the two pairs of palps are carried forwards.

The prothorax is small and ring-like; the mesothorax is the largest segment and the metathorax is somewhat shorter. The legs are long and slender with large, strong coxae: a meron is present in relation to the two hind pairs of coxae, but is less completely developed than in most Lepidoptera. The tibiae are often furnished with spines and movable spurs, the tarsi are 5-segmented, and between the claws there is either a pair of pulvilli or a cushion-like empodium. The wings are almost always fully developed, but the females of *Enoicyla* and *Philopotamus distinctus* are practically apterous. In *Anomalopteryx* (male) and *Thamastes* (both sexes) the hind wings are reduced to scale-like rudiments. The extremely hairy nature of the wings, which is especially characteristic of the order, is due to the presence of macrotrichia both on the veins and wing-membrane. Certain genera, however, exhibit a tendency to a reduction of this clothing, and in some forms there is an almost general absence of hairs. Scattered scales of a primitive type are found on the wings of certain Trichoptera, but are narrow and acuminate, with few striae, and do not assume the broadened form so characteristic of Lepidoptera. The fore wings are denser than the hind wings, and are often slightly more coriaceous. The wing-coupling apparatus (see Riek, 1970, for details) is exhibited in a primitive condition in *Rhyacophila* in which there is a jugal lobe on the fore wing resting on the costa of the hind wing. There are rarely jugal bristles or a frenulum, as in *Plectrotarsus*, and the humeral lobe is suppressed or vestigial. In the majority of genera the jugal lobe is rudimentary or wanting, and an amplexiform type of coupling apparatus is developed. This is brought about by a fold along the whole length of the anal area of the fore wing engaging the costa of the hind wing. In some forms a row of costal hooks along the hind wing grapple the anal margin of the fore wing, and thus securely interlock the two wings of the side. The venation, as exemplified by *Rhyacophila fuscula*, is of an extremely generalized type (Fig. 536) and closely resembles that of the most primitive Lepidoptera. Almost all the veins are longitudinal, not more than two veinlets in the costal series are retained, and the cross-veins are reduced in number. Unlike the Lepidoptera, M_4 of the fore wing is not fused with Cu_{1a}. Near the fork of vein$_{4+5}$ on both pairs of wings there is, ordinarily, a semi-transparent whitish spot generally devoid of hairs and known as the *thyridium*. It is possibly due to the presence of a gland or sensory organ and is wanting in Lepidoptera. The usual number of abdominal segments is apparently 10. The genitalia in the male (Nielsen, 1975 and 1970 in Tuxen, see

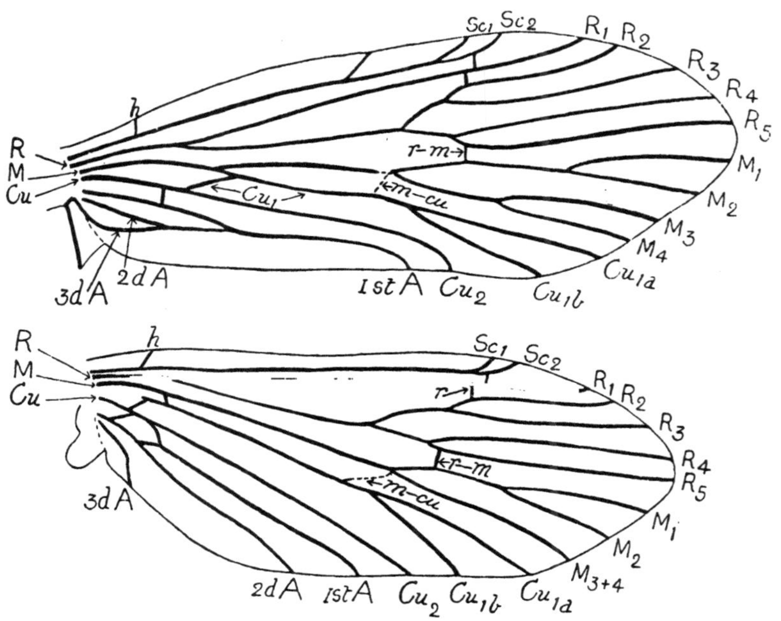

FIG. 536 *Rhyacophila fuscula*, venation
After Comstock, with legend slightly altered.

Lepidoptera) consist in more primitive forms of a pair of claspers and two lobes (parameres?) of the aedeagus, but in such genera as *Limnephilus* the claspers are absent: in the female the terminal segments are sometimes retractile and tubular, thus functioning as an ovipositor. The genital aperture is usually on the 9th sternite. The sperm are sometimes transferred to the female in a spermatophore (Khalifa, 1949).

In the males of species of *Hydroptila* there is an elaborate apparatus of scent-brushes and scent-scales situated at the hinder region of the head and attached to tubes or membranes which are capable of being everted, presumably by means of blood pressure. When not in use these organs are withdrawn into the head (Eltringham, 1920).

The internal anatomy of Trichoptera has been very little investigated and only fragmentary accounts exist. The alimentary canal is relatively short with a small crop, a tubular and slightly coiled mid gut, and an expanded rectal chamber: six Malpighian tubes are present. The crop and mid gut may be separated by a sphincter (Mortimer, 1965). The nervous system, in addition to the usual cephalic centres, consists of 3 thoracic and 7 abdominal ganglia, but the 3rd thoracic and true 1st abdominal ganglia are indistinguishably fused (Glasgow, 1936). The testes are simple ovoid sacs and the ovaries consist of numerous polytrophic ovarioles (Stitz, 1904); Unzicker (1968) describes the spermatheca, accessory glands and bursa copulatrix of a number of genera.

Biology and Metamorphoses

The early stages of Trichoptera, almost without exception, are passed in fresh water. One or two species develop in brackish or salt water (e.g. *Philanisus*) while the larva of *Enoicyla* is terrestrial, living among moss at the bases of trees in woods (Rathjen, 1939; Kelner-Pillault, 1960). The eggs of caddis flies are laid in water, on aquatic vegetation, or on overhanging trees. They are deposited in masses covered by a mucilage which rapidly swells when wetted. The larvae are the familiar objects known as 'caddis worms' and those of the greater number of species form cases or shelters within which they reside. These structures are composed of a basis of silk to which various foreign materials are added. They are commonly tubular in form with an opening at either end. The anterior aperture is wide and through it the head and legs of the contained insect can be protruded. The posterior aperture is usually smaller and is frequently protected by a perforated silken

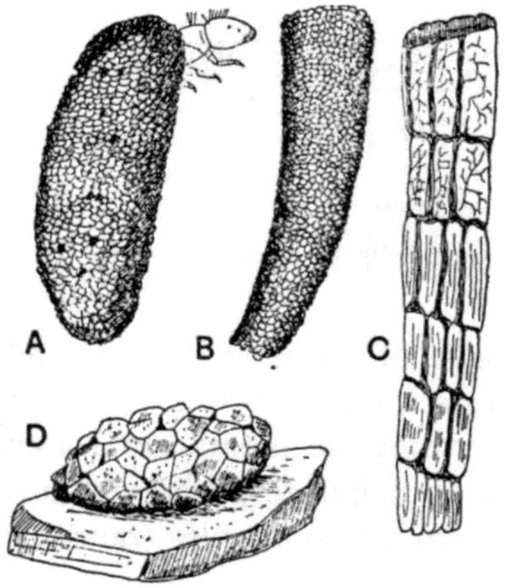

FIG. 537
Cases of Trichoptera, magnified

A, *Hydroptila maclachlani*, case with larva. *After* Klapalek. B, *Odontocerum*, larval case. C, *Phryganea*, larva case. D, *Hydropsyche*, pupal shelter.

plate. As a rule the larva performs undulatory movements with the abdomen which maintain a current of water in contact with the body, flowing out through the posterior opening of the case. At its hinder extremity the larva is provided with a pair of grappling hooks and it is by means of these organs that it is able to maintain a firm hold of its case, dragging the latter along with it while it crawls about. The variety of cases made by caddis larvae is very great (Fig. 537) and their form and the materials used in their construction are in some cases characteristic of particular species, in others of genera or families. Almost all kinds of material which can be found in the water are utilized by one or other of the species. Leaves, pieces of leaves or stalks, straws, pieces of stick, etc., are often employed while other species select seeds, sand, particles of gravel or the shells of small molluscs. It is held (Ross, 1964) that primitive caddises, like the modern *Rhyacophila*, were free-living, without cases. One line, suborder Annulipalpia, developed the habit of building retreats with nearby nets to capture prey, as in *Hydropsyche*,

Philopotamus, *Plectrocnemis*, etc. (Wesenberg-Lund, 1911; Noyes, 1914). Another line developed through the ancestors of the saddle-makers (*Glossosoma*) and the purse-makers (Hydroptilidae) to the higher tube-making caddises. The saddle-cases are short, flattened with the ventral side cut away at pupation and the purse is a small, wide structure, open only at one end. A peculiar variant of the case is the spiral constructions of *Helicopsyche* built of small grains of sand (Moretti and Vigano, 1960). The nets of the Annulipalpia are composed of strong silken threads which are supported on some available framework such as fragments of leaves or twigs. Water flows freely through the net, but the latter holds back the organisms which serve as food for the caddis larvae.

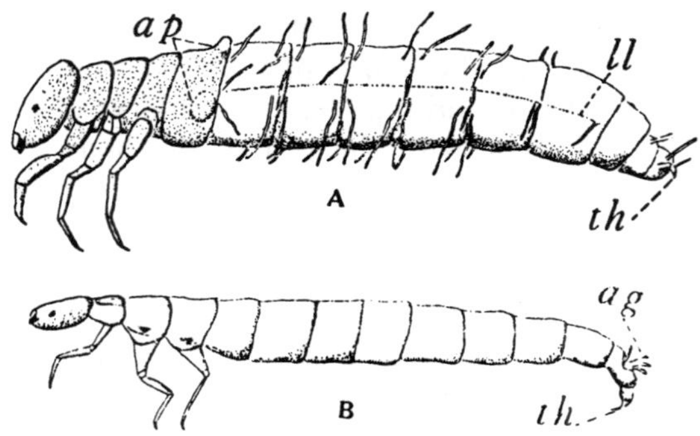

FIG. 538 Trichopterous larvae

A, Eruciform or case-bearing type. B, Campodeoid or non-case-bearing. *ag*, anal gills; *ap*, abdominal papillae; *ll*, lateral line; *th*, terminal hooks.

A typical Trichopterous larva has a well-developed sclerotized head (cf. Badcock, 1961) and very short antennae (Fig. 538A). Biting mandibles are present and the maxillae are single-lobed with short 4- or 5-segmented palpi. The labium bears a small terminal median lobe and very much abbreviated palpi. The thoracic terga vary with regard to their degree of sclerotization and, in case-bearing larvae, one or more of the segments bear sclerotized dorsal plates. The legs are long and well developed with 1-segmented tarsi, each being terminated by a single claw. The abdomen is typically 10-segmented and generally covered with a membranous cuticle. The first segment, in many species, carries three prominent retractile papillae, one being dorsal and the remaining two lateral in position. They serve to maintain the insect in position in its case and thereby allow of an even flow of water through the latter. The anal segment in all larvae bears a pair of short and sometimes jointed appendages: each is terminated by a strong grappling hook and long flexible setae. There may also be anal papillae which take part in both respiration and ion transport (Nüske and Wichard, 1971). Some species also have 10 oval areas on the tergites and sternites of abdominal segments 3–7 of which are concerned with chloride-transport and osmoregulation (Wichard and Komnick, 1973). The larvae are apneustic and live submerged, breathing, in most cases, by means of filamentous tracheal gills. The latter are arranged in segmental groups which are commonly disposed in dorsal, lateral and ventral series along either side of the abdomen. Gills are wanting in newly hatched larvae and are not acquired until later. More rarely gills are absent throughout life and respiration is cutaneous: in some genera a tuft of anal blood gills is present. Most case-bearing larvae bear a delicate longitudinal cuticular fold on either side of the abdomen: it is beset with fine hairs and is known as the lateral line.

Trichopterous larvae are divisible into two general types. In the first type of larva (campodeoid larva of Ulmer) the body is compressed and the head not inclined at an angle. These larvae seldom construct transportable cases and both the lateral line and abdominal papillae are wanting: tracheal gills are seldom present. In the second type (eruciform larva of Ulmer) the head is inclined at a marked angle with the rest of the body. Such larvae are cylindrical in form (Fig. 538B) and construct portable cases. Papillae are developed on the 1st abdominal segment and the lateral line and tracheal gills are present.

The digestive system in Trichopterous larvae forms a straight tube from the mouth to the anus (Betten, 1902; Russ, 1908; Korboot, 1964). The oesophagus leads into a muscular crop which is followed by the stomach: the latter is the most extensive region of the gut and extends from the metathorax into the 6th abdominal segment. The hind intestine is extremely short and is divided into two successive, more or less globular chambers: six Malpighian tubes are present. There are two pairs of salivary glands belonging to the mandibular and maxillary segments respectively (Lucas, 1893): a pair of silk glands open on to the labium and these alone persist in the imago, becoming modified during pupation into salivary glands. According to Gilson (1894) the silk glands and associated structures closely resemble those of Lepidopterous larvae and the silk is produced in a similar manner. Metameric thoracic glands, known as Gilson's glands, occur in many larvae (Quennedey, 1968). In the Phryganeidae they take the form of a pair of branched tubes in each segment of the thorax: the ducts of a pair unite and open by means of a cannula-like papilla on the mid-ventral line of their segment (Fig. 539). In the Limnephilidae and Sericostomatidae (Goerinae and Sericostomatinae) there is a single pair of unbranched glands in the prothorax, opening on a central horn, those of the other segments being wanting. The thoracic glands have been variously homologized with coxal glands and with nephridia: their real function is unknown. The nervous system is very simple: there are 3 thoracic ganglia and 6 to 8 ganglia are mentioned as being found in the abdominal nerve-cord.

Two distinct types of pupal shelter are prevalent (Hickin, 1949). Before pupation a case-bearing larva shortens its habitation when necessary and fixes it to some object in the water. A silken wall is constructed across either end and these partitions are sometimes strengthened by the addition of minute stones or plant fragments. Due provision is always made for the ingress and egress of the water. The pupa lies free within the case, no cocoon being formed. Most caseless larvae (*Rhyacophila*, etc.) construct special pupal shelters which take the form of oval cavern-like structures constructed of small stones, sand or other particles. The pupae in these instances are enclosed in brownish cocoons.

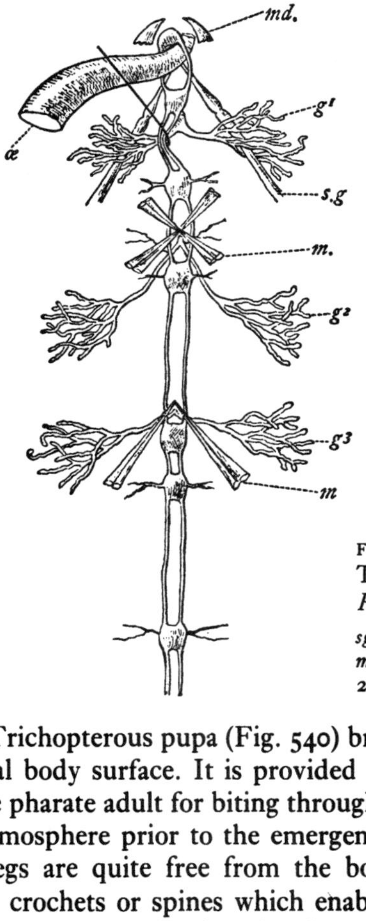

FIG. 539
Thoracic glands (g^1–g^3) of the larva of *Phryganea grandis*

sg, silk gland; *m*, muscles; *oe*, oesophagus; *md*, mandibles. *After* Gilson, *J. Linn. Soc.*, 25.

A Trichopterous pupa (Fig. 540) breathes by means of gills or through the general body surface. It is provided with strong mandibles which are used by the pharate adult for biting through the case to allow of the pupa reaching the atmosphere prior to the emergence of the imago. The antennae, wings and legs are quite free from the body, and the abdomen is armed with dorsal crochets or spines which enable the pupa to work its way out of its habitation. When the time for the emergence of the imago approaches, the pupa makes an upward passage through the water either by crawling or by swimming. In the former method the legs are clawed and the pupa is enabled to cling to vegetation or other objects. In the case of swimming pupae a degree of mobility is exhibited which is not attained by the pupae of any other insects. The motive power is provided by the pharate adult. The middle pair of legs form oars and are provided with hair fringes adapting them to that usage. In some species the pupae are able to swim freely about at the surface until they find suitable objects to crawl out upon: with the inhabitants of swift streams the imago emerges almost as soon as the pupa reaches the surface.

Certain of the more important features in the biology of the different families may be summarized as follows:

A. Larvae of the first type (campodeoid)

RHYACOPHILIDAE. Larvae in swiftly flowing water: those of *Rhyacophila* live free beneath stones and are often provided with tracheal gills. In *Glossosoma* gills are wanting and the larvae live in transportable cases of small stones. Larvae mostly carnivorous but some are vegetarians (Thut, 1969). The pupae in this family are enclosed in cocoons protected by a shelter composed of gravel or sand particles.

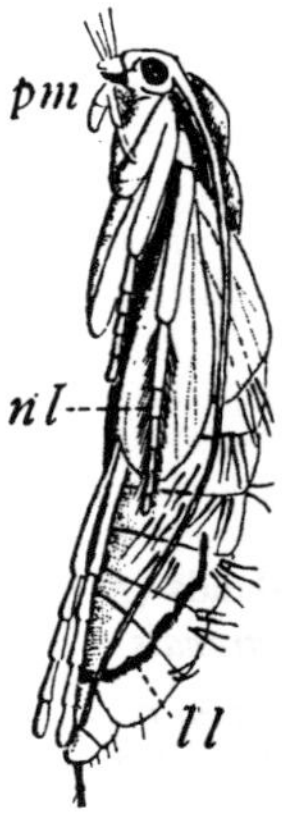

FIG. 540
A typical Trichopterous Pupa
ll, lateral line; *nl*, swimming leg; *pm*, provisional mandibles.

HYDROPTILIDAE. Larvae devoid of tracheal gills and living in standing or flowing water, feeding with modified mouthparts on the juices of *Spirogyra*, etc. There is no case in the first four instars, but the last instar which is of very different appearance, constructs a transportable, seed-like, silken case, sometimes with sand or plant particles attached (Nielsen, 1948).

PHILOPOTAMIDAE, POLYCENTROPIDAE, PSYCHOMYIIDAE, HYDROPSYCHIDAE. In these families, the larvae live in silken non-portable retreats. Tracheal gills are wanting, but anal blood-gills are commonly present. Certain of these larvae are carnivorous and construct silken snares to secure their prey. Johnstone (1964) shows that *Hydropsyche* larvae can stridulate by scraping a point on the fore femur against a file on the side of the head. The sound is apparently a response to mechanical disturbance. The pupae are protected by cavern-like shelters composed of gravel or sand particles. These all belong to the **Hydropsychoidea**.

B. Larvae of the second type (eruciform)

These all belong to the Limnephiloidea.

PHRYGANEIDAE. Larvae mostly in standing water. Cases long and cylindrical, formed of fragments of leaves or fibres arranged in a spiral manner, and open at both ends.

MOLANNIDAE. Larvae in ponds, lakes or streams, living in shield-like or conical cases composed of sand particles.

LEPTOCERIDAE. Larvae in standing or running water, living in straight or slightly curved cylindrical cases of fine sand, vegetable debris, etc.

ODONTOCERIDAE. Larvae in mountain streams, living in slightly curved cylindrical cases of sand. Hind extremity of case closed by a blackish membrane with a central slit: before pupation the mouth is closed by a single stone.

LIMNEPHILIDAE. Larvae of varied habits, living in both standing and running water. Cases of sand, sticks, leaves or shells, or of a mixture of several materials.

SERICOSTOMATIDAE. Larvae chiefly in running water: in cases of sand or stones.

Among the chief writings on the metamorphoses of Trichoptera are papers by Thienemann (1905), Siltala (1907), Lübben (1907), Barnard (1934), Ulmer (1901), Lestage (1921), Hickin (1967) and Hildrew and Morgan (1975). The last four papers give keys for the identification of the early stages while Flint (1960) and Wiggins (1960) deal with the larvae of North American Limnephilidae and Phryganeidae.

Classification

The Trichoptera are usually divided into two suborders, three or four superfamilies and 23–34 families. The basis of this classification is discussed by Ross (1967) and their more detailed classification will be found in Mosely (1939), Ross (1944) whose scheme is extended to the Australian species by Riek (1970). Schmid (1955) deals with the large family Limnephilidae. F. C. J. Fischer (1960–72) has catalogued the species.

Suborder ANNULIPALPIA

Last segment of maxillary palps annulated, flexible, generally much longer than the others. Dorsal part of the tentorium not developed. Larvae making nets.

Superfamily **Hydropsychoidea. Hydropsychidae** and allied families. Some **Dipseudopsidae** make a burrow in silt or wood. Water is pumped by body-movements through a net across the burrow (Gibbs, 1968).

Suborder **INTEGRIPALPIA**

Last segment of maxillary palps not annulated, generally less elongate. Larval ninth abdominal tergite with a sclerotized plate.

Superfamily **Rhyacophiloidea**

Ocelli present except in some **Hydroptilidae.** Maxillary palps with 5 or 6 segments, first two short and stout. This includes the **Rhyacophilidae** with free-living larvae and adult with a preapical spur on the fore tibia; the **Glossosomatidae** with the larva making a flattened, saddle-shaped case with its ventral surface developed only centrally, adult with no preapical spur on fore tibia; and the **Hydroptilidae** which are small, hairy, narrow-winged insects; larva with a form of hypermetamorphosis; early instars free-living, later instars in a purse or tube-like case and usually with a very modified shape. Unspecialized members of all three families are very similar.

Superfamily **Limnephiloidea**

Either ocelli present and male maxillary palps with 3 or 4 segments *or* ocelli absent and maxillary palps with 5 segments. Larvae with tube-like cases. There are two large groups of families. The first, including such families as the **Phryganeidae** and the **Limnephilidae** have preserved the ocelli and the dorsal part of the tentorium but the male has lost one or two segments of the maxillary palps and vein M is missing in the fore wing. The other group, including such families as the **Sericostomatidae, Philanisidae, Helicopsychidae** and **Leptoceridae,** lacks ocelli, the dorsal part of the tentorium is reduced but the male has 5 segments in the maxillary palps, the venation of the fore wing is normal. The African fauna (Marlier, 1962) is peculiar in entirely lacking the Phryganeidae and Limnephilidae.

Literature on Trichoptera

The papers of Betten (1934), Hickin (1946; 1949) and Ross (1967) provide a large number of further references.

BADCOCK, R. M. (1961), The morphology of some parts of the head and maxillolabium in larval Trichoptera, with special reference to the Hydropsychidae, *Trans. R. ent. Soc. Lond.*, **113**, 217–248.

BARNARD, K. H. (1934), South African Caddis-flies (Trichoptera), *Trans. R. Soc. Afr.*, **21**, 291–394.

BETTEN, C. (1902), The larva of the Caddis-fly, *Molanna cinerea* Hagen, *J. New York ent. Soc.*, **10**, 147–154.

—— (1934), The Caddis-flies or Trichoptera of New York State, *Bull. New York State Mus.*, **292**, 576 pp.

CRICHTON, M. I. (1957), The structure and function of the mouthparts of the adult caddis flies (Trichoptera), *Phil. Trans. R. Soc. Lond.* (B), **241**, 45–91.

ELTRINGHAM, H. (1920), On the histology of the scent-organs of the genus *Hydroptila, Trans. ent. Soc. Lond.*, **1919**, 420–430.

FISCHER, F. C. J. (1960–72), *Trichopterorum catalogus*, Parts 1–13, Amsterdam, Nederlandsche ent. Ver.

FLINT, O. S. (1960), Taxonomy and biology of Nearctic Limnephilid larvae (Trichoptera), with special reference to species in eastern United States, *Entomologica am.* (N.S.), **40**, 1–117.

GIBBS, D. G. (1968), The dwelling-tube and feeding of a species of *Protodipseudopsis* (Trichoptera: Dipseudopsidae), *Proc. R. ent. Soc. Lond.* (A), **43**, 73–79.

GILSON, G. (1894), Recherches sur les cellules sécrétantes. La soie et les appareils séricigènes, *La Cellule*, **10**, 37–63.

GLASGOW, J. P. (1936), Internal anatomy of a caddis (*Hydropsyche colonica*), *Quart. J. micr. Sci.*, **79**, 151–179.

HICKIN, N. E. (1949), Pupae of British Trichoptera, *Trans. R. ent. Soc. Lond.*, **100**, 275–289.

—— (1967), *Caddis larvae. Larvae of the British Trichoptera*, George Hutchinson & Co., London.

HILDREW, A. G. AND MORGAN, J. C. (1975), The taxonomy of the British Hydropsychidae (Trichoptera), *J. Ent.* (B), (1974) **43**, 217–229.

JOHNSTONE, G. W. (1964), Stridulation by larval Hydropsychidae (Trichoptera), *Proc. R. ent. Soc. Lond.* (A), **39**, 146–150.

KELNER-PILLAULT, S. (1960), Biologie, écologie d'*Enoicycla pusilla* Burm. (Trichoptères Limnophilides), *Année biol. Paris*, (3) **36**, 51–99.

KHALIFA, A. (1949), Spermatophore production in Trichoptera and some other insects, *Trans. R. ent. Soc. Lond.*, **100**, 449–471.

KORBOOT, K. (1964), Comparative studies of the external and internal anatomy of three species of caddis flies (Trichoptera), *Pap. Dep. Ent. Univ. Queensland*, **2**, no. 1, 1–44.

LESTAGE, J. A. (1921), In: Rousseau, *Les larves et nymphes aquatiques.* **1**, part 9, 343–964, Lebègue et Cie, Bruxelles.

LÜBBEN, H. (1907), Ueber die innere Metamorphose der Trichopteren, *Zool. Jb., Anat.*, **24**, 71–122.

LUCAS, R. (1893), Beiträge zur Kenntnis der Mundwerkzeuge der Trichopteren, *Arch. Naturg.*, **59** Heft 1, 285–330.

MARLIER, G. (1962), Genera des Trichoptères de l'Afrique, *Ann. Mus. Afr. centr. Tervuren (8). Sci. Zool.*, **109**, 1–263.

MCLACHLAN, R. (1874–84), *A Monographic Revision and Synopsis of the Trichoptera of the European Fauna*, J. van Voorst, London.

MORETTI, G. AND VIGANO, A. (1960), L'habitat e la biologia di *Helicopsyche sperata* McL. in Toscana, *Boll. Zool. Turin* (1959) **26**, 573–589.

MORTIMER, T. J. (1965), The alimentary canals of some adult Lepidoptera and Trichoptera, *Trans. R. ent. Soc. Lond.*, **117**, 67–93.

MOSELY, M. E. (1939), *The British Caddis-flies (Trichoptera): a collector's handbook*, Routledge, London.

MOSELY, M. E. AND KIMMINS, D. E. (1953), *The Trichoptera (Caddis-flies) of Australia and New Zealand*, Brit. Mus. (Nat. Hist.), London.

NIELSEN, A. (1948), Postembryonic development and biology of the Hydroptilidae, *K. Danske Vidensk. Selsk. biol. Skrift*, 5 no. 1, 200 pp.

—— (1957), A comparative study of the genital segments and their appendages in male Trichoptera, *Biol. Skr.*, 8 no. 5, 1–159.

NOYES, A. A. (1914), The biology of the net-spinning Trichoptera of Cascadilla Creek, *Ann. ent. Soc. Am.*, 7, 251–271.

NÜSKE, H. AND WICHARD, W. (1971), Die Analpapillen der Köcherfliegenlarven. I. Feinstruktur und histochemischer Nachweis von Natrium und Chlorid bei *Philopotamus montanus* Donov., *Cytobiologie*, 4, 480–486.

PICTET, F. J. (1834), *Recherches pour servir à l'histoire et l'anatomie des Phryganides*, Chezbulièz, Geneva.

QUENNEDEY, A. (1968), Anatomie comparée des glandes de Gilson des larves de Trichoptères, *Bull. Soc. zool. France*, 93, 467–477.

RATHJEN, W. (1939), Experimentelle Untersuchungen zur Biologie und Oekologie von *Enoicyla pusilla* Burm., *Z. Morph. Oekol. Tiere*, 35, 14–83.

RIEK, E. F. (1970), Trichoptera, *The Insects of Australia*. Chapter 35, C.S.I.R.O., Divn. Entomology, Melbourne.

ROSS, H. H. (1944), The Caddis-flies, or Trichoptera of Illinois, *Bull. Illin. nat. Hist. Survey*, 23, no. 1, 326 pp.

—— (1964), Evolution of caddisworm cases and nets, *Amer. Zool.*, 4, 209–220.

—— (1967), The evolution and past dispersal of the Trichoptera, *A. Rev. Ent.*, 12, 169–206.

RUSS, E. A. L. (1908), Die postembryonale Entwicklung des Darmkanals bei den Trichopteren (*Anabolia laevis* Zett.), *Zool. Jb., Anat.*, 25, 675–770.

SCHMID, F. (1955), Contribution à l'étude des Limnophilidae (Trichoptera), *Mitt. schweiz. ent. Ges.*, 28 (Beiheft), 245 pp.

SILTALA, A. J. (1907), Trichopterologische Studien II. Ueber die postembryonale Entwicklung der Trichopterenlarven, *Zool. Jb., Suppl.*, 9, 309–626.

STITZ, H. (1904), Zur Kenntnis des Genitalapparats der Trichopteren, *Zool. Jb., Anat.*, 20, 277–314.

THIENEMANN, A. (1905), Biologie der Trichopterenpuppe, *Zool. Jb., Syst.*, 22, 489–574.

THUT, R. N. (1969), Feeding habits of seven *Rhyacophila* (Trichoptera: Rhyacophilidae) species with notes on other life-history features, *Ann. ent. Soc. Am.*, 62, 894–898.

ULMER, G. (1901), Beiträge zur Metamorphose der deutschen Trichopteren, *Allg. Zs. Ent.*, 6, 115–119, 134–136, 166–168, 200–202, 223–226, 309–311.

—— (1909), Trichoptera, In: Brauer, *Süsswasserfauna Deutschlands*, 5–6, 326 pp., Verlag Gustav Fischer, Jena.

UNZICKER, J. D. (1968), The comparative morphology and evolution of the internal reproductive system of Trichoptera, *Illinois biol. Monogr.*, 40, 1–72.

WESENBERG-LUND, C. J. (1911), Biologische Studien über netzspinnende campodeoide Trichopterenlarven, *Internat. Rev. Hydrobiol. Hydrog.* (Biol. Suppl. 3), 4, 64 pp.

WICHARD, W. AND KOMNICK, H. (1973), Fine structure and function of the abdominal chloride epithelia in caddisfly larvae, *Z. Zellforsch.*, 136, 579–590.

WIGGINS, G. B. (1960), A preliminary systematic study of the North American larvae of the caddisfly family Phryganeidae (Trichoptera), *Can. J. Zool.*, 38, 1153–1170.

Order 29

HYMENOPTERA (ANTS, BEES, WASPS, ICHNEUMON FLIES, SAWFLIES ETC.)

Insects with 2 pairs of membranous wings, often with the venation greatly reduced; the hind wings smaller than the fore pair and interlocked with the latter by means of hooklets. Mouthparts primarily adapted for biting and often for lapping or sucking also. The abdomen usually basally constricted and its first segment fused with the metathorax; an ovipositor always present and modified for sawing, piercing or stinging. Metamorphosis complete; larva generally apodous with a more or less well-developed head, more rarely eruciform with locomotory appendages; tracheal system usually holopneustic or peripneustic throughout life, or at least in the final instar. Pupae adecticous, exarate (rarely obtect) and a cocoon generally present.

This order is one of enormous extent comprising more than 100 000 described species and many thousands of forms still await discovery. If the Hymenoptera be judged by their behaviour, they must be regarded as including the highest members of their class. Structurally the majority of their species have attained an advanced degree of specialization which is only surpassed by the Diptera. In certain species of the order the individuals have acquired the habit of living together in great societies, as in the case of the ants, wasps of the family Vespidae and bees of the family Apidae. A large proportion of the females of these societies have undergone structural changes, in some cases slight, in others more pronounced, so that they constitute a separate caste or type of individual known as the worker whose power of reproduction is either in abeyance or usually limited to the laying of male-producing eggs. Their functions include those of nest-building, feeding and tending the brood and the defence of the colony. The normal reproduction of the species in the social Hymenoptera is either performed, as in certain wasps, by many of the female members of a colony or more usually by a single individual often of large size known as the queen. The sole function of the males is that of impregnating the females, an act which often comparatively few succeed in consummating.

Indications of what, in the higher Hymenoptera, constitutes social behaviour are found among solitary wasps and bees (Wheeler, 1928). Most solitary bees and wasps practise 'mass provisioning' – i.e. they store their cells with sufficient food to satisfy their developing offspring and close them down before the eggs hatch. There are, however, species which feed their larvae from time to time ('progressive provisioning'), thus becoming acquainted with their offspring. Among tropical Vespidae of the tribes Ropalidiini and Polybiini many colonies are perennial and contain numerous fecundated females; their larvae are reared by progressive provisioning. Workers are often hardly differentiated and sometimes numerically weak. Such colonies, when fully developed, emit swarms consisting of fecundated females, usually accompanied by workers. This pleometrotic state is sometimes considered more primitive than what obtains among the Vespidae of temperate zones, whose colonies are haplometrotic, i.e. dominated by a single fecundated female or queen: such colonies are seasonal only and the worker caste is usually clearly differentiated. Among the social bees (Michener, 1974) the most primitive are the species of *Halictus* (s.l.) p. 1255) and the Bombinae. Humble bees construct no true comb but the larvae are reared in waxen pockets. They are at first fed by mass provisioning but in some species the older larvae are fed periodically. Their colonies are haplometrotic and last only for a season. Among the Meliponinae and Apinae the colonies are perennial, haplometrotic and give off swarms. *Melipona* and *Trigona* practise mass provisioning and close their cells; apart from *Halictus*, they are the only social Hymenoptera where there is no contact between parent and larva; in many cases the three castes, which appear to be genetically determined, are all reared in identical cells on a similar diet. In *Apis* the cells are open throughout larval development: the castes are reared in differentiated cells, at least in *A. mellifera*, and queen-producing larvae are fed on a specialized diet. Among ants the castes exhibit their maximum differentiation: the larvae are reared in clusters, there being no cells, and there is a more intimate relation between the workers and the brood than in other social Hymenoptera.

Wheeler attributed great importance to the phenomenon of *trophallaxis*, or the mutual exchange of food between imagines and their larvae. Ant larvae seem to produce a secretion highly acceptable to their nurses. In some species it is saliva, in others an exudation of the integument, while in the Pseudomyrmicinae it is a product of special papillae known as exudatoria. It appears that avidity for these larval secretions helps to sustain the bond between ants and their brood and it further accounts for the relations which ants have acquired with alien insects and other arthropods (Wheeler, 1923). Trophallaxis also occurs in the Vespidae but in them its function is less certain; it may be merely the disposal of excess water produced by the larvae (Brian and Brian, 1952) but in some Vespinae (pp. 1247–48) there is an important exchange of food. Among bees the phenomenon seems to be wanting. An excellent modern summary of social insects will be found in Wilson (1971).

Hymenoptera are also remarkable on account of the highly evolved condition which parasitism has reached in the order, and it has been independently acquired among species belonging to very diverse superfamilies. The Symphyta are essentially phytophagous, nevertheless *Orussus* is parasitic in its larval stage, but its habits have been very little studied. Among the Apocrita, about one half the known species of Cynipoidea are parasites, and this same habit occurs in the whole of the Ichneumonoidea and Proctotrupoid groups, and in almost all the Chalcidoidea. Associated with parasitism is the phenomenon of polyembryony (see p. 305) which is known to occur in a few of the Chalcidoidea, Scelionoidea and Braconidae. Among the aculeate families true parasitism is much rarer and, in the majority of cases of this kind, their larvae devour the provisions accumulated by the host for its own progeny. This involves the destruction of the latter but it is not parasitism in the strict sense. For a general discussion of parasitism in its different phases, and the more important literature thereon, reference should be made to papers by Wheeler (1919) and Clausen (1940).

The effects of Hymenopterous parasites upon their hosts vary in different cases. Certain of the Chalcid parasites of coccids are bivoltine. One generation attacks the young hosts who fail to reach maturity and succumb to the parasitism. The following generation of parasites attacks the older hosts and, in this case, the females of the latter are usually able to lay some or even all their ova prior to being overcome by the parasites. Wheeler has shown that the ectoparasite *Orasema* (p. 1223) produces abortion, or malformation, of certain parts in the ants which it attacks, and none of the latter become imagines. Certain of the Dryinidae are known to parasitize nymphal Homoptera and may modify or otherwise inhibit the development of the secondary sexual characters of their hosts. Lists of parasites and their hosts are given in the catalogue of Thompson (1943–71). Such lists have to be used with caution owing to the difficulties of identification.

Parthenogenesis (see also p. 303) is more frequent among Hymenoptera than in any other order of the animal kingdom, and this method of reproduction is prevalent in a number of widely separated families. In many it is not an occasional phenomenon, but plays an important part in the continuity of the species, and may also be accompanied by an alternation of generations. The best known instance of parthenogenesis is found in the honey bee, in which unfertilized eggs, whether laid by the queens or by fertile workers, normally produce males and the same applies to *Vespa*. Among ants parthenogenesis has been less thoroughly investigated, and it has been claimed that the unfertilized eggs similarly only give rise to males, but Reichenbach, Donisthorpe and others have shown that the workers are capable of laying unfertilized eggs which develop into other workers. In the Cynipidae both sexes may be produced parthenogenetically and the generations, which arise in this way, alternate with those produced by the sexual method. In other species heterogony is absent, and females are produced parthenogenetically generation after generation; in some cases males are absent and in others rare. Among the Tenthredinidae parthenogenesis is also prevalent; in certain species only males arise from the unfertilized eggs, in others only females, or both males and females may be produced. In some Chalcidoidea parthenogenesis is the usual method of reproduction as in *Aphelinus mytilaspidis* and *Tetramesa* (= *Harmolita*) *grandis*, in which examples males are very rare. Many other parasitic Hymenoptera are capable of both sexual and parthenogenetic reproduction and, in these cases, the latter process generally gives rise to males.

General Structure of the Imago

The general structure of the Hymenoptera has been well investigated in comparatively few types. The work of Snodgrass (1956) on the anatomy of the honey bee will serve as an introduction to the general morphology of the order. For the Formicidae the numerous papers by Janet should be consulted. Other useful papers are: Ichneumonoidea, *Doryctes* (Seurat, 1899), *Microbracon* (Soliman, 1941); Chalcidoidea, *Euchalcidia* (Hanna, 1935), *Blastophaga* and *Philotrypesis* (Grandi, 1929 and 1930), *Monodontomerus* (Bucker, 1948); Vespoidea, *Vespula* (Duncan, 1939); Apoidea, Michener (1944).

A. EXTERNAL ANATOMY

The Head is free from the thorax and often extremely mobile. It varies considerably in form and, as a rule, the long axis is the longitudinal one. The cranial capsule is very completely consolidated but both clypeus and labrum are usually distinct; an epipharynx is well developed and trilobed in the higher forms, the median lobe being pointed and projecting. Acuteness of vision is a characteristic of the order and the compound eyes are therefore almost always large; in the male they are sometimes strongly convergent or holoptic. In certain species of ants belonging to the genera *Dorylus* and *Eciton* the eyes have atrophied, and in other species of the latter genus, they are reduced to a single facet on either side. Three *ocelli* are commonly present but, in some cases, they are aborted, as in the Bembicini and in the workers of many of the ants. The *antennae* are extremely variable in character in the Symphyta and among the parasitic families of the Apocrita. As a rule, they are longer in the males than in the females, and frequently exhibit pronounced sexual dimorphism. The latter feature attains its greatest development among the Proctotrupoids and Chalcidoidea, where these organs in the male may be either filiform, clavate, pectinate, branched or verticillate. The number of segments present is singularly inconstant in the lower superfamilies: thus among the Ichneumonidae, for example, it may be as low as 14, or as high as 70. In the sawfly *Arge* there are only three segments, and four are present in some of the ants. In the Sphecoidea, Vespoidea and Apoidea the number for the most part is fixed, there being usually 13 segments in the males and 12 in the females.

The Mouthparts exhibit a wide range of differentiation from the generalized biting, orthopterous type found among the Symphyta to the highly modified sucking type of *Apis*, *Euglossa* and other bees. Mandibles are universally present throughout the order but, except in the predacious members of the Tenthredinidae, their principal function is industrial rather than trophic. They are used to enable the imagines to cut their way through the walls of their cocoons in the case of the parasitic superfamilies, while among the Sphecoidea, Vespoidea and Apoidea their principal functions are the gathering of material and nest-building. If the mouthparts of *Nematus*, or

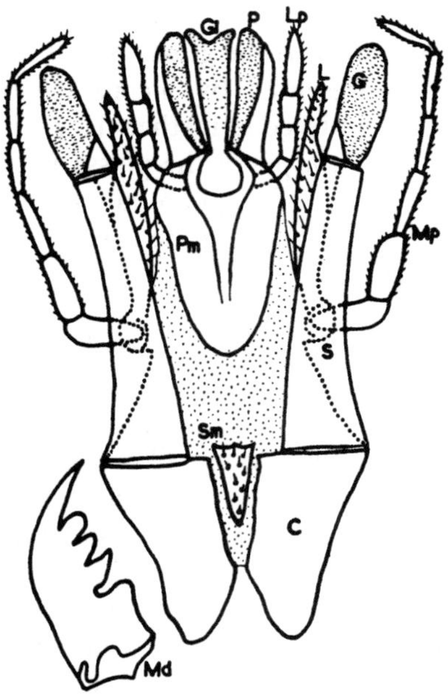

FIG. 541
Mouthparts of *Tenthredo* (Symphyta), ventral view

C, cardo; *G*, galea; *Gl*, glossa; *L*, lacinia; *Lp*, labial palp; *Md*, left mandible (dorsal view); *Mp*, maxillary palp; *P*, paraglossa; *Pm*, prementum; *S*, stipes; *Sm*, submentum.

other typical sawfly (Fig. 541), be examined it will be observed that well-developed dentate mandibles are present; the complete number of parts are evident in the maxillae, and their palpi are 6-segmented. In the labium both prementum and submentum are developed, the labial palpi are 4-segmented, and the ligula appears deeply cleft into three nearly equal lobes – a median glossa and lateral paraglossae. Among the Apocrita this same type of mouthparts is retained in the parasitic group, but it has undergone a variable amount of specialization. The labial and maxillary palpi usually have a reduced number of segments, particularly in the Chalcidoidea. The maxillae are frequently single-lobed, and the ligula is commonly formed by the broadened glossa, the paraglossae being either vestigial or absent. In the higher superfamilies, the glossa becomes increasingly prominent, in conformity with the habit of feeding upon and collecting nectar. This organ becomes progressively lengthened, the associated mouthparts become attenuated accordingly and the result of these modifications is the formation of a proboscis. The latter organ is an adaptation which is necessary in order to extract the juices from the deeply seated nectaries of many flowers.

It is possible to trace the evolution of the proboscis in different genera of the Apoidea, from the simple condition found in the Colletidae, up to the highly specialized apparatus seen in *Apis*, *Euglossa*, etc. In the Colletidae the glossa is extremely short and broad with a bifid extremity; the labial palpi are non-sheathing and 4-segmented, and the maxillary palpi are 6-segmented. In *Andrena* the glossa, although still short, is acuminate, while in *Panurgus* and *Nomada* it is appreciably lengthened, as are also the labial palpi and the maxillary lobes. In *Melecta* the first two segments of the labial palpi ensheath the greatly drawn out glossa, and the maxillary palpi are reduced to small 4-segmented organs. In *Psithyrus* and *Bombus* the glossa is still further elongated, and the maxillary palpi are represented by inconspicuous 2-segmented organs, while in *Apis* they have undergone further degeneration and are in the form of minute papillae. In *Anthophora* the glossa is longer than in any other British bees, but the two pairs of palpi are not specialized to a correspondingly high degree. In the tropical *Euglossa* the maxillary palpi are single segmented, the labial palpi 2-segmented and the glossa attains a length exceeding that of the whole insect.

In Figs. 541, 542 and 543 the mouthparts of a saw-fly, *Vespula* and *Apis* are represented. In the case of the first mentioned type the essentially biting nature of their component parts is evident. In *Vespula* these organs are adapted both for biting (and mastication) and licking. The maxillae are comparatively little modified; the cardines and stipites are well developed, and the palpi are 6-segmented. The laciniae are reduced to small scales, while the galeae assume the form of broad membranous lobes. The labium is composed of a large shield-shaped prementum, the ligula is represented by the curious elongated paraglossae and a wide bilobed glossa, while the palpi are slender 4-segmented organs. The primitive bee *Hylaeus* (Fig. 543A, B) is not very different from *Vespula* but in *Apis* the mouthparts are highly modified to form a proboscis and the glossa has become a sucking organ. The chief basal plate of the maxilla is the stipes and at its proximal end it is articulated with the stalk-like cardo, and near its apex on the outer border is a minute peg-like maxillary palp. Articulating with the distal extremity of the stipes is a large blade-like lobe or galea; a reduced lacinia is present though often overlooked. In the labium, the large strongly sclerotized plate is the prementum, and the latter articulates with a small triangular sclerite or mentum. The base of the latter is supported by a flexible transverse band, the submentum (lorum of some authors), whose extremities are attached to the distal ends of the cardines. The labial palpi are conspicuous 4-segmented organs, each being carried by a basal palpiger. The elongate central organ of the proboscis is the glossa, and at the base of the latter are two small concealed lobes or paraglossae. The glossa is invested with long hairs and at its apex is a small spoon-shaped lobe – the *flabellum* or bouton. The side walls of the glossa are inclined downwards and inwards, until they almost meet along the mid-ventral line, and thereby form the boundaries of a central cavity. Embedded in the roof of the latter is a longitudinal rod which is grooved along its entire length, and this groove is converted into an imperfect tube by means of two rows of hairs which converge from its margins. The dorsal rod is flexible and becomes continuous basally with the

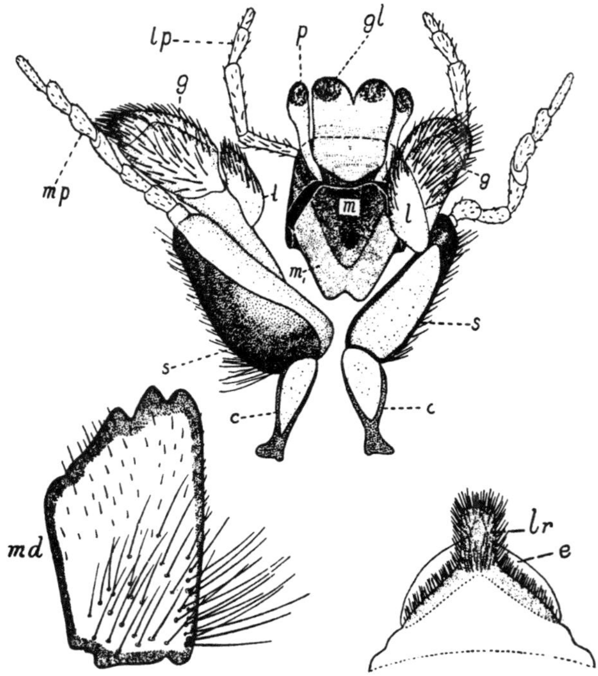

FIG. 542 Mouthparts of *Vespula germanica* (worker), one maxilla is shown extended and the other in its natural position

c, cardo; *e*, epipharynx; *g*, galea; *gl*, glossa; *l*, lacinia; *lp*, labial palp; *lr*, labrum; *m*, m_1, prementum; *md*, mandible; *mp*, maxillary palp; *p*, paraglossa; *s*, stipes.

ventral supporting plate of the ligula. The lining of the cavity of the glossa and its rod can be evaginated through the cleft, a process which admits of the cleansing of the parts in question. In transverse sections, the space between the outer and inner walls of the glossa is seen to contain blood and is in communication with the head cavity. The complete extension of the organ is due to blood pressure. Its retraction is partly due to the release of that pressure, and partly to the contraction of muscles inserted into the base of the dorsal rod. The latter, when drawn backwards, shortens the glossa which, as Snodgrass remarks, become bushy just as does a squirrel's tail if one attempts to pull out the bone at the base. When at rest, the mouthparts are folded down beneath the head against the stipites and mentum. During feeding they are straightened out with the two modified proximal segments of the labial palpi closely applied to the glossa, and partly embraced by the ensheathing laciniae. The glossa is very active while food is being imbibed: not only is the whole ligula alternately retracted into and protruded from the base of the mentum, but the glossa itself alters its length in the manner just described. The liquid food ascends by means of capillary action in the central channel of the glossa, and the effect of the shortening of the latter organ is to squeeze the nectar backwards, until it enters the space between

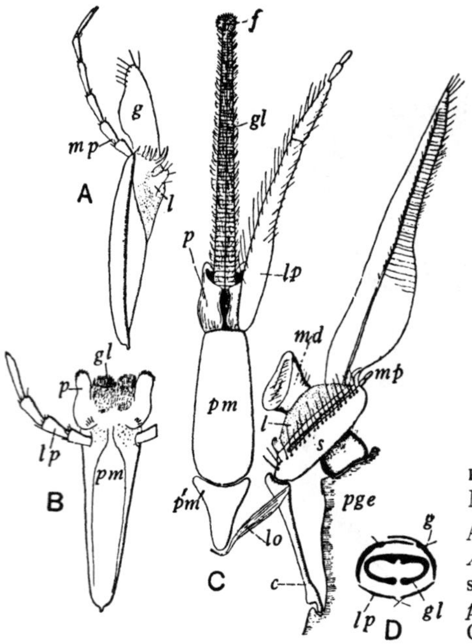

FIG 543.
Mouthparts of bees

A, *Hylaeus*, right maxilla and B, labium (ventral). *Apis*. C, mouthparts (ventral). D, cross-section of same when feeding; *f*, flabellum; *lo*, submentum; *pge*, postgena; *pm*, prementum; *p'm'*, mentum. Other lettering as in Fig. 541.

the paraglossae, and so on into the mouth. Its passage onwards is probably ensured by means of a sucking action exerted by the pharynx. For a detailed investigation of the structure and mode of action of the proboscis, and its musculature, reference should be made to the memoirs by Snodgrass (1925; 1942): the anatomy of this organ in different genera of bees is described by Saunders (1890) and Demoll (1908).

The Thorax of Hymenoptera (Fig. 544) is principally characterized by the fusion of the first abdominal segment with the metathorax, and its complete incorporation in the latter region. The transferred abdominal segment is termed the *propodeum* which was first described by Latreille as the 'median segment'. Among the Symphyta the latter is still evidently part of the abdomen and has undergone but little specialization. In the Apocrita (Fig. 544, *Hylaeus*) it has become transferred to the thorax and fused up with the metapostnotum and metapleura. Its existence in all cases, however, may be ascertained by the fact that it bears the first pair of abdominal spiracles.

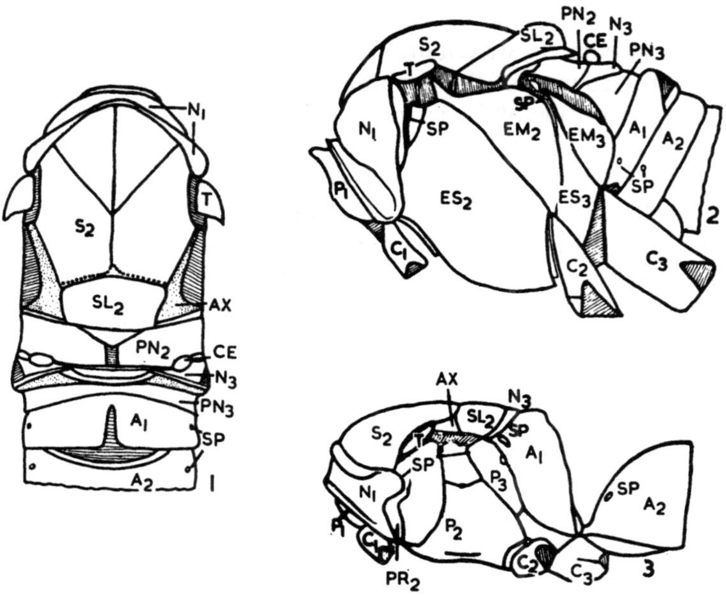

FIG. 544 1. *Tenthredo*, Dorsal aspect of thorax. 2. The same, lateral aspect. 3. *Hylaeus*, lateral aspect

A, abdominal segment; *AX*, axilla; *C*, coxa; *CE*, cenchri; *EM*, epimeron; *ES*, episternum; *N*, notum; *P*, pleuron; *PN*, postnotum; *PR*, prepectus; *S*, scutum; *SL*, scutellum; *SP*, spiracle; *T*, tegula. Numerical suffixes indicate thoracic or abdominal segments.

The study of the thorax in the order, as a whole, indicates that a progressive series of modifications has taken place in the higher forms (Snodgrass, 1910; Daly, 1964). The *pronotum* is separated from the pleuron and attached to the front of the mesothorax. The propleuron and cervical sclerite of each side unite to form what may be conveniently called the *propleuron*; the sternum is sunk beneath the pleura and is only visible without dissection in some lower forms. The *mesonotum* is divided by a transverse sulcus into an anterior plate or *scutum* and a posterior one or *scutellum*. Areas at the sides of the scutellum known as the *axillae* belong however to the scutum. In some Parasitica (Fig. 545) longitudinal sulci demarcate *parapsides* (cf. Tulloch, 1929) at the sides of the scutum. A somewhat similar pair of sulci nearer the mid-line, the *notaulices*, are also sometimes present. Tegulae are present throughout the order. The *mesopostnotum* and its phragma are invaginated and concealed within the thorax; the phragma is often extensive, and may extend backwards into the base of the abdomen, as in *Aphelinus* and some Mymarids. The *metanotum* is usually reduced to a single plate carrying the hind wings, while the *metapostnotum*, in all the higher members of the order, is indistinguishably merged into the front margin of the propodeum.

The Wings – No insects have deviated so far from the primitive venational type as the Hymenoptera, and even the most generalized members of the order are highly specialized as regards the wing-veins. Great difficulties confront any attempt to determine their homologies and, as Comstock has pointed out, the courses of the tracheae do not afford a reliable clue in this respect. An examination of the young pupae of the honey bee reveals the fact that the venation is already foreshadowed before the tracheae develop, and that the latter are formed after the vein cavities are laid down. We have, therefore, to depend very largely upon comparative studies within the order and also with members of related orders. A dominant feature is the extensive fusion of the principal veins and the tendency of their branches to assume a transverse course. The venation is so difficult to interpret that a number of

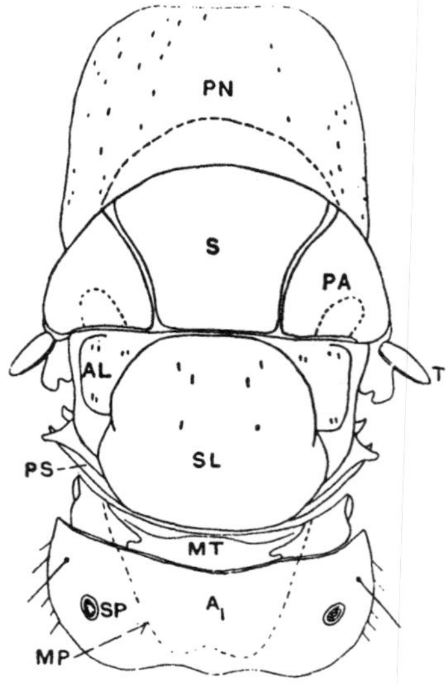

FIG. 545
Dorsal aspect of the thorax of a Torymid (*Philotrypesis caricae*)

AL, axilla; *MP*, mesophragma; *MT*, metanotum; *PA*, parapsides; *PN*, pronotum; *PS*, postscutellum of mesothorax. Other lettering as in Fig. 544. *After* Grandi, *Boll. Lab. Zool. Portici*, 14, 1921.

schemes has been proposed: two only are referred to here. That of Ross (1936), based on a comparison with such forms as *Sialis*, has recently been accepted by a number of morphologists. It is illustrated in Figs. 546, 547. An alternative terminology (that largely used by Cresson: Figs. 548, 549) is one of a number of more or less similar schemes, differing in detail according to the family, but widely used in the systematic literature. The term *cubital* (better *submarginal*) cells of Fig. 549 can be conveniently used in conjunction with Ross's terminology. Specialization by reduction and fusion is evident throughout the Apocrita and attains its maximum development among certain of the Evaniidae and the Chalcidoidea, where there is a solitary compound vein, running near the costa of the fore wing, and the hind wing is veinless; in the Platygasteridae both pairs of wings may be devoid of veins. In the Chalcids, the proximal part of the vein is the *submarginal*; more distally it runs along the front margin as the *marginal vein* and gives off a short *stigmal* vein. Throughout the order the wings of each side are held together by a row of hooks or *hamuli* along the costal margin of the hind pair: these hooks catch on a fold along the posterior margin of the fore wing, so that the wings of a side become interlocked. Among the Chalcids the hamuli are reduced to a localized group of two or three hooks and, in the Mymaridae, the latter may be totally wanting. Apterous forms are a common feature in the order, and are the rule among the workers of all species of ants, and occasionally also among the males of these insects and of many Agaonidae. Wingless females are present in the Mutillidae, Thynninae and Myrmosinae, in which groups the males alone are winged. Similarly apterous females occur frequently in the Proctotrupoids and in certain of the Ichneumonidae and Braconidae. Apterous members of both sexes of the

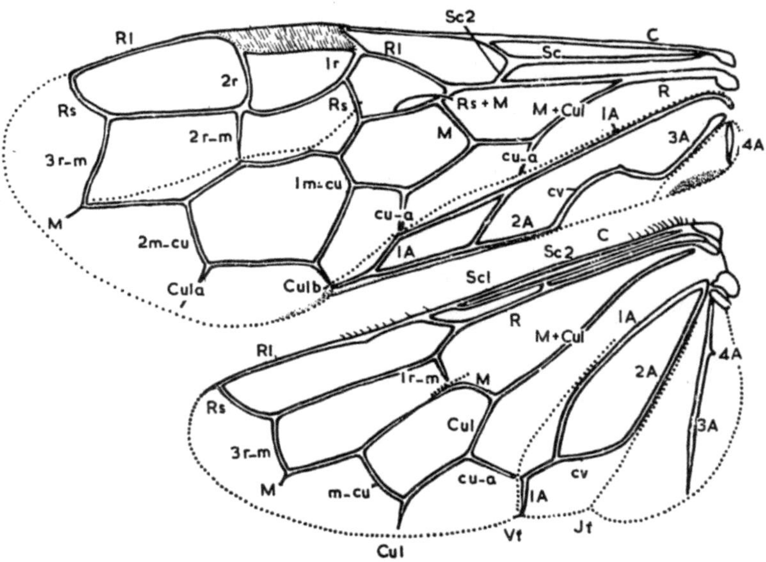

FIG. 546 Left wings of *Pamphilius* (Symphyta), with the veins lettered *cv*, anal cross-vein; *vf*, vannal fold; *jf*, jugal fold.

same species are rare but are known, for example, in the Diapriid *Platymischus*, and in certain members of the Ichneumonid subfamily Hemitelinae.

The Legs exhibit various modifications: in all the parasitic groups, excepting the Pelecinidae, the trochanters are commonly said to be two-segmented, though the second piece, known as the trochantellus, belongs to the femur. In the Apocrita the spur or calcar at the apex of the fore tibia is knife-like in character, and fits against a semicircular emargination of the basitarsus. This cavity is beset with fine comb-like teeth, and the antennae are repeatedly passed through the apparatus, which functions as a preening organ. The Sphecoidea, together with a number of other solitary wasps, are often termed the 'Fossores', and their legs are adapted for digging and running, or for nest-building. In the Apoidea, the legs are comparatively simple in certain primitive genera but, in the higher forms, the posterior pair

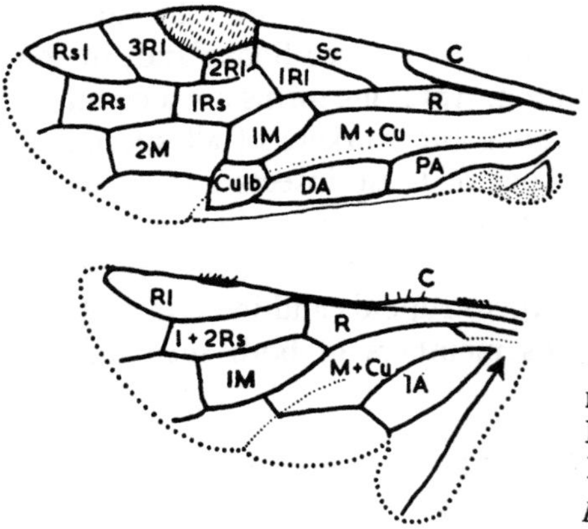

FIG. 547
Left wings of *Xyela* (Symphyta) with the cells lettered

DA, *PA*, distal and proximal anal.

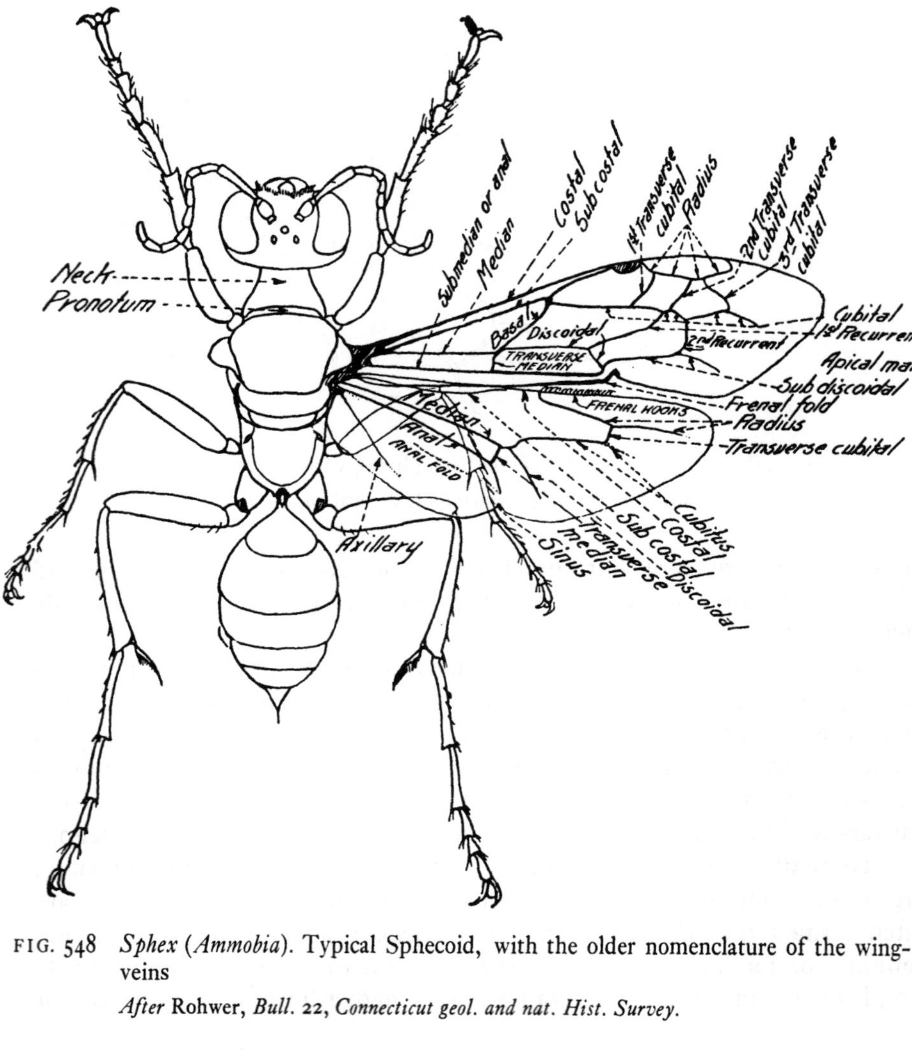

FIG. 548 *Sphex* (*Ammobia*). Typical Sphecoid, with the older nomenclature of the wing-veins

After Rohwer, *Bull.* 22, *Connecticut geol. and nat. Hist. Survey.*

is adapted for pollen-carrying. The posterior tibia is more or less dilated and either bears a large pollen brush or *scopa* or is margined with long hairs, being thus modified to form a *corbicula* or pollen basket. The basitarsus is flattened on its inner aspect, and provided with several rows of short stiff spines which form a brush; by means of the latter the bee gathers the pollen adhering to the hairs of its body. When a sufficient quantity has accumulated on the brushes, it is scraped off over the edge of the hind tibia of the opposite side and stored in the pollen basket. As a rule the tarsi of Hymenoptera are 5-segmented, and an arolium is present between the claws.

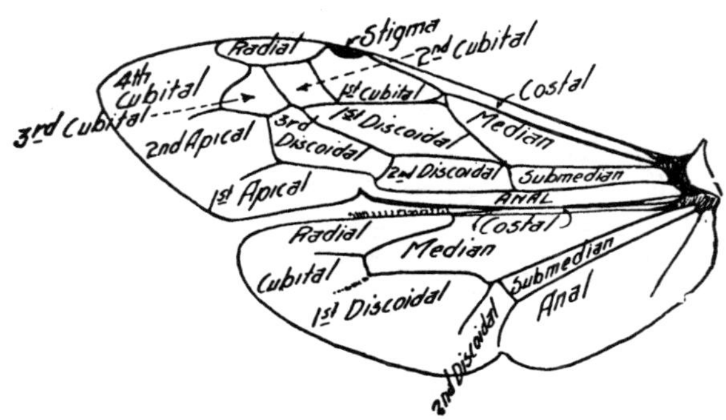

FIG. 549 *Sphex*. Left wings with the cells named according to the older nomenclature

After Rohwer, *loc. cit.*

The Abdomen is restricted physiologically to the region which commences with the 2nd segment, the 1st abdominal segment being the propodeum already referred to. The region behind the propodeum may be called the *gaster* or *metasoma*. The number of segments that can be identified in the gaster varies very greatly: the maximum number of nine are present in the Tenthredinidae. As a rule, in the higher groups, there are six exposed segments in the females and seven or eight in the males. In a number of groups, the 9th gastral tergite bears small, lateral, setigerous processes, the *pygostyles*. In the Symphyta the 1st gastral (2nd abdominal) segment is always unmodified and forms a broad base of attachment. In the Apocrita this region is wholly or partially constricted to form a narrow neck-like zone, which is termed the *petiole*. In the honey bee the latter is so short as to be visible only when the abdomen is deflexed. Almost every transition can be found between this condition and the extremely attenuated petiole of *Sphex*, *Sceliphron* and other genera. Domenichini (1953) illustrates the abdomen of many Chalcids. A ventral pollen brush or *scopa* occurs in some Megachilidae. The *ovipositor* (Fig. 550) is a well-developed organ with considerable uniformity of structure although modified in different groups for sawing, boring, piercing, stinging or, in some groups (e.g. *Formica*) reduced. Its general anatomy is well illustrated in the hive bee (Snodgrass, 1925), though Scudder (1961) has proposed somewhat different homologies and nomenclature. Oeser

(1961) has studied its structure in a range of genera. Morphologically, the ovipositor is composed of three pairs of gonapophyses, which have been shown by Zander to arise from a similar number of abdominal processes in the larva – one pair on the 8th segment and two pairs on the 9th. In the adult, the 7th sternite is usually exposed and little modified. The 8th sternite is reduced but on each side is found a *first valvifer* or *gonangulum* (triangular plate) to which is attached the long *first valvula* or *gonapophysis*. The latter is the active part of the ovipositor, forming the saw in the Symphyta, the effective ovipositor in the Parasitica, and the lancets of the sting in the Aculeata. The sternite of the 9th segment is also reduced and on each side is found a *second valvifer* or *gonocoxa* (oblong plate) to the anterior end of

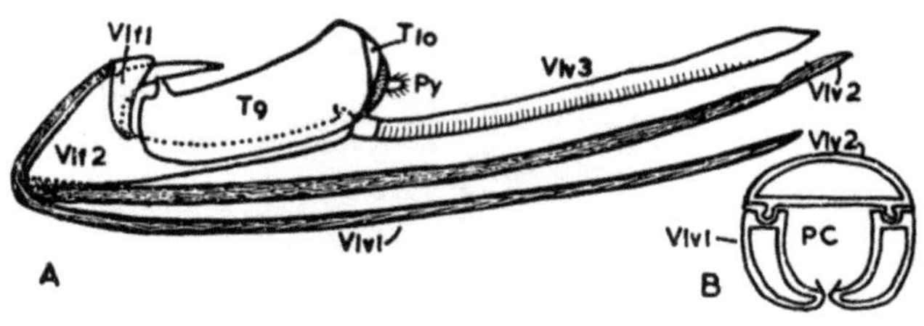

FIG. 550 A. Ovipositor of *Venturia* (Ichneumonidae), from the left side. B. Section of the terebra of *Apis*

PC, poison canal; *Py*, pygostyle; *T9*, *10*, abdominal tergites; *Vlf 1*, *2*, 1st and 2nd valvifers; *Vlv 1*, *2*, *3*, 1st, 2nd and 3rd valvulae.

which is attached the *second valvula* or *gonapophysis*. The 2nd valvulae of each side are immovably joined to one another and form the sting sheath, enclosing the 1st valvulae. The fused 2nd valvulae have along most of their length a pair of ridges projecting into corresponding grooves in the 1st valvulae. Thus the latter can move in and out without being detached from their sheath. These two pairs of valvulae together constitute the *terebra*. In the sawflies, both pairs of valvulae are provided with transverse ridges which terminate below in the serrations of the saw. In the Apocrita, the 2nd valvulae normally bear analogous 'barbs' but the 1st valvulae are smooth. The barbs are specially well developed in *Apis* so that the sting normally remains behind in the wound which it causes. The *third valvulae* or *gonoplacs* (gonostyli, sting palps) each arise from the posterior end of the 2nd valvifer. They are usually less sclerotized than the other valvulae and more or less covered in bristles. The 9th abdominal tergite in all more specialized forms is more or less retracted and desclerotized. In the Aculeata, for instance, it is divided into two lateral plates (quadrate plates) by a membranous area.

The two pairs of valvifers and, in higher forms, the quadrate plates function as levers, and can be moved by powerful muscles. By means of the rotation of the valvifers, the terebra is driven through the tissues of the victim when oviposition or stinging take place. A pair (Fig. 551) of filiform *acid glands* opens, either separately or by means of a common duct, into a large poison-sac. Their secretion is a complex mixture of which the most important constituents are, probably, a protein and certain enzymes. The latter act on the tissues of the victim to release histamine which is responsible for many of the resulting symptoms (summary in Wigglesworth). In *Formica* and its allies the active agent is chiefly formic acid. The poison-sac discharges into the bulb formed by the expanded bases of the 2nd valvulae and, situated close to its opening, is the aperture of an unpaired *alkaline gland*, the function of whose secretion is not clear. Bordas (1897) has studied

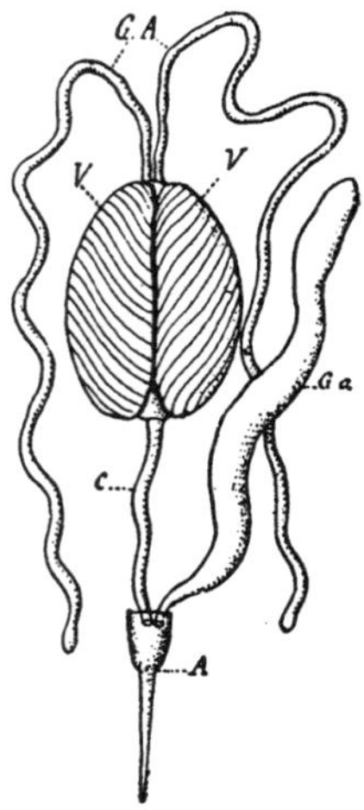

FIG. 551
Vespula germanica, poison glands

G.A, acid gland; *V*, poison sac with canal *c*; *G.a*, alkaline gland; *A*, terebra. *After* Bordas, 1897.

this glandular system in different groups of Hymenoptera and Robertson (1968) both the glands and the ovipositor. Bordas finds that in the Ichneumonidae the acid gland consists of numerous filiform tubes and that an accessory poison gland is present as it is, also, in the Crabroninae (Sphecidae). The Vespoidea and the Scolioidea have the poison-sac strongly musculated (Robertson).

In the male, the 9th abdominal sternite is always developed though concealed and the genitalia lie above it. The latter consist of a basal ring often partly divided longitudinally: a pair of two-segmented forceps to whose inner edge is attached a complicated plate, the *volsella*; and a pair of central penis valves, connected by membranes to form a tube and acting as the intromittent organ. Morphologists are not agreed as to the homologies of these structures. Snodgrass (1941), who has made a valuable survey of the male genitalia in the order, holds that all the structures are secondary differentiations of an aedeagus. Others, such as Michener (1944), hold that the outer forceps are derived from true abdominal appendages.

B. INTERNAL ANATOMY

The Alimentary Canal (Fig. 552) is of a tolerably uniform character throughout the order and presents but few notable deviations in its morphology (Bordas, 1894). In ants there is an *infrabuccal pocket* below the floor of the mouth: it takes the form of a spheroidal sac and opens into the mouth-cavity by means of a short narrow canal. According to Wheeler

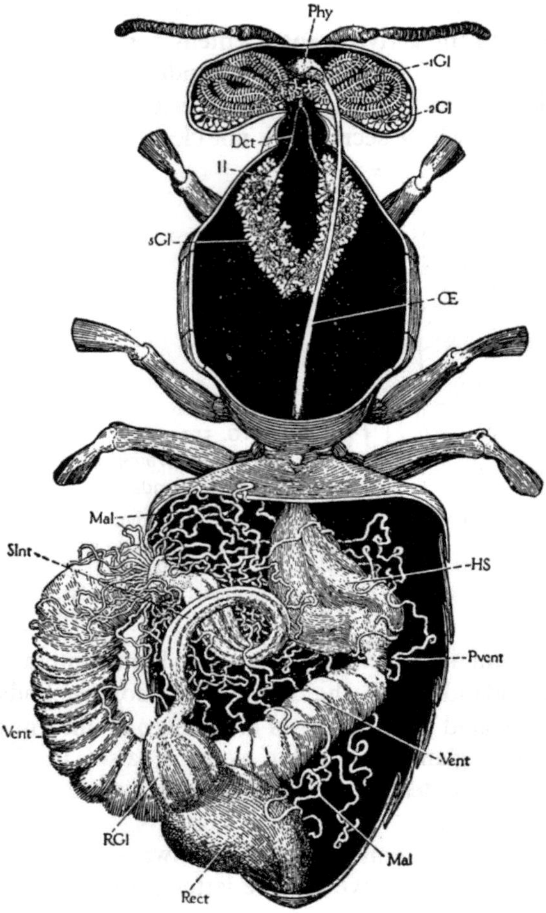

FIG. 552 Alimentary canal of worker bee

1*GL*, lateral pharyngeal gland; 2*GL*, cephalic salivary gland; 3*GL*, thoracic salivary gland with *ll* its reservoir and *Dct* its duct; *Phy*, pharynx; *OE*, oesophagus; *HS*, honey stomach; *Pvent*, proventriculus; *Mal*, Malpighian tubes; *Vent*, ventriculus; *Slnt*, small intestine; *Rect*, rectum and *RGl*, its papillae. *After* Snodgrass, *U.S. Bur. Entom. Tech. Ser. Bull.* 18 (reduced).

(1910) this chamber is used by the ant as a receptacle for the fine particles of solid and viscous food, rasped off or licked up by the tongue. Any juices that may be contained in this nutriment are sucked back into the pharynx, and the solid residue thrown out as a pellet, which retains the form of the chamber in which it was moulded. The mouth-cavity leads into the *pharynx*, which is an organ of suction, and is moved by powerful dilator muscles. The *oesophagus* is a long narrow tube, especially in forms with an elongate petiole, but is relatively short in *Apis* and *Vespula*. Among the Aculeata the oesophagous dilates in the anterior portion of the abdomen into a thin-walled crop or *honey-stomach*. The latter is lined with a sclerotized membrane and its walls contain muscle fibres: it serves as a reservoir for the liquid that has been imbibed, regurgitating it when required. In replete honey ants, the crop is remarkably distensible and, when full, largely determines the shape of the gaster. The crop is succeeded by the *proventriculus*, which is a very characteristic part of the gut in Hymenoptera and forms the neck-like region between the crop and true stomach. In *Apis* it is invaginated into the posterior wall of the crop, and has a X-shaped aperture provided with four triangular lips. The posterior opening of the proventriculus into the stomach is guarded by a well-developed valve. The function of the proventriculus, and its method of action, have given rise to discussion: it apparently serves to pump food from the crop into the stomach and, when closed, to prevent its regurgitation. The *stomach* or *ventriculus* is the largest part of the alimentary canal in *Apis* and *Vespula*, and is bent into a U-shaped loop. In some Sphecoids, Formicids and the Parasitica it is reduced to a small elliptical chamber. In the female of *Doryctes*, which lives but a short time and takes no nourishment, its anterior portion has undergone atrophy (Seurat). A peritrophic membrane is present and consists of a number of thin concentric lamellae. In most Hymenoptera, the *ileum* is a short simple tube but, in *Apis*, its length is much increased, and this region of the gut is looped upon itself. The *rectum* forms an enlarged terminal chamber, and its walls are furnished with three rectal papillae in ants, four in *Doryctes* and six in *Apis* and most other Hymenoptera.

The *Malpighian tubules* are extremely variable in number and, in the Aculeata, they vary from 100 to 125 in the Vespidae; from 20 to 30 in *Megachile* and its allies; and from 6 to 20 among ants. They all open separately into the ileum, and are often disposed in groups. Thus, in *Bombus* and *Apis* there are about 100 of these tubuli and, in the former genus, they are arranged in four bundles; in the Chrysididae there are about 40 Malpighian tubules arranged in three bundles; and in the Eumenidae they number from 40 to 70, and are disposed in two groups. Among the Parasitica, these organs are often much less numerous: in *Blastophaga* they number from 8 to 14 (Grandi), in *Doryctes* 9, in the Ichneumonidae there are generally from 50 to 60, and in the Tenthredinidae 20 to 25. Among Hymenopterous larvae there are four Malpighian tubules in *Apis* and the Formicidae, but in most of the parasitic families there is only a single pair.

Salivary Glands (Bordas, 1894, etc.) are well developed in the bee and consist of two pairs – one situated in the head and the other in the thorax (Figs. 552, 553). Their four ducts unite to form a common canal which opens on the hypopharynx. The *cephalic salivary glands* (postcerebral glands of Bordas; system No. 2 of Cheshire) lie against the posterior wall of the head. The *thoracic salivary glands* (system No. 3 of Cheshire) correspond with the ordinary salivary glands of most other insects. The contents of each gland are discharged into a reservoir, whose duct unites with its fellow to form the main salivary duct which, also, receives those of the cephalic glands. In the drones and queen there is a mass of gland cells situated just above the ocelli. These are the postocellar glands of Bordas but, according to Snodgrass, they are detached lobes of the cephalic glands. In addition to the foregoing, there is a pair of large *lateral pharyngeal glands* (supracerebral glands of Bordas; system No. 1 of Cheshire) which are the source of the royal jelly, which is fed to the larval and

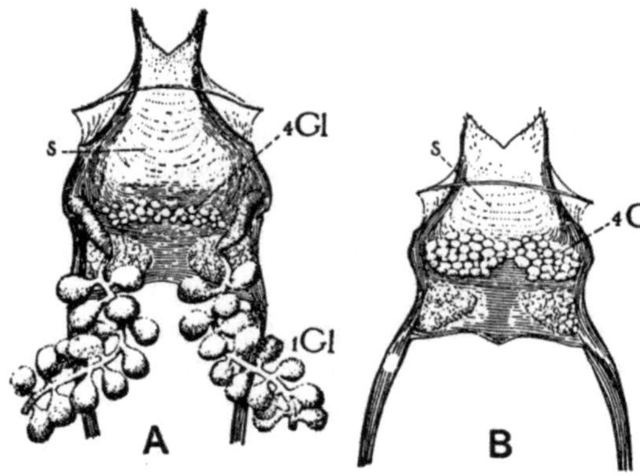

FIG. 553 Pharyngeal plate (*s*) and associated glands of A, worker and B, drone of hive bee

1*GL*, base of lateral pharyngeal gland; 4*GL*, ventral pharyngeal gland. *After* Snodgrass, *loc. cit*

adult queens and drones by the workers. Each is in the form of a long coiled chain of follicles packed away in the antero-dorsal region of the head; these glands are absent in the drone and rudimentary in the queen. Opening into the floor of the pharynx, between the ducts of the lateral pharyngeal glands, is a transverse row of cells which forms the *ventral pharyngeal gland* of Snodgrass (sublingual gland of Bordas). A sac-like *mandibular gland* opens at the inner angle of each jaw: it is larger in the queen than in the worker, and poorly developed in the drone. A second or *internal mandibular gland* has been described by Bordas in the worker of *Apis*, and also found in *Bombus* and *Vespa*; it is a delicate racemose mass, opening near the posterior inner edge of the mandible. The mandibular glands produce a number of pheromones, in particular, 'queen-substance' or 9-oxodec-*trans*-2-enoic acid which inhibits the construction of queen cells. Other substances are attractive to drones and workers (Butler and Fairey, 1964).

The Heart is well developed, and is usually composed of four or five chambers, with a corresponding number of pairs of alary muscles. In *Apis* the chambers are situated in the 3rd to 6th abdominal segments and, in ants, in the 4th to 8th segments. The heart is continued forwards as the aorta which, in the bee, is folded into about eighteen loops in the region of the petiole. In the latter insect both dorsal and ventral diaphragms are well developed; for a comparative study of the structure of the heart in bees, see Wille (1958).

The Respiratory System opens to the exterior by two thoracic (posterior one not always functional) and 2–7 abdominal spiracles. The longitudinal tracheae are often expanded into large sacs (Tonapi, 1958–60).

The Muscular System is one of great complexity and the reader is referred to articles by Janet (1897, etc.) for ants, Duncan (1939) for *Vespula*, Snodgrass (1942) for *Apis* and Alam (1951–53) for *Stenobracon*. In the dealated queens, among ants, the wing muscles are broken down by phagocytes, which take up and convert their substance, and somewhat later discharge it in the form of fat and albuminoid globules into the blood. In this manner the histolysis of the muscles provides nutrient material which contributes to the growth of the eggs (Janet).

The Nervous System – The brain has been studied among the higher members of the order and more especially by von Alten (1910), Jonescu (1909), Kenyon (1896), Thompson (1913) and Viallanes (1886). It is principally characterized by the high degree of differentiation of the mushroom bodies and their related fibre-tracts. In ants, for example, there is considerable variation in their development, not only among different species, but also in different castes of the same species. According to Viallanes, the highest type of brain is found in *Vespula* where the calyces are complexly folded.

The ventral nerve-cord is considerably less specialized than in the cyclorrhaphous Diptera. According to Brandt (1879) the most generalized condition is exhibited in the Tenthredinoidea where there are three thoracic and nine abdominal ganglia. Among the Apocrita the majority of the species similarly possess three thoracic ganglia, but among the Crabroninae and Apoidea there are only two thoracic centres. The first is the prothoracic ganglion and the second is a complex formed by the fusion of the meso- and metathoracic and one, or more, of the abdominal ganglia. The second thoracic centre innervates the 2nd and 3rd pairs of legs, the wings, propodeum, and 2nd abdominal segment. Six abdominal centres are present in many Ichneumonoidea and Formicidae, also in *Ammophila, Cerceris, Odynerus* and others. In most other Apocrita there are fewer abdominal ganglia, and the latter may be reduced to two centres as in *Cynips quercusfolii*, or to a single centre, as in certain Chalcids. In the females of many Aculeata the last two abdominal ganglia are more or less fused: thus in *Mutilla europaea* and *Megachile* there are five such ganglia in the latter sex and four in the male. In *Bombus* the worker and queen have six ganglia and the male five. In the worker of the hive bee there are five ganglia, while the queen as well as the male has but four. In *Vespula* the worker similarly has five ganglia, but the male and queen are exceptional in having six. In *Blastophaga* there are two abdominal centres in the female, while in the male they are fused into a common mass (Grandi).

The Male Reproductive System – The testes are separate in the Symphyta and also in *Apis* and *Bombus*. According to Bordas (1894) they are in close contact in *Vespula* and fused together in other Hymenoptera studied by him. Each testis is enclosed in a double membrane and may consist of 250–300 follicles as in *Vespula, Bombus* and *Apis*; these follicles are much less numerous in ants, and are usually reduced to three in other Hymenoptera. The vasa deferentia enlarge to form vesiculae seminales which are usually cylindrical or sac-like in form. In *Vespula* and *Apis* they are particularly voluminous, while they are tubular and convoluted in *Athalia, Cimbex* and *Bombus*. The two ejaculatory canals, which leave the vesiculae, receive the ducts of a pair of accessory glands. The latter are large and sac-like in almost all members of the order. In *Apis* the ejaculatory canals are rudimentary, and the accessory glands open into the common ejaculatory duct.

The Female Reproductive System (Fig. 554) – The ovaries are composed of polytrophic ovarioles; in *Apis* the latter are very numerous but their number is inconstant. In *Blastophaga* the ovarioles are very attentuated and closely packed together; according to Grandi there are 130–182 to each ovary. In *Cimbex* there are usually 20–30 ovarioles in each ovary; in *Aphelinus* there are five, while in other Chalcids and in the Ichneumonoidea

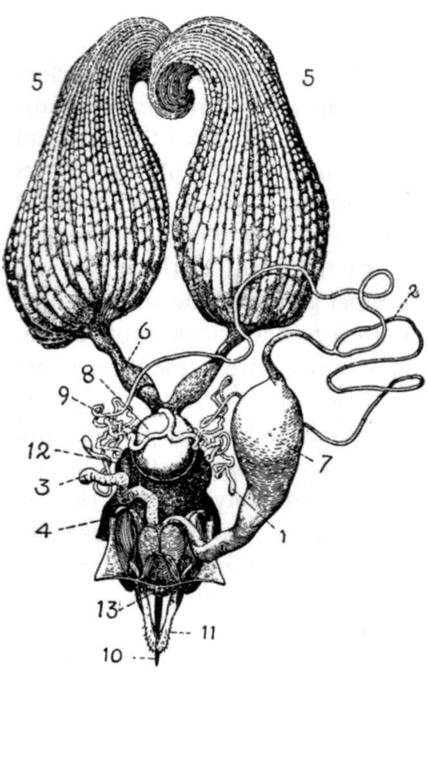

FIG. 554
Reproductive organs, etc., of queen bee
1, acid gland and 2, its duct; 3, alkaline gland; 4, bursa copulatrix; 5, ovary; 6, oviduct; 7, poison sac; 8, spermatheca and 9, its gland; 10, terebra; 11, sting palp; 12, vagina; 13, 9th sternum. Adapted from Snodgrass, *loc. cit.*

there are commonly four. In *Doryctes*, however, each ovary is greatly developed and consists of a single pair of ovarioles; in *Aphidius* the latter are wanting and the follicles are simply enclosed in a sac-like membrane. Among ants the number varies, in different genera and species, between two (*Leptothorax emersoni*) and about 250 (*Neivamyrmex nigrescens* = *Eciton schmitti*): in the workers, however, the number is very much lower, there is often a single ovariole to each ovary and rarely there are as many as twelve. The two oviducts unite to form the vagina and, in *Apis*, the latter is dilated posteriorly as the bursa copulatrix. A median spermatheca is generally present together with a pair of colleterial glands: the latter may open into a median reservoir as in *Cimbex* (Severin) or into the duct of the spermatheca as in *Apis*.

Metamorphoses

THE EGG

The eggs of Hymenoptera are usually ovoid or sausage-shaped and, in the parasitic groups, they are frequently provided with a pedicel. The latter structure may arise from either pole of the egg (Adler) and is of very general occurrence among the Cynipoidea. In the gall-forming species of the latter group it may be five or six times the length of the egg itself. Stalked eggs are also found among the Chalcidoidea and Proctotrupoids: in *Blastophaga* the pedicel may measure more than twice the length of the egg. In the majority of cases the function of this appendage is obscure, but in *Blastothrix* it protrudes through the body-wall of the host, and functions as a kind of respiratory funnel, which enables the newly hatched larva to breathe the outside air (see p.1220). In *Schedius kuvanae* the eggs are deposited within those of the gipsy moth, with their pedicels protruding to the exterior (Howard and Fiske), and it is probable that the latter organs fulfil a similar function in this instance also. A reduced pedicel is found in other Chalcids as well as in certain of the Ichneumonoidea; it is met with both in the case of eggs which are laid externally to their hosts, and in those which are laid within the latter. Iwata has described the eggs (and ovarioles) of many species of Hymenoptera, (e.g. in 1960*a*, *b*).

THE LARVA

A typical Hymenopterous larva is composed of a well-developed head, three thoracic and usually nine or ten abdominal segments. With some exceptions the tracheal system is peripneustic or holopneustic, either throughout life or in the later instars. Among the Symphyta the head is strongly sclerotized and there are powerful biting mouthparts. Three pairs of thoracic limbs and six or eight pairs of abdominal feet are generally present. Such larvae feed upon plant tissues, and are peripneustic or holopneustic throughout life, with nine or ten pairs of spiracles. Larvae which bore into stems or wood have lost the abdominal feet, but retain the thoracic limbs usually in a more or less reduced condition. Among the Apocrita, the larvae are apodous: evanescent thoracic appendages are present, however, in *Eucoila*, and a single pair is found in larvae of the Platygasteridae, and in these instances they are possibly modified survivals of true appendages. As a general rule, the larvae of the Apocrita (Fig. 555) (Clausen, 1940; Short, 1952), are maggot-like in form; the head is less strongly sclerotized than in the Symphyta, and in the parasitic forms it is often greatly reduced and sunk into the prothorax. Degeneration of the organs of special sense is very evident and, in most cases, the larvae are sluggish and move but little. These features are associated with the fact that their possessors live in darkness, and are supplied with an abundance of nutriment in their immediate vicinity, there being no necessity to seek for it. Ocelli are wanting, and the antennae are reduced to

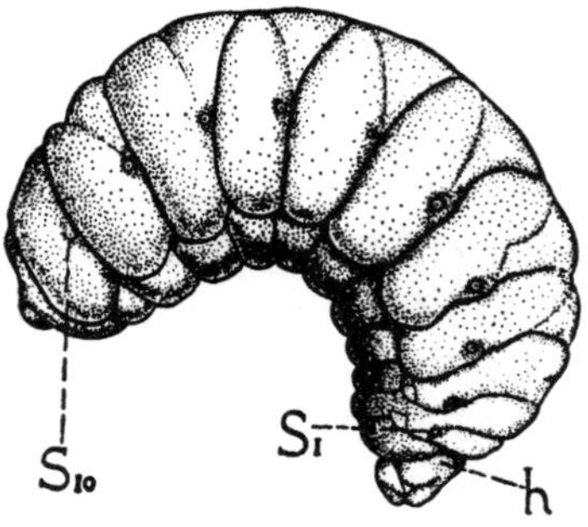

FIG. 555
Larva of a bee: enlarged
h, head; *s*, spiracles. From Nelson.

short sensory processes, small papillae, or may be atrophied. The mandibles may be either dentate, sickle-shaped or simple pointed spines with broad flattened bases. The labrum, maxillae and labium are fleshy lobes, and the two last-mentioned organs exhibit little or no differentiation into separate sclerites. Both the maxillary and labial palpi are usually represented by small papillae or are totally wanting. In almost all the larvae of the Apocrita the stomach is a blind sac and does not communicate with the hind intestine until the final instar, the faecal contents only being evacuated at the conclusion of the larval stage. Well-developed salivary glands are present, often of considerable length, and the ganglia of the ventral nerve-cord are often undifferentiated. In the Aculeata the tracheal system is holopneustic throughout life and generally ten pairs of spiracles are present. In the Parasitica the respiratory system undergoes profound modifications in correlation with varying modes of life (Seurat, 1899). Thus among the ectoparasitic species (Fig. 556) the larvae are hatched with a peripneustic tracheal system. The typical number of spiracles is nine pairs but they are not always borne on the same segments in different species. The Chalcid *Aphelinus* has eight pairs of spiracles and the Ichneumon *Pimpla pomorum* has ten pairs. In the Proctotrupoid *Lygocerus* the larva is hatched with two pairs and there are seven pairs in the last instar. Among the endoparasitic forms the young larvae are commonly apneustic, but this condition is rarely retained throughout life. In the apneustic condition the cuticle is extremely thin and admits of the interchange of gases by means of diffusion. At this stage the larva is haematophagous but its subsequently becomes carnivorous, devouring the various internal organs of its host. When it assumes this mode of life, a certain number of spiracles open on the surface of the body, and in the final instar there are usually nine pairs present (Thorpe, 1932). Fisher (1971) has reviewed the physiology of endoparasitic larvae, stressing their respiratory problems.

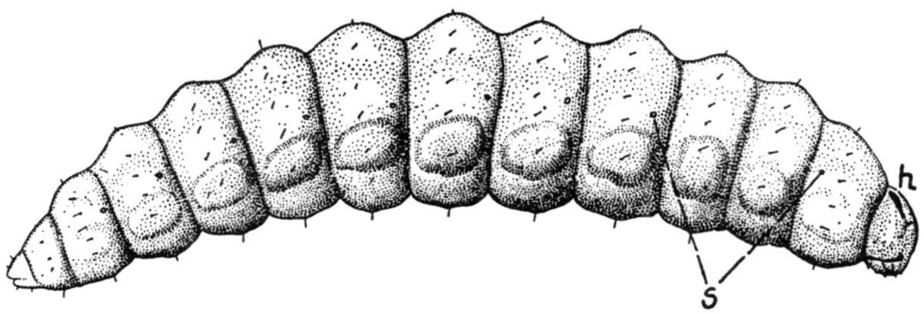

FIG. 556 Fully-grown larva of an ectoparasitic Ichneumon, *Pimpla pomorum*: enlarged

h, head; *s*, spiracles.

Hypermetamorphosis occurs among many of the Parasitica and examples of this type of development are known in five of the major divisions of the order (Richardson, 1913). At least twelve distinct primary larval forms are known in addition to the usual larval type already described (Clausen, 1940). Any attempt at the classification of these forms at present can only be a tentative one pending the growth of more detailed knowledge. The principal types of primary larvae are as follows (Figs. 557, 558). (1) The PLANIDIUM (Fig. 557) is an active larva invested with strongly sclerotized imbricated segmental plates and provided with spine-like locomotory processes. It develops from an egg which is laid away from the host and is a migratory form adapted to seek out the latter. This type is known in the Chalcid families Eucharitidae and Perilampidae. (2) The CAUDATE TYPE is well exhibited in certain Ichneumonidae, Braconidae and in a few of the Chalcidoidea, notably *Aphidencyrtus aphidivorus*. It is somewhat vermiform in shape with a caudal outgrowth of variable length (see also p. 229). (3) The CYCLOPIFORM or NAUPLIIFORM TYPE occurs in certain of the Proctotrupoids. It is characterized by the large swollen cephalothorax, very large sickle-like mandibles and a pair of bifurcate caudal processes of variable form. In its general facies it bears a resemblance to the nauplius of Crustacea. (4) The TELEAFORM TYPE is found in certain other Proctotrupoids and in several of the Chalcidoidea; it derives its name from the primary larva of *Teleas*. The cephalic extremity is prominently hooked or curved; posteriorly the body is prolonged into a caudal process, and the trunk is armed with one or more girdles of setae. Apparently modified examples of this larval type have been described by McColloch in *Eumicrosoma* and by Silvestri in the Chalcids *Poropoea* and *Patasson* (= *Anaphoidea*). (5) The VESICULATE TYPE occurs in *Apanteles* and *Microgaster* and is characterized by the proctodaeum being everted to form a swollen anal vesicle. (6) The EUCOILIFORM TYPE is known in the Eucoilinae: it differs from the teleaform type in possessing three pairs of long thoracic appendages, and in the absence of

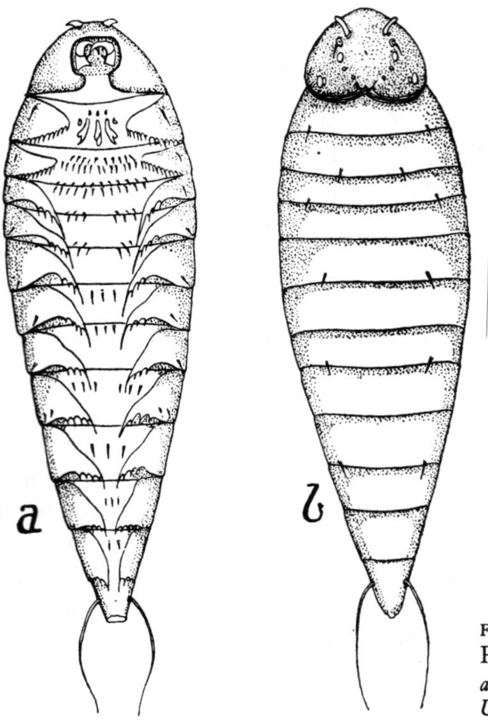

FIG. 557
Planidium of *Perilampus*
a, ventral; *b*, dorsal. *After* H. S. Smith, *U.S. Bur. Ent. Tech. Ser.* 19, pt. 4.

the cephalic process and the girdles of setae. (7) The POLYPODEIFORM TYPE, with 8 to 12 pairs of trunk appendages, occurs in *Ibalia*, the Braconid *Microdus* and in *Phaenoserphus*: in *Eucoila* and allies it follows the protopod stage. The subsequent stages in development in those species in which hypermetamorphosis occurs exhibit wide variation: thus the second larval instar of *Teleas* is of the cyclopoid type, but the final instar in all cases is the ovoid maggot-like type of larva characteristic of the Apocrita.

The presence of a trophic membrane or trophamnion (Fig. 559), enclosing the embryo in certain endoparasitic Hymenoptera, has been already alluded to (p. 304). It has been found in diverse species, comprising members of each of the main parasitic groups, but is evidently not homologous in all cases, and very different methods of formation have been described. This

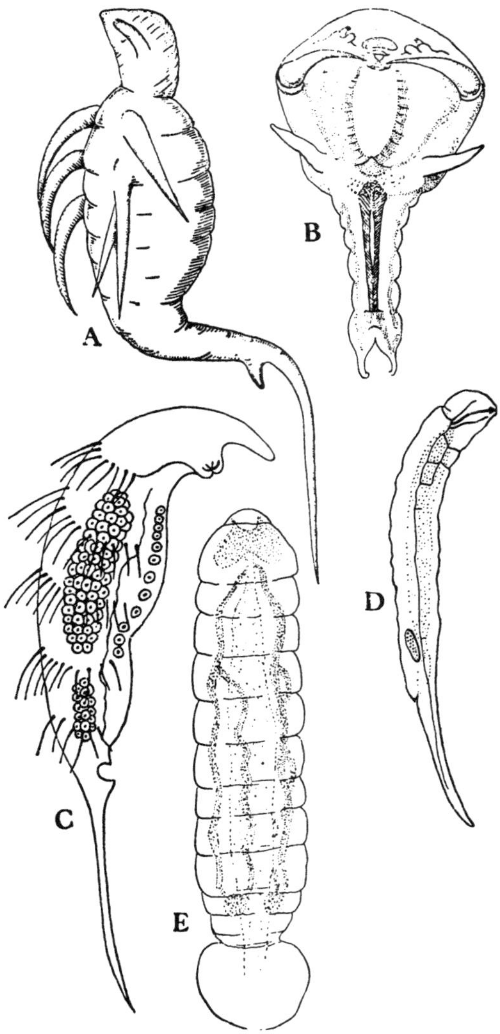

FIG. 558
Primary larvae of various parasitic Hymenoptera

A, eucoiliform (*Eucoila*), *after* Keilin and Pluvinel. B, cyclopiform (*Trichacis*), *after* Marchal. C, teleaform (*Teleas*), *after* Ayers. D, caudate (*Mesochorus*), *after* Seurat. E, vesiculate (*Microgaster*), original. All highly magnified.

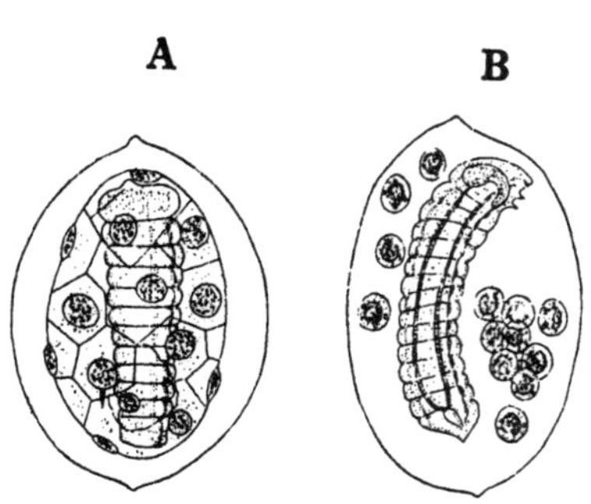

FIG. 559
Chalcis (*Smicra*) *sispes* (= *clavipes*). A, Egg with embryo surrounded by trophamnion. B, young larva and disintegrated trophamnion

After Henneguy, *Les Insectes.*

membrane is believed to play an important part in the nutrition of the embryo.

Michener (1953) gives a partial key to the larvae of the superfamilies.

THE PUPA

In the Apocrita the prepupa (pharate pupa) is intermediate in some characters between the larva and pupa (Fig. 190). The prothoracic segment is distended by the developing pupal head, the wings and legs have assumed the form of those of the pupa, and it is in this stage that the first abdominal segment or propodeum becomes incorporated with the thorax. After the final larval moult, the prepupa passes into the pupa and the latter is in nearly all genera of the exarate type, in which the wings and appendages are free and not soldered to the surface of the body (Fig. 189). With the exception of the Cynipidae and Chalcidoidea a cocoon, though often slight, is of general occurrence in the order. In many Tenthredinoidea it is parchment-like; in others it is formed of agglutinated soil particles; while in *Cimbex* the cocoon is formed of an outer and inner coat, and attains a higher degree of development than in other Hymenoptera. In many of the Aculeata the cocoon is little more than a silken lining to the larval cell, and in some of the ants it is totally wanting. Among the Braconidae dense masses of silken cocoons are often formed by the members of a species which issue from a single individual host.

Classification of Hymenoptera

The Hymenoptera are divisible into the suborders Symphyta and Apocrita. It has long been customary to separate the Apocrita into two main divisions, viz. the Aculeata or stinging forms and the Parasitica (Terebrantia) which are parasites of other insects. The distinctions between these divisions are so difficult to draw and subject to so many exceptions that it is more practical and probably more correct to recognize rather a series of superfamilies. Biologically, there is also no clear distinction. A number of the Parasitica are plant feeders and many Aculeata are parasites. In the Aculeate families Sapygidae, Dryinidae and Chrysididae, the ovipositor retains its egg-laying function as in the Parasitica: in other Aculeates, the ovipositor is converted into a sting and the egg reaches the exterior at its base.

The standard works on European Hymenoptera are those of André (1879–1913) and Schmiedeknecht (1930). Dalla Torre (1892–1902) published a catalogue of the species of the world but this is now very out of date. A valuable catalogue of the North American species was published by Muesebeck *et al.* (1951; supplements 1958, 1967). A completely new edition is due to appear shortly. A new world catalogue, edited by Van der Vecht and Shenfeldt (1969–), has begun to appear.

The two suborders may be recognized:

Suborder SYMPHYTA (CHALASTOGASTRA)

Abdomen broadly attached to the thorax, no marked constriction between 1st and 2nd abdominal segments. Cenchri present except in the Cephidae. Fore tibia nearly always with 2 spurs. Larva (except *Orussus*) with thoracic and generally abdominal legs, excreta voided throughout life.

Suborder APOCRITA (CLISTOGASTRA)

Abdomen deeply constricted between the 1st abdominal segment (propodeum) and the 2nd. Cenchri absent. Fore tibia with one spur (except Ceraphronoidea). Larva apodous, excreta normally voided only just before pupation. A few of the very small Chalcidoids (*Alaptus, Trichogramma*) have the abdomen broadly attached to the propodeum (cf. Soyka, 1949) but this is probably secondary.

Suborder I. SYMPHYTA

Included in this division are all the more primitive members of the Hymenoptera which are recognized by the broadly sessile abdomen and the fact that its first segment is only partially amalgamated with the thorax. The imagines do not exhibit the highly specialized habits and instincts so prevalent among the Apocrita and the ovipositor is adapted for sawing or boring: except in *Orussus* parasitism is wanting. The peculiar structures known as the *cenchri* are raised bosses on the metanotum which engage with a scaly area on the underside of the fore wings and keep them in place when at rest (Zirngiebl, 1936). They are found in all families except the Cephidae. The fore tibia normally has two apical spurs, the smaller of them being sometimes absent (Cephidae, some Siricoidea). The larvae (Yuasa, 1922) have a well-developed head and 13 trunk segments: three pairs of thoracic legs and frequently six or more pairs of abdominal limbs are present. The tarsus and claw of each thoracic leg are fused into a single piece while the abdominal limbs are devoid of crochets. A single pair of ocelli is present and the maxillary and labial palpi are usually 4- and 3-segmented respectively. Spiracles are always present on the prothorax and first eight abdominal segments: metathoracic spiracles are also present in the Cephidae, and in *Sirex* and *Tremex*, but are vestigial or wanting in the larvae of other Symphyta.

The Symphyta fall into two groups, the Orthandria including the first five superfamilies, and the Strophandria with the Tenthredinoidea. The first group show the male genitalia in normal orientation except in a few Xyelidae; in the second they are inverted with the ventral surface uppermost. The British species have been revised by Benson (1951–58).

Superfamily Xyeloidea

FAM. XYELIDAE. This family has the most generalized venation among Hymenoptera (Fig. 547) with vein Rs generally forked in the fore wing. The larvae are also noteworthy since they have feet on all the abdominal segments. The imagines may be easily recognized by the greatly elongated 3rd antennal segment which is followed by a flagellum; the ovipositor is moderately or very long. The hind margin of the pronotum is nearly straight. *Xyela* and *Macroxyela* are typical genera: the only common British species, *X. julii*, is found where *Pinus* and *Betula* grow together. The larva lives in the staminate flowers of the former, whereas the adults visit the flowers of *Betula*. Rasnitsyn (1966, 1969) has described numerous fossils from the Lias upwards of this and the next four superfamilies.

Superfamily Megalodontoidea

Antennae with the 3rd segment not very long; pronotum with hind margin almost straight. Fore wing with cross-vein *2r* present. Abdomen flattened, ovipositor not exserted. Fore tibia with two apical spurs.

FAM. PAMPHILIIDAE. These are robustly-built insects with a short ovipositor and a primitive venation though vein Sc is absent in the fore wing (Fig. 546). The larvae (Fig. 560), on the other hand, have no abdominal feet: they are sometimes gregarious and often live in webs or rolled leaves. *Neurotoma* and *Pamphilius* are British genera: their species are mostly rather infrequent in occurrence.

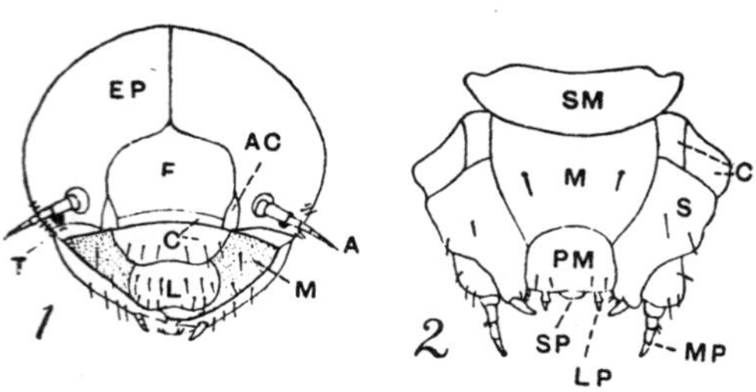

FIG. 560
Pamphilius dentatus, larva. 1, frontal view of head; 2, maxillae and labium

A, antenna; *AC*, antecoxal piece of mandible; *C*, cardo; *EP*, epicranial plate; *F*, frons; *L*, labrum; *LP*, labial palp; *M*, mentum; *MP*, maxillary palp; *S*, stipes; *SM*, submentum; *SP*, spinneret; *T*, mandibular sclerite. Adapted from MacGillivray.

The **MEGALODONTIDAE** are rather similar insects but have flabellate antennae in the adult and shorter mouthparts. The larvae live gregariously in webs spun on herbaceous plants.

Superfamily Siricoidea

Fore tibia with smaller spur very small or often absent.

FAM. SIRICIDAE (Wood-wasps or Horn-tails). A family of large-sized insects with conspicuous coloration, being often black and yellow or metallic blue. The abdomen terminates in a spine or horn, which is short and triangular in the males and lanceolate in the females. Pronotum with hind margin deeply emarginate. Postgenae fusing behind the oral fossa. Maxillary palp with four segments. The ovipositor is exceedingly strong and, when at rest, projects backwards in the horizontal plane, and has the appearance of a powerful sting. This instrument is used

for boring and drilling, and not for sawing as in the Tenthredinidae. Holes are made through the bark into the new wood of various forest and shade trees and a single egg is deposited in each hole. The larvae on hatching burrow into the heart wood and often cause considerable damage. Pupation takes place in the larval gallery and a cocoon of silk and gnawed wood is constructed. The larva has a tolerably large head and three pairs of reduced thoracic limbs: the last trunk segment terminates in a horny process which aids in locomotion. The best known species in the British Isles is *Urocerus gigas* which lives in Pinaceae, and its life-history appears seldom to occupy less than two years. It usually only attacks trees which have passed their full vigour and are not perfectly healthy, but sound felled trees are sometimes selected. Great damage has been done to introduced pines by species which have reached Australia. The metallic blue *Sirex noctilio* is also not infrequently met with, but it is not truly indigenous in Britain. For further information on the British species of the genus, see Chrystal (1928). In the allied genus *Tremex* the larva affects broad-leaved trees in N. America.

The species of the family **Xiphydriidae** lack the spine at the apex of the abdomen and the maxillary palps have four segments; their larvae bore into the wood of deciduous trees; two species of *Xiphydria* are rather uncommon in Britain (Chrystal and Skinner, 1932). The genera of the family were revised by Benson (1954). The family **Syntexidae** includes a single N. American species with a wood-boring larva. In the adult the hind margin of the pronotum is scarcely emarginate.

Superfamily **Orussoidea**

Venation weak, hind wing without any enclosed cells. Antennae inserted between the eyes but below the apparent clypeus but head otherwise like a Siricid. Ovipositor very long and when retracted it is looped within the abdomen.

FAM. ORUSSIDAE. This very small family is evidently a relic of an ancient group, and is distributed over most parts of the world. It is represented in Europe and N. America by the single genus *Orussus* and the species *O. abietinus* has been supposed to occur in Britain. Structurally the members of this family show many peculiar features. The wings have a reduced venation (Fig. 561) quite unlike that of any Symphyta, since there are no closed submarginal cells in the hind wing. In the position of the antennae, and the form and resting position of the long ovipositor when at rest, the family is unique among Hymenoptera. The only known larva is apodous and probably an ectoparasite of Buprestidae though the genus appears also to attack Siricids. It is described by Rohwer and Cushman (1917), who also emphasize the unique features of the family. On account of its adult characters, and the form and habits of the larva, they place the family in a separate suborder, the Idiogastra – intermediate between the Symphyta and Apocrita. Apart from its specializations however, it shares many characters with the Siricidae (cf. Cooper, 1953).

Superfamily **Cephoidea**

Cenchri absent. Fore tibia with one apical spur. Abdomen constricted between the first and second segments.

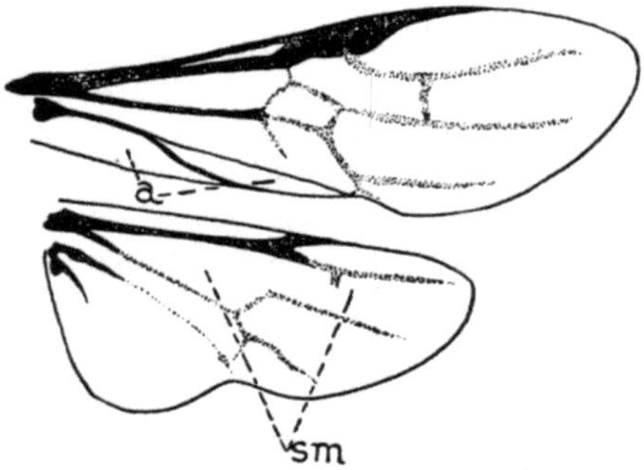

FIG. 561
Right wings of *Orussus abietinus*
a, divided anal cell; *sm*, submarginal area. Adapted from MacGillivray.

FAM. CEPHIDAE (Stem Saw-flies). The Cephidae are a small family of slender, narrow-bodied insects with a thin integument (Fig. 562). The prothorax is exceptionally large and movably articulated with the following segment and its hind margin is nearly straight. They are mostly black or darkly coloured, either with or without narrow yellow bands. In length they seldom measure more than 18 mm and are usually smaller. The larvae bore into the stems and shoots of various plants and are apodous, with the exception of three pairs of reduced tubercle-like thoracic

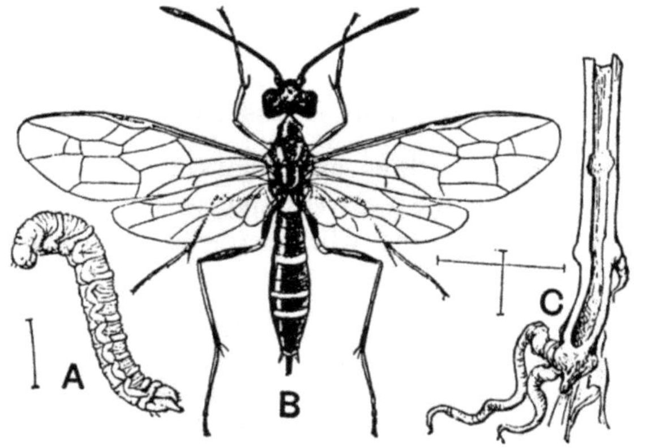

FIG. 562 *Cephus cinctus* (= *occidentalis*)
A, larva; B, female; C, larval gallery in grass-stem. *After* Marlatt, *U.S. Dept. Agric.*

limbs. They are also characterized by the vestigial ocelli, the well developed metathoracic spiracles and the presence of vestigial subanal appendages. The abdomen terminates in a small retractile point or spine which arises from a fleshy protuberance on the last segment, above the anus. The pupae are usually enclosed in transparent cocoons within the stems of the food-plant. For an enumeration of the larval characters in different genera see Middleton (1918). Less than a dozen species occur in the British Isles, the best known being *Cephus pygmaeus*, the Wheat-stem Borer. Although destructive in many parts of Europe, and introduced into N. America, it is rarely injurious in Britain. The eggs of this species are laid in the stem of the wheat plant, and the larva bores its way upwards through the latter, ultimately weakening it below the ear: for an account of its biology and parasites see Salt (1931). *Janus integer* (= *flaviventris*) lays its eggs in the centre of the pith of the shoots of currants and its larvae bore through the stems: its metamorphoses are figured by Marlatt (1894). For a systematic account of the group see Benson (1946).

Superfamily **Tenthredinoidea**

Cenchri present. Fore tibia with two apical spurs. Male genitalia inverted.

FAM. ARGIDAE. *Antennae with 3 segments, third long and sometimes bifid. Fore wings without cross-vein 2r. Mid and hind tibiae often with preapical spurs.* A family of more than 400 species of world-wide distribution, easily recognized by the structure of their antennae. The larvae have 6–8 abdominal legs and feed freely, mostly on woody Angiosperms. In the S. American *Dielocerus* the female sits over and protects her young which spin their cocoon in a communal covering.

FAM. BLASTICOTOMIDAE. *Antennae with a long third segment and a short, terminal fourth. Fore wing with cross-vein 2r present. Mid and hind tibiae without preapical spurs.* The single European species, lately found in Britain, is *Blasticotoma filiceti*. The larva, which has no abdominal legs, bores in the stems of various ferns.

FAM. CIMBICIDAE. *Antennae clubbed with 3rd segment not long. Fore wing with cross-vein 2r present. Pronotum deeply emarginate behind. Sides of abdomen carinate.* A small family of stout, often large insects with strongly clubbed antennae. The larger species are mostly attached to various trees and the larvae, which sit partly curled up and are covered with a waxy powder, are characteristic.

FAM. DIPRIONIDAE. *Antennae with more than 9 more or less serrate or pectinate segments, 3rd not long. Fore wing with cross-vein 2r absent; hind wing with cross-vein m-cu enclosing a median cell, anal cell closed. Diprion* (= *Lophyrus*) and its allies include a number of important pests of coniferous trees. *Gilpinia hercyniae*, attached to *Picea*, has been introduced into Canada where large areas of forest have been defoliated. Males are very rare in this species. *Diprion pini* is the common British species on *Pinus*. The characteristic spotted larvae live gregariously and are very conspicuous.

FAM. PERGIDAE. *Antennae often serrate or clubbed, 3rd segment short. Cross-vein 2r absent. Hind wing with no enclosed median or anal cells.* There is great diversity of structure within this family which occurs mainly in Australia and S. America. The females of *Perga* brood over their young. The larvae have no abdominal legs and feed gregariously on the leaves of *Eucalyptus*. The larvae of *Phylacteophaga* mine the leaves of the same plant.

FAM. TENTHREDINIDAE. *Antennae with 9 segments (rarely 1 or 2 more), 3rd short, not serrate or clubbed. Pronotum deeply emarginate behind. Fore wing with cross-vein 2r present or absent. Postgenae not meeting behind oral fossa. Scutellum with a defined 'postscutellum'. Fore tibia with 2 apical spurs.* This family with about 4000 species (nearly 400 British) includes the bulk of the suborder and exhibits considerable diversity of structure and habit. The adults are most often obtained by sweeping or shaking the vegetation; many frequent flowers and some are carnivorous, preying upon small flies and beetles. Great variation exists as to the proportion of individuals of the sexes and in only a few species are the males as numerous as the females. Cameron estimated that males were unknown in one-third of the British species. Parthenogenesis is common in the family and in some species males, in others females, and in a third group individuals of both sexes are produced from unfertilized eggs. Thus in *Nematus ribesii* only males have been reared from the unfertilized eggs. The impregnated females give rise to individuals of both sexes, but females predominate. In such species as *Croesus varus* or *Monostegia abdominalis* the parthenogenetic eggs produce females and no males are known (Benson, 1950).

The eggs are usually laid in young shoots or in leaves and the saw, or cutting instrument, of the ovipositor is toothed in various ways in conformity with the nature of the oviposition and also as a specific character. Its serrations are large and stout in species which lay their eggs in woody twigs; very fine in those which oviposit in leaf-tissue; or scarcely evident at all in *Nematus ribesii*, which simply attaches its eggs each by means of a small flange into a minute slit on the underside of a leaf. In most species, during oviposition the blades of the ovipositor move alternately, one being thrust forward while the other is withdrawn, until an incision or pocket of the required depth is formed (cf. Keir, 1936). Both the first and second valvulae are more or less complexly serrated towards their apices while the third valvulae serve to protect the whole terebra. The larvae often bear a close general resemblance to those of the Lepidoptera. They are exclusively phytophagous in habit and affect almost all orders of Phanerogamia and certain of the Filices. Trees and bushes, however, support a larger number of species than herbaceous plants. The larvae (Yuasa, 1922; Lorenz and Kraus, 1957) exhibit much diversity of habit and a large number are nocturnal feeders: many are solitary while others are gregarious. The vast majority live exposed, but some live internally in stems, fruit or galls and a certain number are leaf-miners. Many closely simulate their environment and are cryptically coloured, while others are very conspicuous with bright colours. In numerous species the larvae are covered with a whitish powdery exudation: in *Caliroa* they are sluglike and the body is obscured by a darkly coloured slime or exudation and species of *Monophadnoides* are invested with bifurcate spines. The body-segments of saw-fly larvae are usually subdivided, by means of transverse folds, into annulets whose number appears to be constant for each species (Fig. 563). Three pairs of thoracic limbs are present and almost all species carry abdominal feet also. Unlike those of the Lepidoptera there are usually more than five pairs of the latter organs and they are devoid of crochets. The number of these appendages varies from 6 to 8. In many cases the larvae emit secretions which are produced by special glands. *Caliroa* has a pair of ventral digit-like glands opening between the head and prothorax: many larvae are provided with glands resembling osmeteria,

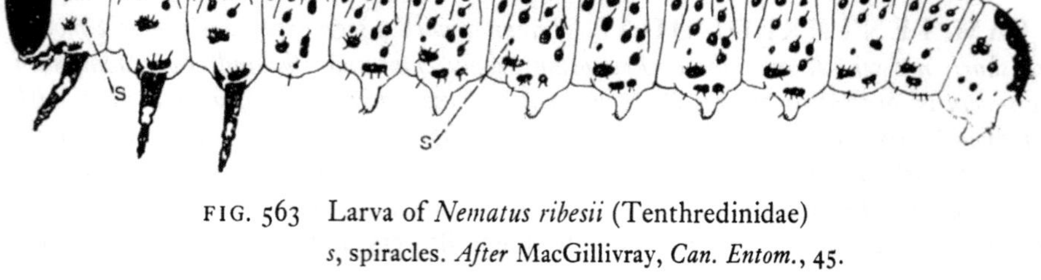

FIG. 563 Larva of *Nematus ribesii* (Tenthredinidae)
s, spiracles. *After* MacGillivray, *Can. Entom.*, 45.

which open by means of a slit-like aperture on the sternum of each of the first 7 abdominal segments. Pupation, as a rule, takes place in an elongate-oval silken cocoon which may or may not be mixed with soil particles; in other cases an earthen cell is constructed.

Suborder II. APOCRITA

Included in this suborder is the vast majority of Hymenoptera, all of which are recognizable by the abdomen being basally constricted or petiolate. The imagines are almost always highly specialized in their habits and some are social, living in large communities. The ovipositor is adapted for piercing in the Parasitica and usually for stinging in the Aculeata. The larvae are apodous, and the head is generally well developed but, among certain of the parasitic families, it is greatly reduced. The larval habits are extremely diverse. Thus many of the Cynipoidea and a few of the Chalcidoidea are phytophagous. Others of the Cynipoidea, all the Ichneumonoidea, and almost all the Chalcidoidea, are carnivorous, being either ecto- or endoparasites. The Sphecoidea and Vespoidea are largely predacious, and the Apoidea are nourished upon nectar and pollen. Various genera of the normally nest-making groups have secondarily become parasites of the nests of their allies.

In recent years, various Cretaceous Aculeate fossils have been described (Sphecoids, Bethyloids, etc.) (Evans, 1973).

The British Aculeates are described and figured in the work of Saunders (1896). Among the Parasitica, the Cynipoidea are dealt with by Cameron (1882–92) and the Ichneumonidae by Morley (1903–14), the Evanioidea by Crosskey (1951), but no monographic works exist on the remaining British parasitic groups. All the older works are now very out of date but modern revisions are very incomplete.

The literature on the biology of the Aculeata has assumed enormous proportions. Among the more important works are those of Fabre (1879–1903), Ferton (1901–21), Friese (1922–23), G. W. and E. G. Peckham (1898), Roubaud (1916), Grandi (1926–31), Williams (1919), Olberg (1959), Krombein (1967) and Wheeler (1910). Wheeler (1923) has given an admirable annotated bibliography of the subject to which the reader is referred.

The first five superfamilies are unambiguous Parasitica and the last six Aculeata, but annectent forms occur in the Proctotrupoid group and Bethyloidea.

Superfamily **Trigonaloidea**

Hind femur with a trochantellus. Fore wing with a pterostigma, Rs usually two branched; hind wing with enclosed cells, with an axillary or anal lobe. Head large, mandibles with 4 strong teeth, antennae with more than 20 segments. Gaster normally attached to propodeum, not petiolate, last visible tergite and sternite of female opposed, ovipositor vestigial.

FAM. TRIGONALIDAE. A small family of rare but widely distributed insects. The multiarticulate antennae, the presence of the trochantellus and the ovipositor (Oeser, 1962) ally them to the Parasitica, but the wing-venation is more of an Aculeate type. They appear to be an archaic but also highly specialized group (Fig. 564). In some species very numerous eggs are laid on leaves and are then eaten by caterpillars or saw-fly larvae within which they attack other Hymenopterous or Dipterous parasites. *Trigonalys maculatus*, however, is a direct parasite of *Perga* and some species are parasites of social wasps (Clausen, 1940). The family has been monographed by Schulz (1907) and the anatomy of the single, rare British species *Pseudogonalos hahni*, is discussed by Bugnion (1910).

Superfamily **Ichneumonoidea** (Fig. 565)

This is probably the largest superfamily of the order. Hind femur with a trochantellus. Fore wing with a pterostigma at distal end of costa, venation usually relatively complete, costal cell usually narrow. Antennae usually

FIG. 564 *Poecilogonalos thwaitesii* (Trigonalidae)
From Clausen, C. P. Entomophagous insects, 1940. fig. 24

long, usually with more than 16 segments. Gaster attached at bottom of propodeum, a little above only in some Braconidae, spiracles on segments 1–7, part of tergite 8 exposed and sclerotized, tergite 9 separately developed, pygostyles present. Without exception all are parasites preying upon some stage in the life-history of other insects, or occasionally upon other

Arthropoda. It will, therefore, be readily appreciated that the group, as a whole, is of the greatest importance, not only on account of the role which it plays in the economy of nature, but also from the fact that the majority of the species are beneficial to man.

FAM. ICHNEUMONIDAE (Ichneumon Flies). *Costal cell narrow or obliterated, cross-vein 2 m-cu present or* (Neorhacodes) *indicated* (*except for a few genera in which the gaster is 3 times as long as head + thorax and the propodeum prolonged beyond the insertion of the hind coxa*). *Hind wing with cross-vein r-m meeting Rs after it leaves Sc + R. Gastral sternites often partly membranous* (Fig. 565). The vast majority of these insects are parasites, or less frequently hyperparasites, of Lepidoptera. After the latter come the Hymenoptera, and more especially the family Tenthredinidae, but all groups including the Parasitica may be attacked. A considerable number of Ichneumonidae are known to utilize Coleoptera as their hosts, but

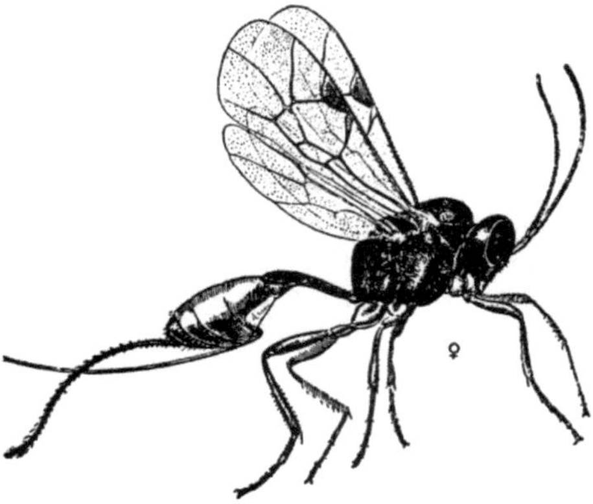

FIG. 565
Tersilochus conotracheli, female : enlarged (*Ichneumonidae*)
After Cushman, *J. agric. Res.*, 6, 1916.

Diptera are much less frequently selected. A still smaller number parasitize Arachnida, and a few attack Aphids and also *Hemerobius, Chrysopa* and *Raphidia*. Most species of the family are probably seldom restricted to any individual specific host, and those so accredited are becoming reduced in number with increasing knowledge. In their behaviour, the Ichneumons are among the most highly evolved of all solitary insects. The remarkable instincts exhibited in the discovery of their hosts and in providing for their offspring, their mating habits, behaviour in captivity, etc., afford a wide field for investigation. The imagines are most active on warm sunny days and are partial to flowers, especially Umbelliferae. In many species at least the females hibernate though the males perish before the advent of winter. Apterous and brachypterous forms are comparatively frequent in the subfamily Cryptinae (= Hemitelinae), and it is often a matter of difficulty to discriminate them from similarly wingless Braconidae. In the Cryptinae, however, the abdominal segments are soft and telescopic whereas, among the Braconidae, the middle segments do not overlap one another. Ichneumon larvae (Short 1959–70) are composed of a variably shaped head and usually 13 body segments. Spiracles, when present, consist typically of nine pairs, which are situated on the pro- or mesothorax, and first eight abdominal segments. Among endoparasitic larvae, there are frequently striking differences between the earlier and later instars. One of the most characteristic features of the newly hatched larvae of many species is the presence of a prominent caudal prolongation or tail. Owing to the fact that it disappears when the tracheal system becomes open to the exterior, this appendage has been regarded as an accessory respiratory organ, but this seems to be incorrect (Thorpe, 1932). Seurat, on the other hand, ascribes to it a locomotory function. The head in the young larva is large, and often strongly sclerotized, the segments between that region and the caudal appendage are sometimes greatly compressed, and the respiratory system is apneustic. The second instar is usually of a transitional nature between the first and third. The tail, though greatly reduced, is still evident, and the head has also undergone reduction and is less strongly sclerotized. In the third instar the larva generally becomes maggot-like, with a greatly abbreviated head, and the tail, as a rule, has disappeared or is vestigial. Towards the end of this stadium Timberlake states that, in *Olesicampe* (= *Limnerium*), the tracheal system communicates with the exterior by the spiracles. The number of instars present is obviously extremely difficult to determine: according to Cushman there are five in *Tersilochus*, and the same number is stated by Smith to be present in *Calliephialtes*. Ectophagous larvae are always devoid of the caudal appendage, the head is well developed and sclerotized, a variable growth of body hairs is evident, and the tracheal system is peripneustic from an early stage. When fully fed, Ichneumon larvae construct silken cocoons often composed of iridescent strands. Some of the most remarkable members of the family belong to the genera *Thalessa* and *Rhyssa* whose larvae are ectoparasites of those of the Siricidae. The adults are notable on account of the great length of the ovipositor and for their specialized habits of egg-laying. *Thalessa* has an ovipositor which may attain a length of six inches, with which it pierces or drills the wood of trees in order to reach the burrows occupied by *Tremex*. The British *Rhyssa persuasoria* similarly parasitizes *Urocerus*, and it has been recorded to reach its host by inserting the terebra along the burrows of the latter and also by passing it through the bark and solid wood. An interesting account of the habits of both genera is given by Riley (1889). The familiar reddish-brown species of *Ophion*, so often attracted to lights, are common parasites of Noctuid larvae. *Hemiteles areator* has been bred from a remarkable range of hosts comprising many Lepidoptera, various Hymenoptera including other Ichneumonidae, and also from several Coleoptera and Diptera. *Agriotypus* is an endoparasite of Trichopterous larvae, and the adults have been observed to dive and swim beneath the water while seeking their host (see Clausen, 1931). Ashmead places this genus in a family of its own on account of the hardened abdominal sterna and the spined scutellum and Mason (1971) supports this because of the structure of the abdominal petiole and of the larval head. He would transfer it to the Proctotrupoids. Among the more important life-history studies of individual species of Ichneumonidae the reader should consult the old though important work of Ratzeburg (1844), particularly for the larval development of *Anomalon*; among others, the papers of Cushman on *Calliephialtes* and *Tersilochus*, Newport (1852–53) on *Netelia*, Timberlake (1912) on *Olesicampe* (= *Limnerium*), and Imms (1918*b*) on *Pimpla* may be mentioned. Morley (1903–14) has monographed the British species, and Schmiedeknecht (1902–11) has produced a general systematic treatise on the family but both accounts are now out of date. Townes and his associates (1959–70) have done a considerable part of a revision of the world fauna, chiefly at the generic level (cf. Townes, 1969).

FAM. BRACONIDAE. *Fore wing with costal cell narrow or obliterated, cross-vein 2 m-cu absent. Hind wing with cross-vein r-m meeting Sc + R before Rs leaves it.*

Sternites of gaster often partly membranous. These insects are closely related in structure and habits to the Ichneumonidae but are readily separated by the absence of 2 *m-cu* in the fore wings. A further distinction is the general presence of the first sector of M + Rs which is absent in the Ichneumonidae (Fig. 566). Also, with the exception of the subfamily Aphidiinae, there is no articulation between the 2nd and 3rd gastral segments. Braconidae are easily distinguished from the Stephanidae by the absence of the costal cell. With regard to their hosts a great variety of insects are selected; the Lepidoptera are the most commonly parasitized, and more than one hundred examples of an individual species of Braconid may issue from a single caterpillar. Braconid larvae are composed of thirteen body segments and, in the first instar, the head is often large and sclerotized (Short, 1952). As in the preceding

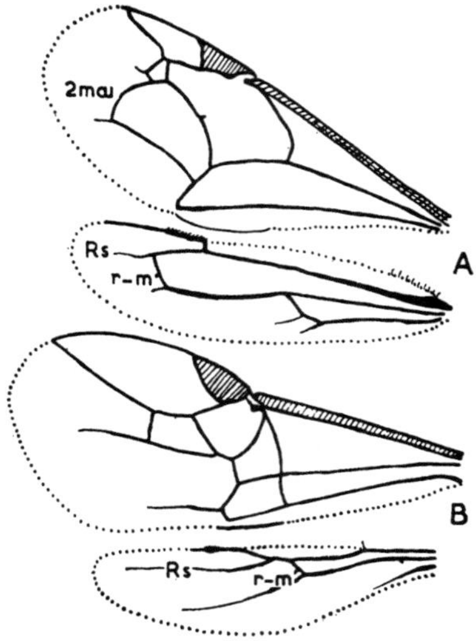

FIG. 566
Left wings of A an Ichneumonid and B a Braconid, to show the absence of 2 *m-cu* in fore wing of the latter, and the different positions of *r-m* in the hind wing

family, the most frequent number of spiracles in the adult larva is nine pairs, of which the first is placed as a rule on the mesothorax, and the remainder on the first eight abdominal segments. The endoparasitic forms are often provided with a caudal appendage similar to that found in Ichneumonid larvae. In *Apanteles, Microplitis, Microgaster*, and probably in other closely allied genera, this appendage is wanting, and the proctodaeum is evaginated to form a swollen anal vesicle, perhaps an accessory respiratory organ. Weissenberg (1908), from the analogy of the hind gut of other parasitic larvae, considers its most important function to be that of excretion. Pupation may occur within the host as in *Rhogas* and *Aphidius* or, more usually, externally as in *Apanteles* and many other genera. The pupa is enclosed in a cocoon which, in the last-named genus, is composed of fine threads of white, yellow or buff-coloured silk. In *Microgaster* the cocoon is of a glistening papyraceous nature. Very frequently members of a species emerging from the same host construct their cocoons in a mass, often enveloped by a common web. They may be closely compacted to form a cake, the individual cocoons being regularly arranged so as to resemble honeycomb. The biology of *Apanteles* has been studied by many observers, notably Grandori (1911), Seurat (1899), Muesebeck (1918) and others. It is a common endoparasite of Lepidopterous larvae, *A. glomeratus* being an abundant enemy of *Pieris*, a single larva of which may support nearly 150 examples. When mature the larval parasites gnaw their way through the skin of the host, and construct sulphur-yellow cocoons, irregularly heaped together. The biology of *Microgaster* is very similar and *M. connexus* is a common parasite of *Euproctis* (= *Porthesia*) *similis* (Gatenby, 1919). The species of *Alysia* and their allies are distinguished by the peculiar attachment of the mandibles, the apices of the latter being directed outwards and not meeting when closed. Those Braconids which exhibit this curious feature are separated by Ashmead into a distinct family – the Alysiidae. Almost all their species are parasitic upon Dipterous larvae and the biology of *Alysia manducator* has been followed by Altson (1920). It is a common endoparasite of *Calliphora, Lucilia*, etc.; the young larva has a caudal appendage, and becomes maggot-like with nine pairs of open spiracles in the last instar. The abundant species of *Dacnusa* and *Chorebus* attack leaf-miners, especially Agromyzidae (Griffiths, 1964–8). One of the most remarkable Braconids is *Sycosoter lavagnei* which is an ectoparasite of the Scolytine *Hypoborus ficus* (Lichtenstein and Picard, 1917): both sexes are dimorphic, having winged and apterous forms, but in the male the alate forms are the commoner.

The Aphidiinae are parasites of aphides, more specially of the apterous viviparous females and, as a general rule, only a single larval parasite develops within the body of an individual host (cf. Vevai, 1942). The life-history of *Aphidius testaceipes*, which is a common enemy of *Toxoptera graminum*, has been followed by Webster and Phillips (1912). These observers state that the aphid may be attacked in any of its instars but if parasitized before the second ecdysis the host fails to reach to maturity. On the other hand, if the *Aphidius* deposits its egg in an aphid which has passed the second ecdysis the parasitism does not prevent its attaining the adult stage. If the aphid has passed the third ecdysis before becoming parasitized, it is capable in all cases of producing a small number of young before succumbing. When about to pupate the *Aphidius* larva makes a ventral fissure in the body-wall of its host, and cements the latter down to the object upon which it finally rests. The dead parasitized aphids are familiar straw-coloured objects and each bears a circular hole through which the adult parasite issued. Species of *Praon* leave their host prior to pupation and spin for themselves a separate shelter, which is usually surmounted by the empty body of its victim (Fig. 567).

Čapek (1969) has discussed the classification of the Braconidae. Nixon (1972–76) is monographing the European *Apanteles*.

FAM. STEPHANIDAE. *Fore wing with costal cell broad. Sternites of gaster evenly sclerotized. Anterior thoracic spiracle concealed by a prothoracic lobe. Head*

FIG. 567
Cocoon of *Praon* beneath the body of its dead host (an aphid) (Braconidae)
After Howard.

tuberculate above. Hind femora with teeth. Ovipositor long. The members of this family have very slender antennae composed of 30 or more segments and the hind wings are usually without enclosed cells. About 100 species have been described (Elliott, 1922). The European *Stephanus serrator* is a parasite of *Xylotrechus* (Cerambycidae) (Blüthgen, 1953) and the Californian *Schlettererius cinctipes* of *Sirex noctilio* (Taylor, 1967).

FAM. MEGALYRIDAE. *Fore wing with costal cell broad. Sternites of gaster evenly sclerotized. Anterior spiracle entirely in prothorax. Mesoscutum with a central furrow but no notauli. Ovipositor long.* The more than 20 Australian species of *Megalyra* are parasites of wood-boring beetles. A second subfamily, the Dinapsinae, are chiefly found in S. Africa and Madagascar with one genus in Java (Hedquist, 1967).

Superfamily **Evanioidea**

The three families placed in this group agree in the abnormal attachment of the petiole, at a point just behind the scutellum, but are otherwise rather diverse. Crosskey (1951) deals with the British species and Kieffer (1912) has monographed the group. Short (1960) gives a brief account of their larvae.

Antennae with 13 or 14 segments. Fore wing with costal cell wide, costa and pterostigma developed; hind wing with a costal vein, anal lobe sometimes developed. Abdomen with spiracles on segments 1 and 8 only, pygostyles present.

Rasnitsyn (1972) has described more primitive members of the group (Praeaulacidae) from the Upper Jurassic.

FAM. EVANIIDAE. *Fore wing with not more than one m-cu cross-vein and 2 r-m present. Hind wing with an anal lobe. Gaster short with a long abrupt petiole and short, unexserted ovipositor. Propleuron short.* A few hundred species of this family are known from all the main geographical regions. All seem to be parasites of the oothecae of cockroaches, *Evania appendigaster* on *Periplaneta* (Cameron, 1957), *Brachygaster minuta* on *Ectobius* and *Zeuxevania splendidula* (Genieys, 1924) on *Loboptera*. In some genera the wing-venation becomes almost completely lost.

FAM. AULACIDAE. *Fore wing with 2 m-cu cross-veins, 2 r-m present. Antennae inserted just above the clypeus. Hind wing with no anal lobe. Ovipositor long.* There are rather more than 100 species of this widespread but generally uncommon family. *Aulacus* attacks the larva of *Xiphydria* but other members of the family attack Coleoptera (Buprestidae, Cerambycidae).

FAM. GASTERUPTIIDAE. *Fore wing with not more than 1 m-cu cross-vein, 2 r-m absent. Hind wing with no anal lobe. Gaster long, gradually clavate, ovipositor often long. Propleuron forming a long 'neck'.* The insects of this family are distinguished by a neck-like prolongation of the propleuron, swollen hind tibia, and long gaster, and sometimes by a long ovipositor. There are about 300 species found in all parts of the world. They are parasites of solitary bees (Höppner, 1904). Crosskey (1962) discusses their classification.

Superfamily **Cynipoidea**

Included in this superfamily are about 1600 species of small, and often minute, insects which are usually black or darkly coloured. Biologically, they are of great interest as the various species are either gall-makers, inquilines or parasites, the first-mentioned exhibiting the phenomena of heterogony and agamogenesis. The eggs are provided with a usually elongate pedicel, the larvae are apodous and maggot-like, and there is no cocoon. In the great majority of the imagines the second gastral tergum is larger than the remainder and in many cases forms almost the whole of the dorsal surface of the gaster. The trochantellus is more or less indicated but, as Kieffer points out, it is often no more than the contracted base of the femur. Authorities differ with regard to the division of the group into families; four families are dealt with in the pages which follow. Kieffer (1914) has written an admirable short account, which is accompanied by a full bibliography; Eady and Quinlan (1963) deal with the British species of gall-makers. For a catalogue of the world's species see Dalla Torre and Kieffer (1910) and the more recent revisional treatise of Weld (1952) who gives a key to the families and genera.

Characters of the superfamily. Antennae in female usually with 13 but up to 19, male with 14–15 segments, not elbowed. Pronotum latero-ventrally pointed, closely co-adapted to mesepisternum, posteriorly running back to the tegulae. Wings, when present, with fore wing without pterostigma and proximal part of costa, venation often reduced but cell R_1 more or less complete; hind wing with no anal lobe. Hind femur with a trochantellus. Female with last visible abdominal tergite and sternite not apposed, ovipositor exposed though not far protruded, tergite 8 exposed and resembling 7, tergite 9 retracted, desclerotized, abdomen more or less compressed; spiracles on segments 1 and 8 only.

FAM. CYNIPIDAE (Gall-wasps). *Second and third gastral tergites usually forming at least one-half of the gaster.* The greater number of species belongs to the subfamily Cynipinae, all of which produce galls for the purpose of providing shelter and nutriment for their offspring or are inquilines in such galls. Their larvae are consequently internal feeders and are maggot-like in form, with well-sclerotized dentate mandibles. The head is small and is followed by twelve body segments, and there are nine pairs of spiracles. The antennae and both pairs of palps are vestigial. Pupation takes place within the larval cell and a cocoon is wanting. The forms of galls produced by these insects are almost endless and all parts of plants may be affected, from the roots to the flowers. In every case the female insect lays an egg or eggs in the tissues of the growing plant, in the interior of which the subsequent development takes place. As a rule this mode of life is accompanied by the production of a gall. Many theories have been advanced to account for the phenomena of gall-formation, but the problem appears to be still far from being solved, largely on account of difficulties attending the experimental side of the subject. A full discussion of the various views which are or have been held is given in the works of

Kieffer. The irritation of the tissues produced by the insertion of the ovipositor is not the initial cause. There also appears to be no evidence that the fluid injected by the female during oviposition is anything more than of the nature of a lubricant. The mere presence of the Cynipid egg in the tissues is not in itself sufficient to produce the gall as, ordinarily, the latter does not commence to develop until the larva has hatched; many months may elapse between the date of oviposition and that of eclosion. All that can be said is that the galls are produced as the result of reactions of the cambium and other meristematic tissues of the plant in response to the stimulus induced by the presence of the living larva. It is probable also that the latter exudes a secretion which exercises an influence upon the growth of the cells of the plant (Triggerson, 1914). The formation and structure of the galls have been studied by Beyerinck (1882), Cook (1902–04), Cosens (1912), Hough (1953) and others. Viewed in section, a gall is usually seen to be composed of the following layers of tissue passing from without inwards (Fig. 568). The outermost coat is the

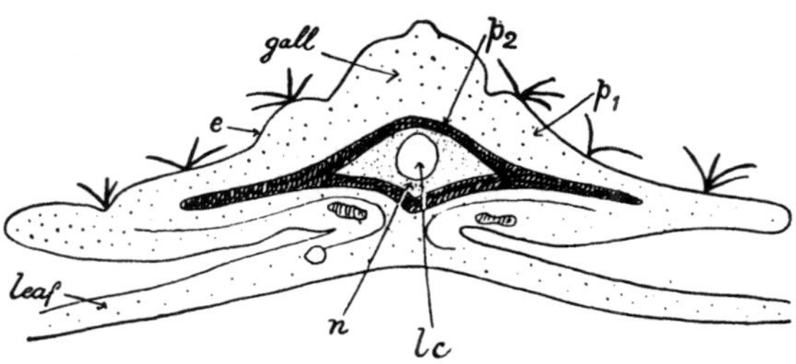

FIG. 568 Diagrammatic section of gall of *Neuroterus lenticularis*, according to Fockeu

e, epidermis; p_1, parenchyma; p_2, protective layer; *n*, nutritive layer; *lc*, larval cell.

epidermis and beneath the latter is an extensive development of parenchymatous tissue. The third layer is protective in function and is usually of a hard consistency but is sometimes wanting, while the innermost layer is nutritive and surrounds the cell containing the larva. Cook concludes that the morphology of a gall is dependent, as a rule, upon the insect which produces it rather than upon the plant upon which it is produced. Galls formed by the same genus of insects exhibit great similarity even though produced on widely different plants. Furthermore, those produced on a particular genus of plants by different insects are very dissimilar. In addition to the species which actually forms the gall, the latter frequently supports a definite biological association of other insects. A large number are inquilines, which comprise not only other Cynipidae but also larvae of Diptera, Coleoptera and Lepidoptera. Furthermore, the larvae and pupae of the true gall-maker, and of the inquilines, are very subject to the attacks of hymenopterous parasites, more particularly Chalcids. Kinsey (1920) estimates that 86 per cent of the known species of gall-wasps produce galls on *Quercus* and are confined to that genus. Another 7 per cent are restricted to species of *Rosa* and the remaining 7 per cent are found on plants belonging to 35 genera of Angiosperms, more especially the Compositae. The reason for this very marked selection of a single genus of plants is hard to understand, particularly as oaks have a limited distribution in the present age. It is true that the galls on this genus are more conspicuous than on other plants where they are more liable to become overlooked. According to Kinsey the tribe Aylacini is, in many respects, the most primitive of the gall-wasps. Its members are not confined to a particular plant family but select those belonging to many genera. The various species of *Aulacidea* either induce extremely simple gall formation or live in stems, producing no gall at all. They have not acquired agamic reproduction, the sexes are produced in about equal numbers, and the alternation of generations so characteristic of the higher Cynipinae is absent. Almost every transition may be observed from the simple condition prevalent in *Aulacidea* to the many types of highly complex galls, and the alternation of morphologically and physiologically different generations found in many other genera. The most highly evolved galls are to be looked upon as almost entirely separate organisms, which are only connected to the host plant by means of a narrow neck of tissue. In some cases the galls develop in size and form new tissue after separation from the parent plant. The galls of two successive generations, produced on different parts of the same plant, often present entirely different forms; and the insects of the two generations are frequently so divergent in characters that they have often been allocated to separate genera until their relationships have been detected. Heterogony among Cynipidae is of an exceptionally remarkable nature. In many species males have never been seen at all, out of many thousands of the insects which have been reared, and there appears to be little doubt of their non-existence. In these very highly specialized cases the successive generations are all similar and agamic and a secondary simplification of the life-cycle results. The majority of the Cynipinae have only the alternate generations alike: each agamic generation is followed by a bisexual generation which, in its turn, produces the agamic one. The latter is the overwintering stage while the bisexual generation is produced during summer. A few of the commoner species, which are prevalent in Britain, may be selected as illustrating the principal biological phenomena already referred to. *Neuroterus lenticularis* is a very abundant gall-wasp in England. The galls (Fig. 568) from which the spring (agamic) generation emerges are lenticular growths found on the lower surface of oak leaves in October. The insects remain in the galls all the winter, and appear as adults early in April. They consist entirely of parthenogenetic females which deposit their eggs deep down among the catkins and young leaves. The resulting galls occur in May and June and are quite different from those preceding being spherical and sappy in character. The summer generation which emerges from them was originally referred to a different species, i.e. *Spathegaster baccarum*. Both males and females are produced but the latter largely predominate in numbers. After copulation the eggs are laid at the sides of the veins in the tissues of the young leaves, and the resulting galls are of the lenticular kind found in October. The most conspicuous difference in the females of the two generations is seen in the ovipositor, which is much larger in the agamic than in the summer individuals. *Biorrhiza pallida* is another very characteristic oak species. In the bisexual generation the males are winged, and the females are either apterous or have vestigial wings. This generation emerges from the 'oak apple' galls and the eggs are laid in the roots of that tree. In this situation other galls are produced from which, in spring, the agamic generation (known as *B. aptera*) is produced. The individuals of this brood consist exclusively of apterous females which migrate up the tree and produce the 'oak apple' galls in due course. The genus *Rhodites* is confined to the Rosaceae – *Rosa* and *Rubus* being most usually selected. The familiar and striking bedeguar or 'pin-cushion' galls are produced on the former genus by *Rhodites rosae*. These galls consist of a mass of moss-like filaments surrounding a cluster of hard cells containing the *Rhodites* larvae. There is no alternation of generations in this species, males are much less frequent than females, and the eggs are known to be capable of parthenogenetic development. The hard spherical 'marble' galls of what is usually known as *Adleria kollari* on oak produce the agamic generation of that species. Marsden-Jones (1953) has now proved what Beyerinck suspected that the bisexual generation is *Andricus circulans*, bred from galls on *Quercus cerris*. The species was introduced into Devonshire about 1830 and soon became abundant: it should now be known as *Andricus kollari*.

The genus *Synergus* and its allies are inquilines though they are easily mistaken for true gall-makers to which they frequently bear an extremely close resemblance. They mostly lay their eggs in cynipid galls found on oak.

The subfamily Charipinae contains species which are hyperparasites of aphids through *Aphidius* or other parasites: less frequently they have similar relations with coccids. An account of the biology of *Charips* is given by Haviland (1921).

The subfamily Eucoilinae includes insects which are parasites of Diptera and the primary larvae are eucoiliform. In *Kleidotoma* the 2nd instar larva is polypodeiform with 10 pairs of trunk limbs (James, 1928). *Eucoila eucera* (Fig. 569) is an important parasite of *Oscinella frit* and *Cothonaspis rapae* attacks *Erioischia brassicae*.

FAM. FIGITIDE. *Second gastral tergite, though large, not forming half the length of the gaster.* *Aspicera* and its allies are parasites of Syrphidae, while the Anacharitinae have been bred from the Neuroptera. The Figitinae are parasites of Diptera. The primary larva is eucoiliform and is followed by a polypodeiform instar with 10 pairs of trunk limbs (James, 1928).

FAM. IBALIIDAE. *Sixth gastral tergite the largest. Cell R_1 closed in fore wing and at least 9 times as long as broad. Hind basitarsus twice as long as the other segments put together.* The members of this small family are among the largest in the Cynipoidea. The genus *Ibalia* is widely distributed in Europe and N. America and its members are endoparasites of Siricidae. The single British species, *I. leucospoides*, has been studied by Chrystal (1930). The egg is laid in the *Sirex* larva and the primary larva is of the polypod type with 12 pairs of trunk limbs: the whole larval period appears to last 3 years.

FAM. LIOPTERIDAE. *Gaster attached to top of propodeum, one of tergites 3–6 the largest. Fore wing with cell R_1 closed, not twice as long as broad.* This family includes a few, mostly rather large insects of S. America, Africa and the Far East (Hedicke and Kerrich, 1940); their habits are not known.

Superfamily **Chalcidoidea** (Fig. 570)

The superfamily is probably one of the largest in the order as regards number of species and it also includes some of the smallest members of the Insecta. The bulk of its species are either parasites or hyperparasites of other insects, and are of even greater economic importance than those of the Ichneumonoidea as a natural means of control. Non-parasitic vegetable-feeding forms are found in the families Agaonidae, Torymidae, Pteromalidae and Eurytomidae: in the majority of cases they infest seeds but certain members of the last two families are gall-producers. The parasitic species, in a relatively small number of instances, are indirectly injurious from the fact that they destroy beneficial insects such as *Laccifer lacca*, or are hyperparasites of other insects which, in their turn, are destroyers of harmful species. The orders most commonly parasitized are the Lepidoptera, Hemiptera-Homoptera and Diptera. Lepidoptera are more frequently selected than any other major group, enormous numbers of their eggs and larvae succumbing to infestation by various Chalcids: on the other hand their pupae are rarely affected. Certain Pteromalidae, however, prefer to

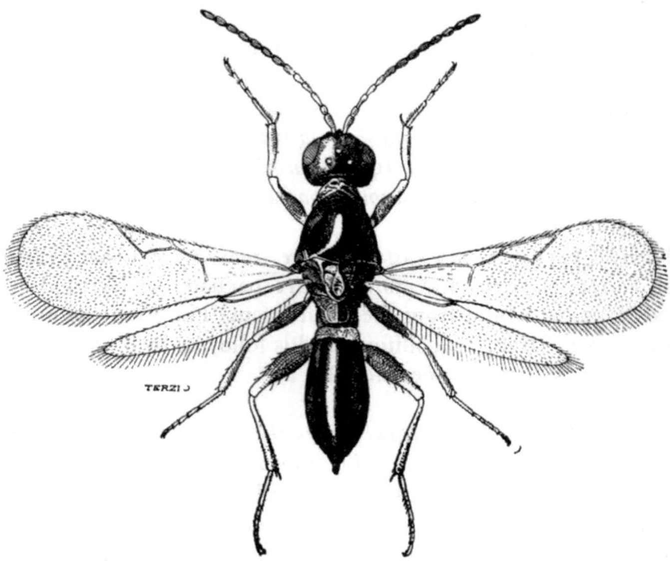

FIG. 569 *Eucoila eucera*, female (Cynipidae). Britain

oviposit in larvae just about to pupate or in newly transformed pupae. The Coccoidea are the most universally attacked of any group of insects, and some species (*Eulecanium coryli*, etc.) are so freely infested that it is often rare to find an unparasitized individual. In temperate regions Chalcids seem to pass through from one to three generations in the year – a higher number is apparently rare. One of the shortest life-cycles occurs in *Euplectrus comstockii*, which develops from the egg to the adult in seven days (Schwarz, 1881), and an equally rapid development is found in *Trichogramma minutum* (= *pretiosa*) which has been reared from the eggs of *Aletia xylina* (Hubbard). Chalcid larvae are composed, as a rule, of a reduced head and thirteen trunk segments. In the ectophagous forms open spiracles are evident at the time of hatching: thus in *Aphytis mytilaspidis* the full number of eight pairs are present at this stage. In other cases, as in *Torymus propinquus*, a reduced number of spiracles is present at the time of eclosion from the egg, additional pairs being acquired subsequently. Among endophagous species the younger larvae are usually apneustic, open spiracles developing later when the destruction of their hosts reaches an advanced stage. In *Blastothrix* and other genera the newly hatched larva is exceptional in being metapneustic (Figs. 570, 571). This condition is an adaptation which allows of the respiration of atmospheric air along the pedicel of the egg, which protrudes externally through the body-wall of the host and functions as a kind of respiratory tube. In *Blastophaga psenes* the tracheal system is apneustic throughout life (Grandi). Hypermetamorphosis is common in the superfamily, and at least five types of primary larvae are known, but probably

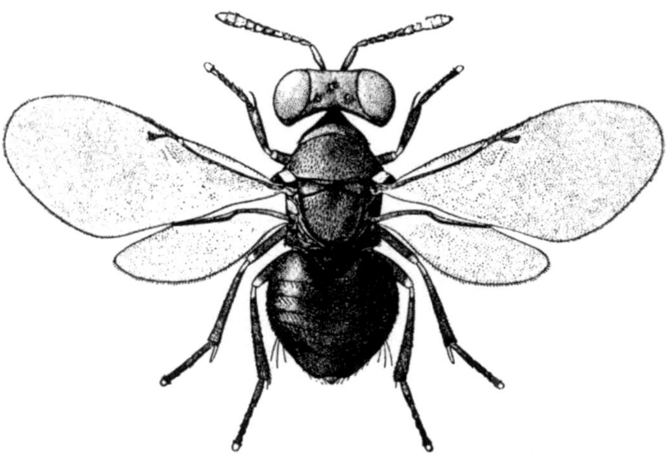

FIG. 570 A typical Chalcid, *Blastothrix sericea*, female: magnified

others await discovery. In the later instars all these types assume an ovoid maggot-like form and, in the majority of species, the latter kind of larva is retained throughout life. Chalcid larvae do not construct any cocoons (except in *Euplectrus*, see p. 1225), and pupation usually occurs either within or in close proximity to the remains of their hosts.

For the classification of the Chalcidoidea the paper by Peck *et al.* (1964) is a useful introduction but the composition of several of the families is still far from settled. A general account of the biology of the group is given by Clausen (1940). Life-history studies of individual species are numerous; in Europe a series of papers has been contributed by Silvestri (*Boll. Lab. zool. Portici*) and a large number by other authors will be found in various American journals and bulletins. Some of the more important of these papers are referred to under the families concerned. A monograph on the British species is greatly to be desired and about 1400 species have been listed.

Characters of the superfamily: Antennae with less than 14 segments, elbowed. Pronotum ventrally rounded, not closely co-adapted to mesepisternum, usually separated from tegulae. Wings, when present, with fore wing with pterostigma and proximal part of costa absent, venation reduced, usually very much; hind wing with no anal lobe. Hind femur with a trochantellus. Female with last visible abdominal tergite and sternite not apposed, so that the ovipositor is at least partly exposed, tergite 8 resembling 7, 10 fused with 9 which is pigmented, pygostyles present; spiracles on segments 1 and 8 only.

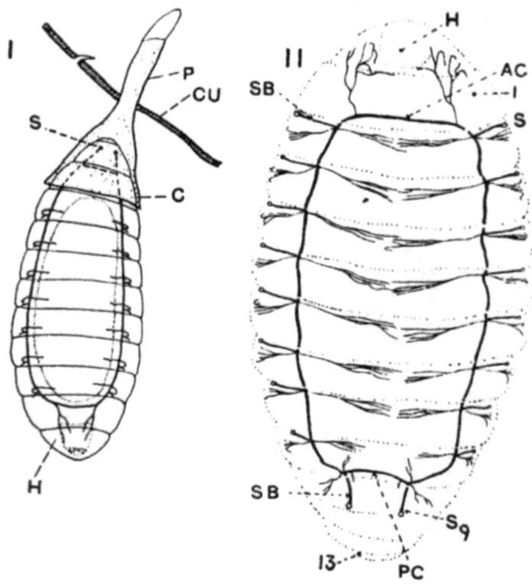

FIG. 571

Blastothrix sericea. I, Newly-hatched larva respiring by means of the pedicel (*P*) of the egg

II, Larva in last instar

C, remains of chorion; *CU*, body-wall of host; *H*, head; *S*, spiracles. *AC*, *PC*, anterior and posterior tracheal commissures; 1, 13, 1st and 13th trunk segments. *S*, S_9, 1st and 9th spiracles; *SB*, spiracular tracheae.

FAM. AGAONIDAE (AGAONTIDAE) (Fig insects). *Head of female long, oblong, with a deep longitudinal groove above. Front and hind legs very stout, the middle ones much more slender. Males nearly always wingless, with stout, 3–9 segmented antennae.* A family which includes some of the most remarkable of all Chalcids both as regards their structure and biology. Sexual dimorphism has reached a very highly specialized condition, the males being wingless and greatly modified in other respects, bearing no resemblance to the members of the opposite sex. The species are caprifiers that live within the receptacles and pollinate, or fructify, the flowers of various species of *Ficus*. The number of known species and varieties of fig is said to reach 500 and, in certain of these, the caprification phenomena are known to vary widely, and many of the insects involved are apparently confined to certain definite species of figs. The investigation of the symbiotic relationship between plant and insect offers, therefore, an extremely wide field for investigation. The best known species is *Blastophaga psenes*, which exists in a state of symbiosis within the fruit of *Ficus carica*. It is well known that, in the Smyrna variety of fig, the receptacles contain only female flowers, and pollination is brought about by the agency of this Chalcid. On the other hand, the caprifigs, or varieties which contain male flowers, are the natural hosts of the *Blastophaga*. Caprification, or the process of hanging caprifigs in the Smyrna trees, is an old custom based upon the belief that the figs would not mature unless it were carried out. Much discussion has arisen with reference to whether caprification is essential or not. In California it is agreed that the culture of the Smyrna fig necessitates the simultaneous cultivation of caprifying varieties in which the *Blastophaga* lives. If the latter insect fails to pollinate the Smyrna figs, the fruit falls without maturing (Condit, 1947). The eggs of this

Chalcid are laid in the ovaries of the caprifig and give rise to galls therein. The male imago emerges first and, on finding a gall containing a female, commences to gnaw a hole through the wall of the ovary and fertilizes the female while the latter is still *in situ* (Fig. 572). The female leaves the receptacle through the opening at its apex and, laden with adherent pollen, flies to a neighbouring fruit. If the latter be in the right condition she seeks the opening and gains admission into the interior of the receptacle, where she commences oviposition. Should the caprifig, from which she has emerged, be suspended in a tree of the Smyrna variety she enters a fruit of the latter, but subsequently discovers that she has selected a wrong host, as the flowers are of such a shape that they do not allow of oviposition within them. After wandering about for a while, she usually crawls out of the receptacle and incidentally pollinates the flowers. The males mostly die without ever leaving the receptacles in which their development took place.

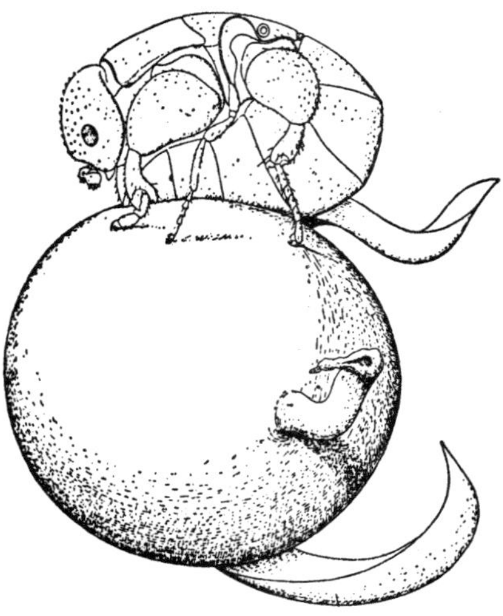

FIG. 572
Male *Blastophaga psenes* fertilizing the female, the latter within a galled flower of the fig (Agaonidae)
After Grandi, *Boll. Lab. zool. Portici*, 14, 1929.

Hill (1967*b*) gives an interesting introduction to the biology of the figs, fig-insects and their parasites and commensals in Hong Kong. Grandi (1929) has investigated the structure and biology of *Blastophaga psenes* and gives a full bibliography. The taxonomic writings of the latter author (1916 *et seq.*) also deal with the external morphology of many genera. Papers by Mayr and Saunders, S. S. and Hill (1967*a*) should also be consulted.

FAM. TORYMIDAE. *Mandibles stout with 3–4 teeth. Notauli present. Axillae scarcely advanced in front of fore margin of scutellum; mesepisternum not enlarged. Hind coxa 5–6 times larger than the front one, hind femur sometimes with 1 or more teeth; tarsi with 5 segments. Ovipositor usually long and exserted.* A very large family whose affinities lie more closely with the Agaonidae than any other group. Members of the subfamily Idarninae are found associated with fig-insects either as parasites or inquilines. Their males are often apterous, but the abdomen is short and not tubularly lengthened or broadened at the apex as in the Agaonidae. The great majority of Torymidae are parasites of gallicolous insects, but a certain number have been reared from the nests of bees and wasps. Species of *Megastigmus* have been bred from Hymenopterous and Dipterous gall-makers while others are phytophagous, attacking the seeds of Coniferae, *Rosa*, etc. *Syntomaspis druparum* is the apple-seed Chalcid of Europe and N. America. *Monodontomerus* parasitizes many insects, *M. obsoletus* having been bred from both Lepidoptera and Hymenoptera. *Torymus* (*Callimome*) chiefly attacks gall-making Diptera and Hymenoptera (Hoffmeyer, 1930–31).

The **Podagrionidae** have been placed here, probably correctly. The hind femora have a row of teeth. The species are parasites of the eggs of Mantids.

FAM. CHALCIDIDAE. *Labrum concealed. Mandibles stout with 3–4 teeth. Prepectus not developed, mesepisternum not enlarged, axillae not advanced in front of scutellum. Hind coxa 5–6 times as large as front one, hind femora with a row of teeth. Fore wings not longitudinally folded. Ovipositor usually short and straight.* This family is especially well represented in S. America and is less common in cool climates. Eight species are British (Ferrière and Kerrich, 1958). Many of its members are primary or secondary parasites of Lepidopterous larvae or pupae, or of Dipterous puparia. *Chalcis* attacks the eggs of *Stratiomys.* The Leucospidae are now usually regarded as a separate family of parasites of bees (Bouček, 1974). The labrum is exposed, the wings longitudinally folded, the prepectus indicated and the abdomen much modified. *Leucospis gigas* was observed by Fabre (1886); it undergoes hypermetamorphosis and is an ectoparasite of *Megachile* (*Chalicodoma*) *muraria.*

The **Chalcidectidae,** mainly American parasites of Coleoptera, have toothed femora like a Chalcidid but have been regarded as nearer to the Cleonyminae (Bouček, 1959). The axillae are considerably advanced.

FAM. EURYTOMIDAE. *Characters of the Chalcididae but hind coxae not enlarged, hind femur not toothed. Pronotum wide, quadrate, scarcely narrower than mesoscutum which is often coarsely punctate. Last segments of ♀ produced.* Few other families of Chalcids exhibit so wide a diversity of habits as is met with among the members of this group. *Tetramesa* (*Harmolita*) produces galls on the stems of wheat, rye, barley and various grasses (Phillips, 1920). *T. grandis* exhibits alternation of generations: the apterous form, *minutum*, occurs in spring, laying its eggs at the base of the young wheat plant, and the larva destroys the tiller affected, or may kill the entire plant. The alate form, *grandis*, is the summer generation which lays its eggs slightly above the nodes. Males are rare and have only been found in the case of the spring brood: *T. orchidearum* is exceptional in that it produces galls on the stems and leaves of certain orchids (*Cattleya*). *Bruchophagus platyptera* is likewise phytophagous and passes its developmental stages in the seeds of clover and alfalfa. Species of *Eurytoma* and other genera attack the seeds of plum, grape, *Ampelopsis* etc. Other members of the family live in the nests of bees and wasps or are parasites of gall-forming Diptera and Hymenoptera; a few are egg-parasites of Orthoptera, while several species of *Eurytoma* attack a wide selection of hosts.

FAM. PERILAMPIDAE. A small family distinguishable from the preceding by the large thorax and small triangular abdomen. The biology of *Perilampus hyalinus*, a hyperparasite of the larva of *Hyphantria* and other hosts, has been studied by Smith (1912; 1917). The newly hatched larva is an active planidium which bores its way into the *Hyphantria*, in whose body-cavity it remains until it meets with either the larva of the Tachinid *Varichaeta* or of the Ichneumon *Olesicambe* (= *Limnerium*), which are primary parasites. Upon discovering one or other of the latter hosts it becomes endoparasitic: subsequently it makes its way out, undergoes hypermetamorphosis into a white maggot-like larva and becomes an ectoparasite of the same host. British species have been bred from various parasites (Ferrière and Kerrich, 1963).

FAM. EUCHARITIDAE. *Mandibles sickle-shaped, usually with only one or two teeth. Thorax very convex, scutellum produced backwards. Gaster compressed, usually with a long petiole.* Included herewith are certain remarkable metallic blue or green Chalcids characterized by the configuration of the scutellum which is frequently produced backwards in the form of powerful spines. So far as known they attack ants, usually as ectoparasites, and are mainly found in the tropics. *Orasema* attacks members of the genera *Pheidole* and *Solenopsis*. According to Wheeler (1907) it parasitizes the prepupa just after the last exuviae have been stripped off by the worker ants. Its newly-hatched larva is a planidium and is found attached near to the head of the host. As a result of the parasitism the hosts undergo degeneration and fail to become imagines. For the biology of *Schizaspidia* see Clausen (1923).

FAM. PTEROMALIDAE. *Very diverse in appearance but lacking the specializations of the other families. Pronotum narrow, transverse, thorax usually not coarsely punctured, axillae not advanced in front of scutellum. Hind coxae not enlarged, hind femora not toothed, tarsi with five segments.* This family is the largest among Chalcids and its members, like those of the Encyrtidae, affect almost all orders of insects either as parasites or hyperparasites. The north-west European species have been revised by Graham (1969). *Pteromalus puparum* is common and widely distributed: it especially parasitizes pupae of *Pieris rapae* and *brassicae* and an account of its biology is given by Martelli (1907). *P. deplanatus* has been recorded as occurring in great swarms in buildings but there is no satisfactory explanation of the habit (Scott, 1919). *Nasonia brevicornis* (*Mormoniella vitripennis*) is a common pupal parasite of *Musca domestica, Calliphora* and other Calyptratae (Altson, 1920). *Spalangia muscidarum* is likewise a pupal parasite of *Musca, Stomoxys* and *Haematobia*, its larva undergoing hypermetamorphosis (Richardson, 1913). *Asaphes* and *Pachyneuron* parasitize aphides, etc., and the Eunotinae mainly affect coccids, *Scutellista cyanea* being an important factor in the control of the Black Scale (*Saissetia oleae*) in California and of *Ceroplastes rusci* in Italy.

The Miscogasterinae are not now treated as a distinct family, since they intergrade with other Pteromalids. Many of the species are Dipterous parasites. The Cleonyminae (Ferrière and Kerrich, 1958) have also been reduced to a subfamily. The species are often relatively large and attack wood-boring beetles. The **Leptofoenidae** are perhaps no more than specialized members of the last subfamily; they are some of the most extraordinary of all Chalcids in appearance and are found on dead wood in tropical America and Australia. The Ormyrinae have often been united with the Torymidae but seem to fit better here. The gaster usually has rows of coarse punctures. The species are usually parasites of gall-making insects.

FAM. ENCYRTIDAE. *Mesepisternum greatly enlarged. Mid tibial spur usually much enlarged.* The Chalcids comprised in this extensive family live as parasites of the ova, larvae or pupae of various insects. Although the Hemiptera-Homoptera and Lepidoptera are most frequently selected hardly a single order of insects is immune from their attacks. Certain genera are definitely restricted with reference to their selection of hosts. Thus *Aphycus* is an ecto- or endoparasite of Coccoidea, particularly of *Eulecanium*; *Blastothrix* almost exclusively parasitizes *Eulecanium* and *Pulvinaria* while *Ageniaspis* is mainly confined to the Lepidopterous genera *Phyllonorycter* and *Yponomeuta*. The family is of more than ordinary interest and importance from the fact that certain species of *Ageniaspis*, *Litomastix* and *Copidosoma*, which parasitize Lepidoptera, are known to exhibit polyembryony. They deposit their eggs in those of the hosts but the larvae of the latter emerge in the normal manner and contain the developing parasites in their body-cavity where embryonic fission takes place (see p. 305). Several members of the family have been the subject of detailed biological studies and reference should be made to papers on *Ageniaspis* by Bugnion (1891) and Marchal (1904), on *Encyrtus* (*Comys*) by Embleton (1904), on *Aphycus* and *Blastothrix* by Imms (1918*a*), on *Copidosoma* by Leiby (1922) and Doutt (1947) and *Litomastix* by Silvestri (1906). The work of Mercet (1921) on the Spanish species is well illustrated and is a valuable taxonomic monograph.

Several smaller groups have more recently been segregated from the Encyrtidae and are regarded as separate families. The **Eupelmidae** includes a few genera with a very wide host-range, having been bred from the eggs of Lepidoptera (*Anastatus*), from Cecidomyidae, Coccoidea and various Coleoptera (*Eupelmus*), and from Arachnids (*Arachnophaga*). The **Thysanidae**, of which two are British, includes the genus *Thysanus* (*Signiphora*), very curious small insects with bristly tibiae which are parasites of Coccoidea, or more rarely Diptera. The **Tanaostigmatidae** include a few Encyrtid-like species, some of which attack Coleoptera.

FAM. EULOPHIDAE. *Axillae advanced strongly in front of anterior margin of scutellum and usually in front of tegulae. Fore wing not broad, pubescence not in rows. Tarsi with 4 segments; 5 only in the Aphelininae and a few other females.* A very large family consisting for the most part of very small species. The Aphelininae (Ferrière, 1965) are important parasites of the Diaspine coccids and of aphids, their larvae being either ectophagous or endophagous. *Aphytis mytilaspidis* is a common ectoparasite of the Mussel Scale and its structure and biology have been fully studied (Imms, 1916). *Prospaltella berlesei* has been introduced into Italy for the purpose of controlling *Diaspis pentagona* and there is now a voluminous literature on the subject. The species of this subfamily exhibit extraordinary specializations in parasitic behaviour (summary of Flanders' work in Clausen, 1940, also Flanders, 1953). The female larva is the direct and almost always internal parasite of a scale insect. The male larva is an obligatory hyperparasite, either of other scale-parasites or frequently on its own species. The male larva is most often an external feeder, though sometimes internal. Fertilized and unfertilized females behave differently, the latter showing interest only in scales which are already parasitized. The Tetrastichinae (Delucchi and Remaudière, 1966) affect nearly all orders of insects either as primary or secondary parasites: the majority parasitize gallicolous Diptera; Hymenoptera and Coleoptera. *Tetrastichus asparagi* is an egg-parasite of *Crioceris asparagi* and, according to Johnston (1915), from 1 to 10 larvae occur in a single egg. The beetle larvae emerge from the infested eggs but fail to pupate, although a pupal cell is constructed, and the adult parasites issue from the latter. *Melittobia* is a common ectoparasite of the pupae of *Bombus*, *Osmia* and other Aculeata (Balfour-Browne,

1922) as well as of certain Diptera, more especially *Calliphora*. Members of the subfamily Eulophinae are principally primary or secondary parasites of leaf-mining Lepidoptera. *Euplectrus* is exceptional in that a cocoon is present. Thomsen (1947) has shown that it is constructed from the products of the Malpighian tubes which are modified in the hind gut and discharged through the anus.

Askew has revised the British species with the exception of the Tetrastichinae (1968).

FAM. TRICHOGRAMMATIDAE. *Axillae as in previous family. Fore wing broad, pubescence in lines or rows, marginal and stigmal veins in a single curve. Tarsi with 3 segments.* The 3-segmented tarsi separate this family from all others and, according to Ashmead, it is related to the Eulophidae, connecting the latter with the Mymaridae. Over 100 species are known, all are egg-parasites, and they include some of the most minute insects (Doutt and Viggiani, 1968). *Trichogramma* usually parasitizes Lepidoptera and Howard mentions that as many as 20 individuals will develop within a single egg. In Europe *T. evanescens* (correctly *T. semblidis*) is a parasite of *Sialis* (misidentified as *Donacia*) (Gatenby, 1917) and of some Lepidoptera. *T. minutum*, in America, has been extensively used in connection with biological control. *Prestwichia aquatica* has been reared from the eggs of *Notonecta*, *Ranatra*, *Dytiscus* and *Hygrobia*, while *Hydrophylax aquivolans* parasitizes those of *Ischnura*. The last-named Chalcid swims beneath the water by the aid of its wings, *Poropoea stollwerckii* affects the eggs of *Attelabus* and, according to Silvestri, it passes through five larval forms (see also Bakkendorf, 1934).

FAM. ELASMIDAE. *Axillae as in Eulophidae. Fore wing narrow, marginal vein long. Hind coxae enlarged, femur compressed, tarsi with 4 segments.* A few dozen small black species with enlarged, compressed hind legs. The group has a wide distribution and parasitizes either Lepidoptera or secondarily their parasites.

FAM. MYMARIDAE (Fairy Flies). *Hind wings linear, base forming a stalk, wings with long fringes. Antennae with short scape and no ring-segments. Ovipositor issuing near tip of gaster.* The species of this family are all exceedingly minute and are exclusively egg-parasites. They are mostly black or yellowish and devoid of metallic colours. Some authorities have placed them among the Proctotrupoids but Ashmead holds their position to be in the present superfamily. One of the most remarkable is *Caraphractus cinctus* (= *Polynema natans*) which parasitizes eggs of Dytiscids; both sexes swim readily beneath the water by means of their wings. *Alaptus* includes probably the smallest of all insects, *A. magnanimus* measuring only 0·21 mm in length. *Anaphes conotracheli* has been reared from the eggs of weevils and *Litus krygeri* from those of *Ocypus olens*.

The genera have been revised by Annecke and Doutt (1961) and the European species by Debauche (1948), the early stages are reviewed by Jackson (1961).

The PROCTOTRUPOID GROUP (Serphoidea)

The old Proctotrupoidea has now been divided by Masner (1956) and Masner and Dessart (1967) into three superfamilies. They are slender insects, mostly of small size and all are parasites. Many attack the eggs of other insects, other species are endoparasites of larvae or pupae, some are hyperparasites and a few are inquilines. The majority of species form a cocoon of a silky or parchment-like nature but in the aphid-infesting genera the pupa is protected by the body of the host. The wings exhibit the greatest diversity of venation and in many forms they are almost veinless, while apterous species are frequent. By some authorities these insects are considered to be closely allied to the Chalcids but Ashmead's view that they are closer to his Vespoids (e.g. Bethyloids) is probably unfounded. The whole group is in need of study at all levels. Few British species have been investigated but a standard work is that by Kieffer (1914–26). The families all share the following characters: Pronotum obtuse at sides below but with a narrow bridge joining the two sides at mesosternal margin. Hind femur without or with very feebly marked trochantellus. Hind wing with no anal lobe. Last visible gastral tergite and sternite apposed in female, ovipositor normally completely concealed, tergite 7 retracted and partly desclerotized, tergite 6 without a spiracle, pygostyles present.

Superfamily **Proctotrupoidea**

Front tibia with one spur. Sides of gaster rounded. Antennae with 11–15 segments, normally inserted near middle of face. Radial cell of fore wing normally closed. Gaster without spiracles or one on tergite 7 which is sclerotized, 8 more or less desclerotized.

FAM. PELECINIDAE. *Large insects. First segment of gaster as long as head and thorax together, gaster in female very long and narrow, segments of equal length, in male clavate. Hind basitarsus much shorter than second segment.* American insects and though their structure is in many ways specialized, they are closely allied to the next family (though Masner does not think so), especially in the structure of the anterior thoracic spiracle which in both groups lies entirely in the prothorax. *Pelecinus polyturator*, which is rather common in temperate N. America, is 50–60 mm long in the female and has been bred from the larvae of *Lachnosterna* (Scarabaeidae).

FAM. PROCTOTRUPIDAE. *Antennae with 13 segments, inserted at middle of face, not on a shelf or prominence. Fore wing with a small or very small enclosed radial cell; hind wing with no enclosed cell.* This relatively small family may be recognized by the sheath (apparently composed of the fused pygostyles) which terminates the female abdomen; in the male, a pair of spine-like processes, the parameres, are often visible. Their larvae appear to be mainly parasitic upon those of Coleoptera (cf. Eastham, 1929). The British species were revised by Nixon (1938).

FAM. DIAPRIIDAE. *Antennae with 11–15 segments, inserted far above the clypeus on a frontal shelf or prominence (except* Ismarus*). Hind wing with one enclosed cell only in the Belytinae.* A family of moderate size, primarily parasites of Diptera. *Diapria conica* attacks *Eristalis tenax* (Sanders, 1911) and *Cinetus* (= *Belyta*) *fulvus* has been bred from *Arachnocampa luminosa. Ismarus*, however, attacks Dryinidae (*Anteon*, Nixon, 1957).

FAM. VANHORNIIDAE. The single North American species is known by its short mandibles with 3 large, outwardly directed teeth, not meeting when closed; gaster with 2 (female) or 4 (male) visible segments, ovipositor long and directed forwards, under the gaster. It has been bred from the larva of an Eucnemid beetle.

FAM. HELORIDAE. *Antennae with 15 segments, not inserted on a prominence. Fore wings with several enclosed cells, radial cell triangular; hind wing with an enclosed cell. Gaster not compressed.* The few but widely distributed species of *Helorus* are parasites of Chrysopidae (Clancy, 1946). *Monomachus*, often put in a separate family, occurs in Australia and S. America and attacks a Stratiomyid fly (Riek, 1970). The **Austroserphidae** with a few Australian species (Riek, 1970) have some resemblance to the Heloridae but the antennal scape is produced dorsally into a long point. The host is not known.

FAM. ROPRONIIDAE. *Antennae with 14 segments, not inserted on a prominence. Fore wing with several enclosed cells, radial cell triangular. Hind wing with no enclosed cell. Gaster compressed.* Three N. American species of unknown host-relationships.

Superfamily **Scelionoidea**

Front tibia with one spur. Sides of gaster with a sharp margin. Antennae with 10–12 segments, sometimes 4–5 fused into a club. Wings with no enclosed cells. Gastral tergites 7–8 membranous in female, more or less exposed and sclerotized in male, spiracles on first 4 tergites.

FAM. SCELIONIDAE. *Antennae with 11 or 12 segments, or 7–8 and an unsegmented club; if with 10, the stigmal vein is present.* The species are numerous and widely distributed and are all parasites of eggs, most often of Lepidoptera, Hemiptera or Orthoptera (s.l.), occasionally of spiders. *Scelio* parasitizes locust egg-pods. The biology of *Telenomus* has been studied by Balduf (1926), Costa Lima (1928) and Jones (1937), and that of *Eumicrosoma* by McColloch and Yuasa (1914–15). Species of the former genus lay in the eggs of Lepidoptera or Heteroptera, of the latter in those of *Blissus leucopterus*. *Asolcus* is an important parasite of the eggs of Pentatomids, including those of the Sun pest, *Eurygaster integriceps* (Javahery, 1968). In *Mantibaria* (= *Rielia*) *manticida*, an exceptionally advanced type of parasitism is presented (Chopard, 1922; Hervé, 1945). Its development takes place in the eggs of *Mantis religiosa* and the adult parasites make their way to the imagines of the host upon whose bodies they settle down. In this situation they cast off their wings and lead an ectoparasitic life. Where the mantis is a female, and has commenced oviposition, the *Mantibaria* migrates to the genital region in order to lay its eggs in the viscid mass of the ootheca while the latter is being formed. Parasites which settle upon male mantids are short-lived and perish along with their hosts. Examples of similar or less obligatory phoresy are recorded in other Scelionids, as *Lepidoscelio* (Channa Basavanna, 1953).

FAM. PLATYGASTERIDAE. *Antennae with 10 or, rarely, fewer segments. Fore wing without marginal or stigmal veins, rarely a submarginal.* This (Fig. 573) forms a very large family whose species mainly parasitize Cecidomyidae (Marchal, 1906). Their eggs are usually laid within those of their host, but the development of the latter is not arrested since the larval parasite does not develop until after the eclosion of the larval Cecidomyid. Such species as *Platygaster herrickii* exhibit polyembryony (Hill and Emery, 1937). The localization of the parasitic larvae in the host is variable. Thus, *Synopeas rhanis* lives free in the body-cavity of the larva of *Dasyneura ulmariae*: *Platygaster minutus* lives in the gut of *Mayetiola destructor* while *Trichacis remulus* forms cysts in the ventral nerve-cord of the same host-species. On the other hand, larvae of *Platygaster dryomyiae* (Figs. 573, 574) and *Inostemma*

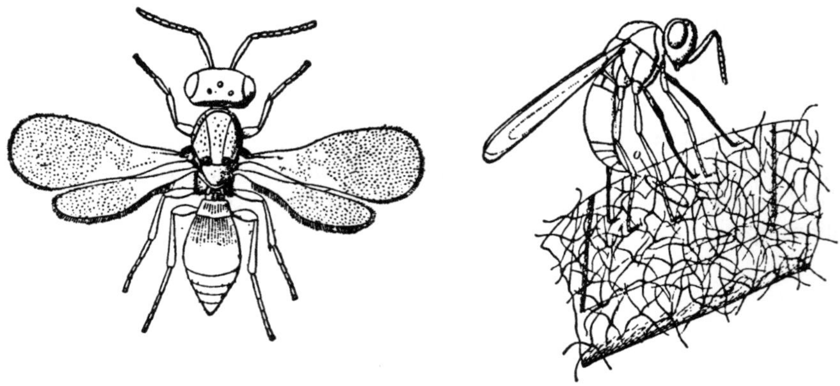

FIG. 573 I, *Platygaster dryomyiae*, female (Platygasteridae); II, the same in the act of oviposition in an egg O, of *Dryomyia* (Cecidomyidae) on a leaf

After Silvestri, *Boll. Lab. zool. Portici*, 11, 1921.

piricola live in cysts in the brain of their hosts. The females of the last-mentioned genus possess on the 1st segment of the gaster a long horn-like growth which curves forwards over the thorax. This peculiar projection lodges the greatly elongate ovipositor with which eggs are inserted within those of *Contarinia pirivora* in the blossom-buds of the pear.

Superfamily **Ceraphronoidea**

Fore tibia with 2 spurs, mid tibia with 1 or 2. Sides of gaster rounded, without spiracles. Antennae inserted very low down, at clypeal margin, with 9–11 segments. Radial cell in fore wing not enclosed. The **Ceraphronidae** are small insects which attack hosts of several orders but are probably most generally secondary parasites of aphids or coccids through various Braconid and Chalcid primaries. The biology of *Lygocerus* has been followed by Haviland (1920), the species observed being ectoparasites of the larvae and pupae of *Aphidius* which lives as an internal parasite of various aphids. There appear to be four larval instars and the first-stage larva is ovoid, with a reduced head and no tail-like appendage. Only two pairs of spiracles are present and these are placed between the 1st and 2nd segments and on the 4th segment respectively: at a later stage seven pairs of spiracles are present.

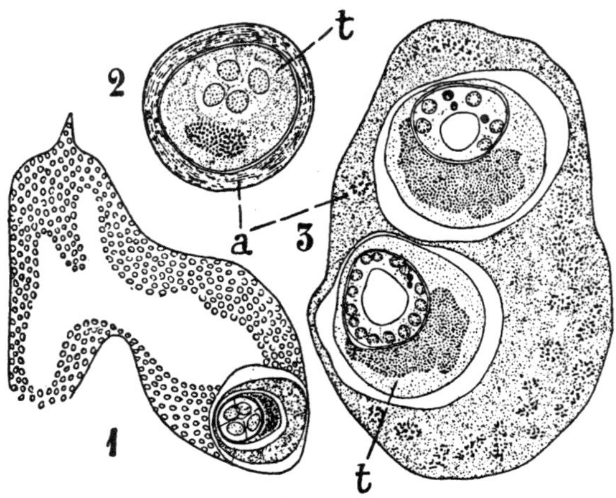

FIG. 574 *Platygaster dryomiae*. 1, Embryo in morula stage in the brain of larva of *Dryomyia* (sagittal section). 2, An embryo with trophamnion *t* and adventitious layer *a*. 3, Two embryos in the blastula stage enclosed in a common adventitious layer

After Silvestri, *loc. cit.*

Superfamily **Bethyloidea**

At first sight this assemblage of families seems very diverse but, except for the Dryinidae which are difficult to place anywhere, there are annectent forms between the extreme types. In particular, the metallic coloured Mesitiinae in the Bethylidae are difficult to separate from the Chrysididae (cf. Reid, 1941). In the whole group, sexual dimorphism in number of antennal segments is absent, in contrast to typical Aculeates. A note on the affinities of the Dryinidae will be found under the Pompiloidea (p. 1245). The British species have partly been described by Richards (1939*a*) and for the Chrysididae the works of Berland and Bernard (1938) and Linsenmaier (1959) should be consulted.

Hind femur with no trochantellus. Hind wing with an anal lobe or if this unclear the integument is metallic. Antennae with 10–13 or with more than 19 segments. Hind wings very rarely with an enclosed cell. Pronotum obtuse below, not closely attached to mesepisternum. Gaster with spiracles on segments 2–6 or (Dryinidae) 7.

FAM. LOBOSCELIDIIDAE. *Antennae with 13 segments, inserted below a ledge. Occiput with a broad flange. Tegula very large. Femora and tibiae with membranous flanges. Fore wing with reduced venation, no pterostigma but usually one enclosed basal cell; anal lobe of hind wing large.* These peculiar insects have been placed in the Cynipoids and in the Diapriidae but Maa and Yoshimoto (1961) seem better advised in placing them in the Bethyloids though they lack the spiracle of tergite 6. There are 19 species, found from Vietnam to New Britain. Their habits are unknown.

FAM. SCLEROGIBBIDAE. *Antennae with more than 19 segments. Female apterous, male winged.* This small group has a wide distribution. All are parasites of Embiids, within whose colonies the females live (Richards, 1939*b*) (see also p. 1231).

FAM. DRYINIDAE. *Antennae with 10 segments, inserted on a prominence near dorsal margin of clypeus. Female with chelate fore tarsus except in* Aphelopinae *and often brachypterous or apterous.* Most species are recognized in the female by the chelate fore tarsi (Fig. 575); females of many species (*Gonatopus*, etc.) are apterous. All are parasitic upon the nymphs of Homoptera and more especially of the families Fulgoridae, Cercopidae, Membracidae and Cicadellidae. Their biology is of exceptional interest and most of what is known thereon will be found in writings of

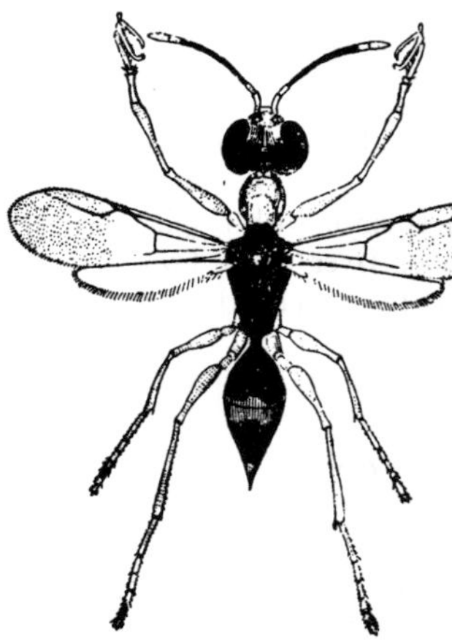

FIG. 575
Echthrodelphax fairchildi, female (Dryinidae)

After Perkins, *Ent. Bull.*, 11 *Hawaiian Sugar Pl. Assn.*

Perkins (1905–07), Fenton (1918), Kornhauser (1919) and Keilin and Thompson (1915). During the larval stages they are endoparasites in the abdomen of the host and, sooner or later, an external gall-like cyst or thylacium containing the parasite is developed as a rule from the moulted larval skins. This cyst protrudes after the manner of a hernia, its position on the host is variable, and one or several may be present on a single individual. In many cases the cyst may be as large as the abdomen of the host, and is usually black or yellow in colour. Pupation takes place either on the foodplant of the Homopteron or in the soil. The effect of the parasitism is often very marked and the changes induced are regarded by Giard as an instance of 'castration parasitaire'. They vary according to the species of insect attacked and, in some cases, they are evident externally owing to the imperfect development of the genitalia of the host. Over 300 species of Dryinidae are known and the greatest number has been described from Europe: in Britain *Aphelopus* frequently parasitizes species of Typhlocybinae. In this genus the cyst is partly formed from a proliferation of the host tissues.

FAM. EMBOLEMIDAE. *Antennae with 10 segments, inserted on a prominence high above clypeus. Female apterous, with simple fore tarsus; head pyriform, narrowed below. Male winged, venation relatively complete.* There are a very few Holarctic species. The N. American *Ampulicomorpha* has been bred from *Epiptera* (Achilidae) (Bridwell, 1958).

FAM. BETHYLIDAE. *Antennae with 11–13 segments, inserted low down on face. Sexual dimorphism slight. Gaster with 7–8 exposed segments. Head prognathous. Pronotum usually parallel-sided. Anal lobe of hind wing conspicuous but apterous species not rare. Colours rarely metallic.* The Bethylidae are a family of mostly small black insects, to some extent forming a link between the Proctotrupoids and the Scolioidea, though nearer the latter group. Reduction of the wings is rather frequent, especially in the females. A few forms, e.g. *Sclerodermus*, are dimorphic in each sex. So far as is known the species parasitize Lepidopterous and Coleopterous larvae and their biology has chiefly been observed by Bridwell (1918; 1919*b*, 1920) and Williams (1919*b*). *Epyris* stings Tenebrionid larvae and lays a single egg on each. *Sclerodermus* utilizes various Coleopterous larvae distributing her eggs over the prey, while *Goniozus* and *Bethylus* attack concealed Lepidopterous larvae. Parthenogenesis has been recorded, the unfertilized eggs producing males. Evans (1964*a*) has published a synopsis of the American species.

FAM. CHRYSIDIDAE (Cuckoo- or Ruby-tailed wasps). *Antennae with 12–13 segments. Sexual dimorphism usually slight. Gaster with 2–6 exposed segments, gaster usually concave beneath. Pronotum short but wide. Head orthognathous. Colours metallic. Most species fully winged.* These insects are usually of a brilliant metallic coloration, generally green, green and ruby, or blue, with a very hard coarsely-sculptured integument. They are easily recognizable by the structure of the gaster, which is peculiar in several ways, and very little longer than the head and thorax. It is convex above, and flat or concave beneath, so that it is capable of being readily turned under the thorax and closely applied to the latter. In this manner the insect rolls itself into a ball when attacked, leaving only the wings projecting. There are, with few exceptions, only three or four segments visible dorsally. The terminal segments in the female are modified to form a retractile tube within which the ovipositor is concealed (Fig. 576). The imagines only fly during hot sunshine, and are usually seen in the neighbourhood of the nests of various solitary bees and wasps within which their transformations take place. The family is very widely distributed and the genus *Chrysis* includes over 1000 species. About 21 species of Chrysididae are British, one of the commonest being *Chrysis ignita*. So far as known all the species are parasites, and the Eumenidae and Megachilidae are especially subject to their attacks. As a rule their larvae prey on those of the host but Chapman (1869) has observed the larva of *Chrysis ignita* feeding upon a caterpillar stored by its host (*Odynerus*). *Chrysis shanghaiensis* is exceptional in that it lays its egg on the mature caterpillar of *Monema* (*Limacodidae*) inside its cocoon (Parker, 1936).

The Cleptinae and Amiseginae (Krombein, 1956, 1960) are more primitive, with more gastral segments exposed, the ventral side not concave and the sculpture often weaker. The larvae of *Cleptes* develops on mature sawfly larvae in their cocoons but little is known of their habits in detail. The Amiseginae, many of which have apterous females, all parasitize the eggs of Phasmids, e.g. *Mesitiopterus* (Milliron, 1950).

Two small families, the **Scolebythidae** (Evans, 1963) and the **Plumariidae** (Evans, 1966*a*), are probably best placed in the Bethyloidea, if only because they have 13 antennal segments in the female, but they also have some marked Scolioid features. They may well represent the remains of the stock from which the two superfamilies were derived. Both are found in S. America and S. Africa or Madagascar and the first-named is associated with burrows in wood.

The **Sclerogibbidae** (see also p. 1230), on the other hand, are always placed near or in the Bethylidae though they are the only Aculeata with more than 13 antennal segments (actually 20–30). The female is apterous with short, stout legs and a fore tarsus with a row of pegs on the posterior side. The male is fully winged and more like an ordinary Bethylid.

Superfamily **Scolioidea**

Head more or less globular or long axis vertical. Antennae with 12 segments in female, 13 in male except in various Formicidae, antennae often elbowed in female. Pronotum often extending back to the tegula, with no conspicuous lobe concealing the spiracle, usually acute below and co-adapted to the mesepisternum. Many females wingless, but the fore wings with a pterostigma, hind wings with at least one enclosed cell and a distinct anal lobe. Legs often short and stout, hind coxae not enlarged, trochantellus not developed. Gaster with spiracles on tergites 1–7. Last visible tergite and sternite apposed, former often with a terminal spike in the male, latter often desclerotized in the female; gaster sometimes with the first two or three segments node-like.

Within this group are found the most primitive members of the true Aculeata. While many of them are parasites, usually of Coleopterous larvae or of other Aculeates, a primitive form of nesting activity is seen in the Scoliids. Some species of *Tiphia* are the only Aculeates in which more than one egg is laid on a single victim. There is a clear affinity with the Bethylidae, some of which also exhibit rudimentary nidification. Sexual dimorphism is usually marked and in the Thynninae is carried to very great lengths. The ants or Formicidae which are now usually placed in this superfamily, are more specialized and in behaviour include some of the highest developments in the Insecta.

FAM. SCOLIIDAE. *Meso- and metasternum together forming a plate overlying the bases of the mid and hind coxae. Male last sternite produced into 3 spines. Female fully winged, wing membrane striolate.* This extensive family includes some of the largest members of the Hymenoptera. They are hairy insects whose prevailing colour is black marked with spots or bands of yellow or red, and the wings are often dark with a metallic iridescence. They are mainly inhabitants of warm countries, and the larvae are ectoparasites of larval Scarabaeidae or much more rarely Curculionidae. The habits of some European species were observed by Fabre who found that the females penetrate the soil in order to discover the larvae upon which they deposit their eggs. Thus *Scolia hirta* (= *bifasciata*) selects those of species of *Cetonia*, *S. flavifrons* is confined to *Oryctes*, and *Elis sexmaculata* (= *interrupta*) chooses *Anoxia*. *Scolia manilae* has been successfully used to control *Anomala orientalis* in Hawaii.

FAM. TIPHIIDAE. *Metasternum not overlying the coxae. Male last sternite rarely 3-spined. Thorax and propodeum never fused dorsally into a single plate. Gastral tergite 2 without lateral 'felt lines'. 1st and 2nd gastral segments separated by a deep constriction or mesosternum with two laminae which overlie the bases of the mid coxae.*

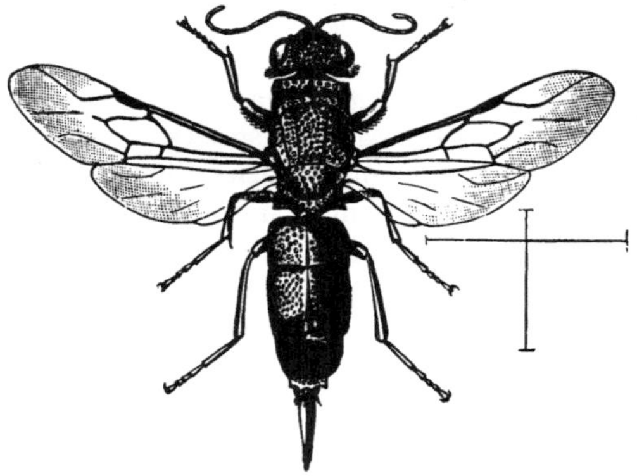

FIG. 576
Chrysis ignita, female. Britain
After Sharp, *Camb. Nat. Hist.*

Females often apterous. This is another large family whose members vary much in structure and habits and perhaps include survivors from the ancestral stock from which several of the higher groups of Aculeates arose. *Tiphia* (Fig. 577) includes two British species and exotic species have been much studied with a view to their use in biological control (Clausen *et al.*, 1932). *Pterombrus* in America and the widespread *Methoca* attack the larvae of Cicindelinae. The latter genus, with one British species, *M. ichneumonides*, has ant-like apterous females and the males are rarely seen (Champion, 1914; Pagden, 1926). *Myrmosa* and its allies also have apterous females but are superficially more like Mutillids. They parasitize various ground-nesting Aculeata.

The Thynninae is a subfamily of some size in S. (one species in N.) America and Australia (Given, 1954). The majority of species lay their eggs on the larvae of Scarabaeidae, but *Diamma* attacks *Gryllotalpa*. The females are apterous and in most species are carried about by the male in a prolonged mating flight, during which the female obtains food from flowers. An illustrated account of the biology of several Chilean species of *Elaphroptera* has been published by Janvier (1933). The family **Sierolomorphidae** includes a few N. American and one Hawaiian species of unknown habits. The lack of an anal lobe in the hind wing has led some authors to place them in a separate family between the Tiphiidae and Mutillidae (Evans, 1961).

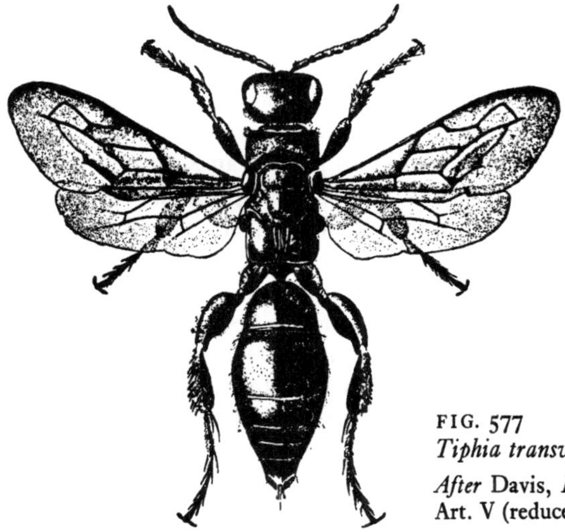

FIG. 577
Tiphia transversa, N. America. Female
After Davis, *Bull. Illin. nat. Hist. Survey*, 13, Art. V (reduced).

FAM. MUTILLIDAE. *Female apterous, male nearly always winged and usually without an anal lobe. Second gastral tergite with lateral felt-lines except in* Rhopalomutilla *in which the dorsum of the female thorax and propodeum is fused into a single plate and the male lacks an anal lobe.* Some thousands of species are placed in this family, most of them with a characteristic common facies. The female is apterous with the thorax and propodeum usually fused into a single plate. The male is nearly always winged and usually rather different structurally from the female. Both sexes are black or reddish in colour, often with spots or bands of silvery pubescence and a generally velvety appearance. The hosts are mostly aculeate Hymenoptera but various other orders are also attacked and one species has been bred from *Glossina* (Mickel, 1928). There are two British species, *Mutilla europaea* a parasite of *Bombus*, and *Smicromyrme rufipes* a parasite of Pompilids and Sphecids.

FAM. SAPYGIDAE. *Both sexes winged. First and second gastral segments not separated by a constriction. Mesosternum without laminae, not overlying the coxae.* The species of this family are parasites of Aculeates, particularly bees. The first-stage larva has big mandibles and destroys the host larva; in later instars the mandibles are smaller and the stored food is devoured (Fabre). *Polochrum* attacks *Xylocopa*, and in Britain, *Sapyga clavicornis* is mainly a parasite of *Chelostoma florisomne* (Megachilidae).

Fedtschenkia which is rather distinct and has been put in a separate family is a parasite of *Pterocheilus* (Eumenidae) (Bohart and Schuster, 1972).

FAM. FORMICIDAE (Ants). *Male and fertile female (queen) nearly always winged though female wings soon shed, workers wingless. 1st or 1st and 2nd segments of gaster scale-like or nodiform and well separated from the part behind. Male gaster without an upturned terminal spine; worker with at least some posterior thoracic sutures and spur of fore tibia not much curved and not externally pectinate. Antennae elbowed but less clearly so in male.*

The ants constitute a single very large family with about 4600 species (Wilson, 1971). They are all social and, with the exception of a few parasitic forms, have a well-differentiated worker caste (Fig. 578). The demarcation between the head, thorax and abdomen is highly accentuated. Myrmecologists usually term the narrow (often scale-like) segment or two segments following the propodeum, the *pedicel* and the more posterior broad segments, the *gaster*; it is more convenient in the Hymenoptera generally to keep the latter term for the whole post-propodeal abdomen and to speak of the differentiation of a petiole of one or two segments. The head varies enormously in shape, and the mandibles present an almost bewildering variety of form, and are subjected to many uses. The labrum is vestigial, the maxillae are composed of the usual sclerites, and their palps are 1- to 6-segmented. The laciniae are membranous and toothless, thereby indicating a soft or liquid diet. In the labium both submentum and prementum are evident, together with a median glossa: at the base of the latter is a pair of small paraglossae beset with rows of setae. The labial palps are 1- to 4-segmented. The antennae are composed of 4 to 13 segments and usually the male has one more segment than the female or worker. Compound eyes and three ocelli are well developed in the males, but in the females, and especially the workers, the eyes are usually reduced or vestigial. The abdomen is the seat of a stridulating organ which consists of an area of extremely fine parallel striae on the mid-dorsal integument of the first broad segment of the gaster (Fig.

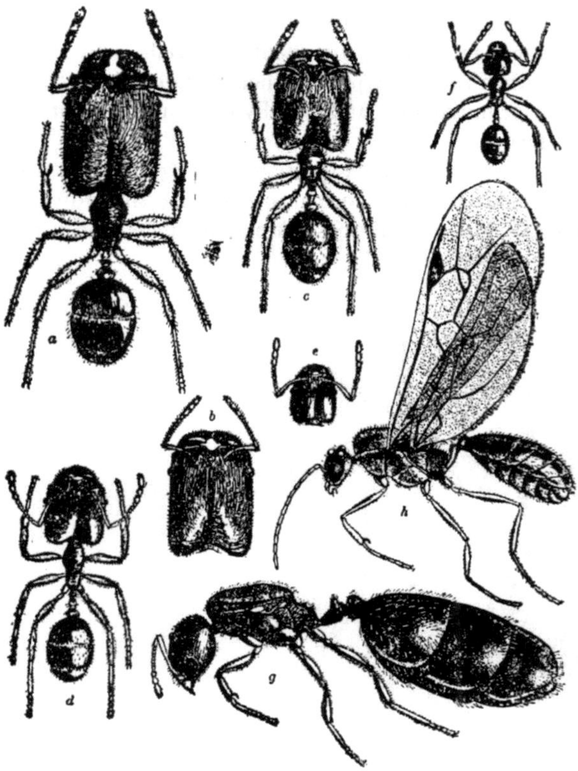

FIG. 578 *Pheidole tepicana* (= *instabilis*)

a, Soldier; *b–e*, intermediate workers; *f*, typical worker (micrergate); *g*, deälated female; *h*, male. *After* Wheeler, *Ants*.

90). The sharp edge of the preceding segment overlaps this area, and is deflexed, so that it may scrape backwards and forwards when the segments are moved on each other, thereby producing a highly pitched sound. A large and well developed sting is present in the females and workers of the subfamilies Ponerinae, Dorylinae and most Myrmicinae, but is vestigial or absent in the remainder.

Ants, as Wheeler observes, have acquired an extensive and uniform experience with all developmental stages of their progeny which they not only feed and clean, but also transport from place to place as conditions may demand. The eggs are small, hardly more than 0·5 mm long even in the largest species, but the popular expression of 'ants' eggs' is often applied to the cocoons or even to the larvae or pupae. The larva (Wheeler and Wheeler, 1970, and earlier papers) consists of a head and 13 trunk segments: eyes are wanting and in a few cases vestigial antennae are present. There are ten pairs of spiracles, situated on the meso- and metathorax and the first eight abdominal segments. The body is almost always invested with hairs which assume many forms and are most abundant in the first instar. In many genera of the Ponerinae there are girdles of large segmentally repeated tubercles. Different species adopt very different methods of nourishing their larvae. Many feed them only on regurgitated liquid while carnivorous species give them portions of other insects, the harvesting ants utilize fragments of seeds, and the fungus growers nourish their larvae with fungus-hyphae. Wheeler states that a cocoon is constantly present in the more primitive ants and equally constantly absent in large groups of highly specialized forms.

Polymorphism attains its highest expression among ants and a large number of types have been recognized, some normal, others, like the individual illustrated in Fig. 579, pathological. The normal phases are male, queen and worker. The male is

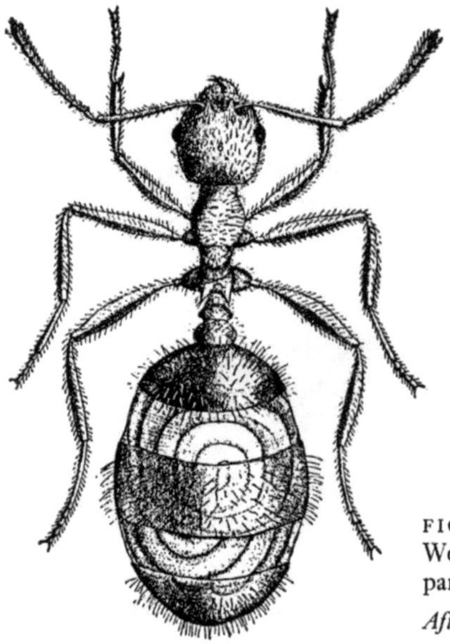

FIG. 579
Worker of *Pheidole dentata* (=*commutata*) parasitized by *Mermis*
After Wheeler.

the least variable of the three castes and retains many scolioid features. The sense-organs, wings and genitalia are highly developed but the mandibles are often weak. The head is smaller and rounder than in the females and workers of the same species, and the antennae longer and more slender. A remarkable type of male occurs in the Dorylinae, characterized by its large peculiar mandibles, long cylindrical abdomen and the specialized genitalia. In a few species (e.g. of *Ponera*), the male is apterous.

The queen is a female characterized by her large stature and well-developed reproductive organs. She is usually larger than the male and worker of the same species and attains a great size in certain forms. The antennae and legs are often shorter and stouter than in the male, the mandibles are well developed and the gaster large. Where the queen is relatively small (e.g. *Formica rufa*), some special method of colony-foundation, such as temporary social parasitism (p.1243), is often adopted.

The worker is a female characterized by the absence of wings, the reduced thorax and small gaster. The eyes are small and the ocelli either absent or minute. A receptaculum seminis is usually wanting and the ovarioles are greatly diminished in number. In some species, no eggs are laid by workers, in others eggs are laid but rarely hatch and are used as food for the young larvae (Brian, 1953). In *Oecophylla*, according to Ledoux (1950) two kinds of egg are laid, small ones giving rise to queens and workers, large ones to males, both without fertilization. Workers are usually variable in size and sometimes in colour or structure. Where they are dimorphic without intermediates, the larger type with large head and mandibles is termed a *soldier*. These are often adapted to particular functions such as fighting, guarding the nest, crushing seeds and other food particles, etc.

A number of other specialized types of worker and queen have been recognized in particular genera, either as the only type or in coexistence with the normal form. There is also a range of types modified by parasitism by *Mermis* or *Orasema* (p. 1223), or by the presence in the nest of certain guest beetles. In *Formica sanguinea*, the influence of the Staphylinid *Lomechusa* leads in the worker to a hypertrophy of the labial glands and a consequent modification of the head and thorax (Novák, 1948). These so-called 'pseudogynes' may develop in the absence of the beetle provided there are pseudogynes in the nest.

When both sexes are winged mating nearly always takes place during what is termed the nuptial flight. As a rule one or other sex predominates in any particular colony and, since the nuptial flight for the colonies of a particular species in the same neighbourhood takes place synchronously, means are thus afforded for intercrossing with individuals of different colonies. Prior to the marriage flight, the workers become much excited and direct the operation, preventing the males and females from leaving the nest until the right time. There is good reason to believe that meteorological conditions exercise an important influence in this matter. There exist in the literature many references to great nuptial swarms of ants which sometimes cloud the air like smoke. On descending to earth the impregnated female sheds her wings, and the deälated individual commences to excavate a small chamber, within which she remains in seclusion until her eggs are mature, and ready to be laid. During the whole of this period, which may extend for months, the chamber is sealed up, and the female draws entirely upon the nutriment afforded by her fat-body and degenerating flight muscles. An exception however is seen in a few of the Myrmeciines and Ponerines in which the queen forages for food. When the first larvae appear, they are fed by the secretion of her salivary glands until they pupate. As soon as the workers are mature they break through the soil and establish a communication between the brood chamber and the outer world. They then go abroad and forage for food, and share it with their parent. The latter is now relieved from the care of the brood, and she limits her activities to egg laying, imbibing liquid food directly from the mouths of the attendant workers. In this capacity, she lives on solely for the purpose of egg-production, sometimes to an age of fifteen years. The number of ants in a fully developed colony appears to vary between wide limits. Yung has made actual counts in the case of *Formica rufa* and found that the numbers vary between about 19 900 and 93 700; in *Formica pratensis* Forel estimated that the largest mound may contain as many as 500 000 individuals, a figure which Yung regards as excessive.

The nests or formicaries present an almost bewildering variety of architecture. Not only has every species its own plan of construction, but this plan may be modified in various ways in adaptation to special local conditions. The Dorylinae can hardly be stated to construct nests, and usually have their abode in some available recess beneath a stone, or log, or they may even temporarily occupy nests of other ants. A large number of ants construct nests in the soil and these consist of a number of more or less irregular excavations, either with or without a definite superstructure around the entrances. The excavations are divided into galleries, or passages of communication, and chambers leading off from the latter. The chambers are used as nurseries for the brood, as granaries for storing seeds, as fungus-gardens and for other purposes. In some species there is nothing to indicate the situation of a subterranean nest, the excavated soil being carried some distance away and scattered irregularly. In others it is heaped up in the vicinity of the entrance, or entrances, to the nests, to form a crater, which varies in size and construction among different ants. Such craters are often difficult to distinguish from mound or hill nests. The latter are usually much larger and are formed not only of excavated soil but also of straws, twigs, pine-needles, leaves, etc., and are perforated with galleries and chambers. Such mound nests are well exhibited in the European *Formica rufa*. Perhaps the largest number of ants' nests is excavated beneath stones or logs. In the tropics many ants take advantage of cavities found in stems, petioles, thorns and bulbs, or even in the spaces enclosed by the overlapping leaves of certain epiphytes. The thorns of Bull's Horn Acacia are inhabited by species of *Pseudomyrmex* and Janzen (1966) has shown the mutual advantage of this symbiosis. Cavities in the bark, the dead wood and stumps of trees are also frequently utilized, and even old deserted Cynipid galls. Suspended nests are of frequent occurrence, hanging from trees in tropical and subtropical forests. These are constructed of earth, carton, or silk and contain anastomosing galleries and chambers. *Oecophylla smaragdina* forms leaf nests, the leaves being fastened together by means of a silken web. The observations of Doflein and others have proved that the silk is provided by the larvae of the species concerned. They are held by the workers in their jaws and used, as it were, as shuttles in weaving the silken tissue of the nest.

Eleven (Wheeler and Wheeler, 1972) subfamilies of the Formicidae are recognized and of these the most primitive are the Ponerinae (Fig. 580) and the Myrmeciinae. The former are characteristic of the tropics and are an important group of ants in Australia. In Britain there are two Ponerines, *P. coarctata* and *Hypoponera punctatissima* in which the males are ergatoid. The nests of the Ponerinae are subterranean and are usually only occupied by a few dozen individuals. The three castes differ very little in size; the workers are monomorphic, they feed their larvae with portions of other insects, and the pupae are enclosed in cocoons. The 'bull-dog' ants (*Myrmecia*) of Australia attain a length of 2–2·5 cm and, as Wheeler remarks, they bite and sting with such ferocity that few observers care to study them at close quarters. A probable ancestral stock of *Myrmecia* is seen in *Sphecomyrma*, an Upper Cretaceous amber fossil (Wilson, Carpenter and Brown, 1967).

The Dorylinae include the driver and legionary ants (Borgmeier, 1958) of the tropics; they are likewise carnivorous, but the workers are blind and highly polymorphic, varying from large soldiers with toothed mandibles, through intermediates, to the smallest workers. The females are very little known and seldom found: they are very large, blind and wingless like the workers. Schneirla (1971) has shown that the queen periodically lays large numbers of eggs and each time this happens her abdomen becomes enormously swollen. The males are likewise very large, with sickle-shaped mandibles, and peculiar genitalia. These insects do not construct permanent nests but merely bivouac in temporary quarters, and wander from place to place in long files. Their sorties are only made on sunless days or at night and are for predatory and migratory purposes. The periodicity of the migrations depends on

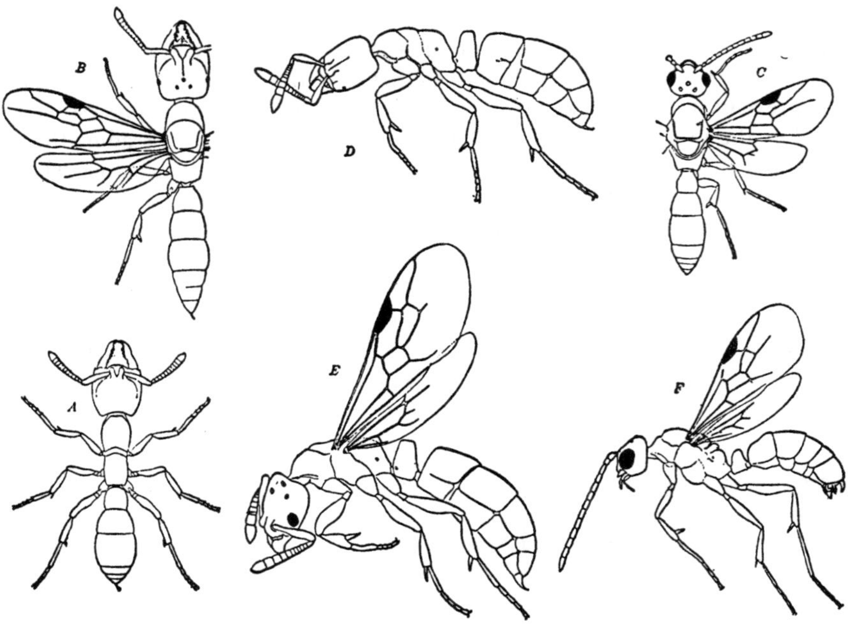

FIG. 580 Ponerine ants. A, B and C, worker, female and male of *Stigmatomma pallipes*; D, E, F, worker, female and male of *Ponera coarctata pennsylvanica*
After Wheeler, *Ants*.

the periodic oviposition and occurs when many larvae require feeding. Belt mentions columns of *Eciton* which he followed for two or three hundred yards without coming to the end. Their prey consists of insects and spiders of various kinds, but Savage records that *Anomma* will succeed in killing large animals if the latter be penned up, and he mentions having lost monkeys, pigs and birds by these insects.

The Myrmeciinae, Ponerinae and Dorylinae, together with the small subfamily Cerapachyinae and many of the lower Myrmicinae are carnivorous ants which represent the savage or hunting stage in the evolution of those insects. The remaining groups of ants have partly abandoned this habit and adopted a vegetarian diet. Wheeler has called attention to the fact that an abundance of food is necessary for the maintenance and fullest development of social life. In warm arid countries insect food is either very scarce, or competition to secure it very keen among ants and other animals. A number of the former have become vegetarians as their last resource in the struggle for existence. Under such circumstances, the seeds of herbaceous plants provide an accessible nutritious food, and the outcome of this is the harvesting habit which is prevalent in many species. The four higher subfamilies, however, have an extremely varied diet, since they not only imbibe the secretions of nectaries, the honey-dew or products of aphids and other Homoptera (Way, 1963), but also feed upon fungi, fruit, seeds and other substances. The harvesting habit appears to have arisen sporadically, and often in distantly related genera, but all of which pertain to the subfamily Myrmicinae. In species of *Messor*, for example, the ants have been observed to gather the seeds both from the ground and from the plants, remove their envelopes, and cast the chaff and empty capsules on the kitchen middens outside the nest. In confirmation of Pliny and Plutarch, Moggridge mentions that the ants bite off the radicle to prevent germination but Schildknecht and Koob (1971) show that they and the Attini secrete a herbicide which prevents the germination of seeds and spores. The latter process is also arrested by the ants bringing the seeds when damp to the surface, spreading them in the sun, and then carrying them back to the special chambers or granaries wherein they are stored. Went, Wheeler and Wheeler (1972) suggest that in some of these ants the adults prepare the seeds but that the larvae digest them and then regurgitate prepared food for their nurses, as occurs in *Vespa* (p. 1247).

The Myrmicine tribe of the Attini, which is peculiar to tropical and subtropical America, are all fungus growers and fungus eaters, and number about 200 species. The fungi are cultivated in special chambers of the nest termed fungus-gardens (Fig. 581) and, according to Moeller (1893), these gardens are practically pure

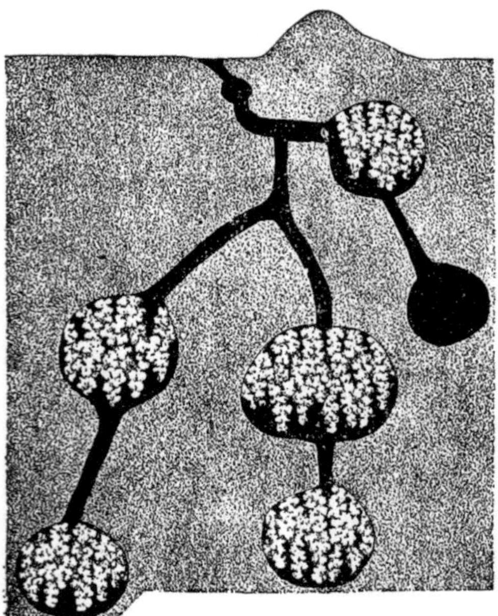

FIG. 581
Diagram of a large nest of *Trachymyrmex septentrionalis* showing, near surface, small original chamber of queen, chambers with pendent fungus-gardens and newly excavated chamber *After* Wheeler, *Ants*.

cultures of the fungi concerned, being assiduously 'weeded' and tended by the ants. Neither free aerial hyphae, nor any form of fruit body develop, but whether this is due to their elimination by the ants, or to environmental conditions, is uncertain. A fungus-garden is a sponge-like mass of comminuted leaf fragments or, in some cases, of insect excrement. The fungi grow rapidly on this substratum and produce numerous swellings or bromatia. The latter form the food of the ants and their larvae and have never been produced in cultures. The systematic position of these fungi is unsettled: several genera have been described which have been referred to the Ascomycetes and Basidiomycetes. The formation of a new fungus-garden is undertaken by the queen who, before departing for the mating flight, fills her infrabuccal pocket (p. 1190) with fungal hyphae. This pellet is expelled within the newly made nest chamber and the growing hyphae are nourished, at first, by the faeces of the insect, who may even sacrifice some of her eggs for the same purposes (see also the papers of Weber, 1937, etc.). The workers cut leaves to provide a substrate for the fungi and are often serious agricultural pests. Tumlinson *et al.* (1971) have shown that a specific pheromone is laid down to define their trail.

The small tropical subfamily Pseudomyrmecinae is notable for its highly specialized larvae. The head in these larvae is surrounded by the hood-like thorax and lies far back on the ventral surface, where it is in contact with the first abdominal segment. Just behind the head there is a pocket, or trophothylax, and food received from the workers is deposited in this pouch, from which it is gradually drawn into the mouth and swallowed. As previously mentioned (p.1176) trophallaxis is highly developed, the larvae supplying their nurses with the secretions of their remarkable exudatoria.

Ants have become associated with a large number of phytophagous insects which possess the habit of excreting liquid of a kind which is exceedingly palatable. In return, the ants render many of such insects certain services, and the relations thus established may be regarded as a kind of symbiosis (Nixon, 1951, Bünzli, 1935). The insects most concerned belong to various families of the Homoptera (Way, 1963), viz. aphids, coccids, Membracidae, Psyllidae, etc., together with the larvae of the Lycaenidae (Hinton, 1951). In the case of many aphids and coccids, for example, they are afforded protection by the ants, who construct tents or shelters for housing them. With aphids the ants frequently betray their sense of ownership by at once carrying them away to safety should the nest be disturbed. This solicitude on the part of the ants may extend to the eggs of the aphids also, and numerous observers have noted ants collecting and storing aphid eggs in autumn and tending the nymphs when they emerge. The latter are carried and placed upon stems or roots which may be situated either within the nest or at some distance outside the latter. Nearly 70 species, representing 29 genera, of Lycaenidae are mentioned as having larvae that are attended by ants, and the relationship in some cases may be exceedingly intimate.

In the subfamilies Formicinae and Dolichoderinae the habit of collecting nectar and honey-dew has become highly developed. The workers of these insects have a pliable integument, which often allows of great distension when their crops are gorged with food. In a few species, the gorging may take place to such an extent that the inflated crop may cause the sclerites of the gaster to be so far forced apart that they appear as islands upon the tense intersegmental membranes: individuals which exhibit this habit are known as *repletes* (Fig. 582), and the species possessing this physiological caste are termed honey ants. In *Prenolepis* all the workers are thus able to distend themselves, and regurgitate the sweet substance which they collect to their larvae or their sister ants. The true or perfect repletes are developed only in the nest, where they remain and store the sweets brought to them by the foragers, thus functioning as living casks or bottles. The contents of the latter are regurgitated when required for feeding the community. Repletes occur among the ants of N. America, S. Africa and Australia but the caste is wanting in the British species.

The relations of ants to aphids and Lycaenid larvae represent only one of the many phases of symbiosis. The others are extremely diversified and the ants are, as a rule, passive or indifferent, and the other insects associated with them are mostly of the nature of inquilines. When the latter regularly inhabit the ants' nests, either throughout life, or during some stage in their development, they are known as myrmecophiles or ant-guests. Our knowledge of these organisms is due to the efforts of many workers, notably Wasmann, Escherich, Janet, Silvestri and others. Was-

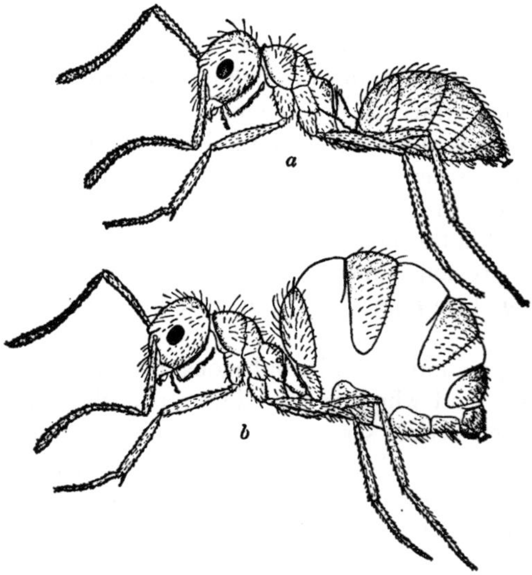

FIG. 582
Prenolepis imparis

a, worker in ordinary condition; *b*, replete. *After* Wheeler, *Ants*.

mann, in 1894, enumerated 1246 species of myrmecophilous Arthropoda, the greater number being insects, and more especially Coleoptera. Since that time many more species have been brought to light, and we are now acquainted with probably over 2000 species, including at least 1200 different Coleoptera. In Britain there are about 300 species, upwards of 70 being Coleoptera (Donisthorpe, 1927*b*). The relations of these myrmecophiles to the ants are extremely diversified, and the following classes are recognized by Wheeler. 1. The *synechthrans*, which live in the nests as scavengers or predators and are treated with marked hostility. They have to elude the ants in order to obtain their food, which usually consists of dead or diseased ants, the brood, or refuse of various kinds. They constitute rather a small group, comprising a number of agile carnivorous Staphylinidae belonging to the genera *Myrmedonia, Quedius, Xantholinus, Myrmoecia, Lamprinus*, etc. The first-mentioned genus is represented by numerous species on all the continents. 2. The *synoeketes*, or indifferently tolerated guests, live in the nests without attracting the notice of the ants, or without arousing any obvious animosity. They are either too small, or too slow of movement, to attract attention, or have no specific odour which differentiates them. Among this large and heterogeneous assembly the most regular synoeketes are the curiously flattened larvae of the Syrphid genus *Microdon*. Verhoeff has observed the fly ovipositing in the nest and it was repeatedly driven away by the ants (*Formica sanguinea*), but kept returning until the eggs were finally laid. In addition to *Microdon*, synoeketes of British ants include Collembola of the genus *Cyphoderus*, larvae of the Chrysomelid beetle *Clytra*, species of *Dinarda*, various Phoridae, etc. A very large number of these guests is associated with the Doryline ants, accompanying the latter from place to place on their wanderings, and some of the Staphylinids, for example, exhibit an extraordinarily close mimetic resemblance to their hosts. The curious Lepismid *Atelura* is common in the nests of various European ants and, according to Janet, its members obtain most of their food by running up and imbibing some of the liquid while it is being regurgitated by one ant to another. The remarkable wingless crickets of the genus *Myrmecophila*, and the diminutive cockroaches of the genus *Attaphila*, lick the ants in order to imbibe the cutaneous secretions of the latter, and often mount the bodies of their hosts in the process. 3. The *symphiles*, or true guests, are species which are amicably treated, licked, fed and even reared by the ants. They are much less numerous than the synoeketes, and consist largely of Coleoptera. Although they belong to many different families of the latter order, they exhibit marked adaptive convergence which is shown in the similarity of coloration, antennal characters, mouthparts and gland structure. These features are developed in order to solicit food from the ants, and to ingratiate themselves by means of special exudations. These true guests are assiduously licked by the ants, and it has long been known that they usually bear tufts of reddish or golden-yellow hairs. The latter are regarded by Wasmann as being the most characteristic organs of the symphiles, and he has shown that they are situated on various regions of the integument, where numerous glands open, and that they have the function of diffusing some aromatic secretion. It is thus evident that the symphiles repay their hosts for their hospitality by secreting a substance which is highly attractive to them. Some of the most remarkable among the ant-guests are the members of the *Lomechusa* group of the Staphylinidae. These insects are tended with the greatest fidelity by the ants, who also rear the *Lomechusa* larvae like their own brood notwithstanding the fact that the guest larvae devour large numbers of both the eggs and young of their hosts. *Atemeles*, a symphile of the same family, spends the winter in nests of *Myrmica* and the summer in those of *Formica* (Hölldobler, 1970). The Paussinae and Clavigeridae, which are remarkable for the bizarre forms assumed by their antennae, also include among their ranks various symphiles. 4. The remaining groups of myrmecophilous insects are *parasites*. The latter include various larval Chalcids such as *Orasema* and other Eucharitid genera, the Phorid *Metopina* and the Gamasid mite *Antennophorus*. The endoparasites include members of all the great groups of parasitic Hymenoptera, the Strepsipteran *Myrmecolax* (p. 1243), several Phoridae and Conopidae, and the Nematode *Mermis*.

So far reference has only been made to the relations of ants to other organisms. There are, however, many instances of social symbiosis between different species of ants. Thus two species of ants belonging to different genera may occupy a compound nest and live amicably together though keeping their broods separate. Other cases have been brought to light by Forel in which small ants nest in close proximity to larger species, and either feed upon the refuse food of the latter, or waylay its workers and compel them to deliver up their booty. True inquiline species are also known which can only live in association with a host of another species and share its nest. Social symbiosis leads us to the condition termed temporary social parasitism. In the latter type of existence the queen seeks adoption in the colony of another species and trusts to the alien workers to rear her first brood of young. The full benefits of this form of parasitism can only be secured by elimination of the queen of the host species. The workers of the latter gradually die out and the nest is ultimately entirely peopled by the parasitic ants. Parasite and host are always members of the same or closely allied genera. From temporary social parasitism the next step is exhibited by slavery. Slave-making ants are confined to the northern hemisphere and are members of four genera only. One of the best known species is the blood-red ant (*Formica sanguinea*) of Europe and N. America, which utilizes as its slave certain other species of its genus, viz. *F. fusca* and its allies. An army of *sanguinea* workers start out and, having found a suitable nest, they do not kill the workers of the slave species unless they should offer resistance, their main object being to capture the pupae and bring the latter back to their nest. It appears probable that a number of the captured pupae is eaten since the number of slaves in a *sanguinea* nest is smaller than the number of cocoons captured. The survivors from the latter emerge and become slaves in the colony of the captors. Wasmann regards this species as a facultative slave-maker, since independent slaveless nests do occur, and there is nothing to show that the slaves are anything more than auxiliary rather than essential workers, in the colony which has adopted them. Obligatory slave-makers or 'amazons' are members of the genus *Polyergus*. The European *P. rufescens* is one of the best known, and its normal slaves belong to the same species as those selected by *sanguinea*. The *Polyergus* never excavates its nest, or cares for its young, and is entirely dependent on the slaves hatched from the worker cocoons pillaged from the alien colonies. The European ant *Anergates atratulus* is a highly specialized social parasite; it possesses no workers, and selects as its host *Tetramorium caespitum*. The *Anergates* queen enters the nest of the latter, and the eggs which she lays gives rise to a progeny which is tended and fed by the host workers. Ants which exhibit this parasitic habit are known to eliminate the queens of their host species which accept the alien substitutes. This mode of life is associated with degeneration; the males of *Anergates*, for example, are sluggish, wingless worker-like individuals, and even more or less pupa-like; the females are also modifed and have rather poorly developed eyes. It is extraordinary that in a small area of the Austrian-Swiss border, one host-ant, *Leptothorax acervorum*, has 4 social parasites, *Harpagoxenus, Doronomyrmex* and two species of *Leptothorax* (Buschinger, 1971).

Ants are frequently a great nuisance to man and many species are serious agricultural pests in the tropics owing to their cultivation of coccids. One of the best known of the noxious species is the Argentine ant (*Iridomyrmex humilis*) which has overrun the warmer parts of the United States and become a serious household pest.

The literature on ants is voluminous but much of what is known concerning these insects will be found in the works of Wheeler (1910, 1922, 1923) which are accompanied by very full bibliographies. The two masters of European myrmecology are Forel (see especially 1921–23) and Emery (1910–25), a mere list of whose writings would occupy several pages. Almost equally numerous are the various papers of Wasmann, particularly with reference to myrmecophilous insects. British ants number 42 species and these are mostly dealt with by Donisthorpe (1927*a*): Wilson (1971) provides an excellent summary.

Superfamily **Pompiloidea**

Pronotum produced back to tegulae, forming a lobe over the anterior thoracic spiracle, obtuse below, loosely overlapping the mesopleuron, which has an oblique groove down from the postero-dorsal corner. Legs long, especially the hind coxa and femur, no trochantellus. Fore wing not longitudinally folded, hind wing with an anal lobe and one or more enclosed cells; a few brachypterous species. Gaster with first segment not scale-like or nodiform, segments 2–7 with spiracles, tergite 7 retracted and partly desclerotized, ovipositor largely hidden.

This is a large and in most respects very homogeneous group. Several thousands of species are already described but this is clearly only a fraction of those awaiting discovery, especially in the tropics. The Pompilids all prey on spiders, usually storing them in simple subterranean nests or, less commonly, building mud-cells. A few have become parasitic on other Pompilids and the Ceropalinae, all of which have this peculiarity, have diverged rather widely from the remainder. In 1941, Reid showed that the anomalous genus *Olixon* is structurally very close to certain brachypterous Pompilids described by Arnold from Africa. Later specimens of the genus *Harpagocryptus* have become available and these are closely allied to *Olixon*. Perkins bred *Harpagocryptus*, now synonymized with *Olixon*, from external sac-like structures on the body of an Australian *Trigonidium* (Gryllidae). Very recently Krombein has suggested that the anomalous *Rhopalosoma* (p. 1246) also has affinities with this group on the grounds of resemblances in the male genitalia. In other respects, however, it is widely different. Finally, the Dryinidae (p. 1230) which developed in sacs on Homoptera might conceivably have some affinity with this superfamily; again there are certain resemblances in the male genitalia but little resemblance in any other part of the body.

FAM. POMPILIDAE (PSAMMOCHARIDAE). The Pompilidae are distributed over almost the whole world; fourteen genera, including about 38 species, are found in the British Isles. In these insects the abdomen is devoid of a definite petiole, the hind pair of legs is very long and the males are more slenderly built and usually smaller than the females (Fig. 583). All are fossorial and predatory

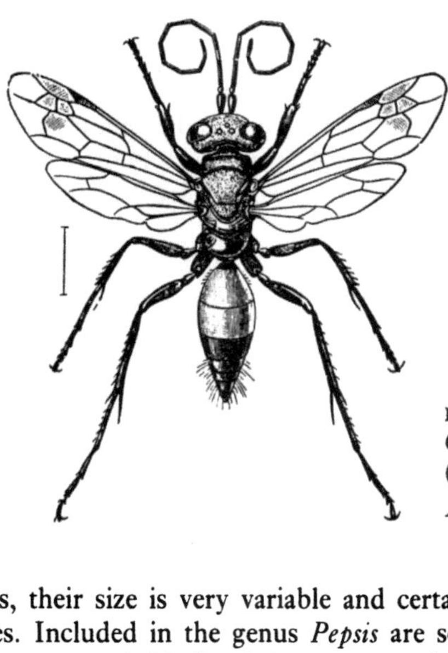

FIG. 583
Calicurgus fasciatellus, female, Britain (Pompilidae)
After Sharp, *Camb. Nat. Hist.*

wasps, their size is very variable and certain species may attain a length of three inches. Included in the genus *Pepsis* are some of the largest of all Hymenoptera. They are remarkable for their extreme activity and possess great powers of running. The nests of these insects are usually burrows in the ground, but *Auplopus* (= *Pseudagenia*) constructs earthen vase-like receptacles which are attached to walls or stones. Their prey consists of spiders and some of them, by means of their highly developed stinging powers, are able to overcome even the largest of these Arachnids.

One of the giants of the family is *Pepsis formosa* which stores its burrows with the great Tarantula spiders (Theraphosidae). The habits of the Pompilidae have been observed by Fabre (1891), the Peckhams (1898), Ferton (1901–21), Williams (1919*a*) and others (for a summary of British species, see Richards and Hamm, 1939; for a wider survey, see Iwata, 1942; Evans, 1953). Evans (1959*a*) deals with the larvae. There appears to be a good deal of variation in the degree of perfection in the art of stinging among different species. Fabre states that *Calicurgus*, correctly *Cryptocheilus*, first stings its prey between the poison fangs of the latter, and subsequently near the junction of the cephalothorax and abdomen, thereby producing complete immobility. The observations of the Peckhams on various Pompilids indicate that stinging is often a much less refined process: in some cases the spider is stung in such a way that it is killed outright, while in others it may live for 40 or more days, but in all cases it is reduced to a sufficiently helpless condition to afford a safe repository for the egg of the wasp.

FAM. RHOPALOSOMATIDAE. A few species of nocturnal habit, superficially somewhat like the Ichneumonid genus *Ophion*, found in all except the Australian regions. The Nearctic species of *Rhopalosoma* are external parasites of crickets (Gurney, 1953) and no nest is made. The affinities of the group have always been a puzzle but they seem to fall either here or among the Scolioids.

Superfamily Vespoidea

Eyes usually emarginate on inner side. Antennae with 11–13 segments. Pronotum produced back to the tegulae. Fore wing usually longitudinally plaited, with first discoidal cell very long; hind wing with an anal cell and enclosed cells. No trochantellus. Gaster with spiracles on segments 1–7, last visible tergite and sternite apposed, segment 1 not scale-like, tergite 7 of female retracted and desclerotized. Larvae with spinneret opening slit-like, maxilla with 2 papillae, mandibles tridentate in lower forms, bi- or unidentate in higher ones.

The wasps are now usually placed in three closely allied families. Exceptions to the characters of the superfamily are especially found in the very specialized Masaridae. The solitary Eumenidae form a very large family, found in all regions. The social Vespidae (recent reviews; Richards, 1971; Spradbery, 1973) are placed in three subfamilies and most species are tropical or subtropical. H. de Saussures' monograph (1852–58) is still valuable.

FAM. EUMENIDAE. *Claws bifid. Inner side of tegula usually emarginate to receive a process of the mesoscutum. Fore wings nearly always with 3 submarginal cells.* This family includes most of the solitary wasps of the temperate regions. They exhibit many variations in nest-building habits: certain species dig tunnels in the ground, and others construct tubular nests in wood or stems, partitioning the tunnels into cells divided by mud-walls. There is, furthermore, a number of species which are mason or potter wasps, constructing oval or globular vase-like nests of mud or clay, fastened to twigs and other objects. The latter types are often of the daintiest description and are said to have served as models for early Indian pottery.

The species of *Odynerus* (*s.l.*) construct varied kinds of nests, while some regularly take advantage of a deserted nest of another wasp, or of a nail-hole or key-hole, rather than build cells of their own. All the species of the family are predacious upon small Lepidopterous larvae, or more rarely, upon those of the Tenthredinidae and Chrysomelidae. The prey when captured is stated by Fabre to be stung into insensibility and a dozen or more larvae may be stored in a single cell. The wasp deposits each egg by means of a suspensory filament from the roof of the cell where it hangs in close proximity to the food thus collected and, after the chamber is sealed, the parent betrays no further care for its offspring. The group is a large one well represented in most regions of the globe and its habits are discussed by Roubaud (1916), Williams (1919), Iwata (1942) and others. Six genera occur in Britain, including *Eumenes* and *Odynerus* (*s.l.*). In the former the first abdominal segment is very long, and narrowed into a petiole (Fig. 584) while in *Odynerus* the petiole is scarcely evident.

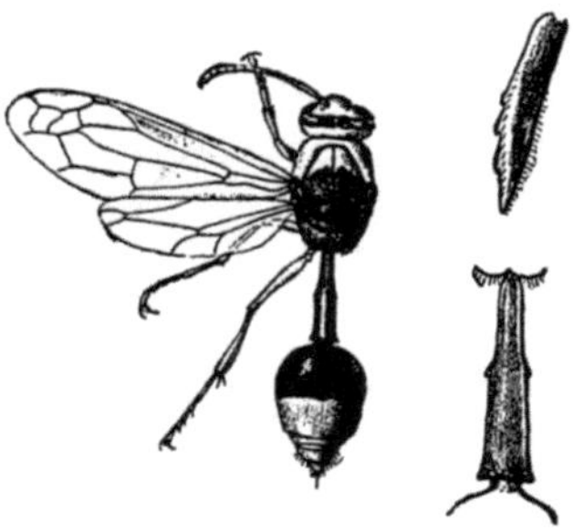

FIG. 584
Eumenes petiolatus (Eumenidae). Female × $\frac{5}{6}$. India
After Bingham (F.B.I.).

FAM. VESPIDAE. *Claws simple or, in the Stenogastrinae, toothed but they have elongate mandibles lying beside the lengthened clypeus; mandibles otherwise short and broad. Tegula not emarginate.* The Vespidae are Asiatic (to New Guinea) and Holarctic and are all fully social wasps except some Stenogastrinae but the position of that group is disputed. Most of them build paper nests but a few partly employ mud and others are parasitic on their allies. They live in large communities each composed of a fertilized female or 'queen', workers and males. In cool climates, the colonies exist for a single season only, the males and workers perishing during autumn, while the impregnated females hibernate and each founds a new colony the next spring. The three forms of individuals are generally alike in coloration, but the queens are considerably larger than the workers and males: the males may be readily distinguished by having seven evident gastral segments and thirteen segments to the antennae, whereas only six gastral segments and twelve antennal segments are found in the queens and workers. Vespinae are largely predacious in habit and feed their larvae upon other insects, portions of which they previously masticate: both fresh and decaying meat and fish are also utilized. The adult wasps are very partial to nectar, ripe fruits and honeydew and this same diet is given to the very young larvae for a short period. Their mouthparts have not attained the length and perfection found among bees, and hence wasps are unable to obtain the secretions of deeply seated nectaries. Although, at times, they cause injury to fruit they render service as scavengers and in reducing the numbers of other insects, more especially Diptera and Lepidopterous larvae. According to Ishay and Ikan (1968), the adults of *Vespa orientalis* lack the enzymes for digesting proteins. These they feed to the larvae which after breaking them down, give back a part of the amino acids in trophallaxis (cf. p. 1176). The British species of the family belong to the genus *Vespa* (*V. crabro*, the hornet) and *Vespula*, and our knowledge of these insects has been greatly extended by the researches of Janet (1893, etc.), Marchal (1896), Roubaud (1916), Bequaert (1918) and Weyrauch (1936, etc.), Duncan (1939) and Spradbery (1973). Great variety of nest construction is found in this genus and the British forms exhibit three distinct types of nidification. Thus *Vespula vulgaris*, *germanica*

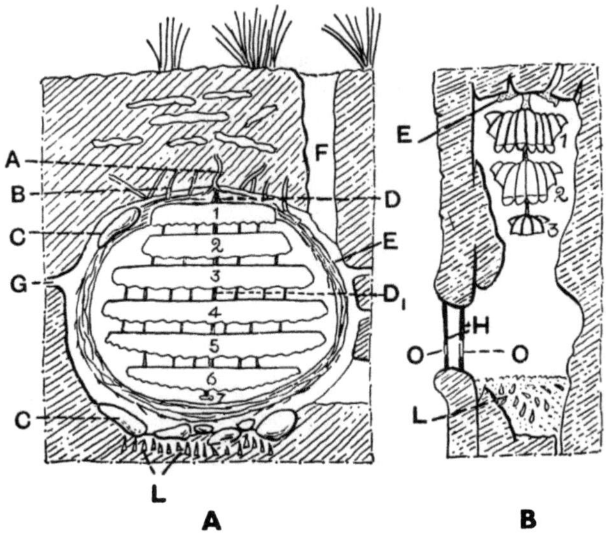

FIG. 585 A, section of subterranean nest of *Vespula germanica*; B, section of nest of *Vespa crabro* in a tree hollow

A, root to which first attachment *D* was made; *B*, secondary attachment; *C*, pieces of flint; D_1, suspensory pillar; *E*, envelope, in B its vestiges; *F*, entrance; *G*, side gallery; *H*, lamellae closing opening to tree hollow; *O*, entrance orifices in lamellae; *L*, saprophagous dipterous larvae. The numerals refer to layers of comb in order of construction. Adapted from Janet.

and *rufa* make underground nest; *V. norvegica* suspends its nests in bushes, etc., while *Vespa crabro* usually nests in hollow trees (Fig. 585B). *Vespula austriaca*, sometimes put in a third genus, is a parasite of *V. rufa* and lacks the worker caste. The stings of these insects are always painful and many of the tropical species are very fierce and easily roused, their stings sometimes involving dangerous consequences to animals and human beings. One of the largest species of the genus is the Himalayan *Vespa ducalis* whose queens attain a length of 40 mm with a wing-expanse of over 80 mm.

After hibernation the female wasps are roused into activity by the warmth of early spring, and commence to seek out likely situations for their nests. Having discovered suitable places they proceed to gather the material for nest-construction. This consists of weather-worn but sound wood, particles of which are rasped off by means of the mandibles, and worked up with the aid of saliva to form a substance known as 'wasp-paper'. In the case of *V. germanica* and *vulgaris*, layers of the substance are applied to the roof of the cavity in the ground destined to hold the nest. From the centre of the disc thus formed, a pedicel is hung with its lower end widened out. Upon the latter the first cells, up to about thirty in number, are constructed: they are hexagonal in form, open below and closed above. An umbrella-like covering is suspended from the roof of the cavity to protect the comb and, in the angle of each cell nearest the centre of the comb, an egg is deposited, being fixed by means of a cement-like substance. In a few days, according to temperature, the larvae hatch and are fed by the parent until ready to pupate. Prior to transforming, the larva spins a cocoon within the cell and closes the mouth of the latter with a tough floor of silk. The contents of the gut are now evacuated for the first time, and transformation into the pupa takes place. After a period of four to six weeks from the time of egg-laying, the adult wasps bite their way through the floors of their cells and emerge. These individuals are always workers, and very soon the entire care of the young and the nest-building is taken over by them, the parent female devoting herself solely to egg-laying. When the nest is fully formed (Fig. 585) it is more or less spherical in form externally, and is invested by several layers of coverings which protect it from rain and also serve to prevent loss of heat from within. New cells are added at the periphery of those already formed and, when one layer of comb has attained a suitable size, new tiers or layers are built below and interconnected with vertical pillars. This goes on until about seven or more combs are constructed and the spaces between the several combs, and between these and the innermost covering of the nest, are just sufficient to allow of the free movement of the occupants of the nest. Each cell of the comb is used for rearing the brood twice or perhaps three times, and it will therefore be seen that the number of cells does not accurately represent the total population of a colony. Janet found in a nest containing seven combs 11 500 cells of which 11 000 had been used twice and the remainder thrice. An average-sized nest probably has a population of about 5000 individuals towards the end of the season. Near the end of summer larger cells are constructed and these 'royal' cells are destined for the females or 'queens' of the next generation. The fertilized eggs produce either females or workers, but the mechanism of caste-determination is unknown. The workers, though probably never fertilized, usually lay some eggs and in the absence of the queen large numbers. Males are always produced from unfertilized eggs whether the latter be laid by the female or workers.

In addition to their normal occupants a large number of other insects has been observed in wasps' nests, either as parasites or inquilines. In the soil beneath the nest, which contains excreta and other organic matter, larvae of *Achanthiptera rohrelliformis* (= *inanis*) are often abundant. Larvae of *Volucella inanis, zonaria* and *pellucens* appear to act as scavengers, devouring excreta, etc., but sometimes also the occupants of the cells: among true parasites the most important is *Metoecus paradoxus* which destroys the larvae. Newstead (1891) has recorded a number of other insects and several species of Acari from nests of *V. germanica* and *V. vulgaris*. *Polistes* (subfamily Polistinae) is the only other type of European social wasp. The colonies are much smaller than those of *Vespula*, and each nest is composed of a single tier of cells suspended by means of a central pedicel without any external envelope. The cycle is essentially like that of *Vespula* and a few species have also become parasites (Weyrauch, 1937). The genus is very large and found in all regions. Of the extra-European social groups the Stenogastrinae, of the Orient and Australia, are the most primitive. Some species are solitary and others social: the latter make fragile naked combs, sometimes using mud, the colony comprises but few individuals, and the larvae are fed by progressive provisioning. The Polybiini are the largest social group: they are tropical and largely S. American (Richards and Richards, 1951). In species of *Belonogaster* in Africa the colonies are small and there is little differentiation of caste, the older females being the egg-layers and the younger individuals are foragers: new colonies are provided for by rudimentary swarming. In other genera true queens and workers can usually be recognized: the colonies are often immensely populous and pleometrotic (with more than one egg-laying queen), while swarming is a regular phase in their life. In *Brachygastra*, nectar forms part of the nest provisions and may be stored. The Ropalidiini are a small group found in the tropics of the Old World: they construct combs (usually naked) and there is little or no distinction into queens and workers.

FAM. MASARIDAE. *Only two submarginal cells (except* Euparagia). *Claws bifid, toothed or simple. Antennae usually clubbed. Eyes often hardly emarginate. Nests built of mud or in burrows in the ground and, except in* Euparagia *stored with nectar and pollen which the female carries in her crop. Mouthparts often very elongate and retractile.* There are about 150 species found in the warm dry regions of all continents except N. Zealand. *Euparagia* (S. U.S.A., Mexico) nests in the ground and stores weevil larvae. The other species are all vegetarian, often rather oligolectic to such flowers as *Heliotropium* or *Penstemon*. In specialized genera the mouthparts are retracted into a sac which protrudes between the fore coxae (Richards, 1962).

Superfamily **Sphecoidea**

Antennae with 13 segments in male, 12 in female. Hind wing with an anal lobe and enclosed cells. Pronotum not produced back as far as the tegulae, with a lobe concealing the anterior spiracle. Trochantellus of hind leg very weak or (usually) not developed, hind basitarsus not broadened. Gaster with spiracles on segments 1–7, tergite 7 partly desclerotized but complete, last visible tergite and sternite apposed, concealing the ovipositor. Pubescence not branched or pectinate. Larva with spinneret openings usually paired, maxillae with two papillae, mandibles usually with more than two teeth.

This superfamily is composed of solitary (except *Microstigmus*) wasps, mostly fossorial but occasionally constructing free mud-cells. Much more rarely resin is used as a building material. They are predacious and store their nests with Lepidopterous larvae, Hemiptera, Orthoptera, Arachnida, etc., or very rarely are parasitic on their allies. Parental care for their larvae occurs in species of *Bembix* and *Ammophila* but in most genera, once the cells have been provisioned, and an egg deposited in each, they are sealed down and the parent exhibits no further concern for her offspring. The prey is probably always stung and the result in most cases is to induce rapid paralysis of the motor centres, thereby eliminating all or almost all power of movement. The assertion that the prey is stung in the ganglionic nerve centres is not often demonstrated, but is an inference drawn from the effects of stinging, and the positions in which the sting is inserted into the bodies of the victims. In a number of cases the prey is stated to be killed outright, but it retains its fresh condition for a variable period up to several weeks, a fact which suggests that the venom may have an antiseptic influence. Many interesting observations on the habits and instincts of the European species of the group are detailed in the writings of Fabre, Ferton, Adlerz and Olberg (1959); a number of American species have been studied by G. W. and E. G. Peckham, Rau (1918) and Evans (1966*b*) and of S. American ones by Janvier.

Opinions have differed as to whether the group should be treated as a single family or should be divided into a considerable number. There is little doubt that at the moment such division is impracticable if the fauna of the whole world is considered. Even the Ampulicinae which are unusually distinct seem to be only a specialized offshoot from the Sphecinae and show affinities with such genera as *Podium*. Evans (1956–59, 1959*c*, 1964*b*) has described the larvae. In the last quoted paper he considers that the larvae of the subfamilies Larrinae, Trypoxylinae and Crabroninae are so alike that they might well be united, but the assemblage of adults which would be so produced is very diverse and impossible to define concisely.

FAM. SPHECIDAE. The subfamily Astatinae includes a small number of species in most of the main regions; they nest in the soil and prey on immature Heteroptera. For the European species, see Verhoeff (1951). The Larrinae form a large, very widely distributed subfamily, most of whose species make small nests in the earth though in the large genus *Motes* (*Notogonia*) mud-cells are affixed to stones, etc. The majority of species prey on Orthoptera or Dictyoptera but *Miscophus* stores small spiders and some genera take Hemiptera. The subfamily Trypoxylinae is mainly made up of the two large genera *Trypoxylon* and *Pison*. Both build with mud – *Trypoxylon* sometimes subdividing natural crevices or, like *Pison*, building mud-cells on tree-trunks, leaves, etc. The prey is always Arachnida of which a considerable number is stored in each cell. There are three British species of *Trypoxylon* but the genus is universally distributed and there are more than 150 species in S. America. The principal genera of the Crabroninae are *Crabro* (*s.l.*) and *Oxybelus*. The former has now been split into many genera with very varied habits (Leclercq, 1954). They construct burrows either in the soil, or in rotten wood or in plant stems. The most usual prey is Diptera, but some of the species attack several other orders, such as Hymenoptera, adult Lepidoptera, Coleoptera or Ephemeroptera. Kohl's monograph of the European species (1915) is a classic and the biology of the British species, which number forty, has been summarized by Hamm and Richards (1926). *Oxybelus* is another large and widespread genus with characteristic spines on the postscutellum. The nest is a burrow in the soil and the prey are Diptera which are often brought in impaled upon the sting. There are three British species of which *O. uniglumis* is abundant. The Pemphredoninae include a number of small black wasps, most of which prey on Homoptera though some *Spilomena* take Thysanopterous nymphs. Most of the nests are in wood or in hollow stems, only a few burrow in the soil. The S. American *Microstigmus* (Myers, 1934) is unique in making a nest of plant wool suspended on a long thread and in storing Collembola. Some species lead a simple form of social life (Matthews, 1968), *Psen* and its allies have longer petioles and prey on various Homoptera. There are about 30 British species of the subfamily. The extensive subfamily Sphecinae includes large but usually slender insects with the propodeum and often the petiole elongate

(Fig. 586). The legs are adapted for digging and running and their methods of stinging are highly specialized. The best known genera are *Sphex* and *Ammophila*, which, as the Peckhams remark, include some of the most graceful and attractive of all wasps – not only on account of their form but also owing to their intelligence and individuality. The above-mentioned observers, and also Fabre, have studied their habits in detail and the records of their observations form some of the most remarkable chapters in insect biology. Stated very briefly, the prey consists respectively of Orthoptera and Lepidopterous larvae which are stored in a single cell situated at the termination of a vertical tunnel in the ground. The method adopted by these insects in stinging their prey is the most complex known and has been observed by Fabre in

FIG. 586
Ammophila sabulosa, female. × 2. Britain (Sphecidae)

the case of *A.* (*Podalonia*) *hirsuta* and by the Peckhams in *A. urnaria*. It is a multiple process but there is some variation with regard to the number of stings administered. In one instance Fabre mentions that stinging took place at twelve different points, beginning between the 1st and 2nd segments and progressing backwards. In his second example, the 3rd, 2nd and 1st segments were stung in the order given and thereafter the remaining segments up to the 9th. In other cases he noted that usually all the segments were stung. After stinging had been accomplished the prey, in some instances, was subjected to a further process known as malaxation, which consists in repeatedly compressing the neck of the victim with the mandibles. The Peckhams' observations largely confirm those of Fabre with the exception that the middle segments of the prey, upon which the egg is deposited, were never touched, while in Fabre's observations they invariably were. They also noted that malaxation was most severe in the case of a caterpillar which was only stung once. It is evident from the various observations which have been recorded that the order in which the segments are stung, the number stung, and the subsequent malaxation which may occur are all somewhat inconstant. The poison introduced during stinging either paralyses or kills the prey and also acts as an antiseptic, keeping the tissues fresh for many weeks. Some species, e.g. *A. aberti*, use a small stone as a tool in tamping down the closure of the burrow (Evans, 1959*b*). In 1941 Baerends published a remarkable account of *Ammophila pubescens*, well known in Britain under the name *A. campestris*. This species practises progressive provisioning and maintains two or three nests at once, each in a different state of development as does *A. azteca* (Evans, 1965). *Sceliphron* includes the 'mud-daubers' whose nests are constructed of kneaded mud or clay and are composed of about 10 to 50 cells. These insects occur in most of the warmer regions of the globe and are very fond of building their nests in human habitations. Their prey consists of spiders and it appears to be a matter of indifference whether the latter be killed or only paralysed and either event may follow as the result of being stung. An examination made by the Peckhams of cells recently provisioned show that while most of the spiders were dead, many clearly exhibited indications of being still alive. The latter died off from day to day and the dead Arachnids remained in good condition for a period of ten or twelve days (cf. also Shafer, 1949).

The Podiini are mostly South American and nearly always prey on cockroaches. *Trigonopsis* constructs free mud-cells attached to the trunks or leaves. The Ampulicinae are rare in individuals and few in species. The prothorax is narrow and elongate and the base of the abdomen is constricted to form a short petiole. So far as known they are predacious upon Blattaria and Bingham mentions that in Burma they enter houses and search for their prey in likely situations. They do not form definite nests and, after having stung their prey into submission, the latter are dragged away and stored in any suitable hole or crevice (Williams, 1919*a*). The family ranges into both hemispheres but is unrepresented in Britain. *Dolichurus* and *Ampulex* occur in France.

The subfamily Nyssoninae is large and very varied in habits and structure. The genus *Mellinus* includes a small number of species which prey on Diptera. The nest is a shallow burrow in the soil and *M. arvensis* is abundant in Britain in the late summer. *Alysson* is a small genus of which the species are rarely common. As far as is known, they nest in the soil and store up Homoptera. The genus *Nysson*, known by the projecting teeth on the propodeum, includes species which are parasites of allied genera. They enter the burrows of *Gorytes* (*s.l.*) and, without destroying the hosts' egg, deposit one of their own between the abdomen and the folded wings of one of the Homopterous prey (Reinhard, 1929). *Gorytes* and its allies nest in the soil and as far as is known prey on Homoptera. *Argogorytes mystaceus*, a common British species, takes the nymphs of *Aphrophora spumaria*. *Sphecius* includes some very large species which prey upon Cicadas. *Stizus* preys on Orthoptera and the allied *Stizoides unicinctus* has become a parasite of *Sphex atratus* which has similar prey. *Bembix* and its allies are known by their elongate labrum and mouthparts. The genus itself is of world-wide distribution but there are numerous allied genera in the south-western United States. The species of *Bembix* are gregarious, a number of individuals occupying a limited area of ground but each one has a separate nest. Thus Wesenberg-Lund (1891) states that fifty *Bembix* will occupy an area about equal to that of an ordinary room. *Bembix* differs from almost all other solitary wasps in that the cells containing its larvae are left unsealed and the latter are fed from day to day. The difference in maternal care entails very great industry on the part of the parent wasps and results in a much less numerous progeny. The prey consists of Diptera and among the genera recorded as serving this purpose are such relatively large forms as *Echinomyia*, *Eristalis* and *Tabanus*. In *Bembix rostrata* a single female supports five or six larvae and each of the latter requires 50 to 80 flies during the fourteen or fifteen days spent in that stage. Parker (1917) discusses the biology of the family and believes that the parent wasps find their skilfully concealed burrows by olfactory sense. He mentions several instances in which the surface of the burrow was disturbed, and even water was poured over it, without causing the wasp to lose track of its nest. In addition to this work and that of Marchal the reader should also consult the writings of Fabre and G. W. and E. G. Peckham and the monographs by Nielsen (1945), Evans (1957) and Evans and Matthews (1973). The

Philanthinae include two tribes which may perhaps be less closely allied than the present classification indicates. *Philanthus* includes a large number of species which burrow in the ground and store bees. The victim is stung on or near the under surface of the mentum and death rapidly supervenes. *Philanthus triangulum* is the 'Bienenwolf' of the Germans, and is occasionally a serious enemy of the hive bee. According to Fabre, after the bee has been stung it is subjected to vigorous malaxation for the purpose of forcing out the contained honey. The latter is imbibed by the captor and its extraction is stated to be necessary before the bee can be safely used as food by the larval *Philanthus*. The species has also been the subject of a valuable study of orientation by Tinbergen (1932). In Britain, the species is established in the Isle of Wight but becomes temporarily established elsewhere after immigration from the continent. *Cerceris* (Fig. 587) is the largest genus of Sphecoids (Arnold) and includes

FIG. 587
Cerceris arenaria, female. × 2·5. Britain (Sphecidae)

more than 600 species. Like *Philanthus*, they make deep burrows in the ground and some of them prey on solitary bees in a way very similar to that genus. The majority, however, catch Coleoptera, especially weevils. The allied genus *Aphilanthops* stores queen ants (Wheeler, 1913).

Superfamily **Apoidea**

Characters of the Sphecoidea but pubescence, at least in part, e.g. near the anterior thoracic spiracle, plumose or branched. Hind basitarsus more or less broadened and often densely pubescent. Gastral tergite 7 in female complete but divided. Larva with spinneret opening slit-like, maxilla with 1 papilla, mandibles uni- or bidentate.

Included in this superfamily are the social and solitary bees. The truly social species, which have evolved a worker caste, are confined to the families Halictidae, Xylocopidae and Apidae, the great majority of forms being solitary. The adults are most important agents for pollinating flowers, the pollen adhering to the plumose body-hairs. The glossa is always well developed, generally pointed, and often exceedingly long. The food consists of nectar and pollen, the former supplying the carbohydrate ingredients and the latter the protein. The larvae are fed upon a similar diet, except that the nectar is regurgitated as honey before being served to them. These substances are stored in the cells, and the latter are never provisioned with animal food (but see

Melipona). The females are provided with corbiculae consisting of special pollen-collecting hairs which are situated either on the abdominal sterna, or on the posterior tibiae and tarsi, or on the femora. Certain genera, however, notably *Nomada*, *Coelioxys* and *Psithyrus*, are inquilines in the nests of other species, and corbiculae are wanting in these instances. A useful account of the structure and biology of certain of the solitary bees is given by Semichon (1906). For the habits of these insects see also Friese (1922–23), Malyshev (1936), Ferton, Fabre and the literature quoted under the different families. The classification adopted is that of Michener (1944) with slight modifications. Useful taxonomic works are those of Mitchell (1960–62) (Eastern U.S.A.) and Michener (1965) for Australia and the S. Pacific. The biology of social bees is discussed by Michener (1974).

FAM. COLLETIDAE. *Labial palp with 4 similar, subcylindrical segments. Part of galea beyond the palp much shorter than the stipes but longer than portion before the palp. Glossa broad, emarginate or truncate, pointed only in a few males, shorter than prementum. Mentum and submentum usually sclerotized but sometimes only weakly so. Fore wing with basal vein (1st sector M) not strongly curved. Mesepisternum with complete antero-dorsal groove. Exposed part of mid coxa shorter than distance from its base to that of hind wing.* The large genera *Hylaeus* (= *Prosopis*) and *Colletes* are Holarctic and also more or less numerous in other regions. Nearly all the other genera are S. African, S. American and especially Australian. The group includes the most primitive Apoidea and the nests are relatively simple structures either in the soil, in hollow stems or in holes in wood. They are often lined with a salivary secretion which dries into a thin transparent pellicle.

FAM. HALICTIDAE. *Labial palpi with 4 similar, subcylindrical segments. Part of galea beyond the palp much shorter than stipes, usually about the same length as portion before the palp. Glossa apically pointed. Mentum and submentum quite desclerotized. Fore wing with basal vein (1st sector of M) usually strongly curved. Mesepisternum with complete antero-dorsal groove. Exposed part of mid coxa shorter than distance from its base to that of hind wing.* This very large family of bees is, with the preceding, the only one to be well represented in Australia. *Halictus* (s.l.), with 36 British species, is very widespread and the species numerous and often difficult to discriminate. The researches of Stöckhert (1923) and Noll (1931) have established that some species are social. In *H. malachurus*, for instance, only the fertilized female survives the winter. In the spring, she produces a brood of workers which greatly enlarge the subterranean nest, producing a comb-like structure by excavating soil surrounding the cells. Many such nests are illustrated by Claude-Joseph (1926). The old female now stays in the nest to lay eggs and to guard the entrance. The next brood consists of both sexes but, after mating, only the females seek winter quarters. The workers are smaller and slightly different structurally from the females and probably lay some of the unfertilized eggs from which the males develop. It is now known (Michener, 1969; Michener and Sakagami, 1962; Plateaux-Quénu, 1972) that every transition occurs from completely solitary to fully social species with all sorts of intermediate stages. In one Australian species of *Lasioglossum*, there is a special male caste with very large head and mandibles, small eyes and reduced wings. It seems not to leave the nest (unlike the ordinary males) but the gonads are fully developed (Houston, 1970). The genus *Sphecodes* includes species which are usually red and black in colour and are very similar in structure but lack the scopa and live as parasites or inquilines in the nests of *Halictus* and *Andrena*. In warmer countries, *Halictus* is partly replaced by the genus *Nomia* which judging by the meagre records probably has a rather similar life-history. Also included in the family are such genera as *Dufourea* and *Systropha* which are much more oligolectic, i.e. visit only a few species of flowers.

FAM. ANDRENIDAE. *Antennal socket connected to epistomal suture by two grooves. Glossa acute, labial palpi with all segments similar or with the 1st (rarely 1st and 2nd) segments elongate and flattened. Mid coxa externally much shorter than distance from its base to that of hind wing. Females and many males with a defined area on last visible gastral tergite.* The principal genus in this family is *Andrena*, the characteristic solitary bee of the Holarctic region, poorly represented elsewhere. There are more than 60 British species (Perkins, 1919), but there are several hundred in the United States, each often gathering pollen and nectar from a very few species of flower. Though solitary bees, the nests are often in large, sometimes compact colonies; rarely (*A. bucephala*, etc.) several females use a common entrance gallery out of which it is presumed the individual burrows diverge. Many species act as hosts for particular species of the parasitic genus *Nomada* and they are also often attacked by Strepsiptera. The latter parasites may induce marked changes in the colour and structure of the host (p. 927). Another well-known genus in the family is *Panurgus*, which includes rather small black bees which visit yellow Compositae; two species are British. *Perdita* is a very large American genus, some of the species of which are minute in size.

FAM. MELITTIDAE. *Labial palpi with all segments similar, cylindrical; glossa acute, often elongate; mentum and V-shaped submentum present. No dorsal mesepisternal groove. Mid coxa (except in* Macropis*) much shorter than distance from its base to that of hind wing.* A small and rather diverse family of bees which are usually markedly oligolectic. British genera include *Melitta*, much resembling *Andrena*, *Dasypoda*, with an enormous tibial scopa in the female, and *Macropis*, which obtains pollen almost exclusively from *Lysimachia vulgaris* and lines its small burrow with a paste made from the sap (Malyshev, 1929).

FAM. MEGACHILIDAE. *Labrum longer than broad, widened to articulation with clypeus. Galea very elongate. Labial palps with first two segments very elongate, sheathing. A groove connects outer side of antennal socket with epistomal suture. Mid coxa usually two-thirds as long as distance from its base to that of hind wing. Two submarginal cells in fore wing of about equal length. Scopa when present on gastral sternites, gaster very rarely with a defined area on last visible tergite.* An enormous family of bees sometimes called 'gastrilegid' because of the ventral scopa of the female. This structure, however, is absent in the many parasitic species. *Megachile* includes the leaf-cutter bees, most of which construct cells either in rotten wood or in the soil out of cut fragments of green leaves; the allied genus *Chalicodoma* builds nests of mud, resin, beeswax, chewed paper or leaves. Fabre's observations on *C. muraria*, the Mason bee, make up one of his most fascinating chapters. The bee is a densely hairy insect, rather larger than a hive bee, black in the female and brown in the male. The nest is built of exceptionally hard masonry constructed with soil particles mixed with salivary secretion and with many small pebbles included and cemented in position. It is attached to walls or to large stones. After eight or nine cells have been built, the whole is then plastered over with the same substance, and the completed nest assumes a dome-like form about the size of half an orange. Notwithstanding the great hardness of these nests, their inmates are very much subject to the attacks of such parasites as Bombyliids, *Leucospis* and *Stelis*. *Osmia* and its allies include a large number of species of Holarctic distribution and of very varied habits. They generally choose hollow places already existing whether they be in wood, stones, mortar, in empty snail shells, Cynipid galls or elsewhere. Most of them build their cells with earth, but some use resin, or chewed-up leaves. Usually about 10–20 cells are found in a nest and each is stored with a mixture of pollen and honey. Smith recorded a nest (doubtless the combined nests of several females) attached beneath a large stone and composed of 230 cells. *O. tridentata* nests in bramble stems while the very common *O. rufa* will form its nest in almost any convenient hollow, whether it be in the ground or in wood, or it may take advantage of a keyhole, snail shell or other object (see summary by Poulton, 1916). In *Anthidium* (Fig. 588) the males are exceptional in being larger than the

FIG. 588
Anthidium manicatum, male. Britain (Megachilidae)

After F. Smith: reproduced by permission of the Trustees of the British Museum.

females, and like *Osmia* the species usually nest in ready-made cavities. Some of the species form a large mass of 'cotton wool' which they strip with their mandibles from the leaves of various plants. The single British species *A. manicatum* chiefly occurs in the southern counties, and within the cottony mass the cells are made of a delicate membrane which serves to retain the pollen and nectar. Other species have been observed by Fabre to use resin in place of cotton for their nest material.

Parasitic genera in this family are *Coelioxys* parasite on *Megachile* and *Anthophora*, *Dioxys* on *Chalicodoma* and *Osmia*, and *Stelis* on *Osmia*, *Chalicodoma* and *Anthidium*.

FAM. FIDELIIDAE. *Similar to Megachilidae but with 3 submarginal cells in fore wing. Dorsal mesepisternal groove present. Nest in the soil, cells completely unlined. Larva like that of Megachilidae.* Two genera and a few species in S. Africa and S. America (Rozen, 1970). The family is of interest for its mixture of primitive and specialized characters.

FAM. ANTHOPHORIDAE. *Glossa long, some labial palp segments sheathing. Labrum broader than long. A groove connects inner side of antennal socket with epistomal suture. Fore wing usually with 3 submarginal cells. Hind tibia with a scopa but not a corbicula, with two spurs. Pygidial plate present on last gastral tergite in nearly all females and most males. Clypeus protuberant, lateral portions (seen from below) bent posteriorly, lying parallel to long axis. Fore coxa a little broader than long. Solitary species.*

The very large genus *Anthophora* (with smaller, closely allied genera) includes stout, often very hairy bees which nest in the soil, sometimes in large colonies. The large black *A. acervorum* is common in Britain and is one of the earliest spring bees, often visiting the flowers of crocus. *Eucera* and its allies are rather similar to *Anthophora* but are known by the elongate male antennae. *Hemisia* (*Centris*) is a large American genus of big, hairy, often banded bees which make very deep burrows in the soil. Of the numerous parasitic genera, the largest is probably *Nomada* which contains almost bare, usually black and yellow, wasp-like bees. They are mostly parasitic on different species of *Andrena*, but *Halictus, Panurgus* and *Eucera* are also attacked. *Melecta* includes black species often marked with patches of white tomentum and their hosts are species of *Anthophora*. *Epeolus* includes small black and reddish species with white, tomentose spots which parasitize *Colletes*. The British species of *Nomada* are described by Perkins (1919) and the interesting problem of the evolution of parasitic bees has been discussed by Wheeler (1919 – parasitic Aculeates generally) and Grütte (1935). There are many more parasitic genera in N. America.

FAM. XYLOCOPIDAE. *Like the Anthophoridae but pygidial plate absent or rarely represented by an inconspicuous flattened spine. Clypeus not protuberant, lateral portions (seen from beneath) transverse rather than longitudinal. Fore coxae transverse. Solitary species.* The genus *Xylocopa* (Hurd and Moure, 1963) or carpenter bees are large, almost Humble Bee-like, though the gaster is usually bare dorsally. They excavate long galleries in timber or make nests in large plant stems. *X. violacea* extends as far north as Paris and its habits attracted the attention of Réaumur: both sexes hibernate and reappear in the following spring. According to Bingham *X. rufescens* is nocturnal and its loud buzzing may be heard throughout moonlit nights in Burma. In the tropical species of the subgenus *Mesotrichia* the propodeum and the first gastral tergite are each produced to form an almost closed cavity in which mites of the genus *Dinogamasus* live. The species of mite is specific to each species of bee (Perkins, 1899; LeVeque, 1928–31). The small, usually metallic bees of the genus *Ceratina* are almost bare except for the legs and nest in the pith of plant-stems. Though only *C. cyanea* is British, it is a very large genus with numerous species in Africa and elsewhere. *Allodape* is a somewhat similar genus with many species in warmer climates. Some S. African species are social (Brauns, 1926; Michener, 1972) and the genus *Eucondylops* has evolved from, and is parasitic on, *Allodape*. There are altogether 11 parasitic species, in several genera; some have their mouthparts so reduced that they must feed only in the nest of the host (Michener, 1970).

FAM. APIDAE. *Generally like the Anthophoridae but hind tibia with a corbicle (except in males and some queens). Hind tibia with no spurs except in Bombinae. Pygidial plate absent. Mostly social species with a worker caste.* The S. American Euglossinae are solitary bees of moderate to very large size, either metallic and bare or dark and densely hairy. In a number of species the males visit the flowers of orchids, attracted by specific chemicals; they store these substances in their modified hind tibia, possibly as part of their courtship behaviour (Zucchi *et al.*, 1969). The hind tibia is developed into a corbicle, much as in *Bombus*, and the nest is built either of mud or of resin. It is possible that some species are subsocial but there is need for further study; at least two parasitic genera have evolved from this stock and are parasitic on *Euglossa*.

The Bombinae (Bumble or Humble Bees) include some of the most familiar insects in temperate climates. They are abundant in the Holarctic region but generally confined to the mountains in tropical countries. They are absent from almost the whole of Africa, the plains of India and none are indigenous to Australia and New Zealand. The species of *Bombus* exhibit, in temperate regions, a social life which resembles that found in *Vespa* much more closely than that which obtains in the hive bee. The societies come to an end in autumn and a certain number of the females hibernate to reappear in spring when they form new colonies. The most abundant caste is that of the workers but the latter are not clearly distinguishable

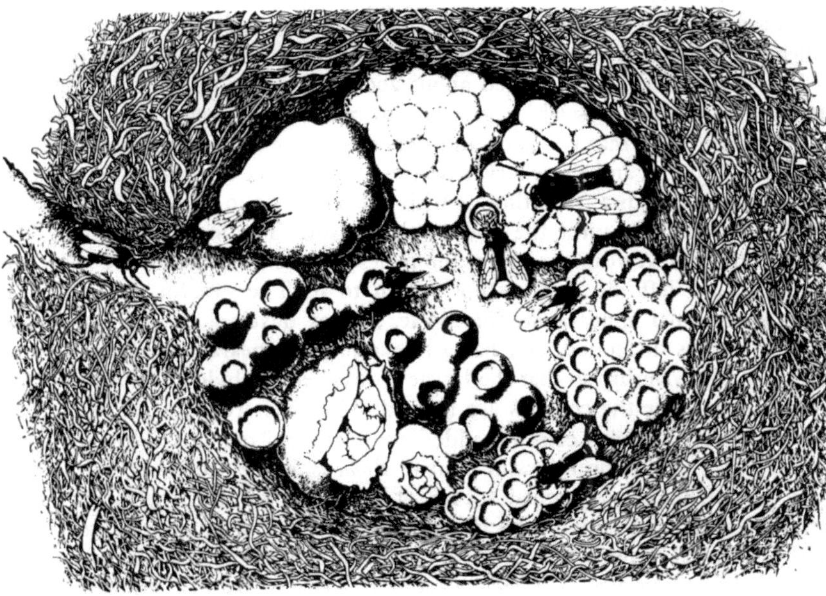

FIG. 589 Nest of *Bombus lapidarius*
After Sladen.

from the queens or females except by their smaller size. Soon after fertilization the females hibernate and this phase may be passed either in the ground, or in thatch, rubbish, moss, etc. In Britain the period of torpor lasts about nine months and according to Sladen (1912) it may commence as early as July, as is in the case of *Bombus pratorum*. The latter species is astir again in March or April while other species often wait until May or even June. Each queen seeks out a situation for her future nest: the latter may be underground and consists of fine grass or moss formed into a hollow ball (Fig. 589). Access to the nest is obtained by means of a tunnel which averages about two feet in length. Other Bombi, known as 'carder bees', form surface nests hidden away among grass, ivy or other herbage. They derive their name from their habit of collecting moss and other material used in nest formation and plaiting it with the aid of their legs and mandibles. Having formed the nest the next act of the queen is to collect a mass of pollen which is formed into a paste. Upon the top of this substance she constructs a circular wall of wax and, in the cell thus formed, she lays her first batch of eggs, capping the latter over with a covering of wax. She also constructs a waxen receptacle, or honey pot, which is filled with a store of honey for her own consumption. This store is drawn upon during inclement weather and while the queen is occupied in incubating her eggs. The larvae hatch in about four days and lie immersed in their food-bed of pollen: the queen further supplies them with regurgitated pollen and nectar which is passed to the brood through a hole which she forms in the upper part of the cell. About the 10th day the larvae spin tough pale yellow cocoons and on the 22nd or 23rd day after oviposition the first adults appear and are always workers. New cells are added to the nest as the season advances, and each cell contains on an average about a dozen eggs. The workers convert their old cocoons into honey pots and, in some species, additional waxen vessels are also constructed. When sufficient workers have emerged, the work of pollen-collecting devolves upon them and the queen becomes restricted to the nest. After the queen has deposited about 200–400 worker eggs, according to the species, she lays other eggs which give rise to males and queens. Both Huber and Schmiedeknecht state that the male and queen cells are not provisioned before the eggs are laid in them, and those larvae destined to produce queens do not appear to receive any different diet from those which will give rise to males. The males and females do not appear until the end of the season. The survivors among these females form the next year's colonies: the males, on the other hand, are short-lived and having once left the nest do not return to it. The nest of *Bombus* usually presents an irregular appearance: the larvae, as they develop, increase in size, and their cell becomes distended, and has a mammilated appearance. The queen adds more wax to the covering so that the larvae always remain hidden, but much of the wax is removed after the cocoons are formed. The cells are only utilized once for rearing purposes and fresh cells are added above the old remains. The members of the genus *Psithyrus* are inquilines in nests of *Bombus*, each species generally sharing the food and shelter of a particular species of host. Furthermore, the colour and size resemblance of the inquiline to the *Bombus* with which it is commonly associated is especially striking. This is very evident in two abundant British Psithyri, i.e.: *P. rupestris* closely resembles *B. lapidarius* and *P. vestalis* likewise simulates *B. terrestris*. According to Sladen the above-mentioned species of *Psithyrus* sting the *Bombus* queens to death and usurp their places in the nests, the *Bombus* workers rearing the *Psithyrus* offspring; other workers, however, have found the two queens subsisting side by side. In such affected nests the population of the host species is naturally greatly reduced in numbers. From the nature of its life *Psithyrus* produces no workers and its males and females differ from those of *Bombus* in their more resistant and shinier integument which, insofar as the abdomen is concerned, is less densely clothed with hair. Owing to the absence of any polliniferous apparatus, the outer surface of the hind tibia of the female *Psithyrus* is convex and uniformly hairy, whereas in *Bombus* it is more or less concave, bare and shiny but marginally clothed with long hairs. In *Psithyrus* also, the female lacks both the comb at the apex of the hind tibia and the auricle at the base of the basitarsus.

There are 19 British species of *Bombus* and 6 of *Psithyrus* (Richards, 1927). The biology of these species is described in the well-illustrated manual of Sladen (1912): the works of Hoffer (1882), Plath (1934) and Cumber (1949) are also important.

The Meliponinae (see also p. 590) include about 250 species which are mainly Neotropical with a certain number of members found in the tropics of the Old World. They nest in hollows in trees and rocks, or in walls, and their colonies include enormous numbers of often minute individuals (sometimes less than 3 mm. long) known as 'mosquito' or 'stingless' bees: the latter expression, however, is a misnomer, since a vestigial sting is present (von Ihering, 1904). The production of queens seems to be determined genetically (Kerr, 1950), excess queens often being destroyed. The workers secrete wax which is produced between the abdominal terga: it is usually mixed with earth or resin forming a dark material called 'cerumen'. The nest consists of a part containing the brood which is separate from that devoted to storing honey and pollen. The entrance to the nest usually projects as a conspicuous funnel which is often guarded by workers during the day and closed with cerumen at night. For an account of the American species and of their habits see Schwarz (1932; 1948), Salt (1929), Nogueira-Neto (1970) and Wille and Michener (1973).

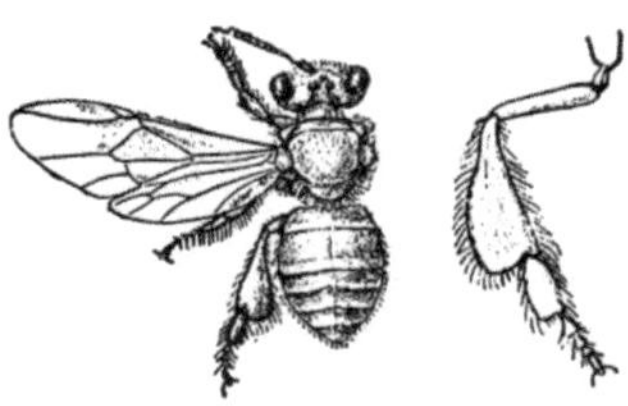

FIG. 590
Trigona lutea worker. × 2. India (Apidae)
After Bingham (F.B.I.).

The best known member of the Apinae is the hive or honey bee, *Apis mellifera*. It has been studied as much as any other species of insect, its habits having attracted attention from very early times. The structure and biology of this insect have been discussed in many volumes dating from the Renaissance onwards, and the details of its economy are so readily accessible that only the more important features will be referred to here. The insect is rarely found wild in Britain, and has been introduced into almost every country of the globe. It is usually regarded as the highest member of the Apoidea, and differentiation into the three forms male, female and worker is more pronounced than among other bees. The male, or drone, is larger and stouter than the worker, and is readily distinguishable from the latter caste by the large holoptic eyes, whose great development is accompanied by a corresponding reduction of the frontal region of the head. Although most males are haploid, diploid males can be produced in inbred strains but the young larvae are normally eaten by the workers (Woyke and Adamska, 1972). The female, or queen, has a particularly long abdomen extending some distance behind the closed wings. She performs none of the functions of next building, food gathering, or brood care and lacks the special organs adapted for these purposes. As already noted (p. 1192), pheromones derived from the queen's mandibular glands play a large part in colony-life. Large prosperous colonies contain 50 000 to 80 000 workers, besides a queen and a variable number of males. The queen is able to survive for several seasons, but the males and workers are relatively short lived. Summer-hatched workers, owing to continuous toil, seldom appear to survive longer than six or seven weeks, but those hatched in autumn live to perform the labours of early spring. The colonies of this species are, therefore, not merely seasonal but are maintained from year to year, and are stored with provisions for winter consumption. When the population increases beyond the capacity of the hive, swarms are emitted which consist of the old queen and a number of workers. In this way the new community is fully prepared for both nest building and reproduction. The original colony is dominated by a new queen and, prior to her emergence, the old queen is prevented from destroying her by the workers. The latter, as far as possible, only allow new queens to develop when it is desirable to emit a swarm. The virgin queen takes what is known as the marriage flight, and is followed by a number of males. Copulation occurs in mid-air and the fertilized queen then returns to the nest. If a second swarm be emitted the same season, a new virgin queen accompanies the workers and, as the swarm usually journeys further from the nest than the previous swarm, an opportunity is afforded for the queen to be fertilized by a male from another colony. At the end of the summer, the workers always eject the males from the hive, since they have no further part to play in the life of the community.

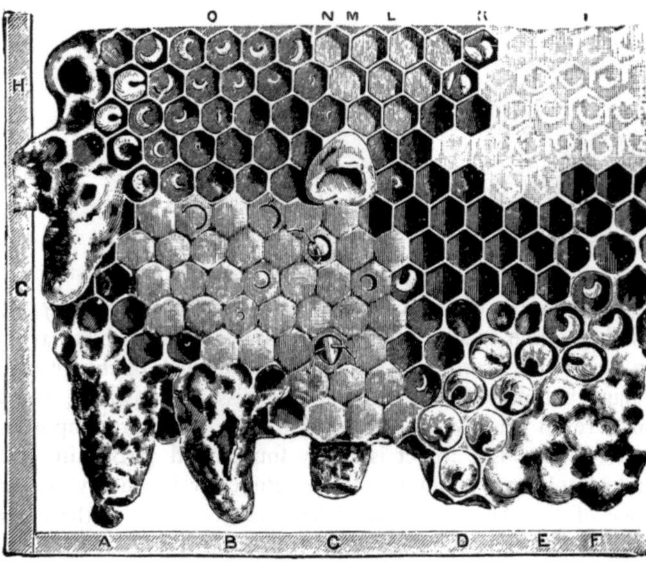

FIG. 591 Comb of hive bee (natural size)

A, empty queen cell; *B*, do, torn open; *C*, do, cut down; *D*, drone larva; *E*, *F*, sealed drone cells; *G*, sealed worker cells; *H*, old queen cell; *I*, sealed honey; *K*, pollen masses; *L*, pollen cells; *M*, abortive queen cell; *N*, emerging bee; *O*, eggs and larvae. *After* Cheshire.

The honeycomb, or structural basis of the nest (Fig. 591), is composed of cells which are mostly hexagonal in form, and arranged in two series, placed back to back. The separate combs hang vertically downwards and the long axes of the cells are almost horizontal. The material used in construction is wax which is secreted by the younger workers. It is a product of epidermal glands situated on the ventral aspect of the 4th to 7th abdominal sterna. The wax is secreted as a fluid and, according to Dreyling (1903), it is exuded through extremely fine cuticular pores, subsequently accumulating and hardening in the form of thin plates. The latter project from pockets situated between adjacent sterna, and the bee removes the wax plates by impaling them on the spines of the distal end of the first tarsal segment of the hind leg (Casteel, 1912). The leg is then flexed forwards, and the wax seized by the mandibles and kneaded into the required condition to form the cells. The cells in which workers develop are smaller than those destined for rearing the males, while the royal cells in which the queens are produced are the largest of all and irregularly ovoid in form. Other of the cells are devoted to the storing of pollen and honey. In addition to wax, the workers also utilize a resinous substance which they collect from the buds and other portions of various trees. The material is termed propolis, and is used as a kind of glue to fasten loose portions of the comb and to fill up crevices, etc. The queen lays a single egg in each brood cell, and the incubation period is about three days. When the larvae are fully grown, the workers seal up the cells by means of a cover of wax and pollen: thus enclosed the larvae form the so-called cocoons in which pupation takes place. The complete development of the queen occupies approximately 15½ days, the worker three weeks, and the male 24 days. The young larvae are at first uniformly nourished on a diet rich in protein (40–43 per cent) which is provided by the workers. This food is a secretion of the lateral pharyngeal glands. The larvae of the queens are fed upon this diet throughout life, while those destined to produce workers and males are nourished upon honey and digested pollen from the fourth day onwards. The subject of sex-determination is a highly complex one and it may be said that it is generally agreed that the virgin eggs produce the drones and the fertilized eggs the queens and workers. In rare cases, however, workers may produce queens and other workers from unfertilized eggs (Jack, 1917).

A variety of flowers is visited by bees in order to gather nectar, the most important being Dutch clover: heather, lime, other clovers, the blossoms of fruit trees and bushes, buckwheat, white mustard, etc., are also largely resorted to. The nectar, when gathered, largely consists of cane sugar which, in its conversion into honey, becomes inverted into glucose and laevulose. The work of von Frisch showing that the dances of foragers returning to the hive convey information as to the distance and direction of food-sources is now widely accepted though there has been some criticism of detail (von Frisch *et al.*, 1967). In order to supplement the above account the reader is referred to the work of Snodgrass (1956) for anatomical details, and for general information to the writings of Ribbands (1953), Butler (1949) and many others.

Other species of *Apis* are the three Indian representatives, *A. dorsata, indica* and *florea*. *Apis indica* is perhaps a subspecies of *mellifera* while *A. florea* is the smallest member of the genus and in some respects transitional between *dorsata* and *indica*. *A. dorsata* constructs a single huge comb sometimes 3–4 feet in diameter. It is suspended quite exposed from rocks, branches, or from buildings. This species is easily irritated and readily attacks man or domestic animals, sometimes with fatal results. The African race of the honeybee, accidentally introduced into Brazil, has there also proved a serious menace to man and his domestic animals.

Literature on Hymenoptera

ADLER, H. (1894), *Alternating Generations, A biological study of oak galls and gall flies.* (trans. C. R. Stratton), Clarendon Press, Oxford.

ALAM, S. M. (1951–53), The skeleto-muscular mechanism of *Stenobracon deesae* Cam. (Brac. Hymenoptera), Parts I and II, *Aligarh Muslim Univ. Publ.* (Zool. Ser.), 3, 74 pp., 75 pp.

ALTEN, H. von (1910), Zur Phylogenie des Hymenopterengehirns, *Jenaische Z. Natw.*, **46**, 511–590.

ALTSON, A. M. (1920), The life-history and habits of two parasites of blow flies, etc., *Proc. zool. Soc. Lond.*, **1920**, 196–243.

ANDRÉ, E. AND E. (1879–1913), *Species des Hyménoptères d'Europe et d'Algérie*, etc. 11 vols. The author, Beaune, later volumes, A. Hermann, Paris.

ANNECKE, D. P. AND DOUTT, R. L. (1961), The genera of the Mymaridae (Hymenoptera: Chalcidoidea), *Dept. Agric. Techn. Adv. Serv., Ent. Mem. Rep. S. Africa*, **5**, 71 pp.

ASKEW, R. R. (1968), Hymenoptera. 2. Chalcidoidea (b). Elasmidae and Eulophidae (Elachertinae, Eulophinae, Euderinae), *R. ent. Soc. Lond., Hndbks. Ident. Brit. Ins.*, **8 (2b)**, 39 pp.

BAERENDS, G. P. (1941), Fortpflanzungsverhalten und Orientierung der Grabwespe *Ammophila campestris* Jur., *Tijdschr. Ent.*, **84**, 68–275.

BAKKENDORF, P. (1934), Biological observations on some Danish Hymenopterous egg parasites, especially in Homopterous eggs, with taxonomic remarks and descriptions of new species, *Ent. Meddr*, **19**, 1–34.

BALDUF, W. V. (1926), *Telenomus cosmopeplae* Gahan, an egg parasite of *Cosmopepla bimaculata* Thomas, *J. econ. Ent.*, **19**, 829–841.

BALFOUR-BROWNE, F. (1922), On the life-history of *Melittobia acasta* Walker, a Chalcid parasite of bees and wasps, *Parasitology*, **14**, 349–370.

BENSON, R. B. (1938), On the classification of sawflies (Hymenoptera Symphyta), *Trans. R. ent. Soc. Lond.*, **87**, 353–384.

—— (1946), Classification of the Cephidae (Hymenoptera Symphyta), *Trans. R. ent. Soc. Lond.*, **96**, 89–108.

—— (1950), An introduction to the natural history of British sawflies (Hymenoptera Symphyta), *Trans. Soc. Brit. Ent.*, **10**, 45–142.

—— (1951–58), Hymenoptera, Symphyta, *R. ent. Soc. Lond., Hndbks. Ident. Brit. Ins.*, **6**, pts. (2a), (2b), (2c), 251 pp.

—— (1954), Classification of the Xiphydriidae (Hymenoptera), *Trans. R. ent. Soc. Lond.*, **105**, 151–162.

BEQUAERT, J. (1918), A revision of the Vespidae of the Belgian Congo based on the collection of the American Museum Congo expedition, with a list of Ethiopian diplopterous wasps, *Bull. Amer. Mus. nat. Hist.*, **39**, 1–384.

BERLAND, L. AND BERNARD, F. (1938), *Faune de France*, **34**, *Hyménoptères vespiformes*, 3 (Cleptidae, Chrysidae, Trigonalidae), Paul Lechevalier, Paris.

BEYERINCK, M. W. (1882), Beobachtungen über die ersten Entwicklungsphasen einiger Cynipidengallen, *Verh. K. Akad. Amsterdam*, **22**, 198 pp.

BISCHOFF, H. (1927), *Biologie der Hymenopteren*, Verlag Julius Springer, Berlin.

BLÜTHGEN, P. (1953), Zur Biologie von *Stephanus serrator* F. (Hym., Stephanidae), *Zool. Anz.*, **150**, 229–234.

BOHART, R. M. AND SCHUSTER, R. O. (1972), A host record for *Fedtschenkia* (Hymenoptera: Sapygidae), *Pan-Pacific Ent.*, **48**, 149.

BORDAS, L. (1894), Appareil glandulaire des Hyménoptères, *Ann. Sci. nat. Zool.*, (7), **19**, 1–362.

—— (1897), *Description anatomique et étude histologique des glandes à venin des insectes Hyménoptères*, Carre & Naud, Paris.

BORGMEIER, T. (1958), Nachträge zu meiner Monographie der neotropischen Wanderameisen (Hym. Formicidae), *Stud. Ent. Petropolis*, n.s. **1**, 197–208.

BOUČEK, Z. (1959), *Chalcedectus sinaiticus* (Masi) from the near East and *Ch. guaraniticus* (Strand) from Paraguay, and new synonymy (Hym. Chalcidoidea), *Acta ent. Mus. nat. Pragae*, **33**, 483–486.

—— (1974), A revision of the Leucospidae (Hymenoptera: Chalcidoidea) of the world, *Bull. Brit. Mus. nat. Hist.* (*Ent.*), Suppl. **23**, 1–241.

BRANDT, E. (1879), Vergleichend-anatomische Untersuchungen über das Nervensystem der Hymenopteren, *Hor. Soc. ent. Ross.*, **15**, 31–50.

BRAUNS, H. (1926), A contribution to the knowledge of the genus *Allodape* St. Farg. et Serv., etc., *Ann. S. Afr. Mus.*, **23**, 417–434.

BRIAN, M. V. (1953), Oviposition by workers of the ant *Myrmica, Phys. comp. oecol.*, **3**, 25–36.

BRIAN, M. V. AND A. D. (1952), The wasp, *Vespula sylvestris* Scopoli: feeding, foraging and colony development, *Trans. R. ent. Soc. Lond.*, **103**, 1–26.

BRIDWELL, J. C. (1918), Notes on the Bruchidae and their parasites in the Hawaiian Islands, *Proc. Hawaiian ent. Soc.*, **3**, 465–500.

—— (1919*a*), Miscellaneous notes on Hymenoptera, *Proc. Hawaiian ent. Soc.*, **4**, 109–165.

—— (1919*b*), Some notes on Hawaiian and other Bethylidae (Hymenoptera) with descriptions of new species, *Proc. Hawaiian ent. Soc.*, **4**, 21–38.

—— (1920), Some notes on Hawaiian and other Bethylidae (Hymenoptera) with the description of a new genus and species, *Proc. Hawaiian ent. Soc.*, **4**, 291–314.

—— (1958), Biological notes on *Ampulicomorpha confusa* Ashmead and its Fulgoroid host (Hymenoptera: Dryinidae and Homoptera: Achilidae), *Proc. ent. Soc. Washington*, **60**, 23–26.

BRUES, C. T. AND MELANDER, A. L. (1945), *Classification of insects*, The authors, Cambridge, Mass.

BUCKER, G. E. (1948), The anatomy of *Monodontomerus dentipes* Boh., an entomophagous Chalcid, *Can. J. Res.*, (C), **26**, 230–281.

BUGNION, D. (1891), Recherches sur le développement post-embryonnaire, l'anatomie et les moeurs de l'*Encyrtus fuscicollis, Rec. zool. Suisse*, **5**, 435–534.

—— (1910), La structure anatomique du *Trigonalys hahni* Spin., *Mitt. Schweiz. ent. Ges.*, **12**, 14–20.

BÜNZLI, G. H. (1935), Untersuchungen über coccidophile Ameisen aus den Kaffeefeldern von Surinam, *Mitt. Schweiz. ent. Ges.*, **16**, 455–593.

BUSCHINGER, A. (1971), Zur Verbreitung der Sozialparasiten von *Leptothorax acervorum* (Fabr.) (Hym. Formicidae), *Bonn. Zool. Beitr.*, **22**, 322–331.

BUTLER, C. G. (1949), *The honeybee. An introduction to bee sense-physiology and behaviour*, Univ. Press, Oxford.

BUTLER, C. G., CALLOW, R. K. AND JOHNSTON, N. C. (1962), The isolation and synthesis of queen-substance, 9-oxodec-*trans*-2-enoic acid, a honeybee pheromone, *Proc. roy. Soc. Lond.*, (B) **158**, 417–432.

BUTLER, C. G. AND FAIREY, E. M. (1964), Pheromones of the honeybee: biological studies of the mandibular gland secretion of the queen, *J. apicult. Res.*, 3 (2), 65–76.

CAMERON, E. (1957), On the parasites and predators of the cockroach, II, *Evania appendigaster* (L.), *Bull. ent. Res.*, **48**, 199–209.

CAMERON, P. (1882–92), *Monograph of the British phytophagous Hymenoptera*, 4 vols., Ray Soc., London.

ČAPEK, M. (1969), An attempt at a natural classification of the family Braconidae based on various unconventional characters. (Hym.), *Proc. ent. Soc. Washington*, **71**, 304–312.

CASTEEL, D. B. (1912), The manipulation of the wax scales of the honeybee, *U.S. Bur. Ent. Circ.*, **161**, 1–13.

CHAMPION, H. G. AND R. J. (1914), Observations on the life-history of *Methoca ichneumonides* Latr., *Entomologist's mon. Mag.*, **50**, 266–270.

CHANNA BASAVANNA, G. P. (1953), Phoresy exhibited by *Lepidoscelio viatrix* Brues (Scelionidae, Hymenoptera), *Indian J. Ent.*, **15**, 264–266.

CHAPMAN, T. A. (1869), On the economy of the Chrysides parasitic on *Odynerus spinipes*, *Entomologist's mon. Mag.*, **6**, 153–158.

CHOPARD, L. (1922), Les parasites de la Mante religieuse, *Ann. Soc. ent. France*, **91**, 249–274.

CHRYSTAL, R. N. (1928), The *Sirex* wood-wasps and their importance in forestry, *Bull. ent. Res.*, **19**, 219–247.

—— (1930), Studies of the *Sirex* parasites, *Oxford For. Mem.*, **11**, 63 pp.

CHRYSTAL, R. N. AND MYERS, J. G. (1928), Natural enemies of *Sirex cyaneus*, Fabr. in England and their life-histories, *Bull. ent. Res.*,**19**, 67–77.

CHRYSTAL, R. N. AND SKINNER, E. R. (1932), Studies in the biology of the wood-wasp *Xiphydria prolongata* Geoffr. (*dromedarius* F.) and its parasite *Thalessa curvipes* Grav., *Scot. For. J.*, **46**, 36–51.

CLANCY, D. W. (1946), The insect parasites of the Chrysopidae (Neuroptera), *Univ. Calif. Publ. Ent.*, **7**, 403–496.

CLAUDE-JOSEPH, F. (= JANVIER, H.) (1926), Recherches biologiques sur les Hyménoptères du Chili (Mellifères), *Ann. Sci. nat. Zool.*, (10), **9**, 113–268.

CLAUSEN, C. P. (1923), The biology of *Schizaspidia tenuicornis*, Ashm., a Eucharid parasite of *Camponotus*, *Ann. ent. Soc. Am.*, **16**, 195–217.

—— (1931), Biological observations on *Agriotypus* (Hymenoptera), *Proc. ent. Soc. Washington*, **33**, 29–37.

—— (1940), *Entomophagous Insects*, McGraw-Hill Book Co., New York and London.

CLAUSEN, C. P. *et al.* (1932), Biology of some Japanese and Chosenese grub parasites (Scoliidae), *U.S. Dept. Agric. Tech. Bull.*, **308**, 26 pp.

COMSTOCK, J. H. (1929), *An Introduction to Entomology*, 3rd edn. of Part I. The Cornstock Publishing Co., Ithaca, N.Y.

CONDIT, I. J. (1947), *The Fig*, Chronica Botanica, Waltham, Mass.

CONNOLD, E. T. (1908), *British Oak Galls*, Hutchinson & Co., London.

COOK, M. T. (1902–04), Galls and insects producing them, *Ohio Nat.*, **2**, 263–278; **4**, 115–147.

COOPER, K. W. (1953), Egg gigantism, oviposition and genital anatomy: their bearing on the biology and phylogenetic position of *Orussus* (Hymenoptera: Siricoidea), *Proc. Rochester Acad. Sci.*, **10**, 38–68.

COSENS, A. (1912), A contribution to the morphology and biology of insect galls, *Trans. Can. Inst.*, **9**, 297–387.

COSTA LIMA, A. DA (1928), Notas sobre a biologia do *Telenomus fariari* Lima, parasito dos ovos de *Triatoma*, *Mem. Inst. Oswaldo Cruz.*, **21**, 201–218.

CROSSKEY, R. W. (1951), The morphology, taxonomy and biology of the British Evanioidea (Hymenoptera), *Trans. R. ent. Soc. Lond.*, **102**, 247–301.

—— (1962), The classification of the Gasteruptiidae (Hymenoptera), *Trans. R. ent. Soc. Lond.*, **114**, 377–402.

CUMBER, R. A. (1949), The biology of humble-bees, with special reference to the production of the worker caste, *Trans. R. ent. Soc. Lond.*, **100**, 1–45.

CUSHMAN, R. A. (1913), The *Calliephialtes* parasite of the Codling moth, *J. agric. Res.*, **1**, 211–237.

DALLA TORRE, K. W. von (1892–1902), *Catalogus Hymenopterorum*, 10 vols., Verlag N. Engelmann, Leipzig.

DALLA TORRE, K. W. von AND KIEFFER, J. J. (1910), Cynipidae, In: *Das Tierreich*, **24**, Verlag R. Friedländer & Sohn, Berlin.

DALY, H. V. (1964), Skeleto-muscular morphogenesis of the thorax and wings of the honeybee *Apis mellifera* (Hymenoptera: Apidae), *Univ. Calif. Publ. Ent.*, **39**, 1–77.

DEBAUCHE, H. R. (1948), Étude sur les Mymarommidae et les Mymaridae de la Belgique (Hymenoptera Chalcidoidea), *Mém. Mus. Hist. nat. Belg.*, **108**, 248 pp.

DELUCCHI, V. AND REMAUDIÈRE, G. (1966), *Hym. Eulophidae. Palearctic Tetrastichinae*, Le François, Paris.

DEMOLL, R. (1908), Die Mundteile der solitären Apiden, *Z. wiss. Zool.*, **91**, 1–51.

DOMENCHINI, G. (1953), Studio sulla morfologia dell'addome degli Hymenoptera Chalcidoidea, *Boll. Zool. agric. Bachic. Milan*, **19**, 183–298.

DONISTHORPE, H. ST. J. K. (1927*a*), *British Ants, their Life-history and Classification*, 2nd edn., London.

—— (1927*b*), *The Guests of British Ants, their Habits and Life-histories*, Routledge, London.

DOUTT, R. L. (1947), Polyembryony in *Copidosoma koehleri* Blanchard, *Am. Nat.*, **81**, 435–463.

DOUTT, R. L. AND VIGGIANI, G. (1968), The classification of the Trichogrammatidae (Hymenoptera: Chalcidoidea), *Proc. Calif. Acad. Sci.*, **35**, 477–586.

DREYLING, L. (1903), Über die wachsbereitenden Organe der Honigbiene, *Zool. Anz.*, **26**, 710–715.

DUNCAN, C. D. (1939), A contribution to the biology of North American Vespine wasps, *Stanford Univ. Publ. Biol. Sci.*, **8**, 272 pp.

EADY, R. D. AND QUINLAN, J. (1963), Hymenoptera Cynipoidea. Key to the families and subfamilies and Cynipinae (including galls), *R. ent. Soc. Lond., Hndbks. Ident. Brit. Ins.*, 8 (1a), 81 pp.

EASTHAM, L. E. S. (1929), The post-embryonic development of *Phaenoserphus viator* Hal. (Proctotrypoidea), a parasite of the larva of *Pterostichus niger* (Carabidae), with notes on the anatomy of the larva, *Parasitology*, **21**, 1–21.

ELLIOTT, E. A. (1922), Monograph of the Hymenopterous family Stephanidae, *Proc. zool. Soc. Lond.*, **1922**, 705–831.

EMBLETON, A. L. (1904), On the anatomy and development of *Comys infelix* Embleton, a hymenopterous parasite of *Lecanium hemisphaericum, Trans. Linn. Soc. Lond. Zool.*, (2), **9**, 231–254.

EMERY, C. (1910–25), Hymenoptera Formicidae, *Gen. Ins.*, **102**, **118**, **137**, **174** (two parts), **183**, Desnet-Verteneuil, Brussels.

EVANS, H. E. (1953), Comparative ethology and the systematics of spider wasps, *Syst. Zool.*, **2**, 155–172.

—— (1956–59), Studies on the larvae of digger wasps (Hymenoptera, Sphecidae), *Trans. Am. ent. Soc.*, Pt. 1 with C. E. Lin, **81**, 131–153; Pt. 2 with C. E. Lin, **82**, 35–66; Pt. 3, **83**, 79–117; Pt. 4, **84**, 109–139; Pt. 5, **85**, 137–191.

EVANS, H. E. (1957), *Studies on the comparative ethology of digger wasps of the genus Bembix*, Comstock Publ. Assoc., Ithaca, N.Y.

—— (1959*a*), The larvae of Pompilidae (Hymenoptera), *Ann. ent. Soc. Am.*, **52**, 430–444.

—— (1959*b*), Observations on the nesting behavior of Digger Wasps of the genus *Ammophila*, *Amer. midl. Nat.*, **62**, 449–473.

—— (1959*c*), The larvae of the Ampulicidae (Hymenoptera), *Ent. News*, **70**, 56–61.

—— (1961), A preliminary review of the nearctic species of *Sierolomorpha* (Hymenoptera), *Breviora*, **140**, 12 pp.

—— (1963), A new family of wasps, *Psyche*, Cambridge, **70**, 7–16.

—— (1964*a*), A synopsis of the American Bethylidae (Hymenoptera, Aculeata), *Bull. Mus. comp. Zool.*, **132**, 1–222.

—— (1964*b*), The classification and evolution of digger wasps as suggested by larval characters. (Hymenoptera: Sphecoidea), *Ent. News*, **75**, 225–237.

—— (1965), Simultaneous care of more than one nest by *Ammophila azteca* Cameron (Hymenoptera, Sphecidae), *Psyche*, Cambridge, **72**, 8–23.

—— (1966*a*), Discovery of the female *Plumarius* (Hymenoptera, Plumariidae), *Psyche*, Cambridge, **73**, 229–237.

—— (1966*b*), *The comparative ethology and evolution of the sand wasps*, Harvard Univ. Press, Cambridge, Mass.

—— (1973), Cretaceous aculeate wasps from Taimyr, Siberia (Hymenoptera), *Psyche*, Cambridge, **80**, 166–178.

EVANS, H. E. AND MATTHEWS, R. W. (1973), Systematics and nesting behavior of Australian *Bembix* sand wasps (Hym. Sphecidae), *Mem. Amer. ent. Inst.*, **20**, 1–386.

FABRE, J. H. (1879–1903), *Souvenirs Entomologiques*, 5 vols., Libr. Ch. Delagrave, Paris.

—— (1886), *Souvenirs Entomologiques*, 3me sér. Paris.

—— (1891), *Souvenirs Entomologiques*, 4me sér. Paris.

FENTON, F. A. (1918), The parasites of leaf-hoppers, with special reference to the biology of the Anteonidae, *Ohio J. Sci.*, **28**, 177–212, 243–278, 285–296.

FERRIÈRE, C. (1965), *Faune de l'Europe et du bassin méditerranéen, I. Hymenoptera Aphelinidae*, Masson et Cie, Paris.

FERRIÈRE, C. AND KERRICH, G. J. (1958), Hymenoptera 2 Chalcidoidea (a), *R. ent. Soc. Lond., Hndbks. Ident. Brit. Ins.*, **8** (2a), 40 pp.

FERTON, C. (1901–21), Notes détachées sur l'instinct des Hyménoptères mellifères et ravisseurs, *Annls. Soc. ent. France*, **70–89**, nine series of notes.

FISHER, R. C. (1971), Aspects of the physiology of endoparasitic Hymenoptera, *Biol. Rev.*, **46**, 243–278.

FLANDERS, S. E. (1953), Aphelinid biologies with implications for taxonomy, *Ann. ent. Soc. Am.*, **46**, 84–94.

FOREL, A. (1921–23), *Le Monde social des Fourmis du Globe*, Librairie Kundig, Geneva, 5 vols.

FRIESE, H. (1922–23), *Die europäischen Bienen*, Walter de Gruyter & Co., Berlin.

FRISCH, K. von, WENNER, A. M. AND JOHNSON, D. L. (1967), Honeybees: Do they use direction and distance information provided by their dancers? *Science*, **158**, 1072–1077.

GATENBY, J. B. (1917), The embryonic development of *Trichogramma evanescens*, Westw., a monoembryonic egg parasite of *Donacia simplex*, Fab., *Q. Jl micr. Sci.*, **62**, 149–187.

—— (1919), Notes on the bionomics, embryology, and anatomy of certain Hymenoptera parasitica, especially of *Microgaster connexus* (Nees), *J. Linn. Soc. Zool.*, **33**, 387–416.

GENIEYS, P. (1924), Contribution à l'étude des Evaniidae. *Zeuxevania splendidula* Costa, *Bull. biol. Fr. Belg.*, **58**, 482–494.

GIVEN, B. B. (1954), A catalogue of the Thynninae (Tiphiidae, Hymenoptera) of Australia and adjacent areas, *Bull. Dep. Sci. industr. Res. N.Z.*, **109**, 1–89.

GRAHAM, M. W. R. DE V. (1969), The Pteromalidae of Northwestern Europe. (Hymenoptera: Chalcidoidea), *Bull. Mus.* (*Nat. Hist.*) *Entom. Suppl.*, **16**, 908 pp.

GRANDI, G. (1916), Gli Agaonini (Hymenoptera) Chalcididae raccolti nell'Africa occidentale dal Prof. F. Silvestri, *Boll. Lab. Portici*, **10**, 121–286 (and other papers in same Journal).

—— (1929), Studio morfologico e biologico della *Blastophaga psenes* (L.), 2nd edn., *Boll. Lab. ent. Bologna*, **2**, 1–147.

—— (1930), Monografia del gen. *Philotrypesis* Först., *Boll. Lab. ent. Bologna*, **3**, 181 pp.

—— (1925–41), Contributi alla conoscenza degli imenotteri melliferi e predatori (21 contributions), Index to 1–15 in *Boll. Ist. Ent. Bologna*, **8**, 122–140 (1935). **16** in same journal; **17–21** in *Mem. Accad. Sci. Ist. Bologna*, (9) **6** to (10) **4**.

—— (1952), Catalogo ragionato delle Agaonine di tutto il mondo descritte fino as oggi. (4d. edizione), *Boll. Ist. Ent. Univ. Bologna*, **19**, 69–96.

GRANDORI, R. (1911), Contributo alla embriologia e alla biologia dell'*Apanteles glomeratus* (L.) Reinh., etc., *Redia*, **7**, 363–428.

GRIFFITHS, G. C. D. (1964–68), The Alysiinae (Hym. Braconidae) parasites of the Agromyzidae (Diptera), Six parts, *Beitr. Ent.*, **14**, **16**, **17**, **18**, 409 pp.

GRÜTTE, E. (1935), Zur Abstammung der Kuckucksbienen (Hymenopt. Apid.), *Arch. Naturg.*, (n.f.) **4** Heft 4, 449–534.

GURNEY, A. B. (1953), Notes on the biology and immature stages of a cricket parasite of the genus *Rhopalosoma*, *Proc. U.S. Nat. Mus.*, **103**, 19–34.

HAMM, A. H. AND RICHARDS, O. W. (1926), The biology of the British Crabronidae, *Trans. ent. Soc. Lond.*, **1926**, 297–331.

HANNA, D. D. (1935), The morphology and anatomy of *Euchalcidia carybori* Hanna (Hymenoptera-Chalcidinae), *Bull. ent. Soc. Egypte*, **19**, 326–364.

HAVILAND, M. D. (1920), On the bionomics and development of *Lygocerus testaceimanus*, Kieffer, and *Lygocerus cameroni*, Kieffer (Proctotrypoidea – Ceraphronidae), parasites of *Aphidius* (Braconidae), *Q. Jl microsc. Sci.*, **65**, 101–127.

—— (1921), On the bionomics and post-embryonic development of certain Cynipid hyperparasites of Aphids, *Q. Jl microsc. Sci.*, **65**, 451–478.

HEDICKE, H. AND KERRICH, G. J. (1940), A revision of the family Liopteridae (Hymenopt. Cynipoidea), *Trans. R. ent. Soc. Lond.*, **90**, 177–225.

HEDQUIST, K.-J. (1967), Notes on Megalyridae (Hym., Ichneumonoidea) and description of new species from Madagascar, *Annls. Soc. ent. France*, **3** (N.S.), 239–246.

HERVÉ, P. (1945), Observations sur un parasite des Mantidae, *Mantibaria* Kirby

(= *Rielia* Kieff) *manticida* Kieff. (Hymn., Scelionidae), *Ann. Soc. Hist. nat. Toulon*, no. **24**, 8–10.

HILL, C. C. AND EMERY, W. T. (1937), The biology of *Platygaster herrickii*, a parasite of the Hessian fly, *J. agric. Res.*, **55**, 199–213.

HILL, D. S. (1967*a*), *The figs* (Ficus *spp.*) *of Hong Kong*, Hong Kong Univ. Press, Hong Kong, 1–128 pp.

—— (1967*b*), Figs (*Ficus* spp.) and fig-wasps (Chalcidoidea), *J. nat. Hist.*, **1**, 413–434.

HINTON, H. E. (1951), Myrmecophilous Lycaenidae and other Lepidoptera – a summary, *Proc. Trans. S. London ent. nat. Hist. Soc.*, **1949–50**, 111–175.

HOFFER, E. (1882), *Die Hummeln Steiermarks*, Leuschner & Kubensky, K.K. Universitäts-Buchhandlung, Graz.

HOFFMEYER, E. B. (1930–31), Beiträge zur Kenntnis der dänischen Callimomiden, mit Bestimmungstabellen der europäischen Arten (Hym. Chalc.), *Ent. Meddr*, **17**, 232–260; **17**, 261–285.

HÖLLDOBLER, B. (1970), Zur Physiologie der Gast-Wirt-Beziehungen (Myrmecophilie) bei Ameisen, 2. Das Gastverhältnis des imaginalen *Atemeles pubicollis* Bris. (Col. Staphylinidae) zu *Myrmica* und *Formica* (Hym. Formicidae), *Z. vergl. Physiol.*, **66**, 215–250.

HÖPPNER, H. (1904), Zur Biologie der Rubus-Bewohner I–III, *Allg. Z. Ent.*, **9**, 97–103, 129–134, 161–171.

HOUGH, J. S. (1953), Studies on the common spangle gall of oak. 1. The developmental history, *New Phytologist*, **52**, 149–177.

HOUSTON, T. F. (1970), Discovery of an apparent male soldier caste in a nest of a Halictine bee (Hymenoptera: Halictidae), with notes on the nest, *Austr. J. Zool.*, **18**, 345–351.

HURD, P. H. AND MOURE, J. S. (1963), A classification of the large carpenter bees (Xylocopini) (Hymenoptera: Apoidea), *Univ. Calif. Publ. Ent.*, **29**, 343 pp.

IHERING, H. von (1904), Biologie der stachellosen Honigbienen Brasiliens, *Zool. Jb., Syst.*, **19**, 179–283.

IMMS, A. D. (1916), Observations on the insect parasites of some Coccidae I, *Q. Jl microsc. Sci.*, **61**, 217–274.

—— (1918*a*), Observations on the insect parasites of some Coccidae. II, *Q. Jl microsc. Sci.*, **63**, 293–374.

—— (1918*b*), Observations on *Pimpla pomorum* Ratz., a parasite of the Apple Blossom weevil, *Ann. appl. Biol.*, **4**, 211–227.

ISHAY, J. AND IKAN, R. (1968), Food exchange between adults and larvae in *Vespa orientalis* F., *Anim. Behav.*, **16**, 298–303.

IWATA, K. (1942), Comparative studies on the habits of solitary wasps, *Tenthredo*, **4**, 1–146.

—— (1960*a*), The comparative study of the ovary in Hymenoptera. Part V. Ichneumonidae, *Acta hym. Fukuoka*, **1**, 115–169.

—— (1960*b*), The comparative study of the ovary in Hymenoptera. Supplement on Aculeata with descriptions of ovarian eggs of certain species, *Acta hym. Fukuoka*, 205–211.

JACK, R. W. (1917), Parthenogenesis amongst workers of the Cape Honey-bee, *Trans. ent. Soc. Lond.*, **1916**, 396–403.

JACKSON, D. J. (1961), Observations on the biology of *Caraphractus cinctus* Walker (Hymenoptera: Mymaridae), a parasitoid of the eggs of Dytiscidae (Coleoptera).

2. Immature stages and seasonal history with a review of Mymarid larvae, *Parasitology*, **51**, 269–294.

JAMES, H. C. (1928), On the life-histories and economic status of certain Cynipid parasites of Dipterous larvae, with descriptions of some new larval forms, *Ann. appl. Biol.*, **15**, 287–316.

JANET, C. (1893–1912), *Études sur les Fourmis, les Guêpes et les Abeilles*, Paris, etc.

JANVIER, H. (= CLAUDE JOSEPH, F.) (1933), Étude biologique de quelques Hyménoptères du Chili, *Annls. Sci. nat. Zool.*, (10) **16**, 289–356.

JANZEN, D. H. (1966), Coevolution of mutualism between ants and acacias in Central America, *Evolution*, **20**, 249–275.

JAVAHERY, M. (1968), The egg parasite complex of British Pentatomoidea (Hemiptera): Taxonomy of Telenominae (Hymenoptera: Scelionidae), *Trans. R. ent. Soc. Lond.*, **120**, 417–436.

JOHNSTON, F. A. (1915), Asparagus-beetle egg parasite, *J. agric. Res.*, **4**, 303–312.

JONES, E. P. (1937), The egg parasites of the Cotton Boll Worm, *Heliothis armigera* Hübn. (*obsoleta* Fabr.) in Southern Rhodesia, *Publ. Mazoe Citrus exp. Sta.*, **6a**, 41–105.

JONESCU, C. N. (1909), Vergleichende Untersuchungen über das Gehirn der Honigbiene, *Jenaische Z. Natw.*, **45**, 111–180.

KEILIN, D. AND THOMPSON, W. R. (1915), Sur le cycle évolutif des Dryinidae, Hyménoptères parasites des hemiptères homoptères, *C.R. Soc. biol. Paris*, **78**, 83–87.

KEIR, W. (1936), The mechanism and manner of action of the sawfly terebrae, *Entomologist*, **69**, 25–31.

KENYON, F. C. (1896), The meaning and structure of the so-called 'Mushroom bodies' of the Hexapod brain, *Am. Nat.*, **30**, 643–650.

KERR, W. E. (1950), Genetic determination of castes in the genus *Melipona*, *Genetics*, **35**, 143–152.

KIEFFER, J. J. (1912), Evaniidae, In: *Das Tierreich*, **30**, Verlag Friedländer & Sohn, Berlin.

—— (1914*a*), Die Gallwespen (Cynipidae), In: *Die Insekten Mitteleuropas*, ed. Schröder, Verlag Quelle & Mayer, Berlin.

—— (1914*b*), Serphidae, In: *Das Tierreich*, **41**, Verlag Friedländer & Sohn, Berlin.

—— (1914*c*), Bethylidae, In: *Das Tierreich*, **42**, Verlag Friedländer & Sohn, Berlin.

—— (1916), Diapriidae, In: *Das Tierreich*, **44**, Verlag Friedländer & Sohn, Berlin.

—— (1926), Scelionidae, In: *Das Tierreich*, **48**, Verlag Friedländer & Sohn, Berlin.

KINSEY, A. C. (1920), Phylogeny of Cynipid genera and biological characteristics, *Bull. Amer. Mus. nat. Hist.*, **42**, 357–402.

—— (1930), The Gall wasps of the genus *Cynips*, *Indiana Univ. Studies*, **16** (1929), 529 pp.

KOHL, F. F. (1896), Die Gattungen der Sphegiden, *Ann. k.k. Hof. Mus. Wien*, **11**, 233–516.

—— (1915), Die Crabronen der paläarktischen Region, *Ann. k.k. Hof. Mus. Wien*, **29**, 1–453.

KORNHAUSER, S. J. (1919), The sexual characteristics of the Membracid *Thelia bimaculata* (Fab.). I. External changes induced by *Aphelopus theliae* Gahan, *J. Morph.*, **32**, 531–636.

KROMBEIN, K. V. (1956), Generic review of the Amiseginae, a group of Phasmatid egg parasites, and notes on the Adelphinae (Hymenoptera, Bethyloidea, Chrysididae), *Trans. Amer. ent. Soc.*, **82**, 147–215.
—— (1960), Additions to the Amiseginae and Adelphinae (Hymenoptera, Chrysididae), *Trans. Amer. ent. Soc.*, **86**, 27–39.
—— (1967), *Trap-nesting wasps and bees: life histories, nests and associates*, Smithsonian Press, Washington, D.C.
LECLERCQ, J. (1954), *Monographie systématique, phylogénetique et zoogéographique des Hyménoptères Crabroniens*, Lejeunia Presse, Liège.
LEDOUX, A. (1950), Recherche sur la biologie de la fourmi fileuse (*Oecophylla longinoda* Latr.), *Ann. Sci. nat. Zool.*, (11), **12**, 314–461.
LEIBY, R. W. (1922), The polyembryonic development of *Copidosoma gelechiae* with notes on its biology, *J. Morph.*, **37**, 195–285.
LeVEQUE, N. (1928), Carpenter bees of the genus *Mesotrichia* obtained by the American Museum Congo Expedition, 1909–1915, *Amer. Mus. Novit.*, **300**, 1–23.
—— (1930*a*), Two new species of *Dinogamasus*, mites found on Carpenter bees of the oriental tropics, *Amer. Mus. Novit.*, **432**, 1–6.
—— (1930*b*), Mites of the genus *Dinogamasus* (*Dolaea*) found in the abdominal pouch of African bees known as *Mesotrichia* or *Koptorthosoma* (Xylocopidae), *Amer. Mus. Novit.*, **434**, 1–17.
—— (1931), New species of *Dinogamasus* (*Dolaea*), symbiotic mites of Carpenter bees from the Oriental tropics, *Amer. Mus. Novit.*, **479**, 1–14.
LICHTENSTEIN, J. L. AND PICARD, F. (1917), Étude morphologique et biologique du *Sycosoter lavagnei* Picard et J. L. Licht., Hecabolide parasite de l'*Hypoborus ficus* Ev., *Bull. biol. Fr. Belg.*, **51**, 440–474.
LINSENMAIER, W. (1959), Revision der Familie Chrysididae (Hymenoptera) mit besonderer Berücksichtigung der europäischen Spezies, *Mitt. schweiz. ent. Ges.*, **32**, 1–232; Nachtrag, *Ibid.*, 233–240; Zweiter Nachtrag, *Ibid.*, **41**, 1–144 (1968).
LORENZ, H. AND KRAUS, M. (1957), Die Larvalsystematik der Blattwespen. (Tenthredinoidea und Megalodontoidea), *Abh. Larvalsystematik*, **1**, 339 pp.
MAA, T. C. AND YOSHIMOTO, C. M. (1961), Loboscelidiidae, a new family of Hymenoptera, *Pacific Insects*, **3**, 523–548.
MALYSHEV, S. I. (1929), The nesting habits of *Macropis* Pz. (Hymen. Apoidea), *Eos*, **5**, 97–109.
—— (1936), The nesting habits of solitary bees, *Eos*, **11**, 201–309.
MARCHAL, P. (1896), La reproduction et l'évolution des guêpes sociales, *Archs Zool. exp. gén.*, **4**, 1–100.
—— (1904), Recherches sur la biologie et la développement des Hyménoptères parasites. I, La polyembryonie spécifique ou germinogonie, *Archs Zool. exp. gén.*, (4), **2**, 257–335.
—— (1906), Recherches sur la biologie et la développement des Hyménoptères parasites. II. Les Platygastres, *Archs Zool. exp. gén.*, **4**, 485–640.
MARLATT, C. L. (1894), The currant stem-girdler, *Ins. Life*, **7**, 387–390.
MARSDEN-JONES, E. M. (1953), A study of the life-cycle of *Adleria kollari* Hartig, the Marble or Devonshire Gall, *Trans. R. ent. Soc. Lond.*, **104**, 195–222.
MARTELLI, G. (1907), Contribuzioni alla biologia della *Pieris brassicae* L. e di alcuni suoi parassiti ed iperparassiti, *Boll. Lab. Portici*, **1**, 170–224.
MASNER, L. (1956), First preliminary report on the occurrence of genera of the

group Proctotrupoidea (Hym.) in CSR. (First part – Scelionidae), *Acta fauna ent. Mus. nat. Prague*, **1**, 99–126.

MASNER, L. AND DESSART, P. (1967), Le reclassification des catégories taxonomiques supérieures des Ceraphronoidea (Hymenoptera), *Bull. Inst. r. Sci. nat. Belg.*, **43** (22), 1–33.

MASON, W. R. M. (1971), An Indian *Agriotypus* (Hym., Agriotypidae), *Can. Ent.*, **103**, 1521–1524.

MATTHEWS, R. W. (1968), Nesting biology of the social wasp *Microstigmus comes* (Hymenoptera: Sphecidae, Pemphredoninae), *Psyche*, Cambridge, **75**, 23–45.

MCCOLLOCH, J. W. AND YUASA, H. (1914), A parasite of the Chinch bug egg, *J. econ. Ent.*, **7**, 219–227.

—— (1915), Further data on the life economy of the Chinch bug egg parasite, *J. econ. Ent.*, **8**, 248–261.

MERCET, R. G (1921), *Fauna Ibérica. Himénopteros*, Fam. *Encirtidos*, Museo nacional de ciencias naturales, Madrid.

MICHENER, C. D. (1944), Comparative external morphology, phylogeny, and a classification of the bees (Hymenoptera), *Bull. Am. Mus. nat. Hist.*, **82**, 157–326.

—— (1953), Comparative morphological and systematic studies of bee larvae with a key to the families of Hymenopterous larvae, *Univ. Kansas Sci. Bull.*, **35**, 987–1102.

—— (1965), A classification of the bees of the Australian and South Pacific regions, *Bull. Am. Mus. nat. Hist.*, **130**, 362 pp.

—— (1969), Comparative social behaviour of bees, *A. Rev. Ent.*, **14**, 299–342.

—— (1970), Social parasites among African Allodapine bees (Hym., Anthophoridae, Ceratinini), *J. linn. Soc. Lond. Zool.*, **49**, 199–215.

—— (1972), Biologies of African Allodapine bees, *Bull. Am. Mus. nat. Hist.*, **145**, 219–302.

—— (1974), *The social behavior of the bees, A comparative study*, Harvard Univ. Press, Cambridge, Mass.

MICHENER, C. D. AND SAKAGAMI, S. F. (1962), *The nest architecture of the sweat bees (Halictinae). A comparative study of behavior*, Univ. Kansas Press, Lawrence.

MICKEL, C. E. (1928), Biological and taxonomic investigations on the Mutillid wasps, *Bull. U.S. natn. Mus.*, **143**, 351 pp.

MIDDLETON, W. (1918), Notes on the larvae of some Cephidae, *Proc. ent. Soc. Washington*, **19**, 175–179.

MILLIRON, H. E. (1950), The identity of a Cleptid egg parasite of the common walking stick, *Diapheromera femorata* Say (Hymenoptera, Cleptidae), *Proc. ent. Soc. Washington*, **52**, 47.

MITCHELL, T. B. (1960–62), Bees of the eastern United States, 1 & 2, *North Carolina agric. Exp. Sta., Tech. Bull.*, **141**, **152**, 538 pp., 557 pp.

MOELLER, A. (1893), Die Pilzgärten einiger südamerikanischer Ameisen, In: Schimper, A. F. W., *Botanische Mittheilungen aus den Tropen*, **6**, 127 pp. Jena.

MORLEY, C. (1903–1914), *Ichneumonologia Britannica, The Ichneumons of Great Britain*, 5 vols, J. H. Keys, Plymouth.

MUESEBECK, C. F. W. (1918), Two important introduced parasites of the Brown-tail Moth, *J. agric. Res.*, **14**, 191–206.

MUESEBECK, C. F. W., KROMBEIN, K. V. AND TOWNES, H. F. (1951), Hymenoptera of America north of Mexico. Synoptic catalog, *U.S. Dept. Agric., Agric. Mongr.*, **2**, 1420 pp. First supplement, Krombein, K. V., 1958; second supplement, Krombein, K. V. and Burks, B. D., 1967.

MYERS, J. G. (1934), Two Collembola-collecting Crabronids in Trinidad, *Trans. R. ent. Soc. Lond.*, **82**, 23–26.

NEWPORT, G. (1852–53), The anatomy and development of certain Chalcididae and Ichneumonidae, etc., *Trans. Linn. Soc. Lond.*, **21**, 61–77, 85–93.

NEWSTEAD, R. (1891), Insects, etc. taken in the nests of British Vespidae, *Entomologist's mon. Mag.*, (2), **2**, 39–41.

NIELSEN, E. T. (1945), Moeurs des *Bembex*, Monographie biologique avec quelques considérations sur la variabilité des habitudes, *Spol. zool. Mus. Hauniensis*, **7**, 174 pp.

NIXON, G. E. J. (1938), A preliminary revision of British Proctotrupinae, *Trans. R. ent. Soc. Lond.*, **87**, 431–466.

—— (1951), *The association of ants with aphids and Coccids*, Cmnwlth Inst. Ent., London.

—— (1957), Hymenoptera Proctotrupoidea Diapriidae (Belytinae), *R. ent. Soc. Lond., Hndbks. Ident. Brit. Ins.*, 8 (3 d ii), 107 pp.

—— (1972–76), A revision of the north-western European species of *Apanteles* Förster (Hymenoptera, Braconidae), *Bull. ent. Res.*, **61**, 701–743; **63**, 169, 228; **64**, 453–524; **65**, 687–732.

NOGUEIRA-NETO, P. (1970), *A criação de abelhas indígenas sem ferrão* (*Meliponinae*), Chácaras e Quintais, S. Paulo.

NOLL, J. (1931), Untersuchungen über die Zeugung und Staatenbildung des *Halictus malachurus* Kirby, *Z. Morph. Ökol. Tiere*, **23**, 285–368.

NOVÁK, V. (1948), On the question of the origin of pathological creatures (pseudogynes) in ants of the genus *Formica*, *Věstnik. Čsl. zool. Spol.*, **12**, 97–131. (Czech, English Summary.)

OESER, R. (1961), Vergleichend-morphologische Untersuchungen über den Ovipositor der Hymenopteren, *Mitt. Zool. Mus. Berlin*, **37**, 3–119.

—— (1962), Der reduzierte Ovipositor von *Pseudogonalos hahni* (Spin.) nebst Bemerkungen über die systematische Stellung der Trigonalidae, *Wandersamml. dtsch. Ent.*, **9** (1961), no. 45, 153–157.

OLBERG, G. (1959), *Das Verhalten der solitären Wespen Mitteleuropas* (*Vespidae, Pompilidae, Sphecidae*), 401 pp., V.E.B. dtsch. Verlag Wiss., Berlin.

PAGDEN, H. (1926), Parthenogenesis in *Methoca*, *Nature*, **117**, 199.

PARKER, D. E. (1936), *Chrysis shanghaiensis* Smith, a parasite of the Oriental moth, *J. agric. Res.*, **52**, 449–458.

PARKER, J. B. (1917), A revision of the Bembicine wasps of America north of Mexico, *Proc. U.S. nat. Mus.*, **52**, 1–155.

PECK, O., BOUČEK, Z. AND HOFFER, A. (1964), Keys to the Chalcidoidea of Czechoslovakia (Insecta: Hymenoptera), *Mem. ent. Soc. Can.*, **34**, 121 pp.

PECKHAM, E. G. AND G. W. (1898), On the instincts and habits of the solitary wasps, *Wisconsin geol. nat. Hist. Survey, Bull.*, **2**, 1–245.

PERKINS, R. C. L. (1899), On a special Acarid chamber formed within the basal abdominal segment of bees of the genus *Koptorthosoma* (Xylocopinae), *Entomologist's mon. Mag.*, **35**, 37–39.

—— (1905–07), Leaf-hoppers and their natural enemies. Dryinidae, *Rep. Expt. Sta.*

Hawaiian Sugar Plantrs. Assoc. Bull., 1, Part 1: 1–69; Part 10: 483–494; **4**, 5–59.
—— (1919), The British species of *Andrena* and *Nomada*, *Trans. ent. Soc. Lond.*, **1919**, 218–319.

PHILLIPS, W. J. (1920), Studies on the life history and habits of the jointworm flies of the genus *Harmolita* (*Isosoma*) with recommendation for control, *U.S. Dept. Agric. Bull.*, **808**, 27 pp.

PLATH, O. A. (1934), *Bumblebees and their Ways*, Macmillan & Co., New York.

PLATEAUX-QUÉNU, C. (1972), *La biologie des Abeilles primitives*, Masson & Cie, Paris.

POULTON, E. B. (1916), Nest-building instincts of bees of the genera *Osmia* and *Anthidium*, *Trans. ent. Soc. Lond.*, **1916**, xxviii–xlvi.

RASNITSŸN, A. P. (1966), New Xyelidae (Hymenoptera) from the Mesozoic of Asia, *Paleont. Zh.*, **1966** (4), 69–85. (In Russian, translated in *Int. Geol. Rev.*, **9**, 723–737 (1967).)
—— (1969), *Origin and evolution of the lower Hymenoptera*, Acad. Sci. S.S.S.R., Moscow.
—— (1972), Praeaulacidae (Hym.) from the late Jurassic of Karatau [Kazakstan, U.S.S.R.], *Paleont. Zh.*, 1, 70–72. (In Russian, translated in *Paleont. J.*, **6**, 62–77 (1972).)

RATZEBURG, J. T. C. (1844–52), *Die Ichneumonen der Forstinsekten*, 3 vols, Nicolaische Buchhandlung, Berlin.

RAU, P. AND N. (1918), *Wasp Studies Afield*, Princeton Univ. Press, Princeton.

REID, J. A. (1941), The thorax of the wingless and short-winged Hymenoptera, *Trans. R. ent. Soc. Lond.*, **91**, 367–446.

REINHARD, E. G. (1929), *The Witchery of Wasps*, The Century Company, New York.

RIBBANDS, C. R. (1953), *The Behaviour and Social Life of Honeybees*, The Bee Res. Ass. Ltd., London.

RICHARDS, O. W. (1927), The specific characters of the British Humblebees (Hymenoptera), *Trans. ent. Soc. Lond.*, **1927**, 233–268.
—— (1939*a*), The British Bethylidae (*s.l.*) (Hymenoptera), *Trans. R. ent. Soc. Lond.*, **89**, 185–344.
—— (1939*b*), The Bethylidae subfamily Sclerogibbinae (Hymenoptera), *Proc. R. ent. Soc. Lond.*, (B), **8**, 211–223.
—— (1956), Hymenoptera. Introduction and keys to families, *R. ent. Soc. Lond., Hndbks. Ident. Brit. Ins.*, **6** Pt. 1, 94 pp.
—— (1962), *A revisional study of the Masarid wasps*, British Museum (Natural History), London.
—— (1971), The biology of social wasps (Hymenoptera, Vespidae), *Biol. Rev.*, **46**, 483–528.

RICHARDS, O. W. AND HAMM, A. H. (1939), The biology of British Pompilidae (Hymenoptera), *Trans. Soc. Brit. Entom.*, **6**, 51–114.

RICHARDS, O. W. AND M. J. (1951), Observations on the social wasps of South America (Hymenoptera Vespidae), *Trans. R. ent. Soc. Lond.*, **102**, 1–170.

RICHARDSON, C. H. (1913), Studies on the habits and development of a Hymenopterous parasite, *Spalangia muscidarum* Rich., *J. Morph.*, **24**, 513–557.

RIEK, E. F. (1970), *Hymenoptera*. In: *Insects of Australia*, Chapter 37, Divn. Entomology, C.S.I.R.O., Melbourne.

RILEY, C. V. (1889), The habits of *Thalessa* and *Tremex*, *Insect Life*, **1**, 168–179.

ROBERTSON, P. L. (1968), A morphological and functional study of the venom apparatus in representatives of some major groups of Hymenoptera, *Austr. J. Zool.*, **16**, 133–166.

ROHWER, S. A. AND CUSHMAN, R. A. (1917), Idiogastra, a new suborder of Hymenoptera, with notes on the immature stages of *Oryssus*, *Proc. ent. Soc. Washington*, **19**, 89–98.

ROHWER, S. A. AND GAHAN, A. B. (1916), Horismology of the Hymenopterous wing, *Proc. ent. Soc. Washington*, **18**, 20–76.

ROSS, H. H. (1936), The ancestry and wing venation of the Hymenoptera, *Ann. ent. Soc. Am.*, **29**, 99–109.

ROUBAUD, E. (1916), Recherches biologiques sur les guêpes solitaires et sociales d'Afrique, *Ann. Sci. nat. Zool.*, (9), **1**, 1–160.

ROZEN, J. G. JR. (1970), Biology, immature stages, and phylogenetic relationships of fideliine bees, with the description of new species of *Neofidelia* (Hymenoptera, Apoidea), *Am. Mus. Novit.*, **2427**, 1–25.

SALT, G. (1929), A contribution to the ethology of the Meliponidae, *Trans. ent. Soc. Lond.*, **77**, 431–470.

—— (1931), Parasites of the Wheat-stem sawfly, *Cephus pygmaeus*, Linnaeus, in England, *Bull. ent Res.*, **22**, 479–545.

SANDERS, G. E. (1911), Notes on the breeding of *Tropidopria conica* Fabr., *Can. Ent.*, **43**, 48–50.

SAUNDERS, E. (1890), On the tongues of the British Anthophila, *J. Linn. Soc. Zool.*, **23**, 410–432.

—— (1896), *The Hymenoptera Aculeata of the British Islands*, L. Reeve & Co., London.

SAUSSURE, H. DE (1852–58), *Études sur la famille des Vespides*, 3 vols, Paris.

SCHILDKNECHT, H. AND KOOB, K. (1971), Myrmicacin, das erste Insekten-Herbicid, *Angew. Chem.*, **83**, 110.

SCHMIEDEKNECHT, O. (1902–36), *Opuscula Ichneumonologica*, The author, Blankenburg.

—— (1930), *Die Hymenopteren Nord- und Mitteleuropas*, 2nd edn, Verlag Gustav Fischer, Jena.

SCHNEIRLA, T. C. (1971), *Army ants. A study in social organisation* (H. R. Topoff, ed.), W. H. Freeman & Co., San Francisco.

SCHULZ, W. A. (1907), Trigonalidae, *Genera insectorum*, **61**, Desnet-Verteneuil, Brussels.

SCHWARZ, E. A. (1881), Biological note on *Euplectrus comstockii* Howard, *Am. Nat.*, **15**, 61–63.

SCHWARZ, H. F. (1932), The genus *Melipona*, *Bull. Amer. Mus. nat. Hist.*, **63**, 231–460.

—— (1948), Stingless bees (Meliponidae) of the Western hemisphere, *Bull. Am. Mus. nat. Hist.*, **90**, 546 pp.

SCOTT, H. (1919), The swarming of the Chalcid *Pteromalus deplanatus* Nees in buildings, *Entomologist's mon. Mag.*, **55**, 13–16.

SCUDDER, G. C. E. (1961), The comparative morphology of the insect ovipositor, *Trans. R. ent. Soc. Lond.*, **113**, 25–40.

SEMICHON, L. (1906), Recherches morphologiques et biologiques sur quelques mellifères solitaires, *Bull. sci. Fr. Belg.*, **40**, 281–412.

SEURAT, L. (1899), Contributions à l'étude des Hyménoptères entomophages, *Ann. Sci. nat. Zool.*, (8) **10**, 1–159.

SHAFER, G. D. (1949), *The Ways of a Mud Dauber*, Stanford Univ. Press.

SHORT, J. R. T. (1952), The morphology of the head of larval Hymenoptera with special reference to the head of Ichneumonoidea, etc., *Trans. ent. Soc. Lond.*, **103**, 27–84.

—— (1959), A description and classification of the final instar larvae of the Ichneumonidae (Insecta: Hymenoptera), *Proc. U.S. natn. Mus.*, **110**, 391–511. Supplement, *Trans. R. ent. Soc. Lond.*, **122**, 185–210 (1970).

—— (1960), The final instar larvae of *Aulacus striatus* Jurine (Hym. Aulacidae) – a correction, *Entomologist's mon. Mag.*, (1959), **95**, 217–219.

SILVESTRI, F. (1906–08), Contribuzioni alla conoscenza biologica degli Imenotteri parassiti, *Boll. Lab. zool. Portici*, **1–4**.

SLADEN, F. W. L. (1912), *The Humble Bee*, Macmillan & Co., London.

SMITH, H. S. (1912), The Chalcidoid genus *Perilampus* and its relations to the problem of parasite introduction, *U.S. Dept. Agric. Bur. Ent. Tech. Ser.*, **19** (4), 33–69.

—— (1917), The habit of leaf-oviposition among the parasitic Hymenoptera, *Psyche*, Cambridge, **24**, 63–68.

SNODGRASS, R. E. (1910), The thorax of the Hymenoptera, *Proc. U.S. natn. Mus.*, **39**, 37–91.

—— (1925), *Anatomy and Physiology of the Honeybee*, McGraw-Hill book Co., New York.

—— (1941), The male genitalia of the Hymenoptera, *Smithson. misc. Coll.*, **99** (14), 86 pp.

—— (1942), The skeleto-muscular mechanisms of the honey-bee, *Smithson. misc. Coll.*, **103** (2), 120 pp.

—— (1956), *Anatomy of the honey bee*, Ithaca, Comstock Publ. Assoc., New York.

SOLIMAN, H. S. (1941), Studies in the structure of *Microbracon hebetor* Say (Hymenoptera Braconidae), *Bull. Soc. Fouad Ier ent.*, **25**, 1–96.

SOYKA, W. (1949), Die systematische Stellung der Familie der 'Mymaridae' und deren Aufstellung (Chalcidoidea, Hymenoptera), *Ent. Nachricht Bl. Burgdorf*, **3**, 12–15.

SPRADBERY, J. P. (1973), *Wasps*, Sidgwick & Jackson, London.

STÖCKHERT, E. (1923), Ueber Entwicklung und Lebensweise der Bienengattung *Halictus* Latr. und ihrer Schmarotzer (Hym.), etc., *Konowia*, **2**, 48–64, 146–165, 216–247.

TAYLOR, K. L. (1967), Parasitism of *Sirex noctilio* F. by *Schlettererius cinctipes* (Cresson) (Hymenoptera: Stephanidae), *J. Austr. ent. Soc.*, **6**, 13–19.

THOMPSON, C. B. (1913), A comparative study of the brains of three genera of ants, with special reference to the mushroom bodies, *J. Comp. Neur.*, **23**, 515–572.

THOMPSON, W. R. (continued by F. J. SIMMONDS AND B. HERTING) (1943–71), *A Catalogue of the Parasites and Predators of Insect Pests*, Cmnwlth. Agric. Bureau, Belleville, Ont. (Sect. 1, pts. 1–11).

THOMSEN, M. (1947), Some observations on the biology and anatomy of a cocoon-making chalcid larva, *Euplectrus bicolor* Swed., *Vid. dansk. naturh. For.*, **84**, 73–89.

THORPE, W. H. (1932), Experiments upon respiration in the larvae of certain parasitic Hymenoptera, *Proc. Roy. Soc.*, (B), **109**, 450–471.

TIMBERLAKE, P. H. (1912), Experimental parasitism: A study of the biology of

Limnerium validum (Cresson), *U.S. Dept. Agric. Bur. Ent., Tech. Ser.*, **19**, 5, 71–92.

TINBERGEN, N. (1932), Ueber die Orientierung des Bienenwolfes (*Philanthus triangulum* Fabr.), *Z. vergl. Phys.*, **16**, 305–334.

TONAPI, G. T. (1958*a*), A comparative study of spiracular structure and mechanism in some Hymenoptera, *Trans. R. ent. Soc. Lond.*, **110**, 489–520.

—— (1958*b*), A comparative study of the respiratory system of some Hymenoptera, Part I. Symphyta. Part II. Apocrita parasitica, *Indian J. Ent.*, **20**, 108–120, 203–220.

—— (1960), A comparative study of the respiratory system of some Hymenoptera, Part III. Apocrita-Aculeata, *Indian J. Ent.*, **20** (1958), 245–269.

TOWNES, H. (1969), The genera of Ichneumonidae, Pts. 1, 2, 3, *Mem. Am. ent. Inst.*, **11**, 1–300; **12**, 1–557; **13**, 1–307.

TRIGGERSON, C. J. (1914), A study of *Dryophanta erinacei* (Mayr) and its gall, *Ann. ent. Soc. Am.*, **7**, 1–34.

TULLOCH, G. S. (1929), The proper use of the terms parapsides and parapsidal furrows, *Psyche*, Cambridge, **36**, 376–382.

TUMLINSON, J. H., SILVERSTEIN, R. M., MOSER, J. C., BROWNLEE, R. G. AND RUTH, J. M. (1971), Identification of the trail pheromone of a leaf-cutting ant, *Atta texana* (Hym., Formicidae), *Nature*, **234**, 348–349.

VECHT, J. van der AND SHENFELDT, R. D. (originally VECHT, J. van der AND FERRIÈRE, C.) (1969–), *Hymenopterorum Catalogus* (nova editio), W. Junk, 's-Gravenhage.

VERHOEFF, P. M. F. (1951), Notes on *Astata* Latreille (Hymenoptera Sphecoidea), *Zool. Meded.*, **31**, 149–164.

VEVAI, E. J. (1942), On the bionomics of *Aphidius matricarius* Hal., a Braconid parasite, of *Myzus persicae* Sulz, *Parasitology*, **34**, 141–151.

VIALLANES, H. (1886), La structure du cerveau des Hyménoptères (Guêpe), *Bull. Soc. philom.*, (7), **10**, 82–83.

WASMANN, E. (1894), *Kritisches Verzeichnis der myrmekophilen und termitophilen Arthropoden*, Verlag von Felix L. Dames, Berlin.

WAY, M. J. (1963), Mutualism between ants and honey-dew producing Homoptera, *A. Rev. Ent.*, **8**, 307–344.

WEBER, N. A. (1937), The biology of the fungus-growing ants. Part I. New forms, *Rev. Ent. Rio de Janeiro*, **7**, 378–409.

WEBSTER, F. M. AND PHILLIPS, W. J. (1912), The spring grain-aphis or 'Green Bug', *Bull. U.S. Dept. Agric. Bur. Ent.*, **110**, 153 pp.

WEISSENBERG, R. (1908), Zur Biologie und Morphologie einer in der Kohlweisslingsraupe parasitisch lebenden Wespenlarve (*Apanteles glomeratus* (L.) Reinh.), *SitzBer. Ges. naturf. Freunde, Berlin*, **1908**, 1–18.

WELD, L. H. (1952), *Cynipoidea* (*Hym.*) *1905–1950*, Ann Arbor, Michigan (privately printed).

WENT, F. W., WHEELER, J. AND WHEELER, G. C. (1972), Feeding and digestion in some ants (*Veromessor* and *Manica*) (Hym., Formicidae), *Bioscience*, **22**, 82–88.

WESENBERG-LUND, C. (1891), *Bembex rostrata*, dens Liv og Instinkter, *Ent. Meddr*, **3**, 19–43.

WEYRAUCH, W. (1936), *Dolichovespula* und *Vespa*, Vergleichende Uebersicht über zwei wesentliche Lebenstypen bei sozialen Wespen, etc., *Biol. Zbl.*, **56**, 287–301.

—— (1937), Zur Systematik und Biologie der Kuckuckswespen *Pseudovespa*, *Pseudovespula* and *Pseudopolistes*, *Zool. Jb.*, *Syst.*, **70**, 243–290.

WHEELER, G. C. AND J. (1970), The ant larvae of the subfamily Formicinae; second supplement, *Ann. ent. Soc. Am.*, **63**, 648–656.

—— (1972), The subfamilies of Formicidae, *Proc. ent. Soc. Washington*, **74**, 35–45.

WHEELER, W. M. (1907), The polymorphism of ants, with an account of some singular abnormalities due to parasitism, *Bull. Am. Mus. nat. Hist.*, **23**, 1–93.

—— (1910), *Ants, Their Structure, Development and Behaviour*, Columbia Univ. Press, New York.

—— (1913), A solitary wasp (*Aphilanthops frigidus* F. Smith) that provisions its nest with queen ants, *J. Anim. Behav.*, **3**, 374–387.

—— (1919), The parasitic Aculeata, a study in evolution, *Proc. Am. philos. Soc.*, **58**, 1–40.

—— (1922), Ants of the American Museum Congo expedition, *Bull. Am. Mus. nat. Hist.*, **45**, 13–1055.

—— (1923), *Social Life among the Insects*, Harcourt, Brace & Co., London.

—— (1928), *The Social Insects*, Kegan Paul, London.

—— (1933), *Colony-founding among Ants with an account of some primitive Australian species*, Harvard Univ. Press, Cambridge, Mass.

WILLE, A. (1958), A comparative study of the dorsal vessels of bees, *Ann. ent. Soc. Am.*, **51**, 538–546.

WILLE, A. AND MICHENER, C. D. (1973), The nest architecture of stingless bees with special reference to those of Costa Rica (Hymenoptera: Apidae), *Rvta. Biol. Tropic.*, **21**, suppl. 1, 278 pp.

WILLIAMS, F. X. (1919*a*), Philippine wasp studies, *Rep. Work Expt. Sta. Hawaiian Sugar Plntrs. Assoc.*, *Ent. Ser.*, *Bull.*, **14**, 186 pp.

—— (1919*b*), *Epyris extraneus* Bridwell (Bethylidae), etc., *Proc. Hawaiian ent. Soc.*, **4** (1918), 55–63.

—— (1928), Studies in tropical wasps – their hosts and associates, etc., *Haiwaiian Sugar Plntrs. Expt. Sta.*, *Ent. Ser.*, *Bull.*, **19**, 179 pp.

WILSON, E. O. (1971), *The Insect Societies*, Harvard Univ. Press, Cambridge, Mass.

WILSON, E. O., CARPENTER, F. M. AND BROWN, W. L. JR. (1967), The first Mesozoic ants, with description of a new family, *Psyche*, Cambridge, **74**, 1–19.

WOYKE, J. AND ADAMSKA, Z. (1972), The biparental origin of adult honey bee drones provided by mutant genes, *J. agric. Res.*, **11**, 41–49.

YUASA, H. (1922), A classification of the larvae of the Tenthredinoidea, *Illinois biol. Monogr.*, **7**, 172 pp.

ZIRNGIEBL, L. (1936), Experimentelle Untersuchungen über die Bedeutung der Cenchri bei den Blattwespen, *Beitr. naturh. Forsch. Südwestdeutschl.*, **1**, 37–41.

ZUCCHI, R., SAKAGAMI, S. F. AND CAMARGO, J. M. F. DE (1969), Biological observations on a neotropical parasocial bee, *Eulema nigrita*, with a review on the biology of the Euglossinae (Hymenoptera, Apidae). A comparative study, *J. Fac. Sci. Hokkaido Univ.* (*Ser. 6 Zool.*), **17**, 271–380.

INDEX

Figures in italics indicate the prime taxonomic reference. Figures in bold type indicate the page on which there is a figure.

MIX
Papier aus verantwortungsvollen Quellen
Paper from responsible sources
FSC® C105338

If you have any concerns about our products,
you can contact us on
ProductSafety@springernature.com

In case Publisher is established outside the EU,
the EU authorized representative is:
Springer Nature Customer Service Center GmbH
Europaplatz 3, 69115 Heidelberg, Germany

Printed by Libri Plureos GmbH
in Hamburg, Germany